Ebook Set up and Use Instructions

Use of the ebook is subject to the single user licence at the back of this book.

Electronic access to your price book is now provided as an ebook on the VitalSource® Bookshelf platform. You can access it online or offline on your PC/Mac, smartphone or tablet. You can browse and search the content across all the books you've got, make notes and highlights and share these notes with other users.

Setting up

1. Create a VitalSource Bookshelf account at https://online.vitalsource.com/user/new (or log into your existing account if you already have one).

2. Retrieve the code by scratching off the security-protected label inside the front cover of this book. Log in to Bookshelf and click the **Redeem Codes link in the Tools** menu at the top right of the screen. Enter the code in the **Redeem code** box and press **Redeem**. Once the code has been redeemed your Spon's Price Book will download and appear in your library. N.B. the code in the scratch-off panel can only be used once.

When you have created a Bookshelf account and redeemed the code you will be able to access the ebook online or offline on your smartphone, tablet or PC/Mac. Your notes and highlights will automatically stay in sync no matter where you make them.

Use ONLINE

1. Log in to your Bookshelf account at https://online.vitalsource.com.
2. Double-click on the title in your **library** to open the ebook.

Use OFFLINE

Download BookShelf to your PC, Mac, iOS device, Android device or Kindle Fire, and log in to your Bookshelf account to access your ebook, as follows:

On your PC/Mac
Go to https://support.vitalsource.com/hc/en-us/categories/360001056774-Bookshelf#360002383594 and follow the instructions to download the free Vital-Source Bookshelf app to your PC or Mac. Double-click the VitalSource Bookshelf icon that appears on your desktop and log into your Bookshelf account. Select **All Titles** from the menu on the left – you should see your price book on your Bookshelf. If your Price Book does not appear, select **Update Booklist** from the **Account** menu. Double-click the price book to open it.

On your iPhone/iPod Touch/iPad
Download the free Bookshelf App available via the iTunes App Store. Open the Bookshelf app and log into your Bookshelf account. Select **All Titles** – you should see your price book on your Bookshelf. Select the price book to open it. You can find more information at https://support.vitalsource.com/hc/en-us/articles/360014105113-iOS

On your Android™ smartphone or tablet
Download the free Bookshelf App available via Google Play. Open the Bookshelf app and log into your Bookshelf account. You should see your price book on your Bookshelf. Select the price book to open it. You can find more information at https://support.vitalsource.com/hc/en-us/articles/360014007294-Android

On your Kindle Fire
Download the free Bookshelf App from Amazon. Open the Bookshelf app and log into your Bookshelf account. Select All Titles - you should see your price book on your Bookshelf. Select the price book to open it. You can find more information at https://support.vitalsource.com/hc/en-us/articles/360014107953-Kindle-Fire

Support

If you have any questions about downloading Bookshelf, creating your account, or accessing and using your ebook edition, please visit http://support.vitalsource.com/

For questions or comments on content, please contact us on sponsonline@tandf.co.uk

Free Updates

with three easy steps…

1. Register today on www.pricebooks.co.uk/updates

2. We'll alert you by email when new updates are posted on our website

3. Then go to www.pricebooks.co.uk/updates
    and download the update.

All four Spon Price Books – *Architects' and Builders'*, *Civil Engineering and Highway Works*, *External Works and Landscape* and *Mechanical and Electrical Services* – are supported by an updating service. Two or three updates are loaded on our website during the year, typically in November, February and May. Each gives details of changes in prices of materials, wage rates and other significant items, with regional price level adjustments for Northern Ireland, Scotland and Wales and regions of England. The updates terminate with the publication of the next annual edition.

As a purchaser of a Spon Price Book you are entitled to this updating service for this 2023 edition – free of charge. Simply register via the website www.pricebooks.co.uk/updates and we will send you an email when each update becomes available.

If you haven't got internet access or if you've some questions about the updates please write to us at Spon Price Book Updates, Spon Press Marketing Department, 4 Park Square, Milton Park, Abingdon, Oxfordshire, OX14 4RN.

Find out more about Spon books
Visit www.pricebooks.co.uk for more details.

Spon's
Mechanical and
Electrical Services
Price Book

2023

Spon's Mechanical and Electrical Services Price Book

Edited by

A=COM

2023

Fifty-fourth edition

First edition 1968
Fifty-fourth edition published 2023
by CRC Press
2 Park Square, Milton Park, Abingdon, Oxon, OX14 4RN

and by CRC Press
Taylor & Francis, 6000 Broken Sound Parkway, NW, Suite 300, Boca Raton, FL 33487

CRC Press is an imprint of the Taylor & Francis Group, an informa business

© 2023 Taylor & Francis

The right of AECOM Ltd to be identified as the Author of this Work has been asserted by them in accordance with the Copyright, Designs and Patents Act 1988.

Trademark notice: Product or corporate names may be trademarks or registered trademarks, and are used only for identification and explanation without intent to infringe.

The publisher makes no representation, express or implied, with regard to the accuracy of the information contained in this book and cannot accept legal responsibility or liability for any errors or omissions that may be made.

British Library Cataloguing in Publication Data
A catalogue record for this book is available from the British Library

ISBN: 978-1-032-33177-5
Ebook: 978-1-003-31892-7

ISSN: 0305-4543

Typeset in Arial by Taylor & Francis Books

Printed and bound by CPI Group (UK) Ltd, Croydon CR0 4YY

Contents

PART 4: MATERIAL COSTS/MEASURED WORK PRICES – ELECTRICAL INSTALLATIONS

PART 5: RATES OF WAGES

PART 6: DAYWORK

PART 7: TABLES AND MEMORANDA

INDEX

Preface

The Fifty-Fourth edition of *Spon's Mechanical and Electrical Services Price Book* continues to cover the widest range and depth of engineering services, reflecting the many alternative systems and products that are commonly used in the industry as well as current industry trends.

AECOM's baseline forecast for tender prices are a 5.9% increase from Q2 2022 to Q2 2023 and 4.2% from Q2 2023 to Q2 2024.

Key Influences

- Impact of the Ukraine war on energy and materials prices
- Ongoing supply chain disruption as a result of the pandemic
- UK Sterling exchange rate against the Euro and US Dollar
- Continued rise in price of metals commodities
- Labour shortages

The wage increases for 2022 With both Mechanical and Electrical trades have been agreed.

Before referring to prices or other information in the book, readers are advised to study the 'Directions' which precede each section of the Materials Costs/Measured Work Prices. As before, no allowance has been made in any of the sections for Value Added Tax.

The order of the book reflects the order of the estimating process, from broad outline costs through to detailed unit rate items.

The approximate estimating section has been revised to provide up to date key data in terms of square metre rates, all-in-rates for key elements and selected specialist activities and elemental analyses on a comprehensive range of building types.

The prime purpose of the Materials Costs/Measured Work Prices part is to provide industry average prices for mechanical and electrical services, giving a reasonably accurate indication of their likely cost. Supplementary information is included which will enable readers to make adjustments to suit their own requirements. It cannot be emphasized too strongly that it is not intended that these prices are used in the preparation of an actual tender without adjustment for the circumstances of the particular project in terms of productivity, locality, project size and current market conditions. Adjustments should be made to standard rates for time, location, local conditions, site constraints and any other factor likely to affect the costs of a specific scheme. Readers are referred to the build up of the gang rates, where allowances are included for supervision, labour related insurances, and where the percentage allowances for overhead, profit and preliminaries are defined.

Readers are reminded of the service available on the Spon's website detailing significant changes to the published information.

www.pricebooks.co.uk/updates

As with previous editions the Editors invite the views of readers, critical or otherwise, which might usefully be considered when preparing future editions of this work.

Whilst every effort is made to ensure the accuracy of the information given in this publication, neither the Editors nor Publishers in any way accept liability for loss of any kind resulting from the use made by any person of such information.

In conclusion, the Editors record their appreciation of the indispensable assistance received from the many individuals and organizations in compiling this book.

AECOM Limited
Aldgate Tower
2 Leman Street
London E1 8FA

Special Acknowledgements

The Editors wish to record their appreciation of the special assistance given by the following organizations in the compilation of this edition.

HOTCHKISS AIR SUPPLY

Hampden Park Industrial Estate
Eastbourne
East Sussex
BN22 9AX
Tel: 01323 501234
Email: info@Hotchkiss.co.uk
www.Hotchkiss.co.uk

23–24, Riverside House
Lower Southend Road
Wickford
Essex
SS11 8BB
Tel: 01268 572116
Email: general@abbeythermal.com

T.Clarke
BUILDING SERVICES GROUP

45 Moorfields
London
EC2Y 9AE
Tel: 020 7997 7400
Email: info@tclarke.co.uk
www.tclarke.co.uk

Dornan Engineering Ltd
114a Cromwell Road
Kensington
London
SW7 4ES
Tel: 020 7340 1030
Email: info@dornangroup.com
www.dornangroup.com

MICHAEL
J.LONSDALE
Creating the right environment

Michael J Lonsdale Ltd
Unit 1 Langley Quay
Waterside Drive
Langley
Berkshire
SL3 6EY
Tel: 01753 588750
www.michaellonsdale.com

Acknowledgements

The editors wish to record their appreciation of the assistance given by many individuals and organizations in the compilation of this edition.

Manufacturers, Distributors and Subcontractors who have contributed this year include:

Aquilar Ltd
Unit 30, Lawson Hunt Industrial Park
Broadbridge Heath
Horsham
West Sussex RH12 3JR
Tel: 01403 216100
Email: info@aquilar.co.uk
www.aquilar.co.uk
Leak Detection

Balmoral Tanks
Rathbone Square Office 5B
24 Tanfield Road
Croydon
Surrey CR0 1AL
Tel: 01208 665410
www.balmoral-group.com
Sprinkler Tanks

Baltimore Aircoil Company
c/o Balticare Ltd
Church House
18–20 Church Street
Staines-on-Thames
TW18 4EP
Tel: 01784 460316
Email: info@balticare.co.uk
www.baltimoreaircoil.eu; www.balticare.eu
Heat Rejection

Biddle
St Mary's Road
Nuneaton
Warwickshire
CV11 5AU
Tel: 02476 384233
Email: sales@biddle-air.co.uk
Air Curtains

Bodet Ltd
4 Sovereign Park
Cleveland Way
Industrial Estate
Hemel Hempstead HP2 7DA
Tel: 01442 418800
www.bodet.co.uk
Clocks

Caice Acoustic Air Movement Ltd
Riverside House
3 Winnersh Fields
Gazelle Close
Winnersh
Wokingham
RG41 5QS
Tel: 0118 918 6470
Email: enquiries@caice.co.uk
Silencers & Acoustic Treatment

DMS Flow Measurement and Controls Ltd
X-Cel House
Chrysalis Way
Langley Bridge
Eastwood
Nottinghamshire
NG16 3RY
Tel: 01773 534555
Email: sales@dmsltd.com
www.dmsltd.com
Chilled Water Plant and Energy Meters

Dormakaba Ltd
Lower Moor Way
Tiverton
Devon EX16 6SS
Tel: 01884 256464
Email: info.gb@dormakaba.com
https://www.dormakaba.com
Access Control Barriers, Revolving Doors

EA-RS Fire Engineering Ltd
4 Swanbridge Industrial Park
Black Croft Road
Witham
Essex, CM8 3YN
Tel: 01376 503680
Email: onesolution@ea-rsgroup.com
www.ea-rsgroup.com
Fire Detection and Alarms

Fire Protection Ltd
Flamebar House
South Road
Templefields
Harlow
Essex
CM20 2AR
Tel: 01279 634 230
Email: info@fireprotection.co.uk
www.fireprotection.co.uk
Fire-Rated Ductwork

Harlequin
Harlequin Manufacturing Ltd
21 Clarehill Road
Moira, County Armagh
Northern Ireland
BT67 0PB
Tel: 028 9261 1077
Email: info@harlequin-mfg.com
www.harlequinplastics.co.uk/
Fuel Oil Storage Tanks

Hoval Ltd
Northgate
Newark
Notts NG24 1JN
Tel: 01636 672711
www.hoval.co.uk
Boilers/CHP/HIU

HRS Hevac Ltd
10–12 Caxton Way
Watford Business Park
Watford
Herts WD18 8JY
Tel: 01923 232335
Email: mail@hrshevac.co.uk
www.hrshevac.co.uk
Heat Exchangers

Hudevad
Record Hall Business Centre,
Rm 215 Hatton Garden
London, EC1N 7RJ
Tel: 02476 881200
https://hudevad.com/
Heat Emitters & Radiators

Hydrotec (UK) Ltd
Hydrotec House
5 Mannor Courtyard
Hughenden Avenue
High Wycombe HP13 5RE
Tel: 01494 796040
www.hydrotec.com
Cleaning and Chemical Treatment

Mitsubishi Electric
Wolvers Home Farm
Ironsbottom
Sidlow
Reigate
RH2 8QG
Tel : 01737387170
https://les.mitsubishielectric.co.uk/
Local Cooling Units/FCU

Safelincs Ltd
33 West Street, Alford
Lincolnshire
LN13 9FX
Tel: 01507 464179
Email: support@safelincs.co.uk
www.safelincs.co.uk
Fire Protection Equipment

Schneider Electric
Cardinal Place
80 Victoria Street
London
SW1E 5JL
Tel: 0870 608 8608
Email: gb-customerservices@gb.schneider-electric.com
www.se.com/uk/en
LV Switch & Distribution Boards/Breakers & Fusers

Socomec Ltd
Knowl Piece
Wilbury Way
Hitchin
Hertfordshire SG4 0TY
Tel: 01462 440033
www.socomec.com
Automatic Transfer Switches

Breathing new life into urban spaces

Copr Bay is a vibrant new neighbourhood for Swansea, part of the city's ambitious regeneration programme.

Our cost management and procurement teams are working behind the scenes to bring this vision to life, so that the people of Swansea can enjoy world-class amenities for generations to come.

 aecom.com

Copr Bay, Swansea – Phase One
Developer: Swansea Council
Image courtesy of: AFL Architects

AECOM

Delivering a better world

Building Performance Simulation for Design and Operation, 2nd edition

Jan Hensen *et al.*

When used appropriately, building performance simulation has the potential to reduce the environmental impact of the built environment, to improve indoor quality and productivity, as well as to facilitate future innovation and technological progress in construction. Since publication of the first edition of *Building Performance Simulation for Design and Operation*, the discussion has shifted from a focus on software features to a new agenda, which centres on the effectiveness of building performance simulation in building life cycle processes.

This new edition provides a unique and comprehensive overview of building performance simulation for the complete building life cycle from conception to demolition, and from a single building to district level. It contains new chapters on building information modelling, occupant behaviour modelling, urban physics modelling, urban building energy modelling and renewable energy systems modelling. This new edition keeps the same chapter structure throughout including learning objectives, chapter summaries and assignments.

May 2019: 792pp
ISBN: 9781138392199

To Order
Tel: +44 (0) 1235 400524
Email: book.orders@tandf.co.uk

For a complete listing of all our titles
visit: www.tandf.co.uk

PART 1

Engineering Features

This section on Engineering Features, deals with current issues and/or technical advancements within the industry. These shall be complimented by cost models and/or itemized prices for items that form part of such.

The intention is that the book provides more than just a schedule of prices to assist the user in the preparation and evaluation of costs but can inform the reader on current and future trends in the industry.

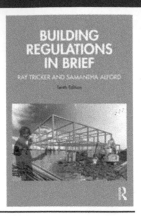

Building Regulations in Brief, 10th edition

By Ray Tricker and Samantha Alford

This tenth edition of the most popular and trusted guide reflects all the latest amendments to the Building Regulations, planning permission and the Approved Documents in England and Wales. This includes coverage of the recent changes to use classes, updated sections on planning permission, permitted development and application fees. We have included the revisions to Approved Document B (as a result of the Hackitt Review), as well as the latest changes to Approved Documents F and L, and the new documents O (overheating) and S (electric vehicle charging points), which come into effect in June 2022.

Giving practical information throughout on how to work with (and within) the Regulations, this book enables compliance in the simplest and most cost-effective manner possible. The no-nonsense approach of *Building Regulations in Brief* cuts through any confusion and explains the meaning of the Regulations. Consequently, it has become a favourite for anyone working in or studying the building industry, as well as those planning to have work carried out on their home. It is essential reading for all building contractors and subcontractors, site engineers, building engineers, building control officers, building surveyors, architects, construction site managers and DIYers.

May 2022: 1146 pp
ISBN: 9780367774233

To Order
Tel:+44 (0) 1235 400524
Email: tandf@bookpoint.co.uk

For a complete listing of all our titles visit:
www.tandf.co.uk

Market Update

After rising strongly last year, construction output is now levelling off as the Ukraine war and new coronavirus lockdowns in China hit investors' confidence. Meanwhile, the same factors are adding inflationary pressures to an already challenging cost environment for the construction sector.

Output and new orders

Construction new work output increased by 0.1% between January and February 2022, according to the Office for National Statistics' construction output data release. Similarly, all work output recorded negligible change between the two months. On a year-to-year basis, the change measure is better for new work and all work output, seeing rises of 5.7% and 6.1% respectively at February 2022. Reasonable momentum is therefore evident across the industry, at an aggregate level, over the year to Q1 2022.

For balance, comparing these February 2022 output values against the same month in 2019, shows a −6.7% shortfall for new work, and −2.4% for all work output. The difference between the two measures of output is influenced by the good levels of repair and maintenance activity across the country, which are included in the all work output data series.

Construction output trends at subsector level have moved at different speeds for some time now, and the latest data continues to underline this variance. Private industrial and infrastructure are still the busiest subsectors, with rates of expansion over the last 12 months of 35% and 18% respectively. The infrastructure output growth is crucial to aggregate measures of industry output because of its relative subsector size.

Private housebuilding contributed vital industry output too, growing by 6% over the year. Again, this subsector's contribution is crucial to the industry's overall output and momentum because it is the largest industry subsector. Private commercial and public non-housing sectors both contracted over the 12 months to February 2022.

Building costs

Input costs are high by nominal and relative measures. AECOM's composite building cost index – a measure of materials and labour costs – increased by 9.6% over the 12 months to March 2022. A wide range of inflation rates is evident still within the basket comprising the index, with almost half of the categories rising between 8 and 15% over the year. The top end of the inflation range has dropped, with the highest item's percentage change over the year not exceeding 35% and fewer categories at these extreme levels than last year. Analysis of the quarterly movement from Q4 2021 to Q1 2022 sees the largest movers in aggregates-based products, finishes and plasterboard work categories.

Latest data indicates some easing of the input cost pressure at the upper end of the range, and a narrowing spread of inflation rates. Nevertheless, a larger core of index categories will continue to post elevated rates of inflation into the medium-term, particularly given the broader geopolitical issues affecting economies and supply chains. The war in Ukraine and its spillover effects will renew inflationary pressures. Energy and metals markets will act as a primary source of inflation, with the inevitable and eventual flow through effects into manufacturing and production processes. Associated lagged effects on the inflation landscape will emerge later this year. The complication for the construction sector is that it is energy intensive. Its activities and operations, and also the processes to create or manufacture the inputs used within the construction industry, are highly exposed to energy-related and metals cost inflation.

Tender prices

Many of the industry's different subsectors are competing with each other to secure resources. This maintains existing industry-wide capacity constraints. Infrastructure is a large demand pull on labour and materials presently, reflecting the notable increase in workload across this subsector over the last 12 months. These long-standing industry constraints, combined with resource competition, add time and cost risks to project delivery. Procurement risks are also amplified because of the current mix of resource constraints and inflation volatility. A rising number of supply chain firms and subcontractors are unable or unwilling to fix prices for typical time frames, and quotation validity periods are often notably shorter.

Contract protections to cover excess and volatile cost inflation movements are more commonly proposed in pre-contract stages at the moment. Current market and pricing volatility is helping to shift ideas on procurement approach, especially within the two-stage procurement segment. It is less certain that a hybrid two-stage approach is the default approach, and conventional two-stage tendering can often be the starting point now when reviewing procurement options. Pressure on preliminaries costs shows little sign of abating, especially as energy costs and staff wage inflation remains strong. Similarly, there can be a need to revisit prelims pricing more frequently in pre-contract stages, to ensure there is appropriate cost coverage in the fast-moving market pricing environment. Improved workloads also helped to support overheads and profit levels over the period.

Outlook

Higher inflation, lower growth. This is the concerning spectre facing many economies as inflation rates on almost all measures continue their record-breaking run. The relentless rises in costs, prices and delivery times are eye-catching enough, but many of the acute inflationary trends and pressures pre-date the Ukraine war and the recent coronavirus lockdowns in China. Input cost pressure is expected to rise further as a result of these significant world events and their associated disruption.

The war in Ukraine introduces direct raw material and commodities supply impacts. Disruption will persist into the medium-term, maintaining inflationary pressures across supply chains and production operations. Global energy and commodity prices are two channels where cost impacts and supply volatility will increase, because of Russia's supply contribution to global markets. Most supply chain firms, irrespective of location, have exposure to commodities in some form, whether processed or raw, and the higher volatility introduces significant risks on their operations.

The number of UK firms with direct reliance on materials and component supplies from Ukraine and Russia is relatively small, yet indirect exposure will hit economies and firms through extended supply chain networks. The lagged effect from disruption will ensure that the supply and inflationary fallout endures for a considerable period of time. The first year of the pandemic in 2020 bears this out, when the flow through inflationary effects to the broader economy became more pronounced in 2021. The UK also has the added complication from the structural changes to its border and trade operations, which introduce additional disruptions, time delays and costs to businesses and consumers.

A series of notable factors will play out this year, all with the potential to bring impacts to the economy and the construction sector, either on their own or in combination. The pervasiveness of inflation introduces a greater risk of an 'inflation creating inflation' psychology taking hold. This applies to cost and pricing trends specific to construction as much as consumer price inflation. Many forecasts for consumer price inflation are further rises in 2022 from the current 7% rate recorded in March.

The sticky inflation environment is now a major dilemma for central banks and their monetary policy decisions. Initially believing higher inflation trends would be transitory, many now view elevated inflation as a likely scenario into the medium-term. The challenge for central banks is navigating the need for a rising interest rate course to temper acute inflationary trends, without damaging economic activity and demand – the narrow path between growth and inflation as the Bank of England Governor recently outlined. The risk is of tightening monetary policy, via higher interest rates, just as major economic headwinds blow onto the UK economy. The potential for a hard landing has grown.

These two factors – inflation psychology and monetary policy dilemmas – raise the risk of damage to aggregate demand across the economy, made worse by recent supply constraints and inflationary trends. In other words, higher costs and prices, combined with higher interest rates, may increase to such an extent that consumers or buyers delay or decide against spending, procuring, or investing. This could result in lower overall demand in the economy, and with attendant impacts on the UK's economic output. The construction sector is unlikely to be immune from spillover effects in this scenario, should it occur, given its historical cyclicality and responsiveness to broader economy-wide trends.

Renewed coronavirus lockdowns in China are set to affect availability of imported goods and components once again, as production and manufacturing disruption brings delays to supplies of finished goods and components for export.

Global shipping and logistics operations are also expected to be disrupted, given China's export volumes and its demand for sea freight. Accompanying dislocations arising from initial delays and reduced throughput capacity at Chinese ports would likely extend into the second half of 2022. Whether the disruption is to the extent seen in 2020 is less likely though, because there will not be a synchronized series of lockdowns across the globe this time. The Chinese lockdowns also link to global demand for energy and commodities by its economy. Lockdowns in China may temporarily alleviate high energy and commodity prices for example. Recently, iron ore, steel, and various metals futures, along with oil prices, all fell at various times in response to news of further lockdowns in major Chinese cities.

Inflationary pressures will endure throughout 2022, as supply constraints, on-going operational challenges and price volatility continue. AECOM's baseline forecast for tender prices is a 6% increase from Q2 2022 to Q2 2023, and 4% from Q2 2023 to Q2 2024. Existing workload momentum and acute input cost inflation is expected to move tender prices higher over the first forecast period. Risks to pricing remain to the upside across the next 12 months due to the persistently acute inflationary environment. The baseline forecast's core assumptions are an inconsistent and increasingly patchy recovery across the economy but without a recession, some variability in construction sector pricing as a result of differing output trends and related competition levels, and enduring operational disruption arising from the permanent changes to the UK's internal and external trading status.

Building Regulations Pocket Book

Ray Tricker and Samantha Alford

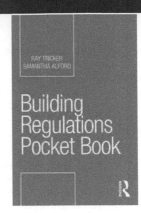

This handy guide provides you with all the information you need to comply with the UK Building Regulations and Approved Documents. On site, in the van, in the office, wherever you are, this is the book you'll refer to time and time again to double check the regulations on your current job. The *Building Regulations Pocket Book* is the must have reliable and portable guide to compliance with the Building Regulations.

Part 1 provides an overview of the Building Act

Part 2 offers a handy guide to the dos and don'ts of gaining the Local Council's approval for Planning Permission and Building Regulations Approval

Part 3 presents an overview of the requirements of the Approved Documents associated with the Building Regulations

Part 4 is an easy to read explanation of the essential requirements of the Building Regulations that any architect, builder or DIYer needs to know to keep their work safe and compliant on both domestic or non-domestic jobs

This book is essential reading for all building contractors and sub-contractors, site engineers, building engineers, building control officers, building surveyors, architects, construction site managers and DIYers. Homeowners will also find it useful to understand what they are responsible for when they have work done on their home (ignorance of the regulations is no defence when it comes to compliance!).

February 2018: 456 pp
ISBN: 9780815368380

To Order
Tel:+44 (0) 1235 400524
Email: tandf@bookpoint.co.uk

For a complete listing of all our titles visit:
www.tandf.co.uk

Building Energy Management Systems (BEMS) or Building Automation Control System (BACS)

A Building Energy Management System (BEMS) is a networked computer based control system installed in buildings that controls and monitors the building's mechanical, electrical and public health (MEP) systems and equipment as well as other ancillary plant and building systems and equipment in the most operational and energy efficient way.

Benefits

- Switches plant on and off automatically according to time, time of day, occupancy/demand and environmental conditions (e.g. senses conditions and activates devices to correct settings)
- Optimizes plant operation and services to meet building occupancy and demand requirements
- Monitors plant status and environmental conditions and thus improves standards of operation and maintenance
- Allows remote access, control and monitoring (e.g. using graphical information and alarms) via web-browser connected PC, tablet or smart phone
- Provides real-time and historical operational and energy performance data for continual building performance analysis in conjunction with Energy Management and Analytics software tools

Types of BMS points

- Is the physical connection of a BMS point to a controller or outstation I/O terminals or 'soft' points obtained through integration by various means with 3rd party systems and equipment? The points schedule and methods of integration are critical when designing the system
- Input – gathers data/monitor; Output – provide command/control
- Analogue points – has a variable value 0% to 100% (e.g. temperature sensor, modulating control valves)
- Digital points – binary I/O (0 or 1) (e.g. on/off control, run/off/fault status)
- Network integration points (NI) or High Level Interface (HLI) – allows to communicate or integrate to stand-alone controller such as chiller, meters, boilers, generators

BMS components – hardware and software/programming

Hardware

- **Primary network/BMS backbone**
  - o Central Virtual BEMS Server located in MER or Data Centre with distributed Web-client PCs or local stand-alone Head end equipment (desktop computer, laptop, monitors, keyboard, network printer (if required), UPS supply, etc.)
  - o Ethernet network cabling (normally Cat 5e or Cat 6 cables)
  - o Ethernet switch (located inside the control panels) or
  - o Part of converged building or client managed MEPS network
  - o Building Enterprise Level system interfaces e.g. Energy Management Reporting, etc.
- 'Middleware' Integration layer providing systems integration, interoperability and smart building operational features

- **Secondary network**
  - o   Control enclosure (CE) or Control panel (CP) or BMS outstation complete with I/O
  - o   Motor control centre (MCC) incorporating BEMS outstations in control section (Typically Form 2, 3 or 4)
  - o   Local operator interfaces or touch screen display panels
  - o   Unitary controller for FCU/VAV/chilled beams
  - o   Power cables and isolators (from MCC panel to mechanical equipment) and associated containment
  - o   Control cables (from MCC/CE panel to field devices) and field network cabling (to 3rd party equipment e.g. Inverters, packaged plant, chillers, etc.)
  - o   Field devices (control valves, meters, sensors/transducers, VSD/Inverter, Soft-start, DOL starter, star-delta starter, actuators)
  - o   Consider options for Wireless controllers, sensors, switches, occupancy PIRs, etc.

Software/programming

- Management level software residing on central server or head-end (typically licensed per user or per point)
- BMS communications protocol – Refers to the data communication within the various layers of the of BMS from PC/server software to the different type of outstation/controller/sensors, meters and other 'edge' devices (e.g. Modbus, M-Bus, LonWorks, BACnet (IP/MSTP), KNX or proprietary)
- Graphical User Interface – Provide true real-time dynamic representation of the installed plant and connected systems. Provide visualization of all points and necessary overrides and adjustments. Colour graphics shall incorporate automatic updating of real-time field data.
- Setting up of system network security and operator access regime
- Configure alarm/reporting strategy and historical data logging
- Software system integration (e.g. energy management system, chillers, boiler, CHP, generator, water treatment plant, closed control units, blind control, lighting control, security and access controls)
- Application of additional Building Performance Analytics software to provide enhanced intelligent alarming and Energy Management reporting, etc.

BMS Pricing – Quick Estimate for a typical medium sized commercial office building:

Shell and Core:

1.	Head end equipment (Workstation) *Extra cost:* • *Additional remote workstation*	£10,000–£20,000 £5,000–£10,000
2.	Network connection (BMS backbone – Copper)	£15,000–£30,000
3.	Main plant – cost per BMS points *Factors that affect the cost:* • *UPS and touch pad display for each control panel* • *Panels enclosure – fire-rated or GRP enclosure*	£550–£900 per pts
4.	Control enclosure (CE) for future tenants. Additional I/O modules will be part of fitting out cost *Factors that affect the cost:* • *UPS and touch pad display for each control panel* • *Panels enclosure – fire-rated or GRP enclosure*	£5,000–£10,000 per panel
5.	Landlord FCUs – depending on points per FCU controller *Factors that affect the cost:* • *Local controller for cellular office, meeting rooms, etc.*	£450–£850 per unit
6.	Software, graphics, commissioning, engineering	Included in BMS pts rate
7.	Trade contract preliminaries	20% to 30%

Category A, B or C Office Fit-out:

1.	Tenant plants – cost per points utilizing existing MCC/CE panel but require additional I/O modules	£330–£550 per pts

2.	Energy meters
	
	  a. LTHW/CHW meters
	  b. Electricity meters (meters by Electrical TC)
	Factors that affect the cost:
	 • *Size of heat meter or type of meter such as ultrasonic flow meter*
		£850–£1,750
		£250–£350

3.	Intelligent unitary controller (IUC) – depends on points per/ IUC controller and the air conditioning solution such as 4-pipe FCU, 2-pipe w/electric heating, VAV, chilled beams, chilled ceiling *Factors that affect the cost:* • *Local controller for cellular office/meeting room* • *For chilled beam and chilled ceiling, controls depend on the zoning* • *Depend also on zone controls*	£650–£850 per unit
4.	Software, graphics, commissioning *Extra over cost:* • *Integration to other stand-alone system (e.g. lighting control, AV, fire alarm, security, central battery)*	Included in pts rate
5.	Trade contract preliminaries	20% to 30%

Building Services Design for Energy Efficient Buildings, 2nd edition

Paul Tymkow *et al.*

The role and influence of building services engineers are undergoing rapid change and are pivotal to achieving low-carbon buildings.

The essential conceptual design issues for planning the principal building services systems that influence energy efficiency are examined in detail, namely HVAC and electrical systems. In addition, the following issues are addressed:

- background issues on climate change, whole-life performance and design collaboration
- generic strategies for energy efficient, low-carbon design
- health and wellbeing and post occupancy evaluation
- building ventilation
- air conditioning and HVAC system selection
- thermal energy generation and distribution systems
- low-energy approaches for thermal control
- electrical systems, data collection, controls and monitoring
- building thermal load assessment
- building electric power load assessment
- space planning and design integration with other disciplines.

In order to deliver buildings that help mitigate climate change impacts, a new perspective is required for building services engineers, from the initial conceptual design and throughout the design collaboration with other disciplines. This book provides a contemporary introduction and guide to this new approach, for students and practitioners alike.

July 2020: 376 pp
ISBN: 9780815365617

To Order
Tel: +44 (0) 1235 400524
Email: book.orders@tandf.co.uk

For a complete listing of all our titles
visit: www.tandf.co.uk

Capital Allowances

Introduction

Capital Allowances provide tax relief by prescribing a statutory rate of depreciation for tax purposes in place of that used for accounting purposes. They are utilized by government to provide an incentive to invest in capital equipment, including assets within commercial property, by allowing the majority of taxpayers a deduction from taxable profits for certain types of capital expenditure, thereby reducing, or deferring, tax liabilities.

The capital allowances most commonly applicable to real estate are those given for plant and machinery in all buildings, other than residential dwellings, except for common areas and Structures and Buildings Allowances (SBA) for the residual construction spend on non-residential property.

Enterprise Zone Allowances are also available for capital expenditure within designated areas only where there is a focus on high value manufacturing. Enhanced rates of allowances were also available on certain types of energy and water saving plant and machinery assets, until their withdrawal from April 2020.

Freeports, introduced in 2020 are specifically designed to incentivize investment and job creation by including lower customs duties, simplified planning rules and tax breaks, such as an enhanced level of capital allowances, suspension of SDLT, business rates, customs and NIC obligations for a five-year period until 2026.

The Act

The primary legislation is contained in the Capital Allowances Act 2001. Major changes to the system were introduced in 2008, 2014, 2018 and 2020 affecting the treatment of tax relief to include plant and machinery allowances, SBA and the incentivized tax relief for the newly introduced Freeports.

Plant and Machinery

Qualifying expenditure on plant and machinery is allocated to either the main pool, or special rate pool, which provide for different rates of recovery and these are set out later.

Various legislative changes and case law precedents in recent years have introduced major changes to the availability of Capital Allowances on property expenditure. The Capital Allowances Act 2001 excludes expenditure on the provision of a building from qualifying for plant and machinery, with prescribed exceptions.

List A in Section 21 of the 2001 Act sets out those assets treated as parts of buildings:

- *Walls, floors, ceilings, doors, gates, shutters, windows and stairs.*
- *Mains services, and systems, for water, electricity and gas.*
- *Waste disposal systems.*
- *Sewerage and drainage systems.*
- *Shafts, or other structures, in which lifts, hoists, escalators and moving walkways are installed.*
- *Fire safety systems.*

Similarly, List B in Section 22 identifies excluded structures and other assets.

Both sections are, however, subject to Section 23. This section sets out expenditure, which, although being part of a building, may still be expenditure on the provision of Plant and Machinery.

List C in Section 23 is reproduced below:

Sections 21 and 22 do not affect the question whether expenditure on any item in List C is expenditure on the provision of Plant or Machinery

1. Machinery (including devices for providing motive power) not within any other item in this list.
2. Gas and sewerage systems provided mainly –
   a. to meet the particular requirements of the qualifying activity, or
   b. to serve particular plant or machinery used for the purposes of the qualifying activity.
3. Omitted.
4. Manufacturing or processing equipment; storage equipment (including cold rooms); display equipment; and counters, checkouts and similar equipment.
5. Cookers, washing machines, dishwashers, refrigerators and similar equipment; washbasins, sinks, baths, showers, sanitaryware and similar equipment; and furniture and furnishings.
6. Hoists.
7. Sound insulation provided mainly to meet the particular requirements of the qualifying activity.
8. Computer, telecommunication and surveillance systems (including their wiring or other links).
9. Refrigeration or cooling equipment.
10. Fire alarm systems; sprinkler and other equipment for extinguishing or containing fires.
11. Burglar alarm systems.
12. Strong rooms in bank or building society premises; safes.
13. Partition walls, where moveable and intended to be moved in the course of the qualifying activity.
14. Decorative assets provided for the enjoyment of the public in hotel, restaurant or similar trades.
15. Advertising hoardings; signs, displays and similar assets.
16. Swimming pools (including diving boards, slides & structures on which such boards or slides are mounted).
17. Any glasshouse constructed so that the required environment (namely, air, heat, light, irrigation and temperature) for the growing of plants is provided automatically by means of devices forming an integral part of its structure.
18. Cold stores.
19. Caravans provided mainly for holiday lettings.
20. Buildings provided for testing aircraft engines run within the buildings.
21. Moveable buildings intended to be moved in the course of the qualifying activity.
22. The alteration of land for the purpose only of installing Plant or Machinery.
23. The provision of dry docks.
24. The provision of any jetty or similar structure provided mainly to carry Plant or Machinery.
25. The provision of pipelines or underground ducts or tunnels with a primary purpose of carrying utility conduits.
26. The provision of towers to support floodlights.
27. The provision of –
    a. any reservoir incorporated into a water treatment works, or
    b. any service reservoir of treated water for supply within any housing estate or other particular locality.
28. The provision of –
    a. silos provided for temporary storage, or
    b. storage tanks.
29. The provision of slurry pits or silage clamps.
30. The provision of fish tanks or fish ponds.
31. The provision of rails, sleepers and ballast for a railway or tramway.
32. The provision of structures and other assets for providing the setting for any ride at an amusement park or exhibition.
33. The provision of fixed zoo cages.

Case Law

The fact that an item appears in List C does not automatically mean that it will qualify for capital allowances. It only means that it may potentially qualify.

Guidance about what can qualify as plant is found in case law dating back to 1887 (*Yarmouth v France*). The case of *Wimpy International Ltd and Associated Restaurants Ltd v Warland* in the late 1980s is one of the most important case law references for determining what can qualify as plant.

The Judge in that case applied three tests when considering whether, or not, an item is plant.
1.   Is the item stock in trade? If the answer is yes, then the item is not plant.
2.   Is the item used for carrying on the business? In order to pass the business use test the item must be employed in carrying on the business; it is not enough for the asset to be simply used in the business. For example, product display lighting in a retail store may be plant but general lighting in a warehouse would fail the test. *(Please note, this case law relates to the pre-Integral Feature Legislation, which introduced lighting as an eligible asset.)*
3.   Is the item the business premises, or part of the business premises? An item cannot be plant if it fails the premises test, i.e. if the business use is the premises itself, or part of the premises, or a place in which the business is conducted. The meaning of 'part of the premises' in this context should not be confused with real property law. HMRC's internal manuals suggest there are four general factors to be considered, each of which is a question of fact and degree:
   •   Does the item appear visually to retain a separate identity
   •   With what degree of permanence has it been attached to the building
   •   To what extent is the structure complete without it
   •   To what extent is it intended to be permanent, or alternatively, is it likely to be replaced within a short period

Certain assets will qualify as plant in most cases. However, many others need to be considered on a case-by-case basis. For example, decorative assets in a hotel restaurant may be plant, but similar assets in an office reception area may be ineligible.

Main Pool Plant and Machinery

Capital Allowances on main pool plant and machinery are currently given in the form of writing down allowances at the current rate of 18% per annum on a reducing balance basis. For every £100 of qualifying expenditure £18 is claimable in year 1, £14.76 in year 2 and so on until either all of the allowances have been claimed, or the asset is sold.

Special Rate Pool Plant and Machinery

Capital Allowances on special rate pool plant and machinery are currently given in the form of writing down allowances at the current rate of 6% per annum on a reducing balance basis. For every £100 of qualifying expenditure £6 is claimable in year 1, £5.64 in year 2 and so on until either all of the allowances have been claimed, or the asset is sold.

The special rate pool specifically covers expenditure on integral features, thermal insulation, solar panels and long-life assets.

Integral Features

The category of qualifying expenditure on 'integral features' was introduced with effect from April 2008. The following items are integral features:
   •   An electrical system (including a lighting system)
   •   A cold water system
   •   A space or water heating system, a powered system of ventilation, air cooling or air purification, and any floor or ceiling comprised in such a system
   •   A lift, an escalator or a moving walkway
   •   External solar shading

Thermal Insulation

The installation of thermal insulation to buildings is allowable where it is provided to an existing building owned or occupied by the taxpayer. The allowance is not available for expenditure incurred on dwellings.

Long-Life Assets

A long-life asset is defined as plant and machinery that has an expected useful economic life of at least 25 years. The useful economic life is taken as the period from of first use until it is likely to cease to be used as a fixed asset of any business. It is important to note that this is likely to be a shorter period than an asset's physical life.

Plant and machinery provided for use in a building used wholly, or mainly, as a dwelling house, showroom, hotel, office, retail shop, or similar premises, or for purposes ancillary to such use, cannot be classified as a long-life asset.

In contrast certain plant and machinery assets in buildings such as factories, cinemas, hospitals etc. could potentially be treated as long-life assets.

Temporary Super-Deduction and Enhanced First Year Allowance for Plant and Machinery Assets

Following Budget 2021, temporary special first year allowances were introduced for new and unused plant and machinery to stimulate investment. The relief will only apply to companies within the charge to corporation tax for expenditure incurred from 1 April 2021 up to and including 31 March 2023. This applies to expenditure contracted after 3 March 2021.

This temporary first year allowance will apply as follows:

- 130% 'super-deduction' for main pool plant and machinery (currently 18% per annum on a reducing balance basis); and
- 50% first year allowance for special rate plant and machinery.

In addition, unincorporated businesses will not benefit because qualifying expenditure must be incurred by a company within the charge to corporation tax.

Refurbishment Schemes

Building refurbishment projects will typically be a mixture of capital costs and revenue expenses, unless the works are so extensive that they are more appropriately classified as a redevelopment. A straightforward repair or a 'like for like' replacement of part of an asset would be a revenue expense, meaning that the entire amount can be deducted from taxable profits in the same year.

Where capital expenditure is incurred which is incidental to the installation of plant or machinery then Section 25 of the Capital Allowances Act 2001 allows it to be treated as part of the expenditure on the qualifying item. Incidental expenditure will often include parts of the building that would be otherwise disallowed, as shown in the Lists reproduced above. For example, the cost of forming a lift shaft inside an existing building would be deemed to be part of the expenditure on the provision of the new lift.

The extent of the application of section 25 Was reviewed for the first time by the Special Commissioners in the case of JD Wetherspoon. The key areas of expenditure considered were overheads and preliminaries where it was held that such costs could be allocated on a pro-rata basis; decorative timber panelling which was found to be part of the premises and so ineligible for allowances; toilet lighting which was considered to provide an attractive ambience and qualified for allowances; and incidental building alterations of which enclosing walls to toilets and kitchens and floor finishes did not qualify except for tiled splashbacks, bespoke toilet cubicles and drainage did qualify, along with the related sanitary fittings and kitchen equipment.

Structures and Building Allowances

The 2018 Autumn Budget introduced the Structures and Buildings Allowance (SBA) for expenditure incurred on non-residential structures and buildings.

Initially, the SBA was available at a flat rate of 2% per annum over a 50-year period, but this increased to 3% per annum over a 33⅓ year period from April 2020. This relates to expenditure incurred on new commercial structures and buildings, including costs for new conversions and renovations, applicable where contracts, for the main construction works, were entered into on, or after 29 October 2018. A temporary 10% rate (over 10 years) applies for freeports between 1 April 2021 and 30 September 2026, as noted earlier.

The SBA extends to qualifying expenditure incurred under capital contributions – for example, a landlord contributing towards a tenant's fitting-out costs.

The relief will be available when the building, or structure, first comes into qualifying use and is available for both UK and overseas assets, provided the business is within the charge to UK tax. This relief is not applicable for expenditure on assets used for residential purposes but does include nursing and care homes.

The SBA expenditure will not qualify for the Annual Investment Allowance (AIA).

The relief will cease to be available if the building, or structure, is brought into residential use, or if it is demolished. However, SBA continues to be available for periods of temporary disuse.

If the building, or structure, is sold, the new owner would benefit from the unutilized residue of the SBA for the remaining period, via an Allowances Statement in accordance with s170IA CAA2001.

The costs of construction will include only the net direct costs related to the construction expenditure of the building, or structure, after any discounts, refunds, or other adjustments. This will include demolition costs, or any land alterations necessary for construction and direct costs required to bring the building, or structure, into existence. Excluded from the above expenditure are indirect costs not associated with the physical construction, such as finance, planning, legal and marketing costs, landscaping etc.

SBA qualifying expenditure will also be applicable to purchases of new and second-hand buildings, or structures. The basis of claim will be dependent upon the vendor's holding structure and whether the property has been used, or unused, at the time of acquisition, as well as being entitled to claim this tax relief.

Where a building, or structure, is acquired and subsequently altered, or renovated, this will trigger a new SBA qualification period. Similar additional expenditure on existing properties will also trigger new SBA streams of qualifying expenditure.

It is important to note that expenditure on plant and machinery does not qualify for SBA. A purchaser will be unable to reclassify expenditure on plant and machinery previously claimed as SBA to plant and machinery at a later date.

Expenditure on qualifying land remediation will also not qualify for SBA.

Annual Investment Allowance

The AIA is available to all businesses of any size and allows a deduction for the whole AIA of qualifying expenditure on plant and machinery, including integral features and long life assets. The AIA rates have fluctuated over the years:

- 1 April 2014 to 31 December 2015 – £500,000
- 1 January 2016 to 31 December 2018 – £200,000
- Currently, the AIA rate from 1 January 2019 to 31 March 2023 is £1,000,000.

For accounting periods of less or greater than 12 months, or if claiming in periods where the rates have changed, time apportionment rules will apply to calculate hybrid rates applicable to the period of claim.

Enterprise Zones

Enterprise zones benefit from a number of reliefs, including a 100% first year allowance for new and unused non-leased plant and machinery assets, where there is a focus on high-value manufacturing.

Freeports

Following a government bid process launched late in 2020, the first eight freeports in England were confirmed as follows:

- East Midlands Airport
- Felixstowe and Harwich
- Humber
- Liverpool City
- Plymouth and South Devon
- Solent
- Teesside
- Thames

Further sites are expected to follow including those in devolved administrations in Scotland, Wales and Northern Ireland.

Dedicated 'tax sites' in freeports will benefit from generous tax allowances to 30 September 2026 including:

- An enhanced 10% rate of SBA, meaning that the relief can be claimed over 10 years, rather than the current 33⅓ years;
- An enhanced capital allowance of 100% for plant and machinery allowances, providing a full deduction, rather than the current rates of 18% (MP) and 6% (SRP) on a reducing balance basis. The allowance is only available to occupiers and not property owners or developers leasing premises;
- Full relief from Stamp Duty Land Tax (SDLT) on land and property purchased for a qualifying commercial purpose;
- Full Business Rates relief for new businesses and existing businesses that expand operations;
- Relief against employer National Insurance contributions, subject to Parliamentary process and approval.

Other Capital Allowances

Other types of allowances include those available for capital expenditure on Mineral Extraction, Research and Development, Know-How, Patents, Dredging and Assured Tenancy.

International Tax Depreciation

The UK is not the only tax regime offering investors, owners and occupiers valuable incentives to invest in plant and machinery and environmentally friendly equipment. Ireland, Australia, Malaysia, Hong Kong and Singapore also have capital allowances regimes which are broadly similar to the UK and provide comparable levels of tax relief to businesses.

Many other overseas countries have tax depreciation regimes based on accounting treatment, instead of capital allowances. Some use a systematic basis over the useful life of the asset and others have prescribed methods spreading the cost over a statutory period, not always equating to the asset's useful life. Some regimes have prescribed statutory rates, whilst others have rates which have become acceptable to the tax authorities through practice.

Contact

For further information or assistance, please contact one of the Fiscal Incentives team:

help.capitalallowances@aecom.com

Data Centre Cooling

Introduction

Cooling of data centres has evolved from cooling small clusters of servers to giant server farms. While these modern large data centres are vital components of the services economy, they consume a formidable amount of energy worldwide. It has been widely documented that the cooling systems – the chiller, humidifier and CRAC units – account for 45% of the total energy consumption of a data centre, while the IT equipment accounts for 30%. This means that 1 kWh consumed by the IT equipment requires another 1 kWh of energy to drive the cooling and auxiliary systems.

From environmental and cost-efficiency perspectives, selecting a cooling method that can reduce this energy demand is clearly beneficial. Traditionally, data centres have been air cooled with chillers, humidifiers and CRAC units, in a variety of 'cold aisle/hot aisle' approaches that aren't terribly efficient and that can result in hot spots within the data hall.

Over the years, server equipment has become more resilient and can now tolerate a greater range in temperature and humidity levels than older technology allowed. These days, legitimate alternatives to cooling are routinely considered in most data centre projects looking for greener and more efficient strategies. Currently, in the northern hemisphere, air-cooling solutions are looking more towards free cooling, which works by using air from outside combined with reclaimed heat (winter) and evaporative cooling (summer) to provide the total cooling solution throughout the year.

Water-based cooling options are a more modern approach and cool the inside of the servers by pumping cold water through pipes or plates. Water-cooled rack systems work well but have an inherent risk of leaks. Understanding the cost drivers and benefits of each are crucial to advising clients effectively.

There are many factors that drive the selection of any option, not least capital and life-cycle costs, but also the location of the data centre and the feasibility of incorporating innovations such as free cooling or aquifer thermal energy storage (ATES). Parameters that drive these decisions include the requirement for power usage efficiency (PUE) levels to hit planning stipulations, and for acoustics and total cost of ownership (TCO) levels to be optimized. PUE measures how efficiently the data centre uses input power – the larger the number the less efficient the solution. In addition, the selected method of supplying power and power resilience play a role in the PUE calculations – and, hence the overall cooling and power solution combined is what is considered against the criteria to finalize the preferred solution for the client.

In this article, we are looking solely at the merits of three cooling solutions that are currently being used on projects, to ascertain the cost drivers of each and understand the cooling-only related costs. These are:

- Air cooling by chiller and CRAC units
- Air cooling by indirect air cooling (IAC) air handling plant
- Chilled-water cooling derived from free-cooling, hybrid cooling towers with chiller assist.

Any power-supply solution, associated building works and main contractor prelims are excluded.

Air cooling by chiller and CRAC units

Air cooling by IAC air handling plant

Chilled-water cooling derived from free cooling hybrid cooling towers with chiller assist

Air cooling by chiller and CRAC units

This chilled-water solution serves CRAC downflow units typically serving cold air to the data hall white space through a floor void. CRAC units normally include humidification elements to control the static within the air and all hot air is redirected back into the CRAC to remove the heat for redistribution into the white space. The source of the cooling water is via a traditional refrigeration chiller located externally, usually on the roof. There is no free cooling and chillers are sized for full peak load.

Air cooling by IAC air handling plant

This 'all air' based cooling solution incorporates an air handling plant mounted externally to the white space. Treated air is distributed to the white space via ductwork or through a plenum. Supply air is relatively low velocity to the cold aisle, giving more control than traditional floor void distribution.

The hot air is returned to the IAC via ductwork and is cooled by the outdoor ambient air at a plate heat exchanger. To assist the cooling process during warm months, the ambient air is adiabatically cooled (water evaporation), which then cools the warm air at the plate heat exchanger in the IAC unit. The water used for adiabatic cooling is bulk stored in the event of a mains supply outage. The process water is distributed from a central pump plantroom to the IAC units.

Chilled-water cooling derived from free cooling, hybrid cooling towers with chiller assist

This chilled-water solution serves CRAC downflow units typically supplying cold air to the white space through a floor void. The source of the cooling water is via 'free cooling' cooling towers located externally, usually on the roof. Ambient air is used to cool the warm return water from the CRAC units, with adiabatic cooling added during the warmer months. At peak times, when approaching the towers cooling-load limits, refrigeration chillers are used to run in parallel with the cooling towers.

Table 1 is a summary of the pros and cons of each system. Bear in mind that numerous factors work in tandem within any given solution; for example, net to gross area, efficiency, power load, capital expenditure cost, and TCO combine to determine the best solution. Always make sure defined parameters are set to allow measurement of any solution against these critical factors. This will ensure the best fit solution can be determined.

Table 1: Benefits and drawbacks of the three data centre cooling systems

* All based on 1500 W/m² IT Load Density requirement; Tier III certification

	1 Traditional	2 Indirect Air Cooling	3 Cool Water with Chiller Assist
Method	Air cooling with Chiller and CRAC units	Air cooling	CHW cooling with free cooling
PUE range (Cooling only)	1.4 to 2.0	1.1 to 1.3	1.1 to 1.3
Pros	• Tried and tested • Simple controls methodology	• Proven technology • No risk of leaks in white space • Limited maintenance needed • Cost-effective for low PUE • Better control of supply air distribution	• Flexibility for distribution • Higher achieved white space ratio • Simple controls methodology
Cons	• Hot spots, inefficient • Highest PUE range • 24/7 operation required • No free cooling	• Acoustic challenges • Impact on net white space for ducts • Reduced flexibility in distribution • Space for bulk water storage required • IAC plant to be located close to DH	• Large plant space required • Increased risk of water leaks • Higher PUE • Great bulk water storage required • Higher maintenance costs

Most data centres use air- or water-based cooling solutions, and this is where our cost comparison has focused. The future is already in place, however, with some clients opting for immersion cooling, by which servers are immersed into a liquid coolant for direct cooling of the electronic components. Immersing servers has been shown to improve rack density, cooling capacity and other design-critical factors. Test projects where data centres are in the sea could result in some significant changes in this industry in the future. We are carrying out advisory work with Atlantis, which proposes to build a data centre on the site of its tidal-energy centre, off the coast of Scotland. It demonstrates how high-power-demand data centres could help fund the emerging tidal power sector, thereby contributing to the future decarbonization of the data centre sector.

Cost Model: Cost Comparison of Cooling Methods

Net Technical Space (m²):	3,000	(4 nr Data Halls)	
Load Density (W/m²):	1,500		
IT Load (kW):	4,500		

Cost/kW of Cooling Installation	CRAC Units and Chiller	Indirect Air Cooling	CRAC Units and Hybrid Cooler with Chiller Assist
	(£/kW)	(£/kW)	(£/kW)
CRAC Units (N+2)	250		250
Air cooled chillers (N+1) including CHW pipework distribution, pumps and other associated plant	940		
IAC Units (N+2)		810	
Process water installation to IAC units including water storage tanks, pumps, water treatment. Supply and return ductwork to IAC units including attenuation.		240	
Hybrid cooler (N+1), water cooled chillers (N+1), including CHW pipework distribution, process water storage tank and pipework distribution, pumps and others associated plant			1200
BMS controls and power supplies to above mechanical plant	350	300	380
Total £/kW of Cooling Installation	**1540**	**1350**	**1830**

Notes:
– Hot/cold aisle containment is excluded
– Main Contractor Prelims and OHP are excluded
– Building/structural/architectural works, dedicated fresh air systems, and electrical infrastructure are excluded

Exposed Services

Background

'Traditional' Category A space has always been offered by developers within the market as designs often required minimal investment and the basic provisions has enabled them to target a wide range of potential tenants – each with varying needs. However, this norm has begun to be seen by many as inflexible and outdated.

The evolution of the office has been driven by reactions to the behavioural changes in the way people work based on the needs of a growing millennial dominated workforce and a shift towards different workplace preferences, for example, activity based working (ABW). Offices are becoming more collaborative, open and place greater emphasis on health and wellbeing. Research has shown that spaces with high ceilings positively promotes the wellbeing of their occupants, even if there is limited natural light as it triggers a psychological sense of freedom. The spaces within now need to be multi-functional, adaptable to changes in density and population whilst also maintaining a sense or privacy when required.

Fuelled by the growth of companies such as Google, Twitter, Amazon, all of whom champion the exposed services, prospective tenants are now placing greater emphasis on their office space. They look to use the space to attract the best talent joining the industry and to differentiate themselves from their competition. Couple this with the explosion in companies offering co-working space, such as WeWork and Fora (which represents around 4% of total office floor area within London and this is rising…). Co-working helps resolve the disconnect between a start-up's initial business plan and the 'crippling' commercial property leases which often start at five years. Start-ups and smaller businesses simply cannot justify the costs and risks associated with a lease so the need for flexible short-term space is an easy decision.

As a result, developers have had to begin to think outside the box in order to differentiate themselves in the market in order to attract new tenants. Whilst exposed services are not a new thing in CAT B it is still in its infancy in speculative CAT A developments. Prospective tenants are also looking to get as much value out of their spaces and it would be incredibly foolish to let a space where the majority if not all of the high level services (including ceilings etc.) are required to be stripped out. Developers have warmed to this and are offering 'exposed' CAT A spaces designed with flexibility and future proofing at the forefront.

Cost Drivers

The largest misconception individuals have about 'exposed services' is that they are less expensive than installing a traditional CAT A utilizing a suspended ceiling as straight away you are removing the costs of a ceiling grid or tiles. It is important to realize that creating this aesthetic is not as straightforward (if done correctly) and involves a lot of additional work to achieve the perfect 'exposed' look. Traditionally, the ceiling grid and tiles were there to hide the nastiness above, ductwork, pipework and cabling, i.e. the services! As a result the following items will need to be considered, all of which represent additional cost.

- **Enhanced treatment to the ductwork**, this could be achieved by a metal wrapping over traditional lagging to provide a clean finish. Alternatively, additional time and labour would improve the final installation.
- **Installation of suspended or track mounted lighting,** including pendants in lieu of ceiling mounted lighting tiles and downlights – fittings are generally more expensive, take longer to install and require additional design coordination.
- **Painting of exposed ceiling and services in a uniform colour**, which is a far more complicated job than simply painting walls. This could also later tie in with a company's overall colour scheme and branding, so in reality the 'casual' exposed look is actually the result of a lot of skilled labour.

- **Increased on-site design coordination and labour intensive tasks**, including the careful and considered routing of electrical conduit to create a uniform and consistent finish and appearance.

Another invisible but inevitable cost consideration of exposed services is acoustics. Ceiling tiles are effective at absorbing sound and keep ambient noise levels from being disruptive. The hard surfaces associated with exposed ceilings, e.g. concrete slabs amplify sounds and easily distract occupants who are trying to concentrate. This does not just include the sounds of people talking; it also includes ambient noise created by things like the HVAC system. Therefore it is important to consider the additional acoustical control options necessary to minimize noise, such as suspended baffles, acoustic panels, and spray-on solutions. The added cost for this varies depending on the options you choose and the level of design coordination required (for lighting, fire alarms, sprinklers etc.). White noise systems can be installed to help; however, some people find the white noise itself to be disruptive and annoying.

Exposed services also come with other hidden costs besides construction costs, most notably, energy costs. By removing a suspended ceiling will mean that the HVAC system will have an increased volume and area of space to heat and cool. Research carried out by the Ceilings & Interiors Systems Association (CISCA) found that offices with suspended ceilings achieved 9–10.3% greater energy savings compared to those with exposed ceilings and services. In addition to this the research found that exposed ceilings require more frequent cleaning, stricter cleaning regimes and periodic repainting that is not necessary with suspended ceilings.

Cost Model

The cost model is based on a central London office development. The model is for the Category A fit-out to one floor occupied by a single tenant, with a net lettable area of 1,800 m². It is purely for mechanical and electrical installations and takes no account of changes to building elements. The exposed services option does not require a suspended ceiling, which would give a saving of about £6/square feet (NIA).

This is only part of the picture, however, with items required to be added back into the exposed services costs. For example, any exposed concrete surface would need dust sealing/painting at £1/square feet NIA. Acoustic panelling would be required to walls, suspended panels and shrouds above FCUs, at a cost of around £3.75–£5/square feet NIA.

Category A Works	Traditional CAT A		
	Based on a NIA 1,800 m²		
	Total (£)	£/m² on NIA	£/ft² on NIA
5.6 Space Heating and Air Treatment Installation of a four pipe fan coil units to the perimeter zones and two pipe fan coil units to the internal zones. Including CHW and LTHW distribution pipework, ductwork distribution, plenums, diffusers, grilles and insulation. Installation of condensate drainage to fan coil units. Includes for testing and commissioning and subcontractor preliminaries.	£459,000	255.00 m²	23.69 ft²
5.8 Electrical Installations Installation of one tenant distribution board, LED lighting including controls based on a modular wiring system. Power to the mechanical services and general earthing and bonding. Includes for testing and commissioning and subcontractor preliminaries.	£358,000	198.89 m²	18.48 ft²

Category A Works	Traditional CAT A		
	Based on a NIA 1,800 m²		
	Total (£)	£/m² on NIA	£/ft² on NIA
5.11 Fire and Lightning Protection			
Sprinkler protection to floorplate. Includes for testing and commissioning and subcontractor pre-liminaries.	£61,000	33.89 m²	3.15 ft²
5.12 Communication, Security and Control Systems			
Fire detection system and BMS to fan coil units. Includes for testing and commissioning and subcontractor pre-liminaries.	£95,000	52.78 m²	4.90 ft²
Traditional Category A MEP Total Costs	**£973,000**	**540.56 m²**	**50.22 ft²**

Category A Works	Exposed Services CAT A		
	Based on a NIA 1,800 m²		
	Total (£)	£/m² on NIA	£/ft² on NIA
5.6 Space Heating and Air Treatment			
Installation of a four pipe fan coil units to the perimeter zones and two pipe fan coil units to the internal zones. Including CHW and LTHW distribution pipework, ductwork distribution, plenums, diffusers, grilles and insulation. Installation of condensate drainage to fan coil units. Includes for testing and commissioning and subcontractor preliminaries.			
Extra over for exposed services; including enhanced finishes to ductwork via metal cladding and enhanced grilles/diffusers.	£550,000	305.56 m²	28.39 ft²
5.8 Electrical Installations			
Installation of one tenant distribution board, LED lighting including controls based on a modular wiring system. Power to the mechanical services and general earthing and bonding. Includes for testing and commissioning and subcontractor preliminaries.			
Extra over for exposed services; including suspended fittings in lieu of ceiling mounted and increased co-ordination of wiring to minimize the visual impact.	£448,000	248.89 m²	23.12 ft²
5.11 Fire and Lightning Protection			
Sprinkler protection to floorplate. Includes for testing and commissioning and subcontractor preliminaries.	£61,000	33.89 m²	3.15 ft²

	Exposed Services CAT A		
Category A Works	Based on a NIA 1,800 m²		
	Total (£)	£/m² on NIA	£/ft² on NIA
5.12 Communication, Security and Control Systems			
Fire detection system and BMS to fan coil units. Includes for testing and commissioning and subcontractor preliminaries.			
Extra over for exposed services; increased coordination of wiring to minimize the visual impact.	£101,000	56.11 m²	5.21 ft²
Exposed Services Category A MEP Total Costs	**£1,160,000**	**644.44 m²**	**59.87 ft²**

Groundwater Cooling

The use of groundwater cooling systems considering the technical and cost implications of this renewable energy technology

The application of groundwater cooling systems is quickly becoming an established technology in the UK with numerous installations having been completed for a wide range of building types, both new build and existing (refurbished).

Buildings in the UK are significant users of energy, accounting for 60% of UK carbon emissions in relation to their construction and occupation. The drivers for considering renewable technologies such as groundwater cooling are well documented and can briefly be summarized as follows:

- **The UK Climate Change Act (2008)** established a legally binding target for the UK to cut CO_2 emissions by 80% by 2050 (against a 1990 baseline), with interim targets of 34% by 2020 and 60% by 2030. It is this primary legislation that is driving national, regional and local policy on climate change. These requirements were further enhanced on 27 June 2019, when the UK committed to a legally binding target of net-zero emissions by 2050.
- **Approved Document Part L2 2021 (England)** came into force on 15 June 2022. Similar regulations apply in Wales, Scotland and Northern Ireland as well as in EU member states. Part L 2021 for England presents an average 27% reduction in carbon emissions compared to the previous Building Regulations Part L 2013.
- In January 2021, MHCLG (now DLUHC) published the **Future Buildings Standard** Consultation, this builds on the Future Homes Standard setting out energy and ventilation standards for non-domestic buildings. The Future Buildings Standard will set out the UK governments ambition for non-domestic buildings to be implemented in 2025. It is designed to transition non-domestic buildings to low-carbon heat sources for space heating and hot water, in turn meaning that new buildings using this standard will become zero carbon over time as the national electricity grid and heat networks decarbonize.
- **The UK Heat Strategy (2013)** set out pathways for the transition to a low carbon heat supply. It set out the commitments of former Department of Energy and Climate Change (DECC), now replaced by the Department for Business Energy and Industrial Strategy (BEIS), to support local authorities in the development of heat networks in their areas through the establishment of the Heat Networks Delivery Unit (HNDU), support for technological innovation, provision of funding for feasibility work, exploration of potential additional financial incentives and government funding for heat networks, and provision of a consumer protection scheme.
- The **CIBSE Heat Networks: Code of Practice** for the UK was produced in July 2015 in collaboration between the Association for Decentralized Energy (ADE), CIBSE and industry partners including AECOM, to establish common standards for the development of district heating. Its aim was to establish minimum standards to ensure the conception, design, build and operation of efficient, cost-effective district heating for all users. While the guidance in the CoP is not legally binding, HNDU and many local authorities are using it as a reference standard in design, procurement and implementation of heat networks in their areas, and it is frequently referenced in tender documentation.
- The **Clean Growth Strategy (2017)** sets out proposals for decarbonizing all sectors of the UK economy through the 2020s. It presents how the UK can benefit from low carbon opportunities while meeting national and international commitments to tackle climate change. Part of this plan is to roll out low carbon heating by building and extending heat networks across the country, underpinned with public funding to 2021. To meet these overall targets, the UK is committed to meeting 15% of its energy demand from renewable sources by 2020.
- The **National Planning Policy Framework (NPPF)** sets out the Government's planning policies for England and stipulates how they should be applied through local planning policies. Local planning authorities are required to design policies which increase the use and supply of low carbon energy, have a positive

strategy to promote energy from renewable and low carbon sources, support community-led initiatives for low carbon energy and identify suitable areas for low carbon energy sources.

- **National Planning Practice Guidance** sets out that Local Plans should promote low carbon design approaches to reduce energy consumption. This includes approaches such as passive solar design.
- Company policies of building developers and end users often require developments to minimize detrimental impact to the environment.

The Ground as a Heat Source/Sink

The thermal capacity of the ground can provide an efficient means of tempering the internal climate of buildings. Whereas the annual swing in mean air temperature in the UK is around 20 K, the temperature of the ground is far more stable. At the modest depth of 2 m, the swing in temperature reduces to 8 K, while at a depth of 50 m the temperature of the ground is stable at 11–13°C. This stability and ambient temperature therefore makes groundwater a useful source of renewable energy for heating and cooling systems in buildings.

Furthermore, former industrial cities like Nottingham, Birmingham, Liverpool and London have a particular problem with rising groundwater levels due to a reduction in groundwater abstraction for use in manufacturing, particularly since the 1970s. The use of groundwater for cooling is therefore encouraged by the Environment Agency in areas with rising groundwater levels as a means of combating this problem.

System Types

Ground and groundwater cooling systems may be defined as either open or closed loop systems.

Open loop systems

Open loop systems generally involve the direct abstraction and use of groundwater, typically from aquifers (porous water bearing rock), although there also are systems which utilize water in former underground mineral workings. Water is abstracted via one or more boreholes and passed through a heat exchanger and then is returned to the aquifer via a separate borehole or boreholes, discharged to foul water drainage or released into a suitable available source such as a river. Typical groundwater supply temperatures are in the range 8–12°C and typical reinjection temperatures for a cooling mode operation is 12–18°C.

The properties of the aquifer are important in ensuring the necessary flow rates to achieve the energy requirements. Whilst reasonable flow rates normally are obtained from the major aquifers, such as the Chalk and the Sherwood Sandstones, there is a small risk that boreholes in these strata will not provide the required yield. In addition, reinjection of the abstracted water tends to be more difficult than abstraction and it often is necessary to install more than one recharge borehole to accept the discharge from a single abstraction borehole. This has obvious cost implications. This issue is often site-specific and an assessment of the hydrogeological conditions, in particular the depth to groundwater, should be obtained before progressing with the scheme.

All open loop abstractions from groundwater require an abstraction licence. If the water is being returned to the aquifer, it is designated as non-consumptive use. The system also requires an environmental permit to discharge the water back into the aquifer. Where it is a simple abstraction, it is regarded as a consumptive use and the abstraction charges are higher. In order to protect groundwater from the discharges from open loop GSHC systems, the Environment Agency restrict the change in temperature between the abstracted and discharged waters to no greater than +/- 8°C and the outflow water temperature must not exceed 25°C.

The efficiency of an open loop system can deteriorate if a large proportion of the reinjected water is contained within the water pumped from the abstraction borehole. This can occur if the abstraction and recharge boreholes are positioned too close to each other, allowing recirculation of the recharged water. This is not an uncommon event and in many circumstances cannot be avoided due to the size of the development. In order to minimize this effect, the abstraction and recharge boreholes should be separated by as large a distance as possible.

Open loop systems fed by groundwater at 8°C, can typically cool water to 12°C on the secondary side of the heat exchanger to serve conventional cooling systems.

Open loop systems are thermally efficient but over time can suffer from blockages caused by silt, and corrosion due to dissolved salts. As a result, additional cost may be incurred in having to provide filtration or water treatment, before the water can be used in the building.

There is also a risk that minerals can precipitate out of the groundwater, especially where the water is iron and/or manganese rich and there is either and/or a change in pressure or exposure of the water to air. This could require cleaning of the pipework and the borehole lining or some form of filtration as indicated above. Several water supply boreholes suffer from this problem, which can cause a reduction in borehole yield.

An abstraction licence and environmental permit/discharge consent need to be obtained for each installation, and this together with the maintenance and durability issues can significantly affect whole life operating costs, making this system less attractive.

A final point to consider is that access by plant will be required to the boreholes for pump replacement, maintenance and hence the boreholes should be located where access for plant is easily achievable.

Closed loop systems

Closed loop systems do not rely on the direct abstraction of water, but instead comprise a continuous pipework loop buried in the ground. Water circulates in the pipework and provides the means of heat transfer with the ground. Since groundwater is not being directly used, closed loop systems therefore suffer fewer of the operational problems of open loop systems, being designed to be virtually maintenance free, but do not contribute to the control of groundwater levels.

There are two types of closed loop system:

Vertical Boreholes – Vertical loops are inserted as U tubes into pre-drilled boreholes, typically less than 150 mm in diameter. These are backfilled with a high conductivity grout to seal the bore, prevent any cross contamination and to ensure good thermal conductivity between the pipe wall and surrounding ground. Vertical boreholes have the highest performance and means of heat rejection, but also have the highest cost due to associated drilling and excavation requirements.

Alterative systems to the U tubes are available on the market with larger diameter propriety pipework systems with flow and return pipes. The larger pipework can have better heat transfer properties, reduced resistance and therefore achieve better efficiencies or less pipework and therefore less ground area requirement than a traditional U tube system which may outweigh the additional cost per metre length of the pipework.

As an alternative to having a separate borehole housing the pipe loop, it can also be integrated with the piling, where the loop is encased within the structural piles. This obviously saves on the costs of drilling and excavation since these would be carried out as part of the piling installation. The feasibility of this option would depend on marrying up the piling layout with the load requirement, and hence the number of loops, for the building.

Horizontal Loops – These are single (or pairs) of pipes laid in 2 m deep trenches, which are backfilled with fine aggregate. These obviously require a greater physical area than vertical loops but are cheaper to install. As they are located closer to the surface where ground temperatures are less stable, efficiency is lower compared to open systems. Alternatively, coiled pipework can also be used where excavation is more straightforward and a large amount of land is available. Although performance may be reduced with this system as the pipe overlaps itself, it does represent a cost-effective way of maximizing the length of pipe installed and hence overall system capacity.

The Case for Heat Pumps

Instead of using the groundwater source directly in the building, referred to as passive cooling, when coupled to a reverse cycle heat pump, substantially increased cooling loads can be achieved.

Heat is extracted from the building and transferred by the heat pump into the water circulating through the loop. As it circulates, it gives up heat to the cooler earth, with the cooler water returning to the heat pump to pick up more heat. In heating mode the cycle is reversed, with the heat being extracted from the earth and being delivered to the HVAC system.

The use of heat pumps provides greater flexibility for heating and cooling applications within the building than passive systems. Ground source heat pumps are inherently more efficient than air source heat pumps, their energy requirement is therefore lower and their associated CO_2 emissions are also reduced, so they are well suited for connection to a groundwater source.

Closed loop systems can typically achieve outputs of 50 W/m (of bore length), although this will vary with geology and borehole construction. When coupled to a reverse cycle heat pump, 1 m of vertical borehole will typically deliver 140 kWh of useful heating and 110 kWh of cooling per annum, although this will depend on hours run and length of heating and cooling seasons.

Key factors Affecting Cost

- The cost is obviously dependent on the type of system used. Deciding on what system is best suited to a particular project is dependent on the geological conditions, the peak cooling and heating loads of the building and its likely load profile. This in turn determines the performance required from the ground loop, in terms of area of coverage in the case of the horizontal looped system, and in the case of vertical boreholes, the depth and number or bores. The cost of the system is therefore a function of the building load.

- In the case of vertical boreholes for a closed loop system, drilling costs are a significant factor, as ground conditions can be variable, and there are potential problems in drilling through sand layers, pebble beds, gravels and clay, which may mean additional costs through having to drill additional holes or the provision of sleeving. The costs of drilling make the vertical borehole solution significantly more expensive than the equivalent horizontal loop.

- The thermal efficiency of the building is also a factor. The higher load associated with a thermally inefficient building obviously results in the requirement for a greater number of boreholes or greater area of horizontal loop coverage, however in the case of boreholes the associated cost differential between a thermally inefficient building and a thermally efficient one is substantially greater than the equivalent increase in the cost of conventional plant. Reducing the energy consumption of the building is cheaper than producing the energy from renewables and the use of renewable energy only becomes cost-effective, and indeed should only be considered, when a building is energy efficient.

- With open loop systems, the principal risk in terms of operation is that the user is not in control of the quantity or quality of the water being taken out of the ground, this being dependent on the local ground and groundwater conditions. Reduced performance due to blockage (silting, etc.) may lead to the system not delivering the design duties whilst bacteriological contamination may lead to the expensive water treatment or the system being taken temporarily out of operation. In order to mitigate the above risk, it may be decided to provide additional means of heat rejection and heating by mechanical means as a back-up to the borehole system, in the event of operational problems. This obviously carries a significant cost. If this additional plant were not provided, then there are space savings to be had over conventional systems due to the absence of heating, heat rejection and possibly refrigeration plant.

- With the design of open loop systems close analysis can be undertaken of existing boreholes in the area and also detailed desktop studies can be undertaken by a Hydrogeologist to determine the feasibility of the scheme and predict the yield for the proposed borehole. As noted above the yield of the borehole is subject to the local hydrogeological conditions in the location where the borehole is located and as such the delivered yield will only be validated after the borehole is completed and tested. There is a risk that you can get less (you could get more) than was predicted requiring additional boreholes (and cost) to be delivered to obtain the required yield.

- Open loop systems may lend themselves particularly well to certain applications increasing their cost effectiveness, i.e. in the case of a leisure centre, the removal of heat from the air-conditioned parts of the centre and the supply of fresh water to the swimming pool.

- In terms of the requirements for abstraction and disposal of the water for open loop systems, there are risks associated with the future availability and cost of the necessary licences; particularly in areas of high forecast energy consumption, such as the South East of England, which needs to be borne in mind when selecting a suitable system. Provided that the system is designed to be non-consumptive, a licence restriction is unlikely.

- Closed loop systems and non-consumptive open loop systems generally require balanced heating and cooling loads (if there is moving groundwater this is not as critical) in order to not cause the ground to heat up or cool down over time. If the heating and cooling loads (kWh) over the year are not balanced this can cause the ground to heat up or cool down over time causing a reduction in the efficiency of the system over time which will have a detrimental effect to the seasonal COPs and operational running costs of the system over time.

- Heat pump systems use electricity to power them and winter triad electrical costs should be considered in lifecycle analysis for commercial and industrial users and appropriate controls storage systems put in place to mitigate costs.

Whilst open loop systems would suit certain applications or end user clients, for commercial buildings the risks associated with this system tend to mean that closed loop applications are the system of choice. When coupled to a reversible heat pump, the borehole acts simply as a heat sink or heat source so the problems associated with open loop systems do not arise.

Typical Costs

Table 1 gives details of the typical borehole cost to an existing site in Central London, using one 140 m deep borehole working on the open loop principle, providing heat rejection for the 600 kW of cooling provided to the building. The borehole passes through rubble, river gravel terraces, clay and finally the Chalk, and is lined above the Chalk to prevent the hole collapsing. The breakdown includes all costs associated with the provision of a working borehole up to the well head, including the manhole chamber and manhole. The costs of any plant or equipment to deliver the water from the borehole are not included.

Heat is drawn out of the cooling circuit and the water is discharged into the Thames at an elevated temperature. In this instance, although the boreholes are more expensive than the dry air cooler alternative, the operating cost is significantly reduced as the system can operate at around three times the efficiency of conventional dry air coolers, so the payback period is a reasonable one. Additionally, the borehole system does not generate any noise, does not require rooftop space and does not require as much maintenance.

This is representative of a typical cost of providing a borehole for an open loop scheme within the London basin. There are obviously economies of scale to be had in drilling more than one well at the same time, with two wells saving approximately 10% of the comparative cost of two separate wells and four wells typically saving 15%.

Table 2 provides a summary of the typical range of costs that could be expected for the different types of system based on current prices.

Table 1: Breakdown of the Cost of a Typical Open Loop Borehole System

Description	Cost £
General Items	
• Mobilization, insurances, demobilization on completion	21,000
• Fencing around working area for the duration of drilling and testing	2,000
• Modifications to existing LV panel and installation of new power supplies for borehole installation	15,000
Trial Hole	
• Allowance for breakout access to nearest walkway (Existing borehole on site used for trial purposes, hence no drilling costs included)	3,000
Construct Borehole	
• Drilling, using temporary casing where required, permanent casing and grouting	32,000
Borehole Cap and Chamber	
• Cap borehole with PN16 flange, construct manhole chamber in roadway, rising main, header pipework, valves, flow meter	12,000
• Permanent pump	14,000
Samples	
• Water samples	1,000
Acidization	
• Mobilization, set up and removal of equipment for acidization of borehole, carry out acidization	12,000
Development and Test Pumping	
• Mobilize pumping equipment and materials and remove on completion of testing	4,000
• Calibration test, pretest monitoring, step testing	4,000
• Constant rate testing and monitoring	20,000
• Waste removal and disposal	3,000
Reinstatement	
• Reinstatement and making good	2,000
Total	**145,000**

Table 2: Summary of the Range of Costs for Different Systems

System	Range Small – 4 kWth	Medium – 50 kWth	Large – 400 kWth	Notes
Heat pump (per unit)	£3,000–5,000	£30,000–42,000	£145,000–175,000	
Slinky pipe (per installation) including excavation	£3,000–4,000	£42,000–52,000	£360,000–390,000 [1]	[1] Based on 90 nr 50 m lengths
Vertical, closed (per installation) using structural piles	N/A	£42,000–63,000	N/A	Based on 50 nr piles. Includes borehole cap and header pipework but excludes connection to pump room and heat pumps
Vertical, closed (per installation) including excavation	£2,000–3,000	£63,000–85,000	£370,000–400,000	Includes borehole cap and header pipework but excludes connection to pump room and heat pumps
Vertical, open (per installation) including excavation	£2,000–3,000	£45,000–65,000	£335,000–370,000	Excludes connection to pump room and heat exchangers

Construction 4.0:
An Innovation Platform for
the Built Environment

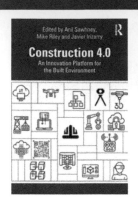

Edited By Anil Sawhney, Michael Riley and Javier Irizarry

Modelled on the concept of Industry 4.0, the idea of Construction 4.0 is based on a confluence of trends and technologies that promise to reshape the way built environment assets are designed, constructed, and operated.

With the pervasive use of Building Information Modelling (BIM), lean principles, digital technologies, and offsite construction, the industry is at the cusp of this transformation. The critical challenge is the fragmented state of teaching, research, and professional practice in the built environment sector. This handbook aims to overcome this fragmentation by describing Construction 4.0 in the context of its current state, emerging trends and technologies, and the people and process issues that surround the coming transformation. Construction 4.0 is a framework that is a confluence and convergence of the following broad themes discussed in this book:

- Industrial production (prefabrication, 3D printing and assembly, offsite manufacture)
- Cyber-physical systems (actuators, sensors, IoT, robots, cobots, drones)
- Digital and computing technologies (BIM, video and laser scanning, AI and cloud computing, big data and data analytics, reality capture, Blockchain, simulation, augmented reality, data standards and interoperability, and vertical and horizontal integration)

The aim of this handbook is to describe the Construction 4.0 framework and consequently highlight the resultant processes and practices that allow us to plan, design, deliver, and operate built environment assets more effectively and efficiently by focusing on the physical-to-digital transformation and then digital-to-physical transformation. This book is essential reading for all built environment and AEC stakeholders who need to get to grips with the technological transformations currently shaping their industry, research, and teaching.

February 2020: 526 pp
ISBN: 9780367027308

To Order
Tel:+44 (0) 1235 400524
Email: tandf@bookpoint.co.uk

For a complete listing of all our titles visit:
www.tandf.co.uk

Large Scale Heat Pump

Introduction

With an increasing drive to demonstrate green credentials from developers and a demand from tenants for greener buildings the use of heat pumps to generate heating and cooling has been increasing not only as part of the heating/cooling installation but completely replacing the traditional approach of boilers and chillers.

The Drive Towards Low-carbon Fuels

There is a drive and demand from governments, developers, tenants and employees to be greener and reduce carbon. Commercial developments for offices and residential buildings are increasingly competing for tenants, sales and investors. One way to achieve the commercial edge is be greener and demonstrate the drive to reduce carbon.

Legislation is driving the interest in the built environment to switch from traditional fuel sources to the use of lower carbon technologies which utilize renewable energy such as solar, geothermal, tidal, water and wind. Alternative heating and cooling systems are being developed to combat the ever growing issue of carbon and as a result large scale heat pumps are increasingly becoming a viable solution.

A Green Grid

Whilst heat pumps demand more electricity than the traditional approach of boilers and chillers, the progressive movement of the UK electrical grid to provide 100% of its energy via renewable sources means that large heat pumps are becoming far more attractive than in the past.

Installing a heat pump that is run on electricity will see its carbon footprint reduced over its working life due to the ongoing decarbonization of the grid. The grid is going green by investing in more renewable sources of energy such as solar and wind.

This is a word of caution with demanding more and more electricity from the grid. There has been a drive in recent years to make office developments taller and increase the density of populations on floor plates to make projects economically viable. From a residential perspective, the growing population is driving the requirement for both low- and high-rise accommodation. In certain locations, electricity is becoming increasingly difficult to deliver at a cost that makes the use of heat pumps viable in terms of capex costs.

Other Benefits

Despite the potential for heat pumps to lead to a higher capital cost than traditional systems, this can be offset during the lifecycle of the equipment due to higher efficiencies. The energy-saving characteristics of heat pumps make them eligible to be included within the ECA (Enhanced Capital Allowance) which means that businesses are able to set 100% of the asset costs against their taxable profits in a single tax year.

There is the potential to tap into the Renewable Heat Incentive – a government environmental programme that provides financial incentives to promote and increase the uptake of heating from renewable sources. It has two schemes (Non-Domestic and Domestic) which have separate tariffs, joining conditions, rules and application processes. Quarterly payments are given over a number of years (depending on the scheme) to eligible installations based on the amount of heat generated.

Systems comparison – Traditional vs Heat Pump

As with any alternative solution to the traditional approach there are numerous and interlinked advantages and disadvantages to be reviewed and taken account of when considering a change in strategy. The primary advantages and disadvantages are listed below

Central Plant (Chiller/Boiler)

Advantages
- More stable heating/cooling source
- Fast heating (boiler)
- Lower initial cost
- High leaving water temperature (boiler)
- Large capacity available
- Traditional technology so known and understood by all parties
- Natural gas is still very cheap (boiler)
- Scalable

Disadvantages
- Less efficient conversion of energy
- High operating cost
- Burns fossil fuel (boiler)

Large Scale Heat Pump (ASHP)

Advantages
- Lower running cost
- Less maintenance
- Safer to use than combustion based system
- Low carbon technology
- Efficient conversion of energy
- Longer lifespan
- No flue or impact on local air quality
- Industry compliant (LEED, ASHRAE)
- Provides both heating and cooling
- Does not contribute adversely to air quality

Disadvantages
- High initial cost
- Noise
- Limited heating output
- Capacity is reduced at low ambient temperature
- Relatively low flow temperature
- Requires careful design/application
- Requires significant space to allow free air flow

Many of the highlighted disadvantages of the heat pump approach can be overcome. The initial high capex cost of the heat pump is often stated as the primary reason for not considering them, however with more units and manufacturers coming to market, the capital cost is reducing. Noise issues can be overcome with careful selection of plant and well-designed attenuation. The issues of limited heat output, reduced capacity at low ambient temperature, and low flow temperatures can be dealt with at early design stages. Heating issues can be dealt with via a small make up boiler and whilst this would introduce a gas driven heat source into a scheme, the makeup boiler is considerably smaller than a traditional approach and is used intimately.

Cost Model

The cost model comparison between an ASHP system and a traditional boiler & chiller central plant is based on a central London office development arranged over 14 floors, with one basement level. The scheme has a GIFA of 15,000 m^2, with an approximate net lettable area of 11,250 m^2 – resulting in a net to gross efficiency of about 75%. The office has been designed to comply with Part L and to achieve a BREEAM Excellent rating. The occupational density (total NIA divided by the total number of workplaces in the building) of the scheme has been assumed to be 1:8.

The comparative costs between the traditional approach to boilers and chillers versus heat pump technology excludes any capital costs associated with upstream reinforcement of the electrical infrastructure as this is site and area specific and could negate any potential decision. However it does take account of additional shell and core infrastructure to support the associated electrical load, such as additional substations, distribution boards and cabling.

	Central Plant				Heat Pump					Cost Variance
	Air Cooled Chiller + Gas Fired Boiler				Air Source Heat Pump (ASHP)					ASHP vs Central Plant
Description	Qty	Unit	Rate	Total	Description	Qty	Unit	Rate	Total	
Heat Source					**Heat Source**					
Boiler	1,170	kW	£56	£65,520	Boiler		kW		not required	(£65,520)
Flue	100	m	£753	£75,300	Flue		m		not required	(£75,300)
Primary pump set / Pressurization units	1	item	£68,265	£68,265	Primary pump set / Pressurization units	1	item	£102,342.00	£102,342	£34,077
Primary distribution	1	item	£102,000	£102,000	Primary distribution	1	item	£102,000.00	£102,000	£0
Space Heating and Air Treatment					**Space Heating and Air Treatment**					
Chiller	1,450	kW	£312	£452,400	Chiller		kW		not required	(£452,400)
					ASHP	1,450	kW	£400.00	£580,000	£580,000
CHW primary and secondary distribution	15,000	m²	£26	£390,000	CHW primary and secondary distribution	15,000	m²	£30.00	£450,000	£60,000
LTHW secondary distribution	15,000	m²	£20	£300,000	LTHW secondary distribution	15,000	m²	£24.00	£360,000	£60,000
Gas Installation					**Gas Installation**					
Gas distribution	15,000	m²	£4	£60,000	Gas distribution				not required	(£60,000)
Electrical Installation					**Electrical Installation**					
Mechanical equipment power supply	1	item	£58,000	£58,000	Mechanical equipment power supply	1	item	£58,000.00	£58,000	£0
Communications, Security and Control Systems					**Communications, Security and Control Systems**					
Controls	1	item	£57,000	£57,000	Controls	1	item	£57,000.00	£57,000	£0
Total				**£1,328,485**	**Total**				**£1,709,342**	**£80,857**

Notes on the above cost:

1. Based on Shell and Core only, excludes all CAT A fit-out costs
2. Base date of Q2 2022, and the prices used are reflective of a project procured through a competitive two-stage tender
3. Unit rates are inclusive of MEP subcontractor testing and commissioning and MEP subcontractor preliminaries
4. Excludes builder's work, main contractor's preliminaries, design reserve, contingencies and inflation allowances

The model compares a 1,450 kW cooling and 1,150 kW heating output from a boiler/chiller combination against the equivalent air source heat pump using 4 pipe water distribution systems.

The aim of the model is to present the costs differences associated with traditional plant compared to an ASHP. It should be noted that the cost variance could vary from project to project depending on but not limited to the following:

a. Project location
b. Project size, shape and configuration
c. Procurement route
d. Programme
e. Design criteria (flow rates, material and performance specifications, etc.)

Later Living

Introduction

This cost model focuses upon a generic luxury later living development in the South of England formed of 50 number apartments comprising 20 one bed units, 25 two bed units and 5 two bed 'close care units' that provide care and facilities for persons with dementia. The type of amenities provided can be but not limited to a restaurant, hair salon, gym, spa, club lounge and parlour and the like, all with 24-hour concierge and on-site nursing staff.

The approach of these style of developments is not to compromise on the level of facilities and location just because the age of the population is higher, in fact there can often be an enhancement in the type of amenities and facilities provided.

Later living developments are a lifestyle concept providing luxury developments to over 65s. This is achieved by offering the best in design, technology, services and care, the goal is to provide on their doorstep everything the later generation need to live life to the fullest.

This article focuses on the high end later living developments located within the South of England.

Innovative Technologies in the Apartments

It is possible within apartments to include a number of elements that are not consider the norm in care apartments or developments. These technologies are often incorporated within the 'close care' apartments.

The apartments are designed to ensure optional functionality incorporating innovative technologies such as 'smart' floors. The smart technology sits under the chosen floor covering within the apartment and has the ability to learn the resident's movements and set parameters so that alarms are not falsely triggered but still offers a high level of monitoring that can detect falls and alert the on-site staff to a resident in difficulties.

There has been an awful lot of research into the benefits of using lighting to assist with dementia patients and their care. Within the 'close care' apartments additional lighting may be incorporated within fixed furniture to establish zones within the apartments, low-level night-time lighting within bathrooms and kitchen and even way finder lighting incorporated into floor finishes. The next step within lighting is the introduction of lighting control to provide circadian lighting. The control of the lighting within rooms and zones is pre-set remotely to provide the optimal level and type of lighting for individual apartments/residents.

The 'close care' units could incorporate aspirating smoke systems within apartments to detect fires at the earliest opportunity, fire suppression systems within cooker hoods that isolate the heat source and discharge a fine water mist that requires limited clean up in the event of discharge. Leak detection and flow monitor devices can be incorporated into bathrooms, kitchens and other wet areas to prevent flooding and accidental overflows from baths and sinks.

Health Care

Health care is provided on site via 24 hour trained medical staff that are provided with sleeping accommodation. Within the later living developments residents can select their desired level of care that best suits their needs and medical requirements. On larger developments it is not uncommon to see a small medical facility that provides a physiotherapist, later living specialist and a doctor, all on a visiting basis.

A limited number of apartments will have nurse call points and medical gases provided to them in order to meet additional care requirements of certain residents.

Concierge Service and Security

The development of this nature will include for in-house concierge service that is manned 24 hours a day aiming to answer the needs of all residents. Dealing with tasks such as making restaurant reservations, organizing dry cleaning or booking a taxi. Security is an important factor for residents. Access control and CCTV within later living developments are at an enhanced level and will often include CCTV within lifts, corridors and stair wells to assist in the tracking and monitoring of residents. The level of cameras and security whilst being enhanced is designed so that it is non-intrusive and discreet.

Amenity

A priority for these developments in not just to provide a place for residents to reside but to provide space to socialize and within later care developments this is critical to both the health and wellbeing of the residents.

The key operational driver is community, therefore developments that encourage interaction and engagement between residents are vital, as it helps tackle loneliness, creates social value and ensures residents can be independent, consequently, numerous amenity spaces are provided. These spaces can be quite large and numerous, thus affecting the efficiency of the development, and as a result their incorporation within the development will come at a cost premium.

The focal point for socialization is through the Drawing rooms which are traditionally based on a library or music room.

High end luxury restaurants are provided within later life developments, furthermore a bespoke wine room is often on site enabling residents to explore and keep fine wine. The level of offering and experience is not compromised within developments such as this and can even be enhanced with feature lighting and music systems.

A fully equipped gym is specially designed for the residents with options for personal training and studios which provide personal or group Yoga and Pilates classes.

This style of development will often include a spa comprising a swimming pool and sauna. Treatment rooms are available for spa styled massages or alternative therapist sessions such as physiotherapists.

As well as treatment rooms, beauty services are also available to residents including an in-house hair and beauty salon.

A high-quality private cinema room is available.

Private and secure gardens that incorporate sensory areas for residents.

Requirements of the extended family are often considered with guest suites available for overnight stays, ability to stay with a resident within their apartment, have in-room catering and a play area for small children within the grounds.

Development Details

This cost model is based on a high end later living development located in the South of England. The basis of the development is a 4-level apartment block with 2 cores from Ground to Level 3, totalling a GIA of 9,000 m^2 and an Apartment NIA of 5,000 m^2. There is a total of 50 apartments located from level 1 and above with the unit mix being based on 20 one-bedroom apartments at an average NIA of 80 m^2, 25 two-bedroom apartments at an average NIA of 110 m and five two-bed 'close care units' at an average NIA of 130 m^2.

Please note the cost model excludes:
1. Lifts
2. External works
3. Utilities
4. Inflation beyond 4th quarter 2019
5. Main contractor's oncosts
6. Fees
7. VAT

COST MODEL

Shell and Core Rates	Based on a GIA 9,000 m²		
	Total (£)	£/m² on GIA	£/ft² on GIA
5.1 Sanitaryware Sanitaryware to communal areas including changing rooms, restaurants, etc. Includes for subcontractor preliminaries.	£30,000	£3.33	£0.31
5.3 Disposal Installations Rainwater and soil and waste drainage all in HDPE pipework and insulated. Includes for testing and commissioning and subcontractor preliminaries.	£341,000	£37.89	£3.52
5.4 Water Installations Boosted cold water to apartments including plant. Boosted cold water distribution and hot water generation to communal areas. Includes for testing and commissioning and subcontractor preliminaries.	£354,000	£39.33	£3.65
5.5 Heat Source Heat generation plant and pipework, including boilers, CHP, pumps etc. Flues to atmosphere. Includes for testing and commissioning and subcontractor preliminaries.	£207,000	£23.00	£2.14
5.6 Space Heating and Air Treatment LTHW distribution to apartments and communal areas. Air cooled chillers and associated plant generating chilled water. CHW distribution to apartments and communal areas. Includes for testing and commissioning and subcontractor preliminaries.	£743,000	£82.56	£7.67
5.7 Ventilation Core Smoke Extract (Colt System) with environmental boost to corridors. Central kitchen extract system. Includes for testing and commissioning and subcontractor preliminaries.	£302,000	£33.56	£3.12

Shell and Core Rates	Based on a GIA 9,000 m²		
	Total (£)	£/m² on GIA	£/ft² on GIA
5.8 Electrical Installations General LV to shell and core areas and ryefield installation to apartments. General small power, containment and earthing and bonding. Lighting to basement and feature lighting above ground. Car Charging Points. Includes for testing and commissioning and subcontractor preliminaries.	£1,058,000	£117.56	£10.92
5.9 Gas Installation Gas supply to plant rooms and kitchens. Includes for testing and commissioning and subcontractor preliminaries.	£33,000	£3.67	£0.34
5.10 Lift Installation Excluded.	£0	£0.00	£0.00
5.11 Protective Installation Sprinkler installation to ground floor communal space and apartments. Dry risers. Lightning protection. Includes for testing and commissioning and subcontractor preliminaries.	£192,000	£21.33	£1.98
5.12 Communication, Security and Control Systems Fire alarm installation and disabled refuge call system. Door entry system, IRS system, fibre installation, telephone installation and data. Access control and closed circuit television. General containment. Medical alert system within shell and core areas. BMS system including energy monitoring system. Includes for testing and commissioning and subcontractor preliminaries.	£924,000	£102.67	£9.54
5.13 Specialist Installation Not applicable.	£0	£0.00	£0.00

Shell and Core Rates	Based on a GIA 9,000 m²		
	Total (£)	£/m² on GIA	£/ft² on GIA
5.14 Builder's work in connection with services BWIC.	£126,000	£14.00	£1.30
Utilities Excluded.	£0	£0.00	£0.00
MEP related External Works Excluded.	£0	£0.00	£0.00
Shell and Core Total Costs	**£4,309,000**	**£478.78**	**£44.48**

Communal Areas Fit-Out Rates	Based on an NIA 1,350 m²		
	Total (£)	£/m² on NIA	£/ft² on NIA
1 Reception Includes for heating and cooling, ventilation, small power, lighting, fire detection and IT/AV. Excludes pendant feature lighting. Includes for testing and commissioning and subcontractor preliminaries.	£97,000	£71.85	£6.68
2 Restaurant Includes for heating and cooling, ventilation, small power, lighting, fire detection, sound system and IT/AV. Excludes pendant feature lighting. Includes for testing and commissioning and subcontractor preliminaries.	£202,000	£149.63	£13.90
3 Restaurant Kitchen	Shell and Core		
4 Wine Room Includes for temperature control, ventilation, small power, lighting, fire detection, sound system and IT/AV. Excludes pendant feature lighting. Includes for testing and commissioning and subcontractor preliminaries.	£74,000	£54.81	£5.09
5 Gym and Studios Includes for heating and cooling, ventilation, small power, lighting, fire detection, sound system and IT/AV. Includes for testing and commissioning and subcontractor preliminaries.	£96,000	£71.11	£6.61

Communal Areas Fit-Out Rates	Based on an NIA 1,350 m²		
	Total (£)	£/m² on NIA	£/ft² on NIA
6 Sauna, Swimming Pool and Treatment Rooms Includes for heating and cooling, ventilation, small power, lighting, fire detection, sound system and IT/AV. Excludes plant for sauna and swimming pool. Includes for testing and commissioning and subcontractor preliminaries.	£363,000	£268.89	£24.98
7 Beauty/Hair Salon Includes for heating and cooling, ventilation, small power, lighting, fire detection, sound system and IT/AV. Includes for testing and commissioning and subcontractor preliminaries.	£49,000	£36.30	£3.37
8 Cinema Room Includes for heating and cooling, ventilation, small power, lighting and fire detection. Includes IT/AV Equipment for cinema system. Includes for testing and commissioning and subcontractor preliminaries.	£148,000	£109.63	£10.18
9 Drawing Rooms Includes for heating and cooling, ventilation, small power, lighting, fire detection, sound system and IT/AV. Excludes pendant feature lighting. Includes for testing and commissioning and subcontractor preliminaries.	£193,000	£142.96	£13.28
Communal Areas Fit-Out Total Costs	**£1,221,000**	**£904.44**	**£84.03**

Extra over for close care apartments (Cost per 2-bed apartment) Extra over for uplifts to dementia care apartments including; Stand-alone flood/leak detectors to all bathrooms and kitchens, Aspirating fire detection system, Full home automation system including video entry panel, medical alert signalling nurse call within each room, link between leak and temperature detectors with the extra care team, heating/cooling controls and Circadian lighting control.	£29,000	£5.80	£0.54
Fit-Out Uplift Per 2-Bed Extra Care Apartment	**£29,000**	**£223.08**	**£20.72**

LED Lighting

Background

LED lighting technology is now sufficiently developed to be widely accepted as the light source of choice for most lighting applications. The technology offers benefits including energy efficacy, high quality light appearance, light colour adjustability, device connectivity and controllability. This makes LED luminaires suitable for applications ranging from street lighting to office and residential lighting.

Currently available types of LED luminaries

Light emitting diodes (LEDs) are small high efficacy light sources that can be packaged into a range of products suitable to both new build and retrofit applications. In new build applications LEDs are often built-in to the luminaire and form an integral part of it, which can offer some technical benefits, whereas in retrofit applications LEDs are packaged into lamp formats that can easily be inserted into existing light fixtures to replace tradition light sources. There are many types of retrofit LED lamps from opalescent and filament LED bulbs that are an alternative to halogen and compact fluorescent bulbs, to linear LED tubes that are an alternative to linear fluorescent lamps.

LED efficacy

LEDs are commercially available with a light source efficacy of over 120 lumens per watt and delivered light output of from luminaires of over 100 lumens per watt. This makes them far more efficient than halogen lamps that have a typical efficacy range of 5–25 lumens per watt and are now banned in the UK, and more efficient than both fluorescent and discharge technology lamps which have an efficacy range of 40–100 lumens per watt.

LED light colour, appearance, and colour rendering

LEDs are available in a wide range of light colour appearances. Very warm white with a colour appearance of 1800 K to very cool white with a colour appearance of over 10,000 K is available. In addition, many saturated colour options are available that can be blended together to create millions of colour options.

Typical LED colour rendering characteristics vary from CRI 70–95+ according to a product's priority for efficacy or colour appearance fidelity. In addition, specialize LED products are available with light spectral properties tuned to specific applications including human centric circadian lighting and retail lighting requirements.

LED binning

The Colour consistency or just discernible differences of chromaticity across manufactured batches of LED is quantified by the MacAdam's ellipse scale – Standard deviation colour matching – SDCM. These can have a significant bearing on LED costings with 1/2 SDCM being excellent quality with no discernible difference in batches and therefore more expensive, 3–4 SDCM being good and 4–7 SDCM acceptable in some applications.

LED control

In both new build and retrofit applications a wide range of LED control options are available, both wired and wireless. This makes it possible to dim LEDs and create programmable colour change effects in both new build and retrofit applications.

Cost implications

Manufacturers and suppliers are keen to highlight potential energy saving from LED lighting technology. Although the principle of realizable savings is completely valid, a detailed study of potential energy savings is essential to ensure the most beneficial and appropriate LED technologies and design approaches can be implemented.

Payback

LED lamps/luminaires are typically slightly more expensive than traditional lighting alternatives across most types. However, the cost variance and energy saving potential for different LED lamps/luminaires types are significantly varied. Therefore, the return on investment from different LED lighting typologies and implementation approaches is not constant and requires detailed study to forecast in detail.

The following provides a return on investment overview for a sample LED Luminaire typology and excludes the cost of installation labour which is assumed to be consistent

Power cost (£)/kWh	0.14
Days in use/year	260
Hours in use/day	12

Note: Range of costs £/kWh varies greatly across suppliers/contracts, currently in 2022 average ranges are 0.11–0.21 (£)/kWh, with some commercial bulk-buy deals going as low as 0.09 (£)/kWh.

	Typical LED Downlight	Compact Fluorescent Downlight	LV Halogen (IRC) Downlight
Input Power (watts)	6	12	35
Lifetime (hours)	50,000	12,000	4,000
Replacement Lamp Cost including labour (£)	–	15.00	15.00
Annual Energy Cost per lamp (£)	2.62	5.24	15.28
Total Energy Cost for 25 luminaires per year (£)	65.52	131.04	382.20
Luminaire Unit Cost (£)	90.00	60.00	40.00

	Typical LED Downlight	Compact Fluorescent Downlight	LV Halogen (IRC) Downlight
Supply only cost of 25 luminaires (£)	2,250.00	1,500.00	1,000.00
Allowance for 4 nr emergency battery packs (£)	320.00	320.00	320.00
Lamps (£)	Included	375.00	375.00
Total cost of luminaries/emergencies/lamps (£)	2,570.00	2,195.00	1,695.00

	Typical LED Downlight	Compact Fluorescent Downlight	LV Halogen (IRC) Downlight
Extra over cost of LED luminaires and emergency luminaires compared to traditional luminaires is (£)		375.00	875.00
Yearly cost of energy and yearly re-lamping allowance (compact fluorescent @ every 4 years; halogen @ every 1.5 year) (£)	65.52	224.79	632.20
The calculated yearly energy and re-lamping saving of using LED luminaires when compared to traditional luminaires is (£)		159.27	566.68
Therefore, time taken in years to 'pay' for the additional cost of the LED luminaires based on the energy and re-lamping costs of 'traditional' luminaires in **years** is		**2.35**	**1.54**

Power cost (£)/kWh	0.14
Days in use/year	260
Hours in use/day	12

	Recessed LED Office Luminaire	Recessed Fluorescent Office Luminaire
Input Power (watts)	32	42
Typical Lifetime (hours)	50,000	12,000
Replacement Lamp Cost including labour (£)	–	20.00
Annual Energy Cost per lamp (£)	13.97	18.34
Total Energy Cost for 25 luminaires per year (£)	349.44	458.64
Luminaire Unit Cost including lamps (£)	185.00	155.00

Lamps (£)	Included	500
Supply only cost of 25 luminaires including lamps (£)	4,625.00	3,875.00
Allowance for 4 nr emergency battery packs (£)	480.00	400.00
Total cost of luminaries/emergencies/lamps (£)	5,105.00	4,775.00

Extra over cost of LED luminaires and emergency luminaires compared to fluorescent luminaires is (£)		330.00
Yearly cost of energy and yearly re-lamping allowance (fluorescent @ every 4 years) (£)	349.44	583.64
The calculated yearly energy and re-lamping saving of using LED luminaires when compared to fluorescent luminaires is (£)		234.20
Therefore time taken in years to 'pay' for the additional cost of the LED luminaires based on the energy and re-lamping costs of fluorescent luminaires in **years** is		**1.41**

Notes and conclusions

1. LED luminaires do save energy.
2. LED luminaires are generally 1.5x the price of a 'traditional' downlight but LED luminaire prices are becoming more competitive as the technology develops.
3. Some LED luminaires may not be converted to emergency so 'additional' luminaires may be required.
4. LED replacement lamps may have completely different light emitting characteristics. And, the luminaire may require modification to accept the retrofit LED lamp.
5. LV Halogen (IRC) downlight utilized for comparison purposes as lamp type now banned in the UK.
6. Recessed office luminaire, based on 600 × 600 modular ceiling and achieving 300 lux on working plane (CIBSE & BS).

Office Refurbishment

Introduction

The demand for a more carbon neutral approach to office development is seeing a move away from new builds and towards the refurbishment of existing buildings. This is in response to the draft London Plan's commitment to circular economy principles and the London Energy Transformation Initiative's (LETI) aim for a net-zero carbon-built environment. This industry professional and government focus, coupled with office tenants' desire to understand their properties' environmental impact, means that buildings need to perform better and still make financial sense for the landlord.

Reusing building elements such as the structure and existing fabric can reduce demolition work and waste and save cost. The speed of a refurbishment means buildings can be brought to market much quicker than new builds.

A refurb can ensure a property is adapted to meet the changing needs of tenants that are driving towards smaller spaces, serviced offices and interconnection with other organizations within the building. Design trends need to be considered to ensure a functional space is produced that is flexible and can accommodate a range of fit-out options.

With more stringent minimum energy efficiency legislation, there is a need for existing buildings to implement new and efficient technology that will uplift their credentials, while addressing the demand for decarbonization. Landlords are seeking to 'sweat their asset' to achieve energy savings without breaking the bank.

Refurbishment projects can range from minor upgrades of plant and superficial replacement of existing fit-out installations, to wholescale replanning and optimization of the space, potentially increasing the net lettable floor area.

Minor refurbishments offer largely cosmetic upgrades, retaining much of the engineering installations where there is adequate remaining lifespan and capacity. The reuse of pipework risers, main plant and life-safety systems is assumed, with only light touches to landlord and toilet areas, and new Cat B fit-out provision.

Medium refurbishments aim to extend the life of the building and provide innovative spaces. They also offer an opportunity to enhance the energy and sustainability performance of a property. The retention of the facade often means the existing thermal properties of the envelope are fixed, so energy improvements will need to come largely through upgrades to the building engineering services. This includes the replacement of MEP central plant and the addition of intelligent controls systems. Thermal and visual comfort will be reassessed to determine whether plant capacities need upgrading; however, in this type of scheme, no major structural alterations are envisaged and there are only minor repairs on the facade. The net lettable space will probably be stripped out and floor layouts replanned to make room for additional functional areas.

Refurb cost drivers

The retention of riser pipework requires intrusive surveys to understand their internal condition, and testing to determine water contamination risks, especially if any parts of the system are to remain operational during the project. Many systems – such as lighting control building management and fire alarms – are often at their original design capacity, and either cannot be extended or the components are no longer manufactured to facilitate their maintenance. Existing buildings are usually based on a lower occupancy for ventilation, lifts, fresh air and cooling/power loads, so these systems need to be reviewed.

The introduction of energy efficient lighting, coupled with controls upgrades and lifts with destination control and regenerative drives, can have a big impact on energy in-use performance of the building, so the lifecycle cost of this type of upgrade indicates significant added value. In addition, the drive for energy efficiency, and legislation aimed at decarbonizing the energy grid, have resulted in a sharp turn away from gas-fired boilers and district heating networks powered by gas-fired CHP. The favoured space-heating approach now uses ground, water or air source heat pump solutions. However, these are not easily introduced in refurbishments because of space and location restrictions.

The lack of sufficient external space for the heat-collection arrays required for ground source heat pumps means air source is a better option. However, noise needs to be balanced out against the desire for recreational space, usually most easily provisioned on rooftop terraces. Restricted floor-to-ceiling heights can often mean that four-pipe fan coil solutions are not feasible and VRF systems are necessary.

Having multiple tenants drives engineering services cost, as electrical panels, plate heat exchangers, riser design and metering increase as a factor of the number of tenancies. Converged networks to run voice, data, security and management systems are becoming more common, but it has a capital-cost uplift. A decision to introduce hydraulic separation on a floor-by-floor basis is also a cost driver but gives future flexibility and control. The IET Regulations require improved electrical protection on all designs after December 2018, which can result in an uplift for existing LV panels and local distribution boards. Refurbs now need to target BREEAM Excellent as a minimum, and Well accreditation. One step further is the use of smart building design in which systems are connected and controlled via a single operating system.

Cost model

This cost model captures the engineering services costs of a medium refurbishment project within a central London location, and is based on a seven-storey building and basement level at 21,700 m² GIA and 17,320 m² NIA.

It assumes the retention of all existing plant and services where feasible, with extended Cat A installations to suit revised office space layout, enhanced reception and new cycle shower facilities. The rates are current in the second quarter of 2020, based on a central London location. Main contractor OHP and preliminaries, professional fees and VAT are excluded. The building is designed to meet BCO 2019 and BREEAM Excellent standards.

Shell and Core Works	Unit	Rate £
Enabling works		
Strip-out works including validation of existing services		33.98
Allowance to make good existing defective services	m²	0.87
Sanitary installations		
New sanitaryware throughout. Relocate DDA toilet as per reconfigured layout		29.74
Additional cycle shower facilities provided	m²	8.65
Disposal installations		
New drainage riser to allow for new tea point for each tenancy, incl. new foul pumps		4.40
Disposal installations for new cycle showers	m²	1.23
Water installations		
Water installations for WC upgrade; upgraded booster pumps for new tea point for each tenancy and water treatment		10.10
Water installations for new cycle showers	m²	4.59
Heat Source	m²	0.0

Shell and Core Works	Unit	Rate £
Space heating and air treatment	m²	
New space heating and cooling to upgraded Reception		8.90
Maintenance related items to extend life of plant – clean fins on air coolers, repair leaks on primary LTHW connection to PHX		0.44
Decommission and remove redundant systems, i.e. CRAC units, condenser water system to make space for tenant cooling equipment for Comms Room		6.65
LTHW/CHW connections to multiple tenancies		12.70
LTHW to serve outside air supply frost coil to basement AHU (previously served by condenser water system)		16.82
Remedial Chemical clean, i.e. flushing and treatment to CHW and LTHW systems		19.31
Cooling and air treatment to Comms Room and Storage rooms		4.12
Ventilation	m²	
Ventilation to new cycle showers		5.41
Cleaning, rebalancing of existing installations including grilles		2.27
Storeroom ventilation added		1.65
WC ventilation extended		0.44
New air handling unit to reception area including fresh air and extract ductwork		5.00
Electrical installations	m²	
Modify DRUPS unit into a life safety generator at reduced capacity of 400 kVA including cabling		14.59
Modify existing Main LV switch panel to accommodate new Life Safety Switchboard including re-cabling of life safety loads		9.61
Modify LV distribution for 2 tenants per floor including reuse of tenant distribution boards to allow tenant metering in accordance with Part L Bldg Reg		5.25
Reconfiguration and enhancement of exposed soffit lighting		1.18
New and replacement lighting to suit reconfigured office plan		3.20
Relamping of existing lighting		2.17
Feature lighting to new Reception		11.90
External lighting		10.82
New lighting to lift lobbies and upgrades to WC areas		4.96
Lighting to Cycle Shower facilities		2.51
Lighting to basement storage		3.47
Gas Installations	m²	0.0
Protective installations	m²	
Replace stainless steel sprinkler tanks (80 m³)		5.33
Decommission and remove gas suppression to data centres and replace with pre-action sprinkler (assume 1 nr pre-action valve)		1.86
Reconfiguration of sprinklers to exposed soffit areas and extension to suit layouts		1.74
New sprinklers to Comms room and storage areas		3.00
Make good existing installations		0.64
Communications	m²	
New entrance CCTV and Reception speed gates		12.22
Fire & voice alarm to existing and new landlord spaces including WC alarms		5.32
New structure cabling network throughout incl Wireless LAN, Wireless Voice		10.58
Integrated Communications Network (ICN) with core switches in basement level B3 and distribution switches throughout the building		8.13
New security system throughout		20.04
Dual telecommunications intake rooms at basement level B3 (or an alternative position) to allow for tenant diverse communication services to their tenancies. One of the intake rooms to act as building Main Equipment Room (MER)		19.55

Shell and Core Works	Unit	Rate £
Lift and conveyor installations – excluded Modernization of lift car finishes	m²	excl.
Special Installations Modifications to the BMS to suit modified systems Automatic metering and billing system	m²	6.21 8.62
Builder's work in connection with services		1.5%
MEP Testing and Commissioning		2.5%
MEP Subcontract Prelims		8.3%
Shell and Core Total Cost		**£395.27/m²**

Category A Fit-Out Works	Unit	Rate £
Enabling works Strip-out and allowance to make good existing defective services	m²	7.58
Space heating and air treatment Allowance for replacing perimeter heating grilles Allowance to clean chilled beams and rebalance mechanical systems; modify pipework zones to suit new tenancy splits Chemical flushing and treatment	m²	0.49 17.56 4.44
Ventilation On floor storeroom ventilation added	m²	1.23
Electrical installations Reconfiguration and enhancement of exposed soffit lighting New and replacement lighting to suit reconfigured office plan Relamping of existing lighting	m²	9.57 16.65 9.39
Protective installations Modification to existing sprinkler system to suit CAT A fit-out space	m²	8.85
Builder's work in connection with services		4.1%
MEP Subcontract Prelims		15%
MEP Testing and Commissioning		5%
Category A Fit-Out Total Cost		**£95.21/m²**

Note: Assuming no occupation during the project.

Shell and Core Works	Unit	Extra Over Rate £
Sanitaryware Sensor taps in lieu of push button, including power supplies Sensor flush plates in lieu of manual, including power supplies	m^2	Included in model above
Space heating and air treatment Activated carbon filtration to air handling units Increase volume of air from 16 l/s person to 20 l/s per person from existing primary air handling plant	m^2 GIA	111.00 to 16.00 3.00 to 5.00
UV lamps to AHUs (excluding lamp-replacement cost every 9,000 hours)	m^2 GIA	3.00 to 5.00
Electrical installations Antimicrobial UV lights to WC and refuse stores	m^2 GIA	6.00 to 7.00
Communications Air quality monitors installed in landlord core areas linked to BMS Full-body arch temperature scanner and associated EWIC Wall-mounted infrared laser temperature scanner and read-out screen Touchless intercom systems incorporating infrared beam to activate bell extra over	m^2 nr nr nr	3.00 31,000.00 1,500.00 to 2,000.00 1,500.00 to 2,000.00
Vertical transportation Lift air purifiers Lift touchless app Lilt touchless operating buttons 9 based on seven floors) Programming of doors to hold open when not in use, for improved ventilation	per lift	3,500.00 2,000.00 5,000.00 1,000.00
Fit-out Works		
Space heating and air treatment Enhanced filtration to fan coil units – activated carbon Enhanced filtration to fan coil units – glass fibre H11 HEPA filters	m^2 NIA	18.00 to 18.50 79.00 to 11.00
Communications Air-quality monitors installed across all occupied spaces linked to BMS	m^2	2.50 to 3.00

Above are just a few examples of the more common type of approaches to dealing with Covid-19 and achieving Well standards and are by no means an exhaustive list. There will no doubt be many refurbishment projects in the near and medium future that will need to ensure working environments adapt to the challenges that Covid-19 presents. It is vital engineering cost consultants are up to speed with how this can be achieved and the options available, and that they work with designers and clients to deliver solutions that are cost-effective and practical.

Other considerations

Where projects need to go ahead in phases while maintaining operational continuity, considerations – which can add £2.20 to £5.70 per square foot GIA to the cost of services for refurbs – include:

- Design of temporary works to facilitate repurposing of space or replacement of critical plant during occupation.
- Introduction of multiple phases that require isolations, drain downs and temporary services modifications.
- Out-of-hours working for tight programmes and to do work that is impossible during working hours.
- Preparatory surveys to identify and ratify existing installations, to ensure critical services are maintained.
- Establishing the condition of existing pipework to ensure reuse is feasible and there is no internal pitting and corrosion.

Future considerations

As electric vehicles become more commonplace, air quality will improve in London and building design will start to incorporate more mixed-mode ventilation strategies. As a consequence of Covid-19, a Building Council for Offices paper has recommended that short- and long-term changes are made to office design to ensure workspaces reduce the transmission of the virus. This is likely to affect the density of floorplates, a move away from hot desking, increased cycle parking, and strict limits on lift and meeting room use. HVAC systems will need to operate differently, including removal of thermal wheels and recirculation systems. Increased air supply and extract ventilation is likely, as is natural ventilation where practical.

Bathroom design will probably move towards individual pods with hands-free flush, washing facilities and door operation. Even the method of construction will be affected, with off-site manufacture helping to mitigate the impact of on-site social distancing.

Research & Development (R&D) Tax Relief

Introduction

Research and Development (R&D) tax relief, or credit, is a Corporation Tax benefit that can either reduce a company's tax bill or, in some circumstances, provide a cash sum. It is based on the company's expenditure on R&D. R&D tax relief/credits are aimed at encouraging expenditure by companies on innovative activities that advance the overall knowledge or capability in the fields of science and technology. The tax credits consist of additional tax relief beyond the amount of the expenditure incurred on qualifying R&D activities and are given against Corporation Tax. In some situations, where the company is making losses, the tax credits can be surrendered for a cash payment.

For there to be R&D for the purpose of the tax relief, a company must be undertaking activity as part of a specific project that seeks to make an advance in science or technology. It is necessary to be able to state what the intended advance is, and to show how, through the resolution of scientific or technological uncertainty, the activity seeks to achieve this.

HMRC has confirmed that qualifying activities include those typically undertaken in the construction and property sector, such as engineering design and developing new construction techniques.

Definition of R&D for Tax Purposes

The primary legislation is contained in Part 13 of the Corporation Tax Act 2009 (CTA 2009) and covers two separate schemes:

- R&D expenditure credits (s.1040 A), commonly referred to as RDEC and is available to large companies.
- Additional tax relief for small or medium sized enterprises (s.1044), commonly referred to as the SME scheme.

R&D is defined in section 1138 of the Corporation Tax Act 2010 as 'activities that fall to be treated as research and development in accordance with generally accepted accounting practice'. This is subject to regulations made by HM Treasury on the definition of R&D, which refer to the detailed guidelines published by the Secretary of State for Business, Energy and Industrial Strategy (BEIS). The BEIS guidelines define R&D for the purpose of obtaining relief as follows:

- 'R&D for tax purposes takes place when a project seeks to achieve an advance in science or technology. The activities which directly contribute to achieving this advance in science or technology through the resolution of scientific or technological uncertainty are R&D. Certain qualifying indirect activities related to the project are also R&D. Activities other than qualifying indirect activities which do not directly contribute to the resolution of the project's scientific or technological uncertainty are not R&D.'

Guidance is provided by HM Revenue & Customs (HMRC) on what constitutes R&D for the purposes of the tax relief and can be found within the Corporate Intangibles R&D Manual at CIRD81900. The work that qualifies for R&D tax relief must be part of a specific project to make an advance in science or technology. It cannot be an advance within arts, humanities, a social science such as economics or a theoretical field such as pure maths. The project must relate to the company's trade.

Advance in Science or Technology

An advance in science or technology means an advance in overall knowledge or capability in a field of science or technology (not a company's own state of knowledge or capability alone). This includes the adaptation of knowledge or capability from another field of science or technology in order to make such an advance where this adaptation was not readily deducible. A process, material, device, product, service, or source of knowledge does not become an advance in science or technology simply because science or technology is used in its creation.

A project which seeks to either

- Extend overall knowledge or capability in a field of science or technology; or
- Create a process, material, device, product, or service which incorporates or represents an increase in overall knowledge or capability in a field of science or technology; or
- Make an appreciable improvement (for example, changing or adapting the scientific or technological characteristics of something to the point where it is 'better' than the original and beyond minor or routine upgrading) to an existing process, material, device, product, or service through scientific or technological changes; or
- Use science or technology to duplicate the effect of an existing process, material, device, product, or service in a new or appreciably improved way (for example, a product which has exactly the same performance characteristics as existing models, but is built in a fundamentally different manner) will therefore be R&D.

Even if the advance in science or technology sought by a project is not achieved or not fully realized, R&D still takes place and the project could qualify for relief. Moreover, if a particular advance in science or technology has already been made or attempted but details are not readily available (for example, if it is a trade secret), work to achieve such an advance can still be an advance in science or technology.

Scientific or Technological Uncertainty

Scientific or technological uncertainty exists when knowledge of whether something is scientifically possible or technologically feasible, or how to achieve it in practice, is not readily available or deducible by a competent professional (an individual who is knowledgeable about the relevant scientific and technological principles involved, aware of the current state of knowledge and has accumulated experience) working in the field. This includes system uncertainty, or a scientific or technological uncertainty that results from the complexity of a system rather than uncertainty about how its individual components behave. Scientific or technological uncertainty will often arise from turning something that has already been established as scientifically feasible into a cost-effective, reliable, and reproducible process, material, device, product, or service.

Uncertainties that can readily be resolved by a competent professional working in the field are not scientific or technological uncertainties. Similarly, improvements, optimizations and fine-tuning which do not materially affect the underlying science or technology do not constitute work to resolve scientific or technological uncertainty.

Qualifying Indirect Activity

These are activities which form part of a project but do not directly contribute to the resolution of the scientific or technological uncertainty.

This includes the following:

- Scientific and technical information services, insofar as they are conducted for the purpose of R&D support (such as the preparation of the original report of R&D findings);
- Indirect supporting activities such as maintenance, security, administration and clerical activities, and finance and personnel activities, insofar as undertaken for R&D;
- Ancillary activities essential to the undertaking of R&D (for example, taking on and paying staff, leasing laboratories, and maintaining R&D equipment including computers used for R&D purposes);
- Training required to directly support an R&D project;
- Research by students and researchers carried out at universities;
- Research (including related data collection) to devise new scientific or technological testing, survey, or sampling methods, where this research is not R&D in its own right; and
- Feasibility studies to inform the strategic direction of a specific R&D activity.

Start and End of R&D for Tax Purposes

R&D begins when work to resolve the scientific or technological uncertainty starts, and ends when that uncertainty is resolved or work to resolve it ceases. For example, R&D could end once a working prototype or a material, device, product, or process is ready for testing, production or if the project does not continue. However, R&D can take place even after production starts. For example, if any new problems arise involving scientific or technological uncertainty after the product has been put into production or into use, then the R&D process may start again.

Types of R&D Tax Relief

There are different types of R&D relief, depending on the size of a company and if the project has been subcontracted to the company or not.

The two principal tax reliefs available to companies undertaking R&D in the UK are as follows:

- SME R&D Tax Relief.
- Research and Development Expenditure Credit (RDEC).

SME R&D Tax Relief

An SME for R&D tax relief purposes is a company or organization with less than 500 staff and either of the following:

- Annual turnover not exceeding €100 million.
- Balance sheet not exceeding €86 million.

A company may need to take into account its own data, a proportion of a partner enterprise's data or the data of a linked enterprise. A company may not be considered an SME if it is part of a larger enterprise that would fail these tests if taken as a whole.

SME R&D Tax relief allows companies to:

- Deduct an extra 130% of their qualifying costs from their yearly profit, as well as the normal 100% deduction, to make a total 230% deduction. For profit making SMEs, R&D tax credits reduce the Corporation Tax bill and businesses can receive up to £24.70 for every £100 spent on qualifying R&D activities.
- Claim a tax credit if the company is loss making, worth up to 14.5% of the surrenderable loss. Loss-making SMEs can receive up to £33.35 for every £100 spent on qualifying R&D activities.

A company can only claim R&D tax relief as an SME if the company is a going concern and not in administration or liquidation when making the claim.

In some circumstances, an SME company may be allowed to claim only the RDEC in respect of particular expenditure (for example, for work subcontracted to it or that is otherwise subsidized due to a grant or subsidy). The SME scheme is a notifiable State Aid, and a company can't get the SME Tax relief if it is receiving any other notifiable State Aids for the same R&D project. If an SME company has received a grant which is a notifiable State Aid, for an R&D project, it cannot claim R&D tax relief under the SME scheme, but eligible expenditure will qualify under the RDEC scheme.

The UK government has recently announced that SME R&D tax credit claims will be subject to a cap in order to deter abuse of the scheme. This measure limits the amount of payable R&D tax credit which an SME can claim to £20,000 plus 300% of its total Pay As You Earn (PAYE) and National Insurance contributions (NICs) liability for the period. The measure will have effect for accounting periods beginning on or after 1 April 2021.

The measure amends CTA 2009 Part 13 Chapter 2. It adds a new condition at section 1058 (which sets the amount of payable tax credit) limiting the amount that can be claimed to the level of the new cap, which is defined as £20,000 plus three times the company's relevant expenditure on workers.

A company is however exempt from the cap if:

- Its employees are creating, preparing to create or managing Intellectual Property (IP); and
- It does not spend more than 15% of its qualifying R&D expenditure on subcontracting R&D to, or the provision of externally provided workers by, connected persons.

RDEC

Large companies (a company which does not meet the SME test for R&D tax relief purposes) can claim RDEC for working on R&D projects. It can also be claimed by SMEs who have been subcontracted to do R&D work by a large company or who have received a grant or subsidy for their R&D project.

The RDEC is an 'above the line' tax credit equal to 13% of qualifying R&D costs. RDEC is a taxable credit. This means that large companies receive up to £10.53 for every £100 spent on qualifying R&D activities. For profit making companies, the credit discharges the Corporation Tax liability that the company would have to pay. Companies with no Corporation Tax liability will benefit from the RDEC either through a cash payment or a reduction of tax or other duties due. Companies in groups can also surrender the RDEC against another group companies' Corporation Tax liability.

Categories of Qualifying Expenditure

- Staffing costs – This includes salaries, wages, Class 1 NICs, pension fund contributions and certain reimbursed expenses for staff directly and actively engaged in the R&D project. Support staff costs, for example administrative or clerical staff, do not qualify, except when they relate to qualifying indirect activities. These can be activities like maintenance, clerical, administrative and security work.
- Externally provided workers – These are the staff costs paid to an external agency for staff who are directly and actively engaged in the R&D project. These are not employees and subcontractors. Relief is usually given on 65% of the payments made to the staff provider. Special rules apply if the company and staff provider are connected or elect to be connected.
- Consumable items
  - Cost of items that are directly employed and consumed in qualifying R&D projects. These include materials and the proportion of water, fuel and power of any kind consumed in the R&D process.
  - From 1 April 2015, the costs of materials incorporated in products that are sold are not eligible for relief.
  - Where a prototype is created to test the R&D being undertaken, the design, construction and testing costs will normally be qualifying expenses. However, if the prototype is also built with a view to selling the prototype itself (for example, the construction of a bespoke machine), HMRC considers that to be production and outside the R&D scheme, even if R&D was undertaken to create the prototype. In this case, the R&D expenditure and production costs will need to be split. For example, the construction costs and materials consumed would not be qualifying expenses, but design, modelling and testing costs could still qualify for R&D tax relief.
- Software – A company may claim for the cost of software that is directly employed in the R&D activity. Where software is only partly employed in direct R&D, an appropriate apportionment should be made.
- Subcontracted expenditure
  - SME Scheme: A company can generally claim for 65% of the payments made to unconnected parties. The subcontracted work may be further subcontracted to any third party. Special rules apply where the parties are connected or elect to be connected. Under the SME Scheme, the subcontractor does not need to be a UK resident and there is no requirement for the subcontracted R&D to be performed in the UK.
  - RDEC Scheme: R&D expenditure subcontracted to other persons is generally not allowable unless it is directly undertaken by a charity, higher education institute, scientific research organization, health service body, or by an individual or a partnership of individuals.

Qualifying expenditure must be allowable as a deduction in calculating the profits of the trade. Capital expenditure is therefore excluded; it may however qualify for R&D allowances.

Costs that Cannot be Claimed for R&D Tax Relief

Companies cannot receive R&D tax relief for the following:

- The production and distribution of goods and services.
- Capital expenditure under either the SME tax relief or RDEC scheme. However, a 100% Research and Development Allowance may be due on capital assets, such as plant, machinery and buildings used for R&D activity.

- The cost of land.
- Cost of patents and trademarks.
- Rent or rates.

Examples of Qualifying R&D Activities in the Construction Industry

Design

- Integrating new or improved technology into buildings.
- Redesigning to improve performance, reliability, quality, safety and lifecycle costs.
- Designing prefabricated components for off-site assembly of large and complex systems.
- Developing innovative sustainable solutions to achieve net zero carbon.
- Developing advanced Buildings Information Management (BIM) capabilities.

On-site

- Solving technical problems that arise during or after construction.
- Developing new techniques to remove contaminants and remediate difficult ground conditions.
- Complex diversions of utilities such as gas, water and electricity.
- Developing new, more efficient construction techniques.
- Developing bespoke temporary solutions on challenging sites.

Research and Development Allowances

Research and Development Allowance (RDA) gives relief for capital expenditure incurred on R&D by a business. RDA is only due if the research and development expenditure is related to the trade being carried on or about to be carried on. Qualifying RDA expenditure is capital expenditure that a business incurs for providing facilities for carrying out R&D, for example, a laboratory in which to carry on R&D. RDA is only available to traders and is not available to a person carrying on a profession or vocation. The rate of RDA is 100% of the qualifying expenditure. However, if a disposal value is brought into account on expenditure in a chargeable period in which an RDA is made, the RDA is 100% of the expenditure less the disposal value. RDA is made in the chargeable period in which the qualifying expenditure was incurred unless it was incurred before the trade began, in which case RDA is made in the chargeable period in which trade began.

How to Claim R&D Tax Relief from HMRC

A company can make a claim for R&D tax relief up to two years after the end of the accounting period it relates to. To make a claim, a company must enter their total qualifying expenditure on the full Company Tax Return form, CT600, to HMRC. If claiming under the SME scheme, a company enters the enhanced expenditure figure into the CT600, and the payable tax credit figure if applicable. If claiming under the RDEC scheme, a company must enter both the total qualifying expenditure and RDEC figures into the CT600. After submitting their full CT600 form, a claimant company can use the HMRC online service to support their R&D tax relief claim and send any additional details, as required.

Contact

For further information or assistance, please contact one of the Fiscal Incentives team:

help.r&d@aecom.com

Construction Project Manager's Pocket Book, 2nd edition

Duncan Cartlidge

The second edition of the *Construction Project Manager's Pocket Book* maintains its coverage of a broad range of project management skills, from technical expertise to leadership, negotiation, team building and communication. However, this new edition has been updated to include:

*   revisions to the CDM regulations,
*   changes to the standard forms of contract and other documentation used by the project manager,
*   the impact of BIM and emerging technologies,
*   implications of Brexit on EU public procurement,
*   other new procurement trends, and
*   ethics and the project manager.

Construction project management activities are tackled in the order they occur on real projects, with reference made to the RIBA Plan of Work throughout. This is the ideal concise reference which no project manager, construction manager, architect or quantity surveyor should be without.

May 2020: 286pp
ISBN: 9780367435936

To Order
Tel: +44 (0) 1235 400524
Email: book.orders@tandf.co.uk

For a complete listing of all our titles
visit: www.tandf.co.uk

RICS SKA Rating

SKA Rating is a Royal Institute of Chartered Surveyors (RICS) sustainability assessment method that helps landlords and tenants assess fit-out projects against a set of sustainability good practice criteria. Assessing existing buildings was previously a blind spot in the industry. It is estimated that 11% of UK construction spending is on fit-outs and that buildings have multiple fit-outs during their lifecycle. Therefore, the creation of SKA Rating in 2009 looked to address this gap in the market. Since then it has become an adopted assessment method and can be used to assess the following building types:

- Offices
- Retail
- Higher Education

Fit-out projects are typically of a short duration and vary widely in scope from a quick refresh to a minor refurbishment or a comprehensive fit-out programme. SKA Rating was designed by the industry to be aligned with these relatively short timeframes for both the design and construction phases of a project. It is a flexible scheme that only assesses what's within the control of the project team and is not penalized by the limitations of the base build. This means that the project team can get rewarded for implementing a good practice fit-out, even if the building has poor transport links. Unlike other tools it provides the assessor with more ownership of what's in scope compared to other schemes that are more rigid. It is a more granular scheme that rewards incremental improvement within the fit-out and looks to avoid complicated calculation tools or additional specialists that would not usually be appointed to a project team, which is to a benefit to the client.

SKA is a freely accessible tool that is entirely online with the guidance, assessment methodology and certification generated through the tool. If a project is certified then there is a fee to pay to RICS. The aim is to promote good practice in the fit-out marketplace, so the online tool and the associated good practice guidance can be accessed by anyone at (ska-tool.rics.org). Also, anyone can train as an assessor and RICS provides a one-day Foundation course via the RICS Online Academy. Then there are separate online modules for the Offices, Retail and Higher Education Schemes.

There are currently nearly 200 accredited SKA assessors. Qualified SKA assessors can assess projects and can generate certificates once they have demonstrated that the project meets the SKA criteria.

SKA covers the following topics and issues:
- Reducing energy and water use by selecting efficient equipment and promoting metering etc.
- Selecting materials with lower environmental impact.
- Reducing construction site impacts, including waste.
- Reducing pollutants such as refrigerant leakage.
- Promoting health and wellbeing in the working environment, including improving internal air quality, daylight, etc.
- Rewarding more sustainable project delivery, such as registration to the Considerate Constructors Scheme and seasonal commissioning.

These issues are addressed by providing a list of good practice measures that are rewarded if the project demonstrates implementation. For example, SKA aims to reduce energy use in the completed fit-out by promoting the use of energy efficient equipment. These are assessed over three project stages:

1. **Design/Planning**. At this stage the measures in scope are identified with the project team and a route to the target rating is established. This will be dependent on the impacts of the SKA to the programme, budget and priorities of the client. Criteria will be embedded into the design specification documentation and contractor preliminaries. No certificate is generated at this stage, just confirmation that the criteria has been incorporated into the design.
2. **Delivery/Construction**. This stage involves the gathering of evidence from as-built information and other sources to prove that what has been specified has actually been delivered, and that the performance and waste benchmarks have been achieved. A Handover Certificate is produced once the SKA assessor has reviewed and approved all the available evidence and it has been uploaded to the SKA online tool. Note there is a fee to pay before the certification can be processed by RICS.
2. **Occupancy Stage Assessment**. Finally, there is the option to review how well a fit-out has performed in use against its original brief from a year after completion. A separate certificate can be issued at this Occupancy Stage if the evidence has been discharged by the SKA Assessor.

Scoring

The below scoring is used for SKA and is based upon the as-built evidence provided at the Handover Stage by the project team:

Benefits of SKA

SKA rating is designed to be of particular use for occupiers but has benefits for other property stakeholders, including landlords, developers, consultants, fit-out contractors, and members of the supply chain.

For occupiers and tenants:
- Manage the bottom line. A decision-making tool which ranks the different aspects of a fit-out based on what will have the biggest impact on sustainability. It can be used to help the design team achieve maximum sustainability for a given budget.
- Legal/statutory compliance. Help to ensure company directors, in line with Section 172 of the Companies Act 2006, to promote the success of the company and have a 'regard to the impact of the company's operations on the community and the environment'.

- Are you ISO 14001 accredited? A SKA certificate helps you ensure that a fit-out project is rated accurately and can be used to support your environmental management system.
- Customer, investor and stakeholder perception. A SKA rating certificate demonstrates that your company takes sustainability seriously and has achieved a standard with your property. This can feature in annual reports, environmental policy or tender documentation, or be displayed in a building.
- Staff engagement. Many employees increasingly place a value on a company's green credentials and wish to work in high-quality spaces. A SKA rating can be used to reinforce staff morale and complements other CSR activity.

For developers and landlords:

- Use SKA rating to set targets, use the formal assessment process to assure yourself that target performance standards are met, and finally use the certificate to report performance to stakeholders.
- Benchmark the sustainability of your fit-out across a portfolio of buildings or against your peers.
- Include the use of the RICS SKA rating in 'Green Lease' provisions to help drive up the sustainability of properties or protect a building already certified under a whole building assessment method.
- Sustainability increasingly makes good business sense. Research demonstrates an expected increase in the asset value of labelled low-carbon buildings compared with standard speculative buildings.

For consultants:

- Embed SKA rating in your standard processes to demonstrate that you follow a sustainable specification and procurement process for fit-outs.
- Use the SKA rating tool and datasheets to support the delivery of professional advice to clients on good practice in sustainable fit-outs, and to assess their designs and projects.
- Offer SKA rating certificated assessments to clients who wish to demonstrate they have achieved a more sustainable fit-out through the SKA rating accredited assessors' scheme.

For contractors:

- Use the assessment process and related guidance to make your design, specification, procurement and construction practices more environmentally sustainable.
- Demonstrate your sustainability credentials to clients and consultants by making the targeting of standards under SKA rating part of your standard tender process.

Product labelling

Product selection is a key part of SKA. This includes everything from procuring energy efficient technologies through to using furniture that have low environmental impacts during manufacture and high levels of recycled material. Grigoriou Interiors, a SKA development partner, has a label and directory for products that are proven to be SKA-compliant. This allows designers to find and specify products much more quickly and it will make the assessment process even simpler than it is now. For more information see: https://www.rics.org/uk/about-rics/responsible-business/ska-rating/ska-rating-product-compliance-label/.

Conclusion

SKA has been a great success story to date and has helped to promote sustainable good practice in fit-out and refurbishment projects. It can provide tangible and reputational benefits to occupiers and other property stakeholders and the ultimate aim of SKA is to get it widely adopted to the point that it is embedded in all projects as part of the established process.

For more information see: https: //www.rics.org/uk/about-rics/responsible-business/ska-rating/

CPD in the Built Environment

Greg Watt and Norman Watts

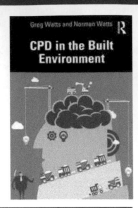

The aim of this book is to provide a single source of information to support continuing professional development (CPD) in the built environment sector.

The book offers a comprehensive introduction to the concept of CPD and provides robust guidance on the methods and benefits of identifying, planning, monitoring, actioning, and recording CPD activities. It brings together theories, standards, professional and industry requirements, and contemporary arguments around individual personal and professional development. Practical techniques and real-life best practice examples outlined from within and outside of the industry empower the reader to take control of their own built environment-related development, whilst also providing information on how to develop fellow staff members. The contents covered in this book align with the requirements of numerous professional bodies, such as the Royal Institution of Chartered Surveyors (RICS), the Institution of Civil Engineers (ICE), and the Chartered Institute of Builders (CIOB).

The chapters are supported by case studies, templates, practical advice, and guidance. The book is designed to help all current and future built environment professionals manage their own CPD as well as managing the CPD of others. This includes helping undergraduate and postgraduate students complete CPD requirements for modules as part of a wide range of built environment university degree courses and current built environment professionals of all levels and disciplines who wish to enhance their careers through personal and professional development, whether due to professional body requirements or by taking control of identifying and achieving their own educational needs.

April 2021: 182 pp
ISBN: 9780367372156

To Order
Tel:+44 (0) 1235 400524
Email: tandf@bookpoint.co.uk

For a complete listing of all our titles visit:
www.tandf.co.uk

Value Added Tax

Introduction

Value Added Tax (VAT) is a tax on the consumption of goods and services. The UK introduced a domestic VAT regime when it joined the European Community in 1973. The principal source of European law in relation to VAT is Council Directive 2006/112/EC, a recast of Directive 77/388/EEC, which is currently restated and consolidated in the UK through the VAT Act 1994 and various Statutory Instruments, as amended by subsequent Finance Acts.

VAT Notice 708: Buildings and construction (December 2020) provides HMRC's interpretation of the VAT law in connection with construction works, however, the UK VAT legislation should always be referred to in conjunction with the publication. Recent VAT tribunals and court decisions since the date of this publication will affect the application of the VAT law in certain instances. The Notice is available on the HM Revenue & Customs website at www.hmrc.gov.uk.

The Scope of VAT

VAT is payable on:

- Supplies of goods and services made in the UK;
- By a taxable person;
- In the course or furtherance of business; and
- Which are not specifically exempted or zero-rated.

Rates of VAT

There are three rates of VAT:

- A standard rate, currently 20%;
- A reduced rate, currently 5%; and
- A zero rate of 0%.

Additionally some supplies are exempt from VAT and others are considered outside the scope of VAT.

Recovery of VAT

When a taxpayer makes taxable supplies he must account for VAT, known as output VAT, at the appropriate rate of either 20%, 5% or 0%. Any VAT due then has to be declared and submitted on a VAT submission to HM Revenue & Customs and will normally be charged to the taxpayer's customers.

As a VAT registered person, the taxpayer is entitled to reclaim from HM Revenue & Customs, commonly referred to as input VAT, the VAT incurred on their purchases and expenses directly related to its business activities in respect of standard-rated, reduced-rated and zero-rated supplies. A taxable person cannot, however, reclaim VAT that relates to any non-business activities (but see below) or, depending on the amount of exempt supplies they made, input VAT may be restricted or not recoverable.

At predetermined intervals the taxpayer will pay to HM Revenue & Customs the excess of VAT collected over the VAT they can reclaim. However, if the VAT reclaimed is more than the VAT collected, the taxpayer, who will be in a net repayment position, can reclaim the difference from HM Revenue & Customs.

Example

X Ltd constructs a block of flats. It sells long leases to buyers for a premium. X Ltd has constructed a new building designed as a dwelling and will have granted a long lease. This first sale of a long lease is a VAT zero-rated supply. This means any VAT incurred in connection with the development, which X Ltd will have properly paid (e.g. payments for consultants and certain preliminary services) will be recoverable. For reasons detailed below, the contractor employed by X Ltd will have charged VAT on his construction services at the zero rate of VAT.

Use for Business and Non-Business Activities

Where a supply relates partly to business use and partly to non-business use, then the basic rule is that it must be apportioned on a fair and reasonable basis so that only the business element is potentially recoverable. In some cases, VAT on land, buildings and certain construction services, purchased for both business and non-business use, could be recovered in full by applying what is known as 'Lennartz' accounting, to reclaim VAT relating to the non-business use and account for VAT on the non-business use over a maximum period of 10 years. Following an ECJ case restricting the scope of this approach, its application to immovable property was removed completely in January 2011 by HMRC (business brief 53/10), when UK VAT law was amended to comply with EU Directive 2009/162/EU.

Taxable Persons

A taxable person is an individual, firm, company, etc., who is required to be registered for VAT. A person who makes taxable supplies above certain turnover limits is compulsorily required to be VAT registered. The current registration limit (since 1 April 2017), known as the VAT threshold, is £85,000. If the threshold is exceeded in any 12 month rolling period, or there is an expectation that the value of the taxable supplies in a single 30 day period, or goods are received into the UK from the EU worth more than the £85,000, then registration for UK VAT is compulsory.

A person who makes taxable supplies below the limit is still entitled to be registered on a voluntary basis if they wish, for example, in order to recover input VAT incurred in relation to those taxable supplies, however output VAT will then become due on the sales and must be accounted for.

VAT Exempt Supplies

Where a supply is exempt from VAT this means that no output VAT is payable – but equally the person making the exempt supply cannot normally recover any of the input VAT on their own costs relating to that exempt supply.

Generally commercial property transactions such as leasing of land and buildings are exempt unless a landlord chooses to standard-rate its interest in the property by a applying for an option to tax. This means that VAT is added to rental income and also that VAT incurred, on say, an expensive refurbishment, is recoverable.

Supplies outside the scope of VAT

Supplies are outside the scope of VAT if they are:

- Made by someone who is not a taxable person
- Made outside the UK; or
- Not made in the course or furtherance of business

In the course or furtherance of business

VAT must be accounted for on all taxable supplies made in the course or furtherance of business, with the corresponding recovery of VAT on expenditure incurred.

If a taxpayer also carries out non-business activities, then VAT incurred in relation to such supplies is generally not recoverable.

In VAT terms, business means any activity continuously performed which is mainly concerned with making supplies for a consideration. This includes:

- Anyone carrying on a trade, vocation or profession;
- The provision of membership benefits by clubs, associations and similar bodies in return for a subscription or other consideration; and
- Admission to premises for a charge.

It may also include the activities of other bodies including charities and non-profit making organizations.

Examples of non-business activities are:

- Providing free services or information.
- Maintaining some museums or particular historic sites.
- Publishing religious or political views.

Construction Services

In general the provision of construction services by a contractor will be VAT standard rated at 20%, however, there are a number of exceptions for construction services provided in relation to certain relevant residential properties and charitable buildings.

The supply of building materials is VAT standard rated at 20%, however, where these materials are supplied and installed as part of the construction services, the VAT liability of those materials follows that of the construction services supplied.

Zero-rated construction services

The following construction services are VAT zero-rated, including the supply of related building materials.

The construction of new dwellings

The supply of services in the course of the construction of a new building designed for use as a dwelling or number of dwellings is zero-rated, other than the services of an architect, surveyor or any other person acting as a consultant or in a supervisory capacity.

The following basic conditions must ALL be satisfied in order for the works to qualify for zero-rating:

1. A qualifying building has been, is being, or will be constructed;
2. Services are made 'in the course of the construction' of that building;
3. Where necessary, you hold a valid certificate; and
4. Your services are not specifically excluded from zero-rating.

The construction of a new building for 'relevant residential or charitable' use

The supply of services in the course of the construction of a building designed for use as a relevant residential Purpose (RRP), or relevant charitable purpose (RCP), is zero-rated, other than the services of an architect, surveyor or any other person acting as a consultant or in a supervisory capacity.

A 'relevant residential' use building means:

1. A home or other institution providing residential accommodation for children;
2. A home or other institution providing residential accommodation with personal care for persons in need of personal care by reason of old age, disablement, past or present dependence on alcohol or drugs or past or present mental disorder;
3. A hospice;
4. Residential accommodation for students or school pupils;
5. Residential accommodation for members of any of the armed forces;
6. A monastery, nunnery, or similar establishment; or
7. An institution which is the sole or main residence of at least 90% of its residents.

A 'relevant residential' purpose building does not include use as a hospital, a prison, or similar institution, or as a hotel, inn, or similar establishment.

A 'relevant charitable' purpose means use by a charity in either, or both of the following ways:

1.   Otherwise than in the course or furtherance of a business; or
2.   As a village hall, or similarly in providing social or recreational facilities for a local community.

Non-qualifying use, which is not expected to exceed 10% of the time the building is normally available for use, can be ignored. The calculation of business use can be based on time, floor area, or head count, subject to approval being acquired from HM Revenue & Customs.

The construction services can only be zero-rated if a certificate is given by the end user to the contractor carrying out the works, confirming that the building is to be used for a qualifying purpose, i.e. for a 'relevant residential or charitable' purpose. It follows that such services can only be zero-rated when supplied to the end user and, unlike supplies relating to dwellings, supplies by subcontractors cannot be zero-rated.

The construction of an annex used for a 'relevant charitable' purpose

Construction services provided in the course of construction of an annexe for use entirely, or partly for a 'relevant charitable' purpose, can be zero-rated.

In order to qualify, the annexe must:

1.   Be capable of functioning independently from the existing building;
2.   Have its own main entrance; and
3.   Be covered by a qualifying use certificate.

The conversion of a non-residential building into dwellings, or the conversion of a building from non-residential use to 'relevant residential' use, where the supply is to a 'relevant' housing association

The supply to a 'relevant' housing association in the course of conversion of a non-residential building, or non-residential part of a building, into:

1.   A new eligible dwelling designed as a dwelling, or number of dwellings; or
2.   A building, or part of a building, for use solely for a relevant residential purpose.

Any services related to the conversion, other than the services of an architect, surveyor or any person acting as a consultant or in a supervisory capacity, are zero-rated.

A 'relevant' housing association is defined as:

1.   A private registered provider of social housing;
2.   A registered social landlord within the meaning of Part I of the Housing Act 1996 (Welsh registered social landlords);
3.   A registered social landlord within the meaning of the Housing (Scotland) Act 2001 (Scottish registered social landlords), or
4.   A registered housing association within the meaning of Part II of the Housing (Northern Ireland) Order 1992 (Northern Irish registered housing associations).

If the building is to be used for a 'relevant residential' purpose, the housing association should issue a qualifying use certificate to the contractor completing the works. Subcontractors' services that are not made directly to a relevant housing association are standard-rated.

The development of a residential caravan park

The supply in the course of the construction of any civil engineering work 'necessary for' the development of a permanent park for residential caravans, and any services related to the construction, other than the services of an architect, surveyor or any person acting as a consultant or in a supervisory capacity, are zero-rated when a new permanent park is being developed, the civil engineering works are necessary for the development of the park and the services are not specifically excluded from zero-rating. This includes access roads, paths, drainage, sewerage and the installation of mains water, power and gas supplies.

Certain building alterations for 'disabled' persons

Certain goods and services supplied to a 'disabled' person, or a charity making these items and services available to 'disabled' persons, can be zero-rated. The recipient of these goods or services needs to give the supplier an appropriate written declaration that they are entitled to benefit from zero rating.

The following services (amongst others) are zero-rated:

1. The installation of specialist lifts and hoists and their repair and maintenance;
2. The construction of ramps, widening doorways or passageways including any preparatory work and making good work;
3. The provision, extension and adaptation of a bathroom, washroom or lavatory; and
4. Emergency alarm call systems.

Sale of Reconstructed Buildings

A protected building is not to be regarded as substantially reconstructed unless, when the reconstruction is completed, the reconstructed building incorporates no more of the original building than the external walls, together with other external features of architectural or historical interest.

DIY Builders and Converters

Private individuals who decide to construct their own home are able to reclaim VAT they pay on goods they use to construct their home by use of a special refund mechanism made by way of an application to HM Revenue & Customs. This also applies to services provided in the conversion of an existing non-residential building to form a new dwelling.

The scheme is meant to ensure that private individuals do not suffer the burden of VAT if they decide to construct their own home.

Charities may also qualify for a refund on the purchase of materials incorporated into a building used for non-business purposes where they provide their own free labour for the construction of a 'relevant charitable' use building.

Installation of Energy Saving Materials (ESM) to Domestic Properties

At the Spring Statement 2022, it was announced that the supply and installation of certain energy saving materials including insulation, draught stripping, central heating, hot water controls and solar panels in a residential building, would benefit from a zero rate until 31 March 2027. From 1 April 2027, ESM will revert to the reduced rate.

The scope of ESM has also been extended to add backwind and water turbines, which had previously been excluded.

Reduced-rated Construction Services

The following construction services are subject to the reduced rate of VAT of 5%, including the supply of related building materials.

Conversion – changing the number of dwellings

In order to qualify for the 5% rate, there must be a different number of 'single household dwellings' within a building than there were before commencement of the conversion works. A 'single household dwelling' is defined as a dwelling that is designed for occupation by a single household.

These conversions can be from 'relevant residential' purpose buildings, non-residential buildings and houses in multiple occupation.

A house in multiple occupation conversion

This relates to construction services provided in the course of converting a 'single household dwelling', a number of 'single household dwellings', a non-residential building or a 'relevant residential' purpose building into a house for multiple occupation, such as bed sit accommodation.

A special residential conversion

A special residential conversion involves the conversion of a 'single household dwelling', a house in multiple occupation, or a non-residential building into a 'relevant residential' purpose building. such as student accommodation or a care home.

Renovation of derelict dwellings

The provision of renovation services in connection with a dwelling or 'relevant residential' purpose building that has been empty for two or more years prior to the date of commencement of construction works can be carried out at a reduced rate of VAT of 5%.

Installation of energy saving materials

The supply and installation of certain energy saving materials including insulation, draught stripping, central heating, hot water controls and solar panels in a residential building, or a building used for a relevant charitable purpose.

Buildings that are used by charities for non-business purposes, and/or as village halls, are not eligible for the reduced rate for the supply of energy saving materials.

Grant-funded installation of heating equipment or connection of a gas supply

The grant-funded supply and installation of heating appliances, connection of a mains gas supply, supply, installation, maintenance and repair of central heating systems, and supply and installation of renewable source heating systems, to qualifying persons. A qualifying person is someone aged 60 or over or is in receipt of various specified benefits.

Grant-funded installation of security goods

The grant-funded supply and installation of security goods to a qualifying person.

Housing alterations for the elderly

Certain home adaptations that support the needs of elderly people are reduced rated.

Building contracts

Design and build contracts

If a contractor provides a design and build service relating to works to which the reduced or zero rate of VAT is applicable, then any design costs incurred by the contractor will follow the VAT liability of the principal supply of construction services.

Management contracts

A management contractor acts as a main contractor for VAT purposes and the VAT liability of his services will follow that of the construction services provided. If the management contractor only provides advice without engaging trade contractors, his services will be VAT standard rated.

Construction Management and Project Management

The project manager or construction manager is appointed by the client to plan, manage and coordinate a construction project. This will involve establishing competitive bids for all the elements of the work and the appointment of trade contractors. The trade contractors are engaged directly by the client for their services. The VAT liability of the trade contractors will be determined by the nature of the construction services they provide and the building being constructed.

The fees of the construction manager or project manager will be VAT standard rated. If the construction manager also provides some construction services, these works may be zero or reduced rated if the works qualify.

Liquidated and Ascertained Damages

Liquidated damages are outside of the scope of VAT as compensation. The employer should not reduce the VAT amount due on a payment under a building contract on account of a deduction of damages. In contrast, an agreed reduction in the contract price will reduce the VAT amount.

Similarly, in certain circumstances, HM Revenue & Customs may agree that a claim by a contractor under a JCT, or other form of contract, is also compensation payment and outside the scope of VAT.

Reverse Charge for Building and Construction Services

A VAT reverse charge was introduced on 1 March 2021 for certain building and construction services. This measure has been introduced to combat missing trader fraud in the construction industry, removing the opportunity for fraudsters to charge VAT and then go missing, before paying it over to the HMRC.

For certain supplies of construction services, the customer will now be liable to account to HMRC for the VAT in respect of those purchases rather than the supplier. This reverse charge will apply through the supply chain where payments are required to be reported through the Construction Industry Scheme (CIS) up to the point where the customer receiving the supply is no longer a business that makes supplies of specified services (end users).

The introduction of a reverse charge does not change the liability for the supply of the specified services: this results from the way in which the VAT on those supplies is accounted for. Rather than the supplier charging and accounting for the VAT, the recipient of those supplies accounts for the VAT in its return, instead of paying the VAT amount to its supplier. It will be able to reclaim that VAT amount as input tax, subject to the normal rules. The supplier will need to issue a VAT invoice that indicates that the supplies are subject to the reverse charge.

The types of construction services covered by the reverse charge are based on the definition of 'construction operations' used in CIS under section 74 of the Finance Act 2004.

As well as excluding supplies of specified services to end users, the reverse charge does not capture supplies of specified services where the supplier and customer are connected in a particular way, and for supplies between landlords and tenants.

Contact

For further information or assistance, please contact one of the Fiscal Incentives team:

help.vat@aecom.com

Estimator's Pocket Book, 2nd edition

Duncan Cartlidge

The *Estimator's Pocket Book, Second Edition* is a concise and practical reference covering the main pricing approaches, as well as useful information such as how to process sub-contractor quotations, tender settlement and adjudication. It is fully up to date with NRM2 throughout, features a look ahead to NRM3 and describes the implications of BIM for estimators.
It includes instructions on how to handle:

*   the NRM order of cost estimate;
*   unit-rate pricing for different trades;
*   pro-rata pricing and dayworks;
*   builders' quantities;
*   approximate quantities.

Worked examples show how each of these techniques should be carried out in clear, easy-to-follow steps. This is the indispensable estimating reference for all quantity surveyors, cost managers, project managers and anybody else with estimating responsibilities. Particular attention is given to NRM2, but the overall focus is on the core estimating skills needed in practice. Updates to this edition include a greater reference to BIM, an update on the current state of the construction industry as well as up-to-date wage rates, legislative changes and guidance notes.

February 2019: 292pp
ISBN: 9781138366701

To Order
Tel: +44 (0) 1235 400524
Email: book.orders@tandf.co.uk

For a complete listing of all our titles
visit: www.tandf.co.uk

Water Network Charges

This article gives guidance on water suppliers, new connections, infrastructure charges and network charges.

New connections and charging methodology

The Water Industry Act 1991 (the Act) (as amended by the Water Act 2014) allows the Water Services Regulation Authority (Ofwat) to set rules about the charges that a water and/or wastewater company can impose on its customers and on a water supply and/or sewerage licensee (Retailer). It also allows Ofwat to make amendments to these rules following a formal public consultation.

The methodology for charging for new connections has changed effective from April 2022, based upon the Ofwat document 'Charging Rules for New Connection Services (English Undertakers)' as Updated relative to the March 2020 Version.

The principles of who can undertake the work remains as the previous rules, but the calculation of connection charges, both for water and sewerage, have been revised.

Existing appointees' and new appointments & variations (NAVs)

Most customers in England and Wales currently receive their water and sewerage services from one of 22 appointed 'water' and 'sewerage and water only' suppliers that were in existence when the sectors were privatized. These suppliers are known as 'existing appointees' or statutory providers.

New Appointments and Variations or (NAVs) refers to small competitor water companies that are not subject to full price controls like the statutory providers. The regulator 'Ofwat' recognizes that NAVs can provide services, such as their own water infrastructure and that these services can create competition in the market place when customers are purchasing new connections.

New appointments provide challenge to existing appointees. This drives efficiencies, stimulates innovation and reveals information. They have the potential to benefit all customers, through:
- Lower prices;
- Improved service;
- Greater choice of supplier for developers and large user customers.

Providing new connections

The term 'new connection' is used to describe where a customer requires either or both:
- Access to the existing public water supply or sewerage system by means of a service pipe or lateral drain.
- A new water main or public sewer.

A customer may choose their own contractor to do the work, which is then known as 'self-lay'. The statutory provider will take over responsibility for (adopt) all self-laid infrastructure that meets the terms of its agreement with the owner, developer or self-lay organization that carries out the work.

The Water Industry Act 1991 (WIA91) places a number of duties on statutory 'water only' and 'water and sewerage' companies in providing or enabling new connections for an individual property or development site.

If a property requires a new water main, sewer, service pipe or lateral drain for domestic purposes (cooking, cleaning, central heating or sanitary facilities), the owner or developer may ask the local statutory provider to install the infrastructure. For water mains and public sewers this is often referred to as 'requisitioning' the infrastructure.

Self-lay

If a development requires a new water main or sewer, you may ask the water or sewerage company to install the pipework. Alternatively, you may choose your own contractor to do the work, which is known as self-lay. The water company will take over responsibility for (adopt) self-laid pipes that meet the terms of its agreement with the developer or self-lay organization (SLO) that carries out the work.

Accreditation

Before a self-lay organization can carry out work, they must be approved by the relevant water company. Self-lay organizations can avoid having to comply with the 22 separate water companies' requirements by becoming accredited under the Water Industry Registration Scheme (WIRS) which is recognized by all the water companies and Water UK.

Under WIRS, Lloyd's Register carries out technical assessments of the service providers who elect to be accredited for contestable works associated with installing water infrastructure.

Areas covered by WIRS include:
- design
- construction
- connections
- commissioning
- project management

Lloyd's Register offers an independent assessment and registration process that includes assessing the self-lay organization's management procedures and processes.

WIRS plays an important part in supporting and expanding the self-lay market, as it promotes and ensures compliance with the highest standards of water infrastructure installation and the essential protection and maintenance of water quality.

The Water Industry Registration Scheme Advisory Panel (WIRSAP) manages the scheme, its members include representatives from water companies, self-lay organizations and the regulator Ofwat.

Other charges to consider

On 20 December 2018 Ofwat set out its final position on Wholesale Charging Rules and Charging Rules for New Connection Services (English Undertakers) in relation to network and infrastructure charging.

With the exception of Small Companies, each Water Company must present its Charging Arrangements using the six worked examples for the following scenarios, and in the standard format set out below.

Scenarios summary
1. Single connection to a house from an existing main
2. Single connection to a block of flats from an existing main
3. Medium housing development requiring new mains and communication pipes (by others)
4. Medium housing development requiring new mains and communication pipes (by Water Company)
5. Large housing development requiring new mains and communication pipes (by others)
6. Large housing development requiring new mains and communication pipes (by Water Company)

Example of 2022 Methodology Water Services Connection Charges – Thames Water Scenario 4

This assumes a medium sized housing development of 50 properties off a new mains. Each with 3 m of pipelaying in customer excavation. Plus a total of 300 m of New Mains of which 10 m of 125 mm pipelaying in road, 50 m of 125 mm pipe laying in the footpath, 140 m of 125 mm pipe and 100 m of 90 mm pipe in unmade ground assuming Barrier Pipe. Prices below include for External Meter and first metre of pipelaying is included in connection Charge.

Application Fee		£50
Design Fee (£50 for first property and £30 for each additional property)		£1,000
Construction Charges	Connection of 125 mm in road (1 × £6,670)	£6,670
	9 m of 125 mm pipe laying in road (9 × £850)	£7,650
	50 m of 125 mm pipe laying in footpath (50 × £800)	£40,000
	140 m of 125 mm pipe laying in unmade (140 × £330)	£46,200
	100 m of 80/90 mm pipe laying in unmade (100 × £270)	£27,000
		£127,520
Connection Charges		£56,500
Consent to Connect (Surface water)		£330
Infrastructure Charges (50 × £445)		£38,250
Income Offset		£0
Total (Excl. VAT)		**£223,650**
Environmental Discount Water (50 × £200)		**−£10,000**

Infrastructure charges

If a supplier is providing new water supplies for domestic purposes (such as to new housing developments) it is legally entitled to levy infrastructure charges. These can be raised where premises are connected to the water company's water supply (or to the sewerage company's sewers) for the first time. These charges are intended to provide a contribution towards the costs of developing or enhancing local networks to serve new customers.

A customer will have to pay infrastructure charges when a property is connected to the water and/or wastewater networks for the first time. This is in addition to the charges for making the actual physical connection to the water main and/or public sewer. When a customer makes a connection to a network the operator is entitled to charge the customer in accordance with its published Charging Arrangements for the connection works in addition to raising infrastructure charges. Infrastructure charges apply for premises where the supply of water or provision of sewerage services is intended for domestic purposes. Water for domestic purposes refers to usage for drinking, washing, cooking, central heating and sanitary purposes for which water is supplied to the premises. All other purposes (including supplies for a laundry business or for a food or drink takeaway business), are regarded as non-domestic purposes. *The definition is about the usage of the water and not the type of property being supplied.* Domestic sewerage purposes refers to the removal of the contents of lavatories, water which has been used for cooking or washing or surface water from the premises and associated land, with the exception of laundries and takeaway restaurants. *As with a potable water connection, the definition is not about the type of property.* Please note that a single development may include a combination of supplies for domestic and non-domestic purposes. For example, a development may include a number of flats (expected to be for domestic purposes), retail units (expected to be for domestic purposes), a fire-fighting supply (non-domestic purposes) and landlord supplies (non-domestic purposes). Similarly a separate supply requested for a swimming pool or a garden tap is considered to be for non-domestic purposes even if within a residential property.

The relevant multiplier

A Relevant Multiplier is used to calculate the amount of an infrastructure charge for the connection of premises other than dwelling houses or flats, if water is provided by a supply larger than the standard size used for new connections of houses – 25 mm or 32 mm external diameter pipe. For other properties, such as student housing, offices or care homes the regulator has introduced a pricing calculator known as a relevant multiplier. This tool is utilized to measure a cost that reflects the increased impact on a provider's network. The wastewater infrastructure charge is calculated on the same basis as the water infrastructure charge unless the customer is able to show that waste and surface water flows are not being discharged to the public sewer.

The Relevant Multiplier (RM) is a way of working out infrastructure charges for the following types of property:
• Residential properties with a single, shared supply pipe and which are subject to a 'common billing agreement'; this includes sheltered housing, student accommodation and high-rise flats.
• Non-residential properties where the supply pipe is larger than the standard size, such as office blocks.

How the relevant multiplier is calculated

Each water fitting (wash basin, bath, shower, etc.) is given a 'loading unit' based on the amount of water it uses. The average number of units per property is taken as 24, equal to an RM of 1.00. The operator can use this as the basis for calculating the RM for each property on a development where the RM applies. The operator will do this by adding up the loading units for all the water fittings on a development. It divides this by the number of properties to give the average loading units per property. It divides this again by 24 (the average loading units) to give the RM for each property. Details of the number of loading units assigned to each water fitting are shown below. For properties subject to a common billing agreement, the RM can be more or less than 1.00. For other properties the minimum is 1.00.

Using the relevant multiplier to calculate infrastructure charge

Operator will utilize the RM multiplied by the standard charge to provide the infrastructure charge for that property.

Example of a residential development

• The development consists of 90 apartments with a common billing agreement and the total loading units are 2,100.
• The total loading units (2,100) are divided by the number of properties (90) and again by the average loading unit (24). This gives an RM for each flat of 0.97 (2,100 ÷ 90 ÷ 24 = 0.97)
• The infrastructure charge for each flat is the RM of 0.97 multiplied by the standard charges.
• The infrastructure charge for the whole development is the RM multiplied by the standard charge multiplied by the number of properties.

The water infrastructure charge for the development is therefore RM 0.97 × 90 properties × £400 standard charge Water charges will differ from each water and sewerage undertaker = £34,900

In 2022 the Income Offset is no longer offered. Instead, an Environmental Discount is available.

Environmental discounts

Environmental Discounts are only applicable to quotes produced from 1 April 2022, and evidence demonstrating compliance with the criteria must be provided alongside the application for New Connection Services. If the development does qualify, details of the discount will be included within the quotation.

Wastewater connections charges

Example of 2022 Methodology Wastewater Services Connection Charges – Thames Water

Application Fee	(Connection to trunk sewer)	£10,700
	(Connection to other public sewer)	£3,600
Connection Charge (In road)		£23,800
Pipe Laying (In road) (300 mm dia. × 6 m length)		£10,860
Infrastructure Charges (50 × £320)		£16,000
Total (Excl. VAT)		**£54,260**
Environmental Discount Water (50 × £25)		**−£1,250**

Choosing your water supplier

Under the current legislation, household customers are not able to change their water supplier or sewerage service provider. The water or sewerage company that supplies your property will depend on where you live.

If you are a business customer in England, you may be able to choose the retailer for your water or sewerage services. With a choice of water retailers (there may be a minimum capacity supply that apply) you may benefit from:

- better standards of service
- tailored service offerings
- advice on saving water
- lower prices
- you will still benefit from the same water quality.

Since April 2017, most non-household customers based wholly or mainly in England are now able to choose both their water and wastewater retailer. They now do not have to use a minimum amount of water, as was the requirement previously. Customers will also be able to apply for a licence (WSSL) to supply themselves retail services. A licensee limited to self-supply ('self-supplier') can only supply its own premises and the premises of people or companies associated with it. Licensees limited to self-supply are not able to supply retail services to premises of third parties, such as supplying other business customers unrelated to their own businesses.

Regulatory information within this document has been referenced directly from the regulator Ofwat. This information relates to its current guidance on infrastructure and network charging and is correct as of December 2018.

Further Information on water distribution network areas can be found at https://www.ofwat.gov.uk/regulated-companies/ofwat-industry-overview/licences/

Grenfell and Construction Industry Reform:
A Guide for the
Construction Professional

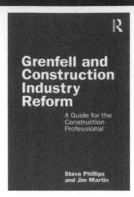

Steve Phillips and Jim Martin

In the wake of the tragic events of the fire at Grenfell Tower, the inquiry into the fire and the independent Hackitt Review revealed deep-rooted and unpalatable truths about the current state of the UK construction industry. Dame Judith Hackitt was scathing in her assessment of the construction industry denouncing it as "an industry that has not reflected and learned for itself, nor looked to other sectors" and defining the key issues as ignorance, indifference, lack of clarity on roles and responsibilities and inadequate regulatory oversight and enforcement tools.

There is an urgent need to change practices and behaviours to prevent a similar tragedy from reoccurring. This book sets out the changes required, why they are required, how they are to be achieved and the progress towards them to date.

Implementation of these major safety reforms will move the construction industry from the conditions that allowed the fire at Grenfell Tower to occur, to a system where construction professionals take greater responsibility for the safety of residents in their buildings. This book provides an overview of how the movement towards implementing a new building safety regime has unfolded over the last three years and details what still needs to be done if residents are to feel safe and be safe in their own homes.

September 2021: 120 pp
ISBN: 9780367552855

To Order
Tel:+44 (0) 1235 400524
Email: tandf@bookpoint.co.uk

For a complete listing of all our titles visit:
www.tandf.co.uk

Building Regulations

England's Building Regulations provide an insight into how building design is evolving over time – the key challenges and preoccupations at a given moment. As a template for what a safe, healthy building should be, they are also an important indicator of the direction of travel the government wants building design and the wider construction industry to move in.

These changes are reflective of a time where cutting carbon and creating well-ventilated spaces are paramount. Introduced by the Department for Levelling-Up, Housing and Communities (DLUHC), the new regulations are primarily driven by the pursuit of the UK's legally binding net zero 2050 goal and reflects the global need to decarbonize.

While 2050 may seem far distant, buildings will increasingly need to meet stringent carbon emission standards if the UK is to meet its environmental goals.

The updated rules have also been influenced by the Coronavirus pandemic. This has made clear the need for effective ventilation in buildings, particularly in high density shared spaces such as offices. Furthermore, office tenants are more aware than ever of a building's health elements.

Perhaps one of the most important things to recognize is that these updates are setting the stage for larger changes in three years' time. These uplifts are an interim – and preparatory – measure ahead of the much wider-reaching Future Homes Standard and Future Buildings Standard, which are due in 2025 and will apply to new homes and new non-domestic buildings, respectively. The 2022 changes are intended to help make the transition to the more stringent new standards easier for the industry.

Current situation in the sector

The affected regulations are Part L and Part F. Broadly speaking, Part L addresses carbon emissions and energy efficiency, while Part F is concerned with delivering good indoor air quality by delivering sufficient ventilation and minimizing the ingress of external pollutants. The change also introduces a new regulation, Part O. This regulation is about reducing the risk of overheating, primarily in domestic buildings.

The previous major revision to UK building regulations was in 2013. Under the 2022 updates, a key focus is on improving the building fabric and building services to make them more energy efficient. The headline element is the requirement for a 27% average cut in operational carbon for new non-domestic buildings (with a 31% average cut for new domestic buildings).

Thanks to increasing recognition of the need to reduce carbon emissions, some parts of the industry are already well prepared for the changes. Some developers are already designing buildings in ways that the new regulations will encourage. This is particularly true in London, as the Greater London Authority has been requiring higher standards for several years. The London Plan, for example, means that similar requirements to many of the changes introduced by the 2022 regulations are already in force. In the capital, it may well feel like business as usual for developers.

Design implications

Setting an effective regulatory environment, which incentivizes positive changes while minimizing the risk of unintended consequences, is extremely challenging given the technical complexity of the built environment and wide range of externalities faced on individual projects. A flexible approach is essential to ensure the most effective solutions are found based on individual project constraints.

In England, the notional building sits at the heart of the standard-setting process. It is used to set a feasible target across different building types. The CO_2 emissions and primary energy of the proposed building should be equal to or lower than that of the notional building, which has the same size and shape as the proposed building, but built to a standard recipe of U-values and building service efficiencies.

The specifications used to define the notional building are tougher this time, typically demonstrating a 27% improvement in operational carbon performance compared with its 2013 counterpart. This includes introducing the use of renewable technologies into the notional building.

The developer has the option to vary the specifications used in the notional building. This means it can relax some parameters relative to the notional specifications (but remaining at least as good as specified backstop values), as long as other parameters are improved beyond the notional specification, so that ultimately the CO_2 emissions and primary energy are as good or better than the notional building. This is intended to allow for flexibility and choice in design – a 'mix and match' approach, enabling certain design choices to be offset by other elements of the building.

Costs will be impacted – but not to a major degree

Much has evolved in the time between 2013, when the regulations were last updated, and 2022.

Technological advancements and increases in supply chain capacity for more energy efficient products have eased pressure on the cost of meeting the new regulations. For example, the premium for triple-glazed windows has dropped in recent years as uptake has increased. The difference in cost versus double-glazed units is now approximately £20–£30/m^2 – assuming that the same frame can accommodate both double and triple-glazed units. Frames are key to pricing. For some projects there is an uplift on the frame cost, which means that the cost differential per square metre can rise to £50–£100/m^2 .

LED lighting was not widely undertaken as an option nine years ago. Technology has improved, and prices have dropped to make LED lighting a mainstream and accessible choice – in fact, the default option. Similarly, lighting and daylight control systems are now generally a standard feature in commercial office specifications throughout England unlike back in 2013.

Under the new regulations, there is a requirement to monitor indoor air quality in new build offices, and higher risk areas in other new non-domestic buildings, to reduce the transmission of airborne illnesses. Carbon dioxide monitors are expected to be installed to flag if CO_2 has risen beyond acceptable levels and trigger an increase in ventilation. Typically, this can be expected to add around £4/m^2 to construction costs. This requirement applies (with a few caveats) to several types of space, including offices, gyms, theatres, nightclubs, pubs and chilled food processing areas.

The attractiveness of electricity as a power and heat source has also changed. The carbon intensity of electricity dropped significantly over the past decade, in large part due to the rapid increase of renewable electricity feeding into the UK energy system. Fossil fuels now account for less than 50% of total UK electricity generation, making electricity a more attractive source of power from a carbon reduction perspective.

Building regulations previously favoured the use of gas as an alternative to electricity – a situation which the rule change reverses. Heat pumps powered by electricity or even direct sources of electricity for heating water are now favoured over gas-fired boilers or combined heat and power (CHP).

Switching to electric power has clear environmental benefits but there can be cost implications. Electric heat pump technology in lieu of traditional gas-fired boilers and chillers can add up to £50/m^2 to construction costs. In addition, increasing the electricity supply increases the maximum demand level which can put in pressure on the grid, especially in densely populated urban areas and increase cost. Often local substations are unable to meet demand and connections to larger substations are required which increases complexity and can impact programme length and cost. Early engagement with the distribution network operator (DNO) is the key to minimizing the risk of potential impacts – addressing how loads can be secured and what is the best procurement approach.

Location and building type will strongly influence compliance choices

Because developers can choose the means by which their building design meets the requirement to match the operational carbon emission standards of a 'notional building' of the same shape and size, there are myriad ways to comply with the 27% average carbon reduction in the new Part L.

One part of a design solution can be to cover a building's roof in photovoltaic panels. This option, however, is only likely to be suitable for projects with large, simple footprints. In a London office project with a tight footprint, in contrast, it is far more likely that the roof has a terrace event space, or other area, to support occupant wellbeing or increase commercial potential, over installing solar panels. That said, solar panels might be the best choice for a large, out-of-town distribution centre building with swathes of available roof space.

As the public becomes increasingly aware of the shift towards net zero, clear indicators of a building's efforts to reduce its carbon output can be a selling point for potential tenants. Lighting controls to prevent unnecessary energy use are a good example. Installing solar PV makes a very clear statement of a building's green credentials – but it is important to note that as the UK's electricity system decarbonizes, the carbon saved by photovoltaics is reduced.

Less obvious or public-facing carbon cutting choices may derive a better carbon reduction, such as installing a more efficient heat pump. Another solution for an office building for which photovoltaics are not suitable could be to improve the U-value of a building's walls. The requirements may need to be met by multiple combined improvements. Hence it may need both improved U-value and low carbon heating. In particular, improving the U-value on its own is unlikely to comply with the demands of the new regulations.

There is likely to be a higher cost impact to meet requirements for out-of-London projects compared with those in London. This is because developers are less likely to have already established a policy of incorporating major carbon reduction measures, such as a shift from gas boilers to heat pumps, into their projects. This difference is primarily because planning requirements outside London are generally more relaxed on energy and carbon. One of the many reasons that developments are built in out-of-town locations is because the land is cheaper, and planning is quicker and simpler.

Ensuring the timing is right

There is a one-year transition period from June 2022. This is to provide developers with assurance about the standards to which they must build. They should not have to make material amendments to work which is already underway when new regulations come into force.

Effectively, developers have one year from when the regulations come into effect to commence work on each individual building on site if they still wish to adopt the current (2013) regulations. To do so, developer would also have needed to submit a building or initial notice or deposit plans.

Looking to the future

We have looked at three standard building types with basic specifications. In particular for larger, more complex projects we advise that developers should go through the regulations in close detail and commission indicative, early-stage modelling to identify appropriate solutions in order to build a picture of relative costs and benefits.

The most important takeaway is that the new regulations are part of a wider shift towards highly efficient buildings where low levels of carbon – both from an emissions and an embodied carbon perspective – is the rule, rather than the exception. In adapting to the new regulations, our key advice is:

• Timing is important. The rules do not change overnight – this year, developers will need to assess whether they will fall under the old regulations or must comply with the new.

• If developing a project that falls under the new Part L then it is vital that the impact of this is considered at the earliest opportunity.

• Our analysis suggests that Greater London Authority-compliant buildings may not need to change very much to meet the new regulations, but this is not a blanket rule.

• The regulations are likely to drive a national shift away from fossil-fuelled heating towards heat pumps and other low carbon sources.

In the coming years, standards for carbon reduction and building efficiencies will only become higher. With the far more ambitious Future Homes and Future Buildings Standards set to go into technical consultation next year ahead of their introduction in 2025, we're advising the industry to aim as high as possible when designing out carbon, and to drive up efficiencies and invest in innovation in order to remain compliant. As ever, it is best to be ahead of the curve rather than behind it.

New challenges and adaptability

The covid pandemic started whilst work was already underway to revise the building regulations Parts L and F. As the ways the virus was spread became better understood, steps were taken to incorporate some of this learning into the regulations so that the impacts of future outbreaks might be managed more easily.

Part F is particularly relevant to this as it is concerned with the ventilation of buildings. Several changes have been made to this part of the regulations; some of these changes seek to future-proof some building types where the spread of air-borne viruses might be reduced through improved ventilation standards.

For example there is a new requirement relating to recirculating ventilation systems in offices:

Recirculation of air within ventilation systems in offices 1.37 Ventilation systems that, under normal operation, recirculate air between more than one space, room or zone should also be able to operate in a mode that reduces the risk of the transmission of airborne infection. This can be achieved by one or more of the following. a. Systems capable of providing 100% outdoor air to the levels specified in paragraphs 1.32 to 1.34 to all occupiable rooms and common spaces, without recirculating air. b. Systems incorporating a UV-C germicidal irradiation system that is able to disinfect the air that is being recirculated. This type of system is commonly located within the heating, ventilation and air conditioning (HVAC) system or ductwork. c. Systems designed so that they can incorporate HEPA filters, if required, which are able to provide filtration of the recirculated air. Note: For some system types some recirculation is necessary or desirable in normal operation. Use of any full outdoor air mode, UV-C germicidal irradiation of HEPA filtration may not be necessary under normal conditions of operation.

Source: Statutory guidance. Ventilation: Approved Document F. Department for Levelling Up, Housing and Communities

Recirculating ventilation systems can be an effective means of reducing energy demands compared with systems that provide 100% fresh air. This new clause in Part F seeks to strike a balance by allowing this system type to be used but requiring the system to have the functionality to mitigate spreading infections. This might be done by switching to a mode with no recirculation or incorporating a means of treating or filtering air before it is recirculated. Some of these features may require a system to be recalibrated or recommissioned when switching to a mode with an increased resistance to the ventilation flow.

Costing Net Zero Homes

The built environment is a major contributor to greenhouse gas emissions, accounting for an estimated 49% of the UK's total output.

Nevertheless, new homes must be built. The government wants 300,000 homes built each year to meet demand but construction is falling far short of this goal.

The industry is tasked with meeting the dual challenge of delivering new residential buildings while reducing carbon output to neutral levels, in order to fulfil the UK's legal requirement of producing net zero carbon emissions by 2050. Building homes to net zero standards will therefore be critical to providing much-needed housing stock while keeping the sector's carbon emissions to a minimum.

Current guidance

As yet, there are very few completed net zero carbon housing developments of significant commercial value and scale to use as standard bearers and templates for future projects. Existing low or zero carbon developments tend to be one-off experiments, such as small projects built to Passivhaus standards.

The London Energy Transformation Initiative (LETI) current guidance is that for a residential building to be whole-life net zero, it must have a zero carbon balance for its operational activities, and be 100% circular – meaning each and every one of the building's materials and products 'are made up of reused materials, and the building is designed for disassembly such that 100% of its materials and products can be reused in future buildings'.

Achieving these standards will seem to many like an impossible goal. Nevertheless, it is a goal that will need to be met in the coming years for the industry to remain viable and new-build projects to be greenlit.

The industry is at a unique point in time where we know we must provide net zero developments in order for the construction sector to meet regulatory requirements and keep up with the rest of the country's efforts to decarbonize; yet a single or an agreed set of overarching guidance, principles and rules for creating net zero homes has not yet been fully established. Perhaps most importantly, if net zero buildings cannot be made affordable – at least not right now, at this early, experimental stage – then we also need to consider what the premium is on delivering these projects.

For a new-build project targeting net zero status in 2021, the goal therefore is to strive to meet operational and embodied carbon targets from trusted industry networks and institutions such as LETI and the RIBA, and to offset any remaining carbon.

The future

There is an obvious need for accurate, reliable data to enable standardization of net zero – with the goal being that the carbon value of each element of design, construction, operation and decommissioning can be easily quantified. Our Scope X™ process addresses these gaps and helps measure critical elements of building design.

The uncertainty around what exactly constitutes a net zero homes project will soon fade. For example, residential buildings will need to be constructed to meet future Part L regulations – updates to which are in consultation, with new guidance due next year. These updated rules will help set the standard for the energy performance and carbon output of new dwellings.

Changes to Part L of the Building Regulations will also inform the upcoming Future Homes Standard, which is due to be implemented in 2025. That policy will require all new homes to be 'zero carbon ready', which will involve a huge step change in how we design and heat our homes. In the meantime, developers, contractors, suppliers and designers are tasked with meeting the net zero challenge using their own metrics and definitions – which will be influenced by existing cost models.

It is worth noting that institution-specific targets and standards are constantly evolving, further adding to the complexity of designing and building for net zero – LETI, the RIBA and the Greater London Authority are all in the process of updating and aligning their embodied carbon targets.

Regardless of which set of principles eventually becomes standardized, building net zero homes will undoubtedly require a massive shift in our collective mindset. For example, LETI advises that we begin to regard new buildings as 'material resource banks' – that is, as a source of materials that can and should be used decades after the building has been decommissioned.

It is also likely that older buildings that become available for redevelopment will be treated as 'donor buildings' for the future, and that we may also see specialist salvage contractors create a market through this. These concepts require us all to start thinking more imaginatively.

Net zero concepts in some ways represent a return to traditional ideas – of sourcing locally and frugally and of considering natural materials and reducing waste – yet they also demand understanding and being ready to deploy forward-looking technology and carbon-quantifying techniques to achieve and measure a successful project.

For net zero housing to move from ideas, plans and goals to a commonplace reality in the UK, adaptability and a willingness to update design and construction techniques will be required. The pay-off for those who do apply net zero principles to their residential projects is to be working at the vanguard of design, construction and building philosophy – and to be actively contributing to meeting the pressing need for a lower carbon building industry.

Clearly, for new innovation that delivers zero carbon to become more cost-effective, it needs to be implemented as mainstream practice and regulation – and we must get beyond the age-old chicken and egg question.

How to Cost-Effectively Manage Evolving Data Centre Design Demands

With 4.66 billion active internet users globally, data centres are one of the fastest growing and most important components of the global economy. By 2025 data usage is projected to have increased tenfold on 2018 levels. To cope, more data storage space is needed. This is leading to a boom in data centre new-build schemes, with construction set to expand by nearly 10% per annum between 2018 and 2025 – creating major opportunities for investors in this alternative, rapidly expanding asset class.

Demand for new data centre construction is primarily being driven by the global move to cloud computing and storage. This has created a need for the hyperscale data centre as owned and developed by the likes of AWS, Microsoft and Google. Co-location facilities continue to provide variable amounts of data centre white space to other client tenants, as well as providing these cloud giants with facilities on a wholesale basis. Enterprise (owner-operated) data centres still exist, but a high proportion of development is within the hyperscale market as most companies and individuals use this for computing and storage.

Although essentially industrial buildings, data centres are highly functional, and their design requirements are changing quickly – not least when it comes to meeting strict carbon reduction requirements and providing adequate cybersecurity. In this article – an abridged version of one that first appeared in *Building* magazine – we focus on three evolving areas of data centre design.

Design considerations

The data centre industry is under pressure to bring resilient, efficient and secure projects to market quickly. In addition, clients are increasingly asking to complete site selection, acquisition, design and construction work as energy-efficiently and innovatively as possible. When designing a data centre, it is critical that the following three elements are considered from the outset for successful delivery.

1. Scalability: The flexibility to accommodate change without major works to the physical or IT infrastructure is key to a strong data centre design. The need to refresh IT servers, often on a three-year cycle, introduces a design requirement for high-frequency equipment change. As a data centre site tends to operate for at least two decades, it must be designed to handle the significant technology upgrades and updates that will occur many times throughout its operational life.

In addition, potential future load requirements are unknown, which adds further demand for adaptability. It makes sense, therefore, to design in a modular UPS system, so that as a data centre's load grows, additional power modules can easily be added. In turn, as power consumption increases, the design should offer accommodation for swift upgrades to the cooling systems that support it. This means carefully balancing current requirements against future demand.

Given potential future load requirements are unknown, it makes sense to design in a modular UPS system, so that as a data centre's load grows, additional power modules can easily be added. In turn, as power consumption increases, the design should offer accommodation for swift upgrades to the cooling systems that support it. This means carefully balancing current requirements against future demand.

2. Resilience: A data centre's critical load comprises all the hardware components that make up the IT business architecture. Any new design must build in sufficient resilience to ensure any component failure or required maintenance activities does not compromise the critical load.

Typically, resilience is measured and defined by the data centre's tier rating, a standardized, industry-accepted mechanism devised by the Uptime Institute. The system ranges from tier 1 to tier 4 – a tier 1 project will have a simple design with minimal backup equipment and a single path for power and cooling, whereas a tier 4 data centre will be designed to be fully fault-tolerant and with redundancy for every single component.

3. Connectivity: Selecting systems that use cloud-based, mobile-friendly data centre infrastructure management (DCIM) software, which can be shared and managed remotely, allows users to leverage information beyond that of their data centre, enabling benefits such as predictive maintenance. This helps to maximize the life of components while ensuring data centre availability and performance, as components such as batteries can be replaced before they fail.

In addition, there has been a significant increase in the use of AI to monitor and manage power and cooling at data centre sites. This information can be used for detailed reporting and analysis and to improve energy efficiency. However, depending on the level of integration and the extent of the information that is being generated and managed, incorporating this kind of software can come at a significant cost.

The question of cooling

Keeping data centres cool is one of the most critical elements of their design. There are many factors that drive the selection of any cooling option, not least capital and lifecycle costs, but also the location of the data centre and the feasibility of incorporating innovations. Understanding the cost drivers and benefits of each are crucial to advising clients effectively.

Solutions include passive cooling, which uses natural ventilation to remove heat from the building, and immersive liquid cooling, where servers are immersed in a rack filled with coolant that can have more than 1,000 times the heat capacity of air. Many operators are now integrating AI and machine learning into their cooling systems. The AI learns to apply the optimum amount of cooling at any given time, driving up efficiency and reducing energy use.

Location helps, too. Data centres can also simply be built in colder climates – such as Iceland, where BMW's data is stored – or the Baltic region, which hosts a number of the world's global hyperscale data centre facilities. Another approach, which is generally required by local authorities in the Baltic region, is to recycle waste heat and use it for district heating schemes.

As data centre growth continues, finding innovative ways to use heat for nearby homes and businesses is an important way that data centres can be better integrated into communities and contribute to wider decarbonization efforts.

Upgrading existing data centres

As data centres near the end of their planned operational life, they can suffer from an increased risk of failure and reduced efficiency. Yet they can also represent an excellent opportunity. An upgrade or refurbishment of an existing facility can cost significantly less than a new-build, and can be achieved in a far shorter time frame, since there is no need for planning approvals or additional utility connections. Upgrade schemes also have significantly reduced embodied carbon compared with a new-build project.

Upgrading legacy installations is an effective way to increase capacity without space and carbon footprint increases. It also offers the opportunity to design in crucial long-term benefits, such as strengthening competitiveness, reliability, safety, flexibility and environmental integration, as well as security and monitoring.

A key component of redesigning the centre will be studying and analysing factors such as air flow, heat propagation, audible noise, and electromagnetic compatibility. Other areas to consider in extending the life of an ageing facility include the need to elevate the data centre operating temperature, upgrade servers and systems, improve the system layout and rack layout for power and cooling efficiency, consider supplemental or alternative cooling schemes, the availability and reliability issues in power distribution and finally the availability of data centre power, including the potential for alternative power sources.

Current challenges and trends – and the road ahead

As the data centre sector has matured, its needs and demands have become clearer. End users are now better placed to choose between construction of their own data centre or engaging with a third-party provider, whether through co-location, managed or cloud services, or engaging with the upgrade of an existing centre.

These decisions are underpinned by a better understanding of the total cost and time frame of ownership, which typically covers a period of five to 10 years. Institutions and funds are therefore now able to invest with greater certainty, as unknown costs can be minimized with the right level of management capability. However, this also means that data centres are becoming more commoditized. Competition is fierce as third-party and outsourcing organizations strive to lower costs and differentiate themselves from the competition.

Europe is attracting strong data centre investment as data creation continues to grow. More capital is available, and competition for deals is increasing. But the availability of power from renewable sources is constraining growth in Europe. In less mature markets, creating and accessing sufficient fibre connectivity is also a key consideration. In the Middle East alone, rapid digitization of 'smart cities' is spurring the data centre market, which is expected to grow annually by 7% between 2018 and 2024.

There are many hurdles for data centres to overcome in the coming decades, not least meeting the net-zero carbon challenge, protecting data from malicious attacks and preventable leaks, and addressing geographical and geopolitical challenges. The challenges facing the wider technology industry are also complex: maintaining talent, minimizing supply chain disruption, and implementing sustainable, environmentally-sound solutions.

For those designing data centres, security, decarbonization and innovation are the watchwords to ensure that projects are fit for purpose in this extremely fast-moving sector.

A Handbook of Sustainable Building Design and Engineering, 2nd edition

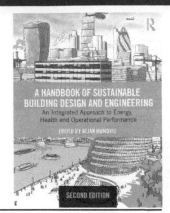

Dejan Mumovic *et al.*

The second edition of this authoritative textbook equips students with the tools they will need to tackle the challenges of sustainable building design and engineering. The book looks at how to design, engineer and monitor energy efficient buildings, how to adapt buildings to climate change, and how to make buildings healthy, comfortable and secure. New material for this edition includes sections on environmental masterplanning, renewable technologies, retrofitting, passive house design, thermal comfort and indoor air quality. With chapters and case studies from a range of international, interdisciplinary authors, the book is essential reading for students and professionals in building engineering, environmental design, construction and architecture.

October 2018: 604pp
ISBN: 9781138215474

To Order
Tel: +44 (0) 1235 400524
Email: book.orders@tandf.co.uk

For a complete listing of all our titles
visit: www.tandf.co.uk

Taylor & Francis
Taylor & Francis Group

Reducing Operational Carbon on Commercial and Educational Campuses

Organizations, educational institutions, and companies with large campuses must decarbonize their real estate. In many cases, this is a complex challenge as each individual building – whether recently-built or 100 years' old – must be made more energy efficient to reduce the amount of operational carbon used.

Historically, this might have been done on a project-by-project basis. To meet publicly stated climate emergency declarations and net zero carbon commitments however, organizations, institutions and companies will need to take a different approach, or risk reputational damage.

We have recently completed the development of an estate strategy for the University of Greenwich, London. Central to this strategy is the University's commitment to reaching net zero carbon emissions by 2030.

In this article, we argue the case for the adoption of a comprehensive decarbonization plan on commercial and institutional campuses as the best way to reduce operational carbon in the most time and cost-efficient way. Furthermore, a comprehensive plan can also reduce reliance on increasingly expensive and bureaucratic carbon offsetting schemes.

1) A piecemeal approach won't cut it
Currently, there is a common knowledge gap between setting net zero targets and understanding what role decarbonization plays on that journey, and how it can be achieved. Decarbonization is not as simple as replacing gas

boilers with heat pumps (as doing so greatly increases running costs). Instead, we strongly advocate that any action to reduce operational carbon by moving away from fossil fuels should be considered in the context of all ongoing maintenance and must touch all capital works.

2) If you can't measure it, you can't manage it

Establishing a baseline for how well an estate is performing is the first step, along with general condition of the plant and fabric. If not already available (as part of either planned maintenance or maintenance backlog information), then an extensive survey will be required. Our Operational Carbon & Energy ANalysis tool (or OCEAN for short) gathers data on carbon and energy performance of individual buildings so that holders of large asset portfolios can understand how their entire portfolio is performing.

If not already installed, adequate metering is a good investment as it is essential to understand how an estate is performing to determine where or when to make interventions. Furthermore, once a solution has been implemented, it is equally important to then monitor performance.

3) Make better use of space

Once a baseline has been established, the buildings and systems should be evaluated to determine whether space could be better used and/or operational hours consolidated.

In universities for example, where the pandemic has accelerated the trend to online study at home complemented by collaborative in-person sessions, the way space is used on campus has altered dramatically. Managed correctly, this is an opportunity to reduce energy demand by delivering learning in selected spaces that are shared between departments and faculties, and thus used more intensively. Similarly, the creation of night hubs allows parts of an estate to remain open out-of-hours while the majority closes.

4) Develop a decarbonization plan

The next step is to develop a decarbonization plan for each building that considers fabric, lighting, controls, and plugged-in load. The plan should consider the opportunities for reducing operating temperatures to allow heat pumps to be used, and the possibility of using thermal storage. A decarbonization plan needs to be built into the planned maintenance and plant replacement schedule along with an adequate budget commitment over the identified decarbonization period.

Finally, each project needs to be delivered in sequence to ensure an institution can decarbonize in line with its publicly stated goal. Having switched to a lower energy approach with all electric services, organizations should subscribe to a certified zero carbon electricity tariff to complete the process.

Essential and achievable

Adopting a comprehensive plan to reduce operational carbon on a commercial or educational campus is essential – and achievable. In our experience decarbonizing commercial and public real estate across the region, the most effective campus-level plans are informed by an organizational-level net zero transition plan. It's the best way for firms and educational institutions to identify opportunities that save carbon, time, and money in the short and longer term, while providing assurance to stakeholders, customers and students.

Typical Engineering Details

In addition to the Engineering Features, Typical Engineering Details are included. These are indicative schematics to assist in the compilation of costing exercises. The user should note that these are only examples and cannot be construed to reflect the design for each and every situation. They are merely provided to assist the user with gaining an understanding of the Engineering concepts and elements making up such.

Electrical

- Urban Network Mainly Underground
- Urban Network Mainly Underground with Reinforcement
- Urban Network Mainly Underground with Substation Reinforcement
- Typical Simple 11 kV Network Connection For LV Intakes Up To 1000 kVA
- Typical 11 kV Network Connections For MV Intakes 1000 kVA To 6000 kVA
- Static UPS System – Simplified Single Line Schematic For a Single Module
- Typical Data Transmission (Structured Cabling)
- Typical Networked Lighting Control System
- Typical Standby Power System, Single Line Schematic
- Typical Fire Detection and Alarm Schematic
- Typical Block Diagram – Access Control System (ACS)
- Typical Block Diagram – Intruder Detection System (IDS)
- Typical Block Diagram – Digital CCTV

Mechanical

- Fan Coil Unit System
- Displacement Ventilation System
- Chilled Ceiling System (Passive System)
- Chilled Beam System (Passive or Active System)
- Variable Air Volume (VAV)
- Variable Refrigerant Volume System (VRV)
- Reverse cycle heat pump

Urban Network Mainly Underground

Details: Connection to small housing development 10 houses, 60 m of LV cable from local 11 kV / 400 V substation route in footpath and verge, 10 m of service cable to each plot in verge

Supply Capacity: 200 kVA

Connection Voltage: LV

3 phase supply

Breakdown of detailed cost information

	Labour	Plant	Materials	Overheads	Total
Mains Cable	£330	£110	£ 1,320	£350	£2,110
Service Cable	£330	£110	£550	£200	£1,190
Jointing	£1,320	£440	£1,650	£680	£4.090
Termination	Incl.	Incl.	Incl.	Incl.	Incl.
Trench/Reinstate	£1,980	£660	£1,650	£860	£5,150
Substation	–	–	–	–	–
HV Trench & Joint	–	–	–	–	–
HV Cable Install	–	–	–	–	–
Total Calculated Price	£3,960	£1,320	£5,170	£2,090	£12,540
Total Non-Contestable Elements and Associated Charges					£3,300
Grand Total Calculated Price Excl. VAT					£15,840

Urban Network Mainly Underground with Reinforcement

Details: Connection to small housing development 10 houses, 60 m of LV cable from local 11 kV / 400 V substation route in footpath and verge, 10 m of service cable to each plot in unmade ground. Scheme includes reinforcement of LV distribution board at substation

Supply Capacity: 200 kVA

Connection Voltage: LV

Single phase supply

Breakdown of detailed cost information

	Labour	Plant	Materials	Overheads	Total
Mains Cable	£330	£110	£ 1,320	£350	£2,110
Service Cable	£330	£110	£550	£200	£1,190
Jointing	£1,320	£440	£1,650	£680	£4.090
Termination	Incl.	Incl.	Incl.	Incl.	Incl.
Trench/Reinstate	£1,980	£660	£1,650	£860	£5,150
Substation	–	–	–	–	–
HV Trench & Joint	–	–	–	–	–
HV Cable Install	–	–	–	–	–
Total Calculated Price	£3,960	£1,320	£5,170	£2,090	£12,540
Total Non-Contestable Elements and Associated Charges|					£14,300
Grand Total Calculated Price Excl. VAT					£26,840

Urban Network Mainly Underground with Substation Reinforcement

Details: Connection to small housing development 10 houses, 60 m of LV cable from local 11 kV / 400 V substation route in footpath and verge, 10 m of service cable to each plot in verge. Scheme includes reinforcement of LV distribution board and new substation and 20 m of HV cable

Supply Capacity: 200 kVA

Connection Voltage: LV

Single phase

Breakdown of detailed cost information

	Labour	Plant	Materials	Overheads	Total
Mains Cable	£330	£110	£ 1,650	£420	£2,510
Service Cable	£330	£110	£550	£200	£1,190
Jointing	£1,320	£440	£1,650	£680	£4.090
Termination	Incl.	Incl.	Incl.	Incl.	Incl.
Trench/Reinstate	£1,980	£660	£1,650	£860	£5,150
Substation	£11,000	£3,300	£88,000	£20,460	£122,760
HV Trench & Joint	£1,980	£660	£1,100	£750	£4,490
HV Cable Install	£330	£110	£1,240	£340	£2,020
Total Calculated Price	£17,270	£5,390	£95,840	£23,710	£142,210
Total Non-Contestable Elements and Associated Charges					£16,500
Grand Total Calculated Price Excl. VAT					£158,710

Client LV Intake Switchboard (400 V)

DNO metering air circuit breaker (ACB)

DNO Transformer (typically iol filled, 500 to 1000 kVA, 11 kV/400 V)

DNO Ring main (RMU)

DNO 11 kV network

Client Demise

DNO* Demise

Note: *DNO – Distribution Network Operator

Typical Simple 11 kV Network Connection for LV Intakes up to 1000 kVA

Client substation / LV Switchboard (400 V)

Client Transformers (typically cast resin 100 kVA to 3000 kVA, 11 kVA / 400 V)

Client MV Intake Switchboard (11 kV)

DNO metering circuit breakers

DNO 11 kV network

Client Demise

DNO* Demise

Note: *DNO - Distribution Network Operator

Typical 11 kV Network Connection for MV Intakes 1000 kVA up to 6000 kVA

94

External Bypass

UPS Module

Internal Maintenance Bypass

Bypass Supply

Main Supply

UPS Input
Switchgear

Batteries

UPS Output
Switchgear

Critical Loads
(IT Equiptment, etc.)

Static UPS System – Simplified Schematic For Single Module

95

Typical Data Transmission (Structured Cabling)

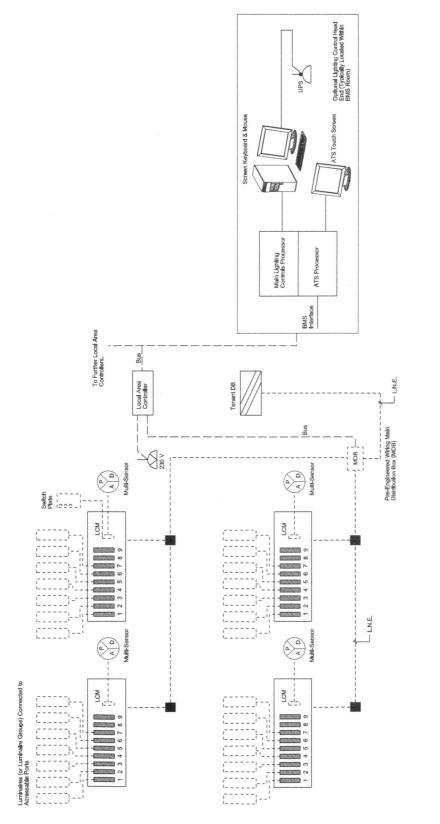

Typical Networked Lighting Control System

97

98

Typical Fire Detection and Alarm Schematic

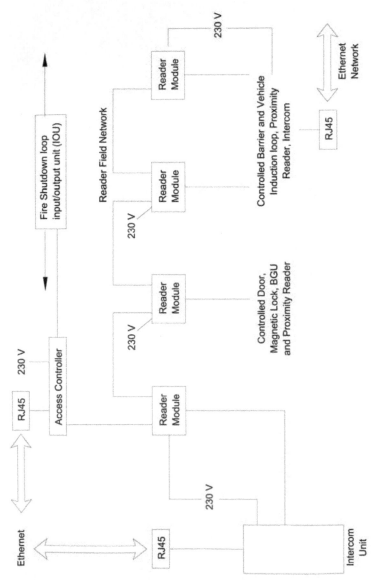

Typical Block Diagram – Access Control System

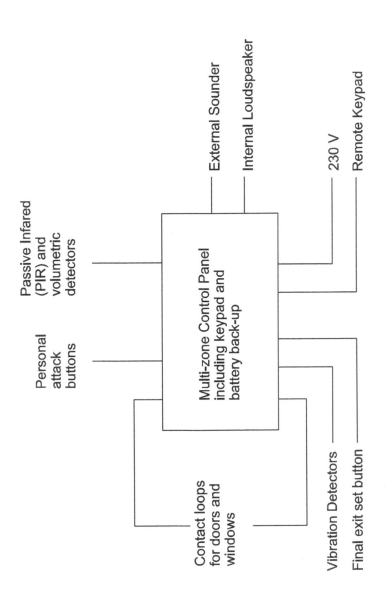

Passive Infared (PIR) and volumetric detectors

Personal attack buttons

External Sounder

Internal Loudspeaker

230 V

Remote Keypad

Multi-zone Control Panel including keypad and battery back-up

Contact loops for doors and windows

Vibration Detectors

Final exit set button

Typical Block Diagram – Intruder Detection System (IDS)

External Cameras

Internal Cameras

Sounders/speakers

Detectors (PIRs)

Microphone

Monitor

Mouse

Keyboard

Multi Channel Digital Video Recorder

Remote workstation

WAN

Network Switch

Automatic Number Plate Recognition

PC 1

Server

Keyboard/ mouse switch

UPS

Microphone

Keyboard/mouse Extender

keyboard

Mouse

Typical Block Diagram – Digital CCTV

FAN COIL UNIT SYSTEM

AIR COOLED CHILLER OR WATER COOLED CHILLER WITH COOLING TOWER

BOILER

CEILING VOID RETURN AIR PELNUM

FAN COIL UNIT

CONDENSATE DRAIN

EXTRACT BELLMOUTH

SUPPLY AIR DIFFUSER

EXTRACT AIR GRILLE

FLOOR VOID

SUPPLY AND EXTRACT AIR HANDLING UNIT

HEAT RECOVERY DEVICE

FRESH AIR IN

EXHAUST AIR OUT

DISPLACEMENT VENTILATION SYSTEM

AIR COOLED CHILLER OR WATER COOLED CHILLER WITH COOLING TOWER

VARIABLE LOW TEMPERATURE HOT WATER

CEILING VOID

CROSS TALK ATTENUATOR

EXTRACT AIR GRILLE

SUPPLY AIR DIFFUSER

PERIMETER TRENCH HEATING

FAN COIL UNIT FOR CELLULAR / HIGHLOAD AREAS

EXTRACT AIR GRILLE

SUPPLY AIR DIFFUSER

EXTRACT BELLMOUTH

SUPPLY AIR INTO FLOOR VOID

BOILER

SUPPLY AND EXTRACT AIR HANDLING UNIT

FRESH AIR IN

EXHAUST AIR OUT

DP

104

CHILLED CEILING SYSTEM

AIR COOLED CHILLER OR WATER COOLED CHILLER WITH COOLING TOWER

ELEVATED / VARIABLE TEMPERATURE CHILLED WATER

CEILING VOID

CHILLED CEILING

SUPPLY AIR DIFFUSER

CHILLED CEILING

RETURN AIR GRILLE

FLOOR VOID

VARIABLE LOW TEMPERATURE HOT WATER

EXTRACT BELLMOUTH

CHILLED CEILING

PERIMETER TRENCH HEATING

SUPPLY AND EXTRACT AIR HANDLING UNIT

HEAT RECOVERY DEVICE

BOILER

FRESH AIR IN

EXHAUST AIR OUT

PASSIVE OR ACTIVE CHILLED BEAM SYSTEM

AIR COOLED CHILLER OR WATER COOLED CHILLER WITH COOLING TOWER

ELEVATED / VARIABLE TEMPERATURE CHILLED WATER

PASSIVE OR ACTIVE CHILLED BEAM WITH INTEGRATED FRESH AIR TERMINAL (ACTIVE BEAM SHOWN)

CEILING VOID

EXTRACT AIR GIRLLE

FLOOR VOID

VARIABLE LOW TEMPERATURE HOT WATER

PERIMETER TRENCH HEATING

SUPPLY AND EXTRACT AIR HANDLING UNIT

HEAT RECOVERY DEVICE

BOILER

FRESH AIR IN

EXHAUST AIR OUT

VARIABLE AIR VOLUME SYSTEM

AIR COOLED CHILLER OR WATER COOLED CHILLER WITH COOLING TOWER

CEILING VOID RETURN AIR PELNUM

SUPPLY AIR DIFFUSER

TRIMINAL VAV UNIT (OPTIONAL HEATING COIL)

EXTRACT AIR GIRLLE

VARIABLE LOW TEMPERATURE HOT WATER

FLOOR VOID

EXTRACT BELLMOUTH

PERIMETER TRENCH HEATING

DP

DP

SUPPLY AND EXTRACT AIR HANDLING UNIT

DP

DP

DP

DP

MIXING ARRANGEMENT

BOILER

DP

DP

FRESH AIR IN

EXHAUST AIR OUT

VARIABLE REFRIGERANT VOLUME SYSTEM

AIR COOLED
CHILLER
OR WATER
COOLED
CHILLER WITH
COOLING
TOWER

VRF HEAT
PUMP UNIT

CEILING VOID
RETURN AIR
PELNUM

SUPPLY AIR
DIFFUSER

FAN COIL UNIT

EXTRACT AIR
GIRLLE

CONDENSATE DRAIN

FLOOR VOID

EXTRACT BELLMOUTH

SUPPLY AND EXTRACT
AIR HANDLING UNIT

HEAT
RECOVERY
DEVICE

BOILER

FRESH AIR IN

EXHAUST AIR
OUT

REVERSE CYCLE HEAT PUMP

PACKAGE HEAT REJECTION UNIT FOR AHU DX UNIT

COOLING TOWER DRY AIR COOLER OR ADIABATIC COOLER

CONSTANT TEMPERATURE CONDENSER PIPEWORK

BOILER

HEAT INPUT WHEN REQUIRED

COMPRESSOR

CONDENSATE DRAIN

HEAT PUMP UNIT

CEILING VOID RETURN AIR PELNUM

FALSE CEILING

OPTION FOR HEAT PUMP UNITS TO PERIMETER WALL

How to Become a Chartered Surveyor

Jen Lemen

Thinking about a career in property or construction? Thinking of becoming of Chartered Surveyor? *How to Become a Chartered Surveyor* demystifies the process and provides a clear road map for candidates to follow. The book outlines potential pathways and practice areas within the profession and includes the breadth and depth of surveying, from commercial, residential and project management, to geomatics and quantity surveying. Experienced APC assessor and trainer, Jen Lemen BSc (Hons) FRICS, provides invaluable guidance, covering:

* routes to becoming a Chartered Surveyor, including t-levels, apprenticeships and alternative APC routes such as the Senior Professional, Academic and Specialist assessments
* areas of professional practice
* advice for the AssocRICS, APC (MRICS), FRICS and Registered Valuer assessments, including both written and interview elements
* advice on referrals and appeals
* how to support candidates, including the role of the Counsellor and Supervisor
* opportunities   for further career progression, including further qualifications and setting up in practice   as an RICS regulated firm
* global perspectives
* professional ethics for surveyors

Written in clear, concise and simple terms and providing practical advice throughout, this book will help candidates to decode and understand the RICS guidance, plan their career and be successful in their journey to become a Chartered Surveyor. It will also be of relevance to academic institutions, employers, school leavers, apprentices, senior professionals, APC Counsellors/Supervisors and careers advisors.

August 2021: 168 pp
ISBN: 9780367742195

To Order
Tel:+44 (0) 1235 400524
Email: tandf@bookpoint.co.uk

For a complete listing of all our titles visit:
www.tandf.co.uk

PART 2

Approximate Estimating

DIRECTIONS

The prices shown in this section of the book are average prices on a fixed price basis for typical buildings tendered during the second quarter of 2022. Unless otherwise noted, they exclude external services and professional fees.

The information in this section has been arranged to follow more closely the order in which estimates may be developed, in accordance with RIBA work stages.

a)   Cost Indices and Regional Variations – These provide information regarding the adjustments to be made to estimates taking into account current pricing levels for different locations in the UK.

b)   Feasibility Costs – These provide a range of data (based on a rate per square metre) for all-in engineering costs, excluding lifts, associated with a wide variety of building types. These would typically be used at work stage 0/1 (feasibility) of a project.

c)   Elemental Rates – The outline costs for offices have been developed further to provide rates for the alternative solutions for each of the services elements. These would typically be used at work stage 2, outline proposal.
     Where applicable, costs have been identified as Shell and Core and Fit-Out to reflect projects where the choice of procurement has dictated that the project is divided into two distinctive contractual parts.
     Such detail would typically be required at work stage 3, detailed proposals.

d)   All-in Rates – These are provided for a number of items and complete parts of a system, i.e. boiler plant, ductwork, pipework, electrical switchgear and small power distribution, together with lifts and escalators. Refer to the relevant section for further guidance notes.

e)   Elemental Costs – These are provided for a diverse range of building types; offices, laboratory, shopping mall, airport terminal building, supermarket, performing arts centre, sports hall, luxury hotel, hospital and secondary school. Also included is a separate analysis of a building management system for an office block. In each case, a full analysis of engineering services costs is given to show the division between all elements and their relative costs to the total building area. A regional variation factor has been applied to bring these analyses to a common London base.

Prices should be applied to the total floor area of all storeys of the building under consideration. The area should be measured between the external walls without deduction for internal walls and staircases/lift shafts, i.e. GIA (Gross Internal Area).

Although prices are reviewed in the light of recent tenders it has only been possible to provide a range of prices for each building type. This should serve to emphasize that these can only be average prices for typical requirements and that such prices can vary widely depending on variations in size, location, phasing, specification, site conditions, procurement route, programme, market conditions and net to gross area efficiencies. Rates per square metre should not therefore be used indiscriminately and each case needs to be assessed on its own merits.

The prices do not include for incidental builder's work nor for profit and attendance by a Main Contractor where the work is executed as a subcontract: they do however include for preliminaries, profit and overheads for the services contractor. Capital contributions to statutory authorities and public undertakings and the cost of work carried out by them have been excluded.

Where services works are procured indirectly, i.e. ductwork via a mechanical Subcontractor, the reader should make due allowance for the addition of a further level of profit etc.

COST INDICES

The following tables reflect the major changes in cost to contractors but do not necessarily reflect changes in tender levels. In addition to changes in labour and materials costs, tenders are affected by other factors such as the degree of competition in the particular industry, the area where the work is to be carried out, the availability of labour and the prevailing economic conditions. This has meant in years when there has been an abundance of work, tender levels have tended to increase at a greater rate than can be accounted for solely by increases in basic labour and material costs and, conversely, when there is a shortage of work this has tended to result in keener tenders. Allowances for these factors are impossible to assess on a general basis and can only be based on experience and knowledge of the particular circumstances.

In compiling the tables, the cost of labour has been calculated on the basis of a notional gang as set out elsewhere in the book. The proportion of labour to materials has been assumed as follows:

Mechanical Services – 30:70, Electrical Services – 50:50, (2015 = 100)

Mechanical Services

Year	First Quarter	Second Quarter	Third Quarter	Fourth Quarter	Annual Average	Annual Change
2021	118.5	121.4	123.4	125.7	122.2	6.5
2022	129.1(P)	132.0(F)	134.9	138.4	133.6	9.3
2023	141.2	142.7	144.3	146.4	143.7	7.5
2024	148.6	150.3	152.0	154.2	151.3	5.3

Electrical Services

Year	First Quarter	Second Quarter	Third Quarter	Fourth Quarter	Annual Average	Annual Change
2021	114.7	118.2	120.4	122.0	118.8	7.0%
2022	125.4(P)	128.0(F)	130.8	133.7	129.5	9.0%
2023	136.9	138.4	139.9	141.4	139.2	7.5%
2024	144.1	145.7	147.3	149.0	146.5	5.3%

(P = Provisional)

(F = Forecast)

COST INDICES

Regional Variations

Prices throughout this Book apply to work in the London area (see Directions at the beginning of the Mechanical Installations and Electrical Installations sections). However, prices for mechanical and electrical services installations will of course vary from region to region, largely as a result of differing labour costs but also depending on the degree of accessibility, urbanization and local market conditions.

The following table of regional factors is intended to provide readers with indicative adjustments that may be made to the prices in the Book for locations outside of London. The figures are of necessity averages for regions and further adjustments should be considered for city centre or very isolated locations, or other known local factors.

Inner London	1.05	North East	0.83
Outer London	1.00	North West	0.88
South East	0.96 (Excl. GL)	Yorkshire and Humberside	0.86
South West	0.90	Scotland	0.88
East of England	0.93	Wales	0.88
East Midlands	0.88	Northern Ireland	0.76
West Midlands	0.87		

RIBA STAGE 0/1 PREPARATION

Item	Unit	Range £
Typical Square Metre Rates for Engineering Services The following examples indicate the range of rates within each building type for engineering services, excluding lifts etc., utilities services and professional fees. Based on Gross Internal Area (GIA).		
Industrial Buildings Factories		
Owner occupation: Includes for rainwater, soil/waste, sprinklers, LTHW heating via HL Electric heaters, local air conditioning, BMS, HV/LV installations, lighting, fire alarms, security, earthing	m²	210.00 to 260.00
Warehouses & Distribution Centres		
High bay for owner occupation: Includes for rainwater, soil/waste, sprinklers, LTHW heating via HL radiant heaters, local air conditioning, BMS, HV/LV installations, lighting, fire alarms, security	m²	290.00 to 310.00
Offices see Stage 2/3		
Health and Welfare Facilities District General Hospitals		
Natural ventilation: Includes for rainwater, soil/waste, cold water, hot water, dry risers, sprinklers, medical gases, LTHW heating, WC/kitchen extract, BMS, LV installations, standby generation, lighting, small power, fire alarms, earthing/lightning protection, nurse call systems, security, IT wireways	m²	1350.00 to 1450.00
Natural ventilation: Includes for rainwater, soil/waste, cold water, hot water, dry risers, sprinklers, medical gases, LTHW heating, localized VAV air conditioning and mechanical ventilation, WC/kitchen extract, BMS, LV installations, standby generation, lighting, small power, fire alarms, earthing/lightning protection, nurse call systems, security, IT wireways	m²	1500.00 to 1550.00
Private Hospitals		
Natural ventilation: Includes for rainwater, soil/waste, cold water, hot water, dry risers, sprinklers, medical gases, LTHW heating, localized VAV air conditioning, WC/kitchen extract, BMS, LV installations, standby generation, lighting, small power, fire alarms, earthing/lightning protection, nurse call systems, security, IT wireways	m²	1500.00 to 1600.00
Air-conditioned: Includes for rainwater, soil/waste, cold water, hot water, dry risers, sprinklers, medical gases, LTHW heating, supply and extract ventilation, 4 pipe air conditioning, WC/kitchen extract, BMS, LV installations, standby generation, lighting, small power, fire alarms, earthing/lightning protection, nurse call systems, security, IT wireways	m²	1725.00 to 1800.00
GP Surgery		
Natural ventilation: Includes for rainwater, soil/waste, cold water, hot water, LTHW heating, WC/kitchen extract, BMS, LV installations, lighting, small power, fire alarms, earthing/lightning protection, nurse call systems, security, IT wireways	m²	960.00 to 1050.00
Air-conditioned: Includes for rainwater, soil/waste, cold water, hot water, LTHW heating, supply and extract, BMS, LV installations, lighting, small power, fire alarms, earthing/lightning protection, nurse call systems, security, IT wireways	m²	1050.00 to 1300.00

Approximate Estimating

RIBA STAGE 0/1 PREPARATION

Item	Unit	Range £
Entertainment and Recreation Buildings		
Performing Arts (with Theatre)		
Air-Conditioned: Includes for rainwater, soil/waste, cold water, central hot water, sprinklers/dry risers, LTHW heating, air conditioning, kitchen/toilet extract, BMS, LV installations, lighting including enhanced dimming/scene setting, small power, fire alarms, earthing, security, IT wireways including production, audio and video recording	m²	990.00 to 1275.00
Sports Halls		
Comfort cooled: Includes for rainwater, soil/waste, cold water, hot water via LTHW heat exchangers, LTHW heating, air conditioning via AHUs to limited areas, toilet extract, BMS, LV installations, lighting, small power, fire alarms, earthing, security	m²	420.00 to 520.00
Multi-Purpose Leisure Centre		
Comfort cooled: Includes for rainwater, soil/waste, cold water, hot water LTHW heat exchangers, LTHW heating, air conditioning via AHUs to limited areas, pool hall supply/extract, kitchen/toilet extract, BMS, LV installations, lighting, small power, fire alarms, earthing, security, IT wireways	m²	740.00 to 910.00
Retail Buildings		
Open Arcade		
Natural ventilation: Includes for rainwater, soil/waste, cold water, hot water, sprinklers/dry risers, LTHW heating, toilet extract, smoke extract, BMS, LV installations, life safety standby generators, lighting, small power, fire alarms, public address, earthing/lightning protection, security, IT wireways	m²	410.00 to 510.00
Department Stores		
Air-conditioned: Includes for sanitaryware, soil/waste, cold water, hot water, sprinklers/dry risers, LTHW heating, air conditioning via AHUs, toilet extract, smoke extract, BMS, LV installations, life safety standby generators, lighting, small power, fire alarms, public address, earthing, lightning protection, CCTV/security, IT installation wireways	m²	480.00 to 600.00
Food Stores		
Air-conditioned: Includes for rainwater, soil/waste, cold water, hot water, sprinklers, LTHW heating, air conditioning via AHUs, toilet extract, BMS, LV installations, lighting, small power, fire alarms, earthing, lightning protection, security, IT wireways, refrigeration	m²	680.00 to 820.00
Educational Buildings		
Secondary Schools (Academy)		
Natural Ventilation: Includes for rainwater, soil/waste, cold water, central hot water, LTHW heating, toilet extract, BMS, LV installations, lighting, small power, fire alarms, earthing, security, IT wireways	m²	510.00 to 570.00
Natural vent with comfort cooling to selected areas (BB93 compliant): Includes for rainwater, soil/waste, cold water, central hot water, LTHW heating, DX air conditioning, general supply/extract, toilet extract, BMS, LV installations, lighting, small power, fire alarms, earthing, security, IT wireways	m²	580.00 to 650.00

RIBA STAGE 0/1 PREPARATION

Item	Unit	Range £
Scientific Buildings		
Educational Research		
Comfort cooled: Includes for rainwater, soil/waste, cold water, central hot water, dry risers, compressed air, medical gasses, LTHW heating, 4 pipe air conditioning, toilet extract, fume, BMS, LV installations, lighting, small power, fire alarms, earthing, lightning protection, security, IT wireways	m^2	910.00 to 1075.00
Air-conditioned: Includes for rainwater, soil/waste, laboratory waste, cold water, central hot water, specialist water, dry risers, compressed air, medical gasses, steam, LTHW heating, VAV air conditioning, Comm's room cooling, toilet extract, fume extract, BMS, LV installations, UPS, standby generators, lighting, small power, fire alarms, earthing, lightning protection, security, IT wireways	m^2	1825.00 to 1950.00
Commercial Research		
Air-conditioned; Includes for rainwater, soil/waste, laboratory waste, cold water, central hot water, specialist water, dry risers, compressed air, medical gasses, steam, LTHW heating, VAV air conditioning, Comm room cooling, toilet extract, fume extract, BMS, LV installations, UPS, standby generators, lighting, small power, fire alarms, earthing, lightning protection, security, IT wireways	m^2	1725.00 to 2075.00
Residential Facilities see Stage 2/3		

Approximate Estimating

RIBA STAGE 2/3 DESIGN

Item	Unit	Range £	
Elemental Rates for Alternative Engineering Services Solutions – Offices			
The following examples of building types indicate the range of rates for alternative design solutions for each of the engineering services elements based on Gross Internal Area for the Shell and Core and Net Internal Area for the Fit-Out. Fit-Out is assumed to be to Cat A Standard.			
Consideration should be made for the size of the building height, sales point (£/ft²) and location, which may affect the economies of scale for rates, i.e. the larger the building the lower the rates.			
5 Services			
Shell & Core			
5.1 Sanitary Installations	m² GIA	16.00 to	29.00
5.3 Disposal Installations	m² GIA	27.00 to	43.00
5.4 Water Installations	m² GIA	49.00 to	67.00
5.6 Space Heating and Air Conditioning			
Comfort cooling;			
2 pipe variable refrigerant volume (VRV)	m² GIA	87.00 to	110.00
Full airconditioning, all electric ASHPs	m² GIA	175.00 to	215.00
5.7 Ventilation Systems	m² GIA	87.00 to	105.00
5.8 Electrical Installations			
LV distribution	m² GIA	54.00 to	65.00
Lighting installations (including lighting controls and luminaires)	m² GIA	87.00 to	110.00
Small power	m² GIA	22.00 to	33.00
Electrical installations for mechanical plant	m² GIA	3.00 to	8.00
5.11 Fire and Lightning Protection			
Sprinkler installation	m² GIA	24.00 to	32.00
Wet risers	m² GIA	11.00 to	16.00
Earthing	m² GIA	2.00 to	3.00
Lightning protection	m² GIA	2.00 to	3.00
5.12 Communication, Security and Control Systems			
Fire alarms	m² GIA	13.00 to	20.00
Security	m² GIA	9.00 to	12.00
BMS controls; including MCC panels and control cabling	m² GIA	27.00 to	33.00
Fit-Out			
5.6 Space Heating and Air Conditioning			
Comfort cooling;			
2 pipe variable refrigerant volume (VRV)	m² NIA	87.00 to	130.00
Full air conditioning;			
4 pipe fan coil – concealed	m² NIA	195.00 to	315.00
4 pipe fan coil – exposed	m² NIA	220.00 to	345.00
4 pipe variable refrigerant volume	m² NIA	175.00 to	215.00
Mixed mode	m² NIA	315.00 to	380.00
Chilled ceiling and active chilled beam	m² NIA	250.00 to	380.00
Active chilled beam	m² NIA	275.00 to	400.00
Displacement	m² NIA	175.00 to	195.00

RIBA STAGE 2/3 DESIGN

Item	Unit	Range £	
5.8 Electrical Installations			
Lighting installations (including lighting controls and luminaires)	m² NIA	125.00 to	165.00
Electrical installations for mechanical plant	m² NIA	5.00 to	11.00
5.11 Fire and Lightning Protection			
Sprinkler installation	m² NIA	27.00 to	33.00
Earthing	m² NIA	2.00 to	3.00
5.12 Communication, Security and Control Systems			
Fire alarms	m² NIA	25.00 to	33.00
BMS controls; including MCC panels and control cabling	m² NIA	27.00 to	38.00

Approximate Estimating

RIBA STAGE 2/3 DESIGN

Item	Unit	Range £	
Elemental Rates for Alternative Engineering Services Solutions – Hotels			
The following examples of building types indicate the range of rates for alternative design solutions for each of the engineering services elements based on Gross Internal Area for the Shell and Core and Net Internal Area for the Fit-Out. Fit-Out is assumed to be to Cat A Standard.			
Consideration should be made for the size of the building, which may affect the economies of scale for rates, i.e. the larger the building the lower the rates.			
5 Services			
Shell & Core			
5.1 Sanitary Installations			
2 to 3 star	m² GIA	34.00 to	42.00
4 to 5 star	m² GIA	50.00 to	69.00
5.3 Disposal Installations			
2 to 3 star	m² GIA	34.00 to	42.00
4 to 5 star	m² GIA	37.00 to	52.00
5.4 Water Installations			
2 to 3 star	m² GIA	48.00 to	61.00
4 to 5 star	m² GIA	68.00 to	92.00
5.6 Space Heating and Air Conditioning			
LPHW heating installation; including gas installations			
2 to 3 star	m² GIA	71.00 to	83.00
4 to 5 star	m² GIA	71.00 to	83.00
Air conditioning; including ventilation			
2 to 3 star – 4 pipe fan coil	m² GIA	275.00 to	370.00
4 to 5 star – 4 pipe fan coil	m² GIA	290.00 to	375.00
2 to 3 star – 3 pipe variable refrigerant volume	m² GIA	175.00 to	230.00
4 to 5 star – 3 pipe variable refrigerant volume	m² GIA	175.00 to	230.00
5.8 Electrical Installations			
LV installations			
Standby generators (life safety only)			
2 to 3 star	m² GIA	20.00 to	24.00
4 to 5 star	m² GIA	20.00 to	28.00
LV distribution			
2 to 3 star	m² GIA	41.00 to	52.00
4 to 5 star	m² GIA	57.00 to	72.00
Lighting installations			
2 to 3 star	m² GIA	29.00 to	37.00
4 to 5 star	m² GIA	45.00 to	60.00
Small power			
2 to 3 star	m² GIA	12.00 to	16.00
4 to 5 star	m² GIA	17.00 to	25.00
Electrical installations for mechanical plant			
2 to 3 star	m² GIA	11.00 to	14.00
4 to 5 star	m² GIA	11.00 to	14.00

RIBA STAGE 2/3 DESIGN

Item	Unit	Range £	
5.11 Fire and Lightning Protection			
Fire Protection			
2 to 3 star – Dry risers	m² GIA	13.00 to	17.00
4 to 5 star – Dry risers	m² GIA	13.00 to	17.00
2 to 3 star – Sprinkler installation	m² GIA	29.00 to	37.00
4 to 5 star – Sprinkler installation	m² GIA	33.00 to	43.00
Earthing			
2 to 3 star	m² GIA	3.00 to	4.00
4 to 5 star	m² GIA	3.00 to	4.00
Lightning Protection			
2 to 3 star	m² GIA	3.00 to	4.00
4 to 5 star	m² GIA	3.00 to	4.00
5.12 Communication, Security and Control Systems			
Fire Alarms			
2 to 3 star	m² GIA	20.00 to	27.00
4 to 5 star	m² GIA	20.00 to	30.00
IT / AV / Containment/TV systems			
2 to 3 star	m² GIA	22.00 to	28.00
4 to 5 star	m² GIA	75.00 to	95.00
Security			
2 to 3 star	m² GIA	22.00 to	31.00
4 to 5 star	m² GIA	25.00 to	32.00
BMS Controls; including MCC panels and control cabling			
2 to 3 star	m² GIA	15.00 to	20.00
4 to 5 star	m² GIA	37.00 to	46.00

Approximate Estimating

RIBA STAGE 2/3 DESIGN

Item	Unit	Range £	
Elemental Rates for Alternative Engineering Services Solutions – Apartments			
The following examples of building types indicate the range of rates for alternative design solutions for each of the engineering services elements based on Gross Internal Area for the Shell and Core and Net Internal Area for the Fit-Out. Fit-Out is assumed to be to Cat A Standard.			
Consideration should be made for the size of the building, which may affect the economies of scale for rates, i.e. the larger the building the lower the rates.			
5 Services			
Shell & Core			
5.1 Sanitary Installations			
Affordable	m² GIA	1.00 to	1.00
Private	m² GIA	3.00 to	4.00
5.3 Disposal Installations			
Affordable	m² GIA	23.00 to	30.00
Private	m² GIA	29.00 to	38.00
5.4 Water Installations			
Affordable	m² GIA	23.00 to	29.00
Private	m² GIA	36.00 to	48.00
5.5 Heat Source (ASHP)			
Affordable	m² GIA	33.00 to	40.00
Private	m² GIA	33.00 to	41.00
5.6 Space Heating and Air Conditioning			
Affordable	m² GIA	33.00 to	38.00
Private	m² GIA	92.00 to	125.00
5.7 Ventilation Systems			
Affordable	m² GIA	23.00 to	25.00
Private	m² GIA	27.00 to	33.00
5.8 Electrical Installations			
Affordable	m² GIA	50.00 to	65.00
Private	m² GIA	75.00 to	105.00
5.11 Fire and Lightning Protection			
Affordable	m² GIA	35.00 to	40.00
Private	m² GIA	35.00 to	55.00
5.12 Communication, Security and Control Systems			
Affordable	m² GIA	40.00 to	50.00
Private	m² GIA	55.00 to	75.00
5.13 Special Installations			
Affordable	m² GIA	15.00 to	18.00
Private	m² GIA	46.00 to	65.00
Fit-Out			
5.1 Sanitary Installations			
Affordable	m² NIA	40.00 to	45.00
Private	m² NIA	125.00 to	150.00

RIBA STAGE 2/3 DESIGN

Item	Unit	Range £	
5.3 Disposal Installations			
Affordable	m² NIA	15.00 to	20.00
Private	m² NIA	30.00 to	40.00
5.4 Water Installation			
Affordable	m² NIA	35.00 to	41.00
Private	m² NIA	60.00 to	80.00
5.6 Space Heating and Air Conditioning			
Affordable	m² NIA	90.00 to	115.00
Private	m² NIA	250.00 to	305.00
5.7 Ventilating Systems			
Affordable (whole house vent)	m² NIA	50.00 to	60.00
Private (whole house vent)	m² NIA	60.00 to	72.00
5.8 Electrical Installations			
Affordable	m² NIA	70.00 to	85.00
Private	m² NIA	150.00 to	185.00
5.11 Fire and Lightning Protection			
Affordable	m² NIA	30.00 to	35.00
Private	m² NIA	35.00 to	43.00
5.12 Communication, Security and Control Systems			
Affordable	m² NIA	25.00 to	30.00
Private	m² NIA	120.00 to	145.00
5.13 Special Installations			
Affordable	m² NIA	15.00 to	18.00
Private	m² NIA	45.00 to	50.00

Note: The range in cost differs due to the vast diversity in services strategies available. The lower end of the scale reflects no comfort cooling, radiator heating in lieu of underfloor, etc. The high end of the scale is based on central plant installations with good quality apartments, which includes comfort cooling, sprinklers, home network installations, video entry, higher quality of sanitaryware and good quality LED lighting.

APPROXIMATE ESTIMATING RATES – 5 SERVICES

Item	Unit	Range £	
5.3 Disposal Installations			
Above Ground Drainage			
Soil and waste	Point	520.00 to	670.00
5.4 Water Installations			
Water Installations			
Cold water	Point	520.00 to	670.00
Hot water	Point	670.00 to	800.00
PIPEWORK			
Hot and Cold Water			
Excludes insulation, valves and ancillaries, etc.			
Light gauge copper tube to EN1057 R250 (TX) formerly BS 2871 Part 1 Table X			
with joints as described including allowance for waste, fittings and supports			
assuming average runs with capillary joints up to 54 mm and bronze welded			
thereafter			
Horizontal High Level Distribution			
15 mm	m	42.00 to	51.00
22 mm	m	43.00 to	52.00
28 mm	m	59.00 to	72.00
35 mm	m	71.00 to	86.00
42 mm	m	84.00 to	105.00
54 mm	m	110.00 to	135.00
67 mm	m	135.00 to	165.00
76 mm	m	155.00 to	190.00
108 mm	m	235.00 to	295.00
Risers			
15 mm	m	24.00 to	32.00
22 mm	m	30.00 to	35.00
28 mm	m	43.00 to	52.00
35 mm	m	52.00 to	62.00
42 mm	m	64.00 to	78.00
54 mm	m	71.00 to	92.00
67 mm	m	130.00 to	150.00
76 mm	m	145.00 to	175.00
108 mm	m	185.00 to	215.00
Toilet Areas, etc., at Low Level			
15 mm	m	70.00 to	82.00
22 mm	m	83.00 to	99.00
28 mm	m	120.00 to	145.00

APPROXIMATE ESTIMATING RATES – 5 SERVICES

Item	Unit	Range £	
5.5 Heat Source			
Heat Source			
Gas fired boilers including gas train and controls	kW	42.00 to	52.00
Gas fired boilers including gas train, controls, flue, plantroom pipework, valves and insulation, pumps and pressurization unit	kW	125.00 to	155.00
5.6 Space Heating and Air Conditioning			
Chilled Water			
Air cooled R134a refrigerant chiller including control panel, anti-vibration mountings	kW	190.00 to	250.00
Air cooled R134a refrigerant chiller including control panel, anti-vibration mountings, plantroom pipework, valves, insulation, pumps and pressurization units	kW	285.00 to	350.00
Water cooled R134a refrigerant chiller including control panel, anti-vibration mountings	kW	110.00 to	160.00
Water cooled R134a refrigerant chiller including control panel, anti-vibration mountings, plantroom pipework, valves, insulation, pumps and pressurization units	kW	215.00 to	250.00
Absorption steam medium chiller including control panel, anti-vibration mountings, plantroom pipework, valves, insulation and pressurization units	kW	375.00 to	450.00
Heat Rejection			
Open circuit, forced draft cooling tower	kW	125.00 to	155.00
Closed circuit, forced draft cooling tower	kW	120.00 to	145.00
Dry air cooler	kW	73.00 to	79.00
Pumps			
Pumps including flexible connections, anti-vibration mountings	kPa	49.00 to	61.00
Pumps including flexible connections, anti-vibration mountings, plantroom pipework, valves, insulation and accessories	kPa	110.00 to	145.00
Ductwork			
The rates below allow for ductwork and for all other labour and material in fabrication, fittings, supports and jointing to equipment, stop and capped ends, elbows, bends, diminishing and transition pieces, regular and reducing couplings, volume control dampers, branch diffuser and 'snap on' grille connections, ties, 'Ys', crossover spigots, etc., turning vanes, regulating dampers, access doors and openings, hand-holes, test holes and covers, blanking plates, flanges, stiffeners, tie rods and all supports and brackets fixed to structure.			
Rectangular galvanized mild steel ductwork as HVCA DW 144 up to 1000 mm longest side	m^2	70.00 to	83.00
Rectangular galvanized mild steel ductwork as HVCA DW 144 up to 2500 mm longest side	m^2	75.00 to	92.00
Rectangular galvanized mild steel ductwork as HVCA DW 144 up to 3000 mm longest side and above	m^2	105.00 to	125.00
Circular galvanized mild steel ductwork as HVCA DW 144	m^2	82.00 to	92.00
Flat oval galvanized mild steel ductwork as HVCA DW 144 up to 545 mm wide	m^2	74.00 to	92.00
Flat oval galvanized mild steel ductwork as HVCA DW 144 up to 880 mm wide	m^2	81.00 to	100.00
Flat oval galvanized mild steel ductwork as HVCA DW 144 up to 1785 mm wide	m^2	96.00 to	120.00

APPROXIMATE ESTIMATING RATES – 5 SERVICES

Item	Unit	Range £	
5.6 Space Heating and Air Conditioning – cont			
Packaged Air Handling Units			
Air handling unit including LPHW preheater coil, pre-filter panel, LPHW heater coils, chilled water coil, filter panels, inverter drive, motorized volume control dampers, sound attenuation, flexible connections to ductwork and all anti-vibration mountings	m³/s	9900.00 to	12500.00
LTHW AND CHILLED WATER			
Excludes insulation, valves and ancillaries, etc.			
Black heavy weight mild steel tube to BS 1387 With joints in the running length, allowance for waste, fittings and supports assuming average runs			
Horizontal Distribution – Basements, etc.			
15 mm	m	69.00 to	82.00
20 mm	m	72.00 to	86.00
25 mm	m	82.00 to	100.00
32 mm	m	96.00 to	115.00
40 mm	m	110.00 to	135.00
50 mm	m	135.00 to	175.00
65 mm	m	135.00 to	180.00
80 mm	m	180.00 to	220.00
100 mm	m	225.00 to	270.00
125 mm	m	300.00 to	365.00
150 mm	m	355.00 to	415.00
200 mm	m	530.00 to	640.00
250 mm	m	670.00 to	810.00
300 mm	m	980.00 to	1175.00
Risers			
15 mm	m	47.00 to	58.00
20 mm	m	49.00 to	59.00
25 mm	m	56.00 to	68.00
32 mm	m	65.00 to	78.00
40 mm	m	74.00 to	86.00
50 mm	m	96.00 to	115.00
65 mm	m	110.00 to	130.00
80 mm	m	125.00 to	150.00
100 mm	m	160.00 to	195.00
125 mm	m	200.00 to	235.00
150 mm	m	250.00 to	310.00
200 mm	m	390.00 to	485.00
250 mm	m	480.00 to	580.00
300 mm	m	740.00 to	930.00
On Floor Distribution			
15 mm	m	76.00 to	91.00
20 mm	m	82.00 to	96.00
25 mm	m	90.00 to	110.00
32 mm	m	105.00 to	125.00
40 mm	m	115.00 to	140.00
50 mm	m	155.00 to	190.00

APPROXIMATE ESTIMATING RATES – 5 SERVICES

Item	Unit	Range £	
Plantroom Areas, etc.			
15 mm	m	75.00 to	88.00
20 mm	m	78.00 to	93.00
25 mm	m	85.00 to	100.00
32 mm	m	97.00 to	120.00
40 mm	m	110.00 to	130.00
50 mm	m	145.00 to	170.00
65 mm	m	160.00 to	185.00
80 mm	m	210.00 to	255.00
100 mm	m	265.00 to	320.00
125 mm	m	345.00 to	420.00
150 mm	m	405.00 to	485.00
200 mm	m	600.00 to	730.00
250 mm	m	740.00 to	880.00
300 mm	m	1075.00 to	1300.00
5.7 Ventilating Systems			
Extract Fans			
Extract fan including inverter drive, sound attenuation, flexible connections to ductwork and all anti vibration mountings	m³/s	3500.00 to	4025.00
5.8 Electrical Installations			
HV/LV INSTALLATIONS			
The cost of HV/LV equipment will vary according to the electricity supplier's requirements, the duty required and the actual location of the site. For estimating purposes, the items indicated below are typical of the equipment required in a HV substation incorporated into a building			
Ring main unit			
Ring main unit, 11 kV including electrical terminations	Point	19500.00 to	23500.00
Transformers			
Oil filled transformers, 11 kV to 415 kV including electrical terminations	kVA	26.00 to	32.00
Cast resin transformers, 11 kV to 415 kV including electrical terminations	kVA	29.00 to	33.00
Midal filled transformers, 11 kV to 415 kV including electrical terminations	kVA	30.00 to	36.00
HV Switchgear			
Cubicle section HV switchpanel, Form 4 type 6 including air circuit breakers, meters and electrical terminations	Section	25000.00 to	36500.00
LV Switchgear			
LV switchpanel, Form 3 including all isolators, fuses, meters and electrical terminations	Isolator	3825.00 to	4775.00
LV switchpanel, Form 4 type 5 including all isolators, fuses, meters and electrical terminations	Isolator	5900.00 to	7300.00
External Packaged Substation			
Extra over cost for prefabricated packaged substation housing, excludes base and protective security fencing	Each	40000.00 to	48500.00

APPROXIMATE ESTIMATING RATES – 5 SERVICES

Item	Unit	Range £	
5.8 Electrical Installations – cont			
STANDBY GENERATING SETS			
Diesel powered including control panel, flue, oil day tank and attenuation			
Approximate installed cost, LV	kVA	340.00 to	410.00
Approximate installed cost, HV	kVA	395.00 to	470.00
UNINTERRUPTIBLE POWER SUPPLY			
Rotary UPS including control panel and choke transformer (excludes distribution)			
Approximate installed cost (range 1000 kVA to 2500 kVA)	kVA	530.00 to	640.00
Static UPS including control panel, automatic bypass, DC isolator and batteries for 30 minutes standby (excludes distribution)			
Approximate installed cost (range 500 kVA to 1000 kVA)	kVA	225.00 to	310.00
SMALL POWER			
Approximate prices for wiring of power points of length not exceeding 20 m, including accessories, wireways but excluding distribution boards			
13 amp Accessories			
Wired in PVC insulated twin and earth cable in ring main circuit			
Domestic properties	Point	70.00 to	86.00
Commercial properties	Point	94.00 to	115.00
Industrial properties	Point	94.00 to	115.00
Wired in PVC insulated twin and earth cable in radial circuit			
Domestic properties	Point	94.00 to	115.00
Commercial properties	Point	115.00 to	140.00
Industrial properties	Point	115.00 to	140.00
Wired in LSF insulated single cable in ring main circuit			
Commercial properties	Point	115.00 to	140.00
Industrial properties	Point	115.00 to	140.00
Wired in LSF insulated single cable in radial circuit			
Commercial properties	Point	135.00 to	165.00
Industrial properties	Point	135.00 to	165.00
45 amp wired in PVC insulated twin and earth cable			
Domestic properties	Point	125.00 to	160.00
Low Voltage Power Circuits			
Three phase four wire radial circuit feeding an individual load, wired in LSF insulated single cable including all wireways, isolator, not exceeding 10 metres; in commercial properties			
$1.5\,mm^2$	Point	225.00 to	290.00
$2.5\,mm^2$	Point	260.00 to	305.00
$4\,mm^2$	Point	270.00 to	310.00
$6\,mm^2$	Point	295.00 to	365.00
$10\,mm^2$	Point	345.00 to	430.00
$16\,mm^2$	Point	380.00 to	460.00

APPROXIMATE ESTIMATING RATES – 5 SERVICES

Item	Unit	Range £	
Three phase four core radial circuit feeding an individual load item, wired in LSF/SWA/XLPE insulated cable including terminations, isolator, clipped to surface, not exceeding 10 metres; in commercial properties			
1.5 mm^2	Point	175.00 to	215.00
2.5 mm^2	Point	195.00 to	225.00
4 mm^2	Point	205.00 to	260.00
6 mm^2	Point	225.00 to	280.00
10 mm^2	Point	345.00 to	420.00
16 mm^2	Point	430.00 to	530.00
LIGHTING			
Approximate prices for wiring of lighting points including rose, wireways but excluding distribution boards, luminaires and switches			
Final Circuits			
Wired in PVC insulated twin and earth cable			
Domestic properties	Point	51.00 to	62.00
Commercial properties	Point	64.00 to	79.00
Industrial properties	Point	64.00 to	79.00
Wired in LSF insulated single cable			
Commercial properties	Point	86.00 to	105.00
Industrial properties	Point	86.00 to	105.00
ELECTRICAL WORKS IN CONNECTION WITH MECHANICAL SERVICES			
The cost of electrical connections to mechanical services equipment will vary depending on the type of building and complexity of the equipment			
Typical rate for power wiring, isolators and associated wireways	m^2	9.00 to	13.00
FIRE ALARMS			
Cost per point for two core FP200 Wired system including all terminations, supports and wireways			
Call point	Point	310.00 to	365.00
Smoke detector	Point	255.00 to	305.00
Smoke/heat detector	Point	295.00 to	355.00
Heat detector	Point	275.00 to	335.00
Heat detector and sounder	Point	255.00 to	290.00
Input/output/relay units	Point	355.00 to	415.00
Alarm sounder	Point	275.00 to	335.00
Alarm sounder/beacon	Point	335.00 to	400.00
Speakers/voice sounders	Point	335.00 to	400.00
Speakers/voice sounders (weatherproof)	Point	355.00 to	415.00
Beacon/strobe	Point	255.00 to	305.00
Beacon/strobe (weatherproof)	Point	365.00 to	445.00
Door release units	Point	365.00 to	445.00
Beam detector	Point	1125.00 to	1225.00
For costs for zone control panel, battery chargers and batteries, see 'Prices for Measured Work' section.			

APPROXIMATE ESTIMATING RATES – 5 SERVICES

Item	Unit	Range £	
5.8 Electrical Installations – cont			
EXTERNAL LIGHTING			
Estate Road Lighting			
Post type road lighting lantern 70 Watt CDM-T 3000k complete with 5 m high column with hinged lockable door, control gear and cut-out including 2.5 mm two core butyl cable internal wiring, interconnections and earthing fed by 16 mm² four core XLPE/SWA/LSF cable and terminations. Approximate installed price per metre road length (based on 300 metres run) including time switch but excluding builder's work in connection			
Columns erected at 30 m intervals along road (cost per metre of road)	m	92.00 to	115.00
Bollard Lighting			
Bollard lighting fitting 26 Watt TC-D 3500k including control gear, all internal wiring, interconnections, earthing and 25 metres of 2.5 mm² three core XLPE/SWA/LSF cable			
Approximate installed price excluding builder's work in connection	Each	1350.00 to	1675.00
Outdoor Flood Lighting			
Wall mounted outdoor flood light fitting complete with tungsten halogen lamp, mounting bracket, wire guard and all internal wiring and containment, fixed to brickwork or concrete and connected			
Installed price 500 Watt	Point	285.00 to	355.00
Installed price 1000 Watt	Point	345.00 to	415.00
Pedestal mounted outdoor flood light fitting complete with 1000 Watt MBF/U lamp, control gear, contained in weatherproof steel box, all internal wiring and containment, interconnections and earthing, fixed to brickwork or concrete and connected			
Approximate installed price excluding builder's work in connection	Each	1325.00 to	1650.00
BUILDING MANAGEMENT INSTALLATIONS			
Category A Fit-Out			
Option 1 – 185 nr four pipe fan coil – 740 points			
1.0 Field Equipment			
Network devices; Valves/actuators; Sensing devices	Point	85.00 to	87.00
2.0 Cabling			
Power – from local isolator to DDC controller; Control – from DDC controller to field equipment	Point	48.00 to	50.00
3.0 Programming			
Software – central facility; Software – network devices; Graphics	Point	25.00 to	28.00
4.0 On-site testing and commissioning			
Equipment; Programming/graphics; Power and control cabling	Point	25.00 to	28.00
Total Option 1 – Four pipefan coil	Point	180.00 to	195.00
Cost/FCU	Each	730.00 to	780.00

APPROXIMATE ESTIMATING RATES – 5 SERVICES

Item	Unit	Range £	
Option 2 – 185 nr two pipe fan coil system with electric heating – 740 points			
1.0 Field Equipment			
Network devices; Valves/actuators/thyristors; Sensing devices	Point	110.00 to	115.00
2.0 Cabling			
Power – from local isolator to DDC controller; Control – from DDC controller to field equipment	Point	48.00 to	50.00
3.0 Programming			
Software – central facility; Software – network devices; Graphics	Point	26.00 to	29.00
4.0 On-site testing and commissioning			
Equipment; Programming/graphics; Power and control cabling	Point	27.00 to	30.00
Total Option 2 – Two pipe fan coil with electric heating	Point	210.00 to	225.00
Cost/FCU	Each	850.00 to	900.00
Option 3 – 180 nr chilled beams with perimeter heating – 567 points			
1.0 Field Equipment			
Network devices; Valves/actuators; Sensing devices	Point	105.00 to	105.00
2.0 Cabling			
Power – from local isolator to DDC controller; Control – from DDC controller to field equipment	Point	67.00 to	69.00
3.0 Programming			
Cost/Point	Point	37.00 to	39.00
4.0 On-site testing and commissioning			
Equipment; Programming/Graphics; Power and control cabling	Point	38.00 to	40.00
Total Option 3 – Chilled beams with perimeter heating	Point	245.00 to	255.00
Cost/Chilled Beam	Each	780.00 to	810.00
Shell & Core Only			
Main Plant – Cost/BMS Points	Each	770.00 to	780.00
Landlord FCUs – Cost/Terminal Units	Each	770.00 to	780.00
Trade Contract Preliminaries	%	20.00 to	30.00
Notes: The following are included in points, rates			
- DDC Controllers/Control Enclosures/Control Panels			
- Motor Control Centre (MCC)			
- Field Devices			
- Control and Power Cabling from DDC Controllers/MCC			
- Programming			
- On-Site Testing and Commissioning			

APPROXIMATE ESTIMATING RATES – 5 SERVICES

Item	Unit	Range £
5.10 Lift and Conveyor Systems		
LIFT INSTALLATIONS		
The cost of lift installations will vary depending upon a variety of circumstances. The following prices assume a car height of 2.2 metres, manufacturer's standard car finish, brushed stainless steel 2 panel centre opening doors to BSEN81–20 and Lift Regulations 2016.		
Passenger Lifts, machine room above		
Electrically operated AC drive serving 2 levels with directional collective controls and a speed of 1.0 m/s		
8 Person	Item	79000.00 to 96000.00
10 Person	Item	85000.00 to 100000.00
13 Person	Item	89000.00 to 110000.00
17 Person	Item	100000.00 to 120000.00
21 Person	Item	110000.00 to 135000.00
26 Person	Item	125000.00 to 155000.00
Electrically operated AC drive serving 4 levels and a speed of 1.0 m/s		
8 Person	Item	92000.00 to 110000.00
10 Person	Item	98000.00 to 115000.00
13 Person	Item	100000.00 to 125000.00
17 Person	Item	115000.00 to 135000.00
21 Person	Item	125000.00 to 160000.00
26 Person	Item	150000.00 to 185000.00
Electrically operated AC drive serving 6 levels and a speed of 1.0 m/s		
8 Person	Item	105000.00 to 125000.00
10 Person	Item	110000.00 to 130000.00
13 Person	Item	115000.00 to 135000.00
17 Person	Item	125000.00 to 160000.00
21 Person	Item	150000.00 to 185000.00
26 Person	Item	165000.00 to 195000.00
Electrically operated AC drive serving 8 levels and a speed of 1.0 m/s		
8 Person	Item	115000.00 to 145000.00
10 Person	Item	120000.00 to 145000.00
13 Person	Item	125000.00 to 160000.00
17 Person	Item	150000.00 to 185000.00
21 Person	Item	155000.00 to 190000.00
26 Person	Item	180000.00 to 225000.00
Electrically operated AC drive serving 10 levels and a speed of 1.0 m/s		
8 Person	Item	125000.00 to 160000.00
10 Person	Item	130000.00 to 165000.00
13 Person	Item	130000.00 to 165000.00
17 Person	Item	155000.00 to 190000.00
21 Person	Item	175000.00 to 210000.00
26 Person	Item	205000.00 to 250000.00
Electrically operated AC drive serving 12 levels and a speed of 1.0 m/s		
8 Person	Item	130000.00 to 165000.00
10 Person	Item	150000.00 to 185000.00
13 Person	Item	150000.00 to 185000.00
17 Person	Item	170000.00 to 210000.00
21 Person	Item	190000.00 to 235000.00
26 Person	Item	215000.00 to 260000.00

APPROXIMATE ESTIMATING RATES – 5 SERVICES

Item	Unit	Range £
Electrically operated AC drive serving 14 levels and a speed of 1.0 m/s		
8 Person	Item	155000.00 to 190000.00
10 Person	Item	155000.00 to 190000.00
13 Person	Item	170000.00 to 210000.00
17 Person	Item	190000.00 to 235000.00
21 Person	Item	205000.00 to 250000.00
26 Person	Item	235000.00 to 285000.00
Add to above for		
Increase speed from 1.0 m/s to 1.6 m/s		
8 Person	Item	5300.00 to 6500.00
10 Person	Item	5500.00 to 6800.00
13 Person	Item	5700.00 to 6900.00
17 Person	Item	5700.00 to 6900.00
21 Person	Item	5700.00 to 6900.00
26 Person	Item	5700.00 to 6900.00
Increase speed from 1.6 m/s to 2.0 m/s		
8 Person	Item	7500.00 to 9000.00
10 Person	Item	7900.00 to 9900.00
13 Person	Item	8800.00 to 10500.00
17 Person	Item	4000.00 to 4700.00
21 Person	Item	4000.00 to 4900.00
26 Person	Item	4000.00 to 4900.00
Increase speed from 2.0 m/s to 2.5 m/s		
8 Person	Item	2800.00 to 3300.00
10 Person	Item	2800.00 to 3300.00
13 Person	Item	3275.00 to 4000.00
17 Person	Item	3275.00 to 4000.00
21 Person	Item	4000.00 to 4700.00
26 Person	Item	4000.00 to 4700.00
Enhanced finish to car – Centre mirror, flat ceiling, carpet		
8 Person	Item	3900.00 to 4575.00
10 Person	Item	4125.00 to 5000.00
13 Person	Item	3975.00 to 5100.00
17 Person	Item	4600.00 to 5600.00
21 Person	Item	5400.00 to 6600.00
26 Person	Item	6200.00 to 7600.00
Bottom motor room		
8 Person	Item	9200.00 to 11500.00
10 Person	Item	9200.00 to 11500.00
13 Person	Item	9200.00 to 11500.00
17 Person	Item	11500.00 to 13500.00
21 Person	Item	11500.00 to 13500.00
26 Person	Item	11500.00 to 14500.00
Fire fighting control		
8 Person	Item	7500.00 to 9000.00
10 Person	Item	7500.00 to 9000.00
13 Person	Item	7500.00 to 9000.00
17 Person	Item	7500.00 to 9000.00
21 Person	Item	7500.00 to 9000.00
26 Person	Item	7500.00 to 9000.00

APPROXIMATE ESTIMATING RATES – 5 SERVICES

Item	Unit	Range £	
5.10 Lift and Conveyor Systems – cont			
Add to above for – cont			
Glass back			
8 Person	Item	3425.00 to	4100.00
10 Person	Item	3900.00 to	4575.00
13 Person	Item	4425.00 to	5500.00
17 Person	Item	5400.00 to	6600.00
21 Person	Item	5400.00 to	6600.00
26 Person	Item	5400.00 to	6600.00
Glass doors			
8 Person	Item	26000.00 to	31500.00
10 Person	Item	26000.00 to	31500.00
13 Person	Item	28000.00 to	34000.00
17 Person	Item	30000.00 to	37500.00
21 Person	Item	28000.00 to	34000.00
26 Person	Item	28000.00 to	34000.00
Painting to entire pit			
8 Person	Item	2700.00 to	3300.00
10 Person	Item	2700.00 to	3300.00
13 Person	Item	2700.00 to	3300.00
17 Person	Item	2700.00 to	3300.00
21 Person	Item	2700.00 to	3300.00
26 Person	Item	2700.00 to	3300.00
Dual seal shaft			
8 Person	Item	5400.00 to	6600.00
10 Person	Item	5400.00 to	6600.00
13 Person	Item	5400.00 to	6600.00
17 Person	Item	6400.00 to	8000.00
21 Person	Item	6400.00 to	8000.00
26 Person	Item	6400.00 to	8000.00
Dust sealing machine room			
8 Person	Item	1050.00 to	1300.00
10 Person	Item	1050.00 to	1300.00
13 Person	Item	1775.00 to	2100.00
17 Person	Item	1775.00 to	2100.00
21 Person	Item	1775.00 to	2100.00
26 Person	Item	1775.00 to	2100.00
Intercom to reception desk and security room			
8 Person	Item	530.00 to	650.00
10 Person	Item	530.00 to	650.00
13 Person	Item	530.00 to	650.00
17 Person	Item	530.00 to	650.00
21 Person	Item	530.00 to	650.00
26 Person	Item	530.00 to	650.00
Heating, cooling and ventilation to machine room			
8 Person	Item	1325.00 to	1675.00
10 Person	Item	1325.00 to	1675.00
13 Person	Item	1325.00 to	1675.00
17 Person	Item	1325.00 to	1675.00
21 Person	Item	1325.00 to	1675.00
26 Person	Item	1325.00 to	1675.00

APPROXIMATE ESTIMATING RATES – 5 SERVICES

Item	Unit	Range £	
Shaft lighting/small power			
8 Person	Item	4625.00 to	5700.00
10 Person	Item	4625.00 to	5700.00
13 Person	Item	4625.00 to	5700.00
17 Person	Item	4625.00 to	5700.00
21 Person	Item	4625.00 to	5700.00
26 Person	Item	4625.00 to	5700.00
Motor room lighting/small power			
8 Person	Item	1700.00 to	2050.00
10 Person	Item	1700.00 to	2050.00
13 Person	Item	2125.00 to	2575.00
17 Person	Item	2125.00 to	2575.00
21 Person	Item	2125.00 to	2575.00
26 Person	Item	2125.00 to	2575.00
Lifting beams			
8 Person	Item	2025.00 to	2450.00
10 Person	Item	2025.00 to	2450.00
13 Person	Item	2025.00 to	2450.00
17 Person	Item	2250.00 to	2775.00
21 Person	Item	2250.00 to	2775.00
26 Person	Item	2250.00 to	2775.00
10 mm equipotential bonding of all entrance metalwork			
8 Person	Item	1050.00 to	1300.00
10 Person	Item	1050.00 to	1300.00
13 Person	Item	1050.00 to	1300.00
17 Person	Item	1050.00 to	1300.00
21 Person	Item	1050.00 to	1300.00
26 Person	Item	1050.00 to	1300.00
Shaft secondary steelwork			
8 Person	Item	7700.00 to	9600.00
10 Person	Item	8000.00 to	9700.00
13 Person	Item	8200.00 to	10000.00
17 Person	Item	8400.00 to	10500.00
21 Person	Item	8400.00 to	10500.00
26 Person	Item	8400.00 to	10500.00
Independent insurance inspection			
8 Person	Item	2525.00 to	3100.00
10 Person	Item	2525.00 to	3100.00
13 Person	Item	2525.00 to	3100.00
17 Person	Item	2525.00 to	3100.00
21 Person	Item	2525.00 to	3100.00
26 Person	Item	2525.00 to	3100.00
12 month warranty service			
8 Person	Item	1375.00 to	1675.00
10 Person	Item	1375.00 to	1675.00
13 Person	Item	1375.00 to	1675.00
17 Person	Item	1375.00 to	1675.00
21 Person	Item	1375.00 to	1675.00
26 Person	Item	1375.00 to	1675.00

APPROXIMATE ESTIMATING RATES – 5 SERVICES

Item	Unit	Range £
5.10 Lift and Conveyor Systems – cont		
Passenger Lifts, machine room less		
Electrically operated AC drive serving 2 levels with directional collective controls and a speed of 1.0 m/s		
8 Person	Item	70000.00 to 86000.00
10 Person	Item	78000.00 to 98000.00
13 Person	Item	82000.00 to 100000.00
17 Person	Item	99000.00 to 120000.00
21 Person	Item	110000.00 to 130000.00
26 Person	Item	120000.00 to 145000.00
Electrically operated AC drive serving 4 levels and a speed of 1.0 m/s		
8 Person	Item	81000.00 to 99000.00
10 Person	Item	89000.00 to 110000.00
13 Person	Item	94000.00 to 115000.00
17 Person	Item	110000.00 to 135000.00
21 Person	Item	120000.00 to 145000.00
26 Person	Item	130000.00 to 165000.00
Electrically operated AC drive serving 6 levels and a speed of 1.0 m/s		
8 Person	Item	94000.00 to 115000.00
10 Person	Item	100000.00 to 120000.00
13 Person	Item	105000.00 to 130000.00
17 Person	Item	120000.00 to 145000.00
21 Person	Item	130000.00 to 165000.00
26 Person	Item	145000.00 to 170000.00
Electrically operated AC drive serving 8 levels and a speed of 1.0 m/s		
8 Person	Item	105000.00 to 130000.00
10 Person	Item	110000.00 to 130000.00
13 Person	Item	120000.00 to 145000.00
17 Person	Item	130000.00 to 165000.00
21 Person	Item	145000.00 to 170000.00
26 Person	Item	155000.00 to 195000.00
Electrically operated AC drive serving 10 levels and a speed of 1.0 m/s		
8 Person	Item	115000.00 to 145000.00
10 Person	Item	120000.00 to 145000.00
13 Person	Item	130000.00 to 155000.00
17 Person	Item	145000.00 to 170000.00
21 Person	Item	155000.00 to 195000.00
26 Person	Item	180000.00 to 225000.00
Electrically operated AC drive serving 12 levels and a speed of 1.0 m/s		
8 Person	Item	130000.00 to 155000.00
10 Person	Item	130000.00 to 155000.00
13 Person	Item	145000.00 to 170000.00
17 Person	Item	155000.00 to 195000.00
21 Person	Item	170000.00 to 210000.00
26 Person	Item	190000.00 to 235000.00

APPROXIMATE ESTIMATING RATES – 5 SERVICES

Item	Unit	Range £	
Electrically operated AC drive serving 14 levels and a speed of 1.0 m/s			
8 Person	Item	145000.00 to	170000.00
10 Person	Item	145000.00 to	170000.00
13 Person	Item	150000.00 to	185000.00
17 Person	Item	180000.00 to	225000.00
21 Person	Item	190000.00 to	235000.00
26 Person	Item	215000.00 to	260000.00
Add to above for			
Increase speed from 1.0 m/s to 1.6 m/s			
8 Person	Item	3450.00 to	4225.00
10 Person	Item	3450.00 to	4225.00
13 Person	Item	3650.00 to	4450.00
17 Person	Item	5600.00 to	6800.00
21 Person	Item	6500.00 to	7900.00
26 Person	Item	8200.00 to	10000.00
Enhanced finish to car – Centre mirror, flat ceiling, carpet			
8 Person	Item	3750.00 to	4575.00
10 Person	Item	4000.00 to	4875.00
13 Person	Item	4000.00 to	4875.00
17 Person	Item	4575.00 to	5600.00
21 Person	Item	5300.00 to	6500.00
26 Person	Item	6300.00 to	7500.00
Fire fighting control			
8 Person	Item	7300.00 to	8900.00
10 Person	Item	7300.00 to	8900.00
13 Person	Item	7300.00 to	8900.00
17 Person	Item	7300.00 to	8900.00
21 Person	Item	7300.00 to	8900.00
26 Person	Item	7300.00 to	8900.00
Painting to entire pit			
8 Person	Item	2650.00 to	3225.00
10 Person	Item	2650.00 to	3225.00
13 Person	Item	2650.00 to	3225.00
17 Person	Item	2650.00 to	3225.00
21 Person	Item	2650.00 to	3225.00
26 Person	Item	2650.00 to	3225.00
Dual seal shaft			
8 Person	Item	5300.00 to	6500.00
10 Person	Item	5300.00 to	6500.00
13 Person	Item	5300.00 to	6500.00
17 Person	Item	6500.00 to	7900.00
21 Person	Item	6500.00 to	7900.00
26 Person	Item	6500.00 to	7900.00
Shaft lighting/small power			
8 Person	Item	4625.00 to	5700.00
10 Person	Item	4625.00 to	5700.00
13 Person	Item	4625.00 to	5700.00
17 Person	Item	4625.00 to	5700.00
21 Person	Item	4625.00 to	5700.00
26 Person	Item	4625.00 to	5700.00

APPROXIMATE ESTIMATING RATES – 5 SERVICES

Item	Unit	Range £	
5.10 Lift and Conveyor Systems – cont			
Add to above for – cont			
Intercom to reception desk and security room			
8 Person	Item	530.00 to	650.00
10 Person	Item	530.00 to	650.00
13 Person	Item	530.00 to	650.00
17 Person	Item	530.00 to	650.00
21 Person	Item	530.00 to	650.00
26 Person	Item	530.00 to	650.00
Lifting beams			
8 Person	Item	2000.00 to	2450.00
10 Person	Item	2000.00 to	2450.00
13 Person	Item	2000.00 to	2450.00
17 Person	Item	2250.00 to	2775.00
21 Person	Item	2250.00 to	2775.00
26 Person	Item	2250.00 to	2775.00
10 mm equipotential bonding of all entrance metalwork			
8 Person	Item	1050.00 to	1300.00
10 Person	Item	1050.00 to	1300.00
13 Person	Item	1050.00 to	1300.00
17 Person	Item	1050.00 to	1300.00
21 Person	Item	1050.00 to	1300.00
26 Person	Item	1050.00 to	1300.00
Shaft secondary steelwork			
8 Person	Item	7600.00 to	9100.00
10 Person	Item	7700.00 to	9500.00
13 Person	Item	8000.00 to	9800.00
17 Person	Item	8200.00 to	10000.00
21 Person	Item	8200.00 to	10000.00
26 Person	Item	8200.00 to	10000.00
Independent insurance inspection			
8 Person	Item	2450.00 to	2975.00
10 Person	Item	2450.00 to	2975.00
13 Person	Item	2450.00 to	2975.00
17 Person	Item	2450.00 to	2975.00
21 Person	Item	2450.00 to	2975.00
26 Person	Item	2450.00 to	2975.00
12 month warranty service			
8 Person	Item	1300.00 to	1575.00
10 Person	Item	1300.00 to	1575.00
13 Person	Item	1300.00 to	1575.00
17 Person	Item	1300.00 to	1575.00
21 Person	Item	1300.00 to	1575.00
26 Person	Item	1300.00 to	1575.00

APPROXIMATE ESTIMATING RATES – 5 SERVICES

Item	Unit	Range £
Goods Lifts, machine room above		
Electrically operated two speed serving 2 levels to take 1000 kg load, prime coated internal finish and a speed of 1.0 m/s		
2000 kg	Item	130000.00 to 165000.00
2250 kg	Item	145000.00 to 170000.00
2500 kg	Item	145000.00 to 170000.00
3000 kg	Item	155000.00 to 195000.00
Electrically operated two speed serving 4 levels and a speed of 1.0 m/s		
2000 kg	Item	145000.00 to 170000.00
2250 kg	Item	155000.00 to 195000.00
2500 kg	Item	165000.00 to 195000.00
3000 kg	Item	200000.00 to 250000.00
Electrically operated two speed serving 6 levels and a speed of 1.0 m/s		
2000 kg	Item	165000.00 to 195000.00
2250 kg	Item	180000.00 to 225000.00
2500 kg	Item	180000.00 to 225000.00
3000 kg	Item	205000.00 to 250000.00
Electrically operated two speed serving 8 levels and a speed of 1.0 m/s		
2000 kg	Item	180000.00 to 225000.00
2250 kg	Item	190000.00 to 235000.00
2500 kg	Item	205000.00 to 250000.00
3000 kg	Item	230000.00 to 280000.00
Electrically operated two speed serving 10 levels and a speed of 1.0 m/s		
2000 kg	Item	205000.00 to 250000.00
2250 kg	Item	215000.00 to 260000.00
2500 kg	Item	215000.00 to 260000.00
3000 kg	Item	245000.00 to 300000.00
Electrically operated two speed serving 12 levels and a speed of 1.0 m/s		
2000 kg	Item	215000.00 to 260000.00
2250 kg	Item	230000.00 to 280000.00
2500 kg	Item	230000.00 to 280000.00
3000 kg	Item	275000.00 to 325000.00
Electrically operated two speed serving 14 levels and a speed of 1.0 m/s		
2000 kg	Item	235000.00 to 285000.00
2250 kg	Item	255000.00 to 310000.00
2500 kg	Item	255000.00 to 310000.00
3000 kg	Item	290000.00 to 350000.00
Add to above for		
Increase speed of travel from 1.0 m/s to 1.6 m/s		
2000 kg	Item	1700.00 to 2050.00
Enhanced finish to car – Centre mirror, flat ceiling, carpet		
2000 kg	Item	4625.00 to 5700.00
2250 kg	Item	4625.00 to 5700.00
2500 kg	Item	4625.00 to 5700.00
Bottom motor room		
2000 kg	Item	11500.00 to 13500.00

APPROXIMATE ESTIMATING RATES – 5 SERVICES

Item	Unit	Range £	
5.10 Lift and Conveyor Systems – cont			
Add to above for – cont			
Painting to entire pit			
2000 kg	Item	2650.00 to	3225.00
2250 kg	Item	2650.00 to	3225.00
2500 kg	Item	2650.00 to	3225.00
3000 kg	Item	2950.00 to	3575.00
Dual seal shaft			
2000 kg	Item	5300.00 to	6500.00
2250 kg	Item	5300.00 to	6500.00
2500 kg	Item	5300.00 to	6500.00
3000 kg	Item	5300.00 to	6500.00
Intercom to reception desk and security room			
2000 kg	Item	540.00 to	660.00
2250 kg	Item	540.00 to	660.00
2500 kg	Item	540.00 to	660.00
3000 kg	Item	540.00 to	660.00
Heating, cooling and ventilation to machine room			
2000 kg	Item	1475.00 to	1800.00
2250 kg	Item	1475.00 to	1800.00
2500 kg	Item	1475.00 to	1800.00
3000 kg	Item	1475.00 to	1800.00
Lifting beams			
2000 kg	Item	1950.00 to	2425.00
2250 kg	Item	1950.00 to	2425.00
2500 kg	Item	1950.00 to	2425.00
3000 kg	Item	1950.00 to	2425.00
10 mm equipotential bonding of all entrance metalwork			
2000 kg	Item	1050.00 to	1300.00
2250 kg	Item	1050.00 to	1300.00
2500 kg	Item	1050.00 to	1300.00
3000 kg	Item	1050.00 to	1300.00
Independent insurance inspection			
2000 kg	Item	2525.00 to	3075.00
2250 kg	Item	2525.00 to	3075.00
2500 kg	Item	2525.00 to	3075.00
3000 kg	Item	2525.00 to	3075.00
12 month warranty service			
2000 kg	Item	1375.00 to	1675.00
2250 kg	Item	1600.00 to	1925.00
2500 kg	Item	1600.00 to	1925.00
3000 kg	Item	1600.00 to	1925.00
Goods Lift, machine room less			
Electrically operated two speed serving 2 levels to take 1000 kg load, prime coated internal finish and a speed of 1.0 m/s			
2000 kg	Item	115000.00 to	145000.00
2250 kg	Item	120000.00 to	145000.00
2500 kg	Item	130000.00 to	165000.00

APPROXIMATE ESTIMATING RATES – 5 SERVICES

Item	Unit	Range £
Electrically operated two speed serving 4 levels and a speed of 1.0 m/s		
2000 kg	Item	130000.00 to 165000.00
2250 kg	Item	135000.00 to 170000.00
2500 kg	Item	150000.00 to 185000.00
Electrically operated two speed serving 6 levels and a speed of 1.0 m/s		
2000 kg	Item	150000.00 to 185000.00
2250 kg	Item	150000.00 to 185000.00
2500 kg	Item	170000.00 to 210000.00
Electrically operated two speed serving 8 levels and a speed of 1.0 m/s		
2000 kg	Item	155000.00 to 195000.00
2250 kg	Item	170000.00 to 210000.00
2500 kg	Item	180000.00 to 225000.00
Electrically operated two speed serving 10 levels and a speed of 1.0 m/s		
2000 kg	Item	175000.00 to 215000.00
2250 kg	Item	180000.00 to 225000.00
2500 kg	Item	190000.00 to 235000.00
Electrically operated two speed serving 12 levels and a speed of 1.0 m/s		
2000 kg	Item	155000.00 to 195000.00
2250 kg	Item	205000.00 to 250000.00
2500 kg	Item	215000.00 to 260000.00
Electrically operated two speed serving 14 levels and a speed of 1.0 m/s		
2000 kg	Item	205000.00 to 250000.00
2250 kg	Item	215000.00 to 260000.00
2500 kg	Item	230000.00 to 280000.00
Add to above for		
Increase speed of travel from 1.0 m/s to 1.6 m/s		
2000 kg	Item	6400.00 to 8000.00
2250 kg	Item	10500.00 to 12500.00
Enhanced finish to car – Centre mirror, flat ceiling, carpet		
2000 kg	Item	5500.00 to 6700.00
2250 kg	Item	7200.00 to 8800.00
2500 kg	Item	10500.00 to 13000.00
Painting to entire pit		
2000 kg	Item	2675.00 to 3275.00
2250 kg	Item	2675.00 to 3275.00
2500 kg	Item	2675.00 to 3275.00
Dual seal shaft		
2000 kg	Item	5300.00 to 6500.00
2250 kg	Item	5300.00 to 6500.00
2500 kg	Item	5300.00 to 6500.00
Intercom to reception desk and security room		
2000 kg	Item	540.00 to 660.00
2250 kg	Item	540.00 to 660.00
2500 kg	Item	540.00 to 660.00
Lifting beams		
2000 kg	Item	2025.00 to 2450.00
2250 kg	Item	2025.00 to 2450.00
2500 kg	Item	2025.00 to 2450.00

APPROXIMATE ESTIMATING RATES – 5 SERVICES

Item	Unit	Range £
5.10 Lift and Conveyor Systems – cont		
Add to above for – cont		
10 mm equipotential bonding of all entrance metalwork		
2000 kg	Item	1050.00 to 1300.00
2250 kg	Item	1050.00 to 1300.00
2500 kg	Item	1050.00 to 1300.00
Independent insurance inspection		
2000 kg	Item	2525.00 to 3100.00
2250 kg	Item	2525.00 to 3100.00
2500 kg	Item	2525.00 to 3100.00
12 month warranty service		
2000 kg	Item	1375.00 to 1675.00
2250 kg	Item	1375.00 to 1675.00
2500 kg	Item	1375.00 to 1675.00
ESCALATOR INSTALLATIONS		
30Ø Pitch escalator with a rise of 3 to 6 metres with standard balustrades		
1000 mm step width	Item	105000.00 to 130000.00
Add to above for		
Balustrade lighting	Item	3300.00 to 4100.00
Skirting lighting	Item	13000.00 to 16500.00
Emergency stop button pedestals	Item	5800.00 to 7300.00
Truss cladding – stainless steel	Item	33000.00 to 40500.00
Truss cladding – spray painted steel	Item	29500.00 to 36000.00

APPROXIMATE ESTIMATING RATES – 5 SERVICES

Item	Unit	Range £
5.11 Fire and Lightning Protection		
SPRINKLER INSTALLATION		
Recommended maximum area coverage per sprinkler head		
Extra light hazard, 21 m² of floor area		
Ordinary hazard, 12 m² of floor area		
Extra high hazard, 9 m² of floor area		
Equipment		
Sprinkler equipment installation, pipework, valve sets, booster pumps and water storage	Item	92000.00 to 105000.00
Price per sprinkler head; including pipework, valves and supports	Point	240.00 to 265.00
HOSE REELS AND DRY RISERS		
Wall mounted concealed hose reel with 36 metre hose including approximately 15 metres of pipework and isolating valve:		
Price per hose reel	Point	2150.00 to 2675.00
100 mm dry riser main including 2 Way breeching valve and box, 65 mm landing valve, complete with padlock and leather strap and automatic air vent and drain valve:		
Price per landing	Point	2600.00 to 3100.00
5.12 Communications and Security Installations		
ACCESS CONTROL SYSTEMS		
Door mounted access control unit inclusive of door furniture, lock plus software; including up to 50 metres of cable and termination; including documentation testing and commissioning		
Internal single leaf door	Point	1325.00 to 1600.00
Internal double door	Point	1450.00 to 1775.00
External single leaf door	Point	1550.00 to 1900.00
External double leaf door	Point	1775.00 to 2225.00
Management control PC with printer software and commissioning up to 1000 users	Point	19000.00 to 22000.00
CCTV INSTALLATIONS		
CCTV equipment inclusive of 50 m of cable including testing and commissioning		
Internal camera with bracket	Point	1150.00 to 1375.00
Internal camera with housing	Point	1400.00 to 1725.00
Internal PTZ camera with bracket	Point	2175.00 to 2650.00
External fixed camera with housing	Point	1550.00 to 1900.00
External PTZ camera dome	Point	3050.00 to 3650.00
External PTZ camera dome with power	Point	3725.00 to 4475.00

APPROXIMATE ESTIMATING RATES – 5 SERVICES

Item	Unit	Range £
5.12 Communications and Security Installations – cont		
TURNSTILES		
Physical Access Control Barrier system – standard security level comprising unit caseworks in stainless steel finish, standard lane width – 650 mm, restricting panels standard 900 mm high, standard level detection sensor system, including provision to integrate Access Control Card Readers (issued by others), including LED Pictogram – Green Arrow/Red Cross, closed base for ease of installation and cable management		
Including delivery, installation and commissioning to a site in London; budget cost land configurations		
Single lane	Item	11500.00 to 12500.00
Double lane	Item	18500.00 to 19500.00
Triple lane	Item	25500.00 to 26500.00
Physical Access Control Barrier system – medium security level comprising unit caseworks in stainless steel finish, standard lane width – 650 mm, restricting panels standard 900 mm high, medium level detection sensor system, including provision to integrate Access Control Card Readers (issued by others), including LED Pictogram – Green Arrow/Red Cross, closed base for ease of installation and cable management		
Including delivery, installation and commissioning to a site in London; budget cost land configurations		
Single lane	Item	13500.00 to 15000.00
Double lane	Item	20000.00 to 21500.00
Triple lane	Item	28500.00 to 29500.00
Physical Access Control Barrier system – high security level comprising unit caseworks in stainless steel finish, standard lane width – 650 mm, restricting panels standard 1600 mm or 1800 mm high (same cost), higher level detection sensor system, including provision to integrate Access Control Card Readers (issued by others), including LED Pictogram – Green Arrow/Red Cross, closed base for ease of installation and cable management		
Including delivery, installation and commissioning to a site in London; budget cost land configurations		
Single lane	Item	13500.00 to 15000.00
Double lane	Item	23000.00 to 25000.00
Triple lane	Item	33000.00 to 34000.00
High security full height security revolving door – 2300 mm high, four wing T25 sections, no centre column, positioning drive with horizontal and vertical safety strips at door leaves, controlled via electronic Access Control Card Readers (supplied by others), sensor system in ceiling monitors door segments, secure simultaneous bi-directional use possible, finish – standard finish – PPC to RAL colour, card reader mounting boxes, 4 LED lights in ceiling, rubber matting		
Including delivery, installation and commissioning to a site in London		
Buget cost – 1800 mm dia., per door	Item	30000.00 to 32000.00
Buget cost – 2000 mm dia., per door	Item	30000.00 to 32000.00
Additional cost – stainless steel finish	Item	2150.00 to 2150.00

APPROXIMATE ESTIMATING RATES – 5 SERVICES

Item	Unit	Range £	
IT INSTALLATIONS			
Data Cabling			
Complete channel link including patch leads, cable, panels, testing and documentation (excludes cabinets and/or frames, patch cords, backbone/ harness connectivity as well as containment)			
Low Level			
Cat 5e (up to 5,000 outlets)	Point	64.00 to	75.00
Cat 5e (5,000 to 15,000 outlets)	Point	42.00 to	53.00
Cat 6 (up to 5,000 outlets)	Point	74.00 to	86.00
Cat 6 (5,000 to 15,000 outlets)	Point	64.00 to	75.00
Cat 6a (up to 5,000 outlets)	Point	86.00 to	105.00
Cat 6a (5,000 to 15,000 outlets)	Point	85.00 to	96.00
Cat 7 (up to 5,000 outlets)	Point	105.00 to	130.00
Cat 7 (5,000 to 15,000 outlets)	Point	96.00 to	115.00
Note: LSZH cable based on average of 50 metres false floor low level installation assuming 1 Workstation in 2.5 m × 2.5 m (to 3.2 m × 3.2 m) density, with 4 data points per workstation. Not applicable for installations with less than 250 No. outlets			
High Level			
Cat 5e (up to 500 outlets)	Point	74.00 to	86.00
Cat 5e (over 500 outlets)	Point	64.00 to	75.00
Cat 6 (up to 500 outlets)	Point	85.00 to	96.00
Cat 6 (over 500 outlets)	Point	74.00 to	97.00
Cat 6a (up to 500 outlets)	Point	96.00 to	115.00
Cat 6a (over 500 outlets)	Point	96.00 to	115.00
Cat 7 (up to 500 outlets)	Point	120.00 to	140.00
Cat 7 (over 500 outlets)	Point	120.00 to	140.00

Note: High level at 10 × 10 m grid

BUILDING MODELS – ELEMENTAL COST SUMMARIES

Item	Unit	Range £	

AIRPORT TERMINAL BUILDING

New build airport terminal building, premium quality, located in the South East of England, handling both domestic and international flights with a gross internal floor area (GIFA) of 25,000 m². These costs exclude baggage handling, check-in systems, pre-check in and boarding security systems, vertical transportation and travellators, pre-conditioned air systems to aircraft, services to stands and visual docking systems. Costs assume that the works are undertaken under landside access/logistics environment.

Item	Unit	Range £	
5 Services			
5.1 Sanitary Installations	m²	8.00 to	11.00
5.3 Disposal Installations			
rainwater	m²	8.00 to	10.00
soil and waste	m²	10.00 to	15.00
condensate	m²	2.00 to	3.00
5.4 Water Installations			
domestic hot and cold water services	m²	22.00 to	25.00
5.5 Heat Source	m²	8.00 to	10.00
5.6 Space Heating and Air Conditioning			
LTHW heating system	m²	97.00 to	110.00
chilled water system	m²	105.00 to	120.00
supply and extract air conditioning system	m²	130.00 to	160.00
local cooling; DX systems to IT rooms	m²	7.00 to	10.00
5.7 Ventilation Systems			
mechanical ventilation to baggage handling and plantrooms	m²	33.00 to	39.00
toilet extract ventilation	m²	11.00 to	14.00
smoke extract installation	m²	14.00 to	21.00
kitchen extract system	m²	3.00 to	6.00
5.8 Electrical Installations			
main HV/MV Installations including switchgear, transformers (Cast Resin)	m²	77.00 to	99.00
low voltage (LV) – incoming LV switchgear, distribution boards and distribution systems	m²	54.00 to	67.00
generators, life safety – containerized standby diesel generators and 24 hour capacity belly tanks	m²	42.00 to	54.00
small power installation	m²	54.00 to	67.00
lighting installations; warehouse, office, ancillary and plant spaces	m²	145.00 to	180.00
emergency lighting installation	m²	14.00 to	23.00
power to mechanical services	m²	8.00 to	11.00
5.9 Fuel Installations (gas)	m²	3.00 to	4.00
5.11 Fire and Lightning Protection			
lightning protection	m²	1.00 to	2.00
earthing and bonding	m²	1.00 to	2.00
sprinkler installations	m²	42.00 to	53.00
dry riser and hosereels installation	m²	8.00 to	10.00
fire suppression to IT rooms	m²	6.00 to	7.00
5.12 Communication, Security and Control Systems			
fire alarm systems; DDI	m²	42.00 to	53.00
voice/public address systems	m²	26.00 to	31.00
other alarm systems, e.g. disabled refuge	m²	2.00 to	4.00

BUILDING MODELS – ELEMENTAL COST SUMMARIES

Item	Unit	Range £	
security installations; CCTV, access control and intruder detection	m^2	45.00 to	55.00
wireways for IT, comms and FA systems	m^2	17.00 to	22.00
structured IT cabling; fibre optic backbone and copper Cat 6 to office and warehouse	m^2	14.00 to	18.00
BMS systems	m^2	69.00 to	78.00
flight information display system	m^2	34.00 to	43.00
Total Cost/m² (based on GIFA of 25,000 m²)	m^2	1150.00 to	1450.00

BUILDING MODELS – ELEMENTAL COST SUMMARIES

Item	Unit	Range £	
SHOPPING MALL (TENANT'S FIT-OUT EXCLUDED)			
Natural ventilation shopping mall with approximately 33,000 m² two storey retail area and a 13,000 m² above ground, mechanically ventilated, covered car park, situated in a town centre in South East England.			
RETAIL BUILDING			
5 Services			
5.1 Sanitary Installations	m²	1.00 to	2.00
5.3 Disposal Installations			
rainwater	m²	8.00 to	9.00
soil, waste and vent	m²	8.00 to	9.00
5.4 Water Installations			
cold water installation	m²	8.00 to	9.00
hot water installation	m²	7.00 to	8.00
5.6 Space Heating and Air Conditioning			
condenser water system	m²	40.00 to	42.00
LTHW installation	m²	6.00 to	7.00
air conditioning system	m²	40.00 to	42.00
over door heaters at entrances	m²	1.00 to	2.00
5.7 Ventilation Systems			
public toilet ventilation	m²	1.00 to	2.00
plantroom ventilation	m²	6.00 to	7.00
supply and extract systems to shop units	m²	18.00 to	18.00
toilet extract systems to shop units	m²	2.00 to	3.00
smoke ventilation to Mall area	m²	13.00 to	14.00
service corridor ventilation	m²	2.00 to	3.00
other miscellaneous ventilation	m²	23.00 to	24.00
5.8 Electrical Installations			
LV distribution	m²	30.00 to	31.00
standby power	m²	6.00 to	7.00
general lighting	m²	81.00 to	85.00
external lighting	m²	7.00 to	8.00
emergency lighting	m²	17.00 to	17.00
small power	m²	13.00 to	14.00
mechanical services power supplies	m²	4.00 to	4.00
general earthing	m²	1.00 to	2.00
UPS for security and CCTV equipment	m²	1.00 to	2.00
5.9 Fuel Installations (gas)			
gas supply and boilers	m²	2.00 to	3.00
gas supplies to anchor (major) stores	m²	1.00 to	2.00
5.11 Fire and Lightning Protection			
lightning protection	m²	1.00 to	2.00
sprinkler installations	m²	18.00 to	20.00
5.12 Communication, Security and Control Systems			
fire alarm installation	m²	11.00 to	11.00
public address/voice alarm	m²	7.00 to	8.00
security installation	m²	14.00 to	15.00
general containment	m²	15.00 to	17.00

BUILDING MODELS – ELEMENTAL COST SUMMARIES

Item	Unit	Range £	
5.13 Specialist Installations			
BMS/Controls	m^2	23.00 to	24.00
Total Cost/m^2 (based on GIFA of 33,000 m^2)	m^2	435.00 to	475.00
CAR PARK – 13,000 m^2			
5 Services			
5.3 Disposal Intallations			
car park drainage	m^2	6.00 to	7.00
5.7 Ventilation Systems			
car park ventilation (impulse fans)	m^2	40.00 to	41.00
5.8 Electrical Installations			
LV distribution	m^2	13.00 to	14.00
general lighting	m^2	28.00 to	29.00
emergency lighting	m^2	7.00 to	8.00
small power	m^2	4.00 to	4.00
mechanical services power supplies	m^2	7.00 to	8.00
general earthing	m^2	1.00 to	2.00
ramp frost protection	m^2	2.00 to	3.00
5.11 Fire and Lightning Protection			
sprinkler installation	m^2	26.00 to	29.00
fire alarm installations	m^2	26.00 to	29.00
5.12 Communication, Security and Control Systems			
security installations	m^2	12.00 to	12.00
BMS/Controls	m^2	8.00 to	9.00
5.13 Specialist Installations			
entry/exit barriers, pay stations	m^2	8.00 to	9.00
Total Cost/m^2 (based on GIFA of 13,000 m^2)	m^2	190.00 to	205.00

BUILDING MODELS – ELEMENTAL COST SUMMARIES

Item	Unit	Range £	
SUPERMARKET			
Supermarket located in the South East with a total gross floor area of 4,000 m², including a sales area of 2,350 m². The building is on one level and incorporates a main sales, coffee shop, bakery, offices and amenities areas and warehouse.			
5 Services			
5.1 Sanitary Installations	m²	2.00 to	3.00
5.3 Disposal Installations	m²	4.00 to	6.00
5.4 Water Installations			
hot and cold water services	m²	33.00 to	34.00
5.6 Space Heating and Air Conditioning			
heating & ventilation with cooling via DX units	m²	13.00 to	14.00
5.7 Ventilation Systems			
supply and extract sytems	m²	3.00 to	4.00
5.8 Electrical Installations			
panels/boards	m²	37.00 to	40.00
containment	m²	2.00 to	3.00
general lighting	m²	25.00 to	26.00
small power	m²	13.00 to	14.00
mechanical services wiring	m²	2.00 to	4.00
5.11 Fire and Lightning Protection			
lightning protection	m²	1.00 to	2.00
5.12 Communication, Security and Control Systems			
fire alarms, detection and public address	m²	3.00 to	4.00
CCTV	m²	3.00 to	4.00
intruder alarm, detection and store security	m²	3.00 to	4.00
telecom and structured cabling	m²	1.00 to	2.00
BMS	m²	9.00 to	10.00
data cabinet	m²	3.00 to	4.00
controls wiring	m²	4.00 to	6.00
5.13 Specialist Installations			
refrigeration			
installation	m²	36.00 to	39.00
plant	m²	36.00 to	39.00
cold store	m²	13.00 to	14.00
cabinets	m²	100.00 to	105.00
Total Cost/m² (based on GIFA of 4,000 m²)	m²	345.00 to	380.00

BUILDING MODELS – ELEMENTAL COST SUMMARIES

Item	Unit	Range £
OFFICE BUILDING		
Speculative 15 storey office in Central London for multiple tenant occupancy with a gross mounted water cooled chillers, located in basement. 1 level of basement of approx. 1,600 m²		
SHELL & CORE – 20,000 m²		
5 Services		
5.1 Sanitary Installations	m²	18.00 to 20.00
5.3 Disposal Installations		
rainwater/soil and waste	m²	30.00 to 33.00
condensate	m²	2.00 to 3.00
5.4 Water Installations		
hot and cold water services	m²	60.00 to 65.00
5.5 Heat Source – ASHP	m²	33.00 to 35.00
5.6 Space Heating and Air Conditioning	m²	135.00 to 145.00
5.7 Ventilation Systems	m²	55.00 to 65.00
5.8 Electrical Installations	m²	195.00 to 210.00
5.11 Fire and Lighting Protection	m²	35.00 to 40.00
5.12 Communication, Security and Control Systems	m²	140.00 to 160.00
Total Cost/m² (based on GIFA of 20,000 m²)	m²	700.00 to 780.00
CATEGORY 'A' FIT-OUT – 13,000 m² NIA		
5 Services		
5.6 Space Heating and Air Conditioning	m²	220.00 to 230.00
5.8 Electrical Installations	m²	165.00 to 170.00
5.11 Fire and Lightning Protection	m²	27.00 to 33.00
5.12 Communication, Security and Control Systems	m²	45.00 to 55.00
Total Cost/m² (based on NIFA of 13,000 m²)	m²	450.00 to 490.00

BUILDING MODELS – ELEMENTAL COST SUMMARIES

Item	Unit	Range £	
BUSINESS PARK			
New build office in South East within M25, part of a speculative business park with a gross floor area of 10,000 m². A full air displacement system with roof mounted air cooled chillers, gas fired boilers and air handling plant. Total M&E services value approximately £2,500,000 (shell & core) and £1,000,000 (Cat A Fit-out). Excluding lifts.			
SHELL & CORE – 10,000 m² GIA			
5 Services			
5.1 Sanitary Installations	m²	9.00 to	12.00
5.3 Disposal Installations			
condensate	m²	2.00 to	4.00
rainwater, soil and waste	m²	13.00 to	17.00
5.4 Water Installations			
hot and cold water services	m²	13.00 to	18.00
5.5 Heat Source	m²	9.00 to	12.00
5.6 Space Heating and Air Conditioning			
LTHW heating; plantroom and risers	m²	13.00 to	18.00
chilled water; plantroom and risers	m²	37.00 to	42.00
ductwork; plantroom and risers	m²	70.00 to	80.00
5.7 Ventilation Systems			
toilet and miscellaneous ventilation	m²	13.00 to	18.00
5.8 Electrical Installations			
LV supply/distribution	m²	30.00 to	37.00
general lighting	m²	24.00 to	30.00
general power	m²	5.00 to	7.00
electrical services in connection with mechanical services	m²	3.00 to	5.00
security (wireways)	m²	2.00 to	4.00
voice and data (wireways)	m²	2.00 to	4.00
5.9 Fuel Installations (gas)	m²	2.00 to	4.00
5.11 Fire and Lightning Protection			
earthing and bonding	m²	3.00 to	5.00
lightning protection	m²	3.00 to	5.00
dry risers	m²	2.00 to	4.00
5.12 Communication, Security and Control Systems			
fire alarms	m²	11.00 to	14.00
BMS	m²	25.00 to	29.00
Total Cost/m² (based on GIFA of 10,000 m²)	m²	290.00 to	370.00

BUILDING MODELS – ELEMENTAL COST SUMMARIES

Item	Unit	Range £	
CATEGORY 'A' FIT-OUT – 8,000 m² NIA			
5 Services			
5.6 Space Heating and Air Conditioning			
LTHW heating and perimeter heaters	m²	41.00 to	52.00
floor swirl diffusers and supply ductwork	m²	26.00 to	31.00
5.8 Electrical Installations			
distribution boards	m²	2.00 to	4.00
general lightining, recessed including lighting controls	m²	56.00 to	68.00
5.11 Fire and Lightning Protection			
earthing and bonding	m²	2.00 to	3.00
5.12 Communication, Security and Control Systems			
fire alarms	m²	8.00 to	10.00
BMS	m²	6.00 to	8.00
Total Cost/m² (based on NIFA of 8,000 m²)	m²	145.00 to	175.00

BUILDING MODELS – ELEMENTAL COST SUMMARIES

Item	Unit	Range £	
PERFORMING ARTS CENTRE (MEDIUM SPECIFICATION)			
Performing Arts centre with a Gross Internal Area (GIA) of approximately 8,000 m², based on a medium specification with cooling to the Auditorium. The development comprises dance studios and a theatre auditorium in the outer London area. The theatre would require all the necessary stage lighting, machinery and equipment installed in a modern professional theatre (these are excluded from the model, as assumed to be FF&E, but the containment and power wiring is included).			
5 Services			
5.1 Sanitary Installations	m²	11.00 to	14.00
5.3 Disposal Installations			
soil, waste and rainwater	m²	18.00 to	23.00
5.4 Water Installations			
cold water services	m²	17.00 to	23.00
hot water services	m²	14.00 to	18.00
5.5 Heat Source	m²	13.00 to	17.00
5.6 Space Heating and Air Conditioning			
heating	m²	85.00 to	91.00
chilled water system	m²	73.00 to	98.00
supply and extract air systems	m²	130.00 to	160.00
5.7 Ventilation Systems			
ventilation and extract systems to toilets, kitchen and workshop	m²	25.00 to	32.00
5.8 Electrical Installations			
LV supply/distribution	m²	48.00 to	54.00
general lighting	m²	120.00 to	145.00
small power	m²	37.00 to	43.00
power to mechanical plant	m²	10.00 to	11.00
5.9 Fuel Installations (gas)	m²	4.00 to	7.00
5.11 Fire and Lightning Protection			
lightning protection	m²	4.00 to	6.00
5.12 Communication, Security and Control Systems			
fire alarms and detection	m²	32.00 to	37.00
voice and data complete installation (excluding active equipment)	m²	37.00 to	48.00
security; access; control; disabled alarms; staff paging	m²	35.00 to	48.00
BMS	m²	54.00 to	67.00
5.13 Specialist Installations			
theatre systems includes for containment and power wiring	m²	32.00 to	37.00
Total Cost/m² (based on NIFA of 8,000 m²)	m²	800.00 to	980.00

BUILDING MODELS – ELEMENTAL COST SUMMARIES

Item	Unit	Range £	
SPORTS HALL			
Single storey sports hall, located in the South East, with a gross internal area of 1,200 m² (40 m × 30 m).			
5 Services			
5.1 Sanitary Installations	m²	13.00 to	14.00
5.3 Disposal Installations			
rainwater	m²	4.00 to	6.00
soil and waste	m²	9.00 to	10.00
5.4 Water Installations			
hot and cold water services	m²	18.00 to	19.00
5.5 Heat Source			
boilers, flues, pumps and controls	m²	18.00 to	19.00
5.6 Space Heating and Air Conditioning			
warm air heating to sports hall area	m²	23.00 to	24.00
radiator heating to ancillary areas	m²	26.00 to	27.00
5.7 Ventilation Systems			
ventilation to changing, fitness and sports hall areas	m²	25.00 to	26.00
5.8 Electrical Installations			
main switchgear and sub-mains	m²	15.00 to	16.00
small power	m²	13.00 to	14.00
lighting and luminaires to sports areas	m²	30.00 to	32.00
lighting and luminaires to ancillary areas	m²	36.00 to	37.00
5.11 Fire and Lightning Protection			
lightning protection	m²	6.00 to	7.00
5.12 Communication, Security and Control Systems			
fire, smoke detection and alarm system, intruder detection	m²	15.00 to	16.00
CCTV installation	m²	17.00 to	18.00
public address and music systems	m²	9.00 to	10.00
wireways for voice and data	m²	4.00 to	6.00
Total Cost/m² (based on GIFA of 1,200 m²)	m²	280.00 to	300.00

BUILDING MODELS – ELEMENTAL COST SUMMARIES

Item	Unit	Range £	
STADIUM – NEW			
A three storey stadium, located in Greater London with a gross internal area of 85,000 m² and incorporating 60,000 spectator seats.			
5 Services			
5.1 Sanitary Installations	m²	13.00 to	14.00
5.3 Disposal Installations			
rainwater	m²	6.00 to	7.00
above ground drainage	m²	14.00 to	15.00
5.4 Water Installations			
hot and cold water services	m²	26.00 to	28.00
5.5 Heat Source	m²	12.00 to	13.00
5.6 Space Heating and Air Conditioning			
heating	m²	8.00 to	9.00
cooling	m²	18.00 to	20.00
5.7 Ventilation Systems	m²	74.00 to	77.00
5.8 Electrical Installations			
HV/LV Supply	m²	13.00 to	14.00
LV distribution	m²	32.00 to	33.00
general lighting	m²	77.00 to	80.00
small power	m²	25.00 to	26.00
earthing and bonding	m²	1.00 to	2.00
power supply to mechanical equipment	m²	1.00 to	2.00
pitch lighting	m²	13.00 to	13.00
5.9 Fuel Installations (gas)	m²	6.00 to	7.00
5.11 Fire and Lightning Protection			
lightning protection	m²	1.00 to	2.00
hydrants	m²	2.00 to	3.00
5.12 Communication, Security and Control Systems			
wireways for data, TV, telecom and PA	m²	10.00 to	11.00
5.7 Ventilation Systems	m²	74.00 to	77.00
public address	m²	21.00 to	22.00
security	m²	18.00 to	20.00
data voice installations	m²	40.00 to	42.00
fire alarms	m²	10.00 to	11.00
disabled/refuse alarm/call systems	m²	4.00 to	5.00
BMS	m²	18.00 to	20.00
Total Cost/m² (based on GIFA of 85,000 m²)	m²	465.00 to	495.00
Total Cost/seat (based on 60,000 seats)	Each	660.00 to	700.00

BUILDING MODELS – ELEMENTAL COST SUMMARIES

Item	Unit	Range £	
HOTELS			
200 Bedroom, four star hotel, situated in Central London, with a gross internal floor area of 16,500 m².			
The development comprises a 10 storey building with large suites on each guest floor, together with banqueting, meeting rooms and leisure facilities.			
5 Services			
5.1 Sanitary Installations	m²	52.00 to	64.00
5.3 Disposal Installations			
rainwater	m²	6.00 to	8.00
soil and waste	m²	33.00 to	41.00
5.4 Water Installations			
hot and cold water services	m²	68.00 to	87.00
5.5 Heat Source			
air source heat pumps	m²	68.00 to	83.00
5.6 Space Heating and Air Conditioning			
air conditioning and space heating system; chillers, pumps, CHW and LTHW pipework, insulation, 4 pipe FCU, ductwork, grilles and diffusers, to guest rooms, public areas, meeting and banquet rooms	m²	200.00 to	275.00
5.7 Ventilation Systems			
general bathroom extract from guest suites, ventilation to kitchens and bathrooms, etc.	m²	55.00 to	72.00
smoke extract	m²	17.00 to	22.00
5.8 Electrical Installations			
HV/LV Installation, standby power, lighting, emergency lighting and small power to guest rooms and public areas, including earthing and bonding	m²	150.00 to	195.00
5.11 Fire and Lightning Protection			
dry risers and sprinkler installation	m²	47.00 to	59.00
lightning protection	m²	3.00 to	4.00
5.12 Communication, Security and Control Systems			
fire/smoke detection and fire alarm system	m²	20.00 to	30.00
security/access control	m²	24.00 to	32.00
integrated sound and AV system	m²	25.00 to	30.00
telephone and data and TV installation	m²	19.00 to	22.00
containment	m²	18.00 to	22.00
wire ways for voice and data (no hotel management and head end equipment)	m²	12.00 to	14.00
BMS	m²	40.00 to	50.00
Total Cost/m² (based on GIFA of 16,500 m²)	m²	860.00 to	1100.00

BUILDING MODELS – ELEMENTAL COST SUMMARIES

Item	Unit	Range £	
PRIVATE HOSPITAL			
New build project building. The works consist of a new 80 bed hospital of approximately 15,000 m², eight storey with a plant room.			
All heat is provided from existing steam boiler plant, medical gases are also served from existing plant. The project includes the provision of additional standby electrical generation to serve the wider site requirements.			
This hospital has six operating theatres, ITU/HDU department, pathology facilities, diagnostic imaging, out patient facilities and physiotherapy.			
5 Services			
5.1 Sanitary Installations	m²	40.00 to	44.00
5.3 Disposal Installations			
rainwater	m²	4.00 to	8.00
soil and waste	m²	44.00 to	45.00
5.4 Water Installations			
hot and cold water services	m²	100.00 to	105.00
5.5 Heat Source (included in 5.6)	N/A		
5.6 Space Heating and Air Conditioning			
LPHW heating	m²	84.00 to	86.00
chilled water	m²	61.00 to	65.00
steam and condensate	m²	40.00 to	44.00
ventilation, comfort cooling and air conditioning	m²	205.00 to	240.00
5.7 Ventilation Systems (included in 5.6)	N/A		
5.8 Electrical Installations			
HV distribution	m²	54.00 to	60.00
LV supply/distribution	m²	80.00 to	87.00
standby power	m²	48.00 to	52.00
UPS	m²	35.00 to	39.00
general lighting	m²	91.00 to	105.00
general power	m²	71.00 to	74.00
emergency lighting	m²	27.00 to	29.00
theatre lighting	m²	21.00 to	23.00
specialist lighting	m²	21.00 to	23.00
external lighting	m²	4.00 to	7.00
electrical supplies for mechanical services	m²	17.00 to	22.00
earthing and bonding	m²	5.00 to	8.00
5.9 Fuel Installations			
gas installations	m²	7.00 to	9.00
oil installations	m²	10.00 to	12.00
5.11 Fire and Lightning Protection			
dry risers	m²	4.00 to	7.00
sprinkler systems/gaseous fire suppression	m²	68.00 to	78.00
lightning protection	m²	2.00 to	4.00

BUILDING MODELS – ELEMENTAL COST SUMMARIES

Item	Unit	Range £	
5.12 Communication, Security and Control Systems			
fire alarms and detection	m^2	40.00 to	43.00
voice and data	m^2	62.00 to	64.00
data containment	m^2	12.00 to	15.00
security and CCTV	m^2	35.00 to	37.00
nurse call and cardiac alarm system	m^2	53.00 to	55.00
TV systems	m^2	22.00 to	25.00
disabled WC alarm	m^2	5.00 to	9.00
BMS	m^2	93.00 to	94.00
5.13 Specialist Installations			
pneumatic tube conveying system	m^2	8.00 to	12.00
medical gases	m^2	96.00 to	100.00
Total Cost/m² (based on GIFA of 15,000 m²)	m^2	1550.00 to	1725.00

BUILDING MODELS – ELEMENTAL COST SUMMARIES

Item	Unit	Range £	
SCHOOL			
New build secondary school (Academy) located in Southern England, with a gross internal floor area of 10,000 m². Total M&E services value approximately £5,000,000 and extensive use of offsite manufacturing.			
The building comprises a three storey teaching block, including provision for music, drama, catering, sports hall, science laboratories, food technology, workshops and reception/admin (BB93 compliant). Excludes IT active systems, renewable energy and sprinkler protection.			
5 Services			
5.1 Sanitary Installations			
toilet cores and changing facilities and lab sinks	m²	16.00 to	18.00
5.3 Disposal Installations			
rainwater installations	m²	5.00 to	7.00
soil and waste	m²	17.00 to	20.00
5.4 Water Installations			
potable hot and cold water services	m²	36.00 to	40.00
non-potable hot and cold water services to labs and art rooms	m²	12.00 to	17.00
5.5 Heat Source			
gas fired boiler installation	m²	13.00 to	17.00
5.6 Space Heating and Air Conditioning			
LTHW heating system (primary)	m²	43.00 to	48.00
LTHW heating system (secondary)	m²	11.00 to	14.00
DX cooling system to ICT server rooms	m²	10.00 to	13.00
mechanical supply and extract ventilation including DX type cooling to Music, Drama, Kitchen/Dining and Sports Hall	m²	66.00 to	84.00
5.7 Ventilation Systems			
toilet extract systems	m²	11.00 to	15.00
changing area extract systems	m²	7.00 to	9.00
extract ventilation from design/food technology and science labs	m²	13.00 to	20.00
5.8 Electrical Installations			
mains and sub-mains distribution	m²	44.00 to	54.00
lighting and luminaires including emergency fittings	m²	82.00 to	100.00
small power installation	m²	42.00 to	49.00
earthing and bonding	m²	2.00 to	3.00
5.9 Fuel Installations (gas)	m²	10.00 to	15.00
5.11 Fire and Lightning Protection			
lightning protection	m²	4.00 to	5.00
5.12 Communication, Security and Control Systems			
containment for telephone, IT data, AV and security systems	m²	7.00 to	13.00
fire, smoke detection and alarm system	m²	22.00 to	29.00
security installations including CCTV, access control and intruder alarm	m²	24.00 to	27.00
disabled toilet, refuge and induction loop systems	m²	3.00 to	5.00
BMS – to plant	m²	26.00 to	33.00
BMS – to opening vents/windows	m²	11.00 to	13.00
Total Cost/m² (based on GIFA of 10,000 m²)	m²	540.00 to	670.00

BUILDING MODELS – ELEMENTAL COST SUMMARIES

Item	Unit	Range £	
AFFORDABLE RESIDENTIAL DEVELOPMENT			
An 8 storey, 117 affordable residential development with a gross internal area of 11,400 m² and a net internal area of 8,400 m², situated within the Central London area. The development has no car park or communal facilities and achieves a net to gross efficiency of 74%.			
Based upon radiator LTHW heating within each apartment, with plate heat exchanger, whole house ventilation, plastic cold and hot water services pipework, sprinkler protection to apartments. Excludes remote metering. Based upon 2 bed units with 1 bathroom. Heating provided by air source heat pumps			
SHELL & CORE			
5 Services			
5.1 Sanitary Installations	m²	1.00 to	1.00
5.3 Disposal Installations	m²	23.00 to	30.00
5.4 Water Installations	m²	23.00 to	29.00
5.5 Heat Source	m²	33.00 to	40.00
5.6 Space Heating and Air Conditioning	m²	30.00 to	35.00
5.7 Ventilation Systems	m²	23.00 to	25.00
5.8 Electrical Installations	m²	50.00 to	65.00
5.9 Fuel Installations – Gas	N/A		
5.11 Fire and Lightning Protection	m²	35.00 to	40.00
5.12 Communication, Security and Control Systems	m²	40.00 to	50.00
5.13 Specialist Installations	m²	18.00 to	18.00
Total Cost/m² (based on GIFA of 11,400 m²)	m²	275.00 to	330.00
FITTING OUT			
5 Services			
5.1 Sanitary Installations	m²	40.00 to	45.00
5.3 Disposal Installations	m²	15.00 to	20.00
5.4 Water Installations	m²	35.00 to	41.00
5.5 Heat Source	m²	40.00 to	52.00
5.6 Space Heating and Air Conditioning	m²	50.00 to	63.00
5.7 Ventilation Systems	m²	50.00 to	60.00
5.8 Electrical Installations	m²	70.00 to	85.00
5.11 Fire and Lightning Protection	m²	30.00 to	35.00
5.12 Communication, Security and Control Systems	m²	25.00 to	30.00
5.13 Specialist Installations	m²	18.00 to	18.00
Total Cost/m² (based on a NIA of 8,400 m²)	m²	375.00 to	450.00
Total cost per apartment – Shell & Core and Fitting Out	Each	54000.00 to	64000.00

BUILDING MODELS – ELEMENTAL COST SUMMARIES

Item	Unit	Range £	
PRIVATE RESIDENTIAL DEVELOPMENT			
A 24 storey, 203 apartment private residential development with a gross internal area of 27,000 m² and a net internal area of 18,000 m², situated within the London's prime residential area. The development has limited car park facilities, residents' communal areas such as gym, cinema room, self-stimulated, etc. Based upon LTHW underfloor heating, 4 pipe fan coils to principle rooms, with plate heat and cooling exchangers, whole house ventilation with summer house facility, copper hot and cold water services pipework, sprinkler protection to apartments, lighting control, TV/data outlets. Flood wiring for apartment, home automation and sound system only. Based upon a combination of 2/3 bedroom apartments with family bathroom and ensuite.			
SHELL & CORE			
5 Services			
5.1 Sanitary Installations	m²	3.00 to	4.00
5.3 Disposal Installations	m²	29.00 to	38.00
5.4 Water Installations	m²	37.00 to	48.00
5.5 Heat Source	m²	33.00 to	41.00
5.6 Space Heating and Air Conditioning	m²	92.00 to	125.00
5.7 Ventilation Systems	m²	27.00 to	33.00
5.8 Electrical Installations	m²	75.00 to	105.00
5.9 Fuel Installations – Gas	N/A		
5.11 Fire and Lightning Protection	m²	35.00 to	55.00
5.12 Communication, Security and Control Systems	m²	55.00 to	75.00
5.13 Specialist Installations	m²	46.00 to	65.00
Total Cost/m² (based on GIFA of 27,000 m²)	m²	435.00 to	590.00
FITTING OUT			
5 Services			
5.1 Sanitary Installations	m²	125.00 to	150.00
5.3 Disposal Installations	m²	30.00 to	40.00
5.4 Water Installations	m²	60.00 to	80.00
5.5 Heat Source	m²	70.00 to	90.00
5.6 Space Heating and Air Conditioning	m²	180.00 to	215.00
5.7 Ventilation Systems	m²	60.00 to	72.00
5.8 Electrical Installations	m²	150.00 to	185.00
5.11 Fire and Lightning Protection	m²	35.00 to	43.00
5.12 Communication, Security and Control Systems	m²	120.00 to	145.00
5.13 Specialist Installations	m²	45.00 to	50.00
Total Cost/m² (based on NIA of 18,000 m²)	m²	880.00 to	1075.00
Total cost per apartment – Shell & Core and Fit-Out	Each	135000.00 to	170000.00

BUILDING MODELS – ELEMENTAL COST SUMMARIES

Item	Unit	Range £	
DISTRIBUTION CENTRE New build distribution centre located in the South East of England with a total gross floor area of 75,000 m², including a refrigerated cold box of 17,500 m². The building is on one level (no mezzanine decks) and incorporates a small office area at ground level, vehicle recovery unit, electric fork lift truck docking bay, gate house and associated plantrooms.			
5 Services			
5.3 Disposal Installations			
soil and waste	m²	8.00 to	9.00
rainwater	m²	3.00 to	4.00
5.4 Water Installations			
domestic hot and cold water services	m²	3.00 to	4.00
emergency drench showers & Cat 5 vehicle wash downs	m²	1.00 to	2.00
5.6 Space Heating and Air Conditioning			
heating with ventilation to offices, displacement system to main warehouse	m²	33.00 to	43.00
local cooling; DX systems to refrigerated cold box	m²	44.00 to	52.00
5.7 Ventilation Systems			
smoke extract system	m²	10.00 to	11.00
specialist extract to battery charging and maintenance areas	m²	2.00 to	3.00
5.8 Electrical Installations			
generator, life safety – containerized standby diesel generator and 24 hour capacity belly tanks	m²	6.00 to	7.00
main HV/MV installations including switchgear, transformers (Cast Resin)	m²	26.00 to	28.00
low voltage (LV) – incoming LV switchgear, distribution boards and distribution systems	m²	21.00 to	22.00
lighting installations; warehouse, office, ancillary and plant spaces	m²	17.00 to	22.00
small power installations inc. mech/HVAC plant power	m²	13.00 to	17.00
5.9 Fuel Installations (gas)			
gas mains services to plantroom	m²	1.00 to	2.00
5.11 Fire and Lightning Protection			
sprinklers (ESFR) including rack protection	m²	58.00 to	63.00
lightning protection	m²	1.00 to	2.00
earthing and bonding	m²	1.00 to	2.00
5.12 Communication, Security and Control Systems			
fire alarm systems; DDI	m²	14.00 to	17.00
security installations; CCTV, access control and intruder detection	m²	9.00 to	10.00
BMS systems	m²	7.00 to	8.00
other alarm systems, e.g. disabled refuge	m²	1.00 to	2.00
structured IT cabling; fibre optic backbone ad copper Cat 6 to office and warehouse	m²	14.00 to	18.00
5.13 Specialist Installations			
testing and commissioning	m²	3.00 to	4.00
Total Cost/m² (based on GIFA of 75,000 m²)	m²	295.00 to	355.00

BUILDING MODELS – ELEMENTAL COST SUMMARIES

Item	Unit	Range £	
DATA CENTRE			
New build data centre in a single storey 'warehouse' type construction in the Greater London area/home counties proximity to the M25. Total Net Technical Area (NTA) of 2,000 m² (2 nr data halls of 1,000 m² NTA each) with typically other areas of 300 m² for NOC/Security Room, office space, 350 m² ancillary space and 1,000 m² of internal plant areas. Total GIA of 3,650 m².			
Power and cooling to Technical Spaces designed to 1,500w/m² of NTA = 3,000 kW IT load.			
Critical infrastructure built to UTI Tier III resilience standard (HVAC generally N+1 and Electrical on a dual 'A' and 'B' string supplies from MV/LV switchgear to IT cab (Critical UPS's N+1 on each string) with all heat rejection and standby generators (N+1) located in a secure, external plant farm.			
Total MEP services value of approximately £18,520,000 to £23,090,000.			
All costs expressed as £/m² against Nett Technical Area – NTA.			
5 Services			
5.3 Disposal Installations			
soil and waste	m²	17.00 to	20.00
rainwater	m²	12.00 to	14.00
condensate	m²	12.00 to	17.00
5.4 Water Installations			
domestic hot and cold water services	m²	26.00 to	35.00
5.5 Heat Source			
for office and ancillary spaces only	m²	20.00 to	30.00
5.6 Space Heating and Air Conditioning			
chilled water plant with redundancy of N+1 to technical space; packaged air-cooled Turbocor chillers with free cooling capability	m²	590.00 to	780.00
chilled water distribution systems; plant, header and primary distribution to technical space CRAC units, office and ancillary space cooling units; single distribution ring with redundancy N+1 on pumps, pressurization, dosing and buffer vessels	m²	1200.00 to	1375.00
chilled water final connections to technical space CRAC office/ancillary space cooling units	m²	155.00 to	195.00
computer room air conditioning (CRAC) units; 100 kW single coil units with 30% of units with humidification; N+2 redundancy	m²	290.00 to	360.00
CRAC units to critical ancillary space, e.g. UPS module and battery rooms	m²	51.00 to	73.00
cooling units to office, support and other spaces; single coil FCUs	m²	41.00 to	64.00
5.7 Ventilation Systems			
supply and extract ventilation systems to data halls, switchrooms and ancillary spaces including dedicated gas extract to gaseous fire suppression protected spaces	m²	110.00 to	160.00
computer room floor grilles; heavy duty 600x600 mm metal grilles with dampers	m²	155.00 to	185.00
general supply and extract to office, support and ancillary spaces	m²	62.00 to	99.00

BUILDING MODELS – ELEMENTAL COST SUMMARIES

Item	Unit	Range £	
5.8 Electrical Installations			
main HV/MV installations including switchgear, transformers (Cast Resin); dual MV supply from local authority (REC supply cost excluded)	m²	225.00 to	285.00
low voltage (LV) switchgear – incoming LV switchgear, UPS Input and Output LV boards, mechanical services and support/ancillary area panels and distribution boards	m²	1175.00 to	1275.00
LV distribution including busbars, cabling and containment systems to provide full dual 'A' and 'B' supplies to data halls and supplies to mechanical services, office and ancillary areas	m²	880.00 to	1025.00
generators – containerized standby diesel generators, DC rated with acoustic treatment and exhaust systems; On an N+1 redundancy for critical IT electrical and cooling loads including synchronization panel and 48 hour capacity belly tanks	m²	1100.00 to	1175.00
uninterruptible power supplies (UPS) – static UPS systems to provide 2 × (N+1) redundancy; critical IT loads – 600 kVA modules with 10 minute battery autonomy, 0.9 PFC, harmonic filtration and static by-pass; 500 kVA Static UPS with 10 minute battery autonomy for mechanical loads on N load basis	m²	910.00 to	1075.00
critical power distribution units (PDUs) to data halls; Dual 'A' and 'B' units to provide diverse supplies to each IT cabinet	m²	580.00 to	900.00
final critical IT power supplies. 2 nr supplies to each cabinet from 'A' and 'B' PDUs	m²	485.00 to	660.00
lighting installations; data halls, office, support, ancillary and plant supplies	m²	155.00 to	195.00
5.9 Fuel Installations			
manual fill point and distribution pipework to generator belly tanks	m²	48.00 to	73.00
5.10 Lift and Conveyor Systems			
dock levellers and DDA access platform lifts only	m²	46.00 to	64.00
5.11 Fire and Lightning Protection			
gaseous fire suppression to technical, UPS and switchgear rooms	m²	360.00 to	405.00
lightning protection	m²	12.00 to	25.00
earthing and clean earth	m²	25.00 to	37.00
leak detection; to chilled water circuits within data hall service corridor	m²	64.00 to	87.00
5.12 Communication, Security and Control Systems			
fire alarm systems; DDI	m²	64.00 to	87.00
very early smoke detection alarm (VESDA) to technical areas	m²	145.00 to	185.00
other alarm systems, e.g. disabled refuge	m²	11.00 to	23.00
structured IT cabling; fibre optic backbone and copper Cat 6 to office and support spaces only. IT cabling to data halls EXCLUDED (Client fit-out item).	m²	145.00 to	200.00
security installations; CCTV, Access Control and Intruder Detection	m²	610.00 to	840.00
BMS/EMS systems	N/A		
5.13 Specialist Installations			
factory acceptance testing, site acceptance testing, integrated systems testing and 'black building' testing @ 2% of MEP value	m²	210.00 to	260.00
Total Cost/m² (based on NTA of 2,000 m²)	m²	10500.00 to	13500.00
Total Cost/kW IT Load (based om 3,000 kW IT Load)	m²	7100.00 to	8900.00

Notes: The above includes; MEP preliminaries and OH&P allowances
The above EXCLUDES; Main contractor's preliminaries and OH&P; Utility connections; Electric, water, drainage and Fibre Connections; Contingency/Risk Allowances; Foreign Exchange fluctuations; VAT at prevailing rates

BUILDING MODELS – ELEMENTAL COST SUMMARIES

Item	Unit	Range £
GYM		
Gym located in Central London on the ground floor of a mixed use development. This model is based on a total of 2,000 m², with costs based on connection to central plant within plant rooms, water source heat pump. The model includes for an AHU dedicated to providing ventilation to the gym. Services provided also include for 4 Pipe Fan Coil Units providing Heating and Cooling, additional underfloor heating to changing rooms. Excluded from the gym fit-out costs are FFE, active IT equipment and gym equipment.		
5 Services		
5.1 Sanitary Installations		
sanitary appliances including WCs, wash hand basins, cleaner's sinks, water fountains, urinals, showers and the provision of disabled toilets and accessible showers	m²	44.00 to 51.00
5.3 Disposal Installations		
soil, waste and vent installation to all sanitaryware points, provision of stub stacks, condensate drainage for fan coil units including insulation, rainwater disposal excluded as part of base build installations	m²	21.00 to 28.00
5.4 Water Installations		
installation of mains cold water services including meter, storage tanks, pumps, electromagnetic water conditioner and connections to sanitaryware, hot water including connection to base built plate heat exchanger for hot water generation, bulk storage, distribution, pump sets and connections to sanitaryware, miscellaneous water points to drinking fountains and plant supplies	m²	31.00 to 37.00
5.5 Heat Source		
connection to base build plate heat exchanger, energy meter, buffer vessels, associated pump sets and primary distribution pipework and insulation and water source heat pump	m²	61.00 to 72.00
5.6 Space Heating and Air Conditioning		
CHW distribution pipework to AHU and FCUs, LTHW distribution to underfloor heating, AHU and FCUs, intake and exhaust air ductwork to AHU. Air handling unit and associated supply and extract ductwork distribution	m²	280.00 to 340.00
5.7 Ventilation Systems		
intake and exhaust air ductwork to fans, toilet and shower supply and extract system. Extract to stores and cleaner's cupboards. MVHR unit to office/staff rest room.	m²	40.00 to 52.00
5.8 Electrical Installations		
LV distribution system including switchgear, containment and cabling, small power and lighting including emergency lighting and controls, floor boxes to gym equipment, enhanced lighting to lobbies, reception areas and toilet/ changing areas, earthing and bonding	m²	230.00 to 270.00
5.11 Fire and Lightning Protection		
connection to base build installations, zone valve, concealed sprinkler heads throughout	m²	33.00 to 41.00

BUILDING MODELS – ELEMENTAL COST SUMMARIES

Item	Unit	Range £
5.12 Communication, Security and Control Systems fire alarm system, PA/VA installation, access control including card reader turnstiles at reception controlling entrance to the gym, security system comprising of security cameras, access control and intruder alarms to back of houses, TV and data installations to gym, Wi-Fi installation, complete sound system including speakers, cabling and central music generation and smart device docks	m²	245.00 to 285.00
Total Cost/m² (based on GIFA of 2,000 m²)	m²	980.00 to 1175.00

BUILDING MODELS – ELEMENTAL COST SUMMARIES

Item	Unit	Range £	
BAR Bar located in Central London on the 15th floor of a mixed use development. This model is based on a total GIA of 500 m², with costs based on for complete standalone plant. The model includes for a displacement ventilation installation, heating and cooling provided by a VRF installation via terminal units and an AHU. Excluded from the cost model is the bar and back bar unit, lifts with connections only provided for electrical, comms, water and drainage, FFE and active IT equipment.			
5 Services			
5.1 Sanitary Installations			
sanitary appliances including WCs, wash hand basins, urinals and the provision of disabled toilets	m²	72.00 to	83.00
5.3 Disposal Installations			
soil, waste and vent installation to all sanitaryware points, provision of stub stacks capped connections to bar areas, condensate drainage for fan coil units including insulation, rainwater gullies to terrace areas	m²	31.00 to	35.00
5.4 Water Installations			
installation of mains cold water services including meter, storage tanks, pumps, electromagnetic water conditioner and connections to sanitaryware, hot water generation, plant bulk storage, distribution, pump sets and connections to sanitaryware, capped connections to bar areas, miscellaneous water points and plant supplies	m²	160.00 to	180.00
5.6 Space Heating and Air Conditioning			
VRF installation serving the AHU and terminal units provide heating and cooling to bar areas standalone cooling only VRF units to bar cellar and stores, specialist temperature control and humidity control units to wine cellar, Intake and exhaust air ductwork to AHU. Air handling unit and associated supply and extract ductwork distribution	m²	300.00 to	345.00
5.7 Ventilation Systems			
supply and extract installations to WC areas and back of house areas. Extract to cleaner's cupboard, stores and the like.	m²	68.00 to	83.00
5.8 Electrical Installations			
LV distribution system including switchgear, containment and cabling small power and lighting including emergency lighting and controls, provision of enhanced lighting to lobbies, bar areas and toilet areas. Allowance for statement lighting, full lighting control and scene setting, capped electrical supply to bar area, earthing and bonding	m²	530.00 to	710.00
5.11 Fire and Lightning Protection			
connection to base build installations, zone valve, concealed sprinkler heads throughout	m²	37.00 to	44.00
5.12 Communication, Security and Control Systems			
fire alarm system, PA/VA installation, security system comprising of security cameras, access control and intruder alarms to back of houses, TV and data installations to gym, Wi-Fi installation, complete sound system including speakers, cabling and central music generation and smart device docks	m²	400.00 to	450.00
Total Cost/m² (based on GIFA of 500 m²)	m²	1600.00 to	1925.00

BUILDING MODELS – ELEMENTAL COST SUMMARIES

Item	Unit	Range £	
SPA High end luxury spa located in Central London within the basement level of a mixed use development. This model is based on a total GIA of 1,050 m², with costs based on connection to central plant within plant rooms. The model includes for a water source heat pump an AHU which is only providing ventilation to the spa area and swimming pool hall. Services provided also include for 4 pipe fan coil units providing heating and cooling, displacement ventilation to the pool hall, additional underfloor heating to changing rooms. Included within the cost plan are costs for swimming pool plant, steam rooms and a sauna. Excluded from the spa fit-out costs are FFE and active IT equipment. Locating elements such as spas within the basements of developments will attract a cost premium due to the difficulty associated with bringing services into these areas. A holistic approach to costing must be considered for spa areas, as basement digs to accommodate swimming pools and the associated structural and water proofing required.			
5 Services			
5.1 Sanitary Installations			
sanitary appliances including WCs, wash hand basins, cleaner's sinks, water fountains, urinals, showers, ice fountain and the provision of disabled toilets and accessible showers	m²	115.00 to	140.00
5.3 Disposal Installations			
soil, waste and vent installation to all sanitaryware points, provision of stub stacks, condensate drainage for fan coil units including insulation, back wash drainage to pool plant, rainwater gullies to terrace areas	m²	34.00 to	41.00
5.4 Water Installations			
installation of mains cold water services including meter, storage tanks, pumps, electromagnetic water conditioner and connections to sanitaryware, hot water including connection to base built plate heat exchanger for hot water generation, bulk storage, distribution, pump sets and connections to sanitaryware, softened water to experience shower, saunas and steam rooms, miscellaneous water points to drinking fountains and plant supplies	m²	70.00 to	85.00
5.5 Heat Source			
connection to base build plate heat exchanger, energy meter, buffer vessels, associated pump sets and primary distribution pipework and insulation	m²	110.00 to	135.00
5.6 Space Heating and Air Conditioning			
CHW distribution pipework to AHU and FCUs, VRF installation to comms/media rooms, LTHW distribution to underfloor heating, AHU and FCUs, supply and extract installation to pool hall, intake and exhaust air ductwork to AHU. Air handling unit and associated supply and extract ductwork distribution.	m²	740.00 to	840.00
5.7 Ventilation Systems			
intake and exhaust air ductwork to fans, toilet and shower supply and extract system. Extract to stores and cleaner's cupboards. MVHR unit to office/staff rest room. Extra over for smoke extract	m²	150.00 to	175.00
5.8 Electrical Installations			
LV distribution system including switchgear, containment and cabling, small power and lighting including emergency lighting and controls, floor boxes to gym equipment, enhanced lighting to lobbies, reception areas and toilet/changing areas, allowance for statement lighting, full lighting control, fibre optic lighting with pool and scene setting, earthing and bonding	m²	390.00 to	475.00

BUILDING MODELS – ELEMENTAL COST SUMMARIES

Item	Unit	Range £
SPA – cont		
5 Services – cont		
5.11 Fire and Lightning Protection		
connection to base build installations, zone valve, concealed sprinkler heads throughout	m^2	35.00 to 40.00
5.12 Communication, Security and Control Systems		
fire alarm system, PA/VA installation, access control including card reader turnstiles at reception controlling entrance to the gym, security system comprising of security cameras, access control and intruder alarms to back of houses, TV and data installations, Wi-Fi installation, complete sound system including speakers, cabling and central music generation and smart device docks, installation of central building management system including central control panels and BMS to plant and equipment	m^2	385.00 to 470.00
5.13 Specialist Installations		
saunas and steam rooms. Pool plant, including all pumps, water treatment, filters and controls	m^2	630.00 to 700.00
Total Cost/m² (based on GIFA of 1,050 m²)	m^2	2675.00 to 3075.00

Material Costs/Measured Work Prices – Mechanical Installations

JCT Contract Administration Pocket Book, 2nd edition

Andy Atkinson

This book is quite simply about contract administration using the JCT contracts. The key features of the new and updated edition continue to be its brevity, readability and relevance to everyday practice. It provides a succinct guide written from the point of view of a construction practitioner, rather than a lawyer, to the traditional form of contract with bills of quantities SBC/Q2016, the design and build form DB2016 and the minor works form MWD2016. The book broadly follows the sequence of producing a building from the initial decision to build through to completion. Chapters cover:

- Procurement and tendering
- Payments, scheduling, progress and claims
- Contract termination and insolvency
- Indemnity and insurance
- Supply chain problems, defects and subcontracting issues
- Quality, dealing with disputes and adjudication
- How to administer contracts for BIM-compliant projects

JCT contracts are administered by a variety of professionals including project managers, architects, engineers, quantity surveyors and construction managers. It is individuals in these groups, whether experienced practitioner or student, who will benefit most from this clear, concise and highly relevant book.

December 2020: 248 pp
ISBN: 9780367632786

To Order
Tel:+44 (0) 1235 400524
Email: tandf@bookpoint.co.uk

For a complete listing of all our titles visit:
www.tandf.co.uk

Taylor & Francis
Taylor & Francis Group

Material Costs/Measured Work Prices

DIRECTIONS

The following explanations are given for each of the column headings and letter codes.

Unit	Prices for each unit are given as singular (i.e. 1 metre, 1 nr) unless stated otherwise.
Net price	Industry tender prices, plus nominal allowance for fixings (unless measured separately), waste and applicable trade discounts.
Material cost	Net price plus percentage allowance for overheads (7%), profit (5%) and preliminaries (13%).
Labour norms	In man-hours for each operation.
Labour cost	Labour constant multiplied by the appropriate all-in man-hour cost based on gang rate (See also relevant Rates of Wages Section) plus percentage allowance for overheads, profit and preliminaries.
Measured work Price (total rate)	Material cost plus Labour cost.

MATERIAL COSTS

The Material Costs given are based at Second Quarter 2022 but exclude any charges in respect of VAT. The average rate of copper during this quarter is US $9,610 / UK £7,677 per tonne. Users of the book are advised to register on the Spon's website www.pricebooks.co.uk/updates to receive the free quarterly updates – alerts will then be provided by email as changes arise.

MEASURED WORK PRICES

These prices are intended to apply to new work in the London area. The prices are for reasonable quantities of work and the user should make suitable adjustments if the quantities are especially small or especially large. Adjustments may also be required for locality (e.g. outside London – refer to cost indices in approximate esti-mating section for details of adjustment factors) and for the market conditions (e.g. volume of work secured or being tendered) at the time of use.

MECHANICAL INSTALLATIONS

The labour rate has been based on average gang rates per man hour effective from 4 October 2021. To this rate has been added 13% and 7% to cover preliminary items, site and head office overheads together with 5% for profit, resulting in an inclusive rate of £34.55 per man hour. The rate has been calculated on a working year of 2,016 hours; a detailed build-up of the rate is given at the end of these directions.

The rates and allowances will be subject to increases as covered by the National Agreements. Future changes will be published in the free Spon's quarterly update by registering on their website.

DUCTWORK INSTALLATIONS

The labour rate basis is as per Mechanical above and to this rate has been added 13% plus 7% to cover site and head office overheads only (factory overhead is included in the material rate) and preliminary items together with 5% for profit, resulting in an inclusive rate of £33.10 per man hour. The rate has been calculated on a working year of 2,016 hours; a detailed build-up of the rate is given at the end of these directions.

In calculating the 'Measured Work Prices' the following assumptions have been made:

(a)   That the work is carried out as a subcontract under the Standard Form of Building Contract.
(b)   That, unless otherwise stated, the work is being carried out in open areas at a height which would not require more than simple scaffolding.
(c)   That the building in which the work is being carried out is no more than six storey's high.

Where these assumptions are not valid, as for example where work is carried out in ducts and similar confined spaces or in multi-storey structures when additional time is needed to get to and from upper floors, then an appropriate adjustment must be made to the prices. Such adjustment will normally be to the labour element only.

Note: The rates do not include for any uplift applied if the ductwork package is procured via the Mechanical SubContractor

LABOUR RATE – MECHANICAL & PUBLIC HEALTH

The annual cost of a notional 12 man gang

	FOREMAN	SENIOR CRAFTSMAN (+2 Welding skill)	SENIOR CRAFTSMAN	CRAFTSMAN	INSTALLER	MATE (Over 18)	SUB TOTALS
	1 NR	1 NR	2 NR	4 NR	2 NR	2 NR	
Hourly Rate from 4 October 2021	**18.05**	**16.11**	**14.92**	**13.74**	**12.40**	**10.46**	
Working hours per annum per man	1,680.00	1,680.00	1,680.00	1,680.00	1,680.00	1,680.00	
x Hourly rate × nr of men = £ per annum	**30,324.00**	**27,064.80**	**50,131.20**	**92,332.80**	**41,664.00**	**35,145.60**	**276,662.40**
Overtime Rate	25.43	22.74	21.04	19.36	17.49	14.73	
Overtime hours per annum per man	336.00	336.00	336.00	336.00	336.00	336.00	
x Hourly rate × nr of men = £ per annum	**8,544.48**	**7,640.64**	**14,138.88**	**26,019.84**	**11,753.28**	**9,898.56**	**77,995.68**
Total	**38,868.48**	**34,705.44**	**64,270.08**	**118,352.64**	**53,417.28**	**45,044.16**	**354,658.08**
Incentive schemes 0.00%	**0.00**	**0.00**	**0.00**	**0.00**	**0.00**	**0.00**	**0.00**
Daily Travel Time Allowance (15–20 miles each way)	10.72	10.51	10.51	10.51	10.51	10.31	
Days per annum per man	224.00	224.00	224.00	224.00	224.00	224.00	
x nr of men = £ per annum	**2,401.28**	**2,354.24**	**4,708.48**	**9,416.96**	**4,708.48**	**4,618.88**	**28,208.32**
Daily Travel Fare (15–20 miles each way)	17.00	17.00	17.00	17.00	17.00	17.00	
Days per annum per man	224.00	224.00	224.00	224.00	224.00	224.00	
x nr of men = £ per annum	**3,808.00**	**3,808.00**	**7,616.00**	**15,232.00**	**7,616.00**	**7,616.00**	**45,696.00**
Employer's Pension contributions at 4 October 2021:							
£ Contributions/annum	**1,759.68**	**1,570.92**	**2,909.92**	**5,819.84**	**2,418.00**	**2,039.44**	**16,517.80**
National Insurance Contributions:							
Gross pay – subject to NI	50,975.84	46,330.03	86,527.00	164,258.36	74,172.25	64,550.75	
% of NI Contributions	15.05	15.05	15.05	15.05	15.05	15.05	
£ Contributions/annum	**5,549.48**	**4,908.35**	**8,970.35**	**16,726.90**	**7,265.39**	**5,937.63**	**49,358.09**
Holiday Credit and Welfare contributions:							
Number of weeks	52	52	52	52	52	52	
Total weekly £ contribution each	92.41	86.23	78.10	86.23	66.61	57.72	
X nr of men = £ Contributions/annum	**4,805.32**	**4,483.96**	**8,122.40**	**17,935.84**	**6,927.44**	**6,002.88**	**48,277.84**
Holiday Top-up Funding including overtime	21.01	18.82	17.40	15.97	14.45	12.20	
Cost	**1,092.76**	**978.39**	**1,810.04**	**3,320.92**	**1,503.05**	**1,268.83**	**9,973.99**

SUBTOTAL		**552,690.12**
TRAINING (INCLUDING ANY TRADE REGISTRATIONS) – SAY	1.00%	**5,526.90**
SEVERANCE PAY AND SUNDRY COSTS – SAY	1.50%	**8,373.26**
EMPLOYER'S LIABILITY AND THIRD PARTY INSURANCE – SAY	2.00%	**11,331.81**
ANNUAL COST OF NOTIONAL GANG		**577,922.08**
THEREFORE ANNUAL COST PER PRODUCTIVE MAN		**55,040.20**
AVERAGE NR OF HOURS WORKED PER MAN = 2016		
THEREFORE ALL-IN MAN-HOURS		**27.30**
PRELIMINARY ITEMS – SAY	13%	**3.55**
SITE AND HEAD OFFICE OVERHEADS AND PROFIT (7% & 5% RESPECTIVELY) – SAY	12%	**3.70**
THEREFORE INCLUSIVE MAN-HOUR RATE		**34.55**

MEN ACTUALLY WORKING = 10.5

Notes:

1)   The following assumptions have been made in the above calculations:
         a)   Increase in Hourly Rate from 4 October 2021, in accordance with the latest BESA Wage Agreement.
         b)   The working week of 37.5 hours, i.e. the normal working week as defined by the National Agreement.
         c)   The actual hours worked are five days of 9 hours each.
         d)   A working year of 2,016 hours.
         e)   Five days in the year are lost through sickness or similar reason.
2)   National insurance contributions are those effective from 1 April 2022.
3)   Weekly Holiday Credit/Welfare Stamp values are those effective from 7 October 2019.
4)   Rates are based from 4 October 2021.
5)   Overtime rates are based on Premium Rate 1.
6)   Fares with Oyster Card (New Malden to Waterloo (underground) + Zone 1 – Anytime fare) current at June 2022 (TfL 0343–2221–234).

LABOUR RATE – DUCTWORK

The annual cost of notional eight man gang

		FOREMAN	SENIOR CRAFTSMAN	CRAFTSMAN	INSTALLER	SUB TOTALS
		1 NR	1 NR	4 NR	2 NR	
Hourly Rate from 4 October 2021		18.05	14.92	13.74	12.40	
Working hours per annum per man		1,680.00	1,680.00	1,680.00	1,680.00	
x Hourly rate × nr of men = £ per annum		30,324.00	25,065.60	92,332.80	41,664.00	189,386.40
Overtime Rate		25.43	21.04	19.36	17.49	
Overtime hours per annum per man		336.00	336.00	336.00	336.00	
x hourly rate × nr of men = £ per annum		8,544.48	7,069.44	26,019.84	11,753.28	53,387.04
Total		38,868.48	32,135.04	118,352.64	53,417.28	242,773.44
Incentive schemes	0.00%	0.00	0.00	0.00	0.00	0.00
Daily Travel Time Allowance (15–20 miles each way)		10.72	10.51	10.51	10.51	
Days per annum per man		224.00	224.00	224.00	224.00	
x nr of men = £ per annum		2,401.28	2,354.24	9,416.96	4,708.48	18,880.96
Daily Travel Fare (15–20 miles each way)		17.00	17.00	17.00	17.00	
Days per annum per man		224.00	224.00	224.00	224.00	
x nr of men = £ per annum		3,808.00	3,808.00	15,232.00	7,616.00	30,464.00
Employer's Pension Contributions from 4 October 2021						
£ Contributions/annum		1,759.68	1,454.96	5,819.84	2,418.00	11,452.48
National Insurance Contributions:						
Gross pay – subject to NI		50,975.84	43,263.50	164,258.36	74,172.25	
% of NI Contributions		15.05	15.05	15.05	15.05	
£ Contributions/annum		5,549.48	4,485.17	16,726.90	7,265.39	34,026.94
Holiday Credit and Welfare contributions:						
Number of weeks		52	52	52	52	
Total weekly £ contribution each		92.41	78.10	86.23	66.61	
x nr of men = £ Contributions/annum		4,805.32	4,061.20	17,935.84	6,927.44	33,729.80
Holiday Top-up Funding including overtime		21.01	17.40	15.97	14.45	
Cost		1,092.76	905.02	3,320.92	1,503.05	6,821.75

	SUBTOTAL		378,149.37
TRAINING (INCLUDING ANY TRADE REGISTRATIONS) – SAY		1.00%	3,781.49
SEVERANCE PAY AND SUNDRY COSTS – SAY		1.50%	5,728.96
EMPLOYER'S LIABILITY AND THIRD PARTY INSURANCE – SAY		2.00%	7,753.20
ANNUAL COST OF NOTIONAL GANG			395,413.02
THEREFORE ANNUAL COST PER PRODUCTIVE MAN			52,721.74
AVERAGE NR OF HOURS WORKED PER MAN = 2016			
THEREFORE ALL-IN MAN-HOURS			26.15
PRELIMINARY ITEMS – SAY		13%	3.40
SITE AND HEAD OFFICE OVERHEADS AND PROFIT (7% & 5% RESPECTIVELY) – SAY		12%	3.55
THEREFORE INCLUSIVE MAN-HOUR RATE			33.10

MEN ACTUALLY WORKING = 7.5

Notes:

1)   The following assumptions have been made in the above calculations:
        a)   Increase in Hourly Rate from 4 October 2021, in accordance with the latest BESA Wage Agreement.
        b)   The working week of 37.5 hours, i.e. the normal working week as defined by the National Agreement.
        c)   The actual hours worked are five days of 9 hours each.
        d)   A working year of 2,016 hours.
        e)   Five days in the year are lost through sickness or similar reason.
2)   National insurance contributions are those effective from 1 April 2022.
3)   Weekly Holiday Credit/Welfare Stamp values are those effective from 7 October 2019.
4)   Rates are based from 4 October 2021.
5)   Overtime rates are based on Premium Rate 1.
6)   Fares with Oyster Card (New Malden to Waterloo (underground) + Zone 1 – Anytime fare) current at June 2022 (TfL – 0343 222 1234).

33 DRAINAGE ABOVE GROUND

Item	Net Price £	Material £	Labour hours	Labour £	Unit	Total rate £
RAINWATER PIPEWORK/GUTTERS						
PVC-u gutters: push fit joints; fixed with brackets to backgrounds; BS 4576 BS EN 607						
Half round gutter, with brackets measured separately						
75 mm	5.67	7.17	0.69	23.83	m	31.00
100 mm	11.40	14.43	0.64	22.10	m	36.53
150 mm	13.14	16.63	0.82	28.32	m	44.95
Brackets: including fixing to backgrounds. For minimum fixing distances, refer to the Tables and Memoranda at the rear of the book						
75 mm; Fascia	2.10	2.65	0.15	5.18	nr	7.83
100 mm; Jointing	3.34	4.23	0.16	5.53	nr	9.75
100 mm; Support	1.35	1.70	0.16	5.53	nr	7.23
150 mm; Fascia	4.05	5.13	0.16	5.53	nr	10.65
Bracket supports: including fixing to backgrounds. For minimum fixing distances, refer to the Tables and Memoranda at the rear of the book						
Side rafter	5.31	6.72	0.16	5.53	nr	12.25
Top rafter	5.31	6.72	0.16	5.53	nr	12.25
Rise and fall	5.74	7.27	0.16	5.53	nr	12.79
Extra over fittings half round PVC-u gutter						
Union						
75 mm	3.10	3.92	0.19	6.56	nr	10.48
100 mm	3.55	4.49	0.24	8.29	nr	12.78
150 mm	13.30	16.83	0.28	9.67	nr	26.50
Rainwater pipe outlets						
Running: 75 × 53 mm dia.	7.64	9.67	0.12	4.14	nr	13.81
Running: 100 × 68 mm dia.	5.21	6.60	0.12	4.14	nr	10.74
Running: 150 × 110 mm dia.	24.39	30.87	0.12	4.14	nr	35.02
Stop end: 100 × 68 mm dia.	5.85	7.41	0.12	4.14	nr	11.55
Internal stop ends: short						
75 mm	3.10	3.92	0.09	3.11	nr	7.03
100 mm	1.59	2.01	0.09	3.11	nr	5.12
150 mm	8.80	11.14	0.09	3.11	nr	14.25
External stop ends: short						
75 mm	6.30	7.98	0.09	3.11	nr	11.08
100 mm	8.53	10.79	0.09	3.11	nr	13.90
150 mm	8.80	11.14	0.09	3.11	nr	14.25
Angles						
75 mm; 45°	8.46	10.71	0.20	6.91	nr	17.62
75 mm; 90°	8.46	10.71	0.20	6.91	nr	17.62
100 mm; 90°	5.84	7.39	0.20	6.91	nr	14.30
100 mm; 120°	6.45	8.17	0.20	6.91	nr	15.07
100 mm; 135°	6.45	8.17	0.20	6.91	nr	15.07
100 mm; Prefabricated to special angle	36.45	46.13	0.23	7.94	nr	54.07
100 mm; Prefabricated to raked angle	42.04	53.21	0.23	7.94	nr	61.15

33 DRAINAGE ABOVE GROUND

Item	Net Price £	Material £	Labour hours	Labour £	Unit	Total rate £
RAINWATER PIPEWORK/GUTTERS – cont						
Extra over fittings half round PVC-u gutter – cont						
150 mm; 90°	20.94	26.50	0.20	6.91	nr	**33.41**
Gutter adaptors						
100 mm; Stainless steel clip	3.20	4.05	0.16	5.53	nr	**9.57**
100 mm; Cast iron spigot	9.32	11.80	0.23	7.94	nr	**19.74**
100 mm; Cast iron socket	9.32	11.80	0.23	7.94	nr	**19.74**
100 mm; Cast iron 'ogee' spigot	9.47	11.98	0.23	7.94	nr	**19.92**
100 mm; Cast iron 'ogee' socket	9.47	11.98	0.23	7.94	nr	**19.92**
100 mm; Half round to Square PVC-u	15.93	20.16	0.23	7.94	nr	**28.10**
100 mm; Gutter overshoot guard	19.69	24.93	0.58	20.03	nr	**44.95**
Square gutter, with brackets measured separately						
120 mm square	5.39	6.82	0.82	28.32	m	**35.14**
Brackets: including fixing to backgrounds. For minimum fixing distances, refer to the Tables and Memoranda at the rear of the book						
Jointing	3.48	4.41	0.16	5.53	nr	**9.93**
Support	1.51	1.91	0.16	5.53	nr	**7.44**
Bracket support: including fixing to backgrounds. For minimum fixing distances, refer to the Tables and Memoranda at the rear of the book						
Side rafter	5.31	6.72	0.16	5.53	nr	**12.25**
Top rafter	5.31	6.72	0.16	5.53	nr	**12.25**
Rise and fall	5.74	7.27	0.16	5.53	nr	**12.79**
Extra over fittings square PVC-u gutter						
Rainwater pipe outlets						
Running: 62 mm square	5.23	6.62	0.12	4.14	nr	**10.76**
Stop end: 62 mm square	5.93	7.51	0.12	4.14	nr	**11.65**
Stop ends: short						
External	2.54	3.22	0.09	3.11	nr	**6.32**
Angles						
90°	6.32	8.00	0.20	6.91	nr	**14.91**
120°	15.81	20.01	0.20	6.91	nr	**26.91**
135°	7.39	9.35	0.20	6.91	nr	**16.26**
Prefabricated to special angle	37.39	47.32	0.23	7.94	nr	**55.26**
Prefabricated to raked angle	46.08	58.32	0.23	7.94	nr	**66.26**
Gutter adaptors						
Cast iron	27.14	34.35	0.23	7.94	nr	**42.30**

33 DRAINAGE ABOVE GROUND

Item	Net Price £	Material £	Labour hours	Labour £	Unit	Total rate £
High capacity square gutter, with brackets measured separately						
137 mm	11.52	14.58	0.82	28.32	m	**42.90**
Brackets: including fixing to backgrounds. For minimum fixing distances, refer to the Tables and Memoranda at the rear of the book						
Jointing	10.95	13.86	0.16	5.53	nr	**19.39**
Support	4.58	5.79	0.16	5.53	nr	**11.32**
Overslung	4.24	5.37	0.16	5.53	nr	**10.89**
Bracket supports: including fixing to backgrounds. For minimum fixing distances, refer to the Tables and Memoranda at the rear of the book						
Side rafter	6.52	8.25	0.16	5.53	nr	**13.77**
Top rafter	6.52	8.25	0.16	5.53	nr	**13.77**
Rise and fall	10.47	13.25	0.16	5.53	nr	**18.77**
Extra over fittings high capacity square PVC-u						
Rainwater pipe outlets						
Running: 75 mm square	17.95	22.72	0.12	4.14	nr	**26.86**
Running: 82 mm dia.	17.95	22.72	0.12	4.14	nr	**26.86**
Running: 110 mm dia.	15.70	19.87	0.12	4.14	nr	**24.02**
Screwed outlet adaptor						
75 mm square pipe	11.42	14.46	0.23	7.94	nr	**22.40**
Stop ends: short						
External	6.30	7.98	0.09	3.11	nr	**11.08**
Angles						
90°	15.70	19.87	0.20	6.91	nr	**26.78**
135°	18.48	23.38	0.20	6.91	nr	**30.29**
Prefabricated to special angle	53.37	67.55	0.23	7.94	nr	**75.49**
Prefabricated to raked internal angle	92.94	117.62	0.23	7.94	nr	**125.56**
Prefabricated to raked external angle	92.94	117.62	0.23	7.94	nr	**125.56**
Deep eliptical gutter, with brackets measured separately						
137 mm	6.24	7.90	0.82	28.32	m	**36.21**
Brackets: including fixing to backgrounds. For minimum fixing distances, refer to the Tables and Memoranda at the rear of the book						
Jointing	4.95	6.26	0.16	5.53	nr	**11.79**
Support	2.12	2.68	0.16	5.53	nr	**8.21**
Bracket support: including fixing to backgrounds. For minimum fixing distances, refer to the Tables and Memoranda at the rear of the book						
Side rafter	6.84	8.66	0.16	5.53	nr	**14.19**
Top rafter	6.84	8.66	0.16	5.53	nr	**14.19**
Rise and fall	10.99	13.91	0.16	5.53	nr	**19.44**

Material Costs/Prices for Measured Works

33 DRAINAGE ABOVE GROUND

Item	Net Price £	Material £	Labour hours	Labour £	Unit	Total rate £
RAINWATER PIPEWORK/GUTTERS – cont						
Extra over fittings deep eliptical PVC-u gutter						
Rainwater pipe outlets						
Running: 68 mm dia.	7.86	9.94	0.12	4.14	nr	**14.09**
Running: 82 mm dia.	7.46	9.44	0.12	4.14	nr	**13.58**
Stop end: 68 mm dia.	7.98	10.10	0.12	4.14	nr	**14.24**
Stop ends: short						
External	3.85	4.87	0.09	3.11	nr	**7.98**
Angles						
90°	7.92	10.03	0.20	6.91	nr	**16.94**
135°	9.42	11.92	0.20	6.91	nr	**18.83**
Prefabricated to special angle	27.56	34.88	0.23	7.94	nr	**42.82**
Gutter adaptors						
Stainless steel clip	6.77	8.57	0.16	5.53	nr	**14.10**
Marley deepflow	5.48	6.93	0.23	7.94	nr	**14.87**
Ogee profile PVC-u gutter, with brackets measured separately						
122 mm	6.57	8.31	0.82	28.32	m	**36.63**
Brackets: including fixing to backgrounds. For minimum fixing distances, refer to the Tables and Memoranda at the rear of the book						
Jointing	5.22	6.61	0.16	5.53	nr	**12.13**
Support	2.02	2.56	0.16	5.53	nr	**8.08**
Overslung	2.02	2.56	0.16	5.53	nr	**8.08**
Extra over fittings Ogee profile PVC-u gutter						
Rainwater pipe outlets						
Running: 68 mm dia.	7.46	9.44	0.12	4.14	nr	**13.58**
Stop ends: short						
Internal/External: left or right hand	3.79	4.79	0.09	3.11	nr	**7.90**
Angles						
90°: internal or external	7.55	9.55	0.20	6.91	nr	**16.46**
135°: internal or external	7.55	9.55	0.20	6.91	nr	**16.46**
PVC-u rainwater pipe: dry push fit joints; fixed with brackets to backgrounds; BS 4576/BS EN 607						
Pipe: circular, with brackets measured separately						
53 mm	8.49	10.74	0.61	21.06	m	**31.81**
68 mm	8.84	11.19	0.61	21.06	m	**32.26**
Pipe clip: including fixing to backgrounds. For minimum fixing distances, refer to the Tables and Memoranda at the rear of the book						
68 mm	1.98	2.51	0.16	5.53	nr	**8.03**

33 DRAINAGE ABOVE GROUND

Item	Net Price £	Material £	Labour hours	Labour £	Unit	Total rate £
Pipe clip adjustable: including fixing to backgrounds. For minimum fixing distances, refer to the Tables and Memoranda at the rear of the book						
53 mm	3.02	3.83	0.16	5.53	nr	**9.35**
68 mm	4.23	5.35	0.16	5.53	nr	**10.88**
Pipe clip drive in: including fixing to backgrounds. For minimum fixing distances, refer to the Tables and Memoranda at the rear of the book						
68 mm	4.75	6.01	0.16	5.53	nr	**11.54**
Extra over fittings circular pipework PVC-u						
Pipe coupler: PVC-u to PVC-u						
68 mm	2.46	3.12	0.12	4.14	nr	**7.26**
Pipe coupler: PVC-u to Cast Iron						
68 mm: to 3" cast iron	10.95	13.85	0.17	5.87	nr	**19.73**
68 mm: to 3 ¾" cast iron	37.65	47.64	0.17	5.87	nr	**53.51**
Access pipe: single socket						
68 mm	15.44	19.54	0.15	5.18	nr	**24.72**
Bend: short radius						
53 mm: 67.5°	3.60	4.55	0.20	6.91	nr	**11.46**
68 mm: 92.5°	5.29	6.69	0.20	6.91	nr	**13.60**
68 mm: 112.5°	5.29	6.69	0.20	6.91	nr	**13.60**
Bend: long radius						
68 mm: 112°	5.01	6.34	0.20	6.91	nr	**13.24**
Branch						
68 mm: 92°	32.47	41.09	0.23	7.94	nr	**49.04**
68 mm: 112°	32.58	41.23	0.23	7.94	nr	**49.17**
Double branch						
68 mm: 112°	64.27	81.35	0.24	8.29	nr	**89.63**
Shoe						
53 mm	5.32	6.74	0.12	4.14	nr	**10.88**
68 mm	7.86	9.95	0.12	4.14	nr	**14.10**
Rainwater head: including fixing to backgrounds						
68 mm	15.44	19.54	0.29	10.01	nr	**29.55**
Pipe: square, with brackets measured separately						
62 mm	5.41	6.85	0.45	15.54	m	**22.39**
75 mm	10.29	13.02	0.45	15.54	m	**28.56**
Pipe clip: including fixing to backgrounds. For minimum fixing distances, refer to the Tables and Memoranda at the rear of the book						
62 mm	1.80	2.28	0.16	5.53	nr	**7.81**
75 mm	4.01	5.08	0.16	5.53	nr	**10.61**
Pipe clip adjustable: including fixing to backgrounds. For minimum fixing distances, refer to the Tables and Memoranda at the rear of the book						
62 mm	5.60	7.09	0.16	5.53	nr	**12.62**

33 DRAINAGE ABOVE GROUND

Item	Net Price £	Material £	Labour hours	Labour £	Unit	Total rate £
RAINWATER PIPEWORK/GUTTERS – cont						
Extra over fittings square pipework PVC-u						
Pipe coupler: PVC-u to PVC-u						
62 mm	2.66	3.36	0.20	6.91	nr	**10.27**
75 mm	4.12	5.21	0.20	6.91	nr	**12.12**
Square to circular adaptor: single socket						
62 mm to 68 mm	5.17	6.54	0.20	6.91	nr	**13.45**
Square to circular adaptor: single socket						
75 mm to 62 mm	6.68	8.46	0.20	6.91	nr	**15.36**
Access pipe						
62 mm	30.91	39.12	0.16	5.53	nr	**44.65**
75 mm	32.63	41.29	0.16	5.53	nr	**46.82**
Bends						
62 mm: 92.5°	4.78	6.05	0.20	6.91	nr	**12.95**
62 mm: 112.5°	3.48	4.41	0.20	6.91	nr	**11.31**
75 mm: 112.5°	8.72	11.03	0.20	6.91	nr	**17.94**
Bends: prefabricated special angle						
62 mm	31.06	39.31	0.23	7.94	nr	**47.26**
75 mm	43.16	54.62	0.23	7.94	nr	**62.56**
Offset						
62 mm	7.48	9.47	0.20	6.91	nr	**16.38**
75 mm	20.61	26.08	0.20	6.91	nr	**32.99**
Offset: prefabricated special angle						
62 mm	31.06	39.31	0.23	7.94	nr	**47.26**
Shoe						
62 mm	4.40	5.56	0.12	4.14	nr	**9.71**
75 mm	7.52	9.52	0.12	4.14	nr	**13.66**
Branch						
62 mm	10.86	13.75	0.23	7.94	nr	**21.69**
75 mm	52.95	67.02	0.23	7.94	nr	**74.96**
Double branch						
62 mm	58.27	73.74	0.24	8.29	nr	**82.03**
Rainwater head						
62 mm	48.86	61.84	0.29	10.01	nr	**71.86**
75 mm	51.40	65.06	3.45	119.08	nr	**184.14**

33 DRAINAGE ABOVE GROUND

Item	Net Price £	Material £	Labour hours	Labour £	Unit	Total rate £
PVC-u rainwater pipe: solvent welded joints; fixed with brackets to backgrounds; BS 4576/BS EN 607						
Pipe: circular, with brackets measured separately						
82 mm	16.67	21.10	0.35	12.09	m	**33.18**
Pipe clip: galvanized; including fixing to backgrounds. For minimum fixing distances, refer to the Tables and Memoranda at the rear of the book						
82 mm	6.78	8.59	0.58	20.03	nr	**28.62**
Pipe clip: galvanized plastic coated; including fixing to backgrounds. For minimum fixing distances, refer to the Tables and Memoranda at the rear of the book						
82 mm	9.39	11.88	0.58	20.03	nr	**31.91**
Pipe clip: PVC-u including fixing to backgrounds. For minimum fixing distances, refer to the Tables and Memoranda at the rear of the book						
82 mm	5.04	6.38	0.58	20.03	nr	**26.41**
Pipe clip: PVC-u adjustable: including fixing to backgrounds. For minimum fixing distances, refer to the Tables and Memoranda at the rear of the book						
82 mm	9.22	11.67	0.58	20.03	nr	**31.70**
Extra over fittings circular pipework PVC-u						
Pipe coupler: PVC-u to PVC-u						
82 mm	8.45	10.69	0.21	7.25	nr	**17.94**
Access pipe						
82 mm	52.04	65.86	0.23	7.94	nr	**73.80**
Bend						
82 mm: 92, 112.5 and 135°	21.24	26.88	0.29	10.01	nr	**36.90**
Shoe						
82 mm	13.11	16.59	0.29	10.01	nr	**26.61**
110 mm	16.45	20.82	0.32	11.05	nr	**31.87**
Branch						
82 mm: 92, 112.5 and 135°	31.66	40.07	0.35	12.09	nr	**52.16**
Rainwater head						
82 mm	27.79	35.17	0.58	20.03	nr	**55.19**
110 mm	25.30	32.01	0.58	20.03	nr	**52.04**
Roof outlets: 178 dia.; Flat						
50 mm	27.77	35.15	1.15	39.71	nr	**74.86**
82 mm	27.77	35.15	1.15	39.71	nr	**74.86**
Roof outlets: 178 mm dia.; Domed						
50 mm	27.77	35.15	1.15	39.71	nr	**74.86**
82 mm	27.77	35.15	1.15	39.71	nr	**74.86**

33 DRAINAGE ABOVE GROUND

Item	Net Price £	Material £	Labour hours	Labour £	Unit	Total rate £
RAINWATER PIPEWORK/GUTTERS – cont						
Extra over fittings circular pipework PVC-u – cont						
Roof outlets: 406 mm dia.; Flat						
82 mm	54.24	68.64	1.15	39.71	nr	**108.36**
110 mm	54.24	68.64	1.15	39.71	nr	**108.36**
Roof outlets: 406 mm dia.; Domed						
82 mm	54.24	68.64	1.15	39.71	nr	**108.36**
110 mm	54.24	68.64	1.15	39.71	nr	**108.36**
Roof outlets: 406 mm dia.; Inverted						
82 mm	119.44	151.16	1.15	39.71	nr	**190.88**
110 mm	119.44	151.16	1.15	39.71	nr	**190.88**
Roof outlets: 406 mm dia.; Vent Pipe						
82 mm	78.45	99.29	1.15	39.71	nr	**139.00**
110 mm	78.45	99.29	1.15	39.71	nr	**139.00**
Balcony outlets: screed						
82 mm	46.42	58.75	1.15	39.71	nr	**98.47**
Balcony outlets: asphalt						
82 mm	46.18	58.45	1.15	39.71	nr	**98.16**
Adaptors						
82 mm × 62 mm square pipe	6.21	7.86	0.21	7.25	nr	**15.11**
82 mm × 68 mm circular pipe	6.21	7.86	0.21	7.25	nr	**15.11**
For 110 mm dia. pipework and fittings refer to 33 Drainage Above Ground						
Cast iron gutters: mastic and bolted joints; BS 460; fixed with brackets to backgrounds						
Half round gutter, with brackets measured separately						
100 mm	24.27	30.71	0.85	29.35	m	**60.07**
115 mm	25.30	32.02	0.97	33.50	m	**65.52**
125 mm	29.60	37.47	0.97	33.50	m	**70.96**
150 mm	50.61	64.05	1.12	38.68	m	**102.72**
Brackets; fixed to backgrounds. For minimum fixing distances, refer to the Tables and Memoranda at the rear of the book						
Fascia						
100 mm	5.70	7.21	0.16	5.53	nr	**12.73**
115 mm	5.70	7.21	0.16	5.53	nr	**12.73**
125 mm	5.70	7.21	0.16	5.53	nr	**12.73**
150 mm	7.21	9.12	0.16	5.53	nr	**14.65**
Rise and fall						
100 mm	11.31	14.31	0.39	13.47	nr	**27.78**
115 mm	11.31	14.31	0.39	13.47	nr	**27.78**
125 mm	11.62	14.71	0.39	13.47	nr	**28.18**
150 mm	11.82	14.97	0.39	13.47	nr	**28.43**

33 DRAINAGE ABOVE GROUND

Item	Net Price £	Material £	Labour hours	Labour £	Unit	Total rate £
Top rafter						
100 mm	6.96	8.81	0.16	5.53	nr	**14.34**
115 mm	6.96	8.81	0.16	5.53	nr	**14.34**
125 mm	9.42	11.92	0.16	5.53	nr	**17.45**
150 mm	12.80	16.20	0.16	5.53	nr	**21.73**
Side rafter						
100 mm	6.96	8.81	0.16	5.53	nr	**14.34**
115 mm	6.96	8.81	0.16	5.53	nr	**14.34**
125 mm	9.42	11.92	0.16	5.53	nr	**17.45**
150 mm	12.80	16.20	0.16	5.53	nr	**21.73**
Extra over fittings half round gutter cast iron BS 460						
Union						
100 mm	13.35	16.89	0.39	13.47	nr	**30.36**
115 mm	16.66	21.08	0.48	16.58	nr	**37.66**
125 mm	18.82	23.81	0.48	16.58	nr	**40.39**
150 mm	21.12	26.73	0.55	18.99	nr	**45.72**
Stop end; internal						
100 mm	6.82	8.63	0.12	4.14	nr	**12.77**
115 mm	8.82	11.16	0.15	5.18	nr	**16.34**
125 mm	8.82	11.16	0.15	5.18	nr	**16.34**
150 mm	12.24	15.49	0.20	6.91	nr	**22.40**
Stop end; external						
100 mm	6.82	8.63	0.12	4.14	nr	**12.77**
115 mm	8.62	10.91	0.15	5.18	nr	**16.09**
125 mm	8.82	11.16	0.15	5.18	nr	**16.34**
150 mm	12.24	15.49	0.20	6.91	nr	**22.40**
90° angle; single socket						
100 mm	20.24	25.62	0.39	13.47	nr	**39.08**
115 mm	20.82	26.34	0.43	14.85	nr	**41.19**
125 mm	24.56	31.08	0.43	14.85	nr	**45.93**
150 mm	44.87	56.78	0.50	17.27	nr	**74.05**
90° angle; double socket						
100 mm	24.54	31.06	0.39	13.47	nr	**44.53**
115 mm	26.04	32.95	0.43	14.85	nr	**47.80**
125 mm	33.68	42.62	0.43	14.85	nr	**57.47**
135° angle; single socket						
100 mm	20.67	26.16	0.39	13.47	nr	**39.63**
115 mm	20.86	26.40	0.43	14.85	nr	**41.25**
125 mm	30.83	39.01	0.43	14.85	nr	**53.86**
150 mm	45.74	57.89	0.50	17.27	nr	**75.16**
Running outlet						
65 mm outlet						
100 mm	19.75	25.00	0.39	13.47	nr	**38.46**
115 mm	21.49	27.20	0.43	14.85	nr	**42.05**
125 mm	24.56	31.08	0.43	14.85	nr	**45.93**
75 mm outlet						
100 mm	19.75	25.00	0.39	13.47	nr	**38.46**
115 mm	21.49	27.20	0.43	14.85	nr	**42.05**
125 mm	24.56	31.08	0.43	14.85	nr	**45.93**
150 mm	42.52	53.82	0.50	17.27	nr	**71.08**

33 DRAINAGE ABOVE GROUND

Item	Net Price £	Material £	Labour hours	Labour £	Unit	Total rate £
RAINWATER PIPEWORK/GUTTERS – cont						
Extra over fittings half round gutter cast iron BS 460 – cont						
100 mm outlet						
150 mm	42.52	53.82	0.50	17.27	nr	**71.08**
Stop end outlet; socket						
65 mm outlet						
100 mm	23.15	29.29	0.39	13.47	nr	**42.76**
115 mm	25.95	32.84	0.43	14.85	nr	**47.69**
75 mm outlet						
125 mm	23.15	29.29	0.43	14.85	nr	**44.14**
150 mm	48.66	115.41	0.50	17.27	nr	**132.67**
100 mm outlet						
150 mm	42.52	53.82	0.50	17.27	nr	**71.08**
Stop end outlet; spigot						
65 mm outlet						
100 mm	23.15	29.29	0.39	13.47	nr	**42.76**
115 mm	25.95	32.84	0.43	14.85	nr	**47.69**
75 mm outlet						
125 mm	23.15	29.29	0.43	14.85	nr	**44.14**
150 mm	48.66	61.59	0.50	17.27	nr	**78.86**
100 mm outlet						
150 mm	48.66	61.59	0.50	17.27	nr	**78.86**
Half round; 3 mm thick double beaded gutter, with brackets measured separately						
100 mm	24.71	31.28	0.85	29.35	m	**60.63**
115 mm	25.79	32.64	0.85	29.35	m	**62.00**
125 mm	29.59	37.45	0.97	33.50	m	**70.95**
Brackets; fixed to backgrounds. For minimum fixing distances, refer to the Tables and Memoranda at the rear of the book						
Fascia						
100 mm	5.70	7.21	0.16	5.53	nr	**12.73**
115 mm	5.70	7.21	0.16	5.53	nr	**12.73**
125 mm	5.70	7.21	0.16	5.53	nr	**12.73**
Extra over fittings half round 3 mm thick gutter BS 460						
Union						
100 mm	13.35	16.89	0.38	13.12	nr	**30.02**
115 mm	16.27	20.59	0.38	13.12	nr	**33.71**
125 mm	18.82	23.81	0.43	14.85	nr	**38.66**
Stop end; internal						
100 mm	6.82	8.63	0.12	4.14	nr	**12.77**
115 mm	8.82	11.16	0.12	4.14	nr	**15.30**
125 mm	8.85	11.20	0.15	5.18	nr	**16.38**

33 DRAINAGE ABOVE GROUND

Item	Net Price £	Material £	Labour hours	Labour £	Unit	Total rate £
Stop end; external						
100 mm	6.82	8.63	0.12	4.14	nr	**12.77**
115 mm	8.82	11.16	0.12	4.14	nr	**15.30**
125 mm	8.85	11.20	0.15	5.18	nr	**16.38**
90° angle; single socket						
100 mm	21.05	26.63	0.38	13.12	nr	**39.76**
115 mm	21.30	26.96	0.38	13.12	nr	**40.09**
125 mm	24.56	31.08	0.43	14.85	nr	**45.93**
135° angle; single socket						
100 mm	20.67	26.16	0.38	13.12	nr	**39.28**
115 mm	20.82	26.34	0.38	13.12	nr	**39.47**
125 mm	25.96	32.86	0.43	14.85	nr	**47.71**
Running outlet						
65 mm outlet						
100 mm	21.20	26.84	0.38	13.12	nr	**39.96**
115 mm	21.91	27.73	0.38	13.12	nr	**40.85**
125 mm	29.01	36.72	0.43	14.85	nr	**51.57**
75 mm outlet						
115 mm	21.91	27.73	0.38	13.12	nr	**40.85**
125 mm	25.42	32.17	0.43	14.85	nr	**47.02**
Stop end outlet; socket						
65 mm outlet						
100 mm	23.15	29.29	0.38	13.12	nr	**42.42**
115 mm	25.95	32.84	0.38	13.12	nr	**45.97**
125 mm	29.01	36.72	0.43	14.85	nr	**51.57**
75 mm outlet						
125 mm	29.55	37.39	0.43	14.85	nr	**52.24**
Stop end outlet; spigot						
65 mm outlet						
100 mm	23.15	29.29	0.38	13.12	nr	**42.42**
115 mm	25.95	32.84	0.38	13.12	nr	**45.97**
125 mm	29.01	36.72	0.43	14.85	nr	**51.57**
Deep half round gutter, with brackets measured separately						
100 × 75 mm	40.65	51.45	0.85	29.35	m	**80.80**
125 × 75 mm	52.52	66.47	0.97	33.50	m	**99.97**
Brackets; fixed to backgrounds. For minimum fixing distances, refer to the Tables and Memoranda at the rear of the book						
Fascia						
100 × 75 mm	19.20	24.30	0.16	5.53	nr	**29.83**
125 × 75 mm	23.65	29.93	0.16	5.53	nr	**35.46**
Extra over fittings deep half round gutter BS 460						
Union						
100 × 75 mm	22.33	28.26	0.38	13.12	nr	**41.38**
125 × 75 mm	23.65	29.93	0.43	14.85	nr	**44.78**

33 DRAINAGE ABOVE GROUND

Item	Net Price £	Material £	Labour hours	Labour £	Unit	Total rate £
RAINWATER PIPEWORK/GUTTERS – cont						
Extra over fittings deep half round gutter BS 460 – cont						
Stop end; internal						
100 × 75 mm	19.55	24.74	0.12	4.14	nr	**28.89**
125 × 75 mm	24.11	30.51	0.15	5.18	nr	**35.69**
Stop end; external						
100 × 75 mm	19.55	24.74	0.12	4.14	nr	**28.89**
125 × 75 mm	24.11	30.51	0.15	5.18	nr	**35.69**
90° angle; single socket						
100 × 75 mm	55.71	70.51	0.38	13.12	nr	**83.63**
125 × 75 mm	70.75	89.54	0.43	14.85	nr	**104.38**
135° angle; single socket						
100 × 75 mm	55.71	70.51	0.38	13.12	nr	**83.63**
125 × 75 mm	70.75	89.54	0.43	14.85	nr	**104.38**
Running outlet						
65 mm outlet						
100 × 75 mm	55.71	70.51	0.38	13.12	nr	**83.63**
125 × 75 mm	70.75	89.54	0.43	14.85	nr	**104.38**
75 mm outlet						
100 × 75 mm	55.71	70.51	0.38	13.12	nr	**83.63**
125 × 75 mm	51.34	64.98	0.43	14.85	nr	**79.83**
Stop end outlet; socket						
65 mm outlet						
100 × 75 mm	75.26	95.25	0.38	13.12	nr	**108.37**
75 mm outlet						
100 × 75 mm	75.26	95.25	0.38	13.12	nr	**108.37**
125 × 75 mm	75.26	95.25	0.43	14.85	nr	**110.10**
Stop end outlet; spigot						
65 mm outlet						
100 × 75 mm	75.26	95.25	0.38	13.12	nr	**108.37**
75 mm outlet						
100 × 75 mm	75.26	95.25	0.38	13.12	nr	**108.37**
125 × 75 mm	75.26	95.25	0.43	14.85	nr	**110.10**
Ogee gutter, with brackets measured separately						
100 mm	27.07	34.26	0.85	29.35	m	**63.62**
115 mm	29.76	37.67	0.97	33.50	m	**71.16**
125 mm	31.23	39.52	0.97	33.50	m	**73.02**
Brackets; fixed to backgrounds. For minimum fixing distances, refer to the Tables and Memoranda at the rear of the book						
Fascia						
100 mm	6.20	7.85	0.16	5.53	nr	**13.37**
115 mm	6.20	7.85	0.16	5.53	nr	**13.37**
125 mm	6.98	8.83	0.16	5.53	nr	**14.35**

33 DRAINAGE ABOVE GROUND

Item	Net Price £	Material £	Labour hours	Labour £	Unit	Total rate £
Extra over fittings ogee cast iron gutter BS 460						
Union						
100 mm	13.36	16.91	0.38	13.12	nr	**30.04**
115 mm	16.27	20.59	0.43	14.85	nr	**35.44**
125 mm	18.82	23.81	0.43	14.85	nr	**38.66**
Stop end; internal						
100 mm	6.96	8.81	0.12	4.14	nr	**12.96**
115 mm	8.99	11.38	0.15	5.18	nr	**16.56**
125 mm	8.99	11.38	0.15	5.18	nr	**16.56**
Stop end; external						
100 mm	6.96	8.81	0.12	4.14	nr	**12.96**
115 mm	8.99	11.38	0.15	5.18	nr	**16.56**
125 mm	8.99	11.38	0.15	5.18	nr	**16.56**
90° angle; internal						
100 mm	21.12	26.73	0.38	13.12	nr	**39.85**
115 mm	22.89	28.97	0.43	14.85	nr	**43.81**
125 mm	24.99	31.62	0.43	14.85	nr	**46.47**
90° angle; external						
100 mm	21.12	26.73	0.38	13.12	nr	**39.85**
115 mm	22.89	28.97	0.43	14.85	nr	**43.81**
125 mm	24.99	31.62	0.43	14.85	nr	**46.47**
135° angle; internal						
100 mm	21.92	27.75	0.38	13.12	nr	**40.87**
115 mm	23.36	29.57	0.43	14.85	nr	**44.42**
125 mm	30.77	38.94	0.43	14.85	nr	**53.79**
135° angle; external						
100 mm	21.92	27.75	0.38	13.12	nr	**40.87**
115 mm	23.36	29.57	0.43	14.85	nr	**44.42**
125 mm	30.77	38.94	0.43	14.85	nr	**53.79**
Running outlet						
65 mm outlet						
100 mm	21.51	27.22	0.38	13.12	nr	**40.34**
115 mm	22.90	28.98	0.43	14.85	nr	**43.83**
125 mm	24.99	31.62	0.43	14.85	nr	**46.47**
75 mm outlet						
125 mm	24.99	31.62	0.43	14.85	nr	**46.47**
Stop end outlet; socket						
65 mm outlet						
100 mm	33.96	42.98	0.38	13.12	nr	**56.11**
115 mm	33.96	42.98	0.43	14.85	nr	**57.83**
125 mm	33.96	42.98	0.43	14.85	nr	**57.83**
75 mm outlet						
125 mm	33.96	42.98	0.43	14.85	nr	**57.83**
Stop end outlet; spigot						
65 mm outlet						
100 mm	33.96	42.98	0.38	13.12	nr	**56.11**
115 mm	33.96	42.98	0.43	14.85	nr	**57.83**
125 mm	33.96	42.98	0.43	14.85	nr	**57.83**
75 mm outlet						
125 mm	33.96	42.98	0.43	14.85	nr	**57.83**

33 DRAINAGE ABOVE GROUND

Item	Net Price £	Material £	Labour hours	Labour £	Unit	Total rate £
RAINWATER PIPEWORK/GUTTERS – cont						
Extra over fittings ogee cast iron gutter BS 460 – cont						
Notts ogee gutter, with brackets measured separately						
115 mm	48.09	60.86	0.85	29.35	m	**90.21**
Brackets; fixed to backgrounds. For minimum fixing distances, refer to the Tables and Memoranda at the rear of the book						
Fascia						
115 mm	18.76	23.74	0.16	5.53	nr	**29.27**
Extra over fittings Notts ogee cast iron gutter BS 460						
Union						
115 mm	22.77	28.82	0.38	13.12	nr	**41.94**
Stop end; internal						
115 mm	18.76	23.74	0.16	5.53	nr	**29.27**
Stop end; external						
115 mm	18.76	23.74	0.16	5.53	nr	**29.27**
90° angle; internal						
115 mm	54.40	68.85	0.43	14.85	nr	**83.70**
90° angle; external						
115 mm	54.40	68.85	0.43	14.85	nr	**83.70**
135° angle; internal						
115 mm	55.44	70.16	0.43	14.85	nr	**85.01**
135° angle; external						
115 mm	55.44	70.16	0.43	14.85	nr	**85.01**
Running outlet						
65 mm outlet						
115 mm	65.22	82.54	0.43	14.85	nr	**97.39**
75 mm outlet						
115 mm	65.22	82.54	0.43	14.85	nr	**97.39**
Stop end outlet; socket						
65 mm outlet						
115 mm	83.97	106.27	0.43	14.85	nr	**121.12**
Stop end outlet; spigot						
65 mm outlet						
115 mm	83.97	106.27	0.43	14.85	nr	**121.12**
No 46 moulded gutter, with brackets measured separately						
100 × 75 mm	46.51	58.86	0.85	29.35	m	**88.21**
125 × 100 mm	68.18	86.29	0.97	33.50	m	**119.79**
Brackets; fixed to backgrounds. For minimum fixing distances, refer to the Tables and Memoranda at the rear of the book						
Fascia						
100 × 75 mm	10.37	13.13	0.16	5.53	nr	**18.65**
125 × 100 mm	10.37	13.13	0.16	5.53	nr	**18.65**

33 DRAINAGE ABOVE GROUND

Item	Net Price £	Material £	Labour hours	Labour £	Unit	Total rate £
Extra over fittings						
Union						
100 × 75 mm	21.91	27.73	0.38	13.12	nr	**40.85**
125 × 100 mm	25.39	32.13	0.43	14.85	nr	**46.98**
Stop end; internal						
100 × 75 mm	19.62	24.83	0.12	4.14	nr	**28.98**
125 × 100 mm	25.39	32.13	0.15	5.18	nr	**37.31**
Stop end; external						
100 × 75 mm	19.62	24.83	0.12	4.14	nr	**28.98**
125 × 100 mm	25.39	32.13	0.15	5.18	nr	**37.31**
90° angle; internal						
100 × 75 mm	51.37	65.01	0.38	13.12	nr	**78.13**
125 × 100 mm	73.85	93.47	0.43	14.85	nr	**108.32**
90° angle; external						
100 × 75 mm	51.37	65.01	0.38	13.12	nr	**78.13**
125 × 100 mm	73.85	93.47	0.43	14.85	nr	**108.32**
135° angle; internal						
100 × 75 mm	52.36	66.27	0.38	13.12	nr	**79.39**
125 × 100 mm	73.85	93.47	0.43	14.85	nr	**108.32**
135° angle; external						
100 × 75 mm	52.36	66.27	0.38	13.12	nr	**79.39**
125 × 100 mm	73.85	93.47	0.43	14.85	nr	**108.32**
Running outlet						
65 mm outlet						
100 × 75 mm	52.36	66.27	0.38	13.12	nr	**79.39**
125 × 100 mm	73.85	93.47	0.43	14.85	nr	**108.32**
75 mm outlet						
100 × 75 mm	52.36	66.27	0.38	13.12	nr	**79.39**
125 × 100 mm	73.85	93.47	0.43	14.85	nr	**108.32**
100 mm outlet						
100 × 75 mm	52.36	66.27	0.38	13.12	nr	**79.39**
125 × 100 mm	73.85	93.47	0.43	14.85	nr	**108.32**
100 × 75 mm outlet						
125 × 100 mm	52.36	66.27	0.43	14.85	nr	**81.12**
Stop end outlet; socket						
65 mm outlet						
100 × 75 mm	99.24	125.60	0.38	13.12	nr	**138.72**
75 mm outlet						
125 × 100 mm	99.24	125.60	0.43	14.85	nr	**140.45**
Stop end outlet; spigot						
65 mm outlet						
100 × 75 mm	99.24	125.60	0.38	13.12	nr	**138.72**
75 mm outlet						
125 × 100 mm	99.24	125.60	0.43	14.85	nr	**140.45**
Box gutter, with brackets measured separately						
100 × 75 mm	75.42	95.45	0.85	29.35	m	**124.81**

33 DRAINAGE ABOVE GROUND

Item	Net Price £	Material £	Labour hours	Labour £	Unit	Total rate £
RAINWATER PIPEWORK/GUTTERS – cont						
Box gutter, with brackets measured separately – cont						
Brackets; fixed to backgrounds. For minimum fixing distances, refer to the Tables and Memoranda at the rear of the book						
Fascia						
100 × 75 mm	11.95	15.13	0.16	5.53	nr	**20.65**
Extra over fittings box cast iron gutter BS 460						
Union						
100 × 75 mm	15.85	20.06	0.38	13.12	nr	**33.19**
Stop end; external						
100 × 75 mm	11.95	15.13	0.12	4.14	nr	**19.27**
90° angle						
100 × 75 mm	60.27	76.28	0.38	13.12	nr	**89.40**
135° angle						
100 × 75 mm	60.27	76.28	0.38	13.12	nr	**89.40**
Running outlet						
65 mm outlet						
100 × 75 mm	60.27	76.28	0.38	13.12	nr	**89.40**
75 mm outlet						
100 × 75 mm	60.27	76.28	0.38	13.12	nr	**89.40**
100 × 75 mm outlet						
100 × 75 mm	60.27	76.28	0.38	13.12	nr	**89.40**
Cast iron rainwater pipe; dry joints; BS 460; fixed to backgrounds: Circular						
Plain socket pipe, with brackets measured separately						
65 mm	43.64	55.24	0.69	23.83	m	**79.06**
75 mm	43.64	55.24	0.69	23.83	m	**79.06**
100 mm	59.57	75.39	0.69	23.83	m	**99.22**
Bracket; fixed to backgrounds. For minimum fixing distances, refer to the Tables and Memoranda at the rear of the book						
65 mm	15.94	20.17	0.29	10.01	nr	**30.19**
75 mm	16.01	20.26	0.29	10.01	nr	**30.28**
100 mm	16.21	20.52	0.29	10.01	nr	**30.53**
Eared socket pipe, with wall spacers measured separately						
65 mm	46.64	59.02	0.62	21.41	m	**80.43**
75 mm	46.64	59.02	0.62	21.41	m	**80.43**
100 mm	62.59	79.21	0.62	21.41	m	**100.62**
Wall spacer plate; eared pipework						
65 mm	10.30	13.04	0.16	5.53	nr	**18.56**
75 mm	10.47	13.25	0.16	5.53	nr	**18.78**
100 mm	16.21	20.52	0.16	5.53	nr	**26.04**

33 DRAINAGE ABOVE GROUND

Item	Net Price £	Material £	Labour hours	Labour £	Unit	Total rate £
Extra over fittings circular cast iron pipework BS 460						
Loose sockets						
Plain socket						
65 mm	12.62	15.97	0.23	7.94	nr	**23.91**
75 mm	12.62	15.97	0.23	7.94	nr	**23.91**
100 mm	19.49	24.67	0.23	7.94	nr	**32.61**
Eared socket						
65 mm	18.23	23.07	0.29	10.01	nr	**33.08**
75 mm	18.23	23.07	0.29	10.01	nr	**33.08**
100 mm	24.63	31.17	0.29	10.01	nr	**41.18**
Shoe; front projection						
Plain socket						
65 mm	39.23	49.65	0.23	7.94	nr	**57.59**
75 mm	39.23	49.65	0.23	7.94	nr	**57.59**
100 mm	52.89	66.94	0.23	7.94	nr	**74.88**
Eared socket						
65 mm	45.47	57.55	0.29	10.01	nr	**67.56**
75 mm	45.47	57.55	0.29	10.01	nr	**67.56**
100 mm	60.36	76.39	0.29	10.01	nr	**86.41**
Access Pipe						
65 mm	70.83	89.64	0.23	7.94	nr	**97.59**
75 mm	74.36	94.11	0.23	7.94	nr	**102.05**
100 mm	131.90	166.93	0.23	7.94	nr	**174.87**
100 mm; eared	148.93	188.48	0.29	10.01	nr	**198.50**
Bends; any degree						
65 mm	28.34	35.87	0.23	7.94	nr	**43.81**
75 mm	33.76	42.73	0.23	7.94	nr	**50.67**
100 mm	47.72	60.39	0.23	7.94	nr	**68.33**
Branch						
92.5°						
65 mm	54.68	69.20	0.29	10.01	nr	**79.21**
75 mm	60.30	76.32	0.29	10.01	nr	**86.33**
100 mm	70.27	88.93	0.29	10.01	nr	**98.95**
112.5°						
65 mm	54.68	69.20	0.29	10.01	nr	**79.21**
75 mm	60.30	76.32	0.29	10.01	nr	**86.33**
135°						
65 mm	54.68	69.20	0.29	10.01	nr	**79.21**
75 mm	60.30	76.32	0.29	10.01	nr	**86.33**
Offsets						
75 to 150 mm projection						
65 mm	42.58	53.89	0.25	8.63	nr	**62.52**
75 mm	42.58	53.89	0.25	8.63	nr	**62.52**
100 mm	80.34	101.68	0.25	8.63	nr	**110.31**
225 mm projection						
65 mm	43.40	54.93	0.25	8.63	nr	**63.56**
75 mm	43.40	54.93	0.25	8.63	nr	**63.56**
100 mm	80.34	101.68	0.25	8.63	nr	**110.31**

33 DRAINAGE ABOVE GROUND

Item	Net Price £	Material £	Labour hours	Labour £	Unit	Total rate £
RAINWATER PIPEWORK/GUTTERS – cont						
Extra over fittings circular cast iron pipework BS 460 – cont						
305 mm projection						
65 mm	58.07	73.50	0.25	8.63	nr	**82.13**
75 mm	60.96	77.16	0.25	8.63	nr	**85.79**
100 mm	99.21	125.56	0.25	8.63	nr	**134.20**
380 mm projection						
65 mm	115.90	146.68	0.25	8.63	nr	**155.32**
75 mm	115.90	146.68	0.25	8.63	nr	**155.32**
100 mm	158.18	200.19	0.25	8.63	nr	**208.82**
455 mm projection						
65 mm	135.66	171.70	0.25	8.63	nr	**180.33**
75 mm	135.66	171.70	0.25	8.63	nr	**180.33**
100 mm	196.71	248.96	0.25	8.63	nr	**257.60**
Cast iron rainwater pipe; dry joints; BS 460; fixed to backgrounds: Rectangular						
Plain socket						
100 × 75 mm	123.31	156.06	1.04	35.91	m	**191.97**
Bracket; fixed to backgrounds. For minimum fixing distances, refer to the Tables and Memoranda at the rear of the book						
100 × 75 mm; build in holdabat	70.17	88.81	0.35	12.09	nr	**100.89**
100 × 75 mm; trefoil earband	55.91	70.77	0.29	10.01	nr	**80.78**
100 × 75 mm; plain earband	54.07	68.44	0.29	10.01	nr	**78.45**
Eared Socket, with wall spacers measured separately						
100 × 75 mm	125.35	158.64	1.16	40.06	m	**198.70**
Wall spacer plate; eared pipework						
100 × 75	16.21	20.52	0.16	5.53	nr	**26.04**
Extra over fittings rectangular cast iron pipework BS 460						
Loose socket						
100 × 75 mm; plain	52.30	66.20	0.23	7.94	nr	**74.14**
100 × 75 mm; eared	86.64	109.65	0.29	10.01	nr	**119.67**
Shoe; front						
100 × 75 mm; plain	138.73	175.58	0.23	7.94	nr	**183.52**
100 × 75 mm; eared	169.51	214.54	0.29	10.01	nr	**224.55**
Shoe; side						
100 × 75 mm; plain	168.32	213.02	0.23	7.94	nr	**220.97**
100 × 75 mm; eared	210.42	266.31	0.29	10.01	nr	**276.33**
Bends; side; any degree						
100 × 75 mm; plain	127.64	161.54	0.25	8.63	nr	**170.17**
100 × 75 mm; 135°; plain	130.37	165.00	0.25	8.63	nr	**173.63**
Bends; side; any degree						
100 × 75 mm; eared	161.40	204.27	0.25	8.63	nr	**212.90**

33 DRAINAGE ABOVE GROUND

Item	Net Price £	Material £	Labour hours	Labour £	Unit	Total rate £
Bends; front; any degree						
100 × 75 mm; plain	120.91	153.02	0.25	8.63	nr	**161.65**
100 × 75 mm; eared	136.99	173.37	0.25	8.63	nr	**182.01**
Offset; side;						
Plain socket						
75 mm projection	168.19	212.86	0.25	8.63	nr	**221.49**
115 mm projection	174.92	221.38	0.25	8.63	nr	**230.01**
225 mm projection	217.92	275.80	0.25	8.63	nr	**284.43**
305 mm projection	251.18	317.89	0.25	8.63	nr	**326.52**
Offset; front						
Plain socket						
75 mm projection	127.95	161.94	0.25	8.63	nr	**170.57**
150 mm projection	141.40	178.96	0.25	8.63	nr	**187.59**
225 mm projection	175.05	221.54	0.25	8.63	nr	**230.18**
305 mm projection	208.28	263.60	0.25	8.63	nr	**272.23**
Eared socket						
75 mm projection	163.67	207.14	0.25	8.63	nr	**215.78**
150 mm projection	176.50	223.38	0.25	8.63	nr	**232.02**
225 mm projection	208.80	264.26	0.25	8.63	nr	**272.89**
305 mm projection	242.93	307.46	0.25	8.63	nr	**316.09**
Offset; plinth						
115 mm projection; plain	132.07	167.15	0.25	8.63	nr	**175.78**
115 mm projection; eared	170.09	215.26	0.25	8.63	nr	**223.90**
Rainwater heads						
Flat hopper						
210 × 160 × 185 mm; 65 mm outlet	100.48	127.17	0.40	13.81	nr	**140.98**
210 × 160 × 185 mm; 75 mm outlet	100.48	127.17	0.40	13.81	nr	**140.98**
250 × 215 × 215 mm; 100 mm outlet	119.09	150.72	0.40	13.81	nr	**164.54**
Flat rectangular						
225 × 125 × 125 mm; 65 mm outlet	139.71	176.81	0.40	13.81	nr	**190.63**
225 × 125 × 125 mm; 75 mm outlet	139.71	176.81	0.40	13.81	nr	**190.63**
280 × 150 × 130 mm; 100 mm outlet	192.89	244.12	0.40	13.81	nr	**257.93**
Rectangular						
250 × 180 × 175 mm; 75 mm outlet	130.26	164.85	0.40	13.81	nr	**178.67**
250 × 180 × 175 mm; 100 mm outlet	130.26	164.85	0.40	13.81	nr	**178.67**
300 × 250 × 200 mm; 65 mm outlet	181.42	229.61	0.40	13.81	nr	**243.42**
300 × 250 × 200 mm; 75 mm outlet	181.42	229.61	0.40	13.81	nr	**243.42**
300 × 250 × 200 mm; 100 mm outlet	181.42	229.61	0.40	13.81	nr	**243.42**
300 × 250 × 200 mm; 100 × 75 mm outlet	181.42	229.61	0.40	13.81	nr	**243.42**
Castellated rectangular						
250 × 180 × 175 mm; 65 mm outlet	130.26	164.85	0.40	13.81	nr	**178.67**

33 DRAINAGE ABOVE GROUND

Item	Net Price £	Material £	Labour hours	Labour £	Unit	Total rate £
DISPOSAL SYSTEMS						
Pricing note: degree angles are only indicated where material prices differ.						
PVC-u overflow pipe; solvent welded joints; fixed with clips to backgrounds						
Pipe, with brackets measured separately						
19 mm	1.98	2.50	0.21	7.25	m	**9.75**
FIXINGS						
Pipe clip: including fixing to backgrounds. For minimum fixing distances, refer to the Tables and Memoranda at the rear of the book						
19 mm	0.77	0.98	0.18	6.22	nr	**7.19**
Extra over fittings overflow pipework PVC-u						
Straight coupler						
19 mm	2.11	2.67	0.17	5.87	nr	**8.54**
Bend						
19 mm: 91.25°	2.48	3.14	0.17	5.87	nr	**9.01**
19 mm: 135°	2.49	3.15	0.17	5.87	nr	**9.02**
Tee						
19 mm	2.70	3.41	0.18	6.22	nr	**9.63**
Reverse nut connector						
19 mm	0.99	1.25	0.15	5.18	nr	**6.43**
BSP adaptor: solvent welded socket to threaded socket						
19 mm × ¾"	3.63	4.59	0.14	4.83	nr	**9.43**
Straight tank connector						
19 mm	3.29	4.17	0.21	7.25	nr	**11.42**
32 mm	1.60	2.02	0.28	9.67	nr	**11.69**
40 mm	1.60	2.02	0.30	10.36	nr	**12.38**
Bent tank connector						
19 mm	3.87	4.90	0.21	7.25	nr	**12.15**
Tundish						
19 mm	51.48	65.16	0.38	13.12	nr	**78.28**
MuPVC waste pipe; solvent welded joints; fixed with clips to backgrounds; BS 5255						
Pipe, with brackets measured separately						
32 mm	3.51	4.44	0.23	7.94	m	**12.38**
40 mm	4.32	5.47	0.23	7.94	m	**13.42**
50 mm	6.56	8.30	0.26	8.98	m	**17.28**
FIXINGS						
Pipe clip: including fixing to backgrounds. For minimum fixing distances, refer to the Tables and Memoranda at the rear of the book						
32 mm	0.71	0.89	0.13	4.49	nr	**5.38**
40 mm	0.79	1.00	0.13	4.49	nr	**5.49**
50 mm	1.07	1.36	0.13	4.49	nr	**5.85**

33 DRAINAGE ABOVE GROUND

Item	Net Price £	Material £	Labour hours	Labour £	Unit	Total rate £
Pipe clip: expansion: including fixing to backgrounds. For minimum fixing distances, refer to the Tables and Memoranda at the rear of the book						
32 mm	0.84	1.06	0.13	4.49	nr	**5.55**
40 mm	0.87	1.10	0.13	4.49	nr	**5.59**
50 mm	2.04	2.59	0.13	4.49	nr	**7.08**
Pipe clip: metal; including fixing to backgrounds. For minimum fixing distances, refer to the Tables and Memoranda at the rear of the book						
32 mm	3.06	3.87	0.13	4.49	nr	**8.36**
40 mm	3.63	4.60	0.13	4.49	nr	**9.08**
50 mm	4.63	5.86	0.13	4.49	nr	**10.35**
Extra over fittings waste pipework MuPVC						
Screwed access plug						
32 mm	1.98	2.51	0.18	6.22	nr	**8.73**
40 mm	1.98	2.51	0.18	6.22	nr	**8.73**
50 mm	2.85	3.61	0.25	8.63	nr	**12.24**
Straight coupling						
32 mm	2.12	2.68	0.27	9.32	nr	**12.00**
40 mm	2.12	2.68	0.27	9.32	nr	**12.00**
50 mm	3.90	4.93	0.27	9.32	nr	**14.25**
Expansion coupling						
32 mm	3.76	4.76	0.27	9.32	nr	**14.09**
40 mm	4.51	5.71	0.27	9.32	nr	**15.04**
50 mm	6.13	7.76	0.27	9.32	nr	**17.08**
MuPVC to copper coupling						
32 mm	3.76	4.76	0.27	9.32	nr	**14.09**
40 mm	4.51	5.71	0.27	9.32	nr	**15.04**
50 mm	6.13	7.76	0.27	9.32	nr	**17.08**
Spigot and socket coupling						
32 mm	3.76	4.76	0.27	9.32	nr	**14.09**
40 mm	4.51	5.71	0.27	9.32	nr	**15.04**
50 mm	6.13	7.76	0.27	9.32	nr	**17.08**
Union						
32 mm	8.92	11.29	0.28	9.67	nr	**20.96**
40 mm	11.69	14.79	0.28	9.67	nr	**24.46**
50 mm	13.30	16.84	0.28	9.67	nr	**26.51**
Reducer: socket						
32 × 19 mm	3.22	4.07	0.27	9.32	nr	**13.40**
40 × 32 mm	2.12	2.68	0.27	9.32	nr	**12.00**
50 × 32 mm	3.07	3.89	0.27	9.32	nr	**13.21**
50 × 40 mm	3.76	4.76	0.27	9.32	nr	**14.09**
Reducer: level invert						
40 × 32 mm	2.65	3.35	0.27	9.32	nr	**12.67**
50 × 32 mm	3.26	4.13	0.27	9.32	nr	**13.45**
50 × 40 mm	3.26	4.13	0.27	9.32	nr	**13.45**

33 DRAINAGE ABOVE GROUND

Item	Net Price £	Material £	Labour hours	Labour £	Unit	Total rate £
DISPOSAL SYSTEMS – cont						
Extra over fittings waste pipework						
MuPVC – cont						
Swept bend						
32 mm	2.18	2.75	0.27	9.32	nr	**12.08**
32 mm: 165°	2.25	2.85	0.27	9.32	nr	**12.17**
40 mm	2.43	3.07	0.27	9.32	nr	**12.39**
40 mm: 165°	4.25	5.38	0.27	9.32	nr	**14.70**
50 mm	4.20	5.32	0.30	10.36	nr	**15.68**
50 mm: 165°	5.62	7.11	0.30	10.36	nr	**17.47**
Knuckle bend						
32 mm	1.98	2.51	0.27	9.32	nr	**11.84**
40 mm	2.19	2.77	0.27	9.32	nr	**12.10**
Spigot and socket bend						
32 mm	3.54	4.48	0.27	9.32	nr	**13.81**
32 mm: 150°	3.68	4.65	0.27	9.32	nr	**13.97**
40 mm	4.06	5.13	0.27	9.32	nr	**14.46**
50 mm	5.78	7.31	0.30	10.36	nr	**17.67**
Swept tee						
32 mm: 91.25°	2.90	3.67	0.31	10.71	nr	**14.37**
32 mm: 135°	3.48	4.41	0.31	10.71	nr	**15.11**
40 mm: 91.25°	3.69	4.67	0.31	10.71	nr	**15.37**
40 mm: 135°	4.60	5.82	0.31	10.71	nr	**16.53**
50 mm	7.19	9.10	0.31	10.71	nr	**19.80**
Swept cross						
40 mm: 91.25°	8.97	11.35	0.31	10.71	nr	**22.05**
50 mm: 91.25°	9.38	11.87	0.43	14.85	nr	**26.72**
50 mm: 135°	11.83	14.98	0.31	10.71	nr	**25.68**
Male iron adaptor						
32 mm	3.23	4.09	0.28	9.67	nr	**13.76**
40 mm	3.81	4.82	0.28	9.67	nr	**14.49**
Female iron adaptor						
32 mm	3.81	4.82	0.28	9.67	nr	**14.49**
40 mm	3.81	4.82	0.28	9.67	nr	**14.49**
50 mm	5.47	6.92	0.31	10.71	nr	**17.63**
Reverse nut adaptor						
32 mm	4.84	6.12	0.20	6.91	nr	**13.03**
40 mm	4.84	6.12	0.20	6.91	nr	**13.03**
Automatic air admittance valve						
32 mm	27.18	34.40	0.27	9.32	nr	**43.72**
40 mm	27.18	34.40	0.28	9.67	nr	**44.07**
50 mm	27.18	34.40	0.31	10.71	nr	**45.10**
MuPVC to metal adpator: including heat shrunk joint to metal						
50 mm	11.72	14.83	0.38	13.12	nr	**27.95**
Caulking bush: including joint to metal						
32 mm	5.10	6.46	0.31	10.71	nr	**17.16**
40 mm	5.10	6.46	0.31	10.71	nr	**17.16**
50 mm	5.10	6.46	0.32	11.05	nr	**17.51**

33 DRAINAGE ABOVE GROUND

Item	Net Price £	Material £	Labour hours	Labour £	Unit	Total rate £
Weathering apron						
50 mm	4.48	5.67	0.65	22.45	nr	**28.12**
Vent Cowl						
50 mm	5.12	6.47	0.19	6.56	nr	**13.04**
ABS waste pipe; solvent welded joints;						
fixed with clips to backgrounds; BS 5255						
Pipe, with brackets measured separately						
32 mm	2.86	3.62	0.23	7.94	m	**11.56**
40 mm	3.58	4.53	0.23	7.94	m	**12.47**
50 mm	4.49	5.69	0.26	8.98	m	**14.66**
FIXINGS						
Pipe clip: including fixing to backgrounds. For						
minimum fixing distances, refer to the Tables						
and Memoranda at the rear of the book						
32 mm	0.46	0.58	0.17	5.87	nr	**6.45**
40 mm	0.57	0.72	0.17	5.87	nr	**6.59**
50 mm	1.72	2.17	0.17	5.87	nr	**8.04**
Pipe clip: expansion: including fixing to						
backgrounds. For minimum fixing distances,						
refer to the Tables and Memoranda at the						
rear of the book						
32 mm	0.46	0.58	0.17	5.87	nr	**6.45**
40 mm	0.57	0.72	0.17	5.87	nr	**6.59**
50 mm	1.72	2.17	0.17	5.87	nr	**8.04**
Pipe clip: metal; including fixing to						
backgrounds. For minimum fixing distances,						
refer to the Tables and Memoranda at the						
rear of the book						
32 mm	2.60	3.30	0.17	5.87	nr	**9.17**
40 mm	3.09	3.91	0.17	5.87	nr	**9.78**
50 mm	3.94	4.98	0.17	5.87	nr	**10.86**
Extra over fittings waste pipework ABS						
Screwed access plug						
32 mm	1.78	2.25	0.18	6.22	nr	**8.47**
40 mm	1.78	2.25	0.18	6.22	nr	**8.47**
50 mm	3.67	4.65	0.25	8.63	nr	**13.28**
Straight coupling						
32 mm	1.78	2.25	0.27	9.32	nr	**11.57**
40 mm	1.78	2.25	0.27	9.32	nr	**11.57**
50 mm	3.67	4.65	0.27	9.32	nr	**13.97**
Expansion coupling						
32 mm	3.67	4.65	0.27	9.32	nr	**13.97**
40 mm	3.67	4.65	0.27	9.32	nr	**13.97**
50 mm	5.63	7.12	0.27	9.32	nr	**16.45**
ABS to copper coupling						
32 mm	3.48	4.41	0.27	9.32	nr	**13.73**
40 mm	3.48	4.41	0.27	9.32	nr	**13.73**
50 mm	5.32	6.74	0.27	9.32	nr	**16.06**

33 DRAINAGE ABOVE GROUND

Item	Net Price £	Material £	Labour hours	Labour £	Unit	Total rate £
DISPOSAL SYSTEMS – cont						
Extra over fittings waste pipework ABS – cont						
Reducer: socket						
40 × 32 mm	1.78	2.25	0.27	9.32	nr	**11.57**
50 × 32 mm	4.04	5.11	0.27	9.32	nr	**14.44**
50 × 40 mm	4.04	5.11	0.27	9.32	nr	**14.44**
Swept bend						
32 mm	1.78	2.25	0.27	9.32	nr	**11.57**
40 mm	1.78	2.25	0.27	9.32	nr	**11.57**
50 mm	3.67	4.65	0.30	10.36	nr	**15.01**
Knuckle bend						
32 mm	1.78	2.25	0.27	9.32	nr	**11.57**
40 mm	1.78	2.25	0.27	9.32	nr	**11.57**
Swept tee						
32 mm	2.54	3.22	0.31	10.71	nr	**13.92**
40 mm	2.54	3.22	0.31	10.71	nr	**13.92**
50 mm	6.72	8.51	0.31	10.71	nr	**19.21**
Swept cross						
40 mm	8.56	10.84	0.23	7.94	nr	**18.78**
50 mm	9.78	12.38	0.43	14.85	nr	**27.23**
Male iron adaptor						
32 mm	3.67	4.65	0.28	9.67	nr	**14.32**
40 mm	3.67	4.65	0.28	9.67	nr	**14.32**
Female iron adapator						
32 mm	3.67	4.65	0.28	9.67	nr	**14.32**
40 mm	3.67	4.65	0.28	9.67	nr	**14.32**
50 mm	5.50	6.96	0.31	10.71	nr	**17.67**
Tank connectors						
32 mm	2.53	3.20	0.29	10.01	nr	**13.21**
40 mm	2.80	3.54	0.29	10.01	nr	**13.55**
Caulking bush: including joint to pipework						
50 mm	4.35	5.50	0.50	17.27	nr	**22.77**
Polypropylene waste pipe; push fit joints; fixed with clips to backgrounds; BS 5254. Pipe, with brackets measured separately						
32 mm	1.87	2.37	0.21	7.25	m	**9.62**
40 mm	2.28	2.89	0.21	7.25	m	**10.14**
50 mm	3.81	4.82	0.38	13.12	m	**17.94**
FIXINGS						
Pipe clip: saddle; including fixing to backgrounds. For minimum fixing distances, refer to the Tables and Memoranda at the rear of the book						
32 mm	0.62	0.78	0.17	5.87	nr	**6.65**
40 mm	0.62	0.78	0.17	5.87	nr	**6.65**

33 DRAINAGE ABOVE GROUND

Item	Net Price £	Material £	Labour hours	Labour £	Unit	Total rate £
Pipe clip: including fixing to backgrounds. For minimum fixing distances, refer to the Tables and Memoranda at the rear of the book						
50 mm	1.46	1.84	0.17	5.87	nr	**7.71**
Extra over fittings waste pipework polypropylene						
Screwed access plug						
32 mm	1.72	2.18	0.16	5.53	nr	**7.70**
40 mm	1.72	2.18	0.16	5.53	nr	**7.70**
50 mm	2.94	3.72	0.20	6.91	nr	**10.63**
Straight coupling						
32 mm	1.72	2.18	0.19	6.56	nr	**8.74**
40 mm	1.72	2.18	0.19	6.56	nr	**8.74**
50 mm	2.94	3.72	0.20	6.91	nr	**10.63**
Universal waste pipe coupler						
32 mm dia.	2.91	3.68	0.20	6.91	nr	**10.59**
40 mm dia.	3.31	4.19	0.20	6.91	nr	**11.09**
Reducer						
40 × 32 mm	2.94	3.72	0.19	6.56	nr	**10.28**
50 × 32 mm	3.10	3.93	0.19	6.56	nr	**10.49**
50 × 40 mm	3.19	4.04	0.20	6.91	nr	**10.94**
Swept bend						
32 mm	1.72	2.18	0.19	6.56	nr	**8.74**
40 mm	1.72	2.18	0.19	6.56	nr	**8.74**
50 mm	2.94	3.72	0.20	6.91	nr	**10.63**
Knuckle bend						
32 mm	1.72	2.18	0.19	6.56	nr	**8.74**
40 mm	1.72	2.18	0.19	6.56	nr	**8.74**
50 mm	2.94	3.72	0.20	6.91	nr	**10.63**
Spigot and socket bend						
32 mm	1.72	2.18	0.19	6.56	nr	**8.74**
40 mm	1.72	2.18	0.19	6.56	nr	**8.74**
Swept tee						
32 mm	1.90	2.40	0.22	7.60	nr	**10.00**
40 mm	1.90	2.40	0.22	7.60	nr	**10.00**
50 mm	3.19	4.04	0.23	7.94	nr	**11.98**
Male iron adaptor						
32 mm	1.72	2.18	0.13	4.49	nr	**6.67**
40 mm	1.72	2.18	0.19	6.56	nr	**8.74**
50 mm	2.94	3.72	0.15	5.18	nr	**8.90**
Tank connector						
32 mm	1.72	2.18	0.24	8.29	nr	**10.46**
40 mm	1.72	2.18	0.24	8.29	nr	**10.46**
50 mm	2.94	3.72	0.35	12.09	nr	**15.81**

33 DRAINAGE ABOVE GROUND

Item	Net Price £	Material £	Labour hours	Labour £	Unit	Total rate £
DISPOSAL SYSTEMS – cont						
Polypropylene traps; including fixing to appliance and connection to pipework; BS 3943						
Tubular P trap; 75 mm seal						
32 mm dia.	7.01	8.87	0.20	6.91	nr	**15.78**
40 mm dia.	8.09	10.23	0.20	6.91	nr	**17.14**
Tubular S trap; 75 mm seal						
32 mm dia.	8.81	11.14	0.20	6.91	nr	**18.05**
40 mm dia.	10.41	13.17	0.20	6.91	nr	**20.08**
Running tubular P trap; 75 mm seal						
32 mm dia.	10.73	13.58	0.20	6.91	nr	**20.49**
40 mm dia.	11.73	14.85	0.20	6.91	nr	**21.75**
Running tubular S trap; 75 mm seal						
32 mm dia.	12.94	16.37	0.20	6.91	nr	**23.28**
40 mm dia.	13.92	17.62	0.20	6.91	nr	**24.52**
Spigot and socket bend; converter from P to S Trap						
32 mm	2.75	3.48	0.20	6.91	nr	**10.39**
40 mm	2.93	3.70	0.21	7.25	nr	**10.95**
Bottle P trap; 75 mm seal						
32 mm dia.	7.81	9.88	0.20	6.91	nr	**16.79**
40 mm dia.	9.35	11.83	0.20	6.91	nr	**18.74**
Bottle S trap; 75 mm seal						
32 mm dia.	9.42	11.93	0.20	6.91	nr	**18.83**
40 mm dia.	11.44	14.47	0.25	8.63	nr	**23.11**
Bottle P trap; resealing; 75 mm seal						
32 mm dia.	9.72	12.30	0.20	6.91	nr	**19.20**
40 mm dia.	11.35	14.36	0.25	8.63	nr	**23.00**
Bottle S trap; resealing; 75 mm seal						
32 mm dia.	11.11	14.06	0.20	6.91	nr	**20.97**
40 mm dia.	12.91	16.33	0.25	8.63	nr	**24.97**
Bath trap, low level; 38 mm seal						
40 mm dia.	9.75	12.33	0.25	8.63	nr	**20.97**
Bath trap, low level; 38 mm seal complete with overflow hose						
40 mm dia.	15.04	19.03	0.25	8.63	nr	**27.67**
Bath trap; 75 mm seal complete with overlow hose						
40 mm dia.	14.99	18.98	0.25	8.63	nr	**27.61**
Bath trap; 75 mm seal complete with overflow hose and overflow outlet						
40 mm dia.	25.59	32.39	0.20	6.91	nr	**39.30**
Bath trap; 75 mm seal complete with overflow hose, overflow outlet and ABS chrome waste						
40 mm dia.	35.35	44.74	0.20	6.91	nr	**51.65**
Washing machine trap; 75 mm seal including stand pipe						
40 mm dia.	20.34	25.75	0.25	8.63	nr	**34.38**

33 DRAINAGE ABOVE GROUND

Item	Net Price £	Material £	Labour hours	Labour £	Unit	Total rate £
Washing machine standpipe						
40 mm dia.	10.61	13.43	0.25	8.63	nr	**22.07**
Plastic unslotted chrome plated basin/sink waste including plug						
32 mm	12.77	16.17	0.34	11.74	nr	**27.91**
40 mm	17.10	21.64	0.34	11.74	nr	**33.38**
Plastic slotted chrome plated basin/sink waste including plug						
32 mm	10.14	12.84	0.34	11.74	nr	**24.58**
40 mm	16.99	21.51	0.34	11.74	nr	**33.25**
Bath overflow outlet; plastic; white						
42 mm	8.92	11.29	0.37	12.78	nr	**24.07**
Bath overlow outlet; plastic; chrome plated						
42 mm	11.76	14.88	0.37	12.78	nr	**27.66**
Combined cistern and bath overflow outlet; plastic; white						
42 mm	18.27	23.13	0.39	13.47	nr	**36.59**
Combined cistern and bath overlow outlet; plastic; chrome plated						
42 mm	18.27	23.13	0.39	13.47	nr	**36.59**
Cistern overflow outlet; plastic; white						
42 mm	14.73	18.64	0.15	5.18	nr	**23.82**
Cistern overlow outlet; plastic; chrome plated						
42 mm	13.27	16.80	0.15	5.18	nr	**21.98**
PVC-u soil and waste pipe; solvent welded joints; fixed with clips to backgrounds; BS 4514/BS EN 607						
Pipe, with brackets measured separately						
82 mm	13.37	16.92	0.35	12.09	m	**29.00**
110 mm	13.63	17.25	0.41	14.16	m	**31.41**
160 mm	35.32	44.70	0.51	17.61	m	**62.31**
FIXINGS						
Galvanized steel pipe clip: including fixing to backgrounds. For minimum fixing distances, refer to the Tables and Memoranda at the rear of the book						
82 mm	5.55	7.03	0.18	6.22	nr	**13.24**
110 mm	5.76	7.28	0.18	6.22	nr	**13.50**
160 mm	13.87	17.56	0.18	6.22	nr	**23.77**
Plastic coated steel pipe clip: including fixing to backgrounds. For minimum fixing distances, refer to the Tables and Memoranda at the rear of the book						
82 mm	7.78	9.84	0.18	6.22	nr	**16.06**
110 mm	10.58	13.39	0.18	6.22	nr	**19.61**
160 mm	13.35	16.90	0.18	6.22	nr	**23.12**

33 DRAINAGE ABOVE GROUND

Item	Net Price £	Material £	Labour hours	Labour £	Unit	Total rate £
DISPOSAL SYSTEMS – cont						
FIXINGS – cont						
Plastic pipe clip: including fixing to backgrounds. For minimum fixing distances, refer to the Tables and Memoranda at the rear of the book						
82 mm	4.10	5.19	0.18	6.22	nr	**11.41**
110 mm	7.90	10.00	0.18	6.22	nr	**16.22**
Plastic coated steel pipe clip: adjustable; including fixing to backgrounds. For minimum fixing distances, refer to the Tables and Memoranda at the rear of the book						
82 mm	6.26	7.93	0.20	6.91	nr	**14.83**
110 mm	7.84	9.92	0.20	6.91	nr	**16.83**
Galvanized steel pipe clip: drive in; including fixing to backgrounds. For minimum fixing distances, refer to the Tables and Memoranda at the rear of the book						
110 mm	11.97	15.15	0.22	7.60	nr	**22.74**
Extra over fittings solvent welded pipework PVC-u						
Straight coupling						
82 mm	7.09	8.97	0.21	7.25	nr	**16.22**
110 mm	8.87	11.22	0.22	7.60	nr	**18.82**
160 mm	25.56	32.35	0.24	8.29	nr	**40.64**
Expansion coupling						
82 mm	10.61	13.43	0.21	7.25	nr	**20.68**
110 mm	10.86	13.75	0.22	7.60	nr	**21.35**
160 mm	32.65	41.32	0.24	8.29	nr	**49.61**
Slip coupling; double ring socket						
82 mm	22.06	27.91	0.21	7.25	nr	**35.17**
110 mm	27.61	34.94	0.22	7.60	nr	**42.54**
160 mm	43.85	55.49	0.24	8.29	nr	**63.78**
Puddle flanges						
110 mm	215.43	272.64	0.45	15.54	nr	**288.18**
160 mm	348.26	440.75	0.55	18.99	nr	**459.75**
Socket reducer						
82 to 50 mm	10.80	13.67	0.18	6.22	nr	**19.88**
110 to 50 mm	13.45	17.03	0.18	6.22	nr	**23.24**
110 to 82 mm	13.87	17.56	0.22	7.60	nr	**25.16**
160 to 110 mm	28.08	35.54	0.26	8.98	nr	**44.51**
Socket plugs						
82 mm	8.36	10.58	0.15	5.18	nr	**15.76**
110 mm	12.22	15.47	0.20	6.91	nr	**22.37**
160 mm	22.50	28.48	0.27	9.32	nr	**37.80**
Access door; including cutting into pipe						
82 mm	23.71	30.00	0.28	9.67	nr	**39.67**
110 mm	23.71	30.00	0.34	11.74	nr	**41.75**
160 mm	42.35	53.59	0.46	15.88	nr	**69.48**

33 DRAINAGE ABOVE GROUND

Item	Net Price £	Material £	Labour hours	Labour £	Unit	Total rate £
Screwed access cap						
82 mm	16.77	21.22	0.15	5.18	nr	**26.40**
110 mm	19.77	25.02	0.20	6.91	nr	**31.93**
160 mm	37.23	47.11	0.27	9.32	nr	**56.44**
Access pipe: spigot and socket						
110 mm	29.23	37.00	0.22	7.60	nr	**44.60**
Access pipe: double socket						
110 mm	29.23	37.00	0.22	7.60	nr	**44.60**
Swept bend						
82 mm	17.79	22.51	0.29	10.01	nr	**32.53**
110 mm	20.84	26.37	0.32	11.05	nr	**37.42**
160 mm	51.91	65.70	0.49	16.92	nr	**82.62**
Bend; special angle						
82 mm	34.29	43.40	0.29	10.01	nr	**53.41**
110 mm	40.88	51.74	0.32	11.05	nr	**62.79**
160 mm	69.04	87.38	0.49	16.92	nr	**104.30**
Spigot and socket bend						
82 mm	17.22	21.79	0.26	8.98	nr	**30.77**
110 mm	20.18	25.53	0.32	11.05	nr	**36.58**
110 mm: 135°	22.68	28.70	0.32	11.05	nr	**39.75**
160 mm: 135°	50.21	63.55	0.44	15.19	nr	**78.74**
Variable bend: single socket						
110 mm	39.08	49.46	0.33	11.40	nr	**60.86**
Variable bend: double socket						
110 mm	38.99	49.35	0.33	11.40	nr	**60.74**
Access bend						
110 mm	57.85	73.21	0.33	11.40	nr	**84.61**
Single branch: two bosses						
82 mm	24.88	31.48	0.35	12.09	nr	**43.57**
82 mm: 104°	26.53	33.57	0.35	12.09	nr	**45.66**
110 mm	27.57	34.89	0.42	14.50	nr	**49.40**
110 mm: 135°	28.76	36.40	0.42	14.50	nr	**50.91**
160 mm	58.54	74.09	0.50	17.27	nr	**91.36**
160 mm: 135°	59.87	75.77	0.50	17.27	nr	**93.03**
Single branch; four bosses						
110 mm	33.57	42.48	0.42	14.50	nr	**56.99**
Single access branch						
82 mm	80.66	102.09	0.35	12.09	nr	**114.17**
110 mm	47.20	59.74	0.42	14.50	nr	**74.24**
Unequal single branch						
160 × 160 × 110 mm	66.09	83.65	0.50	17.27	nr	**100.91**
160 × 160 × 110 mm: 135°	71.03	89.90	0.50	17.27	nr	**107.17**
Double branch						
110 mm	71.22	90.14	0.42	14.50	nr	**104.65**
110 mm: 135°	68.12	86.22	0.42	14.50	nr	**100.72**
Corner branch						
110 mm	120.35	152.32	0.42	14.50	nr	**166.82**
Unequal double branch						
160 × 160 × 110 mm	123.10	155.79	0.50	17.27	nr	**173.06**

33 DRAINAGE ABOVE GROUND

Item	Net Price £	Material £	Labour hours	Labour £	Unit	Total rate £
DISPOSAL SYSTEMS – cont						
Extra over fittings solvent welded pipework PVC-u – cont						
Single boss pipe; single socket						
110 × 110 × 32 mm	7.45	9.42	0.24	8.29	nr	**17.71**
110 × 110 × 40 mm	7.45	9.42	0.24	8.29	nr	**17.71**
110 × 110 × 50 mm	7.86	9.95	0.24	8.29	nr	**18.24**
Single boss pipe; triple socket						
110 × 110 × 40 mm	12.10	15.31	0.24	8.29	nr	**23.60**
Waste boss; including cutting into pipe						
82 to 32 mm	9.74	12.33	0.29	10.01	nr	**22.35**
82 to 40 mm	9.74	12.33	0.29	10.01	nr	**22.35**
110 to 32 mm	9.74	12.33	0.29	10.01	nr	**22.35**
110 to 40 mm	9.74	12.33	0.29	10.01	nr	**22.35**
110 to 50 mm	10.10	12.78	0.29	10.01	nr	**22.80**
160 to 32 mm	13.77	17.43	0.30	10.36	nr	**27.79**
160 to 40 mm	13.77	17.43	0.35	12.09	nr	**29.52**
160 to 50 mm	13.77	17.43	0.40	13.81	nr	**31.24**
Self-locking waste boss; including cutting into pipe						
110 to 32 mm	13.01	16.47	0.30	10.36	nr	**26.83**
110 to 40 mm	13.59	17.21	0.30	10.36	nr	**27.56**
110 to 50 mm	17.57	22.24	0.30	10.36	nr	**32.60**
Adaptor saddle; including cutting to pipe						
82 to 32 mm	6.48	8.20	0.29	10.01	nr	**18.22**
110 to 40 mm	8.00	10.13	0.29	10.01	nr	**20.14**
160 to 50 mm	14.50	18.35	0.29	10.01	nr	**28.36**
Branch boss adaptor						
32 mm	3.88	4.90	0.26	8.98	nr	**13.88**
40 mm	3.88	4.90	0.26	8.98	nr	**13.88**
50 mm	5.54	7.01	0.26	8.98	nr	**15.99**
Branch boss adaptor bend						
32 mm	5.30	6.71	0.26	8.98	nr	**15.68**
40 mm	5.77	7.30	0.26	8.98	nr	**16.28**
50 mm	6.91	8.75	0.26	8.98	nr	**17.73**
Automatic air admittance valve						
82 to 110 mm	60.77	76.91	0.19	6.56	nr	**83.47**
PVC-u to metal adpator: including heat shrunk joint to metal						
110 mm	16.92	21.42	0.57	19.68	nr	**41.10**
Caulking bush: including joint to pipework						
82 mm	16.66	21.08	0.46	15.88	nr	**36.97**
110 mm	16.66	21.08	0.46	15.88	nr	**36.97**
Vent cowl						
82 mm	5.03	6.37	0.13	4.49	nr	**10.86**
110 mm	5.07	6.42	0.13	4.49	nr	**10.90**
160 mm	13.26	16.79	0.13	4.49	nr	**21.28**

33 DRAINAGE ABOVE GROUND

Item	Net Price £	Material £	Labour hours	Labour £	Unit	Total rate £
Weathering apron; to lead slates						
82 mm	5.03	6.37	1.15	39.71	nr	**46.08**
110 mm	5.76	7.28	1.15	39.71	nr	**47.00**
160 mm	17.32	21.92	1.15	39.71	nr	**61.63**
Weathering apron; to asphalt						
82 mm	21.14	26.76	1.10	37.99	nr	**64.74**
110 mm	21.14	26.76	1.10	37.99	nr	**64.74**
Weathering slate; flat; 406 × 406 mm						
82 mm	59.65	75.49	1.04	35.91	nr	**111.41**
110 mm	59.65	75.49	1.04	35.91	nr	**111.41**
Weathering slate; flat; 457 × 457 mm						
82 mm	61.20	77.45	1.04	35.91	nr	**113.37**
110 mm	61.20	77.45	1.04	35.91	nr	**113.37**
Weathering slate; angled; 610 × 610 mm						
82 mm	82.70	104.66	1.04	35.91	nr	**140.57**
110 mm	82.70	104.66	1.04	35.91	nr	**140.57**
PVC-u soil and waste pipe; ring seal joints; fixed with clips to backgrounds; BS 4514/BS EN 607						
Pipe, with brackets measured separately						
82 mm dia.	11.79	14.92	0.35	12.09	m	**27.01**
110 mm dia.	11.88	15.03	0.41	14.16	m	**29.19**
160 mm dia.	42.60	53.91	0.51	17.61	m	**71.53**
FIXINGS						
Galvanized steel pipe clip: including fixing to backgrounds. For minimum fixing distances, refer to the Tables and Memoranda at the rear of the book						
82 mm	5.55	7.03	0.18	6.22	nr	**13.24**
110 mm	5.76	7.28	0.18	6.22	nr	**13.50**
160 mm	13.87	17.56	0.18	6.22	nr	**23.77**
Plastic coated steel pipe clip: including fixing to backgrounds. For minimum fixing distances, refer to the Tables and Memoranda at the rear of the book						
82 mm	7.78	9.84	0.18	6.22	nr	**16.06**
110 mm	10.58	13.39	0.18	6.22	nr	**19.61**
160 mm	13.35	16.90	0.18	6.22	nr	**23.12**
Plastic pipe clip: including fixing to backgrounds. For minimum fixing distances, refer to the Tables and Memoranda at the rear of the book						
82 mm	4.10	5.19	0.18	6.22	nr	**11.41**
110 mm	7.90	10.00	0.18	6.22	nr	**16.22**
Plastic coated steel pipe clip: adjustable; including fixing to backgrounds. For minimum fixing distances, refer to the Tables and Memoranda at the rear of the book						
82 mm	6.26	7.93	0.20	6.91	nr	**14.83**
110 mm	7.84	9.92	0.20	6.91	nr	**16.83**

33 DRAINAGE ABOVE GROUND

Item	Net Price £	Material £	Labour hours	Labour £	Unit	Total rate £
DISPOSAL SYSTEMS – cont						
FIXINGS – cont						
Galvanized steel pipe clip: drive in; including fixing to backgrounds. For minimum fixing distances, refer to the Tables and Memoranda at the rear of the book						
110 mm	11.97	15.15	0.22	7.60	nr	**22.74**
Extra over fittings ring seal pipework PVC-u						
Straight coupling						
82 mm	7.72	9.78	0.21	7.25	nr	**17.03**
110 mm	8.65	10.95	0.22	7.60	nr	**18.55**
160 mm	29.23	37.00	0.24	8.29	nr	**45.29**
Straight coupling; double socket						
82 mm	13.63	17.25	0.21	7.25	nr	**24.51**
110 mm	12.36	15.65	0.22	7.60	nr	**23.24**
160 mm	29.23	37.00	0.24	8.29	nr	**45.29**
Reducer; socket						
82 to 50 mm	12.71	16.08	0.15	5.18	nr	**21.26**
110 to 50 mm	17.72	22.43	0.15	5.18	nr	**27.61**
110 to 82 mm	17.72	22.43	0.19	6.56	nr	**28.99**
160 to 110 mm	25.54	32.32	0.31	10.71	nr	**43.02**
Access cap						
82 mm	13.84	17.51	0.15	5.18	nr	**22.69**
110 mm	16.30	20.63	0.17	5.87	nr	**26.50**
Access cap; pressure plug						
160 mm	38.33	48.51	0.33	11.40	nr	**59.91**
Access pipe						
82 mm	33.96	42.98	0.22	7.60	nr	**50.58**
110 mm	33.90	42.90	0.22	7.60	nr	**50.50**
160 mm	71.16	90.06	0.24	8.29	nr	**98.35**
Bend						
82 mm	17.27	21.85	0.29	10.01	nr	**31.87**
82 mm; adjustable radius	38.94	49.28	0.29	10.01	nr	**59.30**
110 mm	18.59	23.52	0.32	11.05	nr	**34.57**
110 mm; adjustable radius	34.66	43.86	0.32	11.05	nr	**54.92**
160 mm	55.79	70.60	0.49	16.92	nr	**87.53**
160 mm; adjustable radius	73.18	92.62	0.49	16.92	nr	**109.54**
Bend; spigot and socket						
110 mm	18.75	23.73	0.32	11.05	nr	**34.78**
Bend; offset						
82 mm	15.21	19.25	0.21	7.25	nr	**26.50**
110 mm	24.39	30.87	0.32	11.05	nr	**41.92**
160 mm	53.81	68.10	0.32	11.05	nr	**79.15**
Bend; access						
110 mm	51.18	64.77	0.33	11.40	nr	**76.16**

33 DRAINAGE ABOVE GROUND

Item	Net Price £	Material £	Labour hours	Labour £	Unit	Total rate £
Single branch						
82 mm	27.11	34.31	0.35	12.09	nr	**46.40**
110 mm	25.07	31.72	0.42	14.50	nr	**46.23**
110 mm; 45°	25.79	32.64	0.31	10.71	nr	**43.35**
160 mm	64.64	81.81	0.50	17.27	nr	**99.08**
Single branch; access						
82 mm	40.63	51.42	0.35	12.09	nr	**63.51**
110 mm	58.82	74.45	0.42	14.50	nr	**88.95**
Unequal single branch						
160 × 160 × 110 mm	66.05	83.60	0.50	17.27	nr	**100.86**
160 × 160 × 110 mm; 45°	71.03	89.90	0.50	17.27	nr	**107.17**
Double branch; 4 bosses						
110 mm	65.62	83.05	0.49	16.92	nr	**99.97**
Corner branch; 2 bosses						
110 mm	120.33	152.29	0.49	16.92	nr	**169.21**
Multi-branch; 4 bosses						
110 mm	35.93	45.47	0.52	17.96	nr	**63.43**
Boss branch						
110 × 32 mm	7.43	9.41	0.34	11.74	nr	**21.15**
110 × 40 mm	7.43	9.41	0.34	11.74	nr	**21.15**
Strap on boss						
110 × 32 mm	8.28	10.48	0.30	10.36	nr	**20.84**
110 × 40 mm	8.28	10.48	0.30	10.36	nr	**20.84**
110 × 50 mm	8.35	10.56	0.30	10.36	nr	**20.92**
Patch boss						
82 × 32 mm	9.74	12.33	0.31	10.71	nr	**23.04**
82 × 40 mm	9.74	12.33	0.31	10.71	nr	**23.04**
82 × 50 mm	9.74	12.33	0.31	10.71	nr	**23.04**
Boss pipe; collar 4 boss						
110 mm	12.10	15.31	0.35	12.09	nr	**27.39**
Boss adaptor; rubber; push fit						
32 mm	4.52	5.72	0.26	8.98	nr	**14.70**
40 mm	4.52	5.72	0.26	8.98	nr	**14.70**
50 mm	5.12	6.48	0.26	8.98	nr	**15.46**
WC connector; cap and seal; solvent socket						
110 mm	12.60	15.95	0.23	7.94	nr	**23.89**
110 mm; 90°	20.25	25.63	0.27	9.32	nr	**34.95**
Vent terminal						
82 mm	5.03	6.37	0.13	4.49	nr	**10.86**
110 mm	5.07	6.42	0.13	4.49	nr	**10.90**
160 mm	13.29	16.82	0.13	4.49	nr	**21.31**
Weathering slate; inclined; 610 × 610 mm						
82 mm	82.70	104.66	1.04	35.91	nr	**140.57**
110 mm	82.70	104.66	1.04	35.91	nr	**140.57**
Weathering slate; inclined; 450 × 450 mm						
82 mm	61.20	77.45	1.04	35.91	nr	**113.37**
110 mm	61.20	77.45	1.04	35.91	nr	**113.37**
Weathering slate; flat; 400 × 400 mm						
82 mm	59.68	75.53	1.04	35.91	nr	**111.44**
110 mm	59.68	75.53	1.04	35.91	nr	**111.44**

Material Costs/Prices for Measured Works

33 DRAINAGE ABOVE GROUND

Item	Net Price £	Material £	Labour hours	Labour £	Unit	Total rate £
DISPOSAL SYSTEMS – cont						
Extra over fittings ring seal pipework PVC-u – cont						
Air admittance valve						
82 mm	60.77	76.91	0.19	6.56	nr	**83.47**
110 mm	60.77	76.91	0.19	6.56	nr	**83.47**
Cast iron pipe; nitrile rubber gasket joint with continuity clip BSEN 877; fixed vertically to backgrounds						
Pipe, with brackets and jointing couplings measured separately						
50 mm	34.06	43.11	0.31	10.71	m	**53.82**
75 mm	39.69	50.23	0.34	11.74	m	**61.97**
100 mm	44.09	55.80	0.37	12.78	m	**68.58**
150 mm	92.09	116.55	0.60	20.72	m	**137.27**
FIXINGS						
Brackets; fixed to backgrounds. For minimum fixing distances, refer to the Tables and Memoranda at the rear of the book						
50 mm	12.70	16.08	0.15	5.18	nr	**21.26**
75 mm	13.48	17.06	0.18	6.22	nr	**23.27**
100 mm	13.81	17.48	0.18	6.22	nr	**23.69**
150 mm	25.59	32.39	0.20	6.91	nr	**39.29**
Extra over fittings nitrile gasket cast iron pipework BS 416/6087, with jointing couplings measured separately						
Standard coupling						
50 mm	16.47	20.85	0.17	5.87	nr	**26.72**
75 mm	18.04	22.83	0.17	5.87	nr	**28.70**
100 mm	23.55	29.80	0.17	5.87	nr	**35.67**
150 mm	47.00	59.48	0.17	5.87	nr	**65.35**
Conversion coupling						
65 × 75 mm	19.12	24.20	0.60	20.73	nr	**44.92**
70 × 75 mm	19.12	24.20	0.60	20.73	nr	**44.92**
90 × 100 mm	24.61	31.15	0.67	23.14	nr	**54.29**
Access pipe; round door						
50 mm	48.16	60.95	0.41	14.16	nr	**75.11**
75 mm	69.24	87.62	0.46	15.88	nr	**103.51**
100 mm	72.77	92.10	0.67	23.17	nr	**115.28**
150 mm	121.05	153.20	0.83	28.68	nr	**181.88**
Access pipe; square door						
100 mm	143.40	181.49	0.67	23.17	nr	**204.67**
150 mm	219.52	277.82	0.83	28.68	nr	**306.50**
Taper reducer						
75 mm	39.57	50.08	0.60	20.73	nr	**70.81**
100 mm	50.23	63.57	0.67	23.14	nr	**86.72**
150 mm	97.82	123.80	0.83	28.68	nr	**152.48**

33 DRAINAGE ABOVE GROUND

Item	Net Price £	Material £	Labour hours	Labour £	Unit	Total rate £
Blank cap						
50 mm	10.00	12.65	0.24	8.29	nr	**20.94**
75 mm	12.46	15.77	0.26	8.98	nr	**24.74**
100 mm	12.11	15.33	0.32	11.05	nr	**26.38**
150 mm	17.49	22.14	0.40	13.81	nr	**35.95**
Blank cap; 50 mm screwed tapping						
75 mm	24.60	31.13	0.26	8.98	nr	**40.11**
100 mm	28.54	36.12	0.32	11.05	nr	**47.17**
150 mm	31.69	40.11	0.40	13.81	nr	**53.92**
Universal connector						
50 × 56/48/40 mm	14.86	18.81	0.33	11.40	nr	**30.20**
Change piece; BS416						
100 mm	31.69	40.11	0.47	16.23	nr	**56.34**
WC connector						
100 mm	50.98	64.52	0.49	16.92	nr	**81.44**
Boss pipe; 2" BSPT socket						
50 mm	40.71	51.52	0.58	20.03	nr	**71.55**
75 mm	59.58	75.41	0.65	22.45	nr	**97.85**
100 mm	71.18	90.08	0.79	27.28	nr	**117.36**
150 mm	110.89	140.35	0.86	29.70	nr	**170.05**
Boss pipe; 2" BSPT socket; 135°						
100 mm	85.76	108.54	0.79	27.28	nr	**135.82**
Boss pipe; 2 × 2" BSPT socket; opposed						
75 mm	78.94	99.91	0.65	22.45	nr	**122.36**
100 mm	87.42	110.64	0.79	27.28	nr	**137.92**
Boss pipe; 2 × 2" BSPT socket; in line						
100 mm	92.94	117.63	0.79	27.28	nr	**144.91**
Boss pipe; 2 × 2" BSPT socket; 90°						
100 mm	92.04	116.48	0.79	27.28	nr	**143.76**
Bend; short radius						
50 mm	28.80	36.45	0.50	17.27	nr	**53.71**
75 mm	32.58	41.24	0.60	20.73	nr	**61.96**
100 mm	39.83	50.41	0.67	23.14	nr	**73.56**
100 mm; 11°	34.34	43.46	0.67	23.14	nr	**66.60**
100 mm; 67°	39.83	50.41	0.67	23.14	nr	**73.56**
150 mm	71.18	90.08	0.83	28.68	nr	**118.76**
Access bend; short radius						
50 mm	70.93	89.77	0.50	17.27	nr	**107.04**
75 mm	76.95	97.38	0.60	20.73	nr	**118.11**
100 mm	84.30	106.69	0.67	23.14	nr	**129.83**
100 mm; 45°	84.30	106.69	0.67	23.14	nr	**129.83**
150 mm	119.67	151.45	0.83	28.68	nr	**180.14**
150 mm; 45°	119.67	151.45	0.83	28.68	nr	**180.14**
Long radius bend						
75 mm	54.35	68.78	0.60	20.73	nr	**89.51**
100 mm	64.55	81.69	0.67	23.14	nr	**104.83**
100 mm; 5°	39.83	50.41	0.67	23.14	nr	**73.56**
150 mm	140.66	178.01	0.83	28.68	nr	**206.70**
150 mm; 22.5°	147.46	186.63	0.83	28.68	nr	**215.31**

33 DRAINAGE ABOVE GROUND

Item	Net Price £	Material £	Labour hours	Labour £	Unit	Total rate £
DISPOSAL SYSTEMS – cont						
Extra over fittings nitrile gasket cast iron						
pipework BS 416/6087, with jointing						
couplings measured separately – cont						
Access bend; long radius						
75 mm	96.44	122.05	0.60	20.73	nr	**142.78**
100 mm	109.01	137.96	0.67	23.14	nr	**161.11**
150 mm	191.97	242.95	0.83	28.68	nr	**271.64**
Long tail bend						
100 × 250 mm long	51.47	65.14	0.70	24.18	nr	**89.32**
100 × 815 mm long	163.62	207.07	0.70	24.18	nr	**231.25**
Offset						
75 mm projection						
75 mm	42.48	53.76	0.53	18.30	nr	**72.06**
100 mm	41.92	53.05	0.66	22.79	nr	**75.84**
115 mm projection						
75 mm	46.68	59.08	0.53	18.30	nr	**77.38**
100 mm	52.35	66.25	0.66	22.79	nr	**89.04**
150 mm projection						
75 mm	54.85	69.42	0.53	18.30	nr	**87.72**
100 mm	54.85	69.42	0.66	22.79	nr	**92.21**
225 mm projection						
100 mm	59.97	75.90	0.66	22.79	nr	**98.69**
300 mm projection						
100 mm	64.55	81.69	0.66	22.79	nr	**104.48**
Branch; equal and unequal						
50 mm	43.30	54.80	0.78	26.94	nr	**81.73**
75 mm	49.01	62.03	0.85	29.36	nr	**91.39**
100 mm	61.58	77.94	1.00	34.53	nr	**112.47**
150 mm	152.65	193.20	1.20	41.46	nr	**234.65**
150 × 100 mm; 87.5°	116.82	147.85	1.21	41.61	nr	**189.46**
150 × 100 mm; 45°	171.35	216.87	1.21	41.61	nr	**258.48**
Branch; 2″ BSPT screwed socket						
100 mm	82.50	104.41	1.00	34.53	nr	**138.94**
Branch; long tail						
100 × 915 mm long	170.58	215.88	1.00	34.53	nr	**250.42**
Access branch; equal and unequal						
50 mm	94.32	119.37	0.78	26.94	nr	**146.31**
75 mm	94.32	119.37	0.85	29.36	nr	**148.74**
100 mm	106.05	134.21	1.02	35.24	nr	**169.45**
150 mm	225.72	285.66	1.20	41.46	nr	**327.12**
150 × 100 mm; 87.5°	189.70	240.08	1.20	41.46	nr	**281.53**
150 × 100 mm; 45°	205.63	260.25	1.20	41.46	nr	**301.71**
Parallel branch						
100 mm	64.55	81.69	1.00	34.53	nr	**116.22**
Double branch						
75 mm	72.77	92.10	0.95	32.81	nr	**124.91**
100 mm	76.17	96.40	1.30	44.91	nr	**141.31**
150 × 100 mm	214.49	271.46	1.56	53.87	nr	**325.34**

33 DRAINAGE ABOVE GROUND

Item	Net Price £	Material £	Labour hours	Labour £	Unit	Total rate £
Double access branch						
100 mm	120.62	152.65	1.43	49.40	nr	**202.06**
Corner branch						
100 mm	108.29	137.05	1.30	44.91	nr	**181.96**
Puddle flange; grey epoxy coated						
100 mm	72.41	91.65	1.00	34.53	nr	**126.18**
Roof vent connector; asphalt						
75 mm	88.12	111.53	0.90	31.08	nr	**142.61**
100 mm	68.82	87.10	0.97	33.50	nr	**120.59**
P trap						
100 mm	63.86	80.81	1.00	34.53	nr	**115.35**
P trap with access						
50 mm	97.73	123.69	0.77	26.59	nr	**150.28**
75 mm	97.73	123.69	0.90	31.08	nr	**154.77**
100 mm	108.32	137.09	1.16	40.06	nr	**177.15**
150 mm	189.13	239.37	1.77	61.12	nr	**300.49**
Bellmouth gully inlet						
100 mm	92.42	116.97	1.08	37.30	nr	**154.27**
Balcony gully inlet						
100 mm	279.05	353.17	1.08	37.30	nr	**390.47**
Roof outlet						
Flat grate						
75 mm	172.26	218.01	0.83	28.66	nr	**246.67**
100 mm	242.34	306.71	1.08	37.30	nr	**344.01**
Dome grate						
75 mm	172.26	218.01	0.83	28.66	nr	**246.67**
100 mm	271.83	344.03	1.08	37.30	nr	**381.33**
Top hat						
100 mm	368.01	465.76	1.08	37.30	nr	**503.05**
Cast iron pipe; EPDM rubber gasket joint with continuity clip; BS EN877; fixed to backgrounds						
Pipe, with brackets and jointing couplings measured separately						
50 mm	34.94	44.22	0.31	10.71	m	**54.93**
70 mm	40.84	51.69	0.34	11.74	m	**63.43**
100 mm	48.61	61.52	0.37	12.78	m	**74.29**
125 mm	78.00	98.71	0.65	22.45	m	**121.16**
150 mm	96.32	121.91	0.70	24.17	m	**146.08**
200 mm	160.97	203.72	1.14	39.37	m	**243.09**
250 mm	224.91	284.65	1.25	43.17	m	**327.81**
300 mm	280.08	354.46	1.53	52.83	m	**407.30**

33 DRAINAGE ABOVE GROUND

Item	Net Price £	Material £	Labour hours	Labour £	Unit	Total rate £
DISPOSAL SYSTEMS – cont						
FIXINGS						
Brackets; fixed to backgrounds. For minimum fixing distances, refer to the Tables and Memoranda at the rear of the book						
Ductile iron						
50 mm	14.85	18.79	0.10	3.45	nr	**22.24**
70 mm	14.85	18.79	0.10	3.45	nr	**22.24**
100 mm	17.16	21.72	0.15	5.18	nr	**26.90**
150 mm	31.76	40.20	0.20	6.91	nr	**47.10**
200 mm	117.84	149.14	0.25	8.63	nr	**157.77**
Mild steel; vertical						
125 mm	28.93	36.61	0.15	5.18	nr	**41.79**
Mild steel; stand off						
250 mm	62.96	79.69	0.25	8.63	nr	**88.32**
300 mm	69.36	87.79	0.25	8.63	nr	**96.42**
Stack support; rubber seal						
70 mm	58.89	74.53	0.55	18.99	nr	**93.53**
100 mm	65.52	82.93	0.65	22.45	nr	**105.37**
125 mm	72.67	91.97	0.74	25.55	nr	**117.53**
150 mm	103.70	131.24	0.86	29.70	nr	**160.94**
Wall spacer plate; cast iron (eared sockets)						
50 mm	12.64	16.00	0.10	3.45	nr	**19.46**
70 mm	12.64	16.00	0.10	3.45	nr	**19.46**
100 mm	12.64	16.00	0.10	3.45	nr	**19.46**
Extra over fittings EPDM rubber jointed cast iron pipework BS EN 877, with jointing couplings measured separately						
Coupling						
50 mm	15.87	20.08	0.10	3.45	nr	**23.53**
70 mm	17.46	22.10	0.10	3.45	nr	**25.56**
100 mm	22.74	28.78	0.10	3.45	nr	**32.24**
125 mm	28.25	35.76	0.15	5.18	nr	**40.94**
150 mm	45.57	57.68	0.30	10.36	nr	**68.04**
200 mm	101.95	129.02	0.35	12.09	nr	**141.11**
250 mm	146.04	184.82	0.40	13.81	nr	**198.64**
300 mm	168.98	213.86	0.50	17.27	nr	**231.13**
Plain socket						
50 mm	39.89	50.48	0.10	3.45	nr	**53.94**
70 mm	39.89	50.48	0.10	3.45	nr	**53.94**
100 mm	45.79	57.95	0.10	3.45	nr	**61.40**
150 mm	77.22	97.73	0.10	3.45	nr	**101.18**
Eared socket						
50 mm	41.07	51.98	0.25	8.63	nr	**60.61**
70 mm	41.07	51.98	0.25	8.63	nr	**60.61**
100 mm	49.76	62.97	0.25	8.63	nr	**71.61**
150 mm	81.59	103.26	0.25	8.63	nr	**111.90**

33 DRAINAGE ABOVE GROUND

Item	Net Price £	Material £	Labour hours	Labour £	Unit	Total rate £
Slip socket						
50 mm	51.66	65.38	0.25	8.63	nr	74.01
70 mm	51.66	65.38	0.25	8.63	nr	74.01
100 mm	60.33	76.35	0.25	8.63	nr	84.99
150 mm	90.80	114.91	0.25	8.63	nr	123.55
Stack support pipe						
70 mm	38.11	48.23	0.74	25.55	nr	73.78
100 mm	52.35	66.25	0.88	30.39	nr	96.64
125 mm	55.99	70.86	1.00	34.53	nr	105.39
150 mm	75.90	96.05	1.19	41.09	nr	137.15
Access pipe; round door						
50 mm	73.19	92.63	0.27	9.32	nr	101.95
70 mm	77.43	98.00	0.30	10.36	nr	108.36
100 mm	85.12	107.72	0.32	11.05	nr	118.77
150 mm	154.06	194.98	0.71	24.52	nr	219.50
Access pipe; square door						
100 mm	164.58	208.29	0.32	11.05	nr	219.34
125 mm	171.28	216.77	0.67	23.14	nr	239.91
150 mm	257.69	326.14	0.71	24.52	nr	350.65
200 mm	511.89	647.85	1.21	41.78	nr	689.63
250 mm	805.43	1019.35	1.31	45.24	nr	1064.59
300 mm	1004.71	1271.56	1.41	48.69	nr	1320.25
Taper reducer						
70 mm	42.26	53.49	0.30	10.36	nr	63.85
100 mm	49.69	62.88	0.32	11.05	nr	73.93
125 mm	49.96	63.23	0.64	22.10	nr	85.33
150 mm	95.39	120.72	0.67	23.14	nr	143.86
200 mm	154.90	196.04	1.15	39.71	nr	235.75
250 mm	319.95	404.93	1.25	43.17	nr	448.10
300 mm	439.84	556.66	1.37	47.31	nr	603.97
Blank cap						
50 mm	11.06	14.00	0.24	8.29	nr	22.29
70 mm	11.68	14.78	0.26	8.98	nr	23.76
100 mm	13.57	17.17	0.32	11.05	nr	28.22
125 mm	19.20	24.30	0.35	12.09	nr	36.39
150 mm	19.64	24.85	0.40	13.81	nr	38.66
200 mm	86.68	109.71	0.60	20.72	nr	130.43
250 mm	177.14	224.18	0.65	22.45	nr	246.63
300 mm	252.44	319.49	0.72	24.86	nr	344.35
Blank cap; 50 mm screwed tapping						
70 mm	26.04	32.95	0.26	8.98	nr	41.93
100 mm	28.12	35.59	0.32	11.05	nr	46.64
150 mm	33.78	42.75	0.40	13.81	nr	56.56
Universal connector; EPDM rubber						
50 × 56/48/40 mm	22.58	28.58	0.30	10.36	nr	38.94
Blank end; push fit						
100 × 38/32 mm	25.20	31.90	0.32	11.05	nr	42.95
Boss pipe; 2" BSPT socket						
50 mm	61.88	78.32	0.27	9.32	nr	87.64
75 mm	61.88	78.32	0.30	10.36	nr	88.68
100 mm	75.62	95.71	0.32	11.05	nr	106.76
150 mm	123.31	156.06	0.71	24.52	nr	180.58

33 DRAINAGE ABOVE GROUND

Item	Net Price £	Material £	Labour hours	Labour £	Unit	Total rate £
DISPOSAL SYSTEMS – cont						
Extra over fittings EPDM rubber jointed cast iron pipework BS EN 877, with jointing couplings measured separately – cont						
Boss pipe; 2 × 2" BSPT socket; opposed						
100 mm	97.72	123.67	0.32	11.05	nr	**134.72**
Boss pipe; 2 × 2" BSPT socket; 90°						
100 mm	97.72	123.67	0.32	11.05	nr	**134.72**
Manifold connector						
100 mm	150.86	190.92	0.64	22.10	nr	**213.02**
150 mm	210.21	266.04	1.00	34.53	nr	**300.57**
Bend; short radius						
50 mm	27.48	34.77	0.27	9.32	nr	**44.10**
70 mm	30.93	39.14	0.30	10.36	nr	**49.50**
100 mm	36.60	46.32	0.32	11.05	nr	**57.37**
125 mm	64.93	82.18	0.62	21.41	nr	**103.59**
150 mm	65.74	83.20	0.67	23.14	nr	**106.34**
200 mm; 45°	195.71	247.69	1.21	41.78	nr	**289.47**
250 mm; 45°	381.88	483.31	1.31	45.24	nr	**528.54**
300 mm; 45°	537.19	679.87	1.43	49.38	nr	**729.25**
Access bend; short radius						
70 mm	59.93	75.84	0.30	10.36	nr	**86.20**
100 mm	87.53	110.78	0.32	11.05	nr	**121.83**
150 mm	154.05	194.96	0.67	23.14	nr	**218.10**
Bend; long radius bend						
100 mm; 88°	93.01	117.72	0.32	11.05	nr	**128.77**
100 mm; 22°	68.60	86.82	0.32	11.05	nr	**97.87**
150 mm; 88°	266.67	337.50	0.67	23.14	nr	**360.63**
Access bend; long radius						
100 mm	113.24	143.31	0.32	11.05	nr	**154.37**
150 mm	275.10	348.16	0.32	11.05	nr	**359.22**
Bend; long tail						
100 mm	64.16	81.20	0.32	11.05	nr	**92.25**
Bend; long tail double						
70 mm	99.40	125.80	0.30	10.36	nr	**136.16**
100 mm	109.66	138.78	0.32	11.05	nr	**149.83**
Offset; 75 mm projection						
100 mm	56.26	71.20	0.32	11.05	nr	**82.25**
Offset; 130 mm projection						
50 mm	46.54	58.90	0.27	9.32	nr	**68.22**
70 mm	70.62	89.37	0.30	10.36	nr	**99.73**
100 mm	92.75	117.39	0.32	11.05	nr	**128.44**
125 mm	117.54	148.76	0.67	23.14	nr	**171.90**
Branch; equal and unequal						
50 mm	44.05	55.75	0.37	12.78	nr	**68.52**
70 mm	46.52	58.88	0.40	13.81	nr	**72.69**
100 mm	63.83	80.78	0.42	14.50	nr	**95.28**

33 DRAINAGE ABOVE GROUND

Item	Net Price £	Material £	Labour hours	Labour £	Unit	Total rate £
125 mm	127.93	161.90	0.76	26.24	nr	**188.15**
150 mm	139.19	176.16	0.97	33.50	nr	**209.65**
200 mm	357.05	451.88	1.51	52.14	nr	**504.03**
250 mm	456.29	577.48	1.63	56.29	nr	**633.77**
300 mm	752.90	952.87	1.77	61.12	nr	**1013.99**
Branch; radius; equal and unequal						
70 mm	56.68	71.73	0.40	13.81	nr	**85.54**
100 mm	74.62	94.43	0.42	14.50	nr	**108.94**
150 mm	161.37	204.23	1.37	47.31	nr	**251.54**
200 mm	454.09	574.70	1.51	52.14	nr	**626.84**
Branch; long tail						
100 mm	208.14	263.42	0.52	17.96	nr	**281.37**
Access branch; radius; equal and unequal						
70 mm	83.12	105.19	0.40	13.81	nr	**119.01**
100 mm	112.66	142.59	0.42	14.50	nr	**157.09**
150 mm	268.99	340.43	0.97	33.50	nr	**373.92**
Double branch; equal and unequal						
100 mm	62.83	79.52	0.52	17.96	nr	**97.48**
100 mm; 70°	93.73	118.63	0.52	17.96	nr	**136.59**
150 mm	268.99	340.43	1.37	47.31	nr	**387.74**
200 mm	460.03	582.22	1.51	52.14	nr	**634.36**
Double branch; radius; equal and unequal						
100 mm	81.05	102.57	0.52	17.96	nr	**120.53**
150 mm	331.80	419.93	1.37	47.31	nr	**467.24**
Corner branch						
100 mm	163.95	207.49	0.52	17.96	nr	**225.45**
150 mm	181.27	229.41	0.52	17.96	nr	**247.37**
Corner branch; long tail						
100 mm	244.59	309.55	0.52	17.96		**327.51**
Roof vent connector; asphalt						
100 mm	103.14	130.53	0.32	11.05	nr	**141.59**
Movement connector						
100 mm	131.00	165.80	0.32	11.05	nr	**176.85**
150 mm	242.52	306.93	0.67	23.14	nr	**330.07**
Expansion plugs						
70 mm	29.14	36.88	0.32	11.05	nr	**47.94**
100 mm	36.32	45.97	0.39	13.47	nr	**59.44**
150 mm	65.19	82.51	0.55	18.99	nr	**101.50**
P trap						
100 mm dia.	67.94	85.99	0.32	11.05	nr	**97.04**
P trap with access						
50 mm	103.76	131.32	0.27	9.32	nr	**140.64**
70 mm	103.76	131.32	0.30	10.36	nr	**141.68**
100 mm	112.39	142.24	0.32	11.05	nr	**153.29**
150 mm	200.94	254.31	0.67	23.14	nr	**277.45**
Branch trap						
100 mm	246.26	311.66	0.42	14.50	nr	**326.17**
Stench trap						
100 mm	478.89	606.08	0.42	14.50	nr	**620.59**

33 DRAINAGE ABOVE GROUND

Item	Net Price £	Material £	Labour hours	Labour £	Unit	Total rate £
DISPOSAL SYSTEMS – cont						
Extra over fittings EPDM rubber jointed cast iron pipework BS EN 877, with jointing couplings measured separately – cont						
Balcony gully inlet						
100 mm	271.82	344.01	1.00	34.53	nr	**378.55**
Roof outlet						
Flat grate						
70 mm	167.80	212.37	1.00	34.53	nr	**246.90**
100 mm	236.04	298.74	1.00	34.53	nr	**333.27**
Dome grate						
70 mm	167.80	212.37	1.00	34.53	nr	**246.90**
100 mm	264.74	335.06	1.00	34.53	nr	**369.59**
Top hat						
100 mm	358.45	453.65	1.00	34.53	nr	**488.18**
Floor drains; for cast iron pipework BS 416 and BS EN877						
Adjustable clamp plate body						
100 mm; 165 mm nickel bronze grate and frame	159.79	202.23	0.50	17.27	nr	**219.49**
100 mm; 165 mm nickel bronze rodding eye	185.94	235.33	0.50	17.27	nr	**252.59**
100 mm; 150 × 150 mm nickel bronze grate and frame	179.28	226.90	0.50	17.27	nr	**244.16**
100 mm; 150 × 150 mm nickel bronze rodding eye	185.94	235.33	0.50	17.27	nr	**252.59**
Deck plate body						
100 mm; 165 mm nickel bronze grate and frame	159.79	202.23	0.50	17.27	nr	**219.49**
100 mm; 165 mm nickel bronze rodding eye	185.94	235.33	0.50	17.27	nr	**252.59**
100 mm; 150 × 150 mm nickel bronze grate and frame	179.28	226.90	0.50	17.27	nr	**244.16**
100 mm; 150 × 150 mm nickel bronze rodding eye	185.94	235.33	0.50	17.27	nr	**252.59**
Extra for						
100 mm; Srewed extension piece	62.32	78.87	0.30	10.36	nr	**89.23**
100 mm; Grating extension piece; screwed or spigot	47.27	59.82	0.30	10.36	nr	**70.18**
100 mm; Brewary trap	1719.67	2176.41	2.00	69.07	nr	**2245.48**

33 DRAINAGE ABOVE GROUND

Item	Net Price £	Material £	Labour hours	Labour £	Unit	Total rate £
High density polyethylene (HDPE) pipes and fittings						
Pipe, with brackets and jointing couplings measured separately						
40 mm dia.	4.99	6.31	0.47	16.23	m	**22.54**
50 mm dia.	5.92	7.49	0.53	18.31	m	**25.80**
56 mm dia.	7.10	8.99	0.60	20.73	m	**29.71**
63 mm dia.	8.15	10.32	0.60	20.73	m	**31.05**
75 mm dia.	9.38	11.88	0.90	31.08	m	**42.96**
90 mm dia.	13.95	17.65	0.90	31.08	m	**48.74**
110 mm dia.	17.66	22.35	1.10	37.99	m	**60.33**
125 mm dia.	25.06	31.72	1.20	41.46	m	**73.17**
160 mm dia.	40.45	51.19	1.48	51.11	m	**102.30**
200 mm dia.	52.42	66.34	1.77	61.12	m	**127.46**
250 mm dia.	84.02	106.34	1.75	60.48	m	**166.82**
315 mm dia.	129.83	164.32	1.90	65.61	m	**229.93**
Extra over fittings						
Coupler						
40 mm dia.	7.55	9.56	0.48	16.58	nr	**26.13**
50 mm dia.	7.78	9.84	0.52	17.96	nr	**27.80**
56 mm dia.	7.82	9.90	0.58	20.03	nr	**29.93**
63 mm dia.	10.44	13.21	0.58	20.03	nr	**33.24**
90 mm dia.	10.92	13.82	0.67	23.14	nr	**36.95**
110 mm dia.	11.67	14.77	0.74	25.55	nr	**40.32**
125 mm dia.	20.09	25.43	0.83	28.66	nr	**54.09**
160 mm dia.	27.94	35.36	1.00	34.53	nr	**69.90**
200 mm dia.	194.56	246.24	1.35	46.62	nr	**292.86**
250 mm dia.	321.65	407.08	1.50	51.80	nr	**458.88**
315 mm dia.	364.83	461.73	1.80	62.16	nr	**523.89**
Cap						
40 mm dia.	4.32	5.47	0.24	8.29	nr	**13.76**
50 mm dia.	3.69	4.67	0.26	8.98	nr	**13.65**
56 mm dia.	3.69	4.67	0.30	10.36	nr	**15.03**
63 mm dia.	3.69	4.67	0.32	11.05	nr	**15.73**
75 mm dia.	3.69	4.67	0.35	12.09	nr	**16.76**
90 mm dia.	4.67	5.91	0.37	12.78	nr	**18.69**
110 mm dia.	8.74	11.06	0.40	13.81	nr	**24.87**
125 mm dia.	10.98	13.89	0.46	15.88	nr	**29.78**
160 mm dia.	35.09	44.41	0.50	17.27	nr	**61.68**
200 mm dia.	49.71	62.92	0.68	23.48	nr	**86.40**
250 mm dia.	64.37	81.47	0.75	25.90	nr	**107.37**
315 mm dia.	73.15	92.58	0.90	31.08	nr	**123.66**
Reducer						
50 mm dia.	3.86	4.88	0.54	18.65	nr	**23.53**
56 mm dia.	4.22	5.34	0.60	20.72	nr	**26.06**
63 mm dia.	4.22	5.34	0.67	23.14	nr	**28.48**
75 mm dia.	4.94	6.25	0.67	23.14	nr	**29.39**
90 mm dia.	6.32	8.00	0.67	23.14	nr	**31.14**
110 mm dia.	7.75	9.81	0.74	25.55	nr	**35.36**
125 mm dia.	12.30	15.56	0.83	28.66	nr	**44.23**
160 mm dia.	13.77	17.43	1.10	37.99	nr	**55.41**
200 mm dia.	87.61	110.88	1.40	48.35	nr	**159.23**

Material Costs/Prices for Measured Works

33 DRAINAGE ABOVE GROUND

Item	Net Price £	Material £	Labour hours	Labour £	Unit	Total rate £
DISPOSAL SYSTEMS – cont						
Extra over fittings – cont						
Bend; 45°						
40 mm dia.	3.45	4.37	0.45	15.54	nr	**19.91**
50 mm dia.	3.47	4.39	0.50	17.27	nr	**21.66**
56 mm dia.	4.07	5.15	0.55	18.99	nr	**24.14**
63 mm dia.	5.23	6.61	0.58	20.03	nr	**26.64**
75 mm dia.	6.70	8.48	0.67	23.14	nr	**31.61**
90 mm dia.	9.59	12.14	0.67	23.14	nr	**35.28**
110 mm dia.	9.59	12.14	0.74	25.55	nr	**37.70**
125 mm dia.	15.47	19.57	0.83	28.66	nr	**48.24**
160 mm dia.	40.62	51.40	1.00	34.53	nr	**85.94**
200 mm dia.	112.01	141.76	1.40	48.33	nr	**190.09**
250 mm dia.	182.22	230.62	1.80	62.16	nr	**292.78**
315 mm dia.	252.72	319.84	2.40	82.88	nr	**402.72**
Bend; 90°						
40 mm dia.	3.45	58.30	0.40	13.81	nr	**72.11**
50 mm dia.	4.07	53.92	0.50	17.27	nr	**71.19**
56 mm dia.	4.94	72.61	0.45	15.54	nr	**88.15**
63 mm dia.	5.81	66.35	0.58	20.03	nr	**86.38**
75 mm dia.	8.74	102.43	0.67	23.14	nr	**125.57**
90 mm dia.	11.70	148.78	0.64	22.10	nr	**170.88**
110 mm dia.	11.70	148.78	0.74	25.55	nr	**174.33**
125 mm dia.	18.09	172.63	0.83	28.66	nr	**201.29**
160 mm dia.	39.44	327.40	1.00	34.53	nr	**361.93**
200 mm dia.	192.75	925.38	1.40	48.35	nr	**973.73**
250 mm dia.	321.80	1152.66	1.80	62.16	nr	**1214.82**
315 mm dia.	468.08	1525.52	2.40	82.88	nr	**1608.40**
Equal tee						
40 mm dia.	6.89	8.72	0.60	20.72	nr	**29.44**
50 mm dia.	8.74	11.06	0.70	24.17	nr	**35.23**
56 mm dia.	8.74	11.06	0.70	24.17	nr	**35.23**
63 mm dia.	9.08	11.50	0.75	25.90	nr	**37.40**
75 mm dia.	10.75	13.61	0.83	28.66	nr	**42.27**
90 mm dia.	13.69	17.33	0.87	30.04	nr	**47.37**
110 mm dia.	15.75	19.93	1.00	34.53	nr	**54.47**
125 mm dia.	23.33	29.53	1.08	37.30	nr	**66.83**
160 mm dia.	87.73	111.03	1.35	46.62	nr	**157.65**
200 mm dia.	174.01	220.23	1.90	65.61	nr	**285.84**
250 mm dia.	248.66	314.71	2.70	93.24	nr	**407.95**
315 mm dia.	351.05	444.29	3.60	124.32	nr	**568.61**
Reducing tee						
50 mm dia.	8.74	11.06	0.70	24.17	nr	**35.23**
56 mm dia.	8.74	11.06	0.70	24.17	nr	**35.23**
63 mm dia.	9.08	11.50	0.75	25.90	nr	**37.40**
75 mm dia.	10.75	13.61	0.83	28.66	nr	**42.27**
90 mm dia.	13.69	17.33	0.87	30.04	nr	**47.37**
110 mm dia.	15.75	19.93	1.00	34.53	nr	**54.47**
125 mm dia.	23.33	29.53	1.08	37.30	nr	**66.83**

33 DRAINAGE ABOVE GROUND

Item	Net Price £	Material £	Labour hours	Labour £	Unit	Total rate £
160 mm dia.	87.73	111.03	1.35	46.62	nr	**157.65**
200 mm dia.	174.01	220.23	1.90	65.61	nr	**285.84**
250 mm dia.	248.66	314.71	2.70	93.24	nr	**407.95**
315 mm dia.	351.05	444.29	3.60	124.32	nr	**568.61**
Y-branch 45°						
40 mm dia.	8.42	10.66	0.65	22.45	nr	**33.11**
50 mm dia.	8.74	11.06	0.70	24.17	nr	**35.23**
56 mm dia.	8.74	11.06	0.70	24.17	nr	**35.23**
63 mm dia.	9.08	11.50	0.75	25.90	nr	**37.40**
75 mm dia.	10.20	12.90	0.83	28.66	nr	**41.57**
90 mm dia.	14.56	18.43	0.87	30.04	nr	**48.48**
110 mm dia.	17.49	22.14	1.00	34.53	nr	**56.67**
125 mm dia.	23.33	29.53	1.08	37.30	nr	**66.83**
160 mm dia.	80.72	102.16	1.35	46.62	nr	**148.78**
200 mm dia.	260.35	329.49	1.90	65.61	nr	**395.10**
250 mm dia.	438.83	555.38	2.70	93.24	nr	**648.62**
315 mm dia.	643.62	814.56	3.60	124.32	nr	**938.88**
Reducing Y-branch 45°						
50 mm dia.	8.74	11.06	0.70	24.17	nr	**35.23**
56 mm dia.	8.74	11.06	0.70	24.17	nr	**35.23**
63 mm dia.	9.08	11.50	0.75	25.90	nr	**37.40**
75 mm dia.	10.20	12.90	0.83	28.66	nr	**41.57**
90 mm dia.	15.18	19.21	0.87	30.04	nr	**49.26**
110 mm dia.	17.49	22.14	1.00	34.53	nr	**56.67**
125 mm dia.	23.33	29.53	1.08	37.30	nr	**66.83**
160 mm dia.	108.80	137.70	1.35	46.62	nr	**184.31**
200 mm dia.	174.01	220.23	1.90	65.61	nr	**285.84**
250 mm dia.	321.80	407.27	2.70	93.24	nr	**500.51**
315 mm dia.	541.20	684.94	3.60	124.32	nr	**809.26**

38 PIPED SUPPLY SYSTEMS

Item	Net Price £	Material £	Labour hours	Labour £	Unit	Total rate £
COLD WATER PIPELINES: COPPER PIPEWORK						
Copper pipe; capillary or compression joints in the running length; EN1057 R250 (TX) formerly BS 2871 Table X						
Fixed vertically or at low level, with brackets measured separately						
12 mm dia.	2.33	2.94	0.39	13.47	m	**16.41**
15 mm dia.	2.62	3.31	0.40	13.81	m	**17.12**
22 mm dia.	5.26	6.65	0.47	16.23	m	**22.88**
28 mm dia.	6.66	8.43	0.51	17.61	m	**26.05**
35 mm dia.	15.80	20.00	0.58	20.03	m	**40.02**
42 mm dia.	19.24	24.35	0.66	22.79	m	**47.15**
54 mm dia.	24.74	31.31	0.72	24.86	m	**56.17**
67 mm dia.	32.36	40.96	0.75	25.90	m	**66.85**
76 mm dia.	45.74	57.89	0.76	26.24	m	**84.13**
108 mm dia.	65.83	83.31	0.78	26.94	m	**110.24**
133 mm dia.	85.40	108.08	1.05	36.26	m	**144.34**
159 mm dia.	135.36	171.31	1.15	39.71	m	**211.02**
Fixed horizontally at high level or suspended, with brackets measured separately						
12 mm dia.	2.33	2.94	0.45	15.54	m	**18.48**
15 mm dia.	2.62	3.31	0.46	15.88	m	**19.20**
22 mm dia.	5.26	6.65	0.54	18.65	m	**25.30**
28 mm dia.	13.33	16.87	0.59	20.37	m	**37.24**
35 mm dia.	15.80	20.00	0.67	23.14	m	**43.13**
42 mm dia.	19.24	24.35	0.76	26.24	m	**50.60**
54 mm dia.	24.74	31.31	0.83	28.66	m	**59.97**
67 mm dia.	32.36	40.96	0.86	29.70	m	**70.65**
76 mm dia.	45.74	57.89	0.87	30.04	m	**87.93**
108 mm dia.	65.83	83.31	0.90	31.08	m	**114.39**
133 mm dia.	85.40	108.08	1.21	41.78	m	**149.86**
159 mm dia.	135.36	171.31	1.32	45.58	m	**216.89**
Copper pipe; capillary or compression joints in the running length; EN1057 R250 (TY) formerly BS 2871 Table Y						
Fixed vertically or at low level with brackets measured separately (Refer to Copper Pipe Table X Section)						
12 mm dia.	3.01	3.81	0.41	14.16	m	**17.97**
15 mm dia.	4.30	5.44	0.43	14.85	m	**20.29**
22 mm dia.	7.57	9.58	0.50	17.27	m	**26.85**
28 mm dia.	9.78	12.38	0.54	18.65	m	**31.02**
35 mm dia.	14.27	18.07	0.62	21.41	m	**39.48**
42 mm dia.	17.30	21.89	0.71	24.52	m	**46.41**
54 mm dia.	29.50	37.33	0.78	26.94	m	**64.26**
67 mm dia.	39.18	49.59	0.82	28.32	m	**77.91**
76 mm dia.	57.26	72.46	0.60	20.72	m	**93.18**
108 mm dia.	79.49	100.61	0.88	30.39	m	**131.00**

38 PIPED SUPPLY SYSTEMS

Item	Net Price £	Material £	Labour hours	Labour £	Unit	Total rate £
Copper pipe; capillary or compression joints in the running length; EN1057 R250 (TX) formerly BS 2871 Table X						
Plastic coated gas and cold water service pipe for corrosive environments, fixed vertically or at low level with brackets measured separtely						
15 mm dia. (white)	6.01	7.60	0.59	20.37	m	**27.98**
22 mm dia. (white)	10.66	13.49	0.68	23.49	m	**36.98**
28 mm dia. (white)	11.00	13.92	0.74	25.56	m	**39.48**
FIXINGS						
Saddle band						
6 mm dia.	0.07	0.09	0.05	1.77	nr	**1.85**
8 mm dia.	0.07	0.09	0.07	2.37	nr	**2.46**
10 mm dia.	0.11	0.13	0.09	2.95	nr	**3.09**
12 mm dia.	0.11	0.13	0.11	3.69	nr	**3.83**
15 mm dia.	0.11	0.13	0.13	4.49	nr	**4.62**
22 mm dia.	0.22	0.28	0.13	4.49	nr	**4.77**
28 mm dia.	0.11	0.13	0.16	5.53	nr	**5.66**
35 mm dia.	0.40	0.50	0.18	6.22	nr	**6.72**
42 mm dia.	0.83	1.05	0.21	7.25	nr	**8.30**
54 mm dia.	1.11	1.41	0.21	7.25	nr	**8.66**
Single spacing clip						
15 mm dia.	0.23	0.30	0.14	4.83	nr	**5.13**
22 mm dia.	0.23	0.30	0.15	5.18	nr	**5.48**
28 mm dia.	0.55	0.70	0.17	5.87	nr	**6.57**
Two piece spacing clip						
8 mm dia. Bottom	0.19	0.24	0.11	3.80	nr	**4.04**
8 mm dia. Top	0.19	0.24	0.11	3.80	nr	**4.04**
12 mm dia. Bottom	0.19	0.24	0.13	4.49	nr	**4.73**
12 mm dia. Top	0.19	0.24	0.13	4.49	nr	**4.73**
15 mm dia. Bottom	0.22	0.28	0.13	4.49	nr	**4.77**
15 mm dia. Top	0.22	0.28	0.13	4.49	nr	**4.77**
22 mm dia. Bottom	0.22	0.28	0.14	4.83	nr	**5.12**
22 mm dia. Top	0.23	0.30	0.14	4.83	nr	**5.13**
28 mm dia. Bottom	0.22	0.28	0.16	5.53	nr	**5.81**
28 mm dia. Top	0.39	0.49	0.16	5.53	nr	**6.01**
35 mm dia. Bottom	0.23	0.30	0.21	7.25	nr	**7.55**
35 mm dia. Top	0.54	0.68	0.21	7.25	nr	**7.93**
Single pipe bracket						
15 mm dia.	1.94	2.46	0.14	4.83	nr	**7.29**
22 mm dia.	1.97	2.49	0.14	4.83	nr	**7.32**
28 mm dia.	2.70	3.42	0.17	5.87	nr	**9.29**
Single pipe ring						
15 mm dia.	0.41	0.52	0.26	8.98	nr	**9.50**
22 mm dia.	0.49	0.62	0.26	8.98	nr	**9.60**
28 mm dia.	0.54	0.68	0.31	10.71	nr	**11.39**
35 mm dia.	0.73	0.92	0.32	11.05	nr	**11.97**
42 mm dia.	1.09	1.38	0.32	11.05	nr	**12.43**
54 mm dia.	1.67	2.12	0.34	11.74	nr	**13.86**
67 mm dia.	1.92	2.43	0.35	12.09	nr	**14.51**
76 mm dia.	2.25	2.84	0.42	14.50	nr	**17.35**
108 mm dia.	4.54	5.75	0.42	14.50	nr	**20.25**

38 PIPED SUPPLY SYSTEMS

Item	Net Price £	Material £	Labour hours	Labour £	Unit	Total rate £
COLD WATER PIPELINES: COPPER PIPEWORK – cont						
FIXINGS – cont						
Double pipe ring						
15 mm dia.	0.43	0.55	0.26	8.98	nr	**9.53**
22 mm dia.	0.78	0.99	0.26	8.98	nr	**9.97**
28 mm dia.	0.70	0.89	0.31	10.71	nr	**11.59**
35 mm dia.	0.75	0.95	0.32	11.05	nr	**12.00**
42 mm dia.	1.64	2.07	0.32	11.05	nr	**13.12**
54 mm dia.	1.76	2.22	0.34	11.74	nr	**13.96**
67 mm dia.	2.46	3.11	0.35	12.09	nr	**15.20**
76 mm dia.	3.23	4.09	0.42	14.50	nr	**18.59**
108 mm dia.	5.67	7.18	0.42	14.50	nr	**21.69**
Wall bracket						
15 mm dia.	3.84	4.86	0.05	1.73	nr	**6.58**
22 mm dia.	5.04	6.38	0.05	1.73	nr	**8.11**
28 mm dia.	6.01	7.61	0.05	1.73	nr	**9.34**
35 mm dia.	9.00	11.39	0.05	1.73	nr	**13.11**
42 mm dia.	11.92	15.09	0.05	1.73	nr	**16.82**
54 mm dia.	18.57	23.50	0.05	1.73	nr	**25.23**
Hospital bracket						
15 mm dia.	6.34	8.03	0.26	8.98	nr	**17.00**
22 mm dia.	7.51	9.51	0.26	8.98	nr	**18.48**
28 mm dia.	8.81	11.15	0.31	10.71	nr	**21.86**
35 mm dia.	9.32	11.80	0.32	11.05	nr	**22.85**
42 mm dia.	13.33	16.87	0.32	11.05	nr	**27.92**
54 mm dia.	17.83	22.57	0.34	11.74	nr	**34.31**
Screw on backplate, female						
All sizes 15 mm to 54 mm × 10 mm	2.02	2.56	0.10	3.45	nr	**6.02**
Screw on backplate, male						
All sizes 15 mm to 54 mm × 10 mm	2.71	3.44	0.10	3.45	nr	**6.89**
Pipe joist clips, single						
15 mm dia.	1.29	1.63	0.08	2.76	nr	**4.39**
22 mm dia.	1.29	1.63	0.08	2.76	nr	**4.39**
Pipe joist clips, double						
15 mm dia.	1.79	2.27	0.08	2.76	nr	**5.03**
22 mm dia.	0.97	1.23	0.08	2.76	nr	**3.99**
Extra over channel sections for fabricated hangers and brackets						
Galvanized steel; including inserts, bolts, nuts, washers; fixed to backgrounds						
41 × 21 mm	8.39	10.62	0.29	10.01	m	**20.64**
41 × 41 mm	10.06	12.74	0.29	10.01	m	**22.75**
Threaded rods; metric thread; including nuts, washers etc.						
10 mm dia. × 600 mm long for ring clips up to 54 mm	2.71	3.43	0.18	6.22	nr	**9.64**
12 mm dia. × 600 mm long for ring clips from 54 mm	4.24	5.37	0.18	6.22	nr	**11.58**

38 PIPED SUPPLY SYSTEMS

Item	Net Price £	Material £	Labour hours	Labour £	Unit	Total rate £
Extra over copper pipes; capillary fittings; BS 864						
Stop end						
15 mm dia.	1.88	2.38	0.13	4.49	nr	**6.87**
22 mm dia.	3.47	4.40	0.14	4.83	nr	**9.23**
28 mm dia.	6.24	7.89	0.17	5.87	nr	**13.76**
35 mm dia.	13.75	17.40	0.19	6.56	nr	**23.96**
42 mm dia.	23.69	29.99	0.22	7.60	nr	**37.58**
54 mm dia.	33.08	41.86	0.23	7.94	nr	**49.80**
Straight coupling; copper to copper						
6 mm dia.	1.97	2.49	0.23	7.94	nr	**10.43**
8 mm dia.	2.00	2.53	0.23	7.94	nr	**10.47**
10 mm dia.	1.03	1.30	0.23	7.94	nr	**9.25**
15 mm dia.	0.22	0.28	0.23	7.94	nr	**8.22**
22 mm dia.	0.59	0.74	0.26	8.98	nr	**9.72**
28 mm dia.	1.73	2.19	0.30	10.36	nr	**12.55**
35 mm dia.	5.63	7.12	0.34	11.74	nr	**18.86**
42 mm dia.	9.40	11.89	0.38	13.12	nr	**25.01**
54 mm dia.	17.34	21.94	0.42	14.50	nr	**36.45**
67 mm dia.	51.44	65.11	0.53	18.30	nr	**83.41**
Adaptor coupling; imperial to metric						
½" × 15 mm dia.	4.08	5.17	0.27	9.32	nr	**14.49**
¾" × 22 mm dia.	3.58	4.53	0.31	10.71	nr	**15.24**
1" × 28 mm dia.	7.01	8.87	0.36	12.43	nr	**21.30**
1 ¼" × 35 mm dia.	11.71	14.82	0.41	14.16	nr	**28.98**
1 ½" × 42 mm dia.	14.92	18.88	0.46	15.88	nr	**34.76**
Reducing coupling						
15 × 10 mm dia.	4.28	5.42	0.23	7.94	nr	**13.36**
22 × 10 mm dia.	6.28	7.95	0.26	8.98	nr	**16.93**
22 × 15 mm dia.	6.72	8.50	0.27	9.32	nr	**17.82**
28 × 15 mm dia.	7.72	9.77	0.28	9.67	nr	**19.44**
28 × 22 mm dia.	7.78	9.85	0.30	10.36	nr	**20.21**
35 × 28 mm dia.	11.16	14.13	0.34	11.74	nr	**25.87**
42 × 35 mm dia.	16.38	20.73	0.38	13.12	nr	**33.85**
54 × 35 mm dia.	28.70	36.32	0.42	14.50	nr	**50.83**
54 × 42 mm dia.	31.33	39.65	0.42	14.50	nr	**54.16**
Straight female connector						
15 mm × ½" dia.	4.26	5.39	0.27	9.32	nr	**14.71**
22 mm × ¾" dia.	6.15	7.79	0.31	10.71	nr	**18.49**
28 mm × 1" dia.	11.64	14.73	0.36	12.43	nr	**27.17**
35 mm × 1 ¼" dia.	20.14	25.48	0.41	14.16	nr	**39.64**
42 mm × 1 ½" dia.	26.11	33.05	0.46	15.88	nr	**48.94**
54 mm × 2" dia.	41.44	52.45	0.52	17.96	nr	**70.41**
Straight male connector						
15 mm × ½" dia.	3.63	4.59	0.27	9.32	nr	**13.91**
22 mm × ¾" dia.	6.48	8.20	0.31	10.71	nr	**18.91**
28 mm × 1" dia.	10.44	13.21	0.36	12.43	nr	**25.64**
35 mm × 1 ¼" dia.	18.35	23.22	0.41	14.16	nr	**37.38**
42 mm × 1 ½" dia.	23.63	29.91	0.46	15.88	nr	**45.80**
54 mm × 2" dia.	35.87	45.40	0.52	17.96	nr	**63.36**
67 mm × 2 ½" dia.	57.28	72.50	0.63	21.76	nr	**94.25**

38 PIPED SUPPLY SYSTEMS

Item	Net Price £	Material £	Labour hours	Labour £	Unit	Total rate £
COLD WATER PIPELINES: COPPER PIPEWORK – cont						
Extra over copper pipes – cont						
Female reducing connector						
15 mm × ¾" dia.	10.57	13.37	0.27	9.32	nr	**22.70**
Male reducing connector						
15 mm × ¾" dia.	9.45	11.96	0.27	9.32	nr	**21.29**
22 mm × 1" dia.	14.38	18.20	0.31	10.71	nr	**28.90**
Flanged connector						
28 mm dia.	67.54	85.48	0.36	12.43	nr	**97.92**
35 mm dia.	85.52	108.23	0.41	14.16	nr	**122.39**
42 mm dia.	102.20	129.34	0.46	15.88	nr	**145.23**
54 mm dia.	154.51	195.55	0.52	17.96	nr	**213.51**
67 mm dia.	181.69	229.95	0.61	21.06	nr	**251.01**
Tank connector						
15 mm × ½" dia.	9.06	11.46	0.25	8.63	nr	**20.09**
22 mm × ¾" dia.	13.78	17.44	0.28	9.67	nr	**27.11**
28 mm × 1" dia.	18.12	22.94	0.32	11.05	nr	**33.99**
35 mm × 1 ¼" dia.	23.25	29.42	0.37	12.78	nr	**42.20**
42 mm × 1 ½" dia.	30.48	38.57	0.43	14.85	nr	**53.42**
54 mm × 2" dia.	46.59	58.96	0.46	15.88	nr	**74.85**
Tank connector with long thread						
15 mm × ½" dia.	11.71	14.82	0.30	10.36	nr	**25.18**
22 mm × ¾" dia.	16.70	21.13	0.33	11.40	nr	**32.53**
28 mm × 1" dia.	20.65	26.14	0.39	13.47	nr	**39.60**
Reducer						
15 × 10 mm dia.	1.44	1.82	0.23	7.94	nr	**9.76**
22 × 15 mm dia.	1.02	1.29	0.26	8.98	nr	**10.27**
28 × 15 mm dia.	3.80	4.81	0.28	9.67	nr	**14.48**
28 × 22 mm dia.	2.91	3.69	0.30	10.36	nr	**14.05**
35 × 22 mm dia.	10.82	13.70	0.34	11.74	nr	**25.44**
42 × 22 mm dia.	19.55	24.74	0.36	12.43	nr	**37.18**
42 × 35 mm dia.	15.13	19.15	0.38	13.12	nr	**32.27**
54 × 35 mm dia.	31.75	40.19	0.40	13.81	nr	**54.00**
54 × 42 mm dia.	27.38	34.65	0.42	14.50	nr	**49.15**
67 × 54 mm dia.	37.23	47.12	0.53	18.30	nr	**65.42**
Adaptor; copper to female iron						
15 mm × ½" dia.	7.34	9.28	0.27	9.32	nr	**18.61**
22 mm × ¾" dia.	11.17	14.14	0.31	10.71	nr	**24.85**
28 mm × 1" dia.	15.74	19.92	0.36	12.43	nr	**32.35**
35 mm × 1 ¼" dia.	28.49	36.06	0.41	14.16	nr	**50.21**
42 mm × 1 ½" dia.	35.87	45.40	0.46	15.88	nr	**61.28**
54 mm × 2" dia.	43.15	54.61	0.52	17.96	nr	**72.57**
Adaptor; copper to male iron						
15 mm × ½" dia.	7.48	9.46	0.27	9.32	nr	**18.79**
22 mm × ¾" dia.	9.57	12.11	0.31	10.71	nr	**22.82**
28 mm × 1" dia.	15.99	20.24	0.36	12.43	nr	**32.67**
35 mm × 1 ¼" dia.	23.26	29.44	0.41	14.16	nr	**43.60**
42 mm × 1 ½" dia.	32.14	40.68	0.46	15.88	nr	**56.56**
54 mm × 2" dia.	43.15	54.61	0.52	17.96	nr	**72.57**

38 PIPED SUPPLY SYSTEMS

Item	Net Price £	Material £	Labour hours	Labour £	Unit	Total rate £
Union coupling						
15 mm dia.	10.11	12.79	0.41	14.16	nr	**26.95**
22 mm dia.	16.19	20.49	0.45	15.54	nr	**36.03**
28 mm dia.	23.63	29.91	0.51	17.61	nr	**47.52**
35 mm dia.	31.02	39.25	0.64	22.10	nr	**61.36**
42 mm dia.	45.30	57.33	0.68	23.48	nr	**80.82**
54 mm dia.	86.22	109.12	0.78	26.94	nr	**136.05**
67 mm dia.	145.99	184.77	0.96	33.15	nr	**217.92**
Elbow						
15 mm dia.	0.40	0.50	0.23	7.94	nr	**8.45**
22 mm dia.	1.05	1.33	0.26	8.98	nr	**10.31**
28 mm dia.	2.77	3.51	0.31	10.71	nr	**14.21**
35 mm dia.	12.05	15.25	0.35	12.09	nr	**27.34**
42 mm dia.	19.90	25.19	0.41	14.16	nr	**39.35**
54 mm dia.	41.11	52.03	0.44	15.19	nr	**67.23**
67 mm dia.	106.72	135.06	0.54	18.65	nr	**153.71**
Backplate elbow						
15 mm dia.	7.59	9.61	0.51	17.61	nr	**27.22**
22 mm dia.	16.33	20.67	0.54	18.65	nr	**39.32**
Overflow bend						
22 mm dia.	23.00	29.11	0.26	8.98	nr	**38.09**
Return bend						
15 mm dia.	11.38	14.41	0.23	7.94	nr	**22.35**
22 mm dia.	22.37	28.31	0.26	8.98	nr	**37.29**
28 mm dia.	28.58	36.17	0.31	10.71	nr	**46.88**
Obtuse elbow						
15 mm dia.	1.47	1.87	0.23	7.94	nr	**9.81**
22 mm dia.	3.04	3.85	0.26	8.98	nr	**12.83**
28 mm dia.	5.84	7.39	0.31	10.71	nr	**18.09**
35 mm dia.	18.19	23.03	0.36	12.43	nr	**35.46**
42 mm dia.	32.39	40.99	0.41	14.16	nr	**55.15**
54 mm dia.	58.54	74.08	0.44	15.19	nr	**89.28**
67 mm dia.	106.21	134.42	0.54	18.65	nr	**153.07**
Straight tap connector						
15 mm × ½" dia.	1.95	2.47	0.13	4.49	nr	**6.96**
22 mm × ¾" dia.	2.90	3.67	0.14	4.83	nr	**8.51**
Bent tap connector						
15 mm × ½" dia.	2.89	3.66	0.13	4.49	nr	**8.15**
22 mm × ¾" dia.	8.87	11.22	0.14	4.83	nr	**16.06**
Bent male union connector						
15 mm × ½" dia.	14.82	18.76	0.41	14.16	nr	**32.92**
22 mm × ¾" dia.	19.23	24.34	0.45	15.54	nr	**39.88**
28 mm × 1" dia.	27.55	34.87	0.51	17.61	nr	**52.48**
35 mm × 1 ¼" dia.	44.94	56.88	0.64	22.10	nr	**78.98**
42 mm × 1 ½" dia.	73.04	92.44	0.68	23.48	nr	**115.93**
54 mm × 2" dia.	115.37	146.02	0.78	26.94	nr	**172.95**
Bent female union connector						
15 mm dia.	14.82	18.76	0.41	14.16	nr	**32.92**
22 mm × ¾" dia.	19.23	24.34	0.45	15.54	nr	**39.88**
28 mm × 1" dia.	27.55	34.87	0.51	17.61	nr	**52.48**

38 PIPED SUPPLY SYSTEMS

Item	Net Price £	Material £	Labour hours	Labour £	Unit	Total rate £
COLD WATER PIPELINES: COPPER PIPEWORK – cont						
Extra over copper pipes – cont						
Bent female union connector – cont						
35 mm × 1 ¼" dia.	44.94	56.88	0.64	22.10	nr	**78.98**
42 mm × 1 ½" dia.	73.04	92.44	0.68	23.48	nr	**115.93**
54 mm × 2" dia.	115.37	146.02	0.78	26.94	nr	**172.95**
Straight union adaptor						
15 mm × ¾" dia.	6.33	8.01	0.41	14.16	nr	**22.17**
22 mm × 1" dia.	8.99	11.37	0.45	15.54	nr	**26.91**
28 mm × 1 ¼" dia.	14.53	18.39	0.51	17.61	nr	**36.00**
35 mm × 1 ½" dia.	22.37	28.31	0.64	22.10	nr	**50.41**
42 mm × 2" dia.	28.24	35.75	0.68	23.48	nr	**59.23**
54 mm × 2 ½" dia.	43.61	55.19	0.78	26.94	nr	**82.12**
Straight male union connector						
15 mm × ½" dia.	12.62	15.98	0.41	14.16	nr	**30.14**
22 mm × ¾" dia.	16.38	20.73	0.45	15.54	nr	**36.27**
28 mm × 1" dia.	24.38	30.86	0.51	17.61	nr	**48.47**
35 mm × 1 ¼" dia.	35.12	44.45	0.64	22.10	nr	**66.55**
42 mm × 1 ½" dia.	55.19	69.85	0.68	23.48	nr	**93.33**
54 mm × 2" dia.	79.30	100.37	0.78	26.94	nr	**127.30**
Straight female union connector						
15 mm × ½" dia.	12.62	15.98	0.41	14.16	nr	**30.14**
22 mm × ¾" dia.	16.38	20.73	0.45	15.54	nr	**36.27**
28 mm × 1" dia.	24.38	30.86	0.51	17.61	nr	**48.47**
35 mm × 1 ¼" dia.	35.12	44.45	0.64	22.10	nr	**66.55**
42 mm × 1 ½" dia.	55.19	69.85	0.68	23.48	nr	**93.33**
54 mm × 2" dia.	79.30	100.37	0.78	26.94	nr	**127.30**
Male nipple						
¾ × ½" dia.	2.97	3.76	0.24	8.19	nr	**11.95**
1 × ¾" dia.	3.58	4.53	0.32	11.05	nr	**15.58**
1 ¼ × 1" dia.	3.92	4.96	0.37	12.78	nr	**17.74**
1 ½ × 1 ¼" dia.	14.58	18.45	0.42	14.50	nr	**32.95**
2 × 1 ½" dia.	29.84	37.76	0.46	15.88	nr	**53.64**
2 ½ × 2" dia.	38.87	49.19	0.56	19.34	nr	**68.53**
Female nipple						
¾ × ½" dia.	4.61	5.83	0.19	6.56	nr	**12.40**
1 × ¾" dia.	7.21	9.12	0.32	11.05	nr	**20.17**
1 ¼ × 1" dia.	9.85	12.47	0.37	12.78	nr	**25.25**
1 ½ × 1 ¼" dia.	14.58	18.45	0.42	14.50	nr	**32.95**
2 × 1 ½" dia.	29.84	37.76	0.46	15.88	nr	**53.64**
2 ½ × 2" dia.	38.87	49.19	0.56	19.34	nr	**68.53**
Equal tee						
10 mm dia.	3.95	5.00	0.25	8.63	nr	**13.64**
15 mm dia.	0.40	0.50	0.36	12.43	nr	**12.94**
22 mm dia.	1.05	1.33	0.39	13.47	nr	**14.80**
28 mm dia.	7.70	9.74	0.43	14.85	nr	**24.59**
35 mm dia.	19.62	24.83	0.57	19.68	nr	**44.52**
42 mm dia.	31.47	39.83	0.60	20.72	nr	**60.55**
54 mm dia.	63.46	80.32	0.65	22.45	nr	**102.76**
67 mm dia.	86.35	109.28	0.78	26.94	nr	**136.21**

38 PIPED SUPPLY SYSTEMS

Item	Net Price £	Material £	Labour hours	Labour £	Unit	Total rate £
Female tee, reducing branch FI						
15 × 15 mm × ¼" dia.	9.50	12.02	0.36	12.43	nr	**24.46**
22 × 22 mm × ½" dia.	11.62	14.70	0.39	13.48	nr	**28.19**
28 × 28 mm × ¾" dia.	22.76	28.80	0.43	14.85	nr	**43.65**
35 × 35 mm × ¾" dia.	32.85	41.58	0.47	16.23	nr	**57.81**
42 × 42 mm × ½" dia.	39.46	49.95	0.60	20.72	nr	**70.67**
Backplate tee						
15 × 15 mm × ½" dia.	17.96	22.73	0.62	21.41	nr	**44.14**
Heater tee						
½ × ½" × 15 mm dia.	16.15	20.43	0.36	12.43	nr	**32.87**
Union heater tee						
½ × ½" × 15 mm dia.	16.15	20.43	0.36	12.43	nr	**32.87**
Sweep tee – equal						
15 mm dia.	12.83	16.24	0.36	12.43	nr	**28.68**
22 mm dia.	16.52	20.91	0.39	13.47	nr	**34.38**
28 mm dia.	27.81	35.20	0.43	14.85	nr	**50.05**
35 mm dia.	39.44	49.92	0.57	19.68	nr	**69.60**
42 mm dia.	58.49	74.02	0.60	20.72	nr	**94.74**
54 mm dia.	64.77	81.97	0.65	22.45	nr	**104.42**
67 mm dia.	88.30	111.75	0.78	26.94	nr	**138.69**
Sweep tee – reducing						
22 × 22 × 15 mm dia.	13.84	17.52	0.39	13.47	nr	**30.98**
28 × 28 × 22 mm dia.	23.54	29.79	0.43	14.85	nr	**44.64**
35 × 35 × 22 mm dia.	39.44	49.92	0.57	19.68	nr	**69.60**
Sweep tee – double						
15 mm dia.	14.53	18.39	0.36	12.43	nr	**30.82**
22 mm dia.	19.77	25.02	0.39	13.47	nr	**38.49**
28 mm dia.	30.06	38.04	0.43	14.85	nr	**52.89**
Cross						
15 mm dia.	19.22	24.33	0.48	16.58	nr	**40.90**
22 mm dia.	21.46	27.16	0.53	18.30	nr	**45.46**
28 mm dia.	30.82	39.00	0.61	21.06	nr	**60.07**
Extra over copper pipes; high duty capillary fittings; BS 864						
Stop end						
15 mm dia.	9.00	11.39	0.16	5.53	nr	**16.91**
Straight coupling; copper to copper						
15 mm dia.	4.11	5.20	0.27	9.32	nr	**14.52**
22 mm dia.	6.60	8.35	0.32	11.05	nr	**19.40**
28 mm dia.	8.82	11.16	0.37	12.78	nr	**23.94**
35 mm dia.	16.47	20.85	0.43	14.85	nr	**35.70**
42 mm dia.	18.02	22.80	0.50	17.27	nr	**40.07**
54 mm dia.	26.50	33.54	0.54	18.65	nr	**52.19**
Reducing coupling						
15 × 12 mm dia.	7.76	9.82	0.27	9.32	nr	**19.14**
22 × 15 mm dia.	9.00	11.39	0.32	11.05	nr	**22.44**
28 × 22 mm dia.	12.39	15.68	0.37	12.78	nr	**28.46**

38 PIPED SUPPLY SYSTEMS

Item	Net Price £	Material £	Labour hours	Labour £	Unit	Total rate £
COLD WATER PIPELINES: COPPER PIPEWORK – cont						
Extra over copper pipes – cont						
Straight female connector						
15 mm × ½" dia.	10.11	12.79	0.32	11.05	nr	23.84
22 mm × ¾" dia.	11.42	14.45	0.36	12.43	nr	26.88
28 mm × 1" dia.	16.85	21.32	0.42	14.50	nr	35.83
Straight male connector						
15 mm × ½" dia.	9.85	12.47	0.32	11.05	nr	23.52
22 mm × ¾" dia.	11.42	14.45	0.36	12.43	nr	26.88
28 mm × 1" dia.	16.85	21.32	0.42	14.50	nr	35.83
42 mm × 1 ½" dia.	32.87	41.59	0.53	18.30	nr	59.90
54 mm × 2" dia.	53.42	67.61	0.62	21.41	nr	89.02
Reducer						
15 × 12 mm dia.	5.10	6.46	0.27	9.32	nr	15.78
22 × 15 mm dia.	4.97	6.29	0.32	11.05	nr	17.34
28 × 22 mm dia.	9.00	11.39	0.37	12.78	nr	24.16
35 × 28 mm dia.	11.42	14.45	0.43	14.85	nr	29.30
42 × 35 mm dia.	14.72	18.63	0.50	17.27	nr	35.89
54 × 42 mm dia.	23.72	30.01	0.39	13.47	nr	43.48
Straight union adaptor						
15 mm × ¾" dia.	8.23	10.41	0.27	9.32	nr	19.73
22 mm × 1" dia.	11.15	14.11	0.32	11.05	nr	25.16
28 mm × 1 ¼" dia.	14.72	18.63	0.37	12.78	nr	31.40
35 mm × 1 ½" dia.	26.68	33.76	0.43	14.85	nr	48.61
42 mm × 2" dia.	33.80	42.78	0.50	17.27	nr	60.05
Bent union adaptor						
15 mm × ¾" dia.	21.39	27.07	0.27	9.32	nr	36.39
22 mm × 1" dia.	28.88	36.54	0.32	11.05	nr	47.60
28 mm × 1 ¼" dia.	38.87	49.19	0.37	12.78	nr	61.97
Adaptor; male copper to FI						
15 mm × ½" dia.	9.63	12.19	0.27	9.32	nr	21.51
22 mm × ¾" dia.	16.52	20.91	0.32	11.05	nr	31.96
Union coupling						
15 mm dia.	18.54	23.47	0.54	18.65	nr	42.12
22 mm dia.	23.72	30.01	0.60	20.72	nr	50.73
28 mm dia.	32.96	41.71	0.68	23.48	nr	65.19
35 mm dia.	57.52	72.79	0.83	28.66	nr	101.46
42 mm dia.	67.74	85.74	0.89	30.73	nr	116.47
Elbow						
15 mm dia.	11.93	15.10	0.27	9.32	nr	24.43
22 mm dia.	12.76	16.16	0.32	11.05	nr	27.21
28 mm dia.	18.95	23.99	0.37	12.78	nr	36.77
35 mm dia.	29.64	37.51	0.43	14.85	nr	52.36
42 mm dia.	36.91	46.72	0.50	17.27	nr	63.98
54 mm dia.	64.20	81.25	0.52	17.96	nr	99.21
Return bend						
28 mm dia.	28.58	36.17	0.37	12.78	nr	48.95
35 mm dia.	35.23	44.59	0.43	14.85	nr	59.43

38 PIPED SUPPLY SYSTEMS

Item	Net Price £	Material £	Labour hours	Labour £	Unit	Total rate £
Bent male union connector						
15 mm × ½" dia.	27.68	35.03	0.54	18.65	nr	53.68
22 mm × ¾" dia.	37.29	47.19	0.60	20.72	nr	67.91
28 mm × 1" dia.	67.74	85.74	0.68	23.48	nr	109.22
Composite flange						
35 mm dia.	5.58	7.06	0.38	13.13	nr	20.19
42 mm dia.	7.25	9.18	0.41	14.16	nr	23.34
54 mm dia.	9.25	11.71	0.43	14.85	nr	26.57
Equal tee						
15 mm dia.	13.71	17.35	0.44	15.19	nr	32.55
22 mm dia.	17.26	21.84	0.47	16.23	nr	38.07
28 mm dia.	22.74	28.79	0.53	18.30	nr	47.09
35 mm dia.	38.87	49.19	0.70	24.17	nr	73.36
42 mm dia.	49.49	62.64	0.84	29.01	nr	91.64
54 mm dia.	77.92	98.62	0.79	27.28	nr	125.90
Reducing tee						
15 × 12 mm dia.	18.80	23.80	0.44	15.19	nr	38.99
22 × 15 mm dia.	22.21	28.10	0.47	16.23	nr	44.34
28 × 22 mm dia.	31.73	40.16	0.53	18.30	nr	58.46
35 × 28 mm dia.	50.30	63.66	0.73	25.21	nr	88.87
42 × 28 mm dia.	64.39	81.49	0.84	29.01	nr	110.49
54 × 28 mm dia.	101.68	128.69	1.01	34.88	nr	163.57
Extra over copper pipes; compression fittings; BS 864						
Stop end						
15 mm dia.	1.94	2.45	0.10	3.45	nr	5.90
22 mm dia.	2.30	2.91	0.12	4.14	nr	7.06
28 mm dia.	2.63	3.33	0.15	5.18	nr	8.51
Straight connector; copper to copper						
15 mm dia.	4.01	5.07	0.18	6.22	nr	11.29
22 mm dia.	5.31	6.72	0.21	7.25	nr	13.97
28 mm dia.	7.06	8.94	0.24	8.29	nr	17.23
Straight connector; copper to imperial copper						
22 mm dia.	6.61	8.37	0.21	7.25	nr	15.62
Male coupling; copper to MI (BSP)						
15 mm dia.	1.14	1.44	0.19	6.56	nr	8.00
22 mm dia.	1.80	2.27	0.23	7.94	nr	10.21
28 mm dia.	4.17	5.28	0.26	8.98	nr	14.26
Male coupling with long thread and backnut						
15 mm dia.	7.34	9.29	0.19	6.56	nr	15.85
22 mm dia.	9.32	11.80	0.23	7.94	nr	19.74
Female coupling; copper to FI (BSP)						
15 mm dia.	1.41	1.78	0.19	6.56	nr	8.34
22 mm dia.	2.01	2.54	0.23	7.94	nr	10.49
28 mm dia.	5.83	7.37	0.27	9.32	nr	16.70
Elbow						
15 mm dia.	1.49	1.89	0.18	6.22	nr	8.11
22 mm dia.	2.56	3.24	0.21	7.25	nr	10.49
28 mm dia.	8.62	10.91	0.24	8.29	nr	19.20

Material Costs/Prices for Measured Works

38 PIPED SUPPLY SYSTEMS

Item	Net Price £	Material £	Labour hours	Labour £	Unit	Total rate £
COLD WATER PIPELINES: COPPER PIPEWORK – cont						
Extra over copper pipes – cont						
Male elbow; copper to FI (BSP)						
15 mm × ½" dia.	2.88	3.65	0.19	6.56	nr	**10.21**
22 mm × ¾" dia.	3.74	4.73	0.23	7.94	nr	**12.68**
28 mm × 1" dia.	9.00	11.39	0.27	9.32	nr	**20.71**
Female elbow; copper to FI (BSP)						
15 mm × ½" dia.	4.41	5.58	0.19	6.56	nr	**12.14**
22 mm × ¾" dia.	6.37	8.07	0.23	7.94	nr	**16.01**
28 mm × 1" dia.	11.23	14.22	0.27	9.32	nr	**23.54**
Backplate elbow						
15 mm × ½" dia.	6.37	8.07	0.50	17.27	nr	**25.33**
Tank coupling; long thread						
22 mm dia.	9.32	11.80	0.46	15.88	nr	**27.68**
Tee equal						
15 mm dia.	2.13	2.69	0.28	9.67	nr	**12.36**
22 mm dia.	3.56	4.50	0.30	10.36	nr	**14.86**
28 mm dia.	16.23	20.54	0.34	11.74	nr	**32.28**
Tee reducing						
22 mm dia.	9.18	11.62	0.30	10.36	nr	**21.98**
Backplate tee						
15 mm dia.	22.42	28.38	0.62	21.41	nr	**49.79**
Extra over fittings; silver brazed welded joints						
Reducer						
76 × 67 mm dia.	37.00	46.83	1.40	48.35	nr	**95.17**
108 × 76 mm dia.	77.56	98.16	1.80	62.16	nr	**160.32**
133 × 108 mm dia.	154.66	195.74	2.20	75.97	nr	**271.71**
159 × 133 mm dia.	197.80	250.34	2.60	89.78	nr	**340.12**
90° elbow						
76 mm dia.	89.40	113.14	1.60	55.25	nr	**168.39**
108 mm dia.	164.58	208.30	2.00	69.07	nr	**277.36**
133 mm dia.	347.91	440.32	2.40	82.88	nr	**523.20**
159 mm dia.	431.92	546.64	2.80	96.69	nr	**643.33**
45° elbow						
76 mm dia.	83.77	106.03	1.60	55.25	nr	**161.28**
108 mm dia.	131.38	166.27	2.00	69.07	nr	**235.33**
133 mm dia.	335.94	425.16	2.40	82.88	nr	**508.04**
159 mm dia.	483.55	611.98	2.80	96.69	nr	**708.67**
Equal tee						
76 mm dia.	108.88	137.79	2.40	82.88	nr	**220.67**
108 mm dia.	169.13	214.05	3.00	103.60	nr	**317.65**
133 mm dia.	398.30	504.09	3.60	124.32	nr	**628.40**
159 mm dia.	475.03	601.20	4.20	145.04	nr	**746.24**

38 PIPED SUPPLY SYSTEMS

Item	Net Price £	Material £	Labour hours	Labour £	Unit	Total rate £
Extra over copper pipes; dezincification resistant compression fittings; BS 864						
Stop end						
15 mm dia.	2.77	3.51	0.10	3.45	nr	**6.96**
22 mm dia.	4.01	5.07	0.13	4.49	nr	**9.56**
28 mm dia.	8.59	10.87	0.15	5.18	nr	**16.05**
35 mm dia.	13.45	17.02	0.18	6.22	nr	**23.24**
42 mm dia.	22.42	28.38	0.20	6.91	nr	**35.29**
Straight coupling; copper to copper						
15 mm dia.	2.21	2.80	0.18	6.22	nr	**9.02**
22 mm dia.	3.63	4.60	0.21	7.25	nr	**11.85**
28 mm dia.	8.21	10.39	0.24	8.29	nr	**18.68**
35 mm dia.	17.38	22.00	0.29	10.01	nr	**32.01**
42 mm dia.	22.85	28.92	0.33	11.40	nr	**40.32**
54 mm dia.	34.17	43.25	0.38	13.12	nr	**56.37**
Straight swivel connector; copper to imperial copper						
22 mm dia.	8.18	10.35	0.20	6.91	nr	**17.26**
Male coupling; copper to MI (BSP)						
15 mm × ½" dia.	1.98	2.50	0.19	6.56	nr	**9.06**
22 mm × ¾" dia.	3.00	3.80	0.23	7.94	nr	**11.74**
28 mm × 1" dia.	5.82	7.36	0.26	8.98	nr	**16.34**
35 mm × 1 ¼" dia.	13.20	16.71	0.32	11.05	nr	**27.76**
42 mm × 1 ½" dia.	19.82	25.09	0.37	12.78	nr	**37.87**
54 mm × 2" dia.	29.24	37.01	0.57	19.68	nr	**56.69**
Male coupling with long thread and backnuts						
22 mm dia.	11.23	14.22	0.23	7.94	nr	**22.16**
28 mm dia.	12.45	15.75	0.24	8.29	nr	**24.04**
Female coupling; copper to FI (BSP)						
15 mm × ½" dia.	2.37	2.99	0.19	6.56	nr	**9.55**
22 mm × ¾" dia.	3.46	4.38	0.23	7.94	nr	**12.32**
28 mm × 1" dia.	7.54	9.54	0.27	9.32	nr	**18.86**
35 mm × 1 ¼" dia.	15.88	20.09	0.32	11.05	nr	**31.15**
42 mm × 1 ½" dia.	21.32	26.98	0.37	12.78	nr	**39.76**
54 mm × 2" dia.	31.28	39.59	0.42	14.50	nr	**54.09**
Elbow						
15 mm dia.	2.67	3.37	0.18	6.22	nr	**9.59**
22 mm dia.	4.27	5.40	0.21	7.25	nr	**12.65**
28 mm dia.	10.60	13.41	0.24	8.29	nr	**21.70**
35 mm dia.	23.46	29.69	0.29	10.01	nr	**39.70**
42 mm dia.	31.77	40.20	0.33	11.40	nr	**51.60**
54 mm dia.	54.67	69.20	0.38	13.12	nr	**82.32**
Male elbow; copper to MI (BSP)						
15 mm × ½" dia.	4.63	5.86	0.19	6.56	nr	**12.43**
22 mm × ¾" dia.	5.18	6.56	0.23	7.94	nr	**14.50**
28 mm × 1" dia.	9.71	12.29	0.27	9.32	nr	**21.61**
Female elbow; copper to FI (BSP)						
15 mm × ½" dia.	4.95	6.26	0.19	6.56	nr	**12.82**
22 mm × ¾" dia.	7.14	9.03	0.23	7.94	nr	**16.98**
28 mm × 1" dia.	11.84	14.98	0.27	9.32	nr	**24.30**

38 PIPED SUPPLY SYSTEMS

Item	Net Price £	Material £	Labour hours	Labour £	Unit	Total rate £
COLD WATER PIPELINES: COPPER PIPEWORK – cont						
Extra over copper pipes – cont						
Backplate elbow						
15 mm × ½" dia.	7.22	9.14	0.50	17.27	nr	**26.41**
Straight tap connector						
15 mm dia.	4.10	5.18	0.13	4.49	nr	**9.67**
22 mm dia.	8.88	11.24	0.15	5.18	nr	**16.42**
Tank coupling						
15 mm dia.	5.64	7.14	0.19	6.56	nr	**13.70**
22 mm dia.	6.24	7.89	0.23	7.94	nr	**15.83**
28 mm dia.	13.18	16.68	0.27	9.32	nr	**26.00**
35 mm dia.	23.31	29.50	0.32	11.05	nr	**40.55**
42 mm dia.	37.87	47.93	0.37	12.78	nr	**60.71**
54 mm dia.	48.78	61.74	0.31	10.71	nr	**72.45**
Tee equal						
15 mm dia.	3.75	4.75	0.28	9.67	nr	**14.42**
22 mm dia.	6.19	7.84	0.30	10.36	nr	**18.20**
28 mm dia.	16.90	21.39	0.34	11.74	nr	**33.13**
35 mm dia.	30.51	38.61	0.43	14.85	nr	**53.46**
42 mm dia.	47.98	60.72	0.46	15.88	nr	**76.61**
54 mm dia.	77.07	97.54	0.54	18.65	nr	**116.18**
Tee reducing						
22 mm dia.	9.90	12.53	0.30	10.36	nr	**22.89**
28 mm dia.	16.33	20.67	0.34	11.74	nr	**32.41**
35 mm dia.	29.81	37.73	0.43	14.85	nr	**52.58**
42 mm dia.	46.09	58.33	0.46	15.88	nr	**74.21**
54 mm dia.	77.07	97.54	0.54	18.65	nr	**116.18**
Extra over copper pipes; bronze one piece brazing flanges; metric, including jointing ring and bolts						
Bronze flange; PN6						
15 mm dia.	40.41	51.14	0.27	9.32	nr	**60.46**
22 mm dia.	48.14	60.93	0.32	11.05	nr	**71.98**
28 mm dia.	52.89	66.94	0.36	12.43	nr	**79.38**
35 mm dia.	63.19	79.97	0.47	16.23	nr	**96.20**
42 mm dia.	68.94	87.25	0.54	18.65	nr	**105.90**
54 mm dia.	79.97	101.21	0.63	21.76	nr	**122.96**
67 mm dia.	94.27	119.31	0.77	26.59	nr	**145.90**
76 mm dia.	113.61	143.79	0.93	32.12	nr	**175.91**
108 mm dia.	150.11	189.98	1.14	39.37	nr	**229.35**
133 mm dia.	205.69	260.32	1.41	48.69	nr	**309.01**
159 mm dia.	296.73	375.54	1.74	60.09	nr	**435.63**
Bronze flange; PN10						
15 mm dia.	46.26	58.55	0.27	9.32	nr	**67.87**
22 mm dia.	49.24	62.31	0.32	11.05	nr	**73.36**
28 mm dia.	54.15	68.53	0.38	13.12	nr	**81.65**

38 PIPED SUPPLY SYSTEMS

Item	Net Price £	Material £	Labour hours	Labour £	Unit	Total rate £
35 mm dia.	65.47	82.86	0.47	16.23	nr	**99.09**
42 mm dia.	73.38	92.87	0.54	18.65	nr	**111.52**
54 mm dia.	80.30	101.63	0.63	21.76	nr	**123.39**
67 mm dia.	94.35	119.41	0.77	26.59	nr	**146.00**
76 mm dia.	113.61	143.79	0.93	32.12	nr	**175.91**
108 mm dia.	143.59	181.73	1.14	39.37	nr	**221.10**
133 mm dia.	191.72	242.64	1.41	48.69	nr	**291.33**
159 mm dia.	243.02	307.56	1.74	60.09	nr	**367.65**
Bronze flange; PN16						
15 mm dia.	50.90	64.42	0.27	9.32	nr	**73.74**
22 mm dia.	54.18	68.58	0.32	11.05	nr	**79.63**
28 mm dia.	59.57	75.39	0.38	13.12	nr	**88.51**
35 mm dia.	72.02	91.15	0.47	16.23	nr	**107.38**
42 mm dia.	80.74	102.18	0.54	18.65	nr	**120.83**
54 mm dia.	88.31	111.77	0.63	21.76	nr	**133.53**
67 mm dia.	103.78	131.34	0.77	26.59	nr	**157.93**
76 mm dia.	124.96	158.15	0.93	32.12	nr	**190.27**
108 mm dia.	157.95	199.90	1.14	39.37	nr	**239.27**
133 mm dia.	210.90	266.91	1.41	48.69	nr	**315.60**
159 mm dia.	267.30	338.30	1.74	60.09	nr	**398.38**
Extra over copper pipes; bronze blank flanges; metric, including jointing ring and bolts						
Gunmetal blank flange; PN6						
15 mm dia.	33.92	42.93	0.27	9.32	nr	**52.25**
22 mm dia.	43.27	54.76	0.27	9.32	nr	**64.09**
28 mm dia.	44.47	56.29	0.27	9.32	nr	**65.61**
35 mm dia.	72.79	92.12	0.32	11.05	nr	**103.17**
42 mm dia.	98.20	124.28	0.32	11.05	nr	**135.33**
54 mm dia.	108.27	137.03	0.34	11.74	nr	**148.77**
67 mm dia.	133.22	168.61	0.36	12.43	nr	**181.04**
76 mm dia.	171.62	217.20	0.37	12.78	nr	**229.98**
108 mm dia.	272.67	345.09	0.41	14.16	nr	**359.25**
133 mm dia.	321.66	407.09	0.58	20.03	nr	**427.12**
159 mm dia.	402.05	508.83	0.61	21.06	nr	**529.90**
Gunmetal blank flange; PN10						
15 mm dia.	41.01	51.91	0.27	9.32	nr	**61.23**
22 mm dia.	53.08	67.18	0.27	9.32	nr	**76.50**
28 mm dia.	58.76	74.37	0.27	9.32	nr	**83.69**
35 mm dia.	72.79	92.12	0.32	11.05	nr	**103.17**
42 mm dia.	136.42	172.66	0.32	11.05	nr	**183.71**
54 mm dia.	155.53	196.84	0.34	11.74	nr	**208.58**
67 mm dia.	166.84	211.16	0.46	15.88	nr	**227.04**
76 mm dia.	221.72	280.61	0.47	16.23	nr	**296.84**
108 mm dia.	260.21	329.32	0.51	17.61	nr	**346.93**
133 mm dia.	345.68	437.49	0.58	20.03	nr	**457.52**
159 mm dia.	470.78	595.82	0.71	24.52	nr	**620.34**

38 PIPED SUPPLY SYSTEMS

Item	Net Price £	Material £	Labour hours	Labour £	Unit	Total rate £
COLD WATER PIPELINES: COPPER PIPEWORK – cont						
Extra over copper pipes – cont						
Gunmetal blank flange; PN16						
15 mm dia.	41.01	51.91	0.27	9.32	nr	**61.23**
22 mm dia.	53.89	68.20	0.27	9.32	nr	**77.52**
28 mm dia.	58.76	74.37	0.27	9.32	nr	**83.69**
35 mm dia.	72.79	92.12	0.32	11.05	nr	**103.17**
42 mm dia.	136.42	172.66	0.32	11.05	nr	**183.71**
54 mm dia.	155.53	196.84	0.34	11.74	nr	**208.58**
67 mm dia.	193.79	245.27	0.46	15.88	nr	**261.15**
76 mm dia.	221.72	280.61	0.47	16.23	nr	**296.84**
108 mm dia.	260.21	329.32	0.51	17.61	nr	**346.93**
133 mm dia.	597.65	756.39	0.58	20.03	nr	**776.42**
159 mm dia.	713.87	903.48	0.71	24.52	nr	**927.99**
Extra over copper pipes; bronze screwed flanges; metric, including jointing ring and bolts						
Gunmetal screwed flange; 6 BSP						
15 mm dia.	33.92	42.93	0.35	12.09	nr	**55.01**
22 mm dia.	39.20	49.61	0.47	16.23	nr	**65.85**
28 mm dia.	40.86	51.72	0.52	17.96	nr	**69.68**
35 mm dia.	55.09	69.72	0.62	21.41	nr	**91.13**
42 mm dia.	66.13	83.69	0.70	24.17	nr	**107.86**
54 mm dia.	89.88	113.75	0.84	29.01	nr	**142.76**
67 mm dia.	112.78	142.74	1.03	35.57	nr	**178.31**
76 mm dia.	136.14	172.29	1.22	42.13	nr	**214.42**
108 mm dia.	215.60	272.86	1.41	48.69	nr	**321.55**
133 mm dia.	255.58	323.46	1.75	60.43	nr	**383.89**
159 mm dia.	325.34	411.75	2.21	76.32	nr	**488.07**
Gunmetal screwed flange; 10 BSP						
15 mm dia.	41.08	51.99	0.35	12.09	nr	**64.07**
22 mm dia.	47.83	60.54	0.47	16.23	nr	**76.77**
28 mm dia.	52.82	66.85	0.52	17.96	nr	**84.81**
35 mm dia.	75.28	95.27	0.62	21.41	nr	**116.68**
42 mm dia.	91.92	116.34	0.70	24.17	nr	**140.51**
54 mm dia.	130.70	165.42	0.84	29.01	nr	**194.43**
67 mm dia.	153.02	193.66	1.03	35.57	nr	**229.23**
76 mm dia.	172.74	218.61	1.22	42.13	nr	**260.74**
108 mm dia.	229.29	290.19	1.41	48.69	nr	**338.88**
133 mm dia.	277.89	351.70	1.75	60.43	nr	**412.13**
159 mm dia.	493.13	624.10	2.21	76.32	nr	**700.42**
Gunmetal screwed flange; 16 BSP						
15 mm dia.	32.27	40.84	0.35	12.09	nr	**52.93**
22 mm dia.	40.80	51.64	0.47	16.23	nr	**67.87**
28 mm dia.	47.23	59.77	0.52	17.96	nr	**77.73**
35 mm dia.	72.18	91.35	0.62	21.41	nr	**112.76**
42 mm dia.	86.88	109.95	0.70	24.17	nr	**134.12**
54 mm dia.	111.73	141.40	0.84	29.01	nr	**170.41**
67 mm dia.	161.50	204.39	1.03	35.57	nr	**239.96**

38 PIPED SUPPLY SYSTEMS

Item	Net Price £	Material £	Labour hours	Labour £	Unit	Total rate £
76 mm dia.	185.20	234.39	1.22	42.13	nr	**276.52**
108 mm dia.	216.70	274.26	1.41	48.69	nr	**322.95**
133 mm dia.	351.14	444.40	1.75	60.43	nr	**504.83**
159 mm dia.	434.33	549.69	2.21	76.32	nr	**626.01**
Extra over copper pipes; labour						
Made bend						
15 mm dia.	–	–	0.26	8.98	nr	**8.98**
22 mm dia.	–	–	0.28	9.67	nr	**9.67**
28 mm dia.	–	–	0.31	10.71	nr	**10.71**
35 mm dia.	–	–	0.42	14.50	nr	**14.50**
42 mm dia.	–	–	0.51	17.61	nr	**17.61**
54 mm dia.	–	–	0.58	20.03	nr	**20.03**
67 mm dia.	–	–	0.69	23.83	nr	**23.83**
76 mm dia.	–	–	0.80	27.63	nr	**27.63**
Bronze butt weld						
15 mm dia.	–	–	0.25	8.63	nr	**8.63**
22 mm dia.	–	–	0.31	10.71	nr	**10.71**
28 mm dia.	–	–	0.37	12.78	nr	**12.78**
35 mm dia.	–	–	0.49	16.92	nr	**16.92**
42 mm dia.	–	–	0.58	20.03	nr	**20.03**
54 mm dia.	–	–	0.72	24.86	nr	**24.86**
67 mm dia.	–	–	0.88	30.39	nr	**30.39**
76 mm dia.	–	–	1.08	37.30	nr	**37.30**
108 mm dia.	–	–	1.37	47.31	nr	**47.31**
133 mm dia.	–	–	1.73	59.74	nr	**59.74**
159 mm dia.	–	–	2.03	70.10	nr	**70.10**
Press fit (copper fittings); Mechanical press fit joints; butyl rubber O ring						
Coupler						
15 mm dia.	1.38	1.75	0.36	12.43	nr	**14.18**
22 mm dia.	2.19	2.77	0.36	12.43	nr	**15.20**
28 mm dia.	4.49	5.68	0.44	15.19	nr	**20.87**
35 mm dia.	5.66	7.17	0.44	15.19	nr	**22.36**
42 mm dia.	10.19	12.89	0.52	17.96	nr	**30.85**
54 mm dia.	12.99	16.45	0.60	20.72	nr	**37.17**
Stop end						
22 mm dia.	3.51	4.44	0.18	6.22	nr	**10.66**
28 mm dia.	5.56	7.03	0.22	7.60	nr	**14.63**
35 mm dia.	9.53	12.06	0.22	7.60	nr	**19.66**
42 mm dia.	14.32	18.12	0.26	8.98	nr	**27.10**
54 mm dia.	17.24	21.82	0.30	10.36	nr	**32.18**
Reducer						
22 × 15 mm dia.	1.62	2.05	0.36	12.43	nr	**14.48**
28 × 15 mm dia.	4.27	5.41	0.40	13.81	nr	**19.22**
28 × 22 mm dia.	4.44	5.62	0.40	13.81	nr	**19.43**
35 × 22 mm dia.	5.32	6.73	0.40	13.81	nr	**20.55**
35 × 28 mm dia.	5.87	7.42	0.44	15.19	nr	**22.62**
42 × 22 mm dia.	9.17	11.61	0.44	15.19	nr	**26.81**
42 × 28 mm dia.	8.75	11.07	0.48	16.58	nr	**27.65**

38 PIPED SUPPLY SYSTEMS

Item	Net Price £	Material £	Labour hours	Labour £	Unit	Total rate £
COLD WATER PIPELINES: COPPER PIPEWORK – cont						
Press fit (copper fittings) – cont						
Reducer – cont						
42 × 35 mm dia.	8.75	11.07	0.48	16.58	nr	**27.65**
54 × 35 mm dia.	11.84	14.99	0.52	17.96	nr	**32.94**
54 × 42 mm dia.	11.84	14.99	0.56	19.34	nr	**34.32**
90° elbow						
15 mm dia.	1.51	1.91	0.36	12.43	nr	**14.34**
22 mm dia.	2.52	3.19	0.36	12.43	nr	**15.62**
28 mm dia.	5.41	6.85	0.44	15.19	nr	**22.05**
35 mm dia.	11.19	14.16	0.44	15.19	nr	**29.35**
42 mm dia.	21.02	26.60	0.52	17.96	nr	**44.55**
54 mm dia.	29.18	36.93	0.60	20.72	nr	**57.65**
45° elbow						
15 mm dia.	1.86	2.35	0.36	12.43	nr	**14.78**
22 mm dia.	2.56	3.24	0.36	12.43	nr	**15.67**
28 mm dia.	7.69	9.73	0.44	15.19	nr	**24.92**
35 mm dia.	10.96	13.87	0.44	15.19	nr	**29.07**
42 mm dia.	18.24	23.09	0.52	17.96	nr	**41.04**
54 mm dia.	25.95	32.85	0.60	20.72	nr	**53.57**
Equal tee						
15 mm dia.	2.39	3.03	0.54	18.65	nr	**21.67**
22 mm dia.	4.36	5.51	0.54	18.65	nr	**24.16**
28 mm dia.	7.83	9.91	0.66	22.79	nr	**32.70**
35 mm dia.	13.54	17.14	0.66	22.79	nr	**39.93**
42 mm dia.	26.74	33.84	0.78	26.94	nr	**60.78**
54 mm dia.	33.36	42.21	0.90	31.08	nr	**73.29**
Reducing tee						
22 × 15 mm dia.	3.53	4.47	0.54	18.65	nr	**23.12**
28 × 15 mm dia.	6.89	8.72	0.62	21.41	nr	**30.13**
28 × 22 mm dia.	9.25	11.70	0.62	21.41	nr	**33.11**
35 × 22 mm dia.	12.07	15.27	0.62	21.41	nr	**36.68**
35 × 28 mm dia.	13.42	16.99	0.62	21.41	nr	**38.40**
42 × 28 mm dia.	24.28	30.72	0.70	24.17	nr	**54.90**
42 × 35 mm dia.	24.28	30.72	0.70	24.17	nr	**54.90**
54 × 35 mm dia.	41.01	51.90	0.82	28.32	nr	**80.22**
54 × 42 mm dia.	41.01	51.90	0.82	28.32	nr	**80.22**
Male iron connector; BSP thread						
15 mm dia.	5.06	6.40	0.18	6.22	nr	**12.62**
22 mm dia.	7.52	9.52	0.18	6.22	nr	**15.73**
28 mm dia.	10.04	12.71	0.22	7.60	nr	**20.31**
35 mm dia.	18.17	23.00	0.22	7.60	nr	**30.59**
42 mm dia.	24.37	30.84	0.26	8.98	nr	**39.82**
54 mm dia.	47.01	59.49	0.30	10.36	nr	**69.85**
90° elbow; male iron BSP thread						
15 mm dia.	8.19	10.36	0.36	12.43	nr	**22.79**
22 mm dia.	12.84	16.25	0.36	12.43	nr	**28.68**
28 mm dia.	19.66	24.88	0.44	15.19	nr	**40.07**

38 PIPED SUPPLY SYSTEMS

Item	Net Price £	Material £	Labour hours	Labour £	Unit	Total rate £
35 mm dia.	25.57	32.37	0.44	15.19	nr	**47.56**
42 mm dia.	33.33	42.18	0.52	17.96	nr	**60.14**
54 mm dia.	48.69	61.63	0.60	20.72	nr	**82.35**
Female iron connector; BSP thread						
15 mm dia.	5.78	7.32	0.18	6.22	nr	**13.54**
22 mm dia.	7.64	9.67	0.18	6.22	nr	**15.88**
28 mm dia.	10.27	13.00	0.22	7.60	nr	**20.59**
35 mm dia.	20.11	25.45	0.22	7.60	nr	**33.05**
42 mm dia.	28.74	36.37	0.26	8.98	nr	**45.35**
54 mm dia.	49.30	62.40	0.30	10.36	nr	**72.76**
90° elbow; female iron BSP thread						
15 mm dia.	6.90	8.74	0.36	12.43	nr	**21.17**
22 mm dia.	10.19	12.89	0.36	12.43	nr	**25.32**
28 mm dia.	16.80	21.27	0.44	15.19	nr	**36.46**
35 mm dia.	21.73	27.50	0.44	15.19	nr	**42.70**
42 mm dia.	29.57	37.43	0.52	17.96	nr	**55.38**
54 mm dia.	43.48	55.03	0.60	20.72	nr	**75.75**

38 PIPED SUPPLY SYSTEMS

Item	Net Price £	Material £	Labour hours	Labour £	Unit	Total rate £
COLD WATER PIPELINES: STAINLESS STEEL PIPEWORK						
Stainless steel pipes; capillary or compression joints; BS 4127, vertical or at low level, with brackets measured separately						
Grade 304; satin finish						
15 mm dia.	6.83	8.64	0.41	14.16	m	**22.80**
22 mm dia.	9.58	12.12	0.51	17.62	m	**29.74**
28 mm dia.	13.04	16.50	0.58	20.03	m	**36.53**
35 mm dia.	19.71	24.94	0.65	22.45	m	**47.39**
42 mm dia.	24.99	31.62	0.71	24.53	m	**56.15**
54 mm dia.	34.84	44.09	0.80	27.63	m	**71.72**
Grade 316 satin finish						
15 mm dia.	8.76	11.08	0.61	21.07	m	**32.15**
22 mm dia.	16.41	20.76	0.76	26.26	m	**47.03**
28 mm dia.	19.45	24.61	0.87	30.05	m	**54.67**
35 mm dia.	35.29	44.66	0.98	33.86	m	**78.52**
42 mm dia.	45.69	57.83	1.06	36.62	m	**94.45**
54 mm dia.	53.24	67.38	1.16	40.06	m	**107.45**
FIXINGS						
Single pipe ring						
15 mm dia.	15.85	20.06	0.26	8.98	nr	**29.04**
22 mm dia.	18.47	23.37	0.26	8.98	nr	**32.35**
28 mm dia.	19.34	24.48	0.31	10.71	nr	**35.19**
35 mm dia.	21.97	27.81	0.32	11.05	nr	**38.86**
42 mm dia.	24.97	31.60	0.32	11.05	nr	**42.66**
54 mm dia.	28.50	36.07	0.34	11.74	nr	**47.81**
Screw on backplate, female						
All sizes 15 mm to 54 mm dia.	14.30	18.10	0.10	3.45	nr	**21.56**
Screw on backplate, male						
All sizes 15 mm to 54 mm dia.	16.28	20.61	0.10	3.45	nr	**24.06**
Stainless steel threaded rods; metric thread; including nuts, washers etc.						
10 mm dia. × 600 mm long	17.22	21.79	0.18	6.22	nr	**28.01**
Extra over stainless steel pipes; capillary fittings						
Straight coupling						
15 mm dia.	2.08	2.63	0.25	8.63	nr	**11.27**
22 mm dia.	3.35	4.24	0.28	9.67	nr	**13.91**
28 mm dia.	4.43	5.61	0.33	11.40	nr	**17.00**
35 mm dia.	10.23	12.95	0.37	12.78	nr	**25.73**
42 mm dia.	11.81	14.94	0.42	14.50	nr	**29.45**
54 mm dia.	17.76	22.47	0.45	15.54	nr	**38.01**
45° bend						
15 mm dia.	11.17	14.13	0.25	8.63	nr	**22.76**
22 mm dia.	14.66	18.55	0.30	10.24	nr	**28.80**
28 mm dia.	18.00	22.78	0.33	11.40	nr	**34.17**

38 PIPED SUPPLY SYSTEMS

Item	Net Price £	Material £	Labour hours	Labour £	Unit	Total rate £
35 mm dia.	21.34	27.01	0.37	12.78	nr	**39.79**
42 mm dia.	27.47	34.76	0.42	14.50	nr	**49.27**
54 mm dia.	33.96	42.98	0.45	15.54	nr	**58.52**
90° bend						
15 mm dia.	5.79	7.33	0.28	9.67	nr	**17.00**
22 mm dia.	7.79	9.86	0.28	9.67	nr	**19.53**
28 mm dia.	10.98	13.89	0.33	11.40	nr	**25.29**
35 mm dia.	26.77	33.88	0.37	12.78	nr	**46.67**
42 mm dia.	36.85	46.63	0.42	14.50	nr	**61.13**
54 mm dia.	49.99	63.26	0.45	15.54	nr	**78.80**
Reducer						
22 × 15 mm dia.	13.30	16.83	0.28	9.67	nr	**26.50**
28 × 22 mm dia.	14.83	18.77	0.33	11.40	nr	**30.17**
35 × 28 mm dia.	18.18	23.01	0.37	12.78	nr	**35.79**
42 × 35 mm dia.	19.56	24.75	0.42	14.50	nr	**39.25**
54 × 42 mm dia.	58.04	73.46	0.48	16.60	nr	**90.06**
Tap connector						
15 mm dia.	28.08	35.54	0.13	4.49	nr	**40.03**
22 mm dia.	37.13	46.99	0.14	4.83	nr	**51.82**
28 mm dia.	51.53	65.22	0.17	5.87	nr	**71.09**
Tank connector						
15 mm dia.	36.26	45.89	0.13	4.49	nr	**50.38**
22 mm dia.	53.96	68.29	0.13	4.49	nr	**72.78**
28 mm dia.	66.45	84.09	0.15	5.18	nr	**89.27**
35 mm dia.	96.09	121.61	0.18	6.22	nr	**127.82**
42 mm dia.	126.89	160.59	0.21	7.25	nr	**167.84**
54 mm dia.	192.10	243.12	0.24	8.29	nr	**251.40**
Tee equal						
15 mm dia.	10.34	13.08	0.37	12.78	nr	**25.86**
22 mm dia.	12.86	16.28	0.40	13.81	nr	**30.09**
28 mm dia.	15.55	19.69	0.45	15.54	nr	**35.23**
35 mm dia.	37.34	47.26	0.59	20.38	nr	**67.64**
42 mm dia.	46.10	58.35	0.62	21.42	nr	**79.77**
54 mm dia.	93.07	117.79	0.67	23.14	nr	**140.94**
Unequal tee						
22 × 15 mm dia.	20.96	26.52	0.37	12.78	nr	**39.30**
28 × 15 mm dia.	23.61	29.88	0.45	15.54	nr	**45.42**
28 × 22 mm dia.	23.61	29.88	0.45	15.55	nr	**45.43**
35 × 22 mm dia.	41.26	52.22	0.59	20.38	nr	**72.60**
35 × 28 mm dia.	41.26	52.22	0.59	20.38	nr	**72.60**
42 × 28 mm dia.	50.73	64.21	0.62	21.42	nr	**85.63**
42 × 35 mm dia.	50.73	64.21	0.62	21.42	nr	**85.63**
54 × 35 mm dia.	105.01	132.90	0.67	23.14	nr	**156.05**
54 × 42 mm dia.	105.01	132.90	0.67	23.14	nr	**156.05**
Union, conical seat						
15 mm dia.	46.50	58.85	0.25	8.63	nr	**67.49**
22 mm dia.	73.22	92.67	0.28	9.67	nr	**102.34**
28 mm dia.	94.67	119.82	0.33	11.40	nr	**131.22**
35 mm dia.	124.26	157.27	0.37	12.78	nr	**170.05**
42 mm dia.	156.73	198.36	0.42	14.50	nr	**212.86**
54 mm dia.	207.37	262.45	0.45	15.54	nr	**277.99**

38 PIPED SUPPLY SYSTEMS

Item	Net Price £	Material £	Labour hours	Labour £	Unit	Total rate £
COLD WATER PIPELINES: STAINLESS STEEL PIPEWORK – cont						
Extra over stainless steel pipes – cont						
Union, flat seat						
15 mm dia.	48.56	61.45	0.25	8.63	nr	**70.09**
22 mm dia.	75.60	95.68	0.28	9.67	nr	**105.35**
28 mm dia.	97.73	123.69	0.33	11.40	nr	**135.08**
35 mm dia.	127.70	161.62	0.37	12.78	nr	**174.40**
42 mm dia.	160.89	203.63	0.42	14.50	nr	**218.13**
54 mm dia.	215.72	273.02	0.45	15.54	nr	**288.56**
Extra over stainless steel pipes; compression fittings						
Straight coupling						
15 mm dia.	40.42	51.16	0.18	6.22	nr	**57.37**
22 mm dia.	77.00	97.45	0.22	7.60	nr	**105.05**
28 mm dia.	103.67	131.20	0.25	8.63	nr	**139.83**
35 mm dia.	160.03	202.53	0.30	10.36	nr	**212.89**
42 mm dia.	186.79	236.40	0.40	13.81	nr	**250.21**
90° bend						
15 mm dia.	50.95	64.48	0.18	6.22	nr	**70.69**
22 mm dia.	101.22	128.11	0.22	7.60	nr	**135.71**
28 mm dia.	138.03	174.69	0.25	8.63	nr	**183.32**
35 mm dia.	279.51	353.75	0.33	11.40	nr	**365.15**
42 mm dia.	408.48	516.98	0.35	12.09	nr	**529.06**
Reducer						
22 × 15 mm dia.	73.33	92.81	0.28	9.67	nr	**102.48**
28 × 22 mm dia.	100.45	127.13	0.28	9.67	nr	**136.80**
35 × 28 mm dia.	146.77	185.75	0.30	10.36	nr	**196.11**
42 × 35 mm dia.	195.24	247.10	0.37	12.78	nr	**259.88**
Stud coupling						
15 mm dia.	42.01	53.16	0.42	14.50	nr	**67.67**
22 mm dia.	71.05	89.92	0.25	8.63	nr	**98.55**
28 mm dia.	98.65	124.85	0.25	8.63	nr	**133.48**
35 mm dia.	158.03	200.00	0.37	12.78	nr	**212.78**
42 mm dia.	186.79	236.40	0.42	14.50	nr	**250.90**
Equal tee						
15 mm dia.	71.74	90.80	0.37	12.78	nr	**103.58**
22 mm dia.	148.21	187.57	0.40	13.81	nr	**201.38**
28 mm dia.	202.73	256.57	0.45	15.54	nr	**272.11**
35 mm dia.	403.09	510.16	0.59	20.38	nr	**530.54**
42 mm dia.	558.95	707.40	0.62	21.42	nr	**728.82**
Running tee						
15 mm dia.	88.32	111.78	0.37	12.78	nr	**124.56**
22 mm dia.	159.11	201.36	0.40	13.81	nr	**215.18**
28 mm dia.	269.39	340.94	0.59	20.38	nr	**361.32**

38 PIPED SUPPLY SYSTEMS

Item	Net Price £	Material £	Labour hours	Labour £	Unit	Total rate £
Press fit (stainless steel); Press fit jointing system; butyl rubber O ring mechanical joint						
Pipework						
15 mm dia.	8.45	10.69	0.46	15.88	m	**26.58**
22 mm dia.	13.51	17.09	0.48	16.58	m	**33.67**
28 mm dia.	16.64	21.06	0.52	17.96	m	**39.01**
35 mm dia.	24.52	31.03	0.56	19.34	m	**50.37**
42 mm dia.	30.16	38.17	0.58	20.03	m	**58.20**
54 mm dia.	38.41	48.62	0.66	22.79	m	**71.41**
FIXINGS						
For stainless steel pipes						
Refer to fixings for stainless steel pipes; capillary or compression joints; BS 4127						
Extra over stainless steel pipes; Press fit jointing system						
Coupling						
15 mm dia.	9.92	12.56	0.36	12.43	nr	**24.99**
22 mm dia.	12.50	15.82	0.36	12.43	nr	**28.25**
28 mm dia.	14.06	17.79	0.44	15.19	nr	**32.99**
35 mm dia.	17.46	22.10	0.44	15.19	nr	**37.30**
42 mm dia.	23.85	30.19	0.52	17.96	nr	**48.14**
54 mm dia.	28.70	36.32	0.60	20.72	nr	**57.04**
Stop end						
22 mm dia.	9.42	11.92	0.18	6.22	nr	**18.14**
28 mm dia.	10.98	13.89	0.22	7.60	nr	**21.49**
35 mm dia.	17.95	22.72	0.22	7.60	nr	**30.31**
42 mm dia.	25.23	31.93	0.26	8.98	nr	**40.91**
54 mm dia.	29.19	36.94	0.30	10.36	nr	**47.30**
Reducer						
22 × 15 mm dia.	11.82	14.96	0.36	12.43	nr	**27.39**
28 × 15 mm dia.	13.39	16.95	0.40	13.81	nr	**30.76**
28 × 22 mm dia.	13.86	17.55	0.40	13.81	nr	**31.36**
35 × 22 mm dia.	16.93	21.42	0.40	13.81	nr	**35.24**
35 × 28 mm dia.	20.97	26.53	0.44	15.19	nr	**41.73**
42 × 35 mm dia.	22.12	27.99	0.48	16.58	nr	**44.57**
54 × 42 mm dia.	25.28	31.99	0.56	19.34	nr	**51.33**
90° bend						
15 mm dia.	14.20	17.98	0.36	12.43	nr	**30.41**
22 mm dia.	19.85	25.12	0.36	12.43	nr	**37.55**
28 mm dia.	25.04	31.69	0.44	15.19	nr	**46.88**
35 mm dia.	39.40	49.87	0.44	15.19	nr	**65.06**
42 mm dia.	65.78	83.26	0.52	17.96	nr	**101.21**
54 mm dia.	90.85	114.98	0.60	20.72	nr	**135.70**
45° bend						
15 mm dia.	19.26	24.38	0.36	12.43	nr	**36.81**
22 mm dia.	23.95	30.31	0.36	12.43	nr	**42.74**
28 mm dia.	27.87	35.28	0.44	15.19	nr	**50.47**

38 PIPED SUPPLY SYSTEMS

Item	Net Price £	Material £	Labour hours	Labour £	Unit	Total rate £
COLD WATER PIPELINES: STAINLESS STEEL PIPEWORK – cont						
Extra over stainless steel pipes – cont						
45° bend – cont						
35 mm dia.	32.71	41.39	0.44	15.19	nr	**56.59**
42 mm dia.	52.62	66.59	0.52	17.96	nr	**84.55**
54 mm dia.	68.36	86.52	0.60	20.72	nr	**107.24**
Equal tee						
15 mm dia.	23.28	29.47	0.54	18.65	nr	**48.12**
22 mm dia.	28.57	36.16	0.54	18.65	nr	**54.81**
28 mm dia.	33.35	42.21	0.66	22.79	nr	**65.00**
35 mm dia.	42.27	53.50	0.66	22.79	nr	**76.29**
42 mm dia.	60.00	75.93	0.78	26.94	nr	**102.87**
54 mm dia.	71.78	90.85	0.90	31.08	nr	**121.93**
Reducing tee						
22 × 15 mm dia.	24.42	30.91	0.54	18.65	nr	**49.55**
28 × 15 mm dia.	29.59	37.45	0.62	21.41	nr	**58.86**
28 × 22 mm dia.	32.04	40.55	0.62	21.41	nr	**61.96**
35 × 22 mm dia.	37.98	48.06	0.62	21.41	nr	**69.47**
35 × 28 mm dia.	39.63	50.16	0.62	21.41	nr	**71.57**
42 × 28 mm dia.	56.35	71.31	0.70	24.17	nr	**95.49**
42 × 35 mm dia.	58.05	73.47	0.70	24.17	nr	**97.64**
54 × 35 mm dia.	65.57	82.99	0.82	28.32	nr	**111.31**
54 × 42 mm dia.	67.42	85.33	0.82	28.32	nr	**113.65**
FIXINGS						
For stainless steel pipes						
Refer to fixings for stainless steel pipes; capillary or compression joints; BS 4127						

38 PIPED SUPPLY SYSTEMS

Item	Net Price £	Material £	Labour hours	Labour £	Unit	Total rate £
COLD WATER PIPELINES: MEDIUM DENSITY POLYETHYLENE – BLUE PIPEWORK						
Note: MDPE is sized on Outside Dia. i.e. OD not ID						
Pipes for water distribution; laid underground; electrofusion joints in the running length; BS 6572						
Coiled service pipe						
20 mm dia.	1.17	1.48	0.37	12.78	m	14.26
25 mm dia.	1.52	1.92	0.41	14.16	m	16.08
32 mm dia.	2.54	3.22	0.47	16.23	m	19.45
50 mm dia.	6.12	7.74	0.53	18.31	m	26.05
63 mm dia.	9.59	12.14	0.60	20.73	m	32.86
Mains service pipe						
90 mm dia.	14.50	18.35	0.90	31.08	m	49.44
110 mm dia.	21.68	27.44	1.10	37.99	m	65.43
125 mm dia.	27.44	34.72	1.20	41.46	m	76.18
160 mm dia.	43.72	55.34	1.48	51.11	m	106.44
180 mm dia.	56.94	72.07	1.50	51.85	m	123.92
225 mm dia.	86.53	109.52	1.77	61.12	m	170.64
250 mm dia.	109.22	138.23	1.75	60.48	m	198.71
315 mm dia.	168.64	213.42	1.90	65.61	m	279.04
Extra over fittings; MDPE blue; electrofusion joints						
Coupler						
20 mm dia.	9.86	12.47	0.36	12.43	nr	24.91
25 mm dia.	9.86	12.47	0.40	13.81	nr	26.29
32 mm dia.	9.86	12.47	0.44	15.19	nr	27.67
40 mm dia.	14.54	18.40	0.48	16.58	nr	34.98
50 mm dia.	13.96	17.66	0.52	17.96	nr	35.62
63 mm dia.	18.25	23.10	0.58	20.03	nr	43.13
90 mm dia.	26.83	33.96	0.67	23.14	nr	57.10
110 mm dia.	43.16	54.63	0.74	25.55	nr	80.18
125 mm dia.	48.70	61.63	0.83	28.66	nr	90.30
160 mm dia.	77.45	98.03	1.00	34.53	nr	132.56
180 mm dia.	91.07	115.26	1.25	43.17	nr	158.43
225 mm dia.	145.54	184.19	1.35	46.62	nr	230.81
250 mm dia.	213.11	269.71	1.50	51.80	nr	321.51
315 mm dia.	351.45	444.79	1.80	62.16	nr	506.95
Extra over fittings; MDPE blue; butt fused joints						
Cap						
25 mm dia.	18.69	23.66	0.20	6.91	nr	30.56
32 mm dia.	18.69	23.66	0.22	7.60	nr	31.25
40 mm dia.	19.80	25.06	0.24	8.29	nr	33.34

38 PIPED SUPPLY SYSTEMS

Item	Net Price £	Material £	Labour hours	Labour £	Unit	Total rate £
COLD WATER PIPELINES: MEDIUM DENSITY POLYETHYLENE – BLUE PIPEWORK – cont						
Extra over fittings – cont						
Cap – cont						
50 mm dia.	28.88	36.55	0.26	8.98	nr	**45.52**
63 mm dia.	33.12	41.92	0.32	11.05	nr	**52.97**
90 mm dia.	54.65	69.17	0.37	12.78	nr	**81.94**
110 mm dia.	110.05	139.29	0.40	13.81	nr	**153.10**
125 mm dia.	87.86	111.19	0.46	15.88	nr	**127.08**
160 mm dia.	100.44	127.12	0.50	17.27	nr	**144.38**
180 mm dia.	168.00	212.62	0.60	20.72	nr	**233.34**
225 mm dia.	198.45	251.16	0.68	23.48	nr	**274.64**
250 mm dia.	291.28	368.65	0.75	25.90	nr	**394.55**
315 mm dia.	375.60	475.36	0.90	31.08	nr	**506.44**
Reducer						
63 × 32 mm dia.	25.25	31.96	0.54	18.65	nr	**50.60**
63 × 50 mm dia.	29.53	37.38	0.60	20.72	nr	**58.10**
90 × 63 mm dia.	37.37	47.30	0.67	23.14	nr	**70.43**
110 × 90 mm dia.	51.50	65.18	0.74	25.55	nr	**90.73**
125 × 90 mm dia.	74.90	94.79	0.83	28.66	nr	**123.45**
125 × 110 mm dia.	82.53	104.45	1.00	34.53	nr	**138.98**
160 × 110 mm dia.	126.77	160.45	1.10	37.99	nr	**198.43**
180 × 125 mm dia.	137.70	174.28	1.25	43.17	nr	**217.44**
225 × 160 mm dia.	224.54	284.18	1.40	48.35	nr	**332.52**
250 × 180 mm dia.	172.52	218.34	1.80	62.16	nr	**280.50**
315 × 250 mm dia.	198.46	251.17	2.40	82.88	nr	**334.05**
Bend; 45°						
50 mm dia.	38.54	48.77	0.50	17.27	nr	**66.04**
63 mm dia.	46.62	59.00	0.58	20.03	nr	**79.03**
90 mm dia.	72.20	91.37	0.67	23.14	nr	**114.51**
110 mm dia.	105.86	133.97	0.74	25.55	nr	**159.53**
125 mm dia.	118.31	149.73	0.83	28.66	nr	**178.39**
160 mm dia.	219.25	277.48	1.00	34.53	nr	**312.01**
180 mm dia.	249.58	315.87	1.25	43.17	nr	**359.04**
225 mm dia.	318.13	402.62	1.40	48.33	nr	**450.96**
250 mm dia.	344.26	435.69	1.80	62.16	nr	**497.85**
315 mm dia.	428.90	542.82	2.40	82.88	nr	**625.70**
Bend; 90°						
50 mm dia.	38.54	48.77	0.50	17.27	nr	**66.04**
63 mm dia.	46.62	59.00	0.58	20.03	nr	**79.03**
90 mm dia.	72.20	91.37	0.67	23.14	nr	**114.51**
110 mm dia.	105.86	133.97	0.74	25.55	nr	**159.53**
125 mm dia.	118.31	149.73	0.83	28.66	nr	**178.39**
160 mm dia.	219.25	277.48	1.00	34.53	nr	**312.01**
180 mm dia.	425.85	538.95	1.25	43.17	nr	**582.12**
225 mm dia.	538.43	681.44	1.40	48.35	nr	**729.79**
250 mm dia.	588.96	745.38	1.80	62.16	nr	**807.54**
315 mm dia.	737.30	933.12	2.40	82.88	nr	**1016.00**

38 PIPED SUPPLY SYSTEMS

Item	Net Price £	Material £	Labour hours	Labour £	Unit	Total rate £
Equal tee						
50 mm dia.	42.14	53.33	0.70	24.17	nr	**77.51**
63 mm dia.	45.99	58.20	0.75	25.90	nr	**84.10**
90 mm dia.	83.99	106.30	0.87	30.04	nr	**136.34**
110 mm dia.	125.25	158.52	1.00	34.53	nr	**193.05**
125 mm dia.	161.70	204.65	1.08	37.30	nr	**241.94**
160 mm dia.	265.88	336.49	1.35	46.62	nr	**383.11**
180 mm dia.	272.07	344.33	1.63	56.29	nr	**400.62**
225 mm dia.	329.73	417.30	1.90	65.61	nr	**482.91**
250 mm dia.	454.97	575.81	2.70	93.24	nr	**669.04**
315 mm dia.	1134.22	1435.47	3.60	124.32	nr	**1559.78**
Extra over plastic fittings, compression joints						
Straight connector						
20 mm dia.	5.01	6.34	0.38	13.13	nr	**19.47**
25 mm dia.	5.25	6.64	0.45	15.54	nr	**22.18**
32 mm dia.	12.47	15.79	0.50	17.27	nr	**33.05**
50 mm dia.	28.73	36.36	0.68	23.49	nr	**59.85**
63 mm dia.	43.26	54.75	0.85	29.36	nr	**84.11**
Reducing connector						
25 mm dia.	10.30	13.04	0.38	13.13	nr	**26.17**
32 mm dia.	16.63	21.05	0.45	15.54	nr	**36.59**
50 mm dia.	46.07	58.31	0.50	17.27	nr	**75.58**
63 mm dia.	64.30	81.38	0.62	21.42	nr	**102.80**
Straight connector; polyethylene to MI						
20 mm dia.	4.54	5.74	0.31	10.71	nr	**16.45**
25 mm dia.	7.73	9.78	0.35	12.09	nr	**21.87**
32 mm dia.	8.37	10.60	0.40	13.81	nr	**24.41**
50 mm dia.	21.36	27.04	0.55	19.00	nr	**46.03**
63 mm dia.	30.13	38.13	0.65	22.45	nr	**60.58**
Straight connector; polyethylene to FI						
20 mm dia.	6.10	7.73	0.31	10.71	nr	**18.43**
25 mm dia.	6.61	8.36	0.35	12.09	nr	**20.45**
32 mm dia.	7.90	9.99	0.40	13.81	nr	**23.81**
50 mm dia.	25.10	31.77	0.55	19.00	nr	**50.77**
63 mm dia.	35.15	44.49	0.75	25.91	nr	**70.40**
Elbow						
20 mm dia.	6.68	8.45	0.38	13.13	nr	**21.58**
25 mm dia.	9.87	12.49	0.45	15.54	nr	**28.03**
32 mm dia.	14.39	18.21	0.50	17.27	nr	**35.48**
50 mm dia.	33.38	42.24	0.68	23.49	nr	**65.73**
63 mm dia.	45.37	57.43	0.80	27.63	nr	**85.05**
Elbow; polyethylene to MI						
25 mm dia.	8.50	10.76	0.35	12.09	nr	**22.84**
Elbow; polyethylene to FI						
20 mm dia.	6.08	7.69	0.31	10.71	nr	**18.40**
25 mm dia.	8.26	10.45	0.35	12.09	nr	**22.54**
32 mm dia.	12.38	15.66	0.42	14.50	nr	**30.17**
50 mm dia.	29.39	37.19	0.50	17.27	nr	**54.46**
63 mm dia.	38.51	48.74	0.55	19.00	nr	**67.74**

38 PIPED SUPPLY SYSTEMS

Item	Net Price £	Material £	Labour hours	Labour £	Unit	Total rate £
COLD WATER PIPELINES: MEDIUM DENSITY POLYETHYLENE – BLUE PIPEWORK – cont						
Extra over plastic fittings, compression joints – cont						
Tank coupling						
25 mm dia.	12.75	16.14	0.42	14.50	nr	**30.65**
Equal tee						
20 mm dia.	8.99	11.38	0.53	18.31	nr	**29.68**
25 mm dia.	14.06	17.79	0.55	19.00	nr	**36.79**
32 mm dia.	17.61	22.29	0.64	22.11	nr	**44.40**
50 mm dia.	41.06	51.97	0.75	25.91	nr	**77.87**
63 mm dia.	63.60	80.49	0.87	30.05	nr	**110.55**
Equal tee; FI branch						
20 mm dia.	8.71	11.02	0.45	15.54	nr	**26.56**
25 mm dia.	13.87	17.56	0.50	17.27	nr	**34.83**
32 mm dia.	16.88	21.37	0.60	20.73	nr	**42.09**
50 mm dia.	38.91	49.24	0.68	23.49	nr	**72.73**
63 mm dia.	54.63	69.14	0.81	27.99	nr	**97.12**
Equal tee; MI branch						
25 mm dia.	13.64	17.26	0.50	17.27	nr	**34.52**

38 PIPED SUPPLY SYSTEMS

Item	Net Price £	Material £	Labour hours	Labour £	Unit	Total rate £
COLD WATER PIPELINES: ABS PIPEWORK						
Pipes; solvent welded joints in the running length, brackets measured separately						
Class C (9 bar pressure)						
1" dia.	8.77	11.10	0.30	10.36	m	**21.46**
1 ¼" dia.	14.75	18.67	0.33	11.40	m	**30.07**
1 ½" dia.	18.72	23.69	0.36	12.43	m	**36.12**
2" dia.	25.21	31.91	0.39	13.47	m	**45.38**
3" dia.	51.95	65.75	0.46	15.88	m	**81.63**
4" dia.	85.62	108.36	0.53	18.30	m	**126.66**
6" dia.	169.27	214.23	0.76	26.24	m	**240.47**
8" dia.	290.71	367.93	0.97	33.50	m	**401.42**
Class E (15 bar pressure)						
½" dia.	6.69	8.46	0.24	8.29	m	**16.75**
¾" dia.	10.30	13.03	0.27	9.32	m	**22.36**
1" dia.	13.57	17.17	0.30	10.36	m	**27.53**
1 ¼" dia.	20.25	25.63	0.33	11.40	m	**37.03**
1 ½" dia.	26.70	33.79	0.36	12.43	m	**46.22**
2" dia.	33.43	42.30	0.39	13.47	m	**55.77**
3" dia.	67.23	85.09	0.49	16.92	m	**102.01**
4" dia.	108.15	136.88	0.57	19.68	m	**156.56**
FIXINGS						
Refer to steel pipes; galvanized iron. For minimum fixing dimensions, refer to the Tables and Memoranda at the rear of the book						
Extra over fittings; solvent welded joints						
Cap						
½" dia.	2.69	3.41	0.16	5.53	nr	**8.93**
¾" dia.	3.12	3.94	0.19	6.56	nr	**10.51**
1" dia.	3.58	4.53	0.22	7.60	nr	**12.13**
1 ¼" dia.	5.98	7.56	0.25	8.63	nr	**16.20**
1 ½" dia.	9.21	11.66	0.28	9.67	nr	**21.32**
2" dia.	11.66	14.76	0.31	10.71	nr	**25.46**
3" dia.	35.04	44.35	0.36	12.43	nr	**56.78**
4" dia.	53.59	67.82	0.44	15.19	nr	**83.01**
Elbow 90°						
½" dia.	3.76	4.76	0.29	10.01	nr	**14.77**
¾" dia.	4.50	5.70	0.34	11.74	nr	**17.44**
1" dia.	6.30	7.97	0.40	13.81	nr	**21.78**
1 ¼" dia.	10.63	13.46	0.45	15.54	nr	**29.00**
1 ½" dia.	13.81	17.48	0.51	17.61	nr	**35.09**
2" dia.	21.03	26.62	0.56	19.34	nr	**45.96**
3" dia.	60.42	76.47	0.65	22.45	nr	**98.92**
4" dia.	90.24	114.21	0.80	27.63	nr	**141.84**
6" dia.	363.28	459.77	1.21	41.78	nr	**501.56**
8" dia.	554.42	701.67	1.45	50.07	nr	**751.74**

38 PIPED SUPPLY SYSTEMS

Item	Net Price £	Material £	Labour hours	Labour £	Unit	Total rate £
COLD WATER PIPELINES: ABS PIPEWORK – cont						
Extra over fittings – cont						
Elbow 45°						
½" dia.	7.29	9.22	0.29	10.01	nr	**19.23**
¾" dia.	7.38	9.33	0.34	11.74	nr	**21.07**
1" dia.	9.21	11.66	0.40	13.81	nr	**25.47**
1 ¼" dia.	13.49	17.08	0.45	15.54	nr	**32.62**
1 ½" dia.	16.70	21.13	0.51	17.61	nr	**38.75**
2" dia.	23.20	29.36	0.56	19.34	nr	**48.70**
3" dia.	54.64	69.15	0.65	22.45	nr	**91.60**
4" dia.	113.29	143.38	0.80	27.63	nr	**171.01**
6" dia.	234.83	297.20	1.21	41.78	nr	**338.99**
8" dia.	505.32	639.53	1.45	50.07	nr	**689.60**
Reducing bush						
¾" ×½" dia.	2.77	3.51	0.42	14.50	nr	**18.01**
1" ×½" dia.	3.58	4.53	0.45	15.54	nr	**20.07**
1" × ¾" dia.	3.58	4.53	0.45	15.54	nr	**20.07**
1 ¼" × 1" dia.	4.81	6.09	0.48	16.58	nr	**22.66**
1 ½" × ¾" dia.	6.30	7.97	0.51	17.61	nr	**25.58**
1 ½" × 1" dia.	6.30	7.97	0.51	17.61	nr	**25.58**
1 ½" × 1 ¼" dia.	6.30	7.97	0.51	17.61	nr	**25.58**
2" × 1" dia.	8.26	10.45	0.56	19.34	nr	**29.79**
2" × 1 ¼" dia.	8.26	10.45	0.56	19.34	nr	**29.79**
2" × 1 ½" dia.	8.26	10.45	0.56	19.34	nr	**29.79**
3" × 1 ½" dia.	23.20	29.36	0.65	22.45	nr	**51.81**
3" × 2" dia.	23.20	29.36	0.65	22.45	nr	**51.81**
4" × 3" dia.	32.01	40.52	0.80	27.63	nr	**68.14**
6" × 4" dia.	98.56	124.73	1.21	41.78	nr	**166.52**
Union						
½" dia.	14.96	18.93	0.34	11.74	nr	**30.67**
¾" dia.	16.14	20.42	0.39	13.47	nr	**33.89**
1" dia.	21.73	27.50	0.43	14.85	nr	**42.35**
1 ¼" dia.	26.63	33.70	0.50	17.27	nr	**50.97**
1 ½" dia.	36.68	46.43	0.57	19.68	nr	**66.11**
2" dia.	47.84	60.55	0.62	21.41	nr	**81.96**
Sockets						
½" dia.	2.77	3.51	0.34	11.74	nr	**15.25**
¾" dia.	3.12	3.94	0.39	13.47	nr	**17.41**
1" dia.	3.58	4.53	0.43	14.85	nr	**19.38**
1 ¼" dia.	6.30	7.97	0.50	17.27	nr	**25.24**
1 ½" dia.	7.59	9.61	0.57	19.68	nr	**29.29**
2" dia.	10.63	13.46	0.62	21.41	nr	**34.87**
3" dia.	42.80	54.17	0.70	24.17	nr	**78.34**
4" dia.	60.73	76.86	0.70	24.17	nr	**101.03**
6" dia.	151.77	192.08	1.26	43.51	nr	**235.59**
8" dia.	303.12	383.62	1.55	53.53	nr	**437.15**

38 PIPED SUPPLY SYSTEMS

Item	Net Price £	Material £	Labour hours	Labour £	Unit	Total rate £
Barrel nipple						
½" dia.	5.23	6.62	0.34	11.74	nr	**18.36**
¾" dia.	6.80	8.60	0.39	13.47	nr	**22.07**
1" dia.	8.81	11.15	0.43	14.85	nr	**26.00**
1 ¼" dia.	12.22	15.47	0.50	17.27	nr	**32.74**
1 ½" dia.	14.42	18.25	0.57	19.68	nr	**37.93**
2" dia.	17.44	22.08	0.62	21.41	nr	**43.49**
3" dia.	46.38	58.70	0.70	24.17	nr	**82.87**
Tee, 90°						
½" dia.	4.28	5.42	0.41	14.16	nr	**19.58**
¾" dia.	5.98	7.56	0.47	16.23	nr	**23.79**
1" dia.	8.26	10.45	0.55	18.99	nr	**29.45**
1 ¼" dia.	11.88	15.03	0.64	22.10	nr	**37.13**
1 ½" dia.	17.44	22.08	0.71	24.52	nr	**46.59**
2" dia.	26.63	33.70	0.78	26.94	nr	**60.63**
3" dia.	77.67	98.30	0.91	31.42	nr	**129.73**
4" dia.	114.02	144.31	1.12	38.68	nr	**182.98**
6" dia.	398.52	504.36	1.69	58.36	nr	**562.72**
8" dia.	621.38	786.42	2.03	70.10	nr	**856.52**
Full face flange						
½" dia.	46.42	58.75	0.10	3.45	nr	**62.20**
¾" dia.	47.51	60.13	0.13	4.49	nr	**64.61**
1" dia.	51.46	65.13	0.15	5.18	nr	**70.31**
1 ¼" dia.	57.20	72.40	0.18	6.22	nr	**78.61**
1 ½" dia.	68.79	87.06	0.21	7.25	nr	**94.31**
2" dia.	93.10	117.83	0.29	10.01	nr	**127.85**
3" dia.	159.65	202.05	0.37	12.78	nr	**214.82**
4" dia.	209.23	264.80	0.41	14.16	nr	**278.96**

38 PIPED SUPPLY SYSTEMS

Item	Net Price £	Material £	Labour hours	Labour £	Unit	Total rate £
COLD WATER PIPELINES: PVC-u PIPEWORK						
Pipes; solvent welded joints in the running length, brackets Measured separately						
Class C (9 bar pressure)						
2" dia.	20.79	26.31	0.41	14.16	m	**40.46**
3" dia.	39.84	50.42	0.47	16.23	m	**66.65**
4" dia.	70.68	89.45	0.50	17.27	m	**106.72**
6" dia.	152.95	193.57	1.76	60.78	m	**254.35**
Class D (12 bar pressure)						
1 ¼" dia.	12.11	15.33	0.41	14.16	m	**29.48**
1 ½" dia.	16.64	21.07	0.42	14.50	m	**35.57**
2" dia.	25.81	32.67	0.45	15.54	m	**48.21**
3" dia.	55.31	70.01	0.48	16.58	m	**86.58**
4" dia.	92.57	117.15	0.53	18.30	m	**135.46**
6" dia.	171.80	217.43	0.58	20.03	m	**237.46**
Class E (15 bar pressure)						
½" dia.	5.93	7.50	0.38	13.12	m	**20.62**
¾" dia.	8.47	10.72	0.40	13.81	m	**24.54**
1" dia.	9.85	12.46	0.41	14.16	m	**26.62**
1 ¼" dia.	14.45	18.29	0.41	14.16	m	**32.45**
1 ½" dia.	18.81	23.81	0.42	14.50	m	**38.31**
2" dia.	29.39	37.20	0.45	15.54	m	**52.74**
3" dia.	63.69	80.61	0.47	16.23	m	**96.84**
4" dia.	104.57	132.34	0.50	17.27	m	**149.61**
6" dia.	226.50	286.66	0.53	18.30	m	**304.97**
Class 7						
½" dia.	10.47	13.26	0.32	11.05	m	**24.31**
¾" dia.	14.71	18.62	0.33	11.40	m	**30.01**
1" dia.	22.42	28.38	0.40	13.81	m	**42.19**
1 ¼" dia.	30.80	38.98	0.40	13.81	m	**52.79**
1 ½" dia.	38.13	48.25	0.41	14.16	m	**62.41**
2" dia.	63.38	80.21	0.43	14.85	m	**95.06**
FIXINGS						
Refer to steel pipes; galvanized iron. For minimum fixing dimensions, refer to the Tables and Memoranda at the rear of the book						
Extra over fittings; solvent welded joints						
End cap						
½" dia.	1.90	2.40	0.17	5.87	nr	**8.27**
¾" dia.	2.22	2.81	0.19	6.56	nr	**9.37**
1" dia.	2.51	3.18	0.22	7.60	nr	**10.78**
1 ¼" dia.	3.92	4.97	0.25	8.63	nr	**13.60**
1 ½" dia.	6.55	8.29	0.28	9.67	nr	**17.96**
2" dia.	8.04	10.18	0.31	10.71	nr	**20.88**
3" dia.	24.61	31.15	0.36	12.43	nr	**43.58**
4" dia.	38.00	48.10	0.44	15.19	nr	**63.29**
6" dia.	91.86	116.26	0.67	23.14	nr	**139.39**

38 PIPED SUPPLY SYSTEMS

Item	Net Price £	Material £	Labour hours	Labour £	Unit	Total rate £
Socket						
½" dia.	2.01	2.55	0.31	10.71	nr	**13.25**
¾" dia.	2.22	2.81	0.35	12.09	nr	**14.89**
1" dia.	2.59	3.28	0.42	14.50	nr	**17.78**
1 ¼" dia.	4.67	5.91	0.45	15.54	nr	**21.45**
1 ½" dia.	5.48	6.93	0.51	17.61	nr	**24.54**
2" dia.	7.79	9.85	0.56	19.34	nr	**29.19**
3" dia.	29.76	37.66	0.65	22.45	nr	**60.11**
4" dia.	43.13	54.59	0.80	27.63	nr	**82.22**
6" dia.	108.20	136.94	1.21	41.78	nr	**178.72**
Reducing socket						
¾ × ½" dia.	2.35	2.97	0.31	10.71	nr	**13.68**
1 × ¾" dia.	2.96	3.75	0.35	12.09	nr	**15.84**
1 ¼ × 1" dia.	5.64	7.14	0.42	14.50	nr	**21.65**
1 ½ × 1 ¼" dia.	6.30	7.97	0.45	15.54	nr	**23.51**
2 × 1 ½" dia.	9.50	12.03	0.51	17.61	nr	**29.64**
3 × 2" dia.	28.95	36.64	0.56	19.34	nr	**55.98**
4 × 3" dia.	42.83	54.20	0.65	22.45	nr	**76.65**
6 × 4" dia.	156.14	197.62	0.80	27.63	nr	**225.24**
8 × 6" dia.	241.82	306.05	1.21	41.78	nr	**347.83**
Elbow, 90°						
½" dia.	2.66	3.36	0.31	10.71	nr	**14.07**
¾" dia.	3.19	4.04	0.35	12.09	nr	**16.13**
1" dia.	4.46	5.65	0.42	14.50	nr	**20.15**
1 ¼" dia.	7.79	9.85	0.45	15.54	nr	**25.39**
1 ½" dia.	10.02	12.68	0.45	15.54	nr	**28.22**
2" dia.	14.87	18.81	0.56	19.34	nr	**38.15**
3" dia.	42.83	54.20	0.65	22.45	nr	**76.65**
4" dia.	64.53	81.67	0.80	27.63	nr	**109.29**
6" dia.	255.48	323.34	1.21	41.78	nr	**365.12**
Elbow 45°						
½" dia.	5.08	6.43	0.31	10.71	nr	**17.13**
¾" dia.	5.40	6.83	0.35	12.09	nr	**18.92**
1" dia.	6.55	8.29	0.45	15.54	nr	**23.83**
1 ¼" dia.	9.39	11.88	0.45	15.54	nr	**27.42**
1 ½" dia.	11.79	14.92	0.51	17.61	nr	**32.53**
2" dia.	16.61	21.02	0.56	19.34	nr	**40.36**
3" dia.	39.09	49.48	0.65	22.45	nr	**71.92**
4" dia.	80.33	101.66	0.80	27.63	nr	**129.29**
6" dia.	165.74	209.76	1.21	41.78	nr	**251.54**
Bend 90° (long radius)						
3" dia.	119.44	151.16	0.65	22.45	nr	**173.60**
4" dia.	241.27	305.35	0.80	27.63	nr	**332.98**
6" dia.	530.20	671.02	1.21	41.78	nr	**712.81**
Bend 45° (long radius)						
1 ½" dia.	28.36	35.89	0.51	17.61	nr	**53.50**
2" dia.	46.34	58.65	0.56	19.34	nr	**77.99**
3" dia.	99.06	125.36	0.65	22.45	nr	**147.81**
4" dia.	192.79	243.99	0.80	27.63	nr	**271.62**

38 PIPED SUPPLY SYSTEMS

Item	Net Price £	Material £	Labour hours	Labour £	Unit	Total rate £
COLD WATER PIPELINES: PVC-u PIPEWORK – cont						
Extra over fittings – cont						
Socket union						
½" dia.	10.30	13.03	0.34	11.74	nr	**24.78**
¾" dia.	11.79	14.92	0.39	13.47	nr	**28.39**
1" dia.	15.30	19.37	0.45	15.54	nr	**34.91**
1 ¼" dia.	19.03	24.09	0.50	17.27	nr	**41.36**
1 ½" dia.	26.06	32.98	0.57	19.68	nr	**52.67**
2" dia.	33.73	42.69	0.62	21.41	nr	**64.10**
3" dia.	125.59	158.95	0.70	24.17	nr	**183.12**
4" dia.	170.05	215.21	0.89	30.73	nr	**245.95**
Saddle plain						
2" × 1 ¼" dia.	26.50	33.54	0.42	14.50	nr	**48.04**
3" × 1 ½" dia.	37.25	47.14	0.48	16.58	nr	**63.72**
4" × 2" dia.	41.94	53.08	0.68	23.48	nr	**76.56**
6" × 2" dia.	49.26	62.35	0.91	31.42	nr	**93.77**
Straight tank connector						
½" dia.	6.85	8.67	0.13	4.49	nr	**13.16**
¾" dia.	7.75	9.80	0.14	4.83	nr	**14.64**
1" dia.	16.44	20.81	0.14	4.83	nr	**25.64**
1 ¼" dia.	41.79	52.89	0.16	5.53	nr	**58.41**
1 ½" dia.	45.81	57.98	0.18	6.22	nr	**64.20**
2" dia.	54.87	69.44	0.24	8.29	nr	**77.73**
3" dia.	56.27	71.21	0.29	10.01	nr	**81.23**
Equal tee						
½" dia.	3.09	3.91	0.44	15.19	nr	**19.11**
¾" dia.	3.92	4.97	0.48	16.58	nr	**21.54**
1" dia.	5.87	7.43	0.54	18.65	nr	**26.08**
1 ¼" dia.	8.32	10.53	0.70	24.17	nr	**34.71**
1 ½" dia.	11.99	15.18	0.74	25.55	nr	**40.73**
2" dia.	19.03	24.09	0.80	27.63	nr	**51.72**
3" dia.	55.16	69.82	1.04	35.91	nr	**105.73**
4" dia.	80.87	102.35	1.28	44.20	nr	**146.55**
6" dia.	281.71	356.53	1.93	66.65	nr	**423.18**

38 PIPED SUPPLY SYSTEMS

Item	Net Price £	Material £	Labour hours	Labour £	Unit	Total rate £
COLD WATER PIPELINES: PVC-C PIPEWORK						
Pipes; solvent welded in the running length, brackets measured separately						
Pipe; 3 m long; PN25						
16 × 2.0 mm	6.59	8.34	0.20	6.91	m	**15.25**
20 × 2.3 mm	9.96	12.60	0.20	6.91	m	**19.51**
25 × 2.8 mm	12.91	16.34	0.20	6.91	m	**23.25**
32 × 3.6 mm	19.18	24.27	0.20	6.91	m	**31.18**
Pipe; 5 m long; PN25						
40 × 4.5 mm	23.39	29.60	0.20	6.91	m	**36.51**
50 × 5.6 mm	35.20	44.54	0.20	6.91	m	**51.45**
63 × 7.0 mm	54.35	68.78	0.20	6.91	m	**75.69**
FIXINGS						
Refer to steel pipes; galvanized iron. For minimum fixing dimensions, refer to the Tables and Memoranda at the rear of the book						
Extra over fittings; solvent welded joints						
Straight coupling; PN25						
16 mm	1.02	1.29	0.20	6.91	nr	**8.20**
20 mm	1.44	1.83	0.20	6.91	nr	**8.73**
25 mm	1.80	2.28	0.20	6.91	nr	**9.18**
32 mm	5.48	6.93	0.20	6.91	nr	**13.84**
40 mm	7.03	8.90	0.20	6.91	nr	**15.80**
50 mm	9.41	11.91	0.20	6.91	nr	**18.82**
63 mm	16.63	21.05	0.20	6.91	nr	**27.96**
Elbow; 90°; PN25						
16 mm	1.65	2.09	0.20	6.91	nr	**8.99**
20 mm	2.52	3.19	0.20	6.91	nr	**10.10**
25 mm	3.15	3.98	0.20	6.91	nr	**10.89**
32 mm	6.55	8.29	0.20	6.91	nr	**15.20**
40 mm	10.09	12.77	0.20	6.91	nr	**19.68**
50 mm	13.96	17.67	0.20	6.91	nr	**24.58**
63 mm	23.88	30.22	0.20	6.91	nr	**37.13**
Elbow; 45°; PN25						
20 mm	2.52	3.19	0.20	6.91	nr	**10.10**
25 mm	3.15	3.98	0.20	6.91	nr	**10.89**
32 mm	6.55	8.29	0.20	6.91	nr	**15.20**
40 mm	10.09	12.77	0.20	6.91	nr	**19.68**
50 mm	13.96	17.67	0.20	6.91	nr	**24.58**
63 mm	23.88	30.22	0.20	6.91	nr	**37.13**
Reducer fitting; single stage reduction						
20/16 mm	1.80	2.28	0.20	6.91	nr	**9.18**
25/20 mm	2.15	2.72	0.20	6.91	nr	**9.63**
32/25 mm	4.32	5.46	0.20	6.91	nr	**12.37**
40/32 mm	5.65	7.15	0.20	6.91	nr	**14.06**
50/40 mm	6.55	8.29	0.20	6.91	nr	**15.20**
63/50 mm	9.92	12.55	0.20	6.91	nr	**19.46**

38 PIPED SUPPLY SYSTEMS

Item	Net Price £	Material £	Labour hours	Labour £	Unit	Total rate £
COLD WATER PIPELINES: PVC-C PIPEWORK – cont						
Extra over fittings – cont						
Equal tee; 90°; PN25						
16 mm	2.74	3.46	0.20	6.91	nr	**10.37**
20 mm	3.77	4.78	0.20	6.91	nr	**11.68**
25 mm	4.78	6.05	0.20	6.91	nr	**12.96**
32 mm	7.78	9.84	0.20	6.91	nr	**16.75**
40 mm	13.44	17.01	0.20	6.91	nr	**23.92**
50 mm	20.13	25.48	0.20	6.91	nr	**32.39**
63 mm	33.93	42.94	0.20	6.91	nr	**49.85**
Cap; PN25						
20 mm	1.88	2.38	0.20	6.91	nr	**9.29**
25 mm	2.52	3.19	0.20	6.91	nr	**10.10**
32 mm	3.66	4.64	0.20	6.91	nr	**11.54**
40 mm	5.03	6.36	0.20	6.91	nr	**13.27**
50 mm	7.03	8.90	0.20	6.91	nr	**15.80**
63 mm	11.20	14.17	0.20	6.91	nr	**21.08**

38 PIPED SUPPLY SYSTEMS

Item	Net Price £	Material £	Labour hours	Labour £	Unit	Total rate £
COLD WATER PIPELINES: SCREWED STEEL PIPEWORK						
Galvanized steel pipes; screwed and socketed joints; BS 1387: 1985						
Galvanized; medium, fixed vertically, with brackets measured separately, screwed joints are within the running length, but any flanges are additional						
10 mm dia.	7.12	9.01	0.51	17.61	m	**26.62**
15 mm dia.	6.42	8.12	0.52	17.96	m	**26.08**
20 mm dia.	7.25	9.17	0.55	18.99	m	**28.16**
25 mm dia.	10.13	12.82	0.60	20.72	m	**33.54**
32 mm dia.	12.54	15.88	0.67	23.14	m	**39.01**
40 mm dia.	14.57	18.44	0.75	25.90	m	**44.34**
50 mm dia.	23.84	30.18	0.85	29.35	m	**59.53**
65 mm dia.	32.31	40.89	0.93	32.12	m	**73.01**
80 mm dia.	41.84	52.96	1.07	36.95	m	**89.91**
100 mm dia.	59.22	74.94	1.46	50.42	m	**125.36**
125 mm dia.	94.15	119.16	1.72	59.40	m	**178.56**
150 mm dia.	109.34	138.37	1.96	67.68	m	**206.06**
Galvanized; heavy, fixed vertically, with brackets measured separately, screwed joints are within the running length, but any flanges are additional						
15 mm dia.	7.62	9.64	0.52	17.96	m	**27.60**
20 mm dia.	8.60	10.88	0.55	18.99	m	**29.87**
25 mm dia.	12.28	15.54	0.60	20.72	m	**36.26**
32 mm dia.	15.24	19.28	0.67	23.14	m	**42.42**
40 mm dia.	17.79	22.51	0.75	25.90	m	**48.41**
50 mm dia.	28.77	36.41	0.85	29.35	m	**65.76**
65 mm dia.	39.07	49.45	0.93	32.12	m	**81.57**
80 mm dia.	49.62	62.80	1.07	36.95	m	**99.75**
100 mm dia.	69.18	87.56	1.46	50.42	m	**137.97**
125 mm dia.	100.24	126.86	1.72	59.40	m	**186.26**
150 mm dia.	117.24	148.38	1.96	67.68	m	**216.06**
Galvanized; medium, fixed horizontaly or suspended at high level, with brackets measured separately, screwed joints are within the running length, but any flanges are additional						
10 mm dia.	7.12	9.01	0.51	17.61	m	**26.62**
15 mm dia.	6.42	8.12	0.52	17.96	m	**26.08**
20 mm dia.	7.25	9.17	0.55	18.99	m	**28.16**
25 mm dia.	10.13	12.82	0.60	20.72	m	**33.54**
32 mm dia.	12.54	15.88	0.67	23.14	m	**39.01**
40 mm dia.	14.57	18.44	0.75	25.90	m	**44.34**
50 mm dia.	23.84	30.18	0.85	29.35	m	**59.53**
65 mm dia.	32.31	40.89	0.93	32.12	m	**73.01**
80 mm dia.	41.84	52.96	1.07	36.95	m	**89.91**
100 mm dia.	59.22	74.94	1.46	50.42	m	**125.36**
125 mm dia.	94.15	119.16	1.72	59.40	m	**178.56**
150 mm dia.	109.34	138.37	1.96	67.68	m	**206.06**

38 PIPED SUPPLY SYSTEMS

Item	Net Price £	Material £	Labour hours	Labour £	Unit	Total rate £
COLD WATER PIPELINES: SCREWED STEEL PIPEWORK – cont						
Galvanized; heavy, fixed horizontaly or suspended at high level, with brackets measured separately, screwed joints are within the running length, but any flanges are additional						
15 mm dia.	7.62	9.64	0.52	17.96	m	27.60
20 mm dia.	8.60	10.88	0.55	18.99	m	29.87
25 mm dia.	12.28	15.54	0.60	20.72	m	36.26
32 mm dia.	15.24	19.28	0.67	23.14	m	42.42
40 mm dia.	17.79	22.51	0.75	25.90	m	48.41
50 mm dia.	28.77	36.41	0.85	29.35	m	65.76
65 mm dia.	39.07	49.45	0.93	32.12	m	81.57
80 mm dia.	49.62	62.80	1.07	36.95	m	99.75
100 mm dia.	69.18	87.56	1.46	50.42	m	137.97
125 mm dia.	100.24	126.86	1.72	59.40	m	186.26
150 mm dia.	117.24	148.38	1.96	67.68	m	216.06
FIXINGS						
For steel pipes; galvanized iron. For minimum fixing dimensions, refer to the Tables and Memoranda at the rear of the book						
Single pipe bracket, screw on, galvanized iron; screwed to wood						
15 mm dia.	1.93	2.44	0.14	4.83	nr	7.27
20 mm dia.	2.14	2.71	0.14	4.83	nr	7.54
25 mm dia.	2.50	3.16	0.17	5.87	nr	9.03
32 mm dia.	3.41	4.31	0.19	6.56	nr	10.87
40 mm dia.	5.06	6.41	0.22	7.60	nr	14.00
50 mm dia.	6.72	8.50	0.22	7.60	nr	16.10
65 mm dia.	7.93	10.03	0.28	9.67	nr	19.70
80 mm dia.	12.42	15.72	0.32	11.05	nr	26.77
100 mm dia.	17.95	22.72	0.35	12.09	nr	34.81
Single pipe bracket, screw on, galvanized iron; plugged and screwed						
15 mm dia.	1.93	2.44	0.25	8.63	nr	11.07
20 mm dia.	2.14	2.71	0.25	8.63	nr	11.34
25 mm dia.	2.50	3.16	0.30	10.36	nr	13.52
32 mm dia.	3.41	4.31	0.32	11.05	nr	15.36
40 mm dia.	5.06	6.41	0.32	11.05	nr	17.46
50 mm dia.	6.72	8.50	0.32	11.05	nr	19.55
65 mm dia.	7.93	10.03	0.35	12.09	nr	22.12
80 mm dia.	12.42	15.72	0.42	14.50	nr	30.22
100 mm dia.	17.95	22.72	0.42	14.50	nr	37.23
Single pipe bracket for building in, galvanized iron						
15 mm dia.	2.04	2.58	0.10	3.45	nr	6.03
20 mm dia.	2.30	2.91	0.11	3.80	nr	6.70

38 PIPED SUPPLY SYSTEMS

Item	Net Price £	Material £	Labour hours	Labour £	Unit	Total rate £
25 mm dia.	2.50	3.16	0.12	4.14	nr	**7.30**
32 mm dia.	2.64	3.34	0.14	4.83	nr	**8.17**
40 mm dia.	3.41	4.31	0.15	5.18	nr	**9.49**
50 mm dia.	4.11	5.20	0.16	5.53	nr	**10.72**
Pipe ring, single socket, galvanized iron						
15 mm dia.	2.04	2.58	0.10	3.45	nr	**6.03**
20 mm dia.	2.30	2.91	0.11	3.80	nr	**6.70**
25 mm dia.	2.50	3.16	0.12	4.14	nr	**7.30**
32 mm dia.	2.51	3.18	0.15	5.18	nr	**8.36**
40 mm dia.	3.29	4.17	0.15	5.18	nr	**9.35**
50 mm dia.	4.31	5.45	0.16	5.53	nr	**10.98**
65 mm dia.	5.99	7.58	0.30	10.36	nr	**17.94**
80 mm dia.	7.04	8.92	0.35	12.09	nr	**21.00**
100 mm dia.	11.32	14.33	0.40	13.81	nr	**28.14**
125 mm dia.	22.90	28.98	0.60	20.73	nr	**49.71**
150 mm dia.	27.65	34.99	0.77	26.59	nr	**61.58**
Pipe ring, double socket, galvanized iron						
15 mm dia.	18.02	22.81	0.10	3.45	nr	**26.27**
20 mm dia.	20.73	26.24	0.11	3.80	nr	**30.04**
25 mm dia.	22.86	28.93	0.12	4.14	nr	**33.07**
32 mm dia.	25.20	31.89	0.14	4.83	nr	**36.72**
40 mm dia.	31.83	40.28	0.15	5.18	nr	**45.46**
50 mm dia.	36.22	45.84	0.16	5.53	nr	**51.37**
Screw on backplate (Male), galvanized iron; plugged and screwed						
All sizes 15 mm to 50 mm × M12	1.60	2.02	0.10	3.45	nr	**5.47**
Screw on backplate (Female), galvanized iron; plugged and screwed						
All sizes 15 mm to 50 mm × M12	1.60	2.02	0.10	3.45	nr	**5.47**
Extra over channel sections for fabricated hangers and brackets						
Galvanized steel; including inserts, bolts, nuts, washers; fixed to backgrounds						
41 × 21 mm	8.39	10.62	0.29	10.01	m	**20.64**
41 × 41 mm	10.06	12.74	0.29	10.01	m	**22.75**
Threaded rods; metric thread; including nuts, washers etc.						
10 mm dia. × 600 mm long	2.71	3.43	0.18	6.22	nr	**9.64**
12 mm dia. × 600 mm long	4.24	5.37	0.18	6.22	nr	**11.58**
Extra over steel flanges, screwed and drilled; metric; BS 4504						
Screwed flanges; PN6						
15 mm dia.	22.90	28.99	0.35	12.09	nr	**41.07**
20 mm dia.	22.90	28.99	0.47	16.23	nr	**45.22**
25 mm dia.	22.90	28.99	0.53	18.30	nr	**47.29**
32 mm dia.	22.90	28.99	0.62	21.41	nr	**50.40**
40 mm dia.	22.90	28.99	0.70	24.17	nr	**53.16**
50 mm dia.	24.43	30.91	0.84	29.01	nr	**59.92**
65 mm dia.	33.96	42.98	1.03	35.57	nr	**78.55**

38 PIPED SUPPLY SYSTEMS

Item	Net Price £	Material £	Labour hours	Labour £	Unit	Total rate £
COLD WATER PIPELINES: SCREWED STEEL PIPEWORK – cont						
Extra over steel flanges, screwed and drilled – cont						
Screwed flanges; PN6 – cont						
80 mm dia.	47.97	60.71	1.23	42.48	nr	**103.19**
100 mm dia.	56.72	71.79	1.41	48.69	nr	**120.48**
125 mm dia.	104.04	131.67	1.77	61.12	nr	**192.79**
150 mm dia.	104.04	131.67	2.21	76.32	nr	**207.99**
Screwed flanges; PN16						
15 mm dia.	29.13	36.87	0.35	12.09	nr	**48.96**
20 mm dia.	29.13	36.87	0.47	16.23	nr	**53.10**
25 mm dia.	29.13	36.87	0.53	18.30	nr	**55.18**
32 mm dia.	31.11	39.37	0.62	21.41	nr	**60.78**
40 mm dia.	31.11	39.37	0.70	24.17	nr	**63.54**
50 mm dia.	37.95	48.03	0.84	29.01	nr	**77.04**
65 mm dia.	47.38	59.97	1.03	35.57	nr	**95.54**
80 mm dia.	57.71	73.04	1.23	42.48	nr	**115.51**
100 mm dia.	64.73	81.92	1.41	48.69	nr	**130.61**
125 mm dia.	113.30	143.39	1.77	61.12	nr	**204.51**
150 mm dia.	111.47	141.08	2.21	76.32	nr	**217.39**
Extra over steel flanges, screwed and drilled; imperial; BS 10						
Screwed flanges; Table E						
½" dia.	39.13	49.52	0.35	12.09	nr	**61.60**
¾" dia.	39.13	49.52	0.47	16.23	nr	**65.75**
1" dia.	39.13	49.52	0.53	18.30	nr	**67.82**
1 ¼" dia.	39.13	49.52	0.62	21.41	nr	**70.93**
1 ½" dia.	39.13	49.52	0.70	24.17	nr	**73.69**
2" dia.	39.13	49.52	0.84	29.01	nr	**78.53**
2 ½" dia.	46.58	58.96	1.03	35.57	nr	**94.52**
3" dia.	56.12	71.03	1.23	42.48	nr	**113.50**
4" dia.	71.40	90.36	1.41	48.69	nr	**139.05**
5" dia.	151.33	191.52	1.77	61.12	nr	**252.64**
Extra over steel flange connections						
Bolted connection between pair of flanges; including gasket, bolts, nuts and washers						
50 mm dia.	75.07	95.01	0.53	18.30	nr	**113.31**
65 mm dia.	94.58	119.70	0.53	18.30	nr	**138.00**
80 mm dia.	108.50	137.32	0.53	18.30	nr	**155.62**
100 mm dia.	129.48	163.88	0.53	18.30	nr	**182.18**
125 mm dia.	244.57	309.53	0.61	21.06	nr	**330.59**
150 mm dia.	251.55	318.37	0.90	31.08	nr	**349.45**

38 PIPED SUPPLY SYSTEMS

Item	Net Price £	Material £	Labour hours	Labour £	Unit	Total rate £
Extra over heavy steel tubular fittings; BS 1387						
Long screw connection with socket and backnut						
15 mm dia.	8.39	10.62	0.63	21.76	nr	**32.38**
20 mm dia.	10.54	13.34	0.84	29.01	nr	**42.35**
25 mm dia.	13.78	17.44	0.95	32.81	nr	**50.24**
32 mm dia.	18.14	22.96	1.11	38.33	nr	**61.29**
40 mm dia.	22.08	27.95	1.28	44.20	nr	**72.15**
50 mm dia.	32.59	41.24	1.53	52.83	nr	**94.08**
65 mm dia.	74.86	94.74	1.87	64.58	nr	**159.32**
80 mm dia.	96.81	122.52	2.21	76.32	nr	**198.84**
100 mm dia.	109.83	139.00	3.05	105.32	nr	**244.32**
Running nipple						
15 mm dia.	2.11	2.67	0.50	17.27	nr	**19.93**
20 mm dia.	2.62	3.31	0.68	23.48	nr	**26.79**
25 mm dia.	2.81	3.56	0.77	26.59	nr	**30.15**
32 mm dia.	4.54	5.74	0.90	31.08	nr	**36.82**
40 mm dia.	6.13	7.75	1.03	35.57	nr	**43.32**
50 mm dia.	9.31	11.79	1.23	42.48	nr	**54.26**
65 mm dia.	20.01	25.33	1.50	51.80	nr	**77.13**
80 mm dia.	31.21	39.50	1.78	61.47	nr	**100.97**
100 mm dia.	48.92	61.91	2.38	82.19	nr	**144.10**
Barrel nipple						
15 mm dia.	1.76	2.23	0.50	17.27	nr	**19.49**
20 mm dia.	2.65	3.36	0.68	23.48	nr	**26.84**
25 mm dia.	2.98	3.77	0.77	26.59	nr	**30.36**
32 mm dia.	4.94	6.25	0.90	31.08	nr	**37.33**
40 mm dia.	5.49	6.95	1.03	35.57	nr	**42.52**
50 mm dia.	7.85	9.93	1.23	42.48	nr	**52.41**
65 mm dia.	16.78	21.23	1.50	51.80	nr	**73.03**
80 mm dia.	23.42	29.64	1.78	61.47	nr	**91.11**
100 mm dia.	42.37	53.62	2.38	82.19	nr	**135.81**
125 mm dia.	78.70	99.61	2.87	99.11	nr	**198.71**
150 mm dia.	123.99	156.92	3.39	117.07	nr	**273.98**
Close taper nipple						
15 mm dia.	2.49	3.15	0.50	17.27	nr	**20.42**
20 mm dia.	3.24	4.10	0.68	23.48	nr	**27.58**
25 mm dia.	4.24	5.37	0.77	26.59	nr	**31.96**
32 mm dia.	6.32	8.00	0.90	31.08	nr	**39.08**
40 mm dia.	7.82	9.90	1.03	35.57	nr	**45.47**
50 mm dia.	12.03	15.22	1.23	42.48	nr	**57.70**
65 mm dia.	19.00	24.04	1.50	51.80	nr	**75.84**
80 mm dia.	31.14	39.41	1.78	61.47	nr	**100.87**
100 mm dia.	59.20	74.92	2.38	82.19	nr	**157.11**
90° bend with socket						
15 mm dia.	6.87	8.69	0.64	22.10	nr	**30.80**
20 mm dia.	9.25	11.71	0.85	29.35	nr	**41.06**
25 mm dia.	14.15	17.91	0.97	33.50	nr	**51.40**
32 mm dia.	20.30	25.69	1.12	38.68	nr	**64.37**
40 mm dia.	24.79	31.37	1.29	44.55	nr	**75.92**
50 mm dia.	38.55	48.79	1.55	53.53	nr	**102.32**
65 mm dia.	77.72	98.37	1.89	65.27	nr	**163.63**

38 PIPED SUPPLY SYSTEMS

Item	Net Price £	Material £	Labour hours	Labour £	Unit	Total rate £
COLD WATER PIPELINES: SCREWED STEEL PIPEWORK – cont						
Extra over heavy steel tubular fittings – cont						
80 mm dia.	115.36	146.00	2.24	77.35	nr	**223.35**
100 mm dia.	204.55	258.88	3.09	106.71	nr	**365.58**
125 mm dia.	501.03	634.11	3.92	135.37	nr	**769.48**
150 mm dia.	752.22	952.01	4.74	163.68	nr	**1115.70**
Extra over heavy steel fittings; BS 1740						
Plug						
15 mm dia.	1.90	2.40	0.28	9.67	nr	**12.07**
20 mm dia.	2.96	3.75	0.38	13.12	nr	**16.87**
25 mm dia.	5.21	6.59	0.44	15.19	nr	**21.79**
32 mm dia.	8.08	10.23	0.51	17.61	nr	**27.84**
40 mm dia.	8.92	11.28	0.59	20.37	nr	**31.66**
50 mm dia.	12.75	16.13	0.70	24.17	nr	**40.31**
65 mm dia.	30.48	38.57	0.85	29.35	nr	**67.93**
80 mm dia.	57.03	72.17	1.00	34.53	nr	**106.71**
100 mm dia.	109.48	138.56	1.44	49.73	nr	**188.28**
Socket						
15 mm dia.	2.11	2.67	0.64	22.10	nr	**24.77**
20 mm dia.	2.39	3.03	0.85	29.35	nr	**32.38**
25 mm dia.	3.39	4.28	0.97	33.50	nr	**37.78**
32 mm dia.	4.87	6.17	1.12	38.68	nr	**44.84**
40 mm dia.	5.91	7.49	1.29	44.55	nr	**52.03**
50 mm dia.	9.13	11.55	1.55	53.53	nr	**65.08**
65 mm dia.	18.13	22.94	1.89	65.27	nr	**88.21**
80 mm dia.	23.45	29.68	2.24	77.35	nr	**107.03**
100 mm dia.	44.14	55.87	3.09	106.71	nr	**162.57**
150 mm dia.	105.36	133.35	4.74	163.68	nr	**297.03**
Elbow, female/female						
15 mm dia.	12.11	15.33	0.64	22.10	nr	**37.43**
20 mm dia.	15.80	19.99	0.85	29.35	nr	**49.35**
25 mm dia.	21.44	27.13	0.97	33.50	nr	**60.63**
32 mm dia.	39.92	50.52	1.12	38.68	nr	**89.19**
40 mm dia.	47.60	60.25	1.29	44.55	nr	**104.79**
50 mm dia.	78.03	98.76	1.55	53.53	nr	**152.28**
65 mm dia.	190.66	241.30	1.89	65.27	nr	**306.57**
80 mm dia.	227.34	287.72	2.24	77.35	nr	**365.08**
100 mm dia.	393.71	498.28	3.09	106.71	nr	**604.99**
Equal tee						
15 mm dia.	15.07	19.07	0.91	31.42	nr	**50.49**
20 mm dia.	17.50	22.14	1.22	42.13	nr	**64.27**
25 mm dia.	25.79	32.64	1.40	48.35	nr	**80.99**
32 mm dia.	53.27	67.42	1.62	55.94	nr	**123.36**
40 mm dia.	57.97	73.37	1.86	64.23	nr	**137.60**
50 mm dia.	94.26	119.30	2.21	76.32	nr	**195.62**
65 mm dia.	227.97	288.52	2.72	93.93	nr	**382.45**
80 mm dia.	244.68	309.66	3.21	110.85	nr	**420.51**
100 mm dia.	393.77	498.36	4.44	153.32	nr	**651.69**

38 PIPED SUPPLY SYSTEMS

Item	Net Price £	Material £	Labour hours	Labour £	Unit	Total rate £
Extra over malleable iron fittings; BS 143						
Cap						
15 mm dia.	5.59	7.07	0.32	11.05	nr	18.13
20 mm dia.	5.85	7.40	0.43	14.85	nr	22.25
25 mm dia.	10.58	13.39	0.49	16.92	nr	30.31
32 mm dia.	20.08	25.41	0.58	20.03	nr	45.44
40 mm dia.	21.55	27.27	0.66	22.79	nr	50.06
50 mm dia.	34.25	43.35	0.78	26.94	nr	70.29
65 mm dia.	58.87	74.50	0.96	33.15	nr	107.65
80 mm dia.	97.11	122.90	1.13	39.02	nr	161.92
100 mm dia.	165.54	209.51	1.70	58.71	nr	268.22
Plain plug, hollow						
15 mm dia.	2.72	3.45	0.28	9.67	nr	13.12
20 mm dia.	4.22	5.34	0.38	13.12	nr	18.46
25 mm dia.	7.40	9.37	0.44	15.19	nr	24.56
32 mm dia.	11.47	14.51	0.51	17.61	nr	32.12
40 mm dia.	12.66	16.03	0.59	20.37	nr	36.40
50 mm dia.	18.14	22.96	0.70	24.17	nr	47.13
65 mm dia.	43.31	54.81	0.85	29.35	nr	84.16
80 mm dia.	81.07	102.60	1.00	34.53	nr	137.13
100 mm dia.	155.65	196.99	1.44	49.73	nr	246.71
Plain plug, solid						
15 mm dia.	3.38	4.28	0.29	10.01	nr	14.29
20 mm dia.	3.58	4.53	0.38	13.12	nr	17.65
25 mm dia.	4.76	6.03	0.44	15.19	nr	21.22
32 mm dia.	7.34	9.29	0.51	17.61	nr	26.91
40 mm dia.	9.88	12.51	0.59	20.37	nr	32.88
50 mm dia.	12.99	16.44	0.70	24.17	nr	40.61
Elbow, male/female						
15 mm dia.	1.75	2.22	0.64	22.10	nr	24.32
20 mm dia.	2.37	3.00	0.85	29.35	nr	32.35
25 mm dia.	3.92	4.96	0.97	33.50	nr	38.46
32 mm dia.	8.88	11.24	1.12	38.68	nr	49.92
40 mm dia.	12.28	15.54	1.29	44.55	nr	60.09
50 mm dia.	15.83	20.03	1.55	53.53	nr	73.56
65 mm dia.	35.28	44.65	1.89	65.27	nr	109.92
80 mm dia.	48.26	61.07	2.24	77.35	nr	138.43
100 mm dia.	84.35	106.75	3.09	106.71	nr	213.46
Elbow						
15 mm dia.	1.57	1.99	0.64	22.10	nr	24.09
20 mm dia.	2.15	2.73	0.85	29.35	nr	32.08
25 mm dia.	3.35	4.24	0.97	33.50	nr	37.74
32 mm dia.	6.97	8.83	1.12	38.68	nr	47.50
40 mm dia.	10.42	13.19	1.29	44.55	nr	57.74
50 mm dia.	12.22	15.47	1.55	53.53	nr	68.99
65 mm dia.	27.25	34.49	1.89	65.27	nr	99.76
80 mm dia.	40.01	50.64	2.24	77.35	nr	127.99
100 mm dia.	68.69	86.93	3.09	106.71	nr	193.64
125 mm dia.	164.86	208.65	4.44	153.32	nr	361.97
150 mm dia.	306.93	388.45	5.79	199.94	nr	588.40

Material Costs/Prices for Measured Works

38 PIPED SUPPLY SYSTEMS

Item	Net Price £	Material £	Labour hours	Labour £	Unit	Total rate £
COLD WATER PIPELINES: SCREWED STEEL PIPEWORK – cont						
Extra over malleable iron fittings – cont						
45° elbow						
15 mm dia.	4.05	5.13	0.64	22.10	nr	**27.23**
20 mm dia.	5.02	6.35	0.85	29.35	nr	**35.71**
25 mm dia.	6.90	8.73	0.97	33.50	nr	**42.23**
32 mm dia.	16.06	20.32	1.12	38.68	nr	**59.00**
40 mm dia.	18.89	23.91	1.29	44.55	nr	**68.46**
50 mm dia.	25.88	32.76	1.55	53.53	nr	**86.28**
65 mm dia.	36.38	46.04	1.89	65.27	nr	**111.31**
80 mm dia.	54.74	69.28	2.24	77.35	nr	**146.64**
100 mm dia.	105.52	133.55	3.09	106.71	nr	**240.26**
150 mm dia.	321.14	406.43	5.79	199.94	nr	**606.37**
Bend, male/female						
15 mm dia.	3.09	3.92	0.64	22.10	nr	**26.02**
20 mm dia.	5.09	6.44	0.85	29.35	nr	**35.80**
25 mm dia.	7.17	9.08	0.97	33.50	nr	**42.57**
32 mm dia.	12.12	15.34	1.12	38.68	nr	**54.02**
40 mm dia.	17.77	22.49	1.29	44.55	nr	**67.03**
50 mm dia.	33.40	42.27	1.55	53.53	nr	**95.79**
65 mm dia.	51.11	64.68	1.89	65.27	nr	**129.95**
80 mm dia.	78.72	99.62	2.24	77.35	nr	**176.97**
100 mm dia.	194.92	246.69	3.09	106.71	nr	**353.40**
Bend, male						
15 mm dia.	7.12	9.01	0.64	22.10	nr	**31.11**
20 mm dia.	7.99	10.11	0.85	29.35	nr	**39.46**
25 mm dia.	11.72	14.83	0.97	33.50	nr	**48.33**
32 mm dia.	25.92	32.81	1.12	38.68	nr	**71.49**
40 mm dia.	36.36	46.02	1.29	44.55	nr	**90.57**
50 mm dia.	55.17	69.83	1.55	53.53	nr	**123.35**
Bend, female						
15 mm dia.	3.19	4.04	0.64	22.10	nr	**26.14**
20 mm dia.	4.55	5.76	0.85	29.35	nr	**35.11**
25 mm dia.	6.39	8.09	0.97	33.50	nr	**41.58**
32 mm dia.	12.45	15.76	1.12	38.68	nr	**54.43**
40 mm dia.	14.82	18.75	1.29	44.55	nr	**63.30**
50 mm dia.	23.34	29.54	1.55	53.53	nr	**83.07**
65 mm dia.	51.11	64.68	1.89	65.27	nr	**129.95**
80 mm dia.	75.79	95.92	2.24	77.35	nr	**173.28**
100 mm dia.	159.03	201.27	3.09	106.71	nr	**307.97**
125 mm dia.	320.84	406.05	4.44	153.32	nr	**559.37**
150 mm dia.	704.86	892.07	5.79	199.94	nr	**1092.01**
Return bend						
15 mm dia.	16.27	20.59	0.64	22.10	nr	**42.69**
20 mm dia.	26.31	33.30	0.85	29.35	nr	**62.65**
25 mm dia.	32.83	41.55	0.97	33.50	nr	**75.04**
32 mm dia.	45.58	57.68	1.12	38.68	nr	**96.36**
40 mm dia.	61.73	78.13	1.29	44.55	nr	**122.67**
50 mm dia.	94.26	119.29	1.55	53.53	nr	**172.82**

38 PIPED SUPPLY SYSTEMS

Item	Net Price £	Material £	Labour hours	Labour £	Unit	Total rate £
Equal socket, parallel thread						
15 mm dia.	1.53	1.93	0.64	22.10	nr	**24.03**
20 mm dia.	2.00	2.53	0.85	29.35	nr	**31.88**
25 mm dia.	2.52	3.19	0.97	33.50	nr	**36.69**
32 mm dia.	5.62	7.11	1.12	38.68	nr	**45.79**
40 mm dia.	7.63	9.66	1.29	44.55	nr	**54.20**
50 mm dia.	11.04	13.97	1.55	53.53	nr	**67.49**
65 mm dia.	17.51	22.16	1.89	65.27	nr	**87.43**
80 mm dia.	24.80	31.38	2.24	77.35	nr	**108.74**
100 mm dia.	40.94	51.81	3.09	106.71	nr	**158.52**
Concentric reducing socket						
20 × 15 mm dia.	2.44	3.09	0.76	26.24	nr	**29.33**
25 × 15 mm dia.	3.19	4.04	0.86	29.70	nr	**33.74**
25 × 20 mm dia.	3.01	3.81	0.86	29.70	nr	**33.51**
32 × 25 mm dia.	5.78	7.31	1.01	34.88	nr	**42.19**
40 × 25 mm dia.	7.63	9.66	1.16	40.06	nr	**49.71**
40 × 32 mm dia.	8.46	10.70	1.16	40.06	nr	**50.76**
50 × 25 mm dia.	14.63	18.52	1.38	47.65	nr	**66.17**
50 × 40 mm dia.	11.84	14.98	1.38	47.65	nr	**62.63**
65 × 50 mm dia.	20.66	26.15	1.69	58.36	nr	**84.51**
80 × 50 mm dia.	25.78	32.63	2.00	69.07	nr	**101.69**
100 × 50 mm dia.	51.38	65.03	2.75	94.96	nr	**159.99**
100 × 80 mm dia.	153.71	194.53	2.75	94.96	nr	**289.50**
150 × 100 mm dia.	142.83	180.76	4.10	141.58	nr	**322.35**
Eccentric reducing socket						
20 × 15 mm dia.	5.53	7.00	0.76	26.24	nr	**33.25**
25 × 15 mm dia.	15.69	19.85	0.86	29.70	nr	**49.55**
25 × 20 mm dia.	17.81	22.54	0.86	29.70	nr	**52.24**
32 × 25 mm dia.	23.03	29.15	1.01	34.88	nr	**64.02**
40 × 25 mm dia.	26.38	33.39	1.16	40.06	nr	**73.45**
40 × 32 mm dia.	28.73	36.37	1.16	40.06	nr	**76.42**
50 × 25 mm dia.	28.81	36.46	1.18	40.75	nr	**77.20**
50 × 40 mm dia.	28.88	36.55	1.28	44.20	nr	**80.75**
65 × 50 mm dia.	29.26	37.03	1.69	58.36	nr	**95.39**
80 × 50 mm dia.	31.17	39.45	2.00	69.07	nr	**108.52**
Hexagon bush						
20 × 15 mm dia.	1.37	1.73	0.37	12.78	nr	**14.51**
25 × 15 mm dia.	1.88	2.38	0.43	14.85	nr	**17.23**
25 × 20 mm dia.	1.75	2.22	0.43	14.85	nr	**17.07**
32 × 25 mm dia.	2.38	3.01	0.51	17.61	nr	**20.63**
40 × 25 mm dia.	3.55	4.49	0.58	20.03	nr	**24.52**
40 × 32 mm dia.	3.55	4.49	0.58	20.03	nr	**24.52**
50 × 25 mm dia.	7.52	9.51	0.71	24.52	nr	**34.03**
50 × 40 mm dia.	7.00	8.86	0.71	24.52	nr	**33.38**
65 × 50 mm dia.	12.93	16.37	0.84	29.01	nr	**45.38**
80 × 50 mm dia.	19.52	24.71	1.00	34.53	nr	**59.24**
100 × 50 mm dia.	45.22	57.23	1.52	52.49	nr	**109.72**
100 × 80 mm dia.	37.62	47.61	1.52	52.49	nr	**100.10**
150 × 100 mm dia.	135.16	171.05	2.48	85.64	nr	**256.69**

38 PIPED SUPPLY SYSTEMS

Item	Net Price £	Material £	Labour hours	Labour £	Unit	Total rate £
COLD WATER PIPELINES: SCREWED STEEL PIPEWORK – cont						
Extra over malleable iron fittings – cont						
Hexagon nipple						
15 mm dia.	1.50	1.89	0.28	9.67	nr	**11.56**
20 mm dia.	1.67	2.11	0.38	13.12	nr	**15.23**
25 mm dia.	2.37	3.00	0.44	15.19	nr	**18.19**
32 mm dia.	5.03	6.37	0.51	17.61	nr	**23.98**
40 mm dia.	5.78	7.31	0.59	20.37	nr	**27.68**
50 mm dia.	10.54	13.34	0.70	24.17	nr	**37.51**
65 mm dia.	17.64	22.32	0.85	29.35	nr	**51.68**
80 mm dia.	24.38	30.86	1.00	34.53	nr	**65.39**
100 mm dia.	43.28	54.77	1.44	49.73	nr	**104.50**
150 mm dia.	119.81	151.63	2.32	80.12	nr	**231.75**
Union, male/female						
15 mm dia.	7.40	9.37	0.64	22.10	nr	**31.47**
20 mm dia.	9.08	11.50	0.85	29.35	nr	**40.85**
25 mm dia.	10.54	13.34	0.97	33.50	nr	**46.83**
32 mm dia.	18.62	23.57	1.12	38.68	nr	**62.25**
40 mm dia.	23.83	30.16	1.29	44.55	nr	**74.70**
50 mm dia.	37.46	47.41	1.55	53.53	nr	**100.94**
65 mm dia.	83.83	106.10	1.89	65.27	nr	**171.37**
Union, female						
15 mm dia.	7.00	8.86	0.64	22.10	nr	**30.96**
20 mm dia.	7.70	9.75	0.85	29.35	nr	**39.10**
25 mm dia.	9.03	11.42	0.97	33.50	nr	**44.92**
32 mm dia.	15.57	19.71	1.12	38.68	nr	**58.38**
40 mm dia.	17.58	22.25	1.29	44.55	nr	**66.80**
50 mm dia.	26.21	33.17	1.55	53.53	nr	**86.70**
65 mm dia.	67.12	84.95	1.89	65.27	nr	**150.22**
80 mm dia.	88.73	112.29	2.24	77.35	nr	**189.64**
100 mm dia.	168.95	213.83	3.09	106.71	nr	**320.53**
Union elbow, male/female						
15 mm dia.	11.17	14.13	0.64	22.10	nr	**36.23**
20 mm dia.	14.00	17.72	0.85	29.35	nr	**47.08**
25 mm dia.	19.66	24.89	0.97	33.50	nr	**58.38**
Twin elbow						
15 mm dia.	10.74	13.59	0.91	31.42	nr	**45.01**
20 mm dia.	11.88	15.03	1.22	42.13	nr	**57.16**
25 mm dia.	19.24	24.35	1.39	48.00	nr	**72.35**
32 mm dia.	38.54	48.78	1.62	55.94	nr	**104.73**
40 mm dia.	48.80	61.76	1.86	64.23	nr	**125.99**
50 mm dia.	62.72	79.37	2.21	76.32	nr	**155.69**
65 mm dia.	101.36	128.28	2.72	93.93	nr	**222.21**
80 mm dia.	115.04	145.59	3.21	110.85	nr	**256.44**
Equal tee						
15 mm dia.	2.15	2.73	0.91	31.42	nr	**34.15**
20 mm dia.	3.14	3.97	1.22	42.13	nr	**46.10**
25 mm dia.	4.52	5.72	1.39	48.00	nr	**53.72**

38 PIPED SUPPLY SYSTEMS

Item	Net Price £	Material £	Labour hours	Labour £	Unit	Total rate £
32 mm dia.	9.57	12.11	1.62	55.94	nr	**68.05**
40 mm dia.	13.06	16.53	1.86	64.23	nr	**80.76**
50 mm dia.	18.79	23.79	2.21	76.32	nr	**100.10**
65 mm dia.	44.09	55.80	2.72	93.93	nr	**149.73**
80 mm dia.	51.38	65.03	3.21	110.85	nr	**175.87**
100 mm dia.	93.16	117.90	4.44	153.32	nr	**271.23**
125 mm dia.	172.82	218.72	5.38	185.79	nr	**404.50**
150 mm dia.	413.27	523.03	6.31	217.90	nr	**740.93**
Tee reducing on branch						
20 × 15 mm dia.	3.22	4.08	1.22	42.13	nr	**46.21**
25 × 15 mm dia.	4.41	5.58	1.39	48.00	nr	**53.58**
25 × 20 mm dia.	5.02	6.35	1.39	48.00	nr	**54.35**
32 × 25 mm dia.	9.68	12.25	1.62	55.94	nr	**68.20**
40 × 25 mm dia.	12.28	15.54	1.86	64.23	nr	**79.77**
40 × 32 mm dia.	18.02	22.81	1.86	64.23	nr	**87.04**
50 × 25 mm dia.	16.31	20.65	2.21	76.32	nr	**96.96**
50 × 40 mm dia.	25.37	32.11	2.21	76.32	nr	**108.42**
65 × 50 mm dia.	39.14	49.54	2.72	93.93	nr	**143.47**
80 × 50 mm dia.	52.92	66.97	3.21	110.85	nr	**177.82**
100 × 50 mm dia.	87.60	110.87	4.44	153.32	nr	**264.19**
100 × 80 mm dia.	135.16	171.05	4.44	153.32	nr	**324.38**
150 × 100 mm dia.	219.23	277.46	6.28	216.86	nr	**494.33**
Equal pitcher tee						
15 mm dia.	7.49	9.47	0.91	31.42	nr	**40.90**
20 mm dia.	9.23	11.68	1.22	42.13	nr	**53.81**
25 mm dia.	13.80	17.47	1.39	48.00	nr	**65.47**
32 mm dia.	21.40	27.09	1.62	55.94	nr	**83.03**
40 mm dia.	33.14	41.94	1.86	64.23	nr	**106.17**
50 mm dia.	46.49	58.83	2.21	76.32	nr	**135.15**
65 mm dia.	66.17	83.74	2.72	93.93	nr	**177.67**
80 mm dia.	103.24	130.66	3.21	110.85	nr	**241.51**
100 mm dia.	232.30	293.99	4.44	153.32	nr	**447.32**
Cross						
15 mm dia.	6.20	7.85	1.00	34.53	nr	**42.38**
20 mm dia.	9.73	12.31	1.33	45.93	nr	**58.24**
25 mm dia.	12.31	15.57	1.51	52.14	nr	**67.72**
32 mm dia.	18.31	23.17	1.77	61.12	nr	**84.30**
40 mm dia.	24.66	31.20	2.02	69.76	nr	**100.96**
50 mm dia.	38.33	48.51	2.42	83.57	nr	**132.08**
65 mm dia.	54.74	69.28	2.97	102.56	nr	**171.85**
80 mm dia.	72.77	92.10	3.50	120.86	nr	**212.96**
100 mm dia.	150.19	190.08	4.84	167.14	nr	**357.21**

38 PIPED SUPPLY SYSTEMS

Item	Net Price £	Material £	Labour hours	Labour £	Unit	Total rate £
COLD WATER PIPELINES: PIPEWORK ANCILLARIES – VALVES						
Regulators						
Gunmetal; self-acting two port thermostat; single seat; screwed; normally closed; with adjustable or fixed bleed device						
25 mm dia.	1024.10	1296.10	1.46	50.42	nr	**1346.51**
32 mm dia.	1053.64	1333.49	1.45	50.07	nr	**1383.56**
40 mm dia.	1126.21	1425.34	1.55	53.54	nr	**1478.87**
50 mm dia.	1355.74	1715.83	1.68	58.01	nr	**1773.84**
Self-acting temperature regulator for storage calorifier; integral sensing element and pocket; screwed ends						
15 mm dia.	865.22	1095.02	1.32	45.58	nr	**1140.60**
25 mm dia.	949.72	1201.96	1.52	52.49	nr	**1254.45**
32 mm dia.	1226.41	1552.14	1.79	61.81	nr	**1613.96**
40 mm dia.	1500.28	1898.76	1.99	68.72	nr	**1967.48**
50 mm dia.	1753.73	2219.52	2.26	78.04	nr	**2297.56**
Self-acting temperature regulator for storage calorifier; integral sensing element and pocket; flanged ends; bolted connection						
15 mm dia.	1270.13	1607.48	0.61	21.06	nr	**1628.54**
25 mm dia.	1453.66	1839.76	0.72	24.86	nr	**1864.62**
32 mm dia.	1832.39	2319.08	0.94	32.46	nr	**2351.54**
40 mm dia.	2170.33	2746.76	1.03	35.57	nr	**2782.33**
50 mm dia.	2519.91	3189.20	1.18	40.75	nr	**3229.95**
Chrome plated thermostatic mixing valves including non-return valves and inlet swivel connections with strainers; copper compression fittings						
15 mm dia.	140.19	177.43	0.69	23.83	nr	**201.26**
Chrome plated thermostatic mixing valves including non-return valves and inlet swivel connections with angle pattern combined isolating valves and strainers; copper compression fittings						
15 mm dia.	230.06	291.17	0.69	23.83	nr	**314.99**
Gunmetal thermostatic mixing valves including non-return valves and inlet swivel connections with strainers; copper compression fittings						
15 mm dia.	290.96	368.23	0.69	23.83	nr	**392.06**
Gunmetal thermostatic mixing valves including non-return valves and inlet swivel connections with angle pattern combined isolating valves and strainers; copper compression fittings						
15 mm dia.	305.90	387.14	0.69	23.83	nr	**410.97**

38 PIPED SUPPLY SYSTEMS

Item	Net Price £	Material £	Labour hours	Labour £	Unit	Total rate £
Ball float valves						
Bronze, equilibrium; copper float; working pressure cold services up to 16 bar; flanged ends; BS 4504 Table 16/21; bolted connections						
25 mm dia.	217.24	274.93	1.04	35.91	nr	**310.85**
32 mm dia.	301.36	381.40	1.22	42.13	nr	**423.53**
40 mm dia.	408.75	517.32	1.38	47.65	nr	**564.97**
50 mm dia.	642.30	812.89	1.66	57.32	nr	**870.21**
65 mm dia.	688.78	871.72	1.93	66.65	nr	**938.37**
80 mm dia.	768.02	972.01	2.16	74.59	nr	**1046.60**
Heavy, equilibrium; with long tail and backnut; copper float; screwed for iron						
25 mm dia.	158.68	200.83	1.58	54.56	nr	**255.39**
32 mm dia.	217.85	275.71	1.78	61.47	nr	**337.18**
40 mm dia.	307.13	388.70	1.90	65.61	nr	**454.31**
50 mm dia.	403.32	510.44	2.65	91.51	nr	**601.95**
Brass, ball valve; BS 1212; copper float; screwed						
15 mm dia.	10.29	13.02	0.25	8.63	nr	**21.66**
22 mm dia.	18.70	23.67	0.29	10.01	nr	**33.69**
28 mm dia.	81.47	103.11	0.35	12.09	nr	**115.20**
Gate valves						
DZR copper alloy wedge non-rising stem; capillary joint to copper						
15 mm dia.	21.61	27.35	0.84	29.01	nr	**56.35**
22 mm dia.	26.52	33.56	1.01	34.88	nr	**68.44**
28 mm dia.	35.85	45.38	1.19	41.09	nr	**86.47**
35 mm dia.	64.36	81.45	1.38	47.65	nr	**129.11**
42 mm dia.	109.34	138.38	1.62	55.94	nr	**194.32**
54 mm dia.	153.21	193.90	1.94	66.99	nr	**260.89**
Cocks; capillary joints to copper						
Stopcock; brass head with gun metal body						
15 mm dia.	9.95	12.59	0.45	15.54	nr	**28.13**
22 mm dia.	18.56	23.49	0.46	15.88	nr	**39.37**
28 mm dia.	52.76	66.78	0.54	18.65	nr	**85.42**
Lockshield stop cocks; brass head with gun metal body						
15 mm dia.	11.48	14.52	0.45	15.54	nr	**30.06**
22 mm dia.	11.48	14.52	0.46	15.88	nr	**30.41**
28 mm dia.	11.48	14.52	0.54	18.65	nr	**33.17**
DZR stopcock; brass head with gun metal body						
15 mm dia.	27.23	34.46	0.45	15.54	nr	**50.00**
22 mm dia.	47.16	59.68	0.46	15.88	nr	**75.57**
28 mm dia.	78.60	99.47	0.54	18.65	nr	**118.12**

38 PIPED SUPPLY SYSTEMS

Item	Net Price £	Material £	Labour hours	Labour £	Unit	Total rate £
COLD WATER PIPELINES: PIPEWORK ANCILLARIES – VALVES – cont						
Cocks – cont						
Gunmetal stopcock						
35 mm dia.	102.72	130.00	0.69	23.83	nr	**153.83**
42 mm dia.	136.43	172.66	0.71	24.52	nr	**197.18**
54 mm dia.	203.77	257.89	0.81	27.97	nr	**285.87**
Double union stopcock						
15 mm dia.	29.27	37.04	0.60	20.72	nr	**57.76**
22 mm dia.	41.15	52.09	0.60	20.72	nr	**72.80**
28 mm dia.	73.21	92.66	0.69	23.83	nr	**116.49**
Double union DZR stopcock						
15 mm dia.	47.77	60.46	0.60	20.72	nr	**81.18**
22 mm dia.	58.74	74.34	0.61	21.06	nr	**95.40**
28 mm dia.	108.55	137.38	0.69	23.83	nr	**161.21**
Double union gun metal stopcock						
35 mm dia.	180.86	228.89	0.63	21.76	nr	**250.65**
42 mm dia.	248.22	314.14	0.67	23.14	nr	**337.28**
54 mm dia.	390.49	494.20	0.85	29.35	nr	**523.55**
Double union stopcock with easy clean cover						
15 mm dia.	50.71	64.18	0.60	20.72	nr	**84.90**
22 mm dia.	63.28	80.08	0.61	21.06	nr	**101.15**
28 mm dia.	117.61	148.84	0.69	23.83	nr	**172.67**
Combined stopcock and drain						
15 mm dia.	50.30	63.66	0.67	23.14	nr	**86.80**
22 mm dia.	61.37	77.67	0.68	23.48	nr	**101.15**
Combined DZR stopcock and drain						
15 mm dia.	66.48	84.13	0.67	23.14	nr	**107.27**
Gate valve						
DZR copper alloy wedge non-rising stem; compression joint to copper						
15 mm dia.	21.61	27.35	0.84	29.01	nr	**56.35**
22 mm dia.	26.52	33.56	1.01	34.88	nr	**68.44**
28 mm dia.	35.85	45.38	1.19	41.09	nr	**86.47**
35 mm dia.	64.36	81.45	1.38	47.65	nr	**129.11**
42 mm dia.	109.34	138.38	1.62	55.94	nr	**194.32**
54 mm dia.	153.21	193.90	1.94	66.99	nr	**260.89**
Cocks; compression joints to copper						
Stopcock; brass head gun metal body						
15 mm dia.	9.04	11.44	0.42	14.50	nr	**25.95**
22 mm dia.	16.86	21.34	0.42	14.50	nr	**35.84**
28 mm dia.	47.96	60.70	0.45	15.54	nr	**76.24**
Lockshield stopcock; brass head gun metal body						
15 mm dia.	9.57	12.11	0.42	14.50	nr	**26.61**
22 mm dia.	13.70	17.34	0.42	14.50	nr	**31.84**
28 mm dia.	24.37	30.85	0.45	15.54	nr	**46.39**

38 PIPED SUPPLY SYSTEMS

Item	Net Price £	Material £	Labour hours	Labour £	Unit	Total rate £
DZR stopcock						
15 mm dia.	22.69	28.72	0.38	13.12	nr	41.84
22 mm dia.	39.31	49.75	0.39	13.47	nr	63.22
28 mm dia.	65.48	82.87	0.40	13.81	nr	96.68
35 mm dia.	102.72	130.00	0.52	17.96	nr	147.96
42 mm dia.	136.43	172.66	0.54	18.65	nr	191.31
54 mm dia.	203.77	257.89	0.63	21.76	nr	279.65
DZR lockshield stopcock						
15 mm dia.	9.57	12.11	0.38	13.12	nr	25.23
22 mm dia.	15.10	19.11	0.39	13.47	nr	32.57
Combined stop/draincock						
15 mm dia.	50.30	63.66	0.22	7.60	nr	71.26
22 mm dia.	64.73	81.92	0.45	15.54	nr	97.46
DZR combined stop/draincock						
15 mm dia.	66.48	84.13	0.41	14.16	nr	98.29
22 mm dia.	86.14	109.02	0.42	14.50	nr	123.52
Stopcock to polyethylene						
15 mm dia.	34.76	43.99	0.38	13.12	nr	57.11
20 mm dia.	43.77	55.40	0.39	13.47	nr	68.87
25 mm dia.	58.22	73.69	0.40	13.81	nr	87.50
Draw off coupling						
15 mm dia.	21.11	26.72	0.38	13.12	nr	39.84
DZR draw off coupling						
15 mm dia.	21.11	26.72	0.38	13.12	nr	39.84
22 mm dia.	24.43	30.91	0.39	13.47	nr	44.38
Draw off elbow						
15 mm dia.	23.16	29.32	0.38	13.12	nr	42.44
22 mm dia.	28.53	36.11	0.39	13.47	nr	49.58
Lockshield drain cock						
15 mm dia.	24.23	30.66	0.41	14.16	nr	44.82
Check valves						
DZR copper alloy and bronze, WRC approved cartridge double check valve; BS 6282; working pressure cold services up to 10 bar at 65°C; screwed ends						
32 mm dia.	137.54	174.07	1.38	47.65	nr	221.72
40 mm dia.	156.75	198.38	1.62	55.94	nr	254.33
50 mm dia.	257.00	325.26	1.94	66.99	nr	392.26

38 PIPED SUPPLY SYSTEMS

Item	Net Price £	Material £	Labour hours	Labour £	Unit	Total rate £
COLD WATER PIPELINES: PIPEWORK ANCILLARIES – PUMPS						
Packaged cold water pressure booster set; fully automatic; 3 phase supply; includes fixing in position; electrical work elsewhere.						
Pressure booster set						
0.75 l/s @ 30 m head	6167.86	7806.04	9.38	323.92	nr	**8129.96**
1.5 l/s @ 30 m head	7324.35	9269.70	9.38	323.92	nr	**9593.61**
3 l/s @ 30 m head	8948.90	11325.73	10.38	358.45	nr	**11684.18**
6 l/s @ 30 m head	20210.79	25578.78	10.38	358.45	nr	**25937.23**
12 l/s @ 30 m head	25249.69	31956.01	12.38	427.51	nr	**32383.52**
0.75 l/s @ 50 m head	7021.47	8886.38	9.38	323.92	nr	**9210.29**
1.5 l/s @ 50 m head	8948.90	11325.73	9.38	323.92	nr	**11649.65**
3 l/s @ 50 m head	9885.12	12510.60	10.38	358.45	nr	**12869.05**
6 l/s @ 50 m head	22573.29	28568.76	10.38	358.45	nr	**28927.21**
12 l/s @ 50 m head	27727.84	35092.36	12.38	427.51	nr	**35519.87**
0.75 l/s @ 70 m head	7682.30	9722.72	9.38	323.92	nr	**10046.63**
1.5 l/s @ 70 m head	9879.63	12503.66	9.38	323.92	nr	**12827.57**
3 l/s @ 70 m head	10518.41	13312.10	10.38	358.45	nr	**13670.55**
6 l/s @ 70 m head	24643.93	31189.36	10.38	358.45	nr	**31547.81**
12 l/s @ 70 m head	29958.17	37915.05	12.38	427.51	nr	**38342.57**
Automatic sump pump for clear and drainage water; single stage centrifugal pump, presure tight electric motor; single phase supply; includes fixing in position; electrical work elsewhere						
Single pump						
1 l/s @ 2.68 m total head	336.80	426.25	3.50	120.86	nr	**547.12**
1 l/s @ 4.68 m total head	370.74	469.20	3.50	120.86	nr	**590.07**
1 l/s @ 6.68 m total head	489.44	619.44	3.50	120.86	nr	**740.30**
2 l/s @ 4.38 m total head	489.44	619.44	4.00	138.13	nr	**757.57**
2 l/s @ 6.38 m total head	489.44	619.44	4.00	138.13	nr	**757.57**
2 l/s @ 8.38 m total head	629.97	797.29	4.00	138.13	nr	**935.42**
3 l/s @ 3.7 m total head	489.44	619.44	4.50	155.40	nr	**774.83**
3 l/s @ 5.7 m total head	629.97	797.29	4.50	155.40	nr	**952.69**
4 l/s @ 2.9 m total head	489.44	619.44	5.00	172.66	nr	**792.10**
4 l/s @ 4.9 m total head	629.97	797.29	5.00	172.66	nr	**969.95**
4 l/s @ 6.9 m total head	1746.98	2210.97	5.00	172.66	nr	**2383.64**
Extra for high level alarm box with single float switch, local alarm and volt free contacts for remote alarm.	512.63	648.78		–	nr	**648.78**
Duty/standby pump unit						
1 l/s @ 2.68 m total head	639.66	809.56	5.00	172.66	nr	**982.22**
1 l/s @ 4.68 m total head	702.69	889.33	5.00	172.66	nr	**1061.99**
1 l/s @ 6.68 m total head	949.82	1202.09	5.00	172.66	nr	**1374.75**

38 PIPED SUPPLY SYSTEMS

Item	Net Price £	Material £	Labour hours	Labour £	Unit	Total rate £
2 l/s @ 4.38 m total head	949.82	1202.09	5.50	189.93	nr	**1392.02**
2 l/s @ 6.38 m total head	949.82	1202.09	5.50	189.93	nr	**1392.02**
2 l/s @ 8.38 m total head	1216.36	1539.43	5.50	189.93	nr	**1729.36**
3 l/s @ 3.7 m total head	949.82	1202.09	6.00	207.20	nr	**1409.28**
3 l/s @ 5.7 m total head	1216.36	1539.43	6.00	207.20	nr	**1746.62**
4 l/s @ 2.9 m total head	949.82	1202.09	6.50	224.46	nr	**1426.55**
4 l/s @ 4.9 m total head	1216.36	1539.43	6.50	224.46	nr	**1763.89**
4/s @ 6.9 m total head	3232.27	4090.76	7.00	241.73	nr	**4332.48**
Extra for 4 nr float switches to give pump on, off and high level alarm	535.60	677.86		–	nr	**677.86**
Extra for dual pump control panel, internal wall mounted IP54, including volt free contacts	2550.53	3227.94	4.00	138.13	nr	**3366.08**

38 PIPED SUPPLY SYSTEMS

Item	Net Price £	Material £	Labour hours	Labour £	Unit	Total rate £
COLD WATER PIPELINES: PIPEWORK ANCILLARIES – TANKS						
Cisterns; fibreglass; complete with ball valve, screened fittings, outlets, drain provisions, fixing plate and fitted covers						
Rectangular						
60 litre capacity	485.06	613.89	1.33	45.93	nr	**659.82**
100 litre capacity	552.68	699.48	1.40	48.35	nr	**747.82**
150 litre capacity	631.97	799.82	1.61	55.60	nr	**855.42**
250 litre capacity	680.94	861.80	1.61	55.60	nr	**917.40**
420 litre capacity	843.02	1066.92	1.99	68.72	nr	**1135.64**
730 litre capacity	1190.49	1506.68	3.31	114.30	nr	**1620.98**
800 litre capacity	1228.96	1555.38	3.60	124.32	nr	**1679.69**
1700 litre capacity	1739.67	2201.73	13.32	459.97	nr	**2661.70**
2250 litre capacity	2047.50	2591.31	20.18	696.87	nr	**3288.18**
3400 litre capacity	2770.42	3506.24	24.50	846.05	nr	**4352.29**
4500 litre capacity	3206.50	4058.15	29.91	1032.87	nr	**5091.01**
Cisterns; polypropylene; complete with ball valve, fixing plate and cover; includes placing in position						
Rectangular						
18 litre capacity	91.65	115.99	1.00	34.53	nr	**150.52**
68 litre capacity	123.50	156.31	1.00	34.53	nr	**190.84**
91 litre capacity	124.30	157.31	1.00	34.53	nr	**191.84**
114 litre capacity	138.30	175.03	1.00	34.53	nr	**209.56**
182 litre capacity	182.63	231.14	1.00	34.53	nr	**265.67**
227 litre capacity	232.62	294.40	1.00	34.53	nr	**328.93**
Circular						
114 litre capacity	129.36	163.71	1.00	34.53	nr	**198.25**
227 litre capacity	131.92	166.96	1.00	34.53	nr	**201.49**
318 litre capacity	219.42	277.70	1.00	34.53	nr	**312.23**
455 litre capacity	252.18	319.16	1.00	34.53	nr	**353.69**
Steel sectional water storage tank; hot pressed steel tank to BS 1564 TYPE 1; 5 mm plate; pre-insulated and complete with all connections and fittings to comply with BSEN 13280; 2001 and WRAS water supply (water fittings) regulations 1999; externally flanged base and sides; cost of erection (on prepared base) is included within the net price, labour cost allows for offloading and positioning materials **Note:** Prices are based on the most economical tank size for each volume, and the cost will vary with differing tank dimensions, for the same volume						

38 PIPED SUPPLY SYSTEMS

Item	Net Price £	Material £	Labour hours	Labour £	Unit	Total rate £
Volume, size						
4,900 litres, 3.66 m × 1.22 m × 1,22 m (h)	7480.53	9467.36	6.00	207.20	nr	**9674.56**
20,300 litres, 3.66 m × 2.4 m × 2.4 m (h)	15937.41	20170.39	12.00	414.39	nr	**20584.78**
52,000 litres, 6.1 m × 3.6 m × 2.4 m (h)	28413.95	35960.70	19.00	656.12	nr	**36616.81**
94,000 litres, 7.3 m × 3.6 m × 3.6 m (h)	50027.42	63314.70	28.00	966.91	nr	**64281.61**
140,000 litres, 9.7 m × 6.1 m × 2.44 m (h)	55817.95	70643.19	28.00	966.91	nr	**71610.11**
GRP sectional water storage tank; pre-insulated and complete with twin compartment and basic connections and fittings to comply with BSEN 13280; 2001 and WRAS water supply (water fittings) regulations 1999; externally flanged base and sides; cost of erection (on prepared base) is included within the net price, labour cost allows for offloading and positioning materials						
Note: Prices are based on the most economical tank size for each volume, and the cost will vary with differing tank dimensions, for the same volume						
Volume, size						
4,500 litres, 3 m × 1 m × 1.5 m (h)	8169.00	10338.68	5.00	172.66	nr	**10511.34**
10,000 litres, 2.5 m × 2 m × 2 m (h)	11683.32	14786.41	7.00	241.73	nr	**15028.14**
20,000 litres, 4 m × 2.5 m × 2 m (h)	15123.02	19139.69	10.00	345.33	nr	**19485.02**
30,000 litres 5 m × 3 m × 2 m (h)	18837.90	23841.24	12.00	414.39	nr	**24255.63**
40,000 litres, 5 m × 4 m × 2 m (h)	21988.43	27828.55	12.00	414.39	nr	**28242.94**
50,000 litres, 5 m × 4 m × 2.5 m (h)	26893.79	34036.78	14.00	483.46	nr	**34520.24**
60,000 litres, 6 m × 4 m × 2.5 m (h)	28606.64	36204.57	16.00	552.52	nr	**36757.09**
70,000 litres, 7 m × 4 m × 2.5 m (h)	32330.85	40917.92	16.00	552.52	nr	**41470.44**
80,000 litres, 8 m × 4 m × 2.5 m (h)	34205.78	43290.83	16.00	552.52	nr	**43843.35**
90,000 litres, 6 m × 5 m × 3 m (h)	36856.09	46645.07	16.00	552.52	nr	**47197.59**
105,000 litres, 7 m × 5 m × 3 m (h)	41293.89	52261.55	24.00	828.78	nr	**53090.33**
120,000 litres, 8 m × 5 m × 3 m (h)	43158.32	54621.17	24.00	828.78	nr	**55449.96**
135,000 litres, 9 m × 6 m × 2.5 m (h)	47887.62	60606.57	24.00	828.78	nr	**61435.35**
144,000 litres, 8 m × 6 m × 3 m (h)	47952.92	60689.21	24.00	828.78	nr	**61517.99**
CLEANING AND CHEMICAL TREATMENT						
Electromagnetic water conditioner; WRAS approved; complete with control box; maximum inlet pressure 16 bar; electrical work elsewhere						
Connection size, nominal flow rate at 50 mbar						
20 mm dia., 0.3 l/s	1727.25	2186.01	1.25	43.17	nr	**2229.17**
25 mm dia., 0.6 l/s	2268.00	2870.38	1.45	50.07	nr	**2920.45**
32 mm dia., 1.2 l/s	3234.00	4092.95	1.55	53.53	nr	**4146.48**
40 mm dia., 1.7 l/s	3864.00	4890.28	1.65	56.98	nr	**4947.26**
50 mm dia., 3.4 l/s	5061.00	6405.20	1.75	60.43	nr	**6465.63**
65 mm dia., 5.2 l/s	5538.75	7009.84	1.90	65.61	nr	**7075.45**
100 mm dia., 30.5 l/s, 595 mbar	10342.50	13089.47	3.00	103.60	nr	**13193.07**

38 PIPED SUPPLY SYSTEMS

Item	Net Price £	Material £	Labour hours	Labour £	Unit	Total rate £
COLD WATER PIPELINES: PIPEWORK ANCILLARIES – TANKS – cont						
CLEANING AND CHEMICAL TREATMENT – cont						
Ultraviolet water sterillizing unit; WRAS approved; complete with control unit; UV lamp housed in quartz tube; unit complete with UV intensity sensor, flushing and discharge valve and facilities for remote alarm; electrical work elsewhere						
Maximum flow rate (@ 400 J/m^2 exposure in accordance with BSEN 14897), connection size						
0.8 l/s, 25 mm dia.	3218.25	4073.02	1.98	68.37	nr	**4141.39**
1.25 l/s, 40 mm dia.	4068.75	5149.41	1.98	68.37	nr	**5217.78**
2.57 l/s, 50 mm dia.	4630.50	5860.36	1.98	68.37	nr	**5928.74**
4.46 l/s, 80 mm dia.	10573.50	13381.82	1.98	68.37	nr	**13450.20**
10.5 l/s, 100 mm dia.	13135.50	16624.29	1.98	68.37	nr	**16692.66**
20.1 l/s, 100 mm dia.	14773.50	18697.34	3.60	124.32	nr	**18821.66**
26.9 l/s 125 mm dia.	16590.00	20996.30	3.60	124.32	nr	**21120.62**
33.7 l/s 125 mm dia.	18721.50	23693.93	3.60	124.32	nr	**23818.25**
Base exchange water softener; WRAS approved; complete with resin tank, brine tank and consumption data monitoring facilities						
Capacities of softeners are based on 300 ppm hardness and quoted in m^3 of softened water produced. Design flow rates are recommended for continuous use						
Simplex configuration						
Design flow rate, min–max softenend water produced						
0.97 l/s, 7.7 m^3	3123.75	3953.42	10.00	345.33	nr	**4298.74**
1.25 l/s, 11.7 m^3	3202.50	4053.08	12.00	414.39	nr	**4467.47**
1.8 l/s, 19.8 m^3	4772.25	6039.76	12.00	414.39	nr	**6454.15**
2.23 l/s, 23.4 m^3	5355.00	6777.29	12.00	414.39	nr	**7191.68**
2.92 l/s, 31.6 m^3	5538.75	7009.84	15.00	517.99	nr	**7527.83**
3.75 l/s, 39.8 m^3	6615.00	8371.94	15.00	517.99	nr	**8889.93**
4.72 l/s, 60.3 m^3	7512.75	9508.14	18.00	621.59	nr	**10129.72**
6.66 l/s, 96.6 m^3	8872.50	11229.04	20.00	690.65	nr	**11919.69**
Duplex configuration						
Design flow rate, min–max softenend water produced						
0.97 l/s, 7.7 m^3	4609.50	5833.78	12.00	414.39	nr	**6248.17**
1.25 l/s, 11.7 m^3	4798.50	6072.98	12.00	414.39	nr	**6487.37**
1.81 l/s, 19.8 m^3	8048.25	10185.87	15.00	517.99	nr	**10703.85**

38 PIPED SUPPLY SYSTEMS

Item	Net Price £	Material £	Labour hours	Labour £	Unit	Total rate £
2.23 l/s, 23.4 m³	9224.25	11674.21	15.00	517.99	nr	**12192.20**
2.29 l/s, 31.6 m³	9507.75	12033.01	18.00	621.59	nr	**12654.59**
3.75 l/s, 39.8 m³	11823.00	14963.19	18.00	621.59	nr	**15584.77**
4.72 l/s, 60.3 m³	13492.50	17076.11	18.00	621.59	nr	**17697.69**
6.66/s, 96.6 m³	13492.50	17076.11	23.00	794.25	nr	**17870.36**
THERMAL INSULATION						
Flexible closed cell walled insulation; Class 1/Class O; adhesive joints; including around fittings						
6 mm wall thickness						
15 mm dia.	1.70	2.15	0.15	5.18	m	**7.33**
22 mm dia.	2.01	2.54	0.15	5.18	m	**7.72**
28 mm dia.	2.54	3.22	0.15	5.18	m	**8.40**
9 mm wall thickness						
15 mm dia.	1.95	2.47	0.15	5.18	m	**7.65**
22 mm dia.	2.44	3.09	0.15	5.18	m	**8.27**
28 mm dia.	2.62	3.32	0.15	5.18	m	**8.50**
35 mm dia.	2.96	3.75	0.15	5.18	m	**8.93**
42 mm dia.	3.39	4.28	0.15	5.18	m	**9.46**
54 mm dia.	3.51	4.44	0.15	5.18	m	**9.62**
13 mm wall thickness						
15 mm dia.	2.50	3.16	0.15	5.18	m	**8.34**
22 mm dia.	3.06	3.88	0.15	5.18	m	**9.06**
28 mm dia.	3.64	4.61	0.15	5.18	m	**9.79**
35 mm dia.	3.93	4.97	0.15	5.18	m	**10.15**
42 mm dia.	4.62	5.84	0.15	5.18	m	**11.02**
54 mm dia.	6.44	8.15	0.15	5.18	m	**13.33**
67 mm dia.	9.51	12.04	0.15	5.18	m	**17.22**
76 mm dia.	11.19	14.16	0.15	5.18	m	**19.34**
108 mm dia.	11.73	14.85	0.15	5.18	m	**20.03**
19 mm wall thickness						
15 mm dia.	4.05	5.13	0.15	5.18	m	**10.31**
22 mm dia.	4.94	6.25	0.15	5.18	m	**11.43**
28 mm dia.	6.70	8.49	0.15	5.18	m	**13.66**
35 mm dia.	7.76	9.82	0.15	5.18	m	**15.00**
42 mm dia.	9.16	11.59	0.15	5.18	m	**16.77**
54 mm dia.	11.59	14.67	0.15	5.18	m	**19.85**
67 mm dia.	13.89	17.57	0.15	5.18	m	**22.75**
76 mm dia.	16.05	20.31	0.22	7.60	m	**27.91**
108 mm dia.	21.89	27.70	0.22	7.60	m	**35.30**
25 mm wall thickness						
15 mm dia.	7.96	10.07	0.15	5.18	m	**15.25**
22 mm dia.	8.70	11.01	0.15	5.18	m	**16.19**
28 mm dia.	9.86	12.47	0.15	5.18	m	**17.65**
35 mm dia.	10.92	13.82	0.15	5.18	m	**19.00**
42 mm dia.	11.66	14.75	0.15	5.18	m	**19.93**
54 mm dia.	13.69	17.32	0.15	5.18	m	**22.50**
67 mm dia.	18.45	23.35	0.15	5.18	m	**28.53**
76 mm dia.	20.80	26.33	0.22	7.60	m	**33.92**

Material Costs/Prices for Measured Works

38 PIPED SUPPLY SYSTEMS

Item	Net Price £	Material £	Labour hours	Labour £	Unit	Total rate £
COLD WATER PIPELINES: PIPEWORK ANCILLARIES – TANKS – cont						
THERMAL INSULATION – cont						
Flexible closed cell walled insulation – cont						
32 mm wall thickness						
15 mm dia.	9.96	12.60	0.15	5.18	m	**17.78**
22 mm dia.	10.92	13.82	0.15	5.18	m	**19.00**
28 mm dia.	12.67	16.03	0.15	5.18	m	**21.21**
35 mm dia.	13.08	16.55	0.15	5.18	m	**21.73**
42 mm dia.	15.23	19.27	0.15	5.18	m	**24.45**
54 mm dia.	19.16	24.25	0.15	5.18	m	**29.43**
76 mm dia.	28.65	36.26	0.22	7.60	m	**43.86**
Note: For mineral fibre sectional insulation; bright class O foil faced; bright class O foil taped joints; 19 mm aluminium bands rates, refer to section – Thermal Insulation						
Note: For mineral fibre sectional insulation; bright class O foil faced; bright class O foil taped joints; 22 swg plain/embossed aluminium cladding; pop riveted rates, refer to section – Thermal Insulation						
Note: For mineral fibre sectional insulation; bright class O foil faced; bright class O foil taped joints; 0.8 mm polyisobutylene sheeting; welded joints rates, refer to section – Thermal Insulation						

38 PIPED SUPPLY SYSTEMS

Item	Net Price £	Material £	Labour hours	Labour £	Unit	Total rate £
HOT WATER						
PIPELINES						
Note: For pipework prices refer to section – Cold Water						
PIPELINE ANCILLARIES						
Note: For prices for ancillaries refer to section – Cold Water						
STORAGE CYLINDERS/CALORIFIERS						
Insulated copper storage cylinders; BS 699; includes placing in position						
Grade 3 (maximum 10 m working head)						
BS size 6; 115 litre capacity; 400 mm dia.; 1050 mm high	230.46	291.67	1.50	51.85	nr	**343.52**
BS size 7; 120 litre capacity; 450 mm dia.; 900 mm high	272.48	344.85	2.00	69.07	nr	**413.92**
BS size 8; 144 litre capacity; 450 mm dia.; 1050 mm high	289.66	366.59	2.80	96.73	nr	**463.32**
Grade 4 (maximum 6 m working head)						
BS size 2; 96 litre capacity; 400 mm dia.; 900 mm high	178.44	225.84	1.50	51.85	nr	**277.69**
BS size 7; 120 litre capacity; 450 mm dia.; 900 mm high	187.53	237.34	1.50	51.85	nr	**289.19**
BS size 8; 144 litre capacity; 450 mm dia.; 1050 mm high	223.85	283.30	1.50	51.85	nr	**335.15**
BS size 9; 166 litre capacity; 450 mm dia.; 1200 mm high	327.31	414.24	1.50	51.85	nr	**466.09**
Storage cylinders; brazed copper construction; to BS 699; screwed bosses; includes placing in position						
Tested to 2.2 bar, 15 m maximum head						
144 litres	1014.44	1283.88	3.00	103.70	nr	**1387.58**
160 litres	1146.80	1451.39	3.00	103.70	nr	**1555.09**
200 litres	1182.04	1496.00	3.76	129.82	nr	**1625.82**
255 litres	1345.23	1702.52	3.76	129.82	nr	**1832.34**
290 litres	1817.16	2299.80	3.76	129.82	nr	**2429.62**
370 litres	2081.85	2634.78	4.50	155.55	nr	**2790.34**
450 litres	2835.89	3589.10	5.00	172.66	nr	**3761.76**
Tested to 2.55 bar, 17 m maximum head						
550 litres	3070.68	3886.26	5.00	172.66	nr	**4058.92**
700 litres	3589.94	4543.43	6.02	208.03	nr	**4751.45**
800 litres	4152.21	5255.03	6.54	225.70	nr	**5480.74**
900 litres	4493.11	5686.48	8.00	276.26	nr	**5962.74**
1000 litres	4741.49	6000.83	8.00	276.26	nr	**6277.09**
1250 litres	5193.03	6572.29	13.16	454.38	nr	**7026.67**
1500 litres	7954.07	10066.67	15.15	523.22	nr	**10589.89**
2000 litres	9545.59	12080.90	17.24	595.39	nr	**12676.28**
3000 litres	13408.27	16969.50	24.39	842.26	nr	**17811.76**

38 PIPED SUPPLY SYSTEMS

Item	Net Price £	Material £	Labour hours	Labour £	Unit	Total rate £
HOT WATER – cont						
STORAGE CYLINDERS/CALORIFIERS – cont						
Indirect cylinders; copper; bolted top; up to 5 tappings for connections; BS 1586; includes placing in position						
Grade 3, tested to 1.45 bar, 10 m maximum head						
74 litre capacity	465.12	588.65	1.50	51.85	nr	**640.50**
96 litre capacity	473.52	599.29	1.50	51.85	nr	**651.14**
114 litre capacity	486.22	615.36	1.50	51.85	nr	**667.21**
117 litre capacity	504.66	638.69	2.00	69.07	nr	**707.76**
140 litre capacity	520.02	658.14	2.50	86.33	nr	**744.47**
162 litre capacity	727.22	920.37	3.00	103.70	nr	**1024.07**
190 litre capacity	794.91	1006.04	3.51	121.17	nr	**1127.20**
245 litre capacity	930.20	1177.26	3.80	131.30	nr	**1308.56**
280 litre capacity	1648.93	2086.89	4.00	138.13	nr	**2225.02**
360 litre capacity	1784.26	2258.16	4.50	155.55	nr	**2413.71**
440 litre capacity	2071.76	2622.02	4.50	155.55	nr	**2777.57**
Grade 2, tested to 2.2 bar, 15 m maximum head						
117 litre capacity	672.23	850.78	2.00	69.07	nr	**919.84**
140 litre capacity	731.48	925.76	2.50	86.33	nr	**1012.09**
162 litre capacity	837.15	1059.50	2.80	96.73	nr	**1156.23**
190 litre capacity	972.51	1230.81	3.00	103.70	nr	**1334.52**
245 litre capacity	1175.43	1487.63	4.00	138.13	nr	**1625.76**
280 litre capacity	1877.26	2375.86	4.00	138.13	nr	**2513.99**
360 litre capacity	2071.76	2622.02	4.50	155.55	nr	**2777.57**
440 litre capacity	2452.26	3103.58	4.50	155.55	nr	**3259.13**
Grade 1, tested 3.65 bar, 25 m maximum head						
190 litre capacity	1446.04	1830.11	3.00	103.70	nr	**1933.81**
245 litre capacity	1644.67	2081.49	3.00	103.70	nr	**2185.19**
280 litre capacity	2325.46	2943.10	4.00	138.13	nr	**3081.23**
360 litre capacity	2942.80	3724.40	4.50	155.55	nr	**3879.96**
440 litre capacity	3572.71	4521.62	4.50	155.55	nr	**4677.17**
Indirect cylinders, including manhole; BS 853						
Grade 3, tested to 1.5 bar, 10 m maximum head						
550 litre capacity	3007.36	3806.11	5.21	179.86	nr	**3985.97**
700 litre capacity	3329.61	4213.95	6.02	208.03	nr	**4421.98**
800 litre capacity	3866.62	4893.59	6.54	225.70	nr	**5119.29**
1000 litre capacity	4833.31	6117.04	7.04	243.19	nr	**6360.23**
1500 litre capacity	5585.14	7068.55	10.00	345.33	nr	**7413.88**
2000 litre capacity	7733.31	9787.28	16.13	556.98	nr	**10344.25**

38 PIPED SUPPLY SYSTEMS

Item	Net Price £	Material £	Labour hours	Labour £	Unit	Total rate £
Grade 2, tested to 2.55 bar, 15 m maximum head						
550 litre capacity	3334.22	4219.79	5.21	179.86	nr	**4399.65**
700 litre capacity	4173.01	5281.36	6.02	208.03	nr	**5489.39**
800 litre capacity	4403.67	5573.28	6.54	225.70	nr	**5798.98**
1000 litre capacity	5452.18	6900.28	7.04	243.19	nr	**7143.47**
1500 litre capacity	6710.37	8492.64	10.00	345.33	nr	**8837.96**
2000 litre capacity	8387.98	10615.83	16.13	556.98	nr	**11172.81**
Grade 1, tested to 4 bar, 25 m maximum head						
550 litre capacity	3879.42	4909.80	5.21	179.86	nr	**5089.65**
700 litre capacity	4403.67	5573.28	6.02	208.03	nr	**5781.31**
800 litre capacity	4718.24	5971.41	6.54	225.70	nr	**6197.11**
1000 litre capacity	6291.00	7961.89	7.04	243.19	nr	**8205.08**
1500 litre capacity	7549.13	9554.18	10.00	345.33	nr	**9899.50**
2000 litre capacity	9226.74	11677.37	16.13	556.98	nr	**12234.34**
Storage calorifiers; copper; heater battery capable of raising temperature of contents from 10°C to 65°C in one hour; static head not exceeding 1.35 bar; BS 853; includes fixing in position on cradles or legs						
Horizontal; primary LPHW at 82°C/71°C						
400 litre capacity	4554.84	5764.60	7.04	243.19	nr	**6007.79**
1000 litre capacity	7287.70	9223.31	8.00	276.26	nr	**9499.57**
2000 litre capacity	14575.49	18446.74	14.08	486.37	nr	**18933.11**
3000 litre capacity	17991.59	22770.15	25.00	863.31	nr	**23633.47**
4000 litre capacity	21863.16	27670.02	40.00	1381.30	nr	**29051.32**
4500 litre capacity	24638.84	31182.92	50.00	1726.63	nr	**32909.55**
Vertical; primary LPHW at 82°C/71°C						
400 litre capacity	5133.32	6496.72	7.04	243.19	nr	**6739.91**
1000 litre capacity	8250.00	10441.20	8.00	276.26	nr	**10717.46**
2000 litre capacity	15714.20	19887.89	14.08	486.37	nr	**20374.26**
3000 litre capacity	19642.75	24859.87	25.00	863.31	nr	**25723.18**
4000 litre capacity	24095.02	30494.66	40.00	1381.30	nr	**31875.96**
4500 litre capacity	27237.90	34472.28	50.00	1726.63	nr	**36198.91**
Storage calorifiers; galvanized mild steel; heater battery capable of raising temperature of contents from 10°C to 65°C in one hour; static head not exceeding 1.35 bar; BS 853; includes fixing in position on cradles or legs						
Horizontal; primary LPHW at 82°C/71°C						
400 litre capacity	4554.84	5764.60	7.04	243.19	nr	**6007.79**
1000 litre capacity	7287.70	9223.31	8.00	276.26	nr	**9499.57**
2000 litre capacity	14575.49	18446.74	14.08	486.37	nr	**18933.11**
3000 litre capacity	17991.59	22770.15	25.00	863.31	nr	**23633.47**
4000 litre capacity	21863.16	27670.02	40.00	1381.30	nr	**29051.32**
4500 litre capacity	24638.84	31182.92	50.00	1726.63	nr	**32909.55**

38 PIPED SUPPLY SYSTEMS

Item	Net Price £	Material £	Labour hours	Labour £	Unit	Total rate £
HOT WATER – cont						
STORAGE CYLINDERS/CALORIFIERS – cont						
Storage calorifiers – cont						
Vertical; primary LPHW at 82°C/71°C						
400 litre capacity	5133.32	6496.72	7.04	243.19	nr	**6739.91**
1000 litre capacity	8250.00	10441.20	8.00	276.26	nr	**10717.46**
2000 litre capacity	15714.20	19887.89	14.08	486.37	nr	**20374.26**
3000 litre capacity	19642.75	24859.87	25.00	863.31	nr	**25723.18**
4000 litre capacity	24095.02	30494.66	40.00	1381.30	nr	**31875.96**
4500 litre capacity	27237.90	34472.28	50.00	1726.63	nr	**36198.91**
Indirect cylinders; mild steel, welded throughout, galvanized, with connections. Tested to 4 bar, 95C. Includes sensors, with full insulation and cases, includes delivery						
222 litre nominal content	1254.62	1587.84	2.50	86.33	nr	**1674.17**
278 litre nominal content	1393.37	1763.45	2.80	96.73	nr	**1860.18**
474 litre nominal content	1742.00	2204.68	3.00	103.70	nr	**2308.38**
765 litre nominal content	1950.72	2468.83	3.00	103.67	nr	**2572.50**
956 litre nominal content	2299.35	2910.06	3.00	103.70	nr	**3013.76**
1365 litre nominal content	2981.46	3773.34	4.00	138.13	nr	**3911.47**
2039 litre nominal content	3803.49	4813.70	5.00	172.66	nr	**4986.36**
2361 litre nominal content	4918.19	6224.46	6.02	208.03	nr	**6432.49**
Indirect cylinders; mild steel welded throughout, galvanized, with connections. Tested to 4bar, 95C. Includes sensors. Includes delivery						
4021 litre nominal content	6346.54	8032.18	1.50	51.80	nr	**8083.98**
5897 litre nominal content	7524.20	9522.62	1.50	51.80	nr	**9574.42**
8000 litre nominal content	11841.90	14987.10	1.50	51.80	nr	**15038.90**
10170 litre nominal content	13912.71	17607.93	2.00	69.07	nr	**17676.99**
Local electric hot water heaters						
Unvented multipoint water heater; providing hot water for one or more outlets; used with conventional taps or mixers; factory fitted temperature and pressure relief valve; externally adjustable thermostat; elemental 'on' indicator; fitted with 1 m of 3 core cable; electrical supply and connection excluded						
5 litre capacity; 2.2 kW rating	256.36	324.45	1.50	51.80	nr	**376.24**
10 litre capacity; 2.2 kW rating	317.69	402.07	1.50	51.80	nr	**453.87**
15 litre capacity; 2.2 kW rating	566.20	716.58	1.50	51.80	nr	**768.38**
30 litre capacity; 2.2 kW rating	566.20	716.58	2.00	69.07	nr	**785.65**
50 litre capacity; 2.2 kW rating	596.11	754.43	2.00	69.07	nr	**823.50**
80 litre capacity; 2.2 kW rating	1172.97	1484.51	2.00	69.07	nr	**1553.58**
100 litre capacity; 2.2 kW rating	1259.26	1593.72	2.00	69.07	nr	**1662.78**

38 PIPED SUPPLY SYSTEMS

Item	Net Price £	Material £	Labour hours	Labour £	Unit	Total rate £
Accessories						
Pressure reducing valve and expansion kit	204.73	259.10	2.00	69.07	nr	**328.17**
Thermostatic blending valve	108.95	137.89	1.00	34.53	nr	**172.42**
SOLAR THERMAL PACKAGES						
Packages include solar collectors, mild steel storage vessel, fresh water module (including plate heat exchanger), roof fixing kit, glycol, stratified pump station (including plate heat exchanger), solar explansion vessel, controls, anti-legionella protection; delivery and commissioning						
No. of panels; storage size						
6 solar collectors, 1000 l storage vessel	9628.83	12186.24	11.00	379.86	nr	**12566.10**
8 solar collectors, 1500 l storage vessel	11132.97	14089.88	13.00	448.92	nr	**14538.81**
10 solar collectors, 2000 l storage vessel	11686.82	14790.84	15.00	517.99	nr	**15308.82**
12 solar collectors, 2000l storage vessel	12814.34	16217.83	17.00	587.05	nr	**16804.88**
14 solar collectors, 2000 l storage vessel	13950.02	17655.15	18.00	621.59	nr	**18276.74**
16 solar collectors, 2500 l storage vessel	15114.86	19129.36	20.00	690.65	nr	**19820.01**

38 PIPED SUPPLY SYSTEMS

Item	Net Price £	Material £	Labour hours	Labour £	Unit	Total rate £
NATURAL GAS: PIPELINES: MEDIUM DENSITY POLYETHYLENE – YELLOW						
Pipe; laid underground; electrofusion joints in the running length; BS 6572; BGT PL2 standards						
Coiled service pipe						
20 mm dia.	2.14	2.71	0.37	12.78	m	**15.49**
25 mm dia.	2.78	3.52	0.41	14.16	m	**17.68**
32 mm dia.	4.60	5.82	0.47	16.23	m	**22.06**
63 mm dia.	17.55	22.21	0.60	20.72	m	**42.93**
90 mm dia.	23.00	29.11	0.90	31.08	m	**60.19**
Mains service pipe						
63 mm dia.	17.04	21.56	0.60	20.72	m	**42.28**
90 mm dia.	22.33	28.26	0.90	31.08	m	**59.34**
125 mm dia.	43.13	54.58	1.20	41.44	m	**96.02**
180 mm dia.	89.06	112.72	1.50	51.80	m	**164.52**
250 mm dia.	163.93	207.46	1.75	60.43	m	**267.90**
Extra over fittings, electrofusion joints						
Straight connector						
32 mm dia.	13.35	16.89	0.47	16.23	nr	**33.13**
63 mm dia.	25.06	31.71	0.58	20.03	nr	**51.74**
90 mm dia.	36.96	46.77	0.67	23.14	nr	**69.91**
125 mm dia.	69.22	87.60	0.83	28.66	nr	**116.26**
180 mm dia.	124.66	157.77	1.25	43.17	nr	**200.93**
Reducing connector						
90 × 63 mm dia.	51.62	65.33	0.67	23.14	nr	**88.47**
125 × 90 mm dia.	103.52	131.01	0.83	28.66	nr	**159.68**
180 × 125 mm dia.	189.92	240.36	1.25	43.17	nr	**283.53**
Bend; 45°						
90 mm dia.	99.76	126.26	0.67	23.14	nr	**149.39**
125 mm dia.	163.20	206.54	0.83	28.66	nr	**235.20**
180 mm dia.	370.19	468.51	1.25	43.17	nr	**511.67**
Bend; 90°						
63 mm dia.	64.63	81.79	0.58	20.03	nr	**101.82**
90 mm dia.	99.76	126.26	0.67	23.14	nr	**149.39**
125 mm dia.	163.20	206.54	0.83	28.66	nr	**235.20**
180 mm dia.	370.19	468.51	1.25	43.17	nr	**511.67**
Extra over malleable iron fittings, compression joints						
Straight connector						
20 mm dia.	20.56	26.02	0.38	13.13	nr	**39.14**
25 mm dia.	22.40	28.35	0.45	15.54	nr	**43.89**
32 mm dia.	25.14	31.82	0.50	17.27	nr	**49.09**
63 mm dia.	50.49	63.90	0.85	29.36	nr	**93.26**
Straight connector; polyethylene to MI						
20 mm dia.	17.47	22.11	0.31	10.71	nr	**32.82**
25 mm dia.	19.06	24.12	0.35	12.09	nr	**36.21**
32 mm dia.	21.27	26.92	0.40	13.81	nr	**40.73**
63 mm dia.	35.65	45.11	0.65	22.45	nr	**67.57**

38 PIPED SUPPLY SYSTEMS

Item	Net Price £	Material £	Labour hours	Labour £	Unit	Total rate £
Straight connector; polyethylene to FI						
20 mm dia.	17.47	22.11	0.31	10.71	nr	**32.82**
25 mm dia.	21.27	26.92	0.35	12.09	nr	**39.00**
32 mm dia.	21.27	26.92	0.40	13.81	nr	**40.73**
63 mm dia.	35.65	45.11	0.75	25.91	nr	**71.02**
Elbow						
20 mm dia.	26.73	33.84	0.38	13.13	nr	**46.96**
25 mm dia.	29.17	36.91	0.45	15.54	nr	**52.45**
32 mm dia.	32.66	41.34	0.50	17.27	nr	**58.60**
63 mm dia.	65.64	83.08	0.80	27.63	nr	**110.70**
Equal tee						
20 mm dia.	35.65	45.11	0.53	18.31	nr	**63.42**
25 mm dia.	41.65	52.72	0.55	19.00	nr	**71.71**
32 mm dia.	52.48	66.42	0.64	22.11	nr	**88.53**
GAS BOOSTERS						
Complete skid mounted gas booster set, including AV mounts, flexible connections, low pressure switch, control panel and NRV (for run/standby unit); 3 phase supply; in accordance with IGE/UP/2; includes delivery, offloading and positioning						
Single unit						
Flow, pressure range						
0–200 m³/hour, 0.1–2.6 kPa	4997.75	6325.15	10.00	345.33	nr	**6670.48**
0–200 m³/hour, 0.1–4.0 kPa	5702.57	7217.17	10.00	345.33	nr	**7562.49**
0–200 m³/hour, 0.1–7 kPa	6636.17	8398.74	10.00	345.33	nr	**8744.06**
0–200 m³/hour, 0.1–9.5 kPa	6910.81	8746.32	10.00	345.33	nr	**9091.64**
0–200 m³/hour 0.1–11.0 kPa	7720.90	9771.57	10.00	345.33	nr	**10116.89**
0–400 m³/hour, 0.1–4.0 kPa	8397.70	10628.13	10.00	345.33	nr	**10973.46**
0–1000 m³/hour, 0.1–7.4 kPa	8723.13	11039.99	10.00	345.33	nr	**11385.32**
50–1000 m³/hour, 0.1–16.0 kPa	16464.84	20837.90	20.00	690.65	nr	**21528.55**
50–1000 m³/hour, 0.1–24.5 kPa	18425.67	23319.53	20.00	690.65	nr	**24010.18**
50–1000 m³/hour, 0.1–31.0 kPa	20956.62	26522.70	20.00	690.65	nr	**27213.35**
50–1000 m³/hour, 0.1–41.0 kPa	23130.58	29274.06	20.00	690.65	nr	**29964.71**
50–1000 m³/hour, 0.1–51.0 kPa	24000.06	30374.48	20.00	690.65	nr	**31065.13**
100–1800 m³/hour, 3.5–23.5 kPa	24627.14	31168.11	20.00	690.65	nr	**31858.76**
100–1800 m³/hour, 4.5–27.0 kPa	26563.07	33618.22	20.00	690.65	nr	**34308.87**
100–1800 m³/hour, 6.0–32.5 kPa	29418.90	37232.56	20.00	690.65	nr	**37923.21**
100–1800 m³/hour, 7.2–39.0 kPa	33844.55	42833.66	20.00	690.65	nr	**43524.31**
100–1800 m³/hour, 9.0–42.0 kPa	35574.55	45023.15	20.00	690.65	nr	**45713.80**
Run/Standby unit						
Flow, pressure range						
0–200 m³/hour, 0.1–2.6 kPa	24598.46	31131.81	16.00	552.52	nr	**31684.33**
0–200 m³/hour, 0.1–4.0 kPa	25347.63	32079.97	16.00	552.52	nr	**32632.49**
0–200 m³/hour, 0.1–7 kPa	25934.72	32822.98	16.00	552.52	nr	**33375.50**
0–200 m³/hour, 0.1–9.5 kPa	26577.35	33636.30	16.00	552.52	nr	**34188.82**
0–200 m³/hour 0.1–11.0 kPa	27017.67	34193.56	16.00	552.52	nr	**34746.08**
0–400 m³/hour, 0.1–4.0 kPa	29794.41	37707.80	16.00	552.52	nr	**38260.32**
0–1000 m³/hour, 0.1–7.4 kPa	37977.81	48064.72	25.00	863.31	nr	**48928.04**

38 PIPED SUPPLY SYSTEMS

Item	Net Price £	Material £	Labour hours	Labour £	Unit	Total rate £
NATURAL GAS: PIPELINES: MEDIUM DENSITY POLYETHYLENE – YELLOW – cont						
Complete skid mounted gas booster set, including AV mounts, flexible connections, low pressure switch, control panel and NRV (for run/standby unit) – cont						
50–1000 m³/hour, 0.1–16.0 kPa	52051.91	65876.89	25.00	863.31	nr	**66740.21**
50–1000 m³/hour, 0.1–24.5 kPa	58847.00	74476.77	25.00	863.31	nr	**75340.08**
50–1000 m³/hour, 0.1–31.0 kPa	65626.17	83056.48	25.00	863.31	nr	**83919.79**
50–1000 m³/hour, 0.1–41.0 kPa	72425.23	91661.37	25.00	863.31	nr	**92524.68**
50–1000 m³/hour, 0.1–51.0 kPa	72750.48	92073.00	25.00	863.31	nr	**92936.32**
100–1800 m³/hour, 3.5–23.5 kPa	75691.49	95795.15	25.00	863.31	nr	**96658.47**
100–1800 m³/hour, 4.5–27.0 kPa	77766.69	98421.52	25.00	863.31	nr	**99284.83**
100–1800 m³/hour, 6.0–32.5 kPa	82619.75	104563.56	25.00	863.31	nr	**105426.87**
100–1800 m³/hour, 7.2–39.0 kPa	87816.24	111140.24	25.00	863.31	nr	**112003.55**
100–1800 m³/hour, 9.0–42.0 kPa	92306.65	116823.30	25.00	863.31	nr	**117686.61**

38 PIPED SUPPLY SYSTEMS

Item	Net Price £	Material £	Labour hours	Labour £	Unit	Total rate £
NATURAL GAS: PIPELINES: SCREWED STEEL For prices for steel pipework refer to section – Low Temperature Hot Water Heating						
PIPE IN PIPE **Note:** for pipe in pipe, a sleeve size two pipe sizes bigger than actual pipe size has been allowed. All rates refer to actual pipe size. Black steel pipes – Screwed and socketed joints; BS 1387: 1985 up to 50 mm pipe size. Butt welded joints; BS 1387: 1985 65 mm pipe size and above. Pipe dia.						
25 mm	21.78	27.57	1.73	59.74	m	**87.31**
32 mm	28.93	36.61	1.95	67.34	m	**103.95**
40 mm	37.20	47.08	2.16	74.59	m	**121.67**
50 mm	48.81	61.77	2.44	84.26	m	**146.03**
65 mm	66.95	84.73	2.95	101.87	m	**186.60**
80 mm	80.93	102.42	3.42	118.10	m	**220.52**
100 mm	101.06	127.90	4.00	138.13	m	**266.03**
Extra over black steel pipes – Screwed pipework; black malleable iron fittings; BS 143. Welded pipework; butt welded steel fittings; BS 1965 Bend, 90°						
25 mm	12.57	15.91	2.91	100.49	m	**116.40**
32 mm	19.13	24.21	3.45	119.14	m	**143.35**
40 mm	26.80	33.92	5.34	184.40	m	**218.32**
50 mm	30.21	38.23	6.53	225.50	m	**263.73**
65 mm	40.79	51.63	8.84	305.27	m	**356.90**
80 mm	69.55	88.02	10.73	370.53	m	**458.55**
100 mm	92.36	116.89	12.76	440.64	m	**557.53**
Bend, 45°						
25 mm	15.88	20.10	2.91	100.49	m	**120.59**
32 mm	20.57	26.03	3.45	119.14	m	**145.17**
40 mm	23.00	29.11	5.34	184.40	m	**213.51**
50 mm	27.43	34.72	6.53	225.50	m	**260.22**
65 mm	35.83	45.35	8.84	305.27	m	**350.61**
80 mm	61.28	77.56	10.73	370.53	m	**448.09**
100 mm	78.50	99.35	12.76	440.64	m	**539.98**
Equal tee						
25 mm	105.73	133.82	4.18	144.35	m	**278.16**
32 mm	108.57	137.41	4.94	170.59	m	**308.00**
40 mm	133.56	169.04	7.28	251.40	m	**420.44**
50 mm	137.04	173.43	8.48	292.84	m	**466.27**
65 mm	182.45	230.91	11.47	396.09	m	**627.00**
80 mm	313.03	396.17	14.23	491.40	m	**887.57**
100 mm	354.28	448.37	17.92	618.82	m	**1067.19**

38 PIPED SUPPLY SYSTEMS

Item	Net Price £	Material £	Labour hours	Labour £	Unit	Total rate £
NATURAL GAS: PIPELINES: SCREWED STEEL – cont						
Copper pipe; capillary or compression joints in the running length; EN1057 R250 (TX) formerly BS 2871 Table X						
Plastic coated gas service pipe for corrosive environments, fixed vertically or at low level with brackets measured separtely						
15 mm dia. (yellow)	12.42	15.72	0.85	29.36	m	**45.08**
22 mm dia. (yellow)	24.47	30.97	0.96	33.17	m	**64.15**
28 mm dia. (yellow)	31.09	39.34	1.06	36.60	m	**75.95**
Copper pipe; capillary or compression joints in the running length; EN1057 R250 (TY) formerly BS 2871 Table Y						
Plastic coated gas and cold water service pipe for corrosive environments, fixed vertically or at low level with brackets measured separately (Refer to Copper Pipe Table X Section)						
15 mm dia. (yellow)	14.35	18.17	0.61	21.06	m	**39.23**
22 mm dia. (yellow)	26.08	33.00	0.69	23.83	m	**56.83**
FIXINGS						
Refer to Section – Cold Water						
Extra over copper pipes; capillary fittings; BS 864						
Refer to Section – Cold Water						

38 PIPED SUPPLY SYSTEMS

Item	Net Price £	Material £	Labour hours	Labour £	Unit	Total rate £
FUEL OIL STORAGE/DISTRIBUTION						
PIPELINES						
For pipework prices refer to Section – Low Temperature Hot Water Heating						
TANKS						
Fuel storage tanks; mild steel; with all necessary screwed bosses; oil resistant joint rings; includes placing in position. Rectangular shape						
1360 litre (300 gallon) capacity; 2 mm plate	500.08	632.89	12.03	415.43	nr	**1048.32**
2730 litre (600 gallon) capacity; 2.5 mm plate	668.23	845.71	18.60	642.31	nr	**1488.02**
4550 litre (1000 gallon) capacity; 3 mm plate	1401.01	1773.12	25.00	863.31	nr	**2636.43**
Fuel storage tanks; 5 mm plate mild steel to BS 799 type J; complete with raised neck manhole with bolted cover, screwed connections, vent and fill connections, drain valve, gauge and overfill alarm; includes placing in position; excludes pumps and control panel. Nominal capacity size						
5,600 litres, 2.5 m × 1.5 m × 1.5 m high	3270.05	4138.57	20.00	690.65	nr	**4829.22**
Extra for bund unit (internal use)	1880.52	2379.98	30.00	1035.98	nr	**3415.96**
Extra for external use with bund (watertight)	1190.98	1507.31	2.00	69.07	nr	**1576.37**
10,200 litres, 3.05 m × 1.83 m × 1.83 m high	4064.03	5143.44	30.00	1035.98	nr	**6179.42**
Extra for bund unit (internal use)	2737.21	3464.22	40.00	1381.30	nr	**4845.52**
Extra for external use with bund (watertight)	1462.63	1851.10	2.00	69.07	nr	**1920.17**
15,000 litres, 3.75 m × 2 m × 2 m high	5140.11	6505.33	40.00	1381.30	nr	**7886.63**
Extra for bund unit (internal use)	3677.50	4654.24	55.00	1899.29	nr	**6553.53**
Extra for external use with bund (watertight)	1713.38	2168.46	2.00	69.07	nr	**2237.52**
20,000 litres, 4 m × 2.5 m × 2 m high	6780.37	8581.23	50.00	1726.63	nr	**10307.86**
Extra for bund unit (internal use)	4461.07	5645.93	65.00	2244.61	nr	**7890.55**
Extra for external use with bund (watertight)	2026.79	2565.10	2.00	69.07	nr	**2634.17**
Extra for BMS output (all tank sizes)	710.42	899.11		–	nr	**899.11**

38 PIPED SUPPLY SYSTEMS

Item	Net Price £	Material £	Labour hours	Labour £	Unit	Total rate £
FUEL OIL STORAGE/DISTRIBUTION – cont						
Fuel storage tanks; plastic; with all necessary screwed bosses; oil resistant joint rings; includes placing in position						
Cylindrical; horizontal						
1350 litre (300 gallon) capacity	704.13	891.14	4.30	148.49	nr	**1039.63**
2500 litre (550 gallon) capacity	1094.35	1385.01	4.88	168.52	nr	**1553.53**
Cylindrical; vertical						
1365 litre (300 gallon) capacity	495.58	627.20	3.73	128.81	nr	**756.01**
2600 litre (570 gallon) capacity	538.82	681.93	4.88	168.52	nr	**850.45**
Bunded tanks						
1135 litre (250 gallon) capacity	1524.35	1929.22	4.30	148.49	nr	**2077.71**
1590 litre (350 gallon) capacity	1617.88	2047.58	4.88	168.52	nr	**2216.10**
2500 litre (550 gallon) capacity	1939.30	2454.38	5.95	205.47	nr	**2659.85**

38 PIPED SUPPLY SYSTEMS

Item	Net Price £	Material £	Labour hours	Labour £	Unit	Total rate £
FIRE HOSE REELS: PIPEWORK ANCILLARIES						
PIPELINES						
For pipework prices refer to section – Cold Water						
PIPELINE ANCILLARIES						
For prices for ancillaries refer to section – Cold Water						
Hose reels; automatic; connection to 25 mm screwed joint; reel with 30.5 m, 19 mm rubber hose; suitable for working pressure up to 7 bar						
Reels						
Non-swing pattern	185.62	234.92	3.75	129.50	nr	**364.42**
Swinging pattern	257.99	326.51	3.75	129.50	nr	**456.01**
Hose reels; manual; connection to 25 mm screwed joint; reel with 30.5 m, 19 mm rubber hose; suitable for working pressure up to 7 bar						
Reels						
Non-swing pattern	185.62	234.92	3.25	112.23	nr	**347.15**
Swinging pattern	257.99	326.51	3.25	112.23	nr	**438.74**

38 PIPED SUPPLY SYSTEMS

Item	Net Price £	Material £	Labour hours	Labour £	Unit	Total rate £
DRY RISERS: PIPEWORK ANCILLARIES						
PIPELINES						
For pipework prices refer to section – Cold Water						
VALVES (BS 5041, Parts 2 and 3)						
Bronze/gunmetal inlet breeching for pumping in with 65 mm dia. instantaneous male coupling; with cap, chain and 25 mm drain valve						
Double inlet with back pressure valve, flanged to steel	363.24	459.72	1.75	60.43	nr	**520.15**
Quadruple inlet with back pressure valve, flanged to steel	813.32	1029.34	1.75	60.43	nr	**1089.77**
Bronze/gunmetal gate type outlet valve with 65 mm dia. instantaneous female coupling; cap and chain; wheel head secured by padlock and leather strap						
Flanged to BS 4504 PN6 (bolted connection to counter flanges measured separately)	291.31	368.68	1.75	60.43	nr	**429.11**
Bronze/gunmetal landing type outlet valve, with 65 mm dia. instantaneous female coupling; cap and chain; wheelhead secured by padlock and leather strap; bolted connections to counter flanges measured separately						
Horizontal, flanged to BS 4504 PN6	317.26	401.53	1.50	51.80	nr	**453.33**
Oblique, flanged to BS 4504 PN6	317.26	401.53	1.50	51.80	nr	**453.33**
Air valve, screwed joints to steel						
25 mm dia.	63.06	79.80	0.55	18.99	nr	**98.80**
INLET BOXES (BS 5041, part 5)						
Steel dry riser inlet box with hinged wire glazed door suitably lettered (fixing by others)						
610 × 460 × 325 mm; double inlet	386.62	489.30	3.00	103.60	nr	**592.90**
610 × 610 × 356 mm; quadruple inlet	721.52	913.15	3.00	103.60	nr	**1016.75**
OUTLET BOXES (BS 5041, part 5)						
Steel dry riser outlet box with hinged wire glazed door suitably lettered (fixing by others)						
610 × 460 × 325 mm; single outlet	377.52	477.79	3.00	103.60	nr	**581.39**

38 PIPED SUPPLY SYSTEMS

Item	Net Price £	Material £	Labour hours	Labour £	Unit	Total rate £
SPRINKLERS						
PIPELINES						
Prefabricated black steel pipework; screwed joints, including all coupliings, unions and the like to BS 1387:1985; includes fixing to backgrounds, with brackets measured separately						
Heavy weight						
25 mm dia. – pipe plus allowance for coupling every 6 m	8.68	10.99	0.47	16.23	m	**27.22**
32 mm dia. – pipe plus allowance for coupling every 6 m	10.79	13.65	0.53	18.30	m	**31.95**
40 mm dia. – pipe plus allowance for coupling every 6 m	12.55	15.88	0.58	20.03	m	**35.91**
50 mm dia. – pipe plus allowance for coupling every 6 m	17.43	22.05	0.63	21.76	m	**43.81**
FIXINGS						
For steel pipes; black malleable iron. For minimum fixing distances, refer to the Tables and Memoranda at the rear of the book						
Pipe ring, single socket, black malleable iron						
25 mm dia.	1.75	2.22	0.12	4.14	nr	**6.36**
32 mm dia.	1.87	2.36	0.14	4.83	nr	**7.20**
40 mm dia.	2.40	3.03	0.15	5.18	nr	**8.21**
50 mm dia.	3.02	3.83	0.16	5.53	nr	**9.35**
Extra over channel sections for fabricated hangers and brackets						
Galvanized steel; including inserts, bolts, nuts, washers; fixed to backgrounds						
41 × 21 mm	8.39	10.62	0.29	10.01	m	**20.64**
41 × 41 mm	10.06	12.74	0.29	10.01	m	**22.75**
Threaded rods; metric thread; including nuts, washers etc.						
12 mm dia. × 600 mm long	4.24	5.37	0.18	6.22	nr	**11.58**
Extra over for black malleable iron fittings; BS 143						
Plain plug, solid						
25 mm dia.	3.09	3.91	0.40	13.81	nr	**17.72**
32 mm dia.	4.45	5.63	0.44	15.19	nr	**20.82**
40 mm dia.	6.26	7.93	0.48	16.58	nr	**24.50**
50 mm dia.	9.60	12.15	0.56	19.34	nr	**31.49**

38 PIPED SUPPLY SYSTEMS

Item	Net Price £	Material £	Labour hours	Labour £	Unit	Total rate £
SPRINKLERS – cont						
Extra over for black malleable iron fittings – cont						
Concentric reducing socket						
32 mm dia.	3.67	4.64	0.48	16.58	nr	**21.22**
40 mm dia.	4.78	6.04	0.55	18.99	nr	**25.04**
50 mm dia.	7.84	9.92	0.60	20.72	nr	**30.64**
Elbow; 90° female/female						
25 mm dia.	2.14	2.71	0.44	15.19	nr	**17.91**
32 mm dia.	3.96	5.01	0.53	18.30	nr	**23.31**
40 mm dia.	6.62	8.37	0.60	20.72	nr	**29.09**
50 mm dia.	9.06	11.46	0.65	22.45	nr	**33.91**
Tee						
25 mm dia. equal	2.95	3.73	0.51	17.61	nr	**21.34**
32 mm dia. reducing to 25 mm dia.	5.36	6.78	0.54	18.65	nr	**25.42**
40 mm dia.	8.23	10.41	0.65	22.45	nr	**32.86**
50 mm dia.	13.80	17.47	0.78	26.94	nr	**44.40**
Cross tee						
25 mm dia. equal	8.01	10.14	1.16	40.06	nr	**50.20**
32 mm dia.	11.66	14.75	1.40	48.35	nr	**63.10**
40 mm dia.	14.97	18.94	1.60	55.25	nr	**74.20**
50 mm dia.	28.37	35.91	1.68	58.01	nr	**93.92**
Prefabricated black steel pipework; welded joints, including all coupliings, unions and the like to BS 1387:1985; fixing to backgrounds						
Heavy weight						
65 mm dia. – pipe length only, welded joints on runs assumed	26.59	33.66	0.65	22.45	m	**56.10**
80 mm dia. – pipe length only, welded joints on runs assumed	33.85	42.85	0.70	24.17	m	**67.02**
100 mm dia. – pipe length only, welded joints on runs assumed	47.23	59.78	0.85	29.35	m	**89.13**
150 mm dia. – pipe length only, welded joints on runs assumed	63.90	80.87	1.15	39.71	m	**120.59**
FIXINGS						
For steel pipes; black malleable iron. For minimum fixing distances, refer to the Tables and Memoranda at the rear of the book						
Pipe ring, single socket, black malleable iron						
65 mm dia.	4.38	5.54	0.30	10.36	nr	**15.90**
80 mm dia.	5.25	6.64	0.35	12.09	nr	**18.73**
100 mm dia.	7.94	10.05	0.40	13.81	nr	**23.87**
150 mm dia.	17.98	22.76	0.77	26.59	nr	**49.35**

38 PIPED SUPPLY SYSTEMS

Item	Net Price £	Material £	Labour hours	Labour £	Unit	Total rate £
Extra over fittings						
Reducer (one size down)						
65 mm dia.	13.66	17.28	2.70	93.24	nr	**110.52**
80 mm dia.	17.76	22.47	2.86	98.76	nr	**121.24**
100 mm dia.	35.41	44.82	3.22	111.19	nr	**156.01**
150 mm dia.	98.39	124.52	4.20	145.04	nr	**269.56**
Elbow; 90°						
65 mm dia.	18.76	23.74	3.06	105.67	nr	**129.41**
80 mm dia.	27.58	34.90	3.40	117.41	nr	**152.31**
100 mm dia.	34.49	43.65	3.70	127.77	nr	**171.42**
150 mm dia.	51.69	65.41	5.20	179.57	nr	**244.98**
Branch bend						
65 mm dia.	19.11	24.19	3.60	124.32	nr	**148.50**
80 mm dia.	22.37	28.32	3.80	131.22	nr	**159.54**
100 mm dia.	34.41	43.55	5.10	176.12	nr	**219.67**
150 mm dia.	89.05	112.70	7.50	258.99	nr	**371.70**
Prefabricated black steel pipe; Victaulic Firelock joints; including all couplings and the like to BS 1387: 1985; fixing to backgrounds						
Heavy weight						
65 mm dia. – pipe plus allowance for Victaulic coupling every 6 m	22.92	29.00	0.70	24.17	m	**53.18**
80 mm dia. – pipe plus allowance for Victaulic coupling every 6 m	29.80	37.71	0.78	26.94	m	**64.65**
100 mm dia. – pipe plus allowance for Victaulic coupling every 6 m	43.44	54.98	0.93	32.12	m	**87.09**
150 mm dia. – pipe plus allowance for Victaulic coupling every 6 m	86.48	109.45	1.25	43.17	m	**152.61**
FIXINGS						
For fixings refer to For steel pipes; black malleable iron.						
Extra over fittings						
Coupling						
65 mm dia. – Victaulic coupling	16.79	21.25	0.26	8.98	nr	**30.22**
80 mm dia. – Victaulic coupling	17.99	22.77	0.26	8.98	nr	**31.75**
100 mm dia. – Victaulic coupling	23.58	29.84	0.32	11.05	nr	**40.89**
150 mm dia. – Victaulic coupling	41.50	52.52	0.35	12.09	nr	**64.61**
Reducer						
65 mm dia. – Victaulic excl. couplings	13.66	17.28	0.48	16.58	nr	**33.86**
80 mm dia. – Victaulic excl. couplings	17.76	22.47	0.43	14.85	nr	**37.32**
100 mm dia. – Victaulic excl. couplings	35.41	44.82	0.46	15.88	nr	**60.70**
150 mm dia. – Victaulic excl. couplings	53.13	67.24	0.45	15.54	nr	**82.78**
Elbow; any degree						
65 mm dia. – Victaulic excl. couplings	18.76	23.74	0.56	19.34	nr	**43.08**
80 mm dia. – Victaulic excl. couplings	27.58	34.90	0.63	21.76	nr	**56.66**
100 mm dia. – Victaulic excl. couplings	53.18	67.31	0.71	24.52	nr	**91.83**
150 mm dia. – Victaulic excl. couplings	79.78	100.97	0.80	27.63	nr	**128.59**

38 PIPED SUPPLY SYSTEMS

Item	Net Price £	Material £	Labour hours	Labour £	Unit	Total rate £
SPRINKLERS – cont						
Extra over fittings – cont						
Equal tee						
65 mm dia. – Victaulic excl. couplings	29.09	36.82	0.74	25.55	nr	**62.37**
80 mm dia. – Victaulic excl. couplings	35.02	44.32	0.83	28.66	nr	**72.98**
100 mm dia. – Victaulic excl. couplings	64.18	81.23	0.94	32.46	nr	**113.69**
150 mm dia. – Victaulic excl. couplings	96.30	121.88	1.05	36.26	nr	**158.14**
PIPELINE ANCILLARIES						
SPRINKLER HEADS						
Sprinkler heads; brass body; frangible glass bulb; manufactured to standard operating temperature of 57–141°C; quick response; RTI<50						
Conventional pattern; 15 mm dia.	5.47	6.92	0.15	5.18	nr	**12.10**
Sidewall pattern; 15 mm dia.	8.02	10.15	0.15	5.18	nr	**15.33**
Conventional pattern; 15 mm dia.; satin chrome plated	6.41	8.12	0.15	5.18	nr	**13.30**
Sidewall pattern; 15 mm dia.; satin chrome plated	8.33	10.54	0.15	5.18	nr	**15.72**
Fully concealed; fusible link; 15 mm dia.	20.02	25.34	0.15	5.18	nr	**30.52**
VALVES						
Wet system alarm valves; including internal non-return valve; working pressure up to 12.5 bar; BS4504 PN16 flanged ends; bolted connections						
100 mm dia.	1761.89	2229.85	25.00	863.31	nr	**3093.16**
150 mm dia.	2152.21	2723.83	25.00	863.31	nr	**3587.14**
Wet system bypass alarm valves; including internal non-return valve; working pressure up to 12.5 bar; BS4504 PN16 flanged ends; bolted connections						
100 mm dia.	3105.73	3930.61	25.00	863.31	nr	**4793.92**
150 mm dia.	3926.70	4969.64	25.00	863.31	nr	**5832.95**
Alternate system wet/dry alarm station; including butterfly valve, wet alarm valve, dry pipe differential pressure valve and pressure gauges; working pressure up to 10.5 bar; BS4505 PN16 flanged ends; bolted connections						
100 mm dia.	3849.10	4871.42	40.00	1381.30	nr	**6252.72**
150 mm dia.	4485.67	5677.06	40.00	1381.30	nr	**7058.36**
Alternate system wet/dry alarm station; including electrically operated butterfly valve, water supply accelerator set, wet alarm valve, dry pipe differential pressure valve and pressure gauges; working pressure up to 10.5 bar; BS4505 PN16 flanged ends; bolted connections						
100 mm dia.	4387.49	5552.81	45.00	1553.96	nr	**7106.77**
150 mm dia.	5030.14	6366.14	45.00	1553.96	nr	**7920.10**

38 PIPED SUPPLY SYSTEMS

Item	Net Price £	Material £	Labour hours	Labour £	Unit	Total rate £
ALARM/GONGS						
Water operated motor alarm and gong; stainless steel and aluminum body and gong; screwed connections						
Connection to sprinkler system and drain pipework	537.05	679.69	6.00	207.20	nr	**886.89**
WATER TANKS						
Note: Prices are based on the most economical tank size for each volume, and the cost will vary with differing tank dimensions, for the same volume.						
GRP sectional sprinkler tank; ordinary hazard, life safety classification; two compartment tank, complete with all fittings and accessories to comply with LPCB type A requirements; cost of erection (on prepared supports) is included within net price, labour cost allows for offloading and positioning of materials						
Volume, size						
55 m³, 6 m × 4 m × 3 m (h)	44174.34	55907.04	14.00	483.46	nr	**56390.50**
70 m³, 6 m × 5 m × 3 m (h)	49390.12	62508.14	16.00	552.52	nr	**63060.66**
80 m³, 8 m × 4 m × 3 m (h)	51185.85	64780.81	16.00	552.52	nr	**65333.33**
105 m³, 10 m × 4 m × 3 m (h)	57920.76	73304.52	24.00	828.78	nr	**74133.30**
125 m³, 10 m × 5 m × 3 m (h)	65321.78	82671.25	24.00	828.78	nr	**83500.03**
140 m³, 8 m × 7 m × 3 m (h)	68027.08	86095.07	24.00	828.78	nr	**86923.85**
135 m³, 9 m × 6 m × 3 m (h)	70083.75	88697.99	24.00	828.78	nr	**89526.77**
160 m³, 13 m × 5 m × 3 m (h)	76945.28	97381.94	24.00	828.78	nr	**98210.72**
185 m³, 12 m × 6 m × 3 m (h)	81393.66	103011.82	24.00	828.78	nr	**103840.60**
GRP sectional sprinkler tank; ordinary hazard, property protection classification; single compartment tank, complete with all fittings and accessories to comply with LPCB type A requirements; cost of erection (on prepared supports) is within net price, labour cost allows for offloading and positioning of materials						
Volume, size						
55 m³, 6 m × 4 m × 3 m (h)	33984.32	43010.56	14.00	483.46	nr	**43494.01**
70 m³, 6 m × 5 m × 3 m (h)	38393.77	48591.15	16.00	552.52	nr	**49143.67**
80 m³, 8 m × 4 m × 3 m (h)	40980.21	51864.56	16.00	552.52	nr	**52417.08**
105 m³, 10 m × 4 m × 3 m (h)	48009.25	60760.51	24.00	828.78	nr	**61589.29**
125 m³, 10 m × 5 m × 3 m (h)	54288.42	68707.43	24.00	828.78	nr	**69536.21**
140 m³, 8 m × 7 m × 3 m (h)	56200.99	71127.98	24.00	828.78	nr	**71956.76**
135 m³, 9 m × 6 m × 3 m (h)	59048.73	74732.07	24.00	828.78	nr	**75560.85**
160 m³, 13 m × 5 m × 3 m (h)	66198.22	83780.47	24.00	828.78	nr	**84609.25**
185 m³, 12 m × 6 m × 3 m (h)	69542.32	88012.76	24.00	828.78	nr	**88841.54**

38 PIPED SUPPLY SYSTEMS

Item	Net Price £	Material £	Labour hours	Labour £	Unit	Total rate £
WATERMIST SYSTEM						
PIPELINES						
For pipework prices refer to section – Cold Water: Stainless Steel						
PIPELINE ANCILLARIES						
Nozzles						
Semco CJ nozzles	115.64	146.36	0.75	25.90	nr	**172.26**
Extra over for rosettes (1 per nozzle)	15.17	19.20	0.20	6.91	nr	**26.10**
Extra over for protection caps (1 per nozzle)	1.41	1.78	0.10	3.45	nr	**5.24**
PUMP SETS						
Automatic pump units (APU); includes fixing in position; electrical work elsewhere.						
15 l/s pump unit	2115.38	2677.22	16.00	552.52	nr	**3229.74**
18 l/s pump unit	2467.95	3123.44	16.00	552.52	nr	**3675.96**
Extra over for control module to duty pump	352.00	445.49	2.00	69.07	nr	**514.56**

38 PIPED SUPPLY SYSTEMS

Item	Net Price £	Material £	Labour hours	Labour £	Unit	Total rate £
FIRE EXTINGUISHERS AND HYDRANTS						
EXTINGUISHERS						
Fire extinguishers; hand held; BS 5423; placed in position						
Water type; cartridge operated; for Class A fires						
Water type, 9 litre capacity; 55 gm CO_2 cartridge; Class A fires (fire-rating 13 A)	39.52	50.01	1.00	34.53	nr	**84.55**
Foam type, 9 litre capacity; 75 gm CO_2 cartridge; Class A & B fires (fire-rating 13 A:183B)	51.47	65.14	1.00	34.53	nr	**99.67**
Dry powder type; cartridge operated; for Class A, B & C fires and electrical equipment fires						
Dry powder type, 1 kg capacity; 12 gm CO_2 cartridge; Class A, B & C fires (fire-rating 5 A:34B)	40.19	50.87	1.00	34.53	nr	**85.40**
Dry powder type, 2 kg capacity; 28 gm CO_2 cartridge; Class A, B & C fires (fire-rating 13 A:55B)	54.13	68.50	1.00	34.53	nr	**103.03**
Dry powder type, 4 kg capacity; 90 gm CO_2 cartridge; Class A, B & C fires (fire-rating 21 A:183B)	96.42	122.03	1.00	34.53	nr	**156.56**
Dry powder type, 9 kg capacity; 190 gm CO_2 cartridge; Class A, B & C fires (fire-rating 43 A:233B)	129.19	163.51	1.00	34.53	nr	**198.04**
Dry powder type; stored pressure type; for Class A, B & C fires and electrical equipment fires						
Dry powder type, 1 kg capacity; Class A, B & C fires (fire-rating 5 A:34B)	17.48	22.12	1.00	34.53	nr	**56.65**
Dry powder type, 2 kg capacity; Class A, B & C fires (fire-rating 13 A:55B)	16.04	20.30	1.00	34.53	nr	**54.83**
Dry powder type, 4 kg capacity; Class A, B & C fires (fire-rating 21 A:183B)	33.21	42.03	1.00	34.53	nr	**76.56**
Dry powder type, 9 kg capacity; Class A, B & C fires (fire-rating 43 A:233B)	35.20	44.54	1.00	34.53	nr	**79.08**
Carbon dioxide type; for Class B fires and electrical equipment fires						
CO_2 type with hose and horn, 2 kg capacity, Class B fires (fire-rating 34B)	34.54	43.71	1.00	34.53	nr	**78.25**
CO_2 type with hose and horn, 5 kg capacity, Class B fires (fire-rating 55B)	69.19	87.56	1.00	34.53	nr	**122.10**
Glass fibre blanket, in GRP container						
1100 × 1100 mm	13.28	16.80	0.50	17.27	nr	**34.07**
1200 × 1200 mm	15.49	19.61	0.50	17.27	nr	**36.87**
1800 × 1200 mm	19.48	24.65	0.50	17.27	nr	**41.92**

38 PIPED SUPPLY SYSTEMS

Item	Net Price £	Material £	Labour hours	Labour £	Unit	Total rate £
FIRE EXTINGUISHERS AND HYDRANTS – cont						
HYDRANTS						
Fire hydrants; bolted connections						
Underground hydrants, complete with frost plug to BS 750						
Sluice valve pattern type 1	265.40	335.89	4.50	155.40	nr	**491.28**
Screw down pattern type 2	192.77	243.97	4.50	155.40	nr	**399.36**
Stand pipe for underground hydrant; screwed base; light alloy						
Single outlet	152.64	193.18	1.00	34.53	nr	**227.71**
Double outlet	223.34	282.66	1.00	34.53	nr	**317.19**
64 mm dia. bronze/gunmetal outlet valves						
Oblique flanged landing valve	167.08	211.45	1.00	34.53	nr	**245.98**
Oblique screwed landing valve	167.08	211.45	1.00	34.53	nr	**245.98**
Cast iron surface box; fixing by others						
400 × 200 × 100 mm	146.29	185.14	1.00	34.53	nr	**219.67**
500 × 200 × 150 mm	197.10	249.45	1.00	34.53	nr	**283.98**
Frost plug	35.50	44.92	0.25	8.63	nr	**53.56**

38 MECHANICAL/COOLING/HEATING SYSTEMS

Item	Net Price £	Material £	Labour hours	Labour £	Unit	Total rate £
GAS/OIL FIRED BOILERS – DOMESTIC						
Domestic water boilers; stove enamelled casing; electric controls; placing in position; assembling and connecting; electrical work elsewhere						
Gas fired; floor standing; connected to conventional flue						
9–12 kW	700.68	886.78	8.59	296.63	nr	**1183.42**
12–15 kW	737.11	932.88	8.59	296.63	nr	**1229.52**
15–18 kW	785.09	993.61	8.88	306.65	nr	**1300.26**
18–21 kW	912.12	1154.37	9.92	342.56	nr	**1496.94**
21–23 kW	1011.17	1279.74	10.66	368.12	nr	**1647.85**
23–29 kW	1311.47	1659.80	11.81	407.83	nr	**2067.63**
29–37 kW	1551.15	1963.14	11.81	407.83	nr	**2370.97**
37–41 kW	1613.13	2041.58	12.68	437.87	nr	**2479.45**
Gas fired; wall hung; connected to conventional flue						
9–12 kW	615.84	779.41	8.59	296.63	nr	**1076.04**
12–15 kW	795.06	1006.23	8.59	296.63	nr	**1302.87**
13–18 kW	1009.01	1277.00	8.59	296.63	nr	**1573.63**
Gas fired; floor standing; connected to balanced flue						
9–12 kW	875.26	1107.73	9.16	316.32	nr	**1424.04**
12–15 kW	910.68	1152.55	10.95	378.13	nr	**1530.68**
15–18 kW	980.26	1240.62	11.98	413.70	nr	**1654.32**
18–21 kW	1155.15	1461.96	12.78	441.33	nr	**1903.28**
21–23 kW	1332.30	1686.16	12.78	441.33	nr	**2127.48**
23–29 kW	1698.18	2149.22	15.45	533.53	nr	**2682.75**
29–37 kW	3264.57	4131.64	17.65	609.50	nr	**4741.14**
Gas fired; wall hung; connected to balanced flue						
6–9 kW	660.86	836.39	9.16	316.32	nr	**1152.70**
9–12 kW	756.24	957.10	9.16	316.32	nr	**1273.41**
12–15 kW	856.33	1083.77	9.45	326.33	nr	**1410.11**
15–18 kW	1026.79	1299.50	9.74	336.35	nr	**1635.85**
18–22 kW	1089.90	1379.38	9.74	336.35	nr	**1715.73**
Gas fired; wall hung; connected to fan flue (including flue kit)						
6–9 kW	755.83	956.58	9.16	316.32	nr	**1272.89**
9–12 kW	842.29	1066.01	9.16	316.32	nr	**1382.33**
12–15 kW	914.33	1157.18	10.95	378.13	nr	**1535.31**
15–18 kW	984.89	1246.48	11.98	413.70	nr	**1660.18**
18–23 kW	1283.29	1624.13	12.78	441.33	nr	**2065.45**
23–29 kW	1661.83	2103.22	15.45	533.53	nr	**2636.74**
29–35 kW	2040.25	2582.14	17.65	609.50	nr	**3191.64**

38 MECHANICAL/COOLING/HEATING SYSTEMS

Item	Net Price £	Material £	Labour hours	Labour £	Unit	Total rate £
GAS/OIL FIRED BOILERS – DOMESTIC – cont						
Domestic water boilers – cont						
Oil fired; floor standing; connected to conventional flue						
12–15 kW	1312.34	1660.89	10.38	358.45	nr	**2019.34**
15–19 kW	1368.96	1732.56	12.20	421.30	nr	**2153.86**
21–25 kW	1562.38	1977.35	14.30	493.82	nr	**2471.16**
26–32 kW	1712.94	2167.90	15.80	545.61	nr	**2713.52**
35–50 kW	1933.97	2447.63	20.46	706.54	nr	**3154.17**
Fire place mounted natural gas fire and back boiler; cast iron water boiler; electric control box; fire output 3 kW with wood surround						
10.50 kW	382.86	484.55	8.88	306.65	nr	**791.20**

38 MECHANICAL/COOLING/HEATING SYSTEMS

Item	Net Price £	Material £	Labour hours	Labour £	Unit	Total rate £
GAS/OIL FIRED BOILERS – COMMERCIAL; FORCED DRAFT						
Commercial steel shell floor standing boilers; pressure jet burner; 8 bar max working pressure; including controls, enamelled jacket, insulation, placing in position and commissioning; electrical work elsewhere						
Natural gas; burner and boiler (high/low type), connected to conventional flue						
411–500 kW	15721.65	19897.32	12.00	414.39	nr	**20311.71**
601–750 kW	18558.75	23487.95	12.00	414.39	nr	**23902.34**
1151–1250 kW	25208.48	31903.86	14.00	483.46	nr	**32387.31**
1426–1500 kW	30865.19	39062.99	14.00	483.46	nr	**39546.44**
1625–1760 kW	36369.88	46029.72	14.00	483.46	nr	**46513.17**
Natural gas; burner and boiler (modulating type), connected to conventional flue						
411–500 kW	16850.40	21325.87	12.00	414.39	nr	**21740.26**
601–750 kW	19469.10	24640.09	12.00	414.39	nr	**25054.48**
1151–1250 kW	28490.11	36057.09	12.00	414.39	nr	**36471.48**
1426–1500 kW	34008.58	43041.25	14.00	483.46	nr	**43524.71**
1625–1760 kW	40901.95	51765.51	14.00	483.46	nr	**52248.97**
1900–2050 kW	42915.26	54313.56	14.00	483.46	nr	**54797.01**
2450–2650 kW	50370.43	63748.82	14.00	483.46	nr	**64232.27**
Natural gas; low Nox burner and boiler (modulating type), connected to conventional flue						
420 kW	17665.20	22357.08	12.00	414.39	nr	**22771.47**
530 kW	19041.75	24099.24	12.00	414.39	nr	**24513.63**
620 kW	21967.05	27801.50	12.00	414.39	nr	**28215.89**
750 kW	22821.75	28883.21	14.00	483.46	nr	**29366.66**
1000 kW	28491.75	36059.16	14.00	483.46	nr	**36542.61**
1250 kW	35694.75	45175.28	14.00	483.46	nr	**45658.73**
1500 kW	43424.85	54958.49	14.00	483.46	nr	**55441.95**
1800 kW	51150.75	64736.39	14.00	483.46	nr	**65219.84**
2200 kW	54896.10	69476.50	14.00	483.46	nr	**69959.96**
2700 kW	60732.00	76862.42	14.00	483.46	nr	**77345.87**
Oil; burner and boiler (high/low type), connected to conventional flue						
411–500 kW	15184.05	19216.93	12.00	414.39	nr	**19631.32**
601–750 kW	17139.15	21691.31	12.00	414.39	nr	**22105.70**
1151–1250 kW	23160.23	29311.58	12.00	414.39	nr	**29725.97**
1426–1500 kW	28202.19	35692.69	14.00	483.46	nr	**36176.15**
1625–1760 kW	33596.20	42519.35	14.00	483.46	nr	**43002.80**

Material Costs/Prices for Measured Works

38 MECHANICAL/COOLING/HEATING SYSTEMS

Item	Net Price £	Material £	Labour hours	Labour £	Unit	Total rate £
GAS/OIL FIRED BOILERS – **COMMERCIAL; FORCED DRAFT – cont**						
Commercial steel shell floor standing boilers – cont						
Oil; burner and boiler (modulating type), connected to conventional flue						
420 kW	16478.70	20855.44	12.00	414.39	nr	**21269.83**
530 kW	16600.50	21009.59	12.00	414.39	nr	**21423.98**
620 kW	18247.95	23094.61	12.00	414.39	nr	**23509.00**
750 kW	19829.25	25095.90	14.00	483.46	nr	**25579.35**
1000 kW	24635.10	31178.18	14.00	483.46	nr	**31661.64**
1250 kW	35438.59	44851.08	14.00	483.46	nr	**45334.54**
1500 kW	32428.20	41041.13	14.00	483.46	nr	**41524.59**
1800 kW	42014.70	53173.80	14.00	483.46	nr	**53657.26**
2200 kW	60812.75	76964.61	14.00	483.46	nr	**77448.07**
2700 kW	68186.41	86296.72	14.00	483.46	nr	**86780.18**
Dual fuel; burner and boiler (high/low oil and modulating gast type), connected to conventional flue						
411–500 kW	19469.10	24640.09	12.00	414.39	nr	**25054.48**
601–750 kW	21867.30	27675.25	12.00	414.39	nr	**28089.65**
1151–1250 kW	33136.32	41937.33	12.00	414.39	nr	**42351.72**
1426–1500 kW	39853.63	50438.76	14.00	483.46	nr	**50922.21**
1625–1760 kW	46132.80	58385.67	14.00	483.46	nr	**58869.13**
2450–2650-kW	60037.82	75983.87	14.00	483.46	nr	**76467.33**
Dual fuel; low NOX burner and boiler (high/ low oil and modulating gast type), connected to conventional flue						
420 kW	19160.40	24249.40	12.00	414.39	nr	**24663.79**
530 kW	21090.30	26691.88	12.00	414.39	nr	**27106.27**
620 kW	21816.90	27611.47	12.00	414.39	nr	**28025.86**
750 kW	23701.65	29996.81	14.00	483.46	nr	**30480.26**
1000 kW	33625.20	42556.05	14.00	483.46	nr	**43039.51**
1250 kW	40687.50	51494.10	14.00	483.46	nr	**51977.56**
1500 kW	44215.50	55959.14	14.00	483.46	nr	**56442.59**
1800 kW	57669.15	72986.08	14.00	483.46	nr	**73469.53**
2200 kW	62094.90	78587.31	14.00	483.46	nr	**79070.76**
2700 kW	67420.50	85327.38	14.00	483.46	nr	**85810.84**

38 MECHANICAL/COOLING/HEATING SYSTEMS

Item	Net Price £	Material £	Labour hours	Labour £	Unit	Total rate £
GAS/OIL FIRED BOILERS – COMMERCIAL; CONDENSING						
Gas boiler; Low Nox wall mounted condensing boiler, with high efficiency modulating premix burner; Aluminium heat exchanger; includes integral control panel; delivery included						
Maximum output						
35 kW	3507.84	4439.52	11.00	379.86	nr	**4819.38**
45 kW	4044.60	5118.85	11.00	379.86	nr	**5498.70**
60 kW	5532.45	7001.87	11.00	379.86	nr	**7381.73**
80 kW	4111.80	5203.89	11.00	379.86	nr	**5583.75**
100 kW	4599.00	5820.49	11.00	379.86	nr	**6200.35**
120 kW	4744.95	6005.21	11.00	379.86	nr	**6385.07**
Commissioning rate	262.50	332.22		–	nr	**332.22**
Gas boiler; Low Nox floor standing condensing boiler with high efficiency modulating premix burner; Stainless steel heat exchanger; includes controls; delivery and commissioning						
Maximum output						
50 kW	4921.35	6228.46	8.00	276.26	nr	**6504.72**
70 kW	5436.90	6880.94	8.00	276.26	nr	**7157.20**
100 kW	5884.20	7447.04	8.00	276.26	nr	**7723.30**
125 kW	6076.35	7690.23	10.00	345.33	nr	**8035.55**
150 kW	6777.75	8577.92	10.00	345.33	nr	**8923.25**
200 kW	7801.50	9873.58	10.00	345.33	nr	**10218.90**
250 kW	8943.90	11319.40	10.00	345.33	nr	**11664.73**
300 kW	9973.95	12623.03	10.00	345.33	nr	**12968.36**
350 kW	11252.85	14241.61	10.00	345.33	nr	**14586.93**
400 kW	12660.90	16023.64	10.00	345.33	nr	**16368.96**
450 kW	14057.40	17791.05	10.00	345.33	nr	**18136.37**
500 kW	15342.60	19417.59	10.00	345.33	nr	**19762.92**
575 kW	17259.90	21844.13	10.00	345.33	nr	**22189.45**
650 kW	18414.90	23305.90	10.00	345.33	nr	**23651.22**
720 kW	21480.90	27186.23	10.00	345.33	nr	**27531.55**
850 kW	26533.50	33580.80	12.00	414.39	nr	**33995.19**
1000 kW	29568.00	37421.26	12.00	414.39	nr	**37835.65**
1150 kW	35225.40	44581.27	12.00	414.39	nr	**44995.66**
1300 kW	38201.10	48347.31	14.00	483.46	nr	**48830.77**
1440 kW	43464.40	55008.55	14.00	483.46	nr	**55492.00**
1550 kW	51371.25	65015.45	14.00	483.46	nr	**65498.91**
1700 kW	53997.13	68338.77	14.00	483.46	nr	**68822.23**
2000 kW	60574.82	76663.49	14.00	483.46	nr	**77146.94**
2300 kW	62966.01	79689.78	14.00	483.46	nr	**80173.24**
3100 kW	80691.45	102123.10	14.00	483.46	nr	**102606.55**

38 MECHANICAL/COOLING/HEATING SYSTEMS

Item	Net Price £	Material £	Labour hours	Labour £	Unit	Total rate £
GAS/OIL FIRED BOILERS – **COMMERCIAL; CONDENSING – cont**						
Gas boilers, wall hung cascade; low Nox **frame mounted, hung condensing boiler** **with high efficiency modulating premix** **burner; Aluminium heat exchanger;** **includes integral control panel; pumps** **and headers included, delivery included**						
Maximum cascade output						
160 kW (2 units)	12412.05	15708.69	10.00	345.33	nr	**16054.02**
200 kW (2 units)	14487.90	18335.89	10.00	345.33	nr	**18681.21**
240 kW (3 units)	18076.80	22878.00	10.00	345.33	nr	**23223.32**
300 kW (3 units)	21532.35	27251.34	12.00	414.39	nr	**27665.73**
360 kW (3 units)	21983.85	27822.76	12.00	414.39	nr	**28237.15**
400 kW (4 units)	25173.75	31859.90	12.00	414.39	nr	**32274.29**
480 kW (4 units)	28720.65	36348.85	12.00	414.39	nr	**36763.25**
600 kW (5 units)	36584.10	46300.84	14.00	483.46	nr	**46784.29**
720 kW (6 units)	43878.45	55532.57	14.00	483.46	nr	**56016.02**
Commissioning rate	199.50	252.49		–	nr	**252.49**
Oil boiler; Low Nox floor standing **condensing boiler with high efficiency** **modulating premix burner; Stainless steel** **heat exchanger; includes controls;** **delivery and commissioning**						
50 kW	5788.65	7326.12	8.00	276.26	nr	**7602.38**
80 kW	7662.90	9698.17	8.00	276.26	nr	**9974.43**
110 kW	10384.50	13142.62	8.00	276.26	nr	**13418.88**
130 kW	10697.40	13538.63	10.00	345.33	nr	**13883.95**
160 kW	11269.65	14262.87	10.00	345.33	nr	**14608.19**
200 kW	12024.60	15218.33	10.00	345.33	nr	**15563.66**
250 kW	15128.40	19146.50	10.00	345.33	nr	**19491.83**
300 kW	17026.80	21549.12	10.00	345.33	nr	**21894.44**
320 kW	23927.40	30282.52	10.00	345.33	nr	**30627.84**
400 kW	26581.80	33641.93	12.00	414.39	nr	**34056.32**
500 kW	32911.20	41652.41	12.00	414.39	nr	**42066.81**
600 kW	36708.00	46457.64	12.00	414.39	nr	**46872.04**

38 MECHANICAL/COOLING/HEATING SYSTEMS

Item	Net Price £	Material £	Labour hours	Labour £	Unit	Total rate £
GAS/OIL FIRED BOILERS – FLUE SYSTEMS						
Flues; suitable for domestic, medium sized industrial and commercial oil and gas appliances; stainless steel, twin wall, insulated; for use internally or externally						
Straight length; 120 mm long; including one locking band						
127 mm dia.	51.97	65.77	0.49	16.92	nr	**82.69**
152 mm dia.	58.22	73.68	0.51	17.61	nr	**91.30**
178 mm dia.	67.58	85.53	0.54	18.65	nr	**104.18**
203 mm dia.	76.89	97.31	0.58	20.03	nr	**117.34**
254 mm dia.	90.93	115.08	0.70	24.17	nr	**139.26**
304 mm dia.	113.93	144.19	0.74	25.55	nr	**169.74**
355 mm dia.	163.14	206.47	0.80	27.63	nr	**234.10**
Straight length; 300 mm long; including one locking band						
127 mm dia.	80.16	101.45	0.52	17.96	nr	**119.41**
152 mm dia.	90.61	114.68	0.52	17.96	nr	**132.63**
178 mm dia.	104.12	131.78	0.55	18.99	nr	**150.77**
203 mm dia.	117.82	149.12	0.64	22.10	nr	**171.22**
254 mm dia.	130.67	165.37	0.79	27.28	nr	**192.65**
304 mm dia.	156.75	198.39	0.86	29.70	nr	**228.09**
355 mm dia.	172.01	217.70	0.94	32.46	nr	**250.16**
400 mm dia.	184.07	232.96	1.03	35.57	nr	**268.52**
450 mm dia.	210.56	266.48	1.03	35.57	nr	**302.05**
500 mm dia.	225.83	285.81	1.10	37.99	nr	**323.79**
550 mm dia.	249.21	315.40	1.10	37.99	nr	**353.39**
600 mm dia.	274.99	348.03	1.10	37.99	nr	**386.01**
Straight length; 500 mm long; including one locking band						
127 mm dia.	94.20	119.22	0.55	18.99	nr	**138.22**
152 mm dia.	105.02	132.92	0.55	18.99	nr	**151.91**
178 mm dia.	118.27	149.68	0.63	21.76	nr	**171.43**
203 mm dia.	138.25	174.96	0.63	21.76	nr	**196.72**
254 mm dia.	160.58	203.23	0.86	29.70	nr	**232.93**
304 mm dia.	192.37	243.46	0.95	32.81	nr	**276.27**
355 mm dia.	216.04	273.43	1.03	35.57	nr	**308.99**
400 mm dia.	235.89	298.54	1.12	38.68	nr	**337.22**
450 mm dia.	272.43	344.79	1.12	38.68	nr	**383.46**
500 mm dia.	293.54	371.51	1.19	41.09	nr	**412.60**
550 mm dia.	323.59	409.54	1.19	41.09	nr	**450.63**
600 mm dia.	335.20	424.23	1.19	41.09	nr	**465.33**
Straight length; 1000 mm long; including one locking band						
127 mm dia.	168.39	213.11	0.62	21.41	nr	**234.52**
152 mm dia.	187.69	237.53	0.68	23.48	nr	**261.02**
178 mm dia.	211.17	267.26	0.74	25.55	nr	**292.81**
203 mm dia.	248.73	314.80	0.80	27.63	nr	**342.42**
254 mm dia.	281.88	356.75	0.87	30.04	nr	**386.79**
304 mm dia.	325.37	411.79	1.06	36.60	nr	**448.40**
355 mm dia.	373.37	472.53	1.16	40.06	nr	**512.59**

38 MECHANICAL/COOLING/HEATING SYSTEMS

Item	Net Price £	Material £	Labour hours	Labour £	Unit	Total rate £
GAS/OIL FIRED BOILERS – FLUE SYSTEMS – cont						
Flues – cont						
Straight length; 1000 mm long – cont						
400 mm dia.	399.85	506.04	1.26	43.51	nr	**549.56**
450 mm dia.	421.90	533.96	1.26	43.51	nr	**577.47**
500 mm dia.	457.85	579.46	1.33	45.93	nr	**625.39**
550 mm dia.	503.54	637.28	1.33	45.93	nr	**683.21**
600 mm dia.	529.01	669.51	1.33	45.93	nr	**715.44**
Adjustable length; boiler removal; internal use only; including one locking band						
127 mm dia.	77.20	97.70	0.52	17.96	nr	**115.66**
152 mm dia.	87.21	110.38	0.55	18.99	nr	**129.37**
178 mm dia.	98.86	125.11	0.59	20.37	nr	**145.49**
203 mm dia.	113.37	143.48	0.64	22.10	nr	**165.58**
254 mm dia.	168.09	212.73	0.79	27.28	nr	**240.01**
304 mm dia.	201.33	254.80	0.86	29.70	nr	**284.50**
355 mm dia.	225.76	285.72	0.99	34.19	nr	**319.91**
400 mm dia.	395.06	499.99	0.91	31.42	nr	**531.41**
450 mm dia.	422.87	535.19	0.91	31.42	nr	**566.61**
500 mm dia.	461.22	583.72	0.99	34.19	nr	**617.90**
550 mm dia.	503.15	636.79	0.99	34.19	nr	**670.98**
600 mm dia.	526.73	666.63	0.99	34.19	nr	**700.81**
Inspection length; 500 mm long; including one locking band						
127 mm dia.	198.40	251.10	0.55	18.99	nr	**270.09**
152 mm dia.	205.45	260.02	0.55	18.99	nr	**279.01**
178 mm dia.	216.39	273.86	0.63	21.76	nr	**295.62**
203 mm dia.	229.14	290.01	0.63	21.76	nr	**311.76**
254 mm dia.	296.27	374.96	0.86	29.70	nr	**404.66**
304 mm dia.	321.09	406.37	0.95	32.81	nr	**439.18**
355 mm dia.	362.55	458.84	1.03	35.57	nr	**494.41**
400 mm dia.	581.64	736.13	1.12	38.68	nr	**774.80**
450 mm dia.	597.47	756.16	1.12	38.68	nr	**794.83**
500 mm dia.	655.61	829.74	1.19	41.09	nr	**870.84**
550 mm dia.	685.56	867.65	1.19	41.09	nr	**908.74**
600 mm dia.	697.21	882.38	1.19	41.09	nr	**923.48**
Adapters						
127 mm dia.	16.59	21.00	0.49	16.92	nr	**37.92**
152 mm dia.	18.15	22.97	0.51	17.61	nr	**40.59**
178 mm dia.	19.42	24.58	0.54	18.65	nr	**43.23**
203 mm dia.	22.10	27.98	0.58	20.03	nr	**48.00**
254 mm dia.	24.32	30.78	0.70	24.17	nr	**54.95**
304 mm dia.	30.34	38.40	0.74	25.55	nr	**63.95**
355 mm dia.	36.84	46.63	0.80	27.63	nr	**74.25**
400 mm dia.	41.33	52.30	0.89	30.73	nr	**83.04**
450 mm dia.	44.14	55.87	0.89	30.73	nr	**86.60**
500 mm dia.	46.86	59.30	0.96	33.15	nr	**92.45**
550 mm dia.	54.47	68.94	0.96	33.15	nr	**102.09**
600 mm dia.	65.47	82.85	0.96	33.15	nr	**116.01**

38 MECHANICAL/COOLING/HEATING SYSTEMS

Item	Net Price £	Material £	Labour hours	Labour £	Unit	Total rate £
Fittings for flue system						
90° insulated tee; including two locking bands						
127 mm dia.	192.89	244.13	1.89	65.27	nr	**309.39**
152 mm dia.	222.63	281.76	2.04	70.45	nr	**352.21**
178 mm dia.	243.34	307.97	2.39	82.53	nr	**390.50**
203 mm dia.	284.92	360.59	2.56	88.40	nr	**448.99**
254 mm dia.	288.62	365.28	2.95	101.87	nr	**467.15**
304 mm dia.	358.92	454.25	3.41	117.76	nr	**572.01**
355 mm dia.	457.58	579.11	3.77	130.19	nr	**709.30**
400 mm dia.	602.96	763.10	4.25	146.76	nr	**909.87**
450 mm dia.	628.46	795.37	4.76	164.37	nr	**959.75**
500 mm dia.	711.14	900.02	5.12	176.81	nr	**1076.83**
550 mm dia.	764.43	967.46	5.61	193.73	nr	**1161.19**
600 mm dia.	795.39	1006.65	5.98	206.50	nr	**1213.16**
135° insulated tee; including two locking bands						
127 mm dia.	250.18	316.63	1.89	65.27	nr	**381.89**
152 mm dia.	270.50	342.35	2.04	70.45	nr	**412.80**
178 mm dia.	295.75	374.30	2.39	82.53	nr	**456.83**
203 mm dia.	375.34	475.03	2.56	88.40	nr	**563.43**
254 mm dia.	426.49	539.77	2.95	101.87	nr	**641.64**
304 mm dia.	321.35	406.69	3.41	117.76	nr	**524.45**
355 mm dia.	632.30	800.23	3.77	130.19	nr	**930.42**
400 mm dia.	832.32	1053.39	4.25	146.76	nr	**1200.15**
450 mm dia.	901.05	1140.37	4.76	164.37	nr	**1304.75**
500 mm dia.	1049.70	1328.51	5.12	176.81	nr	**1505.31**
550 mm dia.	1077.01	1363.06	5.61	193.73	nr	**1556.79**
600 mm dia.	1136.66	1438.56	5.98	206.50	nr	**1645.06**
Wall sleeve; for 135° tee through wall						
127 mm dia.	25.60	32.40	1.89	65.27	nr	**97.66**
152 mm dia.	34.36	43.48	2.04	70.45	nr	**113.93**
178 mm dia.	35.97	45.53	2.39	82.53	nr	**128.06**
203 mm dia.	40.48	51.23	2.56	88.40	nr	**139.63**
254 mm dia.	45.14	57.13	2.95	101.87	nr	**159.00**
304 mm dia.	52.61	66.59	3.41	117.76	nr	**184.34**
355 mm dia.	58.44	73.97	3.77	130.19	nr	**204.15**
15° insulated elbow; including two locking bands						
127 mm dia.	133.86	169.41	1.57	54.22	nr	**223.63**
152 mm dia.	148.77	188.29	1.79	61.81	nr	**250.10**
178 mm dia.	159.11	201.38	2.05	70.79	nr	**272.17**
203 mm dia.	168.69	213.49	2.33	80.46	nr	**293.95**
254 mm dia.	173.74	219.88	2.45	84.60	nr	**304.49**
304 mm dia.	219.64	277.98	3.43	118.45	nr	**396.42**
355 mm dia.	293.83	371.87	4.71	162.65	nr	**534.52**

38 MECHANICAL/COOLING/HEATING SYSTEMS

Item	Net Price £	Material £	Labour hours	Labour £	Unit	Total rate £
GAS/OIL FIRED BOILERS – FLUE SYSTEMS – cont						
Fittings for flue system – cont						
30° insulated elbow; including two locking bands						
127 mm dia.	133.86	169.41	1.44	49.73	nr	**219.14**
152 mm dia.	148.77	188.29	1.62	55.94	nr	**244.23**
178 mm dia.	159.11	201.38	1.89	65.27	nr	**266.64**
203 mm dia.	168.69	213.49	2.17	74.94	nr	**288.42**
254 mm dia.	173.74	219.88	2.16	74.59	nr	**294.47**
304 mm dia.	219.64	277.98	2.74	94.62	nr	**372.59**
355 mm dia.	293.28	371.17	3.17	109.47	nr	**480.64**
400 mm dia.	293.83	371.87	3.53	121.90	nr	**493.77**
450 mm dia.	340.54	430.99	3.88	133.99	nr	**564.98**
500 mm dia.	357.03	451.85	4.24	146.42	nr	**598.27**
550 mm dia.	383.16	484.93	4.61	159.20	nr	**644.12**
600 mm dia.	417.80	528.77	4.96	171.28	nr	**700.05**
45° insulated elbow; including two locking bands						
127 mm dia.	133.86	169.41	1.44	49.73	nr	**219.14**
152 mm dia.	148.77	188.29	1.51	52.14	nr	**240.43**
178 mm dia.	159.11	201.38	1.58	54.56	nr	**255.94**
203 mm dia.	168.69	213.49	1.66	57.32	nr	**270.81**
254 mm dia.	173.74	219.88	1.72	59.40	nr	**279.28**
304 mm dia.	219.64	277.98	1.80	62.16	nr	**340.13**
355 mm dia.	293.83	371.87	1.94	66.99	nr	**438.87**
400 mm dia.	374.71	474.24	2.01	69.41	nr	**543.65**
450 mm dia.	392.49	496.73	2.09	72.17	nr	**568.91**
500 mm dia.	421.01	532.84	2.16	74.59	nr	**607.43**
550 mm dia.	459.18	581.14	2.23	77.01	nr	**658.14**
600 mm dia.	470.81	595.86	2.30	79.42	nr	**675.28**
Flue supports						
Wall support, galvanized; including plate and brackets						
127 mm dia.	81.27	102.86	2.24	77.35	nr	**180.21**
152 mm dia.	89.35	113.08	2.44	84.26	nr	**197.34**
178 mm dia.	98.65	124.85	2.52	87.02	nr	**211.87**
203 mm dia.	102.72	130.00	2.77	95.66	nr	**225.66**
254 mm dia.	122.50	155.03	2.98	102.91	nr	**257.94**
304 mm dia.	138.50	175.29	3.46	119.48	nr	**294.77**
355 mm dia.	185.39	234.63	4.08	140.89	nr	**375.53**
400 mm dia.; with 300 mm support length and collar	476.21	602.69	4.80	165.76	nr	**768.45**
450 mm dia.; with 300 mm support length and collar	507.71	642.55	5.62	194.07	nr	**836.63**
500 mm dia.; with 300 mm support length and collar	554.66	701.98	6.24	215.48	nr	**917.47**
550 mm dia.; with 300 mm support length and collar	606.84	768.02	6.97	240.69	nr	**1008.71**
600 mm dia.; with 300 mm support length and collar	640.60	810.74	7.49	258.65	nr	**1069.39**

38 MECHANICAL/COOLING/HEATING SYSTEMS

Item	Net Price £	Material £	Labour hours	Labour £	Unit	Total rate £
Ceiling/floor support						
127 mm dia.	25.16	31.85	1.86	64.23	nr	96.08
152 mm dia.	27.97	35.40	2.14	73.90	nr	109.30
178 mm dia.	33.49	42.39	1.93	66.65	nr	109.03
203 mm dia.	50.42	63.81	2.74	94.62	nr	158.43
254 mm dia.	57.52	72.80	3.21	110.85	nr	183.65
304 mm dia.	65.73	83.19	3.68	127.08	nr	210.27
355 mm dia.	78.34	99.15	4.28	147.80	nr	246.95
400 mm dia.	100.07	126.65	4.86	167.83	nr	294.48
450 mm dia.	108.67	137.54	5.46	188.55	nr	326.08
500 mm dia.	131.94	166.98	6.04	208.58	nr	375.55
550 mm dia.	161.41	204.28	6.65	229.64	nr	433.92
600 mm dia.	231.62	293.13	7.24	250.02	nr	543.15
Ceiling/floor firestop spacer						
127 mm dia.	4.90	6.20	0.66	22.79	nr	28.99
152 mm dia.	5.45	6.90	0.69	23.83	nr	30.73
178 mm dia.	6.15	7.79	0.70	24.17	nr	31.96
203 mm dia.	7.30	9.24	0.87	30.04	nr	39.28
254 mm dia.	7.55	9.55	0.91	31.42	nr	40.97
304 mm dia.	9.06	11.47	0.95	32.81	nr	44.27
355 mm dia.	16.49	20.88	0.99	34.19	nr	55.06
Wall band; internal or external use						
127 mm dia.	29.62	37.48	1.03	35.57	nr	73.05
152 mm dia.	30.93	39.15	1.07	36.95	nr	76.10
178 mm dia.	32.15	40.69	1.11	38.33	nr	79.03
203 mm dia.	33.95	42.96	1.18	40.75	nr	83.71
254 mm dia.	35.33	44.71	1.30	44.89	nr	89.60
304 mm dia.	38.18	48.32	1.45	50.07	nr	98.39
355 mm dia.	40.61	51.40	1.65	56.98	nr	108.38
400 mm dia.	50.37	63.75	1.85	63.89	nr	127.64
450 mm dia.	53.67	67.92	2.39	82.53	nr	150.46
500 mm dia.	64.52	81.66	2.25	77.70	nr	159.36
550 mm dia.	67.58	85.53	2.45	84.60	nr	170.14
600 mm dia.	71.44	90.42	2.66	91.86	nr	182.28
Flashings and terminals						
Insulated top stub; including one locking band						
127 mm dia.	72.98	92.36	1.49	51.45	nr	143.82
152 mm dia.	82.53	104.45	1.90	65.61	nr	170.06
178 mm dia.	88.85	112.45	1.92	66.30	nr	178.75
203 mm dia.	94.55	119.66	2.20	75.97	nr	195.63
254 mm dia.	100.39	127.06	2.49	85.99	nr	213.04
304 mm dia.	137.22	173.67	2.79	96.35	nr	270.01
355 mm dia.	181.16	229.28	3.19	110.16	nr	339.44
400 mm dia.	174.69	221.08	3.59	123.97	nr	345.05
450 mm dia.	189.78	240.18	3.97	137.09	nr	377.28
500 mm dia.	219.54	277.85	4.38	151.25	nr	429.10
550 mm dia.	229.72	290.74	4.78	165.07	nr	455.80
600 mm dia.	237.74	300.88	5.17	178.53	nr	479.41

38 MECHANICAL/COOLING/HEATING SYSTEMS

Item	Net Price £	Material £	Labour hours	Labour £	Unit	Total rate £
GAS/OIL FIRED BOILERS – FLUE SYSTEMS – cont						
Flashings and terminals – cont						
Rain cap; including one locking band						
127 mm dia.	39.02	49.39	1.49	51.45	nr	**100.84**
152 mm dia.	40.78	51.61	1.54	53.18	nr	**104.79**
178 mm dia.	44.90	56.82	1.72	59.40	nr	**116.22**
203 mm dia.	53.69	67.95	2.00	69.07	nr	**137.02**
254 mm dia.	70.60	89.35	2.49	85.99	nr	**175.33**
304 mm dia.	48.34	61.18	2.80	96.69	nr	**157.87**
355 mm dia.	127.53	161.40	3.19	110.16	nr	**271.56**
400 mm dia.	127.23	161.02	3.45	119.14	nr	**280.16**
450 mm dia.	138.37	175.12	3.97	137.09	nr	**312.21**
500 mm dia.	149.60	189.33	4.38	151.25	nr	**340.58**
550 mm dia.	160.52	203.15	4.78	165.07	nr	**368.22**
600 mm dia.	171.62	217.21	5.17	178.53	nr	**395.74**
Round top; including one locking band						
127 mm dia.	74.60	94.42	1.49	51.45	nr	**145.87**
152 mm dia.	81.34	102.94	1.65	56.98	nr	**159.92**
178 mm dia.	92.71	117.34	1.92	66.30	nr	**183.64**
203 mm dia.	109.12	138.10	2.20	75.97	nr	**214.07**
254 mm dia.	129.02	163.29	2.49	85.99	nr	**249.27**
304 mm dia.	169.45	214.46	2.80	96.69	nr	**311.15**
355 mm dia.	226.06	286.10	3.19	110.16	nr	**396.26**
Coping cap; including one locking band						
127 mm dia.	41.78	52.88	1.49	51.45	nr	**104.33**
152 mm dia.	43.76	55.39	1.65	56.98	nr	**112.37**
178 mm dia.	48.15	60.94	1.92	66.30	nr	**127.24**
203 mm dia.	57.79	73.14	2.20	75.97	nr	**149.11**
254 mm dia.	70.60	89.35	2.49	85.99	nr	**175.33**
304 mm dia.	95.18	120.46	2.79	96.35	nr	**216.81**
355 mm dia.	127.53	161.40	3.19	110.16	nr	**271.56**
Storm collar						
127 mm dia.	7.91	10.02	0.52	17.96	nr	**27.97**
152 mm dia.	8.47	10.72	0.55	18.99	nr	**29.71**
178 mm dia.	9.39	11.89	0.57	19.68	nr	**31.57**
203 mm dia.	9.84	12.45	0.66	22.79	nr	**35.24**
254 mm dia.	12.33	15.61	0.66	22.79	nr	**38.40**
304 mm dia.	12.83	16.24	0.72	24.86	nr	**41.10**
355 mm dia.	13.72	17.37	0.77	26.59	nr	**43.96**
400 mm dia.	36.40	46.06	0.82	28.32	nr	**74.38**
450 mm dia.	40.06	50.70	0.87	30.04	nr	**80.74**
500 mm dia.	43.67	55.27	0.92	31.77	nr	**87.04**
550 mm dia.	47.34	59.91	0.98	33.84	nr	**93.75**
600 mm dia.	50.96	64.50	1.03	35.57	nr	**100.07**
Flat flashing; including storm collar and sealant						
127 mm dia.	44.49	56.30	1.49	51.45	nr	**107.76**
152 mm dia.	45.97	58.18	1.65	56.98	nr	**115.15**
178 mm dia.	48.23	61.04	1.92	66.30	nr	**127.34**
203 mm dia.	52.82	66.85	2.20	75.97	nr	**142.82**
254 mm dia.	72.17	91.33	2.49	85.99	nr	**177.32**

38 MECHANICAL/COOLING/HEATING SYSTEMS

Item	Net Price £	Material £	Labour hours	Labour £	Unit	Total rate £
304 mm dia.	86.22	109.13	2.80	96.69	nr	**205.82**
355 mm dia.	135.89	171.98	3.20	110.50	nr	**282.48**
400 mm dia.	189.56	239.90	3.59	123.97	nr	**363.87**
450 mm dia.	217.77	275.61	3.97	137.09	nr	**412.70**
500 mm dia.	235.90	298.56	4.38	151.25	nr	**449.81**
550 mm dia.	250.49	317.02	4.78	165.07	nr	**482.09**
600 mm dia.	259.54	328.47	5.17	178.53	nr	**507.01**
5°–30° rigid adjustable flashing; including storm collar and sealant						
127 mm dia.	75.76	95.88	1.49	51.45	nr	**147.34**
152 mm dia.	79.82	101.03	1.65	56.98	nr	**158.00**
178 mm dia.	84.88	107.42	1.92	66.30	nr	**173.72**
203 mm dia.	89.25	112.96	2.20	75.97	nr	**188.93**
254 mm dia.	93.97	118.93	2.49	85.99	nr	**204.92**
304 mm dia.	116.28	147.16	2.80	96.69	nr	**243.85**
355 mm dia.	132.25	167.37	3.19	110.16	nr	**277.53**
400 mm dia.	426.33	539.57	3.59	123.97	nr	**663.54**
450 mm dia.	497.56	629.71	3.97	137.09	nr	**766.80**
500 mm dia.	531.18	672.26	4.38	151.25	nr	**823.51**
550 mm dia.	555.85	703.49	4.77	164.72	nr	**868.21**
600 mm dia.	604.76	765.38	5.17	178.53	nr	**943.92**
Domestic and small commercial; twin walled gas vent system suitable for gas fired appliances; domestic gas boilers; small commercial boilers with internal or external flues						
152 mm long						
100 mm dia.	8.10	10.25	0.52	17.96	nr	**28.21**
125 mm dia.	9.95	12.59	0.52	17.96	nr	**30.55**
150 mm dia.	10.79	13.65	0.52	17.96	nr	**31.61**
305 mm long						
100 mm dia.	12.31	15.58	0.52	17.96	nr	**33.54**
125 mm dia.	14.44	18.27	0.52	17.96	nr	**36.23**
150 mm dia.	17.12	21.66	0.52	17.96	nr	**39.62**
457 mm long						
100 mm dia.	13.60	17.21	0.55	18.99	nr	**36.21**
125 mm dia.	15.29	19.35	0.55	18.99	nr	**38.35**
150 mm dia.	18.92	23.95	0.55	18.99	nr	**42.94**
914 mm long						
100 mm dia.	24.31	30.76	0.62	21.41	nr	**52.17**
125 mm dia.	28.34	35.86	0.62	21.41	nr	**57.27**
150 mm dia.	32.49	41.12	0.62	21.41	nr	**62.53**
1524 mm long						
100 mm dia.	35.10	44.43	0.82	28.32	nr	**72.74**
125 mm dia.	43.21	54.68	0.84	29.01	nr	**83.69**
150 mm dia.	46.39	58.71	0.84	29.01	nr	**87.72**
Adjustable length 305 mm long						
100 mm dia.	15.57	19.71	0.56	19.34	nr	**39.04**
125 mm dia.	17.47	22.12	0.56	19.34	nr	**41.45**
150 mm dia.	22.03	27.88	0.56	19.34	nr	**47.21**

38 MECHANICAL/COOLING/HEATING SYSTEMS

Item	Net Price £	Material £	Labour hours	Labour £	Unit	Total rate £
GAS/OIL FIRED BOILERS – FLUE SYSTEMS – cont						
Domestic and small commercial – cont						
Adjustable length 457 mm long						
100 mm dia.	20.99	26.57	0.56	19.34	nr	**45.90**
125 mm dia.	25.45	32.21	0.56	19.34	nr	**51.55**
150 mm dia.	28.31	35.84	0.56	19.34	nr	**55.17**
Adjustable elbow 0°–90°						
100 mm dia.	17.75	22.47	0.48	16.58	nr	**39.04**
125 mm dia.	20.99	26.57	0.48	16.58	nr	**43.14**
150 mm dia.	26.28	33.26	0.48	16.58	nr	**49.83**
Draughthood connector						
100 mm dia.	5.48	6.93	0.48	16.58	nr	**23.51**
125 mm dia.	6.17	7.80	0.48	16.58	nr	**24.38**
150 mm dia.	6.70	8.48	0.48	16.58	nr	**25.06**
Adaptor						
100 mm dia.	13.29	16.82	0.48	16.58	nr	**33.39**
125 mm dia.	13.57	17.17	0.48	16.58	nr	**33.75**
150 mm dia.	13.86	17.54	0.48	16.58	nr	**34.11**
Support plate						
100 mm dia.	9.58	12.13	0.48	16.58	nr	**28.70**
125 mm dia.	10.20	12.90	0.48	16.58	nr	**29.48**
150 mm dia.	10.92	13.82	0.48	16.58	nr	**30.39**
Wall band						
100 mm dia.	8.67	10.97	0.48	16.58	nr	**27.55**
125 mm dia.	9.24	11.69	0.48	16.58	nr	**28.27**
150 mm dia.	11.71	14.82	0.48	16.58	nr	**31.39**
Firestop						
100 mm dia.	3.75	4.75	0.48	16.58	nr	**21.32**
125 mm dia.	3.75	4.75	0.48	16.58	nr	**21.32**
150 mm dia.	4.32	5.47	0.48	16.58	nr	**22.04**
Flat flashing						
125 mm dia.	25.54	32.33	0.55	18.99	nr	**51.32**
150 mm dia.	35.56	45.01	0.55	18.99	nr	**64.00**
Adjustable flashing 5°–30°						
100 mm dia.	66.59	84.28	0.55	18.99	nr	**103.27**
125 mm dia.	103.95	131.56	0.55	18.99	nr	**150.56**
Storm collar						
100 mm dia.	5.31	6.72	0.55	18.99	nr	**25.71**
125 mm dia.	5.43	6.87	0.55	18.99	nr	**25.87**
150 mm dia.	5.57	7.04	0.55	18.99	nr	**26.04**
Gas vent terminal						
100 mm dia.	20.27	25.65	0.55	18.99	nr	**44.64**
125 mm dia.	22.27	28.19	0.55	18.99	nr	**47.18**
150 mm dia.	28.57	36.16	0.55	18.99	nr	**55.15**
Twin wall galvanized steel flue box, 125 mm dia.; fitted for gas fire, where no chimney exists						
Free-standing	134.91	170.74	2.15	74.24	nr	**244.98**
Recess	134.91	170.74	2.15	74.24	nr	**244.98**
Back boiler	99.39	125.79	2.40	82.88	nr	**208.67**

38 MECHANICAL/COOLING/HEATING SYSTEMS

Item	Net Price £	Material £	Labour hours	Labour £	Unit	Total rate £
PACKAGED STEAM GENERATORS						
Packaged steam boilers; boiler mountings						
centrifugal water feed pump; insulation;						
and sheet steel wrap around casing;						
plastic coated						
Gas fired						
276 kW	17397.89	22018.76	86.00	2969.80	nr	**24988.56**
1384 kW	66443.34	84090.70	148.00	5110.81	nr	**89201.51**
2940 KW	75106.72	95055.07	207.00	7148.23	nr	**102203.30**
4843 kW	112370.92	142216.63	295.00	10187.10	nr	**152403.73**
5213 kW	130161.75	164732.71	310.00	10705.08	nr	**175437.79**
Gas oil fired						
276 kW	13855.58	17535.62	86.00	2969.80	nr	**20505.42**
1384 kW	69325.70	87738.60	148.00	5110.81	nr	**92849.42**
2940 kW	83472.77	105643.14	207.00	7148.23	nr	**112791.38**
4843 kW	120774.28	152851.93	295.00	10187.10	nr	**163039.02**
5213 kW	145734.84	184442.02	310.00	10705.08	nr	**195147.10**

38 MECHANICAL/COOLING/HEATING SYSTEMS

Item	Net Price £	Material £	Labour hours	Labour £	Unit	Total rate £
LOW TEMPERATURE HOT WATER HEATING; PIPELINE: SCREWED STEEL						
Black steel pipes; screwed and socketed joints; BS 1387: 1985. Fixed vertically, brackets measured separately. Screwed joints are within the running length, but any flanges are additional						
Medium weight						
10 mm dia.	6.42	8.12	0.37	12.78	m	**20.90**
15 mm dia.	4.04	5.12	0.37	12.78	m	**17.89**
20 mm dia.	4.76	6.02	0.37	12.78	m	**18.80**
25 mm dia.	6.85	8.67	0.41	14.16	m	**22.83**
32 mm dia.	8.47	10.72	0.48	16.58	m	**27.29**
40 mm dia.	9.84	12.45	0.52	17.96	m	**30.41**
50 mm dia.	13.83	17.51	0.62	21.41	m	**38.92**
65 mm dia.	18.78	23.77	0.65	22.45	m	**46.21**
80 mm dia.	24.42	30.90	1.10	37.99	m	**68.89**
100 mm dia.	34.57	43.75	1.31	45.24	m	**88.98**
125 mm dia.	44.00	55.68	1.66	57.32	m	**113.01**
150 mm dia.	51.08	64.65	1.88	64.92	m	**129.57**
Heavy weight						
15 mm dia.	4.81	6.09	0.37	12.78	m	**18.86**
20 mm dia.	5.70	7.21	0.37	12.78	m	**19.99**
25 mm dia.	8.36	10.58	0.41	14.16	m	**24.74**
32 mm dia.	10.37	13.13	0.48	16.58	m	**29.70**
40 mm dia.	12.10	15.32	0.52	17.96	m	**33.27**
50 mm dia.	16.78	21.23	0.62	21.41	m	**42.64**
65 mm dia.	22.82	28.88	0.64	22.10	m	**50.98**
80 mm dia.	29.05	36.77	1.10	37.99	m	**74.75**
100 mm dia.	40.55	51.32	1.31	45.24	m	**96.56**
125 mm dia.	46.93	59.39	1.66	57.32	m	**116.72**
150 mm dia.	54.84	69.41	1.88	64.92	m	**134.33**
200 mm dia.	86.45	109.41	2.99	103.25	m	**212.66**
250 mm dia.	108.38	137.17	3.49	120.52	m	**257.69**
300 mm dia.	129.05	163.33	3.91	135.02	m	**298.35**
Black steel pipes; screwed and socketed joints; BS 1387: 1985. Fixed at high level or suspended, brackets measured separately. Screwed joints are within the running length, but any flanges are additional						
Medium weight						
10 mm dia.	6.42	8.12	0.58	20.03	m	**28.15**
15 mm dia.	4.04	5.12	0.58	20.03	m	**25.15**
20 mm dia.	4.76	6.02	0.58	20.03	m	**26.05**
25 mm dia.	6.85	8.67	0.60	20.72	m	**29.39**
32 mm dia.	8.47	10.72	0.68	23.48	m	**34.20**
40 mm dia.	9.84	12.45	0.73	25.21	m	**37.66**
50 mm dia.	13.83	17.51	0.85	29.35	m	**46.86**

38 MECHANICAL/COOLING/HEATING SYSTEMS

Item	Net Price £	Material £	Labour hours	Labour £	Unit	Total rate £
65 mm dia.	18.78	23.77	0.88	30.39	m	54.15
80 mm dia.	24.42	30.90	1.45	50.07	m	80.97
100 mm dia.	34.57	43.75	1.74	60.09	m	103.83
125 mm dia.	44.00	55.68	2.21	76.32	m	132.00
150 mm dia.	51.08	64.65	2.50	86.33	m	150.98
Heavy weight						
15 mm dia.	4.81	6.09	0.58	20.03	m	26.12
20 mm dia.	5.70	7.21	0.58	20.03	m	27.24
25 mm dia.	8.36	10.58	0.60	20.72	m	31.30
32 mm dia.	10.37	13.13	0.68	23.48	m	36.61
40 mm dia.	12.10	15.32	0.73	25.21	m	40.53
50 mm dia.	16.78	21.23	0.85	29.35	m	50.58
65 mm dia.	22.82	28.88	0.88	30.39	m	59.27
80 mm dia.	29.05	36.77	1.45	50.07	m	86.84
100 mm dia.	40.55	51.32	1.74	60.09	m	111.41
125 mm dia.	46.93	59.39	2.21	76.32	m	135.71
150 mm dia.	54.84	69.41	2.50	86.33	m	155.74
200 mm dia.	86.45	109.41	2.99	103.25	m	212.66
250 mm dia.	108.38	137.17	3.49	120.52	m	257.69
300 mm dia.	129.05	163.33	3.91	135.02	m	298.35

FIXINGS

For steel pipes; black malleable iron. For minimum fixing distances, refer to the Tables and Memoranda to the rear of the book

Single pipe bracket, screw on, black malleable iron; screwed to wood

Item	Net Price £	Material £	Labour hours	Labour £	Unit	Total rate £
15 mm dia.	1.21	1.53	0.14	4.83	nr	6.37
20 mm dia.	1.35	1.71	0.14	4.83	nr	6.55
25 mm dia.	1.58	2.00	0.17	5.87	nr	7.87
32 mm dia.	2.88	3.65	0.19	6.56	nr	10.21
40 mm dia.	2.75	3.48	0.22	7.60	nr	11.08
50 mm dia.	3.85	4.87	0.22	7.60	nr	12.47
65 mm dia.	5.05	6.39	0.28	9.67	nr	16.06
80 mm dia.	6.90	8.73	0.32	11.05	nr	19.79
100 mm dia.	10.12	12.81	0.35	12.09	nr	24.90

Single pipe bracket, screw on, black malleable iron; plugged and screwed

Item	Net Price £	Material £	Labour hours	Labour £	Unit	Total rate £
15 mm dia.	1.21	1.53	0.25	8.63	nr	10.17
20 mm dia.	1.35	1.71	0.25	8.63	nr	10.35
25 mm dia.	1.58	2.00	0.30	10.36	nr	12.36
32 mm dia.	2.88	3.65	0.32	11.05	nr	14.70
40 mm dia.	2.75	3.48	0.32	11.05	nr	14.53
50 mm dia.	3.85	4.87	0.32	11.05	nr	15.92
65 mm dia.	5.05	6.39	0.35	12.09	nr	18.48
80 mm dia.	6.90	8.73	0.42	14.50	nr	23.24
100 mm dia.	10.12	12.81	0.42	14.50	nr	27.32

38 MECHANICAL/COOLING/HEATING SYSTEMS

Item	Net Price £	Material £	Labour hours	Labour £	Unit	Total rate £
LOW TEMPERATURE HOT WATER HEATING; PIPELINE: SCREWED STEEL – cont						
For steel pipes – cont						
Single pipe bracket for building in, black malleable iron						
15 mm dia.	3.55	4.49	0.10	3.45	nr	**7.95**
20 mm dia.	3.55	4.49	0.11	3.80	nr	**8.29**
25 mm dia.	3.55	4.49	0.12	4.14	nr	**8.64**
32 mm dia.	4.04	5.11	0.14	4.83	nr	**9.94**
40 mm dia.	4.06	5.14	0.15	5.18	nr	**10.32**
50 mm dia.	4.06	5.14	0.16	5.53	nr	**10.67**
Pipe ring, single socket, black malleable iron						
15 mm dia.	1.41	1.79	0.10	3.45	nr	**5.24**
20 mm dia.	1.57	1.99	0.11	3.80	nr	**5.78**
25 mm dia.	1.75	2.22	0.12	4.14	nr	**6.36**
32 mm dia.	1.87	2.36	0.14	4.83	nr	**7.20**
40 mm dia.	2.40	3.03	0.15	5.18	nr	**8.21**
50 mm dia.	3.02	3.83	0.16	5.53	nr	**9.35**
65 mm dia.	4.38	5.54	0.30	10.36	nr	**15.90**
80 mm dia.	5.25	6.64	0.35	12.09	nr	**18.73**
100 mm dia.	7.94	10.05	0.40	13.81	nr	**23.87**
125 mm dia.	15.74	19.92	0.60	20.73	nr	**40.65**
150 mm dia.	17.98	22.76	0.77	26.59	nr	**49.35**
200 mm dia.	49.72	62.93	0.90	31.08	nr	**94.01**
250 mm dia.	62.16	78.67	1.10	37.99	nr	**116.65**
300 mm dia.	74.77	94.62	1.25	43.17	nr	**137.79**
350 mm dia.	87.04	110.16	1.50	51.80	nr	**161.96**
400 mm dia.	103.00	130.36	1.75	60.43	nr	**190.79**
Pipe ring, double socket, black malleable iron						
15 mm dia.	1.63	2.06	0.10	3.45	nr	**5.51**
20 mm dia.	1.88	2.38	0.11	3.80	nr	**6.18**
25 mm dia.	2.10	2.65	0.12	4.14	nr	**6.80**
32 mm dia.	2.48	3.14	0.14	4.83	nr	**7.97**
40 mm dia.	2.71	3.43	0.15	5.18	nr	**8.61**
50 mm dia.	4.22	5.34	0.16	5.53	nr	**10.87**
Screw on backplate (Male), black malleable iron; plugged and screwed						
M12	0.86	1.08	0.10	3.45	nr	**4.54**
Screw on backplate (Female), black malleable iron; plugged and screwed						
M12	0.86	1.08	0.10	3.45	nr	**4.54**
Extra over channel sections for fabricated hangers and brackets						
Galvanized steel; including inserts, bolts, nuts, washers; fixed to backgrounds						
41 × 21 mm	8.39	10.62	0.29	10.01	m	**20.64**
41 × 41 mm	10.06	12.74	0.29	10.01	m	**22.75**

38 MECHANICAL/COOLING/HEATING SYSTEMS

Item	Net Price £	Material £	Labour hours	Labour £	Unit	Total rate £
Threaded rods; metric thread; including nuts, washers etc.						
12 mm dia. × 600 mm long	4.24	5.37	0.18	6.22	nr	11.58
Pipe roller and chair						
Roller and chair; black malleable						
Up to 50 mm dia.	22.33	28.26	0.20	6.91	nr	35.17
65 mm dia.	23.00	29.11	0.20	6.91	nr	36.02
80 mm dia.	24.99	31.62	0.20	6.91	nr	38.53
100 mm dia.	25.43	32.19	0.20	6.91	nr	39.09
125 mm dia.	35.63	45.09	0.20	6.91	nr	51.99
150 mm dia.	35.63	45.09	0.30	10.36	nr	55.45
175 mm dia.	60.10	76.07	0.30	10.36	nr	86.43
200 mm dia.	92.06	116.51	0.30	10.36	nr	126.87
250 mm dia.	120.08	151.98	0.30	10.36	nr	162.34
300 mm dia.	151.22	191.38	0.30	10.36	nr	201.74
Roller bracket; black malleable						
25 mm dia.	4.58	5.79	0.20	6.91	nr	12.70
32 mm dia.	4.79	6.06	0.20	6.91	nr	12.96
40 mm dia.	5.13	6.50	0.20	6.91	nr	13.40
50 mm dia.	5.41	6.84	0.20	6.91	nr	13.75
65 mm dia.	7.13	9.02	0.20	6.91	nr	15.93
80 mm dia.	10.25	12.98	0.20	6.91	nr	19.89
100 mm dia.	11.42	14.45	0.20	6.91	nr	21.36
125 mm dia.	18.84	23.84	0.20	6.91	nr	30.74
150 mm dia.	18.84	23.84	0.30	10.36	nr	34.20
175 mm dia.	41.97	53.12	0.30	10.36	nr	63.48
200 mm dia.	41.97	53.12	0.30	10.36	nr	63.48
250 mm dia.	55.81	70.64	0.30	10.36	nr	81.00
300 mm dia.	68.53	86.74	0.30	10.36	nr	97.10
350 mm dia.	110.21	139.48	0.30	10.36	nr	149.84
400 mm dia.	125.97	159.43	0.30	10.36	nr	169.79
Extra over black steel screwed pipes; black steel flanges, screwed and drilled; metric; BS 4504						
Screwed flanges; PN6						
15 mm dia.	11.00	13.92	0.35	12.09	nr	26.01
20 mm dia.	11.00	13.92	0.47	16.23	nr	30.15
25 mm dia.	11.00	13.92	0.53	18.30	nr	32.22
32 mm dia.	11.00	13.92	0.62	21.41	nr	35.33
40 mm dia.	11.00	13.92	0.70	24.17	nr	38.09
50 mm dia.	11.71	14.82	0.84	29.01	nr	43.83
65 mm dia.	16.30	20.63	1.03	35.57	nr	56.20
80 mm dia.	22.99	29.10	1.23	42.48	nr	71.57
100 mm dia.	27.18	34.40	1.41	48.69	nr	83.09
125 mm dia.	49.88	63.12	1.77	61.12	nr	124.25
150 mm dia.	49.88	63.12	2.21	76.32	nr	139.44

38 MECHANICAL/COOLING/HEATING SYSTEMS

Item	Net Price £	Material £	Labour hours	Labour £	Unit	Total rate £
LOW TEMPERATURE HOT WATER HEATING; PIPELINE: SCREWED STEEL – cont						
Extra over black steel screwed pipes – cont						
Screwed flanges; PN16						
15 mm dia.	13.96	17.67	0.35	12.09	nr	**29.75**
20 mm dia.	13.96	17.67	0.47	16.23	nr	**33.90**
25 mm dia.	13.96	17.67	0.53	18.30	nr	**35.97**
32 mm dia.	14.92	18.88	0.62	21.41	nr	**40.29**
40 mm dia.	14.92	18.88	0.70	24.17	nr	**43.05**
50 mm dia.	18.19	23.03	0.84	29.01	nr	**52.03**
65 mm dia.	22.73	28.77	1.03	35.57	nr	**64.34**
80 mm dia.	27.68	35.03	1.23	42.48	nr	**77.51**
100 mm dia.	31.03	39.27	1.41	48.69	nr	**87.96**
125 mm dia.	54.30	68.72	1.77	61.12	nr	**129.84**
150 mm dia.	53.43	67.63	2.21	76.32	nr	**143.94**
Extra over black steel screwed pipes; black steel flanges, screwed and drilled; imperial; BS 10						
Screwed flanges; Table E						
½" dia.	20.86	26.40	0.35	12.09	nr	**38.49**
¾" dia.	20.86	26.40	0.47	16.23	nr	**42.63**
1" dia.	20.84	26.37	0.53	18.30	nr	**44.67**
1 ¼" dia.	20.86	26.40	0.62	21.41	nr	**47.81**
1 ½" dia.	20.86	26.40	0.70	24.17	nr	**50.57**
2" dia.	20.86	26.40	0.84	29.01	nr	**55.41**
2 ½" dia.	24.82	31.41	1.03	35.57	nr	**66.98**
3" dia.	28.12	35.58	1.23	42.48	nr	**78.06**
4" dia.	38.05	48.15	1.41	48.69	nr	**96.85**
5" dia.	80.60	102.01	1.77	61.12	nr	**163.13**
6" dia.	80.60	102.01	2.21	76.32	nr	**178.33**
Extra over black steel screwed pipes; black steel flange connections						
Bolted connection between pair of flanges; including gasket, bolts, nuts and washers						
50 mm dia.	12.90	16.33	0.53	18.30	nr	**34.63**
65 mm dia.	15.29	19.35	0.53	18.30	nr	**37.66**
80 mm dia.	22.94	29.03	0.53	18.30	nr	**47.33**
100 mm dia.	24.91	31.53	0.61	21.06	nr	**52.59**
125 mm dia.	29.32	37.11	0.61	21.06	nr	**58.17**
150 mm dia.	41.89	53.02	0.90	31.08	nr	**84.09**

38 MECHANICAL/COOLING/HEATING SYSTEMS

Item	Net Price £	Material £	Labour hours	Labour £	Unit	Total rate £
Extra over black steel screwed pipes;						
black heavy steel tubular fittings; BS 1387						
Long screw connection with socket and						
backnut						
15 mm dia.	8.88	11.24	0.63	21.76	nr	**33.00**
20 mm dia.	11.19	14.16	0.84	29.01	nr	**43.17**
25 mm dia.	14.69	18.59	0.95	32.81	nr	**51.40**
32 mm dia.	19.68	24.91	1.11	38.33	nr	**63.24**
40 mm dia.	23.67	29.96	1.28	44.20	nr	**74.16**
50 mm dia.	34.88	44.14	1.53	52.83	nr	**96.97**
65 mm dia.	48.77	61.73	1.87	64.58	nr	**126.30**
80 mm dia.	111.53	141.15	2.21	76.32	nr	**217.47**
100 mm dia.	180.68	228.66	3.05	105.32	nr	**333.99**
Running nipple						
15 mm dia.	2.30	2.92	0.50	17.27	nr	**20.18**
20 mm dia.	2.91	3.68	0.68	23.48	nr	**27.16**
25 mm dia.	3.56	4.50	0.77	26.59	nr	**31.09**
32 mm dia.	5.03	6.36	0.90	31.08	nr	**37.44**
40 mm dia.	6.77	8.57	1.03	35.57	nr	**44.13**
50 mm dia.	10.30	13.03	1.23	42.48	nr	**55.51**
65 mm dia.	22.12	27.99	1.50	51.80	nr	**79.79**
80 mm dia.	34.52	43.68	1.78	61.47	nr	**105.15**
100 mm dia.	54.04	68.40	2.38	82.19	nr	**150.58**
Barrel nipple						
15 mm dia.	1.31	1.66	0.50	17.27	nr	**18.92**
20 mm dia.	2.09	2.64	0.68	23.48	nr	**26.12**
25 mm dia.	2.76	3.50	0.77	26.59	nr	**30.09**
32 mm dia.	4.10	5.19	0.90	31.08	nr	**36.27**
40 mm dia.	5.07	6.42	1.03	35.57	nr	**41.98**
50 mm dia.	7.24	9.17	1.23	42.48	nr	**51.64**
65 mm dia.	15.47	19.57	1.50	51.80	nr	**71.37**
80 mm dia.	21.57	27.30	1.78	61.47	nr	**88.77**
100 mm dia.	39.07	49.44	2.38	82.19	nr	**131.63**
125 mm dia.	72.49	91.74	2.87	99.11	nr	**190.85**
150 mm dia.	114.25	144.59	3.39	117.07	nr	**261.66**
Close taper nipple						
15 mm dia.	2.76	3.50	0.50	17.27	nr	**20.77**
20 mm dia.	3.56	4.50	0.68	23.48	nr	**27.98**
25 mm dia.	4.65	5.89	0.77	26.59	nr	**32.48**
32 mm dia.	6.97	8.82	0.90	31.08	nr	**39.90**
40 mm dia.	8.64	10.93	1.03	35.57	nr	**46.50**
50 mm dia.	13.33	16.88	1.23	42.48	nr	**59.35**
65 mm dia.	25.85	32.71	1.50	51.80	nr	**84.51**
80 mm dia.	31.85	40.31	1.78	61.47	nr	**101.78**
100 mm dia.	65.46	82.85	2.38	82.19	nr	**165.04**

38 MECHANICAL/COOLING/HEATING SYSTEMS

Item	Net Price £	Material £	Labour hours	Labour £	Unit	Total rate £
LOW TEMPERATURE HOT WATER HEATING; PIPELINE: SCREWED STEEL – cont						
Extra over black steel screwed pipes; black malleable iron fittings; BS 143						
Cap						
15 mm dia.	0.94	1.19	0.32	11.05	nr	**12.24**
20 mm dia.	1.09	1.38	0.43	14.85	nr	**16.23**
25 mm dia.	1.38	1.74	0.49	16.92	nr	**18.66**
32 mm dia.	2.41	3.04	0.58	20.03	nr	**23.07**
40 mm dia.	2.85	3.61	0.66	22.79	nr	**26.40**
50 mm dia.	5.93	7.50	0.78	26.94	nr	**34.44**
65 mm dia.	9.90	12.52	0.96	33.15	nr	**45.67**
80 mm dia.	11.20	14.17	1.13	39.02	nr	**53.19**
100 mm dia.	24.55	31.07	1.70	58.71	nr	**89.78**
Plain plug, hollow						
15 mm dia.	0.67	0.85	0.28	9.67	nr	**10.52**
20 mm dia.	0.86	1.08	0.38	13.12	nr	**14.21**
25 mm dia.	1.17	1.48	0.44	15.19	nr	**16.67**
32 mm dia.	1.61	2.04	0.51	17.61	nr	**19.65**
40 mm dia.	3.05	3.86	0.59	20.37	nr	**24.23**
50 mm dia.	4.28	5.41	0.70	24.17	nr	**29.59**
65 mm dia.	6.97	8.82	0.85	29.35	nr	**38.17**
80 mm dia.	10.89	13.78	1.00	34.53	nr	**48.31**
100 mm dia.	20.03	25.34	1.44	49.73	nr	**75.07**
Plain plug, solid						
15 mm dia.	2.07	2.62	0.28	9.67	nr	**12.29**
20 mm dia.	2.01	2.54	0.38	13.12	nr	**15.66**
25 mm dia.	2.99	3.78	0.44	15.19	nr	**18.98**
32 mm dia.	4.30	5.45	0.51	17.61	nr	**23.06**
40 mm dia.	6.04	7.64	0.59	20.37	nr	**28.02**
50 mm dia.	7.95	10.06	0.70	24.17	nr	**34.23**
90° elbow, male/female						
15 mm dia.	1.09	1.38	0.64	22.10	nr	**23.48**
20 mm dia.	1.48	1.87	0.85	29.35	nr	**31.22**
25 mm dia.	2.46	3.11	0.97	33.50	nr	**36.60**
32 mm dia.	4.50	5.70	1.12	38.68	nr	**44.37**
40 mm dia.	7.54	9.54	1.29	44.55	nr	**54.09**
50 mm dia.	9.70	12.27	1.55	53.53	nr	**65.80**
65 mm dia.	20.93	26.49	1.89	65.27	nr	**91.76**
80 mm dia.	28.61	36.20	2.24	77.35	nr	**113.56**
100 mm dia.	50.03	63.32	3.09	106.71	nr	**170.03**
90° elbow						
15 mm dia.	0.99	1.26	0.64	22.10	nr	**23.36**
20 mm dia.	1.35	1.71	0.85	29.35	nr	**31.06**
25 mm dia.	2.07	2.62	0.97	33.50	nr	**36.12**
32 mm dia.	3.82	4.83	1.12	38.68	nr	**43.51**
40 mm dia.	6.39	8.08	1.29	44.55	nr	**52.63**
50 mm dia.	7.50	9.49	1.55	53.53	nr	**63.02**
65 mm dia.	16.17	20.46	1.89	65.27	nr	**85.73**

38 MECHANICAL/COOLING/HEATING SYSTEMS

Item	Net Price £	Material £	Labour hours	Labour £	Unit	Total rate £
80 mm dia.	23.73	30.04	2.24	77.35	nr	**107.39**
100 mm dia.	45.79	57.96	3.09	106.71	nr	**164.66**
125 mm dia.	111.39	140.97	4.44	153.32	nr	**294.30**
150 mm dia.	207.33	262.39	5.79	199.94	nr	**462.34**
45° elbow						
15 mm dia.	2.37	3.00	0.64	22.10	nr	**25.10**
20 mm dia.	2.89	3.66	0.85	29.35	nr	**33.01**
25 mm dia.	4.30	5.45	0.97	33.50	nr	**38.94**
32 mm dia.	8.79	11.13	1.12	38.68	nr	**49.80**
40 mm dia.	10.78	13.64	1.29	44.55	nr	**58.18**
50 mm dia.	14.77	18.69	1.55	53.53	nr	**72.22**
65 mm dia.	21.61	27.35	1.89	65.27	nr	**92.62**
80 mm dia.	32.46	41.09	2.24	77.35	nr	**118.44**
100 mm dia.	71.19	90.10	3.09	106.71	nr	**196.80**
150 mm dia.	198.81	251.61	5.79	199.94	nr	**451.56**
90° bend, male/female						
15 mm dia.	1.93	2.45	0.64	22.10	nr	**24.55**
20 mm dia.	2.83	3.58	0.85	29.35	nr	**32.93**
25 mm dia.	4.15	5.26	0.97	33.50	nr	**38.75**
32 mm dia.	6.61	8.36	1.12	38.68	nr	**47.04**
40 mm dia.	10.44	13.21	1.29	44.55	nr	**57.76**
50 mm dia.	18.24	23.09	1.55	53.53	nr	**76.61**
65 mm dia.	30.33	38.39	1.89	65.27	nr	**103.65**
80 mm dia.	46.67	59.07	2.24	77.35	nr	**136.42**
100 mm dia.	115.67	146.39	3.09	106.71	nr	**253.09**
90° bend, male						
15 mm dia.	4.45	5.63	0.64	22.10	nr	**27.73**
20 mm dia.	4.98	6.31	0.85	29.35	nr	**35.66**
25 mm dia.	7.30	9.24	0.97	33.50	nr	**42.74**
32 mm dia.	15.86	20.07	1.12	38.68	nr	**58.75**
40 mm dia.	22.27	28.19	1.29	44.55	nr	**72.73**
50 mm dia.	33.81	42.80	1.55	53.53	nr	**96.32**
90° bend, female						
15 mm dia.	1.77	2.24	0.64	22.10	nr	**24.35**
20 mm dia.	2.53	3.20	0.85	29.35	nr	**32.55**
25 mm dia.	3.55	4.49	0.97	33.50	nr	**37.98**
32 mm dia.	6.75	8.54	1.12	38.68	nr	**47.21**
40 mm dia.	8.98	11.36	1.29	44.55	nr	**55.91**
50 mm dia.	12.66	16.02	1.55	53.53	nr	**69.55**
65 mm dia.	27.01	34.18	1.89	65.27	nr	**99.45**
80 mm dia.	43.04	54.47	2.24	77.35	nr	**131.82**
100 mm dia.	94.36	119.43	3.09	106.71	nr	**226.13**
125 mm dia.	222.59	281.71	4.44	153.32	nr	**435.04**
150 mm dia.	404.81	512.33	5.79	199.94	nr	**712.27**
Return bend						
15 mm dia.	10.16	12.85	0.64	22.10	nr	**34.95**
20 mm dia.	16.43	20.79	0.85	29.35	nr	**50.15**
25 mm dia.	20.48	25.93	0.97	33.50	nr	**59.42**
32 mm dia.	31.77	40.21	1.12	38.68	nr	**78.88**
40 mm dia.	37.82	47.86	1.29	44.55	nr	**92.41**
50 mm dia.	57.72	73.05	1.55	53.53	nr	**126.58**

38 MECHANICAL/COOLING/HEATING SYSTEMS

Item	Net Price £	Material £	Labour hours	Labour £	Unit	Total rate £
LOW TEMPERATURE HOT WATER HEATING; PIPELINE: SCREWED STEEL – cont						
Extra over black steel screwed pipes – cont						
Equal socket, parallel thread						
15 mm dia.	0.94	1.19	0.64	22.10	nr	**23.29**
20 mm dia.	1.17	1.48	0.85	29.35	nr	**30.83**
25 mm dia.	1.60	2.02	0.97	33.50	nr	**35.52**
32 mm dia.	3.21	4.06	1.12	38.68	nr	**42.74**
40 mm dia.	4.66	5.90	1.29	44.55	nr	**50.45**
50 mm dia.	6.76	8.55	1.55	53.53	nr	**62.08**
65 mm dia.	10.38	13.14	1.89	65.27	nr	**78.40**
80 mm dia.	14.73	18.64	2.24	77.35	nr	**96.00**
100 mm dia.	24.27	30.71	3.09	106.71	nr	**137.42**
Concentric reducing socket						
20 × 15 mm dia.	1.51	1.91	0.76	26.24	nr	**28.16**
25 × 15 mm dia.	1.77	2.24	0.85	29.35	nr	**31.60**
25 × 20 mm dia.	1.88	2.39	0.86	29.70	nr	**32.08**
32 × 25 mm dia.	3.53	4.47	1.01	34.88	nr	**39.35**
40 × 25 mm dia.	4.45	5.63	1.16	40.06	nr	**45.69**
40 × 32 mm dia.	4.61	5.84	1.16	40.06	nr	**45.90**
50 × 25 mm dia.	6.63	8.40	1.38	47.65	nr	**56.05**
50 × 40 mm dia.	6.46	8.18	1.38	47.65	nr	**55.83**
65 × 50 mm dia.	11.76	14.88	1.69	58.36	nr	**73.24**
80 × 50 mm dia.	15.29	19.35	2.00	69.07	nr	**88.42**
100 × 50 mm dia.	30.52	38.62	2.75	94.96	nr	**133.59**
100 × 80 mm dia.	42.07	53.25	2.75	94.96	nr	**148.21**
150 × 100 mm dia.	84.72	107.22	4.10	141.58	nr	**248.80**
Eccentric reducing socket						
20 × 15 mm dia.	2.68	3.39	0.73	25.21	nr	**28.60**
25 × 15 mm dia.	7.73	9.78	0.85	29.35	nr	**39.13**
25 × 20 mm dia.	8.74	11.06	0.85	29.35	nr	**40.42**
32 × 25 mm dia.	11.92	15.08	1.01	34.88	nr	**49.96**
40 × 25 mm dia.	13.65	17.28	1.16	40.06	nr	**57.34**
40 × 32 mm dia.	6.86	8.68	1.16	40.06	nr	**48.74**
50 × 25 mm dia.	8.85	11.21	1.38	47.65	nr	**58.86**
50 × 40 mm dia.	12.21	15.46	1.38	47.65	nr	**63.11**
65 × 50 mm dia.	14.20	17.97	1.69	58.36	nr	**76.33**
80 × 50 mm dia.	24.85	31.45	2.00	69.07	nr	**100.51**
Hexagon bush						
20 × 15 mm dia.	0.86	1.08	0.37	12.78	nr	**13.86**
25 × 15 mm dia.	1.04	1.32	0.43	14.85	nr	**16.17**
25 × 20 mm dia.	1.09	1.38	0.43	14.85	nr	**16.23**
32 × 25 mm dia.	1.40	1.77	0.51	17.61	nr	**19.38**
40 × 25 mm dia.	2.23	2.82	0.58	20.03	nr	**22.85**
40 × 32 mm dia.	2.05	2.59	0.58	20.03	nr	**22.62**
50 × 25 mm dia.	4.61	5.84	0.71	24.52	nr	**30.36**

38 MECHANICAL/COOLING/HEATING SYSTEMS

Item	Net Price £	Material £	Labour hours	Labour £	Unit	Total rate £
50 × 40 mm dia.	4.28	5.41	0.71	24.52	nr	**29.93**
65 × 50 mm dia.	7.14	9.04	0.85	29.35	nr	**38.39**
80 × 50 mm dia.	11.58	14.66	1.00	34.53	nr	**49.19**
100 × 50 mm dia.	26.80	33.91	1.52	52.49	nr	**86.40**
100 × 80 mm dia.	36.98	46.80	1.52	52.49	nr	**99.29**
150 × 100 mm dia.	80.18	101.47	2.57	88.75	nr	**190.22**
Hexagon nipple						
15 mm dia.	0.92	1.16	0.28	9.67	nr	**10.83**
20 mm dia.	1.04	1.32	0.38	13.12	nr	**14.44**
25 mm dia.	1.48	1.87	0.44	15.19	nr	**17.06**
32 mm dia.	3.05	3.86	0.51	17.61	nr	**21.47**
40 mm dia.	3.53	4.47	0.59	20.37	nr	**24.85**
50 mm dia.	6.42	8.13	0.70	24.17	nr	**32.30**
65 mm dia.	10.47	13.25	0.85	29.35	nr	**42.60**
80 mm dia.	15.13	19.15	1.00	34.53	nr	**53.68**
100 mm dia.	25.67	32.49	1.44	49.73	nr	**82.21**
150 mm dia.	57.30	72.52	2.32	80.12	nr	**152.63**
Union, male/female						
15 mm dia.	4.82	6.10	0.64	22.10	nr	**28.21**
20 mm dia.	5.90	7.47	0.85	29.35	nr	**36.82**
25 mm dia.	7.38	9.34	0.97	33.50	nr	**42.83**
32 mm dia.	11.37	14.39	1.12	38.68	nr	**53.07**
40 mm dia.	14.55	18.41	1.29	44.55	nr	**62.96**
50 mm dia.	22.93	29.02	1.55	53.53	nr	**82.54**
65 mm dia.	43.75	55.37	1.89	65.27	nr	**120.63**
80 mm dia.	60.15	76.13	2.24	77.35	nr	**153.48**
Union, female						
15 mm dia.	3.97	5.02	0.64	22.10	nr	**27.12**
20 mm dia.	4.30	5.45	0.85	29.35	nr	**34.80**
25 mm dia.	5.01	6.34	0.97	33.50	nr	**39.84**
32 mm dia.	9.44	11.94	1.12	38.68	nr	**50.62**
40 mm dia.	10.68	13.51	1.29	44.55	nr	**58.06**
50 mm dia.	17.68	22.38	1.55	53.53	nr	**75.90**
65 mm dia.	38.13	48.26	1.89	65.27	nr	**113.52**
80 mm dia.	52.65	66.63	2.24	77.35	nr	**143.99**
100 mm dia.	100.23	126.85	3.09	106.71	nr	**233.56**
Union elbow, male/female						
15 mm dia.	6.98	8.84	0.55	18.99	nr	**27.83**
20 mm dia.	8.75	11.08	0.85	29.35	nr	**40.43**
25 mm dia.	12.28	15.54	0.97	33.50	nr	**49.03**
Twin elbow						
15 mm dia.	5.65	7.16	0.91	31.42	nr	**38.58**
20 mm dia.	6.27	7.94	1.22	42.13	nr	**50.07**
25 mm dia.	10.13	12.82	1.39	48.00	nr	**60.82**
32 mm dia.	20.79	26.32	1.62	55.94	nr	**82.26**
40 mm dia.	26.33	33.32	1.86	64.23	nr	**97.55**
50 mm dia.	33.85	42.84	2.21	76.32	nr	**119.16**
65 mm dia.	60.13	76.10	2.72	93.93	nr	**170.03**
80 mm dia.	102.47	129.69	3.21	110.85	nr	**240.54**

38 MECHANICAL/COOLING/HEATING SYSTEMS

Item	Net Price £	Material £	Labour hours	Labour £	Unit	Total rate £
LOW TEMPERATURE HOT WATER HEATING; PIPELINE: SCREWED STEEL – cont						
Extra over black steel screwed pipes – cont						
Equal tee						
15 mm dia.	1.35	1.71	0.91	31.42	nr	**33.14**
20 mm dia.	1.97	2.50	1.22	42.13	nr	**44.62**
25 mm dia.	2.84	3.59	1.39	48.00	nr	**51.59**
32 mm dia.	5.16	6.53	1.62	55.94	nr	**62.47**
40 mm dia.	7.95	10.06	1.86	64.23	nr	**74.29**
50 mm dia.	11.43	14.47	2.21	76.32	nr	**90.79**
65 mm dia.	25.04	31.69	2.72	93.93	nr	**125.61**
80 mm dia.	30.52	38.62	3.21	110.85	nr	**149.47**
100 mm dia.	55.29	69.98	4.44	153.32	nr	**223.30**
125 mm dia.	160.58	203.23	5.38	185.79	nr	**389.02**
150 mm dia.	255.85	323.80	6.31	217.90	nr	**541.70**
Tee reducing on branch						
20 × 15 mm dia.	1.77	2.24	1.22	42.13	nr	**44.37**
25 × 15 mm dia.	2.46	3.11	1.39	48.00	nr	**51.11**
25 × 20 mm dia.	2.91	3.69	1.39	48.00	nr	**51.69**
32 × 25 mm dia.	5.07	6.42	1.62	55.94	nr	**62.36**
40 × 25 mm dia.	6.71	8.49	1.86	64.23	nr	**72.72**
40 × 32 mm dia.	8.82	11.16	1.86	64.23	nr	**75.39**
50 × 25 mm dia.	9.97	12.62	2.21	76.32	nr	**88.93**
50 × 40 mm dia.	14.88	18.83	2.21	76.32	nr	**95.15**
65 × 50 mm dia.	23.24	29.41	2.72	93.93	nr	**123.34**
80 × 50 mm dia.	31.40	39.74	3.21	110.85	nr	**150.59**
100 × 50 mm dia.	51.99	65.80	4.44	153.32	nr	**219.13**
100 × 80 mm dia.	80.18	101.47	4.44	153.32	nr	**254.80**
150 × 100 mm dia.	188.47	238.52	6.31	217.90	nr	**456.42**
Equal pitcher tee						
15 mm dia.	4.13	5.23	0.91	31.42	nr	**36.65**
20 mm dia.	5.13	6.50	1.22	42.13	nr	**48.63**
25 mm dia.	7.73	9.78	1.39	48.00	nr	**57.78**
32 mm dia.	11.72	14.83	1.62	55.94	nr	**70.77**
40 mm dia.	18.07	22.87	1.86	64.23	nr	**87.10**
50 mm dia.	25.38	32.12	2.21	76.32	nr	**108.44**
65 mm dia.	39.25	49.67	2.72	93.93	nr	**143.60**
80 mm dia.	61.27	77.54	3.21	110.85	nr	**188.39**
100 mm dia.	137.81	174.42	4.44	153.32	nr	**327.74**
Cross						
15 mm dia.	4.03	5.10	1.00	34.53	nr	**39.63**
20 mm dia.	6.06	7.67	1.33	45.93	nr	**53.60**
25 mm dia.	7.73	9.78	1.51	52.14	nr	**61.92**
32 mm dia.	11.22	14.20	1.77	61.12	nr	**75.33**
40 mm dia.	14.45	18.28	2.02	69.76	nr	**88.04**
50 mm dia.	23.47	29.71	2.42	83.57	nr	**113.28**
65 mm dia.	32.46	41.09	2.97	102.56	nr	**143.65**
80 mm dia.	43.16	54.63	3.50	120.86	nr	**175.49**
100 mm dia.	89.09	112.76	4.84	167.14	nr	**279.89**

38 MECHANICAL/COOLING/HEATING SYSTEMS

Item	Net Price £	Material £	Labour hours	Labour £	Unit	Total rate £
LOW TEMPERATURE HOT WATER HEATING; PIPELINE: BLACK WELDED STEEL						
Black steel pipes; butt welded joints; BS 1387: 1985; including protective painting. Fixed Vertical with brackets measured separately (Refer to Screwed Steel Section). Welded butt joints are within the running length, but any flanges are additional						
Medium weight						
10 mm dia.	3.80	4.80	0.37	12.78	m	**17.58**
15 mm dia.	4.04	5.12	0.37	12.78	m	**17.89**
20 mm dia.	4.76	6.02	0.37	12.78	m	**18.80**
25 mm dia.	6.85	8.67	0.41	14.16	m	**22.83**
32 mm dia.	8.47	10.72	0.48	16.58	m	**27.29**
40 mm dia.	9.84	12.45	0.52	17.96	m	**30.41**
50 mm dia.	13.83	17.51	0.62	21.41	m	**38.92**
65 mm dia.	18.78	23.77	0.64	22.10	m	**45.87**
80 mm dia.	24.42	30.90	1.10	37.99	m	**68.89**
100 mm dia.	34.57	43.75	1.31	45.24	m	**88.98**
125 mm dia.	44.00	55.68	1.66	57.32	m	**113.01**
150 mm dia.	51.08	64.65	1.88	64.92	m	**129.57**
Heavy weight						
15 mm dia.	4.81	6.09	0.37	12.78	m	**18.86**
20 mm dia.	5.70	7.21	0.37	12.78	m	**19.99**
25 mm dia.	8.36	10.58	0.41	14.16	m	**24.74**
32 mm dia.	10.37	13.13	0.48	16.58	m	**29.70**
40 mm dia.	12.10	15.32	0.52	17.96	m	**33.27**
50 mm dia.	16.78	21.23	0.62	21.41	m	**42.64**
65 mm dia.	22.82	28.88	0.64	22.10	m	**50.98**
80 mm dia.	29.05	36.77	1.10	37.99	m	**74.75**
100 mm dia.	40.55	51.32	1.31	45.24	m	**96.56**
125 mm dia.	46.93	59.39	1.66	57.32	m	**116.72**
150 mm dia.	54.84	69.41	1.88	64.92	m	**134.33**
Black steel pipes; butt welded joints; BS 1387: 1985; including protective painting. Fixed at High Level or Suspended with brackets measured separately (Refer to Screwed Steel Section). Welded butt joints are within the running length, but any flanges are additional						
Medium weight						
10 mm dia.	3.80	4.80	0.58	20.03	m	**24.83**
15 mm dia.	4.04	5.12	0.58	20.03	m	**25.15**
20 mm dia.	4.76	6.02	0.58	20.03	m	**26.05**
25 mm dia.	6.85	8.67	0.60	20.72	m	**29.39**
32 mm dia.	8.47	10.72	0.68	23.48	m	**34.20**
40 mm dia.	9.84	12.45	0.73	25.21	m	**37.66**
50 mm dia.	13.83	17.51	0.85	29.35	m	**46.86**

38 MECHANICAL/COOLING/HEATING SYSTEMS

Item	Net Price £	Material £	Labour hours	Labour £	Unit	Total rate £
LOW TEMPERATURE HOT WATER HEATING; PIPELINE: BLACK WELDED STEEL – cont						
Black steel pipes – cont						
Medium weight – cont						
65 mm dia.	18.78	23.77	0.88	30.39	m	**54.15**
80 mm dia.	24.42	30.90	1.45	50.07	m	**80.97**
100 mm dia.	34.57	43.75	1.74	60.09	m	**103.83**
125 mm dia.	44.00	55.68	2.21	76.32	m	**132.00**
150 mm dia.	51.08	64.65	2.50	86.33	m	**150.98**
Heavy weight						
15 mm dia.	4.81	6.09	0.58	20.03	m	**26.12**
20 mm dia.	5.70	7.21	0.58	20.03	m	**27.24**
25 mm dia.	8.36	10.58	0.60	20.72	m	**31.30**
32 mm dia.	10.37	13.13	0.68	23.48	m	**36.61**
40 mm dia.	12.10	15.32	0.73	25.21	m	**40.53**
50 mm dia.	16.78	21.23	0.85	29.35	m	**50.58**
65 mm dia.	22.82	28.88	0.88	30.39	m	**59.27**
80 mm dia.	29.05	36.77	1.45	50.07	m	**86.84**
100 mm dia.	40.55	51.32	1.74	60.09	m	**111.41**
125 mm dia.	46.93	59.39	2.21	76.32	m	**135.71**
150 mm dia.	54.84	69.41	2.50	86.33	m	**155.74**
FIXINGS						
Refer to steel pipes; black malleable iron. For minimum fixing distances, refer to the Tables and Memoranda to the rear of the book						
Extra over black steel butt welded pipes; black steel flanges, welded and drilled; metric; BS 4504						
Welded flanges; PN6						
15 mm dia.	5.60	7.09	0.59	20.37	nr	**27.47**
20 mm dia.	5.60	7.09	0.69	23.83	nr	**30.92**
25 mm dia.	5.60	7.09	0.84	29.01	nr	**36.10**
32 mm dia.	9.29	11.75	1.00	34.53	nr	**46.29**
40 mm dia.	9.92	12.56	1.11	38.33	nr	**50.89**
50 mm dia.	11.90	15.06	1.37	47.31	nr	**62.37**
65 mm dia.	17.48	22.13	1.54	53.18	nr	**75.31**
80 mm dia.	26.28	33.25	1.67	57.67	nr	**90.92**
100 mm dia.	28.69	36.31	2.22	76.66	nr	**112.97**
125 mm dia.	37.59	47.58	2.61	90.13	nr	**137.71**
150 mm dia.	38.00	48.09	2.99	103.25	nr	**151.34**
Welded flanges; PN16						
15 mm dia.	11.22	14.20	0.59	20.37	nr	**34.57**
20 mm dia.	11.22	14.20	0.69	23.83	nr	**38.03**
25 mm dia.	11.22	14.20	0.84	29.01	nr	**43.21**
32 mm dia.	15.12	19.14	1.00	34.53	nr	**53.67**
40 mm dia.	15.12	19.14	1.11	38.33	nr	**57.47**
50 mm dia.	23.18	29.34	1.37	47.31	nr	**76.65**
65 mm dia.	26.76	33.87	1.54	53.18	nr	**87.05**

38 MECHANICAL/COOLING/HEATING SYSTEMS

Item	Net Price £	Material £	Labour hours	Labour £	Unit	Total rate £
80 mm dia.	34.02	43.05	1.67	57.67	nr	**100.72**
100 mm dia.	34.73	43.95	2.22	76.66	nr	**120.61**
125 mm dia.	55.11	69.75	2.61	90.13	nr	**159.88**
150 mm dia.	65.80	83.28	2.99	103.25	nr	**186.53**
Blank flanges, slip on for welding; PN6						
15 mm dia.	3.51	4.44	0.48	16.58	nr	**21.01**
20 mm dia.	3.51	4.44	0.55	18.99	nr	**23.43**
25 mm dia.	3.51	4.44	0.64	22.10	nr	**26.54**
32 mm dia.	7.20	9.12	0.76	26.24	nr	**35.36**
40 mm dia.	7.20	9.12	0.84	29.01	nr	**38.12**
50 mm dia.	7.65	9.68	1.01	34.88	nr	**44.56**
65 mm dia.	15.15	19.17	1.30	44.89	nr	**64.07**
80 mm dia.	15.81	20.01	1.41	48.69	nr	**68.70**
100 mm dia.	16.24	20.56	1.78	61.47	nr	**82.03**
125 mm dia.	32.44	41.06	2.06	71.14	nr	**112.19**
150 mm dia.	30.70	38.86	2.35	81.15	nr	**120.01**
Blank flanges, slip on for welding; PN16						
15 mm dia.	3.19	4.03	0.48	16.58	nr	**20.61**
20 mm dia.	4.89	6.18	0.55	18.99	nr	**25.18**
25 mm dia.	5.95	7.53	0.64	22.10	nr	**29.63**
32 mm dia.	6.09	7.70	0.76	26.24	nr	**33.95**
40 mm dia.	7.88	9.97	0.84	29.01	nr	**38.98**
50 mm dia.	14.25	18.03	1.01	34.88	nr	**52.91**
65 mm dia.	17.48	22.13	1.30	44.89	nr	**67.02**
80 mm dia.	22.68	28.71	1.41	48.69	nr	**77.40**
100 mm dia.	28.11	35.58	1.78	61.47	nr	**97.04**
125 mm dia.	41.26	52.22	2.06	71.14	nr	**123.36**
150 mm dia.	53.24	67.38	2.35	81.15	nr	**148.54**
Extra over black steel butt welded pipes; black steel flanges, welding and drilled; imperial; BS 10						
Welded flanges; Table E						
15 mm dia.	11.51	14.57	0.59	20.37	nr	**34.94**
20 mm dia.	11.51	14.57	0.69	23.83	nr	**38.39**
25 mm dia.	11.51	14.57	0.84	29.01	nr	**43.57**
32 mm dia.	11.51	14.57	1.00	34.53	nr	**49.10**
40 mm dia.	11.81	14.95	1.11	38.33	nr	**53.28**
50 mm dia.	15.11	19.12	1.37	47.31	nr	**66.43**
65 mm dia.	18.38	23.26	1.54	53.18	nr	**76.44**
80 mm dia.	24.16	30.58	1.67	57.67	nr	**88.25**
100 mm dia.	35.69	45.17	2.22	76.66	nr	**121.83**
125 mm dia.	69.44	87.89	2.61	90.13	nr	**178.02**
150 mm dia.	69.44	87.89	2.99	103.25	nr	**191.14**
Blank flanges, slip on for welding; Table E						
15 mm dia.	7.92	10.03	0.48	16.58	nr	**26.60**
20 mm dia.	7.92	10.03	0.55	18.99	nr	**29.02**
25 mm dia.	7.92	10.03	0.64	22.10	nr	**32.13**
32 mm dia.	9.95	12.59	0.76	26.24	nr	**38.84**
40 mm dia.	10.83	13.71	0.84	29.01	nr	**42.72**
50 mm dia.	11.95	15.12	1.01	34.88	nr	**50.00**
65 mm dia.	13.79	17.45	1.30	44.89	nr	**62.34**

38 MECHANICAL/COOLING/HEATING SYSTEMS

Item	Net Price £	Material £	Labour hours	Labour £	Unit	Total rate £
LOW TEMPERATURE HOT WATER HEATING; PIPELINE: BLACK WELDED STEEL – cont						
Extra over black steel butt welded pipes – cont						
Blank flanges – cont						
80 mm dia.	30.07	38.06	1.41	48.69	nr	**86.75**
100 mm dia.	34.16	43.23	1.78	61.47	nr	**104.69**
125 mm dia.	56.18	71.10	2.06	71.14	nr	**142.24**
150 mm dia.	72.23	91.41	2.35	81.15	nr	**172.56**
Extra over black steel butt welded pipes; black steel flange connections						
Bolted connection between pair of flanges; including gasket, bolts, nuts and washers						
50 mm dia.	12.90	16.33	0.50	17.27	nr	**33.60**
65 mm dia.	15.29	19.35	0.50	17.27	nr	**36.62**
80 mm dia.	22.94	29.03	0.50	17.27	nr	**46.30**
100 mm dia.	24.91	31.53	0.50	17.27	nr	**48.80**
125 mm dia.	29.32	37.11	0.50	17.27	nr	**54.37**
150 mm dia.	41.89	53.02	0.88	30.39	nr	**83.40**
Extra over fittings; BS 1965; butt welded						
Cap						
25 mm dia.	32.56	41.20	0.47	16.23	nr	**57.43**
32 mm dia.	32.56	41.20	0.59	20.37	nr	**61.58**
40 mm dia.	32.76	41.46	0.70	24.17	nr	**65.63**
50 mm dia.	38.29	48.46	0.99	34.19	nr	**82.64**
65 mm dia.	44.99	56.94	1.35	46.62	nr	**103.56**
80 mm dia.	45.81	57.97	1.66	57.32	nr	**115.29**
100 mm dia.	59.86	75.76	2.23	77.01	nr	**152.77**
125 mm dia.	84.19	106.55	3.03	104.63	nr	**211.19**
150 mm dia.	97.39	123.26	3.79	130.88	nr	**254.13**
Concentric reducer						
20 × 15 mm dia.	18.13	22.95	0.69	23.83	nr	**46.78**
25 × 15 mm dia.	22.64	28.65	0.87	30.04	nr	**58.69**
25 × 20 mm dia.	17.50	22.15	0.87	30.04	nr	**52.20**
32 × 25 mm dia.	24.11	30.52	1.08	37.30	nr	**67.81**
40 × 25 mm dia.	31.45	39.80	1.38	47.65	nr	**87.45**
40 × 32 mm dia.	21.54	27.27	1.38	47.65	nr	**74.92**
50 × 25 mm dia.	29.91	37.85	1.82	62.85	nr	**100.70**
50 × 40 mm dia.	21.18	26.81	1.82	62.85	nr	**89.65**
65 × 50 mm dia.	29.25	37.01	2.52	87.02	nr	**124.03**
80 × 50 mm dia.	29.74	37.64	3.24	111.89	nr	**149.53**
100 × 50 mm dia.	49.28	62.37	4.08	140.89	nr	**203.26**
100 × 80 mm dia.	33.90	42.90	4.08	140.89	nr	**183.79**
125 × 80 mm dia.	74.04	93.70	4.71	162.65	nr	**256.35**
150 × 100 mm dia.	80.08	101.35	5.33	184.06	nr	**285.41**

38 MECHANICAL/COOLING/HEATING SYSTEMS

Item	Net Price £	Material £	Labour hours	Labour £	Unit	Total rate £
Eccentric reducer						
20 × 15 mm dia.	26.98	34.14	0.69	23.83	nr	**57.97**
25 × 15 mm dia.	35.31	44.68	0.87	30.04	nr	**74.73**
25 × 20 mm dia.	29.64	37.52	0.87	30.04	nr	**67.56**
32 × 25 mm dia.	32.76	41.46	1.08	37.30	nr	**78.75**
40 × 25 mm dia.	40.44	51.18	1.38	47.65	nr	**98.84**
40 × 32 mm dia.	38.62	48.87	1.38	47.65	nr	**96.53**
50 × 25 mm dia.	46.09	58.33	1.82	62.85	nr	**121.18**
50 × 40 mm dia.	33.20	42.02	1.82	62.85	nr	**104.87**
65 × 50 mm dia.	39.18	49.59	2.52	87.02	nr	**136.61**
80 × 50 mm dia.	47.96	60.70	3.24	111.89	nr	**172.58**
100 × 50 mm dia.	81.26	102.84	4.08	140.89	nr	**243.74**
100 × 80 mm dia.	59.95	75.87	4.08	140.89	nr	**216.76**
125 × 80 mm dia.	160.15	202.69	4.71	162.65	nr	**365.34**
150 × 100 mm dia.	121.27	153.48	5.33	184.06	nr	**337.54**
45° elbow, long radius						
15 mm dia.	6.81	8.61	0.56	19.34	nr	**27.95**
20 mm dia.	6.89	8.72	0.75	25.90	nr	**34.62**
25 mm dia.	8.96	11.34	0.93	32.12	nr	**43.45**
32 mm dia.	10.88	13.77	1.17	40.40	nr	**54.17**
40 mm dia.	11.01	13.94	1.46	50.42	nr	**64.35**
50 mm dia.	15.05	19.05	1.97	68.03	nr	**87.08**
65 mm dia.	19.39	24.54	2.70	93.24	nr	**117.78**
80 mm dia.	17.93	22.70	3.32	114.65	nr	**137.35**
100 mm dia.	27.11	34.31	4.09	141.24	nr	**175.55**
125 mm dia.	57.65	72.96	4.94	170.59	nr	**243.55**
150 mm dia.	71.62	90.64	5.78	199.60	nr	**290.24**
90° elbow, long radius						
15 mm dia.	7.04	8.91	0.56	19.34	nr	**28.25**
20 mm dia.	6.87	8.70	0.75	25.90	nr	**34.60**
25 mm dia.	9.14	11.57	0.93	32.12	nr	**43.68**
32 mm dia.	10.91	13.81	1.17	40.40	nr	**54.21**
40 mm dia.	11.03	13.96	1.46	50.42	nr	**64.38**
50 mm dia.	15.05	19.05	1.97	68.03	nr	**87.08**
65 mm dia.	19.39	24.54	2.70	93.24	nr	**117.78**
80 mm dia.	21.06	26.66	3.32	114.65	nr	**141.31**
100 mm dia.	31.89	40.37	4.09	141.24	nr	**181.60**
125 mm dia.	66.37	84.00	4.94	170.59	nr	**254.59**
150 mm dia.	84.22	106.59	5.78	199.60	nr	**306.19**
Branch bend (based on branch and pipe sizes being the same)						
15 mm dia.	28.04	35.48	0.85	29.35	nr	**64.84**
20 mm dia.	27.79	35.17	0.85	29.35	nr	**64.52**
25 mm dia.	28.07	35.52	1.02	35.22	nr	**70.75**
32 mm dia.	26.53	33.58	1.11	38.33	nr	**71.91**
40 mm dia.	26.18	33.14	1.36	46.96	nr	**80.10**
50 mm dia.	25.44	32.19	1.70	58.71	nr	**90.90**
65 mm dia.	37.61	47.60	1.78	61.47	nr	**109.06**
80 mm dia.	58.75	74.36	1.82	62.85	nr	**137.21**
100 mm dia.	76.64	97.00	1.87	64.58	nr	**161.57**
125 mm dia.	135.05	170.92	2.21	76.32	nr	**247.23**
150 mm dia.	207.66	262.82	2.65	91.51	nr	**354.33**

38 MECHANICAL/COOLING/HEATING SYSTEMS

Item	Net Price £	Material £	Labour hours	Labour £	Unit	Total rate £
LOW TEMPERATURE HOT WATER HEATING; PIPELINE: BLACK WELDED STEEL – cont						
Extra over fittings – cont						
Equal tee						
15 mm dia.	66.46	84.11	0.82	28.32	nr	**112.42**
20 mm dia.	66.46	84.11	1.10	37.99	nr	**122.09**
25 mm dia.	66.46	84.11	1.35	46.62	nr	**130.72**
32 mm dia.	66.46	84.11	1.63	56.29	nr	**140.39**
40 mm dia.	66.46	84.11	2.14	73.90	nr	**158.01**
50 mm dia.	70.00	88.59	3.02	104.29	nr	**192.88**
65 mm dia.	102.13	129.25	3.61	124.66	nr	**253.91**
80 mm dia.	102.32	129.50	4.18	144.35	nr	**273.85**
100 mm dia.	127.26	161.06	5.24	180.95	nr	**342.02**
125 mm dia.	291.29	368.66	6.70	231.37	nr	**600.03**
150 mm dia.	318.22	402.74	8.45	291.80	nr	**694.54**
Extra over black steel butt welded pipes; labour						
Made bend						
15 mm dia.	–	–	0.42	14.50	nr	**14.50**
20 mm dia.	–	–	0.42	14.50	nr	**14.50**
25 mm dia.	–	–	0.50	17.27	nr	**17.27**
32 mm dia.	–	–	0.62	21.41	nr	**21.41**
40 mm dia.	–	–	0.74	25.55	nr	**25.55**
50 mm dia.	–	–	0.89	30.73	nr	**30.73**
65 mm dia.	–	–	1.05	36.26	nr	**36.26**
80 mm dia.	–	–	1.13	39.02	nr	**39.02**
100 mm dia.	–	–	2.90	100.14	nr	**100.14**
125 mm dia.	–	–	3.56	122.94	nr	**122.94**
150 mm dia.	–	–	4.18	144.35	nr	**144.35**
Splay cut end						
15 mm dia.	–	–	0.14	4.83	nr	**4.83**
20 mm dia.	–	–	0.16	5.53	nr	**5.53**
25 mm dia.	–	–	0.18	6.22	nr	**6.22**
32 mm dia.	–	–	0.25	8.63	nr	**8.63**
40 mm dia.	–	–	0.27	9.32	nr	**9.32**
50 mm dia.	–	–	0.31	10.71	nr	**10.71**
65 mm dia.	–	–	0.35	12.09	nr	**12.09**
80 mm dia.	–	–	0.40	13.81	nr	**13.81**
100 mm dia.	–	–	0.48	16.58	nr	**16.58**
125 mm dia.	–	–	0.56	19.34	nr	**19.34**
150 mm dia.	–	–	0.64	22.10	nr	**22.10**
Screwed joint to fitting						
15 mm dia.	–	–	0.30	10.36	nr	**10.36**
20 mm dia.	–	–	0.40	13.81	nr	**13.81**
25 mm dia.	–	–	0.46	15.88	nr	**15.88**
32 mm dia.	–	–	0.53	18.30	nr	**18.30**
40 mm dia.	–	–	0.61	21.06	nr	**21.06**
50 mm dia.	–	–	0.73	25.21	nr	**25.21**
65 mm dia.	–	–	0.89	30.73	nr	**30.73**

38 MECHANICAL/COOLING/HEATING SYSTEMS

Item	Net Price £	Material £	Labour hours	Labour £	Unit	Total rate £
80 mm dia.	–	–	1.05	36.26	nr	**36.26**
100 mm dia.	–	–	1.46	50.42	nr	**50.42**
125 mm dia.	–	–	2.10	72.52	nr	**72.52**
150 mm dia.	–	–	2.73	94.27	nr	**94.27**
Straight butt weld						
15 mm dia.	–	–	0.31	10.71	nr	**10.71**
20 mm dia.	–	–	0.42	14.50	nr	**14.50**
25 mm dia.	–	–	0.52	17.96	nr	**17.96**
32 mm dia.	–	–	0.69	23.83	nr	**23.83**
40 mm dia.	–	–	0.83	28.66	nr	**28.66**
50 mm dia.	–	–	1.22	42.13	nr	**42.13**
65 mm dia.	–	–	1.57	54.22	nr	**54.22**
80 mm dia.	–	–	1.95	67.34	nr	**67.34**
100 mm dia.	–	–	2.38	82.19	nr	**82.19**
125 mm dia.	–	–	2.83	97.73	nr	**97.73**
150 mm dia.	–	–	3.27	112.92	nr	**112.92**

38 MECHANICAL/COOLING/HEATING SYSTEMS

Item	Net Price £	Material £	Labour hours	Labour £	Unit	Total rate £
LOW TEMPERATURE HOT WATER HEATING; PIPELINE: CARBON WELDED STEEL						
Hot finished seamless carbon steel pipe; BS 806 and BS 3601; wall thickness to BS 3600; butt welded joints; including protective painting, fixed vertically or at low level, brackets measured separately (Refer to Screwed Pipework Section). Welded butt joints are within the running length, but any flanges are additional						
Pipework						
200 mm dia.	162.90	206.17	2.04	70.45	m	**276.61**
250 mm dia.	146.65	185.61	2.59	89.44	m	**275.05**
300 mm dia.	344.78	436.36	2.99	103.25	m	**539.61**
350 mm dia.	363.88	460.52	3.52	121.55	m	**582.08**
400 mm dia.	634.42	802.92	4.08	140.89	m	**943.82**
Hot finished seamless carbon steel pipe; BS 806 and BS 3601; wall thickness to BS 3600; butt welded joints; including protective painting, fixed at high level or suspended, brackets measured separately (Refer to Screwed Pipework Section). Welded butt joints are within the running length, but any flanges are additional						
Pipework						
200 mm dia.	162.90	206.17	3.70	127.77	m	**333.94**
250 mm dia.	146.65	185.61	4.73	163.34	m	**348.95**
300 mm dia.	344.78	436.36	5.65	195.11	m	**631.46**
350 mm dia.	363.88	460.52	6.68	230.68	m	**691.20**
400 mm dia.	634.42	802.92	7.70	265.90	m	**1068.82**
FIXINGS						
Refer to steel pipes; black malleable iron. For minimum fixing distances, refer to the Tables and Memoranda to the rear of the book						
Extra over fittings; BS 1965 part 1; butt welded						
Cap						
200 mm dia.	133.36	168.78	3.70	127.77	nr	**296.55**
250 mm dia.	255.93	323.90	4.73	163.34	nr	**487.24**
300 mm dia.	279.90	354.24	5.65	195.11	nr	**549.35**
350 mm dia.	431.24	545.78	6.68	230.68	nr	**776.45**
400 mm dia.	498.81	631.29	7.70	265.90	nr	**897.19**
Concentric reducer						
200 mm × 150 mm dia.	145.69	184.38	7.27	251.05	nr	**435.44**
250 mm × 150 mm dia.	211.26	267.38	9.05	312.52	nr	**579.90**
250 mm × 200 mm dia.	129.30	163.64	9.10	314.25	nr	**477.89**
300 mm × 150 mm dia.	244.03	308.84	10.75	371.22	nr	**680.07**

38 MECHANICAL/COOLING/HEATING SYSTEMS

Item	Net Price £	Material £	Labour hours	Labour £	Unit	Total rate £
300 mm × 200 mm dia.	252.72	319.84	10.75	371.22	nr	**691.07**
300 mm × 250 mm dia.	224.66	284.33	11.15	385.04	nr	**669.37**
350 mm × 200 mm dia.	383.24	485.03	12.50	431.66	nr	**916.69**
350 mm × 250 mm dia.	321.31	406.65	12.70	438.56	nr	**845.21**
350 mm × 300 mm dia.	321.31	406.65	13.00	448.92	nr	**855.57**
400 mm × 250 mm dia.	469.83	594.62	14.46	499.34	nr	**1093.96**
400 mm × 300 mm dia.	581.84	736.37	14.51	501.07	nr	**1237.44**
400 mm × 350 mm dia.	507.13	641.83	15.16	523.51	nr	**1165.34**
Eccentric reducer						
200 mm × 150 mm dia.	263.12	333.01	7.27	251.05	nr	**584.06**
250 mm × 150 mm dia.	350.61	443.73	9.05	312.52	nr	**756.25**
250 mm × 200 mm dia.	229.41	290.34	9.10	314.25	nr	**604.59**
300 mm × 150 mm dia.	420.61	532.33	10.75	371.22	nr	**903.55**
300 mm × 200 mm dia.	485.13	613.97	10.75	371.22	nr	**985.20**
300 mm × 250 mm dia.	390.51	494.22	11.15	385.04	nr	**879.26**
350 mm × 200 mm dia.	445.57	563.91	12.50	431.66	nr	**995.57**
350 mm × 250 mm dia.	370.43	468.81	12.70	438.56	nr	**907.38**
350 mm × 300 mm dia.	370.43	468.81	13.00	448.92	nr	**917.74**
400 mm × 250 mm dia.	594.93	752.94	14.46	499.34	nr	**1252.28**
400 mm × 300 mm dia.	557.12	705.10	14.51	501.07	nr	**1206.16**
400 mm × 350 mm dia.	686.46	868.78	15.16	523.51	nr	**1392.29**
45° elbow						
200 mm dia.	139.35	176.36	7.75	267.63	nr	**443.98**
250 mm dia.	261.69	331.19	10.05	347.05	nr	**678.25**
300 mm dia.	382.36	483.91	12.20	421.30	nr	**905.21**
350 mm dia.	309.44	391.63	14.65	505.90	nr	**897.53**
400 mm dia.	514.28	650.88	17.12	591.20	nr	**1242.07**
90° elbow						
200 mm dia.	163.95	207.49	7.75	267.63	nr	**475.12**
250 mm dia.	307.94	389.73	10.05	347.05	nr	**736.79**
300 mm dia.	449.87	569.35	12.20	421.30	nr	**990.65**
350 mm dia.	618.43	782.69	14.65	505.90	nr	**1288.59**
400 mm dia.	789.93	999.74	17.12	591.20	nr	**1590.93**
Equal tee						
200 mm dia.	433.07	548.10	11.25	388.49	nr	**936.59**
250 mm dia.	742.61	939.85	14.53	501.76	nr	**1441.61**
300 mm dia.	921.61	1166.39	17.55	606.05	nr	**1772.44**
350 mm dia.	1236.48	1564.89	20.98	724.49	nr	**2289.38**
400 mm dia.	1413.11	1788.44	24.38	841.90	nr	**2630.34**
Extra over black steel butt welded pipes; labour						
Straight butt weld						
200 mm dia.	–	–	4.08	140.89	nr	**140.89**
250 mm dia.	–	–	5.20	179.57	nr	**179.57**
300 mm dia.	–	–	6.22	214.79	nr	**214.79**
350 mm dia.	–	–	7.33	253.12	nr	**253.12**
400 mm dia.	–	–	8.41	290.42	nr	**290.42**
Branch weld						
100 mm dia.	–	–	3.46	119.48	nr	**119.48**
125 mm dia.	–	–	4.23	146.07	nr	**146.07**
150 mm dia.	–	–	5.00	172.66	nr	**172.66**

38 MECHANICAL/COOLING/HEATING SYSTEMS

Item	Net Price £	Material £	Labour hours	Labour £	Unit	Total rate £
LOW TEMPERATURE HOT WATER HEATING; PIPELINE: CARBON WELDED STEEL – cont						
Extra over black steel butt welded pipes; black steel flanges, welding and drilled; metric; BS 4504						
Welded flanges; PN16						
200 mm dia.	89.16	112.84	4.10	141.58	nr	254.43
250 mm dia.	128.08	162.09	5.33	184.06	nr	346.15
300 mm dia.	179.42	227.07	6.40	221.01	nr	448.08
350 mm dia.	347.00	439.17	7.43	256.58	nr	695.74
400 mm dia.	458.16	579.84	8.45	291.80	nr	871.64
Welded flanges; PN25						
200 mm dia.	296.08	374.72	4.10	141.58	nr	516.30
250 mm dia.	354.91	449.17	5.33	184.06	nr	633.23
300 mm dia.	479.91	607.37	6.40	221.01	nr	828.38
Blank flanges, slip on for welding; PN16						
200 mm dia.	206.18	260.94	2.70	93.24	nr	354.18
250 mm dia.	313.35	396.58	3.48	120.17	nr	516.75
300 mm dia.	334.80	423.72	4.20	145.04	nr	568.76
350 mm dia.	523.96	663.12	4.78	165.07	nr	828.19
400 mm dia.	692.80	876.81	5.35	184.75	nr	1061.56
Blank flanges, slip on for welding; PN25						
200 mm dia.	206.18	260.94	2.70	93.24	nr	354.18
250 mm dia.	313.35	396.58	3.48	120.17	nr	516.75
300 mm dia.	334.80	423.72	4.20	145.04	nr	568.76
Extra over black steel butt welded pipes; black steel flange connections						
Bolted connection between pair of flanges; including gasket, bolts, nuts and washers						
200 mm dia.	173.03	218.98	3.83	132.26	nr	351.24
250 mm dia.	267.78	338.90	4.93	170.25	nr	509.14
300 mm dia.	365.12	462.09	5.90	203.74	nr	665.83

38 MECHANICAL/COOLING/HEATING SYSTEMS

Item	Net Price £	Material £	Labour hours	Labour £	Unit	Total rate £
LOW TEMPERATURE HOT WATER HEATING; PIPELINE: PRESS FIT						
Press fit jointing system; operating temperature −20°C to +120°C; operating pressure 16 bar; butyl rubber 'O' ring mechanical joint. With brackets measured separately (Refer to Screwed Steel Section)						
Carbon steel						
Pipework						
15 mm dia.	2.66	3.36	0.46	15.88	m	**19.25**
20 mm dia.	4.27	5.41	0.48	16.58	m	**21.98**
25 mm dia.	6.04	7.65	0.52	17.96	m	**25.61**
32 mm dia.	7.80	9.87	0.56	19.34	m	**29.21**
40 mm dia.	10.62	13.45	0.58	20.03	m	**33.47**
50 mm dia.	13.80	17.46	0.66	22.79	m	**40.26**
Extra over for carbon steel press fit fittings						
Coupling						
15 mm dia.	1.46	1.85	0.36	12.43	nr	**14.28**
20 mm dia.	1.80	2.28	0.36	12.43	nr	**14.71**
25 mm dia.	2.27	2.87	0.44	15.19	nr	**18.07**
32 mm dia.	3.80	4.80	0.44	15.19	nr	**20.00**
40 mm dia.	5.07	6.42	0.52	17.96	nr	**24.38**
50 mm dia.	5.99	7.58	0.60	20.72	nr	**28.30**
Reducer						
20 × 15 mm dia.	1.36	1.73	0.36	12.43	nr	**14.16**
25 × 15 mm dia.	1.80	2.28	0.40	13.81	nr	**16.09**
25 × 20 mm dia.	1.87	2.37	0.40	13.81	nr	**16.18**
32 × 20 mm dia.	2.10	2.65	0.40	13.81	nr	**16.47**
32 × 25 mm dia.	2.23	2.83	0.44	15.19	nr	**18.02**
40 × 32 mm dia.	4.85	6.14	0.48	16.58	nr	**22.71**
50 × 20 mm dia.	14.03	17.75	0.48	16.58	nr	**34.33**
50 × 25 mm dia.	14.14	17.89	0.52	17.96	nr	**35.85**
50 × 40 mm dia.	14.87	18.82	0.56	19.34	nr	**38.16**
90° elbow						
15 mm dia.	2.12	2.68	0.36	12.43	nr	**15.12**
20 mm dia.	2.80	3.55	0.36	12.43	nr	**15.98**
25 mm dia.	3.84	4.87	0.44	15.19	nr	**20.06**
32 mm dia.	9.60	12.15	0.44	15.19	nr	**27.34**
40 mm dia.	15.34	19.42	0.52	17.96	nr	**37.37**
50 mm dia.	18.32	23.18	0.60	20.72	nr	**43.90**
45° elbow						
15 mm dia.	2.54	3.22	0.36	12.43	nr	**15.65**
20 mm dia.	2.84	3.59	0.36	12.43	nr	**16.03**
25 mm dia.	3.86	4.88	0.44	15.19	nr	**20.08**
32 mm dia.	7.61	9.64	0.44	15.19	nr	**24.83**
40 mm dia.	9.54	12.07	0.52	17.96	nr	**30.03**
50 mm dia.	10.79	13.66	0.60	20.72	nr	**34.38**

38 MECHANICAL/COOLING/HEATING SYSTEMS

Item	Net Price £	Material £	Labour hours	Labour £	Unit	Total rate £
LOW TEMPERATURE HOT WATER HEATING; PIPELINE: PRESS FIT – cont						
Extra over for carbon steel press fit fittings – cont						
Equal tee						
15 mm dia.	4.08	5.16	0.54	18.65	nr	**23.81**
20 mm dia.	4.73	5.98	0.54	18.65	nr	**24.63**
25 mm dia.	6.34	8.02	0.66	22.79	nr	**30.81**
32 mm dia.	9.86	12.48	0.66	22.79	nr	**35.27**
40 mm dia.	14.55	18.41	0.78	26.94	nr	**45.35**
50 mm dia.	17.45	22.08	0.90	31.08	nr	**53.16**
Reducing tee						
20 × 15 mm dia.	4.64	5.87	0.54	18.65	nr	**24.52**
25 × 15 mm dia.	6.28	7.94	0.62	21.41	nr	**29.35**
25 × 20 mm dia.	6.81	8.62	0.62	21.41	nr	**30.03**
32 × 15 mm dia.	9.19	11.63	0.62	21.41	nr	**33.04**
32 × 20 mm dia.	9.92	12.56	0.62	21.41	nr	**33.97**
32 × 25 mm dia.	10.08	12.76	0.62	21.41	nr	**34.17**
40 × 20 mm dia.	13.28	16.81	0.70	24.17	nr	**40.98**
40 × 25 mm dia.	13.78	17.44	0.70	24.17	nr	**41.61**
40 × 32 mm dia.	13.47	17.05	0.70	24.17	nr	**41.22**
50 × 20 mm dia.	15.84	20.04	0.82	28.32	nr	**48.36**
50 × 25 mm dia.	16.16	20.45	0.82	28.32	nr	**48.77**
50 × 32 mm dia.	16.64	21.06	0.82	28.32	nr	**49.38**
50 × 40 mm dia.	17.44	22.07	0.82	28.32	nr	**50.39**

38 MECHANICAL/COOLING/HEATING SYSTEMS

Item	Net Price £	Material £	Labour hours	Labour £	Unit	Total rate £
LOW TEMPERATURE HOT WATER HEATING; PIPELINE: MECHANICAL GROOVED						
MECHANICAL GROOVED						
Mechanical grooved jointing system; working temperature not exceeding 82°C BS 5750; pipework complete with grooved joints; painted finish. With brackets measured separately (Refer to Screwed Steel Section)						
Grooved joints						
65 mm	20.66	26.15	0.58	20.03	m	**46.18**
80 mm	21.66	27.41	0.68	23.48	m	**50.89**
100 mm	28.09	35.55	0.79	27.28	m	**62.83**
125 mm	44.10	55.82	1.02	35.22	m	**91.04**
150 mm	46.04	58.27	1.15	39.71	m	**97.98**
Extra over mechanical grooved system fittings						
Couplings						
65 mm	20.66	26.15	0.41	14.16	nr	**40.31**
80 mm	21.66	27.41	0.41	14.16	nr	**41.57**
100 mm	28.09	35.55	0.66	22.79	nr	**58.34**
125 mm	44.10	55.82	0.68	23.48	nr	**79.30**
150 mm	46.04	58.27	0.80	27.63	nr	**85.90**
Concentric reducers (one size down)						
80 mm	33.17	41.97	0.59	20.37	nr	**62.35**
100 mm	33.24	42.07	0.71	24.52	nr	**66.59**
125 mm	57.29	72.51	0.85	29.35	nr	**101.86**
150 mm	74.29	94.02	0.98	33.84	nr	**127.86**
Short radius elbow; 90°						
65 mm	37.98	48.07	0.53	18.30	nr	**66.37**
80 mm	38.75	49.04	0.61	21.06	nr	**70.10**
100 mm	51.88	65.66	0.80	27.63	nr	**93.28**
125 mm	85.56	108.29	0.90	31.08	nr	**139.37**
150 mm	111.04	140.54	0.94	32.46	nr	**173.00**
Short radius elbow; 45°						
65 mm	32.51	41.14	0.53	18.30	nr	**59.44**
80 mm	36.52	46.22	0.61	21.06	nr	**67.29**
100 mm	45.40	57.45	0.80	27.63	nr	**85.08**
125 mm	76.81	97.21	0.90	31.08	nr	**128.28**
150 mm	84.57	107.03	0.94	32.46	nr	**139.49**
Equal tee						
65 mm	68.36	86.52	0.83	28.66	nr	**115.18**
80 mm	72.41	91.65	0.93	32.12	nr	**123.76**
100 mm	80.98	102.48	1.18	40.75	nr	**143.23**
125 mm	214.78	271.82	1.37	47.31	nr	**319.13**
150 mm	199.14	252.03	1.43	49.38	nr	**301.41**

38 MECHANICAL/COOLING/HEATING SYSTEMS

Item	Net Price £	Material £	Labour hours	Labour £	Unit	Total rate £
LOW TEMPERATURE HOT WATER HEATING; PIPELINE: PLASTIC PIPEWORK						
Polypropylene PP-R 80 pipe, mechanically stabilized by fibre compound mixture in middle layer; suitable for continuous working temperatures of 0–90°C; thermally fused joints in the running length						
Pipe; 4 m long; PN 20						
20 mm dia.	4.85	6.14	0.35	12.09	m	**18.23**
25 mm dia.	7.31	9.25	0.39	13.47	m	**22.72**
32 mm dia.	8.35	10.57	0.43	14.85	m	**25.42**
40 mm dia.	11.14	14.10	0.47	16.23	m	**30.33**
50 mm dia.	16.21	20.52	0.51	17.61	m	**38.13**
63 mm dia.	26.76	33.86	0.52	17.96	m	**51.82**
75 mm dia.	34.62	43.81	0.60	20.72	m	**64.53**
90 mm dia.	53.41	67.59	0.69	23.83	m	**91.42**
110 mm dia.	80.27	101.59	0.69	23.83	m	**125.42**
125 mm dia.	86.00	108.84	0.85	29.35	m	**138.19**
FIXINGS						
Refer to steel pipes; black malleable iron. For minimum fixing distances, refer to the Tables and Memoranda to the rear of the book						
Extra over fittings; thermally fused joints						
Overbridge bow						
20 mm dia.	3.13	3.96	0.51	17.61	nr	**21.57**
25 mm dia.	5.76	7.29	0.56	19.34	nr	**26.63**
32 mm dia.	11.53	14.59	0.65	22.45	nr	**37.04**
Elbow 90°						
20 mm dia.	1.05	1.33	0.44	15.19	nr	**16.53**
25 mm dia.	1.40	1.77	0.52	17.96	nr	**19.72**
32 mm dia.	2.00	2.54	0.59	20.37	nr	**22.91**
40 mm dia.	3.16	4.00	0.66	22.79	nr	**26.79**
50 mm dia.	6.83	8.64	0.73	25.21	nr	**33.85**
63 mm dia.	10.43	13.20	0.85	29.35	nr	**42.56**
75 mm dia.	23.11	29.24	0.85	29.35	nr	**58.60**
90 mm dia.	42.69	54.03	1.04	35.91	nr	**89.95**
110 mm dia.	60.68	76.80	1.04	35.91	nr	**112.71**
125 mm dia.	93.52	118.36	1.30	44.89	nr	**163.25**
Long bend 90°						
20 mm dia.	5.61	7.10	0.48	16.58	nr	**23.67**
25 mm dia.	5.89	7.46	0.57	19.68	nr	**27.14**
32 mm dia.	6.74	8.53	0.65	22.45	nr	**30.97**
40 mm dia.	12.60	15.95	0.73	25.21	nr	**41.15**
Elbow 90°, female/male						
20 mm dia.	1.08	1.37	0.44	15.19	nr	**16.57**
25 mm dia.	1.42	1.80	0.52	17.96	nr	**19.76**
32 mm dia.	2.03	2.57	0.59	20.37	nr	**22.95**

38 MECHANICAL/COOLING/HEATING SYSTEMS

Item	Net Price £	Material £	Labour hours	Labour £	Unit	Total rate £
Elbow 45°						
20 mm dia.	1.08	1.37	0.44	15.19	nr	**16.57**
25 mm dia.	1.42	1.80	0.52	17.96	nr	**19.76**
32 mm dia.	2.03	2.57	0.59	20.37	nr	**22.95**
40 mm dia.	3.18	4.02	0.66	22.79	nr	**26.81**
50 mm dia.	6.83	8.64	0.73	25.21	nr	**33.85**
63 mm dia.	10.46	13.24	0.85	29.35	nr	**42.59**
75 mm dia.	23.12	29.26	0.85	29.35	nr	**58.61**
90 mm dia.	42.69	54.03	1.04	35.91	nr	**89.95**
110 mm dia.	60.71	76.84	1.04	35.91	nr	**112.75**
125 mm dia.	93.54	118.38	1.30	44.89	nr	**163.27**
Elbow 45°, female/male						
20 mm dia.	1.08	1.37	0.44	15.19	nr	**16.57**
25 mm dia.	1.42	1.80	0.52	17.96	nr	**19.76**
32 mm dia.	2.03	2.57	0.59	20.37	nr	**22.95**
T-Piece 90°						
20 mm dia.	1.47	1.86	0.61	21.06	nr	**22.92**
25 mm dia.	1.99	2.52	0.72	24.86	nr	**27.38**
32 mm dia.	2.58	3.27	0.83	28.66	nr	**31.93**
40 mm dia.	4.01	5.07	0.92	31.77	nr	**36.84**
50 mm dia.	11.37	14.39	1.01	34.88	nr	**49.26**
63 mm dia.	16.28	20.60	1.11	38.33	nr	**58.93**
75 mm dia.	27.14	34.35	1.18	40.75	nr	**75.10**
90 mm dia.	49.97	63.24	1.46	50.42	nr	**113.65**
110 mm dia.	77.95	98.66	1.46	50.42	nr	**149.08**
125 mm dia.	103.60	131.11	1.82	62.85	nr	**193.96**
T-Piece 90° reducing						
25 × 20 × 20 mm	2.03	2.57	0.72	24.86	nr	**27.44**
25 × 20 × 25 mm	2.03	2.57	0.72	24.86	nr	**27.44**
32 × 20 × 32 mm	2.58	3.27	0.83	28.66	nr	**31.93**
32 × 25 × 32 mm	2.58	3.27	0.83	28.66	nr	**31.93**
40 × 20 × 40 mm	4.01	5.07	0.92	31.77	nr	**36.84**
40 × 25 × 40 mm	4.01	5.07	0.92	31.77	nr	**36.84**
40 × 32 × 40 mm	4.01	5.07	0.92	31.77	nr	**36.84**
50 × 25 × 50 mm	11.37	14.39	1.01	34.88	nr	**49.26**
50 × 32 × 50 mm	11.37	14.39	1.01	34.88	nr	**49.26**
50 × 40 × 50 mm	11.37	14.39	1.01	34.88	nr	**49.26**
63 × 20 × 63 mm	15.33	19.40	1.11	38.33	nr	**57.73**
63 × 25 × 63 mm	15.33	19.40	1.11	38.33	nr	**57.73**
63 × 32 × 63 mm	15.33	19.40	1.11	38.33	nr	**57.73**
63 × 40 × 63 mm	15.33	19.40	1.01	34.88	nr	**54.28**
63 × 50 × 63 mm	15.33	19.40	1.01	34.88	nr	**54.28**
75 × 20 × 75 mm	24.92	31.53	1.18	40.75	nr	**72.28**
75 × 25 × 75 mm	24.92	31.53	1.18	40.75	nr	**72.28**
75 × 32 × 75 mm	24.92	31.53	1.18	40.75	nr	**72.28**
75 × 40 × 75 mm	24.92	31.53	1.18	40.75	nr	**72.28**
75 × 50 × 75 mm	24.92	31.53	1.18	40.75	nr	**72.28**
75 × 63 × 75 mm	24.92	31.53	1.18	40.75	nr	**72.28**
90 × 63 × 90 mm	49.97	63.24	1.46	50.42	nr	**113.65**
110 × 75 × 110 mm	77.95	98.66	1.46	50.42	nr	**149.08**

Material Costs/Prices for Measured Works

38 MECHANICAL/COOLING/HEATING SYSTEMS

Item	Net Price £	Material £	Labour hours	Labour £	Unit	Total rate £
LOW TEMPERATURE HOT WATER HEATING; PIPELINE: PLASTIC PIPEWORK – cont						
Extra over fittings – cont						
T-Piece 90° reducing – cont						
110 × 90 × 110 mm	77.95	98.66	1.46	50.42	nr	**149.08**
125 × 90 × 125 mm	92.74	117.37	1.82	62.85	nr	**180.21**
125 × 110 × 125 mm	94.63	119.77	1.82	62.85	nr	**182.62**
Reducer						
25 × 20 mm	1.17	1.48	0.59	20.37	nr	**21.86**
32 × 20 mm	1.51	1.92	0.62	21.41	nr	**23.33**
32 × 25 mm	1.51	1.92	0.62	21.41	nr	**23.33**
40 × 20 mm	2.39	3.02	0.66	22.79	nr	**25.82**
40 × 25 mm	2.39	3.02	0.66	22.79	nr	**25.82**
40 × 32 mm	2.39	3.02	0.66	22.79	nr	**25.82**
50 × 20 mm	3.83	4.85	0.73	25.21	nr	**30.05**
50 × 25 mm	3.83	4.85	0.73	25.21	nr	**30.05**
50 × 32 mm	3.83	4.85	0.73	25.21	nr	**30.05**
50 × 40 mm	3.83	4.85	0.73	25.21	nr	**30.05**
63 × 40 mm	7.73	9.79	0.78	26.94	nr	**36.72**
63 × 25 mm	7.73	9.79	0.78	26.94	nr	**36.72**
63 × 32 mm	7.73	9.79	0.78	26.94	nr	**36.72**
63 × 50 mm	7.73	9.79	0.78	26.94	nr	**36.72**
75 × 50 mm	8.64	10.93	0.85	29.35	nr	**40.28**
75 × 63 mm	8.64	10.93	0.85	29.35	nr	**40.28**
90 × 63 mm	19.29	24.42	1.04	35.91	nr	**60.33**
90 × 75 mm	19.29	24.42	1.04	35.91	nr	**60.33**
110 × 90 mm	31.16	39.44	1.17	40.40	nr	**79.84**
125 × 110 mm	48.69	61.62	1.43	49.38	nr	**111.00**
Socket						
20 mm dia.	1.04	1.31	0.51	17.61	nr	**18.93**
25 mm dia.	1.17	1.48	0.56	19.34	nr	**20.82**
32 mm dia.	1.50	1.90	0.65	22.45	nr	**24.34**
40 mm dia.	1.84	2.33	0.74	25.55	nr	**27.88**
50 mm dia.	3.80	4.81	0.81	27.97	nr	**32.78**
63 mm dia.	7.73	9.79	0.86	29.70	nr	**39.48**
75 mm dia.	8.64	10.93	0.91	31.42	nr	**42.36**
90 mm dia.	22.29	28.21	0.91	31.42	nr	**59.63**
110 mm dia.	37.87	47.93	0.91	31.42	nr	**79.36**
125 mm dia.	52.76	66.77	1.30	44.89	nr	**111.66**
End cap						
20 mm dia.	1.62	2.05	0.25	8.63	nr	**10.68**
25 mm dia.	1.99	2.52	0.29	10.01	nr	**12.53**
32 mm dia.	2.46	3.12	0.33	11.40	nr	**14.51**
40 mm dia.	3.95	5.00	0.36	12.43	nr	**17.43**
50 mm dia.	5.46	6.91	0.40	13.81	nr	**20.72**
63 mm dia.	9.19	11.63	0.44	15.19	nr	**26.82**
75 mm dia.	13.21	16.72	0.47	16.23	nr	**32.95**
90 mm dia.	29.93	37.88	0.57	19.68	nr	**57.57**
110 mm dia.	35.99	45.55	0.57	19.68	nr	**65.23**
125 mm dia.	54.82	69.38	0.85	29.35	nr	**98.73**

38 MECHANICAL/COOLING/HEATING SYSTEMS

Item	Net Price £	Material £	Labour hours	Labour £	Unit	Total rate £
Stub flange with gasket						
32 mm dia.	36.36	46.01	0.23	7.94	nr	**53.96**
40 mm dia.	45.75	57.90	0.27	9.32	nr	**67.23**
50 mm dia.	55.34	70.04	0.38	13.12	nr	**83.16**
63 mm dia.	66.42	84.07	0.43	14.85	nr	**98.92**
75 mm dia.	77.95	98.66	0.48	16.58	nr	**115.23**
90 mm dia.	105.50	133.52	0.53	18.30	nr	**151.82**
110 mm dia.	147.75	186.99	0.53	18.30	nr	**205.29**
125 mm dia.	211.77	268.01	0.75	25.90	nr	**293.91**
Weld in saddle with female thread						
40–½"	2.61	3.31	0.36	12.43	nr	**15.74**
50–½"	2.61	3.31	0.36	12.43	nr	**15.74**
63–½"	2.61	3.31	0.40	13.81	nr	**17.12**
75–½"	2.61	3.31	0.40	13.81	nr	**17.12**
90–½"	2.61	3.31	0.46	15.88	nr	**19.19**
110–½"	2.61	3.31	0.46	15.88	nr	**19.19**
Weld in saddle with male thread						
50–½"	2.61	3.31	0.36	12.43	nr	**15.74**
63–½"	2.61	3.31	0.40	13.81	nr	**17.12**
75–½"	2.61	3.31	0.40	13.81	nr	**17.12**
90–½"	2.61	3.31	0.46	15.88	nr	**19.19**
110–½"	2.61	3.31	0.46	15.88	nr	**19.19**
Transition piece, round with female thread						
20 × ½"	6.11	7.74	0.29	10.01	nr	**17.75**
20 × ¾"	8.07	10.22	0.29	10.01	nr	**20.23**
25 × ½"	6.11	7.74	0.33	11.40	nr	**19.13**
25 × ¾"	8.07	10.22	0.33	11.40	nr	**21.61**
Transition piece, hexagon with female thread						
32 × 1"	22.75	28.79	0.36	12.43	nr	**41.22**
40 × 1 ¼"	36.02	45.58	0.36	12.43	nr	**58.01**
50 × 1 ½"	41.80	52.91	0.36	12.43	nr	**65.34**
63 × 2"	64.75	81.94	0.40	13.81	nr	**95.76**
75 × 2"	67.57	85.51	0.40	13.81	nr	**99.33**
125 × 5"	129.42	163.79	0.51	17.61	nr	**181.41**
Stop valve for surface assembly						
20 mm dia.	25.12	31.80	0.25	8.63	nr	**40.43**
25 mm dia.	25.12	31.80	0.29	10.01	nr	**41.81**
32 mm dia.	47.32	59.89	0.33	11.40	nr	**71.29**
Ball valve						
20 mm dia.	88.48	111.98	0.25	8.63	nr	**120.61**
25 mm dia.	94.81	120.00	0.29	10.01	nr	**130.01**
32 mm dia.	113.93	144.19	0.33	11.40	nr	**155.58**
40 mm dia.	145.40	184.02	0.36	12.43	nr	**196.45**
50 mm dia.	199.61	252.63	0.40	13.81	nr	**266.44**
63 mm dia.	225.08	284.86	0.44	15.19	nr	**300.05**

38 MECHANICAL/COOLING/HEATING SYSTEMS

Item	Net Price £	Material £	Labour hours	Labour £	Unit	Total rate £
LOW TEMPERATURE HOT WATER HEATING; PIPELINE: PLASTIC PIPEWORK – cont						
Floor or ceiling cover plates						
Plastic						
15 mm dia.	0.59	0.75	0.16	5.53	nr	6.27
20 mm dia.	0.67	0.84	0.22	7.60	nr	8.44
25 mm dia.	0.71	0.90	0.22	7.60	nr	8.50
32 mm dia.	0.77	0.98	0.24	8.29	nr	9.27
40 mm dia.	1.73	2.19	0.26	8.98	nr	11.17
50 mm dia.	1.89	2.39	0.26	8.98	nr	11.37
Chromium plated						
15 mm dia.	4.01	5.07	0.16	5.53	nr	10.60
20 mm dia.	4.29	5.43	0.17	5.87	nr	11.30
25 mm dia.	4.43	5.61	0.21	7.25	nr	12.86
32 mm dia.	4.54	5.74	0.22	7.60	nr	13.34
40 mm dia.	5.13	6.49	0.26	8.98	nr	15.47
50 mm dia.	6.13	7.75	0.26	8.98	nr	16.73

38 MECHANICAL/COOLING/HEATING SYSTEMS

Item	Net Price £	Material £	Labour hours	Labour £	Unit	Total rate £
LOW TEMPERATURE HOT WATER HEATING; PIPELINE: PIPELINE ANCILLARIES						
EXPANSION JOINTS						
Axial movement bellows expansion joints; stainless steel						
Screwed ends for steel pipework; up to 6 bar G at 100°C						
15 mm dia.	107.04	135.47	0.68	23.48	nr	**158.96**
20 mm dia.	114.34	144.71	0.81	27.97	nr	**172.69**
25 mm dia.	118.78	150.33	0.93	32.12	nr	**182.44**
32 mm dia.	125.76	159.17	1.06	36.60	nr	**195.77**
40 mm dia.	133.20	168.58	1.16	40.06	nr	**208.64**
50 mm dia.	132.70	167.95	1.19	41.09	nr	**209.04**
Screwed ends for steel pipework; aluminium and steel outer sleeves; up to 16 bar G at 120°C						
20 mm dia.	116.61	147.59	1.32	45.58	nr	**193.17**
25 mm dia.	118.78	150.33	1.52	52.49	nr	**202.82**
32 mm dia.	125.76	159.17	1.80	62.16	nr	**221.32**
40 mm dia.	133.20	168.58	2.03	70.10	nr	**238.68**
50 mm dia.	135.35	171.29	2.26	78.04	nr	**249.34**
Flanged ends for steel pipework; aluminium and steel outer sleeves; up to 16 bar G at 120°C						
20 mm dia.	103.18	130.59	0.53	18.30	nr	**148.89**
25 mm dia.	110.54	139.90	0.64	22.10	nr	**162.00**
32 mm dia.	152.32	192.78	0.74	25.55	nr	**218.33**
40 mm dia.	164.60	208.31	0.82	28.32	nr	**236.63**
50 mm dia.	195.30	247.17	0.89	30.73	nr	**277.90**
Flanged ends for steel pipework; up to 16 bar G at 120°C						
65 mm dia.	180.55	228.51	1.10	37.99	nr	**266.50**
80 mm dia.	194.07	245.61	1.31	45.24	nr	**290.85**
100 mm dia.	222.30	281.34	1.78	61.47	nr	**342.81**
150 mm dia.	340.24	430.60	3.08	106.36	nr	**536.96**
Screwed ends for non-ferrous pipework; up to 6 bar G at 100°C						
20 mm dia.	116.61	147.59	0.72	24.86	nr	**172.45**
25 mm dia.	118.78	150.33	0.84	29.01	nr	**179.33**
32 mm dia.	125.76	159.17	1.02	35.22	nr	**194.39**
40 mm dia.	133.20	168.58	1.11	38.33	nr	**206.91**
50 mm dia.	135.35	171.29	1.18	40.75	nr	**212.04**
Flanged ends for steel, copper or non-ferrous pipework; up to 16 bar G at 120°C						
65 mm dia.	232.15	293.81	0.87	30.04	nr	**323.85**
80 mm dia.	239.51	303.12	0.95	32.81	nr	**335.93**
100 mm dia.	276.38	349.78	1.15	39.71	nr	**389.50**
150 mm dia.	453.22	573.60	1.36	46.96	nr	**620.56**

38 MECHANICAL/COOLING/HEATING SYSTEMS

Item	Net Price £	Material £	Labour hours	Labour £	Unit	Total rate £
LOW TEMPERATURE HOT WATER HEATING; PIPELINE: PIPELINE ANCILLARIES – cont						
Angular movement bellows expansion joints; stainless steel						
Flanged ends for steel pipework; up to 16 bar G at 120°C						
50 mm dia.	267.60	338.68	0.71	24.52	nr	**363.20**
65 mm dia.	324.77	411.03	0.83	28.66	nr	**439.69**
80 mm dia.	379.52	480.33	0.91	31.42	nr	**511.75**
100 mm dia.	489.95	620.08	0.97	33.50	nr	**653.58**
125 mm dia.	647.17	819.06	1.16	40.06	nr	**859.12**
150 mm dia.	769.54	973.94	1.18	40.75	nr	**1014.68**

38 MECHANICAL/COOLING/HEATING SYSTEMS

Item	Net Price £	Material £	Labour hours	Labour £	Unit	Total rate £
LOW TEMPERATURE HOT WATER HEATING; PIPELINE: VALVES						
Gate valves						
Bronze gate valve; non-rising stem; BS 5154, series B, PN 16; working pressure up to 16 from −10°C to 100°C; 7 bar for saturated steam; screwed ends to steel						
15 mm dia.	67.76	85.76	1.18	40.75	nr	**126.51**
20 mm dia.	92.56	117.14	1.24	42.82	nr	**159.96**
25 mm dia.	117.80	149.08	1.31	45.24	nr	**194.32**
32 mm dia.	160.46	203.07	1.43	49.38	nr	**252.46**
40 mm dia.	214.22	271.11	1.53	52.83	nr	**323.95**
50 mm dia.	295.57	374.07	1.63	56.29	nr	**430.36**
Bronze gate valve; non-rising stem; BS 5154, series B, PN 32; working pressure up to 14 bar for saturated steam, 32 bar from −10°C to 100°C; screwed ends to steel						
15 mm dia.	59.30	75.05	1.11	38.33	nr	**113.38**
20 mm dia.	76.92	97.35	1.28	44.20	nr	**141.55**
25 mm dia.	104.20	131.87	1.49	51.45	nr	**183.33**
32 mm dia.	143.92	182.14	1.88	64.92	nr	**247.07**
40 mm dia.	193.23	244.55	2.31	79.77	nr	**324.32**
50 mm dia.	271.51	343.62	2.80	96.69	nr	**440.31**
Bronze gate valve; non-rising stem; BS 5154, series B, PN 20; working pressure up to 9 bar for saturated steam, 20 bar from −10°C to 100°C; screwed ends to steel						
15 mm dia.	40.45	51.19	0.84	29.01	nr	**80.20**
20 mm dia.	57.46	72.73	1.01	34.88	nr	**107.60**
25 mm dia.	74.36	94.11	1.19	41.09	nr	**135.20**
32 mm dia.	106.11	134.29	1.38	47.65	nr	**181.94**
40 mm dia.	145.22	183.79	1.62	55.94	nr	**239.74**
50 mm dia.	207.52	262.64	1.94	66.99	nr	**329.64**
Bronze gate valve; non-rising stem; BS 5154, series B, PN 16; working pressure up to 7 bar for saturated steam, 16 bar from −10°C to 100°C; BS4504 flanged ends; bolted connections						
15 mm dia.	182.11	230.48	1.18	40.75	nr	**271.23**
20 mm dia.	200.82	254.15	1.24	42.82	nr	**296.98**
25 mm dia.	264.46	334.70	1.31	45.24	nr	**379.94**
32 mm dia.	331.79	419.92	1.43	49.38	nr	**469.30**
40 mm dia.	401.16	507.70	1.53	52.83	nr	**560.54**
50 mm dia.	554.59	701.88	1.63	56.29	nr	**758.17**
65 mm dia.	861.33	1090.09	1.71	59.05	nr	**1149.14**
80 mm dia.	1219.03	1542.81	1.88	64.92	nr	**1607.73**
100 mm dia.	2215.36	2803.75	2.03	70.10	nr	**2873.85**

38 MECHANICAL/COOLING/HEATING SYSTEMS

Item	Net Price £	Material £	Labour hours	Labour £	Unit	Total rate £
LOW TEMPERATURE HOT WATER HEATING; PIPELINE: VALVES – cont						
Gate valves – cont						
Cast iron gate valve; bronze trim; non-rising stem; BS 5150, PN6; working pressure 6 bar from −10°C to 120°C; BS4504 flanged ends; bolted connections						
50 mm dia.	360.70	456.50	1.85	63.89	nr	**520.38**
65 mm dia.	377.00	477.13	2.00	69.07	nr	**546.20**
80 mm dia.	435.70	551.42	2.27	78.39	nr	**629.81**
100 mm dia.	551.15	697.54	2.76	95.31	nr	**792.85**
125 mm dia.	788.89	998.41	6.05	208.92	nr	**1207.34**
150 mm dia.	907.78	1148.88	8.03	277.30	nr	**1426.18**
200 mm dia.	1729.09	2188.33	9.17	316.66	nr	**2504.99**
250 mm dia.	2660.55	3367.19	10.72	370.19	nr	**3737.38**
300 mm dia.	3155.60	3993.73	11.75	405.76	nr	**4399.49**
Cast iron gate valve; bronze trim; non-rising stem; BS 5150, PN10; working pressure up to 8.4 bar for saturated steam, 10 bar from −10° C to 120°C; BS4504 flanged ends; bolted connections						
50 mm dia.	407.95	516.30	1.85	63.89	nr	**580.19**
65 mm dia.	506.97	641.62	2.00	69.07	nr	**710.69**
80 mm dia.	564.45	714.36	2.27	78.39	nr	**792.75**
100 mm dia.	751.37	950.93	2.76	95.31	nr	**1046.24**
125 mm dia.	1050.49	1329.51	6.05	208.92	nr	**1538.43**
150 mm dia.	1239.42	1568.61	8.03	277.30	nr	**1845.90**
200 mm dia.	2277.17	2881.99	9.17	316.66	nr	**3198.65**
250 mm dia.	3424.68	4334.28	10.72	370.19	nr	**4704.47**
300 mm dia.	3853.77	4877.33	11.75	405.76	nr	**5283.09**
350 mm dia.	4482.37	5672.89	12.67	437.53	nr	**6110.41**
Cast iron gate valve; bronze trim; non-rising stem; BS 5163 series A, PN16; working pressure for cold water services up to 16 bar; BS4504 flanged ends; bolted connections						
50 mm dia.	407.95	516.30	1.85	63.89	nr	**580.19**
65 mm dia.	506.97	641.62	2.00	69.07	nr	**710.69**
80 mm dia.	564.45	714.36	2.27	78.39	nr	**792.75**
100 mm dia.	751.37	950.93	2.76	95.31	nr	**1046.24**
125 mm dia.	1050.49	1329.51	6.05	208.92	nr	**1538.43**
150 mm dia.	1239.42	1568.61	8.03	277.30	nr	**1845.90**
Ball valves						
Malleable iron body; lever operated stainless steel ball and stem; working pressure up to 12 bar; flanged ends to BS 4504 16/11; bolted connections						
40 mm dia.	511.60	647.48	1.54	53.18	nr	**700.66**
50 mm dia.	644.67	815.89	1.64	56.63	nr	**872.52**
80 mm dia.	1042.81	1319.78	1.92	66.30	nr	**1386.08**
100 mm dia.	1078.15	1364.51	2.80	96.69	nr	**1461.20**
150 mm dia.	1992.94	2522.26	12.05	416.12	nr	**2938.38**

38 MECHANICAL/COOLING/HEATING SYSTEMS

Item	Net Price £	Material £	Labour hours	Labour £	Unit	Total rate £
Malleable iron body; lever operated stainless steel ball and stem; working pressure up to 16 bar; screwed ends to steel						
15 mm dia.	45.92	58.12	1.34	46.29	nr	**104.41**
20 mm dia.	60.75	76.89	1.34	46.29	nr	**123.18**
25 mm dia.	62.49	79.09	1.40	48.37	nr	**127.46**
32 mm dia.	86.09	108.96	1.46	50.49	nr	**159.45**
40 mm dia.	86.09	108.96	1.54	53.21	nr	**162.17**
50 mm dia.	102.96	130.30	1.64	56.70	nr	**187.00**
Carbon steel body; lever operated stainless steel ball and stem; Class 150; working pressure up to 19 bar; screwed ends to steel						
15 mm dia.	103.30	130.74	0.84	29.02	nr	**159.76**
20 mm dia.	116.12	146.96	1.14	39.38	nr	**186.34**
25 mm dia.	125.72	159.11	1.30	44.91	nr	**204.02**
32 mm dia.	162.13	205.19	1.42	49.04	nr	**254.23**
40 mm dia.	202.65	256.48	1.56	53.87	nr	**310.35**
50 mm dia.	253.33	320.62	1.68	58.01	nr	**378.63**
Globe valves						
Bronze; rising stem; renewable disc; BS 5154 series B, PN32; working pressure up to 14 bar for saturated steam, 32 bar from −10°C to 100°C; screwed ends to steel						
15 mm dia.	64.37	81.47	0.77	26.59	nr	**108.06**
20 mm dia.	88.03	111.42	1.03	35.57	nr	**146.98**
25 mm dia.	126.31	159.85	1.19	41.09	nr	**200.95**
32 mm dia.	177.82	225.05	1.38	47.65	nr	**272.70**
40 mm dia.	247.62	313.39	1.62	55.94	nr	**369.34**
50 mm dia.	380.09	481.04	1.61	55.60	nr	**536.64**
Bronze; needle valve; rising stem; BS 5154, series B, PN32; working pressure up to 14 bar for saturated steam, 32 bar from −10°C to 100°C; screwed ends to steel						
15 mm dia.	48.95	61.95	1.07	36.95	nr	**98.90**
20 mm dia.	82.86	104.87	1.18	40.75	nr	**145.62**
25 mm dia.	116.83	147.86	1.27	43.86	nr	**191.72**
32 mm dia.	243.18	307.76	1.35	46.62	nr	**354.38**
40 mm dia.	383.10	484.86	1.47	50.76	nr	**535.62**
50 mm dia.	485.25	614.13	1.61	55.60	nr	**669.72**
Bronze; rising stem; renewable disc; BS 5154, series B, PN16; working pressure up to 7 bar for saturated steam, 16 bar from −10°C to 100°C; BS4504 flanged ends; bolted connections						
15 mm dia.	130.08	164.63	1.16	40.06	nr	**204.69**
20 mm dia.	150.47	190.43	1.26	43.51	nr	**233.95**
25 mm dia.	263.85	333.93	1.38	47.65	nr	**381.59**
32 mm dia.	333.08	421.54	1.47	50.76	nr	**472.31**
40 mm dia.	378.60	479.16	1.56	53.87	nr	**533.03**
50 mm dia.	543.19	687.47	1.71	59.05	nr	**746.52**

38 MECHANICAL/COOLING/HEATING SYSTEMS

Item	Net Price £	Material £	Labour hours	Labour £	Unit	Total rate £
LOW TEMPERATURE HOT WATER HEATING; PIPELINE: VALVES – cont						
Globe valves – cont						
Bronze; rising stem; renewable disc; BS 2060, class 250; working pressure up to 24 bar for saturated steam, 38 bar from −10°C to 100°C; flanged ends (BS 10 table H); bolted connections						
15 mm dia.	331.28	419.26	1.16	40.06	nr	459.32
20 mm dia.	385.01	487.27	1.26	43.51	nr	530.78
25 mm dia.	529.07	669.59	1.38	47.65	nr	717.25
32 mm dia.	691.16	874.74	1.47	50.76	nr	925.50
40 mm dia.	827.66	1047.48	1.56	53.87	nr	1101.35
50 mm dia.	1286.97	1628.80	1.71	59.05	nr	1687.85
65 mm dia.	1539.10	1947.88	1.88	64.92	nr	2012.81
80 mm dia.	1621.89	2052.66	2.03	70.10	nr	2122.76
Check valves						
Bronze; swing pattern; BS 5154 series B, PN 25; working pressure up to 10.5 bar for saturated steam, 25 bar from −10°C to 100°C; screwed ends to steel						
15 mm dia.	27.51	34.81	0.77	26.59	nr	61.40
20 mm dia.	32.68	41.36	1.03	35.57	nr	76.93
25 mm dia.	45.29	57.32	1.19	41.09	nr	98.41
32 mm dia.	76.74	97.13	1.38	47.65	nr	144.78
40 mm dia.	95.46	120.82	1.62	55.94	nr	176.76
50 mm dia.	146.42	185.31	1.94	66.99	nr	252.31
65 mm dia.	265.54	336.07	2.45	84.60	nr	420.68
80 mm dia.	375.48	475.20	2.83	97.73	nr	572.93
Bronze; vertical lift pattern; BS 5154 series B, PN32; working pressure up to 14 bar for saturated steam, 32 bar from −10°C to 100°C; screwed ends to steel						
15 mm dia.	35.85	45.37	0.96	33.15	nr	78.52
20 mm dia.	50.98	64.52	1.07	36.95	nr	101.47
25 mm dia.	74.86	94.74	1.17	40.40	nr	135.14
32 mm dia.	114.03	144.32	1.33	45.93	nr	190.24
40 mm dia.	148.58	188.05	1.41	48.69	nr	236.74
50 mm dia.	222.61	281.73	1.55	53.53	nr	335.26
65 mm dia.	579.77	733.76	1.80	62.16	nr	795.92
80 mm dia.	866.76	1096.97	1.99	68.72	nr	1165.69
Bronze; oblique swing pattern; BS 5154 series A, PN32; working pressure up to 14 bar for saturated steam, 32 bar from −10°C to 120°C; screwed connections to steel						
15 mm dia.	45.54	57.63	0.96	33.15	nr	90.79
20 mm dia.	54.09	68.46	1.07	36.95	nr	105.41
25 mm dia.	74.99	94.91	1.17	40.40	nr	135.31

38 MECHANICAL/COOLING/HEATING SYSTEMS

Item	Net Price £	Material £	Labour hours	Labour £	Unit	Total rate £
32 mm dia.	112.14	141.93	1.33	45.93	nr	**187.86**
40 mm dia.	144.54	182.93	1.41	48.69	nr	**231.62**
50 mm dia.	239.45	303.05	1.55	53.53	nr	**356.58**
Cast iron; swing pattern; BS 5153 PN6; working pressure up to 6 bar from −10°C to 120°C; BS 4504 flanged ends; bolted connections						
50 mm dia.	864.77	1094.45	1.86	64.23	nr	**1158.68**
65 mm dia.	946.81	1198.28	2.00	69.07	nr	**1267.34**
80 mm dia.	1065.72	1348.78	2.56	88.40	nr	**1437.18**
100 mm dia.	1365.23	1727.84	2.76	95.31	nr	**1823.15**
125 mm dia.	2036.73	2577.69	6.05	208.92	nr	**2786.61**
150 mm dia.	2290.05	2898.28	8.11	280.06	nr	**3178.34**
200 mm dia.	5064.40	6409.50	9.26	319.77	nr	**6729.27**
250 mm dia.	6330.34	8011.68	10.72	370.19	nr	**8381.86**
300 mm dia.	7596.42	9614.02	11.75	405.76	nr	**10019.78**
Cast iron; horizontal lift pattern; BS 5153 PN16; working pressure up to 13 bar for saturated steam, 16 bar from −10°C to 120°C; BS 4504 flanged ends; bolted connections						
50 mm dia.	780.90	988.31	1.86	64.23	nr	**1052.54**
65 mm dia.	780.90	988.31	2.00	69.07	nr	**1057.38**
80 mm dia.	966.83	1223.62	2.56	88.40	nr	**1312.02**
100 mm dia.	1125.91	1424.95	2.96	102.22	nr	**1527.16**
125 mm dia.	1679.84	2126.00	7.76	267.97	nr	**2393.97**
150 mm dia.	1888.63	2390.25	10.50	362.59	nr	**2752.84**
Cast iron; semi-lugged butterfly valve; BS5155 PN16; working pressure 16 bar from −10°C to 120°C; BS 4504 flanged ends; bolted connections						
50 mm dia.	239.03	302.51	2.20	75.97	nr	**378.48**
65 mm dia.	246.67	312.19	2.31	79.77	nr	**391.96**
80 mm dia.	290.57	367.74	2.88	99.45	nr	**467.19**
100 mm dia.	403.93	511.21	3.11	107.40	nr	**618.60**
125 mm dia.	589.54	746.13	5.02	173.35	nr	**919.48**
150 mm dia.	676.70	856.43	6.98	241.04	nr	**1097.47**
200 mm dia.	1080.24	1367.15	8.25	284.89	nr	**1652.04**
250 mm dia.	1594.39	2017.87	10.47	361.56	nr	**2379.42**
300 mm dia.	2391.65	3026.88	11.48	396.43	nr	**3423.31**
Cast iron; semi-lugged butterfly valve; BS5155 PN20; working pressure up to 20 bar from −10°C to 120°C; BS 4504 flanged ends; bolted connections						
50 mm dia.	2140.52	2709.05	2.20	75.97	nr	**2785.02**
65 mm dia.	2214.61	2802.81	2.31	79.77	nr	**2882.58**
80 mm dia.	2364.35	2992.32	2.88	99.45	nr	**3091.78**
100 mm dia.	2879.16	3643.86	3.11	107.40	nr	**3751.26**
125 mm dia.	3695.56	4677.10	5.02	173.35	nr	**4850.45**
150 mm dia.	4194.04	5307.98	5.02	173.35	nr	**5481.33**
200 mm dia.	4823.73	6104.92	8.25	284.89	nr	**6389.81**
250 mm dia.	7606.18	9626.38	10.47	361.56	nr	**9987.94**
300 mm dia.	9216.38	11664.24	11.48	396.43	nr	**12060.68**

38 MECHANICAL/COOLING/HEATING SYSTEMS

Item	Net Price £	Material £	Labour hours	Labour £	Unit	Total rate £
LOW TEMPERATURE HOT WATER HEATING; PIPELINE: VALVES – cont						
Check valves – cont						
Cast iron; semi-lugged butterfly valve; BS5155 PN20; working pressure up to 30 bar from −10°C to 120°C; BS 4504 flanged ends; bolted connections						
50 mm dia.	2140.52	2709.05	2.20	75.97	nr	2785.02
65 mm dia.	2214.61	2802.81	2.31	79.77	nr	2882.58
80 mm dia.	2364.35	2992.32	2.88	99.45	nr	3091.78
100 mm dia.	2879.16	3643.86	3.11	107.40	nr	3751.26
125 mm dia.	3695.56	4677.10	5.02	173.35	nr	4850.45
150 mm dia.	4194.04	5307.98	5.02	173.35	nr	5481.33
200 mm dia.	4823.73	6104.92	8.25	284.89	nr	6389.81
250 mm dia.	7606.18	9626.38	10.47	361.56	nr	9987.94
300 mm dia.	9216.38	11664.24	11.48	396.43	nr	12060.68
Commissioning valves						
Bronze commissioning set; metering station; double regulating valve; BS5154 PN20 Series B; working pressure 20 bar from −10°C to 100°C; screwed ends to steel						
15 mm dia.	118.96	150.55	1.08	37.30	nr	187.85
20 mm dia.	191.14	241.91	1.46	50.42	nr	292.33
25 mm dia.	227.29	287.66	1.68	58.01	nr	345.67
32 mm dia.	304.74	385.68	1.95	67.34	nr	453.02
40 mm dia.	441.95	559.34	2.27	78.39	nr	637.73
50 mm dia.	667.00	844.15	2.73	94.27	nr	938.42
Cast iron commissioning set; metering station; double regulating valve; BS5152 PN16; working pressure 16 bar from −10°C to 90°C; flanged ends (BS 4504, Part 1, Table 16); bolted connections						
65 mm dia.	1010.50	1278.89	1.80	62.16	nr	1341.05
80 mm dia.	1209.93	1531.29	2.56	88.40	nr	1619.70
100 mm dia.	1684.46	2131.85	2.30	79.42	nr	2211.27
125 mm dia.	2467.29	3122.61	2.44	84.26	nr	3206.87
150 mm dia.	3250.19	4113.45	2.90	100.14	nr	4213.59
200 mm dia.	8319.18	10528.75	8.26	285.24	nr	10813.99
250 mm dia.	10407.20	13171.35	10.49	362.25	nr	13533.60
300 mm dia.	11302.43	14304.35	11.49	396.78	nr	14701.13
Cast iron variable orifice double regulating valve; orifice valve; BS5152 PN16; working pressure 16 bar from −10° to 90°C; flanged ends (BS 4504, Part 1, Table 16); bolted connections						
65 mm dia.	739.84	936.34	2.00	69.07	nr	1005.41
80 mm dia.	905.55	1146.07	2.56	88.40	nr	1234.47
100 mm dia.	1242.70	1572.76	2.96	102.22	nr	1674.98

38 MECHANICAL/COOLING/HEATING SYSTEMS

Item	Net Price £	Material £	Labour hours	Labour £	Unit	Total rate £
125 mm dia.	1860.94	2355.21	7.76	267.97	nr	**2623.18**
150 mm dia.	2390.33	3025.20	10.50	362.59	nr	**3387.79**
200 mm dia.	6264.83	7928.77	8.26	285.24	nr	**8214.01**
250 mm dia.	9525.63	12055.63	10.49	362.25	nr	**12417.88**
300 mm dia.	17087.07	21625.39	11.49	396.78	nr	**22022.17**
Cast iron globe valve with double regulating feature; BS5152 PN16; working pressure 16 bar from −10°C to 120°C; flanged ends (BS 4504, Part 1, Table 16); bolted connections						
65 mm dia.	881.33	1115.41	2.00	69.07	nr	**1184.47**
80 mm dia.	1065.81	1348.88	2.56	88.40	nr	**1437.29**
100 mm dia.	1403.97	1776.87	2.96	102.22	nr	**1879.08**
125 mm dia.	1957.51	2477.42	7.76	267.97	nr	**2745.40**
150 mm dia.	2477.51	3135.54	10.50	362.59	nr	**3498.13**
200 mm dia.	6414.00	8117.56	8.26	285.24	nr	**8402.80**
250 mm dia.	9704.25	12281.70	10.49	362.25	nr	**12643.94**
300 mm dia.	15246.79	19296.34	11.49	396.78	nr	**19693.12**
Bronze autoflow commissioning valve; PN25; working pressure 25 bar up to 100°C; screwed ends to steel						
15 mm dia.	134.89	170.72	0.82	28.32	nr	**199.04**
20 mm dia.	142.82	180.76	1.08	37.30	nr	**218.05**
25 mm dia.	177.84	225.07	1.27	43.86	nr	**268.93**
32 mm dia.	277.76	351.53	1.50	51.80	nr	**403.33**
40 mm dia.	304.13	384.91	1.76	60.78	nr	**445.69**
50 mm dia.	449.69	569.12	2.13	73.55	nr	**642.68**
Ductile iron autoflow commissioning valves; PN16; working pressure 16 bar from −10°C to 120°C; for ANSI 150 flanged ends						
65 mm dia.	1184.22	1498.74	2.31	79.77	nr	**1578.51**
80 mm dia.	1309.28	1657.02	2.88	99.45	nr	**1756.48**
100 mm dia.	1672.59	2116.83	3.11	107.40	nr	**2224.23**
150 mm dia.	3445.55	4360.68	6.98	241.04	nr	**4601.72**
200 mm dia.	5018.73	6351.70	8.26	285.24	nr	**6636.94**
250 mm dia.	6970.29	8821.60	10.49	362.25	nr	**9183.85**
300 mm dia.	8902.13	11266.54	11.49	396.78	nr	**11663.32**
Strainers						
Bronze strainer; Y type; PN32; working pressure 32 bar from −10°C to 100°C; screwed ends to steel						
15 mm dia.	57.33	72.56	0.82	28.32	nr	**100.87**
20 mm dia.	72.81	92.14	1.08	37.30	nr	**129.44**
25 mm dia.	96.60	122.26	1.27	43.86	nr	**166.12**
32 mm dia.	156.02	197.46	1.50	51.80	nr	**249.26**
40 mm dia.	201.80	255.40	1.76	60.78	nr	**316.17**
50 mm dia.	337.97	427.74	2.13	73.55	nr	**501.29**

38 MECHANICAL/COOLING/HEATING SYSTEMS

Item	Net Price £	Material £	Labour hours	Labour £	Unit	Total rate £
LOW TEMPERATURE HOT WATER HEATING; PIPELINE: VALVES – cont						
Strainers – cont						
Cast iron strainer; Y type; PN16; working pressure 16 bar from −10°C to 120°C; BS 4504 flanged ends						
65 mm dia.	356.56	451.26	2.31	79.77	nr	531.03
80 mm dia.	422.02	534.11	2.88	99.45	nr	633.56
100 mm dia.	617.51	781.53	3.11	107.40	nr	888.92
125 mm dia.	1138.35	1440.69	5.02	173.35	nr	1614.05
150 mm dia.	1460.52	1848.43	6.98	241.04	nr	2089.47
200 mm dia.	2451.41	3102.50	8.26	285.24	nr	3387.74
250 mm dia.	3544.86	4486.38	10.49	362.25	nr	4848.62
300 mm dia.	5772.73	7305.97	11.49	396.78	nr	7702.75
Regulators						
Gunmetal; self-acting two port thermostatic regulator; single seat; screwed ends; complete with sensing element, 2 m long capillary tube						
15 mm dia.	1010.69	1279.12	1.37	47.31	nr	1326.43
20 mm dia.	1043.21	1320.28	1.24	42.82	nr	1363.10
25 mm dia.	1070.58	1354.93	1.34	46.27	nr	1401.20
Gunmetal; self-acting two port thermostatic regulator; double seat; flanged ends (BS 4504 PN25); with sensing element, 2 m long capillary tube; steel body						
65 mm dia.	3748.21	4743.74	1.23	48.71	nr	4792.45
80 mm dia.	4423.93	5598.92	1.62	64.16	nr	5663.08
Control Valves						
Electrically operated (electrical work elsewhere)						
Pressure independent control valve (PICV)						
Brass; pressure independent control valve; rotary type; PN 25; maximum working pressure 25 bar at from −10°C to 120°C; screwed ends to steel; electrical work elsewhere						
15 mm dia.	116.03	146.84	1.08	37.30	nr	184.14
20 mm dia.	138.91	175.81	1.18	40.75	nr	216.56
25 mm dia.	204.56	258.89	1.35	46.62	nr	305.51
32 mm dia.	297.69	376.75	1.46	50.42	nr	427.17
40 mm dia.	329.74	417.31	1.53	52.83	nr	470.15
50 mm dia.	575.49	728.34	1.61	55.60	nr	783.94

38 MECHANICAL/COOLING/HEATING SYSTEMS

Item	Net Price £	Material £	Labour hours	Labour £	Unit	Total rate £
Brass; pressure independent control valve; rotary type with electronic actuator; 24 V motor; PN 25; maximum working pressure 25 bar at from −10°C to 120°C; screwed ends to steel; electrical work elsewhere						
15 mm dia.	242.73	307.20	2.06	71.14	nr	**378.33**
20 mm dia.	265.62	336.16	2.15	74.24	nr	**410.41**
25 mm dia.	331.26	419.25	2.27	78.39	nr	**497.63**
32 mm dia.	467.12	591.19	2.35	81.15	nr	**672.34**
40 mm dia.	499.17	631.75	2.47	85.30	nr	**717.05**
50 mm dia.	761.73	964.05	2.55	88.06	nr	**1052.11**
Differential pressure control valves (DPCV)						
Brass; differential control valve; PN 25; maximum working pressure 25 bar at from −20°C to 120°C; screwed ends to steel						
15 mm dia.	319.06	403.80	1.99	68.72	nr	**472.52**
20 mm dia.	343.58	434.84	2.03	70.10	nr	**504.94**
25 mm dia.	381.79	483.19	2.11	72.86	nr	**556.05**
32 mm dia.	432.17	546.95	2.17	74.94	nr	**621.89**
40 m m dia.	744.91	942.76	2.26	78.04	nr	**1020.80**
50 mm dia.	894.43	1132.00	2.31	79.77	nr	**1211.77**
Cast iron; butterfly type; two position electrically controlled 240 V motor and linkage mechanism; for low pressure hot water; maximum pressure 6 bar at 120°C; flanged ends; electrical work elsewhere						
25 mm dia.	1151.22	1456.99	1.47	50.76	nr	**1507.75**
32 mm dia.	1189.54	1505.49	1.52	52.49	nr	**1557.98**
40 mm dia.	1662.20	2103.68	1.61	55.60	nr	**2159.27**
50 mm dia.	1709.42	2163.44	1.71	59.05	nr	**2222.49**
65 mm dia.	1752.23	2217.63	2.51	86.68	nr	**2304.30**
80 mm dia.	1820.75	2304.34	2.69	92.89	nr	**2397.23**
100 mm dia.	1910.74	2418.23	2.81	97.04	nr	**2515.27**
125 mm dia.	2116.37	2678.48	2.94	101.53	nr	**2780.01**
150 mm dia.	2293.43	2902.57	3.33	114.99	nr	**3017.56**
200 mm dia.	2844.63	3600.16	3.67	126.73	nr	**3726.90**
Cast iron; three way 240 V motorized; for low pressure hot water; maximum pressure 6 bar 120°C; flanged ends, drilled (BS 10, Table F)						
25 mm dia.	1286.69	1628.44	1.99	68.72	nr	**1697.15**
40 mm dia.	1338.72	1694.28	2.13	73.55	nr	**1767.84**
50 mm dia.	1355.93	1716.07	3.21	110.85	nr	**1826.92**
65 mm dia.	1488.71	1884.11	3.23	111.54	nr	**1995.65**
80 mm dia.	1668.12	2111.17	3.50	120.86	nr	**2232.04**

Material Costs/Prices for Measured Works

38 MECHANICAL/COOLING/HEATING SYSTEMS

Item	Net Price £	Material £	Labour hours	Labour £	Unit	Total rate £
LOW TEMPERATURE HOT WATER HEATING; PIPELINE: VALVES – cont						
2 port control valves						
Cast iron; 2 port; motorized control valve; normally open for heating applications; PN 16; screwed ends to steel						
15 mm dia.	52.27	66.15	1.24	42.82	nr	108.97
22 mm dia.	79.81	101.01	1.31	45.24	nr	146.24
28 mm dia.	95.44	120.80	1.37	47.31	nr	168.10
32 mm dia.	111.52	141.14	1.45	50.07	nr	191.21
40 mm dia.	139.40	176.42	1.52	52.49	nr	228.91
50 mm dia.	174.24	220.52	1.60	55.25	nr	275.77
Two port normally closed motorized valve; electric actuator; spring return; domestic usage						
22 mm dia.	179.96	227.76	1.18	40.75	nr	268.51
28 mm dia.	253.95	321.40	1.35	46.62	nr	368.02
Two port on/off motorized valve; electric actuator; spring return; domestic usage						
22 mm dia.	179.96	227.76	1.18	40.75	nr	268.51
3 port control valves						
Cast iron; 3 port; motorized control valve with temperatyre control unit; PN 25; screwed ends to steel						
22 mm dia.	81.32	102.92	2.00	69.07	nr	171.99
25 mm dia.	92.19	116.68	2.13	73.55	nr	190.23
32 mm dia.	118.03	149.38	2.19	75.63	nr	225.00
40 mm dia.	147.52	186.71	2.26	78.04	nr	264.75
Three port motorized valve; electric actuator; spring return; domestic usage						
22 mm dia.	273.38	345.99	1.18	40.75	nr	386.73
4 port control valves						
Cast iron; 4 port; fixed orifice double regulating and control valve; PN 16; screwed ends to steel						
22 mm dia.	72.07	91.21	2.89	99.80	nr	191.01
Safety and relief valves						
Bronze relief valve; spring type; side outlet; working pressure up to 20.7 bar at 120°C; screwed ends to steel						
15 mm dia.	219.66	278.00	0.26	8.98	nr	286.98
20 mm dia.	240.90	304.88	0.36	12.43	nr	317.31
Bronze relief valve; spring type; side outlet; working pressure up to 17.2 bar at 120°C; screwed ends to steel						
25 mm dia.	329.47	416.98	0.38	13.12	nr	430.10
32 mm dia.	446.39	564.96	0.48	16.58	nr	581.53

38 MECHANICAL/COOLING/HEATING SYSTEMS

Item	Net Price £	Material £	Labour hours	Labour £	Unit	Total rate £
Bronze relief valve; spring type; side outlet; working pressure up to 13.8 bar at 120°C; screwed ends to steel						
40 mm dia.	582.78	737.57	0.64	22.10	nr	**759.67**
50 mm dia.	759.90	961.73	0.76	26.24	nr	**987.98**
65 mm dia.	1406.99	1780.69	0.94	32.46	nr	**1813.15**
80 mm dia.	1845.95	2336.23	1.10	37.99	nr	**2374.22**
Cocks; screwed joints to steel						
Bronze gland cock; complete with malleable iron lever; working pressure up to 10 bar at 100°C; screwed ends to steel						
15 mm dia.	77.20	97.71	0.77	26.59	nr	**124.30**
20 mm dia.	109.82	138.98	1.03	35.57	nr	**174.55**
25 mm dia.	147.97	187.27	1.19	41.09	nr	**228.36**
32 mm dia.	223.59	282.98	1.38	47.65	nr	**330.63**
40 mm dia.	312.22	395.14	1.62	55.94	nr	**451.09**
50 mm dia.	476.49	603.05	1.94	66.99	nr	**670.04**
Bronze three-way plug cock; complete with malleable iron lever; working pressure up to 10 bar at 100°C; screwed ends to steel						
15 mm dia.	156.48	198.05	0.77	26.59	nr	**224.64**
20 mm dia.	181.13	229.24	1.03	35.57	nr	**264.81**
25 mm dia.	253.19	320.44	1.19	41.09	nr	**361.53**
32 mm dia.	359.10	454.47	1.38	47.65	nr	**502.13**
40 mm dia.	432.97	547.97	1.62	55.94	nr	**603.91**
Air vents; including regulating, adjusting and testing						
Automatic air vent; maximum pressure up to 7 bar at 93°C; screwed ends to steel						
15 mm dia.	16.07	20.34	0.80	27.63	nr	**47.97**
Automatic air vent; maximum pressure up to 7 bar at 93°C; lockhead isolating valve; screwed ends to steel						
15 mm dia.	16.07	20.34	0.83	28.66	nr	**49.00**
Automatic air vent; maximum pressure up to 17 bar at 200°C; flanged ends (BS10, Table H); bolted connections to counter flange (measured separately)						
15 mm dia.	578.62	732.30	0.83	28.66	nr	**760.96**
Radiator valves						
Bronze; wheelhead or lockshield; chromium plated finish; screwed joints to steel						
Straight						
15 mm dia.	22.35	28.28	0.59	20.37	nr	**48.66**
20 mm dia.	34.77	44.01	0.73	25.21	nr	**69.22**
25 mm dia.	43.64	55.23	0.85	29.35	nr	**84.58**

38 MECHANICAL/COOLING/HEATING SYSTEMS

Item	Net Price £	Material £	Labour hours	Labour £	Unit	Total rate £
LOW TEMPERATURE HOT WATER HEATING; PIPELINE: VALVES – cont						
Bronze – cont						
Angled						
15 mm dia.	75.61	95.69	0.59	20.37	nr	**116.06**
20 mm dia.	99.67	126.15	0.73	25.21	nr	**151.35**
25 mm dia.	128.16	162.20	0.85	29.35	nr	**191.55**
Bronze; wheelhead or lockshield; chromium plated finish; compression joints to copper						
Straight						
15 mm dia.	40.03	50.66	0.59	20.37	nr	**71.03**
20 mm dia.	50.44	63.84	0.73	25.21	nr	**89.04**
25 mm dia.	63.38	80.21	0.85	29.35	nr	**109.57**
Angled						
15 mm dia.	30.20	38.22	0.59	20.37	nr	**58.59**
20 mm dia.	31.70	40.11	0.73	25.21	nr	**65.32**
25 mm dia.	39.29	49.72	0.85	29.35	nr	**79.08**
Twin entry						
8 mm dia.	40.44	51.18	0.23	7.94	nr	**59.12**
10 mm dia.	45.89	58.07	0.23	7.94	nr	**66.02**
Bronze; thermostatic head; chromium plated finish; compression joints to copper						
Straight						
15 mm dia.	16.54	20.94	0.59	20.37	nr	**41.31**
20 mm dia.	24.27	30.71	0.73	25.21	nr	**55.92**
Angled						
15 mm dia.	15.39	19.47	0.59	20.37	nr	**39.85**
20 mm dia.	22.56	28.55	0.73	25.21	nr	**53.76**

38 MECHANICAL/COOLING/HEATING SYSTEMS

Item	Net Price £	Material £	Labour hours	Labour £	Unit	Total rate £
LOW TEMPERATURE HOT WATER HEATING; PIPELINE: EQUIPMENT						
GAUGES						
Thermometers and pressure gauges						
Dial thermometer; coated steel case and dial; glass window; brass pocket; BS 5235; pocket length 100 mm; screwed end						
Back/bottom entry						
100 mm dia. face	80.94	102.43	0.81	27.99	nr	**130.42**
150 mm dia. face	132.67	167.90	0.81	27.99	nr	**195.89**
Dial thermometer; coated steel case and dial; glass window; brass pocket; BS 5235; pocket length 100 mm; screwed end						
100 mm dia. face	119.99	151.86	0.81	27.99	nr	**179.85**
150 mm dia. face	134.38	170.07	0.81	27.99	nr	**198.05**
PRESSURIZATION UNITS						
LTHW pressurization unit complete with expansion vessel(s), interconnecting pipework and all necessary isolating and drain valves; includes placing in position; electrical work elsewhere. Selection based on a final working pressure of 4 bar, a 3 m static head and system operating temperatures of 82/71°C						
System volume						
2,400 litres	2628.41	3326.51	15.00	517.99	nr	**3844.50**
6,000–20,000 litres	4768.85	6035.45	22.00	759.72	nr	**6795.17**
25,000 litres	5433.30	6876.39	22.00	759.72	nr	**7636.10**
DIRT SEPARATORS						
Dirt separator; maximum operating temperature and pressure of 110°C and 10 bar; fitted with drain valve						
Bore size, flow rate (at 1.0 m/s velocity); threaded connections						
32 mm dia., 3.7 m³/h	116.27	147.15	2.29	79.08	nr	**226.23**
40 mm dia., 5.0 m³/h	138.42	175.18	2.45	84.60	nr	**259.78**
Bore size, flow rate (at 1.5 m/s velocity); flanged connections to PN16						
50 mm dia., 13.0 m³/h	782.52	990.36	3.00	103.60	nr	**1093.95**
65 mm dia., 21.0 m³/h	815.75	1032.42	3.00	103.60	nr	**1136.02**
80 mm dia., 29.0 m³/h	1147.94	1452.84	3.84	132.60	nr	**1585.44**
100 mm dia., 49.0 m³/h	1214.38	1536.92	4.44	153.32	nr	**1690.25**
125 mm dia., 74.0 m³/h	2327.26	2945.38	11.64	401.96	nr	**3347.34**
150 mm dia., 109.0 m³/h	2426.92	3071.50	15.75	543.89	nr	**3615.39**

38 MECHANICAL/COOLING/HEATING SYSTEMS

Item	Net Price £	Material £	Labour hours	Labour £	Unit	Total rate £
LOW TEMPERATURE HOT WATER HEATING; PIPELINE: EQUIPMENT – cont						
Dirt separator – cont						
Bore size, flow rater – cont						
200 mm dia., 181.0 m³/h	3657.89	4629.43	15.75	543.89	nr	**5173.32**
250 mm dia., 288.0 m³/h	6483.47	8205.48	15.75	543.89	nr	**8749.37**
300 mm dia., 407.0 m³/h	9476.95	11994.02	17.24	595.34	nr	**12589.36**
MICROBUBBLE DEAERATORS						
Microbubble deaerator; maximum operating temperature and pressure of 110°C and 10 bar; fitted with drain valve						
Bore size, flow rate (at 1.0 m/s velocity); threaded connections						
32 mm dia., 3.7 m³/h	107.04	135.47	2.29	79.08	nr	**214.55**
40 mm dia., 5.0 m³/h	127.35	161.18	2.45	84.60	nr	**245.78**
Bore size, flow rate (at 1.5 m/s velocity); flanged connections to PN16						
50 mm dia., 13.0 m³/h	782.52	990.36	3.00	103.60	nr	**1093.95**
65 mm dia., 21.0 m³/h	814.19	1030.43	3.00	103.60	nr	**1134.03**
80 mm dia., 29.0 m³/h	1147.94	1452.84	3.84	132.60	nr	**1585.44**
100 mm dia., 49.0 m³/h	1214.38	1536.92	4.44	153.32	nr	**1690.25**
125 mm dia., 74.0 m³/h	2327.26	2945.38	11.64	401.96	nr	**3347.34**
150 mm dia., 109.0 m³/h	2426.92	3071.50	15.75	543.89	nr	**3615.39**
200 mm dia., 181.0 m³/h	3657.89	4629.43	15.75	543.89	nr	**5173.32**
250 mm dia., 288.0 m³/h	6483.47	8205.48	15.75	543.89	nr	**8749.37**
300 mm dia., 407.0 m³/h	9510.17	12036.07	17.24	595.34	nr	**12631.41**
COMBINED MICROBUBBLE DEAERATORS AND DIRT SEPARATORS						
Combined deaerator and dirt separators; maximum operating temperature and pressure of 110°C and 10 bar; fitted with drain valve						
Bore size, flow rate (at 1.5 m/s velocity); threaded connections						
25 mm dia., 2.0 m³/h	147.65	186.86	2.75	94.96	nr	**281.83**
Bore size, flow rate (at 1.5 m/s velocity); flanged connections to PN16						
50 mm dia., 13.0 m³/h	998.45	1263.63	3.60	124.32	nr	**1387.95**
65 mm dia., 21.0 m³/h	1031.67	1305.68	3.60	124.32	nr	**1430.00**
80 mm dia., 29.0 m³/h	1397.09	1768.16	4.61	159.20	nr	**1927.36**
100 mm dia., 49.0 m³/h	1463.92	1852.74	5.33	184.06	nr	**2036.80**
125 mm dia., 74.0 m³/h	2792.34	3533.98	13.97	482.42	nr	**4016.40**
150 mm dia., 109.0 m³/h	2893.84	3662.45	18.90	652.66	nr	**4315.11**
200 mm dia., 181.0 m³/h	4488.41	5680.53	18.90	652.66	nr	**6333.19**
250 mm dia., 288.0 m³/h	7913.78	10015.68	18.90	652.66	nr	**10668.34**
300 mm dia., 407.0 m³/h	12136.41	15359.83	20.69	714.48	nr	**16074.31**

38 MECHANICAL/COOLING/HEATING SYSTEMS

Item	Net Price £	Material £	Labour hours	Labour £	Unit	Total rate £
LOW TEMPERATURE HOT WATER HEATING; PIPELINE: PUMPS						
Centrifugal heating and chilled water pump; belt drive; 3 phase, 1450 rpm motor; max. pressure 1000 kN/m²; max. temperature 125°C; bed plate; coupling guard; bolted connections; supply only mating flanges;includes fixing on prepared concrete base; electrical work elsewhere						
40 mm pump size; 4.0 l/s at 70 kPa max head; 0.25 kW max motor rating	2048.74	2592.89	7.59	262.10	nr	**2854.99**
40 mm pump size; 4.0 l/s at 130 kPa max head; 1.5 kW max motor rating	2563.98	3244.97	8.09	279.37	nr	**3524.34**
50 mm pump size; 8.5 l/s at 90 kPa max head; 2.2 kW max motor rating	2644.17	3346.47	8.67	299.40	nr	**3645.86**
50 mm pump size; 8.5 l/s at 190 kPa max head; 3 kW max motor rating	4595.70	5816.31	11.20	386.76	nr	**6203.08**
50 mm pump size; 8.5 l/s at 215 kPa max head; 4 kW max motor rating	5083.57	6433.77	11.70	404.03	nr	**6837.80**
65 mm pump size; 14.0 l/s at 90 kPa max head; 3 kW max motor rating	2634.47	3334.19	11.70	404.03	nr	**3738.22**
65 mm pump size; 14.0 l/s at 160 kPa max head; 4 kW max motor rating	2921.23	3697.10	11.70	404.03	nr	**4101.13**
80 mm pump size; 14.5 l/s at 210 kPa max head; 5.5 kW max motor rating	4481.49	5671.77	11.70	404.03	nr	**6075.80**
80 mm pump size; 22.0 l/s at 130 kPa max head; 5.5 kW max motor rating	4481.49	5671.77	13.64	471.02	nr	**6142.79**
80 mm pump size; 22.0 l/s at 200 kPa max head; 7.5 kW max motor rating	4692.92	5939.36	13.64	471.02	nr	**6410.38**
100 mm pump size; 22.0 l/s at 250 kPa max head; 11 kW max motor rating	6586.10	8335.37	13.64	471.02	nr	**8806.40**
100 mm pump size; 30.0 l/s at 100 kPa max head; 4.0 kW max motor rating	3686.78	4665.98	19.15	661.30	nr	**5327.28**
100 mm pump size; 36.0 l/s at 250 kPa max head; 15.0 kW max motor rating	6975.00	8827.56	19.15	661.30	nr	**9488.86**
100 mm pump size; 36.0 l/s at 550 kPa max head; 30.0 kW max motor rating	9466.03	11980.21	19.15	661.30	nr	**12641.51**

38 MECHANICAL/COOLING/HEATING SYSTEMS

Item	Net Price £	Material £	Labour hours	Labour £	Unit	Total rate £
LOW TEMPERATURE HOT WATER HEATING; PIPELINE: PUMPS – cont						
Centrifugal heating and chilled water pump; twin head belt drive; 3 phase, 1450 rpm motor; max. pressure 1000 kN/m²; max. temperature 125°C; bed plate; coupling guard; bolted connections; supply only mating flanges; includes fixing on prepared concrete base; electrical work elsewhere						
40 mm pump size; 4.0 l/s at 70 kPa max head; 0.75 kW max motor rating	4389.14	5554.89	7.59	262.10	nr	**5817.00**
40 mm pump size; 4.0 l/s at 130 kPa max head; 1.5 kW max motor rating	5283.50	6686.79	8.09	279.37	nr	**6966.16**
50 mm pump size; 8.5 l/s at 90 kPa max head; 2.2 kW max motor rating	5397.73	6831.37	8.67	299.40	nr	**7130.76**
50 mm pump size; 8.5 l/s at 190 kPa max head; 4 kW max motor rating	6221.58	7874.03	11.20	386.76	nr	**8260.79**
65 mm pump size; 8.5 l/s at 215 kPa max head; 4 kW max motor rating	8549.75	10820.57	11.70	404.03	nr	**11224.60**
65 mm pump size; 14.0 l/s at 90 kPa max head; 3 kW max motor rating	8977.26	11361.61	11.70	404.03	nr	**11765.65**
65 mm pump size; 14.0 l/s at 160 kPa max head; 4 kW max motor rating	9426.13	11929.71	11.70	404.03	nr	**12333.74**
80 mm pump size; 14.5 l/s at 210 kPa max head; 7.5 kW max motor rating	9747.95	12337.00	13.64	471.02	nr	**12808.02**
Centrifugal heating and chilled water pump; close coupled; 3 phase, 1450 rpm motor; max. pressure 1000 kN/m²; max. temperature 110°C; bed plate; coupling guard; bolted connections; supply only mating flanges; includes fixing on prepared concrete base; electrical work elsewhere						
40 mm pump size; 4.0 l/s at 23 kPa max head; 0.55 kW max motor rating	1373.86	1738.76	7.31	252.43	nr	**1991.19**
50 mm pump size; 4.0 l/s at 75 kPa max head; 0.75 kW max motor rating	1522.89	1927.37	7.31	252.43	nr	**2179.80**
50 mm pump size; 7.0 l/s at 65 kPa max head; 0.75 kW max motor rating	1522.89	1927.37	8.01	276.61	nr	**2203.97**
65 mm pump size; 10.0 l/s at 33 kPa max head; 0.75 kW max motor rating	1726.17	2184.64	8.01	276.61	nr	**2461.25**
50 mm pump size; 4.0 l/s at 120 kPa max head; 1.5 kW max motor rating	2078.38	2630.40	8.01	276.61	nr	**2907.01**
80 mm pump size; 16.0 l/s at 80 kPa max head; 2.2 kW max motor rating	2690.81	3405.49	12.35	426.48	nr	**3831.97**
80 mm pump size; 16.0 l/s at 120 kPa max head; 4.0 kW max motor rating	2696.29	3412.43	12.35	426.48	nr	**3838.91**

38 MECHANICAL/COOLING/HEATING SYSTEMS

Item	Net Price £	Material £	Labour hours	Labour £	Unit	Total rate £
100 mm pump size; 28.0 l/s at 40 kPa max head; 2.2 kW max motor rating	2658.35	3364.41	17.86	616.75	nr	**3981.16**
100 mm pump size; 28.0 l/s at 90 kPa max head; 4.0 kW max motor rating	2934.68	3714.13	17.86	616.75	nr	**4330.88**
125 mm pump size; 40.0 l/s at 50 kPa max head; 3.0 kW max motor rating	2799.22	3542.69	25.85	892.67	nr	**4435.35**
125 mm pump size; 40.0 l/s at 120 kPa max head; 7.5 kW max motor rating	3454.99	4372.63	25.85	892.67	nr	**5265.30**
150 mm pump size; 70.0 l/s at 75 kPa max head; 11 kW max motor rating	5078.14	6426.89	30.43	1050.82	nr	**7477.72**
150 mm pump size; 70.0 l/s at 120 kPa max head; 15.0 kW max motor rating	5427.72	6869.32	30.43	1050.82	nr	**7920.15**
150 mm pump size; 70.0 l/s at 150 kPa max head; 15.0 kW max motor rating	5427.72	6869.32	30.43	1050.82	nr	**7920.15**
Centrifugal heating & chilled water pump; close coupled; 3 phase, variable speed motor; max. system pressure 1000 kN/m²; max. temperature 110°C; bed plate; coupling guard; bolted connections; supply only mating flanges; includes fixing on prepared concrete base; electrical work elsewhere.						
40 mm pump size; 4.0 l/s at 23 kPa max head; 0.55 kW max motor rating	2479.46	3138.01	7.31	252.43	nr	**3390.44**
40 mm pump size; 4.0 l/s at 75 kPa max head; 0.75 kW max motor rating	2736.89	3463.81	7.31	252.43	nr	**3716.25**
50 mm pump size; 7.0 l/s at 65 kPa max head; 1.5 kW max motor rating	3357.43	4249.16	8.01	276.61	nr	**4525.77**
50 mm pump size; 10.0 l/s at 33 kPa max head; 1.5 kW max motor rating	3357.43	4249.16	8.01	276.61	nr	**4525.77**
50 mm pump size; 4.0 l/s at 120 kPa max head; 1.5 kW max motor rating	3357.43	4249.16	8.01	276.61	nr	**4525.77**
80 mm pump size; 16.0 l/s at 80 kPa max head; 2.2 kW max motor rating	4376.32	5538.68	12.35	426.48	nr	**5965.15**
80 mm pump size; 16.0 l/s at 120 kPa max head; 3.0 kW max motor rating	4720.46	5974.21	12.35	426.48	nr	**6400.69**
100 mm pump size; 28.0 l/s at 40 kPa max head; 2.2 kW max motor rating	4484.69	5675.83	17.86	616.75	nr	**6292.58**
100 mm pump size; 28.0 l/s at 90 kPa max head; 4.0.kW max motor rating	5349.13	6769.86	17.86	616.75	nr	**7386.61**
125 mm pump size; 40.0 l/s at 50 kPa max head; 3.0 kW max motor rating	5134.96	6498.80	25.85	892.67	nr	**7391.47**
125 mm pump size; 40.0 l/s at 120 kPa max head; 7.5 kW max motor rating	6337.92	8021.27	25.85	892.67	nr	**8913.94**
150 mm pump size; 70.0 l/s at 75 kPa max head; 7.5 kW max motor rating	8354.29	10573.18	30.43	1050.82	nr	**11624.01**

38 MECHANICAL/COOLING/HEATING SYSTEMS

Item	Net Price £	Material £	Labour hours	Labour £	Unit	Total rate £
LOW TEMPERATURE HOT WATER HEATING; PIPELINE: PUMPS – cont						
Glandless domestic heating pump; for low pressure domestic hot water heating systems; 240 volt; 50 Hz electric motor; max working pressure 1000 N/m² and max temperature of 130°C; includes fixing in position; electrical work elsewhere						
1" BSP unions – 2 speed	197.86	250.41	1.58	54.56	nr	**304.97**
1.25" BSP unions – 3 speed	291.56	369.00	1.58	54.56	nr	**423.56**
Glandless pumps; for hot water secondary supply; silent running; 3 phase; max pressure 1000 kN/m²; max temperature 130°C; bolted connections; supply only mating flanges; including fixing in position; electrical elsewhere						
1" BSP unions – 3 speed	406.27	514.17	1.58	54.56	nr	**568.74**
Pipeline mounted circulator; for heating and chilled water; silent running; 3 phase; 1450 rpm motor; max pressure 1000 kN/m²; max temperature 120°C; bolted connections; supply only mating flanges; includes fixing in position; electrical elsewhere						
32 mm pump size; 2.0 l/s at 17 kPa max head; 0.2 kW max motor rating	812.56	1028.38	6.44	222.39	nr	**1250.77**
50 mm pump size; 3.0 l/s at 20 kPa max head; 0.2 kW max motor rating	812.56	1028.38	6.86	236.89	nr	**1265.27**
65 mm pump size; 5.0 l/s at 30 kPa max head; 0.37 kW max motor rating	1285.39	1626.79	7.48	258.30	nr	**1885.09**
65 mm pump size; 8.0 l/s at 37 kPa max head; 0.75 kW max motor rating	1439.18	1821.43	7.48	258.30	nr	**2079.73**
80 mm pump size; 12.0 l/s at 42 kPa max head; 1.1 kW max motor rating	1673.33	2117.76	8.01	276.61	nr	**2394.37**
100 mm pump size; 25.0 l/s at 37 kPa max head; 2.2 kW max motor rating	2384.88	3018.30	9.11	314.59	nr	**3332.89**

38 MECHANICAL/COOLING/HEATING SYSTEMS

Item	Net Price £	Material £	Labour hours	Labour £	Unit	Total rate £
Dual pipeline mounted circulator; for heating & chilled water; silent running; 3 phase; 1450 rpm motor; max pressure 1000 kN/m²; max temperature 120°C; bolted connections; supply only mating flanges; includes fixing in position; electrical work elsewhere						
40 mm pump size; 2.0 l/s at 17 kPa max head; 0.8 kW max motor rating	1498.88	1896.98	7.88	272.12	nr	**2169.10**
50 mm pump size; 3.0 l/s at 20 kPa max head; 0.2 kW max motor rating	1503.45	1902.77	8.01	276.61	nr	**2179.37**
65 mm pump size; 5.0 l/s at 30 kPa max head; 0.37 kW max motor rating	2439.96	3088.01	9.20	317.70	nr	**3405.71**
65 mm pump size; 8.0 l/s at 37 kPa max head; 0.75 kW max motor rating	2821.00	3570.25	9.20	317.70	nr	**3887.95**
100 mm pump size; 12.0 l/s at 42 kPa max head; 1.1 kW max motor rating	3190.53	4037.93	9.45	326.33	nr	**4364.26**
Glandless accelerator pumps; for low and medium pressure heating services; silent running; 3 phase; 1450 rpm motor; max pressure 1000 kN/m²; max temperature 130°C; bolted connections; supply only mating flanges; includes fixing in position; electrical work elsewhere						
40 mm pump size; 4.0 l/s at 15 kPa max head; 0.35 kW max motor rating	591.99	749.22	6.94	239.66	nr	**988.87**
50 mm pump size; 6.0 l/s at 20 kPa max head; 0.45 kW max motor rating	630.88	798.45	7.35	253.81	nr	**1052.26**
80 mm pump size; 13.0 l/s at 28 kPa max head; 0.58 kW max motor rating	1243.17	1573.35	7.76	267.97	nr	**1841.32**
Air source heat pumps Two stage heat pump (air to water) in monoblock design, cascadable for outdoor installation. Providing heating and cooling. Unit and controller only. Heat output & cooling capacity						
17 kW heating, 18 kW cooling	12273.26	15533.03	12.00	414.39	nr	**15947.43**
24 kW heating, 24 kW cooling	13949.60	17654.62	12.00	414.39	nr	**18069.01**
32 kW heating, 31 kW cooling	15612.23	19758.83	12.00	414.39	nr	**20173.22**
64 kW heating, 62 kW cooling	23120.07	29260.76	12.00	414.39	nr	**29675.15**

38 MECHANICAL/COOLING/HEATING SYSTEMS

Item	Net Price £	Material £	Labour hours	Labour £	Unit	Total rate £
LOW TEMPERATURE HOT WATER HEATING; PIPELINE: HEAT EXCHANGERS						
Plate heat exchanger; for use in LTHW systems; painted carbon steel frame; stainless steel plates, nitrile rubber gaskets; design pressure of 10 bar and operating temperature of 110/135°C						
Primary side; 80°C in, 69°C out; secondary side; 82°C in, 71°C out						
107 kW, 2.38 l/s	2479.40	3137.93	10.00	345.33	nr	**3483.25**
245 kW, 5.46 l/s	3740.95	4734.55	10.00	345.33	nr	**5079.87**
287 kW, 6.38 l/s	4025.00	5094.04	10.00	345.33	nr	**5439.37**
328 kW, 7.31 l/s	4574.70	5789.74	10.00	345.33	nr	**6135.07**
364 kW, 8.11 l/s	4717.30	5970.21	10.00	345.33	nr	**6315.54**
403 kW, 8.96 l/s	5071.50	6418.49	10.00	345.33	nr	**6763.82**
453 kW, 10.09 l/s	5425.70	6866.77	10.00	345.33	nr	**7212.09**
490 kW, 10.89 l/s	5965.05	7549.37	10.00	345.33	nr	**7894.69**
1000 kW, 21.7 l/s	9990.05	12643.41	12.00	414.39	nr	**13057.80**
1500 kW, 32.6 l/s	13073.20	16545.44	12.00	414.39	nr	**16959.83**
2000 kW, 43.4 l/s	15654.95	19812.90	15.00	517.99	nr	**20330.89**
2500 kW, 54.3 l/s	19746.65	24991.36	15.00	517.99	nr	**25509.35**
Note: For temperature conditions different to those above, the cost of the units can vary significantly, and so manufacturers advice should be sought.						
District heating HIUs						
District heating station; for conventional connection to a district or local heating network; offering primary F/R and secondary F/R connections, with stainless steel plate heat exchanger, integrated control panel, heat meter and combination valve to adjust output						
Heat output						
149 kW	6262.97	7926.41	12.00	414.39	nr	**8340.81**
194 kW	6993.63	8851.14	12.00	414.39	nr	**9265.53**
240 kW	7087.64	8970.12	12.00	414.39	nr	**9384.51**
400 kW	8425.20	10662.93	12.00	414.39	nr	**11077.32**
494 kW	10045.98	12714.19	12.00	414.39	nr	**13128.58**
572 kW	11570.37	14643.46	12.00	414.39	nr	**15057.85**
915 kW	14111.02	17858.91	14.00	483.46	nr	**18342.36**
1417 kW	18132.03	22947.90	14.00	483.46	nr	**23431.35**

38 MECHANICAL/COOLING/HEATING SYSTEMS

Item	Net Price £	Material £	Labour hours	Labour £	Unit	Total rate £
LOW TEMPERATURE HOT WATER HEATING; PIPELINE: CALORIFIERS						
Non-storage calorifiers; mild steel; heater battery duty 116°C/90°C to BS 853, maximum test on shell 11.55 bar, tubes 26.25 bar						
Horizontal or vertical; primary water at 116°C on, 90°C off						
40 kW capacity	1308.83	1656.46	3.00	103.60	nr	**1760.06**
88 kW capacity	1524.78	1929.76	5.00	172.66	nr	**2102.43**
176 kW capacity	1747.32	2211.41	7.04	243.19	nr	**2454.60**
293 kW capacity	2423.19	3066.79	9.01	311.10	nr	**3377.90**
586 kW capacity	3610.06	4568.89	22.22	767.39	nr	**5336.28**
879 kW capacity	5027.70	6363.05	28.57	986.64	nr	**7349.69**
1465 kW capacity	8077.28	10222.61	50.00	1726.63	nr	**11949.24**
2000 kW capacity	14701.71	18606.48	60.00	2071.95	nr	**20678.43**

38 MECHANICAL/COOLING/HEATING SYSTEMS

Item	Net Price £	Material £	Labour hours	Labour £	Unit	Total rate £
LOW TEMPERATURE HOT WATER HEATING; PIPELINE: HEAT EMITTERS						
Perimeter convector heating; metal casing with standard finish; aluminium extruded grille; including backplates						
Top/sloping/flat front outlet						
60 × 200 mm	40.44	51.18	2.00	69.07	m	**120.24**
60 × 300 mm	43.99	55.67	2.00	69.07	m	**124.74**
60 × 450 mm	54.53	69.01	2.00	69.07	m	**138.07**
60 × 525 mm	58.05	73.46	2.00	69.07	m	**142.53**
60 × 600 mm	63.32	80.13	2.00	69.07	m	**149.20**
90 × 260 mm	43.99	55.67	2.00	69.07	m	**124.74**
90 × 300 mm	45.73	57.87	2.00	69.07	m	**126.94**
90 × 450 mm	56.31	71.26	2.00	69.07	m	**140.33**
90 × 525 mm	61.57	77.92	2.00	69.07	m	**146.98**
90 × 600 mm	65.10	82.39	2.00	69.07	m	**151.45**
Extra over for dampers						
Damper	18.90	23.92	0.25	8.63	nr	**32.55**
Extra over for fittings						
60 mm end caps	15.13	19.14	0.25	8.63	nr	**27.78**
90 mm end caps	23.80	26.89	0.25	6.82	nr	**38.76**
60 mm corners	32.15	36.33	0.25	6.82	nr	**49.33**
90 mm corners	47.29	53.44	0.25	6.82	nr	**68.48**

38 MECHANICAL/COOLING/HEATING SYSTEMS

Item	Net Price £	Material £	Labour hours	Labour £	Unit	Total rate £
LOW TEMPERATURE HOT WATER HEATING; PIPELINE: RADIATORS						
Radiant strip heaters						
Suitable for connection to hot water system; aluminium sheet panels with steel pipe clamped to upper surface; including insulation, sliding brackets, cover plates, end closures; weld or screwed BSP ends						
One pipe						
1500 mm long	79.94	101.17	3.11	107.40	nr	**208.56**
3000 mm long	126.42	160.00	3.11	107.40	nr	**267.39**
4500 mm long	171.51	217.06	3.11	107.40	nr	**324.45**
6000 mm long	236.41	299.21	3.11	107.40	nr	**406.60**
Two pipe						
1500 mm long	149.23	188.87	4.15	143.31	nr	**332.18**
3000 mm long	235.72	298.32	4.15	143.31	nr	**441.63**
4500 mm long	321.89	407.38	4.15	143.31	nr	**550.69**
6000 mm long	433.66	548.83	4.15	143.31	nr	**692.14**
Pressed steel panel type radiators; fixed with and including brackets; taking down once for decoration; refixing						
300 mm high; single panel						
500 mm length	22.33	28.26	2.03	70.10	nr	**98.36**
1000 mm length	44.62	56.47	2.03	70.10	nr	**126.57**
1500 mm length	58.47	73.99	2.03	70.10	nr	**144.09**
2000 mm length	65.59	83.01	2.47	85.30	nr	**168.31**
2500 mm length	72.70	92.01	2.97	102.56	nr	**194.57**
3000 mm length	87.00	110.11	3.22	111.19	nr	**221.30**
300 mm high; double panel; convector						
500 mm length	42.92	54.32	2.13	73.55	nr	**127.88**
1000 mm length	85.90	108.71	2.13	73.55	nr	**182.27**
1500 mm length	128.83	163.05	2.13	73.55	nr	**236.60**
2000 mm length	171.81	217.44	2.57	88.75	nr	**306.19**
2500 mm length	214.72	271.75	3.07	106.01	nr	**377.76**
3000 mm length	257.69	326.13	3.31	114.30	nr	**440.43**
450 mm high; single panel						
500 mm length	20.82	26.35	2.08	71.83	nr	**98.18**
1000 mm length	41.69	52.76	2.08	71.83	nr	**124.59**
1600 mm length	66.69	84.41	2.53	87.37	nr	**171.77**
2000 mm length	83.39	105.54	2.97	102.56	nr	**208.10**
2400 mm length	100.03	126.60	3.47	119.83	nr	**246.43**
3000 mm length	125.04	158.25	3.82	131.91	nr	**290.16**
450 mm high; double panel; convector						
500 mm length	38.17	48.31	2.18	75.28	nr	**123.59**
1000 mm length	76.33	96.60	2.18	75.28	nr	**171.88**
1600 mm length	139.80	176.93	2.63	90.82	nr	**267.75**
2000 mm length	235.02	297.43	3.06	105.67	nr	**403.10**
2400 mm length	282.04	356.95	3.37	116.37	nr	**473.32**
3000 mm length	352.55	446.19	3.92	135.37	nr	**581.55**

38 MECHANICAL/COOLING/HEATING SYSTEMS

Item	Net Price £	Material £	Labour hours	Labour £	Unit	Total rate £
LOW TEMPERATURE HOT WATER HEATING; PIPELINE: RADIATORS – cont						
Pressed steel panel type radiators – cont						
600 mm high; single panel						
500 mm length	27.92	35.33	2.18	75.28	nr	110.61
1000 mm length	55.83	70.65	2.43	83.91	nr	154.57
1600 mm length	89.31	113.03	3.13	108.09	nr	221.12
2000 mm length	111.64	141.29	3.77	130.19	nr	271.48
2400 mm length	133.98	169.57	4.07	140.55	nr	310.11
3000 mm length	167.49	211.97	5.11	176.46	nr	388.43
600 mm high; double panel; convector						
500 mm length	48.06	60.82	2.28	78.73	nr	139.56
1000 mm length	96.12	121.65	2.28	78.73	nr	200.38
1600 mm length	176.04	222.80	3.23	111.54	nr	334.34
2000 mm length	295.96	374.56	3.87	133.64	nr	508.20
2400 mm length	355.15	449.47	4.17	144.00	nr	593.47
3000 mm length	443.93	561.83	5.24	180.95	nr	742.78
700 mm high; single panel						
500 mm length	32.65	41.32	2.23	77.01	nr	118.33
1000 mm length	65.30	82.64	2.83	97.73	nr	180.37
1600 mm length	104.49	132.24	3.73	128.81	nr	261.05
2000 mm length	130.63	165.32	4.46	154.02	nr	319.33
2400 mm length	156.77	198.41	4.48	154.71	nr	353.12
3000 mm length	195.91	247.94	5.24	180.95	nr	428.89
700 mm high; double panel; convector						
500 mm length	62.46	79.05	2.33	80.46	nr	159.51
1000 mm length	168.06	212.69	3.08	106.36	nr	319.05
1600 mm length	268.86	340.27	3.83	132.26	nr	472.53
2000 mm length	336.08	425.35	4.17	144.00	nr	569.35
2400 mm length	403.32	510.44	4.37	150.91	nr	661.34
3000 mm length	504.14	638.04	4.82	166.45	nr	804.49
Flat panel type steel radiators; fixed with and including brackets; taking down once for decoration; refixing						
300 mm high; single panel (44 mm deep)						
500 mm length	229.87	290.92	2.03	70.10	nr	361.02
1000 mm length	361.03	456.92	2.03	70.10	nr	527.02
1500 mm length	492.21	622.94	2.03	70.10	nr	693.04
2000 mm length	623.38	788.95	2.47	85.30	nr	874.25
2400 mm length	728.32	921.76	2.97	102.56	nr	1024.32
3000 mm length	1047.87	1326.19	3.22	111.19	nr	1437.38
300 mm high; double panel (100 mm deep)						
500 mm length	455.97	577.08	2.03	70.10	nr	647.18
1000 mm length	714.56	904.35	2.03	70.10	nr	974.45
1500 mm length	973.17	1231.64	2.03	70.10	nr	1301.75
2000 mm length	1231.76	1558.91	2.47	85.30	nr	1644.21
2400 mm length	1438.64	1820.74	2.97	102.56	nr	1923.30
3000 mm length	1911.10	2418.68	3.22	111.19	nr	2529.88

38 MECHANICAL/COOLING/HEATING SYSTEMS

Item	Net Price £	Material £	Labour hours	Labour £	Unit	Total rate £
500 mm high; single panel (44 mm deep)						
500 mm length	271.08	343.08	2.13	73.55	nr	416.64
1000 mm length	443.49	561.28	2.13	73.55	nr	634.83
1500 mm length	615.88	779.46	2.13	73.55	nr	853.01
2000 mm length	788.27	997.64	2.57	88.75	nr	1086.38
2400 mm length	926.19	1172.18	3.07	106.01	nr	1278.20
3000 mm length	1295.23	1639.24	3.31	114.30	nr	1753.54
500 mm high; double panel (100 mm deep)						
500 mm length	530.93	671.94	2.08	71.83	nr	743.77
1000 mm length	864.48	1094.08	2.08	71.83	nr	1165.91
1500 mm length	1198.00	1516.19	2.53	87.37	nr	1603.55
2000 mm length	1531.57	1938.36	2.97	102.56	nr	2040.92
2400 mm length	1798.41	2276.07	3.47	119.83	nr	2395.90
3000 mm length	2360.83	2987.87	3.82	131.91	nr	3119.78
600 mm high; single panel (44 mm deep)						
500 mm length	306.06	387.35	2.18	75.28	nr	462.64
1000 mm length	500.95	634.00	2.18	75.28	nr	709.28
1500 mm length	695.83	1514.64	2.63	90.82	nr	1605.46
2000 mm length	890.71	1127.29	3.06	105.67	nr	1232.96
2400 mm length	1046.63	1324.61	3.37	116.37	nr	1440.99
3000 mm length	1442.63	1825.79	3.92	135.37	nr	1961.16
600 mm high; double panel (100 mm deep)						
500 mm length	604.38	764.91	2.18	75.28	nr	840.19
1000 mm length	986.66	1248.71	2.43	83.91	nr	1332.63
1500 mm length	1368.92	1732.50	3.13	108.09	nr	1840.59
2000 mm length	1751.20	2216.32	3.77	130.19	nr	2346.51
2400 mm length	2057.01	2603.35	4.07	140.55	nr	2743.90
3000 mm length	2677.88	3389.13	5.11	176.46	nr	3565.59
700 mm high; single panel (44 mm deep)						
500 mm length	332.30	420.56	2.28	78.73	nr	499.29
1000 mm length	553.42	700.41	2.28	78.73	nr	779.14
1500 mm length	774.53	980.25	3.23	111.54	nr	1091.79
2000 mm length	995.64	1260.09	3.87	133.64	nr	1393.73
2400 mm length	1172.90	1484.42	4.17	144.00	nr	1628.42
3000 mm length	1600.04	2025.01	5.24	180.95	nr	2205.96
700 mm high; double panel (100 mm deep)						
500 mm length	653.11	826.58	2.23	77.01	nr	903.59
1000 mm length	1084.11	1372.04	2.83	97.73	nr	1469.77
1500 mm length	1515.10	1917.51	3.73	128.81	nr	2046.31
2000 mm length	1946.08	2462.95	4.46	154.02	nr	2616.97
2400 mm length	2290.87	2899.33	4.48	154.71	nr	3054.03

38 MECHANICAL/COOLING/HEATING SYSTEMS

Item	Net Price £	Material £	Labour hours	Labour £	Unit	Total rate £
LOW TEMPERATURE HOT WATER HEATING; PIPELINE: RADIATORS – cont						
Fan convector; sheet metal casing with lockable access panel; centrifugal fan; air filter; LPHW heating coil; extruded aluminium grilles; 3 speed; includes fixing in position; electrical work elsewhere						
Free-standing flat top, 695 mm high, medium speed rating						
Entering air temperature, 18°C						
695 mm long, 1 row 1.94 kW, 75 l/s	742.13	939.24	2.73	94.27	nr	**1033.51**
695 mm long, 2 row 2.64 kW, 75 l/s	742.13	939.24	2.73	94.27	nr	**1033.51**
895 mm long, 1 row 4.02 kW, 150 l/s	836.26	1058.37	2.73	94.27	nr	**1152.65**
895 mm long, 2 row 5.62 kW, 150 l/s	836.26	1058.37	2.73	94.27	nr	**1152.65**
1195 mm long, 1 row 6.58 kW, 250 l/s	952.10	1204.98	3.00	103.60	nr	**1308.57**
1195 mm long, 2 row 9.27 kW, 250 l/s	952.10	1204.98	3.00	103.60	nr	**1308.57**
1495 mm long, 1 row 9.04 kW, 340 l/s	1062.53	1344.73	3.26	112.58	nr	**1457.31**
1495 mm long, 2 row 12.73 kW, 340 l/s	1062.53	1344.73	3.26	112.58	nr	**1457.31**
Free-standing flat top, 695 mm high, medium speed rating, c/w floor plinth						
695 mm long, 1 row 1.94 kW, 75 l/s	779.23	986.19	2.73	94.27	nr	**1080.46**
695 mm long, 2 row 2.64 kW, 75 l/s	779.23	986.19	2.73	94.27	nr	**1080.46**
895 mm long, 1 row 4.02 kW, 150 l/s	878.07	1111.29	2.73	94.27	nr	**1205.56**
895 mm long, 2 row 5.62 kW, 150 l/s	878.07	1111.29	2.73	94.27	nr	**1205.56**
1195 mm long, 1 row 6.58 kW, 250 l/s	999.73	1265.25	3.00	103.60	nr	**1368.85**
1195 mm long, 2 row 9.27 kW, 250 l/s	999.73	1265.25	3.00	103.60	nr	**1368.85**
1495 mm long, 1 row 9.04 kW, 340 l/s	1115.65	1411.96	3.26	112.58	nr	**1524.54**
1495 mm long, 2 row 12.73 kW, 340 l/s	1115.65	1411.96	3.26	112.58	nr	**1524.54**
Free-standing sloping top, 695 mm high, medium speed rating, c/w floor plinth						
695 mm long, 1 row 1.94 kW, 75 l/s	806.39	1020.57	2.73	94.27	nr	**1114.84**
695 mm long, 2 row 2.64 kW, 75 l/s	806.39	1020.57	2.73	94.27	nr	**1114.84**
895 mm long, 1 row 4.02 kW, 150 l/s	905.23	1145.65	2.73	94.27	nr	**1239.93**
895 mm long, 2 row 5.62 kW, 150 l/s	905.23	1145.65	2.73	94.27	nr	**1239.93**
1195 mm long, 1 row 6.58 kW, 250 l/s	1026.87	1299.60	3.00	103.60	nr	**1403.20**
1195 mm long, 2 row 9.27 kW, 250 l/s	1026.87	1299.60	3.00	103.60	nr	**1403.20**
1495 mm long, 1 row 9.04 kW, 340 l/s	1142.79	1446.31	3.26	112.58	nr	**1558.89**
1495 mm long, 2 row 12.73 kW, 340 l/s	1142.79	1446.31	3.26	112.58	nr	**1558.89**
Wall mounted high level sloping discharge						
695 mm long, 1 row 1.94 kW, 75 l/s	814.54	1030.88	2.73	94.27	nr	**1125.15**
695 mm long, 2 row 2.64 kW, 75 l/s	814.54	1030.88	2.73	94.27	nr	**1125.15**
895 mm long, 1 row 4.02 kW, 150 l/s	838.09	1060.69	2.73	94.27	nr	**1154.96**
895 mm long, 2 row 5.62 kW, 150 l/s	838.09	1060.69	2.73	94.27	nr	**1154.96**
1195 mm long, 1 row 6.58 kW, 250 l/s	1019.96	1290.86	3.00	103.60	nr	**1394.46**
1195 mm long, 2 row 9.27 kW, 250 l/s	1019.96	1290.86	3.00	103.60	nr	**1394.46**
1495 mm long, 1 row 9.04 kW, 340 l/s	1105.95	1399.70	3.26	112.58	nr	**1512.27**
1495 mm long, 2 row 12.73 kW, 340 l/s	1105.95	1399.70	3.26	112.58	nr	**1512.27**

38 MECHANICAL/COOLING/HEATING SYSTEMS

Item	Net Price £	Material £	Labour hours	Labour £	Unit	Total rate £
Ceiling mounted sloping inlet/outlet 665 mm wide						
895 mm long, 1 row 4.02 kW, 150 l/s	921.32	1166.03	4.15	143.31	nr	**1309.34**
895 mm long, 2 row 5.62 kW, 150 l/s	921.32	1166.03	4.15	143.31	nr	**1309.34**
1195 mm long, 1 row 6.58 kW, 250 l/s	1035.38	1310.38	4.15	143.31	nr	**1453.69**
1195 mm long, 2 row 9.27 kW, 250 l/s	1035.38	1310.38	4.15	143.31	nr	**1453.69**
1495 mm long, 1 row 9.04 kW, 340 l/s	1136.73	1438.65	4.15	143.31	nr	**1581.96**
1495 mm long, 2 row 12.73 kW, 340 l/s	1136.73	1438.65	4.15	143.31	nr	**1581.96**
Free-standing unit, extended height 1700/1900/2100 mm						
895 mm long, 1 row 4.02 kW, 150 l/s	1066.51	1349.77	3.11	107.40	nr	**1457.17**
895 mm long, 2 row 5.62 kW, 150 l/s	1066.14	1349.30	3.11	107.40	nr	**1456.70**
1195 mm long, 1 row 6.58 kW, 250 l/s	1245.34	1576.10	3.11	107.40	nr	**1683.50**
1195 mm long, 2 row 9.27 kW, 250 l/s	1245.34	1576.10	3.11	107.40	nr	**1683.50**
1495 mm long, 1 row 9.04 kW, 340 l/s	1370.25	1734.19	3.11	107.40	nr	**1841.58**
1495 mm long, 2 row 12.73 kW, 340 l/s	1370.25	1734.19	3.11	107.40	nr	**1841.58**
LTHW trench heating; water temperatures 90°C/70°C; room air temperature 20°C; convector with copper tubes and aluminium fins within steel duct; Includes fixing within floor screed; electrical work elsewhere						
Natural convection type						
Normal capacity, 182 mm width, complete with linear, natural anodized aluminium grille (grille also costed separately below)						
90 mm deep						
1300 mm long, 287 W output	320.34	405.42	2.00	69.07	nr	**474.49**
2300 mm long, 576 W output	523.45	662.47	4.00	138.13	nr	**800.60**
3300 mm long, 864 W output	716.53	906.84	5.00	172.66	nr	**1079.50**
4500 mm long, 1210 W output	980.79	1241.29	7.00	241.73	nr	**1483.02**
4900 mm long, 1325 W output	1053.01	1332.69	8.00	276.26	nr	**1608.95**
110 mm deep						
1300 mm long, 331 W output	323.09	408.91	2.00	69.07	nr	**477.97**
2300 mm long, 662 W output	528.04	668.29	4.00	138.13	nr	**806.42**
3300 mm long, 993 W output	722.99	915.01	5.00	172.66	nr	**1087.67**
4500 mm long, 1389 W output	989.79	1252.68	7.00	241.73	nr	**1494.41**
4900 mm long, 1522 W output	1062.34	1344.49	8.00	276.26	nr	**1620.75**
140 mm deep						
1300 mm long, 551 W output	380.40	481.44	2.00	69.07	nr	**550.50**
2300 mm long, 1100 W output	625.45	791.57	4.00	138.13	nr	**929.70**
3300 mm long, 1651 W output	859.10	1087.28	5.00	172.66	nr	**1259.94**
4500 mm long, 2310 W output	1170.93	1481.93	7.00	241.73	nr	**1723.65**
4900 mm long, 2530 W output	1258.93	1593.30	8.00	276.26	nr	**1869.56**
190 mm deep						
1300 mm long, 625 W output	393.24	497.68	2.00	69.07	nr	**566.75**
2300 mm long, 1250 W output	648.58	820.85	4.00	138.13	nr	**958.98**
3300 mm long, 1874 W output	892.53	1129.58	5.00	172.66	nr	**1302.25**
4500 mm long, 2624 W output	1217.54	1540.91	7.00	241.73	nr	**1782.64**
4900 mm long, 2875 W output	1309.44	1657.23	8.00	276.26	nr	**1933.49**

38 MECHANICAL/COOLING/HEATING SYSTEMS

Item	Net Price £	Material £	Labour hours	Labour £	Unit	Total rate £
LOW TEMPERATURE HOT WATER HEATING; PIPELINE: RADIATORS – cont						
Fan assisted type (outputs assume fan at 50%)						
Normal capacity, 182 mm width, complete with Natural anodized aluminium grille 140 mm deep						
1300 mm long, 1439 W output	624.83	790.79	2.00	69.07	nr	**859.85**
2300 mm long, 3331 W output	1033.31	1307.75	4.00	138.13	nr	**1445.88**
3300 mm long, 5225 W output	1426.34	1805.17	5.00	172.66	nr	**1977.83**
4300 mm long, 5753 W output	1674.13	2118.78	7.00	241.73	nr	**2360.51**
4900 mm long, 6070 W output	1805.57	2285.13	8.00	276.26	nr	**2561.39**
Linear grille anodized aluminium, 170 mm width (if supplied as a separate item)	170.08	215.26	–	–	m	**215.26**
Roll up grille, natural anodized aluminium	170.08	215.26	–	–	m	**215.26**
Thermostatic valve with remote regulator (c/w valve body)	70.22	88.88	4.00	138.13	nr	**227.01**
Fan speed controller	34.32	43.44	2.00	69.07	nr	**112.50**
Note: as an alternative to thermostatic control, the system can be controlled via two port valves. Refer to valve section for valve prices.						
LTHW underfloor heating; water flow and return temperatures of 60°C and 70°C; pipework at 300 mm centres; pipe fixings; flow and return manifolds and zone actuators; wiring block; insulation; includes fixing in position; excludes secondary pump, mixing valve, zone thermostats and floor finishes; electrical work elsewhere						
Note: All rates are expressed on a m² basis, for the following example areas						
Screeded floor with 15–25 mm stone/marble finish (producing 80–100 W/m²)						
250 m² area (single zone)	31.05	39.30	0.14	4.83	m²	**44.13**
1000 m² area (single zone)	31.41	39.76	0.12	4.14	m²	**43.90**
5000 m² area (multi-zone)	31.33	39.65	0.10	3.45	m²	**43.11**
Screeded floor with 10 mm carpet tile (producing 80–100 W/m²)						
250 m² area (single zone)	31.05	39.30	0.14	4.83	m²	**44.13**
1000 m² area (single zone)	31.41	39.76	0.12	4.14	m²	**43.90**
5000 m² area (multi-zone)	31.33	39.65	0.10	3.45	m²	**43.11**
Floating timber floor with 20 mm timber finish (producing 70–80 Wm²)						
250 m² are (single zone)	40.62	51.41	0.14	4.83	m²	**56.25**
1000 m² area (single zone)	43.55	55.12	0.12	4.14	m²	**59.26**
5000 m² area (multi-zone)	41.60	52.65	0.10	3.45	m²	**56.11**
Floating timber floor with 10 mm carpet tile (producing 70–80 W/m²)						
250 m² are (single zone)	52.88	66.92	0.16	5.53	m²	**72.45**
1000 m² area (single zone)	49.03	62.05	0.12	4.14	m²	**66.20**
5000 m² area (multi-zone)	46.91	59.37	0.10	3.45	m²	**62.82**

38 MECHANICAL/COOLING/HEATING SYSTEMS

Item	Net Price £	Material £	Labour hours	Labour £	Unit	Total rate £
LOW TEMPERATURE HOT WATER HEATING; PIPELINE: PIPE FREEZING						
Freeze isolation of carbon steel or copper pipelines containing static water, either side of work location, freeze duration not exceeding 4 hours assuming that flow and return circuits are treated concurrently and activities undertaken during normal working hours						
Up to 4 freezes						
50 mm dia.	537.42	680.16	1.00	34.53	nr	**714.70**
65 mm dia.	537.42	680.16	1.00	34.53	nr	**714.70**
80 mm dia.	614.39	777.57	1.00	34.53	nr	**812.10**
100 mm dia.	691.38	875.02	1.00	34.53	nr	**909.55**
150 mm dia.	1150.42	1455.97	1.00	34.53	nr	**1490.50**
200 mm dia.	1763.37	2231.72	1.00	34.53	nr	**2266.25**

38 MECHANICAL/COOLING/HEATING SYSTEMS

Item	Net Price £	Material £	Labour hours	Labour £	Unit	Total rate £
LOW TEMPERATURE HOT WATER HEATING; PIPELINE: ENERGY METERS						
Ultrasonic						
Energy meter for measuring energy use in LTHW systems; includes ultrasonic flow meter (with sensor and signal converter), energy calculator, pair of temperature sensors with brass pockets, and 3 m of interconnecting cable; includes fixing in position; electrical work elsewhere						
Pipe size (flanged connections to PN16); maximum flow rate						
50 mm, 36 m³/hr	1486.64	1881.49	1.80	62.16	nr	**1943.65**
65 mm, 60 m³/hr	1638.71	2073.95	2.32	80.12	nr	**2154.06**
80 mm, 100 m³/hr	1836.87	2324.74	2.56	88.40	nr	**2413.14**
125 mm, 250 m³/hr	2128.19	2693.43	3.60	124.32	nr	**2817.75**
150 mm, 360 m³/hr	2310.28	2923.89	4.80	165.76	nr	**3089.65**
200 mm, 600 m³/hr	2581.25	3266.83	6.24	215.48	nr	**3482.31**
250 mm, 1000 m³/hr	2978.60	3769.72	9.60	331.51	nr	**4101.23**
300 mm, 1500 m³/hr	3501.29	4431.23	10.80	372.95	nr	**4804.18**
350 mm, 2000 m³/hr	4218.89	5339.43	13.20	455.83	nr	**5795.26**
400 mm, 2500 m³/hr	4829.40	6112.09	15.60	538.71	nr	**6650.80**
500 mm, 3000 m³/hr	5480.59	6936.24	24.00	828.78	nr	**7765.02**
600 mm, 3500 m³/hr	6157.50	7792.93	28.00	966.91	nr	**8759.84**

38 MECHANICAL/COOLING/HEATING SYSTEMS

Item	Net Price £	Material £	Labour hours	Labour £	Unit	Total rate £
LOW TEMPERATURE HOT WATER HEATING: CONTROL COMPONENTS MECHANICAL						
Room thermostats; light and medium duty; installed and connected						
Range 3°C to 27°C; 240 volt						
1 amp; on/off type	38.44	48.66	0.30	10.36	nr	**59.02**
Range 0°C to +15°C; 240 volt						
6 amp; frost thermostat	25.92	32.81	0.30	10.36	nr	**43.17**
Range 3°C to 27°C; 250 volt						
2 amp; changeover type; dead zone	64.31	81.39	0.30	10.36	nr	**91.75**
2 amp; changeover type	29.85	37.78	0.30	10.36	nr	**48.14**
2 amp; changeover type; concealed setting	37.61	47.60	0.30	10.36	nr	**57.96**
6 amp; on/off type	23.22	29.38	0.30	10.36	nr	**39.74**
6 amp; temperature setback	48.05	60.81	0.30	10.36	nr	**71.17**
16 amp; on/off type	35.77	45.28	0.30	10.36	nr	**55.64**
16 amp; on/off type; concealed setting	38.98	49.33	0.30	10.36	nr	**59.69**
20 amp; on/off type; concealed setting	33.22	42.04	0.30	10.36	nr	**52.40**
20 amp; indicated 'off' position	41.87	52.99	0.30	10.36	nr	**63.35**
20 amp; manual; double pole on/off and neon indicator	76.23	96.48	0.30	10.36	nr	**106.84**
20 amp; indicated 'off' position	49.88	63.13	0.30	10.36	nr	**73.49**
Range 10°C to 40°C; 240 volt						
20 amp; changeover contacts	45.46	57.54	0.30	10.36	nr	**67.90**
2 amp; 'heating-cooling' switch	98.32	124.43	0.30	10.36	nr	**134.79**
Surface thermostats						
Cylinder thermostat						
6 amp; changeover type; with cable	22.56	28.55	0.25	8.63	nr	**37.18**
Electrical thermostats; installed and connected						
Range 5°C to 30°C; 230 volt standard port single time						
10 amp with sensor	31.31	39.63	0.30	10.36	nr	**49.99**
Range 5°C to 30°C; 230 volt standard port double time						
10 amp with sensor	35.48	44.90	0.30	10.36	nr	**55.26**
10 amp with sensor and on/off switch	54.13	68.50	0.30	10.36	nr	**78.86**
Radiator thermostats						
Angled valve body; thermostatic head; built in sensor						
15 mm; liquid filled	23.37	29.57	0.84	29.02	nr	**58.59**
15 mm; wax filled	23.37	29.57	0.84	29.02	nr	**58.59**

38 MECHANICAL/COOLING/HEATING SYSTEMS

Item	Net Price £	Material £	Labour hours	Labour £	Unit	Total rate £
LOW TEMPERATURE HOT WATER HEATING: CONTROL COMPONENTS MECHANICAL – cont						
Immersion thermostats; stem type; domestic water boilers; fitted; electrical work elsewhere						
Temperature range 0°C to 40°C						
Non-standard; 280 mm stem	13.83	17.51	0.25	8.63	nr	**26.14**
Temperature range 18°C to 88°C						
13 amp; 178 mm stem	9.18	11.62	0.25	8.63	nr	**20.26**
20 amp; 178 mm stem	14.29	18.09	0.25	8.63	nr	**26.72**
Non-standard; pocket clip; 280 mm stem	13.26	16.78	0.25	8.63	nr	**25.41**
Temperature range 40°C to 80°C						
13 amp; 178 mm stem	5.06	6.40	0.25	8.63	nr	**15.04**
20 amp; 178 mm stem	9.55	12.08	0.25	8.63	nr	**20.71**
Non-standard; pocket clip; 280 mm stem	14.87	18.82	0.25	8.63	nr	**27.45**
13 amp; 457 mm stem	5.99	7.57	0.25	8.63	nr	**16.21**
20 amp; 457 mm stem	10.45	13.22	0.25	8.63	nr	**21.85**
Temperature range 50°C to 100°C						
Non-standard; 1780 mm stem	12.50	15.81	0.25	8.63	nr	**24.45**
Non-standard; 280 mm stem	13.03	16.49	0.25	8.63	nr	**25.13**
Pockets for thermostats						
For 178 mm stem	16.37	20.72	0.25	8.63	nr	**29.35**
For 280 mm stem	16.64	21.06	0.25	8.63	nr	**29.70**
Immersion thermostats; stem type; industrial installations; fitted; electrical work elsewhere						
Temperature range 5°C to 105°C						
For 305 mm stem	205.54	260.13	0.50	17.27	nr	**277.40**

38 MECHANICAL/COOLING/HEATING SYSTEMS

Item	Net Price £	Material £	Labour hours	Labour £	Unit	Total rate £
THERMAL INSULATION						
For flexible closed cell insulation see						
Section – Cold Water						
Mineral fibre sectional insulation; bright class						
O foil faced; bright class O foil taped joints;						
19 mm aluminium bands						
Concealed pipework						
20 mm thick						
15 mm dia.	4.71	5.96	0.15	5.18	m	**11.14**
20 mm dia.	5.02	6.35	0.15	5.18	m	**11.53**
25 mm dia.	5.37	6.80	0.15	5.18	m	**11.98**
32 mm dia.	6.02	7.61	0.15	5.18	m	**12.79**
40 mm dia.	6.44	8.15	0.15	5.18	m	**13.33**
50 mm dia.	7.36	9.31	0.15	5.18	m	**14.49**
Extra over for fittings concealed insulation						
Flange/union						
15 mm dia.	2.35	2.98	0.13	4.49	nr	**7.47**
20 mm dia.	2.51	3.17	0.13	4.49	nr	**7.66**
25 mm dia.	2.71	3.43	0.13	4.49	nr	**7.92**
32 mm dia.	3.01	3.81	0.13	4.49	nr	**8.30**
40 mm dia.	3.20	4.05	0.13	4.49	nr	**8.54**
50 mm dia.	3.67	4.65	0.13	4.49	nr	**9.14**
Valves						
15 mm dia.	4.71	5.96	0.15	5.18	nr	**11.14**
20 mm dia.	5.37	6.80	0.15	5.18	nr	**11.98**
25 mm dia.	5.37	6.80	0.15	5.18	nr	**11.98**
32 mm dia.	6.02	7.61	0.15	5.18	nr	**12.79**
40 mm dia.	6.44	8.15	0.15	5.18	nr	**13.33**
50 mm dia.	7.36	9.31	0.15	5.18	nr	**14.49**
Expansion bellows						
15 mm dia.	9.40	11.90	0.22	7.60	nr	**19.50**
20 mm dia.	10.05	12.71	0.22	7.60	nr	**20.31**
25 mm dia.	10.80	13.67	0.22	7.60	nr	**21.27**
32 mm dia.	12.01	15.20	0.22	7.60	nr	**22.80**
40 mm dia.	12.86	16.28	0.22	7.60	nr	**23.88**
50 mm dia.	14.70	18.60	0.22	7.60	nr	**26.20**
25 mm thick						
15 mm dia.	5.17	6.55	0.15	5.18	m	**11.73**
20 mm dia.	5.56	7.04	0.15	5.18	m	**12.22**
25 mm dia.	6.25	7.91	0.15	5.18	m	**13.09**
32 mm dia.	6.80	8.61	0.15	5.18	m	**13.79**
40 mm dia.	7.27	9.20	0.15	5.18	m	**14.38**
50 mm dia.	8.33	10.54	0.15	5.18	m	**15.72**
65 mm dia.	9.52	12.05	0.15	5.18	m	**17.23**
80 mm dia.	10.43	13.21	0.22	7.60	m	**20.80**
100 mm dia.	13.74	17.39	0.22	7.60	m	**24.99**
125 mm dia.	15.90	20.12	0.22	7.60	m	**27.71**
150 mm dia.	18.94	23.97	0.22	7.60	m	**31.56**

38 MECHANICAL/COOLING/HEATING SYSTEMS

Item	Net Price £	Material £	Labour hours	Labour £	Unit	Total rate £
THERMAL INSULATION – cont						
25 mm thick – cont						
200 mm dia.	26.76	33.87	0.25	8.63	m	**42.50**
250 mm dia.	32.03	40.54	0.25	8.63	m	**49.18**
300 mm dia.	34.27	43.37	0.25	8.63	m	**52.00**
Extra over for fittings concealed insulation						
Flange/union						
15 mm dia.	2.59	3.27	0.13	4.49	nr	**7.76**
20 mm dia.	2.78	3.51	0.13	4.49	nr	**8.00**
25 mm dia.	3.12	3.95	0.13	4.49	nr	**8.44**
32 mm dia.	3.40	4.30	0.13	4.49	nr	**8.79**
40 mm dia.	3.62	4.58	0.13	4.49	nr	**9.07**
50 mm dia.	4.15	5.25	0.13	4.49	nr	**9.74**
65 mm dia.	4.75	6.01	0.13	4.49	nr	**10.50**
80 mm dia.	5.23	6.62	0.18	6.22	nr	**12.83**
100 mm dia.	6.88	8.71	0.18	6.22	nr	**14.93**
125 mm dia.	7.95	10.06	0.18	6.22	nr	**16.27**
150 mm dia.	9.48	12.00	0.18	6.22	nr	**18.21**
200 mm dia.	13.38	16.93	0.22	7.60	nr	**24.53**
250 mm dia.	16.03	20.29	0.22	7.60	nr	**27.88**
300 mm dia.	17.12	21.66	0.22	7.60	nr	**29.26**
Valves						
15 mm dia.	5.17	6.55	0.15	5.18	nr	**11.73**
20 mm dia.	5.56	7.04	0.15	5.18	nr	**12.22**
25 mm dia.	6.25	7.91	0.15	5.18	nr	**13.09**
32 mm dia.	6.80	8.61	0.15	5.18	nr	**13.79**
40 mm dia.	7.27	9.20	0.15	5.18	nr	**14.38**
50 mm dia.	8.33	10.54	0.15	5.18	nr	**15.72**
65 mm dia.	9.52	12.05	0.15	5.18	nr	**17.23**
80 mm dia.	10.43	13.21	0.20	6.91	nr	**20.11**
100 mm dia.	13.74	17.39	0.20	6.91	nr	**24.30**
125 mm dia.	15.90	20.12	0.20	6.91	nr	**27.02**
150 mm dia.	18.94	23.97	0.20	6.91	nr	**30.87**
200 mm dia.	26.76	33.87	0.25	8.63	nr	**42.50**
250 mm dia.	32.03	40.54	0.25	8.63	nr	**49.18**
300 mm dia.	34.27	43.37	0.25	8.63	nr	**52.00**
Expansion bellows						
15 mm dia.	10.35	13.09	0.22	7.60	nr	**20.69**
20 mm dia.	11.18	14.15	0.22	7.60	nr	**21.74**
25 mm dia.	12.47	15.78	0.22	7.60	nr	**23.37**
32 mm dia.	13.58	17.18	0.22	7.60	nr	**24.78**
40 mm dia.	14.55	18.42	0.22	7.60	nr	**26.01**
50 mm dia.	16.66	21.09	0.22	7.60	nr	**28.68**
65 mm dia.	19.01	24.06	0.22	7.60	nr	**31.66**
80 mm dia.	20.85	26.38	0.29	10.01	nr	**36.40**
100 mm dia.	27.48	34.78	0.29	10.01	nr	**44.80**
125 mm dia.	31.81	40.26	0.29	10.01	nr	**50.28**
150 mm dia.	37.91	47.97	0.29	10.01	nr	**57.99**
200 mm dia.	53.54	67.75	0.36	12.43	nr	**80.19**
250 mm dia.	64.11	81.14	0.36	12.43	nr	**93.57**
300 mm dia.	68.53	86.73	0.36	12.43	nr	**99.17**

38 MECHANICAL/COOLING/HEATING SYSTEMS

Item	Net Price £	Material £	Labour hours	Labour £	Unit	Total rate £
30 mm thick						
15 mm dia.	6.73	8.51	0.15	5.18	m	**13.69**
20 mm dia.	7.20	9.12	0.15	5.18	m	**14.30**
25 mm dia.	7.61	9.64	0.15	5.18	m	**14.82**
32 mm dia.	8.29	10.49	0.15	5.18	m	**15.67**
40 mm dia.	8.80	11.14	0.15	5.18	m	**16.32**
50 mm dia.	10.06	12.73	0.15	5.18	m	**17.91**
65 mm dia.	11.39	14.41	0.15	5.18	m	**19.59**
80 mm dia.	12.44	15.75	0.22	7.60	m	**23.35**
100 mm dia.	16.06	20.33	0.22	7.60	m	**27.92**
125 mm dia.	18.54	23.46	0.22	7.60	m	**31.06**
150 mm dia.	21.74	27.52	0.22	7.60	m	**35.12**
200 mm dia.	30.38	38.45	0.25	8.63	m	**47.08**
250 mm dia.	36.12	45.71	0.25	8.63	m	**54.35**
300 mm dia.	38.31	48.48	0.25	8.63	m	**57.11**
350 mm dia.	42.09	53.27	0.25	8.63	m	**61.90**
Extra over for fittings concealed insulation						
Flange/union						
15 mm dia.	3.35	4.24	0.13	4.49	nr	**8.73**
20 mm dia.	3.61	4.57	0.13	4.49	nr	**9.05**
25 mm dia.	3.79	4.79	0.13	4.49	nr	**9.28**
32 mm dia.	4.15	5.25	0.13	4.49	nr	**9.74**
40 mm dia.	4.38	5.55	0.13	4.49	nr	**10.04**
50 mm dia.	5.04	6.38	0.13	4.49	nr	**10.87**
65 mm dia.	5.71	7.22	0.13	4.49	nr	**11.71**
80 mm dia.	6.19	7.84	0.18	6.22	nr	**14.05**
100 mm dia.	8.06	10.20	0.18	6.22	nr	**16.41**
125 mm dia.	9.24	11.69	0.18	6.22	nr	**17.90**
150 mm dia.	10.88	13.77	0.18	6.22	nr	**19.98**
200 mm dia.	15.22	19.26	0.22	7.60	nr	**26.86**
250 mm dia.	18.08	22.88	0.22	7.60	nr	**30.48**
300 mm dia.	19.17	24.26	0.22	7.60	nr	**31.86**
350 mm dia.	21.05	26.64	0.22	7.60	nr	**34.23**
Valves						
15 mm dia.	7.20	9.12	0.15	5.18	nr	**14.30**
20 mm dia.	7.20	9.12	0.15	5.18	nr	**14.30**
25 mm dia.	7.61	9.64	0.15	5.18	nr	**14.82**
32 mm dia.	8.29	10.49	0.15	5.18	nr	**15.67**
40 mm dia.	8.80	11.14	0.15	5.18	nr	**16.32**
50 mm dia.	10.06	12.73	0.15	5.18	nr	**17.91**
65 mm dia.	11.39	14.41	0.15	5.18	nr	**19.59**
80 mm dia.	12.44	15.75	0.20	6.91	nr	**22.65**
100 mm dia.	16.06	20.33	0.20	6.91	nr	**27.23**
125 mm dia.	18.54	23.46	0.20	6.91	nr	**30.37**
150 mm dia.	21.74	27.52	0.20	6.91	nr	**34.43**
200 mm dia.	30.38	38.45	0.25	8.63	nr	**47.08**
250 mm dia.	36.12	45.71	0.25	8.63	nr	**54.35**
300 mm dia.	38.31	48.48	0.25	8.63	nr	**57.11**
350 mm dia.	42.09	53.27	0.25	8.63	nr	**61.90**

38 MECHANICAL/COOLING/HEATING SYSTEMS

Item	Net Price £	Material £	Labour hours	Labour £	Unit	Total rate £
THERMAL INSULATION – cont						
Extra over for fittings concealed insulation – cont						
Expansion bellows						
15 mm dia.	13.46	17.04	0.22	7.60	nr	24.64
20 mm dia.	14.39	18.21	0.22	7.60	nr	25.80
25 mm dia.	15.24	19.29	0.22	7.60	nr	26.89
32 mm dia.	16.65	21.07	0.22	7.60	nr	28.67
40 mm dia.	17.62	22.29	0.22	7.60	nr	29.89
50 mm dia.	20.14	25.48	0.22	7.60	nr	33.08
65 mm dia.	22.78	28.83	0.22	7.60	nr	36.42
80 mm dia.	24.84	31.44	0.29	10.01	nr	41.45
100 mm dia.	32.18	40.73	0.29	10.01	nr	50.74
125 mm dia.	37.01	46.84	0.29	10.01	nr	56.85
150 mm dia.	43.48	55.03	0.29	10.01	nr	65.04
200 mm dia.	60.76	76.90	0.36	12.43	nr	89.33
250 mm dia.	72.31	91.51	0.36	12.43	nr	103.94
300 mm dia.	76.66	97.02	0.36	12.43	nr	109.45
350 mm dia.	84.16	106.51	0.36	12.43	nr	118.94
40 mm thick						
15 mm dia.	8.60	10.89	0.15	5.18	m	16.07
20 mm dia.	8.88	11.24	0.15	5.18	m	16.42
25 mm dia.	9.56	12.10	0.15	5.18	m	17.28
32 mm dia.	10.18	12.88	0.15	5.18	m	18.06
40 mm dia.	10.72	13.57	0.15	5.18	m	18.75
50 mm dia.	12.12	15.34	0.15	5.18	m	20.52
65 mm dia.	13.59	17.19	0.15	5.18	m	22.37
80 mm dia.	14.76	18.68	0.22	7.60	m	26.28
100 mm dia.	19.21	24.32	0.22	7.60	m	31.91
125 mm dia.	21.70	27.46	0.22	7.60	m	35.06
150 mm dia.	25.31	32.03	0.22	7.60	m	39.63
200 mm dia.	34.98	44.27	0.25	8.63	m	52.90
250 mm dia.	40.98	51.87	0.25	8.63	m	60.50
300 mm dia.	43.81	55.45	0.25	8.63	m	64.08
350 mm dia.	48.46	61.33	0.25	8.63	m	69.97
400 mm dia.	53.99	68.33	0.25	8.63	m	76.96
Extra over for fittings concealed insulation						
Flange/union						
15 mm dia.	4.33	5.48	0.13	4.49	nr	9.97
20 mm dia.	4.43	5.61	0.13	4.49	nr	10.09
25 mm dia.	4.76	6.03	0.13	4.49	nr	10.52
32 mm dia.	5.08	6.43	0.13	4.49	nr	10.92
40 mm dia.	5.34	6.76	0.13	4.49	nr	11.25
50 mm dia.	6.06	7.67	0.13	4.49	nr	12.16
65 mm dia.	6.80	8.61	0.13	4.49	nr	13.10
80 mm dia.	7.40	9.37	0.18	6.22	nr	15.59
100 mm dia.	9.59	12.14	0.18	6.22	nr	18.35
125 mm dia.	10.83	13.71	0.18	6.22	nr	19.93
150 mm dia.	12.67	16.03	0.18	6.22	nr	22.24

38 MECHANICAL/COOLING/HEATING SYSTEMS

Item	Net Price £	Material £	Labour hours	Labour £	Unit	Total rate £
200 mm dia.	17.49	22.14	0.22	7.60	nr	**29.74**
250 mm dia.	20.48	25.92	0.22	7.60	nr	**33.52**
300 mm dia.	21.88	27.69	0.22	7.60	nr	**35.29**
350 mm dia.	24.23	30.67	0.22	7.60	nr	**38.26**
400 mm dia.	27.03	34.21	0.22	7.60	nr	**41.80**
Valves						
15 mm dia.	8.60	10.89	0.15	5.18	nr	**16.07**
20 mm dia.	8.88	11.24	0.15	5.18	nr	**16.42**
25 mm dia.	9.56	12.10	0.15	5.18	nr	**17.28**
32 mm dia.	10.18	12.88	0.15	5.18	nr	**18.06**
40 mm dia.	10.72	13.57	0.15	5.18	nr	**18.75**
50 mm dia.	12.12	15.34	0.15	5.18	nr	**20.52**
65 mm dia.	13.59	17.19	0.15	5.18	nr	**22.37**
80 mm dia.	14.76	18.68	0.20	6.91	nr	**25.59**
100 mm dia.	19.21	24.32	0.20	6.91	nr	**31.22**
125 mm dia.	21.70	27.46	0.20	6.91	nr	**34.37**
150 mm dia.	25.31	32.03	0.20	6.91	nr	**38.94**
200 mm dia.	34.98	44.27	0.25	8.63	nr	**52.90**
250 mm dia.	40.98	51.87	0.25	8.63	nr	**60.50**
300 mm dia.	43.81	55.45	0.25	8.63	nr	**64.08**
350 mm dia.	48.46	61.33	0.25	8.63	nr	**69.97**
400 mm dia.	53.99	68.33	0.25	8.63	nr	**76.96**
Expansion bellows						
15 mm dia.	17.27	21.86	0.22	7.60	nr	**29.46**
20 mm dia.	17.74	22.45	0.22	7.60	nr	**30.05**
25 mm dia.	19.13	24.20	0.22	7.60	nr	**31.80**
32 mm dia.	20.36	25.76	0.22	7.60	nr	**33.36**
40 mm dia.	21.41	27.10	0.22	7.60	nr	**34.70**
50 mm dia.	24.23	30.67	0.22	7.60	nr	**38.26**
65 mm dia.	27.15	34.36	0.22	7.60	nr	**41.96**
80 mm dia.	29.59	37.45	0.29	10.01	nr	**47.47**
100 mm dia.	38.36	48.55	0.29	10.01	nr	**58.56**
125 mm dia.	43.37	54.89	0.29	10.01	nr	**64.90**
150 mm dia.	50.62	64.06	0.29	10.01	nr	**74.07**
200 mm dia.	69.95	88.53	0.36	12.43	nr	**100.96**
250 mm dia.	81.97	103.75	0.36	12.43	nr	**116.18**
300 mm dia.	87.65	110.92	0.36	12.43	nr	**123.36**
350 mm dia.	48.46	61.33	0.36	12.43	nr	**73.77**
400 mm dia.	108.01	136.70	0.36	12.43	nr	**149.13**
50 mm thick						
15 mm dia.	11.99	15.17	0.15	5.18	m	**20.35**
20 mm dia.	12.60	15.94	0.15	5.18	m	**21.12**
25 mm dia.	13.40	16.96	0.15	5.18	m	**22.14**
32 mm dia.	14.00	17.71	0.15	5.18	m	**22.89**
40 mm dia.	14.71	18.61	0.15	5.18	m	**23.79**
50 mm dia.	16.52	20.90	0.15	5.18	m	**26.08**
65 mm dia.	18.05	22.84	0.15	5.18	m	**28.02**
80 mm dia.	19.30	24.43	0.22	7.60	m	**32.03**
100 mm dia.	24.71	31.27	0.22	7.60	m	**38.87**
125 mm dia.	27.72	35.08	0.22	7.60	m	**42.68**
150 mm dia.	31.99	40.49	0.22	7.60	m	**48.08**

38 MECHANICAL/COOLING/HEATING SYSTEMS

Item	Net Price £	Material £	Labour hours	Labour £	Unit	Total rate £
THERMAL INSULATION – cont						
50 mm thick – cont						
200 mm dia.	43.68	55.28	0.25	8.63	m	**63.91**
250 mm dia.	50.44	63.83	0.25	8.63	m	**72.47**
300 mm dia.	53.45	67.64	0.25	8.63	m	**76.28**
350 mm dia.	59.00	74.67	0.25	8.63	m	**83.30**
400 mm dia.	65.43	82.81	0.25	8.63	m	**91.45**
Extra over for fittings concealed insulation						
Flange/union						
15 mm dia.	6.01	7.60	0.13	4.49	nr	**12.09**
20 mm dia.	6.32	7.99	0.13	4.49	nr	**12.48**
25 mm dia.	6.72	8.50	0.13	4.49	nr	**12.99**
32 mm dia.	7.02	8.88	0.13	4.49	nr	**13.37**
40 mm dia.	7.37	9.33	0.13	4.49	nr	**13.82**
50 mm dia.	8.24	10.42	0.13	4.49	nr	**14.91**
65 mm dia.	9.02	11.42	0.13	4.49	nr	**15.91**
80 mm dia.	9.66	12.22	0.18	6.22	nr	**18.44**
100 mm dia.	12.37	15.65	0.18	6.22	nr	**21.87**
125 mm dia.	13.86	17.55	0.18	6.22	nr	**23.76**
150 mm dia.	16.02	20.27	0.18	6.22	nr	**26.49**
200 mm dia.	21.82	27.62	0.22	7.60	nr	**35.22**
250 mm dia.	25.22	31.92	0.22	7.60	nr	**39.51**
300 mm dia.	26.73	33.83	0.22	7.60	nr	**41.43**
350 mm dia.	29.53	37.37	0.22	7.60	nr	**44.97**
400 mm dia.	32.72	41.41	0.22	7.60	nr	**49.01**
Valves						
15 mm dia.	11.99	15.17	0.15	5.18	nr	**20.35**
20 mm dia.	12.60	15.94	0.15	5.18	nr	**21.12**
25 mm dia.	13.40	16.96	0.15	5.18	nr	**22.14**
32 mm dia.	14.00	17.71	0.15	5.18	nr	**22.89**
40 mm dia.	14.71	18.61	0.15	5.18	nr	**23.79**
50 mm dia.	16.52	20.90	0.15	5.18	nr	**26.08**
65 mm dia.	18.05	22.84	0.15	5.18	nr	**28.02**
80 mm dia.	19.30	24.43	0.20	6.91	nr	**31.34**
100 mm dia.	24.71	31.27	0.20	6.91	nr	**38.18**
125 mm dia.	27.72	35.08	0.20	6.91	nr	**41.98**
150 mm dia.	31.99	40.49	0.20	6.91	nr	**47.39**
200 mm dia.	43.68	55.28	0.25	8.63	nr	**63.91**
250 mm dia.	50.44	63.83	0.25	8.63	nr	**72.47**
300 mm dia.	53.45	67.64	0.25	8.63	nr	**76.28**
350 mm dia.	59.00	74.67	0.25	8.63	nr	**83.30**
400 mm dia.	65.43	82.81	0.25	8.63	nr	**91.45**
Expansion bellows						
15 mm dia.	23.92	30.27	0.22	7.60	nr	**37.87**
20 mm dia.	25.19	31.88	0.22	7.60	nr	**39.47**
25 mm dia.	26.78	33.90	0.22	7.60	nr	**41.50**
32 mm dia.	28.01	35.44	0.22	7.60	nr	**43.04**
40 mm dia.	29.46	37.28	0.22	7.60	nr	**44.88**
50 mm dia.	33.01	41.78	0.22	7.60	nr	**49.38**
65 mm dia.	36.08	45.66	0.22	7.60	nr	**53.25**

38 MECHANICAL/COOLING/HEATING SYSTEMS

Item	Net Price £	Material £	Labour hours	Labour £	Unit	Total rate £
80 mm dia.	38.65	48.92	0.29	10.01	nr	**58.93**
100 mm dia.	49.36	62.47	0.29	10.01	nr	**72.49**
125 mm dia.	55.42	70.14	0.29	10.01	nr	**80.16**
150 mm dia.	63.98	80.97	0.29	10.01	nr	**90.99**
200 mm dia.	87.38	110.59	0.36	12.43	nr	**123.02**
250 mm dia.	100.87	127.66	0.36	12.43	nr	**140.09**
300 mm dia.	106.93	135.33	0.36	12.43	nr	**147.76**
350 mm dia.	118.00	149.35	0.36	12.43	nr	**161.78**
400 mm dia.	130.85	165.60	0.36	12.43	nr	**178.03**
Mineral fibre sectional insulation; bright class O foil faced; bright class O foil taped joints; 22 SWG plain/embossed aluminium cladding; pop riveted						
Plantroom pipework						
20 mm thick						
15 mm dia.	7.68	9.72	0.44	15.19	m	**24.92**
20 mm dia.	8.11	10.27	0.44	15.19	m	**25.46**
25 mm dia.	8.69	11.00	0.44	15.19	m	**26.19**
32 mm dia.	9.41	11.91	0.44	15.19	m	**27.11**
40 mm dia.	9.92	12.56	0.44	15.19	m	**27.75**
50 mm dia.	11.04	13.98	0.44	15.19	m	**29.17**
Extra over for fittings plantroom insulation						
Flange/union						
15 mm dia.	8.58	10.86	0.58	20.03	nr	**30.89**
20 mm dia.	9.11	11.53	0.58	20.03	nr	**31.56**
25 mm dia.	9.76	12.35	0.58	20.03	nr	**32.38**
32 mm dia.	10.66	13.49	0.58	20.03	nr	**33.52**
40 mm dia.	11.26	14.24	0.58	20.03	nr	**34.27**
50 mm dia.	12.63	15.99	0.58	20.03	nr	**36.02**
Bends						
15 mm dia.	4.20	5.31	0.44	15.19	nr	**20.50**
20 mm dia.	4.48	5.68	0.44	15.19	nr	**20.87**
25 mm dia.	4.78	6.05	0.44	15.19	nr	**21.25**
32 mm dia.	5.17	6.55	0.44	15.19	nr	**21.74**
40 mm dia.	5.43	6.87	0.44	15.19	nr	**22.06**
50 mm dia.	6.08	7.70	0.44	15.19	nr	**22.89**
Tees						
15 mm dia.	2.52	3.19	0.44	15.19	nr	**18.38**
20 mm dia.	2.70	3.41	0.44	15.19	nr	**18.61**
25 mm dia.	2.86	3.62	0.44	15.19	nr	**18.82**
32 mm dia.	3.11	3.93	0.44	15.19	nr	**19.13**
40 mm dia.	3.29	4.16	0.44	15.19	nr	**19.35**
50 mm dia.	3.65	4.62	0.44	15.19	nr	**19.82**
Valves						
15 mm dia.	4.71	5.96	0.78	26.94	nr	**32.89**
20 mm dia.	5.37	6.80	0.78	26.94	nr	**33.73**
25 mm dia.	5.37	6.80	0.78	26.94	nr	**33.73**

38 MECHANICAL/COOLING/HEATING SYSTEMS

Item	Net Price £	Material £	Labour hours	Labour £	Unit	Total rate £
THERMAL INSULATION – cont						
Extra over for fittings plantroom insulation – cont						
Valves – cont						
32 mm dia.	6.02	7.61	0.78	26.94	nr	**34.55**
40 mm dia.	6.44	8.15	0.78	26.94	nr	**35.08**
50 mm dia.	7.36	9.31	0.78	26.94	nr	**36.25**
Pumps						
15 mm dia.	25.22	31.92	2.34	80.81	nr	**112.72**
20 mm dia.	26.74	33.84	2.34	80.81	nr	**114.65**
25 mm dia.	28.72	36.34	2.34	80.81	nr	**117.15**
32 mm dia.	31.34	39.66	2.34	80.81	nr	**120.46**
40 mm dia.	33.10	41.89	2.34	80.81	nr	**122.70**
50 mm dia.	37.22	47.10	2.34	80.81	nr	**127.91**
Expansion bellows						
15 mm dia.	20.17	25.53	1.05	36.26	nr	**61.78**
20 mm dia.	21.40	27.08	1.05	36.26	nr	**63.34**
25 mm dia.	22.99	29.09	1.05	36.26	nr	**65.35**
32 mm dia.	25.06	31.72	1.05	36.26	nr	**67.98**
40 mm dia.	26.45	33.48	1.05	36.26	nr	**69.74**
50 mm dia.	29.77	37.68	1.05	36.26	nr	**73.94**
25 mm thick						
15 mm dia.	8.42	10.66	0.44	15.19	m	**25.86**
20 mm dia.	9.11	11.53	0.44	15.19	m	**26.73**
25 mm dia.	9.98	12.63	0.44	15.19	m	**27.82**
32 mm dia.	10.62	13.44	0.44	15.19	m	**28.64**
40 mm dia.	11.43	14.47	0.44	15.19	m	**29.66**
50 mm dia.	12.81	16.21	0.44	15.19	m	**31.41**
65 mm dia.	14.47	18.32	0.44	15.19	m	**33.51**
80 mm dia.	15.68	19.85	0.52	17.96	m	**37.81**
100 mm dia.	19.47	24.64	0.52	17.96	m	**42.60**
125 mm dia.	22.61	28.62	0.52	17.96	m	**46.57**
150 mm dia.	26.32	33.31	0.52	17.96	m	**51.27**
200 mm dia.	35.95	45.50	0.60	20.72	m	**66.22**
250 mm dia.	42.72	54.07	0.60	20.72	m	**74.79**
300 mm dia.	47.83	60.53	0.60	20.72	m	**81.25**
Extra over for fittings plantroom insulation						
Flange/union						
15 mm dia.	9.41	11.91	0.58	20.03	nr	**31.94**
20 mm dia.	10.18	12.88	0.58	20.03	nr	**32.91**
25 mm dia.	11.23	14.22	0.58	20.03	nr	**34.25**
32 mm dia.	12.08	15.28	0.58	20.03	nr	**35.31**
40 mm dia.	12.94	16.38	0.58	20.03	nr	**36.41**
50 mm dia.	14.59	18.46	0.58	20.03	nr	**38.49**
65 mm dia.	16.55	20.95	0.58	20.03	nr	**40.97**
80 mm dia.	17.99	22.77	0.67	23.14	nr	**45.91**
100 mm dia.	22.78	28.83	0.67	23.14	nr	**51.96**
125 mm dia.	26.44	33.46	0.67	23.14	nr	**56.60**
150 mm dia.	30.99	39.22	0.67	23.14	nr	**62.36**

38 MECHANICAL/COOLING/HEATING SYSTEMS

Item	Net Price £	Material £	Labour hours	Labour £	Unit	Total rate £
200 mm dia.	42.82	54.20	0.87	30.04	nr	84.24
250 mm dia.	51.00	64.55	0.87	30.04	nr	94.59
300 mm dia.	56.39	71.36	0.87	30.04	nr	101.41
Bends						
15 mm dia.	4.63	5.86	0.44	15.19	nr	21.05
20 mm dia.	5.02	6.35	0.44	15.19	nr	21.54
25 mm dia.	5.48	6.94	0.44	15.19	nr	22.13
32 mm dia.	5.85	7.40	0.44	15.19	nr	22.60
40 mm dia.	6.32	7.99	0.44	15.19	nr	23.19
50 mm dia.	7.05	8.92	0.44	15.19	nr	24.11
65 mm dia.	7.95	10.06	0.44	15.19	nr	25.25
80 mm dia.	8.61	10.90	0.52	17.96	nr	28.86
100 mm dia.	10.72	13.57	0.52	17.96	nr	31.53
125 mm dia.	12.47	15.78	0.52	17.96	nr	33.73
150 mm dia.	14.47	18.32	0.52	17.96	nr	36.28
200 mm dia.	19.77	25.02	0.60	20.72	nr	45.74
250 mm dia.	23.50	29.74	0.60	20.72	nr	50.46
300 mm dia.	26.31	33.29	0.60	20.72	nr	54.01
Tees						
15 mm dia.	2.78	3.51	0.44	15.19	nr	18.71
20 mm dia.	2.99	3.78	0.44	15.19	nr	18.97
25 mm dia.	3.31	4.19	0.44	15.19	nr	19.38
32 mm dia.	3.51	4.44	0.44	15.19	nr	19.63
40 mm dia.	3.76	4.76	0.44	15.19	nr	19.96
50 mm dia.	4.20	5.31	0.44	15.19	nr	20.50
65 mm dia.	4.76	6.03	0.44	15.19	nr	21.22
80 mm dia.	5.17	6.55	0.52	17.96	nr	24.50
100 mm dia.	6.43	8.13	0.52	17.96	nr	26.09
125 mm dia.	7.48	9.47	0.52	17.96	nr	27.43
150 mm dia.	8.69	11.00	0.52	17.96	nr	28.96
200 mm dia.	11.84	14.99	0.60	20.72	nr	35.71
250 mm dia.	14.12	17.87	0.60	20.72	nr	38.59
300 mm dia.	15.81	20.00	0.60	20.72	nr	40.72
Valves						
15 mm dia.	14.96	18.94	0.78	26.94	nr	45.87
20 mm dia.	16.17	20.47	0.78	26.94	nr	47.40
25 mm dia.	17.86	22.60	0.78	26.94	nr	49.54
32 mm dia.	19.16	24.25	0.78	26.94	nr	51.18
40 mm dia.	20.56	26.02	0.78	26.94	nr	52.95
50 mm dia.	23.13	29.28	0.78	26.94	nr	56.21
65 mm dia.	26.28	33.27	0.78	26.94	nr	60.20
80 mm dia.	28.59	36.19	0.92	31.77	nr	67.96
100 mm dia.	36.14	45.74	0.92	31.77	nr	77.51
125 mm dia.	42.02	53.19	0.92	31.77	nr	84.96
150 mm dia.	49.23	62.30	0.92	31.77	nr	94.07
200 mm dia.	68.02	86.09	1.12	38.68	nr	124.76
250 mm dia.	81.03	102.55	1.12	38.68	nr	141.23
300 mm dia.	89.53	113.31	1.12	38.68	nr	151.99

38 MECHANICAL/COOLING/HEATING SYSTEMS

Item	Net Price £	Material £	Labour hours	Labour £	Unit	Total rate £
THERMAL INSULATION – cont						
Extra over for fittings plantroom insulation – cont						
Pumps						
15 mm dia.	27.71	35.06	2.34	80.81	nr	**115.87**
20 mm dia.	29.94	37.89	2.34	80.81	nr	**118.69**
25 mm dia.	33.08	41.86	2.34	80.81	nr	**122.67**
32 mm dia.	35.44	44.86	2.34	80.81	nr	**125.66**
40 mm dia.	38.05	48.16	2.34	80.81	nr	**128.96**
50 mm dia.	42.89	54.28	2.34	80.81	nr	**135.09**
65 mm dia.	48.65	61.57	2.34	80.81	nr	**142.38**
80 mm dia.	52.94	67.00	2.76	95.31	nr	**162.31**
100 mm dia.	66.98	84.77	2.76	95.31	nr	**180.08**
125 mm dia.	77.79	98.45	2.76	95.31	nr	**193.76**
150 mm dia.	91.15	115.36	2.76	95.31	nr	**210.67**
200 mm dia.	125.96	159.42	3.36	116.03	nr	**275.45**
250 mm dia.	150.05	189.90	3.36	116.03	nr	**305.93**
300 mm dia.	165.77	209.80	3.36	116.03	nr	**325.82**
Expansion bellows						
15 mm dia.	22.16	28.04	1.05	36.26	nr	**64.30**
20 mm dia.	23.93	30.29	1.05	36.26	nr	**66.55**
25 mm dia.	26.44	33.46	1.05	36.26	nr	**69.72**
32 mm dia.	28.38	35.92	1.05	36.26	nr	**72.18**
40 mm dia.	30.46	38.55	1.05	36.26	nr	**74.81**
50 mm dia.	34.32	43.44	1.05	36.26	nr	**79.70**
65 mm dia.	38.94	49.28	1.05	36.26	nr	**85.54**
80 mm dia.	42.37	53.62	1.26	43.51	nr	**97.13**
100 mm dia.	53.57	67.80	1.26	43.51	nr	**111.31**
125 mm dia.	62.19	78.71	1.26	43.51	nr	**122.22**
150 mm dia.	72.90	92.27	1.26	43.51	nr	**135.78**
200 mm dia.	100.78	127.54	1.53	52.83	nr	**180.38**
250 mm dia.	120.06	151.94	1.53	52.83	nr	**204.78**
300 mm dia.	132.60	167.82	1.53	52.83	nr	**220.65**
30 mm thick						
15 mm dia.	10.39	13.15	0.44	15.19	m	**28.34**
20 mm dia.	10.98	13.89	0.44	15.19	m	**29.09**
25 mm dia.	11.63	14.72	0.44	15.19	m	**29.92**
32 mm dia.	12.75	16.14	0.44	15.19	m	**31.34**
40 mm dia.	13.26	16.79	0.44	15.19	m	**31.98**
50 mm dia.	14.71	18.61	0.44	15.19	m	**33.81**
65 mm dia.	16.62	21.03	0.44	15.19	m	**36.22**
80 mm dia.	18.16	22.98	0.52	17.96	m	**40.94**
100 mm dia.	22.31	28.24	0.52	17.96	m	**46.19**
125 mm dia.	25.47	32.24	0.52	17.96	m	**50.20**
150 mm dia.	29.68	37.56	0.52	17.96	m	**55.52**
200 mm dia.	39.94	50.55	0.60	20.72	m	**71.26**
250 mm dia.	47.22	59.76	0.60	20.72	m	**80.48**
300 mm dia.	51.98	65.79	0.60	20.72	m	**86.51**
350 mm dia.	57.88	73.25	0.60	20.72	m	**93.97**

38 MECHANICAL/COOLING/HEATING SYSTEMS

Item	Net Price £	Material £	Labour hours	Labour £	Unit	Total rate £
Extra over for fittings plantroom insulation						
Flange/union						
15 mm dia.	11.80	14.93	0.58	20.03	nr	**34.96**
20 mm dia.	12.53	15.86	0.58	20.03	nr	**35.89**
25 mm dia.	13.29	16.82	0.58	20.03	nr	**36.84**
32 mm dia.	14.55	18.42	0.58	20.03	nr	**38.45**
40 mm dia.	15.23	19.27	0.58	20.03	nr	**39.30**
50 mm dia.	17.04	21.56	0.58	20.03	nr	**41.59**
65 mm dia.	19.26	24.37	0.58	20.03	nr	**44.40**
80 mm dia.	21.05	26.64	0.67	23.14	nr	**49.77**
100 mm dia.	26.28	33.27	0.67	23.14	nr	**56.40**
125 mm dia.	30.08	38.07	0.67	23.14	nr	**61.21**
150 mm dia.	35.19	44.53	0.67	23.14	nr	**67.67**
200 mm dia.	47.92	60.65	0.87	30.04	nr	**90.69**
250 mm dia.	56.81	71.90	0.87	30.04	nr	**101.94**
300 mm dia.	61.80	78.22	0.87	30.04	nr	**108.26**
350 mm dia.	68.58	86.79	0.87	30.04	nr	**116.83**
Bends						
15 mm dia.	5.71	7.22	0.44	15.19	nr	**22.42**
20 mm dia.	6.03	7.63	0.44	15.19	nr	**22.82**
25 mm dia.	6.40	8.11	0.44	15.19	nr	**23.30**
32 mm dia.	6.99	8.85	0.44	15.19	nr	**24.04**
40 mm dia.	7.28	9.22	0.44	15.19	nr	**24.41**
50 mm dia.	8.11	10.27	0.44	15.19	nr	**25.46**
65 mm dia.	9.15	11.58	0.44	15.19	nr	**26.77**
80 mm dia.	9.97	12.62	0.52	17.96	nr	**30.57**
100 mm dia.	12.27	15.52	0.52	17.96	nr	**33.48**
125 mm dia.	14.01	17.73	0.52	17.96	nr	**35.69**
150 mm dia.	16.34	20.68	0.52	17.96	nr	**38.64**
200 mm dia.	21.96	27.79	0.60	20.72	nr	**48.51**
250 mm dia.	26.00	32.90	0.60	20.72	nr	**53.62**
300 mm dia.	28.59	36.19	0.60	20.72	nr	**56.91**
350 mm dia.	31.86	40.32	0.60	20.72	nr	**61.04**
Tees						
15 mm dia.	3.41	4.31	0.44	15.19	nr	**19.51**
20 mm dia.	3.62	4.58	0.44	15.19	nr	**19.77**
25 mm dia.	3.82	4.83	0.44	15.19	nr	**20.03**
32 mm dia.	4.18	5.30	0.44	15.19	nr	**20.49**
40 mm dia.	4.37	5.53	0.44	15.19	nr	**20.73**
50 mm dia.	4.85	6.14	0.44	15.19	nr	**21.33**
65 mm dia.	5.47	6.93	0.44	15.19	nr	**22.12**
80 mm dia.	6.01	7.60	0.52	17.96	nr	**25.56**
100 mm dia.	7.37	9.33	0.52	17.96	nr	**27.28**
125 mm dia.	8.42	10.66	0.52	17.96	nr	**28.62**
150 mm dia.	9.82	12.43	0.52	17.96	nr	**30.39**
200 mm dia.	13.18	16.68	0.60	20.72	nr	**37.39**
250 mm dia.	15.61	19.75	0.60	20.72	nr	**40.47**
300 mm dia.	17.15	21.70	0.60	20.72	nr	**42.42**
350 mm dia.	19.13	24.20	0.60	20.72	nr	**44.92**

Material Costs/Prices for Measured Works

38 MECHANICAL/COOLING/HEATING SYSTEMS

Item	Net Price £	Material £	Labour hours	Labour £	Unit	Total rate £
THERMAL INSULATION – cont						
Extra over for fittings plantroom insulation – cont						
Valves						
15 mm dia.	18.76	23.74	0.78	26.94	nr	**50.68**
20 mm dia.	19.89	25.17	0.78	26.94	nr	**52.11**
25 mm dia.	21.10	26.71	0.78	26.94	nr	**53.64**
32 mm dia.	23.12	29.26	0.78	26.94	nr	**56.20**
40 mm dia.	24.18	30.60	0.78	26.94	nr	**57.53**
50 mm dia.	27.07	34.26	0.78	26.94	nr	**61.20**
65 mm dia.	30.60	38.73	0.78	26.94	nr	**65.67**
80 mm dia.	33.41	42.28	0.92	31.77	nr	**74.05**
100 mm dia.	41.74	52.82	0.92	31.77	nr	**84.59**
125 mm dia.	47.80	60.49	0.92	31.77	nr	**92.26**
150 mm dia.	55.88	70.72	0.92	31.77	nr	**102.49**
200 mm dia.	76.09	96.30	1.12	38.68	nr	**134.98**
250 mm dia.	90.22	114.18	1.12	38.68	nr	**152.86**
300 mm dia.	98.17	124.24	1.12	38.68	nr	**162.92**
350 mm dia.	108.90	137.83	1.12	38.68	nr	**176.50**
Pumps						
15 mm dia.	34.75	43.98	2.34	80.81	nr	**124.79**
20 mm dia.	36.87	46.67	2.34	80.81	nr	**127.47**
25 mm dia.	39.06	49.44	2.34	80.81	nr	**130.24**
32 mm dia.	42.80	54.17	2.34	80.81	nr	**134.98**
40 mm dia.	44.78	56.67	2.34	80.81	nr	**137.48**
50 mm dia.	50.16	63.48	2.34	80.81	nr	**144.29**
65 mm dia.	56.64	71.69	2.34	80.81	nr	**152.49**
80 mm dia.	61.85	78.28	2.76	95.31	nr	**173.59**
100 mm dia.	77.27	97.79	2.76	95.31	nr	**193.10**
125 mm dia.	88.50	112.01	2.76	95.31	nr	**207.32**
150 mm dia.	103.50	130.99	2.76	95.31	nr	**226.29**
200 mm dia.	140.93	178.36	3.36	116.03	nr	**294.38**
250 mm dia.	167.11	211.50	3.36	116.03	nr	**327.52**
300 mm dia.	181.80	230.08	3.36	116.03	nr	**346.11**
350 mm dia.	201.66	255.23	3.36	116.03	nr	**371.26**
Expansion bellows						
15 mm dia.	27.81	35.19	1.05	36.26	nr	**71.45**
20 mm dia.	29.50	37.34	1.05	36.26	nr	**73.60**
25 mm dia.	31.24	39.53	1.05	36.26	nr	**75.79**
32 mm dia.	34.24	43.34	1.05	36.26	nr	**79.60**
40 mm dia.	35.82	45.33	1.05	36.26	nr	**81.59**
50 mm dia.	40.09	50.74	1.05	36.26	nr	**87.00**
65 mm dia.	45.32	57.36	1.05	36.26	nr	**93.62**
80 mm dia.	49.53	62.68	1.26	43.51	nr	**106.19**
100 mm dia.	61.83	78.25	1.26	43.51	nr	**121.76**
125 mm dia.	70.83	89.64	1.26	43.51	nr	**133.15**
150 mm dia.	82.81	104.80	1.26	43.51	nr	**148.31**
200 mm dia.	112.75	142.70	1.53	52.83	nr	**195.54**
250 mm dia.	133.69	169.20	1.53	52.83	nr	**222.03**
300 mm dia.	145.43	184.06	1.53	52.83	nr	**236.89**
350 mm dia.	161.33	204.18	1.53	52.83	nr	**257.01**

38 MECHANICAL/COOLING/HEATING SYSTEMS

Item	Net Price £	Material £	Labour hours	Labour £	Unit	Total rate £
40 mm thick						
15 mm dia.	12.70	16.07	0.44	15.19	m	**31.27**
20 mm dia.	13.33	16.87	0.44	15.19	m	**32.07**
25 mm dia.	14.20	17.97	0.44	15.19	m	**33.16**
32 mm dia.	14.89	18.84	0.44	15.19	m	**34.03**
40 mm dia.	15.82	20.02	0.44	15.19	m	**35.21**
50 mm dia.	17.52	22.17	0.44	15.19	m	**37.36**
65 mm dia.	19.50	24.68	0.44	15.19	m	**39.88**
80 mm dia.	20.99	26.57	0.52	17.96	m	**44.52**
100 mm dia.	30.75	38.91	0.52	17.96	m	**56.87**
125 mm dia.	28.80	36.45	0.52	17.96	m	**54.41**
150 mm dia.	33.59	42.51	0.52	17.96	m	**60.47**
200 mm dia.	44.43	56.23	0.60	20.72	m	**76.95**
250 mm dia.	52.04	65.86	0.60	20.72	m	**86.58**
300 mm dia.	57.52	72.80	0.60	20.72	m	**93.52**
350 mm dia.	64.29	81.37	0.60	20.72	m	**102.09**
400 mm dia.	72.32	91.52	0.60	20.72	m	**112.24**
Extra over for fittings plantroom insulation						
Flange/union						
15 mm dia.	14.66	18.56	0.58	20.03	nr	**38.59**
20 mm dia.	15.31	19.37	0.58	20.03	nr	**39.40**
25 mm dia.	16.35	20.69	0.58	20.03	nr	**40.72**
32 mm dia.	17.24	21.82	0.58	20.03	nr	**41.85**
40 mm dia.	18.28	23.14	0.58	20.03	nr	**43.17**
50 mm dia.	20.36	25.76	0.58	20.03	nr	**45.79**
65 mm dia.	22.76	28.80	0.58	20.03	nr	**48.83**
80 mm dia.	24.54	31.06	0.67	23.14	nr	**54.20**
100 mm dia.	30.75	38.91	0.67	23.14	nr	**62.05**
125 mm dia.	34.43	43.58	0.67	23.14	nr	**66.71**
150 mm dia.	40.16	50.83	0.67	23.14	nr	**73.96**
200 mm dia.	53.89	68.20	0.87	30.04	nr	**98.25**
250 mm dia.	63.17	79.95	0.87	30.04	nr	**109.99**
300 mm dia.	69.11	87.46	0.87	30.04	nr	**117.51**
350 mm dia.	77.06	97.52	0.87	30.04	nr	**127.57**
400 mm dia.	86.39	109.34	0.87	30.04	nr	**139.38**
Bends						
15 mm dia.	6.98	8.84	0.44	15.19	nr	**24.03**
20 mm dia.	7.31	9.26	0.44	15.19	nr	**24.45**
25 mm dia.	7.81	9.89	0.44	15.19	nr	**25.08**
32 mm dia.	8.17	10.34	0.44	15.19	nr	**25.53**
40 mm dia.	8.70	11.01	0.44	15.19	nr	**26.21**
50 mm dia.	9.62	12.18	0.44	15.19	nr	**27.37**
65 mm dia.	10.76	13.61	0.44	15.19	nr	**28.81**
80 mm dia.	11.53	14.60	0.52	17.96	nr	**32.55**
100 mm dia.	14.24	18.02	0.52	17.96	nr	**35.98**
125 mm dia.	15.84	20.05	0.52	17.96	nr	**38.00**
150 mm dia.	18.47	23.38	0.52	17.96	nr	**41.33**
200 mm dia.	24.42	30.91	0.60	20.72	nr	**51.63**
250 mm dia.	28.43	35.98	0.60	20.72	nr	**56.70**
300 mm dia.	31.65	40.05	0.60	20.72	nr	**60.77**
350 mm dia.	35.39	44.79	0.60	20.72	nr	**65.51**
400 mm dia.	39.77	50.33	0.60	20.72	nr	**71.05**

38 MECHANICAL/COOLING/HEATING SYSTEMS

Item	Net Price £	Material £	Labour hours	Labour £	Unit	Total rate £
THERMAL INSULATION – cont						
Extra over for fittings plantroom insulation – cont						
Tees						
15 mm dia.	4.18	5.30	0.44	15.19	nr	20.49
20 mm dia.	4.37	5.53	0.44	15.19	nr	20.73
25 mm dia.	4.68	5.93	0.44	15.19	nr	21.12
32 mm dia.	4.91	6.21	0.44	15.19	nr	21.40
40 mm dia.	5.23	6.62	0.44	15.19	nr	21.81
50 mm dia.	5.76	7.29	0.44	15.19	nr	22.49
65 mm dia.	6.44	8.15	0.44	15.19	nr	23.34
80 mm dia.	6.93	8.77	0.52	17.96	nr	26.72
100 mm dia.	8.55	10.82	0.52	17.96	nr	28.77
125 mm dia.	9.52	12.05	0.52	17.96	nr	30.01
150 mm dia.	11.10	14.05	0.52	17.96	nr	32.01
200 mm dia.	14.66	18.56	0.60	20.72	nr	39.28
250 mm dia.	17.21	21.77	0.60	20.72	nr	42.49
300 mm dia.	18.97	24.01	0.60	20.72	nr	44.73
350 mm dia.	21.23	26.87	0.60	20.72	nr	47.59
400 mm dia.	23.85	30.19	0.60	20.72	nr	50.91
Valves						
15 mm dia.	23.30	29.49	0.78	26.94	nr	56.42
20 mm dia.	24.31	30.77	0.78	26.94	nr	57.70
25 mm dia.	26.00	32.90	0.78	26.94	nr	59.84
32 mm dia.	27.37	34.64	0.78	26.94	nr	61.58
40 mm dia.	29.00	36.71	0.78	26.94	nr	63.64
50 mm dia.	32.33	40.92	0.78	26.94	nr	67.86
65 mm dia.	36.11	45.70	0.78	26.94	nr	72.63
80 mm dia.	38.99	49.35	0.92	31.77	nr	81.12
100 mm dia.	48.86	61.84	0.92	31.77	nr	93.61
125 mm dia.	54.65	69.16	0.92	31.77	nr	100.93
150 mm dia.	63.80	80.75	0.92	31.77	nr	112.52
200 mm dia.	85.58	108.31	1.12	38.68	nr	146.99
250 mm dia.	100.32	126.97	1.12	38.68	nr	165.64
300 mm dia.	109.79	138.95	1.12	38.68	nr	177.63
350 mm dia.	122.38	154.88	1.12	38.68	nr	193.56
400 mm dia.	137.23	173.68	1.12	38.68	nr	212.35
Pumps						
15 mm dia.	43.11	54.56	2.34	80.81	nr	135.37
20 mm dia.	44.99	56.94	2.34	80.81	nr	137.74
25 mm dia.	48.13	60.91	2.34	80.81	nr	141.72
32 mm dia.	50.63	64.07	2.34	80.81	nr	144.88
40 mm dia.	53.74	68:01	2.34	80.81	nr	148.81
50 mm dia.	59.90	75.80	2.34	80.81	nr	156.61
65 mm dia.	66.90	84.67	2.34	80.81	nr	165.47
80 mm dia.	72.22	91.40	2.76	95.31	nr	186.71
100 mm dia.	90.44	114.46	2.76	95.31	nr	209.77
125 mm dia.	101.19	128.06	2.76	95.31	nr	223.37
150 mm dia.	118.14	149.51	2.76	95.31	nr	244.82

38 MECHANICAL/COOLING/HEATING SYSTEMS

Item	Net Price £	Material £	Labour hours	Labour £	Unit	Total rate £
200 mm dia.	158.49	200.58	3.36	116.03	nr	316.61
250 mm dia.	185.79	235.14	3.36	116.03	nr	351.17
300 mm dia.	265.83	336.44	3.36	116.03	nr	452.47
350 mm dia.	226.61	286.79	3.36	116.03	nr	402.82
400 mm dia.	254.11	321.60	3.36	116.03	nr	437.63
Expansion bellows						
15 mm dia.	34.48	43.63	1.05	36.26	nr	79.89
20 mm dia.	36.03	45.60	1.05	36.26	nr	81.86
25 mm dia.	38.49	48.72	1.05	36.26	nr	84.98
32 mm dia.	40.53	51.29	1.05	36.26	nr	87.55
40 mm dia.	42.99	54.41	1.05	36.26	nr	90.67
50 mm dia.	47.92	60.65	1.05	36.26	nr	96.91
65 mm dia.	53.54	67.75	1.05	36.26	nr	104.01
80 mm dia.	57.76	73.11	1.26	43.51	nr	116.62
100 mm dia.	72.36	91.58	1.26	43.51	nr	135.09
125 mm dia.	80.93	102.43	1.26	43.51	nr	145.94
150 mm dia.	94.51	119.61	1.26	43.51	nr	163.12
200 mm dia.	126.76	160.43	1.53	52.83	nr	213.26
250 mm dia.	148.66	188.15	1.53	52.83	nr	240.98
300 mm dia.	162.63	205.82	1.53	52.83	nr	258.65
350 mm dia.	181.27	229.42	1.53	52.83	nr	282.26
400 mm dia.	203.29	257.28	1.53	52.83	nr	310.11
50 mm thick						
15 mm dia.	16.56	20.96	0.44	15.19	m	36.15
20 mm dia.	17.45	22.08	0.44	15.19	m	37.28
25 mm dia.	18.49	23.40	0.44	15.19	m	38.60
32 mm dia.	19.45	24.61	0.44	15.19	m	39.81
40 mm dia.	20.34	25.74	0.44	15.19	m	40.93
50 mm dia.	22.44	28.41	0.44	15.19	m	43.60
65 mm dia.	24.11	30.51	0.44	15.19	m	45.71
80 mm dia.	25.90	32.77	0.52	17.96	m	50.73
100 mm dia.	31.99	40.49	0.52	17.96	m	58.44
125 mm dia.	35.42	44.83	0.52	17.96	m	62.78
150 mm dia.	40.68	51.49	0.52	17.96	m	69.44
200 mm dia.	53.32	67.49	0.60	20.72	m	88.21
250 mm dia.	61.64	78.01	0.60	20.72	m	98.73
300 mm dia.	67.10	84.92	0.60	20.72	m	105.64
350 mm dia.	74.78	94.64	0.60	20.72	m	115.36
400 mm dia.	83.63	105.84	0.60	20.72	m	126.56
Extra over for fittings plantroom insulation						
Flange/union						
15 mm dia.	19.50	24.68	0.58	20.03	nr	44.71
20 mm dia.	20.59	26.06	0.58	20.03	nr	46.09
25 mm dia.	21.81	27.60	0.58	20.03	nr	47.63
32 mm dia.	22.89	28.97	0.58	20.03	nr	49.00
40 mm dia.	23.99	30.36	0.58	20.03	nr	50.39
50 mm dia.	26.63	33.70	0.58	20.03	nr	53.73
65 mm dia.	28.77	36.41	0.58	20.03	nr	56.44

Material Costs/Prices for Measured Works

38 MECHANICAL/COOLING/HEATING SYSTEMS

Item	Net Price £	Material £	Labour hours	Labour £	Unit	Total rate £
THERMAL INSULATION – cont						
Extra over for fittings plantroom insulation – cont						
Flange/union – cont						
80 mm dia.	30.82	39.01	0.67	23.14	nr	**62.15**
100 mm dia.	38.56	48.80	0.67	23.14	nr	**71.94**
125 mm dia.	42.88	54.27	0.67	23.14	nr	**77.40**
150 mm dia.	49.35	62.46	0.67	23.14	nr	**85.59**
200 mm dia.	65.58	83.00	0.87	30.04	nr	**113.04**
250 mm dia.	75.81	95.95	0.87	30.04	nr	**125.99**
300 mm dia.	81.77	103.49	0.87	30.04	nr	**133.54**
350 mm dia.	90.94	115.10	0.87	30.04	nr	**145.14**
400 mm dia.	101.44	128.39	0.87	30.04	nr	**158.43**
Bends						
15 mm dia.	9.14	11.56	0.44	15.19	nr	**26.76**
20 mm dia.	9.60	12.15	0.44	15.19	nr	**27.35**
25 mm dia.	10.18	12.88	0.44	15.19	nr	**28.08**
32 mm dia.	10.72	13.57	0.44	15.19	nr	**28.76**
40 mm dia.	11.18	14.15	0.44	15.19	nr	**29.34**
50 mm dia.	12.35	15.64	0.44	15.19	nr	**30.83**
65 mm dia.	13.26	16.79	0.44	15.19	nr	**31.98**
80 mm dia.	14.24	18.02	0.52	17.96	nr	**35.98**
100 mm dia.	17.59	22.27	0.52	17.96	nr	**40.22**
125 mm dia.	19.47	24.64	0.52	17.96	nr	**42.60**
150 mm dia.	22.39	28.34	0.52	17.96	nr	**46.29**
200 mm dia.	29.33	37.12	0.60	20.72	nr	**57.83**
250 mm dia.	33.90	42.90	0.60	20.72	nr	**63.62**
300 mm dia.	36.90	46.70	0.60	20.72	nr	**67.42**
350 mm dia.	41.13	52.05	0.60	20.72	nr	**72.77**
400 mm dia.	46.00	58.22	0.60	20.72	nr	**78.94**
Tees						
15 mm dia.	5.47	6.93	0.44	15.19	nr	**22.12**
20 mm dia.	5.75	7.28	0.44	15.19	nr	**22.47**
25 mm dia.	6.09	7.71	0.44	15.19	nr	**22.91**
32 mm dia.	6.43	8.13	0.44	15.19	nr	**23.33**
40 mm dia.	6.72	8.50	0.44	15.19	nr	**23.69**
50 mm dia.	7.40	9.37	0.44	15.19	nr	**24.56**
65 mm dia.	7.95	10.06	0.44	15.19	nr	**25.25**
80 mm dia.	8.55	10.82	0.52	17.96	nr	**28.77**
100 mm dia.	10.56	13.36	0.52	17.96	nr	**31.32**
125 mm dia.	11.69	14.79	0.52	17.96	nr	**32.75**
150 mm dia.	13.41	16.97	0.52	17.96	nr	**34.93**
200 mm dia.	17.59	22.27	0.60	20.72	nr	**42.99**
250 mm dia.	20.35	25.75	0.60	20.72	nr	**46.47**
300 mm dia.	22.16	28.04	0.60	20.72	nr	**48.76**
350 mm dia.	24.69	31.24	0.60	20.72	nr	**51.96**
400 mm dia.	27.61	34.94	0.60	20.72	nr	**55.66**

38 MECHANICAL/COOLING/HEATING SYSTEMS

Item	Net Price £	Material £	Labour hours	Labour £	Unit	Total rate £
Valves						
15 mm dia.	31.01	39.25	0.78	26.94	nr	**66.19**
20 mm dia.	32.68	41.36	0.78	26.94	nr	**68.29**
25 mm dia.	34.64	43.84	0.78	26.94	nr	**70.78**
32 mm dia.	36.36	46.02	0.78	26.94	nr	**72.96**
40 mm dia.	38.08	48.20	0.78	26.94	nr	**75.13**
50 mm dia.	42.27	53.50	0.78	26.94	nr	**80.43**
65 mm dia.	45.67	57.79	0.78	26.94	nr	**84.73**
80 mm dia.	48.98	61.99	0.92	31.77	nr	**93.76**
100 mm dia.	61.25	77.52	0.92	31.77	nr	**109.29**
125 mm dia.	68.10	86.19	0.92	31.77	nr	**117.96**
150 mm dia.	78.38	99.19	0.92	31.77	nr	**130.96**
200 mm dia.	104.21	131.88	1.12	38.68	nr	**170.56**
250 mm dia.	120.40	152.38	1.12	38.68	nr	**191.06**
300 mm dia.	130.43	165.07	1.12	38.68	nr	**203.74**
350 mm dia.	144.43	182.79	1.12	38.68	nr	**221.47**
400 mm dia.	161.14	203.94	1.12	38.68	nr	**242.61**
Pumps						
15 mm dia.	57.43	72.69	2.34	80.81	nr	**153.49**
20 mm dia.	60.54	76.62	2.34	80.81	nr	**157.42**
25 mm dia.	64.12	81.16	2.34	80.81	nr	**161.96**
32 mm dia.	67.34	85.23	2.34	80.81	nr	**166.04**
40 mm dia.	70.56	89.30	2.34	80.81	nr	**170.11**
50 mm dia.	78.24	99.03	2.34	80.81	nr	**179.83**
65 mm dia.	111.89	141.61	2.34	80.81	nr	**222.41**
80 mm dia.	90.74	114.84	2.76	95.31	nr	**210.15**
100 mm dia.	113.40	143.52	2.76	95.31	nr	**238.83**
125 mm dia.	126.12	159.62	2.76	95.31	nr	**254.93**
150 mm dia.	145.12	183.67	2.76	95.31	nr	**278.98**
200 mm dia.	192.95	244.20	3.36	116.03	nr	**360.23**
250 mm dia.	222.95	282.17	3.36	116.03	nr	**398.20**
300 mm dia.	240.58	304.48	3.36	116.03	nr	**420.51**
350 mm dia.	267.43	338.46	3.36	116.03	nr	**454.49**
400 mm dia.	298.41	377.67	3.36	116.03	nr	**493.70**
Expansion bellows						
15 mm dia.	45.92	58.12	1.05	36.26	nr	**94.38**
20 mm dia.	48.41	61.26	1.05	36.26	nr	**97.52**
25 mm dia.	51.29	64.92	1.05	36.26	nr	**101.18**
32 mm dia.	53.86	68.16	1.05	36.26	nr	**104.42**
40 mm dia.	56.43	71.42	1.05	36.26	nr	**107.68**
50 mm dia.	62.60	79.23	1.05	36.26	nr	**115.49**
65 mm dia.	67.64	85.61	1.05	36.26	nr	**121.87**
80 mm dia.	72.61	91.89	1.26	43.51	nr	**135.40**
100 mm dia.	90.73	114.83	1.26	43.51	nr	**158.34**
125 mm dia.	100.89	127.68	1.26	43.51	nr	**171.19**
150 mm dia.	116.12	146.96	1.26	43.51	nr	**190.47**
200 mm dia.	154.37	195.37	1.53	52.83	nr	**248.20**
250 mm dia.	178.37	225.74	1.53	52.83	nr	**278.57**
300 mm dia.	192.51	243.64	1.53	52.83	nr	**296.47**
350 mm dia.	213.95	270.78	1.53	52.83	nr	**323.61**
400 mm dia.	238.71	302.11	1.53	52.83	nr	**354.94**

38 MECHANICAL/COOLING/HEATING SYSTEMS

Item	Net Price £	Material £	Labour hours	Labour £	Unit	Total rate £
THERMAL INSULATION – cont						
Mineral fibre sectional insulation; bright class O foil faced; bright class O foil taped joints; 0.8 mm polyisobutylene sheeting; welded joints						
External pipework						
20 mm thick						
15 mm dia.	7.44	9.41	0.30	10.36	m	**19.77**
20 mm dia.	7.91	10.02	0.30	10.36	m	**20.38**
25 mm dia.	8.55	10.82	0.30	10.36	m	**21.18**
32 mm dia.	9.38	11.87	0.30	10.36	m	**22.23**
40 mm dia.	10.01	12.67	0.30	10.36	m	**23.03**
50 mm dia.	11.31	14.32	0.30	10.36	m	**24.67**
Extra over for fittings external insulation						
Flange/union						
15 mm dia.	11.39	14.41	0.75	25.90	nr	**40.31**
20 mm dia.	12.10	15.31	0.75	25.90	nr	**41.21**
25 mm dia.	12.99	16.44	0.75	25.90	nr	**42.34**
32 mm dia.	14.16	17.93	0.75	25.90	nr	**43.82**
40 mm dia.	14.95	18.92	0.75	25.90	nr	**44.82**
50 mm dia.	16.75	21.20	0.75	25.90	nr	**47.10**
Bends						
15 mm dia.	1.89	2.39	0.30	10.36	nr	**12.75**
20 mm dia.	1.99	2.51	0.30	10.36	nr	**12.87**
25 mm dia.	2.14	2.71	0.30	10.36	nr	**13.07**
32 mm dia.	2.35	2.98	0.30	10.36	nr	**13.34**
40 mm dia.	2.51	3.17	0.30	10.36	nr	**13.53**
50 mm dia.	2.82	3.57	0.30	10.36	nr	**13.93**
Tees						
15 mm dia.	1.89	2.39	0.30	10.36	nr	**12.75**
20 mm dia.	1.99	2.51	0.30	10.36	nr	**12.87**
25 mm dia.	2.14	2.71	0.30	10.36	nr	**13.07**
32 mm dia.	2.35	2.98	0.30	10.36	nr	**13.34**
40 mm dia.	2.51	3.17	0.30	10.36	nr	**13.53**
50 mm dia.	2.82	3.57	0.30	10.36	nr	**13.93**
Valves						
15 mm dia.	18.10	22.91	1.03	35.57	nr	**58.48**
20 mm dia.	19.23	24.33	1.03	35.57	nr	**59.90**
25 mm dia.	20.63	26.12	1.03	35.57	nr	**61.68**
32 mm dia.	22.47	28.43	1.03	35.57	nr	**64.00**
40 mm dia.	23.72	30.02	1.03	35.57	nr	**65.59**
50 mm dia.	26.58	33.65	1.03	35.57	nr	**69.21**
Expansion bellows						
15 mm dia.	26.80	33.91	1.42	49.04	nr	**82.95**
20 mm dia.	28.49	36.06	1.42	49.04	nr	**85.10**
25 mm dia.	30.61	38.74	1.42	49.04	nr	**87.78**

38 MECHANICAL/COOLING/HEATING SYSTEMS

Item	Net Price £	Material £	Labour hours	Labour £	Unit	Total rate £
32 mm dia.	33.30	42.14	1.42	49.04	nr	**91.18**
40 mm dia.	35.14	44.48	1.42	49.04	nr	**93.51**
50 mm dia.	39.39	49.86	1.42	49.04	nr	**98.89**
25 mm thick						
15 mm dia.	8.20	10.38	0.30	10.36	m	**20.74**
20 mm dia.	8.80	11.14	0.30	10.36	m	**21.50**
25 mm dia.	9.68	12.25	0.30	10.36	m	**22.61**
32 mm dia.	10.48	13.26	0.30	10.36	m	**23.62**
40 mm dia.	11.16	14.12	0.30	10.36	m	**24.48**
50 mm dia.	12.60	15.94	0.30	10.36	m	**26.30**
65 mm dia.	14.29	18.08	0.30	10.36	m	**28.44**
80 mm dia.	15.64	19.79	0.40	13.81	m	**33.61**
100 mm dia.	19.74	24.98	0.40	13.81	m	**38.79**
125 mm dia.	22.73	28.77	0.40	13.81	m	**42.58**
150 mm dia.	26.66	33.74	0.40	13.81	m	**47.56**
200 mm dia.	36.14	45.74	0.50	17.27	m	**63.01**
250 mm dia.	43.17	54.63	0.50	17.27	m	**71.90**
300 mm dia.	46.94	59.41	0.50	17.27	m	**76.68**
Extra over for fittings external insulation						
Flange/union						
15 mm dia.	12.53	15.86	0.75	25.90	nr	**41.76**
20 mm dia.	13.52	17.11	0.75	25.90	nr	**43.01**
25 mm dia.	14.84	18.78	0.75	25.90	nr	**44.68**
32 mm dia.	15.86	20.07	0.75	25.90	nr	**45.97**
40 mm dia.	16.98	21.49	0.75	25.90	nr	**47.39**
50 mm dia.	19.04	24.09	0.75	25.90	nr	**49.99**
65 mm dia.	21.52	27.24	0.75	25.90	nr	**53.14**
80 mm dia.	23.41	29.63	0.89	30.73	nr	**60.36**
100 mm dia.	29.09	36.82	0.89	30.73	nr	**67.55**
125 mm dia.	33.69	42.64	0.89	30.73	nr	**73.37**
150 mm dia.	39.22	49.63	0.89	30.73	nr	**80.37**
200 mm dia.	53.06	67.15	1.15	39.71	nr	**106.86**
250 mm dia.	63.17	79.95	1.15	39.71	nr	**119.66**
300 mm dia.	70.17	88.81	1.15	39.71	nr	**128.52**
Bends						
15 mm dia.	2.04	2.58	0.30	10.36	nr	**12.94**
20 mm dia.	2.21	2.80	0.30	10.36	nr	**13.16**
25 mm dia.	2.40	3.03	0.30	10.36	nr	**13.39**
32 mm dia.	2.60	3.29	0.30	10.36	nr	**13.65**
40 mm dia.	2.79	3.53	0.30	10.36	nr	**13.89**
50 mm dia.	3.15	3.99	0.30	10.36	nr	**14.35**
65 mm dia.	3.59	4.54	0.30	10.36	nr	**14.90**
80 mm dia.	3.91	4.94	0.40	13.81	nr	**18.76**
100 mm dia.	4.93	6.24	0.40	13.81	nr	**20.05**
125 mm dia.	5.67	7.18	0.40	13.81	nr	**20.99**
150 mm dia.	6.68	8.46	0.40	13.81	nr	**22.27**
200 mm dia.	9.02	11.42	0.50	17.27	nr	**28.69**
250 mm dia.	10.80	13.67	0.50	17.27	nr	**30.94**
300 mm dia.	11.73	14.85	0.50	17.27	nr	**32.12**

38 MECHANICAL/COOLING/HEATING SYSTEMS

Item	Net Price £	Material £	Labour hours	Labour £	Unit	Total rate £
THERMAL INSULATION – cont						
Extra over for fittings external insulation – cont						
Tees						
15 mm dia.	2.04	2.58	0.30	10.36	nr	12.94
20 mm dia.	2.21	2.80	0.30	10.36	nr	13.16
25 mm dia.	2.40	3.03	0.30	10.36	nr	13.39
32 mm dia.	2.60	3.29	0.30	10.36	nr	13.65
40 mm dia.	2.79	3.53	0.30	10.36	nr	13.89
50 mm dia.	3.15	3.99	0.30	10.36	nr	14.35
65 mm dia.	3.59	4.54	0.30	10.36	nr	14.90
80 mm dia.	3.91	4.94	0.40	13.81	nr	18.76
100 mm dia.	4.93	6.24	0.40	13.81	nr	20.05
125 mm dia.	5.67	7.18	0.40	13.81	nr	20.99
150 mm dia.	6.68	8.46	0.40	13.81	nr	22.27
200 mm dia.	9.02	11.42	0.50	17.27	nr	28.69
250 mm dia.	10.80	13.67	0.50	17.27	nr	30.94
300 mm dia.	11.73	14.85	0.50	17.27	nr	32.12
Valves						
15 mm dia.	19.92	25.22	1.03	35.57	nr	60.79
20 mm dia.	21.47	27.17	1.03	35.57	nr	62.74
25 mm dia.	23.54	29.80	1.03	35.57	nr	65.36
32 mm dia.	25.23	31.93	1.03	35.57	nr	67.50
40 mm dia.	26.95	34.11	1.03	35.57	nr	69.68
50 mm dia.	30.27	38.31	1.03	35.57	nr	73.88
65 mm dia.	34.23	43.32	1.03	35.57	nr	78.89
80 mm dia.	37.20	47.08	1.25	43.17	nr	90.24
100 mm dia.	46.24	58.52	1.25	43.17	nr	101.69
125 mm dia.	53.51	67.73	1.25	43.17	nr	110.89
150 mm dia.	62.30	78.85	1.25	43.17	nr	122.02
200 mm dia.	84.30	106.70	1.55	53.53	nr	160.22
250 mm dia.	100.32	126.97	1.55	53.53	nr	180.49
300 mm dia.	111.44	141.04	1.55	53.53	nr	194.57
Expansion bellows						
15 mm dia.	29.53	37.37	1.42	49.04	nr	86.40
20 mm dia.	31.81	40.26	1.42	49.04	nr	89.30
25 mm dia.	34.89	44.15	1.42	49.04	nr	93.19
32 mm dia.	37.42	47.36	1.42	49.04	nr	96.39
40 mm dia.	39.97	50.59	1.42	49.04	nr	99.62
50 mm dia.	44.80	56.70	1.42	49.04	nr	105.73
65 mm dia.	50.68	64.14	1.42	49.04	nr	113.18
80 mm dia.	55.10	69.74	1.75	60.43	nr	130.17
100 mm dia.	68.50	86.69	1.75	60.43	nr	147.12
125 mm dia.	79.28	100.33	1.75	60.43	nr	160.76
150 mm dia.	92.29	116.80	1.75	60.43	nr	177.23
200 mm dia.	124.88	158.04	2.17	74.94	nr	232.98
250 mm dia.	148.62	188.09	2.17	74.94	nr	263.03
300 mm dia.	165.10	208.95	3.17	109.47	nr	318.42

38 MECHANICAL/COOLING/HEATING SYSTEMS

Item	Net Price £	Material £	Labour hours	Labour £	Unit	Total rate £
30 mm thick						
15 mm dia.	11.89	15.05	0.30	10.36	m	**25.41**
20 mm dia.	10.76	13.61	0.30	10.36	m	**23.97**
25 mm dia.	11.39	14.41	0.30	10.36	m	**24.77**
32 mm dia.	12.37	15.65	0.30	10.36	m	**26.01**
40 mm dia.	13.02	16.48	0.30	10.36	m	**26.84**
50 mm dia.	14.66	18.56	0.30	10.36	m	**28.92**
65 mm dia.	16.48	20.86	0.30	10.36	m	**31.22**
80 mm dia.	17.92	22.67	0.40	13.81	m	**36.49**
100 mm dia.	22.41	28.36	0.40	13.81	m	**42.18**
125 mm dia.	25.66	32.48	0.40	13.81	m	**46.29**
150 mm dia.	29.80	37.72	0.40	13.81	m	**51.53**
200 mm dia.	40.14	50.80	0.50	17.27	m	**68.06**
250 mm dia.	47.63	60.28	0.50	17.27	m	**77.55**
300 mm dia.	51.37	65.01	0.50	17.27	m	**82.28**
350 mm dia.	56.17	71.08	0.50	17.27	m	**88.35**
Extra over for fittings external insulation						
Flange/union						
15 mm dia.	15.34	19.41	0.75	25.90	nr	**45.31**
20 mm dia.	16.24	20.55	0.75	25.90	nr	**46.45**
25 mm dia.	17.24	21.82	0.75	25.90	nr	**47.72**
32 mm dia.	18.80	23.80	0.75	25.90	nr	**49.70**
40 mm dia.	19.64	24.85	0.75	25.90	nr	**50.75**
50 mm dia.	21.91	27.73	0.75	25.90	nr	**53.63**
65 mm dia.	24.68	31.23	0.75	25.90	nr	**57.13**
80 mm dia.	26.88	34.02	0.89	30.73	nr	**64.76**
100 mm dia.	33.10	41.89	0.89	30.73	nr	**72.63**
125 mm dia.	37.83	47.88	0.89	30.73	nr	**78.61**
150 mm dia.	43.93	55.60	0.89	30.73	nr	**86.34**
200 mm dia.	58.69	74.27	1.15	39.71	nr	**113.99**
250 mm dia.	69.50	87.96	1.15	39.71	nr	**127.67**
300 mm dia.	76.17	96.40	1.15	39.71	nr	**136.11**
350 mm dia.	84.13	106.47	1.15	39.71	nr	**146.18**
Bends						
15 mm dia.	2.52	3.19	0.30	10.36	nr	**13.55**
20 mm dia.	2.69	3.40	0.30	10.36	nr	**13.76**
25 mm dia.	2.86	3.62	0.30	10.36	nr	**13.98**
32 mm dia.	3.09	3.91	0.30	10.36	nr	**14.27**
40 mm dia.	3.25	4.12	0.30	10.36	nr	**14.48**
50 mm dia.	3.67	4.65	0.30	10.36	nr	**15.01**
65 mm dia.	4.13	5.23	0.30	10.36	nr	**15.59**
80 mm dia.	4.50	5.69	0.40	13.81	nr	**19.50**
100 mm dia.	5.63	7.12	0.40	13.81	nr	**20.94**
125 mm dia.	6.43	8.13	0.40	13.81	nr	**21.95**
150 mm dia.	7.45	9.43	0.40	13.81	nr	**23.24**
200 mm dia.	10.05	12.71	0.50	17.27	nr	**29.98**
250 mm dia.	11.93	15.10	0.50	17.27	nr	**32.37**
300 mm dia.	12.83	16.24	0.50	17.27	nr	**33.51**
350 mm dia.	14.04	17.77	0.50	17.27	nr	**35.04**

38 MECHANICAL/COOLING/HEATING SYSTEMS

Item	Net Price £	Material £	Labour hours	Labour £	Unit	Total rate £
THERMAL INSULATION – cont						
Extra over for fittings external insulation – cont						
Tees						
15 mm dia.	2.52	3.19	0.30	10.36	nr	**13.55**
20 mm dia.	2.69	3.40	0.30	10.36	nr	**13.76**
25 mm dia.	2.86	3.62	0.30	10.36	nr	**13.98**
32 mm dia.	3.09	3.91	0.30	10.36	nr	**14.27**
40 mm dia.	3.25	4.12	0.30	10.36	nr	**14.48**
50 mm dia.	3.67	4.65	0.30	10.36	nr	**15.01**
65 mm dia.	4.13	5.23	0.30	10.36	nr	**15.59**
80 mm dia.	4.50	5.69	0.40	13.81	nr	**19.50**
100 mm dia.	5.63	7.12	0.40	13.81	nr	**20.94**
125 mm dia.	6.43	8.13	0.40	13.81	nr	**21.95**
150 mm dia.	7.45	9.43	0.40	13.81	nr	**23.24**
200 mm dia.	10.05	12.71	0.50	17.27	nr	**29.98**
250 mm dia.	11.93	15.10	0.50	17.27	nr	**32.37**
300 mm dia.	12.83	16.24	0.50	17.27	nr	**33.51**
350 mm dia.	14.04	17.77	0.50	17.27	nr	**35.04**
Valves						
15 mm dia.	24.36	30.84	1.03	35.57	nr	**66.40**
20 mm dia.	25.82	32.68	1.03	35.57	nr	**68.24**
25 mm dia.	27.38	34.66	1.03	35.57	nr	**70.23**
32 mm dia.	29.81	37.73	1.03	35.57	nr	**73.30**
40 mm dia.	31.20	39.49	1.03	35.57	nr	**75.06**
50 mm dia.	34.80	44.04	1.03	35.57	nr	**79.61**
65 mm dia.	39.18	49.59	1.03	35.57	nr	**85.16**
80 mm dia.	42.69	54.03	1.25	43.17	nr	**97.19**
100 mm dia.	52.53	66.48	1.25	43.17	nr	**109.64**
125 mm dia.	60.04	75.99	1.25	43.17	nr	**119.15**
150 mm dia.	69.75	88.28	1.25	43.17	nr	**131.44**
200 mm dia.	93.24	118.00	1.55	53.53	nr	**171.53**
250 mm dia.	110.40	139.72	1.55	53.53	nr	**193.25**
300 mm dia.	120.96	153.08	1.55	53.53	nr	**206.61**
350 mm dia.	133.62	169.11	1.55	53.53	nr	**222.64**
Expansion bellows						
15 mm dia.	36.08	45.66	1.42	49.04	nr	**94.69**
20 mm dia.	38.26	48.42	1.42	49.04	nr	**97.46**
25 mm dia.	40.54	51.30	1.42	49.04	nr	**100.34**
32 mm dia.	44.19	55.93	1.42	49.04	nr	**104.96**
40 mm dia.	46.24	58.52	1.42	49.04	nr	**107.56**
50 mm dia.	51.55	65.24	1.42	49.04	nr	**114.28**
65 mm dia.	58.05	73.47	1.42	49.04	nr	**122.51**
80 mm dia.	63.28	80.09	1.75	60.43	nr	**140.52**
100 mm dia.	77.83	98.51	1.75	60.43	nr	**158.94**
125 mm dia.	88.96	112.58	1.75	60.43	nr	**173.01**
150 mm dia.	103.33	130.77	1.75	60.43	nr	**191.21**
200 mm dia.	138.12	174.80	2.17	74.94	nr	**249.74**
250 mm dia.	163.56	207.00	2.17	74.94	nr	**281.94**
300 mm dia.	179.20	226.79	2.17	74.94	nr	**301.73**
350 mm dia.	197.98	250.56	2.17	74.94	nr	**325.50**

38 MECHANICAL/COOLING/HEATING SYSTEMS

Item	Net Price £	Material £	Labour hours	Labour £	Unit	Total rate £
40 mm thick						
15 mm dia.	12.63	15.99	0.30	10.36	m	**26.35**
20 mm dia.	13.08	16.55	0.30	10.36	m	**26.91**
25 mm dia.	13.96	17.67	0.30	10.36	m	**28.03**
32 mm dia.	14.87	18.82	0.30	10.36	m	**29.18**
40 mm dia.	15.54	19.67	0.30	10.36	m	**30.03**
50 mm dia.	17.33	21.93	0.30	10.36	m	**32.29**
65 mm dia.	19.35	24.49	0.30	10.36	m	**34.85**
80 mm dia.	20.97	26.54	0.40	13.81	m	**40.35**
100 mm dia.	26.16	33.11	0.40	13.81	m	**46.92**
125 mm dia.	29.48	37.31	0.40	13.81	m	**51.12**
150 mm dia.	34.04	43.09	0.40	13.81	m	**56.90**
200 mm dia.	45.41	57.47	0.50	17.27	m	**74.74**
250 mm dia.	53.18	67.30	0.50	17.27	m	**84.57**
300 mm dia.	57.56	72.85	0.50	17.27	m	**90.12**
350 mm dia.	63.28	80.09	0.50	17.27	m	**97.35**
400 mm dia.	70.42	89.12	0.50	17.27	m	**106.39**
Extra over for fittings external insulation						
Flange/union						
15 mm dia.	18.88	23.90	0.75	25.90	nr	**49.80**
20 mm dia.	19.72	24.96	0.75	25.90	nr	**50.86**
25 mm dia.	21.02	26.61	0.75	25.90	nr	**52.51**
32 mm dia.	22.16	28.04	0.75	25.90	nr	**53.94**
40 mm dia.	23.39	29.60	0.75	25.90	nr	**55.50**
50 mm dia.	25.91	32.79	0.75	25.90	nr	**58.69**
65 mm dia.	28.87	36.54	0.75	25.90	nr	**62.44**
80 mm dia.	31.12	39.39	0.89	30.73	nr	**70.13**
100 mm dia.	38.33	48.51	0.89	30.73	nr	**79.24**
125 mm dia.	42.88	54.27	0.89	30.73	nr	**85.00**
150 mm dia.	49.68	62.88	0.89	30.73	nr	**93.61**
200 mm dia.	65.51	82.91	1.15	39.71	nr	**122.62**
250 mm dia.	76.80	97.20	1.15	39.71	nr	**136.91**
300 mm dia.	84.38	106.79	1.15	39.71	nr	**146.51**
350 mm dia.	93.60	118.45	1.15	39.71	nr	**158.17**
400 mm dia.	104.81	132.64	1.15	39.71	nr	**172.36**
Bends						
15 mm dia.	3.15	3.99	0.30	10.36	nr	**14.35**
20 mm dia.	3.27	4.14	0.30	10.36	nr	**14.50**
25 mm dia.	3.50	4.43	0.30	10.36	nr	**14.79**
32 mm dia.	3.71	4.69	0.30	10.36	nr	**15.05**
40 mm dia.	3.91	4.94	0.30	10.36	nr	**15.30**
50 mm dia.	4.36	5.52	0.30	10.36	nr	**15.88**
65 mm dia.	4.85	6.14	0.30	10.36	nr	**16.50**
80 mm dia.	5.26	6.66	0.40	13.81	nr	**20.47**
100 mm dia.	6.54	8.27	0.40	13.81	nr	**22.09**
125 mm dia.	7.37	9.33	0.40	13.81	nr	**23.14**
150 mm dia.	8.52	10.79	0.40	13.81	nr	**24.60**
200 mm dia.	11.36	14.37	0.50	17.27	nr	**31.64**
250 mm dia.	13.29	16.82	0.50	17.27	nr	**34.08**
300 mm dia.	14.39	18.21	0.50	17.27	nr	**35.47**
350 mm dia.	15.81	20.00	0.50	17.27	nr	**37.27**
400 mm dia.	17.62	22.29	0.50	17.27	nr	**39.56**

38 MECHANICAL/COOLING/HEATING SYSTEMS

Item	Net Price £	Material £	Labour hours	Labour £	Unit	Total rate £
THERMAL INSULATION – cont						
Extra over for fittings external insulation – cont						
Tees						
15 mm dia.	3.15	3.99	0.30	10.36	nr	**14.35**
20 mm dia.	3.27	4.14	0.30	10.36	nr	**14.50**
25 mm dia.	3.50	4.43	0.30	10.36	nr	**14.79**
32 mm dia.	3.71	4.69	0.30	10.36	nr	**15.05**
40 mm dia.	3.91	4.94	0.30	10.36	nr	**15.30**
50 mm dia.	4.36	5.52	0.30	10.36	nr	**15.88**
65 mm dia.	4.85	6.14	0.30	10.36	nr	**16.50**
80 mm dia.	5.26	6.66	0.40	13.81	nr	**20.47**
100 mm dia.	6.54	8.27	0.40	13.81	nr	**22.09**
125 mm dia.	7.37	9.33	0.40	13.81	nr	**23.14**
150 mm dia.	8.52	10.79	0.40	13.81	nr	**24.60**
200 mm dia.	11.36	14.37	0.50	17.27	nr	**31.64**
250 mm dia.	13.29	16.82	0.50	17.27	nr	**34.08**
300 mm dia.	14.39	18.21	0.50	17.27	nr	**35.47**
350 mm dia.	15.81	20.00	0.50	17.27	nr	**37.27**
400 mm dia.	17.62	22.29	0.50	17.27	nr	**39.56**
Valves						
15 mm dia.	29.97	37.93	1.03	35.57	nr	**73.50**
20 mm dia.	31.32	39.64	1.03	35.57	nr	**75.21**
25 mm dia.	33.38	42.24	1.03	35.57	nr	**77.81**
32 mm dia.	35.15	44.49	1.03	35.57	nr	**80.06**
40 mm dia.	37.17	47.05	1.03	35.57	nr	**82.62**
50 mm dia.	41.19	52.13	1.03	35.57	nr	**87.70**
65 mm dia.	45.87	58.05	1.03	35.57	nr	**93.62**
80 mm dia.	49.43	62.56	1.25	43.17	nr	**105.72**
100 mm dia.	60.88	77.05	1.25	43.17	nr	**120.22**
125 mm dia.	68.10	86.19	1.25	43.17	nr	**129.35**
150 mm dia.	78.94	99.91	1.25	43.17	nr	**143.08**
200 mm dia.	104.06	131.70	1.55	53.53	nr	**185.23**
250 mm dia.	121.92	154.30	1.55	53.53	nr	**207.83**
300 mm dia.	134.04	169.65	1.55	53.53	nr	**223.17**
350 mm dia.	148.66	188.15	1.55	53.53	nr	**241.67**
400 mm dia.	166.44	210.65	1.55	53.53	nr	**264.18**
Expansion bellows						
15 mm dia.	44.39	56.18	1.42	49.04	nr	**105.21**
20 mm dia.	46.41	58.74	1.42	49.04	nr	**107.77**
25 mm dia.	49.44	62.57	1.42	49.04	nr	**111.61**
32 mm dia.	52.09	65.93	1.42	49.04	nr	**114.96**
40 mm dia.	55.07	69.69	1.42	49.04	nr	**118.73**
50 mm dia.	61.04	77.25	1.42	49.04	nr	**126.29**
65 mm dia.	67.93	85.97	1.42	49.04	nr	**135.01**
80 mm dia.	73.23	92.68	1.75	60.43	nr	**153.11**
100 mm dia.	90.21	114.17	1.75	60.43	nr	**174.60**
125 mm dia.	100.89	127.68	1.75	60.43	nr	**188.12**
150 mm dia.	116.96	148.03	1.75	60.43	nr	**208.46**

38 MECHANICAL/COOLING/HEATING SYSTEMS

Item	Net Price £	Material £	Labour hours	Labour £	Unit	Total rate £
200 mm dia.	154.17	195.11	2.17	74.94	nr	**270.05**
250 mm dia.	180.62	228.59	2.17	74.94	nr	**303.53**
300 mm dia.	198.58	251.32	2.17	74.94	nr	**326.26**
350 mm dia.	220.25	278.74	2.17	74.94	nr	**353.68**
400 mm dia.	246.61	312.11	2.17	74.94	nr	**387.04**
50 mm thick						
15 mm dia.	16.65	21.07	0.30	10.36	m	**31.43**
20 mm dia.	17.47	22.11	0.30	10.36	m	**32.47**
25 mm dia.	18.47	23.38	0.30	10.36	m	**33.74**
32 mm dia.	19.94	25.23	0.30	10.36	m	**35.59**
40 mm dia.	20.26	25.64	0.30	10.36	m	**36.00**
50 mm dia.	22.44	28.41	0.30	10.36	m	**38.77**
65 mm dia.	24.46	30.96	0.30	10.36	m	**41.32**
80 mm dia.	26.16	33.11	0.40	13.81	m	**46.92**
100 mm dia.	32.40	41.01	0.40	13.81	m	**54.82**
125 mm dia.	36.25	45.88	0.40	13.81	m	**59.69**
150 mm dia.	41.45	52.46	0.40	13.81	m	**66.27**
200 mm dia.	54.88	69.45	0.50	17.27	m	**86.72**
250 mm dia.	63.39	80.23	0.50	17.27	m	**97.50**
300 mm dia.	67.99	86.04	0.50	17.27	m	**103.31**
350 mm dia.	74.63	94.45	0.50	17.27	m	**111.71**
400 mm dia.	82.66	104.62	0.50	17.27	m	**121.88**
Extra over for fittings external insulation						
Flange/union						
15 mm dia.	24.54	31.06	0.75	25.90	nr	**56.96**
20 mm dia.	25.82	32.68	0.75	25.90	nr	**58.58**
25 mm dia.	27.28	34.53	0.75	25.90	nr	**60.43**
32 mm dia.	28.64	36.24	0.75	25.90	nr	**62.14**
40 mm dia.	29.97	37.93	0.75	25.90	nr	**63.83**
50 mm dia.	33.03	41.81	0.75	25.90	nr	**67.71**
65 mm dia.	35.72	45.21	0.75	25.90	nr	**71.11**
80 mm dia.	38.30	48.47	0.89	30.73	nr	**79.20**
100 mm dia.	47.09	59.59	0.89	30.73	nr	**90.33**
125 mm dia.	52.34	66.24	0.89	30.73	nr	**96.97**
150 mm dia.	59.94	75.86	0.89	30.73	nr	**106.59**
200 mm dia.	78.40	99.22	1.15	39.71	nr	**138.93**
250 mm dia.	90.59	114.65	1.15	39.71	nr	**154.36**
300 mm dia.	98.29	124.40	1.15	39.71	nr	**164.11**
350 mm dia.	108.77	137.66	1.15	39.71	nr	**177.37**
400 mm dia.	121.20	153.39	1.15	39.71	nr	**193.10**
Bends						
15 mm dia.	4.16	5.27	0.30	10.36	nr	**15.63**
20 mm dia.	4.37	5.53	0.30	10.36	nr	**15.89**
25 mm dia.	4.62	5.84	0.30	10.36	nr	**16.20**
32 mm dia.	4.85	6.14	0.30	10.36	nr	**16.50**
40 mm dia.	5.08	6.43	0.30	10.36	nr	**16.79**
50 mm dia.	5.63	7.12	0.30	10.36	nr	**17.48**
65 mm dia.	6.11	7.73	0.30	10.36	nr	**18.09**

38 MECHANICAL/COOLING/HEATING SYSTEMS

Item	Net Price £	Material £	Labour hours	Labour £	Unit	Total rate £
THERMAL INSULATION – cont						
Extra over for fittings external insulation – cont						
Bends – cont						
80 mm dia.	6.52	8.25	0.40	13.81	nr	**22.06**
100 mm dia.	8.10	10.26	0.40	13.81	nr	**24.07**
125 mm dia.	9.04	11.44	0.40	13.81	nr	**25.25**
150 mm dia.	10.38	13.14	0.40	13.81	nr	**26.95**
200 mm dia.	13.72	17.36	0.50	17.27	nr	**34.63**
250 mm dia.	15.85	20.06	0.50	17.27	nr	**37.33**
300 mm dia.	17.02	21.54	0.50	17.27	nr	**38.80**
350 mm dia.	18.66	23.61	0.50	17.27	nr	**40.88**
400 mm dia.	20.65	26.13	0.50	17.27	nr	**43.40**
Tees						
15 mm dia.	4.16	5.27	0.30	10.36	nr	**15.63**
20 mm dia.	4.37	5.53	0.30	10.36	nr	**15.89**
25 mm dia.	4.62	5.84	0.30	10.36	nr	**16.20**
32 mm dia.	4.85	6.14	0.30	10.36	nr	**16.50**
40 mm dia.	5.08	6.43	0.30	10.36	nr	**16.79**
50 mm dia.	5.63	7.12	0.30	10.36	nr	**17.48**
65 mm dia.	6.11	7.73	0.30	10.36	nr	**18.09**
80 mm dia.	6.52	8.25	0.40	13.81	nr	**22.06**
100 mm dia.	8.10	10.26	0.40	13.81	nr	**24.07**
125 mm dia.	9.04	11.44	0.40	13.81	nr	**25.25**
150 mm dia.	10.38	13.14	0.40	13.81	nr	**26.95**
200 mm dia.	13.72	17.36	0.50	17.27	nr	**34.63**
250 mm dia.	15.85	20.06	0.50	17.27	nr	**37.33**
300 mm dia.	17.02	21.54	0.50	17.27	nr	**38.80**
350 mm dia.	18.66	23.61	0.50	17.27	nr	**40.88**
400 mm dia.	20.65	26.13	0.50	17.27	nr	**43.40**
Valves						
15 mm dia.	38.97	49.32	1.03	35.57	nr	**84.89**
20 mm dia.	40.99	51.88	1.03	35.57	nr	**87.45**
25 mm dia.	43.33	54.84	1.03	35.57	nr	**90.41**
32 mm dia.	45.50	57.58	1.03	35.57	nr	**93.15**
40 mm dia.	47.61	60.25	1.03	35.57	nr	**95.82**
50 mm dia.	52.50	66.45	1.03	35.57	nr	**102.02**
65 mm dia.	56.78	71.86	1.03	35.57	nr	**107.42**
80 mm dia.	60.83	76.98	1.25	43.17	nr	**120.15**
100 mm dia.	74.78	94.64	1.25	43.17	nr	**137.81**
125 mm dia.	83.12	105.19	1.25	43.17	nr	**148.36**
150 mm dia.	95.15	120.42	1.25	43.17	nr	**163.59**
200 mm dia.	124.50	157.56	1.55	53.53	nr	**211.09**
250 mm dia.	143.87	182.08	1.55	53.53	nr	**235.60**
300 mm dia.	156.10	197.56	1.55	53.53	nr	**251.08**
350 mm dia.	172.75	218.63	1.55	53.53	nr	**272.16**
400 mm dia.	192.46	243.58	1.55	53.53	nr	**297.11**
Expansion bellows						
15 mm dia.	57.73	73.06	1.42	49.04	nr	**122.10**
20 mm dia.	60.74	76.87	1.42	49.04	nr	**125.91**
25 mm dia.	64.20	81.25	1.42	49.04	nr	**130.29**

38 MECHANICAL/COOLING/HEATING SYSTEMS

Item	Net Price £	Material £	Labour hours	Labour £	Unit	Total rate £
32 mm dia.	67.44	85.36	1.42	49.04	nr	**134.39**
40 mm dia.	70.52	89.25	1.42	49.04	nr	**138.28**
50 mm dia.	77.73	98.38	1.42	49.04	nr	**147.42**
65 mm dia.	84.09	106.43	1.42	49.04	nr	**155.47**
80 mm dia.	90.13	114.07	1.75	60.43	nr	**174.50**
100 mm dia.	110.79	140.21	1.75	60.43	nr	**200.65**
125 mm dia.	123.14	155.85	1.75	60.43	nr	**216.28**
150 mm dia.	140.98	178.43	1.75	60.43	nr	**238.86**
200 mm dia.	184.46	233.45	2.17	74.94	nr	**308.39**
250 mm dia.	213.15	269.77	2.17	74.94	nr	**344.70**
300 mm dia.	231.26	292.68	2.17	74.94	nr	**367.61**
350 mm dia.	255.92	323.89	2.17	74.94	nr	**398.83**
400 mm dia.	285.16	360.90	2.17	74.94	nr	**435.83**

38 MECHANICAL/COOLING/HEATING SYSTEMS

Item	Net Price £	Material £	Labour hours	Labour £	Unit	Total rate £
STEAM HEATING						
PIPELINES						
For pipework prices refer to Section – Low Temperature Hot Water Heating						
PIPELINE ANCILLARIES						
Steam traps and accessories						
Cast iron; inverted bucket type; steam trap pressure range up to 17 bar at 210°C; screwed ends						
½" dia.	131.80	166.80	0.85	29.35	nr	196.15
¾" dia.	131.80	166.80	1.13	39.02	nr	205.82
1" dia.	222.07	281.05	1.35	46.62	nr	327.67
1½" dia.	685.61	867.70	1.80	62.16	nr	929.86
2" dia.	767.42	971.25	2.18	75.28	nr	1046.53
Cast iron; inverted bucket type; steam trap pressure range up to 17 bar at 210°C; flanged ends to BS 4504 PN16; bolted connections						
15 mm dia.	214.31	271.24	1.15	39.71	nr	310.95
20 mm dia.	266.75	337.60	1.25	43.17	nr	380.77
25 mm dia.	331.16	419.11	1.33	45.93	nr	465.04
40 mm dia.	702.52	889.10	1.46	50.42	nr	939.52
50 mm dia.	784.33	992.65	1.60	55.25	nr	1047.90
Steam traps and strainers						
Stainless steel; thermodynamic trap with pressure range up to 42 bar; temperature range to 400°C; screwed ends to steel						
15 mm dia.	85.67	108.42	0.84	29.02	nr	137.44
20 mm dia.	96.97	122.72	1.14	39.38	nr	162.10
Stainless steel; thermodynamic trap with pressure range up to 24 bar; temperature range to 288°C; flanged ends to DIN 2456 PN64; bolted connections						
15 mm dia.	179.37	227.01	1.24	42.84	nr	269.86
20 mm dia.	196.61	248.83	1.34	46.29	nr	295.12
25 mm dia.	227.29	287.66	1.40	48.37	nr	336.02
Malleable iron pipeline strainer; max steam working pressure 14 bar and temperature range to 230°C; screwed ends to steel						
½" dia.	25.70	32.52	0.84	29.02	nr	61.54
¾" dia.	32.12	40.64	1.14	39.38	nr	80.02
1" dia.	50.12	63.43	1.30	44.91	nr	108.33
1½" dia.	102.92	130.26	1.50	51.85	nr	182.11
2" dia.	144.51	182.89	1.74	60.16	nr	243.05

38 MECHANICAL/COOLING/HEATING SYSTEMS

Item	Net Price £	Material £	Labour hours	Labour £	Unit	Total rate £
Bronze pipeline strainer; max steam working pressure 25 bar; flanged ends to BS 4504 PN25; bolted connections						
15 mm dia.	220.76	279.39	1.24	42.84	nr	**322.23**
20 mm dia.	269.27	340.79	1.34	46.29	nr	**387.08**
25 mm dia.	308.73	390.73	1.40	48.37	nr	**439.09**
32 mm dia.	479.26	606.56	1.46	50.49	nr	**657.04**
40 mm dia.	543.88	688.34	1.54	53.21	nr	**741.55**
50 mm dia.	836.48	1058.66	1.64	56.70	nr	**1115.36**
65 mm dia.	926.20	1172.20	2.50	86.33	nr	**1258.54**
80 mm dia.	1152.38	1458.45	2.91	100.39	nr	**1558.84**
100 mm dia.	1996.03	2526.17	3.51	121.17	nr	**2647.34**
Balanced pressure thermostatic steam trap and strainer; max working pressure up to 13 bar; screwed ends to steel						
½" dia.	177.56	224.73	1.26	43.55	nr	**268.27**
¾" dia.	178.02	225.31	1.71	59.13	nr	**284.44**
Bimetallic thermostatic steam trap and strainer; max working pressure up to 21 bar; flanged ends						
15 mm	230.98	292.33	1.24	42.84	nr	**335.17**
20 mm	253.77	321.17	1.34	46.29	nr	**367.46**
Sight glasses						
Pressed brass; straight; single window; screwed ends to steel						
15 mm dia.	52.00	65.81	0.84	29.02	nr	**94.83**
20 mm dia.	57.71	73.03	1.14	39.38	nr	**112.41**
25 mm dia.	72.12	91.28	1.30	44.91	nr	**136.18**
Gunmetal; straight; double window; screwed ends to steel						
15 mm dia.	83.85	106.12	0.84	29.02	nr	**135.14**
20 mm dia.	92.28	116.79	1.14	39.38	nr	**156.16**
25 mm dia.	114.06	144.35	1.30	44.91	nr	**189.26**
32 mm dia.	187.90	237.81	1.35	46.67	nr	**284.47**
40 mm dia.	187.90	237.81	1.74	60.16	nr	**297.97**
50 mm dia.	228.16	288.76	2.08	71.94	nr	**360.70**
SG Iron flanged; BS 4504, PN 25						
15 mm dia.	171.11	216.55	1.00	34.53	nr	**251.09**
20 mm dia.	201.32	254.78	1.25	43.17	nr	**297.95**
25 mm dia.	256.69	324.86	1.50	51.85	nr	**376.72**
32 mm dia.	283.49	358.78	1.70	58.73	nr	**417.51**
40 mm dia.	370.74	469.20	2.00	69.07	nr	**538.27**
50 mm dia.	447.91	566.87	2.30	79.57	nr	**646.44**
Check valve and sight glass; gun metal; screwed						
15 mm dia.	84.69	107.18	0.84	29.02	nr	**136.20**
20 mm dia.	89.74	113.58	1.14	39.38	nr	**152.95**
25 mm dia.	150.97	191.07	1.30	44.91	nr	**235.98**

38 MECHANICAL/COOLING/HEATING SYSTEMS

Item	Net Price £	Material £	Labour hours	Labour £	Unit	Total rate £
STEAM HEATING – cont						
Pressure reducing valves						
Pressure reducing valve for steam; maximum range of 17 bar and 232°C; screwed ends to steel						
15 mm dia.	648.87	821.21	0.87	30.04	nr	**851.25**
20 mm dia.	702.06	888.53	0.91	31.42	nr	**919.96**
25 mm dia.	757.00	958.06	1.35	46.62	nr	**1004.68**
Pressure reducing valve for steam; maximum range of 17 bar and 232°C; flanged ends to BS 4504 PN 25						
25 mm dia.	911.52	1153.61	1.70	58.71	nr	**1212.32**
32 mm dia.	1035.10	1310.02	1.87	64.58	nr	**1374.60**
40 mm dia.	1235.96	1564.23	2.12	73.21	nr	**1637.44**
50 mm dia.	1426.49	1805.37	2.57	88.75	nr	**1894.12**
Safety and relief valves						
Bronze safety valve; 'pop' type; side outlet; including easing lever; working pressure saturated steam up to 20.7 bar; screwed ends to steel						
15 mm dia.	263.72	333.76	0.32	11.05	nr	**344.81**
20 mm dia.	298.11	377.29	0.40	13.81	nr	**391.10**
Bronze safety valve; 'pop' type; side outlet; including easing lever; working pressure saturated steam up to 17.2 bar; screwed ends to steel						
25 mm dia.	424.23	536.91	0.47	16.23	nr	**553.14**
32 mm dia.	506.83	641.44	0.56	19.34	nr	**660.78**
Bronze safety valve; 'pop' type; side outlet; including easing lever; working pressure saturated steam up to 13.8 bar; screwed ends to steel						
40 mm dia.	655.87	830.06	0.64	22.10	nr	**852.16**
50 mm dia.	912.73	1155.15	0.76	26.24	nr	**1181.39**
65 mm dia.	1298.04	1642.80	0.94	32.46	nr	**1675.26**
80 mm dia.	1683.31	2130.39	1.10	37.99	nr	**2168.38**

38 MECHANICAL/COOLING/HEATING SYSTEMS

Item	Net Price £	Material £	Labour hours	Labour £	Unit	Total rate £
EQUIPMENT						
CALORIFIERS						
Non-storage calorifiers; mild steel shell construction with indirect steam heating for secondary LPHW at 82°C flow and 71°C return to BS 853; maximum test on shell 11 bar, tubes 26 bar						
Horizontal/vertical, for steam at 3 bar–5.5 bar						
88 kW capacity	954.94	1208.58	8.00	276.26	nr	**1484.84**
176 kW capacity	1390.14	1759.37	12.05	416.05	nr	**2175.42**
293 kW capacity	1514.51	1916.76	14.08	486.37	nr	**2403.14**
586 kW capacity	2211.20	2798.50	37.04	1278.98	nr	**4077.48**
879 kW capacity	2753.22	3484.48	40.00	1381.30	nr	**4865.78**
1465 kW capacity	3357.42	4249.15	45.45	1569.66	nr	**5818.81**

38 MECHANICAL/COOLING/HEATING SYSTEMS

Item	Net Price £	Material £	Labour hours	Labour £	Unit	Total rate £
LOCAL HEATING UNITS						
Warm air unit heater for connection to LTHW or steam supplies; suitable for heights up to 3 m; recirculating type; mild steel casing; heating coil; adjustable discharge louvre; axial fan; horizontal or vertical discharge; normal speed; entering air temperature 15°C; complete with enclosures; includes fixing in position; includes connections to primary heating supply; electrical work elsewhere						
Low pressure hot water						
7.5 kW, 265 l/s	874.49	1106.76	6.53	225.50	nr	**1332.26**
15.4 kW, 575 l/s	1058.58	1339.74	7.54	260.38	nr	**1600.11**
26.9 kW, 1040 l/s	1434.43	1815.41	8.65	298.71	nr	**2114.12**
48.0 kW, 1620 l/s	1890.83	2393.03	9.35	322.88	nr	**2715.91**
Steam, 2 bar						
9.2 kW, 265 l/s	1244.88	1575.52	6.53	225.50	nr	**1801.02**
18.8 kW, 575 l/s	1347.99	1706.02	6.82	235.51	nr	**1941.53**
34.8 kW, 1040 l/s	1550.42	1962.21	6.82	235.51	nr	**2197.72**
51.6 kW, 1625 l/s	2100.35	2658.20	7.10	245.18	nr	**2903.38**

38 MECHANICAL/COOLING/HEATING SYSTEMS

Item	Net Price £	Material £	Labour hours	Labour £	Unit	Total rate £
CENTRAL REFRIGERATION PLANT						
CHILLERS; Air cooled						
Selection of air cooled chillers based on chilled water flow and return temperatures 6°C and 12°C, and an outdoor temperature of 35°C						
Air cooled liquid chiller; refrigerant 410 A; scroll compressors; twin circuit; integral controls; includes placing in position; electrical work elsewhere						
Cooling load						
100 kW	21824.99	27621.71	8.00	276.26	nr	**27897.97**
150 kW	27504.97	34810.29	8.00	276.26	nr	**35086.55**
200 kW	34045.07	43087.44	8.00	276.26	nr	**43363.70**
Air cooled liquid chiller; refrigerant 410 A; reciprocating compressors; twin circuit; integral controls; includes placing in position; electrical work elsewhere						
Cooling load						
400 kW	60733.44	76864.24	8.00	276.26	nr	**77140.50**
550 kW	86603.74	109605.69	8.00	276.26	nr	**109881.95**
700 kW	100610.78	127333.01	9.00	310.79	nr	**127643.80**
Air cooled liquid chiller; refrigerant R134a; screw compressors; twin circuit; integral controls; includes placing in position; electrical work elsewhere						
Cooling load						
250 kW	61238.19	77503.06	8.00	276.26	nr	**77779.32**
400 kW	73363.99	92849.46	8.00	276.26	nr	**93125.72**
600 kW	98485.88	124643.72	8.00	276.26	nr	**124919.98**
800 kW	118371.60	149811.09	9.00	310.79	nr	**150121.89**
1000 kW	138794.01	175657.69	9.00	310.79	nr	**175968.49**
1200 kW	176299.85	223125.09	10.00	345.33	nr	**223470.42**
Air cooled liquid chiller; ductable for indoor installation; refrigerant 410 A; scroll compressors; integral controls; includes placing in position; electrical work elsewhere						
Cooling load						
40 kW	12980.12	16427.64	6.00	207.20	nr	**16634.84**
80 kW	20326.68	25725.45	6.00	207.20	nr	**25932.64**
CHILLERS; Higher efficiency air cooled						
Selection of air cooled chillers based on chilled water flow and return temperatures of 6°C and 12°C and an outdoor temperature of 25°C						

38 MECHANICAL/COOLING/HEATING SYSTEMS

Item	Net Price £	Material £	Labour hours	Labour £	Unit	Total rate £
CENTRAL REFRIGERATION PLANT – cont						
CHILLERS – cont						
These machines have significantly higher part load operating efficiencies than conventional air cooled machines						
Air cooled liquid chiller, refrigerant R410 A; scroll compressors; complete with free cooling facility; integral controls; including placing in position; electrical work elsewhere						
Cooling load						
250 kW	55423.30	70143.73	8.00	276.26	nr	**70419.99**
300 kW	61151.16	77392.91	8.00	276.26	nr	**77669.17**
350 kW	70664.71	89433.25	8.00	276.26	nr	**89709.51**
400 kW	79751.79	100933.86	8.00	276.26	nr	**101210.12**
450 kW	82868.82	104878.78	8.00	276.26	nr	**105155.04**
500 kW	90086.28	114013.20	8.00	276.26	nr	**114289.46**
600 kW	92447.63	117001.72	8.00	276.26	nr	**117277.98**
650 kW	107976.19	136654.67	9.00	310.79	nr	**136965.46**
700 kW	114361.09	144735.40	9.00	310.79	nr	**145046.19**
CHILLERS; Water cooled						
Selection of water cooled chillers based on chilled water flow and return temperatures of 6°C and 12°C, and condenser entering and leaving temperatures of 27°C and 33°C						
Water cooled liquid chiller; refrigerant 410 A; reciprocating compressors; twin circuit; integral controls; includes placing in position; electrical work elsewhere						
Cooling load						
200 kw	28054.69	35506.02	8.00	276.26	nr	**35782.28**
350 kW	49647.60	62834.01	8.00	276.26	nr	**63110.27**
500 kW	56332.75	71294.73	8.00	276.26	nr	**71570.99**
650 kW	65292.21	82633.82	9.00	310.79	nr	**82944.61**
750 kW	79529.86	100653.00	9.00	310.79	nr	**100963.79**
Water cooled condenserless liquid chiller; refrigerant 410 A; reciprocating compressors; twin circuit; integral controls; includes placing in position; electrical work elsewhere						
Cooling load						
200 kW	24213.11	30644.11	8.00	276.26	nr	**30920.37**
350 kW	42504.63	53793.86	8.00	276.26	nr	**54070.12**
500 kW	53720.43	67988.58	8.00	276.26	nr	**68264.84**
650 kW	65701.35	83151.63	9.00	310.79	nr	**83462.42**
750 kW	74761.21	94617.79	9.00	310.79	nr	**94928.59**

38 MECHANICAL/COOLING/HEATING SYSTEMS

Item	Net Price £	Material £	Labour hours	Labour £	Unit	Total rate £
Water cooled liquid chiller; refrigerant R134a; screw compressors; twin circuit; integral controls; includes placing in position; electrical work elsewhere						
Cooling load						
300 kW	46781.51	59206.67	8.00	276.26	nr	**59482.93**
500 kW	59625.29	75461.76	8.00	276.26	nr	**75738.02**
700 kW	75187.21	95156.94	9.00	310.79	nr	**95467.73**
900 kW	105390.03	133381.62	9.00	310.79	nr	**133692.41**
1100 kW	116419.29	147340.26	10.00	345.33	nr	**147685.58**
1300 kW	134605.08	170356.19	10.00	345.33	nr	**170701.52**
Water cooled liquid chiller; refrigerant R134a; centrifugal compressors; twin circuit; integral controls; includes placing in position; electrical work elsewhere						
Cooling load						
1600 kW	189893.70	240329.47	11.00	379.86	nr	**240709.32**
1900 kW	231272.65	292698.66	11.00	379.86	nr	**293078.52**
2200 kW	275538.37	348721.36	13.00	448.92	nr	**349170.28**
2500 kW	284680.28	360291.36	13.00	448.92	nr	**360740.28**
3000 kW	346427.76	438438.98	15.00	517.99	nr	**438956.97**
3500 kW	471526.67	596764.15	15.00	517.99	nr	**597282.14**
4000 kW	500802.99	633816.27	20.00	690.65	nr	**634506.92**
4500 kW	534303.86	676214.97	20.00	690.65	nr	**676905.62**
5000 kW	651632.21	824705.72	25.00	863.31	nr	**825569.03**
CHILLERS; Absorption						
Absorption chiller, for operation using low pressure steam; selection based on chilled water flow and return temperatures of 6°C and 12°C, steam at 1 bar gauge and condenser entering and leaving temperatures of 27°C and 33°C; integral controls; includes placing in position; electrical work elsewhere						
Cooling load						
400 kW	81814.71	103544.70	8.00	276.26	nr	**103820.96**
700 kW	103885.04	131476.90	9.00	310.79	nr	**131787.70**
1000 kW	118518.40	149996.88	10.00	345.33	nr	**150342.21**
1300 kW	141438.79	179004.93	12.00	414.39	nr	**179419.32**
1600 kW	167392.01	211851.33	14.00	483.46	nr	**212334.78**
2000 kW	198390.90	251083.52	15.00	517.99	nr	**251601.51**
Absorption chiller, for operation using low pressure hot water; selection based on chilled water flow and return temperatures of 6°C and12°C, cooling water temperatures of 27°C and 33°C and hot water at 90°C; integral controls; includes placing in position; electrical work elsewhere						
Cooling load						
700 kW	118518.40	149996.88	9.00	310.79	nr	**150307.68**
1000 kW	148152.36	187501.62	10.00	345.33	nr	**187846.95**
1300 kW	172310.90	218076.67	12.00	414.39	nr	**218491.06**
1600 kW	198390.90	251083.52	14.00	483.46	nr	**251566.98**

38 MECHANICAL/COOLING/HEATING SYSTEMS

Item	Net Price £	Material £	Labour hours	Labour £	Unit	Total rate £
CENTRAL REFRIGERATION PLANT – cont						
HEAT REJECTION						
Dry air liquid coolers						
Flat coil configuration						
500 kW	38066.30	48176.71	15.00	517.99	nr	48694.70
Extra for inverter panels (factory wired and mounted on units)	2926.29	3703.51	15.00	517.99	nr	4221.50
800 kW	64137.53	81172.46	15.00	517.99	nr	81690.44
Extra for inverter panels (factory wired and mounted on units)	38066.30	48176.71	15.00	517.99	nr	48694.70
1100 kW	83695.25	105924.71	15.00	517.99	nr	106442.69
Extra for inverter panels (factory wired and mounted on units)	5852.57	7407.02	15.00	517.99	nr	7925.01
1400 kW	125756.63	159157.59	15.00	517.99	nr	159675.58
Extra for inverter panels (factory wired and mounted on units)	8108.10	10261.61	15.00	517.99	nr	10779.60
1700 kW	125756.63	159157.59	15.00	517.99	nr	159675.58
Extra for inverter panels (factory wired and mounted on units)	8778.86	11110.53	15.00	517.99	nr	11628.51
2000 kW	153069.87	193725.23	15.00	517.99	nr	194243.22
Extra for inverter panels (factory wired and mounted on units)	10120.38	12808.36	15.00	517.99	nr	13326.34
Note: Heat rejection capacities above 500 kW require multiple units. Prices are therefore for total number of units.						
'Vee' type coil configuration						
500 kW	40141.24	50802.75	15.00	517.99	nr	51320.74
Extra for inverter panels (factory wired and mounted on units)	2702.70	3420.54	15.00	517.99	nr	3938.53
800 kW	65175.61	82486.25	15.00	517.99	nr	83004.24
Extra for inverter panels (factory wired and mounted on units)	3597.05	4552.42	15.00	517.99	nr	5070.41
1100 kW	85671.90	108426.36	15.00	517.99	nr	108944.35
Extra for inverter panels (factory wired and mounted on units)	4044.22	5118.37	15.00	517.99	nr	5636.36
1400 kW	116560.08	147518.44	15.00	517.99	nr	148036.43
Extra for inverter panels (factory wired and mounted on units)	6299.75	7972.96	15.00	517.99	nr	8490.95
1700 kW	132130.09	167223.84	15.00	517.99	nr	167741.83
Extra for inverter panels (factory wired and mounted on units)	7641.27	9670.79	15.00	517.99	nr	10188.78
2000 kW	157243.09	199006.85	15.00	517.99	nr	199524.84
Extra for inverter panels (factory wired and mounted on units)	8088.44	10236.73	15.00	517.99	nr	10754.72
Note: Heat rejection capacities above 1100 kW require multiple units. Prices are for total number of units.						

38 MECHANICAL/COOLING/HEATING SYSTEMS

Item	Net Price £	Material £	Labour hours	Labour £	Unit	Total rate £
Air cooled condensers						
Air cooled condenser; refrigerant 410 A; selection based on condensing temperature of 45°C at 32°C dry bulb ambient; includes placing in position; electrical work elsewhere						
Flat coil configuration						
500 kW	22452.19	28415.49	15.00	517.99	nr	28933.48
Extra for inverter panels (factory wired and mounted on units)	10273.49	13002.13	15.00	517.99	nr	13520.12
800 kW	36835.06	46618.45	15.00	517.99	nr	47136.44
Extra for inverter panels (factory wired and mounted on units)	15727.76	19905.05	15.00	517.99	nr	20423.04
1100 kW	49312.70	62410.16	15.00	517.99	nr	62928.14
Extra for inverter panels (factory wired and mounted on units)	22358.82	28297.33	15.00	517.99	nr	28815.31
1400 kW	61566.15	77918.13	15.00	517.99	nr	78436.11
Extra for inverter panels (factory wired and mounted on units)	27495.57	34798.40	15.00	517.99	nr	35316.39
1700 kW	80170.48	101463.77	15.00	517.99	nr	101981.75
Extra for inverter panels (factory wired and mounted on units)	41243.35	52197.58	15.00	517.99	nr	52715.57
2000 kW	92349.24	116877.20	15.00	517.99	nr	117395.19
Extra for inverter panels (factory wired and mounted on units)	41243.35	52197.58	15.00	517.99	nr	52715.57
Note: Heat rejection capacities above 500 kW require multiple units. Prices are for total number of units						
'Vee' type coil configuration						
500 kW	22041.27	27895.43	15.00	517.99	nr	28413.42
Extra for inverter panels (factory wired and mounted on units)	7863.90	9952.55	15.00	517.99	nr	10470.54
800 kW	33155.30	41961.35	15.00	517.99	nr	42479.33
Extra for inverter panels (factory wired and mounted on units)	16101.35	20377.87	15.00	517.99	nr	20895.85
1100 kW	48658.94	61582.76	15.00	517.99	nr	62100.75
Extra for inverter panels (factory wired and mounted on units)	27383.50	34656.56	15.00	517.99	nr	35174.54
1400 kW	56653.57	71700.75	15.00	517.99	nr	72218.74
Extra for inverter panels (factory wired and mounted on units)	32202.68	40755.72	15.00	517.99	nr	41273.70
1700 kW	72988.39	92374.11	15.00	517.99	nr	92892.10
Extra for inverter panels (factory wired and mounted on units)	41093.91	52008.46	15.00	517.99	nr	52526.45
2000 kW	84980.37	107551.16	15.00	517.99	nr	108069.14
Extra for inverter panels (factory wired and mounted on units)	48304.03	61133.58	15.00	517.99	nr	61651.57
Note: Heat rejection capacities above 1100 kW require multiple units. Prices are for total number of units						

38 MECHANICAL/COOLING/HEATING SYSTEMS

Item	Net Price £	Material £	Labour hours	Labour £	Unit	Total rate £
CENTRAL REFRIGERATION PLANT – cont						
Cooling towers						
Cooling towers; forced draught, centrifugal fan, conterflow design; based on water temperatures of 35°C on and 29°C off at 21°C wet bulb ambient temperature; includes placing in position; electrical work elsewhere						
COOLING TOWERS: Open circuit type						
900 kW	22758.23	28802.82	20.00	690.65	nr	**29493.47**
Extra for stainless steel construction	10696.24	13537.16	–	–	nr	**13537.16**
Extra for intake and discharge sound attenuation	10771.06	13631.85	–	–	nr	**13631.85**
1500 kW	28597.37	36192.83	20.00	690.65	nr	**36883.48**
Extra for stainless steel construction	13440.60	17010.42	–	–	nr	**17010.42**
Extra for intake and discharge sound attenuation	16622.21	21037.07	–	–	nr	**21037.07**
2100 kW	36754.96	46517.08	20.00	690.65	nr	**47207.73**
Extra for stainless steel construction	17274.82	21863.02	–	–	nr	**21863.02**
Extra for intake and discharge sound attenuation	24027.80	30409.59	–	–	nr	**30409.59**
2700 kW	47278.96	59836.25	20.00	690.65	nr	**60526.90**
Extra for stainless steel construction	22221.05	28122.96	–	–	nr	**28122.96**
Extra for intake and discharge sound attenuation	22662.19	28681.27	–	–	nr	**28681.27**
3300 kW	57671.48	72989.02	20.00	690.65	nr	**73679.67**
Extra for stainless steel construction	27105.23	34304.38	–	–	nr	**34304.38**
Extra for intake and discharge sound attenuation	26814.72	33936.70	–	–	nr	**33936.70**
3900 kW	70488.85	89210.69	20.00	690.65	nr	**89901.34**
Extra for stainless steel construction	33129.19	41928.31	–	–	nr	**41928.31**
Extra for intake and discharge sound attenuation	31847.89	40306.69	–	–	nr	**40306.69**
4500 kW	72596.30	91877.88	23.00	794.25	nr	**92672.13**
Extra for stainless steel construction	34120.00	43182.28	–	–	nr	**43182.28**
Extra for intake and discharge sound attenuation	30374.31	38441.72	–	–	nr	**38441.72**
5100 kW	91974.18	116402.53	23.00	794.25	nr	**117196.77**
Extra for stainless steel construction	43227.83	54709.14	–	–	nr	**54709.14**
Extra for intake and discharge sound attenuation	42124.26	53312.46	–	–	nr	**53312.46**
5700 kW	102592.10	129840.56	30.00	1035.98	nr	**130876.53**
Extra for stainless steel construction	48218.26	61025.03	–	–	nr	**61025.03**
Extra for intake and discharge sound attenuation	54528.03	69010.67	–	–	nr	**69010.67**
6300 kW	110879.12	140328.61	30.00	1035.98	nr	**141364.58**
Extra for stainless steel construction	52112.79	65953.94	–	–	nr	**65953.94**
Extra for intake and discharge sound attenuation	54527.65	69010.19	–	–	nr	**69010.19**

38 MECHANICAL/COOLING/HEATING SYSTEMS

Item	Net Price £	Material £	Labour hours	Labour £	Unit	Total rate £
COOLING TOWERS: Closed circuit type						
(includes 20% ethylene glycol)						
900 kW	56551.82	71571.99	20.00	690.65	nr	**72262.64**
Extra for stainless steel construction	35344.71	44732.26	–	–	nr	**44732.26**
Extra for intake and discharge sound attenuation	15852.02	20062.32	–	–	nr	**20062.32**
1500 kW	90957.50	115115.81	20.00	690.65	nr	**115806.46**
Extra for stainless steel construction	56848.06	71946.91	–	–	nr	**71946.91**
Extra for intake and discharge sound attenuation	30689.18	38840.23	–	–	nr	**38840.23**
2100 kW	106470.81	134749.46	20.00	690.65	nr	**135440.11**
Extra for stainless steel construction	66544.00	84218.09	–	–	nr	**84218.09**
Extra for intake and discharge sound attenuation	30689.58	38840.73	–	–	nr	**38840.73**
2700 kW	154518.22	195558.26	25.00	863.31	nr	**196421.58**
Extra for stainless steel construction	96573.16	122222.99	–	–	nr	**122222.99**
Extra for intake and discharge sound attenuation	40562.21	51335.53	–	–	nr	**51335.53**
3300 kW	193228.19	244549.59	25.00	863.31	nr	**245412.90**
Extra for stainless steel construction	120766.92	152842.61	–	–	nr	**152842.61**
Extra for intake and discharge sound attenuation	61147.74	77388.58	–	–	nr	**77388.58**
3900 kW	211454.13	267616.34	25.00	863.31	nr	**268479.65**
Extra for stainless steel construction	132158.18	167259.40	–	–	nr	**167259.40**
Extra for intake and discharge sound attenuation	62008.35	78477.77	–	–	nr	**78477.77**
4500 kW	252377.11	319408.47	40.00	1381.30	nr	**320789.77**
Extra for stainless steel construction	157735.04	199629.47	–	–	nr	**199629.47**
Extra for intake and discharge sound attenuation	91565.60	115885.42	–	–	nr	**115885.42**
5100 kW	286045.62	362019.34	40.00	1381.30	nr	**363400.64**
Extra for stainless steel construction	178778.22	226261.72	–	–	nr	**226261.72**
Extra for intake and discharge sound attenuation	101805.51	128845.06	–	–	nr	**128845.06**
5700 kW	345821.90	437672.20	40.00	1381.30	nr	**439053.50**
Extra for stainless steel construction	216137.43	273543.53	–	–	nr	**273543.53**
Extra for intake and discharge sound attenuation	101805.66	128845.24	–	–	nr	**128845.24**
6300 kW	386455.95	489098.65	40.00	1381.30	nr	**490479.95**
Extra for stainless steel construction	241534.61	305686.20	–	–	nr	**305686.20**
Extra for intake and discharge sound attenuation	122294.87	154776.38	–	–	nr	**154776.38**

38 MECHANICAL/COOLING/HEATING SYSTEMS

Item	Net Price £	Material £	Labour hours	Labour £	Unit	Total rate £
CHILLED WATER						
SCREWED STEEL						
PIPELINES						
For pipework prices refer to Section – Low Temperature Hot Water Heating, with the exception of chilled water blocks within brackets as detailed hereafter. For minimum fixing distances, refer to the Tables and Memoranda to the rear of the book						
FIXINGS						
For steel pipes; black malleable iron						
Oversized pipe clip, to contain 30 mm insulation block for vapour barrier						
15 mm dia.	6.26	7.92	0.10	3.45	nr	**11.38**
20 mm dia.	6.32	8.00	0.11	3.80	nr	**11.79**
25 mm dia.	6.63	8.39	0.12	4.14	nr	**12.54**
32 mm dia.	6.83	8.64	0.14	4.83	nr	**13.48**
40 mm dia.	7.20	9.11	0.15	5.18	nr	**14.29**
50 mm dia.	10.30	13.03	0.16	5.53	nr	**18.56**
65 mm dia.	11.11	14.06	0.30	10.36	nr	**24.42**
80 mm dia.	11.78	14.91	0.35	12.09	nr	**26.99**
100 mm dia.	18.68	23.64	0.40	13.81	nr	**37.46**
125 mm dia.	21.68	27.43	0.60	20.73	nr	**48.16**
150 mm dia.	33.91	42.92	0.77	26.59	nr	**69.51**
200 mm dia.	42.87	54.25	0.90	31.08	nr	**85.33**
250 mm dia.	55.67	70.46	1.10	37.99	nr	**108.44**
300 mm dia.	66.25	83.85	1.25	43.17	nr	**127.01**
350 mm dia.	76.88	97.29	1.50	51.80	nr	**149.09**
400 mm dia.	86.09	108.95	1.75	60.43	nr	**169.38**
Screw on backplate (Male), black malleable iron; plugged and screwed						
M12	0.86	1.08	0.10	3.45	nr	**4.54**
Screw on backplate (Female), black malleable iron; plugged and screwed						
M12	1.71	2.17	0.10	3.45	nr	**5.62**
Extra over channel sections for fabricated hangers and brackets						
Galvanized steel; including inserts, bolts, nuts, washers; fixed to backgrounds						
41 × 21 mm	8.39	10.62	0.29	10.01	m	**20.64**
41 × 41 mm	10.06	12.74	0.29	10.01	m	**22.75**
Threaded rods; metric thread; including nuts, washers etc.						
12 mm dia. × 600 mm long	4.24	5.37	0.18	6.22	nr	**11.58**

38 MECHANICAL/COOLING/HEATING SYSTEMS

Item	Net Price £	Material £	Labour hours	Labour £	Unit	Total rate £
For plastic pipework suitable for chilled water systems, refer to ABS pipework details in Section – Cold Water, with the exception of chilled water blocks within brackets as detailed for the aforementioned steel pipe. For minimum fixing distances refer to the Tables and Memoranda to the rear of the book						
For copper pipework, refer to Section – Cold Water with the exception of chilled water blocks within brackets as detailed hereafter. For minimum fixing distances refer to the Tables and Memoranda to the rear of the book						
FIXINGS						
For copper pipework						
Oversized pipe clip, to contain 30 mm insulation block for vapour barrier						
15 mm dia.	6.26	7.92	0.10	3.45	nr	**11.38**
22 mm dia.	6.32	8.00	0.11	3.80	nr	**11.79**
28 mm dia.	6.63	8.39	0.12	4.14	nr	**12.54**
35 mm dia.	6.83	8.64	0.14	4.83	nr	**13.48**
42 mm dia.	7.20	9.11	0.15	5.18	nr	**14.29**
54 mm dia.	10.30	13.03	0.16	5.53	nr	**18.56**
67 mm dia.	11.11	14.06	0.30	10.36	nr	**24.42**
76 mm dia.	11.78	14.91	0.35	12.09	nr	**26.99**
108 mm dia.	13.02	40.12	0.40	13.81	nr	**53.93**
133 mm dia.	21.68	27.43	0.60	20.73	nr	**48.16**
159 mm dia.	33.91	42.92	0.77	26.59	nr	**69.51**
Screw on backplate, female						
All sizes 15 mm to 54 mm × 10 mm	2.02	2.56	0.10	3.45	nr	**6.02**
Screw on backplate, male						
All sizes 15 mm to 54 mm × 10 mm	2.71	3.44	0.10	3.45	nr	**6.89**
Extra over channel sections for fabricated hangers and brackets						
Galvanized steel; including inserts, bolts, nuts, washers; fixed to backgrounds						
41 × 21 mm	8.39	10.62	0.29	10.01	m	**20.64**
41 × 41 mm	10.06	12.74	0.29	10.01	m	**22.75**
Threaded rods; metric thread; including nuts, washers etc.						
10 mm dia. × 600 mm long for ring clips up to 54 mm	2.71	3.43	0.18	6.22	nr	**9.64**
12 mm dia. × 600 mm long for ring clips from 54 mm	4.24	5.37	0.18	6.22	nr	**11.58**

38 MECHANICAL/COOLING/HEATING SYSTEMS

Item	Net Price £	Material £	Labour hours	Labour £	Unit	Total rate £
CHILLED WATER – cont						
PIPELINE ANCILLARIES						
For prices for ancillaries refer to Section: Low Temperature Hot Water Heating						
HEAT EXCHANGERS						
Plate heat exchanger; for use in CHW systems; painted carbon steel frame, stainless steel plates, nitrile rubber gaskets, design pressure of 10 bar and operating temperature of 110/135°C						
Primary side; 13°C in, 8°C out; secondary side; 6°C in, 11°C out						
264 kW, 12.60 l/s	4595.40	5815.94	10.00	345.33	nr	**6161.26**
290 kW, 13.85 l/s	4857.60	6147.78	12.00	414.39	nr	**6562.17**
316 kW, 15.11 l/s	5118.65	6478.16	12.00	414.39	nr	**6892.55**
350 kW, 16.69 l/s	9271.30	11733.76	16.00	552.52	nr	**12286.28**
395 kW, 18.88 l/s	6701.05	8480.85	16.00	552.52	nr	**9033.37**
454 kW, 21.69 l/s	7299.05	9237.68	16.00	552.52	nr	**9790.20**
475 kW, 22.66 l/s	8135.10	10295.78	16.00	552.52	nr	**10848.30**
527 kW, 25.17 l/s	8853.85	11205.43	16.00	552.52	nr	**11757.95**
554 kW, 26.43 l/s	9212.65	11659.53	16.00	552.52	nr	**12212.05**
580 kW, 27.68 l/s	9511.65	12037.94	16.00	552.52	nr	**12590.46**
633 kW, 30.19 l/s	10408.65	13173.19	16.00	552.52	nr	**13725.71**
661 kW, 31.52 l/s	10887.05	13778.65	16.00	552.52	nr	**14331.17**
713 kW, 34.04 l/s	12651.15	16011.30	18.00	621.59	nr	**16632.88**
740 kW, 35.28 l/s	13161.75	16657.51	18.00	621.59	nr	**17279.10**
804 kW, 38.33 l/s	13877.05	17562.79	18.00	621.59	nr	**18184.38**
1925 kW, 91.82 l/s	24920.50	31539.38	20.00	690.65	nr	**32230.04**
2710 kW, 129.26 l/s	39409.35	49876.47	20.00	690.65	nr	**50567.12**
3100 kW, 147.87 l/s	45050.10	57015.41	20.00	690.65	nr	**57706.06**
Note: For temperature conditions different to those above, the cost of the units can vary significantly, and therefore the manufacturers advice should be sought.						
TRACE HEATING						
Trace heating; for freeze protection or temperature maintenance of pipework; to BS 6351; including fixing to parent structures by plastic pull ties						
Straight laid F-S-C Wintergaurd						
15 mm	29.67	37.55	0.27	9.32	m	**46.87**
25 mm	29.67	37.55	0.27	9.32	m	**46.87**
28 mm	29.67	37.55	0.27	9.32	m	**46.87**
32 mm	29.67	37.55	0.30	10.36	m	**47.91**
35 mm	29.67	37.55	0.31	10.71	m	**48.25**
50 mm	29.67	37.55	0.34	11.74	m	**49.29**
100 mm	29.67	37.55	0.40	13.81	m	**51.36**
150 mm F-S-B Wintergaurd (to larger sizes)	29.67	37.55	0.40	13.81	m	**51.36**

38 MECHANICAL/COOLING/HEATING SYSTEMS

Item	Net Price £	Material £	Labour hours	Labour £	Unit	Total rate £
Helically wound F-S-C Wintergaurd						
15 mm	37.82	47.86	1.00	34.53	m	**82.40**
25 mm	37.82	47.86	1.00	34.53	m	**82.40**
28 mm	37.82	47.86	1.00	34.53	m	**82.40**
32 mm	37.82	47.86	1.00	34.53	m	**82.40**
35 mm	37.82	47.86	1.00	34.53	m	**82.40**
50 mm	37.82	47.86	1.00	34.53	m	**82.40**
100 mm	37.82	47.86	1.00	34.53	m	**82.40**
150 mm F-S-B Wintergaurd (to larger sizes)	37.82	47.86	1.00	34.53	m	**82.40**
Accessories for trace heating; weatherproof; polycarbonate enclosure to IP standards; fully installed						
Connection junction box						
100 × 100 × 75 mm junction box incl. power connecting kit	88.87	112.47	1.40	48.35	nr	**160.82**
Single air thermostat						
150 × 150 × 75 mm AT-TS–13	194.68	246.39	1.42	49.04	nr	**295.42**
Single capillary thermostat						
150 × 150 × 75 mm AT-TS–13	280.10	354.50	1.46	50.42	nr	**404.91**
Twin capillary thermostat						
150 × 150 × 75 mm	502.51	635.97	1.46	50.42	nr	**686.39**
PRESSURIZATION UNITS						
Chilled water packaged pressurization unit complete with expansion vessel(s), interconnecting pipework and necessary isolating and drain valves; includes placing in position; electrical work elsewhere. Selection based on a final working pressure of 4 bar, a 3 m static head and system operating temperatures of 6°/12°C						
System volume						
1800 litres	1442.95	1826.20	8.00	276.26	nr	**2102.46**
4500 litres	1570.97	1988.22	8.00	276.26	nr	**2264.48**
7200 litres	1654.46	2093.89	10.00	345.33	nr	**2439.21**
9900 litres	2218.01	2807.12	10.00	345.33	nr	**3152.44**
15300 litres	2319.60	2935.69	13.00	448.92	nr	**3384.61**
22500 litres	2423.96	3067.76	20.00	690.65	nr	**3758.41**
27000 litres	2423.96	3067.76	20.00	690.65	nr	**3758.41**

38 MECHANICAL/COOLING/HEATING SYSTEMS

Item	Net Price £	Material £	Labour hours	Labour £	Unit	Total rate £
CHILLED WATER – cont						
CHILLED BEAMS						
Static (passive) beams; cooling data based on 9 K waterside cooling dt (e.g. water 14°C flow and 16°C return, room temperature 24°C)						
Static passive beam in perforated metal casing for exposed mounting from solid ceiling or roof soffit, 2 pipe cooling only						
Length/Cooling output						
1200 mm/300 W	433.12	548.15	1.31	45.24	m	593.39
1800 mm/450 W	564.70	714.68	1.53	52.83	m	767.52
2400 mm/600 W	674.35	853.45	1.75	60.43	m	913.89
3000 mm/750 W	800.45	1013.04	2.01	69.41	m	1082.45
3600 mm/900 W	986.85	1248.96	2.18	75.28	m	1324.24
Static passive beam, chassis type for mounting above open grid or perforated metal ceiling or roof soffit, 2 pipe cooling only						
Length/Cooling output						
1200 mm/300 W	307.02	388.56	1.31	45.24	m	433.80
1800 mm/450 W	422.15	534.28	1.53	52.83	m	587.11
2400 mm/600 W	520.84	659.17	1.75	60.43	m	719.60
3000 mm/750 W	630.49	797.94	2.01	69.41	m	867.36
3600 mm/900 W	745.62	943.66	2.18	75.28	m	1018.94
Multi-service beams; for exposed mounting on solid ceiling slabs, with Integrated lighting and lighting control, provision of fascia openings for other services (electrical work and other services elsewhere). Performance criteria equivalent to standard beams as listed above						
2 Way discharge exposed linear multi service climate beam for exposed on solid ceiling; closed type with integrated secondary air circulation, 2 pipe cooling only, with primary air						
Length/Cooling output						
1200 mm/550 W	877.20	1110.18	2.26	78.04	m	1188.23
1800 mm/900 W	1206.15	1526.50	2.49	85.99	m	1612.49
2400 mm/1200 W	1535.10	1942.82	2.67	92.20	m	2035.02
3000 mm/1550 W	1864.05	2359.14	2.84	98.07	m	2457.21
3600 mm/1850 W	2247.83	2844.85	3.08	106.36	m	2951.21
2 Way discharge exposed linear multi service climate beam for exposed on solid ceiling; closed type with integrated secondary air circulation, 4 pipe cooling and heating, with primary air						

38 MECHANICAL/COOLING/HEATING SYSTEMS

Item	Net Price £	Material £	Labour hours	Labour £	Unit	Total rate £
Length/Cooling output/Heating output						
1200 mm/550 W/495 W	910.10	1151.82	2.59	89.44	m	**1241.26**
1800 mm/900 W/810 W	1244.53	1575.07	2.82	97.38	m	**1672.46**
2400 mm/1200 W/1080 W	1578.96	1998.33	3.00	103.60	m	**2101.93**
3000 mm/1550 W/1395 W	1913.39	2421.59	3.18	109.81	m	**2531.40**
3600 mm/1850 W/1665 W	2302.65	2914.23	3.41	117.76	m	**3031.99**
Static passive multi-service beam for exposed mounting on solid ceiling, 2 pipe cooling only Length/Cooling output						
1200 mm/300 W	893.65	1131.00	1.81	62.50	m	**1193.50**
1800 mm/450 W	1222.60	1547.32	2.03	70.10	m	**1617.42**
2400 mm/600 W	1551.55	1963.64	2.25	77.70	m	**2041.34**
3000 mm/750 W	1875.02	2373.02	2.51	86.68	m	**2459.70**
3600 mm/900 W	2258.79	2858.72	2.68	92.55	m	**2951.27**
Ventilated (active) beams; cooling data based on 9 K airside and 9 K waterside cooling dt (e.g. water 14°C flow and 16°C return, primary air 15°C, room temperature 24°C), heating data based on −4 K airside and 20 K waterside heating dt (e.g. water at 45°C flow and 35°C return, primary air 16°C, room temperature 20°C)						
1 Way discharge recessed linear active climate beam mounted within a false ceiling; closed type with integrated secondary air circulation; 300 mm wide, 2 pipe cooling only, 5 l/s/m primary air Length/Cooling output						
1200 mm/340 W	335.53	424.65	1.76	60.78	m	**485.42**
1800 mm/530 W	414.48	524.56	1.99	68.72	m	**593.28**
2400 mm/730 W	487.94	617.54	2.17	74.94	m	**692.48**
3000 mm/900 W	562.50	711.91	2.34	80.81	m	**792.71**
3600 mm/1100 W	607.46	768.80	2.58	89.09	m	**857.90**
1 Way discharge recessed linear active climate beam mounted within a false ceiling; closed type with integrated secondary air circulation; 300 mm wide, 4-pipe cooling and heating, with 5 l/s/m primary air Length/Cooling output/Heating output						
1200 mm/340 W/180 W	350.88	444.07	2.09	72.17	m	**516.25**
1800 mm/530 W/310 W	434.21	549.54	2.32	80.12	m	**629.66**
2400 mm/730 W/450 W	513.16	649.46	2.50	86.33	m	**735.79**
3000 mm/900 W/600 W	582.24	736.88	2.68	92.55	m	**829.43**
3600 mm/1100 W/750 W	628.29	795.17	2.91	100.49	m	**895.66**
2 Way discharge recessed linear active climate beam mounted within a false ceiling; closed type with integrated secondary air circulation, 600 mm wide, 2-pipe cooling only, with 10 l/s/m primary air						

38 MECHANICAL/COOLING/HEATING SYSTEMS

Item	Net Price £	Material £	Labour hours	Labour £	Unit	Total rate £
CHILLED WATER – cont						
Ventilated (active) beams – cont						
Length/Cooling output						
1200 mm/550 W	339.92	430.20	1.76	60.78	m	**490.97**
1800 mm/900 W	466.01	589.79	1.99	68.72	m	**658.51**
2400 mm/1200 W	581.15	735.50	2.17	74.94	m	**810.43**
3000 mm/1550 W	679.83	860.39	2.34	80.81	m	**941.20**
3600 mm/1850 W	778.52	985.29	2.58	89.09	m	**1074.38**
2 Way discharge recessed linear active climate beam mounted within a false ceiling; closed type with integrated secondary air circulation; 600 mm wide, 4-pipe cooling and heating, with 10 l/s/m primary air						
Length/Cooling output/Heating output						
1200 mm/550 W/495 W	372.81	471.83	2.09	72.17	m	**544.00**
1800 mm/900 W/810 W	504.39	638.36	2.32	80.12	m	**718.47**
2400 mm/1200 W/1080 W	625.01	791.01	2.50	86.33	m	**877.34**
3000 mm/1550 W/1395 W	729.17	922.84	2.68	92.55	m	**1015.39**
3600 mm/1850 W/1665 W	833.34	1054.68	2.91	100.49	m	**1155.16**
4 Way discharge recessed modular active climate beam mounted within a false ceiling; closed type with integrated secondary air circulation; 600 mm wide, 2-pipe cooling only, with primary air at 10 l/s/m of beam discharge (600 × 600 mm unit, 12 l/s, 1200 × 600 mm unit 18 l/s)						
Length/Cooling output						
600 × 600 mm/500 W	356.36	451.01	1.56	53.87	m	**504.88**
1200 × 600 mm/850 W	411.19	520.40	1.76	60.78	m	**581.18**
4 Way discharge recessed modular active climate beam mounted within a false ceiling; closed type with integrated secondary air circulation; 600 mm wide, 4-pipe cooling and heating, with primary air at 10 l/s/m of beam discharge (600 × 600 mm unit 12 l/s, 1200 × 600 mm unit 18 l/s)						
Length/Cooling output/Heating output						
600 × 600 mm/500 W/450 W	383.78	485.71	1.89	65.27	m	**550.97**
1200 × 600 mm/850 W/765 W	444.08	562.03	2.09	72.17	m	**634.20**
1 Way sidewall discharge active climate beam mounted within a ceiling bulkhead; closed type with integrated secondary air cisculation, 2-pipe cooling only, with 20 l/s/m primary air						
Length/Cooling output						
900 mm/700 W	394.74	499.58	1.89	65.27	m	**564.85**
1100 mm/900 W	427.64	541.21	1.89	65.27	m	**606.48**
1300 mm/1100 W	504.39	638.36	2.09	72.17	m	**710.53**
1500 mm/1200 W	537.29	679.99	2.09	72.17	m	**752.16**

38 MECHANICAL/COOLING/HEATING SYSTEMS

Item	Net Price £	Material £	Labour hours	Labour £	Unit	Total rate £
1 Way sidewall discharge active climate beam mounted within a ceiling bulkhead; closed type with integrated secondary air circulation, 4-pipe only, with 20 l/s/m primary air Length/Cooling output/Heating output						
900 mm/700 W/400 W	427.64	541.21	1.89	65.27	m	**606.48**
1100 mm/900 W/600 W	460.53	582.85	1.89	65.27	m	**648.11**
1300 mm/1100 W/800 W	537.29	679.99	2.09	72.17	m	**752.16**
1500 mm/1200 W/900 W	570.18	721.62	2.09	72.17	m	**793.79**
CHILLED CEILINGS						
Traditional (60% Radiant Absorption, 40% Convection) with bonded elements in perforated metal ceiling tiles (typically 30% free area and 2.5 mm perforated hole sizes)/ tartan grid arrangement. Maximum waterside cooling effect approx. 65 W/m² at 9 dTK						
Radiant elements and metal ceilings combined	153.51	194.28	1.75	60.43	m	**254.71**
Hybrid (40% Radiant Absorption, 60% Convection) with radiant cooling chilled beam elements positioned above a perforated metal ceiling tiles (typically 30% free area and 2.5 mm perforated hole sizes)/tartan grid arrangement. Maximum waterside cooling effect approx. 100 W/m² at 9 dTK						
Radiant chilled beams	65.79	83.26	0.60	20.72	m	**103.98**
Metal ceilings	38.38	48.57	1.15	39.71	m	**88.28**
Convective beam (5% Radiant Absorption, 95% Convection) with convective cooling fin coil batteries positioned above a perforated metal ceiling tiles (typically 50% free area and 5.0 mm perforated hole sizes)/tartan grid arrangement. Maximum waterside cooling effect approx. 100 W/m² at 9 dTK						
Convective fin coil batteries	71.27	90.20	0.60	20.72	m	**110.92**
Metal ceilings	38.38	48.57	1.15	39.71	m	**88.28**
LEAK DETECTION						
Leak detection system consisting of a central control module connected by a leader cable to water sensing cables						
Control modules						
Alarm only (1 zone)	624.20	789.99	4.00	158.42	nr	**948.41**
Alarm only (8 zone)	1222.44	1547.12	4.00	158.42	nr	**1705.53**
Alarm and location	3219.20	4074.21	8.00	316.84	nr	**4391.05**
Cables						
Sensing – 3 m length	133.19	168.57	4.00	158.42	nr	**326.99**
Sensing – 5 m length	163.67	207.14	4.00	158.42	nr	**365.56**
Sensing – 7.5 m length	232.52	294.28	4.00	158.42	nr	**452.70**

38 MECHANICAL/COOLING/HEATING SYSTEMS

Item	Net Price £	Material £	Labour hours	Labour £	Unit	Total rate £
CHILLED WATER – cont						
LEAK DETECTION – cont						
Leak detection system consisting of a central control module connected by a leader cable to water sensing cables – cont						
Cables – cont						
Sensing – 15 m length	325.08	411.42	8.00	316.84	nr	728.26
Sensing – 25 m length	451.50	571.42	12.00	414.39	nr	985.81
Leader – 3.5 m length	53.75	68.03	2.00	79.21	nr	147.24
End terminal						
End terminal	30.98	39.20	0.05	1.98	nr	41.18
ENERGY METERS						
Ultrasonic						
Energy meter for measuring energy use in chilled water systems; includes ultrasonic flow meter (with sensor and signal converter), energy calculator, pair of temperature sensors with brass pockets, and 3 m of interconnecting cable; includes fixing in position; electrical work elsewhere						
Pipe size (flanged connections to PN16); maximum flow rate						
50 mm, 36 m³/hr	1486.64	1881.49	1.80	62.16	nr	1943.65
65 mm, 60 m³/hr	1638.71	2073.95	2.32	80.12	nr	2154.06
80 mm, 100 m³/hr	1836.87	2324.74	2.56	88.40	nr	2413.14
125 mm, 250 m³/hr	2128.19	2693.43	3.60	124.32	nr	2817.75
150 mm, 360 m³/hr	2310.28	2923.89	4.80	165.76	nr	3089.65
200 mm, 600 m³/hr	2581.25	3266.83	6.24	215.48	nr	3482.31
250 mm, 1000 m³/hr	2978.60	3769.72	9.60	331.51	nr	4101.23
300 mm, 1500 m³/hr	3501.29	4431.23	10.80	372.95	nr	4804.18
350 mm, 2000 m³/hr	4218.89	5339.43	13.20	455.83	nr	5795.26
400 mm, 2500 m³/hr	4829.40	6112.09	15.60	538.71	nr	6650.80
500 mm, 3000 m³/hr	5480.59	6936.24	24.00	828.78	nr	7765.02
600 mm, 3500 m³/hr	6157.50	7792.93	28.00	966.91	nr	8759.84
Electromagnetic						
Energy meter for measuring energy use in chilled water systems; includes electromagnetic flow meter (with sensor and signal converter), energy calculator, pair of temperature sensors with brass pockets, and 3 m of interconnecting cable; includes fixing in position; electrical work elsewhere						

38 MECHANICAL/COOLING/HEATING SYSTEMS

Item	Net Price £	Material £	Labour hours	Labour £	Unit	Total rate £
Pipe size (flanged connections to PN40); maximum flow rate						
25 mm, 17.7 m³/hr	1065.81	1348.88	1.48	51.11	nr	**1399.99**
40 mm, 45 m³/hr	1075.99	1361.78	1.55	53.53	nr	**1415.30**
Pipe size (flanged connections to PN16); maximum flow rate						
50 mm, 70 m³/hr	1088.05	1377.04	1.80	62.16	nr	**1439.20**
65 mm, 120 m³/hr	1092.68	1382.90	2.32	80.12	nr	**1463.02**
80 mm, 180 m³/hr	1098.26	1389.96	2.56	88.40	nr	**1478.36**
125 mm, 450 m³/hr	1195.56	1513.11	3.60	124.32	nr	**1637.42**
150 mm, 625 m³/hr	1268.78	1605.77	4.80	165.76	nr	**1771.53**
200 mm, 1100 m³/hr	1365.17	1727.76	6.24	215.48	nr	**1943.24**
250 mm, 1750 m³/hr	1531.99	1938.89	9.60	331.51	nr	**2270.40**
300 mm, 2550 m³/hr	1952.76	2471.41	10.80	372.95	nr	**2844.37**
350 mm, 3450 m³/hr	2540.34	3215.05	13.20	455.83	nr	**3670.88**
400 mm, 4500 m³/hr	2898.07	3667.80	15.60	538.71	nr	**4206.51**

38 MECHANICAL/COOLING/HEATING SYSTEMS

Item	Net Price £	Material £	Labour hours	Labour £	Unit	Total rate £
LOCAL COOLING UNITS						
Split system with ceiling void evaporator unit and external condensing unit						
Ceiling mounted 4 Way blow cassette heat pump unit with remote fan speed and load control; refrigerant 410 A; includes outdoor unit						
Cooling 3.6 kW, heating 4.1 kW	1914.81	2423.39	35.00	1208.64	nr	**3632.02**
Cooling 4.9 kW, heating 5.5 kW	2110.09	2670.53	35.00	1208.64	nr	**3879.17**
Cooling 7.1 kW, heating 8.2 kW	2554.07	3232.43	35.00	1208.64	nr	**4441.07**
Cooling 10 kW, heating 11.2 kW	3020.96	3823.32	35.00	1208.64	nr	**5031.96**
Cooling 12.20 kW, heating 14.60 kW	3304.74	4182.48	35.00	1208.64	nr	**5391.12**
Ceiling mounted 4 Way blow cooling only unit with remote fan speed and load control; refrigerant 410 A; includes outdoor unit						
Cooling 3.80 kW	1733.24	2193.58	35.00	1208.64	nr	**3402.22**
Cooling 5.20 kW	1936.15	2450.39	35.00	1208.64	nr	**3659.03**
Cooling 7.10 kW	2387.77	3021.97	35.00	1208.64	nr	**4230.60**
Cooling 10 kW	3823.79	4839.38	35.00	1208.64	nr	**6048.02**
Cooling 12.2 kW	4408.26	5579.09	35.00	1208.64	nr	**6787.73**
In ceiling, ducted heat pump unit with remote fan speed and load control; refrigerant 410 A; includes outdoor unit						
Cooling 3.60 kW, heating 4.10 kW	1403.68	1776.50	35.00	1208.64	nr	**2985.14**
Cooling 4.90 kW, heating 5.50 kW	1614.23	2042.97	35.00	1208.64	nr	**3251.61**
Cooling 7.10 kW, heating 8.20 kW	1614.23	2042.97	35.00	1208.64	nr	**3251.61**
Cooling 10 kW, heating 11.20 kW	1614.23	2042.97	35.00	1208.64	nr	**3251.61**
Cooling 12.20 kW, heating 14.50 kW	3254.40	4118.77	35.00	1208.64	nr	**5327.41**
In ceiling, ducted cooling only unit with remote fan speed and load control; refrigerant 410 A; includes outdoor unit						
Cooling 3.70 kW	1266.36	1602.71	35.00	1208.64	nr	**2811.35**
Cooling 4.90 kW	1505.91	1905.88	35.00	1208.64	nr	**3114.52**
Cooling 7.10 kW	2273.36	2877.16	35.00	1208.64	nr	**4085.80**
Cooling 10 kW	2592.23	3280.73	35.00	1208.64	nr	**4489.36**
Cooling 12.3 kW	2889.74	3657.25	35.00	1208.64	nr	**4865.89**
Room units						
Ceiling mounted 4 Way blow cassette heat pump unit with remote fan speed and load control; refrigerant 410 A; excludes outdoor unit						
Cooling 3.6 kW, heating 4.1 kW	1208.38	1529.33	17.00	587.05	nr	**2116.38**
Cooling 4.9 kW, heating 5.5 kW	1237.37	1566.02	17.00	587.05	nr	**2153.07**
Cooling 7.1 kW, heating 8.2 kW	1339.60	1695.40	17.00	587.05	nr	**2282.45**
Cooling 10 kW, heating 11.2 kW	1450.98	1836.37	17.00	587.05	nr	**2423.42**
Cooling 12.20 kW, heating 14.60 kW	1573.05	1990.85	17.00	587.05	nr	**2577.90**

38 MECHANICAL/COOLING/HEATING SYSTEMS

Item	Net Price £	Material £	Labour hours	Labour £	Unit	Total rate £
Ceiling mounted 4 Way blow cooling unit with remote fan speed and load control; refrigerant 410 A; excludes outdoor unit						
Cooling 3.80 kW	1141.23	1444.35	17.00	587.05	nr	2031.40
Cooling 5.20 kW	1150.40	1455.95	17.00	587.05	nr	2043.00
Cooling 7.10 kW	1339.60	1695.40	17.00	587.05	nr	2282.45
Cooling 10 kW	1450.98	1836.37	17.00	587.05	nr	2423.42
Cooling 12.2 kW	1573.05	1990.85	17.00	587.05	nr	2577.90
In ceiling, ducted heat pump unit with remote fan speed and load control; refrigerant 410 A; excludes outdoor unit						
Cooling 3.60 kW, heating 4.10 kW	697.26	882.46	17.00	587.05	nr	1469.51
Cooling 4.90 kW, heating 5.50 kW	741.51	938.46	17.00	587.05	nr	1525.51
Cooling 7.10 kW, heating 8.20 kW	1225.17	1550.58	17.00	587.05	nr	2137.63
Cooling 10 kW, heating 11.20 kW	1269.42	1606.58	17.00	587.05	nr	2193.63
Cooling 12.20 kW, heating 14.50 kW	1522.65	1927.06	17.00	587.05	nr	2514.12
In ceiling , ducted cooling unit only with remote fan speed and load control; refrigerant 410 A; excludes outdoor unit						
Cooling 3.70 kW	674.37	853.49	17.00	587.05	nr	1440.54
Cooling 4.90 kW	720.15	911.43	17.00	587.05	nr	1498.48
Cooling 7.10 kW	1225.17	1550.58	17.00	587.05	nr	2137.63
Cooling 10 kW	1269.42	1606.58	17.00	587.05	nr	2193.63
Cooling 12.3 kW	1522.70	1927.13	17.00	587.05	nr	2514.18
External condensing units suitable for connection to multiple indoor units; inverter driven; refrigerant 410 A						
Cooling only						
9 kW	2638.00	3338.65	17.00	587.05	nr	3925.70
Heat pump						
Cooling 5.20 kW, heating 6.10 kW	1879.73	2378.99	17.00	587.05	nr	2966.04
Cooling 6.80 kW, heating 2.50 kW	2396.93	3033.55	17.00	587.05	nr	3620.61
Cooling 8 kW, heating 9.60 kW	2775.33	3512.46	17.00	587.05	nr	4099.51
Cooling 14.50 kW, heating 16.50 kW	4606.20	5829.61	21.00	725.18	nr	6554.79

38 VENTILATION/AIR CONDITIONING SYSTEMS

Item	Net Price £	Material £	Labour hours	Labour £	Unit	Total rate £
DUCTWORK: CIRCULAR						
AIR DUCTLINES						
Galvanized sheet metal DW144 class B spirally wound circular section ductwork; including all necessary stiffeners, joints, couplers in the running length and duct supports						
Straight duct						
80 mm dia.	6.93	8.77	0.87	28.78	m	**37.55**
100 mm dia.	7.13	9.02	0.87	28.78	m	**37.80**
160 mm dia.	9.74	12.33	0.87	28.78	m	**41.11**
200 mm dia.	12.33	15.60	0.87	28.78	m	**44.38**
250 mm dia.	14.63	18.51	1.21	40.03	m	**58.54**
315 mm dia.	17.88	22.63	1.21	40.03	m	**62.66**
355 mm dia.	26.18	33.13	1.21	40.03	m	**73.16**
400 mm dia.	26.63	33.70	1.21	40.03	m	**73.73**
450 mm dia.	31.75	40.19	1.21	40.03	m	**80.21**
500 mm dia.	34.49	43.65	1.21	40.03	m	**83.68**
630 mm dia.	66.84	84.60	1.39	45.98	m	**130.58**
710 mm dia.	74.57	94.38	1.39	45.98	m	**140.36**
800 mm dia.	79.35	100.42	1.44	47.63	m	**148.06**
900 mm dia.	94.41	119.48	1.46	48.30	m	**167.78**
1000 mm dia.	120.08	151.97	1.65	54.58	m	**206.55**
1120 mm dia.	143.26	181.31	2.43	80.38	m	**261.70**
1250 mm dia.	156.03	197.47	2.43	80.38	m	**277.86**
1400 mm dia.	175.53	222.15	2.77	91.63	m	**313.78**
1600 mm dia.	208.91	264.39	3.06	101.22	m	**365.62**
Extra over fittings; circular duct class B						
End cap						
80 mm dia.	2.43	3.07	0.15	4.96	nr	**8.03**
100 mm dia.	2.64	3.34	0.15	4.96	nr	**8.30**
160 mm dia.	3.94	4.98	0.15	4.96	nr	**9.95**
200 mm dia.	4.99	6.31	0.20	6.62	nr	**12.93**
250 mm dia.	7.88	9.97	0.29	9.59	nr	**19.56**
315 mm dia.	9.19	11.63	0.29	9.59	nr	**21.22**
355 mm dia.	13.90	17.59	0.44	14.56	nr	**32.15**
400 mm dia.	15.26	19.31	0.44	14.56	nr	**33.86**
450 mm dia.	17.17	21.73	0.44	14.56	nr	**36.28**
500 mm dia.	18.05	22.84	0.44	14.56	nr	**37.40**
630 mm dia.	70.64	89.41	0.58	19.19	nr	**108.59**
710 mm dia.	79.97	101.21	0.69	22.82	nr	**124.03**
800 mm dia.	85.71	108.48	0.81	26.79	nr	**135.27**
900 mm dia.	108.42	137.22	0.92	30.43	nr	**167.65**
1000 mm dia.	159.74	202.16	1.04	34.40	nr	**236.57**
1120 mm dia.	189.62	239.98	1.16	38.37	nr	**278.35**
1250 mm dia.	214.03	270.88	1.16	38.37	nr	**309.25**
1400 mm dia.	274.61	347.54	1.16	38.37	nr	**385.91**
1600 mm dia.	304.61	385.51	1.16	38.37	nr	**423.88**

38 VENTILATION/AIR CONDITIONING SYSTEMS

Item	Net Price £	Material £	Labour hours	Labour £	Unit	Total rate £
Reducer						
80 mm dia.	12.60	15.95	0.29	9.59	nr	**25.54**
100 mm dia.	12.97	16.41	0.29	9.59	nr	**26.00**
160 mm dia.	17.28	21.87	0.29	9.59	nr	**31.47**
200 mm dia.	21.02	26.60	0.44	14.56	nr	**41.16**
250 mm dia.	20.88	26.43	0.58	19.19	nr	**45.62**
315 mm dia.	23.94	30.30	0.58	19.19	nr	**49.48**
355 mm dia.	26.05	32.97	0.87	28.78	nr	**61.75**
400 mm dia.	30.45	38.54	0.87	28.78	nr	**67.32**
450 mm dia.	32.76	41.46	0.87	28.78	nr	**70.24**
500 mm dia.	35.01	44.30	0.87	28.78	nr	**73.08**
630 mm dia.	106.59	134.89	0.87	28.78	nr	**163.67**
710 mm dia.	131.37	166.26	0.96	31.76	nr	**198.01**
800 mm dia.	141.48	179.05	1.06	35.06	nr	**214.12**
900 mm dia.	170.40	215.66	1.16	38.37	nr	**254.04**
1000 mm dia.	247.25	312.92	1.25	41.35	nr	**354.27**
1120 mm dia.	248.64	314.68	3.47	114.79	nr	**429.47**
1250 mm dia.	325.15	411.51	3.47	114.79	nr	**526.30**
1400 mm dia.	348.66	441.27	4.05	133.97	nr	**575.24**
1600 mm dia.	447.16	565.93	4.62	152.83	nr	**718.76**
90° segmented radius bend						
80 mm dia.	5.29	6.70	0.29	9.59	nr	**16.29**
100 mm dia.	5.57	7.04	0.29	9.59	nr	**16.64**
160 mm dia.	9.42	11.92	0.29	9.59	nr	**21.51**
200 mm dia.	11.63	14.72	0.44	14.56	nr	**29.28**
250 mm dia.	16.23	20.54	0.58	19.19	nr	**39.73**
315 mm dia.	14.98	18.96	0.58	19.19	nr	**38.15**
355 mm dia.	14.69	18.59	0.87	28.78	nr	**47.37**
400 mm dia.	18.29	23.15	0.87	28.78	nr	**51.93**
450 mm dia.	20.63	26.11	0.87	28.78	nr	**54.89**
500 mm dia.	26.31	33.30	0.87	28.78	nr	**62.08**
630 mm dia.	86.66	109.67	0.87	28.78	nr	**138.45**
710 mm dia.	112.16	141.95	0.96	31.76	nr	**173.71**
800 mm dia.	117.11	148.21	1.06	35.06	nr	**183.27**
900 mm dia.	161.66	204.59	1.16	38.37	nr	**242.97**
1000 mm dia.	252.53	319.60	1.25	41.35	nr	**360.95**
1120 mm dia.	298.38	377.63	3.47	114.79	nr	**492.41**
1250 mm dia.	318.96	403.67	3.47	114.79	nr	**518.46**
1400 mm dia.	657.91	832.65	4.05	133.97	nr	**966.62**
1600 mm dia.	977.62	1237.28	4.62	152.83	nr	**1390.11**
45° radius bend						
80 mm dia.	4.49	5.69	0.29	9.59	nr	**15.28**
100 mm dia.	4.84	6.13	0.29	9.59	nr	**15.72**
160 mm dia.	8.12	10.27	0.29	9.59	nr	**19.87**
200 mm dia.	10.05	12.72	0.40	13.23	nr	**25.95**
250 mm dia.	13.98	17.69	0.58	19.19	nr	**36.87**
315 mm dia.	14.74	18.66	0.58	19.19	nr	**37.84**
355 mm dia.	16.98	21.49	0.87	28.78	nr	**50.27**
400 mm dia.	18.34	23.22	0.87	28.78	nr	**51.99**
450 mm dia.	19.49	24.66	0.87	28.78	nr	**53.44**
500 mm dia.	22.85	28.92	0.87	28.78	nr	**57.70**
630 mm dia.	88.03	111.41	0.87	28.78	nr	**140.19**

38 VENTILATION/AIR CONDITIONING SYSTEMS

Item	Net Price £	Material £	Labour hours	Labour £	Unit	Total rate £
DUCTWORK: CIRCULAR – cont						
Extra over fittings – cont						
45° radius bend – cont						
710 mm dia.	101.57	128.54	0.96	31.76	nr	**160.30**
800 mm dia.	116.54	147.49	1.06	35.06	nr	**182.56**
900 mm dia.	140.52	177.84	1.16	38.37	nr	**216.22**
1000 mm dia.	234.48	296.75	1.25	41.35	nr	**338.10**
1120 mm dia.	282.79	357.89	3.47	114.79	nr	**472.68**
1250 mm dia.	306.00	387.28	3.47	114.79	nr	**502.06**
1400 mm dia.	455.13	576.02	4.05	133.97	nr	**709.99**
1600 mm dia.	470.65	595.66	4.62	152.83	nr	**748.49**
90° equal twin bend						
80 mm dia.	29.64	37.51	0.58	19.19	nr	**56.70**
100 mm dia.	30.15	38.15	0.58	19.19	nr	**57.34**
160 mm dia.	39.67	50.21	0.58	19.19	nr	**69.39**
200 mm dia.	51.58	65.27	0.87	28.78	nr	**94.05**
250 mm dia.	57.34	72.57	1.16	38.37	nr	**110.94**
315 mm dia.	88.26	111.71	1.16	38.37	nr	**150.08**
355 mm dia.	104.33	132.04	1.73	57.23	nr	**189.27**
400 mm dia.	121.11	153.27	1.73	57.23	nr	**210.50**
450 mm dia.	141.95	179.65	1.73	57.23	nr	**236.88**
500 mm dia.	150.09	189.95	1.73	57.23	nr	**247.18**
630 mm dia.	265.46	335.97	1.73	57.23	nr	**393.20**
710 mm dia.	328.03	415.16	1.82	60.20	nr	**475.36**
800 mm dia.	451.02	570.81	1.93	63.84	nr	**634.65**
900 mm dia.	591.53	748.64	2.02	66.82	nr	**815.46**
1000 mm dia.	910.76	1152.66	2.11	69.80	nr	**1222.46**
1120 mm dia.	1318.15	1668.25	4.62	152.83	nr	**1821.08**
1250 mm dia.	1329.86	1683.07	4.62	152.83	nr	**1835.89**
1400 mm dia.	1801.21	2279.61	4.62	152.83	nr	**2432.44**
1600 mm dia.	1965.62	2487.69	4.62	152.83	nr	**2640.52**
Conical branch						
80 mm dia.	24.37	30.84	0.58	19.19	nr	**50.03**
100 mm dia.	24.57	31.10	0.58	19.19	nr	**50.28**
160 mm dia.	27.80	35.19	0.58	19.19	nr	**54.38**
200 mm dia.	32.53	41.17	0.87	28.78	nr	**69.95**
250 mm dia.	35.77	45.27	1.16	38.37	nr	**83.65**
315 mm dia.	56.54	71.56	1.16	38.37	nr	**109.93**
355 mm dia.	66.09	83.64	1.73	57.23	nr	**140.87**
400 mm dia.	78.10	98.84	1.73	57.23	nr	**156.07**
450 mm dia.	86.72	109.75	1.73	57.23	nr	**166.98**
500 mm dia.	95.61	121.01	1.73	57.23	nr	**178.24**
630 mm dia.	142.65	180.54	1.73	57.23	nr	**237.77**
710 mm dia.	206.03	260.75	1.82	60.20	nr	**320.96**
800 mm dia.	235.43	297.96	1.93	63.84	nr	**361.81**
900 mm dia.	252.82	319.97	2.02	66.82	nr	**386.79**
1000 mm dia.	301.01	380.96	2.11	69.80	nr	**450.76**
1120 mm dia.	358.76	454.05	4.62	152.83	nr	**606.88**
1250 mm dia.	385.68	488.11	5.20	172.01	nr	**660.12**
1400 mm dia. .	556.64	704.48	5.20	172.01	nr	**876.49**
1600 mm dia.	703.87	890.81	5.20	172.01	nr	**1062.83**

38 VENTILATION/AIR CONDITIONING SYSTEMS

Item	Net Price £	Material £	Labour hours	Labour £	Unit	Total rate £
45° branch						
80 mm dia.	24.58	31.11	0.58	19.19	nr	**50.30**
100 mm dia.	24.78	31.36	0.58	19.19	nr	**50.55**
160 mm dia.	26.29	33.28	0.58	19.19	nr	**52.46**
200 mm dia.	28.37	35.91	0.87	28.78	nr	**64.69**
250 mm dia.	31.73	40.16	1.16	38.37	nr	**78.53**
315 mm dia.	36.66	46.39	1.16	38.37	nr	**84.76**
355 mm dia.	39.96	50.58	1.73	57.23	nr	**107.80**
400 mm dia.	45.01	56.97	1.73	57.23	nr	**114.20**
450 mm dia.	50.70	64.17	1.73	57.23	nr	**121.40**
500 mm dia.	56.26	71.20	1.73	57.23	nr	**128.43**
630 mm dia.	103.50	130.99	1.73	57.23	nr	**188.22**
710 mm dia.	120.71	152.77	1.82	60.20	nr	**212.97**
800 mm dia.	132.00	167.05	2.13	70.46	nr	**237.51**
900 mm dia.	146.44	185.34	2.31	76.41	nr	**261.75**
1000 mm dia.	216.29	273.74	2.31	76.41	nr	**350.15**
1120 mm dia.	265.65	336.21	4.62	152.83	nr	**489.03**
1250 mm dia.	353.29	447.13	4.62	152.83	nr	**599.96**
1400 mm dia.	396.19	501.41	4.62	152.83	nr	**654.24**
1600 mm dia.	452.72	572.96	4.62	152.83	nr	**725.79**

For galvanized sheet metal DW144 class C rates, refer to galvanized sheet metal DW144 class B

38 VENTILATION/AIR CONDITIONING SYSTEMS

Item	Net Price £	Material £	Labour hours	Labour £	Unit	Total rate £
DUCTWORK: FLAT OVAL						
AIR DUCTLINES						
Galvanized sheet metal DW144 class B spirally wound flat oval section ductwork; including all necessary stiffeners, joints, couplers in the running length and duct supports						
Straight duct						
345 × 102 mm	82.26	104.10	2.71	89.65	m	**193.75**
427 × 102 mm	87.27	110.44	2.99	98.91	m	**209.35**
508 × 102 mm	92.31	116.82	3.14	103.87	m	**220.69**
559 × 152 mm	97.73	123.69	3.43	113.46	m	**237.16**
531 × 203 mm	97.73	123.69	3.43	113.46	m	**237.16**
851 × 203 mm	146.24	185.09	5.72	189.22	m	**374.30**
582 × 254 mm	103.54	131.04	3.62	119.75	m	**250.79**
823 × 254 mm	146.36	185.23	5.80	191.86	m	**377.09**
1303 × 254 mm	269.24	340.75	8.13	268.94	m	**609.69**
632 × 305 mm	108.59	137.43	3.93	130.00	m	**267.44**
1275 × 305 mm	262.58	332.33	8.13	268.94	m	**601.26**
765 × 356 mm	123.91	156.82	5.72	189.22	m	**346.04**
1247 × 356 mm	271.05	343.04	8.13	268.94	m	**611.97**
1727 × 356 mm	323.35	409.23	10.41	344.36	m	**753.59**
737 × 406 mm	119.00	150.60	5.72	189.22	m	**339.82**
818 × 406 mm	144.56	182.96	6.21	205.42	m	**388.38**
978 × 406 mm	189.99	240.45	6.92	228.91	m	**469.36**
1379 × 406 mm	284.40	359.94	8.75	289.45	m	**649.39**
1699 × 406 mm	325.62	412.10	10.41	344.36	m	**756.46**
709 × 457 mm	128.87	163.09	5.72	189.22	m	**352.31**
1189 × 457 mm	248.02	313.89	8.80	291.10	m	**605.00**
1671 × 457 mm	322.37	407.99	10.31	341.05	m	**749.04**
678 × 508 mm	128.58	162.73	5.72	189.22	m	**351.95**
919 × 508 mm	162.04	205.07	7.30	241.48	m	**446.55**
1321 × 508 mm	283.68	359.02	8.75	289.45	m	**648.47**
Extra over fittings; flat oval duct class B						
End cap						
345 × 102 mm	39.92	50.52	0.20	6.62	nr	**57.14**
427 × 102 mm	50.06	63.36	0.20	6.62	nr	**69.98**
508 × 102 mm	53.90	68.21	0.20	6.62	nr	**74.83**
559 × 152 mm	56.03	70.91	0.29	9.59	nr	**80.50**
531 × 203 mm	56.05	70.94	0.29	9.59	nr	**80.53**
851 × 203 mm	69.23	87.61	0.44	14.56	nr	**102.17**
582 × 254 mm	59.68	75.53	0.44	14.56	nr	**90.09**
823 × 254 mm	69.25	87.64	0.44	14.56	nr	**102.19**
1303 × 254 mm	205.12	259.60	0.69	22.82	nr	**282.42**
632 × 305 mm	62.30	78.84	0.69	22.82	nr	**101.67**
1275 × 305 mm	201.23	254.68	0.69	22.82	nr	**277.50**

38 VENTILATION/AIR CONDITIONING SYSTEMS

Item	Net Price £	Material £	Labour hours	Labour £	Unit	Total rate £
765 × 356 mm	69.25	87.64	0.69	22.82	nr	110.46
1247 × 356 mm	201.60	255.14	0.69	22.82	nr	277.97
1727 × 356 mm	318.80	403.47	0.69	22.82	nr	426.30
737 × 406 mm	69.31	87.72	1.04	34.40	nr	122.12
818 × 406 mm	75.65	95.75	0.69	22.82	nr	118.57
978 × 406 mm	102.72	130.00	0.69	22.82	nr	152.83
1379 × 406 mm	239.65	303.30	1.04	34.40	nr	337.71
1699 × 406 mm	319.13	403.89	1.04	34.40	nr	438.29
709 × 457 mm	69.26	87.65	1.04	34.40	nr	122.06
1189 × 457 mm	197.98	250.56	1.04	34.40	nr	284.96
1671 × 457 mm	313.16	396.34	1.04	34.40	nr	430.74
678 × 508 mm	69.25	87.64	1.04	34.40	nr	122.04
919 × 508 mm	102.71	129.99	1.04	34.40	nr	164.39
1321 × 508 mm	239.10	302.60	1.04	34.40	nr	337.00
Reducer						
345 × 102 mm	84.28	106.67	0.95	31.43	nr	138.09
427 × 102 mm	91.40	115.68	1.06	35.06	nr	150.74
508 × 102 mm	105.43	133.43	1.13	37.38	nr	170.81
559 × 152 mm	123.92	156.83	1.26	41.68	nr	198.51
531 × 203 mm	125.32	158.60	1.26	41.68	nr	200.28
851 × 203 mm	137.76	174.35	1.34	44.33	nr	218.68
582 × 254 mm	129.27	163.60	1.34	44.33	nr	207.93
823 × 254 mm	139.81	176.94	1.34	44.33	nr	221.27
1303 × 254 mm	372.10	470.93	1.34	44.33	nr	515.26
632 × 305 mm	132.00	167.05	0.70	23.16	nr	190.21
1275 × 305 mm	372.20	471.06	1.16	38.37	nr	509.43
765 × 356 mm	153.03	193.67	1.16	38.37	nr	232.04
1247 × 356 mm	371.12	469.69	1.16	38.37	nr	508.07
1727 × 356 mm	425.93	539.06	1.25	41.35	nr	580.41
737 × 406 mm	156.75	198.39	1.16	38.37	nr	236.76
818 × 406 mm	170.25	215.46	1.27	42.01	nr	257.48
978 × 406 mm	186.20	235.65	1.44	47.63	nr	283.28
1379 × 406 mm	371.84	470.60	1.44	47.63	nr	518.23
1699 × 406 mm	428.95	542.87	1.44	47.63	nr	590.51
709 × 457 mm	154.71	195.80	1.16	38.37	nr	234.17
1189 × 457 mm	386.14	488.70	1.34	44.33	nr	533.02
1671 × 457 mm	424.07	536.71	1.44	47.63	nr	584.34
678 × 508 mm	156.71	198.34	1.16	38.37	nr	236.71
919 × 508 mm	205.74	260.38	1.26	41.68	nr	302.06
1321 × 508 mm	379.72	480.58	1.44	47.63	nr	528.21
90° radius bend						
345 × 102 mm	216.21	273.63	0.29	9.59	nr	283.22
427 × 102 mm	239.90	303.62	0.58	19.19	nr	322.81
508 × 102 mm	295.86	374.44	0.58	19.19	nr	393.62
559 × 152 mm	343.67	434.94	0.58	19.19	nr	454.13
531 × 203 mm	348.00	440.43	0.87	28.78	nr	469.21
851 × 203 mm	353.91	447.91	0.87	28.78	nr	476.69
582 × 254 mm	365.19	462.18	0.87	28.78	nr	490.96
823 × 254 mm	360.32	456.02	0.87	28.78	nr	484.80
1303 × 254 mm	378.79	479.39	0.96	31.76	nr	511.15
632 × 305 mm	382.64	484.27	0.87	28.78	nr	513.05
1275 × 305 mm	396.59	501.92	0.96	31.76	nr	533.67

38 VENTILATION/AIR CONDITIONING SYSTEMS

Item	Net Price £	Material £	Labour hours	Labour £	Unit	Total rate £
DUCTWORK: FLAT OVAL – cont						
Extra over fittings – cont						
90° radius bend – cont						
765 × 356 mm	401.90	508.64	0.87	28.78	nr	**537.42**
1247 × 356 mm	391.51	495.50	0.96	31.76	nr	**527.26**
1727 × 356 mm	759.35	961.03	1.25	41.35	nr	**1002.38**
737 × 406 mm	413.55	523.39	0.96	31.76	nr	**555.15**
818 × 406 mm	394.73	499.57	0.87	28.78	nr	**528.35**
978 × 406 mm	326.06	412.66	0.96	31.76	nr	**444.41**
1379 × 406 mm	634.61	803.16	1.16	38.37	nr	**841.53**
1699 × 406 mm	767.36	971.17	1.25	41.35	nr	**1012.52**
709 × 457 mm	407.05	515.17	0.87	28.78	nr	**543.95**
1189 × 457 mm	454.99	575.83	0.96	31.76	nr	**607.59**
1671 × 457 mm	1104.84	1398.29	1.25	41.35	nr	**1439.64**
678 × 508 mm	413.42	523.22	0.87	28.78	nr	**552.00**
919 × 508 mm	387.43	490.33	0.96	31.76	nr	**522.09**
1321 × 508 mm	738.61	934.79	1.16	38.37	nr	**973.16**
45° radius bend						
345 × 102 mm	106.92	135.32	0.79	26.13	nr	**161.45**
427 × 102 mm	140.32	177.59	0.85	28.12	nr	**205.71**
508 × 102 mm	172.90	218.83	0.95	31.43	nr	**250.25**
559 × 152 mm	200.78	254.11	0.79	26.13	nr	**280.24**
531 × 203 mm	202.95	256.86	0.85	28.12	nr	**284.98**
851 × 203 mm	215.20	272.35	0.98	32.42	nr	**304.77**
582 × 254 mm	213.90	270.71	0.76	25.14	nr	**295.85**
823 × 254 mm	218.42	276.43	0.95	31.43	nr	**307.86**
1303 × 254 mm	407.99	516.35	1.16	38.37	nr	**554.72**
632 × 305 mm	224.90	284.63	0.58	19.19	nr	**303.82**
1275 × 305 mm	412.81	522.45	1.16	38.37	nr	**560.82**
765 × 356 mm	239.22	302.76	0.87	28.78	nr	**331.54**
1247 × 356 mm	410.69	519.76	1.16	38.37	nr	**558.14**
1727 × 356 mm	726.14	919.00	1.26	41.68	nr	**960.68**
737 × 406 mm	245.05	310.13	0.69	22.82	nr	**332.96**
818 × 406 mm	237.96	301.16	0.78	25.80	nr	**326.97**
978 × 406 mm	207.22	262.25	0.87	28.78	nr	**291.03**
1379 × 406 mm	588.27	744.52	1.16	38.37	nr	**782.89**
1699 × 406 mm	730.49	924.50	1.27	42.01	nr	**966.51**
709 × 457 mm	241.46	305.59	0.81	26.79	nr	**332.38**
1189 × 457 mm	438.61	555.10	0.95	31.43	nr	**586.53**
1671 × 457 mm	773.76	979.26	1.26	41.68	nr	**1020.95**
678 × 508 mm	244.98	310.04	0.92	30.43	nr	**340.47**
919 × 508 mm	237.90	301.08	1.10	36.39	nr	**337.47**
1321 × 508 mm	601.31	761.02	1.25	41.35	nr	**802.37**
90° hard bend with turning vanes						
345 × 102 mm	91.72	116.08	0.55	18.19	nr	**134.27**
427 × 102 mm	157.86	199.78	1.16	38.37	nr	**238.16**
508 × 102 mm	214.25	271.16	1.16	38.37	nr	**309.53**
559 × 152 mm	240.87	304.85	1.16	38.37	nr	**343.22**
531 × 203 mm	241.33	305.43	1.73	57.23	nr	**362.66**
851 × 203 mm	352.60	446.25	1.73	57.23	nr	**503.48**
582 × 254 mm	258.50	327.16	1.73	57.23	nr	**384.38**

38 VENTILATION/AIR CONDITIONING SYSTEMS

Item	Net Price £	Material £	Labour hours	Labour £	Unit	Total rate £
823 × 254 mm	352.60	446.25	1.73	57.23	nr	503.48
1303 × 254 mm	625.35	791.44	1.82	60.20	nr	851.65
632 × 305 mm	275.57	348.76	1.73	57.23	nr	405.99
1275 × 305 mm	621.37	786.40	1.82	60.20	nr	846.61
765 × 356 mm	336.08	425.35	1.73	57.23	nr	482.58
1247 × 356 mm	621.75	786.88	1.82	60.20	nr	847.09
1727 × 356 mm	821.14	1039.24	1.82	60.20	nr	1099.44
737 × 406 mm	336.08	425.35	1.73	57.23	nr	482.58
818 × 406 mm	381.53	482.86	1.73	57.23	nr	540.09
978 × 406 mm	440.43	557.41	1.73	57.23	nr	614.64
1379 × 406 mm	685.27	867.28	1.82	60.20	nr	927.49
1699 × 406 mm	821.46	1039.64	2.11	69.80	nr	1109.43
709 × 457 mm	336.09	425.36	1.73	57.23	nr	482.59
1189 × 457 mm	618.03	782.18	1.82	60.20	nr	842.38
1671 × 457 mm	815.36	1031.92	2.11	69.80	nr	1101.71
678 × 508 mm	336.08	425.35	1.82	60.20	nr	485.55
919 × 508 mm	440.43	557.41	1.82	60.20	nr	617.62
1321 × 508 mm	684.69	866.55	2.11	69.80	nr	936.35
90° branch						
345 × 102 mm	172.94	218.87	0.58	19.19	nr	238.05
427 × 102 mm	175.37	221.95	0.58	19.19	nr	241.14
508 × 102 mm	185.04	234.19	1.16	38.37	nr	272.56
559 × 152 mm	215.81	273.12	1.16	38.37	nr	311.50
531 × 203 mm	218.57	276.62	1.16	38.37	nr	314.99
851 × 203 mm	320.16	405.19	1.73	57.23	nr	462.42
582 × 254 mm	229.59	290.57	1.73	57.23	nr	347.80
823 × 254 mm	324.22	410.33	1.73	57.23	nr	467.56
1303 × 254 mm	444.83	562.98	1.82	60.20	nr	623.19
632 × 305 mm	240.86	304.83	1.73	57.23	nr	362.06
1275 × 305 mm	453.87	574.42	1.82	60.20	nr	634.63
765 × 356 mm	353.03	446.80	1.73	57.23	nr	504.02
1247 × 356 mm	450.03	569.56	1.82	60.20	nr	629.76
1727 × 356 mm	674.79	854.02	2.11	69.80	nr	923.82
737 × 406 mm	360.99	456.87	1.73	57.23	nr	514.10
818 × 406 mm	403.19	510.28	1.73	57.23	nr	567.50
978 × 406 mm	355.08	449.39	1.82	60.20	nr	509.59
1379 × 406 mm	454.31	574.98	1.93	63.84	nr	638.82
1699 × 406 mm	679.89	860.46	2.11	69.80	nr	930.26
709 × 457 mm	355.81	450.32	1.73	57.23	nr	507.55
1189 × 457 mm	490.11	620.28	1.82	60.20	nr	680.49
1671 × 457 mm	945.91	1197.15	2.11	69.80	nr	1266.95
678 × 508 mm	359.84	455.41	1.73	57.23	nr	512.64
919 × 508 mm	397.13	502.61	1.82	60.20	nr	562.81
1321 × 508 mm	575.03	727.76	2.11	69.80	nr	797.56
45° branch						
345 × 102 mm	104.22	131.90	0.58	19.19	nr	151.09
427 × 102 mm	179.70	227.42	0.58	19.19	nr	246.61
508 × 102 mm	244.05	308.87	1.16	38.37	nr	347.24
559 × 152 mm	274.89	347.90	1.73	57.23	nr	405.13
531 × 203 mm	274.90	347.91	1.73	57.23	nr	405.14
851 × 203 mm	382.66	484.30	1.73	57.23	nr	541.52
582 × 254 mm	294.37	372.55	1.73	57.23	nr	429.78

38 VENTILATION/AIR CONDITIONING SYSTEMS

Item	Net Price £	Material £	Labour hours	Labour £	Unit	Total rate £
DUCTWORK: FLAT OVAL – cont						
Extra over fittings – cont						
45° branch – cont						
823 × 254 mm	382.68	484.32	1.82	60.20	nr	**544.53**
1303 × 254 mm	667.54	844.84	1.92	63.51	nr	**908.35**
632 × 305 mm	313.77	397.11	1.73	57.23	nr	**454.34**
1275 × 305 mm	663.75	840.04	1.82	60.20	nr	**900.24**
765 × 356 mm	382.68	484.32	1.73	57.23	nr	**541.55**
1247 × 356 mm	664.11	840.50	1.82	60.20	nr	**900.71**
1727 × 356 mm	878.08	1111.30	1.82	60.20	nr	**1171.51**
737 × 406 mm	382.68	484.32	1.73	57.23	nr	**541.55**
818 × 406 mm	414.07	524.04	1.73	57.23	nr	**581.27**
978 × 406 mm	478.00	604.96	1.73	57.23	nr	**662.19**
1379 × 406 mm	732.64	927.23	1.93	63.84	nr	**991.07**
1699 × 406 mm	878.41	1111.71	2.19	72.44	nr	**1184.16**
709 × 457 mm	382.69	484.34	1.73	57.23	nr	**541.56**
1189 × 457 mm	660.58	836.02	1.82	60.20	nr	**896.23**
1671 × 457 mm	872.58	1104.34	2.11	69.80	nr	**1174.14**
678 × 508 mm	382.68	484.32	1.73	57.23	nr	**541.55**
919 × 508 mm	477.99	604.95	1.82	60.20	nr	**665.15**
1321 × 508 mm	732.08	926.52	1.93	63.84	nr	**990.37**
For rates for access doors refer to ancillaries in Ductwork Ancillaries: Access Hatches						

38 VENTILATION/AIR CONDITIONING SYSTEMS

Item	Net Price £	Material £	Labour hours	Labour £	Unit	Total rate £
DUCTWORK: FLEXIBLE						
AIR DUCTLINES						
Aluminium foil flexible ductwork, DW 144 class B; multiply aluminium polyester laminate fabric, with high tensile steel wire helix						
Duct						
102 mm dia.	1.70	2.15	0.33	10.92	m	**13.07**
152 mm dia.	2.50	3.16	0.33	10.92	m	**14.08**
203 mm dia.	3.28	4.15	0.33	10.92	m	**15.06**
254 mm dia.	4.12	5.21	0.33	10.92	m	**16.13**
304 mm dia.	5.04	6.38	0.33	10.92	m	**17.29**
355 mm dia.	6.86	8.68	0.33	10.92	m	**19.59**
406 mm dia.	7.63	9.66	0.33	10.92	m	**20.58**
Insulated aluminium foil flexible ductwork, DW144 class B; laminate construction of aluminium and polyester multiply inner core with 25 mm insulation; outer layer of multiply aluminium polyester laminate, with high tensile steel wire helix						
Duct						
102 mm dia.	3.81	4.82	0.50	16.54	m	**21.36**
152 mm dia.	4.98	6.30	0.50	16.54	m	**22.84**
203 mm dia.	6.03	7.63	0.50	16.54	m	**24.17**
254 mm dia.	7.41	9.38	0.50	16.54	m	**25.92**
304 mm dia.	9.53	12.07	0.50	16.54	m	**28.61**
355 mm dia.	11.83	14.98	0.50	16.54	m	**31.52**
406 mm dia.	13.10	16.58	0.50	16.54	m	**33.12**

38 VENTILATION/AIR CONDITIONING SYSTEMS

Item	Net Price £	Material £	Labour hours	Labour £	Unit	Total rate £
DUCTWORK: PLASTIC						
AIR DUCTLINES						
Rigid grey PVC DW 154 circular section ductwork; solvent welded or filler rod welded joints; excludes couplers and supports (these are detailed separately); ductwork to conform to curent HSE regulations						
Straight duct (standard length 6 m)						
110 mm	10.51	13.30	1.71	56.57	m	**69.87**
160 mm	16.07	20.34	1.74	57.56	m	**77.90**
200 mm	20.32	25.72	1.79	59.21	m	**84.93**
225 mm	24.60	31.13	1.96	64.84	m	**95.96**
250 mm	27.20	34.42	1.98	65.50	m	**99.92**
315 mm	34.87	44.13	2.23	73.77	m	**117.90**
355 mm	42.55	53.86	2.25	74.43	m	**128.29**
400 mm	49.89	63.14	2.39	79.06	m	**142.20**
450 mm	64.35	81.45	2.95	97.58	m	**179.03**
500 mm	85.69	108.46	2.98	98.58	m	**207.03**
600 mm	165.62	209.61	3.12	103.21	m	**312.81**
Extra for supports (BZP finish); Horizontal – Maximum 2.4 m centres; Vertical – Maximum 4.0 m centres						
Duct size						
110 mm	39.95	50.57	0.59	19.52	m	**70.08**
160 mm	40.09	50.74	0.59	19.52	m	**70.26**
200 mm	40.22	50.90	0.59	19.52	m	**70.42**
225 mm	42.71	54.05	0.63	20.84	m	**74.89**
250 mm	42.72	54.07	0.63	20.84	m	**74.91**
315 mm	43.03	54.46	0.63	20.84	m	**75.30**
355 mm	41.16	52.09	0.81	26.79	m	**78.88**
400 mm	55.87	70.71	0.78	25.80	m	**96.51**
450 mm	57.44	72.69	0.80	26.46	m	**99.16**
500 mm	57.70	73.03	0.80	26.46	m	**99.49**
600 mm	61.28	77.56	0.85	28.12	m	**105.67**
Note: These are maximum figures and may be reduced subject to local conditions (i.e. a high number of changes of direction)						
Extra over fittings; Rigid grey PVC						
90° bend						
110 mm	32.09	40.61	1.11	36.72	m	**77.33**
160 mm	35.41	44.82	1.34	44.33	m	**89.15**
200 mm	29.64	37.51	2.00	66.16	m	**103.67**
225 mm	45.49	57.57	1.79	59.21	m	**116.78**
250 mm	48.42	61.28	2.01	66.49	m	**127.77**
315 mm	83.61	105.82	2.56	84.68	m	**190.50**
355 mm	99.56	126.00	2.71	89.65	m	**215.65**
400 mm	126.46	160.05	3.58	118.43	m	**278.47**

38 VENTILATION/AIR CONDITIONING SYSTEMS

Item	Net Price £	Material £	Labour hours	Labour £	Unit	Total rate £
450 mm	382.24	483.77	4.55	150.51	m	634.28
500 mm	443.65	561.48	5.01	165.73	m	727.21
600 mm	759.43	961.13	5.78	191.20	m	1152.33
45° bend						
110 mm	17.92	22.67	0.71	23.49	m	46.16
160 mm	23.73	30.03	0.93	30.76	m	60.80
200 mm	26.54	33.59	1.14	37.71	m	71.30
225 mm	32.46	41.09	1.41	46.64	m	87.73
250 mm	37.05	46.89	1.62	53.59	m	100.48
315 mm	58.96	74.62	1.93	63.84	m	138.46
355 mm	76.18	96.41	2.22	73.44	m	169.85
400 mm	99.47	125.89	2.85	94.28	m	220.17
450 mm	223.88	283.34	3.77	124.71	m	408.05
500 mm	258.51	327.17	4.16	137.61	m	464.78
600 mm	387.90	490.93	4.81	159.11	m	650.04
Tee						
110 mm	58.83	74.46	1.04	34.40	m	108.86
160 mm	84.28	106.67	1.38	45.65	m	152.32
200 mm	115.45	146.11	1.79	59.21	m	205.32
225 mm	138.06	174.73	2.25	74.43	m	249.16
250 mm	166.27	210.44	2.62	86.67	m	297.11
315 mm	251.05	317.72	3.17	104.86	m	422.59
355 mm	344.76	436.33	3.73	123.39	m	559.72
400 mm	358.79	454.09	4.71	155.80	m	609.89
450 mm	397.80	503.45	5.72	189.22	m	692.67
500 mm	498.71	631.17	6.33	209.39	m	840.56
Coupler						
110 mm	11.63	14.72	0.70	23.16	m	37.87
160 mm	16.83	21.29	0.91	30.10	m	51.40
200 mm	22.04	27.89	1.12	37.05	m	64.94
225 mm	28.52	36.10	1.41	46.64	m	82.74
250 mm	32.98	41.74	1.60	52.93	m	94.67
315 mm	40.09	50.74	1.86	61.53	m	112.27
355 mm	44.05	55.75	2.43	80.38	m	136.13
400 mm	50.17	63.50	2.66	87.99	m	151.49
450 mm	109.44	138.50	3.20	105.85	m	244.36
500 mm	123.80	156.69	3.52	116.44	m	273.13
Damper						
110 mm	73.15	92.57	0.96	31.76	m	124.33
160 mm	77.09	97.56	1.26	41.68	m	139.24
200 mm	94.67	119.81	1.52	50.28	m	170.09
225 mm	101.53	128.49	1.88	62.19	m	190.68
250 mm	102.62	129.87	2.16	71.45	m	201.33
315 mm	121.21	153.40	2.94	97.25	m	250.65
355 mm	134.83	170.64	3.42	113.13	m	283.77
400 mm	141.86	179.54	3.79	125.37	m	304.91
Reducer						
160 × 110 mm	13.36	16.91	0.85	28.12	m	45.03
200 × 110 mm	17.13	21.68	0.98	32.42	m	54.10
200 × 160 mm	32.98	41.74	1.07	35.40	m	77.14
225 × 200 mm	45.98	58.19	1.38	45.65	m	103.84
250 × 160 mm	42.57	53.87	1.38	45.65	m	99.52

38 VENTILATION/AIR CONDITIONING SYSTEMS

Item	Net Price £	Material £	Labour hours	Labour £	Unit	Total rate £
DUCTWORK: PLASTIC – cont						
Extra over fittings – cont						
Reducer – cont						
250 × 200 mm	50.04	63.34	1.48	48.96	m	**112.29**
250 × 225 mm	43.45	54.99	1.58	52.27	m	**107.25**
315 × 200 mm	88.88	112.49	1.63	53.92	m	**166.41**
315 × 250 mm	54.98	69.58	1.84	60.87	m	**130.45**
355 × 200 mm	88.88	112.49	1.99	65.83	m	**178.32**
355 × 250 mm	82.52	104.44	1.97	65.17	m	**169.61**
355 × 315 mm	116.54	147.49	2.16	71.45	m	**218.94**
400 × 315 mm	96.97	122.73	2.75	90.97	m	**213.70**
400 × 355 mm	96.97	122.73	2.73	90.31	m	**213.04**
Flange						
110 mm	16.83	21.29	0.70	23.16	m	**44.45**
160 mm	20.46	25.89	0.91	30.10	m	**56.00**
200 mm	22.72	28.76	1.12	37.05	m	**65.81**
225 mm	23.37	29.57	1.35	44.66	m	**74.23**
250 mm	24.82	31.41	1.55	51.27	m	**82.69**
315 mm	35.90	45.44	1.85	61.20	m	**106.63**
355 mm	38.81	49.12	2.06	68.14	m	**117.26**
400 mm	42.09	53.27	2.62	86.67	m	**139.94**
Polypropylene (PPS) DW154 circular						
section ductwork; filler rod welded joints;						
excludes couplers and supports (these are						
detailed separately); ductwork to conform						
to current HSE regulations						
Straight duct (standard length 5 m)						
110 mm	13.85	17.53	0.65	21.50	m	**39.03**
160 mm	20.79	26.32	0.75	24.81	m	**51.13**
200 mm	26.94	34.10	0.85	28.12	m	**62.22**
225 mm	35.25	44.61	1.07	35.40	m	**80.00**
250 mm	39.10	49.49	1.18	39.03	m	**88.52**
315 mm	74.99	94.91	1.41	46.64	m	**141.55**
355 mm	85.14	107.75	1.53	50.61	m	**158.36**
400 mm	114.58	145.01	1.73	57.23	m	**202.24**
Extra for supports (BZP finish);						
Horizontal – Maximum 2.4 m centre;						
Vertical – Maximum 4.0 m centre						
Duct size						
110 mm	55.37	70.08	0.42	13.89	m	**83.97**
160 mm	55.37	70.08	0.42	13.89	m	**83.97**
200 mm	55.37	70.08	0.42	13.89	m	**83.97**
225 mm	55.37	70.08	0.42	13.89	m	**83.97**
250 mm	55.37	70.08	0.42	13.89	m	**83.97**
315 mm	55.37	70.08	0.42	13.89	m	**83.97**
355 mm	55.37	70.08	0.55	18.19	m	**88.27**
400 mm	78.37	99.19	0.55	18.19	m	**117.38**
Note: These are maximum figures and may						
be reduced subject to local conditions (i.e. a						
high number of changes of direction)						

38 VENTILATION/AIR CONDITIONING SYSTEMS

Item	Net Price £	Material £	Labour hours	Labour £	Unit	Total rate £
Extra over fittings; Polypropylene (DW 154)						
90° bend						
110 mm	24.48	30.99	1.01	33.41	m	**64.40**
160 mm	34.50	43.67	1.30	43.00	m	**86.67**
200 mm	40.47	51.22	1.58	52.27	m	**103.49**
225 mm	54.71	69.24	2.10	69.47	m	**138.71**
250 mm	59.49	75.29	2.36	78.07	m	**153.36**
315 mm	145.94	184.70	3.14	103.87	m	**288.57**
355 mm	176.13	222.91	3.52	116.44	m	**339.35**
400 mm	188.56	238.65	4.39	145.22	m	**383.87**
45° bend						
110 mm	18.03	22.82	0.84	27.79	m	**50.60**
160 mm	29.07	36.79	1.13	37.38	m	**74.17**
200 mm	33.32	42.17	1.39	45.98	m	**88.15**
225 mm	41.06	51.96	1.73	57.23	m	**109.19**
250 mm	46.84	59.29	1.99	65.83	m	**125.11**
315 mm	118.75	150.28	2.57	85.01	m	**235.30**
355 mm	132.39	167.55	2.93	96.92	m	**264.47**
400 mm	139.96	177.13	3.60	119.09	m	**296.22**
Tee						
110 mm	91.54	115.85	1.55	51.27	m	**167.12**
160 mm	127.52	161.39	2.17	71.78	m	**233.17**
200 mm	161.24	204.07	2.68	88.65	m	**292.72**
225 mm	182.46	230.92	3.38	111.81	m	**342.73**
250 mm	238.97	302.44	3.95	130.66	m	**433.11**
315 mm	314.81	398.43	4.69	155.14	m	**553.57**
355 mm	384.03	486.03	5.44	179.95	m	**665.99**
400 mm	433.50	548.64	6.09	201.45	m	**750.10**
Coupler						
110 mm	19.16	24.25	0.84	27.79	m	**52.04**
160 mm	23.30	29.48	1.10	36.39	m	**65.87**
200 mm	28.05	35.50	1.36	44.99	m	**80.49**
225 mm	30.16	38.17	1.67	55.24	m	**93.41**
250 mm	32.18	40.73	1.91	63.18	m	**103.91**
315 mm	43.21	54.69	2.24	74.10	m	**128.79**
355 mm	55.45	70.18	2.58	85.35	m	**155.53**
400 mm	63.20	79.98	3.30	109.16	m	**189.14**
Damper						
110 mm	94.22	119.24	0.79	26.13	m	**145.38**
160 mm	104.00	131.62	1.13	37.38	m	**169.00**
200 mm	114.20	144.54	1.44	47.63	m	**192.17**
225 mm	122.20	154.65	1.81	59.87	m	**214.53**
250 mm	127.62	161.52	2.08	68.81	m	**230.32**
315 mm	144.95	183.45	2.40	79.39	m	**262.84**
355 mm	159.59	201.98	2.75	90.97	m	**292.95**
400 mm	172.40	218.18	3.57	118.09	m	**336.28**
Reducer						
160 × 110 mm	17.38	22.00	0.87	28.78	m	**50.78**
200 × 160 mm	16.73	21.17	1.16	38.37	m	**59.54**
225 × 200 mm	21.82	27.61	1.46	48.30	m	**75.91**

38 VENTILATION/AIR CONDITIONING SYSTEMS

Item	Net Price £	Material £	Labour hours	Labour £	Unit	Total rate £
DUCTWORK: PLASTIC – cont						
Extra over fittings – cont						
Reducer – cont						
250 × 200 mm	24.48	30.99	1.61	53.26	m	**84.25**
250 × 225 mm	39.38	49.84	1.70	56.24	m	**106.08**
315 × 200 mm	54.26	68.68	1.66	54.91	m	**123.59**
315 × 250 mm	34.21	43.30	2.03	67.15	m	**110.45**
355 × 250 mm	44.73	56.62	2.05	67.81	m	**124.43**
355 × 315 mm	70.31	88.98	2.22	73.44	m	**162.42**
400 × 315 mm	68.67	86.91	2.88	95.27	m	**182.18**
400 × 355 mm	64.37	81.47	2.92	96.59	m	**178.06**
Flange						
110 mm	18.54	23.47	0.76	25.14	m	**48.61**
160 mm	24.22	30.65	1.02	33.74	m	**64.39**
200 mm	27.63	34.97	1.27	42.01	m	**76.98**
225 mm	29.22	36.98	1.55	51.27	m	**88.26**
250 mm	32.84	41.56	1.79	59.21	m	**100.78**
315 mm	39.24	49.66	2.08	68.81	m	**118.47**
355 mm	44.85	56.76	2.36	78.07	m	**134.82**
400 mm	49.47	62.61	3.07	101.55	m	**164.17**

38 VENTILATION/AIR CONDITIONING SYSTEMS

Item	Net Price £	Material £	Labour hours	Labour £	Unit	Total rate £
DUCTWORK: RECTANGULAR – CLASS B						
AIR DUCTLINES						
Galvanized sheet metal DW144 class B rectangular section ductwork; including all necessary stiffeners, joints, couplers in the running length and duct supports						
Ductwork up to 400 mm longest side						
Sum of two sides 200 mm	27.96	35.39	1.16	38.37	m	**73.76**
Sum of two sides 300 mm	29.77	37.67	1.16	38.37	m	**76.05**
Sum of two sides 400 mm	26.45	33.47	1.19	39.36	m	**72.84**
Sum of two sides 500 mm	28.42	35.97	1.19	39.36	m	**75.34**
Sum of two sides 600 mm	30.18	38.19	1.27	42.01	m	**80.20**
Sum of two sides 700 mm	31.92	40.40	1.27	42.01	m	**82.41**
Sum of two sides 800 mm	33.78	42.75	1.27	42.01	m	**84.76**
Extra over fittings; Rectangular ductwork class B; up to 400 mm longest side						
End cap						
Sum of two sides 200 mm	16.23	20.54	0.38	12.57	nr	**33.11**
Sum of two sides 300 mm	18.09	22.90	0.38	12.57	nr	**35.47**
Sum of two sides 400 mm	19.94	25.24	0.38	12.57	nr	**37.81**
Sum of two sides 500 mm	21.80	27.59	0.38	12.57	nr	**40.16**
Sum of two sides 600 mm	23.66	29.94	0.38	12.57	nr	**42.51**
Sum of two sides 700 mm	25.50	32.28	0.38	12.57	nr	**44.85**
Sum of two sides 800 mm	27.37	34.64	0.38	12.57	nr	**47.21**
Reducer						
Sum of two sides 200 mm	26.51	33.55	1.40	46.31	nr	**79.87**
Sum of two sides 300 mm	29.94	37.89	1.40	46.31	nr	**84.20**
Sum of two sides 400 mm	49.02	62.05	1.42	46.97	nr	**109.02**
Sum of two sides 500 mm	53.07	67.16	1.42	46.97	nr	**114.13**
Sum of two sides 600 mm	57.11	72.28	1.69	55.90	nr	**128.18**
Sum of two sides 700 mm	61.15	77.39	1.69	55.90	nr	**133.30**
Sum of two sides 800 mm	65.15	82.46	1.92	63.51	nr	**145.97**
Offset						
Sum of two sides 200 mm	39.13	49.53	1.63	53.92	nr	**103.45**
Sum of two sides 300 mm	44.23	55.97	1.63	53.92	nr	**109.89**
Sum of two sides 400 mm	65.42	82.79	1.65	54.58	nr	**137.37**
Sum of two sides 500 mm	71.12	90.01	1.65	54.58	nr	**144.59**
Sum of two sides 600 mm	75.78	95.91	1.92	63.51	nr	**159.42**
Sum of two sides 700 mm	80.98	102.48	1.92	63.51	nr	**166.00**
Sum of two sides 800 mm	85.49	108.20	1.92	63.51	nr	**171.71**
Square to round						
Sum of two sides 200 mm	34.40	43.53	1.63	53.92	nr	**97.45**
Sum of two sides 300 mm	38.66	48.93	1.63	53.92	nr	**102.85**
Sum of two sides 400 mm	51.48	65.15	1.65	54.58	nr	**119.74**
Sum of two sides 500 mm	56.02	70.90	1.65	54.58	nr	**125.48**
Sum of two sides 600 mm	60.56	76.65	1.92	63.51	nr	**140.16**
Sum of two sides 700 mm	65.11	82.40	1.92	63.51	nr	**145.92**
Sum of two sides 800 mm	69.60	88.09	1.92	63.51	nr	**151.60**

38 VENTILATION/AIR CONDITIONING SYSTEMS

Item	Net Price £	Material £	Labour hours	Labour £	Unit	Total rate £
DUCTWORK: RECTANGULAR – CLASS B – cont						
Extra over fittings – cont						
90° radius bend						
Sum of two sides 200 mm	25.89	32.77	1.22	40.36	nr	**73.13**
Sum of two sides 300 mm	27.93	35.35	1.22	40.36	nr	**75.71**
Sum of two sides 400 mm	47.07	59.57	1.25	41.35	nr	**100.92**
Sum of two sides 500 mm	50.09	63.39	1.25	41.35	nr	**104.74**
Sum of two sides 600 mm	54.08	68.44	1.33	44.00	nr	**112.43**
Sum of two sides 700 mm	57.56	72.85	1.33	44.00	nr	**116.85**
Sum of two sides 800 mm	61.49	77.82	1.40	46.31	nr	**124.13**
45° radius bend						
Sum of two sides 200 mm	27.96	35.39	0.89	29.44	nr	**64.83**
Sum of two sides 300 mm	30.69	38.84	1.12	37.05	nr	**75.89**
Sum of two sides 400 mm	49.33	62.43	1.10	36.39	nr	**98.82**
Sum of two sides 500 mm	52.87	66.91	1.10	36.39	nr	**103.30**
Sum of two sides 600 mm	56.90	72.01	1.16	38.37	nr	**110.38**
Sum of two sides 700 mm	60.67	76.78	1.16	38.37	nr	**115.16**
Sum of two sides 800 mm	64.65	81.82	1.22	40.36	nr	**122.18**
90° mitre bend						
Sum of two sides 200 mm	43.27	54.76	1.29	42.67	nr	**97.44**
Sum of two sides 300 mm	47.54	60.17	1.29	42.67	nr	**102.84**
Sum of two sides 400 mm	69.13	87.49	1.29	42.67	nr	**130.17**
Sum of two sides 500 mm	74.66	94.48	1.29	42.67	nr	**137.16**
Sum of two sides 600 mm	81.75	103.47	1.39	45.98	nr	**149.45**
Sum of two sides 700 mm	88.34	111.80	1.39	45.98	nr	**157.78**
Sum of two sides 800 mm	95.53	120.90	1.46	48.30	nr	**169.20**
Branch						
Sum of two sides 200 mm	42.86	54.24	0.92	30.43	nr	**84.68**
Sum of two sides 300 mm	47.63	60.28	0.92	30.43	nr	**90.71**
Sum of two sides 400 mm	59.75	75.61	0.95	31.43	nr	**107.04**
Sum of two sides 500 mm	65.08	82.36	0.95	31.43	nr	**113.79**
Sum of two sides 600 mm	70.26	88.92	1.03	34.07	nr	**122.99**
Sum of two sides 700 mm	75.44	95.48	1.03	34.07	nr	**129.55**
Sum of two sides 800 mm	80.63	102.04	1.03	34.07	nr	**136.12**
Grille neck						
Sum of two sides 200 mm	49.15	62.20	1.10	36.39	nr	**98.59**
Sum of two sides 300 mm	55.04	69.66	1.10	36.39	nr	**106.05**
Sum of two sides 400 mm	60.92	77.10	1.16	38.37	nr	**115.47**
Sum of two sides 500 mm	66.80	84.54	1.16	38.37	nr	**122.92**
Sum of two sides 600 mm	72.70	92.01	1.18	39.03	nr	**131.05**
Sum of two sides 700 mm	78.57	99.44	1.18	39.03	nr	**138.47**
Sum of two sides 800 mm	84.45	106.88	1.18	39.03	nr	**145.92**
Ductwork 401 to 600 mm longest side						
Sum of two sides 600 mm	34.41	43.55	1.27	42.01	m	**85.56**
Sum of two sides 700 mm	36.98	46.80	1.27	42.01	m	**88.81**
Sum of two sides 800 mm	39.45	49.93	1.27	42.01	m	**91.94**

38 VENTILATION/AIR CONDITIONING SYSTEMS

Item	Net Price £	Material £	Labour hours	Labour £	Unit	Total rate £
Sum of two sides 900 mm	41.80	52.90	1.27	42.01	m	**94.91**
Sum of two sides 1000 mm	44.15	55.88	1.37	45.32	m	**101.20**
Sum of two sides 1100 mm	46.87	59.32	1.37	45.32	m	**104.64**
Sum of two sides 1200 mm	49.22	62.30	1.37	45.32	m	**107.62**
Extra over fittings; Ductwork 401 to 600 mm longest side						
End cap						
Sum of two sides 600 mm	24.32	30.78	0.38	12.57	nr	**43.35**
Sum of two sides 700 mm	26.26	33.24	0.38	12.57	nr	**45.81**
Sum of two sides 800 mm	28.18	35.67	0.38	12.57	nr	**48.24**
Sum of two sides 900 mm	30.12	38.13	0.58	19.19	nr	**57.31**
Sum of two sides 1000 mm	32.06	40.57	0.58	19.19	nr	**59.76**
Sum of two sides 1100 mm	33.99	43.02	0.58	19.19	nr	**62.20**
Sum of two sides 1200 mm	35.92	45.46	0.58	19.19	nr	**64.65**
Reducer						
Sum of two sides 600 mm	56.19	71.11	1.69	55.90	nr	**127.01**
Sum of two sides 700 mm	60.24	76.24	1.69	55.90	nr	**132.14**
Sum of two sides 800 mm	64.23	81.29	1.92	63.51	nr	**144.80**
Sum of two sides 900 mm	68.28	86.42	1.92	63.51	nr	**149.93**
Sum of two sides 1000 mm	72.35	91.56	2.18	72.11	nr	**163.67**
Sum of two sides 1100 mm	76.73	97.11	2.18	72.11	nr	**169.23**
Sum of two sides 1200 mm	80.79	102.24	2.18	72.11	nr	**174.36**
Offset						
Sum of two sides 600 mm	78.13	98.88	1.92	63.51	nr	**162.39**
Sum of two sides 700 mm	83.88	106.16	1.92	63.51	nr	**169.68**
Sum of two sides 800 mm	88.44	111.93	1.92	63.51	nr	**175.44**
Sum of two sides 900 mm	93.06	117.78	1.92	63.51	nr	**181.29**
Sum of two sides 1000 mm	98.21	124.29	2.18	72.11	nr	**196.40**
Sum of two sides 1100 mm	102.99	130.35	2.18	72.11	nr	**202.46**
Sum of two sides 1200 mm	107.49	136.04	2.18	72.11	nr	**208.15**
Square to round						
Sum of two sides 600 mm	59.45	75.24	1.33	44.00	nr	**119.24**
Sum of two sides 700 mm	64.02	81.02	1.33	44.00	nr	**125.02**
Sum of two sides 800 mm	68.51	86.71	1.40	46.31	nr	**133.02**
Sum of two sides 900 mm	73.10	92.52	1.40	46.31	nr	**138.83**
Sum of two sides 1000 mm	77.66	98.28	1.82	60.20	nr	**158.49**
Sum of two sides 1100 mm	82.31	104.17	1.82	60.20	nr	**164.38**
Sum of two sides 1200 mm	86.89	109.96	1.82	60.20	nr	**170.17**
90° radius bend						
Sum of two sides 600 mm	54.74	69.27	1.16	38.37	nr	**107.65**
Sum of two sides 700 mm	57.76	73.10	1.16	38.37	nr	**111.47**
Sum of two sides 800 mm	62.25	78.79	1.22	40.36	nr	**119.15**
Sum of two sides 900 mm	66.84	84.60	1.22	40.36	nr	**124.95**
Sum of two sides 1000 mm	70.68	89.45	1.40	46.31	nr	**135.76**
Sum of two sides 1100 mm	75.52	95.57	1.40	46.31	nr	**141.88**
Sum of two sides 1200 mm	80.14	101.42	1.40	46.31	nr	**147.73**
45° bend						
Sum of two sides 600 mm	57.96	73.35	1.16	38.37	nr	**111.73**
Sum of two sides 700 mm	61.54	77.89	1.39	45.98	nr	**123.87**
Sum of two sides 800 mm	65.88	83.37	1.46	48.30	nr	**131.67**

38 VENTILATION/AIR CONDITIONING SYSTEMS

Item	Net Price £	Material £	Labour hours	Labour £	Unit	Total rate £
DUCTWORK: RECTANGULAR – CLASS B – cont						
Extra over fittings – cont						
45° radius bend – cont						
Sum of two sides 900 mm	70.30	88.97	1.46	48.30	nr	**137.26**
Sum of two sides 1000 mm	74.31	94.04	1.88	62.19	nr	**156.23**
Sum of two sides 1100 mm	79.05	100.05	1.88	62.19	nr	**162.24**
Sum of two sides 1200 mm	83.49	105.66	1.88	62.19	nr	**167.85**
90° mitre bend						
Sum of two sides 600 mm	92.05	116.50	1.39	45.98	nr	**162.48**
Sum of two sides 700 mm	97.53	123.44	2.16	71.45	nr	**194.89**
Sum of two sides 800 mm	105.42	133.42	2.26	74.76	nr	**208.18**
Sum of two sides 900 mm	113.43	143.56	2.26	74.76	nr	**218.32**
Sum of two sides 1000 mm	120.63	152.68	3.01	99.57	nr	**252.24**
Sum of two sides 1100 mm	129.02	163.29	3.01	99.57	nr	**262.86**
Sum of two sides 1200 mm	137.24	173.68	3.01	99.57	nr	**273.25**
Branch						
Sum of two sides 600 mm	72.56	91.83	1.03	34.07	nr	**125.90**
Sum of two sides 700 mm	77.97	98.68	1.03	34.07	nr	**132.75**
Sum of two sides 800 mm	83.40	105.55	1.03	34.07	nr	**139.62**
Sum of two sides 900 mm	88.84	112.44	1.03	34.07	nr	**146.51**
Sum of two sides 1000 mm	94.28	119.32	1.29	42.67	nr	**161.99**
Sum of two sides 1100 mm	99.94	126.48	1.29	42.67	nr	**169.16**
Sum of two sides 1200 mm	105.37	133.35	1.29	42.67	nr	**176.03**
Grille neck						
Sum of two sides 600 mm	75.33	95.33	1.18	39.03	nr	**134.37**
Sum of two sides 700 mm	81.55	103.21	1.18	39.03	nr	**142.25**
Sum of two sides 800 mm	87.79	111.11	1.18	39.03	nr	**150.14**
Sum of two sides 900 mm	94.02	118.99	1.18	39.03	nr	**158.02**
Sum of two sides 1000 mm	100.25	126.88	1.44	47.63	nr	**174.52**
Sum of two sides 1100 mm	106.48	134.76	1.44	47.63	nr	**182.40**
Sum of two sides 1200 mm	112.72	142.66	1.44	47.63	nr	**190.29**
Ductwork 601 to 800 mm longest side						
Sum of two sides 900 mm	46.75	59.16	1.27	42.01	m	**101.17**
Sum of two sides 1000 mm	49.10	62.14	1.37	45.32	m	**107.46**
Sum of two sides 1100 mm	51.44	65.10	1.37	45.32	m	**110.42**
Sum of two sides 1200 mm	54.16	68.54	1.37	45.32	m	**113.86**
Sum of two sides 1300 mm	56.51	71.52	1.40	46.31	m	**117.83**
Sum of two sides 1400 mm	58.86	74.50	1.40	46.31	m	**120.81**
Sum of two sides 1500 mm	61.22	77.47	1.48	48.96	m	**126.43**
Sum of two sides 1600 mm	63.57	80.45	1.55	51.27	m	**131.72**
Extra over fittings: Ductwork 601 to 800 mm longest side						
End cap						
Sum of two sides 900 mm	30.12	38.13	0.58	19.19	nr	**57.31**
Sum of two sides 1000 mm	32.06	40.57	0.58	19.19	nr	**59.76**
Sum of two sides 1100 mm	33.99	43.02	0.58	19.19	nr	**62.20**

38 VENTILATION/AIR CONDITIONING SYSTEMS

Item	Net Price £	Material £	Labour hours	Labour £	Unit	Total rate £
Sum of two sides 1200 mm	35.92	45.46	0.58	19.19	nr	64.65
Sum of two sides 1300 mm	37.86	47.92	0.58	19.19	nr	67.11
Sum of two sides 1400 mm	39.52	50.02	0.58	19.19	nr	69.21
Sum of two sides 1500 mm	45.91	58.10	0.58	19.19	nr	77.28
Sum of two sides 1600 mm	52.31	66.20	0.58	19.19	nr	85.39
Reducer						
Sum of two sides 900 mm	68.93	87.24	1.92	63.51	nr	150.75
Sum of two sides 1000 mm	72.99	92.37	2.18	72.11	nr	164.48
Sum of two sides 1100 mm	77.06	97.53	2.18	72.11	nr	169.64
Sum of two sides 1200 mm	81.43	103.05	2.18	72.11	nr	175.17
Sum of two sides 1300 mm	85.48	108.18	2.30	76.08	nr	184.27
Sum of two sides 1400 mm	88.99	112.62	2.30	76.08	nr	188.71
Sum of two sides 1500 mm	101.99	129.07	2.47	81.71	nr	210.78
Sum of two sides 1600 mm	114.96	145.50	2.47	81.71	nr	227.21
Offset						
Sum of two sides 900 mm	95.68	121.09	1.92	63.51	nr	184.60
Sum of two sides 1000 mm	99.96	126.51	2.18	72.11	nr	198.62
Sum of two sides 1100 mm	104.23	131.92	2.18	72.11	nr	204.03
Sum of two sides 1200 mm	108.68	137.54	2.18	72.11	nr	209.65
Sum of two sides 1300 mm	112.85	142.83	2.30	76.08	nr	218.91
Sum of two sides 1400 mm	116.96	148.02	2.30	76.08	nr	224.11
Sum of two sides 1500 mm	134.48	170.20	2.47	81.71	nr	251.91
Sum of two sides 1600 mm	151.96	192.32	2.47	81.71	nr	274.02
Square to round						
Sum of two sides 900 mm	72.49	91.75	1.40	46.31	nr	138.06
Sum of two sides 1000 mm	77.07	97.54	1.82	60.20	nr	157.74
Sum of two sides 1100 mm	81.64	103.32	1.82	60.20	nr	163.53
Sum of two sides 1200 mm	86.29	109.21	1.82	60.20	nr	169.41
Sum of two sides 1300 mm	90.87	115.00	2.15	71.12	nr	186.12
Sum of two sides 1400 mm	94.75	119.92	2.15	71.12	nr	191.04
Sum of two sides 1500 mm	110.47	139.81	2.38	78.73	nr	218.54
Sum of two sides 1600 mm	126.20	159.72	2.38	78.73	nr	238.45
90° radius bend						
Sum of two sides 900 mm	62.60	79.23	1.22	40.36	nr	119.59
Sum of two sides 1000 mm	67.28	85.15	1.40	46.31	nr	131.47
Sum of two sides 1100 mm	71.95	91.05	1.40	46.31	nr	137.37
Sum of two sides 1200 mm	76.81	97.21	1.40	46.31	nr	143.52
Sum of two sides 1300 mm	81.49	103.13	1.91	63.18	nr	166.32
Sum of two sides 1400 mm	84.65	107.13	1.91	63.18	nr	170.32
Sum of two sides 1500 mm	98.28	124.38	2.11	69.80	nr	194.18
Sum of two sides 1600 mm	111.90	141.62	2.11	69.80	nr	211.42
45° bend						
Sum of two sides 900 mm	69.10	87.45	1.22	40.36	nr	127.81
Sum of two sides 1000 mm	73.55	93.09	1.40	46.31	nr	139.40
Sum of two sides 1100 mm	77.99	98.71	1.88	62.19	nr	160.90
Sum of two sides 1200 mm	82.78	104.77	1.88	62.19	nr	166.96
Sum of two sides 1300 mm	21.03	26.62	2.26	74.76	nr	101.38
Sum of two sides 1400 mm	90.63	114.70	2.26	74.76	nr	189.46
Sum of two sides 1500 mm	104.02	131.65	2.49	82.37	nr	214.02
Sum of two sides 1600 mm	110.79	140.21	2.49	82.37	nr	222.58

38 VENTILATION/AIR CONDITIONING SYSTEMS

Item	Net Price £	Material £	Labour hours	Labour £	Unit	Total rate £
DUCTWORK: RECTANGULAR – CLASS B – cont						
Extra over fittings: Ductwork 601 to 800 mm longest side – cont						
90° mitre bend						
Sum of two sides 900 mm	106.82	135.19	1.22	40.36	nr	**175.54**
Sum of two sides 1000 mm	115.59	146.30	1.40	46.31	nr	**192.61**
Sum of two sides 1100 mm	124.36	157.39	3.01	99.57	nr	**256.96**
Sum of two sides 1200 mm	133.30	168.70	3.01	99.57	nr	**268.27**
Sum of two sides 1300 mm	142.08	179.81	3.67	121.40	nr	**301.21**
Sum of two sides 1400 mm	148.95	188.51	3.67	121.40	nr	**309.92**
Sum of two sides 1500 mm	171.30	216.79	4.07	134.63	nr	**351.43**
Sum of two sides 1600 mm	193.65	245.09	4.07	134.63	nr	**379.72**
Branch						
Sum of two sides 900 mm	91.47	115.76	1.22	40.36	nr	**156.12**
Sum of two sides 1000 mm	97.04	122.82	1.40	46.31	nr	**169.13**
Sum of two sides 1100 mm	102.62	129.87	1.29	42.67	nr	**172.54**
Sum of two sides 1200 mm	108.44	137.25	1.29	42.67	nr	**179.92**
Sum of two sides 1300 mm	114.03	144.32	1.39	45.98	nr	**190.30**
Sum of two sides 1400 mm	118.88	150.46	1.39	45.98	nr	**196.44**
Sum of two sides 1500 mm	136.34	172.56	1.64	54.25	nr	**226.81**
Sum of two sides 1600 mm	153.83	194.68	1.64	54.25	nr	**248.93**
Grille neck						
Sum of two sides 900 mm	94.02	118.99	1.22	40.36	nr	**159.35**
Sum of two sides 1000 mm	100.25	126.88	1.40	46.31	nr	**173.19**
Sum of two sides 1100 mm	106.48	134.76	1.44	47.63	nr	**182.40**
Sum of two sides 1200 mm	112.72	142.66	1.44	47.63	nr	**190.29**
Sum of two sides 1300 mm	118.95	150.55	1.69	55.90	nr	**206.45**
Sum of two sides 1400 mm	125.18	158.43	1.69	55.90	nr	**214.33**
Sum of two sides 1500 mm	131.42	166.32	1.79	59.21	nr	**225.54**
Sum of two sides 1600 mm	137.66	174.22	1.79	59.21	nr	**233.43**
Ductwork 801 to 1000 mm longest side						
Sum of two sides 1100 mm	75.71	95.81	1.37	45.32	m	**141.13**
Sum of two sides 1200 mm	80.19	101.49	1.37	45.32	m	**146.81**
Sum of two sides 1300 mm	84.66	107.15	1.40	46.31	m	**153.46**
Sum of two sides 1400 mm	89.51	113.29	1.40	46.31	m	**159.60**
Sum of two sides 1500 mm	94.00	118.96	1.48	48.96	m	**167.92**
Sum of two sides 1600 mm	98.48	124.64	1.55	51.27	m	**175.91**
Sum of two sides 1700 mm	102.96	130.31	1.55	51.27	m	**181.58**
Sum of two sides 1800 mm	107.81	136.45	1.61	53.26	m	**189.71**
Sum of two sides 1900 mm	112.30	142.12	1.61	53.26	m	**195.38**
Sum of two sides 2000 mm	116.77	147.78	1.61	53.26	m	**201.04**
Extra over fittings; Ductwork 801 to 1000 mm longest side						
End cap						
Sum of two sides 1100 mm	33.99	43.02	1.44	47.63	nr	**90.65**
Sum of two sides 1200 mm	35.92	45.46	1.44	47.63	nr	**93.10**
Sum of two sides 1300 mm	37.86	47.92	1.44	47.63	nr	**95.55**

38 VENTILATION/AIR CONDITIONING SYSTEMS

Item	Net Price £	Material £	Labour hours	Labour £	Unit	Total rate £
Sum of two sides 1400 mm	39.52	50.02	1.44	47.63	nr	**97.65**
Sum of two sides 1500 mm	45.91	58.10	1.44	47.63	nr	**105.73**
Sum of two sides 1600 mm	52.31	66.20	1.44	47.63	nr	**113.84**
Sum of two sides 1700 mm	58.71	74.30	1.44	47.63	nr	**121.93**
Sum of two sides 1800 mm	65.11	82.40	1.44	47.63	nr	**130.04**
Sum of two sides 1900 mm	71.51	90.50	1.44	47.63	nr	**138.13**
Sum of two sides 2000 mm	77.90	98.59	1.44	47.63	nr	**146.22**
Reducer						
Sum of two sides 1100 mm	64.09	81.11	1.44	47.63	nr	**128.75**
Sum of two sides 1200 mm	67.17	85.01	1.44	47.63	nr	**132.64**
Sum of two sides 1300 mm	70.26	88.92	1.69	55.90	nr	**144.82**
Sum of two sides 1400 mm	73.07	92.48	1.69	55.90	nr	**148.38**
Sum of two sides 1500 mm	85.07	107.67	2.47	81.71	nr	**189.37**
Sum of two sides 1600 mm	97.08	122.87	2.47	81.71	nr	**204.57**
Sum of two sides 1700 mm	109.10	138.07	2.47	81.71	nr	**219.78**
Sum of two sides 1800 mm	121.38	153.62	2.59	85.68	nr	**239.29**
Sum of two sides 1900 mm	133.38	168.81	2.71	89.65	nr	**258.45**
Sum of two sides 2000 mm	145.40	184.02	2.71	89.65	nr	**273.67**
Offset						
Sum of two sides 1100 mm	98.42	124.56	1.44	47.63	nr	**172.19**
Sum of two sides 1200 mm	100.35	127.00	1.44	47.63	nr	**174.64**
Sum of two sides 1300 mm	102.00	129.09	1.69	55.90	nr	**184.99**
Sum of two sides 1400 mm	102.75	130.04	1.69	55.90	nr	**185.95**
Sum of two sides 1500 mm	117.22	148.36	2.47	81.71	nr	**230.06**
Sum of two sides 1600 mm	131.38	166.27	2.47	81.71	nr	**247.98**
Sum of two sides 1700 mm	145.29	183.88	2.59	85.68	nr	**269.55**
Sum of two sides 1800 mm	159.02	201.26	2.61	86.34	nr	**287.60**
Sum of two sides 1900 mm	174.36	220.67	2.71	89.65	nr	**310.32**
Sum of two sides 2000 mm	187.47	237.26	2.71	89.65	nr	**326.90**
Square to round						
Sum of two sides 1100 mm	68.44	86.62	1.44	47.63	nr	**134.25**
Sum of two sides 1200 mm	72.03	91.16	1.44	47.63	nr	**138.80**
Sum of two sides 1300 mm	75.63	95.72	1.69	55.90	nr	**151.62**
Sum of two sides 1400 mm	78.57	99.44	1.69	55.90	nr	**155.34**
Sum of two sides 1500 mm	93.33	118.12	2.38	78.73	nr	**196.85**
Sum of two sides 1600 mm	108.09	136.79	2.38	78.73	nr	**215.52**
Sum of two sides 1700 mm	122.85	155.48	2.55	84.35	nr	**239.83**
Sum of two sides 1800 mm	137.62	174.18	2.55	84.35	nr	**258.53**
Sum of two sides 1900 mm	152.38	192.85	2.83	93.62	nr	**286.46**
Sum of two sides 2000 mm	167.13	211.52	2.83	93.62	nr	**305.13**
90° radius bend						
Sum of two sides 1100 mm	51.91	65.70	1.44	47.63	nr	**113.33**
Sum of two sides 1200 mm	56.03	70.91	1.44	47.63	nr	**118.54**
Sum of two sides 1300 mm	60.14	76.12	1.69	55.90	nr	**132.02**
Sum of two sides 1400 mm	63.85	80.81	1.69	55.90	nr	**136.71**
Sum of two sides 1500 mm	76.88	97.30	2.11	69.80	nr	**167.10**
Sum of two sides 1600 mm	89.92	113.81	2.11	69.80	nr	**183.60**
Sum of two sides 1700 mm	102.96	130.31	2.26	74.76	nr	**205.07**
Sum of two sides 1800 mm	116.14	146.99	2.26	74.76	nr	**221.75**
Sum of two sides 1900 mm	126.85	160.54	2.48	82.04	nr	**242.58**
Sum of two sides 2000 mm	139.89	177.05	2.48	82.04	nr	**259.08**

38 VENTILATION/AIR CONDITIONING SYSTEMS

Item	Net Price £	Material £	Labour hours	Labour £	Unit	Total rate £
DUCTWORK: RECTANGULAR – CLASS B – cont						
Extra over fittings – cont						
45° bend						
Sum of two sides 1100 mm	67.33	85.21	1.44	47.63	nr	**132.84**
Sum of two sides 1200 mm	71.47	90.46	1.44	47.63	nr	**138.09**
Sum of two sides 1300 mm	75.61	95.69	1.69	55.90	nr	**151.60**
Sum of two sides 1400 mm	79.51	100.62	1.69	55.90	nr	**156.53**
Sum of two sides 1500 mm	92.57	117.15	2.49	82.37	nr	**199.52**
Sum of two sides 1600 mm	105.65	133.71	2.49	82.37	nr	**216.08**
Sum of two sides 1700 mm	118.71	150.24	2.67	88.32	nr	**238.57**
Sum of two sides 1800 mm	132.07	167.15	2.67	88.32	nr	**255.47**
Sum of two sides 1900 mm	143.92	182.15	3.06	101.22	nr	**283.37**
Sum of two sides 2000 mm	157.01	198.71	3.06	101.22	nr	**299.93**
90° mitre bend						
Sum of two sides 1100 mm	107.43	135.96	1.44	47.63	nr	**183.59**
Sum of two sides 1200 mm	115.23	145.83	1.44	47.63	nr	**193.47**
Sum of two sides 1300 mm	123.02	155.69	1.69	55.90	nr	**211.60**
Sum of two sides 1400 mm	130.08	164.63	1.69	55.90	nr	**220.54**
Sum of two sides 1500 mm	151.26	191.44	2.07	68.47	nr	**259.91**
Sum of two sides 1600 mm	172.44	218.24	2.07	68.47	nr	**286.72**
Sum of two sides 1700 mm	193.63	245.06	2.80	92.62	nr	**337.68**
Sum of two sides 1800 mm	214.90	271.98	2.67	88.32	nr	**360.30**
Sum of two sides 1900 mm	233.61	295.66	2.95	97.58	nr	**393.25**
Sum of two sides 2000 mm	254.91	322.61	2.95	97.58	nr	**420.20**
Branch						
Sum of two sides 1100 mm	102.85	130.16	1.44	47.63	nr	**177.80**
Sum of two sides 1200 mm	108.44	137.25	1.44	47.63	nr	**184.88**
Sum of two sides 1300 mm	114.03	144.32	1.64	54.25	nr	**198.57**
Sum of two sides 1400 mm	119.11	150.75	1.64	54.25	nr	**205.00**
Sum of two sides 1500 mm	136.59	172.87	1.64	54.25	nr	**227.12**
Sum of two sides 1600 mm	153.41	194.15	1.64	54.25	nr	**248.40**
Sum of two sides 1700 mm	171.54	217.10	1.69	55.90	nr	**273.00**
Sum of two sides 1800 mm	189.25	239.52	1.69	55.90	nr	**295.42**
Sum of two sides 1900 mm	206.73	261.64	1.85	61.20	nr	**322.84**
Sum of two sides 2000 mm	224.21	283.76	1.85	61.20	nr	**344.95**
Grille neck						
Sum of two sides 1100 mm	106.48	134.76	1.44	47.63	nr	**182.40**
Sum of two sides 1200 mm	112.72	142.66	1.44	47.63	nr	**190.29**
Sum of two sides 1300 mm	118.95	150.55	1.69	55.90	nr	**206.45**
Sum of two sides 1400 mm	124.35	157.38	1.69	55.90	nr	**213.28**
Sum of two sides 1500 mm	143.98	182.22	1.79	59.21	nr	**241.43**
Sum of two sides 1600 mm	163.60	207.05	1.79	59.21	nr	**266.27**
Sum of two sides 1700 mm	183.21	231.88	1.86	61.53	nr	**293.40**
Sum of two sides 1800 mm	202.84	256.71	2.02	66.82	nr	**323.53**
Sum of two sides 1900 mm	222.45	281.54	2.02	66.82	nr	**348.36**
Sum of two sides 2000 mm	242.07	306.36	2.02	66.82	nr	**373.18**

38 VENTILATION/AIR CONDITIONING SYSTEMS

Item	Net Price £	Material £	Labour hours	Labour £	Unit	Total rate £
Ductwork 1001 to 1250 mm longest side						
Sum of two sides 1300 mm	100.66	127.40	1.40	46.31	m	**173.71**
Sum of two sides 1400 mm	106.09	134.27	1.40	46.31	m	**180.58**
Sum of two sides 1500 mm	111.21	140.74	1.48	48.96	m	**189.70**
Sum of two sides 1600 mm	11668.65	14767.84	1.55	51.27	m	**14819.12**
Sum of two sides 1700 mm	121.81	154.16	1.55	51.27	m	**205.44**
Sum of two sides 1800 mm	126.92	160.64	1.61	53.26	m	**213.89**
Sum of two sides 1900 mm	132.04	167.11	1.61	53.26	m	**220.36**
Sum of two sides 2000 mm	137.52	174.04	1.61	53.26	m	**227.30**
Sum of two sides 2100 mm	142.64	180.53	2.17	71.78	m	**252.31**
Sum of two sides 2200 mm	147.76	187.00	2.19	72.44	m	**259.44**
Sum of two sides 2300 mm	153.55	194.34	2.19	72.44	m	**266.78**
Sum of two sides 2400 mm	158.67	200.81	2.38	78.73	m	**279.54**
Sum of two sides 2500 mm	164.15	207.74	2.38	78.73	m	**286.47**
Extra over fittings; Ductwork 1001 to 1250 mm longest side						
End cap						
Sum of two sides 1300 mm	38.63	48.89	1.69	55.90	nr	**104.79**
Sum of two sides 1400 mm	40.35	51.07	1.69	55.90	nr	**106.97**
Sum of two sides 1500 mm	46.81	59.24	1.69	55.90	nr	**115.15**
Sum of two sides 1600 mm	53.28	67.43	1.69	55.90	nr	**123.33**
Sum of two sides 1700 mm	59.72	75.59	1.69	55.90	nr	**131.49**
Sum of two sides 1800 mm	66.18	83.76	1.69	55.90	nr	**139.66**
Sum of two sides 1900 mm	72.65	91.95	1.69	55.90	nr	**147.85**
Sum of two sides 2000 mm	79.09	100.09	1.69	55.90	nr	**156.00**
Sum of two sides 2100 mm	85.54	108.26	1.69	55.90	nr	**164.17**
Sum of two sides 2200 mm	92.01	116.45	1.69	55.90	nr	**172.35**
Sum of two sides 2300 mm	98.47	124.62	1.69	55.90	nr	**180.53**
Sum of two sides 2400 mm	104.92	132.78	1.69	55.90	nr	**188.69**
Sum of two sides 2500 mm	111.38	140.97	1.69	55.90	nr	**196.87**
Reducer						
Sum of two sides 1300 mm	68.10	86.19	1.69	55.90	nr	**142.10**
Sum of two sides 1400 mm	70.63	89.39	1.69	55.90	nr	**145.30**
Sum of two sides 1500 mm	82.61	104.56	2.47	81.71	nr	**186.26**
Sum of two sides 1600 mm	94.87	120.06	2.47	81.71	nr	**201.77**
Sum of two sides 1700 mm	106.85	135.23	2.47	81.71	nr	**216.93**
Sum of two sides 1800 mm	118.83	150.39	2.59	85.68	nr	**236.07**
Sum of two sides 1900 mm	130.81	165.55	2.71	89.65	nr	**255.20**
Sum of two sides 2000 mm	143.07	181.07	2.59	85.68	nr	**266.75**
Sum of two sides 2100 mm	155.05	196.24	2.92	96.59	nr	**292.83**
Sum of two sides 2200 mm	167.04	211.41	2.92	96.59	nr	**308.00**
Sum of two sides 2300 mm	179.09	226.65	2.92	96.59	nr	**323.25**
Sum of two sides 2400 mm	191.29	242.10	3.12	103.21	nr	**345.30**
Sum of two sides 2500 mm	203.54	257.60	3.12	103.21	nr	**360.81**
Offset						
Sum of two sides 1300 mm	133.22	168.61	1.69	55.90	nr	**224.51**
Sum of two sides 1400 mm	137.79	174.39	1.69	55.90	nr	**230.29**
Sum of two sides 1500 mm	154.29	195.27	2.47	81.71	nr	**276.97**

38 VENTILATION/AIR CONDITIONING SYSTEMS

Item	Net Price £	Material £	Labour hours	Labour £	Unit	Total rate £
DUCTWORK: RECTANGULAR – CLASS B – cont						
Extra over fittings – cont						
Offset – cont						
Sum of two sides 1600 mm	170.69	216.02	2.47	81.71	nr	**297.73**
Sum of two sides 1700 mm	186.59	236.14	2.59	85.68	nr	**321.82**
Sum of two sides 1800 mm	202.21	255.92	2.61	86.34	nr	**342.25**
Sum of two sides 1900 mm	217.51	275.28	2.71	89.65	nr	**364.92**
Sum of two sides 2000 mm	232.68	294.48	2.71	89.65	nr	**384.13**
Sum of two sides 2100 mm	247.41	313.12	2.92	96.59	nr	**409.72**
Sum of two sides 2200 mm	264.31	334.51	3.26	107.84	nr	**442.35**
Sum of two sides 2300 mm	283.68	359.02	3.26	107.84	nr	**466.86**
Sum of two sides 2400 mm	302.98	383.45	3.48	115.12	nr	**498.57**
Sum of two sides 2500 mm	322.35	407.97	3.48	115.12	nr	**523.08**
Square to round						
Sum of two sides 1300 mm	72.86	92.21	1.69	55.90	nr	**148.12**
Sum of two sides 1400 mm	75.75	95.87	1.69	55.90	nr	**151.77**
Sum of two sides 1500 mm	90.48	114.51	2.38	78.73	nr	**193.24**
Sum of two sides 1600 mm	105.25	133.21	2.38	78.73	nr	**211.94**
Sum of two sides 1700 mm	119.98	151.85	2.55	84.35	nr	**236.20**
Sum of two sides 1800 mm	134.69	170.47	2.55	84.35	nr	**254.82**
Sum of two sides 1900 mm	149.42	189.10	2.83	93.62	nr	**282.71**
Sum of two sides 2000 mm	164.19	207.80	2.83	93.62	nr	**301.41**
Sum of two sides 2100 mm	178.91	226.43	3.85	127.36	nr	**353.78**
Sum of two sides 2200 mm	193.64	245.07	4.18	138.27	nr	**383.34**
Sum of two sides 2300 mm	208.40	263.76	4.22	139.60	nr	**403.35**
Sum of two sides 2400 mm	223.14	282.40	4.68	154.81	nr	**437.21**
Sum of two sides 2500 mm	237.89	301.07	4.70	155.47	nr	**456.55**
90° radius bend						
Sum of two sides 1300 mm	49.16	62.22	1.69	55.90	nr	**118.12**
Sum of two sides 1400 mm	50.32	63.68	1.69	55.90	nr	**119.58**
Sum of two sides 1500 mm	63.84	80.80	2.11	69.80	nr	**150.59**
Sum of two sides 1600 mm	77.43	97.99	2.11	69.80	nr	**167.79**
Sum of two sides 1700 mm	90.95	115.11	2.19	72.44	nr	**187.55**
Sum of two sides 1800 mm	104.45	132.20	2.19	72.44	nr	**204.64**
Sum of two sides 1900 mm	117.98	149.31	2.48	82.04	nr	**231.35**
Sum of two sides 2000 mm	131.60	166.55	2.26	74.76	nr	**241.31**
Sum of two sides 2100 mm	145.09	183.62	2.48	82.04	nr	**265.66**
Sum of two sides 2200 mm	158.62	200.75	2.48	82.04	nr	**282.79**
Sum of two sides 2300 mm	166.26	210.41	2.48	82.04	nr	**292.45**
Sum of two sides 2400 mm	179.81	227.57	3.90	129.01	nr	**356.58**
Sum of two sides 2500 mm	193.39	244.75	3.90	129.01	nr	**373.76**
45° bend						
Sum of two sides 1300 mm	70.36	89.05	1.69	55.90	nr	**144.95**
Sum of two sides 1400 mm	72.72	92.04	1.69	55.90	nr	**147.94**
Sum of two sides 1500 mm	86.05	108.90	2.49	82.37	nr	**191.27**
Sum of two sides 1600 mm	99.66	126.12	2.49	82.37	nr	**208.49**
Sum of two sides 1700 mm	112.99	143.00	2.67	88.32	nr	**231.32**
Sum of two sides 1800 mm	126.32	159.86	2.67	88.32	nr	**248.19**
Sum of two sides 1900 mm	139.66	176.75	3.06	101.22	nr	**277.98**

38 VENTILATION/AIR CONDITIONING SYSTEMS

Item	Net Price £	Material £	Labour hours	Labour £	Unit	Total rate £
Sum of two sides 2000 mm	153.26	193.96	3.06	101.22	nr	**295.19**
Sum of two sides 2100 mm	166.60	210.85	4.05	133.97	nr	**344.83**
Sum of two sides 2200 mm	179.94	227.73	4.05	133.97	nr	**361.70**
Sum of two sides 2300 mm	190.43	241.01	4.39	145.22	nr	**386.23**
Sum of two sides 2400 mm	203.77	257.90	4.85	160.44	nr	**418.33**
Sum of two sides 2500 mm	217.38	275.12	4.85	160.44	nr	**435.55**
90° mitre bend						
Sum of two sides 1300 mm	145.64	184.32	1.69	55.90	nr	**240.22**
Sum of two sides 1400 mm	151.98	192.34	1.69	55.90	nr	**248.25**
Sum of two sides 1500 mm	176.53	223.41	2.80	92.62	nr	**316.03**
Sum of two sides 1600 mm	201.11	254.52	2.80	92.62	nr	**347.14**
Sum of two sides 1700 mm	225.65	285.58	2.95	97.58	nr	**383.16**
Sum of two sides 1800 mm	250.22	316.67	2.95	97.58	nr	**414.26**
Sum of two sides 1900 mm	274.76	347.74	4.05	133.97	nr	**481.71**
Sum of two sides 2000 mm	299.34	378.85	4.05	133.97	nr	**512.82**
Sum of two sides 2100 mm	323.89	409.92	4.07	134.63	nr	**544.55**
Sum of two sides 2200 mm	348.45	441.00	4.07	134.63	nr	**575.64**
Sum of two sides 2300 mm	366.53	463.89	4.39	145.22	nr	**609.10**
Sum of two sides 2400 mm	391.29	495.22	4.85	160.44	nr	**655.66**
Sum of two sides 2500 mm	416.05	526.56	4.85	160.44	nr	**686.99**
Branch						
Sum of two sides 1300 mm	116.96	148.02	1.44	47.63	nr	**195.66**
Sum of two sides 1400 mm	121.98	154.38	1.44	47.63	nr	**202.01**
Sum of two sides 1500 mm	139.64	176.73	1.64	54.25	nr	**230.98**
Sum of two sides 1600 mm	157.54	199.39	1.64	54.25	nr	**253.64**
Sum of two sides 1700 mm	175.19	221.72	1.64	54.25	nr	**275.97**
Sum of two sides 1800 mm	192.85	244.08	1.64	54.25	nr	**298.33**
Sum of two sides 1900 mm	210.50	266.41	1.69	55.90	nr	**322.32**
Sum of two sides 2000 mm	228.40	289.06	1.69	55.90	nr	**344.96**
Sum of two sides 2100 mm	246.06	311.41	1.85	61.20	nr	**372.61**
Sum of two sides 2200 mm	263.70	333.73	1.85	61.20	nr	**394.93**
Sum of two sides 2300 mm	281.60	356.39	2.61	86.34	nr	**442.73**
Sum of two sides 2400 mm	299.25	378.73	2.61	86.34	nr	**465.07**
Sum of two sides 2500 mm	317.15	401.39	2.61	86.34	nr	**487.73**
Grille neck						
Sum of two sides 1300 mm	122.82	155.44	1.79	59.21	nr	**214.65**
Sum of two sides 1400 mm	128.51	162.64	1.79	59.21	nr	**221.85**
Sum of two sides 1500 mm	148.41	187.82	1.79	59.21	nr	**247.04**
Sum of two sides 1600 mm	168.35	213.06	1.79	59.21	nr	**272.27**
Sum of two sides 1700 mm	188.25	238.25	1.86	61.53	nr	**299.78**
Sum of two sides 1800 mm	208.17	263.46	2.02	66.82	nr	**330.28**
Sum of two sides 1900 mm	228.11	288.70	2.02	66.82	nr	**355.52**
Sum of two sides 2000 mm	248.01	313.88	2.02	66.82	nr	**380.70**
Sum of two sides 2100 mm	267.93	339.09	2.61	86.34	nr	**425.43**
Sum of two sides 2200 mm	287.84	364.29	2.61	86.34	nr	**450.62**
Sum of two sides 2300 mm	307.77	389.51	2.61	86.34	nr	**475.85**
Sum of two sides 2400 mm	327.67	414.70	2.88	95.27	nr	**509.97**
Sum of two sides 2500 mm	347.59	439.91	2.88	95.27	nr	**535.18**

38 VENTILATION/AIR CONDITIONING SYSTEMS

Item	Net Price £	Material £	Labour hours	Labour £	Unit	Total rate £
DUCTWORK: RECTANGULAR – CLASS B – cont						
Ductwork 1251 to 1600 mm longest side						
Sum of two sides 1700 mm	149.10	188.70	1.55	51.27	m	**239.97**
Sum of two sides 1800 mm	154.80	195.92	1.61	53.26	m	**249.18**
Sum of two sides 1900 mm	160.70	203.39	1.61	53.26	m	**256.64**
Sum of two sides 2000 mm	166.24	210.39	1.61	53.26	m	**263.65**
Sum of two sides 2100 mm	171.77	217.39	2.17	71.78	m	**289.17**
Sum of two sides 2200 mm	177.30	224.39	2.19	72.44	m	**296.84**
Sum of two sides 2300 mm	183.20	231.86	2.19	72.44	m	**304.31**
Sum of two sides 2400 mm	188.91	239.08	2.38	78.73	m	**317.81**
Sum of two sides 2500 mm	194.44	246.08	2.38	78.73	m	**324.81**
Sum of two sides 2600 mm	199.97	253.09	2.64	87.33	m	**340.42**
Sum of two sides 2700 mm	205.51	260.09	2.66	87.99	m	**348.08**
Sum of two sides 2800 mm	211.21	267.30	2.95	97.58	m	**364.89**
Sum of two sides 2900 mm	235.86	298.51	2.96	97.92	m	**396.42**
Sum of two sides 3000 mm	241.40	305.51	3.15	104.20	m	**409.71**
Sum of two sides 3100 mm	247.15	312.79	3.15	104.20	m	**416.99**
Sum of two sides 3200 mm	252.68	319.79	3.18	105.19	m	**424.99**
Extra over fittings; Ductwork 1251 to 1600 mm longest side						
End cap						
Sum of two sides 1700 mm	59.72	75.59	0.58	19.19	nr	**94.77**
Sum of two sides 1800 mm	66.18	83.76	0.58	19.19	nr	**102.95**
Sum of two sides 1900 mm	72.65	91.95	0.58	19.19	nr	**111.13**
Sum of two sides 2000 mm	79.09	100.09	0.58	19.19	nr	**119.28**
Sum of two sides 2100 mm	85.54	108.26	0.87	28.78	nr	**137.04**
Sum of two sides 2200 mm	92.01	116.45	0.87	28.78	nr	**145.23**
Sum of two sides 2300 mm	98.47	124.62	0.87	28.78	nr	**153.40**
Sum of two sides 2400 mm	104.92	132.78	0.87	28.78	nr	**161.56**
Sum of two sides 2500 mm	111.38	140.97	0.87	28.78	nr	**169.75**
Sum of two sides 2600 mm	117.84	149.14	0.87	28.78	nr	**177.92**
Sum of two sides 2700 mm	124.29	157.30	0.87	28.78	nr	**186.08**
Sum of two sides 2800 mm	130.76	165.49	1.16	38.37	nr	**203.86**
Sum of two sides 2900 mm	137.21	173.66	1.16	38.37	nr	**212.03**
Sum of two sides 3000 mm	143.67	181.83	1.30	43.00	nr	**224.83**
Sum of two sides 3100 mm	148.81	188.33	1.80	59.54	nr	**247.87**
Sum of two sides 3200 mm	153.46	194.22	1.80	59.54	nr	**253.76**
Reducer						
Sum of two sides 1700 mm	72.93	92.30	2.47	81.71	nr	**174.01**
Sum of two sides 1800 mm	84.14	106.48	2.59	85.68	nr	**192.16**
Sum of two sides 1900 mm	95.47	120.82	2.71	89.65	nr	**210.47**
Sum of two sides 2000 mm	106.59	134.89	2.71	89.65	nr	**224.54**
Sum of two sides 2100 mm	117.72	148.98	2.92	96.59	nr	**245.57**
Sum of two sides 2200 mm	128.85	163.07	2.92	96.59	nr	**259.66**
Sum of two sides 2300 mm	140.16	177.39	2.92	96.59	nr	**273.98**
Sum of two sides 2400 mm	151.36	191.56	3.12	103.21	nr	**294.77**
Sum of two sides 2500 mm	162.49	205.64	3.12	103.21	nr	**308.85**
Sum of two sides 2600 mm	173.62	219.73	3.12	103.21	nr	**322.94**
Sum of two sides 2700 mm	184.73	233.79	3.12	103.21	nr	**337.00**

38 VENTILATION/AIR CONDITIONING SYSTEMS

Item	Net Price £	Material £	Labour hours	Labour £	Unit	Total rate £
Sum of two sides 2800 mm	195.92	247.96	3.95	130.66	nr	**378.62**
Sum of two sides 2900 mm	194.05	245.59	3.97	131.33	nr	**376.92**
Sum of two sides 3000 mm	205.18	259.68	4.52	149.52	nr	**409.20**
Sum of two sides 3100 mm	213.41	270.09	4.52	149.52	nr	**419.61**
Sum of two sides 3200 mm	220.95	279.64	4.52	149.52	nr	**429.16**
Offset						
Sum of two sides 1700 mm	184.54	233.55	2.59	85.68	nr	**319.23**
Sum of two sides 1800 mm	205.03	259.49	2.61	86.34	nr	**345.83**
Sum of two sides 1900 mm	218.83	276.95	2.71	89.65	nr	**366.60**
Sum of two sides 2000 mm	232.03	293.66	2.71	89.65	nr	**383.30**
Sum of two sides 2100 mm	244.83	309.85	2.92	96.59	nr	**406.45**
Sum of two sides 2200 mm	257.25	325.58	3.26	107.84	nr	**433.42**
Sum of two sides 2300 mm	269.17	340.66	3.26	107.84	nr	**448.50**
Sum of two sides 2400 mm	288.98	365.73	3.47	114.79	nr	**480.52**
Sum of two sides 2500 mm	300.44	380.23	3.48	115.12	nr	**495.35**
Sum of two sides 2600 mm	319.74	404.66	3.49	115.45	nr	**520.11**
Sum of two sides 2700 mm	339.02	429.07	3.50	115.78	nr	**544.85**
Sum of two sides 2800 mm	358.43	453.63	4.34	143.57	nr	**597.19**
Sum of two sides 2900 mm	358.12	453.24	4.76	157.46	nr	**610.70**
Sum of two sides 3000 mm	377.42	477.67	5.32	175.98	nr	**653.65**
Sum of two sides 3100 mm	392.39	496.60	5.35	176.98	nr	**673.58**
Sum of two sides 3200 mm	406.29	514.20	5.35	176.98	nr	**691.17**
Square to round						
Sum of two sides 1700 mm	86.07	108.93	2.55	84.35	nr	**193.28**
Sum of two sides 1800 mm	100.00	126.56	2.55	84.35	nr	**210.92**
Sum of two sides 1900 mm	113.83	144.06	2.83	93.62	nr	**237.68**
Sum of two sides 2000 mm	127.70	161.62	2.83	93.62	nr	**255.23**
Sum of two sides 2100 mm	141.58	179.19	3.85	127.36	nr	**306.54**
Sum of two sides 2200 mm	155.44	196.73	4.18	138.27	nr	**335.00**
Sum of two sides 2300 mm	169.27	214.23	4.22	139.60	nr	**353.82**
Sum of two sides 2400 mm	183.20	231.86	4.68	154.81	nr	**386.68**
Sum of two sides 2500 mm	197.06	249.40	4.70	155.47	nr	**404.88**
Sum of two sides 2600 mm	210.93	266.96	4.70	155.47	nr	**422.43**
Sum of two sides 2700 mm	224.82	284.53	4.71	155.80	nr	**440.33**
Sum of two sides 2800 mm	238.75	302.16	8.19	270.92	nr	**573.08**
Sum of two sides 2900 mm	239.38	302.96	8.62	285.15	nr	**588.10**
Sum of two sides 3000 mm	253.24	320.50	8.75	289.45	nr	**609.95**
Sum of two sides 3100 mm	263.56	333.56	8.75	289.45	nr	**623.01**
Sum of two sides 3200 mm	272.93	345.42	8.75	289.45	nr	**634.86**
90° radius bend						
Sum of two sides 1700 mm	183.55	232.30	2.19	72.44	nr	**304.75**
Sum of two sides 1800 mm	200.69	253.99	2.19	72.44	nr	**326.43**
Sum of two sides 1900 mm	226.96	287.24	2.26	74.76	nr	**362.00**
Sum of two sides 2000 mm	252.74	319.86	2.26	74.76	nr	**394.62**
Sum of two sides 2100 mm	278.52	352.50	2.48	82.04	nr	**434.54**
Sum of two sides 2200 mm	304.31	385.14	2.48	82.04	nr	**467.17**
Sum of two sides 2300 mm	330.57	418.37	2.48	82.04	nr	**500.41**
Sum of two sides 2400 mm	347.13	439.33	3.90	129.01	nr	**568.34**
Sum of two sides 2500 mm	372.89	471.93	3.90	129.01	nr	**600.94**
Sum of two sides 2600 mm	398.65	504.54	4.26	140.92	nr	**645.46**
Sum of two sides 2700 mm	424.40	537.12	4.55	150.51	nr	**687.63**

38 VENTILATION/AIR CONDITIONING SYSTEMS

Item	Net Price £	Material £	Labour hours	Labour £	Unit	Total rate £
DUCTWORK: RECTANGULAR – CLASS B – cont						
Extra over fittings – cont						
90° radius bend – cont						
Sum of two sides 2800 mm	440.52	557.52	4.55	150.51	nr	**708.03**
Sum of two sides 2900 mm	441.69	559.01	6.87	227.26	nr	**786.26**
Sum of two sides 3000 mm	467.41	591.55	7.00	231.56	nr	**823.11**
Sum of two sides 3100 mm	487.38	616.83	7.00	231.56	nr	**848.38**
Sum of two sides 3200 mm	505.90	640.27	7.00	231.56	nr	**871.83**
45° bend						
Sum of two sides 1700 mm	88.20	111.63	2.67	88.32	nr	**199.95**
Sum of two sides 1800 mm	96.65	122.32	2.67	88.32	nr	**210.65**
Sum of two sides 1900 mm	109.78	138.93	3.06	101.22	nr	**240.16**
Sum of two sides 2000 mm	122.68	155.27	3.06	101.22	nr	**256.49**
Sum of two sides 2100 mm	135.60	171.61	4.05	133.97	nr	**305.58**
Sum of two sides 2200 mm	148.50	187.94	4.05	133.97	nr	**321.92**
Sum of two sides 2300 mm	161.64	204.57	4.39	145.22	nr	**349.79**
Sum of two sides 2400 mm	169.80	214.89	4.85	160.44	nr	**375.33**
Sum of two sides 2500 mm	182.69	231.21	4.85	160.44	nr	**391.65**
Sum of two sides 2600 mm	195.58	247.53	4.87	161.10	nr	**408.63**
Sum of two sides 2700 mm	208.48	263.85	4.87	161.10	nr	**424.95**
Sum of two sides 2800 mm	216.41	273.88	8.81	291.43	nr	**565.31**
Sum of two sides 2900 mm	216.17	273.59	8.81	291.43	nr	**565.02**
Sum of two sides 3000 mm	229.05	289.88	9.31	307.97	nr	**597.85**
Sum of two sides 3100 mm	239.03	302.52	9.31	307.97	nr	**610.49**
Sum of two sides 3200 mm	248.31	314.27	9.39	310.62	nr	**624.88**
90° mitre bend						
Sum of two sides 1700 mm	197.90	250.47	2.67	88.32	nr	**338.79**
Sum of two sides 1800 mm	211.28	267.40	2.80	92.62	nr	**360.02**
Sum of two sides 1900 mm	236.40	299.18	2.95	97.58	nr	**396.77**
Sum of two sides 2000 mm	261.57	331.04	2.95	97.58	nr	**428.62**
Sum of two sides 2100 mm	286.74	362.90	4.05	133.97	nr	**496.88**
Sum of two sides 2200 mm	311.92	394.77	4.05	133.97	nr	**528.74**
Sum of two sides 2300 mm	226.72	286.93	4.39	145.22	nr	**432.15**
Sum of two sides 2400 mm	350.68	443.82	4.85	160.44	nr	**604.26**
Sum of two sides 2500 mm	376.02	475.89	4.85	160.44	nr	**636.32**
Sum of two sides 2600 mm	401.34	507.94	4.87	161.10	nr	**669.04**
Sum of two sides 2700 mm	426.69	540.02	4.87	161.10	nr	**701.11**
Sum of two sides 2800 mm	440.32	557.27	8.81	291.43	nr	**848.70**
Sum of two sides 2900 mm	444.29	562.29	14.81	489.91	nr	**1052.20**
Sum of two sides 3000 mm	469.79	594.57	15.20	502.81	nr	**1097.38**
Sum of two sides 3100 mm	490.77	621.12	15.60	516.04	nr	**1137.16**
Sum of two sides 3200 mm	510.87	646.55	15.60	516.04	nr	**1162.60**
Branch						
Sum of two sides 1700 mm	175.19	221.72	1.69	55.90	nr	**277.63**
Sum of two sides 1800 mm	192.85	244.08	1.69	55.90	nr	**299.98**
Sum of two sides 1900 mm	210.75	266.72	1.85	61.20	nr	**327.92**
Sum of two sides 2000 mm	228.40	289.06	1.85	61.20	nr	**350.26**
Sum of two sides 2100 mm	246.06	311.41	2.61	86.34	nr	**397.75**
Sum of two sides 2200 mm	263.70	333.73	2.61	86.34	nr	**420.07**
Sum of two sides 2300 mm	281.60	356.39	2.61	86.34	nr	**442.73**

38 VENTILATION/AIR CONDITIONING SYSTEMS

Item	Net Price £	Material £	Labour hours	Labour £	Unit	Total rate £
Sum of two sides 2400 mm	299.25	378.73	2.88	95.27	nr	**474.00**
Sum of two sides 2500 mm	316.91	401.08	2.88	95.27	nr	**496.35**
Sum of two sides 2600 mm	334.56	423.42	2.88	95.27	nr	**518.69**
Sum of two sides 2700 mm	352.21	445.76	2.88	95.27	nr	**541.03**
Sum of two sides 2800 mm	369.87	468.11	3.94	130.33	nr	**598.45**
Sum of two sides 2900 mm	387.77	490.76	3.94	130.33	nr	**621.09**
Sum of two sides 3000 mm	405.42	513.09	4.83	159.77	nr	**672.87**
Sum of two sides 3100 mm	419.53	530.95	4.83	159.77	nr	**690.73**
Sum of two sides 3200 mm	432.38	547.22	4.83	159.77	nr	**706.99**
Grille neck						
Sum of two sides 1700 mm	188.25	238.25	1.86	61.53	nr	**299.78**
Sum of two sides 1800 mm	208.17	263.46	2.02	66.82	nr	**330.28**
Sum of two sides 1900 mm	228.11	288.70	2.02	66.82	nr	**355.52**
Sum of two sides 2000 mm	248.01	313.88	2.02	66.82	nr	**380.70**
Sum of two sides 2100 mm	267.93	339.09	2.61	86.34	nr	**425.43**
Sum of two sides 2200 mm	287.84	364.29	2.61	86.34	nr	**450.62**
Sum of two sides 2300 mm	307.77	389.51	2.61	86.34	nr	**475.85**
Sum of two sides 2400 mm	327.67	414.70	2.88	95.27	nr	**509.97**
Sum of two sides 2500 mm	347.59	439.91	2.88	95.27	nr	**535.18**
Sum of two sides 2600 mm	367.51	465.12	2.88	95.27	nr	**560.39**
Sum of two sides 2700 mm	387.43	490.33	2.88	95.27	nr	**585.60**
Sum of two sides 2800 mm	407.35	515.54	3.94	130.33	nr	**645.87**
Sum of two sides 2900 mm	427.26	540.73	4.12	136.29	nr	**677.02**
Sum of two sides 3000 mm	447.17	565.94	5.00	165.40	nr	**731.34**
Sum of two sides 3100 mm	463.10	586.10	5.00	165.40	nr	**751.50**
Sum of two sides 3200 mm	477.61	604.47	5.00	165.40	nr	**769.87**
Ductwork 1601 to 2000 mm longest side						
Sum of two sides 2100 mm	192.51	243.64	2.17	71.78	m	**315.42**
Sum of two sides 2200 mm	198.50	251.22	2.17	71.78	m	**323.01**
Sum of two sides 2300 mm	204.70	259.07	2.19	72.44	m	**331.51**
Sum of two sides 2400 mm	210.50	266.41	2.38	78.73	m	**345.14**
Sum of two sides 2500 mm	216.70	274.25	2.38	78.73	m	**352.98**
Sum of two sides 2600 mm	222.53	281.63	2.64	87.33	m	**368.96**
Sum of two sides 2700 mm	228.35	289.00	2.66	87.99	m	**377.00**
Sum of two sides 2800 mm	234.18	296.38	2.95	97.58	m	**393.97**
Sum of two sides 2900 mm	240.38	304.22	2.96	97.92	m	**402.14**
Sum of two sides 3000 mm	246.20	311.60	2.96	97.92	m	**409.51**
Sum of two sides 3100 mm	265.73	336.31	2.96	97.92	m	**434.23**
Sum of two sides 3200 mm	271.56	343.69	3.15	104.20	m	**447.89**
Sum of two sides 3300 mm	297.05	375.94	3.15	104.20	m	**480.14**
Sum of two sides 3400 mm	302.87	383.32	3.15	104.20	m	**487.52**
Sum of two sides 3500 mm	308.70	390.69	3.15	104.20	nr	**494.89**
Sum of two sides 3600 mm	314.53	398.07	3.18	105.19	m	**503.26**
Sum of two sides 3700 mm	320.72	405.91	3.18	105.19	m	**511.10**
Sum of two sides 3800 mm	326.55	413.28	3.18	105.19	m	**518.47**
Sum of two sides 3900 mm	332.38	420.66	3.18	105.19	m	**525.85**
Sum of two sides 4000 mm	338.21	428.03	3.18	105.19	m	**533.23**

38 VENTILATION/AIR CONDITIONING SYSTEMS

Item	Net Price £	Material £	Labour hours	Labour £	Unit	Total rate £
DUCTWORK: RECTANGULAR – CLASS B – cont						
Extra over fittings; Ductwork 1601 to 2000 mm longest side						
End cap						
Sum of two sides 2100 mm	85.54	108.26	0.87	28.78	nr	**137.04**
Sum of two sides 2200 mm	92.01	116.45	0.87	28.78	nr	**145.23**
Sum of two sides 2300 mm	98.47	124.62	0.87	28.78	nr	**153.40**
Sum of two sides 2400 mm	104.92	132.78	0.87	28.78	nr	**161.56**
Sum of two sides 2500 mm	111.38	140.97	0.87	28.78	nr	**169.75**
Sum of two sides 2600 mm	117.84	149.14	0.87	28.78	nr	**177.92**
Sum of two sides 2700 mm	124.29	157.30	0.87	28.78	nr	**186.08**
Sum of two sides 2800 mm	130.76	165.49	1.16	38.37	nr	**203.86**
Sum of two sides 2900 mm	137.21	173.66	1.16	38.37	nr	**212.03**
Sum of two sides 3000 mm	143.67	181.83	1.73	57.23	nr	**239.06**
Sum of two sides 3100 mm	148.81	188.33	1.80	59.54	nr	**247.87**
Sum of two sides 3200 mm	153.46	194.22	1.80	59.54	nr	**253.76**
Sum of two sides 3300 mm	158.12	200.12	1.80	59.54	nr	**259.66**
Sum of two sides 3400 mm	162.76	205.99	1.80	59.54	nr	**265.53**
Sum of two sides 3500 mm	167.42	211.89	1.80	59.54	nr	**271.43**
Sum of two sides 3600 mm	172.07	217.78	1.80	59.54	nr	**277.32**
Sum of two sides 3700 mm	176.73	223.66	1.80	59.54	nr	**283.21**
Sum of two sides 3800 mm	181.39	229.56	1.80	59.54	nr	**289.11**
Sum of two sides 3900 mm	186.04	235.45	1.80	59.54	nr	**294.99**
Sum of two sides 4000 mm	190.71	241.36	1.80	59.54	nr	**300.91**
Reducer						
Sum of two sides 2100 mm	117.03	148.12	2.61	86.34	nr	**234.45**
Sum of two sides 2200 mm	128.04	162.04	2.61	86.34	nr	**248.38**
Sum of two sides 2300 mm	139.24	176.22	2.61	86.34	nr	**262.56**
Sum of two sides 2400 mm	153.36	194.10	2.88	95.27	nr	**289.37**
Sum of two sides 2500 mm	161.25	204.08	3.12	103.21	nr	**307.28**
Sum of two sides 2600 mm	172.25	218.00	3.12	103.21	nr	**321.21**
Sum of two sides 2700 mm	183.25	231.92	3.12	103.21	nr	**335.12**
Sum of two sides 2800 mm	194.24	245.83	3.95	130.66	nr	**376.49**
Sum of two sides 2900 mm	205.44	260.01	3.97	131.33	nr	**391.33**
Sum of two sides 3000 mm	216.45	273.94	4.52	149.52	nr	**423.46**
Sum of two sides 3100 mm	214.80	271.85	4.52	149.52	nr	**421.37**
Sum of two sides 3200 mm	222.19	281.20	4.52	149.52	nr	**430.72**
Sum of two sides 3300 mm	216.30	273.75	4.52	149.52	nr	**423.27**
Sum of two sides 3400 mm	223.71	283.13	4.52	149.52	nr	**432.65**
Sum of two sides 3500 mm	231.09	292.47	4.52	149.52	nr	**441.99**
Sum of two sides 3600 mm	238.50	301.84	4.52	149.52	nr	**451.36**
Sum of two sides 3700 mm	257.89	326.39	4.52	149.52	nr	**475.91**
Sum of two sides 3800 mm	265.29	335.75	4.52	149.52	nr	**485.27**
Sum of two sides 3900 mm	272.67	345.10	4.52	149.52	nr	**494.62**
Sum of two sides 4000 mm	280.09	354.48	4.52	149.52	nr	**504.00**
Offset						
Sum of two sides 2100 mm	274.40	347.28	2.61	86.34	nr	**433.61**
Sum of two sides 2200 mm	295.06	373.43	2.61	86.34	nr	**459.77**
Sum of two sides 2300 mm	306.90	388.42	2.61	86.34	nr	**474.76**

38 VENTILATION/AIR CONDITIONING SYSTEMS

Item	Net Price £	Material £	Labour hours	Labour £	Unit	Total rate £
Sum of two sides 2400 mm	336.43	425.79	2.88	95.27	nr	521.06
Sum of two sides 2500 mm	347.87	440.26	3.48	115.12	nr	555.38
Sum of two sides 2600 mm	358.65	453.91	3.49	115.45	nr	569.35
Sum of two sides 2700 mm	369.03	467.05	3.50	115.78	nr	582.83
Sum of two sides 2800 mm	378.97	479.62	4.34	143.57	nr	623.19
Sum of two sides 2900 mm	388.61	491.82	4.76	157.46	nr	649.28
Sum of two sides 3000 mm	397.68	503.30	5.32	175.98	nr	679.28
Sum of two sides 3100 mm	398.22	503.99	5.35	176.98	nr	680.97
Sum of two sides 3200 mm	412.22	521.70	5.35	176.98	nr	698.68
Sum of two sides 3300 mm	406.20	514.09	5.35	176.98	nr	691.07
Sum of two sides 3400 mm	420.22	531.83	5.35	176.98	nr	708.81
Sum of two sides 3500 mm	434.24	549.57	5.35	176.98	nr	726.55
Sum of two sides 3600 mm	448.26	567.31	5.35	176.98	nr	744.29
Sum of two sides 3700 mm	473.05	598.69	5.35	176.98	nr	775.66
Sum of two sides 3800 mm	488.17	617.82	5.35	176.98	nr	794.80
Sum of two sides 3900 mm	502.17	635.55	5.35	176.98	nr	812.53
Sum of two sides 4000 mm	516.20	653.30	5.35	176.98	nr	830.28
Square to round						
Sum of two sides 2100 mm	174.15	220.41	2.61	86.34	nr	306.75
Sum of two sides 2200 mm	190.98	241.71	2.61	86.34	nr	328.05
Sum of two sides 2300 mm	207.76	262.95	2.61	86.34	nr	349.28
Sum of two sides 2400 mm	224.67	284.34	2.88	95.27	nr	379.61
Sum of two sides 2500 mm	241.45	305.58	4.70	155.47	nr	461.05
Sum of two sides 2600 mm	258.24	326.82	4.70	155.47	nr	482.30
Sum of two sides 2700 mm	275.05	348.10	4.71	155.80	nr	503.91
Sum of two sides 2800 mm	291.85	369.36	8.19	270.92	nr	640.28
Sum of two sides 2900 mm	308.62	390.58	8.19	270.92	nr	661.51
Sum of two sides 3000 mm	325.42	411.85	8.19	270.92	nr	682.77
Sum of two sides 3100 mm	328.24	415.42	8.19	270.92	nr	686.34
Sum of two sides 3200 mm	339.63	429.84	8.19	270.92	nr	700.76
Sum of two sides 3300 mm	337.51	427.16	8.19	270.92	nr	698.08
Sum of two sides 3400 mm	348.92	441.59	8.62	285.15	nr	726.73
Sum of two sides 3500 mm	360.30	455.99	8.62	285.15	nr	741.14
Sum of two sides 3600 mm	371.70	470.42	8.62	285.15	nr	755.57
Sum of two sides 3700 mm	388.97	492.28	8.62	285.15	nr	777.43
Sum of two sides 3800 mm	400.38	506.72	8.75	289.45	nr	796.16
Sum of two sides 3900 mm	411.77	521.13	8.75	289.45	nr	810.58
Sum of two sides 4000 mm	423.17	535.57	8.75	289.45	nr	825.01
90° radius bend						
Sum of two sides 2100 mm	253.05	320.26	2.61	86.34	nr	406.60
Sum of two sides 2200 mm	485.00	613.81	2.61	86.34	nr	700.15
Sum of two sides 2300 mm	525.55	665.13	2.61	86.34	nr	751.47
Sum of two sides 2400 mm	537.97	680.85	2.88	95.27	nr	776.12
Sum of two sides 2500 mm	578.47	732.11	3.90	129.01	nr	861.12
Sum of two sides 2600 mm	618.14	782.31	4.26	140.92	nr	923.23
Sum of two sides 2700 mm	657.80	832.52	4.55	150.51	nr	983.03
Sum of two sides 2800 mm	697.47	882.72	4.55	150.51	nr	1033.23
Sum of two sides 2900 mm	737.94	933.94	6.87	227.26	nr	1161.19
Sum of two sides 3000 mm	777.61	984.14	6.87	227.26	nr	1211.40
Sum of two sides 3100 mm	786.48	995.37	6.87	227.26	nr	1222.63

38 VENTILATION/AIR CONDITIONING SYSTEMS

Item	Net Price £	Material £	Labour hours	Labour £	Unit	Total rate £
DUCTWORK: RECTANGULAR – CLASS B – cont						
Extra over fittings – cont						
90° radius bend – cont						
Sum of two sides 3200 mm	816.22	1033.00	6.87	227.26	nr	**1260.26**
Sum of two sides 3300 mm	814.18	1030.43	6.87	227.26	nr	**1257.68**
Sum of two sides 3400 mm	843.05	1066.96	7.00	231.56	nr	**1298.52**
Sum of two sides 3500 mm	871.93	1103.52	7.00	231.56	nr	**1335.07**
Sum of two sides 3600 mm	900.78	1140.03	7.00	231.56	nr	**1371.59**
Sum of two sides 3700 mm	965.87	1222.41	7.00	231.56	nr	**1453.97**
Sum of two sides 3800 mm	994.73	1258.93	7.00	231.56	nr	**1490.49**
Sum of two sides 3900 mm	1023.60	1295.47	7.00	231.56	nr	**1527.03**
Sum of two sides 4000 mm	1052.46	1331.99	7.00	231.56	nr	**1563.55**
45° bend						
Sum of two sides 2100 mm	121.85	154.22	2.61	86.34	nr	**240.55**
Sum of two sides 2200 mm	346.68	438.76	2.61	86.34	nr	**525.09**
Sum of two sides 2300 mm	373.80	473.08	2.61	86.34	nr	**559.42**
Sum of two sides 2400 mm	385.98	488.50	2.88	95.27	nr	**583.77**
Sum of two sides 2500 mm	413.06	522.77	4.85	160.44	nr	**683.20**
Sum of two sides 2600 mm	439.54	556.28	4.87	161.10	nr	**717.38**
Sum of two sides 2700 mm	466.01	589.78	4.87	161.10	nr	**750.88**
Sum of two sides 2800 mm	492.48	623.28	8.81	291.43	nr	**914.72**
Sum of two sides 2900 mm	519.58	657.58	8.81	291.43	nr	**949.01**
Sum of two sides 3000 mm	546.05	691.08	9.31	307.97	nr	**999.06**
Sum of two sides 3100 mm	555.17	702.62	9.31	307.97	nr	**1010.59**
Sum of two sides 3200 mm	574.41	726.98	9.31	307.97	nr	**1034.95**
Sum of two sides 3300 mm	578.06	731.59	9.31	307.97	nr	**1039.56**
Sum of two sides 3400 mm	597.33	755.99	9.31	307.97	nr	**1063.96**
Sum of two sides 3500 mm	616.60	780.37	9.39	310.62	nr	**1090.99**
Sum of two sides 3600 mm	635.87	804.76	9.39	310.62	nr	**1115.37**
Sum of two sides 3700 mm	679.39	859.84	9.39	310.62	nr	**1170.46**
Sum of two sides 3800 mm	698.65	884.21	9.39	310.62	nr	**1194.83**
Sum of two sides 3900 mm	717.92	908.60	9.39	310.62	nr	**1219.21**
Sum of two sides 4000 mm	737.19	932.99	9.39	310.62	nr	**1243.61**
90° mitre bend						
Sum of two sides 2100 mm	563.61	713.30	2.61	86.34	nr	**799.64**
Sum of two sides 2200 mm	595.47	753.62	2.61	86.34	nr	**839.96**
Sum of two sides 2300 mm	643.35	814.22	2.61	86.34	nr	**900.56**
Sum of two sides 2400 mm	657.60	832.26	2.88	95.27	nr	**927.53**
Sum of two sides 2500 mm	705.69	893.13	4.85	160.44	nr	**1053.56**
Sum of two sides 2600 mm	754.07	954.35	4.87	161.10	nr	**1115.45**
Sum of two sides 2700 mm	802.44	1015.57	4.87	161.10	nr	**1176.67**
Sum of two sides 2800 mm	850.82	1076.79	8.81	291.43	nr	**1368.22**
Sum of two sides 2900 mm	898.89	1137.64	14.81	489.91	nr	**1627.55**
Sum of two sides 3000 mm	947.27	1198.86	15.20	502.81	nr	**1701.67**
Sum of two sides 3100 mm	960.17	1215.19	15.20	502.81	nr	**1718.00**
Sum of two sides 3200 mm	997.74	1262.74	15.20	502.81	nr	**1765.55**
Sum of two sides 3300 mm	1008.28	1276.08	15.20	502.81	nr	**1778.89**
Sum of two sides 3400 mm	1045.84	1323.62	15.20	502.81	nr	**1826.43**
Sum of two sides 3500 mm	1083.40	1371.15	15.60	516.04	nr	**1887.19**

38 VENTILATION/AIR CONDITIONING SYSTEMS

Item	Net Price £	Material £	Labour hours	Labour £	Unit	Total rate £
Sum of two sides 3600 mm	1120.96	1418.69	15.60	516.04	nr	**1934.73**
Sum of two sides 3700 mm	1170.07	1480.84	15.60	516.04	nr	**1996.88**
Sum of two sides 3800 mm	1207.63	1528.37	15.60	516.04	nr	**2044.41**
Sum of two sides 3900 mm	1245.20	1575.92	15.60	516.04	nr	**2091.96**
Sum of two sides 4000 mm	1282.75	1623.45	15.60	516.04	nr	**2139.49**
Branch						
Sum of two sides 2100 mm	252.67	319.78	2.61	86.34	nr	**406.12**
Sum of two sides 2200 mm	270.33	342.13	2.61	86.34	nr	**428.47**
Sum of two sides 2300 mm	288.26	364.82	2.61	86.34	nr	**451.16**
Sum of two sides 2400 mm	305.63	386.81	2.88	95.27	nr	**482.08**
Sum of two sides 2500 mm	323.57	409.51	2.88	95.27	nr	**504.78**
Sum of two sides 2600 mm	341.26	431.90	2.88	95.27	nr	**527.17**
Sum of two sides 2700 mm	358.94	454.28	2.88	95.27	nr	**549.55**
Sum of two sides 2800 mm	376.62	476.66	3.94	130.33	nr	**606.99**
Sum of two sides 2900 mm	394.55	499.34	3.94	130.33	nr	**629.67**
Sum of two sides 3000 mm	412.22	521.70	3.94	130.33	nr	**652.04**
Sum of two sides 3100 mm	426.38	539.63	3.94	130.33	nr	**669.97**
Sum of two sides 3200 mm	439.26	555.92	3.94	130.33	nr	**686.26**
Sum of two sides 3300 mm	452.38	572.53	3.94	130.33	nr	**702.87**
Sum of two sides 3400 mm	465.28	588.85	4.83	159.77	nr	**748.63**
Sum of two sides 3500 mm	478.15	605.15	4.83	159.77	nr	**764.92**
Sum of two sides 3600 mm	491.04	621.46	4.83	159.77	nr	**781.24**
Sum of two sides 3700 mm	510.07	645.54	4.83	159.77	nr	**805.32**
Sum of two sides 3800 mm	522.96	661.86	4.83	159.77	nr	**821.64**
Sum of two sides 3900 mm	535.84	678.15	4.83	159.77	nr	**837.93**
Sum of two sides 4000 mm	548.73	694.47	4.83	159.77	nr	**854.25**
Grille neck						
Sum of two sides 2100 mm	267.93	339.09	2.61	86.34	nr	**425.43**
Sum of two sides 2200 mm	287.84	364.29	2.61	86.34	nr	**450.62**
Sum of two sides 2300 mm	307.77	389.51	2.61	86.34	nr	**475.85**
Sum of two sides 2400 mm	327.67	414.70	2.88	95.27	nr	**509.97**
Sum of two sides 2500 mm	347.59	439.91	2.88	95.27	nr	**535.18**
Sum of two sides 2600 mm	367.51	465.12	2.88	95.27	nr	**560.39**
Sum of two sides 2700 mm	387.43	490.33	2.88	95.27	nr	**585.60**
Sum of two sides 2800 mm	407.35	515.54	3.94	130.33	nr	**645.87**
Sum of two sides 2900 mm	427.26	540.73	4.12	136.29	nr	**677.02**
Sum of two sides 3000 mm	447.17	565.94	4.12	136.29	nr	**702.23**
Sum of two sides 3100 mm	463.10	586.10	4.12	136.29	nr	**722.39**
Sum of two sides 3200 mm	477.61	604.47	4.12	136.29	nr	**740.76**
Sum of two sides 3300 mm	492.10	622.81	4.12	136.29	nr	**759.09**
Sum of two sides 3400 mm	506.65	641.21	5.00	165.40	nr	**806.61**
Sum of two sides 3500 mm	521.17	659.59	5.00	165.40	nr	**824.99**
Sum of two sides 3600 mm	535.67	677.94	5.00	165.40	nr	**843.34**
Sum of two sides 3700 mm	550.19	696.32	5.00	165.40	nr	**861.72**
Sum of two sides 3800 mm	564.69	714.67	5.00	165.40	nr	**880.07**
Sum of two sides 3900 mm	579.21	733.05	5.00	165.40	nr	**898.45**
Sum of two sides 4000 mm	593.73	751.43	5.00	165.40	nr	**916.83**
Ductwork 2001 to 2500 mm longest side						
Sum of two sides 2500 mm	280.50	355.00	2.38	78.73	m	**433.73**
Sum of two sides 2600 mm	287.72	364.14	2.64	87.33	m	**451.47**
Sum of two sides 2700 mm	293.95	372.02	2.66	87.99	m	**460.01**

38 VENTILATION/AIR CONDITIONING SYSTEMS

Item	Net Price £	Material £	Labour hours	Labour £	Unit	Total rate £
DUCTWORK: RECTANGULAR – CLASS B – cont						
Ductwork 2001 to 2500 mm longest side – cont						
Sum of two sides 2800 mm	301.63	381.75	2.95	97.58	m	**479.33**
Sum of two sides 2900 mm	307.86	389.63	2.96	97.92	m	**487.54**
Sum of two sides 3000 mm	313.74	397.07	3.15	104.20	m	**501.27**
Sum of two sides 3100 mm	319.84	404.79	3.15	104.20	m	**508.99**
Sum of two sides 3200 mm	325.72	412.23	3.15	104.20	m	**516.43**
Sum of two sides 3300 mm	331.94	420.10	3.15	104.20	m	**524.30**
Sum of two sides 3400 mm	337.82	427.54	2.66	87.99	m	**515.53**
Sum of two sides 3500 mm	360.33	456.03	3.15	104.20	m	**560.23**
Sum of two sides 3600 mm	366.21	463.47	3.18	105.19	m	**568.67**
Sum of two sides 3700 mm	392.26	496.44	3.18	105.19	m	**601.64**
Sum of two sides 3800 mm	398.14	503.88	3.18	105.19	m	**609.08**
Sum of two sides 3900 mm	404.02	511.33	3.18	105.19	m	**616.52**
Sum of two sides 4000 mm	411.73	521.08	3.18	105.19	m	**626.27**
Extra over fittings; Ductwork 2001 to 2500 mm longest side						
End cap						
Sum of two sides 2500 mm	111.38	140.97	0.87	28.78	nr	**169.75**
Sum of two sides 2600 mm	117.84	149.14	0.87	28.78	nr	**177.92**
Sum of two sides 2700 mm	124.29	157.30	0.87	28.78	nr	**186.08**
Sum of two sides 2800 mm	130.76	165.49	1.16	38.37	nr	**203.86**
Sum of two sides 2900 mm	137.21	173.66	1.16	38.37	nr	**212.03**
Sum of two sides 3000 mm	143.67	181.83	1.73	57.23	nr	**239.06**
Sum of two sides 3100 mm	148.81	188.33	1.73	57.23	nr	**245.56**
Sum of two sides 3200 mm	153.46	194.22	1.73	57.23	nr	**251.44**
Sum of two sides 3300 mm	158.12	200.12	1.73	57.23	nr	**257.34**
Sum of two sides 3400 mm	162.76	205.99	1.80	59.54	nr	**265.53**
Sum of two sides 3500 mm	167.42	211.89	1.80	59.54	nr	**271.43**
Sum of two sides 3600 mm	172.07	217.78	1.80	59.54	nr	**277.32**
Sum of two sides 3700 mm	176.73	223.66	1.80	59.54	nr	**283.21**
Sum of two sides 3800 mm	181.39	229.56	1.80	59.54	nr	**289.11**
Sum of two sides 3900 mm	186.04	235.45	1.80	59.54	nr	**294.99**
Sum of two sides 4000 mm	190.71	241.36	1.80	59.54	nr	**300.91**
Reducer						
Sum of two sides 2500 mm	131.42	166.32	3.12	103.21	nr	**269.53**
Sum of two sides 2600 mm	142.36	180.17	3.12	103.21	nr	**283.38**
Sum of two sides 2700 mm	153.55	194.34	3.12	103.21	nr	**297.54**
Sum of two sides 2800 mm	164.26	207.89	3.95	130.66	nr	**338.55**
Sum of two sides 2900 mm	175.44	222.04	3.97	131.33	nr	**353.37**
Sum of two sides 3000 mm	186.42	235.93	3.97	131.33	nr	**367.26**
Sum of two sides 3100 mm	194.52	246.19	3.97	131.33	nr	**377.51**
Sum of two sides 3200 mm	201.87	255.49	3.97	131.33	nr	**386.82**
Sum of two sides 3300 mm	209.49	265.12	3.97	131.33	nr	**396.45**
Sum of two sides 3400 mm	216.84	274.43	4.52	149.52	nr	**423.95**
Sum of two sides 3500 mm	213.41	270.09	4.52	149.52	nr	**419.61**

38 VENTILATION/AIR CONDITIONING SYSTEMS

Item	Net Price £	Material £	Labour hours	Labour £	Unit	Total rate £
Sum of two sides 3600 mm	220.78	279.42	4.52	149.52	nr	428.94
Sum of two sides 3700 mm	214.59	271.58	4.52	149.52	nr	421.10
Sum of two sides 3800 mm	221.97	280.93	4.52	149.52	nr	430.45
Sum of two sides 3900 mm	229.35	290.27	4.52	149.52	nr	439.79
Sum of two sides 4000 mm	236.84	299.74	4.52	149.52	nr	449.26
Offset						
Sum of two sides 2500 mm	330.54	418.33	3.48	115.12	nr	533.45
Sum of two sides 2600 mm	351.17	444.44	3.49	115.45	nr	559.89
Sum of two sides 2700 mm	359.10	454.48	3.50	115.78	nr	570.26
Sum of two sides 2800 mm	392.40	496.62	4.34	143.57	nr	640.18
Sum of two sides 2900 mm	399.92	506.14	4.76	157.46	nr	663.60
Sum of two sides 3000 mm	406.79	514.83	5.32	175.98	nr	690.82
Sum of two sides 3100 mm	408.95	517.57	5.35	176.98	nr	694.55
Sum of two sides 3200 mm	409.54	518.32	5.32	175.98	nr	694.30
Sum of two sides 3300 mm	409.84	518.69	5.32	175.98	nr	694.67
Sum of two sides 3400 mm	409.54	518.32	5.35	176.98	nr	695.29
Sum of two sides 3500 mm	406.80	514.85	5.32	175.98	nr	690.83
Sum of two sides 3600 mm	420.78	532.54	5.35	176.98	nr	709.51
Sum of two sides 3700 mm	414.30	524.34	5.35	176.98	nr	701.31
Sum of two sides 3800 mm	428.27	542.02	5.35	176.98	nr	719.00
Sum of two sides 3900 mm	442.28	559.75	5.35	176.98	nr	736.73
Sum of two sides 4000 mm	456.24	577.41	5.35	176.98	nr	754.39
Square to round						
Sum of two sides 2500 mm	211.02	267.07	4.70	155.47	nr	422.54
Sum of two sides 2600 mm	227.80	288.30	4.70	155.47	nr	443.77
Sum of two sides 2700 mm	244.56	309.51	4.71	155.80	nr	465.31
Sum of two sides 2800 mm	261.37	330.78	8.19	270.92	nr	601.71
Sum of two sides 2900 mm	278.11	351.98	8.19	270.92	nr	622.90
Sum of two sides 3000 mm	294.90	373.23	8.19	270.92	nr	644.15
Sum of two sides 3100 mm	307.45	389.11	8.19	270.92	nr	660.03
Sum of two sides 3200 mm	318.82	403.50	8.62	285.15	nr	688.65
Sum of two sides 3300 mm	330.20	417.91	8.62	285.15	nr	703.05
Sum of two sides 3400 mm	341.55	432.27	8.62	285.15	nr	717.42
Sum of two sides 3500 mm	341.51	432.22	8.62	285.15	nr	717.36
Sum of two sides 3600 mm	352.89	446.62	8.75	289.45	nr	736.07
Sum of two sides 3700 mm	350.47	443.55	8.75	289.45	nr	733.00
Sum of two sides 3800 mm	361.84	457.95	8.75	289.45	nr	747.39
Sum of two sides 3900 mm	373.22	472.35	8.75	289.45	nr	761.80
Sum of two sides 4000 mm	384.51	486.64	8.75	289.45	nr	776.08
90° radius bend						
Sum of two sides 2500 mm	494.27	625.54	3.90	129.01	nr	754.55
Sum of two sides 2600 mm	516.52	653.70	4.26	140.92	nr	794.62
Sum of two sides 2700 mm	556.92	704.84	4.55	150.51	nr	855.35
Sum of two sides 2800 mm	559.76	708.43	4.55	150.51	nr	858.94
Sum of two sides 2900 mm	600.09	759.47	6.87	227.26	nr	986.73
Sum of two sides 3000 mm	639.67	809.57	6.87	227.26	nr	1036.82
Sum of two sides 3100 mm	670.67	848.80	6.87	227.26	nr	1076.05
Sum of two sides 3200 mm	699.45	885.22	6.87	227.26	nr	1112.48
Sum of two sides 3300 mm	728.98	922.60	6.87	227.26	nr	1149.86
Sum of two sides 3400 mm	757.79	959.05	6.87	227.26	nr	1186.31
Sum of two sides 3500 mm	758.01	959.33	7.00	231.56	nr	1190.89

38 VENTILATION/AIR CONDITIONING SYSTEMS

Item	Net Price £	Material £	Labour hours	Labour £	Unit	Total rate £
DUCTWORK: RECTANGULAR – CLASS B – cont						
Extra over fittings – cont						
90° radius bend – cont						
Sum of two sides 3600 mm	786.81	995.78	7.00	231.56	nr	**1227.34**
Sum of two sides 3700 mm	779.43	986.44	7.00	231.56	nr	**1218.00**
Sum of two sides 3800 mm	808.21	1022.87	7.00	231.56	nr	**1254.42**
Sum of two sides 3900 mm	837.00	1059.30	7.00	231.56	nr	**1290.86**
Sum of two sides 4000 mm	845.00	1069.43	7.00	231.56	nr	**1300.99**
45° bend						
Sum of two sides 2500 mm	369.41	467.53	4.85	160.44	nr	**627.96**
Sum of two sides 2600 mm	386.94	489.71	4.87	161.10	nr	**650.80**
Sum of two sides 2700 mm	414.00	523.96	4.87	161.10	nr	**685.06**
Sum of two sides 2800 mm	421.35	533.27	8.81	291.43	nr	**824.70**
Sum of two sides 2900 mm	448.38	567.47	8.81	291.43	nr	**858.90**
Sum of two sides 3000 mm	474.81	600.92	9.31	307.97	nr	**908.89**
Sum of two sides 3100 mm	495.61	627.24	9.31	307.97	nr	**935.22**
Sum of two sides 3200 mm	514.82	651.55	9.31	307.97	nr	**959.52**
Sum of two sides 3300 mm	534.65	676.65	9.31	307.97	nr	**984.62**
Sum of two sides 3400 mm	553.88	700.98	9.31	307.97	nr	**1008.96**
Sum of two sides 3500 mm	558.71	707.10	9.31	307.97	nr	**1015.07**
Sum of two sides 3600 mm	577.94	731.44	9.39	310.62	nr	**1042.06**
Sum of two sides 3700 mm	578.46	732.09	9.39	310.62	nr	**1042.71**
Sum of two sides 3800 mm	597.67	756.41	9.39	310.62	nr	**1067.03**
Sum of two sides 3900 mm	616.89	780.73	9.39	310.62	nr	**1091.35**
Sum of two sides 4000 mm	625.71	791.89	9.39	310.62	nr	**1102.51**
90° mitre bend						
Sum of two sides 2500 mm	599.49	758.71	4.85	160.44	nr	**919.15**
Sum of two sides 2600 mm	635.93	804.84	4.87	161.10	nr	**965.93**
Sum of two sides 2700 mm	684.29	866.03	4.87	161.10	nr	**1027.13**
Sum of two sides 2800 mm	685.70	867.83	8.81	291.43	nr	**1159.26**
Sum of two sides 2900 mm	734.25	929.27	14.81	489.91	nr	**1419.18**
Sum of two sides 3000 mm	782.17	989.91	14.81	489.91	nr	**1479.82**
Sum of two sides 3100 mm	823.34	1042.01	15.20	502.81	nr	**1544.82**
Sum of two sides 3200 mm	861.43	1090.23	15.20	502.81	nr	**1593.04**
Sum of two sides 3300 mm	899.17	1137.99	15.20	502.81	nr	**1640.80**
Sum of two sides 3400 mm	937.28	1186.22	15.20	502.81	nr	**1689.04**
Sum of two sides 3500 mm	937.46	1186.45	15.20	502.81	nr	**1689.26**
Sum of two sides 3600 mm	975.56	1234.66	15.60	516.04	nr	**1750.70**
Sum of two sides 3700 mm	980.93	1241.47	15.60	516.04	nr	**1757.51**
Sum of two sides 3800 mm	1019.04	1289.69	15.60	516.04	nr	**1805.73**
Sum of two sides 3900 mm	1057.13	1337.90	15.60	516.04	nr	**1853.95**
Sum of two sides 4000 mm	1068.42	1352.19	15.60	516.04	nr	**1868.23**
Branch						
Sum of two sides 2500 mm	324.16	410.25	2.88	95.27	nr	**505.52**
Sum of two sides 2600 mm	341.81	432.59	2.88	95.27	nr	**527.86**
Sum of two sides 2700 mm	359.73	455.27	2.88	95.27	nr	**550.54**
Sum of two sides 2800 mm	377.12	477.28	3.94	130.33	nr	**607.61**
Sum of two sides 2900 mm	395.04	499.96	3.94	130.33	nr	**630.30**
Sum of two sides 3000 mm	412.72	522.34	3.94	130.33	nr	**652.68**
Sum of two sides 3100 mm	426.86	540.23	3.94	130.33	nr	**670.56**

38 VENTILATION/AIR CONDITIONING SYSTEMS

Item	Net Price £	Material £	Labour hours	Labour £	Unit	Total rate £
Sum of two sides 3200 mm	439.75	556.55	3.94	130.33	nr	686.88
Sum of two sides 3300 mm	452.88	573.16	3.94	130.33	nr	703.49
Sum of two sides 3400 mm	465.76	589.46	3.94	130.33	nr	719.80
Sum of two sides 3500 mm	479.27	606.57	4.83	159.77	nr	766.34
Sum of two sides 3600 mm	492.16	622.87	4.83	159.77	nr	782.65
Sum of two sides 3700 mm	505.29	639.50	4.83	159.77	nr	799.27
Sum of two sides 3800 mm	518.16	655.79	4.83	159.77	nr	815.56
Sum of two sides 3900 mm	531.06	672.11	4.83	159.77	nr	831.88
Sum of two sides 4000 mm	544.14	688.67	4.83	159.77	nr	848.44
Grille neck						
Sum of two sides 2500 mm	347.59	439.91	2.88	95.27	nr	535.18
Sum of two sides 2600 mm	367.51	465.12	2.88	95.27	nr	560.39
Sum of two sides 2700 mm	387.43	490.33	2.88	95.27	nr	585.60
Sum of two sides 2800 mm	407.35	515.54	3.94	130.33	nr	645.87
Sum of two sides 2900 mm	427.14	540.59	3.94	130.33	nr	670.92
Sum of two sides 3000 mm	447.17	565.94	3.94	130.33	nr	696.28
Sum of two sides 3100 mm	463.10	586.10	4.12	136.29	nr	722.39
Sum of two sides 3200 mm	477.61	604.47	4.12	136.29	nr	740.76
Sum of two sides 3300 mm	492.11	622.82	4.12	136.29	nr	759.11
Sum of two sides 3400 mm	506.65	641.21	4.12	136.29	nr	777.50
Sum of two sides 3500 mm	521.17	659.59	5.00	165.40	nr	824.99
Sum of two sides 3600 mm	535.67	677.94	5.00	165.40	nr	843.34
Sum of two sides 3700 mm	550.19	696.32	5.00	165.40	nr	861.72
Sum of two sides 3800 mm	564.69	714.67	5.00	165.40	nr	880.07
Sum of two sides 3900 mm	579.21	733.05	5.00	165.40	nr	898.45
Sum of two sides 4000 mm	593.73	751.43	5.00	165.40	nr	916.83
Ductwork 2501 to 4000 mm longest side						
Sum of two sides 3000 mm	450.45	570.09	2.38	78.73	m	648.82
Sum of two sides 3100 mm	461.23	583.74	2.38	78.73	m	662.47
Sum of two sides 3200 mm	471.45	596.67	2.38	78.73	m	675.40
Sum of two sides 3300 mm	481.67	609.60	2.38	78.73	m	688.33
Sum of two sides 3400 mm	494.31	625.60	2.64	87.33	m	712.93
Sum of two sides 3500 mm	504.53	638.53	2.66	87.99	m	726.52
Sum of two sides 3600 mm	514.75	651.47	2.95	97.58	m	749.06
Sum of two sides 3700 mm	527.73	667.90	2.96	97.92	m	765.81
Sum of two sides 3800 mm	537.95	680.83	3.15	104.20	m	785.03
Sum of two sides 3900 mm	566.95	717.53	3.15	104.20	m	821.73
Sum of two sides 4000 mm	577.16	730.46	3.15	104.20	m	834.66
Sum of two sides 4100 mm	587.73	743.83	3.35	110.82	m	854.64
Sum of two sides 4200 mm	597.94	756.76	3.35	110.82	m	867.57
Sum of two sides 4300 mm	629.08	796.16	3.60	119.09	m	915.25
Sum of two sides 4400 mm	623.65	789.29	3.60	119.09	m	908.37
Sum of two sides 4500 mm	633.86	802.22	3.60	119.09	m	921.30
Extra over fittings; Ductwork 2501 to 4000 mm longest side						
End cap						
Sum of two sides 3000 mm	145.25	183.82	1.73	57.23	nr	241.05
Sum of two sides 3100 mm	150.42	190.38	1.73	57.23	nr	247.60
Sum of two sides 3200 mm	155.13	196.33	1.73	57.23	nr	253.56

38 VENTILATION/AIR CONDITIONING SYSTEMS

Item	Net Price £	Material £	Labour hours	Labour £	Unit	Total rate £
DUCTWORK: RECTANGULAR – CLASS B – cont						
Extra over fittings – cont						
End cap – cont						
Sum of two sides 3300 mm	159.75	202.18	1.73	57.23	nr	259.40
Sum of two sides 3400 mm	164.55	208.25	1.73	57.23	nr	265.48
Sum of two sides 3500 mm	169.27	214.23	1.73	57.23	nr	271.46
Sum of two sides 3600 mm	173.97	220.18	1.73	57.23	nr	277.41
Sum of two sides 3700 mm	178.69	226.15	1.73	57.23	nr	283.38
Sum of two sides 3800 mm	183.39	232.10	1.73	57.23	nr	289.33
Sum of two sides 3900 mm	188.11	238.07	1.80	59.54	nr	297.61
Sum of two sides 4000 mm	192.81	244.02	1.80	59.54	nr	303.57
Sum of two sides 4100 mm	197.53	249.99	1.80	59.54	nr	309.53
Sum of two sides 4200 mm	202.23	255.94	1.80	59.54	nr	315.49
Sum of two sides 4300 mm	206.93	261.90	1.88	62.19	nr	324.09
Sum of two sides 4400 mm	211.65	267.86	1.88	62.19	nr	330.05
Sum of two sides 4500 mm	216.35	273.82	1.88	62.19	nr	336.01
Reducer						
Sum of two sides 3000 mm	147.76	187.00	3.12	103.21	nr	290.21
Sum of two sides 3100 mm	154.72	195.81	3.12	103.21	nr	299.02
Sum of two sides 3200 mm	160.72	203.41	3.12	103.21	nr	306.62
Sum of two sides 3300 mm	166.76	211.05	3.12	103.21	nr	314.26
Sum of two sides 3400 mm	172.70	218.57	3.12	103.21	nr	321.78
Sum of two sides 3500 mm	178.73	226.20	3.12	103.21	nr	329.41
Sum of two sides 3600 mm	184.75	233.82	3.95	130.66	nr	364.48
Sum of two sides 3700 mm	190.94	241.66	3.97	131.33	nr	372.98
Sum of two sides 3800 mm	196.95	249.26	4.52	149.52	nr	398.78
Sum of two sides 3900 mm	192.11	243.13	4.52	149.52	nr	392.65
Sum of two sides 4000 mm	198.12	250.75	4.52	149.52	nr	400.27
Sum of two sides 4100 mm	204.36	258.64	4.52	149.52	nr	408.16
Sum of two sides 4200 mm	210.39	266.27	4.52	149.52	nr	415.79
Sum of two sides 4300 mm	216.39	273.87	4.92	162.75	nr	436.62
Sum of two sides 4400 mm	222.55	281.66	4.92	162.75	nr	444.41
Sum of two sides 4500 mm	228.56	289.27	5.12	169.37	nr	458.64
Offset						
Sum of two sides 3000 mm	410.59	519.65	3.48	115.12	nr	634.76
Sum of two sides 3100 mm	403.99	511.29	3.48	115.12	nr	626.40
Sum of two sides 3200 mm	394.98	499.88	3.48	115.12	nr	615.00
Sum of two sides 3300 mm	385.00	487.26	3.48	115.12	nr	602.38
Sum of two sides 3400 mm	421.80	533.82	3.50	115.78	nr	649.60
Sum of two sides 3500 mm	410.87	519.99	3.49	115.45	nr	635.44
Sum of two sides 3600 mm	398.96	504.92	4.25	140.59	nr	645.51
Sum of two sides 3700 mm	436.89	552.93	4.76	157.46	nr	710.39
Sum of two sides 3800 mm	424.01	536.63	5.32	175.98	nr	712.61
Sum of two sides 3900 mm	449.59	569.00	5.35	176.98	nr	745.98
Sum of two sides 4000 mm	434.55	549.97	5.35	176.98	nr	726.95
Sum of two sides 4100 mm	418.68	529.88	5.85	193.52	nr	723.39
Sum of two sides 4200 mm	401.70	508.39	5.85	193.52	nr	701.91
Sum of two sides 4300 mm	413.21	522.95	6.15	203.44	nr	726.39
Sum of two sides 4400 mm	909.45	1151.00	12.60	416.80	nr	1567.80
Sum of two sides 4500 mm	872.57	1104.33	12.30	406.88	nr	1511.21

38 VENTILATION/AIR CONDITIONING SYSTEMS

Item	Net Price £	Material £	Labour hours	Labour £	Unit	Total rate £
Square to round						
Sum of two sides 3000 mm	601.69	761.50	9.40	310.95	nr	**1072.45**
Sum of two sides 3100 mm	627.52	794.19	9.40	310.95	nr	**1105.14**
Sum of two sides 3200 mm	650.54	823.32	9.40	310.95	nr	**1134.27**
Sum of two sides 3300 mm	673.58	852.48	9.40	310.95	nr	**1163.43**
Sum of two sides 3400 mm	696.61	881.63	9.42	311.61	nr	**1193.24**
Sum of two sides 3500 mm	719.63	910.76	9.42	311.61	nr	**1222.37**
Sum of two sides 3600 mm	742.64	939.89	16.38	541.84	nr	**1481.73**
Sum of two sides 3700 mm	765.66	969.02	17.24	570.29	nr	**1539.31**
Sum of two sides 3800 mm	788.70	998.17	17.24	570.29	nr	**1568.47**
Sum of two sides 3900 mm	964.22	1220.31	17.24	570.29	nr	**1790.60**
Sum of two sides 4000 mm	991.64	1255.02	17.50	578.89	nr	**1833.91**
Sum of two sides 4100 mm	899.87	1138.88	22.46	742.97	nr	**1881.84**
Sum of two sides 4200 mm	1046.43	1324.36	22.46	742.97	nr	**2067.33**
Sum of two sides 4300 mm	1073.81	1359.02	22.50	744.29	nr	**2103.31**
Sum of two sides 4400 mm	1101.22	1393.70	22.50	744.29	nr	**2137.99**
Sum of two sides 4500 mm	1128.65	1428.41	22.52	744.95	nr	**2173.37**
90° radius bend						
Sum of two sides 3000 mm	1745.71	2209.37	7.80	258.02	nr	**2467.39**
Sum of two sides 3100 mm	1828.70	2314.40	7.80	258.02	nr	**2572.43**
Sum of two sides 3200 mm	1903.17	2408.65	8.52	281.84	nr	**2690.49**
Sum of two sides 3300 mm	1977.63	2502.89	8.52	281.84	nr	**2784.73**
Sum of two sides 3400 mm	1931.29	2444.24	8.52	281.84	nr	**2726.07**
Sum of two sides 3500 mm	2004.77	2537.23	9.10	301.02	nr	**2838.26**
Sum of two sides 3600 mm	2078.27	2630.25	9.10	301.02	nr	**2931.28**
Sum of two sides 3700 mm	2026.42	2564.63	13.74	454.51	nr	**3019.15**
Sum of two sides 3800 mm	2098.95	2656.43	13.74	454.51	nr	**3110.95**
Sum of two sides 3900 mm	1964.59	2486.39	13.74	454.51	nr	**2940.90**
Sum of two sides 4000 mm	2036.20	2577.02	14.00	463.11	nr	**3040.13**
Sum of two sides 4100 mm	2109.85	2670.22	14.40	476.35	nr	**3146.57**
Sum of two sides 4200 mm	2181.44	2760.83	14.40	476.35	nr	**3237.17**
Sum of two sides 4300 mm	2253.03	2851.43	14.82	490.24	nr	**3341.67**
Sum of two sides 4400 mm	2104.85	2663.90	14.82	490.24	nr	**3154.14**
Sum of two sides 4500 mm	2175.05	2752.75	15.10	499.50	nr	**3252.25**
45° bend						
Sum of two sides 3000 mm	846.91	1071.85	9.70	320.87	nr	**1392.72**
Sum of two sides 3100 mm	888.22	1124.13	9.70	320.87	nr	**1445.00**
Sum of two sides 3200 mm	925.30	1171.06	9.70	320.87	nr	**1491.93**
Sum of two sides 3300 mm	949.16	1201.25	9.74	322.20	nr	**1523.45**
Sum of two sides 3400 mm	938.13	1187.30	9.74	322.20	nr	**1509.50**
Sum of two sides 3500 mm	974.78	1233.68	9.74	322.20	nr	**1555.87**
Sum of two sides 3600 mm	1011.40	1280.03	17.62	582.86	nr	**1862.89**
Sum of two sides 3700 mm	984.33	1245.77	17.62	582.86	nr	**1828.63**
Sum of two sides 3800 mm	1020.47	1291.51	18.62	615.94	nr	**1907.45**
Sum of two sides 3900 mm	950.71	1203.22	18.62	615.94	nr	**1819.16**
Sum of two sides 4000 mm	986.37	1248.35	18.62	615.94	nr	**1864.29**
Sum of two sides 4100 mm	1023.04	1294.75	20.02	662.25	nr	**1957.01**
Sum of two sides 4200 mm	1058.72	1339.91	18.62	615.94	nr	**1955.85**
Sum of two sides 4300 mm	1094.39	1385.07	20.02	662.25	nr	**2047.32**
Sum of two sides 4400 mm	1018.67	1289.23	21.04	696.00	nr	**1985.22**
Sum of two sides 4500 mm	1053.65	1333.50	21.04	696.00	nr	**2029.50**

38 VENTILATION/AIR CONDITIONING SYSTEMS

Item	Net Price £	Material £	Labour hours	Labour £	Unit	Total rate £
DUCTWORK: RECTANGULAR – CLASS B – cont						
Extra over fittings – cont						
90° mitre bend						
Sum of two sides 3000 mm	2053.51	2598.92	9.70	320.87	nr	**2919.79**
Sum of two sides 3100 mm	2154.73	2727.02	9.70	320.87	nr	**3047.89**
Sum of two sides 3200 mm	2249.92	2847.50	9.74	322.20	nr	**3169.69**
Sum of two sides 3300 mm	2345.13	2968.00	9.74	322.20	nr	**3290.20**
Sum of two sides 3400 mm	2287.11	2894.57	9.74	322.20	nr	**3216.76**
Sum of two sides 3500 mm	2381.46	3013.98	17.62	582.86	nr	**3596.84**
Sum of two sides 3600 mm	2475.82	3133.39	17.62	582.86	nr	**3716.26**
Sum of two sides 3700 mm	2428.97	3074.10	29.62	979.82	nr	**4053.92**
Sum of two sides 3800 mm	2522.60	3192.61	29.62	979.82	nr	**4172.43**
Sum of two sides 3900 mm	2345.74	2968.77	29.62	979.82	nr	**3948.59**
Sum of two sides 4000 mm	2438.56	3086.24	30.40	1005.62	nr	**4091.86**
Sum of two sides 4100 mm	2530.12	3202.12	32.60	1078.40	nr	**4280.52**
Sum of two sides 4200 mm	2622.92	3319.57	33.00	1091.63	nr	**4411.20**
Sum of two sides 4300 mm	2715.74	3437.04	34.02	1125.37	nr	**4562.41**
Sum of two sides 4400 mm	2545.98	3222.19	34.02	1125.37	nr	**4347.56**
Sum of two sides 4500 mm	2637.62	3338.17	34.02	1125.37	nr	**4463.54**
Branch						
Sum of two sides 3000 mm	881.16	1115.20	5.76	190.54	nr	**1305.73**
Sum of two sides 3100 mm	911.63	1153.76	5.76	190.54	nr	**1344.30**
Sum of two sides 3200 mm	939.14	1188.58	5.76	190.54	nr	**1379.12**
Sum of two sides 3300 mm	966.61	1223.34	5.76	190.54	nr	**1413.88**
Sum of two sides 3400 mm	993.99	1258.00	5.76	190.54	nr	**1448.54**
Sum of two sides 3500 mm	1021.48	1292.79	7.88	260.67	nr	**1553.46**
Sum of two sides 3600 mm	1048.99	1327.60	7.88	260.67	nr	**1588.27**
Sum of two sides 3700 mm	1076.80	1362.79	7.88	260.67	nr	**1623.46**
Sum of two sides 3800 mm	1104.31	1397.61	9.66	319.55	nr	**1717.16**
Sum of two sides 3900 mm	1132.89	1433.78	9.66	319.55	nr	**1753.33**
Sum of two sides 4000 mm	1160.40	1468.60	9.66	319.55	nr	**1788.15**
Sum of two sides 4100 mm	1188.35	1503.97	10.88	359.91	nr	**1863.88**
Sum of two sides 4200 mm	1215.84	1538.76	10.88	359.91	nr	**1898.67**
Sum of two sides 4300 mm	1243.31	1573.53	11.70	387.03	nr	**1960.56**
Sum of two sides 4400 mm	1251.24	1583.57	11.70	387.03	nr	**1970.60**
Sum of two sides 4500 mm	1298.60	1643.51	11.70	387.03	nr	**2030.54**
Grille neck						
Sum of two sides 3000 mm	910.16	1151.90	5.76	190.54	nr	**1342.44**
Sum of two sides 3100 mm	942.54	1192.88	5.76	190.54	nr	**1383.42**
Sum of two sides 3200 mm	972.09	1230.28	5.76	190.54	nr	**1420.82**
Sum of two sides 3300 mm	1001.66	1267.70	5.76	190.54	nr	**1458.24**
Sum of two sides 3400 mm	1031.23	1305.12	7.88	260.67	nr	**1565.79**
Sum of two sides 3500 mm	1060.79	1342.54	7.88	260.67	nr	**1603.21**
Sum of two sides 3600 mm	913.79	1156.50	7.88	260.67	nr	**1417.17**
Sum of two sides 3700 mm	1119.87	1417.30	8.24	272.58	nr	**1689.88**
Sum of two sides 3800 mm	1149.44	1454.72	8.24	272.58	nr	**1727.30**
Sum of two sides 3900 mm	1179.00	1492.15	8.24	272.58	nr	**1764.72**
Sum of two sides 4000 mm	1208.55	1529.54	10.00	330.80	nr	**1860.34**

38 VENTILATION/AIR CONDITIONING SYSTEMS

Item	Net Price £	Material £	Labour hours	Labour £	Unit	Total rate £
Sum of two sides 4100 mm	1238.10	1566.94	10.00	330.80	nr	**1897.73**
Sum of two sides 4200 mm	1267.67	1604.36	10.00	330.80	nr	**1935.15**
Sum of two sides 4300 mm	1297.23	1641.78	10.46	346.01	nr	**1987.79**
Sum of two sides 4400 mm	1326.78	1679.17	10.46	346.01	nr	**2025.19**
Sum of two sides 4500 mm	1356.35	1716.59	10.78	356.60	nr	**2073.19**

38 VENTILATION/AIR CONDITIONING SYSTEMS

Item	Net Price £	Material £	Labour hours	Labour £	Unit	Total rate £
DUCTWORK: RECTANGULAR – CLASS C						
AIR DUCTLINES						
Galvanized sheet metal DW144 class C rectangular section ductwork; including all necessary stiffeners, joints, couplers in the running length and duct supports						
Ductwork up to 400 mm longest side						
Sum of two sides 200 mm	58.17	73.62	2.34	77.41	m	**151.03**
Sum of two sides 300 mm	62.92	79.63	2.34	77.41	m	**157.03**
Sum of two sides 400 mm	57.75	73.09	2.38	78.73	m	**151.82**
Sum of two sides 500 mm	62.90	79.60	2.32	76.74	m	**156.34**
Sum of two sides 600 mm	67.60	85.55	2.38	78.73	m	**164.28**
Sum of two sides 700 mm	72.30	91.51	2.38	78.73	m	**170.24**
Sum of two sides 800 mm	77.24	97.75	2.38	78.73	m	**176.48**
Extra over fittings; Ductwork up to 400 mm longest side						
End cap						
Sum of two sides 200 mm	33.14	41.94	0.76	25.14	nr	**67.08**
Sum of two sides 300 mm	37.02	46.86	0.76	25.14	nr	**72.00**
Sum of two sides 400 mm	40.89	51.75	0.76	25.14	nr	**76.89**
Sum of two sides 500 mm	44.75	56.64	0.76	25.14	nr	**81.78**
Sum of two sides 600 mm	48.64	61.55	0.76	25.14	nr	**86.69**
Sum of two sides 700 mm	52.52	66.47	0.76	25.14	nr	**91.61**
Sum of two sides 800 mm	56.36	71.33	0.76	25.14	nr	**96.47**
Reducer						
Sum of two sides 200 mm	53.82	68.12	2.80	92.62	nr	**160.74**
Sum of two sides 300 mm	60.73	76.86	2.80	92.62	nr	**169.49**
Sum of two sides 400 mm	99.41	125.82	2.84	93.95	nr	**219.76**
Sum of two sides 500 mm	107.58	136.16	2.84	93.95	nr	**230.10**
Sum of two sides 600 mm	115.77	146.52	3.38	111.81	nr	**258.33**
Sum of two sides 700 mm	123.92	156.83	3.38	111.81	nr	**268.64**
Sum of two sides 800 mm	132.01	167.07	3.84	127.03	nr	**294.09**
Offset						
Sum of two sides 200 mm	79.38	100.46	3.26	107.84	nr	**208.30**
Sum of two sides 300 mm	89.73	113.57	3.26	107.84	nr	**221.41**
Sum of two sides 400 mm	132.57	167.78	3.30	109.16	nr	**276.95**
Sum of two sides 500 mm	144.12	182.40	3.30	109.16	nr	**291.56**
Sum of two sides 600 mm	153.51	194.28	3.84	127.03	nr	**321.31**
Sum of two sides 700 mm	163.95	207.49	3.84	127.03	nr	**334.52**
Sum of two sides 800 mm	173.06	219.03	3.84	127.03	nr	**346.05**
Square to round						
Sum of two sides 200 mm	69.83	88.37	2.44	80.71	nr	**169.08**
Sum of two sides 300 mm	78.46	99.29	2.44	80.71	nr	**180.01**
Sum of two sides 400 mm	104.39	132.12	2.50	82.70	nr	**214.82**
Sum of two sides 500 mm	113.59	143.76	2.50	82.70	nr	**226.46**
Sum of two sides 600 mm	122.75	155.35	2.66	87.99	nr	**243.34**
Sum of two sides 700 mm	131.99	167.04	2.66	87.99	nr	**255.03**
Sum of two sides 800 mm	141.08	178.55	2.80	92.62	nr	**271.17**

38 VENTILATION/AIR CONDITIONING SYSTEMS

Item	Net Price £	Material £	Labour hours	Labour £	Unit	Total rate £
90° radius bend						
Sum of two sides 200 mm	52.54	66.50	2.20	72.78	nr	**139.27**
Sum of two sides 300 mm	56.68	71.73	2.20	72.78	nr	**144.51**
Sum of two sides 400 mm	95.38	120.72	2.24	74.10	nr	**194.81**
Sum of two sides 500 mm	101.43	128.37	2.24	74.10	nr	**202.47**
Sum of two sides 600 mm	109.52	138.60	2.32	76.74	nr	**215.35**
Sum of two sides 700 mm	116.53	147.48	2.32	76.74	nr	**224.22**
Sum of two sides 800 mm	124.24	157.23	2.44	80.71	nr	**237.95**
45° radius bend						
Sum of two sides 200 mm	56.72	71.79	2.58	85.35	nr	**157.13**
Sum of two sides 300 mm	62.27	78.80	2.58	85.35	nr	**164.15**
Sum of two sides 400 mm	100.23	126.85	2.58	85.35	nr	**212.20**
Sum of two sides 500 mm	107.18	135.65	2.58	85.35	nr	**221.00**
Sum of two sides 600 mm	115.35	145.99	2.78	91.96	nr	**237.95**
Sum of two sides 700 mm	122.98	155.64	2.78	91.96	nr	**247.60**
Sum of two sides 800 mm	131.04	165.84	2.92	96.59	nr	**262.44**
90° mitre bend						
Sum of two sides 200 mm	87.82	111.15	4.08	134.96	nr	**246.11**
Sum of two sides 300 mm	96.47	122.10	4.08	134.96	nr	**257.06**
Sum of two sides 400 mm	140.15	177.38	4.18	138.27	nr	**315.65**
Sum of two sides 500 mm	151.31	191.49	4.18	138.27	nr	**329.76**
Sum of two sides 600 mm	165.65	209.64	4.30	142.24	nr	**351.89**
Sum of two sides 700 mm	178.94	226.47	4.30	142.24	nr	**368.71**
Sum of two sides 800 mm	193.49	244.89	4.52	149.52	nr	**394.41**
Branch						
Sum of two sides 200 mm	87.28	110.46	1.84	60.87	nr	**171.32**
Sum of two sides 300 mm	97.15	122.95	1.84	60.87	nr	**183.81**
Sum of two sides 400 mm	121.40	153.65	1.90	62.85	nr	**216.50**
Sum of two sides 500 mm	133.10	168.45	1.90	62.85	nr	**231.30**
Sum of two sides 600 mm	143.81	182.00	2.06	68.14	nr	**250.15**
Sum of two sides 700 mm	154.56	195.61	2.06	68.14	nr	**263.76**
Sum of two sides 800 mm	165.29	209.19	2.06	68.14	nr	**277.34**
Grille neck						
Sum of two sides 200 mm	100.78	127.55	2.20	72.78	nr	**200.32**
Sum of two sides 300 mm	113.25	143.33	2.20	72.78	nr	**216.11**
Sum of two sides 400 mm	125.71	159.09	2.32	76.74	nr	**235.84**
Sum of two sides 500 mm	138.18	174.88	2.32	76.74	nr	**251.63**
Sum of two sides 600 mm	150.65	190.67	2.36	78.07	nr	**268.74**
Sum of two sides 700 mm	163.11	206.43	2.36	78.07	nr	**284.50**
Sum of two sides 800 mm	175.58	222.22	2.36	78.07	nr	**300.28**
Ductwork 401 to 600 mm longest side						
Sum of two sides 600 mm	94.37	119.44	2.34	77.41	m	**196.85**
Sum of two sides 700 mm	103.80	131.37	2.34	77.41	m	**208.78**
Sum of two sides 800 mm	112.98	142.99	2.34	77.41	m	**220.39**
Sum of two sides 900 mm	121.95	154.34	2.54	84.02	m	**238.36**
Sum of two sides 1000 mm	130.91	165.68	2.96	97.92	m	**263.60**
Sum of two sides 1100 mm	140.60	177.94	2.98	98.58	m	**276.51**
Sum of two sides 1200 mm	149.56	189.29	2.98	98.58	m	**287.86**

38 VENTILATION/AIR CONDITIONING SYSTEMS

Item	Net Price £	Material £	Labour hours	Labour £	Unit	Total rate £
DUCTWORK: RECTANGULAR – CLASS C – cont						
Extra over fittings: Ductwork 401 to 600 mm longest side						
End cap						
Sum of two sides 600 mm	48.64	61.55	0.76	25.14	nr	**86.69**
Sum of two sides 700 mm	52.52	66.47	0.76	25.14	nr	**91.61**
Sum of two sides 800 mm	56.36	71.33	0.76	25.14	nr	**96.47**
Sum of two sides 900 mm	60.25	76.25	0.76	25.14	nr	**101.39**
Sum of two sides 1000 mm	64.11	81.14	0.76	25.14	nr	**106.28**
Sum of two sides 1100 mm	67.98	86.03	0.76	25.14	nr	**111.17**
Sum of two sides 1200 mm	71.84	90.92	0.76	25.14	nr	**116.06**
Reducer						
Sum of two sides 600 mm	103.87	131.45	3.38	111.81	nr	**243.26**
Sum of two sides 700 mm	110.54	139.90	3.38	111.81	nr	**251.71**
Sum of two sides 800 mm	117.10	148.20	3.84	127.03	nr	**275.22**
Sum of two sides 900 mm	123.84	156.73	3.84	127.03	nr	**283.75**
Sum of two sides 1000 mm	130.52	165.18	4.36	144.23	nr	**309.41**
Sum of two sides 1100 mm	137.87	174.48	4.36	144.23	nr	**318.71**
Sum of two sides 1200 mm	144.54	182.93	4.36	144.23	nr	**327.16**
Offset						
Sum of two sides 600 mm	146.33	185.19	3.84	127.03	nr	**312.22**
Sum of two sides 700 mm	156.22	197.71	3.84	127.03	nr	**324.74**
Sum of two sides 800 mm	161.72	204.67	3.84	127.03	nr	**331.70**
Sum of two sides 900 mm	166.78	211.08	3.84	127.03	nr	**338.11**
Sum of two sides 1000 mm	173.75	219.90	4.36	144.23	nr	**364.13**
Sum of two sides 1100 mm	178.27	225.62	4.36	144.23	nr	**369.84**
Sum of two sides 1200 mm	181.78	230.06	4.36	144.23	nr	**374.28**
Square to round						
Sum of two sides 600 mm	110.40	139.72	2.66	87.99	nr	**227.71**
Sum of two sides 700 mm	118.10	149.47	2.66	87.99	nr	**237.46**
Sum of two sides 800 mm	125.71	159.09	2.80	92.62	nr	**251.72**
Sum of two sides 900 mm	133.41	168.85	2.80	92.62	nr	**261.47**
Sum of two sides 1000 mm	141.16	178.65	3.64	120.41	nr	**299.06**
Sum of two sides 1100 mm	149.04	188.62	3.64	120.41	nr	**309.03**
Sum of two sides 1200 mm	156.77	198.40	3.64	120.41	nr	**318.81**
90° radius bend						
Sum of two sides 600 mm	97.97	123.98	2.32	76.74	nr	**200.73**
Sum of two sides 700 mm	100.32	126.96	2.32	76.74	nr	**203.71**
Sum of two sides 800 mm	107.14	135.60	2.44	80.71	nr	**216.31**
Sum of two sides 900 mm	114.16	144.48	2.44	80.71	nr	**225.19**
Sum of two sides 1000 mm	118.42	149.87	2.80	92.62	nr	**242.49**
Sum of two sides 1100 mm	125.83	159.25	2.80	92.62	nr	**251.88**
Sum of two sides 1200 mm	132.74	168.00	2.80	92.62	nr	**260.62**
45° bend						
Sum of two sides 600 mm	109.20	138.20	2.78	91.96	nr	**230.16**
Sum of two sides 700 mm	114.39	144.77	2.78	91.96	nr	**236.73**
Sum of two sides 800 mm	121.82	154.18	2.92	96.59	nr	**250.77**

38 VENTILATION/AIR CONDITIONING SYSTEMS

Item	Net Price £	Material £	Labour hours	Labour £	Unit	Total rate £
Sum of two sides 900 mm	129.38	163.74	2.92	96.59	nr	**260.34**
Sum of two sides 1000 mm	135.56	171.56	3.76	124.38	nr	**295.94**
Sum of two sides 1100 mm	143.77	181.95	3.76	124.38	nr	**306.33**
Sum of two sides 1200 mm	151.33	191.52	3.76	124.38	nr	**315.90**
90° mitre bend						
Sum of two sides 600 mm	171.82	217.46	4.30	142.24	nr	**359.70**
Sum of two sides 700 mm	178.54	225.96	4.30	142.24	nr	**368.21**
Sum of two sides 800 mm	191.94	242.92	4.52	149.52	nr	**392.44**
Sum of two sides 900 mm	205.61	260.22	4.52	149.52	nr	**409.74**
Sum of two sides 1000 mm	216.07	273.46	6.06	200.46	nr	**473.92**
Sum of two sides 1100 mm	230.33	291.50	6.06	200.46	nr	**491.97**
Sum of two sides 1200 mm	244.23	309.10	6.06	200.46	nr	**509.56**
Branch						
Sum of two sides 600 mm	145.11	183.65	2.06	68.14	nr	**251.80**
Sum of two sides 700 mm	155.95	197.37	2.06	68.14	nr	**265.51**
Sum of two sides 800 mm	166.80	211.11	2.06	68.14	nr	**279.25**
Sum of two sides 900 mm	177.68	224.87	2.06	68.14	nr	**293.02**
Sum of two sides 1000 mm	188.56	238.64	2.58	85.35	nr	**323.99**
Sum of two sides 1100 mm	199.88	252.97	2.58	85.35	nr	**338.31**
Sum of two sides 1200 mm	210.74	266.71	2.58	85.35	nr	**352.05**
Grille neck						
Sum of two sides 600 mm	150.65	190.67	2.36	78.07	nr	**268.74**
Sum of two sides 700 mm	163.11	206.43	2.36	78.07	nr	**284.50**
Sum of two sides 800 mm	175.58	222.22	2.36	78.07	nr	**300.28**
Sum of two sides 900 mm	188.03	237.98	2.36	78.07	nr	**316.04**
Sum of two sides 1000 mm	200.51	253.76	2.88	95.27	nr	**349.03**
Sum of two sides 1100 mm	212.96	269.52	2.88	95.27	nr	**364.79**
Sum of two sides 1200 mm	225.44	285.31	2.88	95.27	nr	**380.58**
Ductwork 601 to 800 mm longest side						
Sum of two sides 900 mm	131.82	166.83	2.54	84.02	m	**250.85**
Sum of two sides 1000 mm	140.78	178.18	2.96	97.92	m	**276.09**
Sum of two sides 1100 mm	149.75	189.52	2.98	98.58	m	**288.10**
Sum of two sides 1200 mm	159.45	201.80	2.98	98.58	m	**300.38**
Sum of two sides 1300 mm	168.42	213.15	3.02	99.90	m	**313.05**
Sum of two sides 1400 mm	177.39	224.50	3.10	102.55	m	**327.05**
Sum of two sides 1500 mm	186.35	235.85	3.22	106.52	m	**342.37**
Sum of two sides 1600 mm	195.30	247.17	3.24	107.18	m	**354.35**
Extra over fittings; Ductwork 601 to 800 mm longest side						
End cap						
Sum of two sides 900 mm	60.25	76.25	0.76	25.14	nr	**101.39**
Sum of two sides 1000 mm	64.11	81.14	0.76	25.14	nr	**106.28**
Sum of two sides 1100 mm	67.98	86.03	0.76	25.14	nr	**111.17**
Sum of two sides 1200 mm	71.84	90.92	0.76	25.14	nr	**116.06**
Sum of two sides 1300 mm	75.73	95.84	0.76	25.14	nr	**120.98**
Sum of two sides 1400 mm	79.04	100.04	0.76	25.14	nr	**125.18**
Sum of two sides 1500 mm	91.77	116.14	0.76	25.14	nr	**141.28**
Sum of two sides 1600 mm	104.62	132.41	0.76	25.14	nr	**157.55**

38 VENTILATION/AIR CONDITIONING SYSTEMS

Item	Net Price £	Material £	Labour hours	Labour £	Unit	Total rate £
DUCTWORK: RECTANGULAR – CLASS C – cont						
Extra over fittings – cont						
Reducer						
Sum of two sides 900 mm	125.14	158.38	3.84	127.03	nr	**285.40**
Sum of two sides 1000 mm	131.84	166.85	4.36	144.23	nr	**311.08**
Sum of two sides 1100 mm	138.52	175.31	4.36	144.23	nr	**319.53**
Sum of two sides 1200 mm	145.85	184.58	4.36	144.23	nr	**328.81**
Sum of two sides 1300 mm	152.54	193.06	4.60	152.17	nr	**345.23**
Sum of two sides 1400 mm	158.15	200.16	4.60	152.17	nr	**352.32**
Sum of two sides 1500 mm	182.72	231.25	4.94	163.41	nr	**394.67**
Sum of two sides 1600 mm	207.27	262.32	4.94	163.41	nr	**425.73**
Offset						
Sum of two sides 900 mm	176.48	223.36	3.84	127.03	nr	**350.38**
Sum of two sides 1000 mm	180.94	228.99	4.36	144.23	nr	**373.22**
Sum of two sides 1100 mm	184.86	233.96	4.36	144.23	nr	**378.19**
Sum of two sides 1200 mm	188.62	238.72	4.36	144.23	nr	**382.95**
Sum of two sides 1300 mm	191.35	242.18	4.94	163.41	nr	**405.59**
Sum of two sides 1400 mm	195.22	247.07	4.94	163.41	nr	**410.48**
Sum of two sides 1500 mm	223.76	283.18	4.94	163.41	nr	**446.60**
Sum of two sides 1600 mm	251.73	318.59	4.94	163.41	nr	**482.00**
Square to round						
Sum of two sides 900 mm	132.24	167.36	2.80	92.62	nr	**259.98**
Sum of two sides 1000 mm	139.97	177.14	3.64	120.41	nr	**297.55**
Sum of two sides 1100 mm	147.69	186.92	3.64	120.41	nr	**307.33**
Sum of two sides 1200 mm	155.59	196.91	3.64	120.41	nr	**317.32**
Sum of two sides 1300 mm	163.32	206.69	4.64	153.49	nr	**360.18**
Sum of two sides 1400 mm	169.66	214.72	4.64	153.49	nr	**368.21**
Sum of two sides 1500 mm	199.69	252.73	5.12	169.37	nr	**422.09**
Sum of two sides 1600 mm	229.74	290.76	5.16	170.69	nr	**461.45**
90° radius bend						
Sum of two sides 900 mm	101.22	128.10	2.44	80.71	nr	**208.82**
Sum of two sides 1000 mm	107.92	136.58	2.80	92.62	nr	**229.21**
Sum of two sides 1100 mm	114.60	145.03	2.80	92.62	nr	**237.66**
Sum of two sides 1200 mm	121.67	153.99	2.80	92.62	nr	**246.61**
Sum of two sides 1300 mm	128.35	162.44	3.82	126.36	nr	**288.81**
Sum of two sides 1400 mm	130.31	164.91	3.82	126.36	nr	**291.28**
Sum of two sides 1500 mm	154.75	195.85	4.22	139.60	nr	**335.45**
Sum of two sides 1600 mm	179.19	226.79	4.22	139.60	nr	**366.38**
45° bend						
Sum of two sides 900 mm	124.80	157.95	2.92	96.59	nr	**254.54**
Sum of two sides 1000 mm	132.22	167.33	3.76	124.38	nr	**291.71**
Sum of two sides 1100 mm	139.65	176.74	3.76	124.38	nr	**301.12**
Sum of two sides 1200 mm	147.69	186.92	3.76	124.38	nr	**311.30**
Sum of two sides 1300 mm	155.11	196.30	4.52	149.52	nr	**345.82**
Sum of two sides 1400 mm	159.56	201.94	4.88	161.43	nr	**363.37**
Sum of two sides 1500 mm	184.74	233.80	4.88	161.43	nr	**395.23**
Sum of two sides 1600 mm	209.98	265.75	5.36	177.31	nr	**443.06**

38 VENTILATION/AIR CONDITIONING SYSTEMS

Item	Net Price £	Material £	Labour hours	Labour £	Unit	Total rate £
90° mitre bend						
Sum of two sides 900 mm	186.71	236.30	4.52	149.52	nr	**385.82**
Sum of two sides 1000 mm	201.24	254.69	6.06	200.46	nr	**455.16**
Sum of two sides 1100 mm	215.84	273.16	6.06	200.46	nr	**473.63**
Sum of two sides 1200 mm	230.69	291.95	6.06	200.46	nr	**492.42**
Sum of two sides 1300 mm	245.24	310.37	7.70	254.71	nr	**565.09**
Sum of two sides 1400 mm	253.81	321.22	7.70	254.71	nr	**575.93**
Sum of two sides 1500 mm	295.39	373.84	8.50	281.18	nr	**655.02**
Sum of two sides 1600 mm	336.92	426.41	8.52	281.84	nr	**708.25**
Branch						
Sum of two sides 900 mm	182.93	231.52	2.06	68.14	nr	**299.66**
Sum of two sides 1000 mm	194.08	245.63	2.58	85.35	nr	**330.98**
Sum of two sides 1100 mm	205.23	259.74	2.58	85.35	nr	**345.09**
Sum of two sides 1200 mm	216.89	274.49	2.58	85.35	nr	**359.84**
Sum of two sides 1300 mm	228.06	288.63	2.78	91.96	nr	**380.59**
Sum of two sides 1400 mm	237.76	300.91	2.78	91.96	nr	**392.87**
Sum of two sides 1500 mm	272.69	345.11	3.28	108.50	nr	**453.61**
Sum of two sides 1600 mm	307.65	389.36	3.28	108.50	nr	**497.86**
Grille neck						
Sum of two sides 900 mm	188.03	237.98	2.36	78.07	nr	**316.04**
Sum of two sides 1000 mm	200.51	253.76	2.88	95.27	nr	**349.03**
Sum of two sides 1100 mm	212.96	269.52	2.88	95.27	nr	**364.79**
Sum of two sides 1200 mm	225.44	285.31	2.88	95.27	nr	**380.58**
Sum of two sides 1300 mm	237.91	301.10	3.38	111.81	nr	**412.91**
Sum of two sides 1400 mm	248.70	314.76	3.38	111.81	nr	**426.57**
Sum of two sides 1500 mm	287.95	364.43	3.58	118.42	nr	**482.86**
Sum of two sides 1600 mm	437.51	553.72	3.58	118.42	nr	**672.14**
Ductwork 801 to 1000 mm longest side						
Sum of two sides 1100 mm	151.41	191.62	2.98	98.58	m	**290.20**
Sum of two sides 1200 mm	160.38	202.97	2.98	98.58	m	**301.55**
Sum of two sides 1300 mm	169.32	214.30	3.02	99.90	m	**314.20**
Sum of two sides 1400 mm	179.03	226.57	3.10	102.55	m	**329.12**
Sum of two sides 1500 mm	187.99	237.92	3.22	106.52	m	**344.44**
Sum of two sides 1600 mm	196.96	249.27	3.24	107.18	m	**356.45**
Sum of two sides 1700 mm	205.93	260.62	3.48	115.12	m	**375.74**
Sum of two sides 1800 mm	215.63	272.90	3.52	116.44	m	**389.34**
Sum of two sides 1900 mm	224.60	284.25	3.62	119.75	m	**404.00**
Sum of two sides 2000 mm	233.54	295.57	3.64	120.41	m	**415.98**
Extra over fittings; Ductwork 801 to 1000 mm longest side						
End cap						
Sum of two sides 1100 mm	67.96	86.01	0.76	25.14	nr	**111.15**
Sum of two sides 1200 mm	71.84	90.92	0.76	25.14	nr	**116.06**
Sum of two sides 1300 mm	75.73	95.84	0.76	25.14	nr	**120.98**
Sum of two sides 1400 mm	79.07	100.06	0.76	25.14	nr	**125.21**
Sum of two sides 1500 mm	91.81	116.20	0.76	25.14	nr	**141.34**
Sum of two sides 1600 mm	104.62	132.41	0.76	25.14	nr	**157.55**
Sum of two sides 1700 mm	117.41	148.60	1.16	38.37	nr	**186.97**

38 VENTILATION/AIR CONDITIONING SYSTEMS

Item	Net Price £	Material £	Labour hours	Labour £	Unit	Total rate £
DUCTWORK: RECTANGULAR – CLASS C – cont						
Extra over fittings – cont						
End cap – cont						
Sum of two sides 1800 mm	130.22	164.81	1.16	38.37	nr	**203.18**
Sum of two sides 1900 mm	143.01	180.99	1.16	38.37	nr	**219.37**
Sum of two sides 2000 mm	155.80	197.18	1.16	38.37	nr	**235.55**
Reducer						
Sum of two sides 1100 mm	128.16	162.20	4.36	144.23	nr	**306.43**
Sum of two sides 1200 mm	134.34	170.02	4.36	144.23	nr	**314.24**
Sum of two sides 1300 mm	140.51	177.83	4.60	152.17	nr	**330.00**
Sum of two sides 1400 mm	146.14	184.95	4.60	152.17	nr	**337.12**
Sum of two sides 1500 mm	170.16	215.36	4.94	163.41	nr	**378.77**
Sum of two sides 1600 mm	194.15	245.71	4.94	163.41	nr	**409.12**
Sum of two sides 1700 mm	218.19	276.14	5.18	171.35	nr	**447.49**
Sum of two sides 1800 mm	242.76	307.24	5.18	171.35	nr	**478.59**
Sum of two sides 1900 mm	266.76	337.62	5.42	179.29	nr	**516.91**
Sum of two sides 2000 mm	290.81	368.05	5.42	179.29	nr	**547.34**
Offset						
Sum of two sides 1100 mm	196.83	249.11	4.36	144.23	nr	**393.34**
Sum of two sides 1200 mm	200.70	254.00	4.36	144.23	nr	**398.23**
Sum of two sides 1300 mm	203.99	258.17	4.94	163.41	nr	**421.59**
Sum of two sides 1400 mm	205.51	260.09	4.94	163.41	nr	**423.50**
Sum of two sides 1500 mm	234.44	296.71	4.94	163.41	nr	**460.13**
Sum of two sides 1600 mm	262.75	332.54	4.94	163.41	nr	**495.95**
Sum of two sides 1700 mm	290.58	367.75	5.22	172.68	nr	**540.43**
Sum of two sides 1800 mm	318.02	402.49	5.22	172.68	nr	**575.17**
Sum of two sides 1900 mm	348.73	441.35	5.42	179.29	nr	**620.64**
Sum of two sides 2000 mm	374.93	474.52	5.42	179.29	nr	**653.81**
Square to round						
Sum of two sides 1100 mm	136.90	173.26	3.64	120.41	nr	**293.67**
Sum of two sides 1200 mm	144.06	182.32	3.64	120.41	nr	**302.73**
Sum of two sides 1300 mm	151.26	191.44	4.64	153.49	nr	**344.93**
Sum of two sides 1400 mm	157.14	198.88	4.64	153.49	nr	**352.37**
Sum of two sides 1500 mm	186.65	236.22	5.12	169.37	nr	**405.59**
Sum of two sides 1600 mm	216.17	273.59	5.16	170.69	nr	**444.28**
Sum of two sides 1700 mm	245.68	310.93	5.68	187.89	nr	**498.82**
Sum of two sides 1800 mm	275.25	348.35	5.68	187.89	nr	**536.24**
Sum of two sides 1900 mm	304.75	385.69	6.26	207.08	nr	**592.77**
Sum of two sides 2000 mm	334.28	423.06	6.26	207.08	nr	**630.14**
90° radius bend						
Sum of two sides 1100 mm	103.82	131.40	2.80	92.62	nr	**224.02**
Sum of two sides 1200 mm	112.06	141.82	2.80	92.62	nr	**234.44**
Sum of two sides 1300 mm	120.31	152.26	3.82	126.36	nr	**278.63**
Sum of two sides 1400 mm	127.70	161.62	3.82	126.36	nr	**287.98**
Sum of two sides 1500 mm	153.76	194.60	4.22	139.60	nr	**334.20**
Sum of two sides 1600 mm	179.84	227.61	4.22	139.60	nr	**367.21**
Sum of two sides 1700 mm	205.93	260.62	5.10	168.71	nr	**429.33**
Sum of two sides 1800 mm	232.28	293.97	5.10	168.71	nr	**462.68**
Sum of two sides 1900 mm	253.68	321.06	5.60	185.25	nr	**506.30**
Sum of two sides 2000 mm	279.74	354.04	5.60	185.25	nr	**539.29**

38 VENTILATION/AIR CONDITIONING SYSTEMS

Item	Net Price £	Material £	Labour hours	Labour £	Unit	Total rate £
45° bend						
Sum of two sides 1100 mm	134.65	170.42	3.76	124.38	nr	**294.79**
Sum of two sides 1200 mm	142.93	180.89	3.76	124.38	nr	**305.27**
Sum of two sides 1300 mm	151.24	191.41	4.52	149.52	nr	**340.93**
Sum of two sides 1400 mm	159.01	201.25	4.88	161.43	nr	**362.67**
Sum of two sides 1500 mm	185.16	234.33	5.36	177.31	nr	**411.64**
Sum of two sides 1600 mm	211.28	267.40	5.38	177.97	nr	**445.37**
Sum of two sides 1700 mm	237.43	300.49	5.92	195.83	nr	**496.32**
Sum of two sides 1800 mm	264.12	334.27	5.92	195.83	nr	**530.10**
Sum of two sides 1900 mm	287.85	364.30	6.52	215.68	nr	**579.98**
Sum of two sides 2000 mm	314.01	397.41	6.52	215.68	nr	**613.09**
90° mitre bend						
Sum of two sides 1100 mm	214.87	271.94	6.06	200.46	nr	**472.40**
Sum of two sides 1200 mm	230.45	291.66	6.06	200.46	nr	**492.13**
Sum of two sides 1300 mm	246.04	311.38	7.70	254.71	nr	**566.10**
Sum of two sides 1400 mm	260.17	329.27	7.70	254.71	nr	**583.98**
Sum of two sides 1500 mm	302.53	382.88	8.50	281.18	nr	**664.05**
Sum of two sides 1600 mm	344.88	436.48	8.52	281.84	nr	**718.32**
Sum of two sides 1700 mm	387.26	490.12	9.36	309.63	nr	**799.74**
Sum of two sides 1800 mm	429.81	543.96	9.36	309.63	nr	**853.59**
Sum of two sides 1900 mm	467.21	591.30	9.74	322.20	nr	**913.49**
Sum of two sides 2000 mm	509.82	645.22	9.74	322.20	nr	**967.42**
Branch						
Sum of two sides 1100 mm	205.72	260.35	2.58	85.35	nr	**345.70**
Sum of two sides 1200 mm	216.89	274.49	2.58	85.35	nr	**359.84**
Sum of two sides 1300 mm	228.06	288.63	2.78	91.96	nr	**380.59**
Sum of two sides 1400 mm	238.20	301.47	2.78	91.96	nr	**393.43**
Sum of two sides 1500 mm	273.19	345.75	3.28	108.50	nr	**454.25**
Sum of two sides 1600 mm	308.11	389.95	3.28	108.50	nr	**498.45**
Sum of two sides 1700 mm	343.08	434.20	3.38	111.81	nr	**546.01**
Sum of two sides 1800 mm	378.53	479.06	3.38	111.81	nr	**590.87**
Sum of two sides 1900 mm	413.47	523.29	3.70	122.39	nr	**645.68**
Sum of two sides 2000 mm	448.41	567.51	3.70	122.39	nr	**689.91**
Grille neck						
Sum of two sides 1100 mm	212.96	269.52	2.88	95.27	nr	**364.79**
Sum of two sides 1200 mm	225.44	285.31	2.88	95.27	nr	**380.58**
Sum of two sides 1300 mm	237.91	301.10	3.38	111.81	nr	**412.91**
Sum of two sides 1400 mm	248.70	314.76	3.38	111.81	nr	**426.57**
Sum of two sides 1500 mm	287.95	364.43	3.58	118.42	nr	**482.86**
Sum of two sides 1600 mm	327.18	414.08	3.58	118.42	nr	**532.50**
Sum of two sides 1700 mm	366.43	463.75	3.72	123.06	nr	**586.81**
Sum of two sides 1800 mm	405.66	513.40	3.72	123.06	nr	**636.46**
Sum of two sides 1900 mm	444.91	563.07	4.04	133.64	nr	**696.71**
Sum of two sides 2000 mm	484.13	612.72	4.04	133.64	nr	**746.36**
Ductwork 1001 to 1250 mm longest side						
Sum of two sides 1300 mm	242.40	306.79	3.02	99.90	m	**406.69**
Sum of two sides 1400 mm	253.95	321.40	3.10	102.55	m	**423.95**
Sum of two sides 1500 mm	265.02	335.41	3.22	106.52	m	**441.93**

38 VENTILATION/AIR CONDITIONING SYSTEMS

Item	Net Price £	Material £	Labour hours	Labour £	Unit	Total rate £
DUCTWORK: RECTANGULAR – CLASS C – cont						
Ductwork 1001 to 1250 mm longest side – cont						
Sum of two sides 1600 mm	276.82	350.35	3.24	107.18	m	**457.52**
Sum of two sides 1700 mm	287.89	364.35	3.48	115.12	m	**479.47**
Sum of two sides 1800 mm	298.96	378.36	3.52	116.44	m	**494.80**
Sum of two sides 1900 mm	310.02	392.37	3.62	119.75	m	**512.11**
Sum of two sides 2000 mm	321.83	407.30	3.64	120.41	m	**527.71**
Sum of two sides 2100 mm	332.89	421.31	5.06	167.38	m	**588.69**
Sum of two sides 2200 mm	380.25	481.24	5.10	168.71	m	**649.95**
Sum of two sides 2300 mm	392.66	496.95	5.12	169.37	m	**666.32**
Sum of two sides 2400 mm	403.73	510.95	5.52	182.60	m	**693.55**
Sum of two sides 2500 mm	415.53	525.89	5.54	183.26	m	**709.15**
Extra over fittings; Ductwork 1001 to 1250 mm longest side						
End cap						
Sum of two sides 1300 mm	77.26	97.78	0.76	25.14	nr	**122.92**
Sum of two sides 1400 mm	80.72	102.16	0.76	25.14	nr	**127.30**
Sum of two sides 1500 mm	93.62	118.48	0.76	25.14	nr	**143.62**
Sum of two sides 1600 mm	106.55	134.85	0.76	25.14	nr	**160.00**
Sum of two sides 1700 mm	119.45	151.17	1.16	38.37	nr	**189.55**
Sum of two sides 1800 mm	132.36	167.52	1.16	38.37	nr	**205.89**
Sum of two sides 1900 mm	145.26	183.84	1.16	38.37	nr	**222.21**
Sum of two sides 2000 mm	158.19	200.21	1.16	38.37	nr	**238.58**
Sum of two sides 2100 mm	171.09	216.53	1.74	57.56	nr	**274.09**
Sum of two sides 2200 mm	184.02	232.90	1.74	57.56	nr	**290.46**
Sum of two sides 2300 mm	196.94	249.24	1.74	57.56	nr	**306.80**
Sum of two sides 2400 mm	209.83	265.56	1.74	57.56	nr	**323.12**
Sum of two sides 2500 mm	222.75	281.91	1.74	57.56	nr	**339.47**
Reducer						
Sum of two sides 1300 mm	109.62	138.74	4.60	152.17	nr	**290.90**
Sum of two sides 1400 mm	113.99	144.26	4.60	152.17	nr	**296.43**
Sum of two sides 1500 mm	137.17	173.60	4.94	163.41	nr	**337.02**
Sum of two sides 1600 mm	160.94	203.69	4.94	163.41	nr	**367.10**
Sum of two sides 1700 mm	184.17	233.09	5.18	171.35	nr	**404.44**
Sum of two sides 1800 mm	207.42	262.51	5.18	171.35	nr	**433.86**
Sum of two sides 1900 mm	230.62	291.88	5.42	179.29	nr	**471.17**
Sum of two sides 2000 mm	254.37	321.93	5.42	179.29	nr	**501.23**
Sum of two sides 2100 mm	277.62	351.36	5.84	193.18	nr	**544.54**
Sum of two sides 2200 mm	281.51	356.27	5.84	193.18	nr	**549.46**
Sum of two sides 2300 mm	305.28	386.36	5.84	193.18	nr	**579.54**
Sum of two sides 2400 mm	328.52	415.78	6.24	206.42	nr	**622.20**
Sum of two sides 2500 mm	352.25	445.81	6.24	206.42	nr	**652.23**
Offset						
Sum of two sides 1300 mm	261.14	330.49	4.94	163.41	nr	**493.91**
Sum of two sides 1400 mm	268.91	340.33	4.94	163.41	nr	**503.74**
Sum of two sides 1500 mm	298.94	378.33	4.94	163.41	nr	**541.75**

38 VENTILATION/AIR CONDITIONING SYSTEMS

Item	Net Price £	Material £	Labour hours	Labour £	Unit	Total rate £
Sum of two sides 1600 mm	328.65	415.94	4.94	163.41	nr	**579.35**
Sum of two sides 1700 mm	357.06	451.90	5.22	172.68	nr	**624.57**
Sum of two sides 1800 mm	384.76	486.95	5.22	172.68	nr	**659.63**
Sum of two sides 1900 mm	411.68	521.03	5.42	179.29	nr	**700.32**
Sum of two sides 2000 mm	438.08	554.44	5.42	179.29	nr	**733.73**
Sum of two sides 2100 mm	463.41	586.49	5.84	193.18	nr	**779.67**
Sum of two sides 2200 mm	456.71	578.01	6.52	215.68	nr	**793.69**
Sum of two sides 2300 mm	495.45	627.05	6.52	215.68	nr	**842.72**
Sum of two sides 2400 mm	534.05	675.89	6.94	229.57	nr	**905.47**
Sum of two sides 2500 mm	572.78	724.90	6.96	230.23	nr	**955.14**
Square to round						
Sum of two sides 1300 mm	119.15	150.80	4.64	153.49	nr	**304.29**
Sum of two sides 1400 mm	124.26	157.26	4.64	153.49	nr	**310.75**
Sum of two sides 1500 mm	152.94	193.56	5.12	169.37	nr	**362.93**
Sum of two sides 1600 mm	181.71	229.98	5.16	170.69	nr	**400.67**
Sum of two sides 1700 mm	210.44	266.33	5.68	187.89	nr	**454.23**
Sum of two sides 1800 mm	239.13	302.64	5.68	187.89	nr	**490.53**
Sum of two sides 1900 mm	267.83	338.97	6.26	207.08	nr	**546.05**
Sum of two sides 2000 mm	296.60	375.38	6.26	207.08	nr	**582.46**
Sum of two sides 2100 mm	325.29	411.69	8.52	281.84	nr	**693.53**
Sum of two sides 2200 mm	334.68	423.57	8.54	282.50	nr	**706.07**
Sum of two sides 2300 mm	363.47	460.01	8.60	284.48	nr	**744.49**
Sum of two sides 2400 mm	392.20	496.36	9.54	315.58	nr	**811.94**
Sum of two sides 2500 mm	420.97	532.77	9.58	316.90	nr	**849.68**
90° radius bend						
Sum of two sides 1300 mm	48.70	61.63	3.82	126.36	nr	**188.00**
Sum of two sides 1400 mm	47.84	60.54	3.82	126.36	nr	**186.91**
Sum of two sides 1500 mm	73.48	93.00	4.22	139.60	nr	**232.59**
Sum of two sides 1600 mm	99.23	125.58	4.22	139.60	nr	**265.18**
Sum of two sides 1700 mm	124.80	157.95	5.10	168.71	nr	**326.66**
Sum of two sides 1800 mm	150.38	190.32	5.10	168.71	nr	**359.03**
Sum of two sides 1900 mm	175.94	222.67	5.60	185.25	nr	**407.91**
Sum of two sides 2000 mm	201.75	255.33	5.60	185.25	nr	**440.58**
Sum of two sides 2100 mm	227.28	287.65	5.60	185.25	nr	**472.90**
Sum of two sides 2200 mm	215.57	272.82	5.60	185.25	nr	**458.06**
Sum of two sides 2300 mm	222.41	281.48	8.70	287.79	nr	**569.28**
Sum of two sides 2400 mm	247.95	313.80	8.70	287.79	nr	**601.59**
Sum of two sides 2500 mm	273.55	346.20	8.70	287.79	nr	**633.99**
45° bend						
Sum of two sides 1300 mm	113.69	143.89	4.52	149.52	nr	**293.41**
Sum of two sides 1400 mm	116.78	147.80	4.88	161.43	nr	**309.23**
Sum of two sides 1500 mm	142.65	180.54	5.36	177.31	nr	**357.85**
Sum of two sides 1600 mm	169.09	214.00	5.38	177.97	nr	**391.97**
Sum of two sides 1700 mm	194.96	246.75	5.92	195.83	nr	**442.58**
Sum of two sides 1800 mm	220.84	279.49	5.92	195.83	nr	**475.32**
Sum of two sides 1900 mm	246.75	312.29	6.52	215.68	nr	**527.97**
Sum of two sides 2000 mm	273.17	345.72	6.52	215.68	nr	**561.40**
Sum of two sides 2100 mm	299.04	378.47	15.00	496.19	nr	**874.66**
Sum of two sides 2200 mm	304.67	385.59	15.00	496.19	nr	**881.78**
Sum of two sides 2300 mm	321.34	406.69	15.10	499.50	nr	**906.19**
Sum of two sides 2400 mm	347.24	439.46	16.26	537.87	nr	**977.34**
Sum of two sides 2500 mm	373.57	472.79	16.60	549.12	nr	**1021.91**

38 VENTILATION/AIR CONDITIONING SYSTEMS

Item	Net Price £	Material £	Labour hours	Labour £	Unit	Total rate £
DUCTWORK: RECTANGULAR – CLASS C – cont						
Extra over fittings – cont						
90° mitre bend						
Sum of two sides 1300 mm	259.67	328.63	7.70	254.71	nr	**583.35**
Sum of two sides 1400 mm	270.54	342.40	7.70	254.71	nr	**597.11**
Sum of two sides 1500 mm	320.00	404.99	8.50	281.18	nr	**686.17**
Sum of two sides 1600 mm	369.43	467.55	8.52	281.84	nr	**749.39**
Sum of two sides 1700 mm	418.87	530.12	9.36	309.63	nr	**839.74**
Sum of two sides 1800 mm	468.30	592.68	9.36	309.63	nr	**902.31**
Sum of two sides 1900 mm	517.71	655.22	9.74	322.20	nr	**977.41**
Sum of two sides 2000 mm	567.19	717.83	9.74	322.20	nr	**1040.03**
Sum of two sides 2100 mm	616.60	780.37	15.00	496.19	nr	**1276.57**
Sum of two sides 2200 mm	622.57	787.92	15.00	496.19	nr	**1284.11**
Sum of two sides 2300 mm	650.16	822.84	15.10	499.50	nr	**1322.34**
Sum of two sides 2400 mm	699.91	885.80	16.26	537.87	nr	**1423.68**
Sum of two sides 2500 mm	749.66	948.77	16.60	549.12	nr	**1497.89**
Branch						
Sum of two sides 1300 mm	233.92	296.05	2.78	91.96	nr	**388.01**
Sum of two sides 1400 mm	243.96	308.75	2.78	91.96	nr	**400.71**
Sum of two sides 1500 mm	279.28	353.46	3.28	108.50	nr	**461.96**
Sum of two sides 1600 mm	315.08	398.77	3.28	108.50	nr	**507.27**
Sum of two sides 1700 mm	350.39	443.45	3.38	111.81	nr	**555.26**
Sum of two sides 1800 mm	385.69	488.12	3.38	111.81	nr	**599.93**
Sum of two sides 1900 mm	421.01	532.83	3.70	122.39	nr	**655.22**
Sum of two sides 2000 mm	456.79	578.12	3.70	122.39	nr	**700.51**
Sum of two sides 2100 mm	492.11	622.82	5.22	172.68	nr	**795.50**
Sum of two sides 2200 mm	527.37	667.44	5.22	172.68	nr	**840.12**
Sum of two sides 2300 mm	563.20	712.78	5.22	172.68	nr	**885.46**
Sum of two sides 2400 mm	598.50	757.46	5.76	190.54	nr	**948.00**
Sum of two sides 2500 mm	634.31	802.78	5.76	190.54	nr	**993.32**
Grille neck						
Sum of two sides 1300 mm	245.64	310.88	3.38	111.81	nr	**422.69**
Sum of two sides 1400 mm	257.02	325.28	3.38	111.81	nr	**437.09**
Sum of two sides 1500 mm	296.88	375.73	3.58	118.42	nr	**494.15**
Sum of two sides 1600 mm	336.69	426.12	3.58	118.42	nr	**544.54**
Sum of two sides 1700 mm	376.51	476.51	3.72	123.06	nr	**599.57**
Sum of two sides 1800 mm	416.35	526.93	3.72	123.06	nr	**649.98**
Sum of two sides 1900 mm	456.20	577.37	4.04	133.64	nr	**711.01**
Sum of two sides 2000 mm	496.02	627.76	4.04	133.64	nr	**761.40**
Sum of two sides 2100 mm	535.86	678.18	5.22	172.68	nr	**850.86**
Sum of two sides 2200 mm	575.69	728.60	5.60	185.25	nr	**913.84**
Sum of two sides 2300 mm	615.53	779.02	5.60	185.25	nr	**964.26**
Sum of two sides 2400 mm	655.43	829.51	6.12	202.45	nr	**1031.96**
Sum of two sides 2500 mm	695.18	879.82	6.12	202.45	nr	**1082.27**
Ductwork 1251 to 1600 mm longest side						
Sum of two sides 1700 mm	308.22	390.08	3.48	115.12	m	**505.20**
Sum of two sides 1800 mm	320.21	405.26	3.52	116.44	m	**521.70**
Sum of two sides 1900 mm	332.60	420.94	3.62	119.75	m	**540.68**

38 VENTILATION/AIR CONDITIONING SYSTEMS

Item	Net Price £	Material £	Labour hours	Labour £	Unit	Total rate £
Sum of two sides 2000 mm	344.25	435.69	3.64	120.41	m	**556.10**
Sum of two sides 2100 mm	355.91	450.44	5.06	167.38	m	**617.82**
Sum of two sides 2200 mm	367.56	465.19	5.10	168.71	m	**633.89**
Sum of two sides 2300 mm	379.95	480.87	5.12	169.37	m	**650.24**
Sum of two sides 2400 mm	391.97	496.07	5.52	182.60	m	**678.67**
Sum of two sides 2500 mm	403.62	510.82	5.54	183.26	m	**694.08**
Sum of two sides 2600 mm	452.51	572.69	5.94	196.49	m	**769.19**
Sum of two sides 2700 mm	464.16	587.44	5.98	197.82	m	**785.26**
Sum of two sides 2800 mm	476.43	602.97	6.60	218.33	m	**821.29**
Sum of two sides 2900 mm	488.82	618.65	6.62	218.99	m	**837.63**
Sum of two sides 3000 mm	500.47	633.40	7.06	233.54	m	**866.94**
Sum of two sides 3100 mm	512.57	648.71	7.10	234.87	m	**883.57**
Sum of two sides 3200 mm	524.22	663.46	7.12	235.53	m	**898.98**
Extra over fittings; Ductwork 1251 to 1600 mm longest side						
End cap						
Sum of two sides 1700 mm	119.45	151.17	1.16	38.37	nr	**189.55**
Sum of two sides 1800 mm	132.74	168.00	1.16	38.37	nr	**206.37**
Sum of two sides 1900 mm	145.26	183.84	1.16	38.37	nr	**222.21**
Sum of two sides 2000 mm	171.09	216.53	1.16	38.37	nr	**254.90**
Sum of two sides 2100 mm	184.02	232.90	1.74	57.56	nr	**290.46**
Sum of two sides 2200 mm	196.94	249.24	1.74	57.56	nr	**306.80**
Sum of two sides 2300 mm	209.83	265.56	1.74	57.56	nr	**323.12**
Sum of two sides 2400 mm	222.75	281.91	1.74	57.56	nr	**339.47**
Sum of two sides 2500 mm	235.70	298.31	1.74	57.56	nr	**355.87**
Sum of two sides 2600 mm	248.58	314.60	1.74	57.56	nr	**372.16**
Sum of two sides 2700 mm	261.51	330.97	1.74	57.56	nr	**388.53**
Sum of two sides 2800 mm	274.43	347.32	2.32	76.74	nr	**424.06**
Sum of two sides 2900 mm	287.34	363.66	2.32	76.74	nr	**440.41**
Sum of two sides 3000 mm	444.91	563.07	3.46	114.46	nr	**677.53**
Sum of two sides 3100 mm	297.61	376.66	3.46	114.46	nr	**491.11**
Sum of two sides 3200 mm	306.92	388.43	3.46	114.46	nr	**502.89**
Reducer						
Sum of two sides 1700 mm	161.01	203.77	5.18	171.35	nr	**375.12**
Sum of two sides 1800 mm	183.02	231.62	5.18	171.35	nr	**402.98**
Sum of two sides 1900 mm	205.40	259.96	5.42	179.29	nr	**439.25**
Sum of two sides 2000 mm	158.19	200.21	5.42	179.29	nr	**379.50**
Sum of two sides 2100 mm	227.37	287.76	5.84	193.18	nr	**480.94**
Sum of two sides 2200 mm	249.40	315.64	5.84	193.18	nr	**508.82**
Sum of two sides 2300 mm	271.38	343.46	5.84	193.18	nr	**536.65**
Sum of two sides 2400 mm	293.77	371.79	6.24	206.42	nr	**578.21**
Sum of two sides 2500 mm	315.82	399.70	6.24	206.42	nr	**606.12**
Sum of two sides 2600 mm	337.81	427.53	6.32	209.06	nr	**636.59**
Sum of two sides 2700 mm	333.52	422.11	6.32	209.06	nr	**631.17**
Sum of two sides 2800 mm	355.53	449.96	8.00	264.64	nr	**714.60**
Sum of two sides 2900 mm	377.41	477.65	8.02	265.30	nr	**742.95**
Sum of two sides 3000 mm	422.90	535.22	9.12	301.69	nr	**836.91**
Sum of two sides 3100 mm	461.12	583.59	9.12	301.69	nr	**885.28**
Sum of two sides 3200 mm	475.92	602.33	9.12	301.69	nr	**904.01**

38 VENTILATION/AIR CONDITIONING SYSTEMS

Item	Net Price £	Material £	Labour hours	Labour £	Unit	Total rate £
DUCTWORK: RECTANGULAR – CLASS C – cont						
Extra over fittings – cont						
Offset						
Sum of two sides 1700 mm	396.31	501.57	5.22	172.68	nr	**674.25**
Sum of two sides 1800 mm	437.66	553.90	5.22	172.68	nr	**726.58**
Sum of two sides 1900 mm	465.02	588.53	5.42	179.29	nr	**767.83**
Sum of two sides 2000 mm	491.04	621.46	5.42	179.29	nr	**800.76**
Sum of two sides 2100 mm	516.26	653.38	5.84	193.18	nr	**846.57**
Sum of two sides 2200 mm	540.67	684.27	6.52	215.68	nr	**899.95**
Sum of two sides 2300 mm	564.40	714.30	6.52	215.68	nr	**929.98**
Sum of two sides 2400 mm	603.69	764.03	6.94	229.57	nr	**993.60**
Sum of two sides 2500 mm	625.93	792.17	6.96	230.23	nr	**1022.41**
Sum of two sides 2600 mm	625.67	791.85	6.98	230.90	nr	**1022.75**
Sum of two sides 2700 mm	664.50	840.99	7.00	231.56	nr	**1072.55**
Sum of two sides 2800 mm	703.19	889.95	8.66	286.47	nr	**1176.42**
Sum of two sides 2900 mm	765.26	968.51	9.48	313.59	nr	**1282.11**
Sum of two sides 3000 mm	804.11	1017.68	10.62	351.31	nr	**1368.99**
Sum of two sides 3100 mm	834.27	1055.85	10.68	353.29	nr	**1409.14**
Sum of two sides 3200 mm	862.32	1091.36	10.70	353.95	nr	**1445.31**
Square to round						
Sum of two sides 1700 mm	176.48	223.36	5.68	187.89	nr	**411.25**
Sum of two sides 1800 mm	203.97	258.15	5.68	187.89	nr	**446.04**
Sum of two sides 1900 mm	231.32	292.75	6.26	207.08	nr	**499.83**
Sum of two sides 2000 mm	258.74	327.46	6.26	207.08	nr	**534.54**
Sum of two sides 2100 mm	286.13	362.12	8.52	281.84	nr	**643.96**
Sum of two sides 2200 mm	313.55	396.83	8.54	282.50	nr	**679.33**
Sum of two sides 2300 mm	340.89	431.43	8.60	284.48	nr	**715.92**
Sum of two sides 2400 mm	368.38	466.22	9.54	315.58	nr	**781.80**
Sum of two sides 2500 mm	395.81	500.93	9.58	316.90	nr	**817.84**
Sum of two sides 2600 mm	396.98	502.42	9.90	327.49	nr	**829.91**
Sum of two sides 2700 mm	424.43	537.16	9.90	327.49	nr	**864.65**
Sum of two sides 2800 mm	451.77	571.76	16.98	561.69	nr	**1133.46**
Sum of two sides 2900 mm	490.62	620.93	17.76	587.49	nr	**1208.43**
Sum of two sides 3000 mm	518.09	655.70	18.04	596.76	nr	**1252.45**
Sum of two sides 3100 mm	538.38	681.37	18.04	596.76	nr	**1278.13**
Sum of two sides 3200 mm	556.79	704.68	18.18	601.39	nr	**1306.07**
90° radius bend						
Sum of two sides 1700 mm	399.90	506.12	5.10	168.71	nr	**674.82**
Sum of two sides 1800 mm	432.87	547.84	5.10	168.71	nr	**716.55**
Sum of two sides 1900 mm	485.08	613.92	5.60	185.25	nr	**799.16**
Sum of two sides 2000 mm	536.24	678.66	5.60	185.25	nr	**863.90**
Sum of two sides 2100 mm	587.41	743.43	5.22	172.68	nr	**916.10**
Sum of two sides 2200 mm	638.59	808.20	5.24	173.34	nr	**981.54**
Sum of two sides 2300 mm	690.73	874.19	5.26	174.00	nr	**1048.19**
Sum of two sides 2400 mm	722.34	914.19	8.68	287.13	nr	**1201.32**
Sum of two sides 2500 mm	773.41	978.83	8.70	287.79	nr	**1266.62**
Sum of two sides 2600 mm	777.40	983.88	9.06	299.70	nr	**1283.58**
Sum of two sides 2700 mm	828.47	1048.51	9.06	299.70	nr	**1348.21**

38 VENTILATION/AIR CONDITIONING SYSTEMS

Item	Net Price £	Material £	Labour hours	Labour £	Unit	Total rate £
Sum of two sides 2800 mm	856.09	1083.46	14.26	471.72	nr	**1555.18**
Sum of two sides 2900 mm	954.14	1207.55	14.34	474.36	nr	**1681.92**
Sum of two sides 3000 mm	1005.12	1272.08	14.52	480.32	nr	**1752.40**
Sum of two sides 3100 mm	1044.60	1322.05	14.52	480.32	nr	**1802.37**
Sum of two sides 3200 mm	1081.21	1368.37	14.62	483.62	nr	**1852.00**
45° bend						
Sum of two sides 1700 mm	192.38	243.48	5.92	195.83	nr	**439.31**
Sum of two sides 1800 mm	208.61	264.02	5.92	195.83	nr	**459.85**
Sum of two sides 1900 mm	234.68	297.00	6.52	215.68	nr	**512.68**
Sum of two sides 2000 mm	260.30	329.43	6.52	215.68	nr	**545.11**
Sum of two sides 2100 mm	285.85	361.77	15.00	496.19	nr	**857.97**
Sum of two sides 2200 mm	311.45	394.17	15.00	496.19	nr	**890.37**
Sum of two sides 2300 mm	337.53	427.18	15.10	499.50	nr	**926.68**
Sum of two sides 2400 mm	353.05	446.82	16.26	537.87	nr	**984.70**
Sum of two sides 2500 mm	378.59	479.14	16.60	549.12	nr	**1028.26**
Sum of two sides 2600 mm	310.15	392.52	17.12	566.32	nr	**958.85**
Sum of two sides 2700 mm	404.52	511.96	17.24	570.29	nr	**1082.26**
Sum of two sides 2800 mm	418.05	529.08	18.18	601.39	nr	**1130.47**
Sum of two sides 2900 mm	467.04	591.09	18.18	601.39	nr	**1192.47**
Sum of two sides 3000 mm	492.58	623.40	19.24	636.45	nr	**1259.86**
Sum of two sides 3100 mm	512.25	648.31	19.24	636.45	nr	**1284.76**
Sum of two sides 3200 mm	530.57	671.48	19.24	636.45	nr	**1307.94**
90° mitre bend						
Sum of two sides 1700 mm	420.82	532.59	9.36	309.63	nr	**842.21**
Sum of two sides 1800 mm	446.90	565.60	9.36	309.63	nr	**875.22**
Sum of two sides 1900 mm	497.41	629.52	9.74	322.20	nr	**951.71**
Sum of two sides 2000 mm	548.04	693.60	9.74	322.20	nr	**1015.79**
Sum of two sides 2100 mm	598.69	757.70	15.00	496.19	nr	**1253.90**
Sum of two sides 2200 mm	649.34	821.81	15.00	496.19	nr	**1318.00**
Sum of two sides 2300 mm	699.83	885.70	15.10	499.50	nr	**1385.20**
Sum of two sides 2400 mm	726.12	918.97	16.26	537.87	nr	**1456.85**
Sum of two sides 2500 mm	777.04	983.42	16.60	549.12	nr	**1532.55**
Sum of two sides 2600 mm	772.17	977.26	17.12	566.32	nr	**1543.58**
Sum of two sides 2700 mm	823.07	1041.68	17.24	570.29	nr	**1611.98**
Sum of two sides 2800 mm	845.52	1070.09	30.40	1005.62	nr	**2075.71**
Sum of two sides 2900 mm	936.12	1184.75	30.40	1005.62	nr	**2190.37**
Sum of two sides 3000 mm	987.29	1249.52	31.20	1032.08	nr	**2281.60**
Sum of two sides 3100 mm	1029.46	1302.89	32.08	1061.19	nr	**2364.08**
Sum of two sides 3200 mm	1069.85	1354.00	32.08	1061.19	nr	**2415.19**
Branch						
Sum of two sides 1700 mm	361.14	457.05	3.38	111.81	nr	**568.86**
Sum of two sides 1800 mm	396.46	501.76	3.38	111.81	nr	**613.57**
Sum of two sides 1900 mm	432.33	547.15	3.70	122.39	nr	**669.55**
Sum of two sides 2000 mm	467.67	591.88	3.70	122.39	nr	**714.28**
Sum of two sides 2100 mm	503.06	636.67	5.22	172.68	nr	**809.34**
Sum of two sides 2200 mm	538.46	681.48	5.22	172.68	nr	**854.15**
Sum of two sides 2300 mm	574.31	726.84	5.22	172.68	nr	**899.52**
Sum of two sides 2400 mm	609.59	771.49	5.76	190.54	nr	**962.03**
Sum of two sides 2500 mm	644.99	816.30	5.76	190.54	nr	**1006.84**
Sum of two sides 2600 mm	680.36	861.06	5.76	190.54	nr	**1051.60**
Sum of two sides 2700 mm	715.74	905.84	5.76	190.54	nr	**1096.38**

38 VENTILATION/AIR CONDITIONING SYSTEMS

Item	Net Price £	Material £	Labour hours	Labour £	Unit	Total rate £
DUCTWORK: RECTANGULAR – CLASS C – cont						
Extra over fittings – cont						
Branch – cont						
Sum of two sides 2800 mm	751.02	950.49	7.88	260.67	nr	**1211.16**
Sum of two sides 2900 mm	798.42	1010.48	7.88	260.67	nr	**1271.15**
Sum of two sides 3000 mm	833.81	1055.26	9.66	319.55	nr	**1374.81**
Sum of two sides 3100 mm	862.09	1091.06	9.66	319.55	nr	**1410.61**
Sum of two sides 3200 mm	887.86	1123.67	9.66	319.55	nr	**1443.22**
Grille neck						
Sum of two sides 1700 mm	376.51	476.51	3.72	123.06	nr	**599.57**
Sum of two sides 1800 mm	416.35	526.93	3.72	123.06	nr	**649.98**
Sum of two sides 1900 mm	456.20	577.37	4.04	133.64	nr	**711.01**
Sum of two sides 2000 mm	496.02	627.76	4.04	133.64	nr	**761.40**
Sum of two sides 2100 mm	535.86	678.18	5.60	185.25	nr	**863.43**
Sum of two sides 2200 mm	575.69	728.60	5.60	185.25	nr	**913.84**
Sum of two sides 2300 mm	615.53	779.02	5.60	185.25	nr	**964.26**
Sum of two sides 2400 mm	655.35	829.41	6.12	202.45	nr	**1031.85**
Sum of two sides 2500 mm	695.18	879.82	6.12	202.45	nr	**1082.27**
Sum of two sides 2600 mm	735.00	930.22	6.16	203.77	nr	**1133.99**
Sum of two sides 2700 mm	774.86	980.66	6.16	203.77	nr	**1184.43**
Sum of two sides 2800 mm	814.65	1031.02	8.26	273.24	nr	**1304.26**
Sum of two sides 2900 mm	854.51	1081.47	8.26	273.24	nr	**1354.71**
Sum of two sides 3000 mm	894.35	1131.89	10.04	332.12	nr	**1464.01**
Sum of two sides 3100 mm	926.23	1172.23	10.04	332.12	nr	**1504.35**
Sum of two sides 3200 mm	955.23	1208.94	10.04	332.12	nr	**1541.05**
Ductwork 1601 to 2000 mm longest side						
Sum of two sides 2100 mm	412.78	522.41	5.06	167.38	m	**689.79**
Sum of two sides 2200 mm	426.11	539.29	5.10	168.71	m	**707.99**
Sum of two sides 2300 mm	439.82	556.64	5.10	168.71	m	**725.35**
Sum of two sides 2400 mm	452.74	572.99	5.12	169.37	m	**742.35**
Sum of two sides 2500 mm	466.45	590.34	5.52	182.60	m	**772.94**
Sum of two sides 2600 mm	479.43	606.77	5.54	183.26	m	**790.03**
Sum of two sides 2700 mm	492.41	623.19	5.94	196.49	m	**819.68**
Sum of two sides 2800 mm	505.39	639.62	5.98	197.82	m	**837.43**
Sum of two sides 2900 mm	519.10	656.97	6.60	218.33	m	**875.30**
Sum of two sides 3000 mm	532.08	673.40	6.62	218.99	m	**892.38**
Sum of two sides 3100 mm	611.04	773.33	7.06	233.54	m	**1006.87**
Sum of two sides 3200 mm	624.02	789.75	7.06	233.54	m	**1023.30**
Sum of two sides 3300 mm	637.73	807.11	7.06	233.54	m	**1040.65**
Sum of two sides 3400 mm	650.71	823.53	7.10	234.87	m	**1058.40**
Sum of two sides 3500 mm	663.68	839.96	7.10	234.87	m	**1074.82**
Sum of two sides 3600 mm	676.66	856.38	7.10	234.87	m	**1091.25**
Sum of two sides 3700 mm	690.38	873.74	7.12	235.53	m	**1109.27**
Sum of two sides 3800 mm	703.35	890.16	7.12	235.53	m	**1125.69**
Sum of two sides 3900 mm	716.33	906.59	7.12	235.53	m	**1142.12**
Sum of two sides 4000 mm	729.31	923.01	7.12	235.53	m	**1158.54**

38 VENTILATION/AIR CONDITIONING SYSTEMS

Item	Net Price £	Material £	Labour hours	Labour £	Unit	Total rate £
Extra over fittings; Ductwork 1601 to 2000 mm longest side						
End cap						
Sum of two sides 2100 mm	173.31	219.34	1.74	57.56	nr	**276.90**
Sum of two sides 2200 mm	186.33	235.82	1.74	57.56	nr	**293.38**
Sum of two sides 2300 mm	199.37	252.33	1.74	57.56	nr	**309.89**
Sum of two sides 2400 mm	212.39	268.81	1.74	57.56	nr	**326.36**
Sum of two sides 2500 mm	225.41	285.28	1.74	57.56	nr	**342.84**
Sum of two sides 2600 mm	238.43	301.76	1.74	57.56	nr	**359.32**
Sum of two sides 2700 mm	251.43	318.21	1.74	57.56	nr	**375.77**
Sum of two sides 2800 mm	264.45	334.69	2.32	76.74	nr	**411.44**
Sum of two sides 2900 mm	277.49	351.20	2.32	76.74	nr	**427.94**
Sum of two sides 3000 mm	290.49	367.65	3.46	114.46	nr	**482.10**
Sum of two sides 3100 mm	300.85	380.75	3.46	114.46	nr	**495.21**
Sum of two sides 3200 mm	310.25	392.66	3.46	114.46	nr	**507.11**
Sum of two sides 3300 mm	319.68	404.59	3.46	114.46	nr	**519.05**
Sum of two sides 3400 mm	329.11	416.52	3.46	114.46	nr	**530.98**
Sum of two sides 3500 mm	338.54	428.46	3.46	114.46	nr	**542.91**
Sum of two sides 3600 mm	347.95	440.36	3.46	114.46	nr	**554.82**
Sum of two sides 3700 mm	357.38	452.30	3.46	114.46	nr	**566.75**
Sum of two sides 3800 mm	366.77	464.18	3.46	114.46	nr	**578.63**
Sum of two sides 3900 mm	376.22	476.14	3.46	114.46	nr	**590.59**
Sum of two sides 4000 mm	385.62	488.04	3.46	114.46	nr	**602.50**
Reducer						
Sum of two sides 2100 mm	232.81	294.64	5.84	193.18	nr	**487.82**
Sum of two sides 2200 mm	254.75	322.41	5.84	193.18	nr	**515.60**
Sum of two sides 2300 mm	277.07	350.66	5.84	193.18	nr	**543.85**
Sum of two sides 2400 mm	298.68	378.01	6.24	206.42	nr	**584.43**
Sum of two sides 2500 mm	321.01	406.27	6.24	206.42	nr	**612.68**
Sum of two sides 2600 mm	342.95	434.04	6.32	209.06	nr	**643.10**
Sum of two sides 2700 mm	364.88	461.79	6.32	209.06	nr	**670.85**
Sum of two sides 2800 mm	386.80	489.53	8.00	264.64	nr	**754.17**
Sum of two sides 2900 mm	409.12	517.78	8.02	265.30	nr	**783.08**
Sum of two sides 3000 mm	431.11	545.61	8.02	265.30	nr	**810.91**
Sum of two sides 3100 mm	400.76	507.21	8.02	265.30	nr	**772.51**
Sum of two sides 3200 mm	415.46	525.81	9.12	301.69	nr	**827.50**
Sum of two sides 3300 mm	454.23	574.87	9.12	301.69	nr	**876.56**
Sum of two sides 3400 mm	468.95	593.50	9.12	301.69	nr	**895.19**
Sum of two sides 3500 mm	483.67	612.14	9.12	301.69	nr	**913.82**
Sum of two sides 3600 mm	498.41	630.79	9.12	301.69	nr	**932.48**
Sum of two sides 3700 mm	513.56	649.96	9.12	301.69	nr	**951.64**
Sum of two sides 3800 mm	528.30	668.61	9.12	301.69	nr	**970.30**
Sum of two sides 3900 mm	543.02	687.24	9.12	301.69	nr	**988.93**
Sum of two sides 4000 mm	557.76	705.90	9.12	301.69	nr	**1007.59**
Offset						
Sum of two sides 2100 mm	547.66	693.12	5.84	193.18	nr	**886.30**
Sum of two sides 2200 mm	588.95	745.37	6.52	215.68	nr	**961.05**
Sum of two sides 2300 mm	612.44	775.11	6.52	215.68	nr	**990.79**
Sum of two sides 2400 mm	671.56	849.93	6.94	229.57	nr	**1079.50**
Sum of two sides 2500 mm	694.22	878.60	6.96	230.23	nr	**1108.84**
Sum of two sides 2600 mm	715.58	905.63	6.98	230.90	nr	**1136.53**
Sum of two sides 2700 mm	736.09	931.60	7.00	231.56	nr	**1163.16**

38 VENTILATION/AIR CONDITIONING SYSTEMS

Item	Net Price £	Material £	Labour hours	Labour £	Unit	Total rate £
DUCTWORK: RECTANGULAR – CLASS C – cont						
Extra over fittings – cont						
Offset – cont						
Sum of two sides 2800 mm	755.77	956.50	8.66	286.47	nr	**1242.97**
Sum of two sides 2900 mm	774.77	980.55	9.48	313.59	nr	**1294.15**
Sum of two sides 3000 mm	792.67	1003.20	10.68	353.29	nr	**1356.49**
Sum of two sides 3100 mm	753.42	953.52	10.70	353.95	nr	**1307.48**
Sum of two sides 3200 mm	781.37	988.90	10.62	351.31	nr	**1340.20**
Sum of two sides 3300 mm	833.05	1054.31	10.70	353.95	nr	**1408.26**
Sum of two sides 3400 mm	861.00	1089.68	10.70	353.95	nr	**1443.63**
Sum of two sides 3500 mm	888.97	1125.08	10.70	353.95	nr	**1479.03**
Sum of two sides 3600 mm	916.88	1160.40	10.70	353.95	nr	**1514.36**
Sum of two sides 3700 mm	944.98	1195.97	10.70	353.95	nr	**1549.92**
Sum of two sides 3800 mm	972.91	1231.31	10.70	353.95	nr	**1585.27**
Sum of two sides 3900 mm	1000.84	1266.66	10.70	353.95	nr	**1620.61**
Sum of two sides 4000 mm	1028.79	1302.04	10.70	353.95	nr	**1655.99**
Square to round						
Sum of two sides 2100 mm	347.03	439.19	8.52	281.84	nr	**721.03**
Sum of two sides 2200 mm	380.67	481.77	8.54	282.50	nr	**764.27**
Sum of two sides 2300 mm	414.12	524.11	8.60	284.48	nr	**808.60**
Sum of two sides 2400 mm	447.91	566.87	9.54	315.58	nr	**882.45**
Sum of two sides 2500 mm	481.38	609.24	9.58	316.90	nr	**926.14**
Sum of two sides 2600 mm	514.92	651.68	9.90	327.49	nr	**979.17**
Sum of two sides 2700 mm	548.46	694.13	9.90	327.49	nr	**1021.62**
Sum of two sides 2800 mm	582.02	736.60	16.98	561.69	nr	**1298.29**
Sum of two sides 2900 mm	615.49	778.96	17.76	587.49	nr	**1366.46**
Sum of two sides 3000 mm	649.01	821.38	18.04	596.76	nr	**1418.14**
Sum of two sides 3100 mm	627.59	794.27	18.04	596.76	nr	**1391.03**
Sum of two sides 3200 mm	650.35	823.08	18.18	601.39	nr	**1424.47**
Sum of two sides 3300 mm	684.83	866.72	18.18	601.39	nr	**1468.11**
Sum of two sides 3400 mm	707.53	895.45	18.18	601.39	nr	**1496.84**
Sum of two sides 3500 mm	730.30	924.26	18.18	601.39	nr	**1525.65**
Sum of two sides 3600 mm	753.04	953.05	18.18	601.39	nr	**1554.43**
Sum of two sides 3700 mm	775.70	981.72	18.18	601.39	nr	**1583.11**
Sum of two sides 3800 mm	798.44	1010.51	18.18	601.39	nr	**1611.89**
Sum of two sides 3900 mm	821.18	1039.29	18.18	601.39	nr	**1640.68**
Sum of two sides 4000 mm	843.91	1068.05	18.18	601.39	nr	**1669.44**
90° radius bend						
Sum of two sides 2100 mm	510.59	646.21	5.22	172.68	nr	**818.88**
Sum of two sides 2200 mm	974.57	1233.41	5.24	173.34	nr	**1406.75**
Sum of two sides 2300 mm	1057.39	1338.24	5.26	174.00	nr	**1512.23**
Sum of two sides 2400 mm	1080.64	1367.66	8.68	287.13	nr	**1654.79**
Sum of two sides 2500 mm	1163.46	1472.48	8.70	287.79	nr	**1760.27**
Sum of two sides 2600 mm	1244.65	1575.23	9.06	299.70	nr	**1874.93**
Sum of two sides 2700 mm	1325.84	1677.98	9.06	299.70	nr	**1977.68**
Sum of two sides 2800 mm	1407.02	1780.73	14.26	471.72	nr	**2252.44**
Sum of two sides 2900 mm	1489.85	1885.55	14.34	474.36	nr	**2359.91**
Sum of two sides 3000 mm	1571.05	1988.32	14.52	480.32	nr	**2468.64**
Sum of two sides 3100 mm	1528.97	1935.06	14.52	480.32	nr	**2415.38**

38 VENTILATION/AIR CONDITIONING SYSTEMS

Item	Net Price £	Material £	Labour hours	Labour £	Unit	Total rate £
Sum of two sides 3200 mm	1588.55	2010.46	14.62	483.62	nr	**2494.09**
Sum of two sides 3300 mm	1720.57	2177.56	14.62	483.62	nr	**2661.18**
Sum of two sides 3400 mm	1780.17	2252.98	14.62	483.62	nr	**2736.61**
Sum of two sides 3500 mm	1839.77	2328.41	14.62	483.62	nr	**2812.03**
Sum of two sides 3600 mm	1899.37	2403.84	14.62	483.62	nr	**2887.46**
Sum of two sides 3700 mm	1960.56	2481.28	14.62	483.62	nr	**2964.91**
Sum of two sides 3800 mm	2020.16	2556.71	14.62	483.62	nr	**3040.34**
Sum of two sides 3900 mm	2079.76	2632.14	14.62	483.62	nr	**3115.76**
Sum of two sides 4000 mm	2139.31	2707.51	14.62	483.62	nr	**3191.14**
45° bend						
Sum of two sides 2100 mm	246.16	311.54	15.00	496.19	nr	**807.74**
Sum of two sides 2200 mm	695.86	880.68	15.00	496.19	nr	**1376.87**
Sum of two sides 2300 mm	751.07	950.55	15.10	499.50	nr	**1450.05**
Sum of two sides 2400 mm	774.50	980.21	16.26	537.87	nr	**1518.08**
Sum of two sides 2500 mm	829.67	1050.03	16.60	549.12	nr	**1599.15**
Sum of two sides 2600 mm	883.64	1118.33	17.12	566.32	nr	**1684.66**
Sum of two sides 2700 mm	937.57	1186.58	17.24	570.29	nr	**1756.88**
Sum of two sides 2800 mm	991.56	1254.91	18.18	601.39	nr	**1856.30**
Sum of two sides 2900 mm	1046.75	1324.76	18.18	601.39	nr	**1926.15**
Sum of two sides 3000 mm	1100.72	1393.06	19.24	636.45	nr	**2029.52**
Sum of two sides 3100 mm	1087.40	1376.21	19.24	636.45	nr	**2012.67**
Sum of two sides 3200 mm	1126.97	1426.29	19.24	636.45	nr	**2062.74**
Sum of two sides 3300 mm	1214.93	1537.62	19.24	636.45	nr	**2174.07**
Sum of two sides 3400 mm	1254.50	1587.69	19.24	636.45	nr	**2224.14**
Sum of two sides 3500 mm	1294.06	1637.76	19.24	636.45	nr	**2274.22**
Sum of two sides 3600 mm	1333.58	1687.78	19.24	636.45	nr	**2324.24**
Sum of two sides 3700 mm	1374.39	1739.42	19.24	636.45	nr	**2375.88**
Sum of two sides 3800 mm	1413.93	1789.47	19.24	636.45	nr	**2425.92**
Sum of two sides 3900 mm	1453.47	1839.52	19.24	636.45	nr	**2475.97**
Sum of two sides 4000 mm	1493.00	1889.53	19.24	636.45	nr	**2525.99**
90° mitre bend						
Sum of two sides 2100 mm	1124.19	1422.78	15.00	496.19	nr	**1918.97**
Sum of two sides 2200 mm	1187.59	1503.02	15.00	496.19	nr	**1999.21**
Sum of two sides 2300 mm	1283.23	1624.05	15.10	499.50	nr	**2123.55**
Sum of two sides 2400 mm	1311.26	1659.53	16.26	537.87	nr	**2197.41**
Sum of two sides 2500 mm	1407.27	1781.04	16.60	549.12	nr	**2330.17**
Sum of two sides 2600 mm	1503.85	1903.28	17.12	566.32	nr	**2469.60**
Sum of two sides 2700 mm	1600.41	2025.48	17.24	570.29	nr	**2595.77**
Sum of two sides 2800 mm	1696.95	2147.66	30.40	1005.62	nr	**3153.28**
Sum of two sides 2900 mm	1792.98	2269.20	30.40	1005.62	nr	**3274.82**
Sum of two sides 3000 mm	1889.56	2391.43	32.08	1061.19	nr	**3452.62**
Sum of two sides 3100 mm	1841.01	2329.98	31.20	1032.08	nr	**3362.06**
Sum of two sides 3200 mm	1915.98	2424.86	32.08	1061.19	nr	**3486.05**
Sum of two sides 3300 mm	2034.69	2575.10	32.08	1061.19	nr	**3636.30**
Sum of two sides 3400 mm	2109.66	2669.99	32.08	1061.19	nr	**3731.18**
Sum of two sides 3500 mm	2184.61	2764.84	32.08	1061.19	nr	**3826.04**
Sum of two sides 3600 mm	2259.58	2859.72	32.08	1061.19	nr	**3920.92**
Sum of two sides 3700 mm	2334.00	2953.91	32.08	1061.19	nr	**4015.11**
Sum of two sides 3800 mm	2408.97	3048.80	32.08	1061.19	nr	**4109.99**
Sum of two sides 3900 mm	2483.92	3143.65	32.08	1061.19	nr	**4204.85**
Sum of two sides 4000 mm	2558.87	3238.51	32.08	1061.19	nr	**4299.70**

38 VENTILATION/AIR CONDITIONING SYSTEMS

Item	Net Price £	Material £	Labour hours	Labour £	Unit	Total rate £
DUCTWORK: RECTANGULAR – CLASS C – cont						
Extra over fittings – cont						
Branch						
Sum of two sides 2100 mm	512.00	647.99	5.22	172.68	nr	**820.66**
Sum of two sides 2200 mm	547.64	693.09	5.22	172.68	nr	**865.77**
Sum of two sides 2300 mm	583.78	738.83	5.22	172.68	nr	**911.51**
Sum of two sides 2400 mm	618.87	783.24	5.76	190.54	nr	**973.78**
Sum of two sides 2500 mm	655.07	829.06	5.76	190.54	nr	**1019.60**
Sum of two sides 2600 mm	690.73	874.19	5.76	190.54	nr	**1064.73**
Sum of two sides 2700 mm	726.39	919.32	5.76	190.54	nr	**1109.86**
Sum of two sides 2800 mm	762.13	964.55	7.88	260.67	nr	**1225.22**
Sum of two sides 2900 mm	798.27	1010.29	7.88	260.67	nr	**1270.96**
Sum of two sides 3000 mm	833.99	1055.50	9.66	319.55	nr	**1375.05**
Sum of two sides 3100 mm	862.58	1091.67	9.66	319.55	nr	**1411.22**
Sum of two sides 3200 mm	888.66	1124.68	9.66	319.55	nr	**1444.23**
Sum of two sides 3300 mm	927.02	1173.24	9.66	319.55	nr	**1492.79**
Sum of two sides 3400 mm	953.13	1206.28	9.66	319.55	nr	**1525.83**
Sum of two sides 3500 mm	979.21	1239.29	9.66	319.55	nr	**1558.84**
Sum of two sides 3600 mm	1005.29	1272.30	9.66	319.55	nr	**1591.85**
Sum of two sides 3700 mm	1031.86	1305.92	9.66	319.55	nr	**1625.47**
Sum of two sides 3800 mm	1057.92	1338.90	9.66	319.55	nr	**1658.45**
Sum of two sides 3900 mm	1084.06	1371.99	9.66	319.55	nr	**1691.54**
Sum of two sides 4000 mm	1110.12	1404.97	9.66	319.55	nr	**1724.52**
Grille neck						
Sum of two sides 2100 mm	546.95	692.21	5.60	185.25	nr	**877.46**
Sum of two sides 2200 mm	587.29	743.27	5.60	185.25	nr	**928.52**
Sum of two sides 2300 mm	627.67	794.38	5.60	185.25	nr	**979.62**
Sum of two sides 2400 mm	668.05	845.49	6.12	202.45	nr	**1047.93**
Sum of two sides 2500 mm	708.35	896.49	6.12	202.45	nr	**1098.94**
Sum of two sides 2600 mm	748.73	947.60	6.16	203.77	nr	**1151.37**
Sum of two sides 2700 mm	789.08	998.65	6.16	203.77	nr	**1202.42**
Sum of two sides 2800 mm	829.44	1049.74	8.26	273.24	nr	**1322.97**
Sum of two sides 2900 mm	869.80	1100.82	8.26	273.24	nr	**1374.06**
Sum of two sides 3000 mm	910.16	1151.90	10.04	332.12	nr	**1484.02**
Sum of two sides 3100 mm	942.52	1192.86	10.04	332.12	nr	**1524.98**
Sum of two sides 3200 mm	972.11	1230.30	10.04	332.12	nr	**1562.42**
Sum of two sides 3300 mm	1001.66	1267.70	10.04	332.12	nr	**1599.82**
Sum of two sides 3400 mm	1031.23	1305.12	10.04	332.12	nr	**1637.24**
Sum of two sides 3500 mm	1060.77	1342.51	10.04	332.12	nr	**1674.63**
Sum of two sides 3600 mm	1090.32	1379.91	10.04	332.12	nr	**1712.03**
Sum of two sides 3700 mm	1119.87	1417.30	10.04	332.12	nr	**1749.42**
Sum of two sides 3800 mm	1149.44	1454.72	10.04	332.12	nr	**1786.84**
Sum of two sides 3900 mm	1178.98	1492.12	10.04	332.12	nr	**1824.24**
Sum of two sides 4000 mm	1208.55	1529.54	10.04	332.12	nr	**1861.66**
Ductwork 2001 to 2500 mm longest side						
Sum of two sides 2500 mm	786.18	994.99	5.54	183.26	m	**1178.25**
Sum of two sides 2600 mm	809.32	1024.27	5.94	196.49	m	**1220.77**
Sum of two sides 2700 mm	830.45	1051.01	5.98	197.82	m	**1248.83**

38 VENTILATION/AIR CONDITIONING SYSTEMS

Item	Net Price £	Material £	Labour hours	Labour £	Unit	Total rate £
Sum of two sides 2800 mm	854.49	1081.44	6.60	218.33	m	**1299.77**
Sum of two sides 2900 mm	875.62	1108.18	6.62	218.99	m	**1327.17**
Sum of two sides 3000 mm	896.05	1134.04	7.06	233.54	m	**1367.58**
Sum of two sides 3100 mm	916.92	1160.46	7.10	234.87	m	**1395.32**
Sum of two sides 3200 mm	937.38	1186.34	7.12	235.53	m	**1421.87**
Sum of two sides 3300 mm	958.48	1213.05	7.12	235.53	m	**1448.58**
Sum of two sides 3400 mm	978.92	1238.91	7.12	235.53	m	**1474.44**
Sum of two sides 3500 mm	1072.28	1357.08	7.12	235.53	m	**1592.61**
Sum of two sides 3600 mm	1092.71	1382.94	7.12	235.53	m	**1618.47**
Sum of two sides 3700 mm	1113.84	1409.68	7.12	235.53	m	**1645.20**
Sum of two sides 3800 mm	1134.27	1435.54	7.12	235.53	m	**1671.06**
Sum of two sides 3900 mm	1154.71	1461.40	7.12	235.53	m	**1696.92**
Sum of two sides 4000 mm	1178.79	1491.88	7.12	235.53	m	**1727.41**
Extra over fittings; Ductwork 2001 to 2500 mm longest side						
End cap						
Sum of two sides 2500 mm	225.41	285.28	1.74	57.56	nr	**342.84**
Sum of two sides 2600 mm	238.43	301.76	1.74	57.56	nr	**359.32**
Sum of two sides 2700 mm	251.43	318.21	1.74	57.56	nr	**375.77**
Sum of two sides 2800 mm	264.45	334.69	2.32	76.74	nr	**411.44**
Sum of two sides 2900 mm	277.49	351.20	2.32	76.74	nr	**427.94**
Sum of two sides 3000 mm	290.49	367.65	3.46	114.46	nr	**482.10**
Sum of two sides 3100 mm	300.85	380.75	3.46	114.46	nr	**495.21**
Sum of two sides 3200 mm	310.25	392.66	3.46	114.46	nr	**507.11**
Sum of two sides 3300 mm	319.68	404.59	3.46	114.46	nr	**519.05**
Sum of two sides 3400 mm	329.11	416.52	3.46	114.46	nr	**530.98**
Sum of two sides 3500 mm	338.54	428.46	3.46	114.46	nr	**542.91**
Sum of two sides 3600 mm	347.95	440.36	3.46	114.46	nr	**554.82**
Sum of two sides 3700 mm	357.38	452.30	3.46	114.46	nr	**566.75**
Sum of two sides 3800 mm	366.77	464.18	3.46	114.46	nr	**578.63**
Sum of two sides 3900 mm	376.22	476.14	3.46	114.46	nr	**590.59**
Sum of two sides 4000 mm	385.62	488.04	3.46	114.46	nr	**602.50**
Reducer						
Sum of two sides 2500 mm	199.19	252.09	6.24	206.42	nr	**458.51**
Sum of two sides 2600 mm	218.38	276.38	6.32	209.06	nr	**485.44**
Sum of two sides 2700 mm	238.06	301.28	6.32	209.06	nr	**510.35**
Sum of two sides 2800 mm	256.73	324.91	8.00	264.64	nr	**589.55**
Sum of two sides 2900 mm	276.40	349.81	8.02	265.30	nr	**615.11**
Sum of two sides 3000 mm	295.64	374.16	9.12	301.69	nr	**675.85**
Sum of two sides 3100 mm	309.10	391.20	9.12	301.69	nr	**692.88**
Sum of two sides 3200 mm	321.15	406.45	9.12	301.69	nr	**708.14**
Sum of two sides 3300 mm	333.63	422.24	9.12	301.69	nr	**723.92**
Sum of two sides 3400 mm	345.66	437.47	9.12	301.69	nr	**739.15**
Sum of two sides 3500 mm	308.53	390.48	9.12	301.69	nr	**692.16**
Sum of two sides 3600 mm	320.57	405.71	9.12	301.69	nr	**707.39**
Sum of two sides 3700 mm	357.15	452.01	9.12	301.69	nr	**753.69**
Sum of two sides 3800 mm	369.22	467.29	9.12	301.69	nr	**768.97**
Sum of two sides 3900 mm	381.26	482.52	9.12	301.69	nr	**784.20**
Sum of two sides 4000 mm	393.65	498.20	9.12	301.69	nr	**799.88**

38 VENTILATION/AIR CONDITIONING SYSTEMS

Item	Net Price £	Material £	Labour hours	Labour £	Unit	Total rate £
DUCTWORK: RECTANGULAR – CLASS C – cont						
Extra over fittings – cont						
Offset						
Sum of two sides 2500 mm	623.01	788.48	6.96	230.23	nr	**1018.71**
Sum of two sides 2600 mm	662.55	838.52	6.98	230.90	nr	**1069.42**
Sum of two sides 2700 mm	661.67	837.41	7.00	231.56	nr	**1068.96**
Sum of two sides 2800 mm	741.62	938.59	8.66	286.47	nr	**1225.06**
Sum of two sides 2900 mm	738.82	935.05	9.48	313.59	nr	**1248.65**
Sum of two sides 3000 mm	733.59	928.44	10.62	351.31	nr	**1279.74**
Sum of two sides 3100 mm	717.95	908.63	10.68	353.29	nr	**1261.93**
Sum of two sides 3200 mm	698.02	883.41	10.70	353.95	nr	**1237.36**
Sum of two sides 3300 mm	676.39	856.04	10.70	353.95	nr	**1209.99**
Sum of two sides 3400 mm	652.58	825.90	10.70	353.95	nr	**1179.85**
Sum of two sides 3500 mm	601.06	760.70	10.70	353.95	nr	**1114.66**
Sum of two sides 3600 mm	624.08	789.83	10.70	353.95	nr	**1143.79**
Sum of two sides 3700 mm	671.37	849.69	10.70	353.95	nr	**1203.64**
Sum of two sides 3800 mm	694.43	878.87	10.70	353.95	nr	**1232.82**
Sum of two sides 3900 mm	717.42	907.97	10.70	353.95	nr	**1261.92**
Sum of two sides 4000 mm	740.50	937.18	10.70	353.95	nr	**1291.13**
Square to round						
Sum of two sides 2500 mm	318.99	403.71	9.58	316.90	nr	**720.62**
Sum of two sides 2600 mm	348.45	441.00	9.90	327.49	nr	**768.49**
Sum of two sides 2700 mm	377.92	478.29	9.90	327.49	nr	**805.78**
Sum of two sides 2800 mm	407.38	515.58	16.98	561.69	nr	**1077.27**
Sum of two sides 2900 mm	436.78	552.79	17.76	587.49	nr	**1140.28**
Sum of two sides 3000 mm	466.26	590.10	18.04	596.76	nr	**1186.86**
Sum of two sides 3100 mm	487.24	616.65	18.04	596.76	nr	**1213.41**
Sum of two sides 3200 mm	505.91	640.28	18.18	601.39	nr	**1241.67**
Sum of two sides 3300 mm	524.48	663.78	18.18	601.39	nr	**1265.16**
Sum of two sides 3400 mm	543.14	687.40	18.18	601.39	nr	**1288.79**
Sum of two sides 3500 mm	511.35	647.16	18.18	601.39	nr	**1248.55**
Sum of two sides 3600 mm	530.00	670.77	18.18	601.39	nr	**1272.15**
Sum of two sides 3700 mm	560.68	709.60	18.18	601.39	nr	**1310.98**
Sum of two sides 3800 mm	579.31	733.17	18.18	601.39	nr	**1334.56**
Sum of two sides 3900 mm	597.95	756.77	18.18	601.39	nr	**1358.16**
Sum of two sides 4000 mm	616.50	780.24	18.18	601.39	nr	**1381.63**
90° radius bend						
Sum of two sides 2500 mm	893.63	1130.98	8.70	287.79	nr	**1418.78**
Sum of two sides 2600 mm	919.19	1163.33	9.06	299.70	nr	**1463.03**
Sum of two sides 2700 mm	996.89	1261.67	9.06	299.70	nr	**1561.37**
Sum of two sides 2800 mm	964.53	1220.71	14.26	471.72	nr	**1692.42**
Sum of two sides 2900 mm	1041.20	1317.74	14.34	474.36	nr	**1792.11**
Sum of two sides 3000 mm	1116.34	1412.84	14.52	480.32	nr	**1893.15**
Sum of two sides 3100 mm	1174.28	1486.17	14.52	480.32	nr	**1966.48**
Sum of two sides 3200 mm	1227.83	1553.94	14.62	483.62	nr	**2037.56**
Sum of two sides 3300 mm	1282.87	1623.60	14.62	483.62	nr	**2107.22**
Sum of two sides 3400 mm	1336.42	1691.37	14.62	483.62	nr	**2175.00**
Sum of two sides 3500 mm	1259.01	1593.41	14.62	483.62	nr	**2077.03**

38 VENTILATION/AIR CONDITIONING SYSTEMS

Item	Net Price £	Material £	Labour hours	Labour £	Unit	Total rate £
Sum of two sides 3600 mm	1312.56	1661.18	14.62	483.62	nr	**2144.80**
Sum of two sides 3700 mm	1439.95	1822.40	14.62	483.62	nr	**2306.02**
Sum of two sides 3800 mm	1493.50	1890.17	14.62	483.62	nr	**2373.80**
Sum of two sides 3900 mm	1547.03	1957.92	14.62	483.62	nr	**2441.54**
Sum of two sides 4000 mm	1536.19	1944.20	14.62	483.62	nr	**2427.83**
45° bend						
Sum of two sides 2500 mm	722.17	913.98	16.60	549.12	nr	**1463.10**
Sum of two sides 2600 mm	748.80	947.68	17.12	566.32	nr	**1514.00**
Sum of two sides 2700 mm	802.47	1015.61	17.24	570.29	nr	**1585.90**
Sum of two sides 2800 mm	799.16	1011.41	18.18	601.39	nr	**1612.80**
Sum of two sides 2900 mm	852.35	1078.73	18.18	601.39	nr	**1680.12**
Sum of two sides 3000 mm	904.34	1144.54	19.24	636.45	nr	**1780.99**
Sum of two sides 3100 mm	945.02	1196.02	19.24	636.45	nr	**1832.47**
Sum of two sides 3200 mm	982.63	1243.62	19.24	636.45	nr	**1880.07**
Sum of two sides 3300 mm	1021.40	1292.68	19.24	636.45	nr	**1929.13**
Sum of two sides 3400 mm	1058.97	1340.23	19.24	636.45	nr	**1976.68**
Sum of two sides 3500 mm	1029.06	1302.38	19.24	636.45	nr	**1938.83**
Sum of two sides 3600 mm	1066.61	1349.90	19.24	636.45	nr	**1986.35**
Sum of two sides 3700 mm	1153.66	1460.07	19.24	636.45	nr	**2096.52**
Sum of two sides 3800 mm	1191.25	1507.64	19.24	636.45	nr	**2144.09**
Sum of two sides 3900 mm	1228.79	1555.16	19.24	636.45	nr	**2191.61**
Sum of two sides 4000 mm	1234.13	1561.91	19.24	636.45	nr	**2198.36**
90° mitre bend						
Sum of two sides 2500 mm	896.72	1134.89	16.60	549.12	nr	**1684.01**
Sum of two sides 2600 mm	939.20	1188.66	17.12	566.32	nr	**1754.98**
Sum of two sides 2700 mm	1023.33	1295.13	17.24	570.29	nr	**1865.42**
Sum of two sides 2800 mm	973.88	1232.54	30.40	1005.62	nr	**2238.16**
Sum of two sides 2900 mm	1056.99	1337.73	30.40	1005.62	nr	**2343.35**
Sum of two sides 3000 mm	1140.80	1443.80	30.40	1005.62	nr	**2449.42**
Sum of two sides 3100 mm	1207.21	1527.84	32.08	1061.19	nr	**2589.03**
Sum of two sides 3200 mm	1269.43	1606.59	32.08	1061.19	nr	**2667.78**
Sum of two sides 3300 mm	1330.96	1684.46	32.08	1061.19	nr	**2745.66**
Sum of two sides 3400 mm	1393.18	1763.21	32.08	1061.19	nr	**2824.41**
Sum of two sides 3500 mm	1289.97	1632.58	32.08	1061.19	nr	**2693.78**
Sum of two sides 3600 mm	1352.19	1711.33	32.08	1061.19	nr	**2772.53**
Sum of two sides 3700 mm	1462.69	1851.18	32.08	1061.19	nr	**2912.38**
Sum of two sides 3800 mm	1524.92	1929.93	32.08	1061.19	nr	**2991.13**
Sum of two sides 3900 mm	1587.14	2008.68	32.08	1061.19	nr	**3069.88**
Sum of two sides 4000 mm	1567.71	1984.10	32.08	1061.19	nr	**3045.29**
Branch						
Sum of two sides 2500 mm	695.58	880.33	5.76	190.54	nr	**1070.87**
Sum of two sides 2600 mm	732.59	927.16	5.76	190.54	nr	**1117.70**
Sum of two sides 2700 mm	770.15	974.71	5.76	190.54	nr	**1165.25**
Sum of two sides 2800 mm	806.63	1020.87	7.88	260.67	nr	**1281.54**
Sum of two sides 2900 mm	844.20	1068.42	7.88	260.67	nr	**1329.09**
Sum of two sides 3000 mm	881.29	1115.36	9.66	319.55	nr	**1434.90**
Sum of two sides 3100 mm	911.30	1153.33	9.66	319.55	nr	**1472.88**
Sum of two sides 3200 mm	938.78	1188.13	9.66	319.55	nr	**1507.67**
Sum of two sides 3300 mm	966.76	1223.53	9.66	319.55	nr	**1543.08**
Sum of two sides 3400 mm	994.22	1258.29	9.66	319.55	nr	**1577.84**
Sum of two sides 3500 mm	1022.97	1294.67	9.66	319.55	nr	**1614.22**

38 VENTILATION/AIR CONDITIONING SYSTEMS

Item	Net Price £	Material £	Labour hours	Labour £	Unit	Total rate £
DUCTWORK: RECTANGULAR – CLASS C – cont						
Extra over fittings – cont						
Branch – cont						
Sum of two sides 3600 mm	1050.44	1329.44	9.66	319.55	nr	**1648.99**
Sum of two sides 3700 mm	1090.49	1380.12	9.66	319.55	nr	**1699.67**
Sum of two sides 3800 mm	1118.00	1414.94	9.66	319.55	nr	**1734.49**
Sum of two sides 3900 mm	1145.47	1449.70	9.66	319.55	nr	**1769.25**
Sum of two sides 4000 mm	1173.48	1485.16	9.66	319.55	nr	**1804.71**
Grille neck						
Sum of two sides 2500 mm	708.35	896.49	6.12	202.45	nr	**1098.94**
Sum of two sides 2600 mm	748.73	947.60	6.16	203.77	nr	**1151.37**
Sum of two sides 2700 mm	789.08	998.65	6.16	203.77	nr	**1202.42**
Sum of two sides 2800 mm	829.44	1049.74	8.26	273.24	nr	**1322.97**
Sum of two sides 2900 mm	869.80	1100.82	8.26	273.24	nr	**1374.06**
Sum of two sides 3000 mm	910.16	1151.90	10.04	332.12	nr	**1484.02**
Sum of two sides 3100 mm	942.52	1192.86	10.04	332.12	nr	**1524.98**
Sum of two sides 3200 mm	972.11	1230.30	10.04	332.12	nr	**1562.42**
Sum of two sides 3300 mm	1001.66	1267.70	10.04	332.12	nr	**1599.82**
Sum of two sides 3400 mm	1031.23	1305.12	10.04	332.12	nr	**1637.24**
Sum of two sides 3500 mm	1060.77	1342.51	10.04	332.12	nr	**1674.63**
Sum of two sides 3600 mm	1090.32	1379.91	10.04	332.12	nr	**1712.03**
Sum of two sides 3700 mm	1119.87	1417.30	10.04	332.12	nr	**1749.42**
Sum of two sides 3800 mm	1149.44	1454.72	10.04	332.12	nr	**1786.84**
Sum of two sides 3900 mm	1178.98	1492.12	10.04	332.12	nr	**1824.24**
Sum of two sides 4000 mm	1208.55	1529.54	10.04	332.12	nr	**1861.66**
Ductwork 2501 to 4000 mm longest side						
Sum of two sides 3000 mm	900.90	1140.18	4.76	157.46	m	**1297.64**
Sum of two sides 3100 mm	922.47	1167.47	4.76	157.46	m	**1324.93**
Sum of two sides 3200 mm	942.90	1193.33	4.76	157.46	m	**1350.79**
Sum of two sides 3300 mm	963.33	1219.19	4.76	157.46	m	**1376.65**
Sum of two sides 3400 mm	988.62	1251.19	5.28	174.66	m	**1425.85**
Sum of two sides 3500 mm	1009.05	1277.05	5.32	175.98	m	**1453.04**
Sum of two sides 3600 mm	1029.50	1302.94	5.90	195.17	m	**1498.11**
Sum of two sides 3700 mm	1055.46	1335.79	5.92	195.83	m	**1531.62**
Sum of two sides 3800 mm	1075.89	1361.65	6.30	208.40	m	**1570.05**
Sum of two sides 3900 mm	1133.90	1435.06	6.30	208.40	m	**1643.46**
Sum of two sides 4000 mm	1154.33	1460.92	6.30	208.40	m	**1669.32**
Sum of two sides 4100 mm	1175.45	1487.65	6.70	221.63	m	**1709.29**
Sum of two sides 4200 mm	1195.89	1513.51	6.70	221.63	m	**1735.15**
Sum of two sides 4300 mm	1258.15	1592.32	7.20	238.17	m	**1830.49**
Sum of two sides 4400 mm	1247.30	1578.58	7.20	238.17	m	**1816.75**
Sum of two sides 4500 mm	1267.73	1604.44	7.20	238.17	m	**1842.61**
Extra over fittings; Ductwork 2501 to 4000 mm longest side						
End cap						
Sum of two sides 3000 mm	290.49	367.65	3.46	114.46	nr	**482.10**
Sum of two sides 3100 mm	300.85	380.75	3.46	114.46	nr	**495.21**
Sum of two sides 3200 mm	310.25	392.66	3.46	114.46	nr	**507.11**

38 VENTILATION/AIR CONDITIONING SYSTEMS

Item	Net Price £	Material £	Labour hours	Labour £	Unit	Total rate £
Sum of two sides 3300 mm	319.49	404.35	3.46	114.46	nr	**518.81**
Sum of two sides 3400 mm	329.09	416.50	3.46	114.46	nr	**530.95**
Sum of two sides 3500 mm	338.54	428.46	3.46	114.46	nr	**542.91**
Sum of two sides 3600 mm	347.95	440.36	3.46	114.46	nr	**554.82**
Sum of two sides 3700 mm	357.38	452.30	3.46	114.46	nr	**566.75**
Sum of two sides 3800 mm	366.79	464.20	3.46	114.46	nr	**578.66**
Sum of two sides 3900 mm	376.22	476.14	3.60	119.09	nr	**595.22**
Sum of two sides 4000 mm	385.62	488.04	3.60	119.09	nr	**607.13**
Sum of two sides 4100 mm	395.05	499.98	3.60	119.09	nr	**619.06**
Sum of two sides 4200 mm	404.46	511.88	3.60	119.09	nr	**630.97**
Sum of two sides 4300 mm	413.87	523.79	3.76	124.38	nr	**648.17**
Sum of two sides 4400 mm	423.30	535.72	3.76	124.38	nr	**660.10**
Sum of two sides 4500 mm	432.71	547.63	3.76	124.38	nr	**672.01**
Reducer						
Sum of two sides 3000 mm	295.51	374.00	6.24	206.42	nr	**580.42**
Sum of two sides 3100 mm	309.44	391.62	6.24	206.42	nr	**598.04**
Sum of two sides 3200 mm	321.45	406.82	6.24	206.42	nr	**613.24**
Sum of two sides 3300 mm	333.52	422.11	6.24	206.42	nr	**628.52**
Sum of two sides 3400 mm	345.41	437.15	6.24	206.42	nr	**643.57**
Sum of two sides 3500 mm	357.46	452.40	6.24	206.42	nr	**658.82**
Sum of two sides 3600 mm	369.50	467.63	7.90	261.33	nr	**728.96**
Sum of two sides 3700 mm	381.89	483.31	7.94	262.65	nr	**745.97**
Sum of two sides 3800 mm	393.90	498.52	9.04	299.04	nr	**797.56**
Sum of two sides 3900 mm	384.22	486.26	9.04	299.04	nr	**785.30**
Sum of two sides 4000 mm	396.25	501.49	9.04	299.04	nr	**800.53**
Sum of two sides 4100 mm	408.72	517.28	9.04	299.04	nr	**816.32**
Sum of two sides 4200 mm	420.78	532.54	9.04	299.04	nr	**831.58**
Sum of two sides 4300 mm	432.79	547.74	9.84	325.50	nr	**873.24**
Sum of two sides 4400 mm	445.10	563.31	9.84	325.50	nr	**888.82**
Sum of two sides 4500 mm	457.13	578.54	10.24	338.74	nr	**917.28**
Offset						
Sum of two sides 3000 mm	821.18	1039.29	6.96	230.23	nr	**1269.52**
Sum of two sides 3100 mm	807.98	1022.57	6.96	230.23	nr	**1252.81**
Sum of two sides 3200 mm	789.96	999.77	6.96	230.23	nr	**1230.00**
Sum of two sides 3300 mm	770.01	974.52	6.96	230.23	nr	**1204.76**
Sum of two sides 3400 mm	843.59	1067.65	7.00	231.56	nr	**1299.21**
Sum of two sides 3500 mm	821.73	1039.98	6.98	230.90	nr	**1270.88**
Sum of two sides 3600 mm	797.92	1009.84	8.50	281.18	nr	**1291.02**
Sum of two sides 3700 mm	873.79	1105.87	9.52	314.92	nr	**1420.79**
Sum of two sides 3800 mm	848.02	1073.26	10.64	351.97	nr	**1425.22**
Sum of two sides 3900 mm	899.18	1138.00	10.70	353.95	nr	**1491.95**
Sum of two sides 4000 mm	869.11	1099.94	10.70	353.95	nr	**1453.89**
Sum of two sides 4100 mm	837.35	1059.76	11.70	387.03	nr	**1446.79**
Sum of two sides 4200 mm	803.40	1016.78	11.70	387.03	nr	**1403.81**
Sum of two sides 4300 mm	826.41	1045.91	12.30	406.88	nr	**1452.79**
Sum of two sides 4400 mm	454.72	575.50	6.30	208.40	nr	**783.90**
Sum of two sides 4500 mm	436.29	552.16	6.15	203.44	nr	**755.60**
Square to round						
Sum of two sides 3000 mm	300.85	380.75	4.70	155.47	nr	**536.23**
Sum of two sides 3100 mm	313.76	397.10	4.70	155.47	nr	**552.57**
Sum of two sides 3200 mm	325.27	411.66	4.70	155.47	nr	**567.13**

38 VENTILATION/AIR CONDITIONING SYSTEMS

Item	Net Price £	Material £	Labour hours	Labour £	Unit	Total rate £
DUCTWORK: RECTANGULAR – CLASS C – cont						
Extra over fittings – cont						
Square to round – cont						
Sum of two sides 3300 mm	336.79	426.24	4.70	155.47	nr	**581.71**
Sum of two sides 3400 mm	348.31	440.82	4.71	155.80	nr	**596.62**
Sum of two sides 3500 mm	359.81	455.38	4.71	155.80	nr	**611.19**
Sum of two sides 3600 mm	371.32	469.95	8.19	270.92	nr	**740.87**
Sum of two sides 3700 mm	382.83	484.51	8.62	285.15	nr	**769.66**
Sum of two sides 3800 mm	394.35	499.09	8.62	285.15	nr	**784.23**
Sum of two sides 3900 mm	482.11	610.16	8.62	285.15	nr	**895.30**
Sum of two sides 4000 mm	495.82	627.51	8.75	289.45	nr	**916.96**
Sum of two sides 4100 mm	449.94	569.44	11.23	371.48	nr	**940.92**
Sum of two sides 4200 mm	523.22	662.18	11.23	371.48	nr	**1033.67**
Sum of two sides 4300 mm	536.91	679.51	11.25	372.15	nr	**1051.66**
Sum of two sides 4400 mm	550.61	696.85	11.25	372.15	nr	**1069.00**
Sum of two sides 4500 mm	564.32	714.21	11.26	372.48	nr	**1086.68**
90° radius bend						
Sum of two sides 3000 mm	872.85	1104.68	3.90	129.01	nr	**1233.70**
Sum of two sides 3100 mm	914.35	1157.20	3.90	129.01	nr	**1286.21**
Sum of two sides 3200 mm	951.58	1204.32	4.26	140.92	nr	**1345.24**
Sum of two sides 3300 mm	988.82	1251.45	4.26	140.92	nr	**1392.37**
Sum of two sides 3400 mm	965.64	1222.12	4.26	140.92	nr	**1363.04**
Sum of two sides 3500 mm	1002.38	1268.62	4.55	150.51	nr	**1419.13**
Sum of two sides 3600 mm	1039.13	1315.13	4.55	150.51	nr	**1465.64**
Sum of two sides 3700 mm	1013.21	1282.32	6.87	227.26	nr	**1509.57**
Sum of two sides 3800 mm	1049.48	1328.22	6.87	227.26	nr	**1555.47**
Sum of two sides 3900 mm	982.30	1243.19	6.87	227.26	nr	**1470.45**
Sum of two sides 4000 mm	1018.10	1288.51	7.00	231.56	nr	**1520.07**
Sum of two sides 4100 mm	1054.92	1335.11	7.20	238.17	nr	**1573.29**
Sum of two sides 4200 mm	1090.72	1380.41	7.20	238.17	nr	**1618.59**
Sum of two sides 4300 mm	1126.51	1425.72	7.41	245.12	nr	**1670.84**
Sum of two sides 4400 mm	1052.43	1331.95	7.41	245.12	nr	**1577.07**
Sum of two sides 4500 mm	1087.53	1376.37	7.55	249.75	nr	**1626.13**
45° bend						
Sum of two sides 3000 mm	423.45	535.92	4.85	160.44	nr	**696.36**
Sum of two sides 3100 mm	444.11	562.06	4.85	160.44	nr	**722.50**
Sum of two sides 3200 mm	462.65	585.53	4.85	160.44	nr	**745.97**
Sum of two sides 3300 mm	474.58	600.63	4.87	161.10	nr	**761.72**
Sum of two sides 3400 mm	469.07	593.65	4.87	161.10	nr	**754.75**
Sum of two sides 3500 mm	487.39	616.84	4.87	161.10	nr	**777.94**
Sum of two sides 3600 mm	505.70	640.02	8.81	291.43	nr	**931.45**
Sum of two sides 3700 mm	492.17	622.89	8.81	291.43	nr	**914.32**
Sum of two sides 3800 mm	510.24	645.76	9.31	307.97	nr	**953.73**
Sum of two sides 3900 mm	475.36	601.61	9.31	307.97	nr	**909.58**
Sum of two sides 4000 mm	493.19	624.17	9.31	307.97	nr	**932.15**
Sum of two sides 4100 mm	511.52	647.38	10.01	331.13	nr	**978.50**
Sum of two sides 4200 mm	529.36	669.95	9.31	307.97	nr	**977.93**
Sum of two sides 4300 mm	547.20	692.53	10.01	331.13	nr	**1023.66**
Sum of two sides 4400 mm	509.33	644.61	10.52	348.00	nr	**992.61**
Sum of two sides 4500 mm	526.83	666.75	10.52	348.00	nr	**1014.75**

38 VENTILATION/AIR CONDITIONING SYSTEMS

Item	Net Price £	Material £	Labour hours	Labour £	Unit	Total rate £
90° mitre bend						
Sum of two sides 3000 mm	1026.75	1299.46	4.85	160.44	nr	**1459.89**
Sum of two sides 3100 mm	1077.36	1363.51	4.85	160.44	nr	**1523.95**
Sum of two sides 3200 mm	1124.96	1423.75	4.87	161.10	nr	**1584.85**
Sum of two sides 3300 mm	1172.57	1484.00	4.87	161.10	nr	**1645.10**
Sum of two sides 3400 mm	1143.56	1447.28	4.87	161.10	nr	**1608.38**
Sum of two sides 3500 mm	1190.73	1506.99	8.81	291.43	nr	**1798.42**
Sum of two sides 3600 mm	1237.91	1566.70	8.81	291.43	nr	**1858.13**
Sum of two sides 3700 mm	1214.48	1537.05	14.81	489.91	nr	**2026.96**
Sum of two sides 3800 mm	1261.30	1596.30	14.81	489.91	nr	**2086.21**
Sum of two sides 3900 mm	1172.87	1484.39	14.81	489.91	nr	**1974.29**
Sum of two sides 4000 mm	1219.28	1543.12	15.20	502.81	nr	**2045.93**
Sum of two sides 4100 mm	1265.06	1601.06	16.30	539.20	nr	**2140.26**
Sum of two sides 4200 mm	1311.46	1659.78	16.50	545.81	nr	**2205.60**
Sum of two sides 4300 mm	1357.87	1718.52	17.01	562.68	nr	**2281.21**
Sum of two sides 4400 mm	1272.99	1611.09	17.01	562.68	nr	**2173.78**
Sum of two sides 4500 mm	1318.81	1669.09	17.01	562.68	nr	**2231.77**
Branch						
Sum of two sides 3000 mm	440.58	557.60	2.88	95.27	nr	**652.87**
Sum of two sides 3100 mm	455.82	576.88	2.88	95.27	nr	**672.15**
Sum of two sides 3200 mm	469.57	594.29	2.88	95.27	nr	**689.56**
Sum of two sides 3300 mm	483.30	611.67	2.88	95.27	nr	**706.94**
Sum of two sides 3400 mm	497.00	629.00	2.88	95.27	nr	**724.27**
Sum of two sides 3500 mm	510.74	646.39	3.94	130.33	nr	**776.73**
Sum of two sides 3600 mm	524.50	663.80	3.94	130.33	nr	**794.14**
Sum of two sides 3700 mm	538.40	681.40	3.94	130.33	nr	**811.73**
Sum of two sides 3800 mm	552.15	698.80	4.83	159.77	nr	**858.58**
Sum of two sides 3900 mm	566.44	716.89	4.83	159.77	nr	**876.67**
Sum of two sides 4000 mm	580.20	734.30	4.83	159.77	nr	**894.07**
Sum of two sides 4100 mm	594.17	751.99	5.44	179.95	nr	**931.94**
Sum of two sides 4200 mm	607.92	769.38	5.44	179.95	nr	**949.33**
Sum of two sides 4300 mm	621.65	786.76	5.85	193.52	nr	**980.28**
Sum of two sides 4400 mm	625.62	791.79	5.85	193.52	nr	**985.30**
Sum of two sides 4500 mm	649.30	821.75	5.85	193.52	nr	**1015.27**
Grille neck						
Sum of two sides 3000 mm	455.08	575.95	2.88	95.27	nr	**671.22**
Sum of two sides 3100 mm	471.27	596.44	2.88	95.27	nr	**691.71**
Sum of two sides 3200 mm	486.05	615.14	2.88	95.27	nr	**710.41**
Sum of two sides 3300 mm	500.83	633.85	2.88	95.27	nr	**729.12**
Sum of two sides 3400 mm	515.61	652.56	3.94	130.33	nr	**782.89**
Sum of two sides 3500 mm	530.40	671.27	3.94	130.33	nr	**801.60**
Sum of two sides 3600 mm	456.90	578.25	3.94	130.33	nr	**708.58**
Sum of two sides 3700 mm	559.93	708.65	4.12	136.29	nr	**844.94**
Sum of two sides 3800 mm	574.72	727.36	4.12	136.29	nr	**863.65**
Sum of two sides 3900 mm	589.50	746.07	4.12	136.29	nr	**882.36**
Sum of two sides 4000 mm	604.28	764.77	5.00	165.40	nr	**930.17**
Sum of two sides 4100 mm	619.05	783.47	5.00	165.40	nr	**948.87**
Sum of two sides 4200 mm	633.83	802.18	5.00	165.40	nr	**967.58**
Sum of two sides 4300 mm	648.62	820.89	5.23	173.01	nr	**993.90**
Sum of two sides 4400 mm	663.39	839.59	5.23	173.01	nr	**1012.59**
Sum of two sides 4500 mm	678.17	858.30	5.39	178.30	nr	**1036.60**

38 VENTILATION/AIR CONDITIONING SYSTEMS

Item	Net Price £	Material £	Labour hours	Labour £	Unit	Total rate £
DUCTWORK: RECTANGULAR – CLASS C – cont						
DUCTWORK ANCILLARIES: ACCESS DOORS Refer to ancillaries in Ductwork Ancillaries: Access Hatches						

38 VENTILATION/AIR CONDITIONING SYSTEMS

Item	Net Price £	Material £	Labour hours	Labour £	Unit	Total rate £
DUCTWORK ANCILLARIES: VOLUME CONTROL AND FIRE DAMPERS						
Volume control damper; opposed blade; galvanized steel casing; aluminium aerofoil blades; manually operated						
Rectangular						
Sum of two sides 200 mm	27.40	34.68	1.60	52.93	nr	87.60
Sum of two sides 300 mm	29.30	37.09	1.60	52.93	nr	90.01
Sum of two sides 400 mm	32.04	40.55	1.60	52.93	nr	93.48
Sum of two sides 500 mm	35.07	44.39	1.60	52.93	nr	97.32
Sum of two sides 600 mm	38.78	49.08	1.70	56.26	nr	105.34
Sum of two sides 700 mm	42.62	53.94	2.10	69.49	nr	123.43
Sum of two sides 800 mm	46.56	58.92	2.15	71.12	nr	130.04
Sum of two sides 900 mm	51.10	64.67	2.30	76.08	nr	140.75
Sum of two sides 1000 mm	55.48	70.22	2.40	79.39	nr	149.61
Sum of two sides 1100 mm	60.44	76.50	2.60	86.01	nr	162.50
Sum of two sides 1200 mm	68.36	86.52	2.80	92.62	nr	179.14
Sum of two sides 1300 mm	73.71	93.29	3.10	102.55	nr	195.84
Sum of two sides 1400 mm	79.47	100.58	3.25	107.51	nr	208.09
Sum of two sides 1500 mm	86.03	108.88	3.40	112.47	nr	221.35
Sum of two sides 1600 mm	92.49	117.05	3.45	114.12	nr	231.17
Sum of two sides 1700 mm	98.65	124.85	3.60	119.09	nr	243.94
Sum of two sides 1800 mm	106.03	134.20	3.90	129.01	nr	263.21
Sum of two sides 1900 mm	112.90	142.88	4.20	138.99	nr	281.87
Sum of two sides 2000 mm	121.10	153.27	4.33	143.23	nr	296.50
Circular						
100 mm dia.	36.58	46.30	0.80	26.46	nr	72.76
150 mm dia.	41.78	52.88	0.80	26.46	nr	79.35
160 mm dia.	43.56	55.13	0.90	29.77	nr	84.90
200 mm dia.	47.26	59.82	1.05	34.75	nr	94.56
250 mm dia.	52.62	66.59	1.20	39.71	nr	106.30
300 mm dia.	58.25	73.72	1.35	44.66	nr	118.38
315 mm dia.	60.84	76.99	1.35	44.70	nr	121.70
350 mm dia.	63.99	80.99	1.65	54.59	nr	135.58
400 mm dia.	69.87	88.43	1.90	62.89	nr	151.31
450 mm dia.	75.48	95.52	2.10	69.49	nr	165.02
500 mm dia.	82.22	104.06	2.95	97.58	nr	201.64
550 mm dia.	88.79	112.38	2.94	97.25	nr	209.63
600 mm dia.	96.18	121.72	2.94	97.25	nr	218.98
650 mm dia.	103.85	131.43	4.55	150.51	nr	281.95
700 mm dia.	112.21	142.01	5.20	172.01	nr	314.03
750 mm dia.	121.10	153.27	5.20	172.01	nr	325.28
800 mm dia.	130.31	164.92	5.80	191.86	nr	356.78
850 mm dia.	139.89	177.04	5.80	191.86	nr	368.91
900 mm dia.	149.89	189.69	6.40	211.71	nr	401.40
950 mm dia.	160.17	202.71	6.40	211.71	nr	414.42
1000 mm dia.	170.57	215.88	7.00	231.56	nr	447.44

38 VENTILATION/AIR CONDITIONING SYSTEMS

Item	Net Price £	Material £	Labour hours	Labour £	Unit	Total rate £
DUCTWORK ANCILLARIES: VOLUME CONTROL AND FIRE DAMPERS – cont						
Flat oval						
345 × 102 mm	61.80	78.21	1.20	39.70	nr	117.91
508 × 102 mm	69.61	88.10	1.60	52.93	nr	141.03
559 × 152 mm	78.66	99.55	1.90	62.89	nr	162.44
531 × 203 mm	85.93	108.75	1.90	62.89	nr	171.64
851 × 203 mm	90.84	114.96	4.55	150.51	nr	265.48
582 × 254 mm	95.38	120.71	2.10	69.49	nr	190.20
823 × 254 mm	104.42	132.15	4.10	135.63	nr	267.78
632 × 305 mm	105.09	133.01	2.95	97.58	nr	230.59
765 × 356 mm	117.55	148.77	4.55	150.51	nr	299.29
737 × 406 mm	125.23	158.49	4.55	150.51	nr	309.00
818 × 406 mm	127.85	161.81	5.20	172.01	nr	333.82
978 × 406 mm	138.25	174.97	5.50	181.94	nr	356.91
709 × 457 mm	130.31	164.92	4.50	149.01	nr	313.93
678 × 508 mm	137.85	174.46	4.55	150.51	nr	324.97
919 × 508 mm	154.69	195.78	6.00	198.48	nr	394.26
Fire damper; galvanized steel casing; stainless steel folding shutter; fusible link with manual reset; BS 476 4 hour fire-rated						
Rectangular						
Sum of two sides 200 mm	50.20	63.54	1.60	52.93	nr	116.46
Sum of two sides 300 mm	51.80	65.56	1.60	52.93	nr	118.49
Sum of two sides 400 mm	54.02	68.37	1.60	52.93	nr	121.30
Sum of two sides 500 mm	59.09	74.79	1.60	52.93	nr	127.72
Sum of two sides 600 mm	64.74	81.93	1.70	56.26	nr	138.19
Sum of two sides 700 mm	70.54	89.28	2.10	69.49	nr	158.77
Sum of two sides 800 mm	76.57	96.90	2.15	71.14	nr	168.04
Sum of two sides 900 mm	82.62	104.56	2.30	76.22	nr	180.78
Sum of two sides 1000 mm	88.79	112.38	2.40	79.52	nr	191.89
Sum of two sides 1100 mm	94.30	119.34	2.60	86.15	nr	205.49
Sum of two sides 1200 mm	100.87	127.66	2.80	92.66	nr	220.32
Sum of two sides 1300 mm	107.59	136.17	3.10	102.55	nr	238.71
Sum of two sides 1400 mm	114.83	145.33	3.25	107.51	nr	252.84
Sum of two sides 1500 mm	121.14	153.31	3.40	112.52	nr	265.83
Sum of two sides 1600 mm	128.09	162.11	3.45	114.12	nr	276.24
Sum of two sides 1700 mm	135.09	170.97	3.60	119.09	nr	290.06
Sum of two sides 1800 mm	142.45	180.29	3.90	129.01	nr	309.30
Sum of two sides 1900 mm	149.43	189.12	4.20	138.99	nr	328.11
Sum of two sides 2000 mm	156.00	197.44	4.33	143.23	nr	340.67
Sum of two sides 2100 mm	166.01	210.10	4.43	146.54	nr	356.65
Sum of two sides 2200 mm	175.64	222.29	4.55	150.51	nr	372.80
Circular						
100 mm dia.	61.16	77.41	0.80	26.46	nr	103.87
150 mm dia.	64.35	81.45	0.80	26.46	nr	107.91
160 mm dia.	64.74	81.93	0.90	29.77	nr	111.70

38 VENTILATION/AIR CONDITIONING SYSTEMS

Item	Net Price £	Material £	Labour hours	Labour £	Unit	Total rate £
200 mm dia.	67.94	85.99	1.05	34.73	nr	120.72
250 mm dia.	75.72	95.83	1.20	39.71	nr	135.54
300 mm dia.	84.16	106.51	1.35	44.66	nr	151.17
315 mm dia.	87.82	111.14	1.35	44.70	nr	155.85
350 mm dia.	93.15	117.89	1.65	54.59	nr	172.47
400 mm dia.	102.91	130.24	1.90	62.89	nr	193.13
450 mm dia.	112.54	142.43	2.10	69.49	nr	211.92
500 mm dia.	123.10	155.80	2.95	97.58	nr	253.38
550 mm dia.	133.69	169.20	2.94	97.25	nr	266.45
600 mm dia.	144.90	183.39	4.55	150.51	nr	333.90
650 mm dia.	156.79	198.44	4.55	150.51	nr	348.95
700 mm dia.	168.98	213.87	5.20	172.01	nr	385.88
750 mm dia.	181.79	230.07	5.20	172.01	nr	402.09
800 mm dia.	194.89	246.65	5.80	191.86	nr	438.51
850 mm dia.	208.65	264.07	5.80	191.86	nr	455.94
900 mm dia.	222.82	282.00	6.40	211.71	nr	493.71
950 mm dia.	237.35	300.40	6.40	211.71	nr	512.11
1000 mm dia.	252.61	319.70	7.00	231.56	nr	551.26
Flat oval						
345 × 102 mm	84.11	106.46	1.20	39.71	nr	146.17
427 × 102 mm	91.34	115.59	1.35	44.70	nr	160.30
508 × 102 mm	94.42	119.50	1.60	52.93	nr	172.43
559 × 152 mm	102.93	130.27	1.90	62.89	nr	193.16
531 × 203 mm	108.27	137.03	1.90	62.89	nr	199.92
851 × 203 mm	136.06	172.19	4.55	150.51	nr	322.71
582 × 254 mm	143.57	181.70	2.10	69.49	nr	251.20
632 × 305 mm	150.65	190.66	2.95	97.58	nr	288.25
765 × 356 mm	173.11	219.08	4.55	150.51	nr	369.59
737 × 406 mm	180.58	228.54	4.55	150.51	nr	379.06
818 × 406 mm	191.68	242.59	5.20	172.01	nr	414.61
978 × 406 mm	208.39	263.74	5.50	181.94	nr	445.67
709 × 457 mm	181.95	230.28	4.50	149.01	nr	379.29
678 × 508 mm	189.29	239.56	4.55	150.51	nr	390.08
Smoke/fire damper; galvanized steel casing; stainless steel folding shutter; fusible link and 24 V DC electromagnetic shutter release mechanism; spring operated; BS 476 4 hour fire-rating						
Rectangular						
Sum of two sides 200 mm	384.53	486.67	1.60	52.93	nr	539.59
Sum of two sides 300 mm	385.64	488.06	1.60	52.93	nr	540.99
Sum of two sides 400 mm	386.75	489.47	1.60	52.93	nr	542.40
Sum of two sides 500 mm	396.02	501.20	1.60	52.93	nr	554.13
Sum of two sides 600 mm	405.71	513.47	1.70	56.26	nr	569.73
Sum of two sides 700 mm	415.84	526.28	2.10	69.49	nr	595.78
Sum of two sides 800 mm	426.20	539.40	2.15	71.14	nr	610.54
Sum of two sides 900 mm	437.02	553.10	2.30	76.22	nr	629.32
Sum of two sides 1000 mm	448.22	567.26	2.40	79.52	nr	646.78
Sum of two sides 1100 mm	459.71	581.81	2.60	86.15	nr	667.95
Sum of two sides 1200 mm	471.61	596.87	2.80	92.66	nr	689.53

38 VENTILATION/AIR CONDITIONING SYSTEMS

Item	Net Price £	Material £	Labour hours	Labour £	Unit	Total rate £
DUCTWORK ANCILLARIES: VOLUME CONTROL AND FIRE DAMPERS – cont						
Smoke/fire damper – cont						
Rectangular – cont						
Sum of two sides 1300 mm	483.94	612.47	3.10	102.55	nr	**715.02**
Sum of two sides 1400 mm	496.52	628.40	3.25	107.51	nr	**735.91**
Sum of two sides 1500 mm	509.52	644.84	3.40	112.52	nr	**757.36**
Sum of two sides 1600 mm	522.95	661.84	3.45	114.12	nr	**775.97**
Sum of two sides 1700 mm	536.66	679.20	3.60	119.09	nr	**798.28**
Sum of two sides 1800 mm	550.78	697.07	3.90	129.01	nr	**826.08**
Sum of two sides 1900 mm	565.15	715.26	4.20	138.99	nr	**854.25**
Sum of two sides 2000 mm	579.95	733.99	4.33	143.23	nr	**877.22**
Circular						
100 mm dia.	187.56	237.38	0.80	26.46	nr	**263.84**
150 mm dia.	187.56	237.38	0.90	29.77	nr	**267.15**
200 mm dia.	187.56	237.38	1.05	34.73	nr	**272.11**
250 mm dia.	197.75	250.28	1.20	39.71	nr	**289.99**
300 mm dia.	208.35	263.69	1.35	44.70	nr	**308.40**
350 mm dia.	237.12	300.10	1.65	54.59	nr	**354.69**
400 mm dia.	260.77	330.03	1.90	62.89	nr	**392.92**
450 mm dia.	271.93	344.15	2.10	69.49	nr	**413.65**
500 mm dia.	296.15	374.80	2.95	97.58	nr	**472.39**
550 mm dia.	321.17	406.47	2.95	97.58	nr	**504.05**
600 mm dia.	336.30	425.62	4.55	150.51	nr	**576.13**
650 mm dia.	363.88	460.53	4.55	150.51	nr	**611.04**
700 mm dia.	393.89	498.51	5.20	172.01	nr	**670.52**
750 mm dia.	412.54	522.11	5.20	172.01	nr	**694.12**
800 mm dia.	448.42	567.53	5.80	191.86	nr	**759.39**
850 mm dia.	480.75	608.43	5.80	191.86	nr	**800.30**
900 mm dia.	498.85	631.34	6.40	211.71	nr	**843.05**
950 mm dia.	532.36	673.76	6.40	211.71	nr	**885.47**
1000 mm dia.	556.37	704.15	7.00	231.56	nr	**935.70**
Flat oval						
531 × 203 mm	332.77	421.15	1.90	62.89	nr	**484.04**
851 × 203 mm	371.49	470.15	4.55	150.51	nr	**620.66**
582 × 254 mm	362.24	458.46	2.10	69.49	nr	**527.95**
632 × 305 mm	386.16	488.72	2.95	97.58	nr	**586.31**
765 × 356 mm	416.60	527.25	4.55	150.51	nr	**677.76**
737 × 406 mm	435.40	551.04	4.55	150.51	nr	**701.55**
818 × 406 mm	441.51	558.77	5.20	172.01	nr	**730.78**
978 × 406 mm	466.39	590.26	5.50	181.94	nr	**772.20**
709 × 457 mm	447.90	566.87	4.50	149.01	nr	**715.87**
678 × 508 mm	466.27	590.11	4.55	150.51	nr	**740.62**

38 VENTILATION/AIR CONDITIONING SYSTEMS

Item	Net Price £	Material £	Labour hours	Labour £	Unit	Total rate £
DUCTWORK ANCILLARIES: ACCESS HATCHES						
Access doors, hollow steel construction; 25 mm mineral wool insulation; removeable or hinged; fixed with cams; including subframe and integral sealing gaskets						
Rectangular duct						
150 × 150 mm	20.16	25.51	1.25	41.35	nr	**66.86**
200 × 200 mm	22.18	28.07	1.25	41.35	nr	**69.42**
300 × 150 mm	22.80	28.86	1.25	41.35	nr	**70.21**
300 × 300 mm	25.65	32.46	1.25	41.35	nr	**73.81**
400 × 400 mm	29.26	37.03	1.35	44.70	nr	**81.74**
450 × 300 mm	29.26	37.03	1.50	49.67	nr	**86.70**
450 × 450 mm	32.24	40.80	1.50	49.67	nr	**90.47**
Access doors, hollow steel construction; 25 mm mineral wool insulation; removeable or hinged; fixed with cams; including subframe and integral sealing gaskets						
Flat oval duct						
235 × 90 mm	40.96	51.84	1.25	41.35	nr	**93.19**
235 × 140 mm	43.65	55.24	1.35	44.70	nr	**99.94**
335 × 235 mm	49.87	63.11	1.50	49.67	nr	**112.78**
535 × 235 mm	56.10	71.01	1.50	49.67	nr	**120.67**

38 VENTILATION/AIR CONDITIONING SYSTEMS

Item	Net Price £	Material £	Labour hours	Labour £	Unit	Total rate £
GRILLES/DIFFUSERS/LOUVRES						
Supply grilles; single deflection; extruded aluminium alloy frame and adjustable horizontal vanes; silver grey polyester powder coated; screw fixed						
Rectangular; for duct, ceiling and sidewall applications						
150 × 100 mm	15.31	19.38	0.60	19.85	nr	**39.22**
150 × 150 mm	18.71	23.68	0.60	19.85	nr	**43.53**
200 × 150 mm	22.25	28.16	0.65	21.50	nr	**49.66**
200 × 200 mm	19.25	24.36	0.72	23.82	nr	**48.18**
300 × 100 mm	16.51	20.90	0.72	23.82	nr	**44.71**
300 × 150 mm	19.16	24.25	0.80	26.46	nr	**50.71**
300 × 200 mm	21.78	27.57	0.88	29.11	nr	**56.68**
300 × 300 mm	27.02	34.19	1.04	34.40	nr	**68.59**
400 × 100 mm	18.50	23.42	0.88	29.11	nr	**52.53**
400 × 150 mm	21.38	27.06	0.94	31.09	nr	**58.16**
400 × 200 mm	24.30	30.75	1.04	34.40	nr	**65.16**
400 × 300 mm	30.13	38.13	1.12	37.05	nr	**75.18**
600 × 200 mm	37.07	46.91	1.26	41.68	nr	**88.59**
600 × 300 mm	59.99	75.92	1.40	46.31	nr	**122.24**
600 × 400 mm	71.44	90.42	1.61	53.26	nr	**143.67**
600 × 500 mm	82.94	104.97	1.76	58.22	nr	**163.19**
600 × 600 mm	94.41	119.49	2.17	71.78	nr	**191.27**
800 × 300 mm	70.20	88.85	1.76	58.22	nr	**147.07**
800 × 400 mm	83.58	105.78	2.17	71.78	nr	**177.56**
800 × 600 mm	110.30	139.60	3.00	99.24	nr	**238.84**
1000 × 300 mm	80.45	101.82	2.60	86.01	nr	**187.83**
1000 × 400 mm	95.71	121.12	3.00	99.24	nr	**220.36**
1000 × 600 mm	126.22	159.74	3.80	125.70	nr	**285.45**
1000 × 800 mm	156.72	198.35	3.80	125.70	nr	**324.05**
1200 × 600 mm	142.10	179.84	4.61	152.50	nr	**332.34**
1200 × 800 mm	176.39	223.24	4.61	152.50	nr	**375.74**
1200 × 1000 mm	210.67	266.63	4.61	152.50	nr	**419.12**
Rectangular; for duct, ceiling and sidewall applications; including opposed blade damper volume regulator						
150 × 100 mm	28.54	36.12	0.72	23.83	nr	**59.96**
150 × 150 mm	33.06	41.84	0.72	23.83	nr	**65.67**
200 × 150 mm	35.48	44.91	0.83	27.48	nr	**72.38**
200 × 200 mm	38.36	48.55	0.90	29.78	nr	**78.33**
300 × 100 mm	40.74	51.56	0.90	29.78	nr	**81.34**
300 × 150 mm	43.42	54.95	0.98	32.42	nr	**87.37**
300 × 200 mm	46.79	59.22	1.06	35.08	nr	**94.30**
300 × 300 mm	53.48	67.69	1.20	39.71	nr	**107.40**
400 × 100 mm	52.95	67.01	1.06	35.08	nr	**102.09**
400 × 150 mm	55.91	70.76	1.13	37.38	nr	**108.14**
400 × 200 mm	56.18	71.10	1.20	39.71	nr	**110.81**

38 VENTILATION/AIR CONDITIONING SYSTEMS

Item	Net Price £	Material £	Labour hours	Labour £	Unit	Total rate £
400 × 300 mm	63.27	80.07	1.34	44.34	nr	124.41
600 × 200 mm	77.10	97.57	1.50	49.67	nr	147.24
600 × 300 mm	102.42	129.62	1.66	54.95	nr	184.57
600 × 400 mm	116.43	147.35	1.80	59.60	nr	206.95
600 × 500 mm	131.86	166.89	2.00	66.16	nr	233.04
600 × 600 mm	145.87	184.61	2.60	86.15	nr	270.75
800 × 300 mm	140.46	177.76	2.00	66.16	nr	243.92
800 × 400 mm	157.49	199.32	2.60	86.15	nr	285.47
800 × 600 mm	193.47	244.86	3.61	119.42	nr	364.28
1000 × 300 mm	163.43	206.83	3.00	99.34	nr	306.17
1000 × 400 mm	183.12	231.76	3.61	119.42	nr	351.18
1000 × 600 mm	224.85	284.57	4.61	152.44	nr	437.01
1000 × 800 mm	332.53	420.85	4.61	152.44	nr	573.29
1200 × 600 mm	249.78	316.12	5.62	185.84	nr	501.96
1200 × 800 mm	366.52	463.87	6.10	201.79	nr	665.65
1200 × 1000 mm	416.75	527.44	6.50	215.02	nr	742.46
Supply grilles; double deflection; extruded aluminium alloy frame and adjustable horizontal and vertical vanes; white polyester powder coated; screw fixed Rectangular; for duct, ceiling and sidewall applications						
150 × 100 mm	15.31	19.38	0.88	29.11	nr	48.49
150 × 150 mm	19.76	25.01	0.88	29.11	nr	54.12
200 × 150 mm	22.25	28.16	1.08	35.73	nr	63.88
200 × 200 mm	24.79	31.37	1.25	41.35	nr	72.72
300 × 100 mm	24.15	30.56	1.25	41.35	nr	71.91
300 × 150 mm	27.26	34.50	1.50	49.62	nr	84.12
300 × 200 mm	30.39	38.46	1.75	57.89	nr	96.35
300 × 300 mm	36.67	46.41	2.15	71.12	nr	117.53
400 × 100 mm	28.57	36.15	1.75	57.89	nr	94.04
400 × 150 mm	32.29	40.86	1.95	64.51	nr	105.37
400 × 200 mm	36.02	45.58	2.15	71.12	nr	116.71
400 × 300 mm	43.46	55.00	2.55	84.35	nr	139.35
600 × 200 mm	80.29	101.62	3.01	99.57	nr	201.19
600 × 300 mm	97.02	122.79	3.36	111.15	nr	233.94
600 × 400 mm	113.78	144.00	3.80	125.70	nr	269.70
600 × 500 mm	130.49	165.14	4.20	138.93	nr	304.08
600 × 600 mm	147.22	186.32	4.51	149.19	nr	335.51
800 × 300 mm	120.17	152.08	4.20	138.93	nr	291.02
800 × 400 mm	140.96	178.40	4.51	149.19	nr	327.59
800 × 600 mm	182.55	231.03	5.10	168.71	nr	399.74
1000 × 300 mm	143.29	181.35	4.80	158.78	nr	340.13
1000 × 400 mm	168.15	212.81	5.10	168.71	nr	381.52
1000 × 600 mm	217.84	275.70	5.72	189.22	nr	464.92
1000 × 800 mm	267.57	338.64	6.33	209.39	nr	548.04
1200 × 600 mm	253.15	320.39	5.72	189.22	nr	509.60
1200 × 800 mm	311.00	393.60	6.33	209.39	nr	602.99
1200 × 1000 mm	368.82	466.78	6.33	209.39	nr	676.17

38 VENTILATION/AIR CONDITIONING SYSTEMS

Item	Net Price £	Material £	Labour hours	Labour £	Unit	Total rate £
GRILLES/DIFFUSERS/LOUVRES – cont						
Supply grilles – cont						
Rectangular; for duct, ceiling and sidewall applications; including opposed blade damper volume regulator						
150 × 100 mm	31.20	39.49	1.00	33.08	nr	**72.57**
150 × 150 mm	37.02	46.85	1.00	33.08	nr	**79.93**
200 × 150 mm	40.83	51.68	1.26	41.68	nr	**93.36**
200 × 200 mm	43.88	55.53	1.43	47.30	nr	**102.83**
300 × 100 mm	48.38	61.23	1.43	47.30	nr	**108.53**
300 × 150 mm	51.56	65.25	1.68	55.57	nr	**120.83**
300 × 200 mm	55.40	70.11	1.93	63.84	nr	**133.95**
300 × 300 mm	63.14	79.91	2.31	76.41	nr	**156.32**
400 × 100 mm	63.02	79.76	1.93	63.84	nr	**143.60**
400 × 150 mm	66.80	84.54	2.14	70.79	nr	**155.33**
400 × 200 mm	67.86	85.89	2.31	76.41	nr	**162.30**
400 × 300 mm	77.19	97.69	2.77	91.63	nr	**189.32**
600 × 200 mm	121.02	153.16	3.25	107.51	nr	**260.67**
600 × 300 mm	140.40	177.69	3.62	119.75	nr	**297.44**
600 × 400 mm	159.81	202.26	3.99	131.99	nr	**334.25**
600 × 500 mm	180.63	228.60	4.44	146.87	nr	**375.47**
600 × 600 mm	200.02	253.14	4.94	163.41	nr	**416.55**
800 × 300 mm	191.70	242.62	4.44	146.87	nr	**389.49**
800 × 400 mm	216.29	273.74	4.94	163.41	nr	**437.15**
800 × 600 mm	267.46	338.49	5.71	188.88	nr	**527.38**
1000 × 300 mm	227.77	288.26	5.20	172.01	nr	**460.28**
1000 × 400 mm	257.22	325.54	5.71	188.88	nr	**514.42**
1000 × 600 mm	318.53	403.13	6.53	216.01	nr	**619.14**
1000 × 800 mm	446.41	564.98	6.53	216.01	nr	**780.99**
1200 × 600 mm	363.09	459.53	7.34	242.80	nr	**702.33**
1200 × 800 mm	504.45	638.43	8.80	291.10	nr	**929.53**
1200 × 1000 mm	578.69	732.38	8.80	291.10	nr	**1023.49**
Floor grille suitable for mounting in raised access floors; heavy duty; extruded alumiinium; standard mill finish; complete with opposed blade volume control damper						
Diffuser						
600 mm × 600 mm	175.87	222.58	0.70	23.16	nr	**245.74**
Extra for nylon coated black finish	26.92	34.06		–	nr	**34.06**
Exhaust grilles; aluminium						
0° fixed blade core						
150 × 150 mm	18.63	23.58	0.60	19.92	nr	**43.51**
200 × 200 mm	21.92	27.74	0.72	23.83	nr	**51.58**
250 × 250 mm	25.51	32.29	0.80	26.46	nr	**58.75**
300 × 300 mm	29.42	37.23	1.00	33.08	nr	**70.31**
350 × 350 mm	33.63	42.56	1.20	39.70	nr	**82.25**

38 VENTILATION/AIR CONDITIONING SYSTEMS

Item	Net Price £	Material £	Labour hours	Labour £	Unit	Total rate £
0° fixed blade core; including opposed blade damper volume regulator						
150 × 150 mm	35.08	44.40	0.62	20.52	nr	64.92
200 × 200 mm	40.10	50.75	0.72	23.83	nr	74.58
250 × 250 mm	45.65	57.77	0.80	26.46	nr	84.24
300 × 300 mm	54.62	69.12	1.00	33.08	nr	102.20
350 × 350 mm	61.15	77.39	1.20	39.70	nr	117.08
45° fixed blade core						
150 × 150 mm	18.63	23.58	0.62	20.52	nr	44.10
200 × 200 mm	21.92	27.74	0.72	23.83	nr	51.58
250 × 250 mm	25.51	32.29	0.80	26.46	nr	58.75
300 × 300 mm	29.42	37.23	1.00	33.08	nr	70.31
350 × 350 mm	33.63	42.56	1.20	39.70	nr	82.25
45° fixed blade core; including opposed blade damper volume regulator						
150 × 150 mm	35.08	44.40	0.62	20.52	nr	64.92
200 × 200 mm	40.10	50.75	0.72	23.83	nr	74.58
250 × 250 mm	45.65	57.77	0.80	26.46	nr	84.24
300 × 300 mm	54.62	69.12	1.00	33.08	nr	102.20
350 × 350 mm	61.15	77.39	1.20	39.70	nr	117.08
Eggcrate core						
150 × 150 mm	12.50	15.82	0.62	20.52	nr	36.34
200 × 200 mm	15.45	19.55	1.00	33.08	nr	52.63
250 × 250 mm	18.82	23.82	0.80	26.46	nr	50.28
300 × 300 mm	23.95	30.31	1.00	33.08	nr	63.39
350 × 350 mm	26.76	33.87	1.20	39.70	nr	73.56
Eggcrate core; including opposed blade damper volume regulator						
150 × 150 mm	28.94	36.63	0.62	20.52	nr	57.15
200 × 200 mm	33.63	42.56	0.72	23.83	nr	66.39
250 × 250 mm	38.92	49.26	0.80	26.46	nr	75.72
300 × 300 mm	49.16	62.22	1.00	33.08	nr	95.29
350 × 350 mm	54.30	68.72	1.20	39.70	nr	108.42
Mesh/perforated plate core						
150 × 150 mm	13.75	17.40	0.62	20.51	nr	37.91
200 × 200 mm	16.99	21.50	0.72	23.82	nr	45.32
250 × 250 mm	20.71	26.21	0.80	26.46	nr	52.67
300 × 300 mm	26.36	33.37	1.00	33.08	nr	66.45
350 × 350 mm	29.46	37.29	1.20	39.70	nr	76.99
Mesh/perforated plate core; including opposed blade damper volume regulator						
150 × 150 mm	30.20	38.22	0.62	20.51	nr	58.73
200 × 200 mm	35.18	44.52	0.72	23.82	nr	68.34
250 × 250 mm	40.82	51.66	0.80	26.46	nr	78.13
300 × 300 mm	51.57	65.27	0.80	26.46	nr	91.73
350 × 350 mm	56.96	72.09	1.20	39.70	nr	111.78

38 VENTILATION/AIR CONDITIONING SYSTEMS

Item	Net Price £	Material £	Labour hours	Labour £	Unit	Total rate £
GRILLES/DIFFUSERS/LOUVRES – cont						
Plastic air diffusion system						
Eggcrate grilles						
150 × 150 mm	5.61	7.10	0.62	21.42	nr	**28.52**
200 × 200 mm	9.82	12.43	0.72	24.88	nr	**37.31**
250 × 250 mm	11.21	14.18	0.80	27.63	nr	**41.81**
300 × 300 mm	11.21	14.18	1.00	34.53	nr	**48.71**
Single deflection grilles						
150 × 150 mm	8.83	11.17	0.62	21.42	nr	**32.59**
200 × 200 mm	10.94	13.84	0.72	24.88	nr	**38.72**
250 × 250 mm	12.03	15.23	0.80	27.63	nr	**42.86**
300 × 300 mm	15.13	19.15	1.00	34.53	nr	**53.69**
Double deflection grilles						
150 × 150 mm	11.00	13.92	0.62	21.42	nr	**35.34**
200 × 200 mm	14.59	18.46	0.72	24.88	nr	**43.34**
250 × 250 mm	16.67	21.10	0.80	27.63	nr	**48.73**
300 × 300 mm	23.54	29.79	1.00	34.53	nr	**64.33**
Door transfer grilles						
150 × 150 mm	16.52	20.91	0.62	21.42	nr	**42.33**
200 × 200 mm	19.19	24.29	0.72	24.88	nr	**49.17**
250 × 250 mm	19.88	25.16	0.80	27.63	nr	**52.79**
300 × 300 mm	21.57	27.30	1.00	34.53	nr	**61.83**
Opposed blade dampers						
150 × 150 mm	9.78	12.37	0.62	21.42	nr	**33.79**
200 × 200 mm	11.50	14.55	0.72	24.88	nr	**39.43**
250 × 250 mm	17.24	21.82	0.80	27.63	nr	**49.45**
300 × 300 mm	18.76	23.74	1.00	34.53	nr	**58.27**
Ceiling mounted diffusers; circular aluminium multi-cone diffuser						
Circular; for ceiling mounting						
141 mm dia. neck	63.89	80.85	0.80	26.46	nr	**107.32**
197 mm dia. neck	93.52	118.37	1.10	36.39	nr	**154.76**
309 mm dia. neck	126.67	160.32	1.40	46.33	nr	**206.65**
365 mm dia. neck	108.38	137.17	1.50	49.67	nr	**186.83**
457 mm dia. neck	170.87	216.25	2.00	66.16	nr	**282.41**
Circular; for ceiling mounting; including louvre damper volume control						
141 mm dia. neck	95.48	120.84	1.00	33.08	nr	**153.92**
197 mm dia. neck	125.12	158.36	1.20	39.71	nr	**198.07**
309 mm dia. neck	165.08	208.93	1.60	52.93	nr	**261.86**
365 mm dia. neck	149.67	189.42	1.90	62.89	nr	**252.31**
457 mm dia. neck	219.44	277.72	2.40	79.52	nr	**357.24**

38 VENTILATION/AIR CONDITIONING SYSTEMS

Item	Net Price £	Material £	Labour hours	Labour £	Unit	Total rate £
Ceiling mounted diffusers; rectangular aluminium multi-cone diffuser; four way flow						
Rectangular; for ceiling mounting						
150 × 150 mm neck	26.29	33.28	1.80	59.60	nr	92.88
300 × 150 mm neck	44.84	56.76	2.30	76.08	nr	132.84
300 × 300 mm neck	40.05	50.69	2.80	92.62	nr	143.31
450 × 150 mm neck	57.15	72.32	2.80	92.62	nr	164.95
450 × 300 mm neck	55.93	70.79	3.20	105.85	nr	176.64
450 × 450 mm neck	55.85	70.69	3.40	112.52	nr	183.20
600 × 150 mm neck	70.00	88.59	3.20	105.85	nr	194.44
600 × 300 mm neck	69.61	88.10	3.50	115.78	nr	203.88
600 × 600 mm neck	77.67	98.30	4.00	132.32	nr	230.61
Rectangular; for ceiling mounting; including opposed blade damper volume regulator						
150 × 150 mm neck	33.74	42.71	1.80	59.60	nr	102.31
300 × 150 mm neck	67.98	86.03	2.30	76.08	nr	162.12
300 × 300 mm neck	50.35	63.72	2.80	92.66	nr	156.38
450 × 150 mm neck	91.13	115.34	2.80	92.62	nr	207.96
450 × 300 mm neck	89.47	113.23	3.30	109.16	nr	222.39
450 × 450 mm neck	70.79	89.59	3.51	116.07	nr	205.66
600 × 150 mm neck	111.75	141.43	3.30	109.16	nr	250.59
600 × 300 mm neck	110.92	140.38	4.00	132.32	nr	272.70
600 × 600 mm neck	114.34	144.71	5.62	185.84	nr	330.55
Slot diffusers; continuous aluminium slot diffuser with flanged frame (1500 mm sections)						
Diffuser						
1 slot	43.21	54.69	3.76	124.36	m	179.05
2 slot	52.60	66.57	3.76	124.33	m	190.89
3 slot	64.74	81.93	3.76	124.33	m	206.26
4 slot	72.86	92.22	4.50	149.01	m	241.22
6 slot	96.00	121.49	4.50	149.01	m	270.50
Diffuser; including equalizing deflector						
1 slot	89.44	113.20	5.26	174.10	m	287.30
2 slot	110.97	140.44	5.26	174.10	m	314.55
3 slot	135.72	171.77	5.26	174.10	m	345.87
4 slot	152.41	192.89	6.33	209.36	m	402.25
6 slot	200.48	253.73	6.33	209.36	m	463.09
Extra over for ends						
1 slot	4.83	6.11	1.00	33.08	nr	39.19
2 slot	5.08	6.43	1.00	33.08	nr	39.51
3 slot	5.39	6.82	1.00	33.08	nr	39.90
4 slot	5.66	7.16	1.30	43.00	nr	50.16
6 slot	6.46	8.18	1.40	46.31	nr	54.49
Plenum boxes; 1.0 m long; circular spigot; including cord operated flap damper						
1 slot	61.58	77.93	2.75	91.13	nr	169.06
2 slot	63.19	79.97	2.75	91.13	nr	171.10
3 slot	64.00	81.00	2.75	91.13	nr	172.13
4 slot	69.82	88.36	3.51	116.07	nr	204.43
6 slot	73.26	92.72	3.51	116.07	nr	208.79

38 VENTILATION/AIR CONDITIONING SYSTEMS

Item	Net Price £	Material £	Labour hours	Labour £	Unit	Total rate £
GRILLES/DIFFUSERS/LOUVRES – cont						
Slot diffusers – cont						
Plenum boxes; 2.0 m long; circular spigot; including cord operated flap damper						
1 slot	68.18	86.28	3.26	107.75	nr	**194.03**
2 slot	71.49	90.47	3.26	107.75	nr	**198.22**
3 slot	72.50	91.76	3.26	107.75	nr	**199.51**
4 slot	80.75	102.19	3.76	124.36	nr	**226.55**
6 slot	86.10	108.96	3.76	124.36	nr	**233.32**
Perforated diffusers; rectangular face aluminium perforated diffuser; quick release face plate; for integration with rectangular ceiling tiles						
Circular spigot; rectangular diffuser						
150 mm dia. spigot; 300 × 300 mm diffuser	62.94	79.66	1.00	33.08	nr	**112.74**
300 mm dia. spigot; 600 × 600 mm diffuser	93.92	118.87	1.40	46.33	nr	**165.20**
Circular spigot; rectangular diffuser; including louvre damper volume regulator						
150 mm dia. spigot; 300 × 300 mm diffuser	94.54	119.65	1.00	33.08	nr	**152.73**
300 mm dia. spigot; 600 × 600 mm diffuser	132.32	167.46	1.60	52.93	nr	**220.39**
Rectangular spigot; rectangular diffuser						
150 × 150 mm dia. spigot; 300 × 300 mm diffuser	62.94	79.66	1.00	34.53	nr	**114.19**
300 × 150 mm dia. spigot; 600 × 300 mm diffuser	84.72	107.22	1.20	41.44	nr	**148.66**
300 × 300 mm dia. spigot; 600 × 600 mm diffuser	93.92	118.87	1.40	48.35	nr	**167.21**
600 × 300 mm dia. spigot; 1200 × 600 mm diffuser	98.10	124.15	1.60	55.25	nr	**179.40**
Rectangular spigot; rectangular diffuser; including opposed blade damper volume regulator						
150 × 150 mm dia. spigot; 300 × 300 mm diffuser	94.54	119.65	1.20	41.44	nr	**161.09**
300 × 150 mm dia. spigot; 600 × 300 mm diffuser	126.03	159.51	1.40	48.37	nr	**207.87**
300 × 300 mm dia. spigot; 600 × 600 mm diffuser	132.32	167.46	1.60	55.25	nr	**222.71**
600 × 300 mm dia. spigot; 1200 × 600 mm diffuser	154.54	195.59	1.80	62.16	nr	**257.75**
Floor swirl diffuser; manual adjustment of air discharge direction; complete with damper and dirt trap						
Plastic diffuser						
150 mm dia.	34.88	44.14	0.50	16.54	nr	**60.68**
200 mm dia.	34.88	44.14	0.50	16.54	nr	**60.68**

38 VENTILATION/AIR CONDITIONING SYSTEMS

Item	Net Price £	Material £	Labour hours	Labour £	Unit	Total rate £
Aluminium diffuser						
150 mm dia.	37.34	47.25	0.50	16.54	nr	**63.79**
200 mm dia.	54.57	69.06	0.50	16.54	nr	**85.60**
Plastic air diffusion system						
Cellular diffusers						
300 × 300 mm	16.03	20.29	2.80	96.73	nr	**117.02**
600 × 600 mm	37.11	46.97	4.00	138.13	nr	**185.10**
Multi-cone diffusers						
300 × 300 mm	12.94	16.38	2.80	96.73	nr	**113.11**
450 × 450 mm	21.29	26.95	3.40	117.46	nr	**144.41**
500 × 500 mm	22.50	28.48	3.80	131.30	nr	**159.78**
600 × 600 mm	31.14	39.42	4.00	138.13	nr	**177.55**
625 × 625 mm	53.58	67.81	4.26	146.95	nr	**214.75**
Opposed blade dampers						
300 × 300 mm	16.86	21.34	1.20	41.46	nr	**62.80**
450 × 450 mm	18.54	23.46	1.50	51.85	nr	**75.31**
600 × 600 mm	39.27	49.70	2.60	89.93	nr	**139.63**
Plenum boxes						
300 mm	10.94	13.84	2.80	96.73	nr	**110.57**
450 mm	16.98	21.49	3.40	117.46	nr	**138.95**
600 mm	27.19	34.41	4.00	138.13	nr	**172.54**
Plenum spigot reducer						
600 mm	21.57	27.30	1.00	34.53	nr	**61.83**
Blanking kits for cellular diffusers						
300 mm	7.64	9.67	0.88	30.40	nr	**40.06**
600 mm	9.07	11.48	1.10	37.99	nr	**49.47**
Blanking kits for multi-cone diffusers						
300 mm	4.03	5.11	0.88	30.40	nr	**35.50**
450 mm	12.08	15.29	0.90	31.08	nr	**46.37**
600 mm	14.10	17.84	1.10	37.99	nr	**55.83**
Acoustic louvres; opening mounted; 300 mm deep steel louvres with blades packed with acoustic infill; 12 mm galvanized mesh birdscreen; screw fixing in opening						
Louvre units; self-finished galvanized steel						
900 mm high × 600 mm wide	157.76	199.66	3.00	103.60	nr	**303.26**
900 mm high × 900 mm wide	195.72	247.71	3.00	103.60	nr	**351.31**
900 mm high × 1200 mm wide	231.73	293.28	3.34	115.34	nr	**408.62**
900 mm high × 1500 mm wide	302.40	382.72	3.34	115.34	nr	**498.06**
900 mm high × 1800 mm wide	339.07	429.13	3.34	115.34	nr	**544.47**
900 mm high × 2100 mm wide	375.74	475.54	3.34	115.34	nr	**590.88**
900 mm high × 2400 mm wide	411.73	521.08	3.68	127.08	nr	**648.16**
900 mm high × 2700 mm wide	465.41	589.02	3.68	127.08	nr	**716.10**
900 mm high × 3000 mm wide	498.13	630.43	3.68	127.08	nr	**757.51**
1200 mm high × 600 mm wide	206.85	261.79	3.00	103.60	nr	**365.38**
1200 mm high × 900 mm wide	254.63	322.26	3.34	115.34	nr	**437.60**

38 VENTILATION/AIR CONDITIONING SYSTEMS

Item	Net Price £	Material £	Labour hours	Labour £	Unit	Total rate £
GRILLES/DIFFUSERS/LOUVRES – cont						
Acoustic louvres – cont						
1200 mm high × 1200 mm wide	301.76	381.91	3.34	115.34	nr	**497.25**
1200 mm high × 1500 mm wide	399.95	506.18	3.34	115.34	nr	**621.52**
1200 mm high × 1800 mm wide	447.07	565.81	3.68	127.08	nr	**692.89**
1200 mm high × 2100 mm wide	495.52	627.12	3.68	127.08	nr	**754.20**
1200 mm high × 2400 mm wide	542.66	686.79	3.68	127.08	nr	**813.87**
1500 mm high × 600 mm wide	256.59	324.74	3.00	103.60	nr	**428.34**
1500 mm high × 900 mm wide	314.88	398.51	3.34	115.34	nr	**513.85**
1500 mm high × 1200 mm wide	373.11	472.21	3.34	115.34	nr	**587.54**
1500 mm high × 1500 mm wide	497.50	629.63	3.68	127.08	nr	**756.71**
1500 mm high × 1800 mm wide	555.75	703.36	3.68	127.08	nr	**830.44**
1500 mm high × 2100 mm wide	614.67	777.92	4.00	138.13	nr	**916.05**
1800 mm high × 600 mm wide	305.03	386.04	3.34	115.34	nr	**501.38**
1800 mm high × 900 mm wide	374.41	473.86	3.34	115.34	nr	**589.20**
1800 mm high × 1200 mm wide	444.44	562.49	3.68	127.08	nr	**689.57**
1800 mm high × 1500 mm wide	595.67	753.89	3.68	127.08	nr	**880.97**
Louvre units; polyester powder coated steel						
900 mm high × 600 mm wide	229.12	289.97	3.00	103.60	nr	**393.57**
900 mm high × 900 mm wide	301.76	381.91	3.00	103.60	nr	**485.51**
900 mm high × 1200 mm wide	373.11	472.21	3.34	115.34	nr	**587.54**
900 mm high × 1500 mm wide	479.16	606.42	3.34	115.34	nr	**721.76**
900 mm high × 1800 mm wide	551.14	697.53	3.34	115.34	nr	**812.87**
900 mm high × 2100 mm wide	623.82	789.51	3.34	115.34	nr	**904.85**
900 mm high × 2400 mm wide	695.15	879.78	3.68	127.08	nr	**1006.86**
900 mm high × 2700 mm wide	783.53	991.63	3.68	127.08	nr	**1118.71**
900 mm high × 3000 mm wide	851.60	1077.78	3.68	127.08	nr	**1204.86**
1200 mm high × 600 mm wide	301.10	381.07	3.00	103.60	nr	**484.66**
1200 mm high × 900 mm wide	396.01	501.19	3.34	115.34	nr	**616.53**
1200 mm high × 1200 mm wide	490.93	621.32	3.34	115.34	nr	**736.66**
1200 mm high × 1500 mm wide	635.59	804.40	3.34	115.34	nr	**919.74**
1200 mm high × 1800 mm wide	729.85	923.69	3.68	127.08	nr	**1050.77**
1200 mm high × 2100 mm wide	825.43	1044.67	3.68	127.08	nr	**1171.75**
1200 mm high × 2400 mm wide	919.67	1163.94	3.68	127.08	nr	**1291.02**
1500 mm high × 600 mm wide	374.41	473.86	3.00	103.60	nr	**577.46**
1500 mm high × 900 mm wide	491.57	622.14	3.34	115.34	nr	**737.47**
1500 mm high × 1200 mm wide	608.76	770.44	3.34	115.34	nr	**885.78**
1500 mm high × 1500 mm wide	792.04	1002.41	3.68	127.08	nr	**1129.49**
1500 mm high × 1800 mm wide	909.86	1151.52	3.68	127.08	nr	**1278.60**
1500 mm high × 2100 mm wide	1027.05	1299.83	4.00	138.13	nr	**1437.96**
1800 mm high × 600 mm wide	446.39	564.95	3.34	115.34	nr	**680.29**
1800 mm high × 900 mm wide	586.50	742.27	3.34	115.34	nr	**857.61**
1800 mm high × 1200 mm wide	727.25	920.40	3.68	127.08	nr	**1047.48**
1800 mm high × 1500 mm wide	949.15	1201.24	3.68	127.08	nr	**1328.32**

38 VENTILATION/AIR CONDITIONING SYSTEMS

Item	Net Price £	Material £	Labour hours	Labour £	Unit	Total rate £
Weather louvres; opening mounted;						
300 mm deep galvanized steel louvres;						
screw fixing in position						
Louvre units; including 12 mm galvanized						
mesh birdscreen						
900 × 600 mm	144.22	182.53	2.25	77.78	nr	**260.31**
900 × 900 mm	208.39	263.74	2.25	77.78	nr	**341.51**
900 × 1200 mm	250.74	317.33	2.50	86.33	nr	**403.66**
900 × 1500 mm	305.14	386.19	2.50	86.33	nr	**472.52**
900 × 1800 mm	359.56	455.06	2.50	86.33	nr	**541.39**
900 × 2100 mm	450.71	570.41	2.50	86.33	nr	**656.74**
900 × 2400 mm	484.16	612.75	2.76	95.39	nr	**708.14**
900 × 2700 mm	546.61	691.79	2.76	95.39	nr	**787.18**
900 × 3000 mm	596.95	755.49	2.76	95.39	nr	**850.89**
1200 × 600 mm	195.34	247.22	2.25	77.78	nr	**325.00**
1200 × 900 mm	272.94	345.43	2.50	86.33	nr	**431.76**
1200 × 1200 mm	351.15	444.42	2.50	86.33	nr	**530.75**
1200 × 1500 mm	409.88	518.75	2.50	86.33	nr	**605.08**
1200 × 1800 mm	536.05	678.42	2.76	95.39	nr	**773.81**
1200 × 2100 mm	609.63	771.55	2.76	95.39	nr	**866.94**
1200 × 2400 mm	636.65	805.74	2.76	95.39	nr	**901.13**
1500 × 600 mm	230.87	292.19	2.25	77.78	nr	**369.96**
1500 × 900 mm	318.14	402.64	2.50	86.33	nr	**488.97**
1500 × 1200 mm	419.05	530.35	2.50	86.33	nr	**616.68**
1500 × 1500 mm	491.31	621.80	2.76	95.39	nr	**717.19**
1500 × 1800 mm	579.32	733.18	2.76	95.39	nr	**828.57**
1500 × 2100 mm	724.54	916.98	3.00	103.70	nr	**1020.68**
1800 × 600 mm	258.25	326.84	2.50	86.33	nr	**413.17**
1800 × 900 mm	357.01	451.83	2.50	86.33	nr	**538.16**
1800 × 1200 mm	468.13	592.46	2.76	95.39	nr	**687.85**
1800 × 1500 mm	553.75	700.82	3.00	103.70	nr	**804.52**

38 VENTILATION/AIR CONDITIONING SYSTEMS

Item	Net Price £	Material £	Labour hours	Labour £	Unit	Total rate £
PLANT/EQUIPMENT: FANS						
Axial flow fan; including ancillaries, anti-vibration mountings, mounting feet, matching flanges, flexible connectors and clips; 415 V, 3 phase, 50 Hz motor; includes fixing in position; electrical work elsewhere						
Aerofoil blade fan unit; short duct case						
315 mm dia.; 0.47 m³/s duty; 147 Pa	946.78	1198.24	4.50	155.40	nr	**1353.64**
500 mm dia.; 1.89 m³/s duty; 500 Pa	1570.23	1987.29	5.00	172.66	nr	**2159.95**
560 mm dia.; 2.36 m³/s duty; 147 Pa	1502.44	1901.48	5.50	189.93	nr	**2091.41**
710 mm dia.; 5.67 m³/s duty; 245 Pa	2017.29	2553.08	6.00	207.20	nr	**2760.28**
Aerofoil blade fan unit; long duct case						
315 mm dia.; 0.47 m³/s duty; 147 Pa	946.78	1198.24	4.50	155.40	nr	**1353.64**
500 mm dia.; 1.89 m³/s duty; 500 Pa	1570.23	1987.29	5.00	172.66	nr	**2159.95**
560 mm dia.; 2.36 m³/s duty; 147 Pa	1502.44	1901.48	5.50	189.93	nr	**2091.41**
710 mm dia.; 5.67 m³/s duty; 245 Pa	2017.29	2553.08	6.00	207.20	nr	**2760.28**
Aerofoil blade fan unit; two stage parallel fan arrangement; long duct case						
315 mm; 0.47 m³/s @ 500 Pa	946.78	1198.24	4.50	155.40	nr	**1353.64**
355 mm; 0.83 m³/s @ 147 Pa	1570.23	1987.29	4.75	164.03	nr	**2151.32**
710 mm; 3.77 m³/s @ 431 Pa	1502.44	1901.48	6.00	207.20	nr	**2108.68**
710 mm; 6.61 m³/s @ 500 Pa	2017.29	2553.08	6.00	207.20	nr	**2760.28**
Axial flow fan; suitable for operation at 300°C for 90 minutes; including ancillaries, anti-vibration mountings, mounting feet, matching flanges, flexible connectors and clips; 415 V, 3 phase, 50 Hz motor; includes fixing in position; price includes air operated damper; electrical work elsewhere						
450 mm; 2.0 m³/s @ 300 Pa	2022.56	2559.75	5.00	172.66	nr	**2732.41**
630 mm; 4.6 m³/s @ 200 Pa	2844.65	3600.19	5.50	189.93	nr	**3790.12**
900 mm; 9.0 m³/s @ 300 Pa	4527.73	5730.29	6.50	224.46	nr	**5954.75**
1000 mm; 15.0 m³/s @ 400 Pa	6978.49	8831.97	7.50	258.99	nr	**9090.97**
Bifurcated fan; suitable for temperature up to 200°C with motor protection to IP55; including ancillaries, anti-vibration mountings, mounting feet, matching flanges, flexible connectors and clips; 415 V, 3 phase, 50 Hz motor; includes fixing in position; electrical work elsewhere						
300 mm; 0.50 m³/s @ 100 Pa	1844.96	2334.98	4.50	155.40	nr	**2490.38**
400 mm; 1.97 m³/s @ 200 Pa	2295.68	2905.41	5.00	172.66	nr	**3078.07**
800 mm; 3.86 m³/s @ 200 Pa	3677.30	4653.99	5.50	189.93	nr	**4843.92**
1000 mm; 6.10 m³/s @ 400 Pa	5176.68	6551.60	6.50	224.46	nr	**6776.06**

38 VENTILATION/AIR CONDITIONING SYSTEMS

Item	Net Price £	Material £	Labour hours	Labour £	Unit	Total rate £
Duct mounted in line fan with backward curved centrifugal impellor; including ancillaries, matching flanges, flexible connectors and clips; 415 V, 3phase, 50 Hz motor; includes fixing in position; electrical work elsewhere (NB: inverter included)						
0.5 m³/s @ 200 Pa	3798.39	4807.24	4.50	155.40	nr	4962.64
1.0 m³/s @ 300 Pa	3798.39	4807.24	5.00	172.66	nr	4979.91
3.0 m³/s @ 500 Pa	3662.19	4634.87	5.50	189.93	nr	4824.80
5.0 m³/s @ 750 Pa	7029.30	8896.28	6.50	224.46	nr	9120.74
7.0 m³/s @ 1000 Pa	8836.92	11184.00	7.00	241.73	nr	11425.73
Twin fan extract unit; belt driven; located internally; complete with anti-vibration mounts and non-return shutter; including ancillaries, matching flanges, flexible connectors and clips; 3 phase, 50 Hz motor; includes fixing in position; electrical work elsewhere (NB: inverter and auto change over included)						
0.25 m³/s @ 150 Pa	5738.18	7262.24	4.50	155.40	nr	7417.64
1.00 m³/s @ 200 Pa	5738.18	7262.24	5.00	172.66	nr	7434.90
2.00 m³/s @ 250 Pa	5999.43	7592.88	6.50	224.46	nr	7817.34
Roof mounted extract fan; including ancillaries, fibreglass cowling, fitted shutters and bird guard; 415 V, 3 phase, 50 Hz motor; includes fixing in position; electrical work elsewhere (NB: inverter included)						
Flat roof installation, fixed to curb						
355 mm;	2250.10	2847.73	4.50	155.40	nr	3003.13
400 mm;	2465.24	3120.01	5.50	189.93	nr	3309.94
450 mm;	2465.24	3120.01	7.00	241.73	nr	3361.74
500 mm;	2907.48	3679.71	8.00	276.26	nr	3955.97
Centrifugal fan; single speed for internal domestic kitchens/utility rooms; fitted with standard overload protection; complete with housing; includes placing in position; electrical work elsewhere						
Window mounted						
245 m³/hr	350.60	443.72	0.50	17.27	nr	460.99
500 m³/hr	678.18	858.31	0.50	17.27	nr	875.57
Wall mounted						
245 m³/hr	443.36	561.11	0.83	28.78	nr	589.89
500 m³/hr	858.73	1086.80	0.83	28.78	nr	1115.58

38 VENTILATION/AIR CONDITIONING SYSTEMS

Item	Net Price £	Material £	Labour hours	Labour £	Unit	Total rate £
PLANT/EQUIPMENT: FANS – cont						
Centrifugal fan; various speeds, simultaneous ventilation from separate areas fitted with standard overload protection; complete with housing; includes placing in position; ducting and electrical work elsewhere						
Fan unit						
147–300 m³/hr	699.51	885.30	1.00	34.53	nr	**919.83**
175–411 m³/hr	699.51	885.30	1.00	34.53	nr	**919.83**
Toilet extract units; centrifugal fan; various speeds for internal domestic bathrooms/WCs, complete with housing; includes placing in position; electrical work elsewhere						
Fan unit; fixed to wall; including shutter						
Single speed 85 m³/hr	123.96	156.88	0.75	25.90	nr	**182.78**
Two speed 60–85 m³/hr	162.93	206.20	0.83	28.78	nr	**234.98**
Humidity controlled; autospeed; fixed to wall; including shutter						
30–60–85 m³/hr	288.07	364.58	1.00	34.53	nr	**399.11**

38 VENTILATION/AIR CONDITIONING SYSTEMS

Item	Net Price £	Material £	Labour hours	Labour £	Unit	Total rate £
PLANT/EQUIPMENT: AIR FILTRATION						
High efficiency duct mounted filters; 99.997% H13 (EU13); tested to BS 3928						
Standard; 1700 m³/hr air volume; continuous rating up to 80°C; sealed wood case, aluminium spacers, neoprene gaskets; water repellant filter media; includes placing in position						
610 × 610 × 292 mm	337.67	427.36	1.00	33.08	nr	**460.44**
Side withdrawal frame	127.92	161.90	2.50	82.70	nr	**244.60**
High capacity; 3400 m³/hr air volume; continuous rating up to 80°C; anti-corrosion coated mild steel frame, polyurethane sealant and neoprene gaskets; water repellant filter media; includes placing in position						
610 × 610 × 292 mm	367.15	464.67	1.00	33.08	nr	**497.74**
Side withdrawal frame	127.92	161.90	2.50	82.70	nr	**244.60**
Bag filters; 40/60% F5 (EU5); tested to BSEN 779						
Duct mounted bag filter; continuous rating up to 60°C; rigid filter assembly; sealed into one piece coated mild steel header with sealed pocket separators; includes placing in position						
6 pocket, 592 × 592 × 25 mm header; pockets 380 mm long; 1690 m³/hr	23.86	30.19	1.00	33.08	nr	**63.27**
Side withdrawal frame	22.14	28.02	2.00	66.16	nr	**94.18**
6 pocket, 592 × 592 × 25 mm header; pockets 500 mm long; 2550 m³/hr	24.24	30.68	1.50	49.62	nr	**80.30**
Side withdrawal frame	22.14	28.02	2.50	82.70	nr	**110.72**
6 pocket, 592 × 592 × 25 mm header; pockets 600 mm long; 3380 m³/hr	28.60	36.20	1.50	49.62	nr	**85.82**
Side withdrawal frame	22.14	28.02	3.00	99.24	nr	**127.26**
Bag filters; 80/90% F7, (EU7); tested to BSEN 779						
Duct mounted bag filter; continuous rating up to 60°C; rigid filter assembly; sealed into one piece coated mild steel header with sealed pocket separators; includes placing in position						
6 pocket, 592 × 592 × 25 mm header; pockets 500 mm long; 1688 m³/hr	24.24	30.68	1.00	33.08	nr	**63.76**
Side withdrawal frame	22.14	28.02	2.00	66.16	nr	**94.18**
6 pocket, 592 × 592 × 25 mm header; pockets 635 mm long; 2047 m³/hr	28.60	36.20	1.50	49.62	nr	**85.82**
Side withdrawal frame	22.14	28.02	2.50	82.70	nr	**110.72**
6 pocket, 592 × 592 × 25 mm header; pockets 762 mm long; 2729 m³/hr	32.97	41.73	1.50	49.62	nr	**91.35**
Side withdrawal frame	22.14	28.02	3.00	99.24	nr	**127.26**

38 VENTILATION/AIR CONDITIONING SYSTEMS

Item	Net Price £	Material £	Labour hours	Labour £	Unit	Total rate £
PLANT/EQUIPMENT: AIR FILTRATION – cont						
Grease filters, washable; minimum 65% Double sided extract unit; lightweight stainless steel construction; demountable composite filter media of woven metal mat and expanded metal mesh supports; for mounting on hood and extract systems (hood not included); includes placing in position						
500 × 686 × 565 mm, 4080 m³/hr	497.68	629.87	2.00	66.16	nr	**696.03**
1000 × 686 × 565 mm, 8160 m³/hr	758.95	960.53	3.00	99.24	nr	**1059.77**
1500 × 686 × 565 mm, 12240 m³/hr	1041.00	1317.50	3.50	115.78	nr	**1433.27**
Panel filters; 82% G3 (EU3); tested to BS EN779 Modular duct mounted filter panels; continuous rating up to 100°C; graduated density media; rigid cardboard frame; includes placing in position						
596 × 596 × 47 mm, 2360 m³/hr	6.03	7.63	1.00	33.08	nr	**40.71**
Side withdrawal frame	43.24	54.72	2.50	82.70	nr	**137.42**
596 × 287 × 47 mm, 1140 m³/hr	4.16	5.27	1.00	33.08	nr	**38.35**
Side withdrawal frame	43.24	54.72	2.50	82.70	nr	**137.42**
Panel filters; 90% G4 (EU4); tested to BS EN779 Modular duct mounted filter panels; continuous rating up to 100°C; pleated media with wire support; rigid cardboard frame; includes placing in position						
596 × 596 × 47 mm, 2560 m³/hr	9.12	11.54	1.00	33.08	nr	**44.62**
side withdrawal frame	54.57	69.06	3.00	99.24	nr	**168.30**
596 × 287 × 47 mm, 1230 m³/hr	6.30	7.97	1.00	33.08	nr	**41.05**
Side withdrawal frame	43.24	54.72	3.00	99.24	nr	**153.96**
Carbon filters; standard duty disposable carbon filters; steel frame with bonded carbon panels; for fixing to ductwork; including placing in position						
12 panels						
597 × 597 × 298 mm, 1460 m³/hr	424.16	536.81	0.33	10.92	nr	**547.73**
597 × 597 × 451 mm, 2200 m³/hr	478.15	605.15	0.33	10.92	nr	**616.07**
597 × 597 × 597 mm, 2930 m³/hr	533.24	674.86	0.33	10.92	nr	**685.78**
8 panels						
451 × 451 × 298 mm, 740 m³/hr	319.76	404.69	0.29	9.59	nr	**414.29**
451 × 451 × 451 mm, 1105 m³/hr	353.92	447.92	0.29	9.59	nr	**457.51**
451 × 451 × 597 mm, 1460 m³/hr	386.49	489.15	0.29	9.59	nr	**498.74**
6 panels						
298 × 298 × 298 mm, 365 m³/hr	226.59	286.77	0.25	8.27	nr	**295.04**
298 × 298 × 451 mm, 550 m³/hr	242.60	307.03	0.25	8.27	nr	**315.30**
298 × 298 × 597 mm, 780 m³/hr	257.92	326.42	0.25	8.27	nr	**334.69**

38 VENTILATION/AIR CONDITIONING SYSTEMS

Item	Net Price £	Material £	Labour hours	Labour £	Unit	Total rate £
PLANT/EQUIPMENT: AIR CURTAINS						
The selection of air curtains requires consideration of the particular conditions involved; climatic conditions, wind influence, construction and position all influence selection; consultation with a specialist manufacturer is therefore advisable.						
Commercial grade air curtains; recessed or exposed units with rigid sheet steel casing; aluminium grilles; high quality motor/centrifugal fan assembly; includes fixing in position; electrical work elsewhere						
Ambient temperature; 240 V single phase supply; mounting height 2.40 m						
1000 × 590 × 270 mm	2914.50	3688.59	12.05	416.05	nr	**4104.65**
1500 × 590 × 270 mm	3730.36	4721.15	12.05	416.05	nr	**5137.20**
2000 × 590 × 270 mm	4512.89	5711.51	12.05	416.12	nr	**6127.63**
2500 × 590 × 270 mm	5016.70	6349.14	13.00	448.92	nr	**6798.06**
Ambient temperature; 240 V single phase supply; mounting height 2.80 m						
1000 × 590 × 270 mm	3367.09	4261.39	16.13	556.98	nr	**4818.36**
1500 × 590 × 270 mm	4298.49	5440.17	16.13	556.98	nr	**5997.15**
2000 × 590 × 270 mm	5358.53	6781.76	16.13	556.98	nr	**7338.73**
2500 × 590 × 270 mm	6195.84	7841.45	17.10	590.51	nr	**8431.96**
Ambient temperature 240 V single phase supply; mounting height 3.30 m						
1000 × 774 × 370 mm	4398.54	5566.80	17.24	595.39	nr	**6162.19**
1500 × 774 × 370 mm	5948.11	7527.92	17.24	595.39	nr	**8123.31**
2000 × 774 × 370 mm	7357.11	9311.16	17.24	595.39	nr	**9906.55**
2500 × 774 × 370 mm	9299.72	11769.72	18.30	631.95	nr	**12401.67**
Ambient temperature; 240 V single phase supply; mounting height 4.00 m						
1000 × 774 × 370 mm	4858.30	6148.66	19.10	659.57	nr	**6808.23**
1500 × 774 × 370 mm	6373.30	8066.05	19.10	659.57	nr	**8725.62**
2000 × 774 × 370 mm	7934.78	10042.26	19.10	659.57	nr	**10701.83**
2500 × 774 × 370 mm	9299.72	11769.72	19.90	687.20	nr	**12456.92**
Water heated; 240 V single phase supply; mounting height 2.40 m						
1000 × 590 × 270 mm; 2.30–9.40 kW output	3429.04	4339.79	12.05	416.05	nr	**4755.84**
1500 × 590 × 270 mm; 3.50–14.20 kW output	4389.03	5554.75	12.05	416.05	nr	**5970.81**
2000 × 590 × 270 mm; 4.70–19.00 kW output	5309.70	6719.96	12.05	416.05	nr	**7136.01**
2500 × 590 × 270 mm; 5.90–23.70 kW output	5902.84	7470.64	13.00	448.92	nr	**7919.56**

38 VENTILATION/AIR CONDITIONING SYSTEMS

Item	Net Price £	Material £	Labour hours	Labour £	Unit	Total rate £
PLANT/EQUIPMENT: AIR CURTAINS – cont						
Commercial grade air curtains – cont						
Water heated; 240 V single phase supply; mounting height 2.80 m						
1000 × 590 × 270 mm; 3.30–11.90 kW output	3962.62	5015.09	16.13	556.98	nr	**5572.07**
1500 × 590 × 270 mm; 5.00–17.90 kW output	5058.39	6401.90	16.13	556.98	nr	**6958.88**
2000 × 590 × 270 mm; 6.70–23.90 kW output	6305.42	7980.14	16.13	556.98	nr	**8537.12**
2500 × 590 × 270 mm; 8.30–29.80 kW output	7290.42	9226.75	17.10	590.51	nr	**9817.26**
Water heated; 240 V single phase supply; mounting height 3.30 m						
1000 × 774 × 370 mm; 6.10–21.80 kW output	5175.11	6549.62	17.24	595.39	nr	**7145.01**
1500 × 774 × 370 mm; 9.20–32.80 kW output	6998.61	8857.44	17.24	595.39	nr	**9452.83**
2000 × 774 × 370 mm; 12.30–43.70 kW output	8656.55	10955.73	17.24	595.39	nr	**11551.12**
2500 × 774 × 370 mm; 15.30–54.60 kW output	10246.60	12968.09	18.30	631.95	nr	**13600.04**
Water heated; 240 V single phase supply; mounting height 4.00 m						
1000 × 774 × 370 mm; 7.20–24.20 kW output	5715.85	7233.98	19.10	659.57	nr	**7893.55**
1500 × 774 × 370 mm; 10.90–36.30 kW output	7498.84	9490.54	19.10	659.57	nr	**10150.11**
2000 × 774 × 370 mm; 14.50–48.40 kW output	9335.45	11814.95	19.10	659.57	nr	**12474.52**
2500 × 774 × 370 mm; 18.10–60.60 kW output	10940.99	13846.92	19.90	687.20	nr	**14534.12**
Electrically heated; 415 V three phase supply; mounting height 2.40 m						
1000 × 590 × 270 mm; 2.30–9.40 kW output	4171.06	5278.89	12.05	416.05	nr	**5694.95**
1500 × 590 × 270 mm; 3.50–14.20 kW output	5189.40	6567.70	12.05	416.05	nr	**6983.76**
2000 × 590 × 270 mm; 4.70–19.00 kW output	6185.12	7827.89	12.05	477.17	nr	**8305.06**
2500 × 590 × 270 mm; 5.90–23.70 kW output	7031.97	8899.66	13.00	514.86	nr	**9414.52**
Electrically heated; 415 V three phase supply; mounting height 2.80 m						
1000 × 590 × 270 mm; 3.30–11.90 kW output	4910.70	6214.99	16.13	556.98	nr	**6771.96**
1500 × 590 × 270 mm; 5.00–17.90 kW output	6038.62	7642.48	16.13	556.98	nr	**8199.45**
2000 × 590 × 270 mm; 6.70–23.90 kW output	7625.10	9650.33	16.13	556.98	nr	**10207.30**
2500 × 590 × 270 mm; 8.30–29.80 kW output	8521.97	10785.40	17.10	590.51	nr	**11375.91**

38 VENTILATION/AIR CONDITIONING SYSTEMS

Item	Net Price £	Material £	Labour hours	Labour £	Unit	Total rate £
Electrically heated; 415 V three phase supply; mounting height 3.30 m						
1000 × 774 × 370 mm; 6.10–21.80 kW output	7461.92	9443.81	17.24	595.34	nr	**10039.15**
1500 × 774 × 370 mm; 9.20–32.80 kW output	9997.68	12653.06	17.24	595.34	nr	**13248.40**
2000 × 774 × 370 mm; 12.30–43.70 kW output	12066.53	15271.40	17.24	595.34	nr	**15866.74**
2500 × 774 × 370 mm; 15.30–54.60 kW output	14284.26	18078.16	18.30	631.95	nr	**18710.11**
Electrically heated; 415 V three phase supply; mounting height 4.00 m						
1000 × 774 × 370 mm; 7.20–24.20 kW output	8234.92	10422.11	19.10	659.57	nr	**11081.68**
1500 × 774 × 370 mm; 10.90–36.30 kW output	10683.72	13521.31	19.10	659.57	nr	**14180.89**
2000 × 774 × 370 mm; 14.50–48.40 kW output	12921.71	16353.71	19.10	659.57	nr	**17013.28**
2500 × 774 × 370 mm; 18.10–60.60 kW output	15306.18	19371.50	19.90	687.20	nr	**20058.70**
Industrial grade air curtains; recessed or exposed units with rigid sheet steel casing; aluminium grilles; high quality motor/centrifugal fan assembly; includes fixing in position; electrical work elsewhere						
Ambient temperature; 230 V single phase supply; including wiring between multiple units; horizontally or vertically mounted; opening maximum 6.00 m						
1500 × 585 × 853 mm; 3.0 A supply	5779.33	7314.32	17.24	595.39	nr	**7909.71**
2000 × 585 × 853 mm; 4.0 A supply	7003.58	8863.73	17.24	595.39	nr	**9459.12**
Water heated; 230 V single phase supply; including wiring between multiple units; horizontally or vertically mounted; opening maximum 6.00 m						
1500 × 585 × 956 mm; 3.0 A supply; 34.80 kW output	6624.41	8383.85	17.24	595.39	nr	**8979.24**
2000 × 585 × 956 mm; 4.0 A supply; 50.70 kW output	8094.01	10243.78	17.24	595.39	nr	**10839.17**
Water heated; 230 V single phase supply; including wiring between multiple units; vertically mounted in single bank for openings maximum 6.00 m wide or opposing twin banks for openings maximum 10.00 m wide						
1500 × 585 mm; 3.0 A supply; 41.1 kW output	6624.41	8383.85	17.24	595.39	nr	**8979.24**
2000 × 585 mm; 4.0 A supply; 57.7 kW output	8094.01	10243.78	17.24	595.39	nr	**10839.17**
Remote mounted electronic controller unit; excluding wiring to units						
0–10 V control	101.62	128.61	5.00	172.66	nr	**301.27**

Material Costs/Prices for Measured Works

38 VENTILATION/AIR CONDITIONING SYSTEMS

Item	Net Price £	Material £	Labour hours	Labour £	Unit	Total rate £
SILENCERS/ACOUSTIC TREATMENT						
Attenuators; DW144 galvanized construction c/w splitters; self-securing; fitted to ductwork						
To suit rectangular ducts; unit length 600 mm						
100 × 100 mm	176.13	222.92	0.75	24.81	nr	247.72
150 × 150 mm	183.85	232.68	0.75	24.81	nr	257.49
200 × 200 mm	191.56	242.44	0.75	24.81	nr	267.25
300 × 300 mm	214.31	271.23	0.75	24.81	nr	296.04
400 × 400 mm	240.58	304.48	1.00	33.08	nr	337.56
500 × 500 mm	280.78	355.36	1.25	41.35	nr	396.71
600 × 300 mm	270.46	342.30	1.25	41.35	nr	383.65
600 × 600 mm	378.53	479.06	1.25	41.35	nr	520.41
700 × 300 mm	286.77	362.94	1.50	49.62	nr	412.56
700 × 700 mm	427.00	540.42	1.50	49.62	nr	590.04
800 × 300 mm	297.06	375.96	2.00	66.16	nr	442.12
800 × 800 mm	492.64	623.49	2.00	66.16	nr	689.65
1000 × 1000 mm	630.60	798.09	3.00	99.24	nr	897.33
To suit rectangular ducts; unit length 1200 mm						
200 × 200 mm	211.25	267.36	1.00	33.08	nr	300.44
300 × 300 mm	261.16	330.52	1.00	33.08	nr	363.60
400 × 400 mm	314.00	397.40	1.33	44.00	nr	441.39
500 × 500 mm	378.99	479.65	1.66	54.91	nr	534.57
600 × 300 mm	361.36	457.34	1.66	54.91	nr	512.26
600 × 600 mm	528.72	669.15	1.66	54.91	nr	724.07
700 × 300 mm	387.37	490.25	2.00	66.16	nr	556.41
700 × 700 mm	605.04	765.74	2.00	66.16	nr	831.90
800 × 300 mm	404.98	512.54	2.66	87.99	nr	600.53
800 × 800 mm	698.14	883.57	2.66	87.99	nr	971.56
1000 × 1000 mm	945.09	1196.10	4.00	132.32	nr	1328.42
1300 × 1300 mm	1622.27	2053.14	8.00	264.64	nr	2317.78
1500 × 1500 mm	1856.88	2350.06	8.00	264.64	nr	2614.70
1800 × 1800 mm	2540.64	3215.44	10.66	352.63	nr	3568.07
2000 × 2000 mm	2850.89	3608.09	13.33	440.95	nr	4049.04
To suit rectangular ducts; unit length 1800 mm						
200 × 200 mm	266.60	337.41	1.00	33.08	nr	370.49
300 × 300 mm	354.25	448.34	1.00	33.08	nr	481.42
400 × 400 mm	423.46	535.93	1.33	44.00	nr	579.93
500 × 500 mm	493.50	624.57	1.66	54.91	nr	679.48
600 × 300 mm	516.13	653.22	1.66	54.91	nr	708.13
600 × 600 mm	721.23	912.79	1.66	54.91	nr	967.70
700 × 300 mm	547.17	692.50	2.00	66.16	nr	758.66
700 × 700 mm	861.70	1090.57	2.00	66.16	nr	1156.73
800 × 300 mm	578.60	732.28	2.66	87.99	nr	820.27
800 × 800 mm	1002.61	1268.90	2.66	87.99	nr	1356.90
1000 × 1000 mm	1347.70	1705.65	4.00	132.32	nr	1837.97
1300 × 1300 mm	2251.36	2849.32	8.00	264.64	nr	3113.96
1500 × 1500 mm	2551.40	3229.05	8.00	264.64	nr	3493.69
1800 × 1800 mm	3542.14	4482.94	13.33	440.95	nr	4923.89
2000 × 2000 mm	4010.98	5076.29	10.66	352.63	nr	5428.92
2300 × 2300 mm	5013.64	6345.26	16.00	529.27	nr	6874.53
2500 × 2500 mm	6993.42	8850.87	18.66	617.27	nr	9468.14

38 VENTILATION/AIR CONDITIONING SYSTEMS

Item	Net Price £	Material £	Labour hours	Labour £	Unit	Total rate £
To suit rectangular ducts; unit length 2400 mm						
500 × 500 mm	638.17	807.67	2.08	68.90	nr	876.58
600 × 300 mm	609.26	771.07	2.08	68.90	nr	839.98
600 × 600 mm	907.43	1148.44	2.08	68.90	nr	1217.35
700 × 300 mm	660.42	835.83	2.50	82.70	nr	918.53
700 × 700 mm	1071.86	1356.54	2.50	82.70	nr	1439.24
800 × 300 mm	698.14	883.57	3.33	110.16	nr	993.73
800 × 800 mm	1240.41	1569.86	3.33	110.16	nr	1680.02
1000 × 1000 mm	1673.55	2118.04	5.00	165.40	nr	2283.44
1300 × 1300 mm	2710.61	3430.55	10.00	330.80	nr	3761.35
1500 × 1500 mm	3093.68	3915.36	10.00	330.80	nr	4246.16
1800 × 1800 mm	4414.09	5586.47	13.33	440.89	nr	6027.36
2000 × 2000 mm	4953.34	6268.95	16.66	551.11	nr	6820.06
2300 × 2300 mm	6128.38	7756.08	20.00	661.59	nr	8417.68
2500 × 2500 mm	8605.16	10890.69	23.32	771.55	nr	11662.24
To suit circular ducts; unit length 600 mm						
100 mm dia.	152.95	193.57	0.75	24.81	nr	218.38
200 mm dia.	179.97	227.76	0.75	24.81	nr	252.57
250 mm dia.	204.00	258.19	0.75	24.81	nr	283.00
315 mm dia.	237.46	300.53	1.00	33.08	nr	333.61
355 mm dia.	274.78	347.76	1.00	33.08	nr	380.84
400 mm dia.	295.37	373.82	1.00	33.08	nr	406.90
450 mm dia.	356.68	451.41	1.00	33.08	nr	484.49
500 mm dia.	392.73	497.04	1.26	41.61	nr	538.65
630 mm dia.	463.02	586.00	1.50	49.62	nr	635.62
710 mm dia.	504.22	638.14	1.50	49.62	nr	687.76
800 mm dia.	560.83	709.78	2.00	66.16	nr	775.94
1000 mm dia.	752.82	952.77	3.00	99.24	nr	1052.01
To suit circular ducts; unit length 1200 mm						
100 mm dia.	225.03	284.80	1.00	33.08	nr	317.88
200 mm dia.	246.88	312.45	1.00	33.08	nr	345.53
250 mm dia.	277.35	351.02	1.00	33.08	nr	384.10
315 mm dia.	317.23	401.49	1.33	44.00	nr	445.49
355 mm dia.	380.28	481.28	1.33	44.00	nr	525.28
400 mm dia.	403.45	510.61	1.33	44.00	nr	554.60
450 mm dia.	467.32	591.44	1.33	44.00	nr	635.43
500 mm dia.	478.94	606.15	1.66	54.91	nr	661.06
630 mm dia.	634.18	802.62	2.00	66.16	nr	868.78
710 mm dia.	701.07	887.28	2.00	66.16	nr	953.44
800 mm dia.	699.79	885.65	2.66	87.99	nr	973.64
1000 mm dia.	939.42	1188.93	4.00	132.32	nr	1321.25
1250 mm dia.	1769.65	2239.67	8.00	264.64	nr	2504.30
1400 mm dia.	1924.01	2435.03	8.00	264.64	nr	2699.67
1600 mm dia.	2133.04	2699.58	10.66	352.63	nr	3052.21
To suit circular ducts; unit length 1800 mm						
200 mm dia.	345.96	437.84	1.25	41.35	nr	479.19
250 mm dia.	387.14	489.97	1.25	41.35	nr	531.31
315 mm dia.	460.07	582.27	1.66	54.91	nr	637.18
355 mm dia.	542.41	686.47	1.66	54.91	nr	741.39
400 mm dia.	561.26	710.33	1.66	54.91	nr	765.25
450 mm dia.	573.30	725.57	1.66	54.91	nr	780.49
500 mm dia.	630.25	797.65	2.08	68.81	nr	866.45

38 VENTILATION/AIR CONDITIONING SYSTEMS

Item	Net Price £	Material £	Labour hours	Labour £	Unit	Total rate £
SILENCERS/ACOUSTIC TREATMENT – **cont**						
Attenuators – cont						
To suit circular ducts – cont						
630 mm dia.	697.21	882.39	2.50	82.70	nr	**965.09**
710 mm dia.	815.58	1032.20	2.50	82.70	nr	**1114.89**
800 mm dia.	843.88	1068.02	3.33	110.16	nr	**1178.17**
1000 mm dia.	2203.24	2788.42	6.66	220.31	nr	**3008.73**
1250 mm dia.	2239.28	2834.04	6.66	220.31	nr	**3054.35**
1400 mm dia.	2474.01	3131.10	5.00	165.40	nr	**3296.50**
1600 mm dia.	2475.50	3133.00	9.90	327.49	nr	**3460.49**
THERMAL INSULATION						
Concealed Ductwork						
Flexible wrap; 20 kg–45 kg Bright Class O aluminium foil faced; Bright Class O foil taped joints; 62 mm metal pins and washers; aluminium bands						
40 mm thick insulation	18.03	22.81	0.40	13.81	m^2	**36.63**
Semi-rigid slab; 45 kg Bright Class O aluminium foil faced mineral fibre; Bright Class O foil taped joints; 62 mm metal pins and washers; aluminium bands						
40 mm thick insulation	23.58	29.84	0.65	22.45	m^2	**52.28**
Plantroom Ductwork						
Semi-rigid slab; 45 kg Bright Class O aluminium foil faced mineral fibre; Bright Class O foil taped joints; 62 mm metal pins and washers; 22 SWG plain/embossed aluminium cladding; pop rivited						
50 mm thick insulation	59.48	75.28	1.50	51.80	m^2	**127.08**
External Ductwork						
Semi-rigid slab; 45 kg Bright Class O aluminium foil faced mineral fibre; Bright Class O foil taped joints; 62 mm metal pins and washers; 0.8 mm polyisobutylene sheeting; welded joints						
50 mm thick insulation	43.39	54.91	1.25	43.17	m^2	**98.08**

38 VENTILATION/AIR CONDITIONING SYSTEMS

Item	Net Price £	Material £	Labour hours	Labour £	Unit	Total rate £
DUCTWORK: FIRE RATED						
DUCTLINES						
The relevant BS requires that the fire-rating of ductwork meets 3 criteria; stability (hours), integrity (hours) and insulation (hours). The least of the 3 periods defines the fire-rating. The BS does however allow stability and integrity to be considered in isolation. Rates are therefore provided for both types of system.						
Care should to be taken when using the rates within this section to ensure that the requiremements for stability, integrity and insulation are known and the appropriate rates are used						
High density single layer mineral wool fire-rated ductwork slab, in accordance with BSEN1366, ducts 'Type A' and 'Type B'; 165 kg class O foil faced mineral fibre; 100 mm wide bright class O foil taped joints; welded pins; includes protection to all supports.						
½ hour stability, integrity and insulation						
25 mm thick, vertical and horizontal ductwork	103.18	130.58	1.25	43.17	m²	**173.74**
1 hour stability, integrity and insulation						
30 mm thick, vertical ductwork	121.12	153.29	1.50	51.80	m²	**205.08**
40 mm thick, horizontal ductwork	142.46	180.29	1.50	51.80	m²	**232.09**
1½ hour stability, integrity and insulation						
50 mm thick, vertical ductwork	170.17	215.37	1.75	60.43	m²	**275.80**
70 mm thick, horizontal ductwork	212.86	269.40	1.75	60.43	m²	**329.83**
2 hour stability, integrity and insulation						
70 mm, vertical ductwork	219.32	277.57	2.00	69.07	m²	**346.63**
90 mm horizontal ductwork	262.00	331.59	2.00	69.07	m²	**400.66**
Kitchen extract, 1 hour stability, integrity and insulation						
90 mm, vertical and horizontal	262.00	331.59	2.00	69.07	m²	**400.66**
Galvanized sheet metal rectangular section ductwork to BSEN1366, ducts 'Type A' and 'Type B'; provides 2 hours stability and 2 hours integrity at 1100°C (no rating for insulation); including all necessary stiffeners, joints and supports in the running length						
Ductwork up to 600 mm longest side						
Sum of two sides 200 mm	131.52	166.45	2.91	96.26	m	**262.71**
Sum of two sides 300 mm	142.79	180.72	2.99	98.91	m	**279.62**
Sum of two sides 400 mm	154.03	194.94	3.17	104.86	m	**299.80**

Material Costs/Prices for Measured Works

38 VENTILATION/AIR CONDITIONING SYSTEMS

Item	Net Price £	Material £	Labour hours	Labour £	Unit	Total rate £
DUCTWORK: FIRE RATED – cont						
Ductwork up to 600 mm longest side – cont						
Sum of two sides 500 mm	165.30	209.20	3.37	111.48	m	**320.68**
Sum of two sides 600 mm	176.53	223.42	3.54	117.10	m	**340.52**
Sum of two sides 700 mm	187.81	237.70	3.72	123.06	m	**360.75**
Sum of two sides 800 mm	199.07	251.95	3.90	129.01	m	**380.96**
Sum of two sides 900 mm	210.31	266.17	5.04	166.72	m	**432.89**
Sum of two sides 1000 mm	221.59	280.44	5.58	184.58	m	**465.03**
Sum of two sides 1100 mm	232.83	294.66	5.84	193.18	m	**487.85**
Sum of two sides 1200 mm	244.06	308.88	6.11	202.12	m	**511.00**
Extra over fittings; Ductwork up to 600 mm longest side						
End cap						
Sum of two sides 200 mm	37.36	47.28	0.81	26.79	nr	**74.07**
Sum of two sides 300 mm	39.63	50.15	0.84	27.79	nr	**77.94**
Sum of two sides 400 mm	41.89	53.01	0.87	28.78	nr	**81.79**
Sum of two sides 500 mm	44.15	55.87	0.90	29.77	nr	**85.64**
Sum of two sides 600 mm	46.41	58.73	0.93	30.76	nr	**89.50**
Sum of two sides 700 mm	48.70	61.63	0.96	31.76	nr	**93.39**
Sum of two sides 800 mm	51.01	64.56	0.98	32.42	nr	**96.98**
Sum of two sides 900 mm	53.30	67.45	1.17	38.70	nr	**106.16**
Sum of two sides 1000 mm	55.52	70.27	1.22	40.36	nr	**110.63**
Sum of two sides 1100 mm	57.83	73.19	1.25	41.35	nr	**114.54**
Sum of two sides 1200 mm	60.08	76.03	1.28	42.34	nr	**118.38**
Reducer						
Sum of two sides 200 mm	148.11	187.45	2.23	73.77	nr	**261.22**
Sum of two sides 300 mm	154.92	196.07	2.37	78.40	nr	**274.47**
Sum of two sides 400 mm	161.70	204.64	2.51	83.03	nr	**287.67**
Sum of two sides 500 mm	168.52	213.28	2.65	87.66	nr	**300.94**
Sum of two sides 600 mm	175.31	221.87	2.79	92.29	nr	**314.17**
Sum of two sides 700 mm	182.11	230.48	2.87	94.94	nr	**325.42**
Sum of two sides 800 mm	188.94	239.12	3.01	99.57	nr	**338.69**
Sum of two sides 900 mm	195.74	247.73	3.06	101.22	nr	**348.95**
Sum of two sides 1000 mm	202.52	256.31	3.17	104.86	nr	**361.17**
Sum of two sides 1100 mm	209.30	264.89	3.22	106.52	nr	**371.41**
Sum of two sides 1200 mm	216.15	273.56	3.28	108.50	nr	**382.06**
Offset						
Sum of two sides 200 mm	357.47	452.41	2.95	97.58	nr	**550.00**
Sum of two sides 300 mm	366.33	463.62	3.11	102.88	nr	**566.50**
Sum of two sides 400 mm	375.18	474.83	3.26	107.84	nr	**582.67**
Sum of two sides 500 mm	384.07	486.08	3.42	113.13	nr	**599.22**
Sum of two sides 600 mm	392.94	497.31	3.57	118.09	nr	**615.40**
Sum of two sides 700 mm	410.69	519.77	3.23	106.85	nr	**626.61**
Sum of two sides 800 mm	401.82	508.54	3.73	123.39	nr	**631.93**
Sum of two sides 900 mm	419.59	531.03	3.45	114.12	nr	**645.16**
Sum of two sides 1000 mm	428.44	542.23	3.67	121.40	nr	**663.63**
Sum of two sides 1100 mm	437.28	553.43	3.78	125.04	nr	**678.47**
Sum of two sides 1200 mm	446.21	564.72	3.89	128.68	nr	**693.40**

38 VENTILATION/AIR CONDITIONING SYSTEMS

Item	Net Price £	Material £	Labour hours	Labour £	Unit	Total rate £
90° radius bend						
Sum of two sides 200 mm	144.93	183.43	2.06	68.14	nr	**251.57**
Sum of two sides 300 mm	155.92	197.33	2.21	73.11	nr	**270.44**
Sum of two sides 400 mm	166.88	211.20	2.36	78.07	nr	**289.27**
Sum of two sides 500 mm	177.90	225.15	2.51	83.03	nr	**308.17**
Sum of two sides 600 mm	188.88	239.05	2.66	87.99	nr	**327.04**
Sum of two sides 700 mm	199.81	252.88	2.81	92.95	nr	**345.83**
Sum of two sides 800 mm	210.83	266.82	2.97	98.25	nr	**365.07**
Sum of two sides 900 mm	221.75	280.65	2.99	98.91	nr	**379.56**
Sum of two sides 1000 mm	232.80	294.64	3.04	100.56	nr	**395.20**
Sum of two sides 1100 mm	243.73	308.46	3.07	101.55	nr	**410.02**
Sum of two sides 1200 mm	254.68	322.33	3.09	102.22	nr	**424.54**
45° radius bend						
Sum of two sides 200 mm	178.71	226.17	1.52	50.28	nr	**276.45**
Sum of two sides 300 mm	183.15	231.80	1.59	52.60	nr	**284.39**
Sum of two sides 400 mm	187.63	237.46	1.66	54.91	nr	**292.38**
Sum of two sides 500 mm	192.05	243.06	1.73	57.23	nr	**300.29**
Sum of two sides 600 mm	196.47	248.65	1.80	59.54	nr	**308.19**
Sum of two sides 700 mm	200.94	254.31	1.87	61.86	nr	**316.17**
Sum of two sides 800 mm	205.33	259.87	1.93	63.84	nr	**323.71**
Sum of two sides 900 mm	209.80	265.52	1.99	65.83	nr	**331.35**
Sum of two sides 1000 mm	214.22	271.12	2.05	67.81	nr	**338.93**
Sum of two sides 1100 mm	218.67	276.75	2.11	69.80	nr	**346.54**
Sum of two sides 1200 mm	223.09	282.34	2.17	71.78	nr	**354.13**
90° mitre bend						
Sum of two sides 200 mm	150.26	190.17	2.47	81.71	nr	**271.88**
Sum of two sides 300 mm	174.15	220.41	2.65	87.66	nr	**308.07**
Sum of two sides 400 mm	198.02	250.62	2.84	93.95	nr	**344.56**
Sum of two sides 500 mm	221.96	280.91	3.02	99.90	nr	**380.81**
Sum of two sides 600 mm	245.83	311.12	3.20	105.85	nr	**416.97**
Sum of two sides 700 mm	269.74	341.38	3.38	111.81	nr	**453.19**
Sum of two sides 800 mm	293.65	371.64	3.56	117.76	nr	**489.41**
Sum of two sides 900 mm	317.53	401.86	3.59	118.76	nr	**520.62**
Sum of two sides 1000 mm	341.47	432.17	3.66	121.07	nr	**553.24**
Sum of two sides 1100 mm	365.31	462.34	3.69	122.06	nr	**584.40**
Sum of two sides 1200 mm	389.28	492.67	3.72	123.06	nr	**615.72**
Branch (Side-on Shoe)						
Sum of two sides 200 mm	55.78	70.60	0.98	32.42	nr	**103.02**
Sum of two sides 300 mm	56.09	70.98	1.06	35.06	nr	**106.05**
Sum of two sides 400 mm	56.40	71.38	1.14	37.71	nr	**109.09**
Sum of two sides 500 mm	56.71	71.78	1.22	40.36	nr	**112.13**
Sum of two sides 600 mm	57.01	72.15	1.30	43.00	nr	**115.15**
Sum of two sides 700 mm	57.30	72.52	1.38	45.65	nr	**118.17**
Sum of two sides 800 mm	58.03	73.45	1.37	45.32	nr	**118.77**
Sum of two sides 900 mm	57.59	72.89	1.46	48.30	nr	**121.18**
Sum of two sides 1000 mm	58.18	73.64	1.46	48.30	nr	**121.93**
Sum of two sides 1100 mm	58.50	74.04	1.50	49.62	nr	**123.65**
Sum of two sides 1200 mm	58.80	74.42	1.55	51.27	nr	**125.69**

38 VENTILATION/AIR CONDITIONING SYSTEMS

Item	Net Price £	Material £	Labour hours	Labour £	Unit	Total rate £
DUCTWORK: FIRE RATED – cont						
Ductwork 601 to 800 mm longest side						
Sum of two sides 900 mm	210.28	266.13	5.04	166.72	m	**432.85**
Sum of two sides 1000 mm	221.57	280.42	5.58	184.58	m	**465.00**
Sum of two sides 1100 mm	232.83	294.66	5.84	193.18	m	**487.85**
Sum of two sides 1200 mm	244.06	308.88	6.11	202.12	m	**511.00**
Sum of two sides 1300 mm	255.40	323.23	6.38	211.05	m	**534.28**
Sum of two sides 1400 mm	266.63	337.45	6.65	219.98	m	**557.43**
Sum of two sides 1500 mm	277.95	351.77	6.91	228.58	m	**580.35**
Sum of two sides 1600 mm	289.20	366.02	7.18	237.51	m	**603.53**
Extra over fittings; Ductwork 601 to 800 mm longest side						
End cap						
Sum of two sides 900 mm	49.49	62.63	1.17	38.70	nr	**101.34**
Sum of two sides 1000 mm	53.04	67.12	1.22	40.36	nr	**107.48**
Sum of two sides 1100 mm	56.59	71.63	1.25	41.35	nr	**112.98**
Sum of two sides 1200 mm	60.12	76.09	1.28	42.34	nr	**118.43**
Sum of two sides 1300 mm	63.65	80.55	1.31	43.33	nr	**123.89**
Sum of two sides 1400 mm	67.19	85.04	1.34	44.33	nr	**129.37**
Sum of two sides 1500 mm	70.72	89.50	1.36	44.99	nr	**134.49**
Sum of two sides 1600 mm	74.27	93.99	1.39	45.98	nr	**139.97**
Reducer						
Sum of two sides 900 mm	185.26	234.47	3.06	101.22	nr	**335.69**
Sum of two sides 1000 mm	195.56	247.50	3.17	104.86	nr	**352.36**
Sum of two sides 1100 mm	205.83	260.50	3.22	106.52	nr	**367.02**
Sum of two sides 1200 mm	216.17	273.59	3.28	108.50	nr	**382.09**
Sum of two sides 1300 mm	226.49	286.64	3.33	110.16	nr	**396.80**
Sum of two sides 1400 mm	236.79	299.69	3.38	111.81	nr	**411.50**
Sum of two sides 1500 mm	247.08	312.70	3.44	113.79	nr	**426.50**
Sum of two sides 1600 mm	257.41	325.78	3.49	115.45	nr	**441.22**
Offset						
Sum of two sides 900 mm	393.98	498.62	3.45	114.12	nr	**612.75**
Sum of two sides 1000 mm	412.73	522.35	3.67	121.40	nr	**643.76**
Sum of two sides 1100 mm	431.51	546.12	3.78	125.04	nr	**671.16**
Sum of two sides 1200 mm	450.26	569.85	3.89	128.68	nr	**698.53**
Sum of two sides 1300 mm	468.98	593.54	4.00	132.32	nr	**725.86**
Sum of two sides 1400 mm	487.76	617.30	4.11	135.96	nr	**753.26**
Sum of two sides 1500 mm	506.51	641.04	4.23	139.93	nr	**780.97**
Sum of two sides 1600 mm	525.85	665.51	4.34	143.57	nr	**809.08**
90° radius bend						
Sum of two sides 900 mm	196.97	249.29	2.99	98.91	nr	**348.20**
Sum of two sides 1000 mm	216.34	273.80	3.04	100.56	nr	**374.37**
Sum of two sides 1100 mm	235.64	298.22	3.07	101.55	nr	**399.78**
Sum of two sides 1200 mm	255.00	322.72	3.09	102.22	nr	**424.94**
Sum of two sides 1300 mm	274.31	347.17	3.12	103.21	nr	**450.38**
Sum of two sides 1400 mm	293.65	371.64	3.15	104.20	nr	**475.84**
Sum of two sides 1500 mm	312.99	396.12	3.17	104.86	nr	**500.98**
Sum of two sides 1600 mm	332.33	420.60	3.20	105.85	nr	**526.46**

38 VENTILATION/AIR CONDITIONING SYSTEMS

Item	Net Price £	Material £	Labour hours	Labour £	Unit	Total rate £
45° bend						
Sum of two sides 900 mm	195.59	247.54	1.74	57.56	nr	**305.10**
Sum of two sides 1000 mm	204.76	259.14	1.85	61.20	nr	**320.34**
Sum of two sides 1100 mm	213.96	270.79	1.91	63.18	nr	**333.97**
Sum of two sides 1200 mm	223.11	282.37	1.96	64.84	nr	**347.21**
Sum of two sides 1300 mm	232.28	293.98	2.02	66.82	nr	**360.80**
Sum of two sides 1400 mm	241.47	305.60	2.07	68.47	nr	**374.07**
Sum of two sides 1500 mm	250.63	317.19	2.13	70.46	nr	**387.65**
Sum of two sides 1600 mm	259.82	328.83	2.18	72.11	nr	**400.94**
90° mitre bend						
Sum of two sides 900 mm	294.91	373.24	3.59	118.76	nr	**492.00**
Sum of two sides 1000 mm	326.38	413.06	3.66	121.07	nr	**534.13**
Sum of two sides 1100 mm	357.80	452.84	3.69	122.06	nr	**574.90**
Sum of two sides 1200 mm	389.28	492.67	3.72	123.06	nr	**615.72**
Sum of two sides 1300 mm	420.66	532.39	3.75	124.05	nr	**656.44**
Sum of two sides 1400 mm	452.12	572.20	3.78	125.04	nr	**697.25**
Sum of two sides 1500 mm	483.54	611.97	3.81	126.03	nr	**738.00**
Sum of two sides 1600 mm	515.00	651.78	3.84	127.03	nr	**778.81**
Branch (Side-on Shoe)						
Sum of two sides 900 mm	54.73	69.27	1.37	45.32	nr	**114.59**
Sum of two sides 1000 mm	56.09	70.98	1.46	48.30	nr	**119.28**
Sum of two sides 1100 mm	57.47	72.74	1.50	49.62	nr	**122.35**
Sum of two sides 1200 mm	58.82	74.45	1.55	51.27	nr	**125.72**
Sum of two sides 1300 mm	60.20	76.18	1.59	52.60	nr	**128.78**
Sum of two sides 1400 mm	59.14	74.84	1.68	55.57	nr	**130.42**
Sum of two sides 1500 mm	61.62	77.99	1.63	53.92	nr	**131.91**
Sum of two sides 1600 mm	64.33	81.41	1.72	56.90	nr	**138.31**
Ductwork 801 to 1000 mm longest side						
Sum of two sides 1100 mm	232.83	294.66	5.84	193.18	m	**487.85**
Sum of two sides 1200 mm	244.06	308.88	6.11	202.12	m	**511.00**
Sum of two sides 1300 mm	255.34	323.16	6.38	211.05	m	**534.21**
Sum of two sides 1400 mm	266.61	337.42	6.65	219.98	m	**557.40**
Sum of two sides 1500 mm	277.87	351.67	6.91	228.58	m	**580.25**
Sum of two sides 1600 mm	289.14	365.93	7.18	237.51	m	**603.45**
Sum of two sides 1700 mm	300.43	380.22	7.45	246.44	m	**626.67**
Sum of two sides 1800 mm	311.70	394.49	7.71	255.04	m	**649.53**
Sum of two sides 1900 mm	322.97	408.75	7.98	263.98	m	**672.72**
Sum of two sides 2000 mm	334.25	423.02	8.25	272.91	m	**695.93**
Extra over fittings; Ductwork 801 to 1000 mm longest side						
End cap						
Sum of two sides 1100 mm	56.59	71.63	1.25	41.35	nr	**112.98**
Sum of two sides 1200 mm	60.12	76.09	1.28	42.34	nr	**118.43**
Sum of two sides 1300 mm	63.65	80.55	1.31	43.33	nr	**123.89**
Sum of two sides 1400 mm	67.19	85.04	1.34	44.33	nr	**129.37**
Sum of two sides 1500 mm	70.72	89.50	1.36	44.99	nr	**134.49**
Sum of two sides 1600 mm	74.23	93.95	1.39	45.98	nr	**139.93**
Sum of two sides 1700 mm	77.77	98.43	1.42	46.97	nr	**145.40**

38 VENTILATION/AIR CONDITIONING SYSTEMS

Item	Net Price £	Material £	Labour hours	Labour £	Unit	Total rate £
DUCTWORK: FIRE RATED – cont						
Extra over fittings – cont						
End cap – cont						
Sum of two sides 1800 mm	81.32	102.92	1.45	47.97	nr	150.88
Sum of two sides 1900 mm	84.83	107.36	1.48	48.96	nr	156.32
Sum of two sides 2000 mm	88.36	111.83	1.51	49.95	nr	161.78
Reducer						
Sum of two sides 1100 mm	205.83	260.50	3.22	106.52	nr	367.02
Sum of two sides 1200 mm	216.21	273.64	3.28	108.50	nr	382.14
Sum of two sides 1300 mm	226.56	286.74	3.33	110.16	nr	396.89
Sum of two sides 1400 mm	236.96	299.89	3.38	111.81	nr	411.70
Sum of two sides 1500 mm	247.27	312.95	3.44	113.79	nr	426.74
Sum of two sides 1600 mm	257.68	326.12	3.49	115.45	nr	441.57
Sum of two sides 1700 mm	268.02	339.20	3.54	117.10	nr	456.30
Sum of two sides 1800 mm	278.38	352.32	3.60	119.09	nr	471.40
Sum of two sides 1900 mm	288.78	365.48	3.65	120.74	nr	486.22
Sum of two sides 2000 mm	304.28	385.10	3.70	122.39	nr	507.49
Offset						
Sum of two sides 1100 mm	428.10	541.80	3.78	125.04	nr	666.85
Sum of two sides 1200 mm	446.21	564.72	3.89	128.68	nr	693.40
Sum of two sides 1300 mm	464.28	587.59	4.00	132.32	nr	719.91
Sum of two sides 1400 mm	482.38	610.50	4.11	135.96	nr	746.46
Sum of two sides 1500 mm	500.49	633.41	4.23	139.93	nr	773.34
Sum of two sides 1600 mm	518.61	656.35	4.34	143.57	nr	799.92
Sum of two sides 1700 mm	536.68	679.23	4.45	147.20	nr	826.43
Sum of two sides 1800 mm	554.79	702.14	4.56	150.84	nr	852.98
Sum of two sides 1900 mm	572.85	725.00	4.67	154.48	nr	879.48
Sum of two sides 2000 mm	600.01	759.37	5.53	182.93	nr	942.30
90° radius bend						
Sum of two sides 1100 mm	235.28	297.77	3.07	101.55	nr	399.33
Sum of two sides 1200 mm	254.65	322.29	3.09	102.22	nr	424.50
Sum of two sides 1300 mm	274.05	346.84	3.12	103.21	nr	450.05
Sum of two sides 1400 mm	293.43	371.37	3.15	104.20	nr	475.57
Sum of two sides 1500 mm	312.81	395.90	3.17	104.86	nr	500.76
Sum of two sides 1600 mm	332.18	420.41	3.20	105.85	nr	526.27
Sum of two sides 1700 mm	351.55	444.92	3.22	106.52	nr	551.44
Sum of two sides 1800 mm	370.93	469.45	3.25	107.51	nr	576.96
Sum of two sides 1900 mm	390.31	493.98	3.28	108.50	nr	602.48
Sum of two sides 2000 mm	409.11	517.77	3.30	109.16	nr	626.93
45° bend						
Sum of two sides 1100 mm	215.36	272.56	1.91	63.18	nr	335.74
Sum of two sides 1200 mm	224.36	283.95	1.96	64.84	nr	348.78
Sum of two sides 1300 mm	233.37	295.35	2.02	66.82	nr	362.17
Sum of two sides 1400 mm	242.40	306.78	2.07	68.47	nr	375.25
Sum of two sides 1500 mm	251.41	318.18	2.13	70.46	nr	388.64
Sum of two sides 1600 mm	260.40	329.57	2.18	72.11	nr	401.68
Sum of two sides 1700 mm	269.40	340.95	2.24	74.10	nr	415.05
Sum of two sides 1800 mm	278.41	352.36	2.30	76.08	nr	428.44
Sum of two sides 1900 mm	287.39	363.72	2.35	77.74	nr	441.45
Sum of two sides 2000 mm	296.45	375.19	2.76	91.30	nr	466.49

38 VENTILATION/AIR CONDITIONING SYSTEMS

Item	Net Price £	Material £	Labour hours	Labour £	Unit	Total rate £
90° mitre bend						
Sum of two sides 1100 mm	357.80	452.84	3.69	122.06	nr	**574.90**
Sum of two sides 1200 mm	389.28	492.67	3.72	123.06	nr	**615.72**
Sum of two sides 1300 mm	420.66	532.39	3.75	124.05	nr	**656.44**
Sum of two sides 1400 mm	452.12	572.20	3.78	125.04	nr	**697.25**
Sum of two sides 1500 mm	483.60	612.05	3.81	126.03	nr	**738.08**
Sum of two sides 1600 mm	515.00	651.78	3.84	127.03	nr	**778.81**
Sum of two sides 1700 mm	546.46	691.60	3.87	128.02	nr	**819.62**
Sum of two sides 1800 mm	577.91	731.40	3.90	129.01	nr	**860.41**
Sum of two sides 1900 mm	609.34	771.18	3.93	130.00	nr	**901.18**
Sum of two sides 2000 mm	640.79	810.98	3.96	131.00	nr	**941.98**
Branch (Side-on Shoe)						
Sum of two sides 1100 mm	57.43	72.68	1.50	49.62	nr	**122.30**
Sum of two sides 1200 mm	58.75	74.35	1.55	51.27	nr	**125.62**
Sum of two sides 1300 mm	60.17	76.16	1.59	52.60	nr	**128.75**
Sum of two sides 1400 mm	61.58	77.94	1.63	53.92	nr	**131.86**
Sum of two sides 1500 mm	62.96	79.69	1.68	55.57	nr	**135.26**
Sum of two sides 1600 mm	64.33	81.41	1.72	56.90	nr	**138.31**
Sum of two sides 1700 mm	65.74	83.21	1.77	58.55	nr	**141.76**
Sum of two sides 1800 mm	67.14	84.97	1.81	59.87	nr	**144.85**
Sum of two sides 1900 mm	68.52	86.72	1.86	61.53	nr	**148.25**
Sum of two sides 2000 mm	70.59	89.34	1.90	62.85	nr	**152.19**
Ductwork 1001 to 1250 mm longest side						
Sum of two sides 1300 mm	347.88	440.27	6.38	211.05	m	**651.32**
Sum of two sides 1400 mm	362.86	459.24	6.65	219.98	m	**679.22**
Sum of two sides 1500 mm	377.89	478.25	6.91	228.58	m	**706.83**
Sum of two sides 1600 mm	389.11	492.46	6.91	228.58	m	**721.04**
Sum of two sides 1700 mm	400.34	506.67	7.45	246.44	m	**753.11**
Sum of two sides 1800 mm	415.27	525.57	7.71	255.04	m	**780.62**
Sum of two sides 1900 mm	430.25	544.53	7.98	263.98	m	**808.50**
Sum of two sides 2000 mm	445.25	563.51	8.25	272.91	m	**836.42**
Sum of two sides 2100 mm	460.22	582.46	9.62	318.23	m	**900.68**
Sum of two sides 2200 mm	471.47	596.69	9.62	318.23	m	**914.92**
Sum of two sides 2300 mm	482.68	610.89	10.07	333.11	m	**944.00**
Sum of two sides 2400 mm	497.66	629.84	10.47	346.34	m	**976.19**
Sum of two sides 2500 mm	512.64	648.80	10.87	359.58	m	**1008.37**
Extra over fittings; Ductwork 1001 to 1250 mm longest side						
End cap						
Sum of two sides 1300 mm	63.67	80.58	1.31	43.33	nr	**123.91**
Sum of two sides 1400 mm	67.19	85.04	1.34	44.33	nr	**129.37**
Sum of two sides 1500 mm	70.68	89.45	1.36	44.99	nr	**134.44**
Sum of two sides 1600 mm	74.23	93.95	1.39	45.98	nr	**139.93**
Sum of two sides 1700 mm	77.77	98.43	1.42	46.97	nr	**145.40**
Sum of two sides 1800 mm	81.26	102.85	1.45	47.97	nr	**150.81**
Sum of two sides 1900 mm	84.83	107.36	1.48	48.96	nr	**156.32**
Sum of two sides 2000 mm	88.36	111.83	1.51	49.95	nr	**161.78**
Sum of two sides 2100 mm	91.90	116.30	2.66	87.99	nr	**204.29**
Sum of two sides 2200 mm	95.42	120.76	2.80	92.62	nr	**213.39**
Sum of two sides 2300 mm	98.97	125.25	2.95	97.58	nr	**222.84**

38 VENTILATION/AIR CONDITIONING SYSTEMS

Item	Net Price £	Material £	Labour hours	Labour £	Unit	Total rate £
DUCTWORK: FIRE RATED – cont						
Extra over fittings – cont						
End cap – cont						
Sum of two sides 2400 mm	102.47	129.69	3.10	102.55	nr	232.24
Sum of two sides 2500 mm	106.00	134.15	3.24	107.18	nr	241.33
Reducer						
Sum of two sides 1300 mm	226.49	286.64	3.33	110.16	nr	396.80
Sum of two sides 1400 mm	236.68	299.54	3.38	111.81	nr	411.35
Sum of two sides 1500 mm	246.87	312.44	3.44	113.79	nr	426.24
Sum of two sides 1600 mm	257.12	325.41	3.49	115.45	nr	440.85
Sum of two sides 1700 mm	267.39	338.41	3.54	117.10	nr	455.51
Sum of two sides 1800 mm	277.65	351.40	3.60	119.09	nr	470.48
Sum of two sides 1900 mm	287.86	364.32	3.65	120.74	nr	485.06
Sum of two sides 2000 mm	298.12	377.29	3.70	122.39	nr	499.69
Sum of two sides 2100 mm	308.35	390.24	3.75	124.05	nr	514.29
Sum of two sides 2200 mm	318.57	403.18	3.80	125.70	nr	528.88
Sum of two sides 2300 mm	328.82	416.15	3.85	127.36	nr	543.51
Sum of two sides 2400 mm	339.06	429.12	3.90	129.01	nr	558.13
Sum of two sides 2500 mm	349.34	442.12	3.95	130.66	nr	572.78
Offset						
Sum of two sides 1300 mm	462.73	585.63	4.00	132.32	nr	717.95
Sum of two sides 1400 mm	481.11	608.89	4.11	135.96	nr	744.84
Sum of two sides 1500 mm	499.45	632.10	4.23	139.93	nr	772.03
Sum of two sides 1600 mm	517.81	655.34	4.34	143.57	nr	798.91
Sum of two sides 1700 mm	536.15	678.56	4.45	147.20	nr	825.76
Sum of two sides 1800 mm	554.50	701.77	4.56	150.84	nr	852.61
Sum of two sides 1900 mm	572.83	724.97	4.67	154.48	nr	879.45
Sum of two sides 2000 mm	591.17	748.18	5.53	182.93	nr	931.11
Sum of two sides 2100 mm	609.50	771.38	5.76	190.54	nr	961.92
Sum of two sides 2200 mm	627.54	794.22	5.99	198.15	nr	992.36
Sum of two sides 2300 mm	646.20	817.83	6.22	205.76	nr	1023.58
Sum of two sides 2400 mm	664.52	841.01	6.45	213.36	nr	1054.38
Sum of two sides 2500 mm	682.90	864.28	6.68	220.97	nr	1085.25
90° radius bend						
Sum of two sides 1300 mm	273.68	346.37	3.12	103.21	nr	449.58
Sum of two sides 1400 mm	304.55	385.44	3.15	104.20	nr	489.64
Sum of two sides 1500 mm	312.59	395.61	3.17	104.86	nr	500.47
Sum of two sides 1600 mm	331.98	420.15	3.20	105.85	nr	526.01
Sum of two sides 1700 mm	351.42	444.76	3.22	106.52	nr	551.28
Sum of two sides 1800 mm	370.85	469.34	3.25	107.51	nr	576.85
Sum of two sides 1900 mm	390.28	493.94	3.28	108.50	nr	602.44
Sum of two sides 2000 mm	409.73	518.55	3.30	109.16	nr	627.71
Sum of two sides 2100 mm	429.16	543.15	3.32	109.82	nr	652.97
Sum of two sides 2200 mm	448.57	567.72	3.34	110.49	nr	678.20
Sum of two sides 2300 mm	468.03	592.34	3.36	111.15	nr	703.49
Sum of two sides 2400 mm	487.48	616.95	3.38	111.81	nr	728.76
Sum of two sides 2500 mm	506.90	641.53	3.40	112.47	nr	754.00
45° bend						
Sum of two sides 1300 mm	221.69	280.57	2.02	66.82	nr	347.39
Sum of two sides 1400 mm	231.67	293.20	2.07	68.47	nr	361.67
Sum of two sides 1500 mm	241.63	305.81	2.13	70.46	nr	376.26

38 VENTILATION/AIR CONDITIONING SYSTEMS

Item	Net Price £	Material £	Labour hours	Labour £	Unit	Total rate £
Sum of two sides 1600 mm	251.62	318.45	2.18	72.11	nr	**390.57**
Sum of two sides 1700 mm	261.58	331.06	2.24	74.10	nr	**405.16**
Sum of two sides 1800 mm	271.59	343.72	2.30	76.08	nr	**419.80**
Sum of two sides 1900 mm	281.55	356.33	2.35	77.74	nr	**434.06**
Sum of two sides 2000 mm	291.55	368.99	2.76	91.30	nr	**460.29**
Sum of two sides 2100 mm	301.54	381.63	2.89	95.60	nr	**477.23**
Sum of two sides 2200 mm	311.54	394.28	3.01	99.57	nr	**493.85**
Sum of two sides 2300 mm	321.50	406.89	3.13	103.54	nr	**510.43**
Sum of two sides 2400 mm	331.51	419.56	3.26	107.84	nr	**527.40**
Sum of two sides 2500 mm	341.47	432.17	3.37	111.48	nr	**543.65**
90° mitre bend						
Sum of two sides 1300 mm	419.71	531.18	3.75	124.05	nr	**655.23**
Sum of two sides 1400 mm	450.87	570.62	3.78	125.04	nr	**695.66**
Sum of two sides 1500 mm	483.26	611.61	3.81	126.03	nr	**737.64**
Sum of two sides 1600 mm	514.98	651.76	3.84	127.03	nr	**778.78**
Sum of two sides 1700 mm	546.69	691.89	3.87	128.02	nr	**819.91**
Sum of two sides 1800 mm	578.40	732.02	3.90	129.01	nr	**861.03**
Sum of two sides 1900 mm	610.14	772.19	3.93	130.00	nr	**902.19**
Sum of two sides 2000 mm	641.90	812.39	3.96	131.00	nr	**943.39**
Sum of two sides 2100 mm	673.61	852.52	3.99	131.99	nr	**984.51**
Sum of two sides 2200 mm	705.35	892.70	4.02	132.98	nr	**1025.68**
Sum of two sides 2300 mm	737.10	932.87	4.05	133.97	nr	**1066.84**
Sum of two sides 2400 mm	768.76	972.95	4.08	134.96	nr	**1107.91**
Sum of two sides 2500 mm	800.49	1013.11	4.11	135.96	nr	**1149.06**
Branch (Side-on Shoe)						
Sum of two sides 1300 mm	60.27	76.28	1.59	52.60	nr	**128.88**
Sum of two sides 1400 mm	61.67	78.05	1.63	53.92	nr	**131.97**
Sum of two sides 1500 mm	63.03	79.77	1.68	55.57	nr	**135.34**
Sum of two sides 1600 mm	64.38	81.48	1.72	56.90	nr	**138.38**
Sum of two sides 1700 mm	65.79	83.26	1.77	58.55	nr	**141.81**
Sum of two sides 1800 mm	67.19	85.04	1.81	59.87	nr	**144.91**
Sum of two sides 1900 mm	68.56	86.76	1.86	61.53	nr	**148.29**
Sum of two sides 2000 mm	69.96	88.54	1.90	62.85	nr	**151.40**
Sum of two sides 2100 mm	71.32	90.27	2.58	85.35	nr	**175.61**
Sum of two sides 2200 mm	72.68	91.98	2.61	86.34	nr	**178.32**
Sum of two sides 2300 mm	74.10	93.79	2.64	87.33	nr	**181.12**
Sum of two sides 2400 mm	75.49	95.54	2.88	95.27	nr	**190.81**
Sum of two sides 2500 mm	76.84	97.25	2.91	96.26	nr	**193.51**
Ductwork 1251 to 2000 mm longest side						
Sum of two sides 1700 mm	425.91	539.03	7.45	246.44	m	**785.47**
Sum of two sides 1800 mm	453.21	573.59	7.71	255.04	m	**828.63**
Sum of two sides 1900 mm	480.50	608.12	7.98	263.98	m	**872.10**
Sum of two sides 2000 mm	507.76	642.63	8.25	272.91	m	**915.53**
Sum of two sides 2100 mm	535.10	677.23	9.62	318.23	m	**995.45**
Sum of two sides 2200 mm	562.36	711.72	9.66	319.55	m	**1031.27**
Sum of two sides 2300 mm	589.61	746.21	10.07	333.11	m	**1079.32**
Sum of two sides 2400 mm	616.90	780.75	10.47	346.34	m	**1127.09**
Sum of two sides 2500 mm	644.18	815.28	10.87	359.58	m	**1174.86**
Sum of two sides 2600 mm	671.54	849.90	11.27	372.81	m	**1222.70**
Sum of two sides 2700 mm	698.79	884.39	11.67	386.04	m	**1270.43**

Material Costs/Prices for Measured Works

38 VENTILATION/AIR CONDITIONING SYSTEMS

Item	Net Price £	Material £	Labour hours	Labour £	Unit	Total rate £
DUCTWORK: FIRE RATED – cont						
Ductwork 1251 to 2000 mm longest side – cont						
Sum of two sides 2800 mm	726.09	918.94	12.08	399.60	m	**1318.54**
Sum of two sides 2900 mm	753.34	953.43	12.48	412.83	m	**1366.26**
Sum of two sides 3000 mm	780.63	987.96	12.88	426.07	m	**1414.03**
Sum of two sides 3100 mm	807.93	1022.52	13.26	438.64	m	**1461.16**
Sum of two sides 3200 mm	835.22	1057.06	13.69	452.86	m	**1509.92**
Sum of two sides 3300 mm	513.86	650.35	14.09	466.09	m	**1116.44**
Sum of two sides 3400 mm	530.09	670.88	14.49	479.32	m	**1150.20**
Sum of two sides 3500 mm	546.33	691.44	14.89	492.56	m	**1183.99**
Sum of two sides 3600 mm	562.56	711.98	15.29	505.79	m	**1217.77**
Sum of two sides 3700 mm	578.87	732.62	15.69	519.02	m	**1251.64**
Sum of two sides 3800 mm	595.10	753.15	16.09	532.25	m	**1285.40**
Sum of two sides 3900 mm	611.36	773.74	16.49	545.48	m	**1319.22**
Sum of two sides 4000 mm	627.61	794.30	16.89	558.71	m	**1353.01**
Extra over fittings; Ductwork 1251 to 2000 mm longest side						
End cap						
Sum of two sides 1700 mm	115.53	146.21	1.42	46.97	nr	**193.18**
Sum of two sides 1800 mm	122.97	155.63	1.45	47.97	nr	**203.59**
Sum of two sides 1900 mm	130.39	165.02	1.48	48.96	nr	**213.97**
Sum of two sides 2000 mm	137.84	174.45	1.51	49.95	nr	**224.40**
Sum of two sides 2100 mm	145.23	183.81	2.66	87.99	nr	**271.80**
Sum of two sides 2200 mm	152.70	193.25	2.80	92.62	nr	**285.88**
Sum of two sides 2300 mm	160.08	202.60	2.95	97.58	nr	**300.19**
Sum of two sides 2400 mm	167.52	212.02	3.10	102.55	nr	**314.57**
Sum of two sides 2500 mm	174.95	221.42	3.24	107.18	nr	**328.60**
Sum of two sides 2600 mm	182.41	230.85	3.39	112.14	nr	**342.99**
Sum of two sides 2700 mm	189.78	240.19	3.54	117.10	nr	**357.29**
Sum of two sides 2800 mm	197.23	249.62	3.68	121.73	nr	**371.35**
Sum of two sides 2900 mm	204.72	259.09	3.83	126.69	nr	**385.78**
Sum of two sides 3000 mm	212.11	268.45	3.98	131.66	nr	**400.11**
Sum of two sides 3100 mm	219.54	277.86	4.12	136.29	nr	**414.14**
Sum of two sides 3200 mm	227.00	287.29	4.27	141.25	nr	**428.54**
Sum of two sides 3300 mm	165.58	209.56	4.42	146.21	nr	**355.77**
Sum of two sides 3400 mm	170.84	216.22	4.57	151.17	nr	**367.39**
Sum of two sides 3500 mm	175.16	221.68	4.72	156.14	nr	**377.82**
Sum of two sides 3600 mm	178.67	226.13	4.87	161.10	nr	**387.23**
Sum of two sides 3700 mm	183.94	232.80	5.02	166.06	nr	**398.86**
Sum of two sides 3800 mm	276.17	349.52	5.17	171.02	nr	**520.54**
Sum of two sides 3900 mm	197.07	249.41	5.32	175.98	nr	**425.40**
Sum of two sides 4000 mm	202.34	256.08	5.47	180.95	nr	**437.02**
Reducer						
Sum of two sides 1700 mm	249.49	315.76	3.54	117.10	nr	**432.86**
Sum of two sides 1800 mm	279.88	354.22	3.60	119.09	nr	**473.30**
Sum of two sides 1900 mm	310.24	392.64	3.65	120.74	nr	**513.38**

38 VENTILATION/AIR CONDITIONING SYSTEMS

Item	Net Price £	Material £	Labour hours	Labour £	Unit	Total rate £
Sum of two sides 2000 mm	371.00	469.53	2.92	96.59	nr	**566.13**
Sum of two sides 2100 mm	340.64	431.11	3.70	122.39	nr	**553.51**
Sum of two sides 2200 mm	401.39	508.00	3.10	102.55	nr	**610.54**
Sum of two sides 2300 mm	431.76	546.43	3.29	108.83	nr	**655.26**
Sum of two sides 2400 mm	462.15	584.89	3.48	115.12	nr	**700.01**
Sum of two sides 2500 mm	492.48	623.29	3.66	121.07	nr	**744.36**
Sum of two sides 2600 mm	522.89	661.78	3.85	127.36	nr	**789.13**
Sum of two sides 2700 mm	553.27	700.22	4.03	133.31	nr	**833.53**
Sum of two sides 2800 mm	583.66	738.69	4.22	139.60	nr	**878.28**
Sum of two sides 2900 mm	614.04	777.13	4.40	145.55	nr	**922.68**
Sum of two sides 3000 mm	659.54	834.72	4.59	151.84	nr	**986.55**
Sum of two sides 3100 mm	659.54	834.72	4.78	158.12	nr	**992.84**
Sum of two sides 3200 mm	689.95	873.21	4.96	164.07	nr	**1037.28**
Sum of two sides 3300 mm	519.62	657.63	5.15	170.36	nr	**827.99**
Sum of two sides 3400 mm	541.07	684.78	5.34	176.65	nr	**861.43**
Sum of two sides 3500 mm	562.54	711.95	5.53	182.93	nr	**894.88**
Sum of two sides 3600 mm	583.98	739.08	5.72	189.22	nr	**928.30**
Sum of two sides 3700 mm	605.46	766.27	5.91	195.50	nr	**961.77**
Sum of two sides 3800 mm	626.90	793.41	6.10	201.79	nr	**995.19**
Sum of two sides 3900 mm	648.39	820.60	6.29	208.07	nr	**1028.68**
Sum of two sides 4000 mm	669.85	847.76	6.48	214.36	nr	**1062.12**
Offset						
Sum of two sides 1700 mm	249.49	315.76	4.45	147.20	nr	**462.96**
Sum of two sides 1800 mm	270.05	341.78	4.56	150.84	nr	**492.62**
Sum of two sides 1900 mm	290.65	367.85	4.67	154.48	nr	**522.33**
Sum of two sides 2000 mm	311.12	393.75	5.53	182.93	nr	**576.68**
Sum of two sides 2100 mm	331.69	419.78	5.76	190.54	nr	**610.32**
Sum of two sides 2200 mm	352.24	445.80	5.99	198.15	nr	**643.95**
Sum of two sides 2300 mm	372.84	471.86	6.22	205.76	nr	**677.62**
Sum of two sides 2400 mm	393.36	497.84	6.45	213.36	nr	**711.20**
Sum of two sides 2500 mm	413.92	523.86	6.68	220.97	nr	**744.83**
Sum of two sides 2600 mm	434.50	549.91	6.91	228.58	nr	**778.49**
Sum of two sides 2700 mm	455.06	575.93	7.14	236.19	nr	**812.12**
Sum of two sides 2800 mm	475.64	601.97	7.37	243.80	nr	**845.77**
Sum of two sides 2900 mm	496.14	627.91	7.60	251.41	nr	**879.32**
Sum of two sides 3000 mm	516.74	653.99	7.83	259.01	nr	**913.00**
Sum of two sides 3100 mm	537.31	680.02	8.06	266.62	nr	**946.64**
Sum of two sides 3200 mm	557.86	706.03	8.29	274.23	nr	**980.26**
Sum of two sides 3300 mm	408.61	517.14	8.52	281.84	nr	**798.98**
Sum of two sides 3400 mm	423.13	535.51	8.75	289.45	nr	**824.96**
Sum of two sides 3500 mm	437.66	553.90	8.98	297.06	nr	**850.96**
Sum of two sides 3600 mm	452.18	572.27	9.21	304.66	nr	**876.94**
Sum of two sides 3700 mm	466.71	590.67	9.44	312.27	nr	**902.94**
Sum of two sides 3800 mm	481.22	609.04	9.67	319.88	nr	**928.92**
Sum of two sides 3900 mm	495.75	627.42	9.90	327.49	nr	**954.91**
Sum of two sides 4000 mm	510.30	645.83	10.13	335.10	nr	**980.93**
90° radius bend						
Sum of two sides 1700 mm	347.22	439.44	3.22	106.52	nr	**545.95**
Sum of two sides 1800 mm	378.59	479.14	3.25	107.51	nr	**586.65**
Sum of two sides 1900 mm	399.57	505.70	3.32	109.82	nr	**615.52**

38 VENTILATION/AIR CONDITIONING SYSTEMS

Item	Net Price £	Material £	Labour hours	Labour £	Unit	Total rate £
DUCTWORK: FIRE RATED – cont						
Extra over fittings – cont						
90° radius bend – cont						
Sum of two sides 2000 mm	409.99	518.88	3.28	108.50	nr	**627.38**
Sum of two sides 2100 mm	441.36	558.59	3.30	109.16	nr	**667.75**
Sum of two sides 2200 mm	504.97	639.10	3.32	109.82	nr	**748.92**
Sum of two sides 2300 mm	535.50	677.73	3.36	111.15	nr	**788.88**
Sum of two sides 2400 mm	566.83	717.37	3.38	111.81	nr	**829.18**
Sum of two sides 2500 mm	598.24	757.14	3.40	112.47	nr	**869.61**
Sum of two sides 2600 mm	629.64	796.87	3.42	113.13	nr	**910.00**
Sum of two sides 2700 mm	661.03	836.61	3.44	113.79	nr	**950.40**
Sum of two sides 2800 mm	692.42	876.33	3.46	114.46	nr	**990.78**
Sum of two sides 2900 mm	723.73	915.95	3.48	115.12	nr	**1031.07**
Sum of two sides 3000 mm	755.14	955.70	3.50	115.78	nr	**1071.48**
Sum of two sides 3100 mm	786.50	995.39	3.52	116.44	nr	**1111.83**
Sum of two sides 3200 mm	790.87	1000.92	3.54	117.10	nr	**1118.03**
Sum of two sides 3300 mm	560.34	709.16	3.56	117.76	nr	**826.92**
Sum of two sides 3400 mm	825.41	1044.64	3.58	118.43	nr	**1163.07**
Sum of two sides 3500 mm	579.90	733.92	3.60	119.09	nr	**853.01**
Sum of two sides 3600 mm	599.83	759.15	3.62	119.75	nr	**878.90**
Sum of two sides 3700 mm	619.76	784.37	3.64	120.41	nr	**904.78**
Sum of two sides 3800 mm	639.74	809.65	3.66	121.07	nr	**930.73**
Sum of two sides 3900 mm	659.70	834.92	3.68	121.73	nr	**956.65**
Sum of two sides 4000 mm	679.65	860.16	3.70	122.39	nr	**982.56**
45° bend						
Sum of two sides 1700 mm	257.41	325.78	2.24	74.10	nr	**399.87**
Sum of two sides 1800 mm	285.86	361.79	2.30	76.08	nr	**437.87**
Sum of two sides 1900 mm	314.22	397.68	2.35	77.74	nr	**475.41**
Sum of two sides 2000 mm	342.71	433.73	2.76	91.30	nr	**525.03**
Sum of two sides 2100 mm	441.36	558.59	3.01	99.57	nr	**658.16**
Sum of two sides 2200 mm	428.05	541.74	3.13	103.54	nr	**645.28**
Sum of two sides 2300 mm	472.72	598.28	2.89	95.60	nr	**693.88**
Sum of two sides 2400 mm	456.44	577.67	3.26	107.84	nr	**685.51**
Sum of two sides 2500 mm	484.88	613.66	3.37	111.48	nr	**725.14**
Sum of two sides 2600 mm	513.27	649.59	3.49	115.45	nr	**765.04**
Sum of two sides 2700 mm	541.77	685.66	3.62	119.75	nr	**805.41**
Sum of two sides 2800 mm	570.19	721.63	3.74	123.72	nr	**845.35**
Sum of two sides 2900 mm	598.61	757.60	3.86	127.69	nr	**885.29**
Sum of two sides 3000 mm	627.04	793.59	3.98	131.66	nr	**925.24**
Sum of two sides 3100 mm	655.49	829.58	4.10	135.63	nr	**965.21**
Sum of two sides 3200 mm	683.89	865.53	4.22	139.60	nr	**1005.12**
Sum of two sides 3300 mm	503.25	636.92	4.34	143.57	nr	**780.48**
Sum of two sides 3400 mm	523.34	662.34	4.46	147.54	nr	**809.87**
Sum of two sides 3500 mm	543.43	687.77	4.58	151.50	nr	**839.27**
Sum of two sides 3600 mm	563.54	713.21	4.70	155.47	nr	**868.69**
Sum of two sides 3700 mm	583.58	738.58	4.82	159.44	nr	**898.02**
Sum of two sides 3800 mm	603.68	764.02	4.94	163.41	nr	**927.43**
Sum of two sides 3900 mm	623.74	789.41	5.06	167.38	nr	**956.79**
Sum of two sides 4000 mm	643.87	814.88	5.18	171.35	nr	**986.24**

38 VENTILATION/AIR CONDITIONING SYSTEMS

Item	Net Price £	Material £	Labour hours	Labour £	Unit	Total rate £
90° mitre bend						
Sum of two sides 1700 mm	520.04	658.16	3.87	128.02	nr	**786.18**
Sum of two sides 1800 mm	609.91	771.90	3.90	129.01	nr	**900.92**
Sum of two sides 1900 mm	699.81	885.67	3.93	130.00	nr	**1015.68**
Sum of two sides 2000 mm	789.61	999.34	3.96	131.00	nr	**1130.33**
Sum of two sides 2100 mm	879.52	1113.12	3.82	126.36	nr	**1239.48**
Sum of two sides 2200 mm	969.38	1226.85	3.83	126.69	nr	**1353.54**
Sum of two sides 2300 mm	1059.28	1340.62	3.83	126.69	nr	**1467.31**
Sum of two sides 2400 mm	1149.15	1454.36	3.84	127.03	nr	**1581.39**
Sum of two sides 2500 mm	1239.01	1568.09	3.85	127.36	nr	**1695.45**
Sum of two sides 2600 mm	1328.83	1681.76	3.85	127.36	nr	**1809.12**
Sum of two sides 2700 mm	1418.71	1795.52	3.86	127.69	nr	**1923.21**
Sum of two sides 2800 mm	1508.58	1909.26	3.86	127.69	nr	**2036.95**
Sum of two sides 2900 mm	1598.46	2023.01	3.87	128.02	nr	**2151.03**
Sum of two sides 3000 mm	1688.32	2136.74	3.87	128.02	nr	**2264.75**
Sum of two sides 3100 mm	1778.22	2250.52	3.88	128.35	nr	**2378.87**
Sum of two sides 3200 mm	1868.06	2364.22	3.88	128.35	nr	**2492.57**
Sum of two sides 3300 mm	1383.23	1750.61	3.89	128.68	nr	**1879.29**
Sum of two sides 3400 mm	1446.71	1830.96	3.90	129.01	nr	**1959.97**
Sum of two sides 3500 mm	1510.21	1911.32	3.91	129.34	nr	**2040.66**
Sum of two sides 3600 mm	1573.70	1991.68	3.92	129.67	nr	**2121.35**
Sum of two sides 3700 mm	1637.20	2072.04	3.93	130.00	nr	**2202.04**
Sum of two sides 3800 mm	1700.68	2152.38	3.94	130.33	nr	**2282.71**
Sum of two sides 3900 mm	1764.19	2232.75	3.95	130.66	nr	**2363.42**
Sum of two sides 4000 mm	1827.67	2313.10	3.96	131.00	nr	**2444.09**
Branch (Side-on Shoe)						
Sum of two sides 1700 mm	58.96	74.62	1.77	58.55	nr	**133.18**
Sum of two sides 1800 mm	64.40	81.51	1.81	59.87	nr	**141.38**
Sum of two sides 1900 mm	69.93	88.50	1.86	61.53	nr	**150.03**
Sum of two sides 2000 mm	75.41	95.44	1.90	62.85	nr	**158.29**
Sum of two sides 2100 mm	80.90	102.38	2.58	85.35	nr	**187.73**
Sum of two sides 2200 mm	86.36	109.29	2.61	86.34	nr	**195.63**
Sum of two sides 2300 mm	91.87	116.27	2.65	87.66	nr	**203.94**
Sum of two sides 2400 mm	97.34	123.19	2.68	88.65	nr	**211.84**
Sum of two sides 2500 mm	102.83	130.14	2.71	89.65	nr	**219.79**
Sum of two sides 2600 mm	108.33	137.11	2.75	90.97	nr	**228.08**
Sum of two sides 2700 mm	113.76	143.98	2.78	91.96	nr	**235.94**
Sum of two sides 2800 mm	119.29	150.97	2.81	92.95	nr	**243.93**
Sum of two sides 2900 mm	124.77	157.91	2.84	93.95	nr	**251.86**
Sum of two sides 3000 mm	130.25	164.84	2.87	94.94	nr	**259.78**
Sum of two sides 3100 mm	135.73	171.78	2.90	95.93	nr	**267.71**
Sum of two sides 3200 mm	141.23	178.74	2.93	96.92	nr	**275.67**
Sum of two sides 3300 mm	103.64	131.17	2.93	96.92	nr	**228.09**
Sum of two sides 3400 mm	107.51	136.07	3.00	99.24	nr	**235.31**
Sum of two sides 3500 mm	111.39	140.98	3.03	100.23	nr	**241.21**
Sum of two sides 3600 mm	115.23	145.84	3.06	101.22	nr	**247.06**
Sum of two sides 3700 mm	119.14	150.78	3.09	102.22	nr	**253.00**
Sum of two sides 3800 mm	123.03	155.71	3.12	103.21	nr	**258.92**
Sum of two sides 3900 mm	126.89	160.60	3.15	104.20	nr	**264.80**
Sum of two sides 4000 mm	130.74	165.47	3.18	105.19	nr	**270.66**

38 VENTILATION/AIR CONDITIONING SYSTEMS

Item	Net Price £	Material £	Labour hours	Labour £	Unit	Total rate £
DUCTWORK: FIRE RATED – cont						
Rectangular section ductwork to BSEN1366 (Duraduct LT), ducts 'Type A' and 'Type B'; manufactured from 6 mm thick laminate fire board consisting of steel circular hole punched facings pressed to a fibre cement core; provides up to 4 hours stability, 4 hours integrity and 32 minutes insulation; including all necessary stiffeners, joints and supports in the running length						
Ductwork up to 600 mm longest side						
Sum of two sides 200 mm	529.68	670.37	4.00	132.32	m	**802.69**
Sum of two sides 400 mm	542.61	686.73	4.00	132.32	m	**819.04**
Sum of two sides 600 mm	567.02	717.62	4.00	132.32	m	**849.94**
Sum of two sides 800 mm	748.24	946.97	5.50	181.94	m	**1128.91**
Sum of two sides 1000 mm	1008.02	1275.75	5.50	181.94	m	**1457.69**
Sum of two sides 1200 mm	1041.93	1318.66	6.00	198.48	m	**1517.14**
Extra over fittings; Ductwork up to 600 mm longest side						
End cap						
Sum of two sides 200 mm	125.71	159.09	0.81	26.79	m	**185.89**
Sum of two sides 400 mm	125.71	159.09	0.87	28.78	m	**187.87**
Sum of two sides 600 mm	125.71	159.09	0.93	30.76	m	**189.86**
Sum of two sides 800 mm	203.05	256.98	0.98	32.42	m	**289.40**
Sum of two sides 1000 mm	203.05	256.98	1.22	40.36	m	**297.34**
Sum of two sides 1200 mm	309.37	391.54	1.28	42.34	m	**433.88**
Reducer						
Sum of two sides 200 mm	111.16	140.69	2.23	73.77	m	**214.46**
Sum of two sides 400 mm	111.16	140.69	2.51	83.03	m	**223.72**
Sum of two sides 600 mm	111.16	140.69	2.79	92.29	m	**232.98**
Sum of two sides 800 mm	217.54	275.32	3.01	99.57	m	**374.89**
Sum of two sides 1000 mm	217.54	275.32	3.17	104.86	m	**380.18**
Sum of two sides 1200 mm	275.56	348.75	3.28	108.50	m	**457.25**
Offset						
Sum of two sides 200 mm	111.16	140.69	2.95	97.58	m	**238.27**
Sum of two sides 400 mm	111.16	140.69	3.26	107.84	m	**248.53**
Sum of two sides 600 mm	111.16	140.69	3.57	118.09	m	**258.78**
Sum of two sides 800 mm	217.54	275.32	3.23	106.85	m	**382.16**
Sum of two sides 1000 mm	217.54	275.32	3.67	121.40	m	**396.72**
Sum of two sides 1200 mm	275.56	348.75	3.89	128.68	m	**477.43**
90° radius bend						
Sum of two sides 200 mm	372.21	471.07	2.06	68.14	m	**539.22**
Sum of two sides 400 mm	372.21	471.07	2.36	78.07	m	**549.14**
Sum of two sides 600 mm	372.21	471.07	2.66	87.99	m	**559.07**
Sum of two sides 800 mm	435.08	550.63	2.97	98.25	m	**648.88**
Sum of two sides 1000 mm	435.08	550.63	3.04	100.56	m	**651.20**
Sum of two sides 1200 mm	493.10	624.07	3.09	102.22	m	**726.28**

38 VENTILATION/AIR CONDITIONING SYSTEMS

Item	Net Price £	Material £	Labour hours	Labour £	Unit	Total rate £
45° radius bend						
Sum of two sides 200 mm	372.21	471.07	1.52	50.28	m	**521.36**
Sum of two sides 400 mm	372.21	471.07	1.66	54.91	m	**525.99**
Sum of two sides 600 mm	372.21	471.07	1.80	59.54	m	**530.62**
Sum of two sides 800 mm	435.08	550.63	1.93	63.84	m	**614.48**
Sum of two sides 1000 mm	435.08	550.63	2.05	67.81	m	**618.45**
Sum of two sides 1200 mm	493.10	624.07	2.17	71.78	m	**695.85**
90° mitre bend						
Sum of two sides 200 mm	372.21	471.07	2.47	81.71	m	**552.78**
Sum of two sides 400 mm	372.21	471.07	2.84	93.95	m	**565.02**
Sum of two sides 600 mm	372.21	471.07	3.20	105.85	m	**576.93**
Sum of two sides 800 mm	435.08	550.63	3.56	117.76	m	**668.40**
Sum of two sides 1000 mm	435.08	550.63	3.66	121.07	m	**671.71**
Sum of two sides 1200 mm	493.10	624.07	3.72	123.06	m	**747.12**
Branch						
Sum of two sides 200 mm	135.37	171.32	0.98	32.42	m	**203.74**
Sum of two sides 400 mm	135.37	171.32	1.14	37.71	m	**209.03**
Sum of two sides 600 mm	135.37	171.32	1.30	43.00	m	**214.32**
Sum of two sides 800 mm	169.19	214.12	1.46	48.30	m	**262.42**
Sum of two sides 1000 mm	184.47	233.47	1.46	48.30	m	**281.77**
Sum of two sides 1200 mm	227.17	287.50	1.55	51.27	m	**338.78**
Ductwork 601 to 1000 mm longest side						
Sum of two sides 1000 mm	748.24	946.97	6.00	198.48	m	**1145.45**
Sum of two sides 1100 mm	1041.93	1318.66	6.00	198.48	m	**1517.14**
Sum of two sides 1300 mm	1059.31	1340.67	6.00	198.48	m	**1539.14**
Sum of two sides 1500 mm	1084.01	1371.92	6.00	198.48	m	**1570.40**
Sum of two sides 1700 mm	1310.11	1658.07	6.00	198.48	m	**1856.55**
Sum of two sides 1900 mm	1501.35	1900.11	6.00	198.48	m	**2098.59**
Extra over fittings; Ductwork 601 to 1000 mm longest side						
End cap						
Sum of two sides 1000 mm	203.05	256.98	1.25	41.35	m	**298.33**
Sum of two sides 1100 mm	309.37	391.54	1.25	41.35	m	**432.89**
Sum of two sides 1300 mm	309.37	391.54	1.31	43.33	m	**434.88**
Sum of two sides 1500 mm	309.59	391.82	1.36	44.99	m	**436.81**
Sum of two sides 1700 mm	546.26	691.35	1.42	46.97	m	**738.32**
Sum of two sides 1900 mm	546.26	691.35	1.48	48.96	m	**740.31**
Reducer						
Sum of two sides 1000 mm	217.54	275.32	3.22	106.52	m	**381.83**
Sum of two sides 1100 mm	275.56	348.75	3.22	106.52	m	**455.27**
Sum of two sides 1300 mm	275.56	348.75	3.33	110.16	m	**458.91**
Sum of two sides 1500 mm	275.56	348.75	3.44	113.79	m	**462.55**
Sum of two sides 1700 mm	338.42	428.30	3.54	117.10	m	**545.40**
Sum of two sides 1900 mm	338.42	428.30	3.65	120.74	m	**549.04**
Offset						
Sum of two sides 1000 mm	217.54	275.32	3.78	125.04	m	**400.36**
Sum of two sides 1100 mm	275.56	348.75	3.78	125.04	m	**473.79**
Sum of two sides 1300 mm	275.56	348.75	4.00	132.32	m	**481.07**

38 VENTILATION/AIR CONDITIONING SYSTEMS

Item	Net Price £	Material £	Labour hours	Labour £	Unit	Total rate £
DUCTWORK: FIRE RATED – cont						
Extra over fittings – cont						
Offset – cont						
Sum of two sides 1500 mm	275.56	348.75	4.23	139.93	m	**488.68**
Sum of two sides 1700 mm	338.42	428.30	4.45	147.20	m	**575.50**
Sum of two sides 1900 mm	338.42	428.30	4.67	154.48	m	**582.78**
90° radius bend						
Sum of two sides 1000 mm	435.08	550.63	3.78	125.04	m	**675.68**
Sum of two sides 1100 mm	493.10	624.07	3.07	101.55	m	**725.62**
Sum of two sides 1300 mm	493.10	624.07	3.12	103.21	m	**727.28**
Sum of two sides 1500 mm	493.10	624.07	3.17	104.86	m	**728.93**
Sum of two sides 1700 mm	647.79	819.84	3.22	106.52	m	**926.36**
Sum of two sides 1900 mm	647.79	819.84	3.28	108.50	m	**928.34**
45° bend						
Sum of two sides 1000 mm	493.10	624.07	1.91	63.18	m	**687.25**
Sum of two sides 1100 mm	493.10	624.07	2.00	66.16	m	**690.23**
Sum of two sides 1300 mm	493.10	624.07	2.02	66.82	m	**690.89**
Sum of two sides 1500 mm	493.10	624.07	2.13	70.46	m	**694.53**
Sum of two sides 1700 mm	647.79	819.84	2.24	74.10	m	**893.94**
Sum of two sides 1900 mm	647.79	819.84	2.35	77.74	m	**897.58**
90° mitre bend						
Sum of two sides 1000 mm	435.08	550.63	3.78	125.04	m	**675.68**
Sum of two sides 1100 mm	493.10	624.07	3.69	122.06	m	**746.13**
Sum of two sides 1300 mm	493.10	624.07	3.75	124.05	m	**748.12**
Sum of two sides 1500 mm	493.10	624.07	3.81	126.03	m	**750.10**
Sum of two sides 1700 mm	647.79	819.84	3.87	128.02	m	**947.86**
Sum of two sides 1900 mm	647.79	819.84	3.93	130.00	m	**949.84**
Branch						
Sum of two sides 1000 mm	175.91	222.63	2.38	78.73	m	**301.36**
Sum of two sides 1100 mm	227.17	287.50	1.50	49.62	m	**337.12**
Sum of two sides 1300 mm	227.17	287.50	1.59	52.60	m	**340.10**
Sum of two sides 1500 mm	227.17	287.50	1.68	55.57	m	**343.08**
Sum of two sides 1700 mm	290.05	367.09	1.77	58.55	m	**425.64**
Sum of two sides 1900 mm	290.05	367.09	1.86	61.53	m	**428.62**
Ductwork 1001 to 1250 mm longest side						
Sum of two sides 1300 mm	1041.93	1318.66	6.00	198.48	m	**1517.14**
Sum of two sides 1500 mm	1107.45	1401.58	6.00	198.48	m	**1600.06**
Sum of two sides 1700 mm	1310.11	1658.07	6.00	198.48	m	**1856.55**
Sum of two sides 1900 mm	1331.94	1685.70	6.00	198.48	m	**1884.18**
Sum of two sides 2100 mm	1745.44	2209.02	6.50	215.02	m	**2424.04**
Sum of two sides 2300 mm	1840.26	2329.03	6.50	215.02	m	**2544.05**
Sum of two sides 2500 mm	2190.75	2772.62	6.50	215.02	m	**2987.63**
Extra over fittings; Ductwork 1001 to 1250 mm longest side						
End cap						
Sum of two sides 1300 mm	309.37	391.54	1.31	43.33	m	**434.88**
Sum of two sides 1500 mm	309.37	391.54	1.36	44.99	m	**436.53**
Sum of two sides 1700 mm	546.26	691.35	1.42	46.97	m	**738.32**

38 VENTILATION/AIR CONDITIONING SYSTEMS

Item	Net Price £	Material £	Labour hours	Labour £	Unit	Total rate £
Sum of two sides 1900 mm	546.26	691.35	1.48	48.96	m	**740.31**
Sum of two sides 2100 mm	744.45	942.18	2.66	87.99	m	**1030.17**
Sum of two sides 2300 mm	744.45	942.18	2.95	97.58	m	**1039.76**
Sum of two sides 2500 mm	744.45	942.18	3.24	107.18	m	**1049.35**
Reducer						
Sum of two sides 1300 mm	275.56	348.75	3.33	110.16	m	**458.91**
Sum of two sides 1500 mm	275.56	348.75	3.44	113.79	m	**462.55**
Sum of two sides 1700 mm	338.42	428.30	3.54	117.10	m	**545.40**
Sum of two sides 1900 mm	338.42	428.30	3.65	120.74	m	**549.04**
Sum of two sides 2100 mm	401.22	507.78	3.75	124.05	m	**631.83**
Sum of two sides 2300 mm	401.22	507.78	3.85	127.36	m	**635.13**
Sum of two sides 2500 mm	401.22	507.78	3.95	130.66	m	**638.44**
Offset						
Sum of two sides 1300 mm	275.56	348.75	4.00	132.32	m	**481.07**
Sum of two sides 1500 mm	275.56	348.75	4.23	139.93	m	**488.68**
Sum of two sides 1700 mm	338.42	428.30	4.45	147.20	m	**575.50**
Sum of two sides 1900 mm	338.42	428.30	4.67	154.48	m	**582.78**
Sum of two sides 2100 mm	401.22	507.78	5.76	190.54	m	**698.32**
Sum of two sides 2300 mm	401.22	507.78	6.22	205.76	m	**713.53**
Sum of two sides 2500 mm	401.22	507.78	6.68	220.97	m	**728.75**
90° radius bend						
Sum of two sides 1300 mm	493.10	624.07	3.12	103.21	m	**727.28**
Sum of two sides 1500 mm	493.10	624.07	3.17	104.86	m	**728.93**
Sum of two sides 1700 mm	647.79	819.84	3.22	106.52	m	**926.36**
Sum of two sides 1900 mm	647.79	819.84	3.28	108.50	m	**928.34**
Sum of two sides 2100 mm	662.28	838.18	3.32	109.82	m	**948.00**
Sum of two sides 2300 mm	662.28	838.18	3.36	111.15	m	**949.33**
Sum of two sides 2500 mm	662.28	838.18	3.40	112.47	m	**950.65**
45° bend						
Sum of two sides 1300 mm	493.10	624.07	2.02	66.82	m	**690.89**
Sum of two sides 1500 mm	493.10	624.07	2.13	70.46	m	**694.53**
Sum of two sides 1700 mm	647.79	819.84	2.24	74.10	m	**893.94**
Sum of two sides 1900 mm	647.79	819.84	2.35	77.74	m	**897.58**
Sum of two sides 2100 mm	662.28	838.18	2.89	95.60	m	**933.78**
Sum of two sides 2300 mm	662.28	838.18	3.13	103.54	m	**941.72**
Sum of two sides 2500 mm	662.28	838.18	3.37	111.48	m	**949.66**
90° mitre bend						
Sum of two sides 1300 mm	493.10	624.07	3.75	124.05	m	**748.12**
Sum of two sides 1500 mm	493.10	624.07	3.81	126.03	m	**750.10**
Sum of two sides 1700 mm	647.79	819.84	3.87	128.02	m	**947.86**
Sum of two sides 1900 mm	647.79	819.84	3.93	130.00	m	**949.84**
Sum of two sides 2100 mm	662.28	838.18	3.99	131.99	m	**970.17**
Sum of two sides 2300 mm	662.28	838.18	4.05	133.97	m	**972.15**
Sum of two sides 2500 mm	662.28	838.18	4.11	135.96	m	**974.14**
Branch						
Sum of two sides 1300 mm	227.17	287.50	1.59	52.60	m	**340.10**
Sum of two sides 1500 mm	227.17	287.50	1.68	55.57	m	**343.08**
Sum of two sides 1700 mm	290.05	367.09	1.77	58.55	m	**425.64**
Sum of two sides 1900 mm	290.05	367.09	1.86	61.53	m	**428.62**
Sum of two sides 2100 mm	304.51	385.39	2.58	85.35	m	**470.73**
Sum of two sides 2300 mm	304.51	385.39	2.64	87.33	m	**472.72**
Sum of two sides 2500 mm	304.51	385.39	2.91	96.26	m	**481.65**

38 VENTILATION/AIR CONDITIONING SYSTEMS

Item	Net Price £	Material £	Labour hours	Labour £	Unit	Total rate £
DUCTWORK: FIRE RATED – cont						
Ductwork 1251 to 2000 mm longest side						
Sum of two sides 1800 mm	1310.11	1658.07	6.00	198.48	m	**1856.55**
Sum of two sides 2000 mm	1528.42	1934.37	6.00	198.48	m	**2132.85**
Sum of two sides 2200 mm	1745.44	2209.02	6.50	215.02	m	**2424.04**
Sum of two sides 2400 mm	1962.83	2484.16	6.50	215.02	m	**2699.17**
Sum of two sides 2600 mm	2067.88	2617.11	6.66	220.31	m	**2837.42**
Sum of two sides 2800 mm	2268.25	2870.70	6.66	220.31	m	**3091.01**
Sum of two sides 3000 mm	2467.40	3122.74	6.66	220.31	m	**3343.05**
Sum of two sides 3200 mm	2485.84	3146.08	9.00	297.72	m	**3443.80**
Sum of two sides 3400 mm	2734.42	3460.68	9.00	297.72	m	**3758.40**
Sum of two sides 3600 mm	2783.79	3523.17	11.70	387.03	m	**3910.20**
Sum of two sides 3800 mm	3031.49	3836.65	11.70	387.03	m	**4223.68**
Sum of two sides 4000 mm	3506.97	4438.42	11.70	387.03	m	**4825.45**
Extra over fittings; Ductwork 1251 to 2000 mm longest sides						
End cap						
Sum of two sides 1800 mm	546.26	691.35	1.45	47.97	m	**739.32**
Sum of two sides 2000 mm	546.26	691.35	1.51	49.95	m	**741.30**
Sum of two sides 2200 mm	744.45	942.18	2.80	92.62	m	**1034.80**
Sum of two sides 2400 mm	744.45	942.18	3.10	102.55	m	**1044.72**
Sum of two sides 2600 mm	1019.95	1290.85	3.39	112.14	m	**1402.99**
Sum of two sides 2800 mm	1019.95	1290.85	3.68	121.73	m	**1412.58**
Sum of two sides 3000 mm	1019.95	1290.85	3.98	131.66	m	**1422.50**
Sum of two sides 3200 mm	1377.67	1743.58	4.27	141.25	m	**1884.83**
Sum of two sides 3400 mm	1377.67	1743.58	4.57	151.17	m	**1894.76**
Sum of two sides 3600 mm	1696.70	2147.34	4.87	161.10	m	**2308.43**
Sum of two sides 3800 mm	1696.70	2147.34	5.17	171.02	m	**2318.36**
Sum of two sides 4000 mm	1696.70	2147.34	5.47	180.95	m	**2328.28**
Reducer						
Sum of two sides 1800 mm	338.42	428.30	3.60	119.09	m	**547.38**
Sum of two sides 2000 mm	338.42	428.30	3.70	122.39	m	**550.69**
Sum of two sides 2200 mm	401.22	507.78	3.10	102.55	m	**610.33**
Sum of two sides 2400 mm	401.22	507.78	3.48	115.12	m	**622.90**
Sum of two sides 2600 mm	555.89	703.54	3.85	127.36	m	**830.89**
Sum of two sides 2800 mm	555.89	703.54	4.22	139.60	m	**843.13**
Sum of two sides 3000 mm	555.89	703.54	4.59	151.84	m	**855.37**
Sum of two sides 3200 mm	618.73	783.07	4.96	164.07	m	**947.14**
Sum of two sides 3400 mm	618.73	783.07	5.34	176.65	m	**959.71**
Sum of two sides 3600 mm	676.77	856.52	5.72	189.22	m	**1045.73**
Sum of two sides 3800 mm	676.77	856.52	6.10	201.79	m	**1058.30**
Sum of two sides 4000 mm	676.77	856.52	6.48	214.36	m	**1070.87**
Offset						
Sum of two sides 1800 mm	338.42	428.30	4.56	150.84	m	**579.14**
Sum of two sides 2000 mm	338.42	428.30	5.53	182.93	m	**611.23**
Sum of two sides 2200 mm	401.22	507.78	5.99	198.15	m	**705.93**
Sum of two sides 2400 mm	401.22	507.78	6.45	213.36	m	**721.14**
Sum of two sides 2600 mm	555.89	703.54	6.91	228.58	m	**932.12**
Sum of two sides 2800 mm	555.89	703.54	7.37	243.80	m	**947.33**
Sum of two sides 3000 mm	555.89	703.54	7.83	259.01	m	**962.55**

38 VENTILATION/AIR CONDITIONING SYSTEMS

Item	Net Price £	Material £	Labour hours	Labour £	Unit	Total rate £
Sum of two sides 3200 mm	618.73	783.07	8.29	274.23	m	**1057.30**
Sum of two sides 3400 mm	618.73	783.07	8.75	289.45	m	**1072.52**
Sum of two sides 3600 mm	676.77	856.52	9.21	304.66	m	**1161.18**
Sum of two sides 3800 mm	676.77	856.52	9.67	319.88	m	**1176.40**
Sum of two sides 4000 mm	676.77	856.52	10.13	335.10	m	**1191.61**
90° radius bend						
Sum of two sides 1800 mm	647.79	819.84	3.25	107.51	m	**927.35**
Sum of two sides 2000 mm	647.79	819.84	3.30	109.16	m	**929.00**
Sum of two sides 2200 mm	662.28	838.18	3.34	110.49	m	**948.66**
Sum of two sides 2400 mm	662.28	838.18	3.38	111.81	m	**949.99**
Sum of two sides 2600 mm	1015.14	1284.76	3.42	113.13	m	**1397.89**
Sum of two sides 2800 mm	1015.14	1284.76	3.46	114.46	m	**1399.22**
Sum of two sides 3000 mm	1015.14	1284.76	3.50	115.78	m	**1400.54**
Sum of two sides 3200 mm	1218.17	1541.71	3.54	117.10	m	**1658.82**
Sum of two sides 3400 mm	1218.17	1541.71	3.58	118.43	m	**1660.14**
Sum of two sides 3600 mm	1425.99	1804.74	3.62	119.75	m	**1924.49**
Sum of two sides 3800 mm	1425.99	1804.74	3.66	121.07	m	**1925.81**
Sum of two sides 4000 mm	1425.99	1804.74	3.70	122.39	m	**1927.13**
45° bend						
Sum of two sides 1800 mm	647.79	819.84	2.30	76.08	m	**895.92**
Sum of two sides 2000 mm	647.79	819.84	2.76	91.30	m	**911.14**
Sum of two sides 2200 mm	662.28	838.18	3.01	99.57	m	**937.75**
Sum of two sides 2400 mm	662.28	838.18	3.26	107.84	m	**946.02**
Sum of two sides 2600 mm	1015.14	1284.76	3.49	115.45	m	**1400.21**
Sum of two sides 2800 mm	1015.14	1284.76	3.74	123.72	m	**1408.48**
Sum of two sides 3000 mm	1015.14	1284.76	3.98	131.66	m	**1416.42**
Sum of two sides 3200 mm	1218.17	1541.71	4.22	139.60	m	**1681.31**
Sum of two sides 3400 mm	1218.17	1541.71	4.46	147.54	m	**1689.25**
Sum of two sides 3600 mm	1425.99	1804.74	4.70	155.47	m	**1960.21**
Sum of two sides 3800 mm	1425.99	1804.74	4.94	163.41	m	**1968.15**
Sum of two sides 4000 mm	1425.99	1804.74	5.18	171.35	m	**1976.09**
90° mitre bend						
Sum of two sides 1800 mm	647.79	819.84	3.90	129.01	m	**948.85**
Sum of two sides 2000 mm	647.79	819.84	3.96	131.00	m	**950.83**
Sum of two sides 2200 mm	662.28	838.18	3.83	126.69	m	**964.87**
Sum of two sides 2400 mm	880.59	1114.48	3.84	127.03	m	**1241.50**
Sum of two sides 2600 mm	1015.14	1284.76	3.85	127.36	m	**1412.12**
Sum of two sides 2800 mm	1015.14	1284.76	3.86	127.69	m	**1412.45**
Sum of two sides 3000 mm	1015.14	1284.76	3.87	128.02	m	**1412.78**
Sum of two sides 3200 mm	1218.17	1541.71	3.88	128.35	m	**1670.06**
Sum of two sides 3400 mm	1218.17	1541.71	3.90	129.01	m	**1670.72**
Sum of two sides 3600 mm	1425.99	1804.74	3.92	129.67	m	**1934.41**
Sum of two sides 3800 mm	1425.99	1804.74	3.94	130.33	m	**1935.07**
Sum of two sides 4000 mm	1425.99	1804.74	3.96	131.00	m	**1935.73**
Branch						
Sum of two sides 1800 mm	290.05	367.09	1.81	59.87	m	**426.96**
Sum of two sides 2000 mm	290.05	367.09	1.90	62.85	m	**429.94**
Sum of two sides 2200 mm	304.51	385.39	2.61	86.34	m	**471.73**
Sum of two sides 2400 mm	304.51	385.39	2.68	88.65	m	**474.04**
Sum of two sides 2600 mm	362.53	458.82	2.75	90.97	m	**549.79**
Sum of two sides 2800 mm	362.53	458.82	2.81	92.95	m	**551.78**
Sum of two sides 3000 mm	362.53	458.82	2.87	94.94	m	**553.76**

38 VENTILATION/AIR CONDITIONING SYSTEMS

Item	Net Price £	Material £	Labour hours	Labour £	Unit	Total rate £
DUCTWORK: FIRE RATED – cont						
Extra over fittings – cont						
Branch – cont						
Sum of two sides 3200 mm	377.08	477.23	2.93	96.92	m	**574.15**
Sum of two sides 3400 mm	377.08	477.23	3.00	99.24	m	**576.47**
Sum of two sides 3600 mm	483.42	611.82	3.06	101.22	m	**713.04**
Sum of two sides 3800 mm	483.42	611.82	3.12	103.21	m	**715.02**
Sum of two sides 4000 mm	483.42	611.82	3.18	105.19	m	**717.01**

38 VENTILATION/AIR CONDITIONING SYSTEMS

Item	Net Price £	Material £	Labour hours	Labour £	Unit	Total rate £
LOW VELOCITY AIR CONDITIONING: AIR HANDLING UNITS						
Supply air handling unit; inlet with motorized damper, LTHW frost coil (at −5°C to +5°C), panel filter (EU4), bag filter (EU6), cooling coil (at 28°C db/20°C wb to 12°C db/11.5°C wb), LTHW heating coil (at 5°C to 21°C), supply fan, outlet plenum; includes access sections; all units located internally; Includes placing in position and fitting of sections together; electrical work elsewhere						
Volume, external pressure						
2 m³/s at 350 Pa	10201.09	12910.50	40.00	1381.30	nr	**14291.80**
2 m³/s at 700 Pa	10898.42	13793.03	40.00	1381.30	nr	**15174.34**
5 m³/s at 350 Pa	17201.08	21769.68	65.00	2244.61	nr	**24014.30**
5 m³/s at 700 Pa	17802.80	22531.23	65.00	2244.61	nr	**24775.84**
8 m³/s at 350 Pa	25763.56	32606.37	77.00	2659.00	nr	**35265.37**
8 m/³s at 700 Pa	26165.19	33114.67	77.00	2659.00	nr	**35773.67**
10 m/³ at 350 Pa	28370.65	35905.89	100.00	3453.25	nr	**39359.14**
10 m³/s at 700 Pa	29725.13	37620.12	100.00	3453.25	nr	**41073.38**
13 m³/s at 350 Pa	36398.75	46066.26	108.00	3729.51	nr	**49795.78**
13 m³/s at 700 Pa	37522.93	47489.02	108.00	3729.51	nr	**51218.53**
15 m³/s at 350 Pa	41684.59	52756.02	120.00	4143.90	nr	**56899.92**
15 m³/s at 700 Pa	42443.06	53715.94	120.00	4143.90	nr	**57859.85**
18 m³/s at 350 Pa	47301.31	59864.54	133.00	4592.83	nr	**64457.36**
18 m³/s at 700 Pa	48170.85	60965.03	133.00	4592.83	nr	**65557.86**
20 m³/s at 350 Pa	54814.18	69372.83	142.00	4903.62	nr	**74276.45**
20 m³/s at 700 Pa	55602.47	70370.49	142.00	4903.62	nr	**75274.11**
Extra for inlet and discharge attenuators at 900 mm long						
2 m³/s at 350 Pa	3151.74	3988.85	5.00	172.66	nr	**4161.51**
5 m³/s at 350 Pa	6614.78	8371.66	10.00	345.33	nr	**8716.99**
10 m³/s at 700 Pa	9411.43	11911.10	13.00	448.92	nr	**12360.03**
15 m³/s at 700 Pa	14271.33	18061.79	16.00	552.52	nr	**18614.31**
20 m³/s at 700 Pa	18417.24	23308.86	20.00	690.65	nr	**23999.51**
Extra for locating units externally						
2 m³/s at 350 Pa	2335.29	2955.54	–	–	nr	**2955.54**
5 m³/s at 350 Pa	2963.03	3750.01	–	–	nr	**3750.01**
10 m³/s at 700 Pa	4740.28	5999.30	–	–	nr	**5999.30**
15 m³/s at 700 Pa	6850.40	8669.87	–	–	nr	**8669.87**
20 m³/s at 700 Pa	11873.96	15027.69	–	–	nr	**15027.69**

38 VENTILATION/AIR CONDITIONING SYSTEMS

Item	Net Price £	Material £	Labour hours	Labour £	Unit	Total rate £
LOW VELOCITY AIR CONDITIONING: AIR HANDLING UNITS – cont						
Modular air handling unit with supply and extract sections. Supply side; inlet with motorized damper, LTHW frost coil (at –5°C to 5°C), panel filter (EU4), bag filter (EU6), cooling coil at 28°Cdb/20°Cwb to 12°Cdb/11.5°C wb), LTHW heating coil (at 5°C to 21°C), supply fan, outlet plenum. Extract side; inlet with motorized damper, extract fan; includes access sections; placing in position and fitting of sections together; electrical work elsewhere						
Volume, external pressure						
2 m³/s at 350 Pa	14250.54	18035.48	50.00	1726.63	nr	**19762.11**
2 m³/s at 700 Pa	14940.07	18908.16	50.00	1726.63	nr	**20634.78**
5 m³/s at 350 Pa	24560.10	31083.26	86.00	2969.80	nr	**34053.06**
5 m³/s at 700 Pa	25021.09	31666.69	86.00	2969.80	nr	**34636.49**
8 m³/s at 350 Pa	34105.23	43163.58	105.00	3625.92	nr	**46789.50**
8 m/3 s at 700 Pa	34923.90	44199.69	105.00	3625.92	nr	**47825.61**
10 m/3 at 350 Pa	39381.37	49841.06	120.00	4143.90	nr	**53984.97**
10 m/s at 700 Pa	40229.76	50914.79	120.00	4143.90	nr	**55058.69**
13 m³/s at 350 Pa	50678.00	64138.07	130.00	4489.23	nr	**68627.30**
13 m³/s at 700 Pa	52267.54	66149.79	130.00	4489.23	nr	**70639.02**
15 m³/s at 350 Pa	56629.10	71669.79	145.00	5007.22	nr	**76677.00**
15 m³/s at 700 Pa	59308.12	75060.36	145.00	5007.22	nr	**80067.58**
18 m³/s at 350 Pa	64733.87	81927.19	160.00	5525.20	nr	**87452.39**
18 m³/s at 700 Pa	66249.45	83845.30	160.00	5525.20	nr	**89370.51**
20 m³/s at 350 Pa	70751.49	89543.09	175.00	6043.19	nr	**95586.28**
20 m³/s at 700 Pa	73211.12	92656.00	175.00	6043.19	nr	**98699.19**
Extra for inlet and discharge attenuators at 900 mm long						
2 m³/s at 350 Pa	5745.80	7271.89	8.00	276.26	nr	**7548.15**
5 m³/s at 350 Pa	11346.32	14359.91	10.00	345.33	nr	**14705.23**
10 m³/s at 700 Pa	16235.64	20547.83	13.00	448.92	nr	**20996.75**
15 m³/s at 700 Pa	27771.79	35147.97	16.00	552.52	nr	**35700.50**
20 m³/s at 700 Pa	32419.09	41029.60	20.00	690.65	nr	**41720.25**
Extra for locating units externally						
2 m³/s at 350 Pa	4599.26	5820.83	–	–	nr	**5820.83**
5 m³/s at 350 Pa	6530.92	8265.53	–	–	nr	**8265.53**
10 m³/s at 700 Pa	9322.44	11798.48	–	–	nr	**11798.48**
15 m³/s at 700 Pa	17700.87	22402.22	–	–	nr	**22402.22**
20 m³/s at 700 Pa	27211.66	34439.08	–	–	nr	**34439.08**
Extra for humidifier, self generating type						
2 m³/s at 350 Pa (10 kg/hr)	4007.94	5072.45	5.00	172.66	nr	**5245.11**
5 m³/s at 350 Pa (18 kg/hr)	4959.90	6277.24	5.00	172.66	nr	**6449.91**
10 m³/s at 700 Pa (30 kg/hr)	5644.12	7143.19	6.00	207.20	nr	**7350.39**
15 m³/s at 700 Pa (60 kg/hr)	10452.55	13228.75	8.00	276.26	nr	**13505.01**
20 m³/s at 700 Pa (90 kg/hr)	15715.35	19889.34	10.00	345.33	nr	**20234.67**

38 VENTILATION/AIR CONDITIONING SYSTEMS

Item	Net Price £	Material £	Labour hours	Labour £	Unit	Total rate £
Extra for mixing box						
2 m³/s at 350 Pa	1924.53	2435.68	4.00	138.13	nr	**2573.81**
5 m³/s at 350 Pa	2712.72	3433.22	4.00	138.13	nr	**3571.35**
10 m³/s at 700 Pa	3800.51	4809.92	5.00	172.66	nr	**4982.58**
15 m³/s at 700 Pa	5072.26	6419.46	6.00	207.20	nr	**6626.65**
20 m³/s at 700 Pa	7732.53	9786.29	6.00	207.20	nr	**9993.48**
Extra for runaround coil, including pump and associated pipework; typical outputs in brackets (based on minimal distance between the supply and extract units)						
2 m³/s at 350 Pa (26 kW)	5761.43	7291.66	30.00	1035.98	nr	**8327.64**
5 m³/s at 350 Pa (37 kW)	10215.13	12928.27	30.00	1035.98	nr	**13964.24**
10 m³/s at 700 Pa (85 kW)	17685.93	22383.32	40.00	1381.30	nr	**23764.62**
15 m³/s at 700 Pa (151 kW)	24884.75	31494.14	50.00	1726.63	nr	**33220.76**
20 m³/s at 700 Pa (158 kW)	33921.44	42930.97	60.00	2071.95	nr	**45002.92**
Extra for thermal wheel (typical outputs in brackets)						
2 m³/s at 350 Pa (37 kW)	11916.74	15081.83	12.00	414.39	nr	**15496.22**
5 m³/s at 350 Pa (65 kW)	17115.33	21661.16	12.00	414.39	nr	**22075.55**
10 m³/s at 700 Pa (127 kW)	22824.54	28886.74	15.00	517.99	nr	**29404.73**
15 m³/s at 700 Pa (160 kW)	42763.50	54121.49	17.00	587.05	nr	**54708.54**
20 m³/s at 700 Pa (262 kW)	49952.75	63220.20	19.00	656.12	nr	**63876.31**
Extra for plate heat exchanger, including additional filtration in extract leg (typical outputs in brackets)						
2 m³/s at 350 Pa (25 kW)	7552.60	9558.57	12.00	414.39	nr	**9972.96**
5 m³/s at 350 Pa (51 kW)	14764.39	18685.82	12.00	414.39	nr	**19100.21**
10 m³/s at 700 Pa (98 kW)	21663.46	27417.28	15.00	517.99	nr	**27935.27**
15 m³/s at 700 Pa (160 kW)	37411.01	47347.37	17.00	587.05	nr	**47934.42**
20 m³/s at 700 Pa (190 kW)	48313.19	61145.17	19.00	656.12	nr	**61801.29**
Extra for electric heating in lieu of LTHW						
2 m³/s at 350 Pa	2617.65	3312.89	–	–	nr	**3312.89**
5 m³/s at 350 Pa	4904.44	6207.06	–	–	nr	**6207.06**
10 m³/s at 700 Pa	6621.77	8380.52	–	–	nr	**8380.52**
15 m³/s at 700 Pa	6943.55	8787.76	–	–	nr	**8787.76**
20 m³/s at 700 Pa	8194.38	10370.81	–	–	nr	**10370.81**

38 VENTILATION/AIR CONDITIONING SYSTEMS

Item	Net Price £	Material £	Labour hours	Labour £	Unit	Total rate £
VAV AIR CONDITIONING						
VAV TERMINAL BOXES						
VAV terminal box; integral acoustic silencer; factory installed and pre-wired control components (excluding electronic controller); selected at 200 Pa at entry to unit; includes fixing in position; electrical work elsewhere						
80 l/s–110 l/s	635.67	804.50	2.00	66.16	nr	**870.66**
Extra for secondary silencer	154.59	195.65	0.50	16.54	nr	**212.19**
Extra for 2 row LTHW heating coil	381.91	483.35	–	–	nr	**483.35**
150 l/s–190 l/s	668.90	846.56	2.00	66.16	nr	**912.72**
Extra for secondary silencer	169.75	214.83	0.50	16.54	nr	**231.37**
Extra for 2 row LTHW heating coil	409.17	517.85	–	–	nr	**517.85**
250 l/s–310 l/s	746.19	944.38	2.00	66.16	nr	**1010.54**
Extra for secondary silencer	224.66	284.34	0.50	16.54	nr	**300.88**
Extra for 2 row LTHW heating coil	463.74	586.91	–	–	nr	**586.91**
420 l/s–520 l/s	808.32	1023.01	2.00	66.16	nr	**1089.17**
Extra for secondary silencer	255.70	323.62	0.50	16.54	nr	**340.16**
Extra for 2 row LTHW heating coil	564.67	714.65	–	–	nr	**714.65**
650 l/s–790 l/s	978.79	1238.75	2.00	66.16	nr	**1304.91**
Extra for secondary silencer	327.23	414.14	0.50	16.54	nr	**430.68**
Extra for 2 row LTHW heating coil	591.95	749.18	–	–	nr	**749.18**
1130 l/s–1370 l/s	1136.99	1438.97	2.00	66.16	nr	**1505.13**
Extra for secondary silencer	457.25	578.69	0.50	16.54	nr	**595.23**
Extra for 2 row LTHW heating coil	619.22	783.68	–	–	nr	**783.68**
Extra for electric heater & thyristor controls, 3 kW/lph (per box)	504.64	638.68	–	–	nr	**638.68**
Extra for fitting free issue box controller	–	–	–	–	nr	**–**
Fan assisted VAV terminal box; factory installed and prewired control components (excluding electronic controller); selected at 40 Pa external static pressure; includes fixing in position, electrical work elsewhere						
100 l/s–175 l/s	1407.16	1780.91	3.00	99.24	nr	**1880.15**
Extra for secondary silencer	167.58	212.09	0.50	16.54	nr	**228.63**
Extra for 1 row LTHW heating coil	450.10	569.65	–	–	nr	**569.65**
170 l/s–360 l/s	1535.00	1942.70	3.00	99.24	nr	**2041.94**
Extra for secondary silencer	212.38	268.78	0.50	16.54	nr	**285.32**
Extra for 1 row LTHW heating coil	477.38	604.18	–	–	nr	**604.18**
300 l/s–640 l/s	1759.66	2227.02	3.00	99.24	nr	**2326.26**
Extra for secondary silencer	325.07	411.41	0.50	16.54	nr	**427.95**
Extra for 1 row LTHW heating coil	518.30	655.96	–	–	nr	**655.96**
620 l/s–850 l/s	1759.66	2227.02	3.00	99.24	nr	**2326.26**
Extra for secondary silencer	325.07	411.41	0.50	16.54	nr	**427.95**
Extra for 1 row LTHW heating coil	559.21	707.74	–	–	nr	**707.74**
Extra for electric heater plus thyristor controls, 3 kW/lph (per box)	518.30	655.96	–	–	nr	**655.96**
Extra for fitting free issue controller	–	–	2.13	70.46	nr	**70.46**

38 VENTILATION/AIR CONDITIONING SYSTEMS

Item	Net Price £	Material £	Labour hours	Labour £	Unit	Total rate £
FAN COIL AIR CONDITIONING						
All selections based on summer return air condition of 23°C @ 50% RH, CHW @ 6°/ 12°C, LTHW @ 82°/71°C (where applicable), medium speed, external resistance of 30 Pa						
All selections are based on heating and cooling units. For waterside control units there is no significant reduction in cost between 4 pipe heating and cooling and 2 pipe cooling only units (excluding controls). For airside control units, there is a marginal reduction (less than 5%) between 4 pipe heating and cooling units and 2 pipe cooling only units (excluding controls).						
Ceiling void mounted horizontal waterside control fan coil unit; cooling coil; LTHW heating coil; multi-tapped speed transformer; fine wire mesh filter; includes fixing in position; electrical work elsewhere						
Total cooling load, heating load						
2800 W, 1000 W	572.90	725.07	4.00	138.13	nr	**863.20**
4000 W, 1700 W	545.26	690.08	4.00	138.13	nr	**828.21**
4500 W, 1900 W	845.71	1070.33	4.00	138.13	nr	**1208.46**
6000 W, 2600 W	1106.13	1399.91	4.00	138.13	nr	**1538.04**
Ceiling void mounted horizontal waterside control fan coil unit; cooling coil ; electric heating coil; multi-tapped speed transformer; fine wire mesh filter; includes fixing in position; electrical work elsewhere + thyristor and 2 No. HTCOs						
Total cooling load, heating load						
2800 W, 1500 W	794.64	1005.70	4.00	138.13	nr	**1143.83**
4000 W, 2000 W	822.18	1040.55	4.00	138.13	nr	**1178.68**
4500 W, 2000 W	1046.87	1324.92	4.00	138.13	nr	**1463.05**
6000 W, 3000 W	1293.34	1636.86	4.00	138.13	nr	**1774.99**
Ceiling void mounted horizontal airside control fan coil unit; cooling coil ; LHTW heating coil; multi-tapped speed transformer; fine wire mesh filter, damper actuator & fixing kit; includes fixing in position; electrical work elsewhere.						
Total cooling load, heating load						
2600 W, 2200 W	900.27	1139.39	4.00	138.13	nr	**1277.52**
3600 W, 3200 W	926.32	1172.35	4.00	138.13	nr	**1310.48**
4000 W, 3600 W	1036.68	1312.03	4.00	138.13	nr	**1450.16**
5400 W, 5000 W	1237.57	1566.27	4.00	138.13	nr	**1704.40**

Material Costs/Prices for Measured Works

38 VENTILATION/AIR CONDITIONING SYSTEMS

Item	Net Price £	Material £	Labour hours	Labour £	Unit	Total rate £
FAN COIL AIR CONDITIONING – cont						
Ceiling void mounted horizontal airside control fan coil unit; cooling coil; electric heating coil; multi-tapped speed transformer; fine wire mesh filter, damper actuator & fixing kit; includes fixing in position; electrical work elsewhere + thyristor and 2 No. HTCOs						
Total cooling load, heating load						
2600 W, 1500 W	958.56	1213.15	4.00	138.13	nr	**1351.28**
3600 W, 2000 W	983.35	1244.53	4.00	138.13	nr	**1382.66**
4000 W, 2000 W	1087.52	1376.37	4.00	138.13	nr	**1514.50**
5400 W, 3000 W	1292.12	1635.31	4.00	138.13	nr	**1773.44**
Ceiling void mounted slimline horizontal waterside control fan coil unit, 170 mm deep; cooilng coil; LTHW heating coil; multi-tapped speed transformer; fine wire mesh filter; includes fixing in position; electrical work elsewhere						
Total cooling load, heating load						
1100 W, 1500 W	706.83	894.56	3.50	120.86	nr	**1015.42**
3200 W, 3700 W	1350.40	1709.07	4.00	138.13	nr	**1847.20**
Ceiling void mounted slimline horizontal waterside control fan coil unit, 170 mm deep; cooilng coil; electric heating coil; multi-tapped speed transformer; fine wire mesh filter; includes fixing in position; electrical work elsewhere						
Total cooling load, heating load						
1100 W, 1000 W	522.06	660.72	3.50	120.86	nr	**781.59**
3400 W, 2000 W	1470.70	1861.32	4.00	138.13	nr	**1999.45**
Ceiling void mounted slimline horizontal airside control fan coil unit, 170 mm deep; cooling coil; multi-tapped speed transformer; fine wire mesh filter, damper actuator & fixing kit; includes fixing in position; electrical work elsewhere						
Total cooling load						
1000 W	917.63	1161.35	3.50	120.86	nr	**1282.22**
3000 W	1223.93	1549.01	4.00	138.13	nr	**1687.14**
Ceiling void mounted slimline horizontal airside control fan coil unit, 170 mm deep; cooilng coil; electric heating coil; multi-tapped speed transformer; fine wire mesh filter, damper actuator & fixing kit; includes fixing in position; electrical work elsewhere						
Total cooling load, heating load						
1000 W, 1000 W	679.55	860.04	3.50	120.86	nr	**980.90**
3000 W, 2000 W	1344.21	1701.23	4.00	138.13	nr	**1839.36**

38 VENTILATION/AIR CONDITIONING SYSTEMS

Item	Net Price £	Material £	Labour hours	Labour £	Unit	Total rate £
Low level perimeter waterside control fan coil unit; cooling coil; LTHW heating coil; multi-tapped speed transformer; fine wire mesh filter; includes fixing in position; electrical work elsewhere						
Total cooling load, heating load						
1700 W, 1400 W	622.51	787.85	3.50	120.86	nr	**908.71**
Extra over for standard cabinet	246.41	311.86	1.00	34.53	nr	**346.39**
2200 W, 1900 W	804.79	1018.54	3.50	120.86	nr	**1139.40**
Extra over for standard cabinet	300.15	379.87	1.00	34.53	nr	**414.40**
2600 W, 2200 W	804.79	1018.54	3.50	120.86	nr	**1139.40**
Extra over for standard cabinet	307.56	389.24	1.00	34.53	nr	**423.78**
3900 W, 3200 W	1025.52	1297.90	3.50	120.86	nr	**1418.76**
Extra over for standard cabinet	335.35	424.42	1.00	34.53	nr	**458.95**
Low level perimeter waterside control fan coil unit; cooling coil; electric heating coil; multi-tapped speed transformer; fine wire mesh filter; includes fixing in position; electrical work elsewhere						
Total cooling load, heating load						
1700 W, 1500 W	740.31	936.93	3.50	120.86	nr	**1057.80**
Extra over for standard cabinet	246.41	311.86	1.00	34.53	nr	**346.39**
2200 W, 2000 W	925.07	1170.77	3.50	120.86	nr	**1291.63**
Extra over for standard cabinet	300.15	379.87	1.00	34.53	nr	**414.40**
2600 W, 2000 W	925.07	1170.77	3.50	120.86	nr	**1291.63**
Extra over for standard cabinet	307.56	389.24	1.00	34.53	nr	**423.78**
3800 W, 3000 W	1148.28	1453.27	3.50	120.86	nr	**1574.13**
Extra over for standard cabinet	335.35	424.42	1.00	34.53	nr	**458.95**

38 VENTILATION/AIR CONDITIONING SYSTEMS

Item	Net Price £	Material £	Labour hours	Labour £	Unit	Total rate £
MECHANICAL VENTILATION WITH HEAT RECOVERY (MVHR) UNITS A ventilation unit with high efficiency (up to 130%) rotary heat exchanger (rotary heat exchangers do not require a condense drain or heater battery installed). The unit recovers hot, cold and humid air. Includes main unit, built-in controller and delivery Air flow rate						
40–200 m³/h, 55 l/s	1570.51	1987.64	3.00	103.60	nr	**2091.23**
50–250 m³/h, 69 l/s	1730.65	2190.31	3.00	103.60	nr	**2293.90**
60–300 m³/h, 83 l/s	1893.49	2396.40	3.00	103.60	nr	**2500.00**
60–350 m³/h, 97 l/s	1986.35	2513.93	3.00	103.60	nr	**2617.52**
70–450 m³/h, 125 l/s	2165.33	2740.44	3.00	103.60	nr	**2844.04**
High performance ventilation unit with enthalpy exchanger, class leading efficiency, automatic modulating true summer by-pass, integrated humidity sensor, four variable speed flow rate set points, dial-a-duty motor Air flow rate						
70–600 m³/h	4718.00	5971.10	3.00	103.60	nr	**6074.70**
Extra over for additional cooling module with 1.5 kW cooling	4698.99	5947.04	–	–	nr	**5947.04**
Extra over for post treatment air to water exchanger						
300 m³/h	743.03	940.38	3.00	103.60	nr	**1043.98**
400 m³/h	961.52	1216.90	3.00	103.60	nr	**1320.50**
500 m³/h	854.48	1081.43	3.00	103.60	nr	**1185.03**
600 m³/h	1128.28	1427.95	3.00	103.60	nr	**1531.55**
Extra over for subsoil heat exchanger	4929.81	6239.17	3.00	103.60	nr	**6342.77**

PART 4

Material Costs/Measured Work Prices – Electrical Installations

Leadership in the Construction Industry

George Ofori and Shamas-ur-Rehman Toor

This book presents a new framework for leadership in the construction industry which draws from the authentic leadership construct. The framework has three major themes: self-leadership, self-transcendent leadership, and sustainable leadership.

Despite its significance, leadership has not been given due importance in the construction industry as focus is placed on managerial functionalism. At the project level, even with the technological advances in the industry in recent years, construction is realized in the form of people undertaking distinct interdependent activities which require effective leadership. The industry faces many challenges including: demanding client requirements and project parameters; more stringent regulations, codes and systems; intense competition in the industry; and threats from disruptive enterprise. In such a complex environment, technology-driven and tool-based project and corporate management is insufficient. It must be complemented by a strategic, genuine, stakeholder-focused and ethical leadership.

Leadership in the Construction Industry is based on a study on authentic leadership and its development in Singapore. Leadership theories and concepts are reviewed; the importance of leadership in the construction industry is discussed; and the grounded theory approach which was applied in the study is explained. Many eminent construction professionals in Singapore were interviewed in the field study. Emerging from the experiences of the leaders documented in this book are three major themes: (1) self-leadership: how leaders engage in various self-related processes such as self-awareness, self-regulation, and role modelling. (2) self-transcendent leadership: how leaders go beyond leading themselves to leading others through servant leadership, shared leadership, spiritual leadership, and socially-responsible leadership; and, finally, (3) sustainable leadership or the strategies leaders employ to make the impact of their leadership lasting. A synthesis of these themes and their implications for leadership development is presented before the book concludes with some recommendations for current and aspiring leaders about how they can engage with them. This book is essential reading for all construction practitioners from all backgrounds; and researchers on leadership and management in construction.

March 2021: 364 pp
ISBN: 9780367482152

To Order
Tel:+44 (0) 1235 400524
Email: tandf@bookpoint.co.uk

For a complete listing of all our titles visit:
www.tandf.co.uk

Taylor & Francis
Taylor & Francis Group

Material Costs/Measured Work Prices

DIRECTIONS

The following explanations are given for each of the column headings and letter codes.

Unit	Prices for each unit are given as singular (i.e. 1 metre, 1 nr) unless stated otherwise.
Net price	Industry tender prices, plus nominal allowance for fixings, waste and applicable trade discounts.
Material cost	Net price plus percentage allowance for overheads (7%), profit (5%) and preliminaries (13%).
Labour norms	In man-hours for each operation.
Labour cost	Labour constant multiplied by the appropriate all-in man-hour cost based on gang rate (See also relevant Rates of Wages Section) plus percentage allowance for overheads, profit and preliminaries.
Measured work Price (total rate)	Material cost plus Labour cost.

MATERIAL COSTS

The Material Costs given are based at Second Quarter 2022 but exclude any charges in respect of VAT. The average rate of copper during this quarter is US $9,610 / UK £7,677 per tonne. Users of the book are advised to register on the Spon's website www.pricebooks.co.uk/updates to receive the free quarterly updates – alerts will then be provided by email as changes arise.

MEASURED WORK PRICES

These prices are intended to apply to new work in the London area. The prices are for reasonable quantities of work and the user should make suitable adjustments if the quantities are especially small or especially large. Adjustments may also be required for locality (e.g. outside London – refer to cost indices in approximate estimating section for details of adjustment factors) and for the market conditions (e.g. volume of work secured or being tendered) at the time of use.

ELECTRICAL INSTALLATIONS

The labour rate has been based on average gang rates per man hour effective from 3 January 2022 including allowances for all other emoluments and expenses. Future changes will be published in the free Spon's quarterly update by registering on their website. To this rate, has been added 13% and 7% to cover site and head office overheads and preliminary items together with a further 5% for profit, resulting in an inclusive rate of £39.54 per man hour. The rate has been calculated on a working year of 2,016 hours; a detailed build-up of the rate is given at the end of these directions.

DIRECTIONS

In calculating the 'Measured Work Prices' the following assumptions have been made:

(a)   That the work is carried out as a subcontract under the Standard Form of Building Contract.
(b)   That, unless otherwise stated, the work is being carried out in open areas at a height which would not require more than simple scaffolding.
(c)   That the building in which the work is being carried out is no more than six storeys high.

Where these assumptions are not valid, as for example where work is carried out in ducts and similar confined spaces or in multi-storey structures when additional time is needed to get to and from upper floors, then an appropriate adjustment must be made to the prices. Such adjustment will normally be to the labour element only.

DIRECTIONS

LABOUR RATE – ELECTRICAL

The annual cost of a notional 11 man gang

	TECHNICIAN	APPROVED ELECTRICIANS	ELECTRICIANS	LABOURERS	SUB-TOTALS
	1 NR	4 NR	4 NR	2 NR	
Hourly Rate from 3 January 2022	22.31	19.82	18.29	14.68	
Working hours per annum per man	1,680.00	1,680.00	1,680.00	1,680.00	
x Hourly rate × nr of men = £ per annum	37,480.80	133,190.40	122,908.80	49,324.80	342,904.80
Overtime Rate	33.47	29.73	27.44	22.02	
Overtime hours per annum per man	336.00	336.00	336.00	336.00	
x Hourly rate × nr of men = £ per annum	11,244.24	39,957.12	36,872.64	14,797.44	102,871.44
Total	48,725.04	173,147.52	159,781.44	64,122.24	445,776.24
Incentive schemes (insert percentage) 0.00%	0.00	0.00	0.00	0.00	0.00
Daily Travel Allowance (15–20 miles each way)	7.04	7.04	7.04	7.04	
Days per annum per man	224.00	224.00	224.00	224.00	
x nr of men = £ per annum	1,576.96	6,307.84	6,307.84	3,153.92	17,346.56
Daily Travel Allowance (20 miles each way)	8.80	8.80	8.80	8.80	
Days per annum per man	224.00	224.00	224.00	224.00	
x nr of men = £ per annum	1,971.20	7,884.80	7,884.80	3,942.40	21,683.20
JIB Pension Scheme @ 3.0%	1,734.76	6,196.13	5,739.66	2,331.31	16,001.85
JIB combined benefits scheme (nr of weeks per man)	52.00	52.00	52.00	52.00	
Benefit Credit effective from 1 October 2018	74.61	67.71	63.27	53.31	
x nr of men = £ per annum	3,879.72	14,083.68	13,160.16	5,544.24	36,667.80
Holiday Top-up Funding	70.07	62.49	58.04	47.02	
x nr of men @ 7.5 hrs per day = £ per annum	3,643.48	12,998.70	12,072.45	4,889.82	33,604.45
National Insurance Contributions:					
Annual gross pay (subject to NI) each	59,796.40	214,422.54	199,206.69	81,652.62	
% of NI Contributions	15.05	15.05	15.05	15.05	
£ Contributions/annum	6,873.14	24,075.28	21,975.49	8,510.55	61,434.46
			SUBTOTAL		632,514.56
TRAINING (INCLUDING ANY TRADE REGISTRATIONS) – SAY				1.00%	6,325.15
SEVERANCE PAY AND SUNDRY COSTS – SAY				1.50%	9,582.60
EMPLOYER'S LIABILITY AND THIRD PARTY INSURANCE – SAY				2.00%	12,968.60
ANNUAL COST OF NOTIONAL GANG					661,390.75
MEN ACTUALLY WORKING = 10.5			THEREFORE ANNUAL COST PER PRODUCTIVE MAN		62,989.59
AVERAGE NR OF HOURS WORKED PER MAN = 2016					
THEREFORE ALL-IN MAN-HOUR					31.24
PRELIMINARY ITEMS – SAY				13%	4.06
SITE AND HEAD OFFICE OVERHEADS AND PROFIT (7% & 5% RESPECTIVELY) – SAY				12%	4.24
THEREFORE INCLUSIVE MAN-HOUR					39.54

DIRECTIONS

Notes:

(1)   Hourly wage rates are those effective from 3 January 2022.
(2)   The following assumptions have been made in the above calculations:
      (a)   Hourly rates are based on London rate and job reporting own transport.
      (b)   The working week of 37.5 hours is made up of 7.5 hours Monday to Friday.
      (c)   Five days in the year are lost through sickness or similar reason.
      (d)   A working year of 2,016 hours.
(3)   The incentive scheme addition of 0% is intended to reflect bonus schemes typically in use.
(4)   National insurance contributions are those effective from 1 April 2022.
(5)   Weekly JIB Combined Benefit Credit Scheme are those effective from 1 October 2018 – Contact JIB for ECI (0330 221 0240).
(6)   Overtime is paid after 37.5 hours.

ELECTRICAL SUPPLY/POWER/LIGHTING

Item	Net Price £	Material £	Labour hours	Labour £	Unit	Total rate £
ELECTRICAL GENERATION PLANT						
STANDBY GENERATORS						
Standby diesel generating sets; supply and installation; fixing to base; all supports and fixings; all necessary connections to equipment						
Three phase, 400 volt, four wire 50 Hz packaged standby diesel generating set, complete with radio and television suppressors, daily service fuel tank and associated piping, 4 metres of exhaust pipe and primary exhaust silencer, control panel, mains failure relay, starting battery with charger, all internal wiring, interconnections, earthing and labels. Rated for standby duty; including UK delivery, installation and commissioning						
60 kVA	20023.00	25341.11	100.00	3960.47	nr	**29301.58**
100 kVA	28070.22	35525.67	100.00	3960.47	nr	**39486.14**
150 kVA	20226.00	25598.03	100.00	3960.47	nr	**29558.50**
300 kVA	30992.00	39223.48	120.00	4752.57	nr	**43976.04**
500 kVA	57331.00	72558.11	120.00	4752.57	nr	**77310.68**
750 kVA	116096.67	146931.95	140.00	5544.66	nr	**152476.61**
1000 kVA	165875.10	209931.52	140.00	5544.66	nr	**215476.18**
1500 kVA	195218.96	247069.12	170.00	6732.81	nr	**253801.92**
2000 kVA	300710.21	380578.84	170.00	6732.81	nr	**387311.65**
2500 kVA	411712.88	521063.82	210.00	8316.99	nr	**529380.82**
Extra for residential silencer; peformance 75 dBA at 1 m; including connection to exhaust pipe						
60 kVA	1182.55	1496.63	10.00	396.05	nr	**1892.68**
100 kVA	1413.03	1788.33	10.00	396.05	nr	**2184.37**
150 kVA	1620.45	2050.85	10.00	396.05	nr	**2446.89**
315 kVA	3152.26	3989.50	15.00	594.07	nr	**4583.57**
500 kVA	4031.64	5102.45	15.00	594.07	nr	**5696.52**
750 kVA	6006.65	7602.01	20.00	792.09	nr	**8394.11**
1000 kVA	8089.86	10238.53	20.00	792.09	nr	**11030.62**
1500 kVA	12887.37	16310.26	20.00	792.09	nr	**17102.36**
2000 kVA	13958.23	17665.54	30.00	1188.14	nr	**18853.68**
2500 kVA	15959.86	20198.80	30.00	1188.14	nr	**21386.94**

ELECTRICAL SUPPLY/POWER/LIGHTING

Item	Net Price £	Material £	Labour hours	Labour £	Unit	Total rate £
ELECTRICAL GENERATION PLANT – cont						
Synchronization panel for paralleling generators – not generators to mains; including interconnecting cables; commissioning and testing; fixing to backgrounds						
2 × 60 kVA	9939.91	12579.95	80.00	3168.38	nr	**15748.33**
2 × 100 kVA	10685.95	13524.13	80.00	3168.38	nr	**16692.51**
2 × 150 kVA	13442.84	17013.26	80.00	3168.38	nr	**20181.64**
2 × 315 kVA	22853.22	28923.03	80.00	3168.38	nr	**32091.41**
2 × 500 kVA	32993.51	41756.59	80.00	3168.38	nr	**44924.97**
2 × 750 kVA	37231.73	47120.48	80.00	3168.38	nr	**50288.86**
2 × 1000 kVA	40126.41	50783.98	120.00	4752.57	nr	**55536.55**
2 × 1500 kVA	47502.66	60119.36	120.00	4752.57	nr	**64871.93**
2 × 2000 kVA	51468.95	65139.10	120.00	4752.57	nr	**69891.67**
2 × 2500 kVA	60004.47	75941.66	120.00	4752.57	nr	**80694.23**
Prefabricated drop-over acoustic housing; performance 85 dBA at 1 m over the range from 60 kVA to 315 kVA, 75 dBA from 500 kVA to 2500 kVA						
100 kVA	4302.11	5444.75	7.00	277.23	nr	**5721.98**
150 kVA	6985.97	8841.44	15.00	594.07	nr	**9435.51**
315 kVA	14905.80	18864.78	25.00	990.12	nr	**19854.90**
500 kVA	40077.35	50721.89	40.00	1584.19	nr	**52306.08**
750 kVA	63158.02	79932.79	40.00	1584.19	nr	**81516.98**
1000 kVA	79903.69	101126.11	40.00	1584.19	nr	**102710.30**
1500 kVA	79903.69	101126.11	40.00	1584.19	nr	**102710.30**
2000 kVA	158495.79	200592.27	60.00	2376.28	nr	**202968.55**
2500 kVA	190509.15	241108.38	70.00	2772.33	nr	**243880.71**
COMBINED HEAT AND POWER (CHP) UNITS						
Gas fired engine; skid-mounted with acoustic enclosure complete with exhaust fan and attenuators; exhaust gas attenuation; includes 6 m long pipe connections; dry air cooler for secondary water circuit to reject excess heat; controls and panel; commissioning						
Electrical output; heat output						
18 kW; 30 kW	48329.66	61166.02	–	–	nr	**61166.02**
18 kW; 34 kW	50097.14	63402.94		–	nr	**63402.94**
18 kW; 43 kW	51981.01	65787.17	–	–	nr	**65787.17**
43 kW; 63 kW	94652.88	119792.68	–	–	nr	**119792.68**
50 kW; 79 kW	97089.86	122876.93	–	–	nr	**122876.93**
70 kW; 113 kW	107014.94	135438.11	–	–	nr	**135438.11**
104 kW, 136 kW	140680.49	178045.23	–	–	nr	**178045.23**
142 kW, 193 kW	141493.16	179073.74	–	–	nr	**179073.74**
166 kW, 248 kW	194195.17	245773.41	–	–	nr	**245773.41**
183 kW, 263 kW	195122.17	246946.62	–	–	nr	**246946.62**
264 kW, 375 kW	222693.21	281840.53	–	–	nr	**281840.53**

ELECTRICAL SUPPLY/POWER/LIGHTING

Item	Net Price £	Material £	Labour hours	Labour £	Unit	Total rate £
281 kW, 426 kW	264600.82	334878.80	–	–	nr	334878.80
324 kW, 497 kW	267966.86	339138.86	–	–	nr	339138.86
349 kW; 516 kW	300879.48	380793.07		–	nr	380793.07
385 kW; 581 kW	301512.93	381594.76		–	nr	381594.76
418 kW, 665 kW	311150.64	393792.25	–	–	nr	393792.25
439 kW, 687 kW	313820.40	397171.10	–	–	nr	397171.10
Heat dump; emergency cooling						
18 kW; 30 kW	3904.73	4941.83		–	nr	4941.83
18 kW; 34 kW	3904.73	4941.83		–	nr	4941.83
18 kW; 43 kW	3904.73	4941.83	–	–	nr	4941.83
43 kW; 63 kW	4063.35	5142.58	–	–	nr	5142.58
50 kW; 79 kW	4372.35	5533.65	–	–	nr	5533.65
70 kW; 113 kW	5523.89	6991.04	–	–	nr	6991.04
104 kW, 136 kW	6102.75	7723.64	–	–	nr	7723.64
142 kW, 193 kW	7252.23	9178.42	–	–	nr	9178.42
166 kW, 248 kW	10394.76	13155.61	–	–	nr	13155.61
183 kW, 263 kW	10394.76	13155.61	–	–	nr	13155.61
264 kW, 375 kW	15401.59	19492.25	–	–	nr	19492.25
281 kW, 426 kW	17123.75	21671.82	–	–	nr	21671.82
324 kW, 497 kW	17123.75	21671.82	–	–	nr	21671.82
349 kW; 516 kW	19069.42	24134.26		–	nr	24134.26
385 kW; 581 kW	19069.42	24134.26		–	nr	24134.26
418 kW, 665 kW	21620.73	27363.20	–	–	nr	27363.20
439 kW, 687 kW	21620.73	27363.20	–	–	nr	27363.20
Note: The costs detailed are based on a specialist subcontract package, as part of the M&E contract works, and include installation						
Upgraded catalytic converter						
43 kW; 63 kW	1254.54	1587.75	–	–	nr	1587.75
50 kW; 79 kW	1254.54	1587.75	–	–	nr	1587.75
70 kW; 113 kW	1254.54	1587.75	–	–	nr	1587.75
142 kW, 193 kW	3723.45	4712.40	–	–	nr	4712.40
264 kW, 375 kW	6633.20	8394.98	–	–	nr	8394.98
Flexible connections						
43 kW; 63 kW	711.73	900.77	–	–	nr	900.77
50 kW; 79 kW	711.73	900.77	–	–	nr	900.77
70 kW; 113 kW	911.55	1153.66	–	–	nr	1153.66
104 kW, 136 kW	1227.76	1553.85	–	–	nr	1553.85
142 kW, 193 kW	1291.62	1634.67	–	–	nr	1634.67
166 kW, 248 kW	1792.20	2268.21	–	–	nr	2268.21
183 kW, 263 kW	1796.32	2273.42	–	–	nr	2273.42
264 kW, 375 kW	1738.64	2200.42	–	–	nr	2200.42
281 kW, 426 kW	1961.12	2481.99	–	–	nr	2481.99
324 kW, 497 kW	1961.12	2481.99	–	–	nr	2481.99
349 kW; 516 kW	2204.20	2789.64		–	nr	2789.64
385 kW; 581 kW	2204.20	2789.64		–	nr	2789.64
418 kW, 665 kW	2528.65	3200.26	–	–	nr	3200.26
439 kW, 687 kW	2528.65	3200.26	–	–	nr	3200.26

ELECTRICAL SUPPLY/POWER/LIGHTING

Item	Net Price £	Material £	Labour hours	Labour £	Unit	Total rate £
HV SUPPLY						
Cable; 6350/11000 volts, 3 core, XLPE; stranded copper conductors; steel wire armoured; LSOH to BS 7835						
Laid in trench/duct including marker tape (cable tiles measured elsewhere)						
95 mm²	43.92	55.59	0.23	9.11	m	**64.70**
120 mm²	58.11	73.55	0.23	9.11	m	**82.66**
150 mm²	65.47	82.86	0.25	9.90	m	**92.76**
185 mm²	77.25	97.76	0.25	9.90	m	**107.66**
240 mm²	106.67	135.00	0.27	10.69	m	**145.70**
300 mm²	127.66	161.56	0.29	11.49	m	**173.05**
Pile tape; ES 1–12–23; 1 m	48.31	61.14	0.01	0.40	m	**61.53**
Clipped direct to backgrounds including cleats						
95 mm²	43.92	55.59	0.47	18.61	m	**74.21**
120 mm²	58.11	73.55	0.50	19.80	m	**93.35**
150 mm²	65.47	82.86	0.53	20.99	m	**103.85**
185 mm²	77.25	97.76	0.55	21.78	m	**119.54**
240 mm²	106.67	135.00	0.60	23.76	m	**158.76**
300 mm²	127.66	161.56	0.68	26.93	m	**188.50**
Terminations for above cables, including heat-shrink kit and glanding off						
95 × 3 Core XLPE SWA LSF 11 kV Termination	1088.06	1377.05	4.75	188.12	nr	**1565.18**
120 × 3 Core XLPE SWA LSF 11 kV Termination	1109.90	1404.69	5.00	198.02	nr	**1602.72**
150 × x3 Core XLPE SWA LSF 11 kV Termination	1176.30	1488.72	6.00	237.63	nr	**1726.35**
185 × 3 Core XLPE SWA LSF 11 kV Termination	1188.15	1503.73	6.90	273.27	nr	**1777.00**
240 × 3 Core XLPE SWA LSF 11 kV Termination	1248.37	1579.94	7.50	297.04	nr	**1876.97**
300 × 3 Core XLPE SWA LSF 11 kV Termination	1383.74	1751.27	8.75	346.54	nr	**2097.81**
Cable tiles; single width; laid in trench above cables on prepared sand bed (cost of excavation excluded); reinforced concrete covers; concave/convex ends						
914 × 152 × 63/38 mm	48.31	61.14	0.11	4.36	m	**65.49**
914 × 229 × 63/38 mm	48.31	61.14	0.11	4.36	m	**65.49**
914 × 305 × 63/38 mm	48.31	61.14	0.11	4.36	m	**65.49**

ELECTRICAL SUPPLY/POWER/LIGHTING

Item	Net Price £	Material £	Labour hours	Labour £	Unit	Total rate £
HV SWITCHGEAR AND TRANSFORMERS						
HV Circuit Breakers; installed on prepared foundations including all supports, fixings and inter panel connections where relevant. Excludes main and multi core cabling and heat shrink cable termination kits						
Three phase 11 kV, 630 amp, Air or SF6 insulated, with fixed pattern vacuum or SF6 circuit breaker panels; hand charged spring closing operation; prospective fault level up to 25 kA for 3 seconds. Feeders include ammeter with selector switch, VIP relays, overcurrent and earth fault relays with necessary current relays with necessary current transformers; incomers include 3 phase VT, voltmeter and phase selector switch; Includes IDMT overcurrent and earth fault relays/CTs						
Single panel with cable chamber	38017.35	48114.76	31.70	1255.47	nr	**49370.23**
Three panel with one incomer and two feeders; with cable chambers	70244.21	88901.08	67.83	2686.39	nr	**91587.46**
Five panel with two incoming, two feeders and a bus section; with cable chambers	125838.56	159261.28	99.17	3927.60	nr	**163188.89**
Tripping batteries						
Battery chargers; switchgear tripping and closing; double wound transfomer and earth screen; including fixing to background, commissioning and testing						
Valve regulated lead acid battery NGTS 3.12.2; BS 6290 Part2; TPS 9/3; IEEE485						
30 volt; 19 Ah; 3 A	3743.57	4737.87	6.50	257.43	nr	**4995.30**
40 volt; 29 Ah; 3 A	5997.49	7590.43	8.50	336.64	nr	**7927.07**
110 volt; 19 Ah; 3 A	6118.04	7742.99	6.50	257.43	nr	**8000.42**
110 volt; 29 Ah; 3 A	6675.50	8448.51	8.50	336.64	nr	**8785.15**
100 volt; 38 Ah; 3 A	7133.38	9028.01	10.00	396.05	nr	**9424.06**
Step down transformers; 11/0.415 kV, Dyn 11, 50 Hz. Complete with lifting lugs, mounting skids, provisions for wheels, undrilled gland plates to air-filled cable boxes, off load tapping facility, including UK delivery						
Oil-filled in free breathing ventialted steel tank						
500 kVA	12892.29	16316.48	30.00	1188.14	nr	**17504.62**
800 kVA	14682.28	18581.90	30.00	1188.14	nr	**19770.04**
1000 kVA	16640.46	21060.16	30.00	1188.14	nr	**22248.31**
1250 kVA	20220.39	25590.92	35.00	1386.17	nr	**26977.09**
1500 kVA	23961.31	30325.44	35.00	1386.17	nr	**31711.60**
2000 kVA	31121.17	39386.95	35.00	1386.17	nr	**40773.11**

ELECTRICAL SUPPLY/POWER/LIGHTING

Item	Net Price £	Material £	Labour hours	Labour £	Unit	Total rate £
HV SUPPLY – cont						
Step down transformers – cont						
MIDEL – filled in gasket-sealed steel tank						
500 kVA	17076.02	21611.40	30.00	1188.14	nr	**22799.55**
800 kVA	19462.65	24631.93	30.00	1188.14	nr	**25820.07**
1000 kVA	22159.49	28045.05	30.00	1188.14	nr	**29233.19**
1250 kVA	26783.54	33897.24	35.00	1386.17	nr	**35283.41**
1500 kVA	31866.98	40330.85	35.00	1386.17	nr	**41717.01**
2000 kVA	41264.27	52224.06	40.00	1584.19	nr	**53808.25**
Extra for						
Fluid temperature indicator with 2 N/O contacts	479.03	606.26	2.00	79.21	nr	**685.47**
Winding temperature indicator with 2 N/O contacts	1173.63	1485.35	2.00	79.21	nr	**1564.56**
Dehydrating breather	119.76	151.56	2.00	79.21	nr	**230.77**
Plain rollers	359.27	454.69	2.00	79.21	nr	**533.90**
Pressure relief device with 1 N/O contact	718.53	909.37	2.00	79.21	nr	**988.58**
Step down transformers; 11/0.415 kV, Dyn 11, 50 Hz. Complete with lifting lugs, mounting skids, provisions for wheels, undrilled gland plates to air-filled cable boxes, off load tapping facility, including UK delivery						
Cast Resin type in ventilated steel encloure, AN – Air Natural including winding temperture indicator with 2 N/O contacts						
500 kVA	18269.32	23121.65	40.00	1584.19	nr	**24705.84**
800 kVA	21252.58	26897.27	40.00	1584.19	nr	**28481.46**
1000 kVA	25709.57	32538.04	40.00	1584.19	nr	**34122.23**
1250 kVA	27827.68	35218.71	45.00	1782.21	nr	**37000.92**
1600 kVA	31568.67	39953.31	45.00	1782.21	nr	**41735.52**
2000 kVA	35745.27	45239.21	50.00	1980.24	nr	**47219.45**
Cast Resin type in ventilated steel enclosure with temperature controlled fans to achieve 40% increase to AN/AF rating. Includes winding temperature indicator with 2 N/O contacts						
500/700 kVA	20059.26	25387.00	42.00	1663.40	nr	**27050.40**
800/1120 kVA	23340.88	29540.22	42.00	1663.40	nr	**31203.61**
1000/1400 kVA	28878.97	36549.23	42.00	1663.40	nr	**38212.62**
12501750 kVA	30072.26	38059.45	47.00	1861.42	nr	**39920.88**
1600/2240 kVA	33974.31	42997.89	47.00	1861.42	nr	**44859.31**
2000/2800 kVA	38747.53	49038.88	52.00	2059.45	nr	**51098.32**

ELECTRICAL SUPPLY/POWER/LIGHTING

Item	Net Price £	Material £	Labour hours	Labour £	Unit	Total rate £
LV DISTRIBUTION: CONDUIT AND CABLE TRUNKING						
Heavy gauge, screwed drawn steel; surface fixed on saddles to backgrounds, with standard pattern boxes and fittings including all fixings and supports. (forming holes, conduit entry, draw wires etc. and components for earth continuity are included)						
Black enamelled						
20 mm dia.	3.80	4.82	0.49	19.41	m	**24.22**
25 mm dia.	5.17	6.54	0.56	22.18	m	**28.72**
32 mm dia.	11.10	14.05	0.64	25.35	m	**39.39**
38 mm dia.	14.38	18.20	0.73	28.91	m	**47.11**
50 mm dia.	27.26	34.51	1.04	41.19	m	**75.70**
Galvanized						
20 mm dia.	3.92	4.96	0.49	19.41	m	**24.36**
25 mm dia.	5.36	6.78	0.56	22.18	m	**28.96**
32 mm dia.	10.26	12.99	0.64	25.35	m	**38.33**
38 mm dia.	13.30	16.83	0.73	28.91	m	**45.75**
50 mm dia.	25.57	32.36	1.04	41.19	m	**73.55**
High impact PVC; surface fixed on saddles to backgrounds; with standard pattern boxes and fittings; including all fixings and supports						
Light gauge						
16 mm dia.	0.96	1.21	0.27	10.69	m	**11.91**
20 mm dia.	1.08	1.36	0.28	11.09	m	**12.45**
25 mm dia.	1.48	1.87	0.33	13.07	m	**14.94**
32 mm dia.	2.35	2.98	0.38	15.05	m	**18.03**
38 mm dia.	3.13	3.96	0.44	17.43	m	**21.38**
50 mm dia.	5.16	6.53	0.48	19.01	m	**25.54**
Heavy gauge						
16 mm dia.	1.53	1.94	0.27	10.69	m	**12.63**
20 mm dia.	1.80	2.27	0.28	11.09	m	**13.36**
25 mm dia.	2.45	3.10	0.33	13.07	m	**16.17**
32 mm dia.	3.68	4.66	0.38	15.05	m	**19.71**
38 mm dia.	5.15	6.51	0.44	17.43	m	**23.94**
50 mm dia.	8.53	10.79	0.48	19.01	m	**29.80**
Flexible conduits; including adaptors and locknuts (for connections to equipment)						
Metallic, PVC covered conduit; not exceeding 1 m long; including zinc plated mild steel adaptors, lock nuts and earth conductor						
16 mm dia.	23.89	30.24	0.42	16.63	nr	**46.87**
20 mm dia.	25.38	32.12	0.46	18.22	nr	**50.33**
25 mm dia.	37.66	47.66	0.43	17.03	nr	**64.69**
32 mm dia.	58.27	73.74	0.51	20.20	nr	**93.94**
38 mm dia.	73.57	93.11	0.56	22.18	nr	**115.29**
50 mm dia.	256.72	324.91	0.82	32.48	nr	**357.39**

ELECTRICAL SUPPLY/POWER/LIGHTING

Item	Net Price £	Material £	Labour hours	Labour £	Unit	Total rate £
LV DISTRIBUTION: CONDUIT AND CABLE TRUNKING – cont						
Flexible conduits – cont						
PVC conduit; not exceeding 1 m long; including nylon adaptors, lock nuts						
16 mm dia.	7.22	9.13	0.46	18.22	nr	**27.35**
20 mm dia.	7.22	9.13	0.48	19.01	nr	**28.14**
25 mm dia.	9.05	11.45	0.50	19.80	nr	**31.25**
32 mm.dia.	13.14	16.63	0.58	22.97	nr	**39.60**
PVC adaptable boxes; fixed to backgrounds; including all supports and fixings (cutting and connecting conduit to boxes is included)						
Square pattern						
75 × 75 × 53 mm	4.32	5.47	0.69	27.33	nr	**32.80**
100 × 100 × 75 mm	7.73	9.78	0.71	28.12	nr	**37.90**
150 × 150 × 75 mm	9.39	11.88	0.80	31.68	nr	**43.57**
Terminal strips to be fixed in metal or polythene adaptable boxes)						
20 amp high density polythene						
2 way	1.36	1.72	0.23	9.11	nr	**10.83**
3 way	1.71	2.16	0.23	9.11	nr	**11.27**
4 way	2.26	2.86	0.23	9.11	nr	**11.97**
5 way	2.83	3.58	0.23	9.11	nr	**12.69**
6 way	3.40	4.30	0.25	9.90	nr	**14.20**
7 way	3.97	5.02	0.25	9.90	nr	**14.92**
8 way	4.52	5.73	0.29	11.49	nr	**17.21**
9 way	5.09	6.44	0.30	11.88	nr	**18.33**
10 way	5.66	7.16	0.34	13.47	nr	**20.63**
11 way	6.23	7.88	0.34	13.47	nr	**21.35**
13 way	7.37	9.32	0.37	14.65	nr	**23.98**
14 way	7.93	10.04	0.37	14.65	nr	**24.70**
15 way	8.49	10.75	0.39	15.45	nr	**26.19**
16 way	9.06	11.47	0.45	17.82	nr	**29.29**
18 way	10.20	12.90	0.45	17.82	nr	**30.73**

ELECTRICAL SUPPLY/POWER/LIGHTING

Item	Net Price £	Material £	Labour hours	Labour £	Unit	Total rate £
LV DISTRIBUTION: TRUNKING						
Galvanized steel trunking; fixed to backgrounds; jointed with standard connectors (including plates for air gap between trunking and background); earth continuity straps included						
Single compartment						
50 × 50 mm	12.06	15.26	0.39	15.45	m	**30.71**
75 × 50 mm	16.58	20.98	0.44	17.43	m	**38.40**
75 × 75 mm	19.24	24.35	0.47	18.61	m	**42.97**
100 × 50 mm	20.58	26.04	0.50	19.80	m	**45.84**
100 × 75 mm	22.24	28.15	0.57	22.57	m	**50.72**
100 × 100 mm	22.57	28.57	0.62	24.55	m	**53.13**
150 × 50 mm	33.30	42.15	0.78	30.89	m	**73.04**
150 × 100 mm	37.94	48.01	0.78	30.89	m	**78.91**
150 × 150 mm	40.51	51.27	0.86	34.06	m	**85.33**
225 × 75 mm	53.57	67.80	0.88	34.85	m	**102.65**
225 × 150 mm	66.97	84.75	0.84	33.27	m	**118.02**
225 × 225 mm	68.48	86.67	0.99	39.21	m	**125.88**
300 × 75 mm	55.09	69.72	0.96	38.02	m	**107.74**
300 × 100 mm	63.42	80.27	0.99	39.21	m	**119.47**
300 × 150 mm	48.54	61.44	0.99	39.21	m	**100.64**
300 × 225 mm	41.30	52.27	1.09	43.17	m	**95.44**
300 × 300 mm	96.66	122.34	1.16	45.94	m	**168.28**
Double compartment						
100 × 50 mm	16.13	20.42	0.54	21.39	m	**41.80**
100 × 75 mm	17.40	22.02	0.62	24.55	m	**46.58**
100 × 100 mm	33.01	41.77	0.66	26.14	m	**67.91**
150 × 50 mm	24.52	31.04	0.70	27.72	m	**58.76**
150 × 100 mm	30.37	38.44	0.83	32.87	m	**71.31**
150 × 150 mm	56.51	71.52	0.92	36.44	m	**107.95**
Triple compartment						
150 × 50 mm	28.84	36.50	0.79	31.29	m	**67.79**
150 × 100 mm	29.60	37.46	0.78	30.89	m	**68.35**
150 × 150 mm	61.19	77.44	1.01	40.00	m	**117.45**
Galvanized steel trunking fittings; cutting and jointing trunking to fittings is included						
Stop end						
50 × 50 mm	1.85	2.35	0.19	7.52	nr	**9.87**
75 × 50 mm	1.96	2.48	0.20	7.92	nr	**10.40**
75 × 75 mm	1.96	2.48	0.21	8.32	nr	**10.80**
100 × 50 mm	2.04	2.58	0.31	12.28	nr	**14.86**
100 × 75 mm	2.20	2.79	0.27	10.69	nr	**13.48**
100 × 100 mm	2.16	2.74	0.27	10.69	nr	**13.43**
150 × 50 mm	2.40	3.04	0.28	11.09	nr	**14.13**
150 × 100 mm	2.62	3.31	0.30	11.88	nr	**15.20**
150 × 150 mm	2.70	3.42	0.32	12.67	nr	**16.09**
225 × 75 mm	2.78	3.52	0.35	13.86	nr	**17.38**
225 × 150 mm	3.09	3.91	0.37	14.65	nr	**18.56**
225 × 225 mm	4.81	6.08	0.38	15.05	nr	**21.13**

ELECTRICAL SUPPLY/POWER/LIGHTING

Item	Net Price £	Material £	Labour hours	Labour £	Unit	Total rate £
LV DISTRIBUTION: TRUNKING – cont						
Galvanized steel trunking fittings – cont						
Stop end – cont						
300 × 75 mm	4.83	6.12	0.42	16.63	nr	**22.75**
300 × 100 mm	4.14	5.23	0.42	16.63	nr	**21.87**
300 × 150 mm	4.83	6.12	0.43	17.03	nr	**23.15**
300 × 225 mm	5.80	7.34	0.45	17.82	nr	**25.16**
300 × 300 mm	6.20	7.85	0.48	19.01	nr	**26.86**
Flanged connector						
50 × 50 mm	1.96	2.48	0.19	7.52	nr	**10.01**
75 × 50 mm	3.01	3.81	0.20	7.92	nr	**11.73**
75 × 75 mm	2.43	3.08	0.21	8.32	nr	**11.39**
100 × 50 mm	3.13	3.96	0.26	10.30	nr	**14.26**
100 × 75 mm	3.22	4.08	0.27	10.69	nr	**14.77**
100 × 100 mm	3.13	3.96	0.27	10.69	nr	**14.65**
150 × 50 mm	3.16	3.99	0.28	11.09	nr	**15.08**
150 × 100 mm	3.37	4.27	0.30	11.88	nr	**16.15**
150 × 150 mm	3.42	4.33	0.32	12.67	nr	**17.01**
225 × 75 mm	2.70	3.42	0.35	13.86	nr	**17.28**
225 × 150 mm	3.30	4.18	0.37	14.65	nr	**18.83**
225 × 225 mm	4.85	6.13	0.38	15.05	nr	**21.18**
300 × 75 mm	4.06	5.13	0.42	16.63	nr	**21.77**
300 × 100 mm	4.69	5.93	0.42	16.63	nr	**22.57**
300 × 150 mm	5.30	6.71	0.43	17.03	nr	**23.74**
300 × 225 mm	5.30	6.71	0.45	17.82	nr	**24.53**
300 × 300 mm	6.08	7.70	0.48	19.01	nr	**26.71**
Bends 90°; single compartment						
50 × 50 mm	8.14	10.30	0.42	16.63	nr	**26.93**
75 × 50 mm	10.51	13.31	0.45	17.82	nr	**31.13**
75 × 75 mm	9.84	12.46	0.48	19.01	nr	**31.47**
100 × 50 mm	10.74	13.60	0.53	20.99	nr	**34.59**
100 × 75 mm	11.01	13.94	0.56	22.18	nr	**36.11**
100 × 100 mm	11.01	13.94	0.58	22.97	nr	**36.91**
150 × 50 mm	13.63	17.25	0.64	25.35	nr	**42.60**
150 × 100 mm	17.39	22.01	0.91	36.04	nr	**58.05**
150 × 150 mm	16.58	20.99	0.89	35.25	nr	**56.24**
225 × 75 mm	25.34	32.07	0.76	30.10	nr	**62.17**
225 × 150 mm	31.49	39.85	0.82	32.48	nr	**72.33**
225 × 225 mm	37.07	46.92	0.83	32.87	nr	**79.79**
300 × 75 mm	33.82	42.81	0.85	33.66	nr	**76.47**
300 × 100 mm	34.23	43.32	0.90	35.64	nr	**78.96**
300 × 150 mm	37.65	47.65	0.96	38.02	nr	**85.67**
300 × 225 mm	41.41	52.41	0.98	38.81	nr	**91.22**
300 × 300 mm	42.34	53.58	1.06	41.98	nr	**95.56**
Bends 90°; double compartment						
100 × 50 mm	9.78	12.37	0.53	20.99	nr	**33.36**
100 × 75 mm	10.30	13.03	0.56	22.18	nr	**35.21**
100 × 100 mm	16.81	21.28	0.58	22.97	nr	**44.25**
150 × 50 mm	10.74	13.60	0.65	25.74	nr	**39.34**
150 × 100 mm	11.32	14.33	0.69	27.33	nr	**41.65**
150 × 150 mm	28.27	35.77	0.73	28.91	nr	**64.68**

ELECTRICAL SUPPLY/POWER/LIGHTING

Item	Net Price £	Material £	Labour hours	Labour £	Unit	Total rate £
Bends 90°; triple compartment						
150 × 50 mm	14.13	17.88	0.68	26.93	nr	**44.81**
150 × 100 mm	19.75	25.00	0.73	28.91	nr	**53.91**
150 × 150 mm	34.46	43.61	0.77	30.50	nr	**74.10**
Tees; single compartment						
50 × 50 mm	9.88	12.51	0.56	22.18	nr	**34.69**
75 × 50 mm	11.76	14.89	0.57	22.57	nr	**37.46**
75 × 75 mm	11.17	14.14	0.60	23.76	nr	**37.90**
100 × 50 mm	13.78	17.44	0.65	25.74	nr	**43.18**
100 × 75 mm	13.86	17.54	0.72	28.52	nr	**46.05**
100 × 100 mm	13.29	16.82	0.71	28.12	nr	**44.94**
150 × 50 mm	15.76	19.95	0.82	32.48	nr	**52.43**
150 × 100 mm	19.58	24.78	0.84	33.27	nr	**58.05**
150 × 150 mm	18.99	24.03	0.91	36.04	nr	**60.07**
225 × 75 mm	33.07	41.86	0.94	37.23	nr	**79.08**
225 × 150 mm	44.67	56.54	1.01	40.00	nr	**96.54**
225 × 225 mm	52.27	66.16	1.02	40.40	nr	**106.55**
300 × 75 mm	47.75	60.43	1.07	42.38	nr	**102.81**
300 × 100 mm	49.47	62.61	1.07	42.38	nr	**104.98**
300 × 150 mm	53.29	67.45	1.14	45.15	nr	**112.60**
300 × 225 mm	57.71	73.04	1.19	47.13	nr	**120.17**
300 × 300 mm	62.42	79.00	1.26	49.90	nr	**128.91**
Tees; double compartment						
100 × 50 mm	9.49	12.01	0.65	25.74	nr	**37.76**
100 × 75 mm	22.36	28.29	0.71	28.12	nr	**56.41**
100 × 100 mm	22.80	28.86	0.72	28.52	nr	**57.37**
150 × 50 mm	12.47	15.79	0.82	32.48	nr	**48.26**
150 × 100 mm	13.07	16.54	0.85	33.66	nr	**50.20**
150 × 150 mm	38.62	48.87	0.91	36.04	nr	**84.91**
Tees; triple compartment						
150 × 50 mm	14.27	18.06	0.87	34.46	nr	**52.52**
150 × 100 mm	15.68	19.85	0.89	35.25	nr	**55.10**
150 × 150 mm	45.75	57.90	0.96	38.02	nr	**95.92**
Crossovers; single compartment						
50 × 50 mm	12.54	15.87	0.65	25.74	nr	**41.62**
75 × 50 mm	17.00	21.51	0.66	26.14	nr	**47.65**
75 × 75 mm	17.39	22.01	0.69	27.33	nr	**49.33**
100 × 50 mm	21.48	27.19	0.74	29.31	nr	**56.50**
100 × 75 mm	15.83	20.04	0.80	31.68	nr	**51.72**
100 × 100 mm	21.66	27.41	0.81	32.08	nr	**59.49**
150 × 50 mm	25.65	32.46	0.91	36.04	nr	**68.50**
150 × 100 mm	26.02	32.93	0.94	37.23	nr	**70.16**
150 × 150 mm	24.93	31.56	0.99	39.21	nr	**70.77**
225 × 75 mm	45.67	57.80	1.01	40.00	nr	**97.80**
225 × 150 mm	49.84	63.08	1.08	42.77	nr	**105.85**
225 × 225 mm	67.81	85.82	1.09	43.17	nr	**128.99**
300 × 75 mm	63.85	80.81	1.14	45.15	nr	**125.96**
300 × 100 mm	65.77	83.24	1.16	45.94	nr	**129.18**
300 × 150 mm	71.00	89.86	1.19	47.13	nr	**136.99**
300 × 225 mm	75.13	95.08	1.21	47.92	nr	**143.00**
300 × 300 mm	80.10	101.37	1.29	51.09	nr	**152.46**

ELECTRICAL SUPPLY/POWER/LIGHTING

Item	Net Price £	Material £	Labour hours	Labour £	Unit	Total rate £
LV DISTRIBUTION: TRUNKING – cont						
Galvanized steel trunking fittings – cont						
100 × 50 mm	12.61	15.96	0.74	29.31	nr	**45.26**
100 × 75 mm	13.23	16.74	0.80	31.68	nr	**48.42**
100 × 100 mm	25.66	32.48	0.81	32.08	nr	**64.56**
150 × 50 mm	14.93	18.90	0.86	34.06	nr	**52.96**
150 × 100 mm	14.99	18.97	0.94	37.23	nr	**56.19**
150 × 150 mm	43.60	55.18	1.00	39.60	nr	**94.78**
Crossovers; triple compartment						
150 × 50 mm	16.34	20.68	0.97	38.42	nr	**59.10**
150 × 100 mm	17.98	22.75	0.99	39.21	nr	**61.96**
150 × 150 mm	52.02	65.83	1.06	41.98	nr	**107.82**
Galvanized steel flush floor trunking; fixed to backgrounds; supports and fixings; standard coupling joints; earth continuity straps included						
Triple compartment						
350 × 60 mm	69.63	88.12	1.32	52.28	m	**140.40**
Four compartment						
350 × 60 mm	72.50	91.75	1.32	52.28	m	**144.03**
Galvanized steel flush floor trunking; fittings (cutting and jointing trunking to fittings is included)						
Stop end; triple compartment						
350 × 60 mm	6.38	8.07	0.53	20.99	nr	**29.06**
Stop end; four compartment						
350 × 60 mm	14.79	18.72	0.53	20.99	nr	**39.71**
Rising bend; standard; triple compartment						
350 × 60 mm	52.60	66.57	1.30	51.49	nr	**118.05**
Rising bend; standard; four compartment						
350 × 60 mm	55.11	69.75	1.30	51.49	nr	**121.24**
Rising bend; skirting; triple compartment						
350 × 60 mm	103.19	130.60	1.33	52.67	nr	**183.27**
Rising bend; skirting; four compartment						
350 × 60 mm	117.11	148.22	1.33	52.67	nr	**200.89**
Junction box; triple compartment						
350 × 60 mm	68.58	86.79	1.16	45.94	nr	**132.73**
Junction box; four compartment						
350 × 60 mm	71.33	90.27	1.16	45.94	nr	**136.22**
Body coupler (pair)						
3 and 4 compartment	3.54	4.48	0.16	6.34	nr	**10.81**
Service outlet module comprising flat lid with flanged carpet trim; twin 13 A outlet and drilled plate for mounting 2 telephone outlets; one blank plate; triple compartment						
3 compartment	85.07	107.67	0.47	18.61	nr	**126.28**

ELECTRICAL SUPPLY/POWER/LIGHTING

Item	Net Price £	Material £	Labour hours	Labour £	Unit	Total rate £
Service outlet module comprising flat lid with flanged carpet trim; twin 13 A outlet and drilled plate for mounting 2 telephone outlets; two blank plates; four compartment						
4 compartment	94.75	119.92	0.47	18.61	nr	**138.53**
Single compartment PVC trunking; grey finish; clip on lid; fixed to backgrounds; including supports and fixings (standard coupling joints)						
Dimensions						
50 × 50 mm	14.87	18.82	0.27	10.69	m	**29.51**
75 × 50 mm	16.12	20.40	0.28	11.09	m	**31.49**
75 × 75 mm	18.26	23.11	0.29	11.49	m	**34.60**
100 × 50 mm	20.68	26.17	0.34	13.47	m	**39.64**
100 × 75 mm	22.69	28.72	0.37	14.65	m	**43.37**
100 × 100 mm	23.95	30.32	0.37	14.65	m	**44.97**
150 × 50 mm	20.79	26.31	0.41	16.24	m	**42.55**
150 × 75 mm	37.05	46.90	0.44	17.43	m	**64.32**
150 × 100 mm	44.59	56.44	0.44	17.43	m	**73.87**
150 × 150 mm	45.62	57.74	0.48	19.01	m	**76.75**
Single compartment PVC trunking; fittings (cutting and jointing trunking to fittings is included)						
Crossover						
50 × 50 mm	24.37	30.84	0.29	11.49	nr	**42.32**
75 × 50 mm	27.01	34.19	0.30	11.88	nr	**46.07**
75 × 75 mm	29.31	37.10	0.31	12.28	nr	**49.38**
100 × 50 mm	39.08	49.45	0.35	13.86	nr	**63.32**
100 × 75 mm	49.29	62.38	0.36	14.26	nr	**76.64**
100 × 100 mm	43.60	55.18	0.40	15.84	nr	**71.02**
150 × 75 mm	55.85	70.68	0.45	17.82	nr	**88.50**
150 × 100 mm	66.99	84.79	0.46	18.22	nr	**103.00**
150 × 150 mm	103.99	131.61	0.47	18.61	nr	**150.23**
Stop end						
50 × 50 mm	1.01	1.28	0.12	4.75	nr	**6.03**
75 × 50 mm	1.90	2.41	0.12	4.75	nr	**7.16**
75 × 75 mm	1.45	1.83	0.13	5.15	nr	**6.98**
100 × 50 mm	2.59	3.28	0.16	6.34	nr	**9.62**
100 × 75 mm	4.06	5.13	0.16	6.34	nr	**11.47**
100 × 100 mm	4.07	5.15	0.18	7.13	nr	**12.28**
150 × 75 mm	10.60	13.42	0.20	7.92	nr	**21.34**
150 × 100 mm	13.13	16.61	0.21	8.32	nr	**24.93**
150 × 150 mm	13.39	16.95	0.22	8.71	nr	**25.66**
Flanged coupling						
50 × 50 mm	6.10	7.73	0.32	12.67	nr	**20.40**
75 × 50 mm	6.98	8.84	0.33	13.07	nr	**21.91**
75 × 75 mm	8.38	10.60	0.34	13.47	nr	**24.07**

ELECTRICAL SUPPLY/POWER/LIGHTING

Item	Net Price £	Material £	Labour hours	Labour £	Unit	Total rate £
LV DISTRIBUTION: TRUNKING – cont						
Single compartment PVC trunking – cont						
Flanged coupling – cont						
100 × 50 mm	9.43	11.93	0.44	17.43	nr	**29.36**
100 × 75 mm	10.68	13.52	0.45	17.82	nr	**31.34**
100 × 100 mm	11.37	14.39	0.46	18.22	nr	**32.61**
150 × 75 mm	12.05	15.25	0.57	22.57	nr	**37.83**
150 × 100 mm	12.70	16.08	0.57	22.57	nr	**38.65**
150 × 150 mm	13.33	16.87	0.59	23.37	nr	**40.23**
Internal coupling						
50 × 50 mm	2.15	2.73	0.07	2.77	nr	**5.50**
75 × 50 mm	2.55	3.23	0.07	2.77	nr	**6.00**
75 × 75 mm	2.57	3.25	0.07	2.77	nr	**6.02**
100 × 50 mm	3.43	4.34	0.08	3.17	nr	**7.51**
100 × 75 mm	3.84	4.86	0.08	3.17	nr	**8.03**
100 × 100 mm	4.24	5.37	0.08	3.17	nr	**8.54**
External coupling						
50 × 50 mm	2.35	2.98	0.09	3.56	nr	**6.54**
75 × 50 mm	2.82	3.57	0.09	3.56	nr	**7.13**
75 × 75 mm	2.79	3.53	0.09	3.56	nr	**7.10**
100 × 50 mm	3.75	4.75	0.10	3.96	nr	**8.71**
100 × 75 mm	4.27	5.40	0.10	3.96	nr	**9.36**
100 × 100 mm	4.66	5.89	0.10	3.96	nr	**9.85**
150 × 75 mm	5.33	6.75	0.11	4.36	nr	**11.11**
150 × 100 mm	5.56	7.04	0.11	4.36	nr	**11.39**
150 × 150 mm	5.75	7.27	0.11	4.36	nr	**11.63**
Angle; flat cover						
50 × 50 mm	8.65	10.94	0.18	7.13	nr	**18.07**
75 × 50 mm	11.52	14.58	0.19	7.52	nr	**22.10**
75 × 75 mm	13.25	16.77	0.20	7.92	nr	**24.69**
100 × 50 mm	19.79	25.05	0.23	9.11	nr	**34.16**
100 × 75 mm	30.14	38.14	0.26	10.30	nr	**48.44**
100 × 100 mm	27.49	34.79	0.26	10.30	nr	**45.09**
150 × 75 mm	37.47	47.42	0.30	11.88	nr	**59.30**
150 × 100 mm	44.57	56.41	0.33	13.07	nr	**69.48**
150 × 150 mm	67.54	85.48	0.34	13.47	nr	**98.94**
Angle; internal or external cover						
50 × 50 mm	10.52	13.31	0.18	7.13	nr	**20.44**
75 × 50 mm	14.50	18.35	0.19	7.52	nr	**25.87**
75 × 75 mm	18.62	23.57	0.20	7.92	nr	**31.49**
100 × 50 mm	20.06	25.38	0.23	9.11	nr	**34.49**
100 × 75 mm	32.25	40.82	0.26	10.30	nr	**51.12**
100 × 100 mm	32.40	41.00	0.26	10.30	nr	**51.30**
150 × 75 mm	39.61	50.13	0.30	11.88	nr	**62.01**
150 × 100 mm	46.68	59.08	0.33	13.07	nr	**72.15**
150 × 150 mm	65.57	82.98	0.34	13.47	nr	**96.45**
Tee; flat cover						
50 × 50 mm	7.24	9.16	0.24	9.51	nr	**18.66**
75 × 50 mm	11.04	13.97	0.25	9.90	nr	**23.87**
75 × 75 mm	11.34	14.36	0.26	10.30	nr	**24.66**

ELECTRICAL SUPPLY/POWER/LIGHTING

Item	Net Price £	Material £	Labour hours	Labour £	Unit	Total rate £
100 × 50 mm	24.51	31.02	0.32	12.67	nr	**43.70**
100 × 75 mm	26.04	32.96	0.33	13.07	nr	**46.03**
100 × 100 mm	33.98	43.01	0.34	13.47	nr	**56.47**
150 × 75 mm	44.94	56.88	0.41	16.24	nr	**73.11**
150 × 100 mm	57.64	72.95	0.42	16.63	nr	**89.59**
150 × 150 mm	78.42	99.24	0.44	17.43	nr	**116.67**
Tee; internal or external cover						
50 × 50 mm	20.51	25.96	0.24	9.51	nr	**35.46**
75 × 50 mm	22.53	28.51	0.25	9.90	nr	**38.42**
75 × 75 mm	25.06	31.71	0.26	10.30	nr	**42.01**
100 × 50 mm	32.16	40.70	0.32	12.67	nr	**53.37**
100 × 75 mm	36.73	46.49	0.33	13.07	nr	**59.56**
100 × 100 mm	41.22	52.16	0.34	13.47	nr	**65.63**
150 × 75 mm	53.40	67.58	0.41	16.24	nr	**83.82**
150 × 100 mm	64.36	81.45	0.42	16.63	nr	**98.09**
150 × 150 mm	85.93	108.75	0.44	17.43	nr	**126.18**
Division strip (1.8 m long)						
50 mm	11.12	14.07	0.07	2.77	nr	**16.84**
75 mm	14.24	18.03	0.07	2.77	nr	**20.80**
100 mm	18.10	22.91	0.08	3.17	nr	**26.08**
PVC miniature trunking; white finish; fixed to backgrounds; including supports and fixing; standard coupling joints						
Single compartment						
16 × 16 mm	2.01	2.54	0.20	7.92	m	**10.46**
25 × 16 mm	2.45	3.10	0.21	8.32	m	**11.41**
38 × 16 mm	3.06	3.87	0.24	9.51	m	**13.38**
38 × 25 mm	3.66	4.63	0.25	9.90	m	**14.53**
Compartmented						
38 × 16 mm	3.58	4.53	0.24	9.51	m	**14.03**
38 × 25 mm	4.27	5.40	0.25	9.90	m	**15.30**
PVC miniature trunking fittings; single compartment; white finish; cutting and jointing trunking to fittings is included						
Coupling						
16 × 16 mm	0.63	0.79	0.10	3.96	nr	**4.75**
25 × 16 mm	0.63	0.79	0.12	4.55	nr	**5.35**
38 × 16 mm	0.63	0.79	0.12	4.75	nr	**5.54**
38 × 25 mm	1.46	1.85	0.14	5.54	nr	**7.40**
Stop end						
16 × 16 mm	0.63	0.79	0.12	4.75	nr	**5.54**
25 × 16 mm	0.63	0.79	0.13	5.03	nr	**5.82**
38 × 16 mm	0.63	0.79	0.15	5.94	nr	**6.73**
38 × 25 mm	1.46	1.85	0.17	6.73	nr	**8.58**
Bend; flat, internal or external						
16 × 16 mm	0.63	0.79	0.18	7.13	nr	**7.92**
25 × 16 mm	0.63	0.79	0.20	7.84	nr	**8.63**
38 × 16 mm	0.63	0.79	0.21	8.32	nr	**9.11**
38 × 25 mm	1.46	1.85	0.23	9.11	nr	**10.96**

ELECTRICAL SUPPLY/POWER/LIGHTING

Item	Net Price £	Material £	Labour hours	Labour £	Unit	Total rate £
LV DISTRIBUTION: TRUNKING – cont						
PVC miniature trunking fittings – cont						
Tee						
16 × 16 mm	1.04	1.31	0.19	7.52	nr	**8.84**
25 × 16 mm	1.04	1.31	0.23	9.11	nr	**10.42**
38 × 16 mm	1.04	1.31	0.26	10.30	nr	**11.61**
38 × 25 mm	1.45	1.83	0.29	11.49	nr	**13.32**
PVC bench trunking; white or grey finish; fixed to backgrounds; including supports and fixings; standard coupling joints						
Trunking						
90 × 90 mm	32.40	41.00	0.33	13.07	m	**54.07**
PVC bench trunking fittings; white or grey finish; cutting and jointing trunking to fittings is included						
Stop end						
90 × 90 mm	7.13	9.02	0.09	3.56	nr	**12.59**
Coupling						
90 × 90 mm	4.42	5.59	0.09	3.56	nr	**9.15**
Internal or external bend						
90 × 90 mm	24.22	30.65	0.28	11.09	nr	**41.74**
Socket plate						
90 × 90 mm – 1 gang	1.15	1.46	0.10	3.96	nr	**5.42**
90 × 90 mm – 2 gang	1.38	1.74	0.10	3.96	nr	**5.70**
PVC underfloor trunking; single compartment; fitted in floor screed; standard coupling joints						
Trunking						
60 × 25 mm	16.52	20.91	0.22	8.71	m	**29.62**
90 × 35 mm	23.79	30.11	0.27	10.69	m	**40.81**
PVC underfloor trunking fittings; single compartment; fitted in floor screed; (cutting and jointing trunking to fittings is included)						
Jointing sleeve						
60 × 25 mm	1.10	1.40	0.08	3.17	nr	**4.57**
90 × 35 mm	1.84	2.32	0.10	3.96	nr	**6.28**
Duct connector						
90 × 35 mm	0.85	1.08	0.17	6.73	nr	**7.81**
Socket reducer						
90 × 35 mm	1.52	1.92	0.12	4.75	nr	**6.67**
Vertical access box; 2 compartment						
Shallow	85.45	108.15	0.37	14.65	nr	**122.80**

ELECTRICAL SUPPLY/POWER/LIGHTING

Item	Net Price £	Material £	Labour hours	Labour £	Unit	Total rate £
Duct bend; vertical						
60 × 25 mm	16.66	21.09	0.27	10.69	nr	**31.78**
90 × 35 mm	18.81	23.80	0.35	13.86	nr	**37.66**
Duct bend; horizontal						
60 × 25 mm	19.66	24.88	0.30	11.88	nr	**36.76**
90 × 35 mm	19.95	25.25	0.37	14.65	nr	**39.90**
Zinc coated steel underfloor ducting; fixed to backgrounds; standard coupling joints; earth continuity straps (including supports and fixing, packing shims where required)						
Double compartment						
150 × 25 mm	11.63	14.71	0.57	22.57	m	**37.29**
Triple compartment						
225 × 25 mm	20.60	26.07	0.93	36.83	m	**62.90**
Zinc coated steel underfloor ducting fittings (cutting and jointing to fittings is included)						
Stop end; double compartment						
150 × 25 mm	3.66	4.63	0.31	12.28	nr	**16.91**
Stop end; triple compartment						
225 × 25 mm	4.19	5.30	0.37	14.65	nr	**19.95**
Rising bend; double compartment; standard trunking						
150 × 25 mm	23.54	29.79	0.71	28.12	nr	**57.91**
Rising bend; triple compartment; standard trunking						
225 × 25 mm	42.30	53.53	0.85	33.66	nr	**87.20**
Rising bend; double compartment; to skirting						
150 × 25 mm	52.39	66.31	0.90	35.65	nr	**101.96**
Rising bend; triple compartment; to skirting						
225 × 25 mm	61.78	78.19	0.95	37.62	nr	**115.81**
horizontal bend; double compartment						
150 × 25 mm	35.79	45.29	0.64	25.35	nr	**70.64**
Horizontal bend; triple compartment						
225 × 25 mm	41.27	52.24	0.77	30.50	nr	**82.73**
Junction or service outlet boxes; terminal; double compartment						
150 mm	37.03	46.87	0.91	36.04	nr	**82.91**
Junction or service outlet boxes; terminal; triple compartment						
225 mm	42.32	53.56	1.11	43.96	nr	**97.52**
Junction or service outlet boxes; through or angle; double compartment						
150 mm	49.17	62.23	0.97	38.42	nr	**100.65**
Junction or service outlet boxes; through or angle; triple compartment						
225 mm	54.63	69.13	1.17	46.34	nr	**115.47**

ELECTRICAL SUPPLY/POWER/LIGHTING

Item	Net Price £	Material £	Labour hours	Labour £	Unit	Total rate £
LV DISTRIBUTION: TRUNKING – cont						
Zinc coated steel underfloor ducting fittings – cont						
Junction or service outlet boxes; tee; double compartment						
150 mm	49.17	62.23	1.02	40.40	nr	**102.63**
Junction or service outlet boxes; tee; triple compartment						
225 mm	54.63	69.13	1.22	48.32	nr	**117.45**
Junction or service outlet boxes; cross; double compartment						
up to 150 mm	49.17	62.23	1.03	40.79	nr	**103.03**
Junction or service outlet boxes; cross;triple compartment						
225 mm	54.63	69.13	1.23	48.71	nr	**117.85**
Plates for junction/inspection boxes; double and triple compartment						
Blank plate	10.74	13.59	0.92	36.44	nr	**50.03**
Conduit entry plate	13.48	17.06	0.86	34.06	nr	**51.12**
Trunking entry plate	13.48	17.06	0.86	34.06	nr	**51.12**
Service outlet box comprising flat lid with flanged carpet trim; twin 13 A outlet and drilled plate for mounting 2 telephone outlets and terminal blocks; terminal outlet box; double compartment						
150 × 25 mm trunking	76.68	97.05	1.68	66.54	nr	**163.58**
Service outlet box comprising flat lid with flanged carpet trim; twin 13 A outlet and drilled plate for mounting 2 telephone outlets and terminal blocks; terminal outlet box; triple compartment						
225 × 25 mm trunking	85.35	108.02	1.93	76.44	nr	**184.46**
PVC skirting/dado modular trunking; white (cutting and jointing trunking to fittings and backplates for fixing to walls is included)						
Main carrier/backplate						
50 × 170 mm	26.40	33.41	0.22	8.71	m	**42.13**
62 × 190 mm	29.31	37.10	0.22	8.71	m	**45.81**
Extension carrier/backplate						
50 × 42 mm	16.04	20.30	0.58	22.97	m	**43.27**
Carrier/backplate						
Including cover seal	9.36	11.85	0.53	20.99	m	**32.84**
Chamfered covers for fixing to backplates						
50 × 42 mm	5.44	6.88	0.33	13.07	m	**19.95**
Square covers for fixing to backplates						
50 × 42 mm	10.91	13.80	0.33	13.07	m	**26.87**
Plain covers for fixing to backplates						
85 mm	5.44	6.88	0.34	13.47	m	**20.35**
Retainers-clip to backplates to hold cables						
For chamfered covers	1.50	1.90	0.07	2.77	m	**4.67**
For square-recessed covers	1.30	1.65	0.07	2.77	m	**4.42**
For plain covers	5.01	6.35	0.07	2.77	m	**9.12**

ELECTRICAL SUPPLY/POWER/LIGHTING

Item	Net Price £	Material £	Labour hours	Labour £	Unit	Total rate £
Prepackaged corner assemblies						
Internal; for 170 × 50 Assy	10.55	13.35	0.51	20.20	nr	**33.55**
Internal; for 190 × 62 Assy	11.72	14.83	0.51	20.20	nr	**35.03**
Internal; for 215 × 50 Assy	13.11	16.60	0.53	20.99	nr	**37.59**
Internal; for 254 × 50 Assy	15.61	19.76	0.53	20.99	nr	**40.75**
External; for 170 × 50 Assy	10.55	13.35	0.56	22.18	nr	**35.53**
External; for 190 × 62 Assy	11.72	14.83	0.56	22.18	nr	**37.01**
External; for 215 × 50 Assy	13.11	16.60	0.58	22.97	nr	**39.57**
External; for 254 × 50 Assy	15.61	19.76	0.58	22.97	nr	**42.73**
Clip on end caps						
170 × 50 Assy	6.32	8.00	0.11	4.36	nr	**12.35**
215 × 50 Assy	7.47	9.46	0.11	4.36	nr	**13.82**
254 × 50 Assy	8.76	11.09	0.11	4.36	nr	**15.45**
190 × 62 Assy	8.83	11.18	0.11	4.36	nr	**15.53**
Outlet box						
1 gang; in horizontal trunking; clip in	5.64	7.14	0.34	13.47	nr	**20.60**
2 gang; in horizontal trunking; clip in	7.04	8.90	0.34	13.47	nr	**22.37**
1 gang; in vertical trunking; clip in	5.64	7.14	0.34	13.47	nr	**20.60**
Sheet steel adaptable boxes; with plain or knockout sides; fixed to backgrounds; including supports and fixings (cutting and connecting conduit to boxes is included)						
Square pattern – black						
75 × 75 × 37 mm	2.54	3.22	0.69	27.33	nr	**30.54**
75 × 75 × 50 mm	3.07	3.89	0.69	27.33	nr	**31.21**
75 × 75 × 75 mm	2.76	3.49	0.69	27.33	nr	**30.82**
100 × 100 × 50 mm	5.20	6.58	0.71	28.12	nr	**34.70**
150 × 150 × 50 mm	3.85	4.88	0.79	31.29	nr	**36.16**
150 × 150 × 75 mm	6.66	8.43	0.80	31.68	nr	**40.11**
150 × 150 × 100 mm	7.53	9.53	0.80	31.68	nr	**41.22**
200 × 200 × 50 mm	6.04	7.64	0.80	31.68	nr	**39.32**
225 × 225 × 50 mm	7.72	9.77	0.93	36.83	nr	**46.60**
225 × 225 × 100 mm	10.24	12.95	0.94	37.23	nr	**50.18**
300 × 300 × 100 mm	10.98	13.90	0.99	39.21	nr	**53.11**
Square pattern – galvanized						
75 × 75 × 37 mm	2.33	2.95	0.69	27.33	nr	**30.28**
75 × 75 × 50 mm	2.73	3.45	0.70	27.72	nr	**31.17**
75 × 75 × 75 mm	3.17	4.02	0.69	27.33	nr	**31.34**
100 × 100 × 50 mm	2.89	3.65	0.71	28.12	nr	**31.77**
150 × 150 × 50 mm	3.37	4.26	0.84	33.27	nr	**37.53**
150 × 150 × 75 mm	4.06	5.14	0.80	31.68	nr	**36.82**
150 × 150 × 100 mm	4.91	6.21	0.80	31.68	nr	**37.90**
225 × 225 × 50 mm	6.35	8.03	0.93	36.83	nr	**44.87**
225 × 225 × 100 mm	7.72	9.77	0.94	37.23	nr	**46.99**
300 × 300 × 100 mm	13.12	16.61	0.96	38.02	nr	**54.63**
Rectangular pattern – black						
100 × 75 × 50 mm	4.30	5.44	0.69	27.33	nr	**32.77**
150 × 75 × 50 mm	4.51	5.71	0.70	27.72	nr	**33.43**
150 × 75 × 75 mm	4.89	6.19	0.71	28.12	nr	**34.31**

ELECTRICAL SUPPLY/POWER/LIGHTING

Item	Net Price £	Material £	Labour hours	Labour £	Unit	Total rate £
LV DISTRIBUTION: TRUNKING – cont						
Sheet steel adaptable boxes – cont						
Rectangular pattern – cont						
150 × 100 × 75 mm	10.89	13.78	0.71	28.12	nr	**41.90**
225 × 75 × 50 mm	9.41	11.91	0.78	30.89	nr	**42.80**
225 × 150 × 75 mm	15.03	19.02	0.81	32.08	nr	**51.10**
225 × 150 × 100 mm	28.21	35.70	0.81	32.08	nr	**67.78**
300 × 150 × 50 mm	28.21	35.70	0.93	36.83	nr	**72.53**
300 × 150 × 75 mm	28.21	35.70	0.94	37.23	nr	**72.93**
300 × 150 × 100 mm	28.21	35.70	0.96	38.02	nr	**73.72**
Rectangular pattern – galvanized						
100 × 75 × 50 mm	4.08	5.17	0.69	27.33	nr	**32.49**
150 × 75 × 50 mm	4.28	5.41	0.70	27.72	nr	**33.14**
150 × 75 × 75 mm	4.65	5.88	0.71	28.12	nr	**34.00**
150 × 100 × 75 mm	10.33	13.07	0.71	28.12	nr	**41.19**
225 × 75 × 50 mm	8.95	11.32	0.89	35.25	nr	**46.57**
225 × 150 × 75 mm	14.27	18.06	0.81	32.08	nr	**50.14**
225 × 150 × 100 mm	26.81	33.93	0.81	32.08	nr	**66.01**
300 × 150 × 50 mm	26.81	33.93	0.93	36.83	nr	**70.76**
300 × 150 × 75 mm	26.81	33.93	0.94	37.23	nr	**71.15**
300 × 150 × 100 mm	26.81	33.93	0.96	38.02	nr	**71.95**

ELECTRICAL SUPPLY/POWER/LIGHTING

Item	Net Price £	Material £	Labour hours	Labour £	Unit	Total rate £
LV DISTRIBUTION: CABLES AND WIRING						
ARMOURED CABLE						
Cable; XLPE insulated; PVC sheathed; copper stranded conductors to BS 5467; laid in trench/duct including marker tape (cable tiles measured elsewhere)						
600/1000 volt grade; single core (aluminium wire armour)						
70 mm²	8.67	10.97	0.18	7.13	m	**18.10**
95 mm²	11.40	14.42	0.20	7.92	m	**22.34**
120 mm²	13.98	17.69	0.22	8.71	m	**26.41**
150 mm²	17.77	22.48	0.24	9.51	m	**31.99**
185 mm²	21.04	26.63	0.26	10.30	m	**36.93**
240 mm²	27.39	34.66	0.30	11.88	m	**46.55**
300 mm²	34.05	43.09	0.31	12.28	m	**55.36**
400 mm²	44.26	56.02	0.38	15.05	m	**71.07**
500 mm²	56.34	71.31	0.44	17.43	m	**88.73**
630 mm²	70.64	89.40	0.52	20.59	m	**110.00**
600/1000 volt grade; two core (galvanized steel wire armour)						
1.5 mm²	1.32	1.67	0.06	2.38	m	**4.04**
2.5 mm²	1.63	2.06	0.06	2.38	m	**4.43**
4 mm²	2.03	2.57	0.08	3.17	m	**5.74**
6 mm²	2.57	3.25	0.08	3.17	m	**6.42**
10 mm²	3.58	4.54	0.10	3.96	m	**8.50**
16 mm²	5.20	6.58	0.10	3.96	m	**10.54**
25 mm²	6.71	8.49	0.15	5.94	m	**14.43**
35 mm²	9.29	11.75	0.15	5.94	m	**17.70**
50 mm²	11.83	14.98	0.17	6.73	m	**21.71**
70 mm²	15.64	19.80	0.18	7.13	m	**26.93**
95 mm²	22.02	27.87	0.20	7.92	m	**35.79**
120 mm²	27.37	34.64	0.22	8.71	m	**43.35**
150 mm²	32.95	41.70	0.24	9.51	m	**51.20**
185 mm²	43.87	55.52	0.26	10.30	m	**65.82**
240 mm²	54.64	69.16	0.30	11.88	m	**81.04**
300 mm²	72.21	91.39	0.31	12.28	m	**103.67**
600/1000 volt grade; three core (galvanized steel wire armour)						
1.5 mm²	1.49	1.88	0.07	2.77	m	**4.65**
2.5 mm²	1.92	2.43	0.07	2.77	m	**5.20**
4 mm²	2.52	3.19	0.09	3.56	m	**6.75**
6 mm²	3.05	3.86	0.10	3.96	m	**7.82**
10 mm²	4.81	6.08	0.11	4.36	m	**10.44**
16 mm²	7.09	8.98	0.11	4.36	m	**13.33**
25 mm²	9.67	12.24	0.16	6.34	m	**18.57**
35 mm²	13.01	16.46	0.16	6.34	m	**22.80**
50 mm²	17.48	22.13	0.19	7.52	m	**29.65**
70 mm²	24.20	30.62	0.21	8.32	m	**38.94**
95 mm²	33.04	41.82	0.23	9.11	m	**50.93**

ELECTRICAL SUPPLY/POWER/LIGHTING

Item	Net Price £	Material £	Labour hours	Labour £	Unit	Total rate £
LV DISTRIBUTION: CABLES AND WIRING – cont						
Cable – cont						
600/1000 volt grade; three core – cont						
120 mm²	40.58	51.36	0.24	9.51	m	**60.87**
150 mm²	55.68	70.47	0.27	10.69	m	**81.16**
185 mm²	62.95	79.67	0.30	11.88	m	**91.55**
240 mm²	81.12	102.66	0.33	13.07	m	**115.73**
300 mm²	102.81	130.11	0.35	13.86	m	**143.98**
600/1000 volt grade; four core (galvanized steel wire armour)						
1.5 mm²	1.45	1.84	0.08	3.17	m	**5.01**
2.5 mm²	2.01	2.55	0.09	3.56	m	**6.11**
4 mm²	2.76	3.49	0.10	3.96	m	**7.45**
6 mm²	3.97	5.03	0.10	3.96	m	**8.99**
10 mm²	6.11	7.73	0.12	4.75	m	**12.48**
16 mm²	9.14	11.57	0.12	4.75	m	**16.32**
25 mm²	12.13	15.36	0.18	7.13	m	**22.48**
35 mm²	16.32	20.66	0.19	7.52	m	**28.18**
50 mm²	21.70	27.47	0.21	8.32	m	**35.78**
70 mm²	31.30	39.61	0.23	9.11	m	**48.72**
95 mm²	42.50	53.79	0.26	10.30	m	**64.09**
120 mm²	53.78	68.06	0.28	11.09	m	**79.15**
150 mm²	65.75	83.21	0.32	12.67	m	**95.88**
185 mm²	82.40	104.29	0.35	13.86	m	**118.15**
240 mm²	106.05	134.22	0.36	14.26	m	**148.48**
300 mm²	133.54	169.01	0.40	15.84	m	**184.85**
600/1000 volt grade; seven core (galvanized steel wire armour)						
1.5 mm²	2.33	2.95	0.10	3.96	m	**6.91**
2.5 mm²	3.35	4.24	0.10	3.96	m	**8.20**
600/1000 volt grade; twelve core (galvanized steel wire armour)						
1.5 mm²	3.67	4.65	0.11	4.36	m	**9.01**
2.5 mm²	5.39	6.82	0.11	4.36	m	**11.18**
600/1000 volt grade; nineteen core (galvanized steel wire armour)						
1.5 mm²	5.27	6.67	0.13	5.15	m	**11.82**
2.5 mm²	8.21	10.39	0.14	5.54	m	**15.93**
600/1000 volt grade; twenty-seven core (galvanized steel wire armour)						
1.5 mm²	7.63	9.65	0.14	5.54	m	**15.20**
2.5 mm²	11.64	14.73	0.16	6.34	m	**21.07**
600/1000 volt grade; thirty-seven core (galvanized steel wire armour)						
1.5 mm²	9.84	12.46	0.15	5.94	m	**18.40**
2.5 mm²	14.46	18.30	0.17	6.73	m	**25.03**

ELECTRICAL SUPPLY/POWER/LIGHTING

Item	Net Price £	Material £	Labour hours	Labour £	Unit	Total rate £
Cable; XLPE insulated; PVC sheathed copper stranded conductors to BS 5467; clipped direct to backgrounds including cleat						
600/1000 volt grade; single core (aluminium wire armour)						
70 mm²	11.12	14.07	0.39	15.45	m	**29.52**
95 mm²	15.29	19.35	0.42	16.63	m	**35.99**
120 mm²	18.16	22.98	0.47	18.61	m	**41.60**
150 mm²	21.91	27.73	0.51	20.20	m	**47.93**
185 mm²	25.16	31.84	0.59	23.37	m	**55.21**
240 mm²	31.66	40.07	0.68	26.93	m	**67.01**
300 mm²	38.26	48.42	0.74	29.31	m	**77.73**
400 mm²	48.07	60.84	0.88	34.85	m	**95.69**
500 mm²	61.93	78.38	0.88	34.85	m	**113.24**
630 mm²	76.77	97.16	1.05	41.58	m	**138.74**
600/1000 volt grade; two core (galvanized steel wire armour)						
1.5 mm²	1.73	2.19	0.20	7.92	m	**10.11**
2.5 mm²	2.03	2.57	0.20	7.92	m	**10.49**
4.0 mm²	2.44	3.09	0.21	8.32	m	**11.40**
6.0 mm²	3.02	3.82	0.22	8.71	m	**12.53**
10.0 mm²	4.26	5.39	0.24	9.51	m	**14.89**
16.0 mm²	5.89	7.45	0.25	9.90	m	**17.35**
25 mm²	7.89	9.98	0.35	13.86	m	**23.84**
35 mm²	10.24	12.96	0.36	14.26	m	**27.22**
50 mm²	12.85	16.26	0.37	14.65	m	**30.92**
70 mm²	16.15	20.43	0.39	15.45	m	**35.88**
95 mm²	22.56	28.55	0.42	16.63	m	**45.18**
120 mm²	27.92	35.33	0.47	18.61	m	**53.95**
150 mm²	33.79	42.77	0.51	20.20	m	**62.96**
185 mm²	44.87	56.78	0.59	23.37	m	**80.15**
240 mm²	56.45	71.45	0.68	26.93	m	**98.38**
300 mm²	74.02	93.68	0.74	29.31	m	**122.99**
600/1000 volt grade; three core (galvanized steel wire armour)						
1.5 mm²	1.92	2.43	0.20	7.92	m	**10.35**
2.5 mm²	2.34	2.97	0.21	8.32	m	**11.28**
4.0 mm²	2.93	3.71	0.22	8.71	m	**12.42**
6.0 mm²	3.72	4.70	0.22	8.71	m	**13.42**
10.0 mm²	5.68	7.19	0.25	9.90	m	**17.09**
16.0 mm²	7.92	10.03	0.26	10.30	m	**20.33**
25 mm²	10.66	13.49	0.37	14.65	m	**28.15**
35 mm²	14.04	17.77	0.39	15.45	m	**33.21**
50 mm²	19.19	24.29	0.40	15.84	m	**40.14**
70 mm²	24.31	30.76	0.42	16.63	m	**47.40**
95 mm²	33.31	42.16	0.45	17.82	m	**59.98**
120 mm²	40.71	51.52	0.52	20.59	m	**72.12**
150 mm²	55.71	70.51	0.55	21.78	m	**92.29**
185 mm²	63.92	80.89	0.63	24.95	m	**105.85**
240 mm²	82.04	103.84	0.71	28.12	m	**131.95**
300 mm²	103.33	130.78	0.78	30.89	m	**161.67**

ELECTRICAL SUPPLY/POWER/LIGHTING

Item	Net Price £	Material £	Labour hours	Labour £	Unit	Total rate £
LV DISTRIBUTION: CABLES AND WIRING – cont						
Cable – cont						
600/1000 volt grade; four core (galvanized steel wire armour)						
1.5 mm²	2.13	2.69	0.21	8.32	m	**11.01**
2.5 mm²	2.67	3.38	0.22	8.71	m	**12.09**
4.0 mm²	3.46	4.37	0.22	8.71	m	**13.09**
6.0 mm²	4.84	6.12	0.23	9.11	m	**15.23**
10.0 mm²	7.23	9.15	0.26	10.30	m	**19.44**
16.0 mm²	10.24	12.95	0.26	10.30	m	**23.25**
25 mm²	13.29	16.82	0.39	15.45	m	**32.26**
35 mm²	17.44	22.07	0.40	15.84	m	**37.91**
50 mm²	22.88	28.96	0.41	16.24	m	**45.19**
70 mm²	31.87	40.34	0.45	17.82	m	**58.16**
95 mm²	42.98	54.39	0.50	19.80	m	**74.20**
120 mm²	54.30	68.73	0.54	21.39	m	**90.11**
150 mm²	67.25	85.12	0.60	23.76	m	**108.88**
185 mm²	84.03	106.35	0.67	26.54	m	**132.88**
240 mm²	107.71	136.32	0.75	29.70	m	**166.03**
300 mm²	135.37	171.32	0.83	32.87	m	**204.19**
600/1000 volt grade; seven core (galvanized steel wire armour)						
1.5 mm²	2.99	3.78	0.20	7.92	m	**11.70**
2.5 mm²	3.99	5.04	0.20	7.92	m	**12.97**
600/1000 volt grade; twelve core (galvanized steel wire armour)						
1.5 mm²	5.29	14.79	0.23	9.11	m	**23.90**
2.5 mm²	6.40	8.10	0.24	9.51	m	**17.60**
600/1000 volt grade; nineteen core (galvanized steel wire armour)						
1.5 mm²	6.40	8.10	0.26	10.30	m	**18.39**
2.5 mm²	9.40	11.89	0.28	11.09	m	**22.98**
600/1000 volt grade; twenty-seven core (galvanized steel wire armour)						
1.5 mm²	8.79	11.13	0.29	11.49	m	**22.61**
2.5 mm²	12.85	16.26	0.30	11.88	m	**28.14**
600/1000 volt grade; thirty-seven core (galvanized steel wire armour)						
1.5 mm²	10.88	13.77	0.32	12.67	m	**26.44**
2.5 mm²	16.10	20.38	0.33	13.07	m	**33.44**

ELECTRICAL SUPPLY/POWER/LIGHTING

Item	Net Price £	Material £	Labour hours	Labour £	Unit	Total rate £
Cable termination; brass weatherproof gland with inner and outer seal, shroud, brass locknut and earth ring (including drilling and cutting mild steel gland plate)						
600/1000 volt grade; single core (aluminium wire armour)						
70 mm²	10.00	12.66	2.12	83.96	nr	96.62
95 mm²	10.06	12.73	2.39	94.66	nr	107.38
120 mm²	10.09	12.77	2.47	97.82	nr	110.59
150 mm²	20.36	25.77	2.73	108.12	nr	133.89
185 mm²	20.68	26.17	3.05	120.79	nr	146.97
240 mm²	20.58	26.04	3.45	136.64	nr	162.68
300 mm²	21.03	26.61	3.84	152.08	nr	178.70
400 mm²	30.04	38.01	4.21	166.74	nr	204.75
500 mm²	32.05	40.56	5.70	225.75	m	266.31
630 mm²	39.08	49.45	6.20	245.55	m	295.00
600/1000 volt grade; two core (galvanized steel wire armour)						
1.5 mm²	8.01	10.14	0.58	22.97	nr	33.11
2.5 mm²	8.01	10.14	0.58	23.01	nr	33.15
4 mm²	8.01	10.14	0.58	22.97	nr	33.11
6 mm²	8.04	10.17	0.67	26.54	nr	36.71
10 mm²	8.93	11.30	1.00	39.60	nr	50.90
16 mm²	9.14	11.56	1.11	43.96	nr	55.53
25 mm²	10.49	13.28	1.70	67.33	nr	80.61
35 mm²	10.41	13.18	1.79	70.89	nr	84.07
50 mm²	20.00	25.31	2.06	81.59	nr	106.89
70 mm²	20.16	25.51	2.12	83.96	nr	109.48
95 mm²	19.19	24.28	2.39	94.66	nr	118.94
120 mm²	19.28	24.40	2.47	97.82	nr	122.23
150 mm²	30.09	38.08	2.73	108.12	nr	146.21
185 mm²	30.77	38.94	3.05	120.79	nr	159.74
240 mm²	32.71	41.40	3.45	136.64	nr	178.04
300 mm²	47.63	60.28	3.84	152.08	nr	212.36
600/1000 volt grade; three core (galvanized steel wire armour)						
1.5 mm²	8.04	10.17	0.61	24.35	nr	34.52
2.5 mm²	8.04	10.17	0.62	24.55	nr	34.73
4 mm²	8.04	10.17	0.62	24.55	nr	34.73
6 mm²	8.95	11.32	0.71	28.12	nr	39.44
10 mm²	10.35	13.10	1.06	41.98	nr	55.08
16 mm²	10.66	13.49	1.19	47.13	nr	60.62
25 mm²	19.85	25.12	1.81	71.68	nr	96.81
35 mm²	19.88	25.16	1.99	78.81	nr	103.98
50 mm²	19.92	25.21	2.23	88.32	nr	113.53
70 mm²	19.09	24.16	2.40	95.05	nr	119.21
95 mm²	19.26	24.38	2.63	104.16	nr	128.54
120 mm²	19.37	24.51	2.83	112.08	nr	136.59
150 mm²	31.18	39.46	3.22	127.53	nr	166.98
185 mm²	32.14	40.68	3.44	136.24	nr	176.92
240 mm²	48.78	61.74	3.83	151.69	nr	213.43
300 mm²	50.16	63.49	4.28	169.51	nr	232.99

ELECTRICAL SUPPLY/POWER/LIGHTING

Item	Net Price £	Material £	Labour hours	Labour £	Unit	Total rate £
LV DISTRIBUTION: CABLES AND WIRING – cont						
Cable termination – cont						
600/1000 volt grade; four core (galvanized steel wire armour)						
1.5 mm²	8.13	10.29	0.67	26.54	nr	**36.82**
2.5 mm²	8.13	10.29	0.67	26.54	nr	**36.82**
4 mm²	8.13	10.29	0.71	28.12	nr	**38.41**
6 mm²	10.43	13.20	0.76	30.10	nr	**43.30**
10 mm²	10.43	13.20	1.14	45.15	nr	**58.35**
16 mm²	10.84	13.72	1.29	51.09	nr	**64.81**
25 mm²	19.92	25.21	1.99	78.81	nr	**104.02**
35 mm²	19.96	25.26	2.16	85.55	nr	**110.81**
50 mm²	18.95	23.98	2.49	98.62	nr	**122.59**
70 mm²	19.27	24.38	2.65	104.95	nr	**129.34**
95 mm²	27.75	35.12	2.98	118.02	nr	**153.14**
120 mm²	27.88	35.29	3.15	124.75	nr	**160.05**
150 mm²	32.49	41.12	3.50	138.62	nr	**179.74**
185 mm²	47.39	59.98	3.72	147.33	nr	**207.31**
240 mm²	51.22	64.82	4.33	171.49	nr	**236.31**
300 mm²	73.99	93.64	4.86	192.48	nr	**286.12**
600/1000 volt grade; seven core (galvanized steel wire armour)						
1.5 mm²	9.51	12.04	0.81	32.08	nr	**44.12**
2.5 mm²	9.51	12.04	0.85	33.66	nr	**45.70**
600/1000 volt grade; twelve core (galvanized steel wire armour)						
1.5 mm²	10.90	13.80	1.14	45.15	nr	**58.95**
2.5 mm²	12.38	15.67	1.13	44.75	nr	**60.42**
600/1000 volt grade; nineteen core (galvanized steel wire armour)						
1.5 mm²	13.34	16.88	1.54	60.99	nr	**77.87**
2.5 mm²	13.34	16.88	1.54	60.99	nr	**77.87**
600/1000 volt grade; twenty-seven core (galvanized steel wire armour)						
1.5 mm²	14.67	18.57	1.94	76.83	nr	**95.40**
2.5 mm²	24.29	30.74	2.31	91.49	nr	**122.23**
600/1000 volt grade; thirty-seven core (galvanized steel wire armour)						
1.5 mm²	23.99	30.36	2.53	100.20	nr	**130.56**
2.5 mm²	23.99	30.36	2.87	113.67	nr	**144.03**

ELECTRICAL SUPPLY/POWER/LIGHTING

Item	Net Price £	Material £	Labour hours	Labour £	Unit	Total rate £
Cable; XLPE insulated; LSOH sheathed (LSF); copper stranded conductors to BS 6724; laid in trench/duct including marker tape (cable tiles measured elsewhere)						
600/1000 volt grade; single core (aluminium wire armour)						
50 mm²	5.59	7.07	0.17	6.73	m	**13.81**
70 mm²	7.70	9.75	0.18	7.13	m	**16.87**
95 mm²	10.12	12.81	0.20	7.92	m	**20.73**
120 mm²	12.13	15.36	0.22	8.71	m	**24.07**
150 mm²	14.49	18.33	0.24	9.51	m	**27.84**
185 mm²	17.45	22.08	0.26	10.30	m	**32.38**
240 mm²	23.31	29.50	0.30	11.88	m	**41.38**
300 mm²	28.31	35.83	0.31	12.28	m	**48.11**
400 mm²	37.00	46.83	0.35	13.86	m	**60.69**
500 mm²	46.82	59.25	0.44	17.43	m	**76.68**
630 mm²	62.19	78.71	0.52	20.59	m	**99.31**
800 mm²	78.22	99.00	0.62	24.55	m	**123.55**
1000 mm²	99.84	126.35	0.65	25.74	m	**152.10**
600/1000 volt grade; two core (galvanized steel wire armour)						
1.5 mm²	1.11	1.41	0.06	2.38	m	**3.78**
2.5 mm²	1.33	1.68	0.06	2.38	m	**4.06**
4 mm²	1.74	2.21	0.08	3.17	m	**5.38**
6 mm²	2.40	3.03	0.08	3.17	m	**6.20**
10 mm²	3.55	4.50	0.10	3.96	m	**8.46**
16 mm²	4.86	6.15	0.10	3.96	m	**10.11**
25 mm²	6.52	8.25	0.15	5.94	m	**14.19**
35 mm²	8.23	10.41	0.17	6.73	m	**17.15**
50 mm²	10.33	13.07	0.15	5.94	m	**19.01**
70 mm²	14.08	17.82	0.18	7.13	m	**24.95**
95 mm²	18.46	23.36	0.20	7.92	m	**31.28**
120 mm²	23.44	29.67	0.22	8.71	m	**38.38**
150 mm²	28.34	35.87	0.24	9.51	m	**45.38**
185 mm²	35.43	44.84	0.26	10.30	m	**55.13**
240 mm²	45.08	57.05	0.30	11.88	m	**68.93**
300 mm²	42.10	53.29	0.31	12.28	m	**65.56**
400 mm²	65.03	82.31	0.35	13.86	m	**96.17**
600/1000 volt grade; three core (galvanized steel wire armour)						
1.5 mm²	1.31	1.66	0.07	2.77	m	**4.43**
2.5 mm²	1.71	2.16	0.07	2.77	m	**4.94**
4 mm²	2.23	2.82	0.09	3.56	m	**6.39**
6 mm²	3.02	3.82	0.10	3.96	m	**7.78**
10 mm²	4.64	5.87	0.11	4.36	m	**10.23**
16 mm²	6.35	8.04	0.11	4.36	m	**12.39**
25 mm²	8.85	11.20	0.16	6.34	m	**17.53**
35 mm²	11.41	14.44	0.16	6.34	m	**20.78**
50 mm²	14.74	18.66	0.19	7.52	m	**26.18**
70 mm²	19.87	25.15	0.21	8.32	m	**33.46**
95 mm²	27.58	34.91	0.23	9.11	m	**44.02**

ELECTRICAL SUPPLY/POWER/LIGHTING

Item	Net Price £	Material £	Labour hours	Labour £	Unit	Total rate £
LV DISTRIBUTION: CABLES AND WIRING – cont						
Cable – cont						
600/1000 volt grade; three core – cont						
120 mm²	33.91	42.92	0.24	9.51	m	**52.42**
150 mm²	41.89	53.02	0.27	10.69	m	**63.71**
185 mm²	51.65	65.37	0.30	11.88	m	**77.25**
240 mm²	67.77	85.76	0.33	13.07	m	**98.83**
300 mm²	83.01	105.05	0.35	13.86	m	**118.91**
400 mm²	116.74	147.74	0.41	16.24	m	**163.98**
600/1000 volt grade; four core (galvanized steel wire armour)						
1.5 mm²	1.47	1.87	0.08	3.17	m	**5.03**
2.5 mm²	1.98	2.51	0.09	3.56	m	**6.07**
4 mm²	2.82	3.56	0.10	3.96	m	**7.52**
6 mm²	4.00	5.07	0.10	3.96	m	**9.03**
10 mm²	5.58	7.06	0.12	4.75	m	**11.81**
16 mm²	7.87	9.95	0.12	4.75	m	**14.71**
25 mm²	11.45	14.49	0.18	7.13	m	**21.62**
35 mm²	15.13	19.14	0.19	7.52	m	**26.67**
50 mm²	19.80	25.06	0.21	8.32	m	**33.38**
70 mm²	25.97	32.87	0.23	9.11	m	**41.98**
95 mm²	34.79	44.03	0.26	10.30	m	**54.33**
120 mm²	43.49	55.05	0.28	11.09	m	**66.14**
150 mm²	59.06	74.75	0.32	12.67	m	**87.42**
185 mm²	72.97	92.36	0.35	13.86	m	**106.22**
240 mm²	94.39	119.46	0.36	14.26	m	**133.72**
300 mm²	117.33	148.49	0.40	15.84	m	**164.33**
400 mm²	156.65	198.26	0.45	17.82	m	**216.08**
600/1000 volt grade; seven core (galvanized steel wire armour)						
1.5 mm²	2.10	2.65	0.10	3.96	m	**6.61**
2.5 mm²	2.83	3.58	0.10	3.96	m	**7.54**
4 mm²	5.23	6.62	0.11	4.36	m	**10.97**
600/1000 volt grade; twelve core (galvanized steel wire armour)						
1.5 mm²	3.54	4.48	0.11	4.36	m	**8.84**
2.5 mm²	5.06	6.40	0.11	4.36	m	**10.76**
600/1000 volt grade; nineteen core (galvanized steel wire armour)						
1.5 mm²	5.19	6.57	0.13	5.15	m	**11.72**
2.5 mm²	7.47	9.45	0.14	5.54	m	**15.00**
600/1000 volt grade; twenty-seven core (galvanized steel wire armour)						
1.5 mm²	7.33	9.28	0.14	5.54	m	**14.82**
2.5 mm²	10.40	13.16	0.16	6.34	m	**19.49**
600/1000 volt grade; thirty-seven core (galvanized steel wire armour)						
1.5 mm²	8.73	11.05	0.15	5.94	m	**16.99**
2.5 mm²	12.41	15.71	0.17	6.73	m	**22.44**

ELECTRICAL SUPPLY/POWER/LIGHTING

Item	Net Price £	Material £	Labour hours	Labour £	Unit	Total rate £
Cable; XLPE insulated; LSOH sheathed (LSF) copper stranded conductors to BS 6724; clipped direct to backgrounds including cleat						
600/1000 volt grade; single core (aluminium wire armour)						
50 mm²	4.43	5.61	0.37	14.65	m	20.26
70 mm²	6.73	8.52	0.39	15.45	m	23.97
95 mm²	9.05	11.46	0.42	16.63	m	28.09
120 mm²	10.79	13.66	0.47	18.61	m	32.27
150 mm²	13.33	16.87	0.51	20.20	m	37.07
185 mm²	16.02	20.27	0.59	23.37	m	43.64
240 mm²	20.65	26.13	0.68	26.93	m	53.06
300 mm²	25.16	31.84	0.74	29.31	m	61.15
400 mm²	32.69	41.37	0.81	32.08	m	73.45
500 mm²	41.94	53.08	0.88	34.85	m	87.94
630 mm²	52.60	66.57	1.05	41.58	m	108.16
800 mm²	67.78	85.78	1.33	52.67	m	138.46
1000 mm²	78.49	99.33	1.40	55.45	m	154.78
600/1000 volt grade; two core (galvanized steel wire armour)						
1.5 mm²	0.94	1.19	0.20	7.92	m	9.11
2.5 mm²	1.08	1.37	0.20	7.92	m	9.29
4.0 mm²	1.40	1.77	0.21	8.32	m	10.09
6.0 mm²	1.70	2.15	0.22	8.71	m	10.87
10.0 mm²	2.28	2.88	0.24	9.51	m	12.39
16.0 mm²	3.43	4.35	0.25	9.90	m	14.25
25 mm²	3.43	4.35	0.35	13.86	m	18.21
35 mm²	6.93	8.78	0.36	14.26	m	23.03
50 mm²	8.59	10.87	0.37	14.65	m	25.53
70 mm²	12.77	16.16	0.39	15.45	m	31.60
95 mm²	17.15	21.71	0.42	16.63	m	38.34
120 mm²	21.00	26.58	0.47	18.61	m	45.20
150 mm²	26.05	32.97	0.51	20.20	m	53.17
185 mm²	33.19	42.01	0.59	23.37	m	65.37
240 mm²	42.50	53.79	0.68	26.93	m	80.72
300 mm²	57.07	72.23	0.74	29.31	m	101.54
400 mm²	75.51	95.57	0.81	32.08	m	127.65
600/1000 volt grade; three core (galvanized steel wire armour)						
1.5 mm²	1.07	1.35	0.20	7.92	m	9.27
2.5 mm²	1.35	1.71	0.21	8.32	m	10.03
4.0 mm²	1.67	2.11	0.22	8.71	m	10.82
6.0 mm²	2.15	2.73	0.22	8.71	m	11.44
10.0 mm²	3.29	4.16	0.25	9.90	m	14.06
16.0 mm²	4.61	5.83	0.26	10.30	m	16.13
25 mm²	6.22	7.87	0.37	14.65	m	22.52
35 mm²	8.65	10.94	0.39	15.45	m	26.39
50 mm²	12.34	15.62	0.40	15.84	m	31.46
70 mm²	18.33	23.19	0.42	16.63	m	39.83
95 mm²	25.23	31.93	0.45	17.82	m	49.75

ELECTRICAL SUPPLY/POWER/LIGHTING

Item	Net Price £	Material £	Labour hours	Labour £	Unit	Total rate £
LV DISTRIBUTION: CABLES AND WIRING – cont						
Cable – cont						
600/1000 volt grade; three core – cont						
120 mm²	31.37	39.70	0.52	20.59	m	**60.29**
150 mm²	38.97	49.32	0.55	21.78	m	**71.11**
185 mm²	48.02	60.77	0.63	24.95	m	**85.72**
240 mm²	61.44	77.76	0.71	28.12	m	**105.88**
300 mm²	77.87	98.55	0.78	30.89	m	**129.44**
400 mm²	92.47	117.04	0.87	34.46	m	**151.49**
600/1000 volt grade; four core (galvanized steel wire armour)						
1.5 mm²	1.24	1.57	0.21	8.32	m	**9.89**
2.5 mm²	1.55	1.96	0.22	8.71	m	**10.68**
4.0 mm²	1.93	2.44	0.22	8.71	m	**11.15**
6.0 mm²	2.83	3.58	0.23	9.11	m	**12.69**
10.0 mm²	3.97	5.03	0.26	10.30	m	**15.32**
16.0 mm²	5.73	7.25	0.26	10.30	m	**17.55**
25 mm²	8.27	10.47	0.39	15.45	m	**25.91**
35 mm²	10.79	13.66	0.40	15.84	m	**29.50**
50 mm²	15.62	19.77	0.41	16.24	m	**36.01**
70 mm²	24.09	30.49	0.45	17.82	m	**48.31**
95 mm²	32.27	40.85	0.50	19.80	m	**60.65**
120 mm²	40.54	51.30	0.54	21.39	m	**72.69**
150 mm²	49.07	62.10	0.60	23.76	m	**85.87**
185 mm²	60.21	76.21	0.67	26.54	m	**102.74**
240 mm²	77.84	98.51	0.75	29.70	m	**128.21**
300 mm²	97.66	123.60	0.83	32.87	m	**156.47**
400 mm²	101.53	128.50	0.91	36.04	m	**164.54**
600/1000 volt grade; seven core (galvanized steel wire armour)						
1.5 mm²	1.75	2.21	0.20	7.92	m	**10.14**
2.5 mm²	2.44	3.09	0.20	7.92	m	**11.01**
4.0 mm²	4.43	5.60	0.23	9.11	m	**14.71**
600/1000 volt grade; twelve core (galvanized steel wire armour)						
1.5 mm²	2.83	3.58	0.23	9.11	m	**12.69**
2.5 mm²	4.25	5.37	0.24	9.51	m	**14.88**
600/1000 volt grade; nineteen core (galvanized steel wire armour)						
1.5 mm²	4.32	5.46	0.26	10.30	m	**15.76**
2.5 mm²	6.47	8.19	0.28	11.09	m	**19.28**
600/1000 volt grade; twenty-seven core (galvanized steel wire armour)						
1.5 mm²	6.09	7.70	0.29	11.49	m	**19.19**
2.5 mm²	8.49	10.75	0.30	11.88	m	**22.63**
600/1000 volt grade; thirty-seven core (galvanized steel wire armour)						
1.5 mm²	7.14	9.03	0.32	12.67	m	**21.71**
2.5 mm²	10.10	12.78	0.33	13.07	m	**25.85**

ELECTRICAL SUPPLY/POWER/LIGHTING

Item	Net Price £	Material £	Labour hours	Labour £	Unit	Total rate £
Cable termination; brass weatherproof gland with inner and outer seal, shroud, brass locknut and earth ring (including drilling and cutting mild steel gland plate)						
600/1000 volt grade; single core (aluminium wire armour)						
25 mm²	4.19	5.31	1.70	67.33	nr	72.63
35 mm²	4.57	5.78	1.79	70.89	nr	76.67
50 mm²	5.06	6.41	2.06	81.59	nr	87.99
70 mm²	7.24	9.17	2.12	83.96	nr	93.13
95 mm²	9.67	12.24	2.39	94.66	nr	106.90
120 mm²	11.46	14.50	2.47	97.82	nr	112.33
150 mm²	14.09	17.83	2.73	108.12	nr	125.95
185 mm²	16.78	21.24	3.05	120.79	nr	142.03
240 mm²	21.99	27.83	3.45	136.64	nr	164.47
300 mm²	26.58	33.64	3.84	152.08	nr	185.72
400 mm²	38.80	49.11	4.21	166.74	nr	215.85
500 mm²	48.90	61.88	5.70	225.75	m	287.63
630 mm²	63.60	80.49	6.20	245.55	m	326.04
800 mm²	79.08	100.08	7.50	297.04	m	397.12
1000 mm²	88.43	111.92	10.00	396.05	m	507.97
600/1000 volt grade; two core (galvanized steel wire armour)						
1.5 mm²	1.28	1.61	0.52	20.48	nr	22.09
2.5 mm²	1.47	1.87	0.58	22.97	nr	24.84
4 mm²	1.86	2.35	0.58	22.97	nr	25.33
6 mm²	2.16	2.74	0.67	26.54	nr	29.27
10 mm²	2.81	3.55	1.00	39.60	nr	43.16
16 mm²	4.07	5.15	1.11	43.96	nr	49.11
25 mm²	3.55	4.49	1.70	67.33	nr	71.81
35 mm²	7.91	10.01	1.79	70.89	nr	80.90
50 mm²	9.73	12.32	2.06	81.59	nr	93.91
70 mm²	14.47	18.32	2.12	83.96	nr	102.28
95 mm²	18.85	23.85	2.39	94.66	nr	118.51
120 mm²	23.19	29.35	2.47	97.82	nr	127.17
150 mm²	28.71	36.34	2.73	108.12	nr	144.46
185 mm²	36.55	46.26	3.05	120.79	nr	167.05
240 mm²	45.49	57.57	3.45	136.64	nr	194.21
300 mm²	63.69	80.61	3.84	152.08	nr	232.69
400 mm²	79.30	100.37	4.21	166.74	nr	267.10
600/1000 volt grade; three core (galvanized steel wire armour)						
1.5 mm²	1.35	1.70	0.62	24.55	nr	26.26
2.5 mm²	1.61	2.04	0.62	24.55	nr	26.59
4 mm²	2.05	2.59	0.62	24.55	nr	27.15
6 mm²	2.53	3.20	0.71	28.12	nr	31.32
10 mm²	3.70	4.69	1.06	41.98	nr	46.67
16 mm²	5.09	6.45	1.19	47.13	nr	53.58
25 mm²	7.14	9.04	1.81	71.68	nr	80.72
35 mm²	9.28	11.75	1.99	78.81	nr	90.56
50 mm²	13.21	16.72	2.23	88.32	nr	105.04
70 mm²	19.18	24.28	2.40	95.05	nr	119.33

ELECTRICAL SUPPLY/POWER/LIGHTING

Item	Net Price £	Material £	Labour hours	Labour £	Unit	Total rate £
LV DISTRIBUTION: CABLES AND WIRING – cont						
Cable termination – cont						
600/1000 volt grade; three core – cont						
95 mm²	26.35	33.34	2.63	104.16	nr	**137.50**
120 mm²	32.71	41.39	2.83	112.08	nr	**153.47**
150 mm²	40.71	51.52	3.22	127.53	nr	**179.05**
185 mm²	50.06	63.36	3.44	136.24	nr	**199.60**
240 mm²	66.46	84.11	3.83	151.69	nr	**235.80**
300 mm²	83.66	105.88	4.28	169.51	nr	**275.39**
400 mm²	99.31	125.69	5.00	198.02	nr	**323.72**
600/1000 volt grade; four core (galvanized steel wire armour)						
1.5 mm²	1.54	1.95	0.67	26.54	nr	**28.48**
2.5 mm²	1.90	2.41	0.67	26.54	nr	**28.94**
4 mm²	2.33	2.94	0.71	28.12	nr	**31.06**
6 mm²	3.25	4.11	0.76	30.10	nr	**34.21**
10 mm²	4.46	5.64	1.14	45.15	nr	**50.79**
16 mm²	6.25	7.91	1.29	51.09	nr	**59.00**
25 mm²	8.90	11.27	1.99	78.81	nr	**90.08**
35 mm²	11.62	14.70	2.16	85.55	nr	**100.25**
50 mm²	16.62	21.04	2.49	98.62	nr	**119.65**
70 mm²	25.17	31.86	2.65	104.95	nr	**136.81**
95 mm²	33.74	42.71	2.98	118.02	nr	**160.73**
120 mm²	42.28	53.51	3.15	124.75	nr	**178.27**
150 mm²	53.35	67.52	3.50	138.62	nr	**206.14**
185 mm²	65.20	82.51	3.72	147.33	nr	**229.84**
240 mm²	84.41	106.82	4.33	171.49	nr	**278.31**
300 mm²	105.93	134.06	4.86	192.48	nr	**326.54**
400 mm²	109.88	139.07	5.46	216.24	nr	**355.31**
600/1000 volt grade; seven core (galvanized steel wire armour)						
1.5 mm²	2.37	3.00	0.81	32.08	nr	**35.08**
2.5 mm²	3.25	4.11	0.85	33.66	nr	**37.78**
4 mm²	5.90	7.46	0.93	36.83	nr	**44.30**
600/1000 volt grade; twelve core (galvanized steel wire armour)						
1.5 mm²	3.33	4.21	1.14	45.15	nr	**49.36**
2.5 mm²	5.00	6.32	1.13	44.75	nr	**51.08**
600/1000 volt grade; nineteen core (galvanized steel wire armour)						
1.5 mm²	4.95	6.27	1.54	60.99	nr	**67.26**
2.5 mm²	7.27	9.20	1.54	60.99	nr	**70.19**
600/1000 volt grade; twenty-seven core (galvanized steel wire armour)						
1.5 mm²	6.68	8.45	1.94	76.83	nr	**85.28**
2.5 mm²	9.29	11.76	2.31	91.49	nr	**103.25**
600/1000 volt grade; thirty-seven core (galvanized steel wire armour)						
1.5 mm²	7.71	9.76	2.53	100.20	nr	**109.96**
2.5 mm²	10.78	13.64	2.87	113.67	nr	**127.31**

ELECTRICAL SUPPLY/POWER/LIGHTING

Item	Net Price £	Material £	Labour hours	Labour £	Unit	Total rate £
UN-ARMOURED CABLE						
Cable: XLPE insulated; PVC sheathed 90c copper to CMA Code 6181e; for internal wiring; clipped to backgrounds; (Supports and fixings included)						
300/500 volt grade; single core						
6.0 mm²	0.82	1.03	0.09	3.56	m	**4.60**
10 mm²	1.33	1.69	0.10	3.96	m	**5.65**
16 mm²	2.10	2.66	0.12	4.75	m	**7.41**
Cable; LSF insulated to CMA Code 6491B; non-sheathed copper; laid/drawn in trunking/conduit						
450/750 volt grade; single core						
1.5 mm²	0.20	0.26	0.03	1.19	m	**1.44**
2.5 mm²	0.31	0.40	0.03	1.19	m	**1.59**
4.0 mm²	0.49	0.62	0.03	1.19	m	**1.81**
6.0 mm²	0.73	0.92	0.04	1.58	m	**2.51**
10.0 mm²	1.20	1.52	0.04	1.58	m	**3.10**
16.0 mm²	1.96	2.48	0.05	1.98	m	**4.46**
25.0 mm²	2.98	3.77	0.06	2.38	m	**6.15**
35.0 mm²	4.05	5.13	0.06	2.38	m	**7.51**
50.0 mm²	5.61	7.10	0.07	2.77	m	**9.87**
70.0 mm²	8.04	10.18	0.08	3.17	m	**13.35**
95.0 mm²	11.03	13.96	0.08	3.17	m	**17.13**
120.0 mm²	13.94	17.65	0.10	3.96	m	**21.61**
150.0 mm²	17.17	21.73	0.13	5.15	m	**26.88**
185.0 mm²	21.59	27.33	0.13	5.15	m	**32.48**
240.0 mm²	28.43	35.98	0.13	5.15	m	**41.12**
300.0 mm²	35.29	44.66	0.13	5.15	m	**49.81**
Cable; twin & earth to CMA code 6242Y; clipped to backgrounds						
300/500 volt grade; PVC/PVC						
1.5 mm² 2C+E	0.46	0.58	0.01	0.40	m	**0.98**
1.5 mm² 3C+E	0.57	0.72	0.02	0.79	m	**1.52**
2.5 mm² 2C+E	0.86	1.09	0.02	0.79	m	**1.88**
4.0 mm² 2C+E	1.25	1.59	0.02	0.79	m	**2.38**
6.0 mm² 2C+E	1.85	2.34	0.02	0.79	m	**3.13**
10.0 mm² 2C+E	3.06	3.87	0.03	1.19	m	**5.06**
16.0 mm² 2C+E	4.73	5.99	0.03	1.19	m	**7.17**
300/500 volt grade; LSF/LSF						
1.5 mm² 2C+E	0.57	0.72	0.01	0.40	m	**1.12**
1.5 mm² 3C+E	0.81	1.03	0.02	0.79	m	**1.82**
2.5 mm² 2C+E	1.01	1.28	0.02	0.79	m	**2.07**
4.0 mm² 2C+E	1.21	1.53	0.02	0.79	m	**2.32**
6.0 mm² 2C+E	1.76	2.23	0.02	0.79	m	**3.02**
10.0 mm² 2C+E	4.03	5.10	0.03	1.19	m	**6.28**
16.0 mm² 2C+E	6.46	8.17	0.03	1.19	m	**9.36**

ELECTRICAL SUPPLY/POWER/LIGHTING

Item	Net Price £	Material £	Labour hours	Labour £	Unit	Total rate £
LV DISTRIBUTION: CABLES AND WIRING – cont						
EARTH CABLE						
Cable; LSF insulated to CMA Code 6491B; non-sheathed copper; laid/drawn in trunking/conduit						
450/750 volt grade; single core						
1.5 mm²	0.16	0.20	0.03	1.19	m	**1.39**
2.5 mm²	0.25	0.31	0.03	1.19	m	**1.50**
4.0 mm²	0.38	0.48	0.03	1.19	m	**1.67**
6.0 mm²	0.54	0.68	0.04	1.58	m	**2.26**
10.0 mm²	0.94	1.19	0.04	1.58	m	**2.77**
16.0 mm²	1.47	1.86	0.05	1.98	m	**3.84**
25.0 mm²	2.27	2.88	0.06	2.38	m	**5.25**
35.0 mm²	3.10	3.93	0.06	2.38	m	**6.30**
50.0 mm²	4.32	5.47	0.07	2.77	m	**8.24**
70.0 mm²	6.12	7.74	0.08	3.17	m	**10.91**
95.0 mm²	8.37	10.59	0.08	3.17	m	**13.76**
120.0 mm²	10.55	13.35	0.10	3.96	m	**17.31**
150.0 mm²	13.07	16.54	0.13	5.15	m	**21.69**
185.0 mm²	16.25	20.57	0.16	6.34	m	**26.90**
240.0 mm²	21.25	26.89	0.20	7.92	m	**34.81**
300.0 mm²	26.51	33.55	0.13	5.15	m	**38.70**
FLEXIBLE CABLE						
Flexible cord; PVC insulated; PVC sheathed; copper stranded to CMA Code 218Y (laid loose)						
300 volt grade; two core						
0.50 mm²	0.18	0.23	0.07	2.77	m	**3.00**
0.75 mm²	0.23	0.29	0.07	2.77	m	**3.06**
300 volt grade; three core						
0.50 mm²	0.26	0.33	0.07	2.77	m	**3.10**
0.75 mm²	0.36	0.45	0.07	2.77	m	**3.22**
1.0 mm²	0.38	0.48	0.07	2.77	m	**3.25**
1.5 mm²	0.60	0.76	0.07	2.77	m	**3.53**
2.5 mm²	0.81	1.02	0.08	3.17	m	**4.19**
Flexible cord; PVC insulated; PVC sheathed; copper stranded to CMA Code 318Y (laid loose)						
300/500 volt grade; two core						
1.0 mm²	0.97	1.22	0.07	2.77	m	**3.99**
1.5 mm²	1.37	1.73	0.07	2.77	m	**4.50**
2.5 mm²	2.89	3.65	0.07	2.77	m	**6.43**

ELECTRICAL SUPPLY/POWER/LIGHTING

Item	Net Price £	Material £	Labour hours	Labour £	Unit	Total rate £
300/500 volt grade; three core						
0.75 mm²	0.86	1.09	0.07	2.77	m	**3.86**
1.5 mm²	1.20	1.51	0.07	2.77	m	**4.29**
2.5 mm²	2.26	2.86	0.08	3.17	m	**6.03**
300/500 volt grade; four core						
0.75 mm²	1.82	2.30	0.08	3.17	m	**5.47**
1.0 mm²	2.13	2.70	0.08	3.17	m	**5.87**
1.5 mm²	3.03	3.84	0.08	3.17	m	**7.01**
2.5 mm²	4.67	5.91	0.09	3.56	m	**9.48**
Flexible cord; PVC insulated; PVC						
sheathed for use in high temperature						
zones; copper stranded to CMA Code						
309Y (laid loose)						
300/500 volt grade; two core						
0.50 mm²	0.91	1.16	0.07	2.77	m	**3.93**
0.75 mm²	1.11	1.41	0.07	2.77	m	**4.18**
300/500 volt grade; three core						
1.0 mm²	1.82	2.30	0.07	2.77	m	**5.07**
1.5 mm²	2.52	3.19	0.07	2.77	m	**5.96**
2.5 mm²	3.79	4.80	0.07	2.77	m	**7.57**
Flexible cord; rubber insulated; rubber						
sheathed; copper stranded to CMA code						
318 (laid loose)						
300/500 volt grade; three core						
0.75 mm²	1.40	1.77	0.07	2.77	m	**4.54**
1.0 mm²	1.79	2.26	0.07	2.77	m	**5.03**
1.5 mm²	2.25	2.84	0.07	2.77	m	**5.62**
2.5 mm²	2.79	3.53	0.07	2.77	m	**6.31**
300/500 volt grade; four core						
0.75 mm²	1.87	2.37	0.08	3.17	m	**5.53**
1.0 mm²	2.16	2.74	0.08	3.17	m	**5.91**
1.5 mm²	2.87	3.63	0.08	3.17	m	**6.80**
2.5 mm²	3.26	4.12	0.08	3.17	m	**7.29**
Flexible cord; rubber insulated; rubber						
sheathed; for 90C operation; copper						
stranded to CMA Code 318 (laid loose)						
450/750 volt grade; two core						
1.0 mm²	0.80	1.01	0.07	2.77	m	**3.78**
1.5 mm²	1.01	1.27	0.07	2.77	m	**4.05**
2.5 mm²	1.47	1.87	0.07	2.77	m	**4.64**
450/750 volt grade; three core						
1.0 mm²	0.90	1.14	0.07	2.77	m	**3.91**
1.5 mm²	1.26	1.60	0.07	2.77	m	**4.37**
2.5 mm²	1.88	2.38	0.07	2.77	m	**5.16**
450/750 volt grade; four core						
0.75 mm²	0.88	1.11	0.07	2.77	m	**3.88**
1.0 mm²	1.15	1.45	0.08	3.17	m	**4.62**
1.5 mm²	1.76	2.22	0.08	3.17	m	**5.39**
2.5 mm²	2.45	3.09	0.08	3.17	m	**6.26**

ELECTRICAL SUPPLY/POWER/LIGHTING

Item	Net Price £	Material £	Labour hours	Labour £	Unit	Total rate £
LV DISTRIBUTION: CABLES AND WIRING – cont						
Flexible cord – cont						
Heavy flexible cable; rubber insulated; rubber sheathed; copper stranded to CMA Code 638P (laid loose)						
450/750 volt grade; two core						
1.0 mm²	0.69	0.87	0.08	3.17	m	**4.04**
1.5 mm²	0.82	1.04	0.08	3.17	m	**4.21**
2.5 mm²	1.12	1.42	0.08	3.17	m	**4.59**
450/750 volt grade; three core						
1.0 mm²	0.82	1.04	0.08	3.17	m	**4.21**
1.5 mm²	0.96	1.21	0.08	3.17	m	**4.38**
2.5 mm²	1.11	1.41	0.08	3.17	m	**4.58**
450/750 volt grade; four core						
1.0 mm²	1.12	1.42	0.08	3.17	m	**4.59**
1.5 mm²	1.23	1.55	0.08	3.17	m	**4.72**
2.5 mm²	1.68	2.13	0.08	3.17	m	**5.30**
FIRE RATED CABLE						
Cable, mineral insulated; copper sheathed with copper conductors; fixed with clips to backgrounds. BASEC approval to BS 6207 Part 1 1995; complies with BS 6387 Category CWZ						
Light duty 500 volt grade; bare						
2L 1.0	5.49	6.95	0.23	9.11	m	**16.06**
2L 1.5	6.26	7.93	0.23	9.11	m	**17.04**
2L 2.5	7.55	9.55	0.25	9.90	m	**19.45**
2L 4.0	10.66	13.50	0.25	9.90	m	**23.40**
3L 1.0	6.49	8.21	0.24	9.51	m	**17.72**
3L 1.5	7.72	9.77	0.25	9.90	m	**19.67**
3L 2.5	11.02	13.95	0.25	9.90	m	**23.85**
4L 1.0	7.41	9.38	0.25	9.90	m	**19.28**
4L 1.5	8.31	10.52	0.25	9.90	m	**20.42**
4L 2.5	13.06	16.53	0.26	10.30	m	**26.82**
7L 1.5	13.17	16.66	0.28	11.09	m	**27.75**
7L 2.5	16.68	21.12	0.27	10.69	m	**31.81**
Light duty 500 volt grade; LSF sheathed						
2L 1.0	4.28	5.42	0.23	9.11	m	**14.53**
2L 1.5	4.89	6.18	0.23	9.11	m	**15.29**
2L 4.0	8.05	10.18	0.25	9.90	m	**20.08**
3L 1.0	5.06	6.40	0.24	9.51	m	**15.90**
3L 1.5	6.05	7.66	0.25	9.90	m	**17.56**
4L 1.0	5.69	7.20	0.25	9.90	m	**17.11**
4L 1.5	6.93	8.77	0.25	9.90	m	**18.67**
4L 2.5	9.72	12.30	0.26	10.30	m	**22.60**
7L type="x02" 1.5	10.05	12.72	0.28	11.09	m	**23.81**
7L 2.5	12.52	15.84	0.27	10.69	m	**26.54**
7L 1.0	8.19	10.37	0.27	10.69	m	**21.06**

ELECTRICAL SUPPLY/POWER/LIGHTING

Item	Net Price £	Material £	Labour hours	Labour £	Unit	Total rate £
Heavy duty 750 volt grade; bare						
1H 10	11.19	14.17	0.25	9.90	m	**24.07**
1H 16	15.15	19.18	0.26	10.30	m	**29.47**
1H 25	21.06	26.65	0.27	10.69	m	**37.35**
1H 35	30.40	38.47	0.32	12.67	m	**51.15**
1H 50	33.28	42.12	0.35	13.86	m	**55.98**
1H 70	44.42	56.21	0.38	15.05	m	**71.26**
1H 95	65.86	83.35	0.41	16.24	m	**99.59**
1H 120	69.28	87.68	0.46	18.22	m	**105.90**
1H 150	84.98	107.55	0.50	19.80	m	**127.35**
1H 185	103.63	131.15	0.56	22.18	m	**153.33**
1H 240	134.82	170.63	0.69	27.33	m	**197.96**
2H 1.5	10.33	13.08	0.25	9.90	m	**22.98**
2H 2.5	12.47	15.78	0.26	10.30	m	**26.08**
2H 4.0	15.26	19.31	0.26	10.30	m	**29.61**
2H 6.0	19.63	24.84	0.29	11.49	m	**36.33**
2H 10.0	25.08	31.74	0.34	13.47	m	**45.21**
2H 16.0	35.88	45.41	0.40	15.84	m	**61.26**
2H 25.0	49.48	62.62	0.44	17.43	m	**80.04**
3H 1.5	11.48	14.53	0.25	9.90	m	**24.43**
3H 2.5	14.21	17.99	0.25	9.90	m	**27.89**
3H 4.0	17.32	21.92	0.27	10.69	m	**32.61**
3H 6.0	21.39	27.07	0.30	11.88	m	**38.95**
3H 10.0	31.72	40.14	0.35	13.86	m	**54.01**
3H 16.0	42.73	54.08	0.41	16.24	m	**70.31**
3H 25.0	62.69	79.34	0.47	18.61	m	**97.95**
4H 1.5	13.93	17.63	0.24	9.51	m	**27.14**
4H 2.5	16.90	21.38	0.26	10.30	m	**31.68**
4H 4.0	20.86	26.40	0.29	11.49	m	**37.89**
4H 6.0	26.39	33.40	0.31	12.28	m	**45.68**
4H 10.0	37.49	47.44	0.37	14.65	m	**62.09**
4H 16.0	53.72	67.98	0.44	17.43	m	**85.41**
4H 25.0	76.69	97.06	0.52	20.59	m	**117.66**
7H 1.5	18.79	23.78	0.30	11.88	m	**35.66**
7H 2.5	25.05	31.70	0.32	12.67	m	**44.37**
12H 2.5	42.73	54.08	0.39	15.45	m	**69.52**
19H 1.5	62.63	79.27	0.42	16.63	m	**95.90**
Heavy duty 750 volt grade; LSF sheathed						
1H 10	8.79	11.12	0.25	9.90	m	**21.02**
1H 16	11.96	15.13	0.26	10.30	m	**25.43**
1H 25	16.43	20.79	0.27	10.69	m	**31.48**
1H 35	22.29	28.21	0.32	12.67	m	**40.88**
1H 50	27.06	34.25	0.35	13.86	m	**48.11**
1H 70	34.38	43.51	0.38	15.05	m	**58.56**
1H 95	48.19	60.99	0.41	16.24	m	**77.23**
2H 1.5	8.21	10.39	0.25	9.90	m	**20.29**
2H 2.5	9.84	12.46	0.26	10.30	m	**22.75**
2H 4.0	12.07	15.27	0.26	10.30	m	**25.57**
2H 6.0	15.45	19.55	0.29	11.49	m	**31.03**
2H 10.0	20.40	25.82	0.34	13.47	m	**39.29**
2H 16.0	28.00	35.44	0.40	15.84	m	**51.28**
2H 25.0	38.97	49.32	0.44	17.43	m	**66.74**

ELECTRICAL SUPPLY/POWER/LIGHTING

Item	Net Price £	Material £	Labour hours	Labour £	Unit	Total rate £
LV DISTRIBUTION: CABLES AND WIRING – cont						
Cable, mineral insulated – cont						
Heavy duty 750 volt grade – cont						
3H 1.5	8.95	11.33	0.25	9.90	m	**21.23**
3H 2.5	11.15	14.11	0.25	9.90	m	**24.01**
3H 4.0	13.72	17.37	0.27	10.69	m	**28.06**
3H 6.0	17.63	22.31	0.30	11.88	m	**34.19**
3H 10.0	23.36	29.56	0.35	13.86	m	**43.42**
3H 16.0	32.68	41.36	0.41	16.24	m	**57.60**
3H 25.0	46.97	59.44	0.47	18.61	m	**78.06**
4H 1.5	10.97	13.88	0.24	9.51	m	**23.38**
4H 2.5	13.34	16.89	0.26	10.30	m	**27.18**
4H 4.0	16.97	21.48	0.29	11.49	m	**32.96**
4H 6.0	21.53	27.25	0.31	12.28	m	**39.52**
4H 10.0	28.35	35.88	0.37	14.65	m	**50.53**
4H 16.0	40.48	51.23	0.44	17.43	m	**68.66**
4H 25.0	57.26	72.47	0.52	20.59	m	**93.06**
7H 1.5	14.15	17.91	0.30	11.88	m	**29.79**
7H 2.5	19.35	24.49	0.32	12.67	m	**37.16**
12H 1.5	24.31	30.76	0.39	15.45	m	**46.21**
12H 2.5	32.34	40.93	0.39	15.45	m	**56.38**
19H 1.5	46.90	59.36	0.42	16.63	m	**75.99**
Cable terminations for MI Cable; Polymeric one piece moulding; containing grey sealing compound; testing; phase marking and connection						
Light duty 500 volt grade; brass gland; polymeric one moulding containing grey sealing compound; coloured conductor sleeving; Earth tag; plastic gland shroud						
2L 1.5	13.53	17.12	0.27	10.69	m	**27.82**
2L 2.5	13.53	17.12	0.27	10.69	m	**27.82**
3L 1.5	13.57	17.18	0.27	10.69	m	**27.87**
4L 1.5	13.57	17.18	0.27	10.69	m	**27.87**
Cable terminations; for MI copper sheathed cable. Certified for installation in potentially explosive atmospheres; testing; phase marking and connection; BS 6207 Part 2 1995						
Light duty 500 volt grade; brass gland; brass pot with earth tail; pot closure; sealing compound; conductor sleving; plastic gland shroud; identification markers						
2L 1.5	13.41	16.97	0.41	16.24	nr	**33.20**
2L 2.5	13.41	16.97	0.44	17.27	nr	**34.23**
2L 4.0	13.41	16.97	0.46	18.22	nr	**35.18**
3L 1.5	13.43	16.99	0.44	17.33	nr	**34.32**
3L 2.5	13.43	16.99	0.44	17.43	nr	**34.42**
4L 1.5	13.43	16.99	0.47	18.79	nr	**35.78**

ELECTRICAL SUPPLY/POWER/LIGHTING

Item	Net Price £	Material £	Labour hours	Labour £	Unit	Total rate £
4L 2.5	13.43	16.99	0.50	19.80	nr	**36.79**
7L 1.5	28.69	36.31	0.70	27.72	nr	**64.04**
7L 2.5	28.69	36.31	0.74	29.31	nr	**65.62**
Heavy duty 750 volt grade; brass gland; brass pot with earth tail; pot closure; sealing compound; conductor sleeving; plastic gland shroud; identification markers						
1H 10	13.46	17.03	0.37	14.65	nr	**31.69**
1H 16	13.46	17.03	0.39	15.45	nr	**32.48**
1H 25	13.46	17.03	0.56	22.18	nr	**39.21**
1H 35	13.46	17.03	0.57	22.57	nr	**39.61**
1H 50	28.82	36.47	0.60	23.76	nr	**60.24**
1H 70	28.82	36.47	0.67	26.54	nr	**63.01**
1H 95	28.82	36.47	0.75	29.70	nr	**66.18**
2H 1.5	13.46	17.03	0.42	16.63	nr	**33.67**
2H 2.5	13.46	17.03	0.44	17.54	nr	**34.58**
2H 4	13.46	17.03	0.47	18.61	nr	**35.65**
2H 6	13.46	17.03	0.54	21.39	nr	**38.42**
2H 10	28.82	36.47	0.58	22.97	nr	**59.44**
2H 16	28.82	36.47	0.69	27.33	nr	**63.80**
2H 25	44.79	56.69	0.77	30.50	nr	**87.19**
3H 1.5	13.47	17.05	0.44	17.43	nr	**34.47**
3H 2.5	13.47	17.05	0.47	18.73	nr	**35.78**
3H 4	13.47	17.05	0.57	22.57	nr	**39.62**
3H 6	28.89	36.56	0.61	24.16	nr	**60.72**
3H 10	28.89	36.56	0.65	25.74	nr	**62.31**
3H 16	28.89	36.56	0.78	30.89	nr	**67.46**
3H 25	69.00	87.33	0.85	33.66	nr	**121.00**
4H 1.5	13.47	17.05	0.52	20.59	nr	**37.64**
4H 2.5	13.47	17.05	0.53	20.99	nr	**38.04**
4H 4	28.89	36.56	0.60	23.76	nr	**60.33**
4H 6	28.89	36.56	0.65	25.74	nr	**62.31**
4H 10	28.89	36.56	0.69	27.33	nr	**63.89**
4H 16	44.79	56.69	0.88	34.85	nr	**91.54**
4H 25	69.00	87.33	0.93	36.83	nr	**124.16**
7H 1.5	28.82	36.47	0.71	28.12	nr	**64.59**
7H 2.5	28.82	36.47	0.74	29.31	nr	**65.78**
12H 1.5	44.79	56.69	0.85	33.66	nr	**90.35**
12H 2.5	44.79	56.69	1.00	39.60	nr	**96.29**
19H 2.5	69.00	87.33	1.11	43.96	nr	**131.29**
Cable; FP100; LOSH insulated; non-sheathed fire-resistant to LPCB Approved to BS 6387 Catergory CWZ; in conduit or trunking including terminations						
450/750 volt grade; single core						
1.5 mm²	0.76	0.97	0.13	5.15	m	**6.11**
2.5 mm²	1.19	1.51	0.13	5.15	m	**6.66**
4.0 mm²	1.89	2.40	0.14	5.43	m	**7.82**
6.0 mm²	2.80	3.55	0.13	5.15	m	**8.70**
10 mm²	4.61	5.84	0.16	6.34	m	**12.17**

ELECTRICAL SUPPLY/POWER/LIGHTING

Item	Net Price £	Material £	Labour hours	Labour £	Unit	Total rate £
LV DISTRIBUTION: CABLES AND WIRING – cont						
Cable; FP200; Insudite insulated; LSOH sheathed screened fire-resistant BASEC Approved to BS 7629; fixed with clips to backgrounds						
300/500 volt grade; two core						
1.0 mm²	3.80	4.81	0.20	7.92	m	**12.73**
1.5 mm²	3.09	3.91	0.20	7.92	m	**11.83**
2.5 mm²	5.12	6.48	0.20	7.92	m	**14.40**
4.0 mm²	8.02	10.16	0.23	9.11	m	**19.26**
300/500 volt grade; three core						
1.0 mm²	4.34	5.49	0.23	9.11	m	**14.60**
1.5 mm²	5.22	6.60	0.23	9.11	m	**15.71**
2.5 mm²	6.32	8.00	0.23	9.11	m	**17.11**
4.0 mm²	10.36	13.11	0.25	9.90	m	**23.01**
300/500 volt grade; four core						
1.0 mm²	5.10	6.45	0.25	9.90	m	**16.35**
1.5 mm²	6.19	7.83	0.25	9.90	m	**17.73**
2.5 mm²	8.53	10.80	0.25	9.90	m	**20.70**
4.0 mm	13.08	16.55	0.28	11.09	m	**27.64**
Terminations; including glanding-off, connection to equipment						
Two core						
1.0 mm²	0.73	0.92	0.35	13.86	nr	**14.78**
1.5 mm²	0.73	0.92	0.35	13.86	nr	**14.78**
2.5 mm²	0.73	0.92	0.35	13.86	nr	**14.78**
4.0 mm²	0.73	0.92	0.35	13.86	nr	**14.78**
Three core						
1.0 mm²	0.73	0.92	0.35	13.86	nr	**14.78**
1.5 mm²	0.73	0.92	0.35	13.86	nr	**14.78**
2.5 mm²	0.73	0.92	0.35	14.00	nr	**14.92**
4.0 mm²	0.73	0.92	0.35	13.86	nr	**14.78**
Four core						
1.0 mm²	0.73	0.92	0.35	13.86	nr	**14.78**
1.5 mm²	0.73	0.92	0.35	13.86	nr	**14.78**
2.5 mm²	0.73	0.92	0.35	13.86	nr	**14.78**
4.0 mm²	0.73	0.92	0.35	13.86	nr	**14.78**
Cable; FP400; polymeric insulated; LSOH sheathed fire-resistant; armoured; with copper stranded copper conductors; BASEC Approved to BS 7846; fixed with clips to backgrounds						
600/1000 volt grade; two core						
1.5 mm²	3.51	4.45	0.20	7.92	m	**12.37**
2.5 mm²	4.07	5.15	0.20	7.92	m	**13.07**
4.0 mm²	4.51	5.70	0.21	8.32	m	**14.02**

ELECTRICAL SUPPLY/POWER/LIGHTING

Item	Net Price £	Material £	Labour hours	Labour £	Unit	Total rate £
6.0 mm²	5.56	7.03	0.22	8.71	m	**15.74**
10 mm²	5.99	7.58	0.24	9.51	m	**17.09**
16 mm²	9.11	11.53	0.25	9.90	m	**21.43**
25 mm²	12.59	15.94	0.35	13.86	m	**29.80**
600/1000 volt grade; three core						
1.5 mm²	3.90	4.93	0.20	7.92	m	**12.85**
2.5 mm²	4.54	5.75	0.21	8.32	m	**14.06**
4.0 mm²	5.36	6.78	0.22	8.71	m	**15.50**
6.0 mm²	5.68	7.18	0.22	8.71	m	**15.90**
10 mm²	6.89	8.72	0.25	9.90	m	**18.62**
16 mm²	11.29	14.28	0.26	10.30	m	**24.58**
25 mm²	14.09	17.83	0.37	14.65	m	**32.49**
600/1000 volt grade; four core						
1.5 mm²	5.20	6.57	0.21	8.32	m	**14.89**
2.5 mm²	3.80	4.81	0.22	8.71	m	**13.52**
4.0 mm²	7.13	9.02	0.22	8.71	m	**17.73**
6.0 mm²	7.55	9.56	0.23	9.11	m	**18.67**
10 mm²	7.68	9.72	0.26	10.30	m	**20.02**
16 mm²	10.92	13.83	0.26	10.30	m	**24.12**
25 mm²	12.63	15.98	0.39	15.45	m	**31.43**
Terminations; including glanding-off, connection to equipment						
Two core						
1.5 mm²	2.63	3.33	0.58	22.97	nr	**26.30**
2.5 mm²	2.79	3.53	0.58	22.97	nr	**26.50**
4.0 mm²	3.05	3.86	0.61	24.28	nr	**28.14**
6.0 mm²	3.37	4.26	0.67	26.54	nr	**30.80**
10 mm²	3.70	4.69	1.00	39.60	nr	**44.29**
16 mm²	4.07	5.15	1.11	43.96	nr	**49.12**
25 mm²	4.47	5.65	1.70	67.33	nr	**72.98**
Three core						
1.5 mm²	3.96	5.01	0.62	24.55	nr	**29.57**
2.5 mm²	4.16	5.26	0.62	24.55	nr	**29.82**
4.0 mm²	4.59	5.81	0.66	26.30	nr	**32.11**
6.0 mm²	5.04	6.37	0.71	28.12	nr	**34.49**
10 mm²	5.53	7.00	1.06	41.98	nr	**48.98**
16 mm²	6.11	7.73	1.19	47.13	nr	**54.86**
25 mm²	6.70	8.48	1.81	71.68	nr	**80.17**
Four core						
1.5 mm²	5.27	6.67	0.67	26.54	nr	**33.20**
2.5 mm²	7.86	9.95	0.69	27.13	nr	**37.08**
4.0 mm²	5.83	7.37	0.71	28.12	nr	**35.49**
6.0 mm²	6.18	7.83	0.76	30.10	nr	**37.93**
10 mm²	8.99	11.37	1.14	45.15	nr	**56.52**
16 mm²	9.44	11.95	1.29	51.09	nr	**63.04**
25 mm²	16.60	21.01	1.35	53.47	nr	**74.48**

ELECTRICAL SUPPLY/POWER/LIGHTING

Item	Net Price £	Material £	Labour hours	Labour £	Unit	Total rate £
LV DISTRIBUTION: CABLES AND WIRING – cont						
Cable; FP600; polymeric insulated; LSOH sheathed fire-resistant; armoured; with copper stranded copper conductors; BASEC Approved to BS 7846; fixed with clips to backgrounds						
600/1000 volt grade; two core						
4.0 mm²	6.06	7.66	0.21	8.32	m	**15.98**
10 mm²	7.04	8.91	0.24	9.51	m	**18.42**
16 mm²	8.32	10.54	0.25	9.90	m	**20.44**
600/1000 volt grade; three core						
4.0 mm²	7.04	8.91	0.22	8.71	m	**17.62**
6.0 mm²	6.94	8.79	0.22	8.71	m	**17.50**
10 mm²	8.82	11.16	0.25	9.90	m	**21.06**
16 mm²	10.59	13.41	0.26	10.30	m	**23.70**
25 mm²	14.10	17.85	0.37	14.65	m	**32.50**
35 mm²	17.00	21.52	0.38	15.05	m	**36.57**
50 mm²	22.60	28.60	0.40	15.84	m	**44.44**
70 mm²	27.64	34.98	0.42	16.63	m	**51.61**
95 mm²	40.84	51.69	0.45	17.82	m	**69.51**
600/1000 volt grade; four core						
4.0 mm²	6.91	8.75	0.22	8.71	nr	**17.46**
6.0 mm²	7.62	9.64	0.23	9.11	nr	**18.75**
10 mm²	8.51	10.77	0.26	10.30	nr	**21.06**
16 mm²	9.77	12.36	0.26	10.30	nr	**22.66**
25 mm²	11.92	15.09	0.39	15.45	nr	**30.53**
35 mm²	15.26	19.31	0.40	15.84	m	**35.15**
50 mm²	18.95	23.98	0.42	16.63	m	**40.62**
70 mm²	24.48	30.98	0.45	17.82	m	**48.81**
95 mm²	32.40	41.00	0.47	18.61	m	**59.62**
120 mm²	40.23	50.92	0.48	19.01	m	**69.93**
150 mm²	48.35	61.20	0.51	20.20	m	**81.39**
185 mm²	59.46	75.25	0.52	20.59	m	**95.85**
240 mm²	76.44	96.74	0.55	21.78	m	**118.53**
300 mm²	93.66	118.53	0.59	23.37	m	**141.90**
400 mm²	117.31	148.47	0.63	24.95	m	**173.42**
Terminations; including glanding-off, connection to equipment						
Two core						
4.0 mm²	4.25	5.38	0.35	13.86	nr	**19.24**
10 mm²	4.94	6.25	0.35	13.86	nr	**20.11**
16 mm²	5.83	7.38	0.35	13.86	nr	**21.24**
Three core						
4.0 mm²	4.94	6.25	0.35	13.86	nr	**20.11**
6.0 mm²	4.86	6.16	0.35	13.86	nr	**20.02**
10 mm²	6.18	7.83	0.35	13.86	nr	**21.69**

ELECTRICAL SUPPLY/POWER/LIGHTING

Item	Net Price £	Material £	Labour hours	Labour £	Unit	Total rate £
16 mm²	7.43	9.40	0.35	13.86	nr	**23.26**
25 mm²	9.89	12.51	0.35	13.86	nr	**26.37**
35 mm²	11.92	15.08	0.35	13.86	m	**28.95**
50 mm²	15.85	20.06	0.35	13.86	m	**33.92**
70 mm²	19.37	24.52	0.35	13.86	m	**38.38**
95 mm²	28.63	36.23	0.35	13.86	m	**50.09**
Four core						
4.0 mm²	4.84	6.13	0.35	13.86	nr	**19.99**
6.0 mm²	5.35	6.77	0.35	13.86	nr	**20.63**
10 mm²	5.97	7.55	0.35	13.86	nr	**21.41**
16 mm²	6.85	8.67	0.35	13.86	nr	**22.53**
25 mm²	8.36	10.58	0.35	13.86	nr	**24.44**
35 mm²	10.70	13.54	0.35	13.86	m	**27.40**
50 mm²	13.29	16.81	0.35	13.86	m	**30.68**
70 mm²	17.17	21.73	0.35	13.86	m	**35.59**
95 mm²	22.72	28.76	0.35	13.86	m	**42.62**
120 mm²	28.21	35.70	0.35	13.86	m	**49.56**
150 mm²	33.90	42.91	0.35	13.86	m	**56.77**
185 mm²	41.70	52.77	0.35	13.86	m	**66.63**
240 mm²	53.59	67.82	0.35	13.86	m	**81.69**
300 mm²	65.65	83.09	0.35	13.86	m	**96.95**
400 mm²	82.24	104.09	0.35	13.86	m	**117.95**
Cable; Firetuff fire-resistant to BS 6387; fixed with clips to backgrounds						
Two core						
1.5 mm²	3.00	3.80	0.20	7.92	m	**11.72**
2.5 mm²	3.63	4.59	0.20	7.92	m	**12.51**
4.0 mm²	4.71	5.96	0.21	8.32	m	**14.28**
Three core						
1.5 mm²	3.67	4.64	0.20	7.92	m	**12.56**
2.5 mm²	4.15	5.25	0.21	8.32	m	**13.56**
4.0 mm²	5.99	7.58	0.22	8.71	m	**16.29**
Four core						
1.5 mm²	4.10	5.18	0.21	8.32	m	**13.50**
2.5 mm²	5.02	6.35	0.22	8.71	m	**15.06**
4.0 mm²	6.84	8.66	0.22	8.71	m	**17.37**

ELECTRICAL SUPPLY/POWER/LIGHTING

Item	Net Price £	Material £	Labour hours	Labour £	Unit	Total rate £
LV DISTRUBITION: MODULAR WIRING						
Modular wiring systems; including commissioning						
Master distribution box; steel; fixed to backgrounds; 6 Port						
4.0 mm 18 core armoured home run cable	199.40	252.36	0.90	35.64	nr	**288.00**
4.0 mm 24 core armoured home run cable	199.40	252.36	0.95	37.62	nr	**289.98**
4.0 mm 18 core armoured home run cable & data cable	211.83	268.10	0.95	37.62	nr	**305.72**
6.0 mm 18 core armoured home run cable	199.40	252.36	1.00	39.60	nr	**291.96**
6.0 mm 24 core armoured home run cable	199.40	252.36	1.10	43.57	nr	**295.92**
6.0 mm 18 core armoured home run cable & data cable	211.83	268.10	1.10	43.57	nr	**311.66**
Master distribution box; steel; fixed to backgrounds; 9 Port						
4.0 mm 27 core armoured home run cable	249.22	315.42	1.30	51.49	nr	**366.90**
4.0 mm 27 core armoured home run cable & data cable	249.22	315.42	1.45	57.43	nr	**372.84**
6.0 mm 27 core armoured home run cable	261.69	331.20	1.45	57.43	nr	**388.63**
6.0 mm 27 core armoured home run cable & data cable	261.69	331.20	1.55	61.39	nr	**392.59**
Metal clad cable; BSEN 60439 Part 2 1993; BASEC approved						
4.0 mm 18 core	20.09	25.42	0.30	11.88	m	**37.30**
4.0 mm 24 core	30.92	39.13	0.32	12.67	m	**51.80**
4.0 mm 27 core	31.16	39.44	0.35	13.86	m	**53.30**
6.0 mm 18 core	26.76	33.87	0.32	12.67	m	**46.55**
6.0 mm 27 core	37.28	47.18	0.35	13.86	m	**61.05**
Metal clad data cable						
Single twisted pair	3.92	4.96	0.18	7.13	m	**12.09**
Twin twisted pair	6.31	7.99	0.18	7.13	m	**15.12**
Distribution cables; armoured; BSEN 60439 Part 2 1993; BASEC approved						
3 Wire; 6.1 metre long	92.40	116.94	0.92	36.44	nr	**153.38**
4 Wire; 6.1 metre long	104.21	131.88	0.96	38.02	nr	**169.90**
Extender cables; armoured; BSEN 60439 Part 2 1993; BASEC approved						
3 wire						
0.9 metre long	57.08	72.24	0.13	5.15	nr	**77.39**
1.5 metre long	62.85	79.54	0.23	9.11	nr	**88.65**
2.1 metre long	68.60	86.82	0.31	12.28	nr	**99.10**
2.7 metre long	74.36	94.11	0.40	15.84	nr	**109.96**
3.4 metre long	81.09	102.63	0.51	20.20	nr	**122.83**
4.6 metre long	92.62	117.22	0.69	27.33	nr	**144.55**
6.1 metre long	107.02	135.45	0.92	36.44	nr	**171.88**

ELECTRICAL SUPPLY/POWER/LIGHTING

Item	Net Price £	Material £	Labour hours	Labour £	Unit	Total rate £
7.6 metre long	121.45	153.71	1.14	45.15	nr	**198.86**
9.1 metre long	135.84	171.92	1.37	54.26	nr	**226.18**
10.7 metre long	151.21	191.38	1.61	63.76	nr	**255.14**
4 wire						
0.9 metre long	61.09	77.32	0.14	5.54	nr	**82.86**
1.5 metre long	67.98	86.04	0.24	9.51	nr	**95.54**
2.1 metre long	74.86	94.75	0.32	12.67	nr	**107.42**
2.7 metre long	81.76	103.47	0.43	17.03	nr	**120.50**
3.4 metre long	89.80	113.65	0.51	20.20	nr	**133.85**
4.6 metre long	103.59	131.10	0.67	26.54	nr	**157.64**
6.1 metre long	121.18	153.37	0.92	36.44	nr	**189.80**
7.6 metre long	138.03	174.69	1.22	48.32	nr	**223.01**
9.1 metre long	155.26	196.50	1.46	57.82	nr	**254.32**
10.7 metre long	173.64	219.75	1.71	67.72	nr	**287.48**
3 wire; including twisted pair						
0.9 metre long	78.23	99.01	0.13	5.15	nr	**104.16**
1.5 metre long	87.91	111.25	0.23	9.11	nr	**120.36**
2.1 metre long	97.55	123.46	0.31	12.28	nr	**135.74**
2.7 metre long	107.20	135.68	0.40	15.84	nr	**151.52**
3.4 metre long	118.48	149.94	0.51	20.20	nr	**170.14**
4.6 metre long	137.79	174.38	0.69	27.33	nr	**201.71**
6.1 metre long	160.49	203.12	0.92	36.44	nr	**239.55**
7.6 metre long	186.04	235.45	1.14	45.15	nr	**280.60**
9.1 metre long	210.20	266.03	1.37	54.26	nr	**320.29**
10.7 metre long	235.93	298.59	1.61	63.76	nr	**362.36**
Extender whip ended cables; armoured; BSEN 60439 Part 2 1993; BASEC approved						
3 Wire; 3.0 metre long	52.90	66.95	0.30	11.88	nr	**78.83**
4 Wire; 3.0 metre long	58.63	74.20	0.30	11.88	nr	**86.08**
T connectors						
3 Wire						
6 pole 20a tee	71.94	91.05	0.10	3.96	nr	**95.01**
Snap fix	36.66	46.39	0.10	3.96	nr	**50.35**
1.5 metre flexible cable	42.52	53.81	0.10	3.96	nr	**57.77**
1.5 metre armoured cable	48.00	60.75	0.15	5.94	nr	**66.69**
1.5 metre armoured cable with twisted pair	61.18	77.43	0.15	5.94	nr	**83.37**
4 wire						
Snap fix	43.94	55.62	0.10	3.96	nr	**59.58**
1.5 metre flexible cable	51.59	65.29	0.10	3.96	nr	**69.25**
1.5 metre armoured cable	58.01	73.41	0.18	7.13	nr	**80.54**
Splitters						
5 Wire	38.47	48.69	0.20	7.92	nr	**56.61**
5 Wire converter	49.59	62.76	0.20	7.92	nr	**70.68**
Switch modules						
3 Wire; 6.1 metre long armoured cable	90.91	115.06	0.75	29.70	nr	**144.76**
4 Wire; 6.1 metre long armoured cable	102.30	129.47	0.80	31.68	nr	**161.15**

ELECTRICAL SUPPLY/POWER/LIGHTING

Item	Net Price £	Material £	Labour hours	Labour £	Unit	Total rate £
LV DISTRUBITION: MODULAR WIRING – cont						
Distribution cables; unarmoured; IEC 998 DIN/VDE 0628						
3 Wire; 6.1 metre long	36.07	45.65	0.70	27.72	nr	73.38
4 Wire; 6.1 metre long	45.12	57.10	0.75	29.70	nr	86.81
Extender cables; unarmoured; IEC 998 DIN/VDE 0628						
3 wire						
0.9 metre long	21.68	27.44	0.07	2.77	nr	30.21
1.5 metre long	23.73	30.03	0.12	4.75	nr	34.78
2.1 metre long	25.80	32.66	0.17	6.73	nr	39.39
2.7 metre long	34.15	43.21	0.27	10.69	nr	53.91
3.4 metre long	30.26	38.30	0.22	8.71	nr	47.01
4.6 metre long	34.41	43.55	0.37	14.65	nr	58.21
6.1 metre long	39.57	50.07	0.49	19.41	nr	69.48
7.6 metre long	44.70	56.58	0.61	24.16	nr	80.73
9.1 metre long	49.85	63.09	0.73	28.91	nr	92.01
10.7 metre long	55.36	70.07	0.86	34.06	nr	104.13
4 wire						
0.9 metre long	26.10	33.03	0.08	3.17	nr	36.20
1.5 metre long	28.79	36.44	0.14	5.54	nr	41.98
2.1 metre long	31.48	39.84	0.19	7.52	nr	47.37
2.7 metre long	34.15	43.21	0.24	9.51	nr	52.72
3.4 metre long	36.55	46.26	0.31	12.28	nr	58.54
4.6 metre long	42.65	53.98	0.41	16.24	nr	70.22
6.1 metre long	49.35	62.46	0.55	21.78	nr	84.24
7.6 metre long	56.04	70.93	0.68	26.93	nr	97.86
9.1 metre long	62.76	79.42	0.82	32.48	nr	111.90
10.7 metre long	69.91	88.47	0.96	38.02	nr	126.49
5 wire						
0.9 metre long	39.59	50.11	0.09	3.56	nr	53.67
1.5 metre long	43.06	54.50	0.15	5.94	nr	60.44
2.1 metre long	46.51	58.86	0.21	8.32	nr	67.18
2.7 metre long	49.99	63.27	0.27	10.69	nr	73.97
3.4 metre long	54.03	68.38	0.34	13.47	nr	81.85
4.6 metre long	60.96	77.15	0.46	18.22	nr	95.37
6.1 metre long	69.64	88.13	0.61	24.16	nr	112.29
7.6 metre long	78.31	99.11	0.76	30.10	nr	129.21
9.1 metre long	86.96	110.05	0.91	36.04	nr	146.10
10.7 metre long	96.21	121.76	1.07	42.38	nr	164.14
Extender whip ended cables; armoured; IEC 998 DIN/VDE 0628						
3 Wire; 2.5 mm; 3.0 metre long	20.94	26.50	0.30	11.88	nr	38.38
4 Wire; 2.5 mm; 3.0 metre long	24.64	31.18	0.30	11.88	nr	43.06

ELECTRICAL SUPPLY/POWER/LIGHTING

Item	Net Price £	Material £	Labour hours	Labour £	Unit	Total rate £
T connectors						
3 wire						
5 pin; direct fix	19.32	24.45	0.10	3.96	nr	**28.41**
5 pin; 1.5 mm flexible cable; 0.3 metre long	27.06	34.25	0.15	5.94	nr	**40.19**
4 wire						
5 pin; direct fix	24.64	31.18	0.20	7.92	nr	**39.10**
5 pin; 1.5 mm flexible cable; 0.3 metre long	30.66	38.80	0.20	7.92	nr	**46.72**
5 wire						
5 pin; direct fix	28.03	35.48	0.20	7.92	nr	**43.40**
Splitters						
3 way; 5 pin	16.14	20.43	0.25	9.90	nr	**30.33**
Switch modules						
3 wire	46.04	58.26	0.20	7.92	nr	**66.18**
4 wire	47.30	59.87	0.22	8.71	nr	**68.58**

ELECTRICAL SUPPLY/POWER/LIGHTING

Item	Net Price £	Material £	Labour hours	Labour £	Unit	Total rate £
LV DISTRUBITION: BUSBAR TRUNKING						
MAINS BUSBAR						
Low impedance busbar trunking; fixed to backgrounds including supports, fixings and connections/jointing to equipment						
Straight copper busbar						
1000 amp TP&N	513.41	649.78	3.41	135.05	m	**784.83**
1350 amp TP&N	631.06	798.67	3.58	141.78	m	**940.45**
2000 amp TP&N	791.50	1001.73	5.00	198.02	m	**1199.75**
2500 amp TP&N	1195.84	1513.45	5.90	233.67	m	**1747.12**
Extra for fittings mains bus bar						
IP54 protection						
1000 amp TP&N	35.01	44.31	2.16	85.55	m	**129.86**
1350 amp TP&N	37.62	47.61	2.61	103.37	m	**150.98**
2000 amp TP&N	52.19	66.05	3.51	139.01	m	**205.07**
2500 amp TP&N	61.97	78.43	3.96	156.83	m	**235.26**
End cover						
1000 amp TP&N	40.27	50.96	0.56	22.18	nr	**73.14**
1350 amp TP&N	41.99	53.15	0.56	22.18	nr	**75.33**
2000 amp TP&N	67.39	85.29	0.66	26.14	nr	**111.43**
2500 amp TP&N	68.26	86.39	0.66	26.14	nr	**112.53**
Edge elbow						
1000 amp TP&N	576.76	729.95	2.01	79.61	nr	**809.55**
1350 amp TP&N	649.10	821.50	2.01	79.61	nr	**901.11**
2000 amp TP&N	1006.76	1274.15	2.40	95.05	nr	**1369.21**
2500 amp TP&N	1322.31	1673.51	2.40	95.05	nr	**1768.56**
Flat elbow						
1000 amp TP&N	500.40	633.30	2.01	79.61	nr	**712.91**
1350 amp TP&N	542.59	686.70	2.01	79.61	nr	**766.30**
2000 amp TP&N	769.66	974.08	2.40	95.05	nr	**1069.13**
2500 amp TP&N	958.57	1213.16	2.40	95.05	nr	**1308.21**
Offset						
1000 amp TP&N	1004.79	1271.66	3.00	118.81	nr	**1390.48**
1350 amp TP&N	1235.87	1564.12	3.00	118.81	nr	**1682.94**
2000 amp TP&N	1947.28	2464.48	3.50	138.62	nr	**2603.10**
2500 amp TP&N	2216.52	2805.23	3.50	138.62	nr	**2943.85**
Edge Z unit						
1000 amp TP&N	1505.16	1904.93	3.00	118.81	nr	**2023.75**
1350 amp TP&N	1927.18	2439.04	3.00	118.81	nr	**2557.85**
2000 amp TP&N	2887.76	3654.75	3.50	138.62	nr	**3793.36**
2500 amp TP&N	3297.70	4173.57	3.50	138.62	nr	**4312.19**
Flat Z unit						
1000 amp TP&N	1627.39	2059.62	3.00	118.81	nr	**2178.44**
1350 amp TP&N	1627.76	2060.09	3.00	118.81	nr	**2178.91**
2000 amp TP&N	2423.54	3067.23	3.50	138.62	nr	**3205.85**
2500 amp TP&N	2913.87	3687.79	3.50	138.62	nr	**3826.41**

ELECTRICAL SUPPLY/POWER/LIGHTING

Item	Net Price £	Material £	Labour hours	Labour £	Unit	Total rate £
Edge tee						
1000 amp TP&N	1505.16	1904.93	2.20	87.13	nr	1992.06
1350 amp TP&N	1927.18	2439.04	2.20	87.13	nr	2526.17
2000 amp TP&N	2887.76	3654.75	2.60	102.97	nr	3757.72
2500 amp TP&N	3298.68	4174.81	2.60	102.97	nr	4277.78
Tap off; TP&N integral contactor/breaker						
18 amp	244.82	309.85	0.82	32.48	nr	342.33
Tap off; TP&N fusable with on-load switch;						
excludes fuses						
32 amp	692.70	876.69	0.82	32.48	nr	909.16
63 amp	707.77	895.76	0.88	34.85	nr	930.61
100 amp	865.57	1095.47	1.18	46.73	nr	1142.20
160 amp	983.93	1245.26	1.41	55.84	nr	1301.10
250 amp	1270.51	1607.96	1.76	69.70	nr	1677.66
315 amp	1500.24	1898.71	2.06	81.59	nr	1980.29
Tap off; TP&N MCCB						
63 amp	876.01	1108.68	0.88	34.85	nr	1143.53
125 amp	1050.06	1328.95	1.18	46.73	nr	1375.69
160 amp	1146.36	1450.84	1.41	55.84	nr	1506.68
250 amp	1474.72	1866.40	1.76	69.70	nr	1936.11
400 amp	1879.66	2378.90	2.06	81.59	nr	2460.48
RISING MAINS BUSBAR						
Rising mains busbar; insulated supports,						
earth continuity bar; including couplers;						
fixed to backgrounds						
Straight aluminium bar						
200 amp TP&N	223.72	283.14	2.13	84.36	m	367.50
315 amp TP&N	250.64	317.21	2.15	85.15	m	402.36
400 amp TP&N	290.27	367.36	2.15	85.15	m	452.51
630 amp TP&N	358.25	453.40	2.47	97.82	m	551.23
800 amp TP&N	538.06	680.97	2.88	114.06	m	795.04
Extra for fittings rising busbar						
End feed unit						
200 amp TP&N	447.46	566.30	2.57	101.78	nr	668.09
315 amp TP&N	448.86	568.07	2.76	109.31	nr	677.38
400 amp TP&N	501.26	634.39	2.76	109.31	nr	743.70
630 amp TP&N	502.67	636.18	3.64	144.16	nr	780.34
800 amp TP&N	557.90	706.08	4.54	179.81	nr	885.89
Top feeder unit						
200 amp TP&N	447.46	566.30	2.57	101.78	nr	668.09
315 amp TP&N	448.86	568.07	2.76	109.31	nr	677.38
400 amp TP&N	501.26	634.39	2.76	109.31	nr	743.70
630 amp TP&N	502.67	636.18	3.64	144.16	nr	780.34
800 amp TP&N	557.90	706.08	4.54	179.81	nr	885.89
End cap						
200 amp TP&N	38.23	48.38	0.18	7.13	nr	55.51
315 amp TP&N	38.23	48.38	0.27	10.69	nr	59.08
400 amp TP&N	42.47	53.75	0.27	10.69	nr	64.44
630 amp TP&N	42.47	53.75	0.41	16.24	nr	69.99
800 Amp TP&N	123.18	155.90	0.41	16.24	nr	172.14

ELECTRICAL SUPPLY/POWER/LIGHTING

Item	Net Price £	Material £	Labour hours	Labour £	Unit	Total rate £
LV DISTRUBITION: BUSBAR TRUNKING – cont						
Extra for fittings rising busbar – cont						
Edge elbow						
200 amp TP&N	52.39	66.30	0.55	21.78	nr	**88.09**
315 amp TP&N	53.81	68.10	0.94	37.23	nr	**105.33**
400 amp TP&N	400.72	507.15	0.94	37.23	nr	**544.38**
630 amp TP&N	402.14	508.95	1.45	57.43	nr	**566.38**
800 amp TP&N	380.90	482.06	1.45	57.43	nr	**539.49**
Flat elbow						
200 amp TP&N	172.75	218.63	0.55	21.78	nr	**240.42**
315 amp TP&N	172.75	218.63	0.94	37.23	nr	**255.86**
400 amp TP&N	233.64	295.69	0.94	37.23	nr	**332.92**
630 amp TP&N	235.05	297.47	1.45	57.43	nr	**354.90**
800 amp TP&N	324.27	410.39	1.45	57.43	nr	**467.82**
Edge tee						
200 amp TP&N	242.12	306.43	0.61	24.16	nr	**330.59**
315 amp TP&N	243.55	308.24	1.02	40.40	nr	**348.64**
400 amp TP&N	341.26	431.90	1.02	40.40	nr	**472.29**
630 amp TP&N	341.26	431.90	1.57	62.18	nr	**494.08**
800 amp TP&N	557.90	706.08	1.57	62.18	nr	**768.26**
Flat tee						
200 amp TP&N	311.51	394.25	0.61	24.16	nr	**418.41**
315 amp TP&N	243.55	308.24	1.02	40.40	nr	**348.64**
400 amp TP&N	489.93	620.05	1.02	40.40	nr	**660.45**
630 amp TP&N	491.35	621.85	1.57	62.18	nr	**684.03**
800 amp TP&N	686.75	869.15	1.57	62.18	nr	**931.33**
Tap off units						
TP&N fusable with on-load switch; excludes fuses						
32 amp	227.98	288.54	0.82	32.48	nr	**321.01**
63 amp	301.60	381.70	0.88	34.85	nr	**416.55**
100 amp	404.96	512.52	1.18	46.73	nr	**559.26**
250 amp	607.46	768.80	1.41	55.84	nr	**824.64**
400 amp	884.99	1120.04	2.06	81.59	nr	**1201.63**
TP&N MCCB						
32 amp	232.23	293.91	0.82	32.48	nr	**326.38**
63 amp	321.43	406.81	0.88	34.85	nr	**441.66**
100 amp	515.41	652.30	1.18	46.73	nr	**699.03**
250 amp	873.66	1105.70	1.41	55.84	nr	**1161.55**
400 amp	1534.93	1942.61	2.06	81.59	nr	**2024.19**

ELECTRICAL SUPPLY/POWER/LIGHTING

Item	Net Price £	Material £	Labour hours	Labour £	Unit	Total rate £
LIGHTING BUSBAR						
Prewired busbar, plug-in trunking for lighting; galvanized sheet steel housing (PE); tin-plated copper conductors with tap-off units at 1 m intervals						
Straight lengths – 25 amp						
2 pole & PE	37.19	47.07	0.16	6.34	m	**53.41**
4 pole & PE	41.32	52.29	0.16	6.34	m	**58.63**
Straight lengths – 40 amp						
2 pole & PE	37.19	47.07	0.16	6.34	m	**53.41**
4 pole & PE	49.57	62.73	0.16	6.34	m	**69.07**
Components for prewired busbars, plug-in trunking for lighting						
Plug-in tap off units						
10 amp 4 pole & PE; 3 m of cable	31.67	40.08	0.10	3.96	nr	**44.04**
16 amp 4 pole & PE; 3 m of cable	34.43	43.57	0.10	3.96	nr	**47.53**
16 amp with phase selection, 2 pole & PE; no cable	26.17	33.12	0.10	3.96	nr	**37.08**
Trunking components						
End feed unit & cover; 4 pole & PE	39.93	50.54	0.23	9.11	nr	**59.64**
Centre feed unit	199.70	252.74	0.29	11.49	nr	**264.22**
Right hand, intermediate terminal box feed unit	42.69	54.03	0.23	9.11	nr	**63.14**
End cover (for R/hand feed)	11.03	13.95	0.06	2.38	nr	**16.33**
Flexible elbow unit	95.03	120.28	0.12	4.75	nr	**125.03**
Fixing bracket – universal	6.90	8.73	0.10	3.96	nr	**12.69**
Suspension bracket – flat	5.50	6.96	0.10	3.96	nr	**10.92**
UNDERFLOOR BUSBAR						
Prewired busbar, plug-in trunking for underfloor power distribution; galvanized sheet steel housing (PE); copper conductors with tap-off units at 300 mm intervals						
Straight lengths – 63 amp						
2 pole & PE	23.40	29.62	0.28	11.09	m	**40.71**
3 pole & PE; clean earth system	30.30	38.35	0.28	11.09	m	**49.44**
Components for prewired busbars, plug-in trunking for underfloor power distribution						
Plug-in tap off units						
32 amp 2 pole & PE; 3 m metal flexible prewired conduit	38.57	48.81	0.25	9.90	nr	**58.71**
32 amp 3 pole & PE; clean earth; 3 m metal flexible prewired conduit	46.83	59.27	0.28	11.09	nr	**70.36**

ELECTRICAL SUPPLY/POWER/LIGHTING

Item	Net Price £	Material £	Labour hours	Labour £	Unit	Total rate £
LV DISTRUBITION: BUSBAR TRUNKING – **cont**						
Components for prewired busbars, plug-in **trunking for underfloor power** **distribution – cont**						
Trunking components						
End feed unit & cover; 2 pole & PE	41.32	52.29	0.35	13.86	nr	**66.15**
End feed unit & cover; 3 pole & PE; clean earth	45.44	57.51	0.38	15.05	nr	**72.56**
End cover; 2 pole & PE	12.40	15.69	0.11	4.36	nr	**20.05**
End cover; 3 pole & PE	13.76	17.42	0.11	4.36	nr	**21.78**
Flexible interlink/corner; 2 pole & PE; 1 m long	71.61	90.63	0.34	13.47	nr	**104.09**
Flexible interlink/corner; 3 pole & PE; 1 m long	81.26	102.84	0.35	13.86	nr	**116.70**
Flexible interlink/corner; 2 pole & PE; 2 m long	88.14	111.54	0.37	14.65	nr	**126.20**
Flexible interlink/corner; 3 pole & PE; 2 m long	96.40	122.00	0.37	14.65	nr	**136.66**

ELECTRICAL SUPPLY/POWER/LIGHTING

Item	Net Price £	Material £	Labour hours	Labour £	Unit	Total rate £
LV DISTRUBITION: CABLE SUPPORTS						
LADDER RACK						
Light duty galvanized steel ladder rack; fixed to backgrounds; including supports, fixings and brackets; earth continuity straps						
Straight lengths						
150 mm wide ladder	38.25	48.40	0.69	27.33	m	**75.73**
300 mm wide ladder	41.44	52.44	0.88	34.85	m	**87.29**
450 mm wide ladder	25.65	32.46	1.26	49.90	m	**82.37**
600 mm wide ladder	46.98	59.46	1.51	59.80	m	**119.26**
750 mm wide ladder	35.42	44.83	1.69	66.93	m	**111.76**
900 mm wide ladder	37.47	47.42	1.75	69.31	m	**116.72**
Extra over (cutting and jointing racking to fittings is included)						
Inside riser bend						
150 mm wide ladder	130.67	165.38	0.33	13.07	nr	**178.45**
300 mm wide ladder	133.30	168.70	0.56	22.18	nr	**190.88**
450 mm wide ladder	112.14	141.92	0.85	33.66	nr	**175.58**
600 mm wide ladder	180.90	228.95	0.99	39.21	nr	**268.16**
750 mm wide ladder	97.27	123.10	1.07	42.38	nr	**165.48**
900 mm wide ladder	101.57	128.55	1.12	44.36	nr	**172.90**
Outside riser bend						
150 mm wide ladder	130.67	165.38	0.43	17.03	nr	**182.41**
300 mm wide ladder	133.30	168.70	0.43	17.03	nr	**185.73**
450 mm wide ladder	88.59	112.12	0.73	28.91	nr	**141.03**
600 mm wide ladder	180.90	228.95	0.86	34.06	nr	**263.01**
750 mm wide ladder	97.27	123.10	0.97	38.42	nr	**161.52**
900 mm wide ladder	101.57	128.55	1.15	45.55	nr	**174.09**
Equal tee						
150 mm wide ladder	108.54	137.37	0.62	24.55	nr	**161.92**
300 mm wide ladder	152.26	192.70	0.62	24.55	nr	**217.26**
450 mm wide ladder	139.95	177.13	1.09	43.17	nr	**220.29**
600 mm wide ladder	125.09	158.32	1.12	44.36	nr	**202.68**
750 mm wide ladder	204.96	259.40	1.16	45.94	nr	**305.34**
900 mm wide ladder	211.99	268.29	1.21	47.92	nr	**316.22**
Unequal tee						
150 mm wide ladder	96.05	121.56	0.57	22.57	nr	**144.13**
300 mm wide ladder	89.11	112.78	0.57	22.57	nr	**135.35**
450 mm wide ladder	113.31	143.40	1.17	46.34	nr	**189.74**
600 mm wide ladder	97.01	122.78	1.17	46.34	nr	**169.11**
750 mm wide ladder	116.89	147.94	1.37	54.26	nr	**202.20**
900 mm wide ladder	117.87	149.17	1.37	54.26	nr	**203.43**
4 way crossovers						
150 mm wide ladder	277.98	351.81	0.72	28.52	nr	**380.33**
300 mm wide ladder	286.06	362.03	0.72	28.52	nr	**390.55**
450 mm wide ladder	204.36	258.63	1.13	44.75	nr	**303.39**
600 mm wide ladder	317.90	402.33	1.29	51.09	nr	**453.42**
750 mm wide ladder	197.21	249.59	1.41	55.84	nr	**305.43**
900 mm wide ladder	202.54	256.33	1.64	64.95	nr	**321.28**

ELECTRICAL SUPPLY/POWER/LIGHTING

Item	Net Price £	Material £	Labour hours	Labour £	Unit	Total rate £
LV DISTRUBITION: CABLE SUPPORTS – cont						
Extra over (cutting and jointing racking to fittings is included) – cont						
Flat bend (light duty)						
150 mm wide ladder	124.95	158.14	0.36	14.26	nr	**172.40**
300 mm wide ladder	130.25	164.85	0.40	15.84	nr	**180.69**
450 mm wide ladder	100.45	127.13	0.42	16.63	nr	**143.76**
600 mm wide ladder	163.10	206.42	0.59	23.37	nr	**229.79**
750 mm wide ladder	133.64	169.13	0.78	30.89	nr	**200.03**
900 mm wide ladder	145.90	184.66	0.86	34.06	nr	**218.72**
Heavy duty galvanized steel ladder rack; fixed to backgrounds; including supports, fixings and brackets; earth continuity straps						
Straight lengths						
150 mm wide ladder	50.83	64.33	0.68	26.93	m	**91.26**
300 mm wide ladder	54.81	69.37	0.79	31.29	m	**100.66**
450 mm wide ladder	58.61	74.18	1.07	42.38	m	**116.55**
600 mm wide ladder	63.41	80.25	1.24	49.11	m	**129.36**
750 mm wide ladder	73.48	93.00	1.49	59.01	m	**152.01**
900 mm wide ladder	63.58	80.47	1.67	66.14	m	**146.61**
Extra over (cutting and jointing racking to fittings is included)						
Flat bend						
150 mm wide ladder	137.31	173.77	0.34	13.47	nr	**187.24**
300 mm wide ladder	144.93	183.42	0.39	15.45	nr	**198.87**
450 mm wide ladder	161.18	203.99	0.43	17.03	nr	**221.02**
600 mm wide ladder	179.59	227.29	0.61	24.16	nr	**251.45**
750 mm wide ladder	202.25	255.97	0.82	32.48	nr	**288.44**
900 mm wide ladder	217.22	274.91	0.97	38.42	nr	**313.33**
Inside riser bend						
150 mm wide ladder	167.59	212.10	0.27	10.69	nr	**222.79**
300 mm wide ladder	169.91	215.03	0.45	17.82	nr	**232.85**
450 mm wide ladder	181.98	230.31	0.65	25.74	nr	**256.06**
600 mm wide ladder	194.22	245.80	0.81	32.08	nr	**277.88**
750 mm wide ladder	201.81	255.40	0.92	36.44	nr	**291.84**
900 mm wide ladder	218.07	275.99	1.06	41.98	nr	**317.97**
Outside riser bend						
150 mm wide ladder	167.59	212.10	0.27	10.69	nr	**222.79**
300 mm wide ladder	169.91	215.03	0.33	13.07	nr	**228.10**
450 mm wide ladder	181.98	230.31	0.61	24.16	nr	**254.47**
600 mm wide ladder	194.22	245.80	0.76	30.10	nr	**275.90**
750 mm wide ladder	201.81	255.40	0.94	37.23	nr	**292.63**
900 mm wide ladder	222.43	281.51	1.05	41.58	nr	**323.09**

ELECTRICAL SUPPLY/POWER/LIGHTING

Item	Net Price £	Material £	Labour hours	Labour £	Unit	Total rate £
Equal tee						
150 mm wide ladder	190.07	240.55	0.37	14.65	nr	**255.20**
300 mm wide ladder	217.58	275.37	0.57	22.57	nr	**297.95**
450 mm wide ladder	234.61	296.92	0.83	32.87	nr	**329.79**
600 mm wide ladder	258.75	327.47	0.92	36.44	nr	**363.91**
750 mm wide ladder	323.71	409.69	1.13	44.75	nr	**454.44**
900 mm wide ladder	340.74	431.24	1.20	47.53	nr	**478.76**
Unequal tee						
300 mm wide ladder	205.12	259.60	0.57	22.57	nr	**282.18**
450 mm wide ladder	219.09	277.28	1.17	46.34	nr	**323.62**
600 mm wide ladder	231.47	292.94	1.17	46.34	nr	**339.28**
750 mm wide ladder	287.38	363.71	1.25	49.51	nr	**413.21**
900 mm wide ladder	292.23	369.84	1.33	52.67	nr	**422.52**
4 way crossovers						
150 mm wide ladder	296.09	374.73	0.50	19.80	nr	**394.53**
300 mm wide ladder	310.79	393.33	0.67	26.54	nr	**419.87**
450 mm wide ladder	383.88	485.84	0.92	36.44	nr	**522.27**
600 mm wide ladder	408.45	516.93	1.07	42.38	nr	**559.31**
750 mm wide ladder	427.62	541.20	1.25	49.51	nr	**590.70**
900 mm wide ladder	457.89	579.50	1.36	53.86	nr	**633.36**
Double set						
150 mm wide ladder	244.09	308.92	0.51	20.20	nr	**329.12**
300 mm wide ladder	257.64	326.06	0.63	24.95	nr	**351.02**
450 mm wide ladder	286.52	362.62	0.86	34.06	nr	**396.68**
600 mm wide ladder	319.23	404.02	0.99	39.21	nr	**443.23**
750 mm wide ladder	359.53	455.02	1.16	45.94	nr	**500.96**
900 mm wide ladder	161.87	204.86	1.31	51.88	nr	**256.74**
Extra heavy duty galvanized steel ladder rack; fixed to backgrounds; including supports, fixings and brackets; earth continuity straps						
Straight lengths						
150 mm wide ladder	64.19	81.24	0.63	24.95	m	**106.19**
300 mm wide ladder	67.37	85.26	0.70	27.72	m	**112.98**
450 mm wide ladder	70.70	89.48	0.83	32.87	m	**122.36**
600 mm wide ladder	71.65	90.69	0.89	35.25	m	**125.93**
750 mm wide ladder	71.85	90.93	1.22	48.32	m	**139.25**
900 mm wide ladder	74.98	94.89	1.44	57.03	m	**151.93**
Extra over (cutting and jointing racking to fittings is included)						
Flat bend						
150 mm wide ladder	154.86	195.99	0.36	14.26	nr	**210.24**
300 mm wide ladder	161.14	203.94	0.39	15.45	nr	**219.39**
450 mm wide ladder	174.00	220.22	0.43	17.03	nr	**237.25**
600 mm wide ladder	196.84	249.12	0.61	24.16	nr	**273.28**
750 mm wide ladder	213.39	270.07	0.82	32.48	nr	**302.55**
900 mm wide ladder	231.12	292.50	0.97	38.42	nr	**330.92**

ELECTRICAL SUPPLY/POWER/LIGHTING

Item	Net Price £	Material £	Labour hours	Labour £	Unit	Total rate £
LV DISTRUBITION: CABLE SUPPORTS – cont						
Extra over (cutting and jointing racking to fittings is included) – cont						
Inside riser bend						
150 mm wide ladder	161.68	204.62	0.36	14.26	nr	**218.88**
300 mm wide ladder	164.30	207.94	0.39	15.45	nr	**223.39**
450 mm wide ladder	173.68	219.80	0.43	17.03	nr	**236.83**
600 mm wide ladder	191.55	242.43	0.61	24.16	nr	**266.59**
750 mm wide ladder	196.52	248.71	0.82	32.48	nr	**281.19**
900 mm wide ladder	211.90	268.18	0.97	38.42	nr	**306.59**
Outside riser bend						
150 mm wide ladder	162.11	205.17	0.36	14.26	nr	**219.42**
300 mm wide ladder	164.30	207.94	0.39	15.45	nr	**223.39**
450 mm wide ladder	173.68	219.80	0.41	16.24	nr	**236.04**
600 mm wide ladder	191.55	242.43	0.57	22.57	nr	**265.00**
750 mm wide ladder	196.52	248.71	0.82	32.48	nr	**281.19**
900 mm wide ladder	211.90	268.18	0.93	36.83	nr	**305.01**
Equal tee						
150 mm wide ladder	209.36	264.96	0.37	14.65	nr	**279.62**
300 mm wide ladder	231.14	292.53	0.57	22.57	nr	**315.11**
450 mm wide ladder	244.82	309.84	0.83	32.87	nr	**342.71**
600 mm wide ladder	275.93	349.22	0.92	36.44	nr	**385.65**
750 mm wide ladder	322.41	408.04	1.13	44.75	nr	**452.80**
900 mm wide ladder	334.08	422.81	1.20	47.53	nr	**470.34**
Unequal tee						
150 mm wide ladder	208.37	263.71	0.37	14.65	nr	**278.37**
300 mm wide ladder	229.47	290.42	0.57	22.57	nr	**312.99**
450 mm wide ladder	242.73	307.20	1.17	46.34	nr	**353.54**
600 mm wide ladder	273.08	345.60	1.17	46.34	nr	**391.94**
750 mm wide ladder	333.52	422.11	1.25	49.51	nr	**471.61**
900 mm wide ladder	335.68	424.84	1.33	52.67	nr	**477.51**
4 way crossovers						
150 mm wide ladder	271.71	343.87	0.50	19.80	nr	**363.67**
300 mm wide ladder	282.37	357.36	0.67	26.54	nr	**383.90**
450 mm wide ladder	330.32	418.06	0.92	36.44	nr	**454.49**
600 mm wide ladder	367.22	464.76	1.07	42.38	nr	**507.13**
750 mm wide ladder	381.73	483.12	1.25	49.51	nr	**532.63**
900 mm wide ladder	398.87	504.80	1.36	53.86	nr	**558.67**

ELECTRICAL SUPPLY/POWER/LIGHTING

Item	Net Price £	Material £	Labour hours	Labour £	Unit	Total rate £
LV DISTRUBITION: CABLE TRAYS						
Galvanized steel cable tray to BS 729; including standard coupling joints, fixings and earth continuity straps (supports and hangers are excluded)						
Light duty tray						
Straight lengths						
50 mm wide	4.94	6.26	0.19	7.52	m	13.78
75 mm wide	6.23	7.88	0.23	9.11	m	16.99
100 mm wide	7.68	9.71	0.31	12.28	m	21.99
150 mm wide	10.05	12.72	0.33	13.07	m	25.79
225 mm wide	19.20	24.29	0.39	15.45	m	39.74
300 mm wide	26.99	34.16	0.49	19.41	m	53.57
450 mm wide	15.20	19.24	0.60	23.76	m	43.00
600 mm wide	38.41	48.62	0.79	31.29	m	79.91
750 mm wide	48.80	61.76	1.04	41.19	m	102.95
900 mm wide	60.69	76.81	1.26	49.90	m	126.71
Extra over (cutting and jointing tray to fittings is included)						
Straight reducer						
75 mm wide	20.12	25.47	0.22	8.71	nr	34.18
100 mm wide	20.52	25.97	0.25	9.90	nr	35.87
150 mm wide	26.82	33.94	0.27	10.69	nr	44.64
225 mm wide	35.21	44.57	0.34	13.47	nr	58.03
300 mm wide	44.83	56.74	0.39	15.45	nr	72.18
450 mm wide	35.11	44.44	0.49	19.41	nr	63.85
600 mm wide	56.68	71.74	0.54	21.39	nr	93.12
750 mm wide	74.47	94.25	0.61	24.16	nr	118.41
900 mm wide	86.52	109.50	0.69	27.33	nr	136.83
Flat bend; 90°						
50 mm wide	12.08	15.28	0.19	7.52	nr	22.81
75 mm wide	12.57	15.91	0.24	9.51	nr	25.41
100 mm wide	14.14	17.90	0.28	11.09	nr	28.98
150 mm wide	15.29	19.35	0.30	11.88	nr	31.23
225 mm wide	19.85	25.12	0.36	14.26	nr	39.38
300 mm wide	29.03	36.74	0.44	17.43	nr	54.17
450 mm wide	21.17	26.80	0.57	22.57	nr	49.37
600 mm wide	46.80	59.22	0.69	27.33	nr	86.55
750 mm wide	66.27	83.88	0.81	32.08	nr	115.96
900 mm wide	97.47	123.36	0.94	37.23	nr	160.59
Adjustable riser						
50 mm wide	24.26	30.71	0.26	10.30	nr	41.00
75 mm wide	26.25	33.23	0.29	11.49	nr	44.71
100 mm wide	25.41	32.16	0.32	12.67	nr	44.84
150 mm wide	29.42	37.23	0.36	14.26	nr	51.49
225 mm wide	38.74	49.02	0.44	17.43	nr	66.45
300 mm wide	41.50	52.53	0.52	20.59	nr	73.12
450 mm wide	35.86	45.38	0.66	26.14	nr	71.52

ELECTRICAL SUPPLY/POWER/LIGHTING

Item	Net Price £	Material £	Labour hours	Labour £	Unit	Total rate £
LV DISTRUBITION: CABLE TRAYS – cont						
Extra over (cutting and jointing tray to fittings is included) – cont						
Adjustable riser – cont						
600 mm wide	55.43	70.16	0.79	31.29	nr	**101.45**
750 mm wide	71.34	90.29	1.03	40.79	nr	**131.08**
900 mm wide	83.96	106.26	1.10	43.57	nr	**149.83**
Inside riser; 90°						
50 mm wide	7.63	9.65	0.28	11.09	nr	**20.74**
75 mm wide	7.97	10.09	0.31	12.28	nr	**22.37**
100 mm wide	8.76	21.18	0.33	13.07	nr	**34.25**
150 mm wide	10.41	13.17	0.37	14.65	nr	**27.83**
225 mm wide	12.79	16.19	0.44	17.43	nr	**33.62**
300 mm wide	13.53	17.13	0.53	20.99	nr	**38.12**
450 mm wide	21.65	27.41	0.67	26.54	nr	**53.94**
600 mm wide	55.30	69.99	0.79	31.29	nr	**101.27**
750 mm wide	68.21	86.33	0.95	37.62	nr	**123.96**
900 mm wide	81.04	102.57	1.11	43.96	nr	**146.53**
Outside riser; 90°						
50 mm wide	7.63	9.65	0.28	11.09	nr	**20.74**
75 mm wide	7.97	10.09	0.31	12.28	nr	**22.37**
100 mm wide	8.76	11.09	0.33	13.07	nr	**24.16**
150 mm wide	10.41	13.17	0.37	14.65	nr	**27.83**
225 mm wide	12.79	16.19	0.44	17.43	nr	**33.62**
300 mm wide	13.53	17.13	0.53	20.99	nr	**38.12**
450 mm wide	21.65	27.41	0.67	26.54	nr	**53.94**
600 mm wide	30.75	38.92	0.79	31.29	nr	**70.21**
750 mm wide	41.75	52.84	0.95	37.62	nr	**90.47**
900 mm wide	41.44	52.45	1.11	43.96	nr	**96.41**
Equal tee						
50 mm wide	18.24	23.09	0.30	11.88	nr	**34.97**
75 mm wide	18.96	24.00	0.31	12.28	nr	**36.27**
100 mm wide	19.53	24.72	0.35	13.86	nr	**38.58**
150 mm wide	23.71	30.00	0.36	14.26	nr	**44.26**
225 mm wide	31.38	39.72	0.44	17.43	nr	**57.14**
300 mm wide	43.59	55.17	0.54	21.39	nr	**76.56**
450 mm wide	31.02	39.26	0.71	28.12	nr	**67.38**
600 mm wide	61.74	78.14	0.92	36.44	nr	**114.57**
750 mm wide	92.55	117.13	1.19	47.13	nr	**164.26**
900 mm wide	131.63	166.60	1.44	57.03	nr	**223.63**
Unequal tee						
75 mm wide	11.75	14.88	0.38	15.05	nr	**29.93**
100 mm wide	11.75	14.88	0.39	15.45	nr	**30.32**
150 mm wide	12.10	15.31	0.43	17.03	nr	**32.34**
225 mm wide	14.04	17.77	0.50	19.80	nr	**37.57**
300 mm wide	17.69	22.38	0.63	24.95	nr	**47.34**
450 mm wide	23.46	29.69	0.80	31.68	nr	**61.37**
600 mm wide	36.80	46.57	1.02	40.40	nr	**86.97**
750 mm wide	105.11	133.03	1.12	44.36	nr	**177.38**
900 mm wide	138.56	175.36	1.35	53.47	nr	**228.82**

ELECTRICAL SUPPLY/POWER/LIGHTING

Item	Net Price £	Material £	Labour hours	Labour £	Unit	Total rate £
4 way crossovers						
50 mm wide	26.77	33.88	0.38	15.05	nr	48.93
75 mm wide	27.28	34.52	0.40	15.84	nr	50.37
100 mm wide	28.70	36.32	0.40	15.84	nr	52.16
150 mm wide	34.00	43.03	0.44	17.43	nr	60.46
225 mm wide	45.81	57.97	0.53	20.99	nr	78.96
300 mm wide	63.05	79.79	0.64	25.35	nr	105.14
450 mm wide	54.83	69.39	0.84	33.27	nr	102.66
600 mm wide	61.74	78.14	1.03	40.79	nr	118.93
750 mm wide	92.55	117.13	1.13	44.75	nr	161.89
900 mm wide	131.63	166.60	1.36	53.86	nr	220.46
Medium duty tray with return flange						
Straight lengths						
50 mm wide	11.00	13.92	0.33	13.07	m	26.99
75 mm wide	10.79	13.66	0.33	13.07	m	26.73
100 mm wide	11.69	14.80	0.35	13.86	m	28.66
150 mm wide	14.18	17.94	0.39	15.45	m	33.39
225 mm wide	18.33	23.20	0.45	17.82	m	41.02
300 mm wide	28.30	35.82	0.57	22.57	m	58.40
450 mm wide	39.55	50.06	0.69	27.33	m	77.38
600 mm wide	52.96	67.03	0.91	36.04	m	103.07
Extra over (cutting and jointing tray to fittings is included)						
Straight reducer						
100 mm wide	38.72	49.01	0.25	9.90	nr	58.91
150 mm wide	42.31	53.55	0.27	10.69	nr	64.24
225 mm wide	48.74	61.68	0.34	13.47	nr	75.15
300 mm wide	56.44	71.43	0.39	15.45	nr	86.87
450 mm wide	72.94	92.31	0.49	19.41	nr	111.71
600 mm wide	90.36	114.36	0.54	21.39	nr	135.75
Flat bend; 90°						
75 mm wide	53.82	68.11	0.24	9.51	nr	77.61
100 mm wide	56.71	71.77	0.28	11.09	nr	82.86
150 mm wide	60.68	76.79	0.30	11.88	nr	88.67
225 mm wide	69.81	88.35	0.36	14.26	nr	102.61
300 mm wide	83.75	106.00	0.44	17.43	nr	123.42
450 mm wide	89.83	113.69	0.57	22.57	nr	136.27
600 mm wide	143.54	181.66	0.69	27.33	nr	208.99
Adjustable bend						
75 mm wide	59.04	74.73	0.29	11.49	nr	86.21
100 mm wide	64.26	81.33	0.32	12.67	nr	94.00
150 mm wide	73.49	93.01	0.36	14.26	nr	107.27
225 mm wide	82.35	104.23	0.44	17.43	nr	121.65
300 mm wide	93.07	117.79	0.52	20.59	nr	138.38
Adjustable riser						
75 mm wide	52.67	66.65	0.29	11.49	nr	78.14
100 mm wide	53.53	67.75	0.32	12.67	nr	80.42
150 mm wide	59.01	74.68	0.36	14.26	nr	88.94
225 mm wide	62.59	79.22	0.44	17.43	nr	96.64
300 mm wide	66.73	84.46	0.52	20.59	nr	105.05
450 mm wide	87.53	110.78	0.66	26.14	nr	136.92
600 mm wide	107.36	135.87	0.79	31.29	nr	167.16

ELECTRICAL SUPPLY/POWER/LIGHTING

Item	Net Price £	Material £	Labour hours	Labour £	Unit	Total rate £
LV DISTRUBITION: CABLE TRAYS – cont						
Extra over (cutting and jointing tray to fittings is included) – cont						
Inside riser; 90°						
75 mm wide	31.64	40.05	0.31	12.28	nr	**52.32**
100 mm wide	31.95	40.44	0.33	13.07	nr	**53.51**
150 mm wide	36.39	46.05	0.37	14.65	nr	**60.71**
225 mm wide	44.46	56.27	0.44	17.43	nr	**73.69**
300 mm wide	54.00	68.34	0.53	20.99	nr	**89.33**
450 mm wide	77.94	98.64	0.67	26.54	nr	**125.18**
600 mm wide	125.08	158.31	0.79	31.29	nr	**189.59**
Outside riser; 90°						
75 mm wide	31.64	40.05	0.31	12.28	nr	**52.32**
100 mm wide	31.95	40.44	0.33	13.07	nr	**53.51**
150 mm wide	36.39	46.05	0.37	14.65	nr	**60.71**
225 mm wide	44.46	56.27	0.44	17.43	nr	**73.69**
300 mm wide	54.00	68.34	0.53	20.99	nr	**89.33**
450 mm wide	77.94	98.64	0.67	26.54	nr	**125.18**
600 mm wide	125.08	158.31	0.79	31.29	nr	**189.59**
Equal tee						
75 mm wide	76.57	96.91	0.31	12.28	nr	**109.18**
100 mm wide	79.67	100.83	0.35	13.86	nr	**114.70**
150 mm wide	84.36	106.76	0.36	14.26	nr	**121.02**
225 mm wide	91.08	115.27	0.74	29.31	nr	**144.58**
300 mm wide	111.85	141.55	0.54	21.39	nr	**162.94**
450 mm wide	144.70	183.13	0.71	28.12	nr	**211.25**
600 mm wide	206.81	261.74	0.92	36.44	nr	**298.17**
Unequal tee						
100 mm wide	74.99	94.90	0.39	15.45	nr	**110.35**
150 mm wide	74.99	94.90	0.43	17.03	nr	**111.94**
225 mm wide	87.71	111.00	0.50	19.80	nr	**130.80**
300 mm wide	106.47	134.75	0.63	24.95	nr	**159.70**
450 mm wide	140.25	177.50	0.80	31.68	nr	**209.18**
600 mm wide	201.48	254.99	1.02	40.40	nr	**295.39**
4 way crossovers						
75 mm wide	99.72	126.21	0.40	15.84	nr	**142.05**
100 mm wide	107.47	136.01	0.40	15.84	nr	**151.86**
150 mm wide	93.40	118.21	0.44	17.43	nr	**135.64**
225 mm wide	135.05	170.91	0.53	20.99	nr	**191.90**
300 mm wide	157.13	198.87	0.64	25.35	nr	**224.21**
450 mm wide	199.98	253.10	0.84	33.27	nr	**286.37**
600 mm wide	291.52	368.95	1.03	40.79	nr	**409.74**
Heavy duty tray with return flange						
Straight lengths						
75 mm	19.59	24.79	0.34	13.47	m	**38.26**
100 mm	19.87	25.15	0.36	14.26	m	**39.41**
150 mm	21.96	27.80	0.40	15.84	m	**43.64**
225 mm	26.65	33.73	0.46	18.22	m	**51.94**
300 mm	30.24	38.28	0.58	22.97	m	**61.25**
450 mm	53.21	67.34	0.70	27.72	m	**95.07**

ELECTRICAL SUPPLY/POWER/LIGHTING

Item	Net Price £	Material £	Labour hours	Labour £	Unit	Total rate £
600 mm	74.15	93.84	0.92	36.44	m	**130.28**
750 mm	102.55	129.79	1.01	40.00	m	**169.79**
900 mm	107.35	135.86	1.14	45.15	m	**181.01**
Extra over (cutting and jointing tray to fittings is included)						
Straight reducer						
100 mm wide	56.74	71.82	0.25	9.90	nr	**81.72**
150 mm wide	58.92	74.57	0.27	10.69	nr	**85.26**
225 mm wide	66.92	84.69	0.34	13.47	nr	**98.16**
300 mm wide	74.59	94.40	0.39	15.45	nr	**109.85**
450 mm wide	106.69	135.03	0.49	19.41	nr	**154.44**
600 mm wide	118.62	150.12	0.54	21.39	nr	**171.51**
750 mm wide	150.43	190.39	0.60	23.76	nr	**214.15**
900 mm wide	165.02	208.85	0.66	26.14	nr	**234.99**
Flat bend; 90°						
75 mm wide	71.30	90.24	0.24	9.51	nr	**99.75**
100 mm wide	77.25	97.77	0.28	11.09	nr	**108.86**
150 mm wide	84.13	106.48	0.30	11.88	nr	**118.36**
225 mm wide	85.41	108.09	0.36	14.26	nr	**122.35**
300 mm wide	96.98	122.73	0.44	17.43	nr	**140.16**
450 mm wide	155.75	197.12	0.57	22.57	nr	**219.69**
600 mm wide	212.00	268.31	0.69	27.33	nr	**295.63**
750 mm wide	270.54	342.39	0.83	32.87	nr	**375.26**
900 mm wide	282.80	357.91	1.01	40.00	nr	**397.91**
Adjustable bend						
75 mm wide	66.01	83.55	0.29	11.49	nr	**95.03**
100 mm wide	72.99	92.37	0.32	12.67	nr	**105.04**
150 mm wide	77.11	97.60	0.36	14.26	nr	**111.85**
225 mm wide	84.88	107.42	0.44	17.43	nr	**124.85**
300 mm wide	100.70	127.44	0.52	20.59	nr	**148.04**
Adjustable riser						
75 mm wide	58.18	73.63	0.29	11.49	nr	**85.12**
100 mm wide	56.20	71.13	0.32	12.67	nr	**83.80**
150 mm wide	66.98	84.77	0.36	14.26	nr	**99.03**
225 mm wide	71.73	90.78	0.44	17.43	nr	**108.20**
300 mm wide	77.68	98.32	0.52	20.59	nr	**118.91**
450 mm wide	94.55	119.67	0.66	26.14	nr	**145.81**
600 mm wide	112.65	142.57	0.79	31.29	nr	**173.86**
750 mm wide	136.42	172.65	1.03	40.79	nr	**213.44**
900 mm wide	158.42	200.49	1.10	43.57	nr	**244.06**
Inside riser; 90°						
75 mm wide	51.89	65.67	0.31	12.28	nr	**77.95**
100 mm wide	52.57	66.53	0.33	13.07	nr	**79.60**
150 mm wide	56.94	72.07	0.37	14.65	nr	**86.72**
225 mm wide	59.53	75.34	0.44	17.43	nr	**92.76**
300 mm wide	61.40	77.71	0.53	20.99	nr	**98.70**
450 mm wide	105.36	133.34	0.67	26.54	nr	**159.87**
600 mm wide	129.26	163.59	0.79	31.29	nr	**194.88**
750 mm wide	159.99	202.48	0.95	37.62	nr	**240.11**
900 mm wide	188.61	238.71	1.11	43.96	nr	**282.67**

ELECTRICAL SUPPLY/POWER/LIGHTING

Item	Net Price £	Material £	Labour hours	Labour £	Unit	Total rate £
LV DISTRUBITION: CABLE TRAYS – cont						
Extra over (cutting and jointing tray to fittings is included) – cont						
Outside riser; 90°						
75 mm wide	51.89	65.67	0.31	12.28	nr	77.95
100 mm wide	52.57	66.53	0.33	13.07	nr	79.60
150 mm wide	56.94	72.07	0.37	14.65	nr	86.72
225 mm wide	59.53	75.34	0.44	17.43	nr	92.76
300 mm wide	61.40	77.71	0.53	20.99	nr	98.70
450 mm wide	105.36	133.34	0.67	26.54	nr	159.87
600 mm wide	129.26	163.59	0.79	31.29	nr	194.88
750 mm wide	159.99	202.48	0.95	37.62	nr	240.11
900 mm wide	188.61	238.71	1.11	43.96	nr	282.67
Equal tee						
75 mm wide	94.48	119.57	0.31	12.28	nr	131.85
100 mm wide	100.73	127.49	0.35	13.86	nr	141.35
150 mm wide	109.37	138.42	0.36	14.26	nr	152.68
225 mm wide	121.39	153.63	0.44	17.43	nr	171.05
300 mm wide	137.70	174.28	0.54	21.39	nr	195.66
450 mm wide	199.94	253.04	0.71	28.12	nr	281.16
600 mm wide	259.84	328.86	0.92	36.44	nr	365.30
750 mm wide	320.54	405.68	1.19	47.13	nr	452.81
900 mm wide	391.18	495.08	1.45	57.43	nr	552.51
Unequal tee						
75 mm wide	90.39	114.40	0.38	15.05	nr	129.45
100 mm wide	100.96	127.77	0.39	15.45	nr	143.22
150 mm wide	110.67	140.07	0.43	17.03	nr	157.10
225 mm wide	127.36	161.18	0.50	19.80	nr	180.99
300 mm wide	137.70	174.28	0.63	24.95	nr	199.23
450 mm wide	199.94	253.04	0.80	31.68	nr	284.72
600 mm wide	263.97	334.08	1.02	40.40	nr	374.48
750 mm wide	348.16	440.63	1.12	44.36	nr	484.98
900 mm wide	421.15	533.01	1.35	53.47	nr	586.48
4 way crossovers						
75 mm wide	91.74	116.10	0.40	15.84	nr	131.94
100 mm wide	131.50	166.42	0.40	15.84	nr	182.27
150 mm wide	131.50	166.42	0.44	17.43	nr	183.85
225 mm wide	182.83	231.39	0.53	20.99	nr	252.38
300 mm wide	201.60	255.15	0.64	25.35	nr	280.50
450 mm wide	290.00	367.03	0.84	33.27	nr	400.30
600 mm wide	394.15	498.83	1.03	40.79	nr	539.63
750 mm wide	480.73	608.41	1.13	44.75	nr	653.17
900 mm wide	579.89	733.91	1.36	53.86	nr	787.78

ELECTRICAL SUPPLY/POWER/LIGHTING

Item	Net Price £	Material £	Labour hours	Labour £	Unit	Total rate £
GRP cable tray including standard coupling joints and fixings (supports and hangers excluded)						
Tray						
100 mm wide	53.71	67.98	0.34	13.47	m	**81.44**
200 mm wide	68.78	87.05	0.39	15.45	m	**102.50**
400 mm wide	107.86	136.50	0.53	20.99	m	**157.49**
Cover						
100 mm wide	30.26	38.30	0.10	3.96	m	**42.26**
200 mm wide	40.09	50.74	0.11	4.36	m	**55.10**
400 mm wide	68.46	86.64	0.14	5.54	m	**92.19**
Extra for (cutting and jointing to fittings included)						
Reducer						
200 mm wide	142.73	180.64	0.23	9.11	nr	**189.75**
400 mm wide	186.47	235.99	0.30	11.88	nr	**247.88**
Reducer cover						
200 mm wide	89.34	113.07	0.25	9.90	nr	**122.97**
400 mm wide	130.56	165.24	0.28	11.09	nr	**176.33**
Bend						
100 mm wide	119.35	151.05	0.34	13.47	nr	**164.52**
200 mm wide	138.54	175.34	0.40	15.84	nr	**191.18**
400 mm wide	178.06	225.35	0.32	12.67	nr	**238.02**
Bend cover						
100 mm wide	59.30	75.06	0.10	3.96	nr	**79.02**
200 mm wide	79.21	100.25	0.10	3.96	nr	**104.21**
400 mm wide	109.43	138.49	0.13	5.15	nr	**143.64**
Tee						
100 mm wide	151.91	192.25	0.37	14.65	nr	**206.91**
200 mm wide	167.50	211.98	0.43	17.03	nr	**229.01**
400 mm wide	207.76	262.94	0.56	22.18	nr	**285.11**
Tee cover						
100 mm wide	76.54	96.86	0.27	10.69	nr	**107.56**
200 mm wide	92.12	116.59	0.31	12.28	nr	**128.87**
400 mm wide	129.95	164.46	0.37	14.65	nr	**179.12**

ELECTRICAL SUPPLY/POWER/LIGHTING

Item	Net Price £	Material £	Labour hours	Labour £	Unit	Total rate £
LV DISTRUBITION: BASKET TRAY						
Mild steel cable basket; zinc plated including standard coupling joints, fixings and earth continuity straps (supports and hangers are excluded)						
Basket 54 mm deep						
100 mm wide	7.19	9.10	0.22	8.71	m	**17.81**
150 mm wide	8.02	10.14	0.25	9.90	m	**20.05**
200 mm wide	8.76	11.09	0.28	11.09	m	**22.18**
300 mm wide	10.28	13.01	0.34	13.47	m	**26.48**
450 mm wide	12.47	15.78	0.44	17.43	m	**33.20**
600 mm wide	15.25	19.30	0.70	27.72	m	**47.02**
Extra for (cutting and jointing to fittings is included)						
Reducer						
150 mm wide	23.53	29.78	0.25	9.90	nr	**39.68**
200 mm wide	27.61	34.95	0.28	11.09	nr	**46.03**
300 mm wide	28.06	35.51	0.38	15.05	nr	**50.56**
450 mm wide	30.80	38.98	0.48	19.01	nr	**57.99**
600 mm wide	38.36	48.55	0.48	19.01	nr	**67.56**
Bend						
100 mm wide	22.47	28.44	0.23	9.11	nr	**37.55**
150 mm wide	26.32	33.32	0.26	10.30	nr	**43.61**
200 mm wide	26.77	33.88	0.30	11.88	nr	**45.76**
300 mm wide	29.31	37.10	0.35	13.86	nr	**50.96**
450 mm wide	36.52	46.22	0.50	19.80	nr	**66.03**
600 mm wide	42.96	54.38	0.58	22.97	nr	**77.35**
Tee						
100 mm wide	28.24	35.75	0.28	11.09	nr	**46.83**
150 mm wide	29.96	37.91	0.30	11.88	nr	**49.80**
200 mm wide	30.59	38.71	0.33	13.07	nr	**51.78**
300 mm wide	38.47	48.68	0.39	15.45	nr	**64.13**
450 mm wide	50.38	63.76	0.56	22.18	nr	**85.94**
600 mm wide	52.08	65.92	0.65	25.74	nr	**91.66**
Crossovers						
100 mm wide	38.54	48.77	0.40	15.84	nr	**64.61**
150 mm wide	39.18	49.59	0.42	16.63	nr	**66.22**
200 mm wide	42.56	53.87	0.46	18.22	nr	**72.08**
300 mm wide	48.35	61.19	0.51	20.20	nr	**81.39**
450 mm wide	56.70	71.75	0.74	29.31	nr	**101.06**
600 mm wide	58.52	74.07	0.82	32.48	nr	**106.54**

ELECTRICAL SUPPLY/POWER/LIGHTING

Item	Net Price £	Material £	Labour hours	Labour £	Unit	Total rate £
Mild steel cable basket; expoxy coated including standard coupling joints, fixings and earth continuity straps (supports and hangers are excluded)						
Basket 54 mm deep						
100 mm wide	16.10	20.38	0.22	8.71	m	**29.09**
150 mm wide	18.20	23.04	0.25	9.90	m	**32.94**
200 mm wide	20.27	25.66	0.28	11.09	m	**36.75**
300 mm wide	23.02	29.14	0.34	13.47	m	**42.60**
450 mm wide	28.42	35.96	0.44	17.43	m	**53.39**
600 mm wide	32.36	40.96	0.70	27.72	m	**68.68**
Extra for (cutting and jointing to fittings is included)						
Reducer						
150 mm wide	33.67	42.62	0.28	11.09	nr	**53.70**
200 mm wide	37.73	47.75	0.28	11.09	nr	**58.84**
300 mm wide	40.25	50.94	0.38	15.05	nr	**65.99**
450 mm wide	47.05	59.54	0.48	19.01	nr	**78.55**
600 mm wide	58.64	74.21	0.48	19.01	nr	**93.22**
Bend						
100 mm wide	32.65	41.32	0.23	9.11	nr	**50.43**
150 mm wide	36.48	46.17	0.26	10.30	nr	**56.46**
200 mm wide	36.94	46.75	0.30	11.88	nr	**58.63**
300 mm wide	41.56	52.60	0.35	13.86	nr	**66.46**
450 mm wide	52.74	66.75	0.50	19.80	nr	**86.55**
600 mm wide	63.33	80.15	0.58	22.97	nr	**103.12**
Tee						
100 mm wide	38.38	48.57	0.28	11.09	nr	**59.66**
150 mm wide	40.15	50.81	0.30	11.88	nr	**62.69**
200 mm wide	40.72	51.54	0.33	13.07	nr	**64.61**
300 mm wide	50.65	64.10	0.39	15.45	nr	**79.54**
450 mm wide	64.54	81.68	0.56	22.18	nr	**103.86**
600 mm wide	72.39	91.62	0.65	25.74	nr	**117.36**
Crossovers						
100 mm wide	48.68	61.61	0.40	15.84	nr	**77.45**
150 mm wide	49.23	62.31	0.42	16.63	nr	**78.94**
200 mm wide	52.74	66.75	0.46	18.22	nr	**84.96**
300 mm wide	60.54	76.61	0.51	20.20	nr	**96.81**
450 mm wide	72.91	92.27	0.74	29.31	nr	**121.58**
600 mm wide	78.78	99.70	0.82	32.48	nr	**132.17**

ELECTRICAL SUPPLY/POWER/LIGHTING

Item	Net Price £	Material £	Labour hours	Labour £	Unit	Total rate £
LV DISTRIBUTION: SWITCHGEAR AND DISTRIBUTION BOARDS						
LV switchboard components, factory-assembled modular construction to IP41; form 4, type 5; 2400 mm high, with front and rear access; top cable entry/exit; includes delivery, offloading, positioning and commissioning (hence separate labour costs are not detailed below); excludes cabling and cable terminations						
Air circuit breakers (ACBs) to BSEN 60947–2, withdrawable type, fitted with adjustable instantaneous and overload protection. Includes enclosure and copper links, assembled into LV switchboard						
ACB–100 kA fault rated.						
4 pole, 6300 A (1600 mm wide)	48722.52	61663.22	–	–	nr	**61663.22**
4 pole, 5000 A (1600 mm wide)	38668.67	48939.07	–	–	nr	**48939.07**
4 pole, 4000 A (1600 mm wide)	22157.53	28042.57	–	–	nr	**28042.57**
4 pole, 3200 A (1600 mm wide)	17726.02	22434.05	–	–	nr	**22434.05**
4 pole, 2500 A (1600 mm wide)	12880.28	16301.28	–	–	nr	**16301.28**
4 pole, 2000 A (1600 mm wide) 85ka	10304.22	13041.02	–	–	nr	**13041.02**
4 pole, 1600 A (1600 mm wide) 85ka	9364.37	11851.55	–	–	nr	**11851.55**
4 pole, 1250 A (1600 mm wide)	7315.93	9259.03	–	–	nr	**9259.03**
4 pole, 1000 A (1600 mm wide)	7178.35	9084.92	–	–	nr	**9084.92**
4 pole, 800 A (1600 mm wide) 85ka	6848.51	8667.48	–	–	nr	**8667.48**
3 pole, 6300 A (1600 mm wide)	40327.63	51038.65	–	–	nr	**51038.65**
3 pole, 5000 A (1600 mm wide)	32006.05	40506.86	–	–	nr	**40506.86**
3 pole, 4000 A (1600 mm wide)	17029.99	21553.16	–	–	nr	**21553.16**
3 pole, 3200 A (1600 mm wide)	13624.00	17242.53	–	–	nr	**17242.53**
3 pole, 2500 A (1600 mm wide)	10643.73	13470.71	–	–	nr	**13470.71**
3 pole, 2000 A (1600 mm wide)	9431.79	11936.87	–	–	nr	**11936.87**
3 pole, 1600 A (1600 mm wide)	7545.42	9549.49	–	–	nr	**9549.49**
3 pole, 1250 A (1600 mm wide)	5894.88	7460.56	–	–	nr	**7460.56**
3 pole, 1000 A (1600 mm wide)	5807.99	7350.59	–	–	nr	**7350.59**
3 pole, 800 A (1600 mm wide)	4646.39	5880.48	–	–	nr	**5880.48**
ACB–65 kA fault rated.						
4 pole, 4000 A (1600 mm wide)	48722.52	61663.22	–	–	nr	**61663.22**
4 pole, 3200 A (1600 mm wide)	16352.42	20695.62	–	–	nr	**20695.62**
4 pole, 2500 A (1600 mm wide)	12247.68	15500.66	–	–	nr	**15500.66**
4 pole, 2000 A (1600 mm wide)	9798.15	12400.54	–	–	nr	**12400.54**
4 pole, 1600 A (1600 mm wide)	8858.31	11211.08	–	–	nr	**11211.08**
4 pole, 1250 A (1600 mm wide)	8354.46	10573.40	–	–	nr	**10573.40**
4 pole, 1000 A (1600 mm wide)	8117.85	10273.95	–	–	nr	**10273.95**
4 pole, 800 A (1600 mm wide)	6494.27	8219.15	–	–	nr	**8219.15**
3 pole, 6300 A (1600 mm wide)	40327.63	51038.65	–	–	nr	**51038.65**
3 pole, 4000 A (1600 mm wide)	25604.83	32405.48	–	–	nr	**32405.48**
3 pole, 3200 A (1600 mm wide)	12611.85	15961.56	–	–	nr	**15961.56**

ELECTRICAL SUPPLY/POWER/LIGHTING

Item	Net Price £	Material £	Labour hours	Labour £	Unit	Total rate £
3 pole, 2500 A (1600 mm wide)	9853.01	12469.97	–	–	nr	**12469.97**
3 pole, 2000 A (1600 mm wide)	8907.67	11273.55	–	–	nr	**11273.55**
3 pole, 1600 A (1600 mm wide)	7126.12	9018.82	–	–	nr	**9018.82**
3 pole, 1250 A (1600 mm wide)	5567.28	7045.95	–	–	nr	**7045.95**
3 pole, 1000 A (1600 mm wide)	5545.94	7018.94	–	–	nr	**7018.94**
3 pole, 800 A (1600 mm wide)	4436.74	5615.14	–	–	nr	**5615.14**
Extra for						
Cable box (one per ACB for form 4, types 6 & 7)	427.54	541.10	–	–	nr	**541.10**
Opening coil (shunt trip)	194.48	246.13	–	–	nr	**246.13**
Closing coil	128.83	163.05	–	–	nr	**163.05**
Undervoltage release	230.62	291.87	–	–	nr	**291.87**
Motor operator	775.99	982.09	–	–	nr	**982.09**
Mechnical interlock (per ACB)	536.43	678.91	–	–	nr	**678.91**
ACB fortress/Castell adaptor kit (one per ACB)	115.68	146.40	–	–	nr	**146.40**
Fortress/Castell ACB lock (one per ACB)	364.57	461.40	–	–	nr	**461.40**
Fortress/Castell key	90.78	114.89	–	–	nr	**114.89**
Moulded case circuit breakers (MCCBs) to BS EN 60947–2; plug-in type, fitted with electronic trip unit. Includes metalwork section and copper links, assembled into LV switchboard						
MCCB–150 kA fault rated						
4 pole, 630 A (800 mm wide, 600 mm high)	3284.85	4157.30	–	–	nr	**4157.30**
4 pole, 400 A (800 mm wide, 400 mm high)	2199.58	2783.79	–	–	nr	**2783.79**
4 pole, 250 A (800 mm wide, 400 mm high)	2031.05	2570.49	–	–	nr	**2570.49**
4 pole, 160 A (800 mm wide, 300 mm high)	1391.26	1760.78	–	–	nr	**1760.78**
4 pole, 100 A (800 mm wide, 200 mm high)	2193.10	2775.58	–	–	nr	**2775.58**
3 pole, 630 A (800 mm wide, 600 mm high)	2595.62	3285.01	–	–	nr	**3285.01**
3 pole, 400 A (800 mm wide, 400 mm high)	1766.71	2235.95	–	–	nr	**2235.95**
3 pole, 250 A (800 mm wide, 400 mm high)	1155.38	1462.25	–	–	nr	**1462.25**
3 pole, 160 A (800 mm wide, 300 mm high)	830.16	1050.65	–	–	nr	**1050.65**
3 pole, 100 A (800 mm wide, 200 mm high)	675.54	854.96	–	–	nr	**854.96**
MCCB–70kA fault rated						
4 pole, 630 A (800 mm wide, 600 mm high)	2743.35	3471.99	–	–	nr	**3471.99**
4 pole, 400 A (800 mm wide, 400 mm high)	1885.77	2386.63	–	–	nr	**2386.63**
4 pole, 250 A (800 mm wide, 400 mm high)	1373.29	1738.03	–	–	nr	**1738.03**
4 pole, 160 A (800 mm wide, 300 mm high)	1035.05	1309.95	–	–	nr	**1309.95**
4 pole, 100 A (800 mm wide, 200 mm high)	852.92	1079.46	–	–	nr	**1079.46**
3 pole, 630 A (800 mm wide, 600 mm high)	2010.58	2544.59	–	–	nr	**2544.59**
3 pole, 400 A (800 mm wide, 400 mm high)	1357.00	1717.42	–	–	nr	**1717.42**
3 pole, 250 A (800 mm wide, 400 mm high)	946.35	1197.70	–	–	nr	**1197.70**
3 pole, 160 A (800 mm wide, 300 mm high)	702.49	889.08	–	–	nr	**889.08**
3 pole, 100 A (800 mm wide, 200 mm high)	556.36	704.13	–	–	nr	**704.13**

ELECTRICAL SUPPLY/POWER/LIGHTING

Item	Net Price £	Material £	Labour hours	Labour £	Unit	Total rate £
LV DISTRIBUTION: SWITCHGEAR AND DISTRIBUTION BOARDS – cont						
Moulded case circuit breakers (MCCBs) to BS EN 60947–2 – cont						
MCCB–45kA fault rated						
4 pole, 630 A (800 mm wide, 600 mm, LI, SSKA high)	2526.57	3197.63	–	–	nr	**3197.63**
4 pole, 400 A (800 mm wide, 400 mm, LI, SSKA high)	1753.96	2219.81	–	–	nr	**2219.81**
3 pole, 630 A (800 mm wide, 600 mm, LI, SSKA high)	1883.52	2383.78	–	–	nr	**2383.78**
3 pole, 400 A (800 mm wide, 400 mm, LI, SSKA high)	1293.33	1636.84	–	–	nr	**1636.84**
MCCB–36kA fault rated						
4 pole, 250 A (800 mm wide, 400 mm high)	2103.39	2662.05	–	–	nr	**2662.05**
4 pole, 160 A (800 mm wide, 300 mm high)	1593.37	2016.56	–	–	nr	**2016.56**
3 pole, 250 A (800 mm wide, 400 mm high)	1195.02	1512.42	–	–	nr	**1512.42**
3 pole, 160 A (800 mm wide, 300 mm high)	1052.42	1331.95	–	–	nr	**1331.95**
Extra for						
Cable box (one per MCCB for form 4, types 6 & 7)	180.82	228.84	–	–	nr	**228.84**
Shunt trip (for ratings 100 A to 630 A)	55.73	70.54	–	–	nr	**70.54**
Undervoltage release (for ratings 100 A to 630 A)	72.30	91.50	–	–	nr	**91.50**
Motor operator for 630 A MCCB	133.03	168.36	–	–	nr	**168.36**
Motor operator for 400 A MCCB	133.03	168.36	–	–	nr	**168.36**
Motor operator for 250 A MCCB	64.57	81.72	–	–	nr	**81.72**
Motor operator for 160 A/100 A MCCB	57.40	72.65	–	–	nr	**72.65**
Door handle for 630/400 A MCCB	113.17	143.23	–	–	nr	**143.23**
Door handle for 250/160/100 A MCCB	89.67	113.49	–	–	nr	**113.49**
MCCB earth fault protection	682.03	863.17	–	–	nr	**863.17**
LV switchboard busbar						
Copper busbar assembled into LV switchboard, ASTA type tested to appropriate fault level. Busbar Length may be estimated by adding the widths of the ACB sections to the width of the MCCB sections. ACBs up to 2000 A rating may be stacked two high; larger ratings are one per section. To determine the number of MCCB sections, add together all the MCCB heights and divide by 1800 mm, rounding up as necessary						
6000 A (6 × 10 mm × 100 mm)	3548.91	4491.50	–	–	nr	**4491.50**
5000 A (4 × 10 mm × 100 mm)	2916.99	3691.74	–	–	nr	**3691.74**
4000 A (4 × 10 mm × 100 mm)	2916.99	3691.74	–	–	nr	**3691.74**
3200 A (3 × 10 mm × 100 mm)	2026.05	2564.16	–	–	nr	**2564.17**
2500 A (2 × 10 mm × 100 mm)	1709.36	2163.36	–	–	nr	**2163.36**
2000 A (2 × 10 mm × 80 mm)	1256.47	1590.19	–	–	nr	**1590.19**
1600 A (2 × 10 mm × 50 mm)	911.64	1153.77	–	–	nr	**1153.77**

ELECTRICAL SUPPLY/POWER/LIGHTING

Item	Net Price £	Material £	Labour hours	Labour £	Unit	Total rate £
1250 A (2 × 10 mm × 40 mm)	714.81	904.67	–	–	nr	**904.67**
1000 A (2 × 10 mm × 30 mm)	714.81	904.67	–	–	nr	**904.67**
800 A (2 × 10 mm × 20 mm)	556.44	704.24	–	–	nr	**704.24**
630 A (2 × 10 mm × 20 mm)	556.44	704.24	–	–	nr	**704.24**
400 A (2 × 10 mm × 10 mm)	476.55	603.12	–	–	nr	**603.12**
Automatic power factor correction (PFC); floor standing steel enclosure to IP 42, complete with microprocessor based relay and status indication; includes delivery, offloading, positioning and commissioning; excludes cabling and cable terminations						
Standard PFC (no detuning)						
100 kVAr	6958.73	8806.97		–	nr	**8806.97**
200 kVAr	9732.15	12317.00		–	nr	**12317.00**
400 kVAr	16613.90	21026.55		–	nr	**21026.55**
600 kVAr	22757.17	28801.47		–	nr	**28801.47**
PFC with detuning reactors						
100 kVAr	11321.61	14328.63		–	nr	**14328.63**
200 kVAr	15822.13	20024.49		–	nr	**20024.49**
400 kVAr	27269.53	34512.32		–	nr	**34512.32**
600 kVAr	40103.64	50755.17		–	nr	**50755.17**

ELECTRICAL SUPPLY/POWER/LIGHTING

Item	Net Price £	Material £	Labour hours	Labour £	Unit	Total rate £
LV DISTRIBUTION: AUTOMATIC TRANSFER SWITCHES						
Automatic transfer switches; steel enclosure; solenoid operating; programmable controller, keypad and LCD display; fixed to backgrounds; including commissioning and testing						
Panel mounting type 4 pole M6 s; non BMS connection						
40 amp	2269.84	2872.71	2.40	95.05	nr	**2967.76**
63 amp	2408.83	3048.61	2.50	99.01	nr	**3147.62**
80 amp	2464.17	3118.66	2.60	102.97	nr	**3221.63**
100 amp	2500.28	3164.36	2.60	102.97	nr	**3267.33**
125 amp	2567.66	3249.63	2.70	106.93	nr	**3356.56**
160 amp	2700.44	3417.68	2.80	110.89	nr	**3528.57**
Panel mounting type 4 pole M6e; BMS connection						
40 amp	3132.21	3964.13	2.50	99.01	nr	**4063.14**
63 amp	3317.57	4198.72	2.60	102.97	nr	**4301.69**
80 amp	3391.34	4292.08	2.70	106.93	nr	**4399.02**
100 amp	3439.49	4353.01	2.80	110.89	nr	**4463.91**
125 amp	3529.33	4466.72	2.90	114.85	nr	**4581.58**
160 amp	3706.34	4690.75	3.00	118.81	nr	**4809.56**
Enclosed type 3 pole or 4 pole						
125 amp	4655.86	5892.45	2.60	102.97	nr	**5995.43**
160 amp	4882.97	6179.89	2.90	114.85	nr	**6294.74**
200 amp	5110.09	6467.33	3.30	130.70	nr	**6598.02**
250 amp	5325.85	6740.40	4.30	170.30	nr	**6910.70**
300 amp	5507.54	6970.35	4.84	191.69	nr	**7162.03**
400 amp	6109.39	7732.05	5.12	202.78	nr	**7934.82**
500 amp	6416.00	8120.09	5.50	217.83	nr	**8337.92**
630 amp	7381.24	9341.69	6.00	237.63	nr	**9579.32**
800 amp	8369.19	10592.05	6.20	245.55	nr	**10837.60**
1000 amp	8971.04	11353.75	6.90	273.27	nr	**11627.02**
1250 amp	9254.94	11713.05	7.70	304.96	nr	**12018.01**
1600 amp	9936.29	12575.36	8.50	336.64	nr	**12912.00**
2000 amp	12491.33	15809.02	–		nr	**15809.02**
Enclosed type 3 pole or 4 pole; with single bypass						
40 amp	10798.80	13666.96	2.60	102.97	nr	**13769.93**
63 amp	11008.43	13932.27	2.60	102.97	nr	**14035.24**
80 amp	11147.84	14108.71	2.60	102.97	nr	**14211.68**
100 amp	11300.66	14302.11	2.60	102.97	nr	**14405.08**
125 amp	11885.19	15041.90	2.90	114.85	nr	**15156.75**
160 amp	12859.40	16274.86	2.90	114.85	nr	**16389.71**
250 amp	16736.62	21181.87	3.30	130.70	nr	**21312.56**
400 amp	17585.94	22256.76	4.30	170.30	nr	**22427.06**
630 amp	24121.76	30528.49	4.84	191.69	nr	**30720.18**
800 amp	25914.08	32796.86	5.12	202.78	nr	**32999.64**
1000 amp	36634.02	46364.01	5.50	217.83	nr	**46581.84**

ELECTRICAL SUPPLY/POWER/LIGHTING

Item	Net Price £	Material £	Labour hours	Labour £	Unit	Total rate £
1250 amp	44685.18	56553.56	6.00	237.63	nr	**56791.19**
1600 amp	54327.03	68756.29	6.20	245.55	nr	**69001.84**
2000 amp	66239.80	83833.09	6.90	273.27	nr	**84106.36**
2500 amp	76972.01	97415.78	7.70	304.96	nr	**97720.74**
3200 amp	91289.42	115535.88	8.50	336.64	nr	**115872.52**
Enclosed type 3 pole or 4 pole; with dual bypass						
40 amp	14139.35	17894.76	2.60	102.97	nr	**17997.74**
63 amp	14677.90	18576.35	2.60	102.97	nr	**18679.32**
80 amp	14859.44	18806.11	2.60	102.97	nr	**18909.08**
100 amp	15067.58	19069.53	2.60	102.97	nr	**19172.51**
125 amp	15846.93	20055.87	2.90	114.85	nr	**20170.72**
160 amp	17145.85	21699.79	2.90	114.85	nr	**21814.64**
250 amp	22315.50	28242.50	3.30	130.70	nr	**28373.20**
400 amp	23447.94	29675.71	4.30	170.30	nr	**29846.01**
630 amp	32162.30	40704.60	4.84	191.69	nr	**40896.29**
800 amp	34552.11	43729.15	5.12	202.78	nr	**43931.92**
1000 amp	48845.37	61818.70	5.50	217.83	nr	**62036.53**
1250 amp	57260.15	72468.45	6.00	237.63	nr	**72706.08**
1600 amp	67062.65	84874.49	6.20	245.55	nr	**85120.04**
2000 amp	79968.38	101207.98	6.90	273.27	nr	**101481.25**
2500 amp	91155.67	115366.61	7.70	304.96	nr	**115671.57**
3200 amp	107729.42	136342.36	8.50	336.64	nr	**136679.00**

ELECTRICAL SUPPLY/POWER/LIGHTING

Item	Net Price £	Material £	Labour hours	Labour £	Unit	Total rate £
LV DISTRIBUTION: BREAKERS AND FUSES						
MCCB panelboards; IP4X construction, 50 kA busbars and fully-rated neutral; fitted with doorlock, removable glandplate; form 3b Type2; BSEN 60439–1; including fixing to backgrounds						
Panelboards cubicle with MCCB incomer						
Up to 250 A						
4 way TPN	1392.31	1762.10	1.00	39.60	nr	**1801.71**
Extra over for integral incomer metering	1426.00	1804.75	1.50	59.41	nr	**1864.15**
Up to 630 A						
6 way TPN	2504.55	3169.76	2.00	79.21	nr	**3248.97**
12 way TPN	2828.65	3579.93	2.50	99.01	nr	**3678.95**
18 way TPN	3292.74	4167.29	3.00	118.81	nr	**4286.10**
Extra over for integral incomer metering	1685.25	2132.86	1.50	59.41	nr	**2192.27**
Up to 800 A						
6 way TPN	4285.75	5424.05	2.00	79.21	nr	**5503.26**
12 way TPN	5141.38	6506.93	2.50	99.01	nr	**6605.94**
18 way TPN	5473.23	6926.93	3.00	118.81	nr	**7045.74**
Extra over for integral incomer metering	1685.25	2132.86	1.50	59.41	nr	**2192.27**
Up to 1200 A						
20 way TPN	11467.61	14513.41	3.50	138.62	nr	**14652.03**
Up to 1600 A						
28 way TPN	14851.07	18795.51	3.50	138.62	nr	**18934.13**
Up to 2000 A						
28 way TPN	16165.58	20459.16	4.00	158.42	nr	**20617.57**
Feeder MCCBs						
Single pole						
32 A	117.60	148.83	0.75	29.70	nr	**178.54**
63 A	120.28	152.23	0.75	29.70	nr	**181.93**
100 A	122.94	155.59	0.75	29.70	nr	**185.29**
160 A	130.97	165.75	1.00	39.60	nr	**205.36**
Double pole						
32 A	176.41	223.26	0.75	29.70	nr	**252.97**
63 A	179.07	226.63	0.75	29.70	nr	**256.33**
100 A	261.93	331.50	0.75	29.70	nr	**361.21**
160 A	326.07	412.68	1.00	39.60	nr	**452.28**
Triple pole						
32 A	235.22	297.70	0.75	29.70	nr	**327.40**
63 A	240.54	304.42	0.75	29.70	nr	**334.13**
100 A	312.70	395.76	0.75	29.70	nr	**425.46**
160 A	403.61	510.81	1.00	39.60	nr	**550.42**
250 A	606.74	767.89	1.00	39.60	nr	**807.49**
400 A	820.57	1038.51	1.25	49.51	nr	**1088.02**
630 A	1347.18	1704.99	1.50	59.41	nr	**1764.40**

ELECTRICAL SUPPLY/POWER/LIGHTING

Item	Net Price £	Material £	Labour hours	Labour £	Unit	Total rate £
MCB distribution boards; IP3X external protection enclosure; removable earth and neutral bars and DIN rail; 125/250 amp incomers; including fixing to backgrounds						
SP&N						
6 way	115.41	146.07	2.00	79.21	nr	**225.28**
8 way	136.87	173.23	2.50	99.01	nr	**272.24**
12 way	157.10	198.83	3.00	118.81	nr	**317.64**
16 way	186.73	236.32	4.00	158.42	nr	**394.74**
24 way	393.36	497.84	5.00	198.02	nr	**695.86**
TP&N						
4 way	812.56	1028.37	3.00	118.81	nr	**1147.18**
6 way	841.11	1064.51	3.50	138.62	nr	**1203.13**
8 way	915.89	1159.15	4.00	158.42	nr	**1317.57**
12 way	976.56	1235.93	4.00	158.42	nr	**1394.35**
16 way	1117.46	1414.26	5.00	198.02	nr	**1612.28**
24 way	1407.25	1781.02	6.40	253.47	nr	**2034.49**
Miniature circuit breakers for distribution boards; BS EN 60 898; DIN rail mounting; including connecting to circuit						
SP&N; including connecting of wiring						
6 amp	14.59	18.46	0.10	3.96	nr	**22.42**
10–40 amp	15.18	19.21	0.10	3.96	nr	**23.17**
50–63 amp	15.89	20.12	0.14	5.54	nr	**25.66**
TP&N; including connecting of wiring						
6 amp	61.86	78.29	0.30	11.88	nr	**90.18**
10–40 amp	64.30	81.38	0.45	17.82	nr	**99.20**
50–63 amp	67.36	85.25	0.45	17.82	nr	**103.07**
Residual current circuit breakers for distribution boards; DIN rail mounting; including connecting to circuit						
SP&N						
10 mA						
6 amp	95.80	121.24	0.21	8.32	nr	**129.56**
10–32 amp	93.97	118.93	0.26	10.30	nr	**129.23**
45 amp	95.51	120.88	0.26	10.30	nr	**131.18**
30 mA						
6 amp	95.80	121.24	0.21	8.32	nr	**129.56**
10–40 amp	93.97	118.93	0.21	8.32	nr	**127.25**
50 –63 amp	97.05	122.83	0.26	10.30	nr	**133.13**
100 mA						
6 amp	177.13	224.17	0.21	8.32	nr	**232.49**
10–40 amp	177.13	224.17	0.23	9.23	nr	**233.40**
50–63 amp	177.13	224.17	0.26	10.30	nr	**234.47**

ELECTRICAL SUPPLY/POWER/LIGHTING

Item	Net Price £	Material £	Labour hours	Labour £	Unit	Total rate £
LV DISTRIBUTION: BREAKERS AND FUSES – cont						
HRC fused distribution boards; IP4X external protection enclosure; including earth and neutral bars; fixing to backgrounds						
SP&N						
20 amp incomer						
4 way	218.01	275.92	1.00	39.60	nr	**315.52**
6 way	263.21	333.11	1.20	47.53	nr	**380.64**
8 way	308.55	390.50	1.40	55.45	nr	**445.95**
12 way	399.28	505.33	1.80	71.29	nr	**576.62**
32 amp incomer						
4 way	262.49	332.20	1.00	39.60	nr	**371.81**
6 way	345.19	436.87	1.20	47.53	nr	**484.40**
8 way	406.13	513.99	1.40	55.45	nr	**569.44**
12 way	522.87	661.74	1.80	71.29	nr	**733.03**
TP&N						
20 amp incomer						
4 way	434.64	550.07	1.50	59.41	nr	**609.48**
6 way	549.71	695.71	2.10	83.17	nr	**778.88**
8 way	649.93	822.55	2.70	106.93	nr	**929.49**
12 way	911.84	1154.02	3.90	154.46	nr	**1308.48**
32 amp incomer						
4 way	519.96	658.06	1.50	59.41	nr	**717.46**
6 way	698.07	883.48	2.10	83.17	nr	**966.65**
8 way	852.72	1079.21	2.70	106.93	nr	**1186.14**
12 way	1182.72	1496.85	3.90	154.46	nr	**1651.31**
63 amp incomer						
4 way	1105.25	1398.81	2.17	85.94	nr	**1484.75**
6 way	1417.48	1793.97	2.83	112.08	nr	**1906.05**
8 way	1706.98	2160.35	2.57	101.78	nr	**2262.13**
100 amp incomer						
4 way	1747.95	2212.21	2.40	95.05	nr	**2307.26**
6 way	2284.73	2891.55	2.73	108.12	nr	**2999.67**
8 way	2793.67	3535.67	3.87	153.27	nr	**3688.94**
200 amp incomer						
4 way	4329.51	5479.42	5.36	212.28	nr	**5691.70**
6 way	5722.38	7242.24	6.17	244.36	nr	**7486.60**
HRC fuse; includes fixing to fuse holder						
2–30 amp	4.20	5.31	0.10	3.96	nr	**9.27**
35–63 amp	9.05	11.45	0.12	4.75	nr	**16.20**
80 amp	13.36	16.91	0.15	5.94	nr	**22.85**
100 amp	16.08	20.36	0.15	5.94	nr	**26.30**
125 amp	24.29	30.74	0.15	5.94	nr	**36.68**
160 amp	25.47	32.24	0.15	5.94	nr	**38.18**
200 amp	26.38	33.39	0.15	5.94	nr	**39.33**

ELECTRICAL SUPPLY/POWER/LIGHTING

Item	Net Price £	Material £	Labour hours	Labour £	Unit	Total rate £
Consumer units; fixed to backgrounds; including supports, fixings, connections/ jointing to equipment.						
Switched and insulated; moulded plastic case, 63 amp 230 volt SP&N; earth and neutral bars; 30 mA RCCB protection; fitted MCBs						
2 way	181.91	230.23	1.67	66.14	nr	**296.37**
4 way	204.11	258.32	1.59	62.97	nr	**321.30**
6 way	221.88	280.81	2.50	99.01	nr	**379.82**
8 way	239.43	303.03	3.00	118.81	nr	**421.84**
12 way	278.14	352.01	4.00	158.42	nr	**510.43**
16 way	336.06	425.32	5.50	217.83	nr	**643.15**
Switched and insulated; moulded plastic case, 100 amp 230 volt SP&N; earth and neutral bars; 30 mA RCCB protection; fitted MCBs						
2 way	181.91	230.23	1.67	66.14	nr	**296.37**
4 way	204.11	258.32	1.59	62.97	nr	**321.30**
6 way	221.88	280.81	2.50	99.01	nr	**379.82**
8 way	239.43	303.03	3.00	118.81	nr	**421.84**
12 way	278.14	352.01	4.00	158.42	nr	**510.43**
16 way	336.06	425.32	5.50	217.83	nr	**643.15**
Extra for						
Residual current device; double pole; 230 volt/30 mA tripping current						
16 amp	86.48	109.45	0.22	8.71	nr	**118.16**
30 amp	87.86	111.19	0.22	8.71	nr	**119.90**
40 amp	89.22	112.92	0.22	8.71	nr	**121.63**
63 amp	110.47	139.81	0.22	8.71	nr	**148.52**
80 amp	122.88	155.51	0.22	8.71	nr	**164.23**
100 amp	151.24	191.41	0.25	9.90	nr	**201.31**
Residual current device; double pole; 230 volt/100 mA tripping current						
63 amp	100.99	127.81	0.22	8.71	nr	**136.52**
80 amp	116.83	147.86	0.22	8.71	nr	**156.57**
100 amp	151.26	191.43	0.25	9.90	nr	**201.34**
Heavy duty fuse switches; with HRC fuses BS 5419; short circuit rating 65kA, 500 volt; including retractable operating switches						
SP&N						
63 amp	421.95	534.02	1.30	51.49	nr	**585.51**
100 amp	616.81	780.63	1.95	77.23	nr	**857.86**
TP&N						
63 amp	531.60	672.79	1.83	72.48	nr	**745.27**
100 amp	747.00	945.40	2.48	98.22	nr	**1043.62**
200 amp	1152.29	1458.34	3.13	123.96	nr	**1582.30**
300 amp	2001.86	2533.55	4.45	176.24	nr	**2709.79**
400 amp	2197.67	2781.37	4.45	176.24	nr	**2957.61**
600 amp	3317.27	4198.34	5.72	226.54	nr	**4424.88**
800 amp	5181.97	6558.30	7.88	312.09	nr	**6870.39**

ELECTRICAL SUPPLY/POWER/LIGHTING

Item	Net Price £	Material £	Labour hours	Labour £	Unit	Total rate £
LV DISTRIBUTION: BREAKERS AND FUSES – cont						
Switch disconnectors to BSEN 60947–3; in sheet steel case; IP41 With door interlock fixed to backgrounds						
Double pole						
20 amp	80.55	101.95	1.02	40.40	nr	142.34
32 amp	97.09	122.88	1.02	40.40	nr	163.28
63 amp	354.29	448.39	1.21	47.92	nr	496.31
100 amp	325.11	411.46	1.86	73.66	nr	485.12
TP&N						
20 amp	101.12	127.98	1.29	51.09	nr	179.07
32 amp	117.67	148.92	1.83	72.48	nr	221.40
63 amp	399.30	505.35	2.48	98.22	nr	603.57
100 amp	402.61	509.54	2.48	98.22	nr	607.76
125 amp	419.92	531.46	2.48	98.22	nr	629.68
160 amp	966.18	1222.79	2.48	98.22	nr	1321.01
Enclosed switch disconnector to BSEN 60947–3; enclosure minimum IP55 rating; complete with earth connection bar; fixed to backgrounds						
TP						
20 amp	80.55	101.95	1.02	40.40	nr	142.34
32 amp	97.09	122.88	1.02	40.40	nr	163.28
63 amp	354.29	448.39	1.21	47.92	nr	496.31
TP&N						
20 amp	101.12	127.98	1.29	51.09	nr	179.07
32 amp	117.67	148.92	1.83	72.48	nr	221.40
63 amp	399.30	505.35	2.48	98.22	nr	603.57
Busbar chambers; fixed to background including all supports, fixings, connections/jointing to equipment						
Sheet steel case enclosing 4 pole 550 volt copper bars, detachable metal end plates						
600 mm long						
200 amp	733.81	928.71	2.62	103.76	nr	1032.47
300 amp	942.97	1193.42	3.03	120.00	nr	1313.42
500 amp	1615.75	2044.89	4.48	177.43	nr	2222.32
900 mm long						
200 amp	1057.00	1337.74	3.04	120.40	nr	1458.13
300 amp	1245.71	1576.57	3.59	142.18	nr	1718.75
500 amp	1842.62	2332.02	4.42	175.05	nr	2507.08
1350 mm long						
200 amp	1443.78	1827.24	3.38	133.86	nr	1961.11
300 amp	1699.30	2150.63	3.94	156.04	nr	2306.68
500 amp	2720.09	3442.54	4.82	190.89	nr	3633.44

ELECTRICAL SUPPLY/POWER/LIGHTING

Item	Net Price £	Material £	Labour hours	Labour £	Unit	Total rate £
Contactor relays; pressed steel enclosure; fixed to backgrounds including supports, fixings, connections/jointing to equipment						
Relays						
6 amp, 415/240 volt, 4 pole N/O	84.98	107.55	0.52	20.59	nr	**128.15**
6 amp, 415/240 volt, 8 pole N/O	103.94	131.55	0.85	33.66	nr	**165.21**
Push button stations; heavy gauge pressed steel enclosure; polycarbonate cover; IP65; fixed to backgrounds including supports, fixings, connections/joining to equipment						
Standard units						
One button (start or stop)	107.57	136.14	0.39	15.45	nr	**151.59**
Two button (start or stop)	114.16	144.49	0.47	18.61	nr	**163.10**
Three button (forward-reverse-stop)	161.80	204.78	0.57	22.57	nr	**227.35**
Weatherproof junction boxes; enclosures with rail mounted terminal blocks; side hung door to receive padlock; fixed to backgrounds, including all supports and fixings (Suitable for cable up to 2.5 mm²; including glandplates and gaskets)						
Sheet steel with zinc spray finish enclosure						
Overall Size 229 × 152; suitable to receive 3 × 20(A) glands per gland plate	108.92	137.85	1.43	56.63	nr	**194.49**
Overall Size 306 × 306; suitable to receive 14 × 20(A) glands per gland plate	146.13	184.94	2.17	85.94	nr	**270.89**
Overall Size 458 × 382; suitable to receive 18 × 20(A) glands per gland plate	212.79	269.30	3.51	139.01	nr	**408.32**
Overall Size 762 x508; suitable to receive 26 × 20(A) glands per gland plate	225.05	284.83	4.85	192.08	nr	**476.91**
Overall Size 914 × 610; suitable to receive 45 × 20(A) glands per gland plate	248.75	314.82	7.01	277.63	nr	**592.45**
Weatherproof junction boxes; enclosures with rail mounted terminal blocks; screw fixed lid; fixed to backgrounds, including all supports and fixings (suitable for cable up to 2.5 mm²; including glandplates and gaskets)						
Glassfibre reinforced polycarbonate enclosure						
Overall Size 190 × 190 × 130 mm	143.25	181.29	1.43	56.63	nr	**237.93**
Overall Size 190 × 190 × 180 mm	209.75	265.45	1.53	60.60	nr	**326.05**
Overall Size 280 × 190 × 130 mm	236.91	299.83	2.17	85.94	nr	**385.77**
Overall Size 280 × 190 × 180 mm	265.40	335.89	2.37	93.86	nr	**429.75**
Overall Size 380 × 190 × 130 mm	296.13	374.78	3.33	131.88	nr	**506.66**
Overall Size 380 × 190 × 180 mm	318.03	402.49	3.30	130.70	nr	**533.19**
Overall Size 380 × 280 × 130 mm	339.98	430.28	4.66	184.56	nr	**614.84**
Overall Size 380 × 280 × 180 mm	366.29	463.58	5.36	212.28	nr	**675.86**
Overall Size 560 × 280 × 130 mm	440.88	557.98	7.01	277.63	nr	**835.61**
Overall Size 560 × 380 × 180 mm	454.03	574.63	7.67	303.77	nr	**878.39**

ELECTRICAL SUPPLY/POWER/LIGHTING

Item	Net Price £	Material £	Labour hours	Labour £	Unit	Total rate £
GENERAL LIGHTING						
LED tube luminaires; surface fixed to backgrounds						
Batten type; surface mounted						
600 mm Single – 10 W	38.12	48.24	0.58	22.97	nr	**71.21**
600 mm Twin – 20 W	51.79	65.55	0.59	23.37	nr	**88.91**
1200 mm Single – 20 W	50.88	64.40	0.76	30.10	nr	**94.50**
1200 mm Twin – 40 W	81.84	103.58	0.84	33.27	nr	**136.85**
1500 mm Single – 26 W	62.64	79.27	0.77	30.50	nr	**109.77**
1500 mm Twin – 50 W	99.42	125.82	0.85	33.66	nr	**159.49**
1800 mm Single – 30 W	71.39	90.35	1.05	41.58	nr	**131.93**
1800 mm Twin – 60 W	128.63	162.80	1.06	41.98	nr	**204.78**
Extra over cost for emergency (remote)	104.80	132.63	–	–	nr	**132.63**
Extra over cost for DALI dimming (remote)	37.88	47.94	–	–	nr	**47.94**
Surface mounted LED batten, non-dimmable, opal diffuser						
600 mm Single – 10 W	75.76	95.88	0.62	24.55	nr	**120.43**
600 mm Twin – 20 W	110.14	139.39	0.62	24.55	nr	**163.94**
1200 mm Single – 20 W	87.27	110.45	0.79	31.29	nr	**141.74**
1200 mm Twin – 40 W	113.79	144.01	0.80	31.68	nr	**175.69**
1500 mm Single – 26 W	99.75	126.24	0.88	34.85	nr	**161.09**
1500 mm Twin – 50 W	121.00	153.13	0.90	35.64	nr	**188.78**
1800 mm Single – 30 W	103.99	131.61	1.09	43.17	nr	**174.78**
1800 mm Twin – 60 W	139.29	176.29	1.10	43.57	nr	**219.85**
Extra over cost for emergency (remote)	104.80	132.63	1.30	51.49	nr	**184.12**
Extra over cost for DALI dimming (remote)	37.88	47.94	1.31	51.88	nr	**99.82**
Surface mounted linear LED batten, non-dimmable, MPO diffuser						
600 mm Single – 10 W	79.54	100.67	1.09	43.17	nr	**143.84**
600 mm Twin – 20 W	116.60	147.57	1.09	43.17	nr	**190.74**
1200 mm Single – 20 W	92.38	116.92	0.79	31.29	nr	**148.21**
1200 mm Twin – 40 W	120.10	152.00	1.25	49.51	nr	**201.50**
1500 mm Single – 26 W	103.53	131.03	0.90	35.64	nr	**166.68**
1500 mm Twin – 50 W	126.22	159.75	0.90	35.64	nr	**195.39**
1800 mm Single – 30 W	110.30	139.60	0.90	35.64	nr	**175.24**
1800 mm Twin – 60 W	148.29	187.68	0.90	35.64	nr	**223.32**
Extra over cost for emergency (remote)	104.80	132.63	–	–	nr	**132.63**
Extra over cost for DALI dimming (remote)	37.88	47.94		–	nr	**47.94**
Modular recessed LED; non-dimmable, low brightness; MPO diffuser fitted to exposed T grid ceiling						
300 × 300 mm, 2 W	44.19	55.93	0.84	33.27	nr	**89.20**
Over cost for emergency	101.01	127.84		–	nr	**127.84**
Over cost for DALI dimming	53.03	67.11		–	nr	**67.11**
600 × 600 mm, 40 W	40.40	51.13	0.87	34.46	nr	**85.59**
Over cost for emergency	101.01	127.84		–	nr	**127.84**
Over cost for DALI dimming	37.88	47.94		–	nr	**47.94**
300 × 1200 mm, 70 W	101.64	128.63	0.87	34.46	nr	**163.09**
Over cost for emergency (remote)	138.89	175.77		–	nr	**175.77**
Over cost for DALI dimming (remote)	56.82	71.91		–	nr	**71.91**

ELECTRICAL SUPPLY/POWER/LIGHTING

Item	Net Price £	Material £	Labour hours	Labour £	Unit	Total rate £
Ceiling recessed round asymetric LED downlighter; for wall washing with non-dimmable gear						
1 × 12 W	90.91	115.05	0.75	29.70	nr	**144.76**
1 × 18 W	123.73	156.60	0.75	29.70	nr	**186.30**
1 × 26 W	155.30	196.55	0.75	29.70	nr	**226.25**
1 × 38 W	145.83	184.56	0.75	29.70	nr	**214.27**
1 × 50 W	183.71	232.50	0.75	29.70	nr	**262.20**
Extra over cost for emergency (remote)	183.08	231.70	–		nr	**231.70**
Extra over cost for DALI dimming (remote)	37.88	47.94	–		nr	**47.94**
Ceiling recessed asymetric LED wall washer; non-dimmable control gear; linear 80 mm × 600 mm						
1 × 35 W	277.77	351.55	0.75	29.70	nr	**381.25**
Extra over cost for emergency (integral)	183.08	231.70	–		nr	**231.70**
Extra over cost for DALI dimming (integral)	37.88	47.94	–		nr	**47.94**
Wall mounted LED uplighter; Integral driver; non-dimmable						
1 × 15 W	138.89	175.77	0.84	33.27	nr	**209.04**
1 × 30 W	183.08	231.70	0.84	33.27	nr	**264.97**
1 × 45 W	220.96	279.64	0.84	33.27	nr	**312.91**
Extra over cost for emergency (remote)	183.08	231.70	–		nr	**231.70**
Extra over cost for DALI dimming (remote)	37.88	47.94	–		nr	**47.94**
Suspended linear LED up/downlight 1.5 m Integral gear; MPO diffuser cut-off; 30% uplight, 70% downlight						
1 × 45 W (downlight), 1 × 30 W (uplight)	303.02	383.51	0.75	29.70	nr	**413.21**
Extra over cost for emergency (remote)	183.08	231.70	–		nr	**231.70**
Extra over cost for DALI dimming (remote)	37.88	47.94	–		nr	**47.94**
Recessed 'architectural' LED, Integral gear, Bat wing, low brightness, delivers direct, ceiling and graduated wall washing illumination						
600 × 600 mm, 40 W	277.77	351.55	0.87	34.46	nr	**386.00**
600 × 600 mm, 50 W	353.53	447.43	0.87	34.46	nr	**481.88**
Extra over cost for emergency (remote)	183.08	231.70	–		nr	**231.70**
Extra over cost for DALI dimming (remote)	50.50	63.92	–		nr	**63.92**
500 × 500 mm, 70 W	252.52	319.59	0.87	34.46	nr	**354.05**
500 × 500 mm, 85 W	277.77	351.55	0.87	34.46	nr	**386.00**
Extra over cost for emergency (remote)	227.27	287.63	–		nr	**287.63**
Extra over cost for DALI dimming (remote)	63.13	79.90	–		nr	**79.90**
Ceiling recessed round fixed LED downlighter with non dimmable gear						
1 × 12 W	78.28	99.07	0.75	29.70	nr	**128.78**
1 × 18 W	111.11	140.62	0.75	29.70	nr	**170.32**
1 × 26 W	141.41	178.97	0.75	29.70	nr	**208.67**
1 × 38 W	132.57	167.78	0.75	29.70	nr	**197.49**
1 × 50 W	170.45	215.72	0.75	29.70	nr	**245.43**
Extra over cost for emergency (remote)	183.08	231.70	–		nr	**231.70**
Extra over cost for DALI dimming (remote)	37.88	47.94	–		nr	**47.94**

ELECTRICAL SUPPLY/POWER/LIGHTING

Item	Net Price £	Material £	Labour hours	Labour £	Unit	Total rate £
GENERAL LIGHTING – cont						
Ceiling recessed round adjustable LED downlighter with non-dimmable gear						
1 × 12 W	94.70	119.85	0.75	29.70	nr	**149.55**
1 × 18 W	116.16	147.01	0.75	29.70	nr	**176.71**
1 × 26 W	146.08	184.88	0.75	29.70	nr	**214.59**
1 × 38 W	165.40	209.33	0.75	29.70	nr	**239.03**
1 × 50 W	208.33	263.66	0.75	29.70	nr	**293.36**
Extra over cost for emergency (remote)	183.08	231.70	–		nr	**231.70**
Extra over cost for DALI dimming (remote)	37.88	47.94	–		nr	**47.94**
LUMINAIRES						
High/low bay, LED non-dimmable; aluminium reflector						
35 W	265.15	335.57	1.50	59.41	nr	**394.98**
45 W	309.34	391.50	1.50	59.41	nr	**450.90**
55 W	334.59	423.46	1.50	59.41	nr	**482.86**
LED batten fitting						
Corrosion resistant GRP body; gasket sealed; acrylic diffuser						
600 mm–35 W	96.76	122.46	0.49	19.41	nr	**141.87**
1200 mm–40 W	129.80	164.27	0.64	25.35	nr	**189.62**
1500 mm–45 W	162.84	206.09	0.72	28.52	nr	**234.61**
1800 mm–50 W	199.42	252.39	0.94	37.23	nr	**289.61**
Extra over cost for emergency (Integral)	183.08	231.70	–		nr	**231.70**
Flameproof to IIA/IIB, I.P. 64; Aluminium Body; BS 229 and 899						
600 mm–35 W	461.38	583.92	1.04	41.19	nr	**625.11**
1200 mm–40 W	552.24	698.91	1.31	51.88	nr	**750.80**
1500 mm–45 W	619.50	784.04	1.64	64.95	nr	**848.99**
1800 mm–50 W	686.76	869.16	1.97	78.02	nr	**947.18**
Extra over cost for emergency (Integral)	183.08	231.70	–		nr	**231.70**
LED downlight/spotlight; track mounted (3ct); adjustable fixing; including driver, polycarbonate optical lenses/diffuser, heat sink and 3000 K colour temperature 100 mm dia.						
20 W	251.26	317.99	0.75	29.70	nr	**347.69**
30 W	277.77	351.55	0.75	29.70	nr	**381.25**
Extra for emergency pack (track mounted remote)	189.39	239.69	0.25	9.90	nr	**249.59**
Extra for DALI lighting control (with DALI track)	56.82	71.91	0.25	9.90	nr	**81.81**
LED square modular spotlight; recessed into ceiling; including driver, polycarbonate optical lenses/diffuser, heat sink and 3000 K colour temperature						
1 × 15 W	195.70	247.68	0.84	33.27	nr	**280.95**
Extra for emergency pack 1 lamp (remote)	189.39	239.69	0.25	9.90	nr	**249.59**
Extra for DALI lighting control	37.88	47.94	0.25	9.90	nr	**57.84**

ELECTRICAL SUPPLY/POWER/LIGHTING

Item	Net Price £	Material £	Labour hours	Labour £	Unit	Total rate £
2 × 15 W	265.15	335.57	0.89	35.25	nr	370.82
Extra for emergency pack 1 lamp (remote)	189.39	239.69	0.25	9.90	nr	249.59
Extra for DALI lighting control	69.44	87.89	0.25	9.90	nr	97.79
3 × 15 W	359.84	455.41	0.89	35.25	nr	490.66
Extra for emergency pack 1 lamp (remote)	189.39	239.69	0.25	9.90	nr	249.59
Extra for DALI lighting control	119.95	151.80	0.25	9.90	nr	161.71
LED downlight/spotlight; track mounted; adjustable fixing; including driver, polycarbonate optical lenses/diffuser, heat sink, RGB colour						
100 mm dia.	356.05	450.62	0.75	29.70	nr	480.32
Internal flexible linear LED, IP20, including driver, extrusion and opal cover white 3000 K colour temperature, per/m						
7.2 W	70.20	88.85	0.25	9.90	nr	98.75
14.4 W	91.66	116.01	0.25	9.90	nr	125.91
17.3 W	108.08	136.78	0.25	9.90	nr	146.69
Extra for DALI lighting control (remote)	63.13	79.90	0.25	9.90	nr	89.80
Internal flexible linear LED, IP20, including DMX driver and opal cover RGB, per/m						
6 W	94.70	119.85	0.25	9.90	nr	129.75
9 W	104.80	132.63	0.25	9.90	nr	142.53
12 W	119.95	151.80	0.25	9.90	nr	161.71
15 W	138.89	175.77	0.25	9.90	nr	185.68
LED in-ground adjustable performance uplighters, IP68, including driver, polycarbonate optical lenses/diffuser, heat sink, 4000 K colour temperature						
15 W	404.03	511.34	0.25	9.90	nr	521.24
30 W	448.22	567.27	0.25	9.90	nr	577.17
45 W	498.73	631.19	0.25	9.90	nr	641.09
Extra for DALI lighting control	63.13	79.90	0.25	9.90	nr	89.80
LED in-ground adjustable performance uplighters, IP68, including driver, polycarbonate optical lenses/diffuser, heat sink, RGB						
15 W	422.97	535.31	0.25	9.90	nr	545.21
30 W	467.16	591.24	0.25	9.90	nr	601.14
45 W	530.29	671.14	0.25	9.90	nr	681.04
LED in-ground fixed decorative uplighters, IP68, including driver, polycarbonate optical lenses/diffuser, heat sink, 4000 K colour, 50 mm dia.						
3 W	233.58	295.62	0.25	9.90	nr	305.52
6 W	252.52	319.59	0.25	9.90	nr	329.49
9 W	277.77	351.55	0.25	9.90	nr	361.45
Extra for DALI lighting control	69.44	87.89	0.25	9.90	nr	97.79

ELECTRICAL SUPPLY/POWER/LIGHTING

Item	Net Price £	Material £	Labour hours	Labour £	Unit	Total rate £
GENERAL LIGHTING – cont						
LUMINAIRES – cont						
LED in-ground fixed decorative uplighters, IP68, including driver, polycarbonate optical lenses/diffuser, heat sink, RGB 50 mm dia.						
3 W	195.70	247.68	0.25	9.90	nr	257.58
6 W	208.33	263.66	0.25	9.90	nr	273.56
9 W	220.96	279.64	0.25	9.90	nr	289.54
External flexible linear LED, IP66, including driver, extrusion and opal cover, 3000 K colour temperature, per/m						
6 W	107.32	135.83	0.25	9.90	nr	145.73
9 W	122.47	155.00	0.25	9.90	nr	164.90
12 W	138.89	175.77	0.25	9.90	nr	185.68
15 W	157.83	199.74	0.25	9.90	nr	209.64
Extra for DALI lighting control	63.13	79.90	0.25	9.90	nr	89.80
External flexible linear LED, IP66, including driver, extrusion and opal cover, RGB, per/m						
6 W	119.95	151.80	0.25	9.90	nr	161.71
9 W	137.62	174.18	0.25	9.90	nr	184.08
12 W	152.77	193.35	0.25	9.90	nr	203.25
15 W	175.50	222.11	0.25	9.90	nr	232.02
Extra for DALI lighting control	63.13	79.90	0.25	9.90	nr	89.80
Handrail puck, LED, IP66, including driver and controller, optical lenses/diffuser 3000 K colour temperature						
9 W Asymetric	183.08	231.70	0.25	9.90	nr	241.60
9 W Down	208.33	263.66	0.25	9.90	nr	273.56
Extra for emergency pack (per unit)	189.39	239.69	0.25	9.90	nr	249.59
Extra for DALI lighting control	63.13	79.90	0.25	9.90	nr	89.80
Handrail puck, LED, IP66, including driver and controller, optical lenses/diffuser, RGB						
9 W Asymetric	202.02	255.67	0.25	9.90	nr	265.57
9 W Down	224.74	284.43	0.25	9.90	nr	294.34
Extra for emergency pack (per unit)	189.39	239.69	0.25	9.90	nr	249.59
Extra for DALI lighting control	63.13	79.90	0.25	9.90	nr	89.80
External lighting						
Bulkhead; aluminium body and polycarbonate bowl; vandal-resistant; IP65						
15 W	239.89	303.61	0.75	29.70	nr	333.31
25 W	265.15	335.57	0.75	29.70	nr	365.27
40 W	296.71	375.52	0.75	29.70	nr	405.22
Extra for emergency pack (Integral)	189.39	239.69	0.25	9.90	nr	249.59
Photocell	31.57	39.95	0.75	29.70	nr	69.65

ELECTRICAL SUPPLY/POWER/LIGHTING

Item	Net Price £	Material £	Labour hours	Labour £	Unit	Total rate £
1500 mm high LED circular bollard; polycarbonate visor; vandal-resistant; IP54						
15 W	391.41	495.36	1.75	69.31	nr	**564.67**
25 W	422.97	535.31	1.75	69.31	nr	**604.62**
40 W	448.22	567.27	1.75	69.31	nr	**636.58**
Floodlight; enclosed high performance LED; integeral control gear; reflector; toughened glass; IP65						
20 W	233.58	295.62	1.25	49.51	nr	**345.13**
35 W	265.15	335.57	1.25	49.51	nr	**385.07**
50 W	309.34	391.50	1.25	49.51	nr	**441.00**
65 W	347.22	439.44	1.25	49.51	nr	**488.94**
80 W	385.09	487.37	1.25	49.51	nr	**536.88**
Extra for photocell	31.57	39.95	0.75	29.70	nr	**69.65**
Lighting track						
Single circuit; 25 A 2 P&E steel trunking; low voltage with copper conductors; including couplers and supports; suspended						
Straight track/m	27.78	35.15	0.50	19.80	m	**54.96**
Live end feed unit complete with end stop	8.21	10.39	0.33	13.07	nr	**23.46**
Flexible couplers 0.5 m	10.76	13.61	0.33	13.07	nr	**26.68**
Tap off complete with 0.8 m of cable	11.81	14.94	0.25	9.90	nr	**24.84**
Three circuit; 25 A 2 P&E steel trunking; low voltage with copper conductors; including couplers and supports incorporating integral twisted pair comms bus bracket; fixed to backgrounds						
Straight track	32.01	40.51	0.75	29.70	nr	**70.21**
Live end feed unit complete with end stop	9.47	11.98	0.50	19.80	nr	**31.79**
Flexible couplers 0.5 m	12.49	15.80	0.45	17.82	nr	**33.63**
Tap off complete with 0.8 m of cable	13.07	16.54	0.25	9.90	nr	**26.44**
LIGHTING ACCESSORIES						
Switches						
6 amp metal clad surface mounted switch, gridswitch; one way						
1 gang	11.14	14.09	0.43	17.03	nr	**31.12**
2 gang	15.20	19.23	0.55	21.78	nr	**41.01**
3 gang	26.68	33.76	0.77	30.50	nr	**64.26**
4 gang	28.28	35.79	0.88	34.85	nr	**70.64**
6 gang	48.72	61.66	1.00	39.60	nr	**101.27**
8 gang	67.09	84.91	1.28	50.69	nr	**135.60**
10 gang	105.83	133.93	1.67	66.14	nr	**200.07**
Extra for						
10 amp – two way switch	3.98	5.03	0.03	1.19	nr	**6.22**
20 amp – two way switch	5.18	6.55	0.04	1.58	nr	**8.14**
20 amp – intermediate	9.72	12.31	0.08	3.17	nr	**15.47**
20 amp – one way SP switch	3.86	4.88	0.08	3.17	nr	**8.05**
Steel blank plate; 1 gang	2.35	2.98	0.07	2.77	nr	**5.75**
Steel blank plate; 2 gang	3.99	5.05	0.08	3.17	nr	**8.22**

ELECTRICAL SUPPLY/POWER/LIGHTING

Item	Net Price £	Material £	Labour hours	Labour £	Unit	Total rate £
GENERAL LIGHTING – cont						
Switches – cont						
6 amp modular type switch; galvanized steel box, bronze or satin chrome coverplate; metalclad switches; flush mounting; one way						
1 gang	30.68	38.83	0.43	17.03	nr	**55.86**
2 gang	34.75	43.98	0.55	21.78	nr	**65.76**
3 gang	50.19	63.53	0.77	30.50	nr	**94.02**
4 gang	72.30	91.50	0.88	34.85	nr	**126.36**
6 gang	121.89	154.26	1.18	46.73	nr	**201.00**
8 gang	145.41	184.03	1.63	64.56	nr	**248.58**
9 gang	182.38	230.82	1.83	72.48	nr	**303.29**
12 gang	217.22	274.92	2.29	90.69	nr	**365.61**
6 amp modular type swtich; galvanized steel box; bronze or satin chrome coverplate; flush mounting; two way						
1 gang	31.90	40.37	0.43	17.03	nr	**57.40**
2 gang	43.71	55.32	0.55	21.78	nr	**77.11**
3 gang	63.15	79.92	0.77	30.50	nr	**110.41**
4 gang	75.19	95.15	0.88	34.85	nr	**130.01**
6 gang	126.76	160.42	1.18	46.73	nr	**207.16**
8 gang	151.21	191.38	1.63	64.56	nr	**255.93**
9 gang	189.69	240.07	1.83	72.48	nr	**312.55**
12 gang	225.91	285.91	2.22	87.92	nr	**373.84**
Plate switches; 10 amp flush mounted, white plastic fronted; 16 mm metal box; fitted brass earth terminal						
1 gang, 1 way, single pole	12.16	15.39	0.28	11.09	nr	**26.48**
1 gang, 2 way, single pole	13.62	17.23	0.33	13.07	nr	**30.30**
2 gang, 2 way, single pole	17.01	21.53	0.44	17.43	nr	**38.96**
3 gang, 2 way, single pole	19.97	25.27	0.56	22.18	nr	**47.45**
1 gang, Intermediate	26.93	34.08	0.43	17.03	nr	**51.11**
1 gang, 1 way, double pole	23.28	29.46	0.33	13.07	nr	**42.53**
1 gang, single pole with bell symbol	18.63	23.57	0.23	9.11	nr	**32.68**
1 gang, single pole marked PRESS	14.90	18.86	0.23	9.11	nr	**27.97**
Time delay switch, suppressed	82.47	104.38	0.49	19.41	nr	**123.79**
Plate switches; 6 amp flush mounted white plastic fronted; 25 mm metal box; fitted brass earth terminal						
4 gang, 2 way, single pole	36.35	46.01	0.42	16.63	nr	**62.64**
6 gang, 2 way, single way	57.73	73.07	0.47	18.61	nr	**91.68**
Architrave plate switches; 6 amp flush mounted, white plastic fronted; 27 mm metal box; brass earth terminal						
1 gang, 2 way, single pole	4.87	6.16	0.30	11.88	nr	**18.04**
2 gang, 2 way, single pole	9.87	12.49	0.36	14.26	nr	**26.74**
Ceiling switches, white moulded plastic, pull cord; standard unit						
6 amp, 1 way, single pole	5.34	6.76	0.32	12.67	nr	**19.43**
6 amp, 2 way, single pole	6.23	7.89	0.34	13.47	nr	**21.35**
16 amp, 1 way, double pole	9.28	11.75	0.37	14.65	nr	**26.40**

ELECTRICAL SUPPLY/POWER/LIGHTING

Item	Net Price £	Material £	Labour hours	Labour £	Unit	Total rate £
45 amp, 1 way, double pole with neon indicator	17.10	21.64	0.47	18.61	nr	**40.25**
10 amp splash proof moulded switch with plain, threaded or PVC entry						
1 gang, 2 way single pole	29.93	37.88	0.34	13.47	nr	**51.34**
2 gang, 1 way single pole	33.82	42.81	0.36	14.26	nr	**57.06**
2 gang, 2 way single pole	42.85	54.23	0.40	15.84	nr	**70.07**
6 amp watertight switch; metalclad; BS 3676; ingress protected to IP65 surface mounted						
1 gang, 2 way; terminal entry	25.51	32.29	0.41	16.24	nr	**48.53**
1 gang, 2 way; through entry	25.51	32.29	0.42	16.63	nr	**48.92**
2 gang, 2 way; terminal entry	31.10	39.37	0.54	21.39	nr	**60.75**
2 gang, 2 way; through entry	31.10	39.37	0.53	20.99	nr	**60.36**
2 way replacement switch	22.94	29.03	0.10	3.96	nr	**32.99**
15 amp watertight switch; metalclad; BS 3676; ingress protected to IP65; surface mounted						
1 gang, 2 way, terminal entry	38.68	48.95	0.42	16.63	nr	**65.59**
1 gang, 2 way, through entry	38.68	48.95	0.43	17.03	nr	**65.98**
2 gang, 2 way, terminal entry	90.79	114.90	0.55	21.78	nr	**136.68**
2 gang, 2 way, through entry	90.79	114.90	0.54	21.39	nr	**136.28**
Intermediate interior only	22.94	29.03	0.11	4.36	nr	**33.39**
2 way interior only	22.94	29.03	0.11	4.36	nr	**33.39**
Double pole interior only	22.94	29.03	0.11	4.36	nr	**33.39**
Electrical accessories; fixed to backgrounds (Including fixings)						
Dimmer switches; rotary action; for individual lights; moulded plastic case; metal backbox; flush mounted						
1 gang, 1 way; 250 watt	23.83	30.16	0.28	11.09	nr	**41.24**
1 gang, 1 way; 400 watt	31.39	39.73	0.28	11.09	nr	**50.82**
Dimmer switches; push on/off action; for individual lights; moulded plastic case; metal backbox; flush mounted						
1 gang, 2 way; 250 watt	36.42	46.10	0.34	13.47	nr	**59.56**
3 gang, 2 way; 250 watt	53.69	67.94	0.48	19.01	nr	**86.95**
4 gang, 2 way; 250 watt	70.64	89.40	0.57	22.57	nr	**111.97**
Dimmer switches; rotary action; metal cald; metal backbox; BS 5518 and BS 800; flush mounted						
1 gang, 1 way; 400 watt	66.16	83.74	0.33	13.07	nr	**96.81**
Ceiling roses						
Ceiling rose: white moulded plastic; flush fixed to conduit box						
Plug in type; ceiling socket with 2 terminals, loop-in and ceiling plug with 3 terminals and cover	13.32	16.86	0.34	13.47	nr	**30.32**

ELECTRICAL SUPPLY/POWER/LIGHTING

Item	Net Price £	Material £	Labour hours	Labour £	Unit	Total rate £
GENERAL LIGHTING – cont						
Ceiling roses – cont						
BC lampholder; white moulded plastic; heat resistent PVC insulated and sheathed cable; flush fixed						
2 core; 0.75 mm²	4.21	5.33	0.33	13.07	nr	**18.40**
Batten holder: white moulded plastic; 3 terminals; BS 5042; fixed to conduit						
Straight pattern; 2 terminals with loop-in and earth	6.17	7.81	0.29	11.49	nr	**19.30**
Angled pattern; looped in terminal	–	–	0.29	11.49	nr	**11.49**
LIGHTING CONTROL MODULES						
SPV.92 Lighting control module; plug in; 9 output, 9 channel – switching						
Base and lid assembly	243.52	308.20	2.05	81.19	nr	**389.39**
SPV.92 Lighting control module; plug in; 9 output, 9 channel – dimming (DALI, DSI, 1–10 V)						
Base and lid assembly	297.81	376.90	2.05	81.19	nr	**458.09**
SPH.12+ Lighting control module; hard wired; 4 circuit switching						
Base and lid assembly	262.78	332.58	1.85	73.27	nr	**405.85**
SPH.27r Lighting control module; hard wired; 2 circuit 40 ballast drive DALI (with relays)						
Base and lid assembly	376.67	476.71	1.85	73.27	nr	**549.98**
SPH.27 Lighting control module; hard wired; 2 circuit 40 ballast drive DALI (without relays)						
Base and lid assembly	359.15	454.54	1.85	73.27	nr	**527.81**
SPV.56 Compact lighting control module; 3 output 18 ballast drive; dimmable (DALI)						
Base and lid assembly	332.84	421.24	1.85	73.27	nr	**494.51**
SPB.82 Blind control module; 8 outputs						
Base and lid assembly	262.78	332.58	1.85	73.27	nr	**405.85**
Interfaces						
SPO.20 DALI to DMX converter	350.36	443.41	1.85	73.27	nr	**516.68**
Dual bus presence detectors (DALI & E-Bus)						
SPU.7-S Dual-bus presence detector and infrared sensor	87.60	110.87	0.60	23.76	nr	**134.63**
SPU.6-S Dual-bus universal sensor	96.36	121.96	0.60	23.76	nr	**145.72**
SPU.6-C Compact Dual-bus Universal Sensor	97.92	245.89	0.60	23.76	nr	**269.65**
SPU.6-C1 Compact Dual-bus Universal Sensor conduit/surface mount	103.75	131.31	0.60	23.76	nr	**155.07**
Scene switch plate; anodized aluminium finish						
SPK.9–4+4 Eight button intelligent switch plate	175.17	221.70	2.00	79.21	nr	**300.91**
SPK.9–4-S Four button intelligent switch plate	105.12	133.04	1.60	63.37	nr	**196.40**
SPK.9–2-S Two button intelligent switch plate	87.60	110.87	1.20	47.53	nr	**158.39**

ELECTRICAL SUPPLY/POWER/LIGHTING

Item	Net Price £	Material £	Labour hours	Labour £	Unit	Total rate £
Input devices/interfaces						
SPF.4d Compact DALI interface for 4 gang retractive switch plate	43.80	55.43	1.20	47.53	nr	**102.96**
SPL.20 External Photocell 20,000 Lux range	210.21	266.04	1.50	59.41	nr	**325.45**
SPF.6-T AV interface; RS232 interface	227.75	288.24	1.85	73.27	nr	**361.51**
Sub-addressing for DALI strings						
Sub-addressing charge per SPH.27R (40 sub-addresses)	171.20	216.67	3.00	118.81	nr	**335.49**
Sub-addressing charge per SPV.56 (20 sub-addresses)	85.59	325.00	3.00	118.81	nr	**443.82**
Commissioning						
Not included – worked out based on complexity of project						

ELECTRICAL SUPPLY/POWER/LIGHTING

Item	Net Price £	Material £	Labour hours	Labour £	Unit	Total rate £
GENERAL LV POWER						
ACCESSORIES						
OUTLETS						
Socket outlet: unswitched; 13 amp metal clad; BS 1363; galvanized steel box and coverplate with white plastic inserts; fixed surface mounted						
1 gang	12.83	16.24	0.41	16.24	nr	**32.48**
2 gang	21.38	27.06	0.41	16.24	nr	**43.30**
Socket outlet: switched; 13 amp metal clad; BS 1363; galvanized steel box and coverplate with white plastic inserts; fixed surface mounted						
1 gang	14.99	18.98	0.43	17.03	nr	**36.01**
2 gang	27.29	34.54	0.45	17.82	nr	**52.37**
Socket outlet: switched with neon indicator; 13 amp metal clad; BS 1363; galvanized steel box and coverplate withwhite plastic inserts; fixed surface mounted						
1 gang	31.73	40.16	0.43	17.03	nr	**57.19**
2 gang	57.70	73.02	0.45	17.82	nr	**90.84**
Socket outlet: unswitched; 13 amp; BS 1363; white moulded plastic box and coverplate; fixed surface mounted						
1 gang	7.46	9.44	0.41	16.24	nr	**25.68**
2 gang	14.72	18.63	0.41	16.24	nr	**34.86**
Socket outlet; switched; 13 amp; BS 1363; white moulded plastic box and coverplate; fixed surface mounted						
1 gang	5.44	6.89	0.43	17.03	nr	**23.92**
2 gang	8.74	11.06	0.45	17.82	nr	**28.88**
Socket outlet: switched with neon indicator; 13 amp; BS 1363; white moulded plastic box and coverplate; fixed surface mounted						
1 gang	24.85	31.45	0.43	17.03	nr	**48.48**
2 gang	33.51	42.41	0.45	17.82	nr	**60.23**
Socket outlet: switched; 13 amp; BS 1363; galvanized steel box, white moulded coverplate; flush fitted						
1 gang	6.78	8.58	0.43	17.03	nr	**25.61**
2 gang	18.70	23.67	0.45	17.82	nr	**41.49**
Socket outlet: switched with neon indicator; 13 amp; BS 1363; galvanized steel box, white moulded coverplate; flush fixed						
1 gang	24.85	31.45	0.43	17.03	nr	**48.48**
2 gang	42.97	54.39	0.45	17.82	nr	**72.21**

ELECTRICAL SUPPLY/POWER/LIGHTING

Item	Net Price £	Material £	Labour hours	Labour £	Unit	Total rate £
Socket outlet: switched; 13 amp; BS 1363; galvanized steel box, satin chrome coverplate; BS 4662; flush fixed						
1 gang	28.55	36.14	0.43	17.03	nr	**53.17**
2 gang	36.11	45.71	0.45	17.82	nr	**63.53**
Socket outlet: switched with neon indicator; 13 amp; BS 1363; steel backbox, satin chrome coverplate; BS 4662; flush fixed						
1 gang	35.96	45.51	0.43	17.03	nr	**62.54**
2 gang	64.84	82.07	0.45	17.82	nr	**99.89**
RCD protected socket outlets, 13 amp, to BS 1363; galvanized steel box, white moulded cover plate; flush fitted						
2 gang, 10 mA tripping (active control)	161.48	204.37	0.45	17.82	nr	**222.19**
2 gang, 30 mA tripping (active control)	143.85	182.06	0.45	17.82	nr	**199.88**
2 gang, 30 mA tripping (passive control)	143.85	182.06	0.45	17.82	nr	**199.88**
Filtered socket outlets, 13 amp, to BS 1363, with separate 'clean earth' terminal; galvanized steel box, white moulded cover plate; flush fitted						
2 gang (spike protected)	121.59	153.89	0.50	19.80	nr	**173.69**
2 gang (spike and RFI protected)	153.91	194.79	0.55	21.78	nr	**216.57**
Replacement filter cassette	42.66	53.99	0.15	5.94	nr	**59.93**
Non-standard socket outlets, 13 amp, to BS 1363, with separate 'clean earth' terminal; for plugs with T-shaped earth pin; galvanized steel box, white moulded cover plate; flush fitted						
1 gang	22.57	28.56	0.43	17.03	nr	**45.59**
2 gang	40.61	51.40	0.43	17.03	nr	**68.43**
2 gang coloured RED	56.40	71.38	0.43	17.03	nr	**88.41**
Weatherproof socket outlet: 40 amp; switched; single gang; RCD protected; water and dust protected to I.P.66; surface mounted						
40 A 30 mA tripping current protecting 1 socket	168.63	213.41	0.52	20.59	nr	**234.01**
40 A 30 mA tripping current protecting 2 sockets	74.95	94.86	0.64	25.35	nr	**120.20**
Plug for weatherproof socket outlet: protected to I.P.66						
13 amp plug	7.38	9.34	0.21	8.32	nr	**17.66**
Floor service outlet box; comprising flat lid with flanged carpet trim; twin 13 A switched socket outlets; punched plate for mounting 2 telephone outlets; one blank plate; triple compartment						
3 compartment	75.90	96.06	0.88	34.85	nr	**130.91**
Floor service outlet box; comprising flat lid with flanged carpet trim; 2 twin 13 A switched socket outlets; punched plate for mounting 1 telephone outlet; one blank plate; triple compartment						
3 compartment	67.74	85.73	0.88	34.85	nr	**120.58**

ELECTRICAL SUPPLY/POWER/LIGHTING

Item	Net Price £	Material £	Labour hours	Labour £	Unit	Total rate £
GENERAL LV POWER – cont						
Floor service outlet box; comprising flat lid with flanged carpet trim; twin 13 A switched socket outlets; punched plate for mounting 2 telephone outlets; two blank plates; four compartment						
4 compartment	89.08	112.75	0.88	34.85	nr	**147.60**
Floor service outlet box; comprising flat lid with flanged carpet trim; single 13 A unswitched socket outlet; single compartment; circular						
1 compartment	95.03	120.27	0.79	31.29	nr	**151.56**
Floor service grommet, comprising flat lid with flanged carpet trim; circular						
Floor grommet	42.99	54.40	0.49	19.41	nr	**73.81**
POWER POSTS/POLES/PILLARS						
Power post						
Power post; aluminium painted body; PVC-u cover; 5 nr outlets	503.02	636.62	4.00	158.42	nr	**795.04**
Power pole						
Power pole; 3.6 metres high; aluminium painted body; PVC-u cover; 6 nr outlets	645.02	816.33	4.00	158.42	nr	**974.75**
Extra for						
Power pole extension bar; 900 mm long	70.48	89.20	1.50	59.41	nr	**148.61**
Vertical multi-compartment pillar; PVC-u; BS 4678 Part4 EN60529; excludes accessories						
Single						
630 mm long	224.67	284.34	2.00	79.21	nr	**363.55**
3000 mm long	647.44	819.40	2.00	79.21	nr	**898.61**
Double						
630 mm long	224.67	284.34	3.00	118.81	nr	**403.15**
3000 mm long	687.15	869.66	3.00	118.81	nr	**988.47**
CONNECTION UNITS						
Connection units: moulded pattern; BS 5733; moulded plastic box; white coverplate; knockout for flex outlet; surface mounted – standard fused						
DP switched	13.06	16.53	0.49	19.41	nr	**35.93**
DP unswitched	12.00	15.19	0.49	19.41	nr	**34.60**
DP switched with neon indicator	16.57	20.98	0.49	19.41	nr	**40.38**
Connection units: moulded pattern; BS 5733; galvanized steel box; white coverplate; knockout for flex outlet; surface mounted						
DP switched	16.45	37.35	0.49	19.41	nr	**56.76**
DP unswitched	15.40	19.49	0.49	19.41	nr	**38.89**
DP switched with neon indicator	19.98	25.29	0.49	19.41	nr	**44.69**

ELECTRICAL SUPPLY/POWER/LIGHTING

Item	Net Price £	Material £	Labour hours	Labour £	Unit	Total rate £
Connection units: galvanized pressed steel pattern; galvanized steel box; satin chrome or satin brass finish; white moulded plastic inserts; flush mounted – standard fused						
DP switched	22.76	28.80	0.49	19.41	nr	**48.21**
DP unswitched	21.39	27.07	0.49	19.41	nr	**46.48**
DP switched with neon indicator	30.71	38.87	0.49	19.41	nr	**58.27**
Connection units: galvanized steel box; satin chrome or satin brass finish; white moulded plastic inserts; flex outlet; flush mounted – standard fused						
DP switched	21.83	27.63	0.49	19.41	nr	**47.03**
DP unswitched	20.71	26.21	0.49	19.41	nr	**45.62**
DP switched with neon indicator	28.09	35.55	0.49	19.41	nr	**54.96**
SHAVER SOCKETS						
Shaver unit: self-setting overload device; 200/250 voltage supply; white moulded plastic faceplate; unswitched						
Surface type with moulded plastic box	42.77	54.13	0.55	21.78	nr	**75.92**
Flush type with galvanized steel box	44.90	56.83	0.57	22.57	nr	**79.40**
Shaver unit: dual voltage supply unit; white moulded plastic faceplate; unswitched						
Surface type with moulded plastic box	51.58	65.27	0.62	24.55	nr	**89.83**
Flush type with galvanized steel box	53.61	67.85	0.64	25.35	nr	**93.20**
COOKER CONTROL UNITS						
Cooker control unit: BS 4177; 45 amp DP main switch; 13 amp switched socket outlet; metal coverplate; plastic inserts; neon indicators						
Surface mounted with mounting box	60.56	76.64	0.61	24.16	nr	**100.80**
Flush mounted with galvanized steel box	57.88	73.25	0.61	24.16	nr	**97.41**
Cooker control unit: BS 4177; 45 amp DP main switch; 13 amp switched socket outlet; moulded plastic box and coverplate; surface mounted						
Standard	42.02	53.18	0.61	24.16	nr	**77.34**
With neon indicators	49.30	62.40	0.61	24.16	nr	**86.56**
CONTROL COMPONENTS						
Connector unit: moulded white plastic cover and block; galvanized steel back box; to immersion heaters						
3 kW up to 915 mm long; fitted to thermostat	45.02	56.98	0.75	29.71	nr	**86.69**
Water heater switch: 20 amp; switched with neon indicator						
DP switched with neon indicator	22.37	28.31	0.45	17.82	nr	**46.13**

ELECTRICAL SUPPLY/POWER/LIGHTING

Item	Net Price £	Material £	Labour hours	Labour £	Unit	Total rate £
GENERAL LV POWER – cont						
SWITCH DISCONNECTORS						
Switch disconnectors; moulded plastic						
enclosure; fixed to backgrounds						
3 pole; IP54; Grey						
16 amp	45.87	58.05	0.80	31.68	nr	89.73
25 amp	54.31	68.73	0.80	31.68	nr	100.41
40 amp	88.43	111.92	0.80	31.68	nr	143.60
63 amp	137.65	174.21	1.00	39.60	nr	213.82
80 amp	238.64	302.02	1.25	49.51	nr	351.53
6 pole; IP54; Grey						
25 amp	76.30	96.56	1.00	39.60	nr	136.17
63 amp	128.84	163.06	1.25	49.51	nr	212.57
80 amp	244.23	309.10	1.80	71.29	nr	380.39
3 pole; IP54; Yellow						
16 amp	50.41	63.80	0.80	31.68	nr	95.49
25 amp	59.60	75.44	0.80	31.68	nr	107.12
40 amp	96.70	122.38	0.80	31.68	nr	154.07
63 amp	150.44	190.40	1.00	39.60	nr	230.00
6 pole; IP54; Yellow						
25 amp	74.84	94.72	1.00	39.60	nr	134.33
INDUSTRIAL SOCKETS/PLUGS						
Plugs; Splashproof; 100–130 volts, 50–60 Hz;						
IP 44 (Yellow)						
2 pole and earth						
16 amp	10.47	13.25	0.55	21.78	nr	35.03
32 amp	18.18	23.01	0.60	23.76	nr	46.77
3 pole and earth						
16 amp	19.70	24.93	0.65	25.74	nr	50.67
32 amp	20.76	26.27	0.72	28.52	nr	54.78
3 pole; neutral and earth						
16 amp	27.60	34.93	0.72	28.52	nr	63.45
32 amp	24.58	31.11	0.78	30.89	nr	62.00
Connectors; Splashproof; 100–130 volts,						
50–60 Hz; IP 44 (Yellow)						
2 pole and earth						
16 amp	10.57	13.37	0.42	16.63	nr	30.01
32 amp	18.35	23.23	0.50	19.80	nr	43.03
3 pole and earth						
16 amp	19.89	25.17	0.48	19.01	nr	44.18
32 amp	21.45	27.15	0.58	22.97	nr	50.12
3 pole; neutral and earth						
16 amp	27.86	35.26	0.52	20.59	nr	55.86
32 amp	28.24	35.74	0.73	28.91	nr	64.66

ELECTRICAL SUPPLY/POWER/LIGHTING

Item	Net Price £	Material £	Labour hours	Labour £	Unit	Total rate £
Angled sockets; surface mounted; Splashproof; 100–130 volts, 50–60 Hz; IP 44 (Yellow)						
2 pole and earth						
16 amp	6.46	8.17	0.55	21.78	nr	**29.96**
32 amp	23.51	29.75	0.60	23.76	nr	**53.51**
3 pole and earth						
16 amp	22.39	28.34	0.65	25.74	nr	**54.08**
32 amp	42.04	53.20	0.72	28.52	nr	**81.72**
3 pole; neutral and earth						
16 amp	31.08	39.34	0.72	28.52	nr	**67.85**
32 amp	40.64	51.43	0.78	30.89	nr	**82.32**
Plugs; Watertight; 100–130 volts, 50–60 Hz; IP67 (Yellow)						
2 pole and earth						
16 amp	16.69	21.13	0.55	21.78	nr	**42.91**
32 amp	29.09	36.81	0.60	23.76	nr	**60.57**
63 amp	79.76	100.94	0.75	29.70	nr	**130.64**
Connectors; Watertight; 100–130 volts, 50–60 Hz; IP 67 (Yellow)						
2 pole and earth						
16 amp	34.10	43.15	0.42	16.63	nr	**59.79**
32 amp	57.78	73.13	0.50	19.80	nr	**92.93**
63 amp	139.85	177.00	0.67	26.54	nr	**203.53**
Angled sockets; surface mounted; Watertight; 100–130 volts, 50–60 Hz; IP 67 (Yellow)						
2 pole and earth						
16 amp	28.17	35.65	0.55	21.78	nr	**57.44**
32 amp	55.10	69.73	0.60	23.76	nr	**93.49**
Plugs; Splashproof; 200–250 volts, 50–60 Hz; IP 44 (Blue)						
2 pole and earth						
16 amp	4.02	5.09	0.55	21.78	nr	**26.88**
32 amp	14.44	18.27	0.60	23.76	nr	**42.03**
63 amp	69.56	88.03	0.75	29.70	nr	**117.74**
3 pole and earth						
16 amp	15.98	20.22	0.65	25.74	nr	**45.97**
32 amp	21.46	27.17	0.72	28.52	nr	**55.68**
63 amp	69.81	88.35	0.83	32.87	nr	**121.22**
3 pole; neutral and earth						
16 amp	17.04	21.56	0.72	28.52	nr	**50.08**
32 amp	25.85	32.71	0.78	30.89	nr	**63.60**

ELECTRICAL SUPPLY/POWER/LIGHTING

Item	Net Price £	Material £	Labour hours	Labour £	Unit	Total rate £
GENERAL LV POWER – cont						
INDUSTRIAL SOCKETS/PLUGS – cont						
Connectors; Splashproof; 200–250 volts, 50–60 Hz; IP 44 (Blue)						
2 pole and earth						
16 amp	9.94	12.58	0.42	16.63	nr	**29.21**
32 amp	24.68	31.24	0.50	19.80	nr	**51.04**
63 amp	87.16	110.32	0.67	26.54	nr	**136.85**
3 pole and earth						
16 amp	26.31	33.30	0.48	19.01	nr	**52.31**
32 amp	34.76	43.99	0.58	22.97	nr	**66.96**
63 amp	71.71	90.75	0.75	29.70	nr	**120.46**
3 pole; neutral and earth						
16 amp	30.57	38.69	0.52	20.59	nr	**59.28**
32 amp	104.71	132.52	0.73	28.91	nr	**161.43**
Angled sockets; surface mounted; Splashproof; 200–250 volts, 50–60 Hz; IP 44 (Blue)						
2 pole and earth						
16 amp	13.06	16.53	0.55	21.78	nr	**38.31**
32 amp	22.60	28.61	0.60	23.76	nr	**52.37**
63 amp	99.37	125.76	0.75	29.70	nr	**155.46**
3 pole and earth						
16 amp	25.41	32.15	0.65	25.74	nr	**57.90**
32 amp	43.96	55.64	0.72	28.52	nr	**84.15**
63 amp	96.78	122.49	0.83	32.87	nr	**155.36**
3 pole; neutral and earth						
16 amp	22.11	27.98	0.72	28.52	nr	**56.49**
32 amp	38.68	48.95	0.78	30.89	nr	**79.84**
Plugs; Watertight; 200–250 volts, 50–60 Hz; IP67 (Blue)						
2 pole and earth						
16 amp	17.04	21.56	0.41	16.24	nr	**37.80**
32 amp	31.28	39.59	0.50	19.80	nr	**59.39**
63 amp	85.85	108.65	0.66	26.14	nr	**134.79**
125 amp	219.96	278.39	0.86	34.06	nr	**312.45**
Connectors; Watertight; 200–250 volts, 50–60 Hz; IP 67 (Blue)						
2 pole and earth						
16 amp	17.04	21.56	0.41	16.24	nr	**37.80**
32 amp	29.04	36.75	0.50	19.80	nr	**56.55**
63 amp	74.56	94.36	0.67	26.54	nr	**120.89**
125 amp	296.38	375.10	0.87	34.46	nr	**409.56**
Angled sockets; surface mounted; Watertight; 200–250 volts, 50–60 Hz; IP 67 (Blue)						
2 pole and earth						
16 amp	28.17	35.65	0.55	21.78	nr	**57.44**
32 amp	55.09	69.72	0.60	23.76	nr	**93.48**
125 amp	401.23	507.79	1.00	39.60	nr	**547.40**

ELECTRICAL SUPPLY/POWER/LIGHTING

Item	Net Price £	Material £	Labour hours	Labour £	Unit	Total rate £
UNINTERRUPTIBLE POWER SUPPLY						
Uninterruptible power supply; sheet steel enclosure; self contained battery pack; including installation, testing and commissioning						
Single phase input and output; 5 year battery life; standard 13 A socket outlet connection						
1.0 kVA (10 minute supply)	1294.86	1638.77	0.30	11.88	nr	1650.65
1.0 kVA (30 minute supply)	1845.37	2335.50	0.50	19.80	nr	2355.30
2.0 kVA (10 minute supply)	2310.14	2923.71	0.50	19.80	nr	2943.51
2.0 kVA (60 minute supply)	3877.08	4906.83	0.50	19.80	nr	4926.63
3.0 kVA (10 minute supply)	3052.64	3863.42	0.50	19.80	nr	3883.22
3.0 kVA (40 minute supply)	3941.69	4988.61	1.00	39.60	nr	5028.21
5.0 kVA (30 minute supply)	4996.61	6323.71	1.00	39.60	nr	6363.31
8.0 kVA (10 minute supply)	6010.35	7606.70	2.00	79.21	nr	7685.91
8.0 kVA (30 minute supply)	8127.38	10286.01	2.00	79.21	nr	10365.22
Uninterruptible power supply; including final connections and testing and commissioning						
Medium sized static; single phase input and output; 10 year battery life; in cubicle						
10.0 kVA (10 minutes supply)	6010.02	7606.28	10.00	396.05	nr	8002.33
10.0 kVA (30 minutes supply)	8127.02	10285.55	15.00	594.07	nr	10879.62
15.0 kVA (10 minutes supply)	9809.83	12415.32	10.00	396.05	nr	12811.36
15.0 kVA (30 minutes supply)	13559.34	17160.70	15.00	594.07	nr	17754.77
20.0 kVA (10 minutes supply)	9809.83	12415.32	10.00	396.05	nr	12811.36
20.0 kVA (30 minutes supply)	17453.59	22089.26	15.00	594.07	nr	22683.33
Medium sized static; three phase input and output; 10 year battery life; in cubicle						
10.0 kVA (10 minutes supply)	8112.07	10266.63	10.00	396.05	nr	10662.68
10.0 kVA (30 minutes supply)	10162.13	12861.19	15.00	594.07	nr	13455.26
15.0 kVA (10 minutes supply)	9251.72	11708.97	15.00	594.07	nr	12303.04
15.0 kVA (30 minutes supply)	11432.85	14469.42	20.00	792.09	nr	15261.51
20.0 kVA (10 minutes supply)	9340.63	11821.51	20.00	792.09	nr	12613.60
20.0 kVA (30 minutes supply)	14132.79	17886.46	25.00	990.12	nr	18876.58
30.0 kVA (10 minutes supply)	13978.13	17690.72	25.00	990.12	nr	18680.84
30.0 kVA (30 minutes supply)	17571.98	22239.09	30.00	1188.14	nr	23427.24
Large sized static; three phase input and output; 10 year battery life; in cubicle						
40 kVA (10 minutes supply)	13978.13	17690.72	30.00	1188.14	nr	18878.86
40 kVA (30 minutes supply)	21113.63	26721.41	30.00	1188.14	nr	27909.55
60 kVA (10 minutes supply)	22484.17	28455.96	35.00	1386.17	nr	29842.13
60 kVA (30 minutes supply)	26187.97	33143.50	35.00	1386.17	nr	34529.66
100 kVA (10 minutes supply)	33043.27	41819.57	40.00	1584.19	nr	43403.76
200 kVA (10 minutes supply)	54532.54	69016.38	40.00	1584.19	nr	70600.57
300 kVA (10 minutes supply)	88747.50	112318.83	40.00	1584.19	nr	113903.02
400 kVA (10 minutes supply)	104129.75	131786.62	50.00	1980.24	nr	133766.85
500 kVA (10 minutes supply)	122204.15	154661.57	60.00	2376.28	nr	157037.86
600 kVA (10 minutes supply)	164407.40	208074.00	70.00	2772.33	nr	210846.34
800 kVA (10 minutes supply)	205324.84	259859.12	80.00	3168.38	nr	263027.50

ELECTRICAL SUPPLY/POWER/LIGHTING

Item	Net Price £	Material £	Labour hours	Labour £	Unit	Total rate £
UNINTERRUPTIBLE POWER SUPPLY – cont						
Uninterruptible power supply – cont						
Integral diesel rotary; 400 V three phase input and output; no break supply; including ventilation and acoustic attenuation oil day tank and interconnecting pipework						
300 kVA	600237.06	759660.03	120.00	4752.57	nr	**764412.60**
500 kVA	618170.70	782356.83	120.00	4752.57	nr	**787109.40**
800 kVA	674960.84	854230.44	140.00	5544.66	nr	**859775.10**
1000 kVA	748825.25	947713.24	140.00	5544.66	nr	**953257.90**
1670 kVA	891968.51	1128875.34	170.00	6732.81	nr	**1135608.15**
2000 kVA	1086633.63	1375243.52	200.00	7920.95	nr	**1383164.46**
2500 kVA	1255789.58	1589327.29	230.00	9109.09	nr	**1598436.38**
Uninterruptible power supply (UPS) consisting of a single cabinet, sheet steel enclosure with single 3 phase input and output. Internally consists of multiple power modules to provide the rated output, with an additional power module for redundancy. 10 or 30 minute back-up, 10 year design life VRLA battery either internal or external (sizes dependent), including delivery, testing and commissioning						
20 kVA N+1 (10 mins)	14523.14	18380.48	10.00	396.05	nr	**18776.53**
20 kVA N+1 (30 mins)	16015.48	20269.19	10.00	396.05	nr	**20665.24**
40 kVA N+1 (10 mins)	19593.25	24797.22	30.00	1188.14	nr	**25985.36**
40 kVA N+1 (30 mins)	24777.90	31358.91	30.00	1188.14	nr	**32547.05**
60 kVA N+1 (10 mins)	26771.47	33881.97	35.00	1386.17	nr	**35268.14**
60 kVA N+1 (30 mins)	32922.29	41666.45	35.00	1386.17	nr	**43052.61**
100 kVA N+1 (10 mins)	38712.49	48994.53	40.00	1584.19	nr	**50578.72**
160 kVA N+1 (10 mins)	56335.99	71298.82	40.00	1584.19	nr	**72883.01**
200 kVA N+1 (10 mins)	71989.72	91110.19	40.00	1584.19	nr	**92694.38**

ELECTRICAL SUPPLY/POWER/LIGHTING

Item	Net Price £	Material £	Labour hours	Labour £	Unit	Total rate £
COMMERCIAL BATTERY STORAGE SYSTEM						
Commercial battery storage systems						
Battery storage system; self-contained compact battery pack; flexible layout; rated for outdoor use; includes labour, materials, civils, design and project management						
100 kW nominal/169 kWh batteries (1.6 hr capacity)	116240.88	147114.46	48.00	1901.03	nr	**149015.48**
500 kW nominal/845 kWh batteries (1.6 hr capacity)	423609.02	536119.57	58.00	2297.07	nr	**538416.65**
1000 kW nominal/1690 kWh batteries (1.6 hr capacity)	801646.58	1014563.91	76.00	3009.96	nr	**1017573.87**

ELECTRICAL SUPPLY/POWER/LIGHTING

Item	Net Price £	Material £	Labour hours	Labour £	Unit	Total rate £
EMERGENCY LIGHTING						
24 volt/50 volt/110 volt fluorescent slave luminaires						
For use with DC central battery systems						
Indoor, 8 watt	58.13	73.57	0.80	31.68	nr	105.25
Indoor, exit sign box	72.41	91.64	0.80	31.68	nr	123.33
Outdoor, 8 watt weatherproof	64.78	81.99	0.80	31.68	nr	113.67
Conversion module AC/DC	72.41	91.64	0.25	9.90	nr	101.54
Self-contained; polycarbonate base and diffuser; LED charging light to European sign directive; 3 hour standby						
Non maintained						
Indoor, 8 watt	64.18	81.22	1.00	39.60	nr	120.83
Outdoor, 8 watt weatherproof, vandal-resistant IP65	92.16	116.63	1.00	39.60	nr	156.24
Maintained						
Indoor, 8 watt	45.95	58.15	1.00	39.60	nr	97.75
Outdoor, 8 watt weatherproof, vandal-resistant IP65	135.93	172.03	1.00	39.60	nr	211.64
Exit signage						
Exit sign; gold effect, pendular including brackets						
Non maintained, 8 watt	174.54	220.90	1.00	39.60	nr	260.50
Maintained, 8 watt	187.86	237.76	1.00	39.60	nr	277.36
Modification kit						
Module & battery for 58 W fluorescent modification from mains fitting to emergency; 3 hour standby	60.16	76.14	0.50	19.80	nr	95.94
Extra for remote box (when fitting is too small for modification)	29.87	37.81	0.50	19.80	nr	57.61
12 volt low voltage lighting; non maintained; 3 hour standby						
2 × 20 watt lamp load	207.29	262.35	1.20	47.53	nr	309.88
1 × 50 watt lamp load	226.30	286.40	1.00	39.60	nr	326.00
Maintained 3 hour standby						
2 × 20 watt lamp load	196.23	248.35	1.20	47.53	nr	295.88
1 × 50 watt lamp load	215.22	272.39	1.00	39.60	nr	311.99
DC central battery systems BS5266 compliant 24/50/110 volt						
DC supply to luminaires on mains failure; metal cubicle with battery charger, changeover device and battery as integral unit; ICEL 1001 compliant; 10 year design life valve regulated lead acid battery; 24 hour recharge; LCD display & LED indication; ICEL alarm pack; Includes on-site commissioning on 110 volt systems only						

ELECTRICAL SUPPLY/POWER/LIGHTING

Item	Net Price £	Material £	Labour hours	Labour £	Unit	Total rate £
24 volt, wall mounted						
300 W maintained, 1 hour	3305.55	4183.50	4.00	158.42	nr	**4341.92**
635 W maintained, 3 hour	4010.24	5075.37	6.00	237.63	nr	**5312.99**
470 W non maintained, 1 hour	2947.02	3729.75	4.00	158.42	nr	**3888.17**
780 W non maintained, 3 hour	3481.72	4406.47	6.00	237.63	nr	**4644.09**
50 volt						
935 W maintained, 3 hour	3166.49	4007.51	8.00	316.84	nr	**4324.34**
1965 W maintained, 3 hour	4407.40	5578.00	8.00	316.84	nr	**5894.84**
1311 W non maintained, 3 hour	4314.68	5460.66	8.00	316.84	nr	**5777.50**
2510 W non maintained 3, hour	6589.47	8339.64	8.00	316.84	nr	**8656.47**
110 volt						
1603 W maintained, 3 hour	6210.86	7860.46	8.00	316.84	nr	**8177.30**
4446 W maintained, 3 hour	7093.27	8977.25	10.00	396.05	nr	**9373.30**
2492 W non maintained, 3 hour	7879.86	9972.75	10.00	396.05	nr	**10368.80**
5429 W non maintained, 3 hour	10755.80	13612.54	12.00	475.26	nr	**14087.80**
DC central battery systems; BS EN 50171 compliant; 24/50/110 volt						
Central power systems						
DC supply to luminaires on mains failure; metal cubicle with battery charger, changeover device and battery as integral unit; 10 year design life valve regulated lead acid battery; 12 hour recharge to 80% of specified duty; low volts discount; LCD display & LED indication; includes on-site commissioning for CPS systems only; battery sized for 'end of life' @ 20°C test pushbutton						
24 volt, floor standing						
400 W non maintained, 1 hour	3104.66	3929.25	4.00	158.42	nr	**4087.67**
600 W maintained, 3 hour	3991.71	5051.91	6.00	237.63	nr	**5289.54**
50 volt						
2133 W non maintained, 3 hour	6978.91	8832.51	8.00	316.84	nr	**9149.34**
1900 W maintained, 3 hour	6669.82	8441.33	8.00	316.84	nr	**8758.17**
110 volt						
2200 W non maintained, 3 hour	6455.02	8169.47	8.00	316.84	nr	**8486.31**
4000 W maintained, 3 hour	7204.53	9118.06	12.00	475.26	nr	**9593.32**
Low power systems						
DC supply to luminaires on mains failure; metal cubicle with battery charger, changeover device and battery as integral unit; 5 Year design life valve regulated lead acid battery; low volts discount; LED display & LED indication; battery sized for 'end of life' @ 20°C test pushbutton						
24 volt, floor standing						
300 W non maintained, 1 hour	3404.46	4308.69	4.00	158.42	nr	**4467.11**
600 W maintained, 3 hour	3769.16	4770.25	6.00	237.63	nr	**5007.88**

ELECTRICAL SUPPLY/POWER/LIGHTING

Item	Net Price £	Material £	Labour hours	Labour £	Unit	Total rate £
EMERGENCY LIGHTING – cont						
AC static inverter system; BS5266 compliant; one hour standby						
Central system supplying AC power on mains failure to mains luminaires; ICEL 1001 compliant metal cubicle(s) with changeover device, battery charger, battery & static inverter; 10 year design LIfe valve regulated lead acid battery; 24 hour recharge; LED indication and LCD display; pure sinewave output						
One hour						
750 VA, 600 W single phase I/P & O/P	3477.18	4400.72	6.00	237.63	nr	4638.35
3 kVA, 2.55 kW single phase I/P & O/P	5704.01	7218.99	8.00	316.84	nr	7535.83
5 kVA, 4.25 kW single phase I/P & O/P	7532.77	9533.47	10.00	396.05	nr	9929.52
8 kVA, 6.80 kW single phase I/P & O/P	8873.92	11230.84	12.00	475.26	nr	11706.09
10 kVA, 8.5 kW single phase I/P & O/P	11456.10	14498.84	14.00	554.47	nr	15053.30
13 kVA, 11.05 kW single phase I/P & O/P	15326.00	19396.58	16.00	633.68	nr	20030.26
15 kVA, 12.75 kW single phase I/P & O/P	15884.38	20103.27	30.00	1188.14	nr	21291.41
20 kVA, 17.0 kW 3 phase I/P & single phase O/P	21938.37	27765.21	40.00	1584.19	nr	29349.39
30 kVA, 25.5 kW 3 phase I/P & O/P	29647.46	37521.82	60.00	2376.28	nr	39898.10
40 kVA, 34.0 kW 3 phase I/P & O/P	36961.13	46778.01	80.00	3168.38	nr	49946.39
50 kVA, 42.5 kW 3 phase I/P & O/P	48206.19	61009.75	90.00	3564.43	nr	64574.18
65 kVA, 55.25 kW 3 phase I/P & O/P	59135.97	74842.48	100.00	3960.47	nr	78802.96
90 kVA, 68.85 kW 3 phase I/P & O/P	78146.20	98901.84	120.00	4752.57	nr	103654.40
120 kVA, 102 kW 3 phase I/P & O/P	87229.84	110398.08	150.00	5940.71	nr	116338.80
Three hour						
750 VA, 600 W single phase I/P & O/P	4068.94	5149.65	6.00	237.63	nr	5387.27
3 kVA, 2.55 kW single phase I/P & O/P	6847.47	8666.15	8.00	316.84	nr	8982.99
5 kVA, 4.25 kW single phase I/P & O/P	10854.98	13738.06	10.00	396.05	nr	14134.11
8 kVA, 6.80 kW single phase I/P & O/P	13505.28	17092.28	12.00	475.26	nr	17567.54
10 kVA, 8.5 kW single phase I/P & O/P	17814.67	22546.25	14.00	554.47	nr	23100.71
13 kVA, 11.05 kW single phase I/P & O/P	21966.43	27800.72	16.00	633.68	nr	28434.39
15 kVA, 12.75 kW single phase I/P & O/P	22327.11	28257.19	30.00	1188.14	nr	29445.33
20 kVA, 17.0 kW 3 phase I/P & single phase O/P	32332.48	40919.99	40.00	1584.19	nr	42504.18
30 kVA, 25.5 kW 3 phase I/P & O/P	48585.55	61489.87	60.00	2376.28	nr	63866.15
40 kVA, 34.0 kW 3 phase I/P & O/P	58087.34	73515.34	80.00	3168.38	nr	76683.72
50 kVA, 42.5 kW 3 phase I/P & O/P	78346.56	99155.41	90.00	3564.43	nr	102719.84
65 kVA, 55.25 kW 3 phase I/P & O/P	94729.21	119889.29	100.00	3960.47	nr	123849.76
90 kVA, 68.85 kW 3 phase I/P & O/P	112112.40	141889.45	120.00	4752.57	nr	146642.02
120 kVA, 102 kW 3 phase I/P & O/P	138181.17	174882.09	150.00	5940.71	nr	180822.80

ELECTRICAL SUPPLY/POWER/LIGHTING

Item	Net Price £	Material £	Labour hours	Labour £	Unit	Total rate £
AC static inverter system; BS EN 50171 compliant; one hour standby; Low power system (typically wall mounted) Central system supplying AC power on mains failure to mains luminaires; metal cubicle(s) with changeover device, battery charger, battery & static inverter; 5 year design life valve regulated lead acid battery; LED indication and LCD display; 12 hour recharge to 80% duty; inverter rated for 120% of load for 100% of duty; battery sized for 'end of life' @ 20°C test pushbutton						
One hour						
300 VA, 240 W single phase I/P & O/P	1554.92	1967.90	3.00	118.81	nr	2086.72
600 VA, 480 W single phase I/P & O/P	1874.17	2371.95	4.00	158.42	nr	2530.37
750 VA, 600 W single phase I/P & O/P	3458.47	4377.04	6.00	237.63	nr	4614.67
Three hour						
150 VA, 120 W single phase I/P & O/P	1713.87	2169.07	3.00	118.81	nr	2287.89
450 VA, 360 W single phase I/P & O/P	2033.14	2573.14	4.00	158.42	nr	2731.56
750 VA, 600 W single phase I/P & O/P	4047.56	5122.59	6.00	237.63	nr	5360.21
AC static inverter system central power system; CPS BS EN 50171 compliant; one hour standby Central system supplying AC power on mains failure to mains luminaires; metal cubicle(s) with changeover device, battery charger, battery & static inverter; LED indication and LCD display; pure sinewave output; 10 year design life valve regulated lead acid battery; 12 hour recharge to 80% duty specified; unverter rated for 120% of load for 100% of duty; battery sized for 'end of life' @ 20°C test push button; includes on-site commissioning						
One hour						
750 VA, 600 W single phase I/P & O/P	3869.90	4897.74	6.00	237.63	nr	5135.37
3 kVA, 2.55 kW single phase I/P & O/P	6166.19	7803.93	8.00	316.84	nr	8120.77
5 kVA, 4.25 kW single phase I/P & O/P	7984.27	10104.89	10.00	396.05	nr	10500.94
8 kVA, 6.80 kW single phase I/P & O/P	9318.77	11793.83	12.00	475.26	nr	12269.09
10 kVA, 8.5 kW single phase I/P & O/P	11966.38	15144.65	14.00	554.47	nr	15699.12
13 kVA, 11.05 kW single phase I/P & O/P	15816.25	20017.05	16.00	633.68	nr	20650.72
15 kVA, 12.75 kW single phase I/P & O/P	16370.63	20718.67	30.00	1188.14	nr	21906.81
20 kVA, 17.0 kW 3 phase I/P & single phase O/P	22551.51	28541.19	40.00	1584.19	nr	30125.38
30 kVA, 25.5 kW 3 phase I/P & O/P	30378.16	38446.60	60.00	2376.28	nr	40822.88
40 kVA, 34.0 kW 3 phase I/P & O/P	37651.77	47652.08	80.00	3168.38	nr	50820.46
50 kVA, 42.5 kW 3 phase I/P & O/P	49035.73	62059.62	90.00	3564.43	nr	65624.04
65 kVA, 55.25 kW 3 phase I/P & O/P	60565.29	76651.43	100.00	3960.47	nr	80611.90
90 kVA, 68.85 kW 3 phase I/P & O/P	79474.00	100582.30	120.00	4752.57	nr	105334.87
120 kVA, 102 kW 3 phase I/P & O/P	88509.57	112017.71	150.00	5940.71	nr	117958.42

ELECTRICAL SUPPLY/POWER/LIGHTING

Item	Net Price £	Material £	Labour hours	Labour £	Unit	Total rate £
EMERGENCY LIGHTING – cont						
AC static inverter system central power system – cont						
Three hour						
750 VA, 600 W single phase I/P & O/P	4477.72	5667.00	6.00	237.63	nr	**5904.63**
3 kVA, 2.55 kW single phase I/P & O/P	7302.99	9242.67	8.00	316.84	nr	**9559.51**
5 kVA, 4.25 kW single phase I/P & O/P	11315.83	14321.32	10.00	396.05	nr	**14717.36**
8 kVA, 6.80 kW single phase I/P & O/P	13926.05	17624.80	12.00	475.26	nr	**18100.06**
10 kVA, 8.5 kW single phase I/P & O/P	18292.89	23151.48	14.00	554.47	nr	**23705.95**
13 kVA, 11.05 kW single phase I/P & O/P	22421.94	28377.21	16.00	633.68	nr	**29010.88**
15 kVA, 12.75 kW single phase I/P & O/P	22778.60	28828.59	30.00	1188.14	nr	**30016.74**
20 kVA, 17.0 kW 3 phase I/P & single phase O/P	33194.10	42010.45	40.00	1584.19	nr	**43594.64**
30 kVA, 25.5 kW 3 phase I/P & O/P	49213.39	62284.46	60.00	2376.28	nr	**64660.75**
40 kVA, 34.0 kW 3 phase I/P & O/P	58665.74	74247.36	80.00	3168.38	nr	**77415.74**
50 kVA, 42.5 kW 3 phase I/P & O/P	79014.48	100000.72	90.00	3564.43	nr	**103565.15**
65 kVA, 55.25 kW 3 phase I/P & O/P	95968.86	121458.20	100.00	3960.47	nr	**125418.67**
90 kVA, 68.85 kW 3 phase I/P & O/P	119311.18	151000.23	120.00	4752.57	nr	**155752.80**
120 kVA, 102 kW 3 phase I/P & O/P	139185.71	176153.43	150.00	5940.71	nr	**182094.14**

COMMUNICATIONS/SECURITY/CONTROL

Item	Net Price £	Material £	Labour hours	Labour £	Unit	Total rate £
TELECOMMUNICATIONS						
CABLES						
Multipair internal telephone cable; BS 6746; loose laid on tray/basket						
0.5 millimetre dia. conductor LSZH insulated and sheathed multipair cables; BT specification CW 1308						
3 pair	0.24	0.30	0.03	1.19	m	**1.49**
4 pair	0.27	0.35	0.03	1.19	m	**1.54**
6 pair	0.42	0.54	0.03	1.19	m	**1.72**
10 pair	0.72	0.91	0.05	1.98	m	**2.89**
15 pair	1.09	1.38	0.06	2.38	m	**3.75**
20 pair + 1 Wire	1.55	1.96	0.06	2.38	m	**4.33**
25 pair	1.96	2.48	0.08	3.17	m	**5.65**
40 pair + earth	2.77	3.51	0.08	3.17	m	**6.67**
50 pair + earth	3.33	4.22	0.10	3.96	m	**8.18**
80 pair + earth	4.45	5.63	0.10	3.96	m	**9.60**
100 pair + earth	7.10	8.98	0.12	4.75	m	**13.73**
Multipair internal telephone cable; BS 6746; installed in conduit/trunking						
0.5 millimetre dia. conductor LSZH insulated and sheathed multipair cables; BT specification CW 1308						
3 pair	0.24	0.30	0.05	1.98	m	**2.28**
4 pair	0.27	0.35	0.06	2.38	m	**2.72**
6 pair	0.42	0.54	0.06	2.38	m	**2.91**
10 pair	0.72	0.91	0.07	2.77	m	**3.68**
15 pair	1.09	1.38	0.07	2.77	m	**4.15**
20 pair + 1 Wire	1.55	1.96	0.09	3.56	m	**5.52**
25 pair	1.96	2.48	0.10	3.96	m	**6.44**
40 pair + earth	2.77	3.51	0.12	4.75	m	**8.26**
50 pair + earth	3.33	4.22	0.14	5.54	m	**9.76**
80 pair + earth	4.45	5.63	0.14	5.54	m	**11.18**
100 pair + earth	7.10	8.98	0.15	5.94	m	**14.92**
Low speed data; unshielded twisted pair; solid copper conductors; LSOH sheath; nominal impedance 100 ohm; Category 3 to ISO IS 1801/EIA/TIA 568B and EN50173/ 50174 standards to current revisions						
Installed in riser						
25 pair 24 AWG	2.15	2.72	0.03	1.19	m	**3.91**
50 pair 24 AWG	3.62	4.58	0.06	2.38	m	**6.95**
100 pair 24 AWG	8.07	10.21	0.10	3.96	m	**14.17**
Installed below floor						
25 pair 24 AWG	2.15	2.72	0.02	0.79	m	**3.52**
50 pair 24 AWG	3.62	4.58	0.05	1.98	m	**6.56**
100 pair 24 AWG	8.07	10.21	0.08	3.17	m	**13.38**

COMMUNICATIONS/SECURITY/CONTROL

Item	Net Price £	Material £	Labour hours	Labour £	Unit	Total rate £
TELECOMMUNICATIONS – cont						
ACCESSORIES						
Telephone outlet: moulded plastic plate with box; fitted and connected; flush or surface mounted						
Single master outlet	9.42	11.92	0.35	13.86	nr	**25.78**
Single secondary outlet	6.96	8.81	0.35	13.86	nr	**22.67**
Telephone outlet: bronze or satin chromeplate; with box; fitted and connected; flush or surface mounted						
Single master outlet	15.32	19.39	0.35	13.86	nr	**33.25**
Single secondary outlet	16.73	21.18	0.35	13.86	nr	**35.04**
Frames and Box Connections						
Provision and Installation of a Dual Vertical Krone 108 A Voice Distribution Frame that can accommodate a total of						
138 × Krone 237 A Strips	318.44	403.01	1.50	59.41	nr	**462.42**
Label Frame (Traffolyte style)	2.01	2.55	0.27	10.69	nr	**13.24**
Provision and Installation of a Box Connection 301 A Voice Termination Unit that can accommodate a total of 10 ×						
Krone 237 A Strips	27.61	34.95	0.25	9.90	nr	**44.85**
Label Frame (Traffolyte style)	0.67	0.85	0.08	3.29	nr	**4.14**
Provision and Installation of a Box Connection 201 Voice Termination Unit						
that can accommodate 20 pairs	14.69	18.59	0.17	6.57	nr	**25.16**
Label Frame (Traffolyte style)	0.67	0.85	0.08	3.29	nr	**4.14**
Terminate, Test and Label Voice Multicore System						
Patch panels						
Voice; 19" wide fully loaded, finished in black including termination and forming of cables (assuming 2 pairs per port)						
25 port – RJ45 UTP – Krone	69.09	87.44	2.60	102.97	nr	**190.41**
50 port – RJ45 UTP – Krone	94.19	119.20	4.65	184.16	nr	**303.36**
900 pair fully loaded Systimax style of frame including forming and termination of						
9 × 100 pair cables	778.28	984.99	25.00	990.12	nr	**1975.11**
Installation and termination of Krone Strip (10 pair block – 237 A) including						
designation label strip	4.65	5.88	0.50	19.80	nr	**25.68**
Patch panel and Outlet labelling per port						
(Traffolyte style)	0.24	0.30	0.02	0.79	nr	**1.09**
Provision and installation of a voice jumper, for cross termination on Krone						
Termination Strips	0.07	0.09	0.06	2.38	nr	**2.47**
CW1308/Cat 3 cable circuit test per pair	–	–	0.05	1.98	nr	**1.98**

COMMUNICATIONS/SECURITY/CONTROL

Item	Net Price £	Material £	Labour hours	Labour £	Unit	Total rate £
RADIO AND TELEVISION						
RADIO						
Cables						
Radio frequency cable; BS 2316; PVC sheathed; laid loose 7/0.41 mm tinned copper inner conductor; solid polyethylene dielectric insulation; bare copper wire braid; PVC sheath; 75 ohm impedance						
Cable	1.46	1.85	0.05	1.98	m	**3.83**
Twin 1/0.58 mm copper covered steel solid core wire conductor; solid polyethylene dielectric insulation; barecopper wire braid; PVC sheath; 75 ohm impedance						
Cable	0.34	0.43	0.05	1.98	m	**2.41**
TELEVISION						
Cables						
Television aerial cable; coaxial; PVC sheathed; fixed to backgrounds General purpose TV aerial downlead; copper stranded inner conductor; cellular polythene insulation; copper braid outer conductor; 75 ohm impedance						
7/0.25 mm	0.33	0.41	0.06	2.38	m	**2.79**
Low loss TV aerial downlead; solid copper inner conductor; cellular polythene insulation; copper braid outer; conductor; 75 ohm impedance						
1/1.12 mm	0.34	0.43	0.06	2.38	m	**2.81**
Low loss air spaced; solid copper inner conductor; air spaced polythene insulation; copper braid outer conductor; 75 ohm impedance						
1/1.00 mm	0.45	0.57	0.06	2.38	m	**2.95**
Satelite aerial downlead; solid copper inner conductor; air spaced polythene insulation; copper tape and braid outer conductor; 75 ohm impedance						
1/1.00 mm	0.39	0.49	0.06	2.38	m	**2.87**
Satelite TV coaxial; solid copper inner conductor; semi air spaced polyethylene dielectric insulation; plain annealed copper foil and copper braid screen in outer conductor; PVC sheath; 75 ohm impedance						
1/1.25 mm	0.64	0.81	0.08	3.17	m	**3.98**

COMMUNICATIONS/SECURITY/CONTROL

Item	Net Price £	Material £	Labour hours	Labour £	Unit	Total rate £
RADIO AND TELEVISION – cont						
Television aerial cable – cont						
Satelite TV coaxial; solid copper inner conductor; air spaced polyethylene dielectric insulation; plain annealed copper foil and copper braid screen in outer conductor; PVC sheath; 75 ohm impedance						
1/1.67 mm	1.06	1.34	0.09	3.56	m	**4.90**
Video cable; PVC flame retardant sheath; laid loose						
7/0.1 mm silver coated copper covered annealed steel wire conductor; polyethylene dielectric insulation with tin coated copper wire braid; 75 ohm impedance						
Cable	0.25	0.32	0.05	1.98	m	**2.30**
ACCESSORIES						
TV co-axial socket outlet: moulded plastic box; flush or surface mounted						
One way Direct Connection	10.54	13.34	0.35	13.86	nr	**27.20**
Two way Direct Connection	14.68	18.57	0.35	13.86	nr	**32.44**
One way Isolated UHF/VHF	18.55	23.47	0.35	13.86	nr	**37.33**
Two way Isolated UHF/VHF	25.25	31.95	0.35	13.86	nr	**45.82**

COMMUNICATIONS/SECURITY/CONTROL

Item	Net Price £	Material £	Labour hours	Labour £	Unit	Total rate £
CLOCKS						
Master clock						
Master clock module 230 V with programming circuits, NTP client and server for synchronization over Ethernet network, sounders Harmonys or Melodys, control of DHF and NTP clock systems						
Rack mounted	1185.03	1499.77	3.00	118.81	nr	**1618.59**
Wall mounted	1065.96	1349.08	5.00	198.02	nr	**1547.10**
Master clock module 230 V with programming circuits for control of wired and DHF clock systems, relays or sounders, NTP client and server for synchronization over Ethernet network						
Rack mounted	1600.07	2025.05	3.00	118.81	nr	**2143.87**
Wall mounted	1299.56	1644.73	5.00	198.02	nr	**1842.75**
Clocks and bells, master clock module 230 V with 3 programming circuits for control of wired and DHF clock systems, relays and sounders						
Rack mounted	1033.07	1307.46	3.00	118.81	nr	**1426.27**
Wall mounted	816.48	1033.34	5.00	198.02	nr	**1231.36**
Clocks only, master clock module 230 V for control of wired and DHF clock systems						
Rack mounted	781.33	988.85	3.00	118.81	nr	**1107.66**
Wall mounted	606.69	767.83	5.00	198.02	nr	**965.85**
Extra over for						
GPS antenna with 20 m cable	238.14	301.39	2.50	99.01	nr	**400.40**
DHF transmitter	350.41	443.47	1.50	59.41	nr	**502.88**
DHF repeater, signal booster up to 200 m	401.44	508.06	3.00	118.81	nr	**626.87**
UPS	1037.61	1313.20	5.00	198.02	nr	**1511.22**
Rack mounted external AC/DC power supply	596.48	754.91	3.00	118.81	nr	**873.72**
Indoor clock, 30 cm dia.						
Stand-alone						
HMS display, quartz movement, battery operated	56.70	71.76	0.80	31.68	nr	**103.44**
DHF wireless						
Hour and minute display, battery operated	129.28	163.61	0.80	31.68	nr	**195.30**
Hour, minute and second display, battery operated	141.75	179.40	0.80	31.68	nr	**211.08**
HM display, TBT	133.81	169.35	0.80	31.68	nr	**201.04**
HMS display, TBT	144.02	182.27	0.80	31.68	nr	**213.95**
Slave wired						
Hour and minute display, 24 V	77.11	97.59	1.00	39.60	nr	**137.20**
Hour, minute and second display, 24 V	92.99	117.69	1.00	39.60	nr	**157.29**
Hour, minute display, TBT, AFNOR	128.14	162.18	1.00	39.60	nr	**201.78**
Hour, minute and second display, TBT AFNOR	141.75	179.40	1.00	39.60	nr	**219.00**
Hour, minute display, POE, NTP	179.17	226.76	1.00	39.60	nr	**266.36**
Hour, minute and second display, POE, NTP	190.51	241.11	1.00	39.60	nr	**280.72**

COMMUNICATIONS/SECURITY/CONTROL

Item	Net Price £	Material £	Labour hours	Labour £	Unit	Total rate £
CLOCKS – cont						
Indoor clock, 40 cm dia.						
Stand-alone						
HMS display, quartz movement, battery operated	107.73	136.34	0.80	31.68	nr	**168.03**
DHF wireless						
Hour and minute display, battery operated	189.38	239.68	0.80	31.68	nr	**271.36**
Hour, minute and second display, battery operated	195.05	246.85	0.80	31.68	nr	**278.54**
HM display, TBT	189.38	239.68	0.80	31.68	nr	**271.36**
HMS display, TBT	202.99	256.90	0.80	31.68	nr	**288.58**
Slave wired						
Hour and minute display, 24 V	142.88	180.83	1.00	39.60	nr	**220.44**
Hour, minute and second display, 24 V	131.54	166.48	1.00	39.60	nr	**206.09**
Hour, minute display, TBT, AFNOR	200.72	254.03	1.00	39.60	nr	**293.63**
Hour, minute and second display, TBT AFNOR	207.52	262.64	1.00	39.60	nr	**302.24**
Hour, minute display, POE, NTP	237.01	299.95	1.00	39.60	nr	**339.56**
Hour, minute and second display, POE, NTP	243.81	308.57	1.00	39.60	nr	**348.17**
Digital clock, 7 cm digit height, 230 V						
Independent quartz movement						
Red display	238.14	301.39	0.80	31.68	nr	**333.07**
Green display	238.14	301.39	0.80	31.68	nr	**333.07**
Yellow display	238.14	301.39	0.80	31.68	nr	**333.07**
Blue display	323.19	409.03	0.80	31.68	nr	**440.71**
White display	323.19	409.03	0.80	31.68	nr	**440.71**
AFNOR, 230 V						
Red display	243.81	308.57	0.80	31.68	nr	**340.25**
Green display	243.81	308.57	0.80	31.68	nr	**340.25**
Yellow display	243.81	308.57	0.80	31.68	nr	**340.25**
Blue display	328.86	416.21	0.80	31.68	nr	**447.89**
White display	328.86	416.21	0.80	31.68	nr	**447.89**
DHF wireless, 230 V						
Red display	260.82	330.09	0.80	31.68	nr	**361.78**
Green display	260.82	330.09	0.80	31.68	nr	**361.78**
Yellow display	260.82	330.09	0.80	31.68	nr	**361.78**
Blue display	345.87	437.73	0.80	31.68	nr	**469.42**
White display	345.87	437.73	0.80	31.68	nr	**469.42**
NTP POE						
Red display	283.50	358.80	0.80	31.68	nr	**390.48**
Green display	283.50	358.80	0.80	31.68	nr	**390.48**
Yellow display	283.50	358.80	0.80	31.68	nr	**390.48**
Blue display	368.55	466.44	0.80	31.68	nr	**498.12**
White display	368.55	466.44	0.80	31.68	nr	**498.12**

COMMUNICATIONS/SECURITY/CONTROL

Item	Net Price £	Material £	Labour hours	Labour £	Unit	Total rate £
Digital clock, 5 cm digit height, 230 V						
Independent quartz movement						
Red display	198.45	251.16	0.80	31.68	nr	**282.84**
Green display	198.45	251.16	0.80	31.68	nr	**282.84**
Yellow display	198.45	251.16	0.80	31.68	nr	**282.84**
Blue display	260.82	330.09	0.80	31.68	nr	**361.78**
White display	260.82	330.09	0.80	31.68	nr	**361.78**
AFNOR, 230 V						
Red display	198.45	251.16	0.80	31.68	nr	**282.84**
Green display	198.45	251.16	0.80	31.68	nr	**282.84**
Yellow display	198.45	251.16	0.80	31.68	nr	**282.84**
Blue display	260.82	330.09	0.80	31.68	nr	**361.78**
White display	260.82	330.09	0.80	31.68	nr	**361.78**
DHF wireless, 230 V						
Red display	232.47	294.21	0.80	31.68	nr	**325.90**
Green display	232.47	294.21	0.80	31.68	nr	**325.90**
Yellow display	232.47	294.21	0.80	31.68	nr	**325.90**
Blue display	294.84	373.15	0.80	31.68	nr	**404.83**
White display	294.84	373.15	0.80	31.68	nr	**404.83**
NTP POE						
Red display	249.48	315.74	0.80	31.68	nr	**347.43**
Green display	249.48	315.74	0.80	31.68	nr	**347.43**
Yellow display	249.48	315.74	0.80	31.68	nr	**347.43**
Blue display	311.85	394.68	0.80	31.68	nr	**426.36**
White display	311.85	394.68	0.80	31.68	nr	**426.36**
Digital clock, 10 cm digit height, 230 V						
Independent quartz movement						
Red display	425.25	538.20	0.80	31.68	nr	**569.88**
Green display	425.25	538.20	0.80	31.68	nr	**569.88**
Yellow display	425.25	538.20	0.80	31.68	nr	**569.88**
Blue display	527.31	667.36	0.80	31.68	nr	**699.05**
White display	527.31	667.36	0.80	31.68	nr	**699.05**
AFNOR, 230 V						
Red display	425.25	538.20	0.80	31.68	nr	**569.88**
Green display	425.25	538.20	0.80	31.68	nr	**569.88**
Yellow display	425.25	538.20	0.80	31.68	nr	**569.88**
Blue display	527.31	667.36	0.80	31.68	nr	**699.05**
White display	527.31	667.36	0.80	31.68	nr	**699.05**
DHF wireless, 230 V						
Red display	442.26	559.72	0.80	31.68	nr	**591.41**
Green display	442.26	559.72	0.80	31.68	nr	**591.41**
Yellow display	442.26	559.72	0.80	31.68	nr	**591.41**
Blue display	544.32	688.89	0.80	31.68	nr	**720.58**
White display	544.32	688.89	0.80	31.68	nr	**720.58**
NTP POE						
Red display	459.27	581.25	0.80	31.68	nr	**612.94**
Green display	459.27	581.25	0.80	31.68	nr	**612.94**
Yellow display	459.27	581.25	0.80	31.68	nr	**612.94**
Blue display	561.33	710.42	0.80	31.68	nr	**742.10**
White display	561.33	710.42	0.80	31.68	nr	**742.10**

COMMUNICATIONS/SECURITY/CONTROL

Item	Net Price £	Material £	Labour hours	Labour £	Unit	Total rate £
DATA TRANSMISSION						
Cabinets						
Floor standing; suitable for 19" patch panels with glass lockable doors, metal rear doors, side panels, vertical cable management, 2 × 4 way PDUs, 4 way fan, earth bonding kit; installed on raised floor						
600 mm wide × 800 mm deep – 18U	927.13	1173.38	3.00	118.81	nr	1292.19
600 mm wide × 800 mm deep – 24U	965.75	1222.25	3.00	118.81	nr	1341.06
600 mm wide × 800 mm deep – 33U	1037.21	1312.70	4.00	158.42	nr	1471.12
600 mm wide × 800 mm deep – 42U	1112.55	1408.04	4.00	158.42	nr	1566.46
600 mm wide × 800 mm deep – 47U	1174.47	1486.41	4.00	158.42	nr	1644.83
800 mm wide × 800 mm deep – 42U	1259.37	1593.86	4.00	158.42	nr	1752.28
800 mm wide × 800 mm deep – 47U	1313.50	1662.36	4.00	158.42	nr	1820.78
Label cabinet	2.86	3.61	0.25	9.90	nr	13.52
Wall mounted; suitable for 19" patch panels with glass lockable doors, side panels, vertical cable management, 2 × 4 way PDUs, 4 way fan, earth bonding kit; fixed to wall						
19 mm wide × 500 mm deep – 9U	618.37	782.61	3.00	118.81	nr	901.42
19 mm wide × 500 mm deep – 12U	633.58	801.86	3.00	118.81	nr	920.67
19 mm wide × 500 mm deep – 15U	656.57	830.96	3.00	118.81	nr	949.77
19 mm wide × 500 mm deep – 18U	772.61	977.82	3.00	118.81	nr	1096.63
19 mm wide × 500 mm deep – 21U	803.73	1017.20	3.00	118.81	nr	1136.01
Label cabinet	2.97	3.76	0.25	9.90	nr	13.66
Frames						
Floor standing; suitable for 19" patch panels with supports, vertical cable management, earth bonding kit; installed on raised floor						
19 mm wide × 500 mm deep – 25U	695.49	880.21	2.50	99.01	nr	979.22
19 mm wide × 500 mm deep – 39U	849.72	1075.40	2.50	99.01	nr	1174.42
19 mm wide × 500 mm deep – 42U	927.19	1173.45	2.50	99.01	nr	1272.46
19 mm wide × 500 mm deep – 47U	1004.66	1271.50	2.50	99.01	nr	1370.51
Label frame	2.97	3.76	0.25	9.90	nr	13.66
Copper data cabling						
Unshielded twisted pair; solid copper conductors; PVC insulation; nominal impedance 100 ohm; Cat 5e to ISO 11801, EIA/TIA 568B and EN 50173/50174 standards to the current revisions						
4 pair 24 AWG; nominal outside dia. 5.6 mm; installed above ceiling	0.28	0.35	0.02	0.79	m	1.14
4 pair 24 AWG; nominal outside dia. 5.6 mm; installed in riser	0.28	0.35	0.02	0.79	m	1.14
4 pair 24 AWG; nominal outside dia. 5.6 mm; installed below floor	0.28	0.35	0.01	0.40	m	0.75
4 pair 24 AWG; nominal outside dia. 5.6 mm; installed in trunking	0.28	0.35	0.02	0.79	m	1.14

COMMUNICATIONS/SECURITY/CONTROL

Item	Net Price £	Material £	Labour hours	Labour £	Unit	Total rate £
Unshielded twisted pair; solid copper conductors; LSOH sheathed; nominal impedance 100 ohm; Cat 5e to ISO 11801, EIA/TIA 568B and EN 50173/50174 standards to the current revisions						
4 pair 24 AWG; nominal outside dia. 5.6 mm; installed above ceiling	0.30	0.38	0.02	0.79	m	**1.17**
4 pair 24 AWG; nominal outside dia. 5.6 mm; installed in riser	0.30	0.38	0.02	0.79	m	**1.17**
4 pair 24 AWG; nominal outside dia. 5.6 mm; installed below floor	0.30	0.38	0.01	0.40	m	**0.77**
4 pair 24 AWG; nominal outside dia. 5.6 mm; installed in trunking	0.30	0.38	0.02	0.79	m	**1.17**
Unshielded twisted pair; solid copper conductors; PVC insulation; nominal impedance 100 ohm; Cat 6 to ISO 11801, EIA/TIA 568B and EN 50173/50174 standards to the current revisions						
4 pair 24 AWG; nominal outside dia. 5.6 mm; installed above ceiling	0.44	0.56	0.02	0.59	m	**1.15**
4 pair 24 AWG; nominal outside dia. 5.6 mm; installed in riser	0.44	0.56	0.02	0.59	m	**1.15**
4 pair 24 AWG; nominal outside dia. 5.6 mm; installed below floor	0.44	0.56	0.01	0.32	m	**0.87**
4 pair 24 AWG; nominal outside dia. 5.6 mm; installed in trunking	0.44	0.56	0.02	0.79	m	**1.35**
Unshielded twisted pair; solid copper conductors; LSOH sheathed; nominal impedance 100 ohm; Cat 6 to ISO 11801, EIA/TIA 568B and EN 50173/50174 standards to the current revisions						
4 pair 24 AWG; nominal outside dia. 5.6 mm; installed above ceiling	0.41	0.52	0.02	0.87	m	**1.39**
4 pair 24 AWG; nominal outside dia. 5.6 mm; installed in riser	0.41	0.52	0.02	0.87	m	**1.39**
4 pair 24 AWG; nominal outside dia. 5.6 mm; installed below floor	0.41	0.52	0.01	0.44	m	**0.96**
4 pair 24 AWG; nominal outside dia. 5.6 mm; installed in trunking	0.41	0.52	0.02	0.87	m	**1.39**
Patch panels						
Category 5e; 19" wide fully loaded, finished in black including termination and forming of cables						
24 port – RJ45 UTP – Krone/110	69.07	87.41	4.75	188.12	nr	**275.53**
48 port – RJ45 UTP – Krone/110	129.55	163.95	9.35	370.30	nr	**534.26**
Patch panel labelling per port	0.31	0.40	0.02	0.79	nr	**1.19**
Category 6; 19" wide fully loaded, finished in black including termination and forming of cables						
24 port – RJ45 UTP – Krone/110	131.91	166.95	5.00	198.02	nr	**364.97**
48 port – RJ45 UTP – Krone/110	240.59	304.50	9.80	388.13	nr	**692.62**
Patch panel labelling per port	0.31	0.40	0.02	0.79	nr	**1.19**

COMMUNICATIONS/SECURITY/CONTROL

Item	Net Price £	Material £	Labour hours	Labour £	Unit	Total rate £
DATA TRANSMISSION – cont						
Workstation						
Category 5e RJ45 data outlet plate and multiway						
outlet boxes for wall, ceiling and below floor						
installations including label to ISO						
11801 standards						
Wall mounted; fully loaded						
One gang LSOH PVC plate	6.37	8.07	0.15	5.94	nr	**14.01**
Two gang LSOH PVC plate	9.13	11.55	0.20	7.92	nr	**19.47**
Four gang LSOH PVC plate	15.63	19.78	0.40	15.84	nr	**35.62**
One gang satin brass plate	20.69	26.18	0.25	9.90	nr	**36.08**
Two gang satin brass plate	23.34	29.54	0.33	13.07	nr	**42.61**
Ceiling mounted; fully loaded						
One gang metal clad plate	10.98	13.90	0.33	13.07	nr	**26.97**
Two gang metal clad plate	14.92	18.89	0.45	17.82	nr	**36.71**
Below floor; fully loaded						
Four way outlet box, 5 m length 20 mm						
flexible conduit with glands and starin						
relief bracket	37.15	47.02	0.80	31.68	nr	**78.70**
Six way outlet box, 5 m length 25 mm						
flexible conduit with glands and strain						
relief bracket	48.03	60.78	1.20	47.53	nr	**108.31**
Eight way outlet box, 5 m length 25 mm						
flexible conduit with glands and strain						
relief bracket	60.47	76.53	1.40	55.45	nr	**131.98**
Installation of outlet boxes to desks	–	–	0.40	15.84	nr	**15.84**
Category 6 RJ45 data outlet plate and						
multiway outlet boxes for wall, ceiling and						
below floor installations including label to ISO						
11801 standards						
Wall mounted; fully loaded						
One gang LSOH PVC plate	8.14	10.30	0.17	6.53	nr	**16.83**
Two gang LSOH PVC plate	11.95	15.12	0.22	8.71	nr	**23.83**
Four gang LSOH PVC plate	21.57	27.30	0.44	17.43	nr	**44.72**
One gang satin brass plate	19.89	25.17	0.28	10.89	nr	**36.06**
Two gang satin brass plate	24.88	31.49	0.36	14.38	nr	**45.87**
Ceiling mounted; fully loaded						
One gang metal clad plate	13.18	16.68	0.36	14.38	nr	**31.06**
Two gang metal clad plate	18.62	23.56	0.50	19.60	nr	**43.16**
Below floor; fully loaded						
Four way outlet box, 5 m length 20 mm						
flexible conduit with glands and starin						
relief bracket	45.53	57.63	0.88	34.85	nr	**92.48**
Six way outlet box, 5 m length 25 mm flexible						
conduit with glands and strain relief bracket	60.56	76.65	1.32	52.28	nr	**128.93**
Eight way outlet box, 5 m length 25 mm						
flexible conduit with glands and strain						
relief bracket	77.13	97.62	1.54	60.99	nr	**158.61**
Installation of outlet boxes to desks	–	–	0.40	15.84	nr	**15.84**
Category 5e cable test	–	–	0.10	3.96	nr	**3.96**
Category 6 cable test	–	–	0.11	4.36	nr	**4.36**

COMMUNICATIONS/SECURITY/CONTROL

Item	Net Price £	Material £	Labour hours	Labour £	Unit	Total rate £
Copper patch leads						
Category 5e; straight through booted RJ45						
UTP – RJ45 UTP						
Patch lead 1 m length	2.24	2.84	0.09	3.56	nr	**6.40**
Patch lead 3 m length	2.67	3.38	0.09	3.56	nr	**6.94**
Patch lead 5 m length	3.78	4.78	0.10	3.96	nr	**8.74**
Patch lead 7 m length	5.54	7.02	0.10	3.96	nr	**10.98**
Category 6; straight through booted RJ45						
UTP – RJ45 UTP						
Patch lead 1 m length	5.95	7.53	0.09	3.56	nr	**11.09**
Patch lead 3 m length	8.97	11.36	0.09	3.56	nr	**14.92**
Patch lead 5 m length	10.52	13.32	0.10	3.96	nr	**17.28**
Patch lead 7 m length	13.34	16.88	0.10	3.96	nr	**20.84**
Note 1 – With an intelligent Patching System, add a 25% uplift to the Cat 5e and Cat 6 System price. This would be dependent on the System and IMS solution chosen, i.e. iPatch, RIT or iTRACs.						
Note 2 – For a Cat 6 augmented solution (10Gbps capable), add 25% to a Cat 6 System price.						
Fibre infrastructure						
Fibre optic cable, tight buffered, internal/ external application, single mode, LSOH sheathed						
4 core fibre-optic cable	1.74	2.20	0.10	3.96	m	**6.16**
8 core fibre-optic cable	2.90	3.67	0.10	3.96	m	**7.63**
12 core fibre-optic cable	3.59	4.54	0.10	3.96	m	**8.50**
16 core fibre-optic cable	4.37	5.53	0.10	3.96	m	**9.49**
24 core fibre-optic cable	5.27	6.67	0.10	3.96	m	**10.63**
Fibre optic cable OM1 and OM2, tight buffered, internal/external application, 62.5/ 125 multimode fibre, LSOH sheathed						
4 core fibre-optic cable	1.85	2.34	0.10	3.96	m	**6.30**
8 core fibre-optic cable	3.07	3.89	0.10	3.96	m	**7.85**
12 core fibre-optic cable	3.93	4.97	0.10	3.96	m	**8.93**
16 core fibre-optic cable	4.86	6.16	0.10	3.96	m	**10.12**
24 core fibre-optic cable	5.64	7.14	0.10	3.96	m	**11.10**
Fibre optic cable OM3, tight buffered, internal only application, 50/125 multimode fibre, LSOH sheathed						
4 core fibre-optic cable	2.30	2.91	0.10	3.96	nr	**6.87**
8 core fibre-optic cable	3.59	4.54	0.10	3.96	nr	**8.50**
12 core fibre-optic cable	5.20	6.59	0.10	3.96	nr	**10.55**
24 core fibre cable	9.94	12.59	0.10	3.96	nr	**16.55**

COMMUNICATIONS/SECURITY/CONTROL

Item	Net Price £	Material £	Labour hours	Labour £	Unit	Total rate £
DATA TRANSMISSION – cont						
Fibre infrastructure – cont						
Fibre-optic single and multimode connectors and couplers including termination						
ST single mode booted connector	8.76	11.09	0.25	9.90	nr	**20.99**
ST multimode booted connector	4.23	5.36	0.25	9.90	nr	**15.26**
SC simplex single mode booted connector	12.42	15.71	0.25	9.90	nr	**25.62**
SC simplex multimode booted connector	4.53	5.73	0.25	9.90	nr	**15.63**
SC duplex multimode booted connector	9.03	11.43	0.25	9.90	nr	**21.33**
ST – SC duplex adaptor	21.94	27.77	0.01	0.40	nr	**28.17**
ST inline bulkhead coupler	4.71	5.96	0.01	0.40	nr	**6.35**
SC duplex coupler	8.78	11.11	0.01	0.40	nr	**11.51**
MTRJ small form factor duplex connector	8.75	11.07	0.25	9.90	nr	**20.97**
LC simplex multimode booted connector	–	–	0.25	9.90	nr	**9.90**
Single mode core test per core	–	–	0.20	7.92	nr	**7.92**
Multimode core test per core	–	–	0.20	7.92	nr	**7.92**
Fibre; 19" wide fully loaded, labled, alluminium alloy c/w couplers, fibre management and glands (excludes termination of fibre cores)						
8 way ST; fixed drawer	94.26	119.30	0.50	19.80	nr	**139.10**
16 way ST; fixed drawer	141.36	178.91	0.50	19.80	nr	**198.71**
24 way ST; fixed drawer	196.40	248.57	0.50	19.80	nr	**268.37**
8 way ST; sliding drawer	119.92	151.77	0.50	19.80	nr	**171.57**
16 way ST; sliding drawer	164.77	208.54	0.50	19.80	nr	**228.34**
24 way ST; sliding drawer	216.86	274.45	0.50	19.80	nr	**294.25**
8 way (4 duplex) SC; fixed drawer	112.37	142.21	0.50	19.80	nr	**162.01**
16 way (8 duplex) SC; fixed drawer	164.45	208.12	0.50	19.80	nr	**227.93**
24 way (12 duplex) SC; fixed drawer	202.14	255.82	0.50	19.80	nr	**275.63**
8 way (4 duplex) SC; sliding drawer	142.18	179.94	0.50	19.80	nr	**199.74**
16 way (8 duplex) SC; sliding drawer	194.60	246.28	0.50	19.80	nr	**266.08**
24 way (12 duplex) SC; sliding drawer	231.94	293.54	0.50	19.80	nr	**313.34**
8 way (4 duplex) MTRJ; fixed drawer	127.45	161.30	0.50	19.80	nr	**181.10**
16 way (8 duplex) MTRJ; fixed drawer	194.60	246.28	0.50	19.80	nr	**266.08**
24 way (12 duplex) MTRJ; fixed drawer	231.94	293.54	0.50	19.80	nr	**313.34**
8 way (4 duplex) FC/PC; fixed drawer	133.61	169.10	0.50	19.80	nr	**188.90**
16 way (8 duplex) FC/PC; fixed drawer	204.19	258.42	0.50	19.80	nr	**278.23**
24 way (12 duplex) FC/PC; fixed drawer	243.60	308.29	0.50	19.80	nr	**328.10**
Patch panel label per way	0.31	0.40	0.02	0.79	nr	**1.19**
Fibre patch leads						
Single mode						
Duplex OS1 LC – LC						
Fibre patch lead 1 m length	28.91	36.59	0.08	3.17	nr	**39.75**
Fibre patch lead 3 m length	27.04	34.23	0.08	3.17	nr	**37.40**
Fibre patch lead 5 m length	28.76	36.40	0.10	3.96	nr	**40.36**

COMMUNICATIONS/SECURITY/CONTROL

Item	Net Price £	Material £	Labour hours	Labour £	Unit	Total rate £
Multimode						
Duplex 50/125 OM3 MTRJ – MTRJ						
Fibre patch lead 1 m length	25.37	32.11	0.08	3.17	nr	**35.28**
Fibre patch lead 3 m length	27.58	34.90	0.08	3.17	nr	**38.07**
Fibre patch lead 5 m length	28.96	36.66	0.10	3.96	nr	**40.62**
Duplex 50/125 OM3 ST – ST						
Fibre patch lead 1 m length	31.64	40.04	0.08	3.17	nr	**43.21**
Fibre patch lead 3 m length	34.42	43.56	0.08	3.17	nr	**46.72**
Fibre patch lead 5 m length	36.14	45.74	0.10	3.96	nr	**49.70**
Duplex 50/125 OM3 SC – SC						
Fibre patch lead 1 m length	28.39	35.93	0.08	3.17	nr	**39.10**
Firbe patch lead 3 m length	30.89	39.10	0.08	3.17	nr	**42.27**
Fibre patch lead 5 m length	32.43	41.04	0.10	3.96	nr	**45.00**
Duplex 50/125 OM3 LC – LC						
Fibre patch lead 1 m length	30.42	38.50	0.08	3.17	nr	**41.67**
Fibre patch lead 3 m length	33.10	41.89	0.08	3.17	nr	**45.05**
Fibre patch lead 5 m length	34.75	43.98	0.10	3.96	nr	**47.95**

COMMUNICATIONS/SECURITY/CONTROL

Item	Net Price £	Material £	Labour hours	Labour £	Unit	Total rate £
ACCESS CONTROL EQUIPMENT						
Equipment to control the movement of personnel into defined spaces; includes fixing to backgrounds, termination of power and data cables; excludes cable containment and cable installation						
Access control						
Proximity reader	129.17	163.48	1.50	59.41	nr	**222.89**
Exit button	23.49	29.73	1.50	59.41	nr	**89.14**
Exit PIR	155.01	196.18	2.00	79.21	nr	**275.39**
Emergency break glass double pole	35.23	44.58	1.00	39.60	nr	**84.19**
Alarm contact flush	5.87	7.43	1.00	39.60	nr	**47.03**
Alarm contact surface	5.87	7.43	1.00	39.60	nr	**47.03**
Reader controller 16 door	5037.97	6376.06	4.00	158.42	nr	**6534.48**
Reader controller 8 door	3343.38	4231.38	4.00	158.42	nr	**4389.80**
Reader controller 2 door	1170.83	1481.80	4.00	158.42	nr	**1640.22**
Reader interface	474.85	600.97	1.50	59.41	nr	**660.38**
Lock power supply 12 volt 3 amp	283.02	358.19	2.00	79.21	nr	**437.40**
Lock power supply 24 volt 3 amp	494.40	625.72	2.00	79.21	nr	**704.93**
Rechargeable battery 12 volt 7 Ahr	14.09	17.83	0.25	9.90	nr	**27.73**
Lock equipment						
Single slimline magnetic lock, monitored	122.13	154.57	2.00	79.21	nr	**233.78**
Single slimline magnetic lock, unmonitored	105.70	133.77	1.75	69.31	nr	**203.08**
Double slimline magnetic lock, monitored	241.91	306.17	4.00	158.42	nr	**464.58**
Double slimline magnetic lock, unmonitored	207.86	263.07	4.50	178.22	nr	**441.29**
Standard single magnetic lock, monitored	155.01	196.18	1.25	49.51	nr	**245.69**
Standard single magnetic lock, unmonitored	124.48	157.54	1.00	39.60	nr	**197.14**
Standard single magnetic lock, double monitored	207.86	263.07	1.50	59.41	nr	**322.47**
Standard double magnetic lock, monitored	310.03	392.38	1.50	59.41	nr	**451.79**
Standard double magnetic lock, unmonitored	243.09	307.65	1.25	49.51	nr	**357.16**
12 V electric release fail safe, monitored	162.07	205.11	1.25	49.51	nr	**254.62**
12 V electric release fail secure, monitored	162.07	205.11	1.00	39.60	nr	**244.72**
Solenoid bolt	306.50	387.91	1.50	59.41	nr	**447.31**
Electric mortice lock	115.09	145.65	1.00	39.60	nr	**185.26**

COMMUNICATIONS/SECURITY/CONTROL

Item	Net Price £	Material £	Labour hours	Labour £	Unit	Total rate £
SECURITY DETECTION AND ALARM						
Detection and alarm systems for the protection of property and persons; includes fixing of equipment to backgrounds and termination of power and data cabling; excludes cable containment and cable installation						
Detection, alarm equipment						
Alarm contact flush	6.96	8.81	1.00	39.60	nr	**48.42**
Alarm contact surface	6.96	8.81	1.00	39.60	nr	**48.42**
Roller shutter contact	22.30	28.23	1.00	39.60	nr	**67.83**
Personal attack button	23.69	29.99	1.00	39.60	nr	**69.59**
Acoustic break glass detectors	32.06	40.57	2.00	79.21	nr	**119.78**
Vibration detectors	41.81	52.91	2.00	79.21	nr	**132.12**
12 metre PIR detector	48.78	61.74	1.50	59.41	nr	**121.15**
15 metre dual detector	80.84	102.31	1.50	59.41	nr	**161.72**
8 zone alarm panel	429.29	543.30	3.00	118.81	nr	**662.12**
8–24 zone end station	421.85	533.89	4.00	158.42	nr	**692.31**
Remote keypad	142.17	179.93	2.00	79.21	nr	**259.13**
8 zone expansion	108.72	137.60	1.50	59.41	nr	**197.00**
Final exit set button	11.59	14.67	1.00	39.60	nr	**54.28**
Self-contained external sounder	104.53	132.30	2.00	79.21	nr	**211.51**
Internal loudspeaker	20.90	26.46	2.00	79.21	nr	**105.67**
Rechargeable battery 12 volt 7 Ahr (ampere hours)	16.73	21.17	0.25	9.90	nr	**31.07**
Surveillance equipment						
Vandal-resistant camera, colour	224.00	283.49	3.00	118.81	nr	**402.31**
External camera, colour	448.00	566.99	5.00	198.02	nr	**765.01**
Auto dome external, colour	1120.00	1417.47	5.00	198.02	nr	**1615.50**
Auto dome external, colour/monochrome	1120.00	1417.47	5.00	198.02	nr	**1615.50**
Auto dome internal, colour	448.00	566.99	4.00	158.42	nr	**725.41**
Auto dome internal colour/monochrome	448.00	566.99	4.00	158.42	nr	**725.41**
Mini internal domes	743.93	941.51	3.00	118.81	nr	**1060.33**
Camera switcher, 32 inputs	1614.50	2043.31	4.00	158.42	nr	**2201.73**
Full function keyboard	303.65	384.29	1.00	39.60	nr	**423.90**
16 CH multiplexors, duplex	894.78	1132.44	2.00	79.21	nr	**1211.65**
16 way DVR, 250 GB harddrive	1120.00	1417.47	3.00	118.81	nr	**1536.29**
Additional 250 GB harddrive	298.67	377.99	3.00	118.81	nr	**496.81**
16 way DVR, 500 GB harddrive	1493.34	3307.44	3.00	118.81	nr	**3426.26**
16 way DVR, 750 GB harddrive	4480.01	5669.90	1.00	39.60	nr	**5709.50**
10" colour monitor	194.13	245.70	2.00	79.21	nr	**324.91**
15" colour monitor	224.00	283.49	2.00	79.21	nr	**362.70**
17" colour monitor, high resolution	268.80	340.19	2.00	79.21	nr	**419.40**
21" colour monitor, high resolution	298.67	377.99	2.00	79.21	nr	**457.20**

COMMUNICATIONS/SECURITY/CONTROL

Item	Net Price £	Material £	Labour hours	Labour £	Unit	Total rate £
SECURITY DETECTION AND ALARM – cont						
Detection and alarm systems for the protection of property and persons – cont						
IP CCTV						
Internal fixed dome IP camera h.264	497.78	629.99	–	–	nr	**629.99**
Internal fixed dome IP dome camera 1MP	497.78	629.99	–	–	nr	**629.99**
External fixed dome IP	560.01	708.74	–	–	nr	**708.74**
External PTZ IP	1991.12	2519.96	–	–	nr	**2519.96**
24 port PoE switch	2177.79	2756.21	–	–	nr	**2756.21**
16 base channel network video recorder with 4TB storage	2488.89	3149.94	–	–	nr	**3149.94**
IP camera channel licence	99.56	126.00	–	–	nr	**126.00**

COMMUNICATIONS/SECURITY/CONTROL

Item	Net Price £	Material £	Labour hours	Labour £	Unit	Total rate £
FIRE DETECTION AND ALARM						
STANDARD FIRE DETECTION						
CONTROL PANEL						
Zone control panel; 16 zone, mild steel case						
Single loop	513.19	649.49	3.00	118.81	nr	**768.30**
Single loop (flush mounted)	540.72	684.34	0.25	9.90	nr	**694.24**
Loop extension card (1 loop)	268.36	339.64	3.51	138.97	nr	**478.60**
Two loop	766.56	970.16	4.00	158.42	nr	**1128.58**
Two loop (flush mounted)	794.10	1005.01	5.00	198.02	nr	**1203.03**
Repeater panels						
Standard network card	445.24	563.50	4.00	158.42	nr	**721.92**
FR – focus repeater	445.24	563.50	4.00	158.42	nr	**721.92**
Focus repeater with controls	491.14	621.59	0.25	9.90	nr	**631.49**
EQUIPMENT						
Manual call point units: plastic covered						
Surface mounted						
Call point	49.01	62.03	0.50	19.80	nr	**81.83**
Call point; weatherproof	177.50	224.65	0.80	31.68	nr	**256.33**
Flush mounted						
Call point	49.01	62.03	0.56	22.19	nr	**84.21**
Call point; weatherproof	177.50	224.65	0.86	34.08	nr	**258.73**
Detectors						
Smoke, ionization type with mounting base	20.01	25.32	0.75	29.71	nr	**55.03**
Smoke, optical type with mounting base	19.28	24.40	0.75	29.71	nr	**54.11**
Fixed temperature heat detector with mounting base (60°C)	22.96	29.06	0.75	29.71	nr	**58.77**
Rate of Rise heat detector with mounting base (90°C)	26.14	33.08	0.75	29.71	nr	**62.79**
Duct detector including optical smoke detector and base	276.36	349.76	2.00	79.21	nr	**428.97**
Remote smoke detector LED indicator with base	6.58	8.33	0.50	19.80	nr	**28.13**
Sounders						
Intelligent wall sounder beacon	103.88	131.46	0.75	29.71	nr	**161.18**
Waterproof intelligent sounder beacon	176.10	222.87	0.75	29.70	nr	**252.57**
Siren; 230 V	86.29	109.21	1.25	49.51	nr	**158.72**
Magnetic door holder; 230 V; surface fixed	81.38	103.00	1.50	59.47	nr	**162.47**

COMMUNICATIONS/SECURITY/CONTROL

Item	Net Price £	Material £	Labour hours	Labour £	Unit	Total rate £
FIRE DETECTION AND ALARM – cont						
ADDRESSABLE FIRE DETECTION						
CONTROL PANEL						
Analogue addressable panel; BS EN54 Part 2 and 4 1998; incorporating 120 addresses per loop (maximum 1–2 km length); sounders wired on loop; sealed lead acid integral battery standby providing 48 hour standby; 24 volt DC; mild steel case; surface fixed						
1 loop; 4 × 12 volt batteries	1717.69	2173.91	6.00	237.63	nr	**2411.54**
Extra for 1 loop panel						
Loop expander card	97.11	122.91	1.00	39.60	nr	**162.51**
Repeater panel	1192.49	1509.22	6.00	237.63	nr	**1746.85**
Network nodes	2040.62	2582.61	6.00	237.63	nr	**2820.24**
Interface unit; for other systems						
Mains powered	611.10	773.41	1.50	59.41	nr	**832.82**
Loop powered	268.23	339.48	1.00	39.60	nr	**379.08**
Single channel I/O	139.40	176.42	1.00	39.60	nr	**216.02**
Zone module	173.37	219.42	1.50	59.41	nr	**278.83**
4 loop; 4 × 12 volt batteries; 24 hour standby; 30 minute alarm	2879.68	3644.52	8.00	316.84	nr	**3961.36**
Extra for 4 loop panel						
Loop card	603.86	764.25	1.00	39.60	nr	**803.85**
Repeater panel	1709.96	2164.12	6.00	237.63	nr	**2401.75**
Mimic panel	4198.15	5313.18	5.00	198.02	nr	**5511.20**
Network nodes	2039.04	2580.61	6.00	237.63	nr	**2818.24**
Interface unit; for other systems						
Mains powered	611.10	773.41	1.50	59.41	nr	**832.82**
Loop powered	268.23	339.48	1.00	39.60	nr	**379.08**
Single channel I/O	139.40	176.42	1.00	39.60	nr	**216.02**
Zone module	173.37	219.42	1.50	59.41	nr	**278.83**
Line modules	31.96	40.45	1.00	39.60	nr	**80.06**
8 loop; 4 × 12 volt batteries; 24 hour standby; 30 minute alarm	5870.36	7429.52	12.00	475.26	nr	**7904.78**
Extra for 8 loop panel						
Loop card	603.86	764.25	1.00	39.60	nr	**803.85**
Repeater panel	1709.96	2164.12	6.00	237.63	nr	**2401.75**
Mimic panel	4198.15	5313.18	5.00	198.02	nr	**5511.20**
Network nodes	2039.04	2580.61	6.00	237.63	nr	**2818.24**
Interface unit; for other systems						
Mains powered	611.10	773.41	1.50	59.41	nr	**832.82**
Loop powered	268.27	339.52	1.00	39.60	nr	**379.12**
Single channel I/O	58.52	74.06	1.00	39.60	nr	**113.66**
Zone module	173.37	219.42	1.50	59.41	nr	**278.83**
Line modules	31.96	40.45	1.00	39.60	nr	**80.06**

COMMUNICATIONS/SECURITY/CONTROL

Item	Net Price £	Material £	Labour hours	Labour £	Unit	Total rate £
EQUIPMENT						
Manual call point						
Surface mounted						
Call point	80.51	101.90	1.00	39.60	nr	**141.50**
Call point; weatherproof	295.04	373.40	1.25	49.51	nr	**422.91**
Flush mounted						
Call point	80.51	101.90	1.00	39.60	nr	**141.50**
Call point; weatherproof	295.04	373.40	1.25	49.51	nr	**422.91**
Detectors						
Smoke, ionization type with mounting base	81.99	103.77	0.75	29.71	nr	**133.48**
Smoke, optical type with mounting base	81.17	102.72	0.75	29.71	nr	**132.44**
Fixed temperature heat detector with mounting base (60°C)	81.58	103.25	0.75	29.71	nr	**132.96**
Rate of rise heat detector with mounting base (90°C)	81.58	103.25	0.75	29.71	nr	**132.96**
Duct detector including optical smoke detector and addressable base	519.88	657.96	2.00	79.21	nr	**737.17**
Beam smoke detector with transmitter and receiver unit	801.48	1014.35	2.00	79.21	nr	**1093.56**
Zone short circuit isolator	51.55	65.24	0.75	29.70	nr	**94.94**
Plant interface unit	37.98	48.07	0.50	19.80	nr	**67.87**
Sounders						
Xenon flasher, 24 volt, conduit box	86.57	109.56	0.50	19.80	nr	**129.36**
Xenon flasher, 24 volt, conduit box; weatherproof	128.86	163.09	0.50	19.80	nr	**182.89**
6" bell, conduit box	207.75	351.81	0.75	29.71	nr	**381.52**
6" bell, conduit box; weatherproof	70.23	88.88	0.75	29.70	nr	**118.59**
Siren; 24 V polarized	87.68	110.96	1.00	39.60	nr	**150.57**
Siren; 240 V	86.29	109.21	1.25	49.51	nr	**158.72**
Magnetic door holder; 240 V; surface fixed	81.38	103.00	1.50	59.47	nr	**162.47**
Addressable wireless fire detection system, BS 5839 and EN54 Part 25 compliant, 868 Mhz operating frequency						
CONTROL PANEL						
Wireless fire alarm panel, complete with back-up battery						
8 zone	574.88	727.57	4.00	158.42	nr	**885.99**
20 zone	775.43	981.38	5.00	198.02	nr	**1179.40**

COMMUNICATIONS/SECURITY/CONTROL

Item	Net Price £	Material £	Labour hours	Labour £	Unit	Total rate £
FIRE DETECTION AND ALARM – cont						
EQUIPMENT						
Smoke detector	153.75	194.59	0.50	19.80	nr	**214.39**
Combined smoke detector and sounder	294.12	372.24	0.50	19.80	nr	**392.04**
Combined smoke detector, sounder and beacon	347.60	439.93	0.50	19.80	nr	**459.73**
Heat detector	147.06	186.12	0.50	19.80	nr	**205.92**
Combined heat detector and sounder	287.45	363.80	0.50	19.80	nr	**383.60**
Combined heat detector, sounder and beacon	334.23	423.00	0.50	19.80	nr	**442.81**
Manual call point	153.75	194.59	0.50	19.80	nr	**214.39**
Manual call point (weatherproof)	360.98	456.85	0.75	29.70	nr	**486.56**
Sounder	207.24	262.28	0.50	19.80	nr	**282.08**
Beacon	254.01	321.48	0.50	19.80	nr	**341.28**
Combined sounder and beacon	260.69	329.94	0.50	19.80	nr	**349.74**
Remote silence button	147.06	186.12	0.50	19.80	nr	**205.92**
Output unit	274.08	346.88	0.50	19.80	nr	**366.68**
Signal booster panel	508.04	642.97	0.50	19.80	nr	**662.78**
Extra over for additional wired antenna	267.39	338.41	1.00	39.60	nr	**378.01**

COMMUNICATIONS/SECURITY/CONTROL

Item	Net Price £	Material £	Labour hours	Labour £	Unit	Total rate £
EARTHING AND BONDING						
Earth bar; polymer insulators and base mounting; including connections						
Non disconnect link						
6 way	204.30	258.57	0.81	32.08	nr	**290.65**
8 way	226.22	286.31	0.81	32.08	nr	**318.39**
10 way	262.58	332.32	0.81	32.08	nr	**364.40**
Disconnect link						
6 way	230.16	291.29	1.01	40.00	nr	**331.29**
8 way	254.46	322.05	1.01	40.00	nr	**362.05**
10 way	282.19	357.14	1.01	40.00	nr	**397.14**
Soild earth bar; including connections						
150 × 50 × 6 mm	62.26	78.80	1.01	40.00	nr	**118.80**
Extra for earthing						
Disconnecting link						
300 × 50 × 6 mm	70.57	89.32	1.16	45.94	nr	**135.26**
500 × 50 × 6 mm	77.65	98.27	1.16	45.94	nr	**144.21**
Crimp lugs; including screws and connections to cable						
25 mm	0.58	0.73	0.31	12.28	nr	**13.01**
35 mm	0.78	0.99	0.31	12.28	nr	**13.27**
50 mm	0.90	1.14	0.32	12.67	nr	**13.82**
70 mm	1.62	2.05	0.32	12.67	nr	**14.73**
95 mm	2.71	3.43	0.46	18.22	nr	**21.65**
120 mm	1.91	2.42	1.25	49.51	nr	**51.93**
Earth clamps; connection to pipework						
15 mm to 32 mm dia.	1.59	2.01	0.15	5.94	nr	**7.95**
32 mm to 50 mm dia.	1.99	2.52	0.18	7.13	nr	**9.65**
50 mm to 75 mm dia.	2.32	2.94	0.20	7.92	nr	**10.86**

COMMUNICATIONS/SECURITY/CONTROL

Item	Net Price £	Material £	Labour hours	Labour £	Unit	Total rate £
LIGHTNING PROTECTION						
CONDUCTOR TAPE						
PVC sheathed copper tape						
25 × 3 mm	25.34	32.07	0.30	11.88	m	**43.95**
25 × 6 mm	47.90	60.63	0.30	11.88	m	**72.51**
50 × 6 mm	97.91	123.91	0.30	11.88	m	**135.79**
PVC sheathed copper solid circular conductor						
8 mm	14.78	18.71	0.50	19.80	m	**38.51**
Bare copper tape						
20 × 3 mm	20.78	26.30	0.30	11.88	m	**38.18**
25 × 3 mm	23.81	30.13	0.30	11.88	m	**42.01**
25 × 6 mm	43.66	55.26	0.40	15.84	m	**71.10**
50 × 6 mm	84.45	106.88	0.50	19.80	m	**126.68**
Bare copper solid circular conductor						
8 mm	13.44	17.01	0.50	19.80	m	**36.81**
Tape fixings; flat; metalic						
PVC sheathed copper						
25 × 3 mm	25.34	32.07	0.33	13.07	nr	**45.14**
25 × 6 mm	47.90	60.63	0.33	13.07	nr	**73.70**
50 × 6 mm	97.91	123.91	0.33	13.07	nr	**136.98**
8 mm	8.49	10.74	0.50	19.80	nr	**30.55**
Bare copper						
20 × 3 mm	21.82	27.61	0.30	11.88	nr	**39.50**
25 × 3 mm	23.81	30.13	0.30	11.88	nr	**42.01**
25 × 6 mm	43.66	55.26	0.40	15.84	nr	**71.10**
50 × 6 mm	84.45	106.88	0.50	19.80	nr	**126.68**
8 mm	14.11	17.86	0.50	19.80	nr	**37.66**
Tape fixings; flat; non-metalic; PVC sheathed copper						
25 × 3 mm	2.96	3.74	0.30	11.88	nr	**15.62**
Tape fixings; flat; non-metalic; bare copper						
20 × 3 mm	1.16	1.47	0.30	11.88	nr	**13.35**
25 × 3 mm	1.16	1.47	0.30	11.88	nr	**13.35**
50 × 6 mm	6.16	7.80	0.30	11.88	nr	**19.68**
Puddle flanges; copper						
600 mm long	134.98	170.83	0.93	36.83	nr	**207.66**
AIR RODS						
Pointed air rod fixed to structure; copper 10 mm dia.						
500 mm long	22.47	28.43	1.00	39.60	nr	**68.04**
1000 mm long	37.87	47.93	1.50	59.41	nr	**107.34**
Extra for						
Air terminal base	28.05	35.50	0.35	13.86	nr	**49.36**
Strike pad	29.79	37.70	0.35	13.86	nr	**51.56**

COMMUNICATIONS/SECURITY/CONTROL

Item	Net Price £	Material £	Labour hours	Labour £	Unit	Total rate £
16 mm dia.						
500 mm long	35.61	45.07	0.91	36.04	nr	81.11
1000 mm long	65.07	82.35	1.75	69.31	nr	151.66
2000 mm long	118.81	150.36	2.50	99.01	nr	249.38
Extra for						
Multiple point	59.82	75.71	0.35	13.86	nr	89.57
Air terminal base	32.87	41.60	0.35	13.86	nr	55.47
Ridge saddle	61.73	78.12	0.35	13.86	nr	91.98
Side mounting bracket	61.43	77.74	0.50	19.80	nr	97.54
Rod to tape coupling	24.78	31.36	0.50	19.80	nr	51.16
Strike pad	29.79	37.70	0.35	13.86	nr	51.56
AIR TERMINALS						
16 mm dia.						
500 mm long	35.61	45.07	0.65	25.74	nr	70.82
1000 mm long	65.07	82.35	0.78	30.89	nr	113.24
2000 mm long	118.81	150.36	1.50	59.41	nr	209.77
Extra for						
Multiple point	59.82	75.71	0.35	13.86	nr	89.57
Flat saddle	61.43	77.74	0.35	13.86	nr	91.60
Side bracket	30.70	38.86	0.50	19.80	nr	58.66
Rod to cable coupling	25.49	32.26	0.50	19.80	nr	52.06
BONDS AND CLAMPS						
Bond to flat surface; copper						
26 mm	5.34	6.76	0.45	17.82	nr	24.58
8 mm dia.	22.02	27.86	0.33	13.07	nr	40.93
Pipe bond						
26 mm	5.34	6.76	0.45	17.82	nr	24.58
8 mm dia.	54.74	69.28	0.33	13.07	nr	82.35
Rod to tape clamp						
26 mm	9.88	12.50	0.45	17.82	nr	30.33
Square clamp; copper						
25 × 3 mm	11.21	14.19	0.33	13.07	nr	27.26
50 × 6 mm	57.42	72.67	0.50	19.80	nr	92.47
8 mm dia.	11.99	15.17	0.33	13.07	nr	28.24
Test clamp; copper						
26 × 8 mm; oblong	16.72	21.16	0.50	19.80	nr	40.96
26 × 8 mm; plate type	50.88	64.39	0.50	19.80	nr	84.20
26 × 8 mm; screw down	45.73	57.88	0.50	19.80	nr	77.68
Cast in earth points						
2 hole	37.16	47.03	0.75	29.70	nr	76.74
4 hole	55.15	69.79	1.00	39.60	nr	109.40
Extra for cast in earth points						
Cover plate; 25 × 3 mm	41.71	52.79	0.25	9.90	nr	62.69
Cover plate; 8 mm	41.71	52.79	0.25	9.90	nr	62.69
Rebar clamp; 8 mm	28.72	36.35	0.25	9.90	nr	46.25
Static earth receptacle	195.42	247.33	0.50	19.80	nr	267.13

COMMUNICATIONS/SECURITY/CONTROL

Item	Net Price £	Material £	Labour hours	Labour £	Unit	Total rate £
LIGHTNING PROTECTION – cont						
BONDS AND CLAMPS – cont						
Copper braided bonds						
25 × 3 mm						
200 mm hole centres	19.30	24.42	0.33	13.07	nr	**37.49**
400 mm holes centres	29.57	37.43	0.40	15.84	nr	**53.27**
U bolt clamps						
16 mm	12.74	16.12	0.33	13.07	nr	**29.19**
20 mm	14.45	18.29	0.33	13.07	nr	**31.36**
25 mm	17.71	22.41	0.33	13.07	nr	**35.48**
EARTH PITS/MATS						
Earth inspection pit; hand to others for fixing						
Concrete	57.31	72.53	1.00	39.60	nr	**112.13**
Polypropylene	56.01	70.88	1.00	39.60	nr	**110.49**
Extra for						
5 hole copper earth bar; concrete pit	111.13	140.65	0.35	13.86	nr	**154.51**
5 hole earth bar; polypropylene	46.55	58.91	0.35	13.86	nr	**72.77**
Waterproof electrode seal						
Single flange	349.27	442.03	0.93	36.83	nr	**478.87**
Double flange	587.70	743.80	0.93	36.83	nr	**780.63**
Earth electrode mat; laid in ground and connected						
Copper tape lattice						
600 × 600 × 3 mm	157.64	199.51	0.93	36.83	nr	**236.34**
900 × 900 × 3 mm	282.68	357.76	0.93	36.83	nr	**394.59**
Copper tape plate						
600 × 600 × 1.5 mm	136.57	172.84	0.93	36.83	nr	**209.67**
600 × 600 × 3 mm	270.82	342.76	0.93	36.83	nr	**379.59**
900 × 900 × 1.5 mm	303.66	384.31	0.93	36.83	nr	**421.14**
900 × 900 × 3 mm	588.93	745.35	0.93	36.83	nr	**782.18**
EARTH RODS						
Solid cored copper earth electrodes driven into ground and connected						
15 mm dia.						
1200 mm long	55.48	70.22	0.93	36.83	nr	**107.05**
Extra for						
Coupling	1.68	2.12	0.06	2.38	nr	**4.50**
Driving stud	2.06	2.61	0.06	2.38	nr	**4.99**
Spike	1.90	2.41	0.06	2.38	nr	**4.78**
Rod clamp; flat tape	9.88	12.50	0.25	9.90	nr	**22.40**
Rod clamp; solid conductor	4.42	5.59	0.25	9.90	nr	**15.49**
20 mm dia.						
1200 mm long	107.62	136.20	0.98	38.81	nr	**175.01**
Extra for						
Coupling	1.68	2.12	0.06	2.38	nr	**4.50**
Driving stud	3.38	4.27	0.06	2.38	nr	**6.65**
Spike	3.11	3.93	0.06	2.38	nr	**6.31**
Rod clamp; flat tape	9.88	12.50	0.25	9.90	nr	**22.40**
Rod clamp; solid conductor	4.89	6.19	0.25	9.90	nr	**16.09**

COMMUNICATIONS/SECURITY/CONTROL

Item	Net Price £	Material £	Labour hours	Labour £	Unit	Total rate £
Stainless steel earth electrodes driven into ground and connected						
16 mm dia.						
1200 mm long	70.87	89.70	0.93	36.83	nr	**126.53**
Extra for						
Coupling	2.16	2.73	0.06	2.38	nr	**5.11**
Driving head	2.06	2.61	0.06	2.38	nr	**4.99**
Spike	1.90	2.41	0.06	2.38	nr	**4.78**
Rod clamp; flat tape	9.88	12.50	0.25	9.90	nr	**22.40**
Rod clamp; solid conductor	4.42	5.59	0.25	9.90	nr	**15.49**
SURGE PROTECTION						
Single phase; including connection to equipment						
90–150 V	405.08	512.67	5.00	198.02	nr	**710.69**
200–280 V	384.02	486.02	5.00	198.02	nr	**684.04**
Three phase; including connection to equipment						
156–260 V	817.59	1034.74	10.00	396.05	nr	**1430.79**
346–484 V	761.84	964.19	10.00	396.05	nr	**1360.23**
349–484 V; remote display	847.33	1072.38	10.00	396.05	nr	**1468.42**
346–484 V; 60 kA	1407.25	1781.02	10.00	396.05	nr	**2177.06**
346–484 V; 120 kA	2691.85	3406.81	10.00	396.05	nr	**3802.86**

COMMUNICATIONS/SECURITY/CONTROL

Item	Net Price £	Material £	Labour hours	Labour £	Unit	Total rate £
CENTRAL CONTROL/BUILDING MANAGEMENT						
EQUIPMENT						
Switches/sensors; includes fixing in position; electrical work elsewhere. **Note:** These are normally free issued to the mechanical contractor for fitting. The labour times applied assume the installation has been prepared for the fitting of the component.						
Pressure devices						
Liquid differential pressure sensor	191.64	242.54	0.50	19.80	nr	262.34
Liquid differential pressure switch	105.02	132.91	0.50	19.80	nr	152.71
Air differential pressure transmitter	138.60	175.41	0.50	19.80	nr	195.21
Air differential pressure switch	18.94	23.97	0.50	19.80	nr	43.77
Liquid level switch	95.63	121.03	0.50	19.80	nr	140.83
Static pressure sensor	108.29	137.05	0.50	19.80	nr	156.86
High pressure switch	95.63	121.03	0.50	19.80	nr	140.83
Low pressure switch	95.63	121.03	0.50	19.80	nr	140.83
Water pressure switch	146.58	185.51	0.50	19.80	nr	205.31
Duct averaging temperature sensor	63.76	80.70	1.00	39.60	nr	120.30
Temperature devices						
Return air sensor (fan coils)	6.24	7.90	1.00	39.60	nr	47.50
Frost thermostat	40.89	51.76	0.50	19.80	nr	71.56
Immersion thermostat	66.18	83.76	0.50	19.80	nr	103.56
Temperature high limit	66.18	83.76	0.50	19.80	nr	103.56
Temperature sensor with averaging element	63.76	80.70	0.50	19.80	nr	100.50
Immersion temperature sensor	27.08	34.27	0.50	19.80	nr	54.07
Space temperature sensor	7.84	9.93	1.00	39.60	nr	49.53
Combined space temperature & humidity sensor	108.29	137.05	1.00	39.60	nr	176.66
Outside air temperature sensor	15.16	19.19	2.00	79.21	nr	98.39
Outside air temperature & humidity sensor	355.15	449.48	2.00	79.21	nr	528.69
Duct humidity sensor	139.69	176.80	0.50	19.80	nr	196.60
Space humidity sensor	108.29	137.05	1.00	39.60	nr	176.66
Immersion water flow sensor	27.08	34.27	0.50	19.80	nr	54.07
Rain sensor	253.38	320.68	2.00	79.21	nr	399.89
Wind speed and direction sensor	801.29	1014.11	2.00	79.21	nr	1093.32
Controllers; includes fixing in position; electrical work elsewhere						
Zone						
Fan coil controller	204.65	259.01	2.00	79.21	nr	338.22
VAV controller	267.62	338.69	2.00	79.21	nr	417.90

COMMUNICATIONS/SECURITY/CONTROL

Item	Net Price £	Material £	Labour hours	Labour £	Unit	Total rate £
Plant						
Controller, 96 I/O points (exact configuration is dependent upon the number of I/O boards added)	3756.23	4753.89	0.50	19.80	nr	**4773.69**
Controller, 48 I/O points (exact configuration is dependent upon the number of I/O boards added)	3135.80	3968.67	0.50	19.80	nr	**3988.47**
Controller, 32 I/O points (exact configuration is dependent upon the number of I/O boards added)	2604.06	3295.70	0.50	19.80	nr	**3315.50**
Additional Digital Input Boards (12 DI)	620.44	785.23	0.20	7.92	nr	**793.15**
Additional Digital Output Boards (6 DO)	530.58	671.50	0.20	7.92	nr	**679.42**
Additional Analogue Input Boards (8 AI)	530.58	671.50	0.20	7.92	nr	**679.42**
Additional Analogue Output Boards (8 AO)	530.58	671.50	0.20	7.92	nr	**679.42**
Outstation Enclosure (fitted in riser with space allowance for controller and network device)	530.58	671.50	5.00	198.02	nr	**869.53**
Damper actuator; electrical work elsewhere						
Damper actuator 0–10 V	112.62	142.53	1.00	39.60	nr	**182.13**
Damper actuator with auxiliary switches	74.71	94.56	1.00	39.60	nr	**134.16**
Frequency inverters: not mounted within MCC; includes fixing in position; electrical work elsewhere						
2.2 kW	856.92	1084.52	2.00	79.21	nr	**1163.73**
3 kW	947.53	1199.19	2.00	79.21	nr	**1278.40**
7.5 kW	1401.99	1774.35	2.00	79.21	nr	**1853.56**
11 kW	1713.83	2169.02	2.00	79.21	nr	**2248.23**
15 kW	2070.99	2621.04	2.00	79.21	nr	**2700.25**
18.5 kW	2420.15	3062.95	2.50	99.01	nr	**3161.96**
20 kW	2732.00	3457.62	2.50	99.01	nr	**3556.64**
30 kW	3310.39	4189.63	2.50	99.01	nr	**4288.64**
55 kW	5978.41	7566.28	3.00	118.81	nr	**7685.09**
Miscellaneous; includes fixing in position; electrical work elsewhere						
1 kW Thyristor	76.30	96.57	2.00	79.21	nr	**175.78**
10 kW Thyristor	146.92	185.94	2.00	79.21	nr	**265.15**
Front end and networking; electrical work elsewhere						
PC/monitor	1132.78	1433.65	2.00	79.21	nr	**1512.86**
Printer	333.17	421.66	2.00	79.21	nr	**500.87**
PC software	–	–	–	–	nr	**–**
Network server software	2724.00	3447.50	–	–	nr	**3447.50**
Router (allows connection to a network)	137.27	173.73	–	–	nr	**173.73**

COMMUNICATIONS/SECURITY/CONTROL

Item	Net Price £	Material £	Labour hours	Labour £	Unit	Total rate £
WI-FI						
Wi-Fi						
Zone director controller, 1000 Mbps, RJ 45, 2 ports, auto MDX and suto sensing						
Up to 2,000 clients	1011.07	1279.61	4.00	158.42	nr	**1438.02**
Up to 5,000 clients, 500 access points	6997.79	8856.40	4.00	158.42	nr	**9014.82**
Up to 20,000 clients, 1,000 access points	8302.29	10507.38	4.00	158.42	nr	**10665.80**
Wi-Fi access points, suitable for large developments (commercial buildings, stadiums, hotels, education)						
Indoor access points, 802.11ac Wi-Fi, dual band, 2.4 GHz and 5 GHz						
Up to 100 connected devices	337.44	427.06	1.20	47.53	nr	**474.59**
Up to 100 connected devices, 867 Mbps	456.97	578.35	1.20	47.53	nr	**625.87**
Up to 300 connected devices, 867 Mbps	595.19	753.27	1.20	47.53	nr	**800.80**
Up to 400 connected devices, 1300 Mbps	733.39	928.18	1.20	47.53	nr	**975.71**
Up to 500 connected devices, 1300 Mbps	917.69	1161.42	1.20	47.53	nr	**1208.95**
Up to 500 connected devices, 1733 Mbps	1207.81	1528.60	1.20	47.53	nr	**1576.13**
Outdoor access points, 802.11ac Wi-Fi, dual band, 2.4 GHz and 5 GHz, IP67, plastic enclosure, with flexible wall or pole mounting						
Standard	1194.10	1511.26	1.80	71.29	nr	**1582.55**
Up to 500 Mbps	2203.94	2789.31	1.80	71.29	nr	**2860.60**
Wi-Fi access points, suitable for small/ medium sized and residential						
Indoor access points, 802.11ac Wi-Fi, dual band, 2.4 GHz and 5 GHz						
Up to 100 connected devices, 867 Mbps	456.97	578.35	1.20	47.53	nr	**625.87**
Up to 300 connected devices, 567 Mbps	595.19	753.27	1.20	47.53	nr	**800.80**
Up to 400 connected devices, 1300 Mbps	733.39	928.18	1.20	47.53	nr	**975.71**
Outdoor access points, 802.11ac Wi-Fi, dual band, 2.4 GHz and 5 GHz, IP67, plastic enclosure, with flexible wall or pole mounting						
Standard	1194.10	1511.26	1.80	71.29	nr	**1582.55**

PART 5

Rates of Wages

Page

Metric Handbook: Planning and Design Data, 7th edition

Edited by Pamela Buxton

- Consistently updated since 2015 by expert authors in the field
- Significantly revised in reference to changing building types and construction standards
- New chapters added on data centres and logistics facilities
- Sustainable design integrated into chapters throughout
- Over 100,000 copies sold to successive generations of architects and designers
- This book belongs in every design office.

The *Metric Handbook* is the major handbook of planning and design data for architects and architecture students. Covering basic design data for all the major building types it is the ideal starting point for any project. For each building type, the book gives the basic design requirements and all the principal dimensional data, and succinct guidance on how to use the information and what regulations the designer needs to be aware of.

As well as buildings, the *Metric Handbook* deals with broader aspects of design such as materials, acoustics and lighting, and general design data on human dimensions and space requirements. The *Metric Handbook* is the unique reference for solving everyday planning problems.

November 2021: 880 pp
ISBN: 9780367511395

To Order
Tel:+44 (0) 1235 400524
Email: tandf@bookpoint.co.uk

For a complete listing of all our titles visit:
www.tandf.co.uk

Taylor & Francis
Taylor & Francis Group

Mechanical Installations

Rates of Wages

HEATING, VENTILATING, AIR CONDITIONING, PIPING AND DOMESTIC ENGINEERING INDUSTRY

For full details of the wage agreement and the Heating Ventilating Air Conditioning Piping and Domestic Engineering Industry's National Working Rule Agreement, contact:

Building & Engineering Services Association
Lincoln House
137 – 143 Hammersmith Road
London W14 0QL
Tel: 020 7313 4900
Internet: www.b-es.org

WAGE RATES, ALLOWANCES AND OTHER PROVISIONS

Hourly rates of wages

All districts of the United Kingdom

Main Grades	From 4 October 2021 p/hr
Foreman	18.05
Senior Craftsman (+2nd welding skill)	16.11
Senior Craftsman	14.92
Craftsman (+2nd welding skill)	14.92
Craftsman	13.74
Operative	12.40
Adult Trainee	10.46
Mate (18 and over)	10.46
Modern Apprentices	
Junior Apprentice	6.78
Intermediate Apprentice	9.62
Senior Apprentice	12.40

Note: Ductwork Erection Operatives are entitled to the same rates and allowances as the parallel Fitter grades shown.

HEATING, VENTILATING, AIR CONDITIONING, PIPING AND DOMESTIC ENGINEERING INDUSTRY

Trainee *Rates of Pay*

Junior Ductwork Trainees (Probationary)

Age at entry	From 4 October 2021 p/hr
17	6.15
18	6.15
19	6.15
20	6.15

Junior Ductwork Erectors (Year of Training)

	From 4 October 2021		
	1 yr	*2 yr*	*3 yr*
Age at entry	p/hr	p/hr	p/hr
17	7.66	9.54	10.81
18	7.66	9.54	10.81
19	7.66	9.54	10.81
20	7.66	9.54	10.81

Responsibility Allowance (Craftsmen)	From 4 October 2021 p/hr
Second welding skill or supervisory responsibility (one unit)	0.59
Second welding skill and supervisory responsibility (two units)	0.59

Responsibility Allowance (Senior Craftsmen)	From 4 October 2021 p/hr
Second welding skill	0.59
Supervising responsibility	1.18
Second welding skill and supervisory responsibility	0.59

Daily travelling allowance – Scale 2

C: Craftsmen including Installers

M&A: Mates, Apprentices and Adult Trainees

Direct distance from centre to job in miles

		From 4 October 2021	
Over	*Not exceeding*	*C p/hr*	*p/hr M&A*
15	20	2.88	2.46
20	30	7.37	6.39
30	40	10.62	9.18
40	50	13.99	11.95

HEATING, VENTILATING, AIR CONDITIONING, PIPING AND DOMESTIC ENGINEERING INDUSTRY

Daily travelling allowance – Scale 1

C: Craftsmen including Installers

M&A: Mates, Apprentices and Adult Trainees

Direct distance from centre to job in miles

		From 4 October 2021	
Over	*Not exceeding*	*C p/hr*	*M&A p/hr*
0	15	7.84	7.84
15	20	10.72	10.31
20	30	15.23	14.22
30	40	18.46	17.04
40	50	21.82	19.83

Weekly Holiday Credit and Welfare Contributions

	From 7 October 2019						
	£	£	£	£	£	£	£
	a	b	c	d	e	f	g
Weekly Holiday Credit	82.43	76.25	73.55	70.84	68.12	65.45	62.78
Combined Weekly/Welfare Holiday Credit and Contribution	92.08	85.90	83.20	80.49	77.77	75.10	72.43

	From 7 October 2019			
	£	£	£	£
	h	i	j	K
Weekly Holiday Credit	56.63	47.74	43.92	30.99
Combined Weekly/Welfare Holiday Credit and Contribution	66.28	57.39	53.57	40.64

HEATING, VENTILATING, AIR CONDITIONING, PIPING AND DOMESTIC ENGINEERING INDUSTRY

The grades of H&V Operatives entitled to the different rates of Weekly Holiday Credit and Welfare Contribution are as follows:

a Foreman	b Senior Craftsman (RAS & RAW)	c Senior Craftsman (RAS)	d Senior Craftsman (RAW)
e Senior Craftsman Craftsman (+2 RA)	f Craftsman (+1 RA)	g Craftsman	h Installer Senior Modern Apprentice
i Adult Trainee Mate (over 18)	j Intermediate Modern Apprentice	k Junior Modern Apprentice	l No grade allocated to this Credit Value Category

Daily abnormal conditions money	*From 4 October 2021*
Per day	3.52

Lodging allowance	*From 1 October 2018*
Per night	40.40

Explanatory Notes

1.   Working Hours

    The normal working week (Monday to Friday) shall be 37.5 hours.

2.   Overtime

    Time worked in excess of 37.5 hours during the normal working week shall be paid at time and a half until 12 hours have been worked since the actual starting time. Thereafter double time shall be paid until normal starting time the following morning. Weekend overtime shall be paid at time and a half for the first 5 hours worked on a Saturday and at double time thereafter until normal starting time on Monday morning.

3.   Lodging Allowance

    At the time of review, the rate of nightly Lodging Allowance for the periods from 4 October 2021 and 3 October 2022 are still subject to HMRC approval.

PLUMBING MECHANICAL ENGINEERING SERVICES INDUSTRY

The Joint Industry Board for Plumbing Mechanical Engineering Services has agreed a two year wage agreement for 2022 and 2023 with effect from 3 January 2022 and subsequently from 2 January 2023.

For full details of this wage agreement and the JIB PMES National Working Rules, contact:

The Joint Industry Board for Plumbing Mechanical Engineering Services in England and Wales
JIB-PMES
PO Box 267
PE19 9DN
Tel: 01480 476925
Email: info@jib-pmes.org.uk

WAGE RATES, ALLOWANCES AND OTHER PROVISIONS

EFFECTIVE FROM 3 January 2022

Basic Rates of Hourly Pay

Applicable in England and Wales

	Hourly rate £
Operatives	
Technical plumber and gas service technician	18.06
Advanced plumber and gas service engineer	16.26
Trained plumber and gas service fitter	13.96
Apprentices	
4th year of training with NVQ level 3	13.50
4th year of training with NVQ level 2	12.23
4th year of training	10.76
3rd year of training with NVQ level 2	10.63
3rd year of training	8.75
2nd year of training	7.76
1st year of training	6.75
Adult Trainees	
3rd 6 months of employment	12.16
2nd 6 months of employment	11.68
1st 6 months of employment	10.39

PLUMBING MECHANICAL ENGINEERING SERVICES INDUSTRY

Major Projects Agreement

Where a job is designated as being a Major Project then the following Major Project Performance Payment hourly rate supplement shall be payable:

Employee Category

	National Payment £	London* Payment £
Technical Plumber and Gas Service Technician	2.20	3.57
Advanced Plumber and Gas Service Engineer	2.20	3.57
Trained Plumber and Gas Service Fitter	2.20	3.57
All 4th year apprentices	1.76	2.86
All 3rd year apprentices	1.32	2.68
2nd year apprentice	1.21	1.96
1st year apprentice	0.88	1.43
All adult trainees	1.76	2.86

* The London Payment Supplement applies only to designated Major Projects that are within the M25 London orbital motorway and are effective from 1 February 2007.

* National payment hourly rates are unchanged for 2007 and will continue to be at the rates shown in Promulgation 138 A issued 14 October 2003.

PLUMBING MECHANICAL ENGINEERING SERVICES INDUSTRY

Allowances

Daily travel time allowance plus return fares

All daily travel allowances are to be paid at the daily rate as follows:

Over	Not exceeding	All Operatives	3rd & 4th Year Apprentices	1st & 2nd Year Apprentices
20	30	£5.11	3.28	2.05
30	40	11.91	7.67	4.92
40	50	13.62	8.13	5.11

Responsibility/Incentive Pay Allowance

As from 3 September 2003, Employers may, in consultation with the employees concerned, enhance the basic graded rates of pay by the payment of an additional amount, as per the bands shown below, where it is agreed that their work involves extra responsibility, productivity or flexibility.

Band 1   an additional rate of up to £ 0.33 per hour
Band 2   an additional rate of up to £ 0.55 per hour
Band 3   an additional rate of up to £ 0.82 per hour
Band 4   an additional rate of up to £ 1.07 per hour

This allowance forms part of an operative's basic rate of pay and shall be used to calculate premium payments.

Mileage allowance	£0.45 per mile
Lodging allowance	£42.00 per night
Subsistence allowance (London only)	£5.87 per night

Plumbers welding supplement

Possession of Gas or Arc Certificate	£0.36 per hour
Possession of Gas and Arc Certificate	£0.57 per hour

PLUMBING MECHANICAL ENGINEERING SERVICES INDUSTRY

JIB-PMES Additional Holiday Pay (AHP) (From 3 January 2022)

All PHMES Operatives, Apprentices etc. who are in current membership of 'Unite the Union' at the time a holiday is taken, shall also be entitled to receive an additional payment of AHP from the JIB-PMES – to be paid via their employer – for each Credit funding their holiday pay. 62 Credits is the maximum number payable in a year.

The amount of AHP payable per credit for all holidays is as follows:

From January 2019

Operatives etc. £2.61 per credit (Max – £172.26)

All Apprentices £1.32 per credit (Max – £87.04)

Explanatory Notes

1.   Working Hours

The normal working week (Monday to Friday) shall be 37½ hours, with 45 hours to be worked in the same period before overtime rates become applicable.

2.   Overtime

Overtime shall be paid at time and a half up to 8.00pm (Monday to Friday) and up to 1.00pm (Saturday). Overtime worked after these times shall be paid at double time.

3.   Major Projects Agreement

Under the Major Projects Agreement the normal working week shall be 38 hours (Monday to Friday) with overtime rates payable for all hours worked in excess of 38 hours in accordance with 2 above. However, it should be noted that the hourly rate supplement shall be paid for each hour worked but does not attract premium time enhancement.

4.   Pension

In addition to their hourly rates of pay, plumbing employees are entitled to inclusion within the Industry Pension Scheme (or one providing equivalent benefits). The current levels of industry scheme contributions are 6½% (employers) and 3¼% (employees).

5.   Additional Holiday Pay (AHP) Contributions

As last year, JIB-PMES Holiday Pay Schemes shall apply for the Holiday Credit Period 2021–2022, with 66 credits being the maximum number payable in a year.

For full details please refer to the Additional Holiday Pay (AHP) section as published by the JIB for Plumbing Mechanical Engineering Services in England and Wales.

Electrical Installations

Rates of Wages

ELECTRICAL CONTRACTING INDUSTRY

For full details of this wage agreement and the Joint Industry Board for the Electrical Contracting Industry's National Working Rules, contact:

Joint Industry Board
PO Box 127
Swanley
Kent
BR8 9BH
Tel: 0333 321 230
Internet: www.jib.org.uk

WAGES (Graded Operatives)

Rates

Since 7 January 2002 two different wage rates have applied to JIB Graded Operatives working on site, depending on whether the Employer transports them to site or whether they provide their own transport. The two categories are:

Job Employed (Transport Provided)

Payable to an Operative who is transported to and from the job by his Employer. The Operative shall also be entitled to payment for Travel Time, when travelling in his own time, as detailed in the appropriate scale.

Job Employed (Own Transport)

Payable to an Operative who travels by his own means to and from the job. The Operative shall be entitled to payment for Travel Allowance and also Travel Time, when travelling in his own time, as detailed in the appropriate scale.

ELECTRICAL CONTRACTING INDUSTRY

The JIB rates of wages are set out below:

From and including 3 January 2022, the JIB hourly rates of wages shall be as set out below:

(i)   **National Standard Rate**

Grade	Transport Provided	Own Transport
Technician (or equivalent specialist grade)	£ 18.97	£ 19.94
Approved Electrician (or equivalent specialist grade)	£ 16.78	£ 17.70
Electrician (or equivalent specialist grade)	£ 15.36	£ 16.32
Senior Graded Electrical Trainee	£ 14.62	£ 15.51
Electrical Improver	£ 13.81	£ 14.71
Labourer	£ 12.20	£ 13.10
Adult Trainee	£ 12.20	£ 13.10

(ii)   **London Rate**

Grade	Transport Provided	Own Transport
Technician (or equivalent specialist grade)	£ 21.27	£ 22.31
Approved Electrician (or equivalent specialist grade)	£ 18.78	£ 19.82
Electrician (or equivalent specialist grade)	£ 17.20	£ 18.29
Senior Graded Electrical Trainee	£ 16.33	£ 17.38
Electrical Improver	£ 15.47	£ 16.47
Labourer	£ 13.68	£ 14.68
Adult Trainee	£ 13.68	£ 14.68

From and including 3 January 2022, the JIB hourly rates for Job Employed apprentices shall be:

(i)   **National Standard Rates**

	At College	At Work
Stage 1	£ 4.92	£ 5.59
Stage 2	£ 7.10	£ 7.89
Stage 3	£ 10.35	£ 11.31
Stage 4	£ 11.02	£ 12.15

(ii)   **London Rate**

	At College	At Work
Stage 1	£ 5.51	£ 6.26
Stage 2	£ 8.02	£ 8.86
Stage 3	£ 11.59	£ 12.66
Stage 4	£ 12.35	£ 13.62

ELECTRICAL CONTRACTING INDUSTRY

Travelling Time and Travel Allowances

From and including 3 January 2022

From Monday 2 January 2023, in line with National Working Rule 11.2 the calculation of the actual distance between the Shop and the Job will be by the fastest route [replacing the shortest route].

Lodging allowances

£42.66 from and including 3 January 2022

Lodgings weekend retention fee, maximum reimbursement

£42.66 from and including 3 January 2022

Annual holiday lodging allowance retention

Maximum £14.04 per night (£ 98.28 per week) from and including 3 January 2022

Responsibility money

From and including 30 March 1998 the minimum payment increased to 10p per hour and the maximum to £1.00 per hour (no change)

From and including 4 January 1992 responsibility payments are enhanced by overtime and shift premiums where appropriate (no change)

Combined JIB Benefits Stamp Value (from week commencing 6 January 2020)

JIB grade	Weekly JIB combined £	Holiday value £
Technician	£ 66.93	£ 59.61
Approved Electrician	£ 59.70	£ 52.71
Electrician	£ 55.43	£ 48.27
Labourer & Adult Trainee	£ 44.93	£ 38.31

Explanatory Notes

1.   Working Hours

The normal working week (Monday to Friday) shall be 37½ hours, with 38 hours to be worked in the same period before overtime rates become applicable.

2.   Overtime

Overtime shall be paid at time and a half for all weekday overtime. Saturday overtime shall be paid at time and a half for the first 6 hours, or up to 3.00pm (whichever comes first). Thereafter double time shall be paid until normal starting time on Monday.

NEC4: 100 Questions and Answers

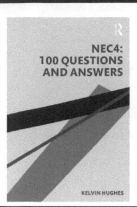

Kelvin Hughes

This book details some of the most important and interesting questions raised about the NEC4 family of contracts and provides clear, comprehensive answers to those questions.

Written by an NEC expert with over 20 years' experience using, advising and training others, the book has several distinctive features:

- It covers the *whole* NEC4 family
- It is written by a very experienced NEC author who explains sometimes complex issues in a simple and accessible style
- The questions and answers range from beginner level up to a masterclass level
- The questions are real life questions asked by actual NEC practitioners on real projects.

The book includes questions and answers relating to tendering, early warnings, programme issues, quality management, payment provisions, compensation events, liabilities, insurances, adjudication, termination and much more. It is essential reading for anyone working with the NEC4 family of contracts, whether professionals or students in construction, architecture, project management and engineering.

May 2019: 246pp
ISBN: 9781138365254

To Order
Tel: +44 (0) 1235 400524
Email: book.orders@tandf.co.uk

For a complete listing of all our titles
visit: www.tandf.co.uk

Taylor & Francis
Taylor & Francis Group

PART 6

Daywork

When work is carried out in connection with a contract that cannot be valued in any other way, it is usual to assess the value on a cost basis with suitable allowances to cover overheads and profit. The basis of costing is a matter for agreement between the parties concerned but definitions of prime cost for the Heating and Ventilating and Electrical Industries have been published jointly by the Royal Institution of Chartered Surveyors and the appropriate bodies of the industries concerned, for those who wish to use them.

These, together with a schedule of basic plant hire charges are reproduced on the following pages, with the kind permission of the Royal Institution of Chartered Surveyors, who own the copyright.

Net Zero Energy Building

Ming Hu

What do we mean by net zero energy? Zero operating energy? Zero energy costs? Zero emissions? There is no one answer: approaches to net zero building vary widely across the globe and are influenced by different environmental and cultural contexts.

Net Zero Energy Building: Predicted and Unintended Consequences presents a comprehensive overview of variations in 'net zero' building practices. Drawing on examples from countries such as the United States, United Kingdom, Germany, Japan, Hong Kong, and China, Ming Hu examines diverse approaches to net zero and reveals their intended and unintended consequences.

Existing approaches often focus on operating energy: how to make buildings more efficient by reducing the energy consumed by climate control, lighting, and appliances. Hu goes beyond this by analyzing overall energy consumption and environmental impact across the entire life cycle of a building—ranging from the manufacture of building materials to transportation, renovation, and demolition. Is net zero building still achievable once we look at these factors?

With clear implications for future practice, this is key reading for professionals in building design, architecture, and construction, as well as students on sustainable and green architecture courses.

April 2019: 162 pp
ISBN: 9780815367802

To Order
Tel: +44 (0) 1235 400524
Email: book.orders@tandf.co.uk

For a complete listing of all our titles
visit: www.tandf.co.uk

Taylor & Francis
Taylor & Francis Group

HEATING AND VENTILATING INDUSTRY

DEFINITION OF PRIME COST OF DAYWORK CARRIED OUT UNDER A HEATING, VENTILATING, AIR CONDITIONING, REFRIGERATION, PIPEWORK AND/OR DOMESTIC ENGINEERING CONTRACT (JULY 1980 EDITION)

This Definition of Prime Cost is published by the Royal Institution of Chartered Surveyors and the Heating and Ventilating Contractors Association for convenience, and for use by people who choose to use it. Members of the Heating and Ventilating Contractors Association are not in any way debarred from defining Prime Cost and rendering accounts for work carried out on that basis in any way they choose. Building owners are advised to reach agreement with contractors on the Definition of Prime Cost to be used prior to entering into a contract or subcontract.

SECTION 1: APPLICATION

1.1 This Definition provides a basis for the valuation of daywork executed under such heating, ventilating, air conditioning, refrigeration, pipework and or domestic engineering contracts as provide for its use.
1.2 It is not applicable in any other circumstances, such as jobbing or other work carried out as a separate or main contract nor in the case of daywork executed after a date of practical completion.
1.3 The terms 'contract' and 'contractor' herein shall be read as 'subcontract' and 'subcontractor' as applicable.

SECTION 2: COMPOSITION OF TOTAL CHARGES

2.1 The Prime Cost of daywork comprises the sum of the following costs:
   (a) Labour as defined in Section 3.
   (b) Materials and goods as defined in Section 4.
   (c) Plant as defined in Section 5.
2.2 Incidental costs, overheads and profit as defined in Section 6, as provided in the contract and expressed therein as percentage adjustments, are applicable to each of 2.1 (a)–(c).

SECTION 3: LABOUR

3.1 The standard wage rates, emoluments and expenses referred to below and the standard working hours referred to in 3.2 are those laid down for the time being in the rules or decisions or agreements of the Joint Conciliation Committee of the Heating, Ventilating and Domestic Engineering Industry applicable to the works (or those of such other body as may be appropriate) and to the grade of operative concerned at the time when and the area where the daywork is executed.
3.2 Hourly base rates for labour are computed by dividing the annual prime cost of labour, based upon the standard working hours and as defined in 3.4, by the number of standard working hours per annum. See example.
3.3 The hourly rates computed in accordance with 3.2 shall be applied in respect of the time spent by operatives directly engaged on daywork, including those operating mechanical plant and transport and erecting and dismantling other plant (unless otherwise expressly provided in the contract) and handling and distributing the materials and goods used in the daywork.
3.4 The annual prime cost of labour comprises the following:
   (a) Standard weekly earnings (i.e. the standard working week as determined at the appropriate rate for the operative concerned).
   (b) Any supplemental payments.
   (c) Any guaranteed minimum payments (unless included in Section 6.1 (a)–(p)).
   (d) Merit money.
   (e) Differentials or extra payments in respect of skill, responsibility, discomfort, inconvenience or risk (excluding those in respect of supervisory responsibility – see 3.5)
   (f) Payments in respect of public holidays.
   (g) Any amounts which may become payable by the contractor to or in respect of operatives arising from the rules etc. referred to in 3.1 which are not provided for in 3.4 (a)–(f) nor in Section 6.1 (a)–(p).
   (h) Employers contributions to the WELPLAN, the HVACR Welfare and Holiday Scheme or payments in lieu thereof.
   (i) Employers National Insurance contributions as applicable to 3.4 (a)–(h).
   (j) Any contribution, levy or tax imposed by Statute, payable by the contractor in his capacity as an employer.

HEATING AND VENTILATING INDUSTRY

3.5   Differentials or extra payments in respect of supervisory responsibility are excluded from the annual prime cost (see Section 6). The time of principals, staff, foremen, chargehands and the like when working manually is admissible under this Section at the rates for the appropriate grades.

SECTION 4: MATERIALS AND GOODS

4.1   The prime cost of materials and goods obtained specifically for the daywork is the invoice cost after deducting all trade discounts and any portion of cash discounts in excess of 5%.
4.2   The prime cost of all other materials and goods used in the daywork is based upon the current market prices plus any appropriate handling charges.
4.3   The prime cost referred to in 4.1 and 4.2 includes the cost of delivery to site.
4.4   Any Value Added Tax which is treated, or is capable of being treated, as input tax (as defined by the Finance Act 1972, or any re-enactment or amendment thereof or substitution therefore) by the contractor is excluded.

SECTION 5: PLANT

5.1   Unless otherwise stated in the contract, the prime cost of plant comprises the cost of the following:
        (a)   Use or hire of mechanically-operated plant and transport for the time employed on and/or provided or retained for the daywork;
        (b)   Use of non-mechanical plant (excluding non-mechanical hand tools) for the time employed on and/or provided or retained for the daywork;
        (c)   Transport to and from the site and erection and dismantling where applicable.
5.2   The use of non-mechanical hand tools and of erected scaffolding, staging, trestles or the like is excluded (see Section 6), unless specifically retained for the daywork.

SECTION 6: INCIDENTAL COSTS, OVERHEADS AND PROFIT

6.1   The percentage adjustments provided in the contract which are applicable to each of the totals of Sections 3, 4 and 5 comprise the following:
        (a)   Head office charges.
        (b)   Site staff including site supervision.
        (c)   The additional cost of overtime (other than that referred to in 6.2).
        (d)   Time lost due to inclement weather.
        (e)   The additional cost of bonuses and all other incentive payments in excess of any included in 3.4.
        (f)   Apprentices' study time.
        (g)   Fares and travelling allowances.
        (h)   Country, lodging and periodic allowances.
        (i)   Sick pay or insurances in respect thereof, other than as included in 3.4.
        (j)   Third party and employers' liability insurance.
        (k)   Liability in respect of redundancy payments to employees.
        (l)   Employer's National Insurance contributions not included in 3.4.
        (m)   Use and maintenance of non-mechanical hand tools.
        (n)   Use of erected scaffolding, staging, trestles or the like (but see 5.2).
        (o)   Use of tarpaulins, protective clothing, artificial lighting, safety and welfare facilities, storage and the like that may be available on site.
        (p)   Any variation to basic rates required by the contractor in cases where the contract provides for the use of a specified schedule of basic plant charges (to the extent that no other provision is made for such variation – see 5.1).
        (q)   In the case of a subcontract which provides that the subcontractor shall allow a cash discount, such provision as is necessary for the allowance of the prescribed rate of discount.
        (r)   All other liabilities and obligations whatsoever not specifically referred to in this Section nor chargeable under any other Section.
        (s)   Profit.
        6.2   The additional cost of overtime where specifically ordered by the Architect/Supervising Officer shall only be chargeable in the terms of a prior written agreement between the parties.

HEATING AND VENTILATING INDUSTRY

MECHANICAL INSTALLATIONS

Calculation of Hourly Base Rate of Labour for Typical Main Grades applicable from 4 October 2021, refer to notes within Section Four – Rates of Wages.

	Foreman	Senior Craftsman (+ 2nd Welding Skill)	Senior Craftsman	Craftsman	Installer	Mate over 18
Hourly Rate from 4 October 2021	18.05	16.11	14.92	13.74	12.40	10.46
Annual standard earnings excluding all holidays, 45.6 weeks × 38 hours	31,000.88	27,668.93	25,625.10	23,598.45	21,297.00	17,965.05
Employers national insurance contributions for this year	3,737.66	3,234.98	2,895.37	2,674.50	2,217.24	1,695.19
Weekly holiday credit and welfare contributions (52 weeks) for this year	4,805.32	4,483.96	4,061.20	4,483.96	3,463.72	3,001.44
Annual prime cost of labour	39,454.12	35,308.84	32,508.33	30,686.90	26,916.20	22,611.15
Hourly base rate	22.97	20.56	18.93	17.87	15.67	13.17

Notes:
(1) Annual industry holiday (4.6 weeks × 37.5 hours) and public holidays (1.6 weeks × 37.5 hours) are paid through weekly holiday credit and welfare stamp scheme.
(2) Where applicable, Merit money and other variables (e.g. daily abnormal conditions money), which attract Employer's National Insurance contribution, should be included.
(3) Contractors in Northern Ireland should add the appropriate amount of CITB Levy to the annual prime cost of labour prior to calculating the hourly base rate.
(4) Hourly rate based on 1,717.50 hours per annum and calculated as follows

52 Weeks @ 37.5 hrs/wk	=		1,950.00
Less			
Public Holiday = 8/5 = 1.6 weeks @ 3.75 hrs/wk	=	60	
Annual holidays = 4.6 weeks @ 37.5 hrs/wk	=	172.50	232.50
Hours	=		1,717.50

(5) For calculation of Holiday Credits and ENI refer to detailed labour rate evaluation.
(6) National Insurance contributions are those effective from 1 April 2022.
(7) Weekly holiday credit/welfare stamp values are those assumed and effective from 7 October 2019
(8) Hourly rates of wages are those assumed and effective from 4 October 2021.

Professional Ethics in Construction and Surveying

Greg Watts, Jason Challender, Anthony Higham and Peter McDermott

This textbook responds to the increasing demand for practical, industry aligned, ethical practices in quantity surveying, construction management and related AEC professions.

Professional Ethics for Construction and Surveying addresses how existing ethical standards can be pragmatically applied to both private and contracting practice, with case studies aligned with the ethical requirements of the main professional bodies. After an introduction to ethics, the authors present real-world situations where the minimum legal and contractual requirements necessitate the combination of professional judgement and ethical decision-making. They outline how such situations arise, then address how decisions can and should be made that are in keeping with the moral, contractual and CSR requirements, with cases covering the building lifecycle from procurement to handover. Consequently, the book brings together ethical theory, existing worldwide ethical standards and the requirements of the RICS, the CIOB and the ICES, with the authors' experiences of examining candidates for entry into the professional bodies.

The result is a professionally focused textbook aimed at vocational learners (at both undergraduate and postgraduate taught levels) and practitioners in construction, engineering, architecture and the wider built environment.

May 2021: 188 pp
ISBN: 9780367354190

To Order
Tel:+44 (0) 1235 400524
Email: tandf@bookpoint.co.uk

For a complete listing of all our titles visit:
www.tandf.co.uk

ELECTRICAL INDUSTRY

DEFINITION OF PRIME COST OF DAYWORK CARRIED OUT UNDER AN ELECTRICAL CONTRACT (MARCH 1981 EDITION)

This Definition of Prime Cost is published by The Royal Institution of Chartered Surveyors and The Electrical Contractors' Associations for convenience and for use by people who choose to use it. Members of The Electrical Contractors' Association are not in any way debarred from defining Prime Cost and rendering accounts for work carried out on that basis in any way they choose. Building owners are advised to reach agreement with contractors on the Definition of Prime Cost to be used prior to entering into a contract or subcontract.

SECTION 1: APPLICATION

1.1 This Definition provides a basis for the valuation of daywork executed under such electrical contracts as provide for its use.

1.2 It is not applicable in any other circumstances, such as jobbing, or other work carried out as a separate or main contract, nor in the case of daywork executed after the date of practical completion.

1.3 The terms 'contract' and 'contractor' herein shall be read as 'subcontract' and 'subcontractor' as the context may require.

SECTION 2: COMPOSITION OF TOTAL CHARGES

2.1 The Prime Cost of daywork comprises the sum of the following costs:
(a) Labour as defined in Section 3.
(b) Materials and goods as defined in Section 4.
(c) Plant as defined in Section 5.

2.2 Incidental costs, overheads and profit as defined in Section 6, as provided in the contract and expressed therein as percentage adjustments, are applicable to each of 2.1 (a)–(c).

SECTION 3: LABOUR

3.1 The standard wage rates, emoluments and expenses referred to below and the standard working hours referred to in 3.2 are those laid down for the time being in the rules and determinations or decisions of the Joint Industry Board or the Scottish Joint Industry Board for the Electrical Contracting Industry (or those of such other body as may be appropriate) applicable to the works and relating to the grade of operative concerned at the time when and in the area where daywork is executed.

3.2 Hourly base rates for labour are computed by dividing the annual prime cost of labour, based upon the standard working hours and as defined in 3.4 by the number of standard working hours per annum. See examples.

3.3 The hourly rates computed in accordance with 3.2 shall be applied in respect of the time spent by operatives directly engaged on daywork, including those operating mechanical plant and transport and erecting and dismantling other plant (unless otherwise expressly provided in the contract) and handling and distributing the materials and goods used in the daywork.

3.4 The annual prime cost of labour comprises the following:
(a) Standard weekly earnings (i.e. the standard working week as determined at the appropriate rate for the operative concerned).
(b) Payments in respect of public holidays.
(c) Any amounts which may become payable by the Contractor to or in respect of operatives arising from operation of the rules etc. referred to in 3.1 Which are not provided for in 3.4(a) and (b) nor in Section 6.
(d) Employer's National Insurance Contributions as applicable to 3.4 (a)–(c).
(e) Employer's contributions to the Joint Industry Board Combined Benefits Scheme or Scottish Joint Industry Board Holiday and Welfare Stamp Scheme, and holiday payments made to apprentices in compliance with the Joint Industry Board National Working Rules and Industrial Determinations as an employer.
(f) Any contribution, levy or tax imposed by Statute, payable by the Contractor in his capacity as an employer.

ELECTRICAL INDUSTRY

3.5   Differentials or extra payments in respect of supervisory responsibility are excluded from the annual prime cost (see Section 6). The time of principals and similar categories, when working manually, is admissible under this Section at the rates for the appropriate grades.

SECTION 4: MATERIALS AND GOODS

4.1   The prime cost of materials and goods obtained specifically for the daywork is the invoice cost after deducting all trade discounts and any portion of cash discounts in excess of 5%.
4.2   The prime cost of all other materials and goods used in the daywork is based upon the current market prices plus any appropriate handling charges.
4.3   The prime cost referred to in 4.1 and 4.2 includes the cost of delivery to site.
4.4   Any Value Added Tax which is treated, or is capable of being treated, as input tax (as defined by the Finance Act 1972, or any re-enactment or amendment thereof or substitution therefore) by the Contractor is excluded.

SECTION 5: PLANT

5.1   Unless otherwise stated in the contract, the prime cost of plant comprises the cost of the following:
       (a)   Use or hire of mechanically-operated plant and transport for the time employed on and/or provided or retained for the daywork;
       (b)   Use of non-mechanical plant (excluding non-mechanical hand tools) for the time employed on and/or provided or retained for the daywork;
       (c)   Transport to and from the site and erection and dismantling where applicable.
5.2   The use of non-mechanical hand tools and of erected scaffolding, staging, trestles or the likes is excluded (see Section 6), unless specifically retained for daywork.
5.3   Note: Where hired or other plant is operated by the Electrical Contractor's operatives, such time is to be included under Section 3 unless otherwise provided in the contract.

SECTION 6: INCIDENTAL COSTS, OVERHEADS AND PROFIT

6.1   The percentage adjustments provided in the contract which are applicable to each of the totals of Sections 3, 4 and 5, compromise the following:
       (a)   Head Office charges.
       (b)   Site staff including site supervision.
       (c)   The additional cost of overtime (other than that referred to in 6.2).
       (d)   Time lost due to inclement weather.
       (e)   The additional cost of bonuses and other incentive payments.
       (f)   Apprentices' study time.
       (g)   Travelling time and fares.
       (h)   Country and lodging allowances.
       (i)   Sick pay or insurance in lieu thereof, in respect of apprentices.
       (j)   Third party and employers' liability insurance.
       (k)   Liability in respect of redundancy payments to employees.
       (l)   Employers' National Insurance Contributions not included in 3.4.
       (m)   Use and maintenance of non-mechanical hand tools.
       (n)   Use of erected scaffolding, staging, trestles or the like (but see 5.2.).
       (o)   Use of tarpaulins, protective clothing, artificial lighting, safety and welfare facilities, storage and the like that may be available on site.
       (p)   Any variation to basic rates required by the Contractor in cases where the contract provides for the use of a specified schedule of basic plant charges (to the extent that no other provision is made for such variation – see 5.1).
       (q)   All other liabilities and obligations whatsoever not specifically referred to in this Section nor chargeable under any other Section.
       (r)   Profit.
       (s)   In the case of a subcontract which provides that the subcontractor shall allow a cash discount, such provision as is necessary for the allowance of the prescribed rate of discount.
6.2   The additional cost of overtime where specifically ordered by the Architect/Supervising Officer shall only be chargeable in the terms of a prior written agreement between the parties.

ELECTRICAL INDUSTRY

ELECTRICAL INSTALLATIONS

Calculation of Hourly Base Rate of Labour for Typical Main Grades applicable from 3 January 2022.

	Technician	Approved Electrician	Electrician	Labourer
Hourly Rate from 3 January 2022 (London Rates)	22.31	19.82	18.29	14.68
Annual standard earnings excluding all holidays, 45.8 Weeks × 37.5 hours	38,317.43	34,040.85	31,413.08	25,212.90
Employers national insurance contributions from 1 April 2022	4,920.05	4,249.99	3,838.26	2,866.81
JIB Combined benefits from 1 October 2018	3,879.72	3,520.92	3,290.04	2,772.12
Holiday top up funding	2,087.36	1,867.23	1,742.39	1,420.98
Annual prime cost of labour	49,204.55	43,678.99	40,283.76	32,272.81
Hourly base rate	28.65	25.43	23.45	18.79

Notes:

(1) Annual industry holiday (4.4 Weeks × 37.5 hours) and public holidays (1.6 Weeks × 37.5 hours).
(2) It should be noted that all labour costs incurred by the Contractor in his capacity as an Employer, other than those contained in the hourly rate above, must be taken into account under Section 6.
(3) Public Holidays are paid through weekly holiday credit and welfare stamp scheme.
(4) Contractors in Northern Ireland should add the appropriate amount of CITB Levy to the annual prime cost of labour prior to calculating the hourly base rate.
(5) Hourly rate based on 1,710 .00 hours per annum and calculated as follows:

52 Weeks @ 37.5 hrs/wk	=		1,950.00
Less			
Public Holiday = 8/5 = 1.6 weeks @ 37.5 hrs/wk	=	60	
Annual holidays = 4.8 weeks @ 37.5 hrs/wk	=	180	240
Hours	=		1,710.00

(6) For calculation of holiday credits and ENI refer to detailed labour rate evaluation.
(7) Hourly wage rates are those effective from 3rd January 2022.
(8) National Insurance contributions are those effective from 1 April 2022.

BUILDING INDUSTRY PLANT HIRE COSTS

SCHEDULE OF BASIC PLANT CHARGES (JULY 2010)

This Schedule is published by the Royal Institution of Chartered Surveyors and is for use in connection with Dayworks under a Building Contract.

EXPLANATORY NOTES

1   The rates in the Schedule are intended to apply solely to daywork carried out under and incidental to a Building Contract. They are NOT intended to apply to:
     (i)   jobbing or any other work carried out as a main or separate contract; or
     (ii)   work carried out after the date of commencement of the Defects Liability Period.
2   The rates apply to plant and machinery already on site, whether hired or owned by the Contractor.
3   The rates, unless otherwise stated, include the cost of fuel and power of every description, lubricating oils, grease, maintenance, sharpening of tools, replacement of spare parts, all consumable stores and for licences and insurances applicable to items of plant.
4   The rates, unless otherwise stated, do not include the costs of drivers and attendants (unless otherwise stated).
5   The rates in the Schedule are base costs and may be subject to an overall adjustment for price movement, overheads and profit, quoted by the Contractor prior to the placing of the Contract.
6   The rates should be applied to the time during which the plant is actually engaged in daywork.
7   Whether or not plant is chargeable on daywork depends on the daywork agreement in use and the inclusion of an item of plant in this schedule does not necessarily indicate that item is chargeable.
8   Rates for plant not included in the Schedule or which is not already on site and is specifically provided or hired for daywork shall be settled at prices which are reasonably related to the rates in the Schedule having regard to any overall adjustment quoted by the Contractor in the Conditions of Contract.
9   The data below is as per the RICS Schedule of Basic Plant Charges 2010 document published on BCIS. In order to adjust the data for current market (or forecast), please follow the steps below:

     –   Log into BCIS Online: https: //service.bcis.co.uk/BCISOnline/Account/LogOn?ReturnUrl=%2fBCI-SOnline
     –   Search for 'Schedule of Basic Plant Charges'
     –   On the Basic parameters tab, click 'Edit' then select 'Adjustment selection'
     –   Then choose the quarter and year you would like to adjust the data to
     –   The results will then appear in the Resources section within the Results tab
     –   Now proceed to download where you can download the data that has been adjusted
     –   However, please note that only 1 section can be downloaded at a time

BUILDING INDUSTRY PLANT HIRE COSTS

MECHANICAL PLANT AND TOOLS

Item of plant	Size/Rating	Unit	Rate per Hour (£)
PUMPS			
Mobile Pumps			
Including pump hoses, values and strainers, etc.			
Diaphragm	50 mm dia.	Each	1.17
Diaphragm	76 mm dia.	Each	1.89
Diaphragm	102 mm dia.	Each	3.54
Submersible	50 mm dia.	Each	0.76
Submersible	76 mm dia.	Each	0.86
Submersible	102 mm dia.	Each	1.03
Induced Flow	50 mm dia.	Each	0.77
Induced Flow	76 mm dia.	Each	1.67
Centrifugal, self-priming	25 mm dia.	Each	1.30
Centrifugal, self-priming	50 mm dia.	Each	1.92
Centrifugal, self-priming	75 mm dia.	Each	2.74
Centrifugal, self-priming	102 mm dia.	Each	3.35
Centrifugal, self-priming	152 mm dia.	Each	4.27
SCAFFOLDING, SHORING, FENCING			
Complete Scaffolding			
Mobile working towers, single width	2.0 m × 0.72 m base × 7.45 m high	Each	3.36
Mobile working towers, single width	2.0 m × 0.72 m base × 8.84 m high	Each	3.79
Mobile working towers, double width	2.0 m × 1.35 m × 7.45 m high	Each	3.79
Mobile working towers, double width	2.0 m × 1.35 m × 15.8 m high	Each	7.13
Chimney scaffold, single unit		Each	1.92
Chimney scaffold, twin unit		Each	3.59
Push along access platform	1.63–3.1 m	Each	5.00
Push along access platform	1.80 m × 0.70 m	Each	1.79
Trestles			
Trestle, adjustable	Any height	Pair	0.41
Trestle, painters	1.8 m high	Pair	0.31
Trestle, painters	2.4 m high	Pair	0.36
Shoring, Planking and Strutting			
'Acrow' adjustable prop	Sizes up to 4.9 m (open)	Each	0.06
'Strong boy' support attachment		Each	0.22
Adjustable trench strut	Sizes up to 1.67 m (open)	Each	0.16

BUILDING INDUSTRY PLANT HIRE COSTS

Item of plant	Size/Rating	Unit	Rate per Hour (£)
Trench sheet		Metre	0.03
Backhoe trench box	Base unit	Each	1.23
Backhoe trench box	Top unit	Each	0.87
Temporary Fencing			
Including block and coupler			
Site fencing steel grid panel	3.5 m × 2.0 m	Each	0.05
Anti-climb site steel grid fence panel	3.5 m × 2.0 m	Each	0.08
Solid panel Heras	2.0 m × 2.0 m	Each	0.09
Pedestrian gate		Each	0.36
Roadway gate		Each	0.60

LIFTING APPLIANCES AND CONVEYORS

Cranes

Mobile Cranes

Rates are inclusive of drivers

Lorry mounted, telescopic jib

Item of plant	Size/Rating	Unit	Rate per Hour (£)
Two wheel drive	5 tonnes	Each	19.00
Two wheel drive	8 tonnes	Each	42.00
Two wheel drive	10 tonnes	Each	50.00
Two wheel drive	12 tonnes	Each	77.00
Two wheel drive	20 tonnes	Each	89.69
Four wheel drive	18 tonnes	Each	46.51
Four wheel drive	25 tonnes	Each	35.90
Four wheel drive	30 tonnes	Each	38.46
Four wheel drive	45 tonnes	Each	46.15
Four wheel drive	50 tonnes	Each	53.85
Four wheel drive	60 tonnes	Each	61.54
Four wheel drive	70 tonnes	Each	71.79

Static Tower Crane

Rates inclusive of driver

Note: Capacity equals maximum lift in tonnes times maximum radius at which it can be lifted

	Capacity (metre/tonnes) Up to	Height under hook above ground (m) Up to	Unit	Rate per Hour (£)
Tower crane	30	22	Each	22.23
Tower crane	40	22	Each	26.62
Tower crane	40	30	Each	33.33

BUILDING INDUSTRY PLANT HIRE COSTS

Item of plant	Size/Rating		Unit	Rate per Hour (£)
Tower crane	50	22	Each	29.16
Tower crane	60	22	Each	35.90
Tower crane	60	36	Each	35.90
Tower crane	70	22	Each	41.03
Tower crane	80	22	Each	39.12
Tower crane	90	42	Each	37.18
Tower crane	110	36	Each	47.62
Tower crane	140	36	Each	55.77
Tower crane	170	36	Each	64.11
Tower crane	200	36	Each	71.95
Tower crane	250	36	Each	84.77
Tower crane with luffing jig	30	25	Each	22.23
Tower crane with luffing jig	40	30	Each	26.62
Tower crane with luffing jig	50	30	Each	29.16
Tower crane with luffing jig	60	36	Each	41.03
Tower crane with luffing jig	65	30	Each	33.13
Tower crane with luffing jig	80	22	Each	48.72
Tower crane with luffing jig	100	45	Each	48.72
Tower crane with luffing jig	125	30	Each	53.85
Tower crane with luffing jig	160	50	Each	53.85
Tower crane with luffing jig	200	50	Each	74.36
Tower crane with luffing jig	300	60	Each	100.00
Crane Equipment				
Muck tipping skip	Up to 200 litres		Each	0.67
Muck tipping skip	500 litres		Each	0.82
Muck tipping skip	750 litres		Each	1.08
Muck tipping skip	1000 litres		Each	1.28
Muck tipping skip	1500 litres		Each	1.41
Muck tipping skip	2000 litres		Each	1.67
Mortar skip	250 litres, plastic		Each	0.41
Mortar skip	350 litres steel		Each	0.77
Boat skip	250 litres		Each	0.92
Boat skip	500 litres		Each	1.08
Boat skip	750 litres		Each	1.23
Boat skip	1000 litres		Each	1.38
Boat skip	1500 litres		Each	1.64
Boat skip	2000 litres		Each	1.90

BUILDING INDUSTRY PLANT HIRE COSTS

Item of plant	Size/Rating	Unit	Rate per Hour (£)
Boat skip	3000 litres	Each	2.82
Boat skip	4000 litres	Each	3.23
Master flow skip	250 litres	Each	0.77
Master flow skip	500 litres	Each	1.03
Master flow skip	750 litres	Each	1.28
Master flow skip	1000 litres	Each	1.44
Master flow skip	1500 litres	Each	1.69
Master flow skip	2000 litres	Each	1.85
Grand master flow skip	500 litres	Each	1.28
Grand master flow skip	750 litres	Each	1.64
Grand master flow skip	1000 litres	Each	1.69
Grand master flow skip	1500 litres	Each	1.95
Grand master flow skip	2000 litres	Each	2.21
Cone flow skip	500 litres	Each	1.33
Cone flow skip	1000 litres	Each	1.69
Geared rollover skip	500 litres	Each	1.28
Geared rollover skip	750 litres	Each	1.64
Geared rollover skip	1000 litres	Each	1.69
Geared rollover skip	1500 litres	Each	1.95
Geared rollover skip	2000 litres	Each	2.21
Multi skip, rope operated	200 mm outlet size, 500 litres	Each	1.49
Multi skip, rope operated	200 mm outlet size, 750 litres	Each	1.64
Multi skip, rope operated	200 mm outlet size, 1000 litres	Each	1.74
Multi skip, rope operated	200 mm outlet size, 1500 litres	Each	2.00
Multi skip, rope operated	200 mm outlet size, 2000 litres	Each	2.26
Multi skip, man riding	200 mm outlet size, 1000 litres	Each	2.00
Multi skip	4 point lifting frame	Each	0.90
Multi skip	Chain brothers	Set	0.87
Crane Accessories			
Multi-purpose crane forks	1.5 and 2 tonnes S.W.L.	Each	1.13
Self-levelling crane forks		Each	1.28
Man cage	1 man, 230 kg S.W.L.	Each	1.90
Man cage	2 man, 500 kg S.W.L.	Each	1.95
Man cage	4 man, 750 kg S.W.L.	Each	2.15
Man cage	8 man, 1000 kg S.W.L.	Each	3.33
Stretcher cage	500 kg, S.W.L.	Each	2.69
Goods carrying cage	1500 kg, S.W.L.	Each	1.33
Goods carrying cage	3000 kg, S.W.L.	Each	1.85

BUILDING INDUSTRY PLANT HIRE COSTS

Item of plant	Size/Rating		Unit	Rate per Hour (£)
Builders' skip lifting cradle	12 tonnes, S.W.L.		Each	2.31
Board/pallet fork	1600 kg, S.W.L.		Each	1.90
Gas bottle carrier	500 kg, S.W.L.		Each	0.92
Hoists				
Scaffold hoist	200 kg		Each	2.46
Rack and pinion (goods only)	500 kg		Each	4.56
Rack and pinion (goods only)	1100 kg		Each	5.90
Rack and pinion (goods and passenger)	8 person, 80 kg		Each	7.44
Rack and pinion (goods and passenger)	14 person, 1400 kg		Each	8.72
Wheelbarrow chain sling			Each	1.67
Conveyors				
Belt Conveyors				
Conveyor	8 m long × 450 mm wide		Each	5.90
Miniveyor, control box and loading hopper	3 m unit		Each	4.49
Other Conveying Equipment				
Wheelbarrow			Each	0.62
Hydraulic superlift			Each	4.56
Pavac slab lifter (tile hoist)			Each	4.49
High lift pallet truck			Each	3.08
Lifting Trucks	Payload	Maximum Lift		
Fork lift, two wheel drive	1100 kg	up to 3.0 m	Each	5.64
Fork lift, two wheel drive	2540 kg	up to 3.7 m	Each	5.64
Fork lift, four wheel drive	1524 kg	up to 6.0 m	Each	5.64
Fork lift, four wheel drive	2600 kg	up to 5.4 m	Each	7.44
Fork life, four wheel drive	4000 kg	up to 17 m	Each	10.77
Lifting Platforms				
Hydraulic platform (Cherry picker)	9 m		Each	4.62
Hydraulic platform (Cherry picker)	12 m		Each	7.56
Hydraulic platform (Cherry picker)	15 m		Each	10.13
Hydraulic platform (Cherry picker)	17 m		Each	15.63
Hydraulic platform (Cherry picker)	20 m		Each	18.13
Hydraulic platform (Cherry picker)	25.6 m		Each	32.38
Scissor lift	7.6 m, electric		Each	3.85
Scissor lift	7.8 m, electric		Each	5.13
Scissor lift	9.7 m, electric		Each	4.23
Scissor lift	10 m, diesel		Each	6.41
Telescopic handler	7 m, 2 tonnes		Each	5.13

BUILDING INDUSTRY PLANT HIRE COSTS

Item of plant	Size/Rating	Unit	Rate per Hour (£)
Telescopic handler	13 m, 3 tonnes	Each	7.18
Lifting and Jacking Gear			
Pipe winch including gantry	1 tonne	Set	1.92
Pipe winch including gantry	3 tonnes	Set	3.21
Chain block	1 tonne	Each	0.35
Chain block	2 tonnes	Each	0.58
Chain block	5 tonnes	Each	1.14
Pull lift (Tirfor winch)	1 tonne	Each	0.64
Pull lift (Tirfor winch)	1.6 tonnes	Each	0.90
Pull lift (Tirfor winch)	3.2 tonnes	Each	1.15
Brother or chain slings, two legs	not exceeding 3.1 tonnes	Set	0.21
Brother or chain slings, two legs	not exceeding 4.25 tonnes	Set	0.31
Brother or chain slings, four legs	not exceeding 11.2 tonnes	Set	1.09
CONSTRUCTION VEHICLES			
Lorries			
Plated lorries (Rates are inclusive of driver)			
Platform lorry	7.5 tonnes	Each	16.21
Platform lorry	17 tonnes	Each	22.90
Platform lorry	24 tonnes	Each	30.68
Extra for lorry with crane attachment	up to 2.5 tonnes	Each	3.25
Extra for lorry with crane attachment	up to 5 tonnes	Each	6.00
Extra for lorry with crane attachment	up to 7.5 tonnes	Each	9.10
Tipper Lorries			
(Rates are inclusive of driver)			
Tipper lorry	up to 11 tonnes	Each	15.78
Tipper lorry	up to 17 tonnes	Each	23.95
Tipper lorry	up to 25 tonnes	Each	31.35
Tipper lorry	up to 31 tonnes	Each	37.79
Dumpers			
Site use only (excl. tax, insurance and extra cost of DERV etc. when operating on highway)	Makers Capacity		
Two wheel drive	1 tonne	Each	1.71
Four wheel drive	2 tonnes	Each	2.43
Four wheel drive	3 tonnes	Each	2.44

BUILDING INDUSTRY PLANT HIRE COSTS

Item of plant	Size/Rating	Unit	Rate per Hour (£)
Four wheel drive	5 tonnes	Each	3.08
Four wheel drive	6 tonnes	Each	3.85
Four wheel drive	9 tonnes	Each	5.65
Tracked	0.5 tonnes	Each	3.33
Tracked	1.5 tonnes	Each	4.23
Tracked	3.0 tonnes	Each	8.33
Tracked	6.0 tonnes	Each	16.03
Dumper Trucks (*Rates are inclusive of drivers*)			
Dumper truck	up to 15 tonnes	Each	28.56
Dumper truck	up to 17 tonnes	Each	32.82
Dumper truck	up to 23 tonnes	Each	54.64
Dumper truck	up to 30 tonnes	Each	63.50
Dumper truck	up to 35 tonnes	Each	73.02
Dumper truck	up to 40 tonnes	Each	87.84
Dumper truck	up to 50 tonnes	Each	133.44
Tractors			
Agricultural Type			
Wheeled, rubber-clad tyred	up to 40 kW	Each	8.63
Wheeled, rubber-clad tyred	up to 90 kW	Each	25.31
Wheeled, rubber-clad tyred	up to 140 kW	Each	36.49
Crawler Tractors			
With bull or angle dozer	up to 70 kW	Each	29.38
With bull or angle dozer	up to 85 kW	Each	38.63
With bull or angle dozer	up to 100 kW	Each	52.59
With bull or angle dozer	up to 115 kW	Each	55.85
With bull or angle dozer	up to 135 kW	Each	60.43
With bull or angle dozer	up to 185 kW	Each	76.44
With bull or angle dozer	up to 200 kW	Each	96.43
With bull or angle dozer	up to 250 kW	Each	117.68
With bull or angle dozer	up to 350 kW	Each	160.03
With bull or angle dozer	up to 450 kW	Each	219.86
With loading shovel	0.8 m³	Each	26.92
With loading shovel	1.0 m³	Each	32.59
With loading shovel	1.2 m³	Each	37.53
With loading shovel	1.4 m³	Each	42.89
With loading shovel	1.8 m³	Each	52.22
With loading shovel	2.0 m³	Each	57.22

BUILDING INDUSTRY PLANT HIRE COSTS

Item of plant	Size/Rating	Unit	Rate per Hour (£)
With loading shovel	2.1 m³	Each	60.12
With loading shovel	3.5 m³	Each	87.26
Light vans			
VW Caddivan or the like		Each	5.26
VW Transport transit or the like	1.0 tonne	Each	6.03
Luton Box Van or the like	1.8 tonnes	Each	9.87
Water/Fuel Storage			
Mobile water container	110 litres	Each	0.62
Water bowser	1100 litres	Each	0.72
Water bowser	3000 litres	Each	0.87
Mobile fuel container	110 litres	Each	0.62
Fuel bowser	1100 litres	Each	1.23
Fuel bowser	3000 litres	Each	1.87
EXCAVATIONS AND LOADERS			
Excavators			
Wheeled, hydraulic	up to 11 tonnes	Each	25.86
Wheeled, hydraulic	up to 14 tonnes	Each	30.82
Wheeled, hydraulic	up to 16 tonnes	Each	34.50
Wheeled, hydraulic	up to 21 tonnes	Each	39.10
Wheeled, hydraulic	up to 25 tonnes	Each	43.81
Wheeled, hydraulic	up to 30 tonnes	Each	55.30
Crawler, hydraulic	up to 11 tonnes	Each	25.86
Crawler, hydraulic	up to 14 tonnes	Each	30.82
Crawler, hydraulic	up to 17 tonnes	Each	34.50
Crawler, hydraulic	up to 23 tonnes	Each	39.10
Crawler, hydraulic	up to 30 tonnes	Each	43.81
Crawler, hydraulic	up to 35 tonnes	Each	55.30
Crawler, hydraulic	up to 38 tonnes	Each	71.73
Crawler, hydraulic	up to 55 tonnes	Each	95.63
Mini excavator	1000/1500 kg	Each	4.87
Mini excavator	2150/2400 kg	Each	6.67
Mini excavator	2700/3500 kg	Each	7.31
Mini excavator	3500/4500 kg	Each	8.21
Mini excavator	4500/6000 kg	Each	9.23
Mini excavator	7000 kg	Each	14.10
Micro excavator	725 mm wide	Each	5.13

BUILDING INDUSTRY PLANT HIRE COSTS

Item of plant	Size/Rating	Unit	Rate per Hour (£)
Loaders			
Shovel loader	0.4 m³	Each	7.69
Shovel loader	1.57 m³	Each	8.97
Shovel loader, four wheel drive	1.7 m³	Each	4.83
Shovel loader, four wheel drive	2.3 m³	Each	4.38
Shovel loader, four wheel drive	3.3 m³	Each	5.06
Skid steer loader wheeled	300/400 kg payload	Each	7.31
Skid steer loader wheeled	625 kg payload	Each	7.67
Tracked skip loader	650 kg	Each	4.42
Excavator Loaders			
Wheeled tractor type with black-hoe excavator			
Four wheel drive			
Four wheel drive, 2 Wheel steer	6 tonnes	Each	6.41
Four wheel drive, 2 Wheel steer	8 tonnes	Each	8.59
Attachments			
Breakers for excavator		Each	8.72
Breakers for mini excavator		Each	1.75
Breakers for back-hoe excavator/loader		Each	5.13
COMPACTION EQUIPMENT			
Rollers			
Vibrating roller	368–420 kg	Each	1.43
Single roller	533 kg	Each	1.94
Single roller	750 kg	Each	3.43
Twin roller	up to 650 kg	Each	6.03
Twin roller	up to 950 kg	Each	6.62
Twin roller with seat end steering wheel	up to 1400 kg	Each	7.68
Twin roller with seat end steering wheel	up to 2500 kg	Each	10.61
Pavement roller	3–4 tonnes dead weight	Each	6.00
Pavement roller	4–6 tonnes	Each	6.86
Pavement roller	6–10 tonnes	Each	7.17
Pavement roller	10–13 tonnes	Each	19.86
Rammers			
Tamper rammer 2 stroke-petrol	225 mm–275 mm	Each	1.52

BUILDING INDUSTRY PLANT HIRE COSTS

Item of plant	Size/Rating	Unit	Rate per Hour (£)
Soil Compactors			
Plate compactor	75 mm–400 mm	Each	1.53
Plate compactor rubber pad	375 mm–1400 mm	Each	1.53
Plate compactor reversible plate-petrol	400 mm	Each	2.44
CONCRETE EQUIPEMENT			
Concrete/Mortar Mixers			
Open drum without hopper	0.09/0.06 m³	Each	0.61
Open drum without hopper	0.12/0.09 m³	Each	1.22
Open drum without hopper	0.15/0.10 m³	Each	0.72
Concrete/Mortar Transport Equipment			
Concrete pump incl. hose, valve and couplers			
Lorry mounted concrete pump	24 m max. distance	Each	50.00
Lorry mounted concrete pump	34 m max. distance	Each	66.00
Lorry mounted concrete pump	42 m max. distance	Each	91.50
Concrete Equipment			
Vibrator, poker, petrol type	up to 75 mm dia.	Each	0.69
Air vibrator (*excluding compressor and hose*)	up to 75 mm dia.	Each	0.64
Extra poker heads	25/36/60 mm dia.	Each	0.76
Vibrating screed unit with beam	5.00 m	Each	2.48
Vibrating screed unit with adjustable beam	3.00 – 5.00 m	Each	3.54
Power float	725 mm – 900 mm	Each	2.56
Power float finishing pan		Each	0.62
Floor grinder	660 × 1016 mm, 110 V electric	Each	4.31
Floor plane	450 × 1100 mm	Each	4.31
TESTING EQUIPMENT			
Pipe Testing Equipment			
Pressure testing pump, electric		Set	2.19
Pressure test pump		Set	0.80

SITE ACCOMODATION AND TEMPORARY SERVICES			Rate per Hour (£)
Heating equipment			
Space heater – propane	80,000 Btu/hr	Each	1.03
Space heater – propane/electric	125,000 Btu/hr	Each	2.09

BUILDING INDUSTRY PLANT HIRE COSTS

Item of plant	Size/Rating	Unit	Rate per Hour (£)
Space heater – propane/electric	250,000 Btu/hr	Each	2.33
Space heater – propane	125,000 Btu/hr	Each	1.54
Space heater – propane	260,000 Btu/hr	Each	1.88
Cabinet heater		Each	0.82
Cabinet heater, catalytic		Each	0.57
Electric halogen heater		Each	1.27
Ceramic heater	3 kW	Each	0.99
Fan heater	3 kW	Each	0.66
Cooling fan		Each	1.92
Mobile cooling unit, small		Each	3.60
Mobile cooling unit, large		Each	4.98
Air conditioning unit		Each	2.81
Site Lighting and Equipment			
Tripod floodlight	500 W	Each	0.48
Tripod floodlight	1000 W	Each	0.62
Towable floodlight	4 × 100 W	Each	3.85
Hand held floodlight	500 W	Each	0.51
Rechargeable light		Each	0.41
Inspection light		Each	0.37
Plasterer's light		Each	0.65
Lighting mast		Each	2.87
Festoon light string	25 m	Each	0.55
Site Electrical Equipment			
Extension leads	240 V/14 m	Each	0.26
Extension leads	110 V/14 m	Each	0.36
Cable reel	25 m 110 V/240 V	Each	0.46
Cable reel	50 m 110 V/240 V	Each	0.88
4 Way junction box	110 V	Each	0.56
Power Generating Units			
Generator – petrol	2 kVA	Each	1.23
Generator – silenced petrol	2 kVA	Each	2.87
Generator – petrol	3 kVA	Each	1.47
Generator – diesel	5 kVA	Each	2.44
Generator – silenced diesel	10 kVA	Each	1.90
Generator – silenced diesel	15 kVA	Each	2.26
Generator – silenced diesel	30 kVA	Each	3.33
Generator – silenced diesel	50 kVA	Each	4.10
Generator – silenced diesel	75 kVA	Each	4.62

BUILDING INDUSTRY PLANT HIRE COSTS

Item of plant	Size/Rating	Unit	Rate per Hour (£)
Generator – silenced diesel	100 kVA	Each	5.64
Generator – silenced diesel	150 kVA	Each	7.18
Generator – silenced diesel	200 kVA	Each	9.74
Generator – silenced diesel	250 kVA	Each	11.28
Generator – silenced diesel	350 kVA	Each	14.36
Generator – silenced diesel	500 kVA	Each	15.38
Tail adaptor	240 V	Each	0.10
Transformers			
Transformer	3 kVA	Each	0.32
Transformer	5 kVA	Each	1.23
Transformer	7.5 kVA	Each	0.59
Transformer	10 kVA	Each	2.00
Rubbish Collection and Disposal Equipment			
Rubbish Chutes			
Standard plastic module	1 m section	Each	0.15
Steel liner insert		Each	0.30
Steel top hopper		Each	0.22
Plastic side entry hopper		Each	0.22
Plastic side entry hopper liner		Each	0.22
Dust Extraction Plant			
Dust extraction unit, light duty		Each	2.97
Dust extraction unit, heavy duty		Each	2.97

SITE EQUIPMENT – Welding Equipment

Arc-(Electric) Complete With Leads			
Welder generator – petrol	200 amp	Each	3.53
Welder generator – diesel	300/350 amp	Each	3.78
Welder generator – diesel	4000 amp	Each	7.92
Extra welding lead sets		Each	0.69
Gas-Oxy Welder			
Welding and cutting set (including oxygen and acetylene, excluding underwater equipment and thermic boring)			
Small		Each	2.24
Large		Each	3.75
Lead burning gun		Each	0.50

BUILDING INDUSTRY PLANT HIRE COSTS

Item of plant	Size/Rating	Unit	Rate per Hour (£)
Mig welder		Each	1.38
Fume extractor		Each	2.46
Road Works Equipment			
Traffic lights, mains/generator	2-way	Set	10.94
Traffic lights, mains/generator	3-way	Set	11.56
Traffic lights, mains/generator	4-way	Set	12.19
Flashing light		Each	0.10
Road safety cone	450 mm	Each	0.08
Safety cone	750 mm	Each	0.10
Safety barrier plank	1.25 m	Each	0.13
Safety barrier plank	2 m	Each	0.15
Safety barrier plank post		Each	0.13
Safety barrier plank post base		Each	0.10
Safety four gate barrier	1 m each gate	Set	0.77
Guard barrier	2 m	Each	0.19
Road sign	750 mm	Each	0.23
Road sign	900 mm	Each	0.31
Road sign	1200 mm	Each	0.42
Speed ramp/cable protection	500 mm section	Each	0.14
Hose ramp open top	3 m section	Each	0.07
DPC Equipment			
Damp-proofing injection machine		Each	2.56
Cleaning Equipment			
Vacuum cleaner (industrial wet) single motor		Each	1.08
Vacuum cleaner (industrial wet) twin motor	30 litre capacity	Each	1.79
Vacuum cleaner (industrial wet) twin motor	70 litre capacity	Each	2.21
Steam cleaner	Diesel/electric 1 phase	Each	3.33
Steam cleaner	Diesel/electric 3 phase	Each	3.85
Pressure washer, light duty electric	1450 PSI	Each	0.72
Pressure washer, heavy duty, diesel	2500 PSI	Each	1.33
Pressure washer, heavy duty, diesel	4000 PSI	Each	2.18
Cold pressure washer, electric		Each	2.39
Hot pressure washer, petrol		Each	4.19
Hot pressure washer, electric		Each	5.13
Cold pressure washer, petrol		Each	2.92

BUILDING INDUSTRY PLANT HIRE COSTS

Item of plant	Size/Rating	Unit	Rate per Hour (£)
Sandblast attachment to last washer		Each	1.23
Drain cleaning attachment to last washer		Each	1.03
Surface Preparation Equipment			
Rotavator	5 h.p.	Each	2.46
Rotavator	9 h.p.	Each	5.00
Scabbler, up to three heads		Each	1.53
Scabbler, pole		Each	2.68
Scrabbler, multi-headed floor		Each	3.89
Floor preparation machine		Each	1.05
Compressors and Equipment			
Portable Compressors			
Compressor – electric	4 cfm	Each	1.36
Compressor – electric	8 cfm lightweight	Each	1.31
Compressor – electric	8 cfm	Each	1.36
Compressor – electric	14 cfm	Each	1.56
Compressor – petrol	24 cfm	Each	2.15
Compressor – electric	25 cfm	Each	2.10
Compressor – electric	30 cfm	Each	2.36
Compressor – diesel	100 cfm	Each	2.56
Compressor – diesel	250 cfm	Each	5.54
Compressor – diesel	400 cfm	Each	8.72
Mobile Compressors			
Lorry mounted compressor			
(machine plus lorry only)	up to 3 m³	Each	41.47
(machine plus lorry only)	up to 5 m³	Each	48.94
Tractor mounted compressor			
(machine plus rubber tyred tractor)	up to 4 m³	Each	21.03
Accessories (Pneumatic Tools) *(with and including up to 15 m of air hose)*			
Demolition pick, medium duty		Each	0.90
Demolition pick, heavy duty		Each	1.03
Breakers (with six steels) light	up to 150 kg	Each	1.19
Breakers (with six steels) medium	295 kg	Each	1.24
Breakers (with six steels) heavy	386 kg	Each	1.44
Rock drill (for use with compressor) hand held		Each	1.18
Additional hoses	15 m	Each	0.09
Breakers			

BUILDING INDUSTRY PLANT HIRE COSTS

Item of plant	Size/Rating	Unit	Rate per Hour (£)
Demolition hammer drill, heavy duty, electric		Each	1.54
Road breaker, electric		Each	2.41
Road breaker, 2 stroke, petrol		Each	4.06
Hydraulic breaker unit, light duty, petrol		Each	3.06
Hydraulic breaker unit, heavy duty, petrol		Each	3.46
Hydraulic breaker unit, heavy duty, diesel		Each	4.62
Quarrying and Tooling Equipment			
Block and stone splitter, hydraulic	600 mm × 600 mm	Each	1.90
Block and stone splitter, manual		Each	1.64
Steel Reinforcement Equipment			
Bar bending machine – manual	up to 13 mm dia. rods	Each	1.03
Bar bending machine – manual	up to 20 mm dia. rods	Each	1.41
Bar shearing machine – electric	up to 38 mm dia. rods	Each	3.08
Bar shearing machine – electric	up to 40 mm dia. rods	Each	4.62
Bar cropper machine – electric	up to 13 mm dia. rods	Each	2.05
Bar cropper machine – electric	up to 20 mm dia. rods	Each	2.56
Bar cropper machine – electric	up to 40 mm dia. rods	Each	4.62
Bar cropper machine – 3 phase	up to 40 mm dia. rods	Each	4.62
Dehumidifiers			
110/240v Water	68 litre extraction per 24 hours	Each	2.46
110/240v Water	90 litre extraction per 24 hours	Each	3.38
SMALL TOOLS			
Saws			
Masonry bench saw	350 mm–500 mm dia.	Each	1.13
Floor saw	125 mm max. cut	Each	1.15
Floor saw	150 mm max. cut	Each	3.83
Floor saw, reversible	350 mm max. cut	Each	3.32
Wall saw, electric		Each	2.05
Chop/cut off saw, electric	350 mm dia.	Each	1.79
Circular saw, electric	230 mm dia.	Each	0.72
Tyrannosaw		Each	1.74
Reciprocating saw		Each	0.79
Door trimmer		Each	1.17
Stone saw	300 mm	Each	1.44
Chainsaw, petrol	500 mm	Each	3.92

BUILDING INDUSTRY PLANT HIRE COSTS

Item of plant	Size/Rating	Unit	Rate per Hour (£)
Full chainsaw safety kit		Each	0.41
Worktop jig		Each	1.08
Pipe Work Equipment			
Pipe bender	15 mm–22 mm	Each	0.92
Pipe bender, hydraulic	50 mm	Each	1.76
Pipe bender, electric	50 mm–150 mm dia.	Each	2.19
Pipe cutter, hydraulic		Each	0.46
Tripod pipe vice		Set	0.75
Ratchet threader	12 mm–32 mm	Each	0.93
Pipe threading machine, electric	12 mm–75 mm	Each	3.07
Pipe threading machine, electric	12 mm–100 mm	Each	4.93
Impact wrench, electric		Each	1.33
Hand-held Drills and equipment			
Impact or hammer drill	up to 25 mm dia.	Each	1.03
Impact or hammer drill	35 mm dia.	Each	1.29
Dry diamond core cutter		Each	0.99
Angle head drill		Each	0.90
Stirrer, mixer drill		Each	1.13
Paint, Insulation Application Equipment			
Airless spray unit		Each	4.13
Portaspray unit		Each	1.16
HPVL turbine spray unit		Each	2.23
Compressor and spray gun		Each	1.91
Other Handtools			
Staple gun		Each	0.96
Air nail gun	110 V	Each	1.01
Cartridge hammer		Each	1.08
Tongue and groove nailer complete with mallet		Each	1.59
Diamond wall chasing machine		Each	2.63
Masonry chain saw	300 mm	Each	5.49
Floor grinder		Each	3.99
Floor plane		Each	1.79
Diamond concrete planer		Each	1.93
Autofeed screwdriver, electric		Each	1.38
Laminate trimmer		Each	0.91
Biscuit jointer		Each	1.49
Random orbital sander		Each	0.97

BUILDING INDUSTRY PLANT HIRE COSTS

Item of plant	Size/Rating	Unit	Rate per Hour (£)
Floor sander		Each	1.54
Palm, delta, flap or belt sander		Each	0.75
Disc cutter, electric	300 mm	Each	1.49
Disc cutter, 2 stroke petrol	300 mm	Each	1.24
Dust suppressor for petrol disc cutter		Each	0.51
Cutter cart for petrol disc cutter		Each	1.21
Grinder, angle or cutter	up to 225 mm	Each	0.50
Grinder, angle or cutter	300 mm	Each	1.41
Mortar raking tool attachment		Each	0.19
Floor/polisher scrubber	325 mm	Each	1.76
Floor tile stripper		Each	2.44
Wallpaper stripper, electric		Each	0.81
Hot air paint stripper		Each	0.50
Electric diamond tile cutter	all sizes	Each	2.42
Hand tile cutter		Each	0.82
Electric needle gun		Each	1.29
Needle chipping gun		Each	1.85
Pedestrian floor sweeper	250 mm dia.	Each	0.82
Pedestrian floor sweeper	petrol	Each	2.20
Diamond tile saw		Each	1.84
Blow lamp equipment and glass		Set	0.50

Quantity Surveyor's Pocket Book, 3rd edition

Duncan Cartlidge

The third edition of the *Quantity Surveyor's Pocket Book* has been updated in line with NRM1, NRM2 and NRM3, and remains a must-have guide for students and qualified practitioners. Its focused coverage of the data, techniques and skills essential to the quantity surveying role makes it an invaluable companion for everything from initial cost advice to the final account stage.

Key features and updates included in this new edition:

*   an up-to-date analysis of NRM1, 2 and 3;
*   measurement and estimating examples in NRM2 format;
*   changes in procurement practice;
*   changes in professional development, guidance notes and schemes of work;
*   the increased use of NEC3 form of contract;
*   the impact of BIM.

This text includes recommended formats for cost plans, developer's budgets, financial reports, financial statements and final accounts. This is the ideal concise reference for quantity surveyors, project and commercial managers, and students of any of the above.

March 2017: 480 pp
ISBN: 9781138698369

To Order
Tel:+44 (0) 1235 400524
Email: tandf@bookpoint.co.uk

For a complete listing of all our titles visit:
www.tandf.co.uk

PART 7

Tables and Memoranda

This part contains the following sections:

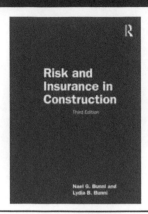

Risk and
Insurance in
Construction
Third Edition

Nael G. Bunni and
Lydia B. Bunni

Risk and Insurance in Construction

Nael G. Bunni and Lydia B. Bunni

The new 3rd edition of Nael Bunni's *Risk and Insurance in Construction,* now co-authored with Lydia Bunni, explains the need for insurance in construction and engineering projects and why it must be incorporated into the Conditions of Contract for such projects. It is unique in bringing together the background of the two topics of "Risk" and "Construction Insurance", explaining the flow and the interaction between them and then dealing with how they have been used to formulate the 2017 FIDIC Suite of Contracts and the 2021 Green Book.

This edition has been fully updated and new chapters deal with the latest definitions of "Risk" outlined in ISO 31000: 2009 and specifically explains the principles embodied in the new Clauses 17, 18 and 19 of the Major Suite of the FIDIC Forms of Contract and how they should be used. An important chapter is included discussing the logical transition from decision making to risk identification to responsibility for those allocated with particular risks, to the potential liability that results. This includes discussing particular liabilities that may arise for parties typically involved in construction and engineering projects, including developers, owners, contractors and designers. This part of the book links insurance to the law and explains the interaction between the two topics. The correlation between liability and the need for indemnity, which can only be provided properly through insurance, is highlighted.

The book is essential reading for practitioners from both the engineering profession and the insurance industry in all types of projects. Engineers who are required to use one of the Major Forms of the FIDIC Suite of Contracts, whether they are designers or contractors, and those involved in the insurance sector, whether brokers, claim consultants or insurance company personnel, will find this book to be an indispensable reference.

November 2021: 512 pp
ISBN: 9781032119830

To Order
Tel:+44 (0) 1235 400524
Email: tandf@bookpoint.co.uk

For a complete listing of all our titles visit:
www.tandf.co.uk

Taylor & Francis
Taylor & Francis Group

CONVERSION TABLES

Length	Unit	Conversion factors			
Millimetre	mm	1 in	= 25.4 mm	1 mm	= 0.0394 in
Centimetre	cm	1 in	= 2.54 cm	1 cm	= 0.3937 in
Metre	m	1 ft	= 0.3048 m	1 m	= 3.2808 ft
		1 yd	= 0.9144 m		= 1.0936 yd
Kilometre	km	1 mile	= 1.6093 km	1 km	= 0.6214 mile

Note:

1 cm	= 10 mm	1 ft	= 12 in
1 m	= 1 000 mm	1 yd	= 3 ft
1 km	= 1 000 m	1 mile	= 1 760 yd

Area	Unit	Conversion factors			
Square Millimetre	mm^2	$1 in^2$	$= 645.2 mm^2$	$1 mm^2$	$= 0.0016 in^2$
Square Centimetre	cm^2	$1 in^2$	$= 6.4516 cm^2$	$1 cm^2$	$= 1.1550 in^2$
Square Metre	m^2	$1 ft^2$	$= 0.0929 m^2$	$1 m^2$	$= 10.764 ft^2$
		$1 yd^2$	$= 0.8361 m^2$	$1 m^2$	$= 1.1960 yd^2$
Square Kilometre	km^2	$1 mile^2$	$= 2.590 km^2$	$1 km^2$	$= 0.3861 mile^2$

Note:

$1 cm^2$	$= 100 mm^2$	$1 ft^2$	$= 144 in^2$
$1 m^2$	$= 10 000 cm^2$	$1 yd^2$	$= 9 ft^2$
$1 km^2$	= 100 hectares	1 acre	$= 4 840 yd^2$
		$1 mile^2$	= 640 acres

Volume	Unit	Conversion factors			
Cubic Centimetre	cm^3	$1 cm^3$	$= 0.0610 in^3$	$1 in^3$	$= 16.387 cm^3$
Cubic Decimetre	dm^3	$1 dm^3$	$= 0.0353 ft^3$	$1 ft^3$	$= 28.329 dm^3$
Cubic Metre	m^3	$1 m^3$	$= 35.3147 ft^3$	$1 ft^3$	$= 0.0283 m^3$
		$1 m^3$	$= 1.3080 yd^3$	$1 yd^3$	$= 0.7646 m^3$
Litre	l	1 l	= 1.76 pint	1 pint	= 0.5683 l
			= 2.113 US pt		= 0.4733 US l

Note:

$1 dm^3$	$= 1 000 cm^3$	$1 ft^3$	$= 1 728 in^3$	1 pint	= 20 fl oz
$1 m^3$	$= 1 000 dm^3$	$1 yd^3$	$= 27 ft^3$	1 gal	= 8 pints
1 l	$= 1 dm^3$				

Neither the Centimetre nor Decimetre are SI units, and as such their use, particularly that of the Decimetre, is not widespread outside educational circles.

Mass	Unit	Conversion factors			
Milligram	mg	1 mg	= 0.0154 grain	1 grain	= 64.935 mg
Gram	g	1 g	= 0.0353 oz	1 oz	= 28.35 g
Kilogram	kg	1 kg	= 2.2046 lb	1 lb	= 0.4536 kg
Tonne	t	1 t	= 0.9842 ton	1 ton	= 1.016 t

Note:

1 g	= 1000 mg	1 oz	= 437.5 grains	1 cwt	= 112 lb
1 kg	= 1000 g	1 lb	= 16 oz	1 ton	= 20 cwt
1 t	= 1000 kg	1 stone	= 14 lb		

Force	Unit	Conversion factors			
Newton	N	1 lbf	= 4.448 N	1 kgf	= 9.807 N
Kilonewton	kN	1 lbf	= 0.004448 kN	1 ton f	= 9.964 kN
Meganewton	MN	100 tonf	= 0.9964 MN		

Tables and Memoranda

CONVERSION TABLES

Pressure and stress	Unit	Conversion factors
Kilonewton per square metre	kN/m^2	1 lbf/in^2 = 6.895 kN/m^2
		1 bar = 100 kN/m^2
Meganewton per square metre	MN/m^2	1 tonf/ft^2 = 107.3 kN/m^2 = 0.1073 MN/m^2
		1 kgf/cm^2 = 98.07 kN/m^2
		1 lbf/ft^2 = 0.04788 kN/m^2

Coefficient of consolidation (Cv) or swelling	Unit	Conversion factors
Square metre per year	m^2/year	1 cm^2/s = 3 154 m^2/year
		1 ft^2/year = 0.0929 m^2/year

Coefficient of permeability	Unit	Conversion factors
Metre per second	m/s	1 cm/s = 0.01 m/s
Metre per year	m/year	1 ft/year = 0.3048 m/year
		= 0.9651 × (10)^{8}m/s

Temperature	Unit	Conversion factors	
Degree Celsius	°C	°C = 5/9 × (°F − 32)	°F = (9 × °C)/ 5 + 32

CONVERSION TABLES

SPEED CONVERSION

km/h	m/min	mph	fpm
1	16.7	0.6	54.7
2	33.3	1.2	109.4
3	50.0	1.9	164.0
4	66.7	2.5	218.7
5	83.3	3.1	273.4
6	100.0	3.7	328.1
7	116.7	4.3	382.8
8	133.3	5.0	437.4
9	150.0	5.6	492.1
10	166.7	6.2	546.8
11	183.3	6.8	601.5
12	200.0	7.5	656.2
13	216.7	8.1	710.8
14	233.3	8.7	765.5
15	250.0	9.3	820.2
16	266.7	9.9	874.9
17	283.3	10.6	929.6
18	300.0	11.2	984.3
19	316.7	11.8	1038.9
20	333.3	12.4	1093.6
21	350.0	13.0	1148.3
22	366.7	13.7	1203.0
23	383.3	14.3	1257.7
24	400.0	14.9	1312.3
25	416.7	15.5	1367.0
26	433.3	16.2	1421.7
27	450.0	16.8	1476.4
28	466.7	17.4	1531.1
29	483.3	18.0	1585.7
30	500.0	18.6	1640.4
31	516.7	19.3	1695.1
32	533.3	19.9	1749.8
33	550.0	20.5	1804.5
34	566.7	21.1	1859.1
35	583.3	21.7	1913.8
36	600.0	22.4	1968.5
37	616.7	23.0	2023.2
38	633.3	23.6	2077.9
39	650.0	24.2	2132.5
40	666.7	24.9	2187.2
41	683.3	25.5	2241.9
42	700.0	26.1	2296.6
43	716.7	26.7	2351.3
44	733.3	27.3	2405.9
45	750.0	28.0	2460.6

Tables and Memoranda

CONVERSION TABLES

km/h	m/min	mph	fpm
46	766.7	28.6	2515.3
47	783.3	29.2	2570.0
48	800.0	29.8	2624.7
49	816.7	30.4	2679.4
50	833.3	31.1	2734.0

GEOMETRY

Two dimensional figures

Figure	Diagram of figure	Surface area	Perimeter
Square		a^2	$4a$
Rectangle		ab	$2(a+b)$
Triangle		$\frac{1}{2}ch$	$a+b+c$
Circle		πr^2 $\frac{1}{4}\pi d^2$ where $2r = d$	$2\pi r$ πd
Parallelogram		ah	$2(a+b)$
Trapezium		$\frac{1}{2}h(a+b)$	$a+b+c+d$
Ellipse		Approximately πab	$\pi(a+b)$
Hexagon		$2.6 \times a^2$	

GEOMETRY

Figure	Diagram of figure	Surface area	Perimeter
Octagon		$4.83 \times a^2$	$6a$
Sector of a circle		$\frac{1}{2} rb$ or $\frac{q}{360} \pi r^2$ note b = angle $\frac{q}{360} \times \pi 2r$	
Segment of a circle		$S - T$ where S = area of sector, T = area of triangle	
Bellmouth		$\frac{3}{14} \times r^2$	

GEOMETRY

Three dimensional figures

Figure	Diagram of figure	Surface area	Volume
Cube		$6a^2$	a^3
Cuboid/ rectangular block		$2(ab + ac + bc)$	abc
Prism/ triangular block		$bd + hc + dc + ad$	$\frac{1}{2}hcd$
Cylinder		$2\pi r^2 + 2\pi h$	$\pi r^2 h$
Sphere		$4\pi r^2$	$\frac{4}{3}\pi r^3$
Segment of sphere		$2\pi Rh$	$\frac{1}{6}\pi h(3r^2 + h^2)$ $\frac{1}{3}\pi h^2(3R - H)$
Pyramid		$(a + b)l + ab$	$\frac{1}{3}abh$

GEOMETRY

Figure	Diagram of figure	Surface area	Volume
Frustum of a pyramid		$l(a + b + c + d) + \sqrt{(ab + cd)}$ [rectangular figure only]	$\frac{h}{3}(ab + cd + \sqrt{abcd})$
Cone		πrl (excluding base) $\pi rl + \pi r^2$ (including base)	$\frac{1}{3}\pi r^2 h$ $\frac{1}{12}\pi d^2 h$
Frustum of a cone		$\pi r^2 + \pi R^2 + \pi l(R + r)$	$\frac{1}{3}\pi(R^2 + Rr + r^2)$

FORMULAE

Formulae

Formula	Description
Pythagoras Theorem	$A^2 = B^2 + C^2$ where A is the hypotenuse of a right-angled triangle and B and C are the two adjacent sides
Simpsons Rule	The Area is divided into an even number of strips of equal width, and therefore has an odd number of ordinates at the division points $$\text{area} = \frac{S(A + 2B + 4C)}{3}$$ where S = common interval (strip width) A = sum of first and last ordinates B = sum of remaining odd ordinates C = sum of the even ordinates The Volume can be calculated by the same formula, but by substituting the area of each coordinate rather than its length
Trapezoidal Rule	A given trench is divided into two equal sections, giving three ordinates, the first, the middle and the last $$\text{volume} = \frac{S \times (A + B + 2C)}{2}$$ where S = width of the strips A = area of the first section B = area of the last section C = area of the rest of the sections
Prismoidal Rule	A given trench is divided into two equal sections, giving three ordinates, the first, the middle and the last $$\text{volume} = \frac{L \times (A + 4B + C)}{6}$$ where L = total length of trench A = area of the first section B = area of the middle section C = area of the last section

TYPICAL THERMAL CONDUCTIVITY OF BUILDING MATERIALS

(Always check manufacturer's details – variation will occur depending on product and nature of materials)

	Thermal conductivity (W/mK)		Thermal conductivity (W/mK)
Acoustic plasterboard	0.25	Oriented strand board	0.13
Aerated concrete slab (500 kg/m^3)	0.16	Outer leaf brick	0.77
Aluminium	237	Plasterboard	0.22
Asphalt (1700 kg/m^3)	0.5	Plaster dense (1300 kg/m^3)	0.5
Bitumen-impregnated fibreboard	0.05	Plaster lightweight (600 kg/m^3)	0.16
Blocks (standard grade 600 kg/m^3)	0.15	Plywood (950 kg/m^3)	0.16
Blocks (solar grade 460 kg/m^3)	0.11	Prefabricated timber wall panels (check manufacturer)	0.12
Brickwork (outer leaf 1700 kg/m^3)	0.84	Screed (1200 kg/m^3)	0.41
Brickwork (inner leaf 1700 kg/m^3)	0.62	Stone chippings (1800 kg/m^3)	0.96
Dense aggregate concrete block 1800 kg/m^3 (exposed)	1.21	Tile hanging (1900 kg/m^3)	0.84
Dense aggregate concrete block 1800 kg/m^3 (protected)	1.13	Timber (650 kg/m^3)	0.14
Calcium silicate board (600 kg/m^3)	0.17	Timber flooring (650 kg/m^3)	0.14
Concrete general	1.28	Timber rafters	0.13
Concrete (heavyweight 2300 kg/m^3)	1.63	Timber roof or floor joists	0.13
Concrete (dense 2100 kg/m^3 typical floor)	1.4	Roof tile (1900 kg/m^3)	0.84
Concrete (dense 2000 kg/m^3 typical floor)	1.13	Timber blocks (650 kg/m^3)	0.14
Concrete (medium 1400 kg/m^3)	0.51	Cellular glass	0.045
Concrete (lightweight 1200 kg/m^3)	0.38	Expanded polystyrene	0.034
Concrete (lightweight 600 kg/m^3)	0.19	Expanded polystyrene slab (25 kg/m^3)	0.035
Concrete slab (aerated 500 kg/m^3)	0.16	Extruded polystyrene	0.035
Copper	390	Glass mineral wool	0.04
External render sand/cement finish	1	Mineral quilt (12 kg/m^3)	0.04
External render (1300 kg/m^3)	0.5	Mineral wool slab (25 kg/m^3)	0.035
Felt – Bitumen layers (1700 kg/m^3)	0.5	Phenolic foam	0.022
Fibreboard (300 kg/m^3)	0.06	Polyisocyanurate	0.025
Glass	0.93	Polyurethane	0.025
Marble	3	Rigid polyurethane	0.025
Metal tray used in wriggly tin concrete floors (7800 kg/m^3)	50	Rock mineral wool	0.038
Mortar (1750 kg/m^3)	0.8		

EARTHWORK

Weights of Typical Materials Handled by Excavators

The weight of the material is that of the state in its natural bed and includes moisture.
Adjustments should be made to allow for loose or compacted states

Material	Mass (kg/m³)	Mass (lb/cu yd)
Ashes, dry	610	1028
Ashes, wet	810	1365
Basalt, broken	1954	3293
Basalt, solid	2933	4943
Bauxite, crushed	1281	2159
Borax, fine	849	1431
Caliche	1440	2427
Cement, clinker	1415	2385
Chalk, fine	1221	2058
Chalk, solid	2406	4055
Cinders, coal, ash	641	1080
Cinders, furnace	913	1538
Clay, compacted	1746	2942
Clay, dry	1073	1808
Clay, wet	1602	2700
Coal, anthracite, solid	1506	2538
Coal, bituminous	1351	2277
Coke	610	1028
Dolomite, lumpy	1522	2565
Dolomite, solid	2886	4864
Earth, dense	2002	3374
Earth, dry, loam	1249	2105
Earth, Fullers, raw	673	1134
Earth, moist	1442	2430
Earth, wet	1602	2700
Felsite	2495	4205
Fieldspar, solid	2613	4404
Fluorite	3093	5213
Gabbro	3093	5213
Gneiss	2696	4544
Granite	2690	4534
Gravel, dry ¼ to 2 inch	1682	2835
Gravel, dry, loose	1522	2565
Gravel, wet ¼ to 2 inch	2002	3374
Gypsum, broken	1450	2444
Gypsum, solid	2787	4697
Hardcore (consolidated)	1928	3249
Lignite, dry	801	1350
Limestone, broken	1554	2619
Limestone, solid	2596	4375
Magnesite, magnesium ore	2993	5044
Marble	2679	4515
Marl, wet	2216	3735
Mica, broken	1602	2700
Mica, solid	2883	4859
Peat, dry	400	674
Peat, moist	700	1179

EARTHWORK

Material	Mass (kg/m³)	Mass (lb/cu yd)
Peat, wet	1121	1889
Potash	1281	2159
Pumice, stone	640	1078
Quarry waste	1438	2423
Quartz sand	1201	2024
Quartz, solid	2584	4355
Rhyolite	2400	4045
Sand and gravel, dry	1650	2781
Sand and gravel, wet	2020	3404
Sand, dry	1602	2700
Sand, wet	1831	3086
Sandstone, solid	2412	4065
Shale, solid	2637	4444
Slag, broken	2114	3563
Slag, furnace granulated	961	1619
Slate, broken	1370	2309
Slate, solid	2667	4495
Snow, compacted	481	810
Snow, freshly fallen	160	269
Taconite	2803	4724
Trachyte	2400	4045
Trap rock, solid	2791	4704
Turf	400	674
Water	1000	1685

Transport Capacities

Type of vehicle	Capacity of vehicle	
	Payload	Heaped capacity
Wheelbarrow	150	0.10
1 tonne dumper	1250	1.00
2.5 tonne dumper	4000	2.50
Articulated dump truck (Volvo A20 6 × 4)	18500	11.00
Articulated dump truck (Volvo A35 6 × 6)	32000	19.00
Large capacity rear dumper (Euclid R35)	35000	22.00
Large capacity rear dumper (Euclid R85)	85000	50.00

EARTHWORK

Machine Volumes for Excavating and Filling

Machine type	Cycles per minute	Volume per minute (m^3)
1.5 tonne excavator	1	0.04
	2	0.08
	3	0.12
3 tonne excavator	1	0.13
	2	0.26
	3	0.39
5 tonne excavator	1	0.28
	2	0.56
	3	0.84
7 tonne excavator	1	0.28
	2	0.56
	3	0.84
21 tonne excavator	1	1.21
	2	2.42
	3	3.63
Backhoe loader JCB3CX excavator Rear bucket capacity 0.28 m^3	1	0.28
	2	0.56
	3	0.84
Backhoe loader JCB3CX loading Front bucket capacity 1.00 m^3	1	1.00
	2	2.00

Machine Volumes for Excavating and Filling

Machine type	Loads per hour	Volume per hour (m^3)
1 tonne high tip skip loader Volume 0.485 m^3	5	2.43
	7	3.40
	10	4.85
3 tonne dumper Max volume 2.40 m^3 Available volume 1.9 m^3	4	7.60
	5	9.50
	7	13.30
	10	19.00
6 tonne dumper Max volume 3.40 m^3 Available volume 3.77 m^3	4	15.08
	5	18.85
	7	26.39
	10	37.70

EARTHWORK

Bulkage of Soils (after excavation)

Type of soil	Approximate bulking of 1 m³ after excavation
Vegetable soil and loam	25–30%
Soft clay	30–40%
Stiff clay	10–20%
Gravel	20–25%
Sand	40–50%
Chalk	40–50%
Rock, weathered	30–40%
Rock, unweathered	50–60%

Shrinkage of Materials (on being deposited)

Type of soil	Approximate bulking of 1 m³ after excavation
Clay	10%
Gravel	8%
Gravel and sand	9%
Loam and light sandy soils	12%
Loose vegetable soils	15%

Voids in Material Used as Subbases or Beddings

Material	m³ of voids/m³
Alluvium	0.37
River grit	0.29
Quarry sand	0.24
Shingle	0.37
Gravel	0.39
Broken stone	0.45
Broken bricks	0.42

Angles of Repose

Type of soil		Degrees
Clay	– dry	30
	– damp, well drained	45
	– wet	15–20
Earth	– dry	30
	– damp	45
Gravel	– moist	48
Sand	– dry or moist	35
	– wet	25
Loam		40

EARTHWORK

Slopes and Angles

Ratio of base to height	Angle in degrees
5:1	11
4:1	14
3:1	18
2:1	27
1½:1	34
1:1	45
1:1½	56
1:2	63
1:3	72
1:4	76
1:5	79

Grades (in Degrees and Percents)

Degrees	Percent	Degrees	Percent
1	1.8	24	44.5
2	3.5	25	46.6
3	5.2	26	48.8
4	7.0	27	51.0
5	8.8	28	53.2
6	10.5	29	55.4
7	12.3	30	57.7
8	14.0	31	60.0
9	15.8	32	62.5
10	17.6	33	64.9
11	19.4	34	67.4
12	21.3	35	70.0
13	23.1	36	72.7
14	24.9	37	75.4
15	26.8	38	78.1
16	28.7	39	81.0
17	30.6	40	83.9
18	32.5	41	86.9
19	34.4	42	90.0
20	36.4	43	93.3
21	38.4	44	96.6
22	40.4	45	100.0
23	42.4		

EARTHWORK

Bearing Powers

Ground conditions		Bearing power		
		kg/m²	lb/in²	Metric t/m²
Rock,	broken	483	70	50
	solid	2415	350	240
Clay,	dry or hard	380	55	40
	medium dry	190	27	20
	soft or wet	100	14	10
Gravel,	cemented	760	110	80
Sand,	compacted	380	55	40
	clean dry	190	27	20
Swamp and alluvial soils		48	7	5

Earthwork Support

Maximum depth of excavation in various soils without the use of earthwork support

Ground conditions	Feet (ft)	Metres (m)
Compact soil	12	3.66
Drained loam	6	1.83
Dry sand	1	0.3
Gravelly earth	2	0.61
Ordinary earth	3	0.91
Stiff clay	10	3.05

It is important to note that the above table should only be used as a guide. Each case must be taken on its merits and, as the limited distances given above are approached, careful watch must be kept for the slightest signs of caving in

CONCRETE WORK

Weights of Concrete and Concrete Elements

Type of material		kg/m³	lb/cu ft
Ordinary concrete (dense aggregates)			
Non-reinforced plain or mass concrete			
Nominal weight		2305	144
Aggregate	– limestone	2162 to 2407	135 to 150
	– gravel	2244 to 2407	140 to 150
	– broken brick	2000 (av)	125 (av)
	– other crushed stone	2326 to 2489	145 to 155
Reinforced concrete			
Nominal weight		2407	150
Reinforcement	– 1%	2305 to 2468	144 to 154
	– 2%	2356 to 2519	147 to 157
	– 4%	2448 to 2703	153 to 163
Special concretes			
Heavy concrete			
Aggregates	– barytes, magnetite	3210 (min)	200 (min)
	– steel shot, punchings	5280	330
Lean mixes			
Dry-lean (gravel aggregate)		2244	140
Soil-cement (normal mix)		1601	100

CONCRETE WORK

Type of material		kg/m^2 per mm thick	lb/sq ft per inch thick
Ordinary concrete (dense aggregates)			
Solid slabs (floors, walls etc.)			
Thickness:	75 mm or 3 in	184	37.5
	100 mm or 4 in	245	50
	150 mm or 6 in	378	75
	250 mm or 10 in	612	125
	300 mm or 12 in	734	150
Ribbed slabs			
Thickness:	125 mm or 5 in	204	42
	150 mm or 6 in	219	45
	225 mm or 9 in	281	57
	300 mm or 12 in	342	70
Special concretes			
Finishes etc.			
	Rendering, screed etc.		
	Granolithic, terrazzo	1928 to 2401	10 to 12.5
	Glass-block (hollow) concrete	1734 (approx)	9 (approx)
Prestressed concrete		Weights as for reinforced concrete (upper limits)	
Air-entrained concrete		Weights as for plain or reinforced concrete	

CONCRETE WORK

Average Weight of Aggregates

Materials	Voids %	Weight kg/m³
Sand	39	1660
Gravel 10–20 mm	45	1440
Gravel 35–75 mm	42	1555
Crushed stone	50	1330
Crushed granite (over 15 mm)	50	1345
(n.e. 15 mm)	47	1440
'All-in' ballast	32	1800–2000

Material	kg/m³	lb/cu yd
Vermiculite (aggregate)	64–80	108–135
All-in aggregate	1999	125

Applications and Mix Design

Site mixed concrete

Recommended mix	Class of work suitable for	Cement (kg)	Sand (kg)	Coarse aggregate (kg)	Nr 25 kg bags cement per m³ of combined aggregate
1:3:6	Roughest type of mass concrete such as footings, road haunching over 300 mm thick	208	905	1509	8.30
1:2.5:5	Mass concrete of better class than 1:3:6 such as bases for machinery, walls below ground etc.	249	881	1474	10.00
1:2:4	Most ordinary uses of concrete, such as mass walls above ground, road slabs etc. and general reinforced concrete work	304	889	1431	12.20
1:1.5:3	Watertight floors, pavements and walls, tanks, pits, steps, paths, surface of 2 course roads, reinforced concrete where extra strength is required	371	801	1336	14.90
1:1:2	Works of thin section such as fence posts and small precast work	511	720	1206	20.40

CONCRETE WORK

Ready mixed concrete

Application	Designated concrete	Standardized prescribed concrete	Recommended consistence (nominal slump class)
Foundations			
Mass concrete fill or blinding	GEN 1	ST2	S3
Strip footings	GEN 1	ST2	S3
Mass concrete foundations			
Single storey buildings	GEN 1	ST2	S3
Double storey buildings	GEN 3	ST4	S3
Trench fill foundations			
Single storey buildings	GEN 1	ST2	S4
Double storey buildings	GEN 3	ST4	S4
General applications			
Kerb bedding and haunching	GEN 0	ST1	S1
Drainage works – immediate support	GEN 1	ST2	S1
Other drainage works	GEN 1	ST2	S3
Oversite below suspended slabs	GEN 1	ST2	S3
Floors			
Garage and house floors with no embedded steel	GEN 3	ST4	S2
Wearing surface: Light foot and trolley traffic	RC30	ST4	S2
Wearing surface: General industrial	RC40	N/A	S2
Wearing surface: Heavy industrial	RC50	N/A	S2
Paving			
House drives, domestic parking and external parking	PAV 1	N/A	S2
Heavy-duty external paving	PAV 2	N/A	S2

CONCRETE WORK

Prescribed Mixes for Ordinary Structural Concrete

Weights of cement and total dry aggregates in kg to produce approximately one cubic metre of fully compacted concrete together with the percentages by weight of fine aggregate in total dry aggregates

Conc. grade	Nominal max size of aggregate (mm)	40		20		14		10	
	Workability	Med.	High	Med.	High	Med.	High	Med.	High
	Limits to slump that may be expected (mm)	50–100	100–150	25–75	75–125	10–50	50–100	10–25	25–50
7	Cement (kg)	180	200	210	230	–	–	–	–
	Total aggregate (kg)	1950	1850	1900	1800	–	–	–	–
	Fine aggregate (%)	30–45	30–45	35–50	35–50	–	–	–	–
10	Cement (kg)	210	230	240	260	–	–	–	–
	Total aggregate (kg)	1900	1850	1850	1800	–	–	–	–
	Fine aggregate (%)	30–45	30–45	35–50	35–50	–	–	–	–
15	Cement (kg)	250	270	280	310	–	–	–	–
	Total aggregate (kg)	1850	1800	1800	1750	–	–	–	–
	Fine aggregate (%)	30–45	30–45	35–50	35–50	–	–	–	–
20	Cement (kg)	300	320	320	350	340	380	360	410
	Total aggregate (kg)	1850	1750	1800	1750	1750	1700	1750	1650
	Sand								
	Zone 1 (%)	35	40	40	45	45	50	50	55
	Zone 2 (%)	30	35	35	40	40	45	45	50
	Zone 3 (%)	30	30	30	35	35	40	40	45
25	Cement (kg)	340	360	360	390	380	420	400	450
	Total aggregate (kg)	1800	1750	1750	1700	1700	1650	1700	1600
	Sand								
	Zone 1 (%)	35	40	40	45	45	50	50	55
	Zone 2 (%)	30	35	35	40	40	45	45	50
	Zone 3 (%)	30	30	30	35	35	40	40	45
30	Cement (kg)	370	390	400	430	430	470	460	510
	Total aggregate (kg)	1750	1700	1700	1650	1700	1600	1650	1550
	Sand								
	Zone 1 (%)	35	40	40	45	45	50	50	55
	Zone 2 (%)	30	35	35	40	40	45	45	50
	Zone 3 (%)	30	30	30	35	35	40	40	45

REINFORCEMENT

Weights of Bar Reinforcement

Nominal sizes (mm)	Cross-sectional area (mm²)	Mass (kg/m)	Length of bar (m/tonne)
6	28.27	0.222	4505
8	50.27	0.395	2534
10	78.54	0.617	1622
12	113.10	0.888	1126
16	201.06	1.578	634
20	314.16	2.466	405
25	490.87	3.853	260
32	804.25	6.313	158
40	1265.64	9.865	101
50	1963.50	15.413	65

Weights of Bars (at specific spacings)

Weights of metric bars in kilogrammes per square metre

Size (mm)	Spacing of bars in millimetres									
	75	100	125	150	175	200	225	250	275	300
6	2.96	2.220	1.776	1.480	1.27	1.110	0.99	0.89	0.81	0.74
8	5.26	3.95	3.16	2.63	2.26	1.97	1.75	1.58	1.44	1.32
10	8.22	6.17	4.93	4.11	3.52	3.08	2.74	2.47	2.24	2.06
12	11.84	8.88	7.10	5.92	5.07	4.44	3.95	3.55	3.23	2.96
16	21.04	15.78	12.63	10.52	9.02	7.89	7.02	6.31	5.74	5.26
20	32.88	24.66	19.73	16.44	14.09	12.33	10.96	9.87	8.97	8.22
25	51.38	38.53	30.83	25.69	22.02	19.27	17.13	15.41	14.01	12.84
32	84.18	63.13	50.51	42.09	36.08	31.57	28.06	25.25	22.96	21.04
40	131.53	98.65	78.92	65.76	56.37	49.32	43.84	39.46	35.87	32.88
50	205.51	154.13	123.31	102.76	88.08	77.07	68.50	61.65	56.05	51.38

Basic weight of steelwork taken as 7850 kg/m³
Basic weight of bar reinforcement per metre run = 0.00785 kg/mm²
The value of π has been taken as 3.141592654

REINFORCEMENT

Fabric Reinforcement

Preferred range of designated fabric types and stock sheet sizes

Fabric reference	Longitudinal wires			Cross wires			
	Nominal wire size (mm)	Pitch (mm)	Area (mm²/m)	Nominal wire size (mm)	Pitch (mm)	Area (mm²/m)	Mass (kg/m²)
Square mesh							
A393	10	200	393	10	200	393	6.16
A252	8	200	252	8	200	252	3.95
A193	7	200	193	7	200	193	3.02
A142	6	200	142	6	200	142	2.22
A98	5	200	98	5	200	98	1.54
Structural mesh							
B1131	12	100	1131	8	200	252	10.90
B785	10	100	785	8	200	252	8.14
B503	8	100	503	8	200	252	5.93
B385	7	100	385	7	200	193	4.53
B283	6	100	283	7	200	193	3.73
B196	5	100	196	7	200	193	3.05
Long mesh							
C785	10	100	785	6	400	70.8	6.72
C636	9	100	636	6	400	70.8	5.55
C503	8	100	503	5	400	49.0	4.34
C385	7	100	385	5	400	49.0	3.41
C283	6	100	283	5	400	49.0	2.61
Wrapping mesh							
D98	5	200	98	5	200	98	1.54
D49	2.5	100	49	2.5	100	49	0.77

Stock sheet size 4.8 m × 2.4 m, Area 11.52 m²

Average weight kg/m³ of steelwork reinforcement in concrete for various building elements

Substructure	kg/m³ concrete	Substructure	kg/m³ concrete
Pile caps	110–150	Plate slab	150–220
Tie beams	130–170	Cant slab	145–210
Ground beams	230–330	Ribbed floors	130–200
Bases	125–180	Topping to block floor	30–40
Footings	100–150	Columns	210–310
Retaining walls	150–210	Beams	250–350
Raft	60–70	Stairs	130–170
Slabs – one way	120–200	Walls – normal	40–100
Slabs – two way	110–220	Walls – wind	70–125

Note: For exposed elements add the following %:
Walls 50%, Beams 100%, Columns 15%

FORMWORK

Formwork Stripping Times – Normal Curing Periods

Conditions under which concrete is maturing	Minimum periods of protection for different types of cement					
	Number of days (where the average surface temperature of the concrete exceeds 10°C during the whole period)			Equivalent maturity (degree hours) calculated as the age of the concrete in hours multiplied by the number of degrees Celsius by which the average surface temperature of the concrete exceeds 10°C		
	Other	SRPC	OPC or RHPC	Other	SRPC	OPC or RHPC
1. Hot weather or drying winds	7	4	3	3500	2000	1500
2. Conditions not covered by 1	4	3	2	2000	1500	1000

KEY
OPC – Ordinary Portland Cement
RHPC – Rapid-hardening Portland Cement
SRPC – Sulphate-resisting Portland Cement

Minimum Period before Striking Formwork

	Minimum period before striking		
	Surface temperature of concrete		
	16°C	17°C	t°C (0–25)
Vertical formwork to columns, walls and large beams	12 hours	18 hours	300 hours t+10
Soffit formwork to slabs	4 days	6 days	100 days t+10
Props to slabs	10 days	15 days	250 days t+10
Soffit formwork to beams	9 days	14 days	230 days t+10
Props to beams	14 days	21 days	360 days t+10

MASONRY

Number of Bricks Required for Various Types of Work per m² of Walling

Description	Brick size	
	215 × 102.5 × 50 mm	215 × 102.5 × 65 mm
Half brick thick		
Stretcher bond	74	59
English bond	108	86
English garden wall bond	90	72
Flemish bond	96	79
Flemish garden wall bond	83	66
One brick thick and cavity wall of two half brick skins		
Stretcher bond	148	119

Quantities of Bricks and Mortar Required per m² of Walling

	Unit	No of bricks required	Mortar required (cubic metres)		
Standard bricks			**No frogs**	**Single frogs**	**Double frogs**
Brick size 215 × 102.5 × 50 mm					
half brick wall (103 mm)	m²	72	0.022	0.027	0.032
2 × half brick cavity wall (270 mm)	m²	144	0.044	0.054	0.064
one brick wall (215 mm)	m²	144	0.052	0.064	0.076
one and a half brick wall (322 mm)	m²	216	0.073	0.091	0.108
Mass brickwork	m³	576	0.347	0.413	0.480
Brick size 215 × 102.5 × 65 mm					
half brick wall (103 mm)	m²	58	0.019	0.022	0.026
2 × half brick cavity wall (270 mm)	m²	116	0.038	0.045	0.055
one brick wall (215 mm)	m²	116	0.046	0.055	0.064
one and a half brick wall (322 mm)	m²	174	0.063	0.074	0.088
Mass brickwork	m³	464	0.307	0.360	0.413
Metric modular bricks			**Perforated**		
Brick size 200 × 100 × 75 mm					
90 mm thick	m²	67	0.016	0.019	
190 mm thick	m²	133	0.042	0.048	
290 mm thick	m²	200	0.068	0.078	
Brick size 200 × 100 × 100 mm					
90 mm thick	m²	50	0.013	0.016	
190 mm thick	m²	100	0.036	0.041	
290 mm thick	m²	150	0.059	0.067	
Brick size 300 × 100 × 75 mm					
90 mm thick	m²	33	–	0.015	
Brick size 300 × 100 × 100 mm					
90 mm thick	m²	44	0.015	0.018	

Note: Assuming 10 mm thick joints

MASONRY

Mortar Required per m² Blockwork (9.88 blocks/m²)

Wall thickness	75	90	100	125	140	190	215
Mortar m³/m²	0.005	0.006	0.007	0.008	0.009	0.013	0.014

Mortar Group	Cement: lime: sand	Masonry cement: sand	Cement: sand with plasticizer
1	1:0–0.25:3		
2	1:0.5:4–4.5	1:2.5-3.5	1:3–4
3	1:1:5–6	1:4–5	1:5–6
4	1:2:8–9	1:5.5–6.5	1:7–8
5	1:3:10–12	1:6.5–7	1:8

Group 1: strong inflexible mortar
Group 5: weak but flexible

All mixes within a group are of approximately similar strength
Frost resistance increases with the use of plasticizers
Cement: lime: sand mixes give the strongest bond and greatest resistance to rain penetration
Masonry cement equals ordinary Portland cement plus a fine neutral mineral filler and an air entraining agent

Calcium Silicate Bricks

Type	Strength	Location
Class 2 crushing strength	14.0 N/mm²	not suitable for walls
Class 3	20.5 N/mm²	walls above dpc
Class 4	27.5 N/mm²	cappings and copings
Class 5	34.5 N/mm²	retaining walls
Class 6	41.5 N/mm²	walls below ground
Class 7	48.5 N/mm²	walls below ground

The Class 7 calcium silicate bricks are therefore equal in strength to Class B bricks
Calcium silicate bricks are not suitable for DPCs

Durability of Bricks	
FL	Frost resistant with low salt content
FN	Frost resistant with normal salt content
ML	Moderately frost resistant with low salt content
MN	Moderately frost resistant with normal salt content

MASONRY

Brickwork Dimensions

No. of horizontal bricks	Dimensions (mm)	No. of vertical courses	Height of vertical courses (mm)
½	112.5	1	75
1	225.0	2	150
1½	337.5	3	225
2	450.0	4	300
2½	562.5	5	375
3	675.0	6	450
3½	787.5	7	525
4	900.0	8	600
4½	1012.5	9	675
5	1125.0	10	750
5½	1237.5	11	825
6	1350.0	12	900
6½	1462.5	13	975
7	1575.0	14	1050
7½	1687.5	15	1125
8	1800.0	16	1200
8½	1912.5	17	1275
9	2025.0	18	1350
9½	2137.5	19	1425
10	2250.0	20	1500
20	4500.0	24	1575
40	9000.0	28	2100
50	11250.0	32	2400
60	13500.0	36	2700
75	16875.0	40	3000

TIMBER

Weights of Timber

Material	kg/m³	lb/cu ft
General	806 (avg)	50 (avg)
Douglas fir	479	30
Yellow pine, spruce	479	30
Pitch pine	673	42
Larch, elm	561	35
Oak (English)	724 to 959	45 to 60
Teak	643 to 877	40 to 55
Jarrah	959	60
Greenheart	1040 to 1204	65 to 75
Quebracho	1285	80
Material	**kg/m² per mm thickness**	**lb/sq ft per inch thickness**
Wooden boarding and blocks		
Softwood	0.48	2.5
Hardwood	0.76	4
Hardboard	1.06	5.5
Chipboard	0.76	4
Plywood	0.62	3.25
Blockboard	0.48	2.5
Fibreboard	0.29	1.5
Wood-wool	0.58	3
Plasterboard	0.96	5
Weather boarding	0.35	1.8

TIMBER

Conversion Tables (for timber only)

Inches	Millimetres	Feet	Metres
1	25	1	0.300
2	50	2	0.600
3	75	3	0.900
4	100	4	1.200
5	125	5	1.500
6	150	6	1.800
7	175	7	2.100
8	200	8	2.400
9	225	9	2.700
10	250	10	3.000
11	275	11	3.300
12	300	12	3.600
13	325	13	3.900
14	350	14	4.200
15	375	15	4.500
16	400	16	4.800
17	425	17	5.100
18	450	18	5.400
19	475	19	5.700
20	500	20	6.000
21	525	21	6.300
22	550	22	6.600
23	575	23	6.900
24	600	24	7.200

Planed Softwood

The finished end section size of planed timber is usually 3/16" less than the original size from which it is produced. This however varies slightly depending upon availability of material and origin of the species used.

Standards (timber) to cubic metres and cubic metres to standards (timber)

Cubic metres	Cubic metres standards	Standards
4.672	1	0.214
9.344	2	0.428
14.017	3	0.642
18.689	4	0.856
23.361	5	1.070
28.033	6	1.284
32.706	7	1.498
37.378	8	1.712
42.050	9	1.926
46.722	10	2.140
93.445	20	4.281
140.167	30	6.421
186.890	40	8.561
233.612	50	10.702
280.335	60	12.842
327.057	70	14.982
373.779	80	17.122

TIMBER

1 cu metre = 35.3148 cu ft = 0.21403 std

1 cu ft = 0.028317 cu metres

1 std = 4.67227 cu metres

Basic sizes of sawn softwood available (cross-sectional areas)

Thickness (mm)	Width (mm)								
	75	100	125	150	175	200	225	250	300
16	X	X	X	X					
19	X	X	X	X					
22	X	X	X	X					
25	X	X	X	X	X	X	X	X	X
32	X	X	X	X	X	X	X	X	X
36	X	X	X	X					
38	X	X	X	X	X	X	X		
44	X	X	X	X	X	X	X	X	X
47*	X	X	X	X	X	X	X	X	X
50	X	X	X	X	X	X	X	X	X
63	X	X	X	X	X	X	X		
75	X	X	X	X	X	X	X	X	
100		X		X		X		X	X
150				X		X			X
200						X			
250								X	
300									X

* This range of widths for 47 mm thickness will usually be found to be available in construction quality only

Note: The smaller sizes below 100 mm thick and 250 mm width are normally but not exclusively of European origin. Sizes beyond this are usually of North and South American origin

Basic lengths of sawn softwood available (metres)

1.80	2.10	3.00	4.20	5.10	6.00	7.20
	2.40	3.30	4.50	5.40	6.30	
	2.70	3.60	4.80	5.70	6.60	
		3.90			6.90	

Note: Lengths of 6.00 m and over will generally only be available from North American species and may have to be recut from larger sizes

TIMBER

Reductions from basic size to finished size by planning of two opposed faces

Purpose	Reductions from basic sizes for timber			
	15–35 mm	36–100 mm	101–150 mm	over 150 mm
a) Constructional timber	3 mm	3 mm	5 mm	6 mm
b) Matching interlocking boards	4 mm	4 mm	6 mm	6 mm
c) Wood trim not specified in BS 584	5 mm	7 mm	7 mm	9 mm
d) Joinery and cabinet work	7 mm	9 mm	11 mm	13 mm

Note: The reduction of width or depth is overall the extreme size and is exclusive of any reduction of the face by the machining of a tongue or lap joints

Maximum Spans for Various Roof Trusses

Maximum permissible spans for rafters for Fink trussed rafters

Basic size	Actual size	Pitch (degrees)								
(mm)	(mm)	15 (m)	17.5 (m)	20 (m)	22.5 (m)	25 (m)	27.5 (m)	30 (m)	32.5 (m)	35 (m)
38 × 75	35 × 72	6.03	6.16	6.29	6.41	6.51	6.60	6.70	6.80	6.90
38 × 100	35 × 97	7.48	7.67	7.83	7.97	8.10	8.22	8.34	8.47	8.61
38 × 125	35 × 120	8.80	9.00	9.20	9.37	9.54	9.68	9.82	9.98	10.16
44 × 75	41 × 72	6.45	6.59	6.71	6.83	6.93	7.03	7.14	7.24	7.35
44 × 100	41 × 97	8.05	8.23	8.40	8.55	8.68	8.81	8.93	9.09	9.22
44 × 125	41 × 120	9.38	9.60	9.81	9.99	10.15	10.31	10.45	10.64	10.81
50 × 75	47 × 72	6.87	7.01	7.13	7.25	7.35	7.45	7.53	7.67	7.78
50 × 100	47 × 97	8.62	8.80	8.97	9.12	9.25	9.38	9.50	9.66	9.80
50 × 125	47 × 120	10.01	10.24	10.44	10.62	10.77	10.94	11.00	11.00	11.00

TIMBER

Sizes of Internal and External Doorsets

Description	Internal size (mm)	Permissible deviation	External size (mm)	Permissible deviation
Coordinating dimension: height of door leaf height sets	2100		2100	
Coordinating dimension: height of ceiling height set	2300 2350 2400 2700 3000		2300 2350 2400 2700 3000	
Coordinating dimension: width of all doorsets S = Single leaf set D = Double leaf set	600 S 700 S 800 S&D 900 S&D 1000 S&D 1200 D 1500 D 1800 D 2100 D		900 S 1000 S 1200 D 1800 D 2100 D	
Work size: height of door leaf height set	2090	± 2.0	2095	± 2.0
Work size: height of ceiling height set	2285 2335 2385 2685 2985	± 2.0	2295 2345 2395 2695 2995	± 2.0
Work size: width of all doorsets S = Single leaf set D = Double leaf set	590 S 690 S 790 S&D 890 S&D 990 S&D 1190 D 1490 D 1790 D 2090 D	± 2.0	895 S 995 S 1195 D 1495 D 1795 D 2095 D	± 2.0
Width of door leaf in single leaf sets F = Flush leaf P = Panel leaf	526 F 626 F 726 F&P 826 F&P 926 F&P	± 1.5	806 F&P 906 F&P	± 1.5
Width of door leaf in double leaf sets F = Flush leaf P = Panel leaf	362 F 412 F 426 F 562 F&P 712 F&P 826 F&P 1012 F&P	± 1.5	552 F&P 702 F&P 852 F&P 1002 F&P	± 1.5
Door leaf height for all doorsets	2040	± 1.5	1994	± 1.5

ROOFING

Total Roof Loadings for Various Types of Tiles/Slates

	Roof load (slope) kg/m^2		
	Slate/Tile	Roofing underlay and battens^2	Total dead load kg/m
Asbestos cement slate (600 × 300)	21.50	3.14	24.64
Clay tile interlocking	67.00	5.50	72.50
plain	43.50	2.87	46.37
Concrete tile interlocking	47.20	2.69	49.89
plain	78.20	5.50	83.70
Natural slate (18" × 10")	35.40	3.40	38.80
	Roof load (plan) kg/m^2		
Asbestos cement slate (600 × 300)	28.45	76.50	104.95
Clay tile interlocking	53.54	76.50	130.04
plain	83.71	76.50	60.21
Concrete tile interlocking	57.60	76.50	134.10
plain	96.64	76.50	173.14

ROOFING

Tiling Data

Product		Lap (mm)	Gauge of battens	No. slates per m^2	Battens (m/m^2)	Weight as laid (kg/m^2)
CEMENT SLATES						
Eternit slates	600 × 300 mm	100	250	13.4	4.00	19.50
(Duracem)		90	255	13.1	3.92	19.20
		80	260	12.9	3.85	19.00
		70	265	12.7	3.77	18.60
	600 × 350 mm	100	250	11.5	4.00	19.50
		90	255	11.2	3.92	19.20
	500 × 250 mm	100	200	20.0	5.00	20.00
		90	205	19.5	4.88	19.50
		80	210	19.1	4.76	19.00
		70	215	18.6	4.65	18.60
	400 × 200 mm	90	155	32.3	6.45	20.80
		80	160	31.3	6.25	20.20
		70	165	30.3	6.06	19.60
CONCRETE TILES/SLATES						
Redland Roofing						
Stonewold slate	430 × 380 mm	75	355	8.2	2.82	51.20
Double Roman tile	418 × 330 mm	75	355	8.2	2.91	45.50
Grovebury pantile	418 × 332 mm	75	343	9.7	2.91	47.90
Norfolk pantile	381 × 227 mm	75	306	16.3	3.26	44.01
		100	281	17.8	3.56	48.06
Renown interlocking tile	418 × 330 mm	75	343	9.7	2.91	46.40
'49' tile	381 × 227 mm	75	306	16.3	3.26	44.80
		100	281	17.8	3.56	48.95
Plain, vertical tiling	265 × 165 mm	35	115	52.7	8.70	62.20
Marley Roofing						
Bold roll tile	420 × 330 mm	75	344	9.7	2.90	47.00
		100	–	10.5	3.20	51.00
Modern roof tile	420 × 330 mm	75	338	10.2	3.00	54.00
		100	–	11.0	3.20	58.00
Ludlow major	420 × 330 mm	75	338	10.2	3.00	45.00
		100	–	11.0	3.20	49.00
Ludlow plus	387 × 229 mm	75	305	16.1	3.30	47.00
		100	–	17.5	3.60	51.00
Mendip tile	420 × 330 mm	75	338	10.2	3.00	47.00
		100	–	11.0	3.20	51.00
Wessex	413 × 330 mm	75	338	10.2	3.00	54.00
		100	–	11.0	3.20	58.00
Plain tile	267 × 165 mm	65	100	60.0	10.00	76.00
		75	95	64.0	10.50	81.00
		85	90	68.0	11.30	86.00
Plain vertical tiles (feature)	267 × 165 mm	35	110	53.0	8.70	67.00
		34	115	56.0	9.10	71.00

ROOFING

Slate Nails, Quantity per Kilogram

Length	Type			
	Plain wire	**Galvanized wire**	**Copper nail**	**Zinc nail**
28.5 mm	325	305	325	415
34.4 mm	286	256	254	292
50.8 mm	242	224	194	200

Metal Sheet Coverings

Thicknesses and weights of sheet metal coverings								
Lead to BS 1178								
BS Code No	3	4	5	6	7	8		
Colour code	Green	Blue	Red	Black	White	Orange		
Thickness (mm)	1.25	1.80	2.24	2.50	3.15	3.55		
Density kg/m^2	14.18	20.41	25.40	30.05	35.72	40.26		
Copper to BS 2870								
Thickness (mm)		0.60	0.70					
Bay width								
Roll (mm)		500	650					
Seam (mm)		525	600					
Standard width to form bay	600	750						
Normal length of sheet	1.80	1.80						
Zinc to BS 849								
Zinc Gauge (Nr)	9	10	11	12	13	14	15	16
Thickness (mm)	0.43	0.48	0.56	0.64	0.71	0.79	0.91	1.04
Density (kg/m^2)	3.1	3.2	3.8	4.3	4.8	5.3	6.2	7.0
Aluminium to BS 4868								
Thickness (mm)	0.5	0.6	0.7	0.8	0.9	1.0	1.2	
Density (kg/m^2)	12.8	15.4	17.9	20.5	23.0	25.6	30.7	

ROOFING

Type of felt	Nominal mass per unit area (kg/10 m)	Nominal mass per unit area of fibre base (g/m^2)	Nominal length of roll (m)
Class 1			
1B fine granule	14	220	10 or 20
surfaced bitumen	18	330	10 or 20
	25	470	10
1E mineral surfaced bitumen	38	470	10
1F reinforced bitumen	15	160 (fibre)	15
		110 (hessian)	
1F reinforced bitumen, aluminium faced	13	160 (fibre)	15
		110 (hessian)	
Class 2			
2B fine granule surfaced bitumen asbestos	18	500	10 or 20
2E mineral surfaced bitumen asbestos	38	600	10
Class 3			
3B fine granule surfaced bitumen glass fibre	18	60	20
3E mineral surfaced bitumen glass fibre	28	60	10
3E venting base layer bitumen glass fibre	32	60*	10
3H venting base layer bitumen glass fibre	17	60*	20

* Excluding effect of perforations

GLAZING

GLAZING

Nominal thickness (mm)	Tolerance on thickness (mm)	Approximate weight (kg/m²)	Normal maximum size (mm)
Float and polished plate glass			
3	+ 0.2	7.50	2140 × 1220
4	+ 0.2	10.00	2760 × 1220
5	+ 0.2	12.50	3180 × 2100
6	+ 0.2	15.00	4600 × 3180
10	+ 0.3	25.00)	6000 × 3300
12	+ 0.3	30.00)	
15	+ 0.5	37.50	3050 × 3000
19	+ 1.0	47.50)	3000 × 2900
25	+ 1.0	63.50)	
Clear sheet glass			
2 *	+ 0.2	5.00	1920 × 1220
3	+ 0.3	7.50	2130 × 1320
4	+ 0.3	10.00	2760 × 1220
5 *	+ 0.3	12.50)	2130 × 2400
6 *	+ 0.3	15.00)	
Cast glass			
3	+ 0.4		
	− 0.2	6.00)	2140 × 1280
4	+ 0.5	7.50)	
5	+ 0.5	9.50	2140 × 1320
6	+ 0.5	11.50)	3700 × 1280
10	+ 0.8	21.50)	
Wired glass			
(Cast wired glass)			
6	+ 0.3	−)	
	− 0.7	)	3700 × 1840
7	+ 0.7	−)	
(Polished wire glass)			
6	+ 1.0	−	330 × 1830

* The 5 mm and 6 mm thickness are known as *thick drawn sheet*. Although 2 mm sheet glass is available it is not recommended for general glazing purposes

METAL

Weights of Metals

Material	kg/m³	lb/cu ft
Metals, steel construction, etc.		
Iron		
– cast	7207	450
– wrought	7687	480
– ore – general	2407	150
– (crushed) Swedish	3682	230
Steel	7854	490
Copper		
– cast	8731	545
– wrought	8945	558
Brass	8497	530
Bronze	8945	558
Aluminium	2774	173
Lead	11322	707
Zinc (rolled)	7140	446

	g/mm² per metre	lb/sq ft per foot
Steel bars	7.85	3.4

Structural steelwork	Net weight of member @ 7854 kg/m³	
riveted	+ 10% for cleats, rivets, bolts, etc.	
welded	+ 1.25% to 2.5% for welds, etc.	
Rolled sections		
beams	+ 2.5%	
stanchions	+ 5% (extra for caps and bases)	
Plate		
web girders	+ 10% for rivets or welds, stiffeners, etc.	

	kg/m	lb/ft
Steel stairs: industrial type		
1 m or 3 ft wide	84	56
Steel tubes		
50 mm or 2 in bore	5 to 6	3 to 4
Gas piping		
20 mm or ¾ in	2	1¼

METAL

Universal Beams BS 4: Part 1: 2005

Designation	Mass (kg/m)	Depth of section (mm)	Width of section (mm)	Thickness		Surface area (m²/m)
				Web (mm)	Flange (mm)	
1016 × 305 × 487	487.0	1036.1	308.5	30.0	54.1	3.20
1016 × 305 × 438	438.0	1025.9	305.4	26.9	49.0	3.17
1016 × 305 × 393	393.0	1016.0	303.0	24.4	43.9	3.15
1016 × 305 × 349	349.0	1008.1	302.0	21.1	40.0	3.13
1016 × 305 × 314	314.0	1000.0	300.0	19.1	35.9	3.11
1016 × 305 × 272	272.0	990.1	300.0	16.5	31.0	3.10
1016 × 305 × 249	249.0	980.2	300.0	16.5	26.0	3.08
1016 × 305 × 222	222.0	970.3	300.0	16.0	21.1	3.06
914 × 419 × 388	388.0	921.0	420.5	21.4	36.6	3.44
914 × 419 × 343	343.3	911.8	418.5	19.4	32.0	3.42
914 × 305 × 289	289.1	926.6	307.7	19.5	32.0	3.01
914 × 305 × 253	253.4	918.4	305.5	17.3	27.9	2.99
914 × 305 × 224	224.2	910.4	304.1	15.9	23.9	2.97
914 × 305 × 201	200.9	903.0	303.3	15.1	20.2	2.96
838 × 292 × 226	226.5	850.9	293.8	16.1	26.8	2.81
838 × 292 × 194	193.8	840.7	292.4	14.7	21.7	2.79
838 × 292 × 176	175.9	834.9	291.7	14.0	18.8	2.78
762 × 267 × 197	196.8	769.8	268.0	15.6	25.4	2.55
762 × 267 × 173	173.0	762.2	266.7	14.3	21.6	2.53
762 × 267 × 147	146.9	754.0	265.2	12.8	17.5	2.51
762 × 267 × 134	133.9	750.0	264.4	12.0	15.5	2.51
686 × 254 × 170	170.2	692.9	255.8	14.5	23.7	2.35
686 × 254 × 152	152.4	687.5	254.5	13.2	21.0	2.34
686 × 254 × 140	140.1	383.5	253.7	12.4	19.0	2.33
686 × 254 × 125	125.2	677.9	253.0	11.7	16.2	2.32
610 × 305 × 238	238.1	635.8	311.4	18.4	31.4	2.45
610 × 305 × 179	179.0	620.2	307.1	14.1	23.6	2.41
610 × 305 × 149	149.1	612.4	304.8	11.8	19.7	2.39
610 × 229 × 140	139.9	617.2	230.2	13.1	22.1	2.11
610 × 229 × 125	125.1	612.2	229.0	11.9	19.6	2.09
610 × 229 × 113	113.0	607.6	228.2	11.1	17.3	2.08
610 × 229 × 101	101.2	602.6	227.6	10.5	14.8	2.07
533 × 210 × 122	122.0	544.5	211.9	12.7	21.3	1.89
533 × 210 × 109	109.0	539.5	210.8	11.6	18.8	1.88
533 × 210 × 101	101.0	536.7	210.0	10.8	17.4	1.87
533 × 210 × 92	92.1	533.1	209.3	10.1	15.6	1.86
533 × 210 × 82	82.2	528.3	208.8	9.6	13.2	1.85
457 × 191 × 98	98.3	467.2	192.8	11.4	19.6	1.67
457 × 191 × 89	89.3	463.4	191.9	10.5	17.7	1.66
457 × 191 × 82	82.0	460.0	191.3	9.9	16.0	1.65
457 × 191 × 74	74.3	457.0	190.4	9.0	14.5	1.64
457 × 191 × 67	67.1	453.4	189.9	8.5	12.7	1.63
457 × 152 × 82	82.1	465.8	155.3	10.5	18.9	1.51
457 × 152 × 74	74.2	462.0	154.4	9.6	17.0	1.50
457 × 152 × 67	67.2	458.0	153.8	9.0	15.0	1.50
457 × 152 × 60	59.8	454.6	152.9	8.1	13.3	1.50
457 × 152 × 52	52.3	449.8	152.4	7.6	10.9	1.48
406 × 178 × 74	74.2	412.8	179.5	9.5	16.0	1.51
406 × 178 × 67	67.1	409.4	178.8	8.8	14.3	1.50
406 × 178 × 60	60.1	406.4	177.9	7.9	12.8	1.49

METAL

Designation	Mass	Depth of section	Width of section	Thickness		Surface area
	(kg/m)	(mm)	(mm)	Web (mm)	Flange (mm)	(m²/m)
406 × 178 × 50	54.1	402.6	177.7	7.7	10.9	1.48
406 × 140 × 46	46.0	403.2	142.2	6.8	11.2	1.34
406 × 140 × 39	39.0	398.0	141.8	6.4	8.6	1.33
356 × 171 × 67	67.1	363.4	173.2	9.1	15.7	1.38
356 × 171 × 57	57.0	358.0	172.2	8.1	13.0	1.37
356 × 171 × 51	51.0	355.0	171.5	7.4	11.5	1.36
356 × 171 × 45	45.0	351.4	171.1	7.0	9.7	1.36
356 × 127 × 39	39.1	353.4	126.0	6.6	10.7	1.18
356 × 127 × 33	33.1	349.0	125.4	6.0	8.5	1.17
305 × 165 × 54	54.0	310.4	166.9	7.9	13.7	1.26
305 × 165 × 46	46.1	306.6	165.7	6.7	11.8	1.25
305 × 165 × 40	40.3	303.4	165.0	6.0	10.2	1.24
305 × 127 × 48	48.1	311.0	125.3	9.0	14.0	1.09
305 × 127 × 42	41.9	307.2	124.3	8.0	12.1	1.08
305 × 127 × 37	37.0	304.4	123.3	7.1	10.7	1.07
305 × 102 × 33	32.8	312.7	102.4	6.6	10.8	1.01
305 × 102 × 28	28.2	308.7	101.8	6.0	8.8	1.00
305 × 102 × 25	24.8	305.1	101.6	5.8	7.0	0.992
254 × 146 × 43	43.0	259.6	147.3	7.2	12.7	1.08
254 × 146 × 37	37.0	256.0	146.4	6.3	10.9	1.07
254 × 146 × 31	31.1	251.4	146.1	6.0	8.6	1.06
254 × 102 × 28	28.3	260.4	102.2	6.3	10.0	0.904
254 × 102 × 25	25.2	257.2	101.9	6.0	8.4	0.897
254 × 102 × 22	22.0	254.0	101.6	5.7	6.8	0.890
203 × 133 × 30	30.0	206.8	133.9	6.4	9.6	0.923
203 × 133 × 25	25.1	203.2	133.2	5.7	7.8	0.915
203 × 102 × 23	23.1	203.2	101.8	5.4	9.3	0.790
178 × 102 × 19	19.0	177.8	101.2	4.8	7.9	0.738
152 × 89 × 16	16.0	152.4	88.7	4.5	7.7	0.638
127 × 76 × 13	13.0	127.0	76.0	4.0	7.6	0.537

METAL

Universal Columns BS 4: Part 1: 2005

Designation	Mass (kg/m)	Depth of section (mm)	Width of section (mm)	Thickness		Surface area (m²/m)
				Web (mm)	Flange (mm)	
356 × 406 × 634	633.9	474.7	424.0	47.6	77.0	2.52
356 × 406 × 551	551.0	455.6	418.5	42.1	67.5	2.47
356 × 406 × 467	467.0	436.6	412.2	35.8	58.0	2.42
356 × 406 × 393	393.0	419.0	407.0	30.6	49.2	2.38
356 × 406 × 340	339.9	406.4	403.0	26.6	42.9	2.35
356 × 406 × 287	287.1	393.6	399.0	22.6	36.5	2.31
356 × 406 × 235	235.1	381.0	384.8	18.4	30.2	2.28
356 × 368 × 202	201.9	374.6	374.7	16.5	27.0	2.19
356 × 368 × 177	177.0	368.2	372.6	14.4	23.8	2.17
356 × 368 × 153	152.9	362.0	370.5	12.3	20.7	2.16
356 × 368 × 129	129.0	355.6	368.6	10.4	17.5	2.14
305 × 305 × 283	282.9	365.3	322.2	26.8	44.1	1.94
305 × 305 × 240	240.0	352.5	318.4	23.0	37.7	1.91
305 × 305 × 198	198.1	339.9	314.5	19.1	31.4	1.87
305 × 305 × 158	158.1	327.1	311.2	15.8	25.0	1.84
305 × 305 × 137	136.9	320.5	309.2	13.8	21.7	1.82
305 × 305 × 118	117.9	314.5	307.4	12.0	18.7	1.81
305 × 305 × 97	96.9	307.9	305.3	9.9	15.4	1.79
254 × 254 × 167	167.1	289.1	265.2	19.2	31.7	1.58
254 × 254 × 132	132.0	276.3	261.3	15.3	25.3	1.55
254 × 254 × 107	107.1	266.7	258.8	12.8	20.5	1.52
254 × 254 × 89	88.9	260.3	256.3	10.3	17.3	1.50
254 × 254 × 73	73.1	254.1	254.6	8.6	14.2	1.49
203 × 203 × 86	86.1	222.2	209.1	12.7	20.5	1.24
203 × 203 × 71	71.0	215.8	206.4	10.0	17.3	1.22
203 × 203 × 60	60.0	209.6	205.8	9.4	14.2	1.21
203 × 203 × 52	52.0	206.2	204.3	7.9	12.5	1.20
203 × 203 × 46	46.1	203.2	203.6	7.2	11.0	1.19
152 × 152 × 37	37.0	161.8	154.4	8.0	11.5	0.912
152 × 152 × 30	30.0	157.6	152.9	6.5	9.4	0.901
152 × 152 × 23	23.0	152.4	152.2	5.8	6.8	0.889

METAL

Joists BS 4: Part 1: 2005 (retained for reference, Corus have ceased manufacture in UK)

Designation	Mass (kg/m)	Depth of section (mm)	Width of section (mm)	Thickness		Surface area (m²/m)
				Web (mm)	Flange (mm)	
254 × 203 × 82	82.0	254.0	203.2	10.2	19.9	1.210
203 × 152 × 52	52.3	203.2	152.4	8.9	16.5	0.932
152 × 127 × 37	37.3	152.4	127.0	10.4	13.2	0.737
127 × 114 × 29	29.3	127.0	114.3	10.2	11.5	0.646
127 × 114 × 27	26.9	127.0	114.3	7.4	11.4	0.650
102 × 102 × 23	23.0	101.6	101.6	9.5	10.3	0.549
102 × 44 × 7	7.5	101.6	44.5	4.3	6.1	0.350
89 × 89 × 19	19.5	88.9	88.9	9.5	9.9	0.476
76 × 76 × 13	12.8	76.2	76.2	5.1	8.4	0.411

Parallel Flange Channels

Designation	Mass (kg/m)	Depth of section (mm)	Width of section (mm)	Thickness		Surface area (m²/m)
				Web (mm)	Flange (mm)	
430 × 100 × 64	64.4	430	100	11.0	19.0	1.23
380 × 100 × 54	54.0	380	100	9.5	17.5	1.13
300 × 100 × 46	45.5	300	100	9.0	16.5	0.969
300 × 90 × 41	41.4	300	90	9.0	15.5	0.932
260 × 90 × 35	34.8	260	90	8.0	14.0	0.854
260 × 75 × 28	27.6	260	75	7.0	12.0	0.79
230 × 90 × 32	32.2	230	90	7.5	14.0	0.795
230 × 75 × 26	25.7	230	75	6.5	12.5	0.737
200 × 90 × 30	29.7	200	90	7.0	14.0	0.736
200 × 75 × 23	23.4	200	75	6.0	12.5	0.678
180 × 90 × 26	26.1	180	90	6.5	12.5	0.697
180 × 75 × 20	20.3	180	75	6.0	10.5	0.638
150 × 90 × 24	23.9	150	90	6.5	12.0	0.637
150 × 75 × 18	17.9	150	75	5.5	10.0	0.579
125 × 65 × 15	14.8	125	65	5.5	9.5	0.489
100 × 50 × 10	10.2	100	50	5.0	8.5	0.382

METAL

Equal Angles BS EN 10056-1

Designation	Mass (kg/m)	Surface area (m²/m)
200 × 200 × 24	71.1	0.790
200 × 200 × 20	59.9	0.790
200 × 200 × 18	54.2	0.790
200 × 200 × 16	48.5	0.790
150 × 150 × 18	40.1	0.59
150 × 150 × 15	33.8	0.59
150 × 150 × 12	27.3	0.59
150 × 150 × 10	23.0	0.59
120 × 120 × 15	26.6	0.47
120 × 120 × 12	21.6	0.47
120 × 120 × 10	18.2	0.47
120 × 120 × 8	14.7	0.47
100 × 100 × 15	21.9	0.39
100 × 100 × 12	17.8	0.39
100 × 100 × 10	15.0	0.39
100 × 100 × 8	12.2	0.39
90 × 90 × 12	15.9	0.35
90 × 90 × 10	13.4	0.35
90 × 90 × 8	10.9	0.35
90 × 90 × 7	9.61	0.35
90 × 90 × 6	8.30	0.35

Unequal Angles BS EN 10056-1

Designation	Mass (kg/m)	Surface area (m²/m)
200 × 150 × 18	47.1	0.69
200 × 150 × 15	39.6	0.69
200 × 150 × 12	32.0	0.69
200 × 100 × 15	33.7	0.59
200 × 100 × 12	27.3	0.59
200 × 100 × 10	23.0	0.59
150 × 90 × 15	26.6	0.47
150 × 90 × 12	21.6	0.47
150 × 90 × 10	18.2	0.47
150 × 75 × 15	24.8	0.44
150 × 75 × 12	20.2	0.44
150 × 75 × 10	17.0	0.44
125 × 75 × 12	17.8	0.40
125 × 75 × 10	15.0	0.40
125 × 75 × 8	12.2	0.40
100 × 75 × 12	15.4	0.34
100 × 75 × 10	13.0	0.34
100 × 75 × 8	10.6	0.34
100 × 65 × 10	12.3	0.32
100 × 65 × 8	9.94	0.32
100 × 65 × 7	8.77	0.32

METAL

Structural Tees Split from Universal Beams BS 4: Part 1: 2005

Designation	Mass (kg/m)	Surface area (m²/m)
305 × 305 × 90	89.5	1.22
305 × 305 × 75	74.6	1.22
254 × 343 × 63	62.6	1.19
229 × 305 × 70	69.9	1.07
229 × 305 × 63	62.5	1.07
229 × 305 × 57	56.5	1.07
229 × 305 × 51	50.6	1.07
210 × 267 × 61	61.0	0.95
210 × 267 × 55	54.5	0.95
210 × 267 × 51	50.5	0.95
210 × 267 × 46	46.1	0.95
210 × 267 × 41	41.1	0.95
191 × 229 × 49	49.2	0.84
191 × 229 × 45	44.6	0.84
191 × 229 × 41	41.0	0.84
191 × 229 × 37	37.1	0.84
191 × 229 × 34	33.6	0.84
152 × 229 × 41	41.0	0.76
152 × 229 × 37	37.1	0.76
152 × 229 × 34	33.6	0.76
152 × 229 × 30	29.9	0.76
152 × 229 × 26	26.2	0.76

Universal Bearing Piles BS 4: Part 1: 2005

Designation	Mass (kg/m)	Depth of Section (mm)	Width of section (mm)	Thickness	
				Web (mm)	Flange (mm)
356 × 368 × 174	173.9	361.4	378.5	20.3	20.4
356 × 368 × 152	152.0	356.4	376.0	17.8	17.9
356 × 368 × 133	133.0	352.0	373.8	15.6	15.7
356 × 368 × 109	108.9	346.4	371.0	12.8	12.9
305 × 305 × 223	222.9	337.9	325.7	30.3	30.4
305 × 305 × 186	186.0	328.3	320.9	25.5	25.6
305 × 305 × 149	149.1	318.5	316.0	20.6	20.7
305 × 305 × 126	126.1	312.3	312.9	17.5	17.6
305 × 305 × 110	110.0	307.9	310.7	15.3	15.4
305 × 305 × 95	94.9	303.7	308.7	13.3	13.3
305 × 305 × 88	88.0	301.7	307.8	12.4	12.3
305 × 305 × 79	78.9	299.3	306.4	11.0	11.1
254 × 254 × 85	85.1	254.3	260.4	14.4	14.3
254 × 254 × 71	71.0	249.7	258.0	12.0	12.0
254 × 254 × 63	63.0	247.1	256.6	10.6	10.7
203 × 203 × 54	53.9	204.0	207.7	11.3	11.4
203 × 203 × 45	44.9	200.2	205.9	9.5	9.5

METAL

Hot Formed Square Hollow Sections EN 10210 S275J2H & S355J2H

Size (mm)	Wall thickness (mm)	Mass (kg/m)	Superficial area (m²/m)
40 × 40	2.5	2.89	0.154
	3.0	3.41	0.152
	3.2	3.61	0.152
	3.6	4.01	0.151
	4.0	4.39	0.150
	5.0	5.28	0.147
50 × 50	2.5	3.68	0.194
	3.0	4.35	0.192
	3.2	4.62	0.192
	3.6	5.14	0.191
	4.0	5.64	0.190
	5.0	6.85	0.187
	6.0	7.99	0.185
	6.3	8.31	0.184
60 × 60	3.0	5.29	0.232
	3.2	5.62	0.232
	3.6	6.27	0.231
	4.0	6.90	0.230
	5.0	8.42	0.227
	6.0	9.87	0.225
	6.3	10.30	0.224
	8.0	12.50	0.219
70 × 70	3.0	6.24	0.272
	3.2	6.63	0.272
	3.6	7.40	0.271
	4.0	8.15	0.270
	5.0	9.99	0.267
	6.0	11.80	0.265
	6.3	12.30	0.264
	8.0	15.00	0.259
80 × 80	3.2	7.63	0.312
	3.6	8.53	0.311
	4.0	9.41	0.310
	5.0	11.60	0.307
	6.0	13.60	0.305
	6.3	14.20	0.304
	8.0	17.50	0.299
90 × 90	3.6	9.66	0.351
	4.0	10.70	0.350
	5.0	13.10	0.347
	6.0	15.50	0.345
	6.3	16.20	0.344
	8.0	20.10	0.339
100 × 100	3.6	10.80	0.391
	4.0	11.90	0.390
	5.0	14.70	0.387
	6.0	17.40	0.385
	6.3	18.20	0.384
	8.0	22.60	0.379
	10.0	27.40	0.374
120 × 120	4.0	14.40	0.470
	5.0	17.80	0.467
	6.0	21.20	0.465

Tables and Memoranda

METAL

Size (mm)	Wall thickness (mm)	Mass (kg/m)	Superficial area (m²/m)
	6.3	22.20	0.464
	8.0	27.60	0.459
	10.0	33.70	0.454
	12.0	39.50	0.449
	12.5	40.90	0.448
140 × 140	5.0	21.00	0.547
	6.0	24.90	0.545
	6.3	26.10	0.544
	8.0	32.60	0.539
	10.0	40.00	0.534
	12.0	47.00	0.529
	12.5	48.70	0.528
150 × 150	5.0	22.60	0.587
	6.0	26.80	0.585
	6.3	28.10	0.584
	8.0	35.10	0.579
	10.0	43.10	0.574
	12.0	50.80	0.569
	12.5	52.70	0.568
Hot formed from seamless hollow	16.0	65.2	0.559
160 × 160	5.0	24.10	0.627
	6.0	28.70	0.625
	6.3	30.10	0.624
	8.0	37.60	0.619
	10.0	46.30	0.614
	12.0	54.60	0.609
	12.5	56.60	0.608
	16.0	70.20	0.599
180 × 180	5.0	27.30	0.707
	6.0	32.50	0.705
	6.3	34.00	0.704
	8.0	42.70	0.699
	10.0	52.50	0.694
	12.0	62.10	0.689
	12.5	64.40	0.688
	16.0	80.20	0.679
200 × 200	5.0	30.40	0.787
	6.0	36.20	0.785
	6.3	38.00	0.784
	8.0	47.70	0.779
	10.0	58.80	0.774
	12.0	69.60	0.769
	12.5	72.30	0.768
	16.0	90.30	0.759
250 × 250	5.0	38.30	0.987
	6.0	45.70	0.985
	6.3	47.90	0.984
	8.0	60.30	0.979
	10.0	74.50	0.974
	12.0	88.50	0.969
	12.5	91.90	0.968
	16.0	115.00	0.959

METAL

Size (mm)	Wall thickness (mm)	Mass (kg/m)	Superficial area (m²/m)
300 × 300	6.0	55.10	1.18
	6.3	57.80	1.18
	8.0	72.80	1.18
	10.0	90.20	1.17
	12.0	107.00	1.17
	12.5	112.00	1.17
	16.0	141.00	1.16
350 × 350	8.0	85.40	1.38
	10.0	106.00	1.37
	12.0	126.00	1.37
	12.5	131.00	1.37
	16.0	166.00	1.36
400 × 400	8.0	97.90	1.58
	10.0	122.00	1.57
	12.0	145.00	1.57
	12.5	151.00	1.57
	16.0	191.00	1.56
(Grade S355J2H only)	20.00*	235.00	1.55

Note: * SAW process

METAL

Hot Formed Square Hollow Sections JUMBO RHS: JIS G3136

Size (mm)	Wall thickness (mm)	Mass (kg/m)	Superficial area (m²/m)
350 × 350	19.0	190.00	1.33
	22.0	217.00	1.32
	25.0	242.00	1.31
400 × 400	22.0	251.00	1.52
	25.0	282.00	1.51
450 × 450	12.0	162.00	1.76
	16.0	213.00	1.75
	19.0	250.00	1.73
	22.0	286.00	1.72
	25.0	321.00	1.71
	28.0 *	355.00	1.70
	32.0 *	399.00	1.69
500 × 500	12.0	181.00	1.96
	16.0	238.00	1.95
	19.0	280.00	1.93
	22.0	320.00	1.92
	25.0	360.00	1.91
	28.0 *	399.00	1.90
	32.0 *	450.00	1.89
	36.0 *	498.00	1.88
550 × 550	16.0	263.00	2.15
	19.0	309.00	2.13
	22.0	355.00	2.12
	25.0	399.00	2.11
	28.0 *	443.00	2.10
	32.0 *	500.00	2.09
	36.0 *	555.00	2.08
	40.0 *	608.00	2.06
600 × 600	25.0 *	439.00	2.31
	28.0 *	487.00	2.30
	32.0 *	550.00	2.29
	36.0 *	611.00	2.28
	40.0 *	671.00	2.26
700 × 700	25.0 *	517.00	2.71
	28.0 *	575.00	2.70
	32.0 *	651.00	2.69
	36.0 *	724.00	2.68
	40.0 *	797.00	2.68

Note: * SAW process

METAL

Hot Formed Rectangular Hollow Sections: EN10210 S275J2h & S355J2H

Size (mm)	Wall thickness (mm)	Mass (kg/m)	Superficial area (m²/m)
50 × 30	2.5	2.89	0.154
	3.0	3.41	0.152
	3.2	3.61	0.152
	3.6	4.01	0.151
	4.0	4.39	0.150
	5.0	5.28	0.147
60 × 40	2.5	3.68	0.194
	3.0	4.35	0.192
	3.2	4.62	0.192
	3.6	5.14	0.191
	4.0	5.64	0.190
	5.0	6.85	0.187
	6.0	7.99	0.185
	6.3	8.31	0.184
80 × 40	3.0	5.29	0.232
	3.2	5.62	0.232
	3.6	6.27	0.231
	4.0	6.90	0.230
	5.0	8.42	0.227
	6.0	9.87	0.225
	6.3	10.30	0.224
	8.0	12.50	0.219
76.2 × 50.8	3.0	5.62	0.246
	3.2	5.97	0.246
	3.6	6.66	0.245
	4.0	7.34	0.244
	5.0	8.97	0.241
	6.0	10.50	0.239
	6.3	11.00	0.238
	8.0	13.40	0.233
90 × 50	3.0	6.24	0.272
	3.2	6.63	0.272
	3.6	7.40	0.271
	4.0	8.15	0.270
	5.0	9.99	0.267
	6.0	11.80	0.265
	6.3	12.30	0.264
	8.0	15.00	0.259
100 × 50	3.0	6.71	0.292
	3.2	7.13	0.292
	3.6	7.96	0.291
	4.0	8.78	0.290
	5.0	10.80	0.287
	6.0	12.70	0.285
	6.3	13.30	0.284
	8.0	16.30	0.279

METAL

Size (mm)	Wall thickness (mm)	Mass (kg/m)	Superficial area (m²/m)
100 × 60	3.0	7.18	0.312
	3.2	7.63	0.312
	3.6	8.53	0.311
	4.0	9.41	0.310
	5.0	11.60	0.307
	6.0	13.60	0.305
	6.3	14.20	0.304
	8.0	17.50	0.299
120 × 60	3.6	9.70	0.351
	4.0	10.70	0.350
	5.0	13.10	0.347
	6.0	15.50	0.345
	6.3	16.20	0.344
	8.0	20.10	0.339
120 × 80	3.6	10.80	0.391
	4.0	11.90	0.390
	5.0	14.70	0.387
	6.0	17.40	0.385
	6.3	18.20	0.384
	8.0	22.60	0.379
	10.0	27.40	0.374
150 × 100	4.0	15.10	0.490
	5.0	18.60	0.487
	6.0	22.10	0.485
	6.3	23.10	0.484
	8.0	28.90	0.479
	10.0	35.30	0.474
	12.0	41.40	0.469
	12.5	42.80	0.468
160 × 80	4.0	14.40	0.470
	5.0	17.80	0.467
	6.0	21.20	0.465
	6.3	22.20	0.464
	8.0	27.60	0.459
	10.0	33.70	0.454
	12.0	39.50	0.449
	12.5	40.90	0.448
200 × 100	5.0	22.60	0.587
	6.0	26.80	0.585
	6.3	28.10	0.584
	8.0	35.10	0.579
	10.0	43.10	0.574
	12.0	50.80	0.569
	12.5	52.70	0.568
	16.0	65.20	0.559
250 × 150	5.0	30.40	0.787
	6.0	36.20	0.785
	6.3	38.00	0.784
	8.0	47.70	0.779
	10.0	58.80	0.774
	12.0	69.60	0.769
	12.5	72.30	0.768
	16.0	90.30	0.759

METAL

Size (mm)	Wall thickness (mm)	Mass (kg/m)	Superficial area (m²/m)
300 × 200	5.0	38.30	0.987
	6.0	45.70	0.985
	6.3	47.90	0.984
	8.0	60.30	0.979
	10.0	74.50	0.974
	12.0	88.50	0.969
	12.5	91.90	0.968
	16.0	115.00	0.959
400 × 200	6.0	55.10	1.18
	6.3	57.80	1.18
	8.0	72.80	1.18
	10.0	90.20	1.17
	12.0	107.00	1.17
	12.5	112.00	1.17
	16.0	141.00	1.16
450 × 250	8.0	85.40	1.38
	10.0	106.00	1.37
	12.0	126.00	1.37
	12.5	131.00	1.37
	16.0	166.00	1.36
500 × 300	8.0	98.00	1.58
	10.0	122.00	1.57
	12.0	145.00	1.57
	12.5	151.00	1.57
	16.0	191.00	1.56
	20.0	235.00	1.55

METAL

Hot Formed Circular Hollow Sections EN 10210 S275J2H & S355J2H

Outside diameter (mm)	Wall thickness (mm)	Mass (kg/m)	Superficial area (m²/m)
21.3	3.2	1.43	0.067
26.9	3.2	1.87	0.085
33.7	3.0	2.27	0.106
	3.2	2.41	0.106
	3.6	2.67	0.106
	4.0	2.93	0.106
42.4	3.0	2.91	0.133
	3.2	3.09	0.133
	3.6	3.44	0.133
	4.0	3.79	0.133
48.3	2.5	2.82	0.152
	3.0	3.35	0.152
	3.2	3.56	0.152
	3.6	3.97	0.152
	4.0	4.37	0.152
	5.0	5.34	0.152
60.3	2.5	3.56	0.189
	3.0	4.24	0.189
	3.2	4.51	0.189
	3.6	5.03	0.189
	4.0	5.55	0.189
	5.0	6.82	0.189
76.1	2.5	4.54	0.239
	3.0	5.41	0.239
	3.2	5.75	0.239
	3.6	6.44	0.239
	4.0	7.11	0.239
	5.0	8.77	0.239
	6.0	10.40	0.239
	6.3	10.80	0.239
88.9	2.5	5.33	0.279
	3.0	6.36	0.279
	3.2	6.76	0.27
	3.6	7.57	0.279
	4.0	8.38	0.279
	5.0	10.30	0.279
	6.0	12.30	0.279
	6.3	12.80	0.279
114.3	3.0	8.23	0.359
	3.2	8.77	0.359
	3.6	9.83	0.359
	4.0	10.09	0.359
	5.0	13.50	0.359
	6.0	16.00	0.359
	6.3	16.80	0.359

METAL

Outside diameter (mm)	Wall thickness (mm)	Mass (kg/m)	Superficial area (m²/m)
139.7	3.2	10.80	0.439
	3.6	12.10	0.439
	4.0	13.40	0.439
	5.0	16.60	0.439
	6.0	19.80	0.439
	6.3	20.70	0.439
	8.0	26.00	0.439
	10.0	32.00	0.439
168.3	3.2	13.00	0.529
	3.6	14.60	0.529
	4.0	16.20	0.529
	5.0	20.10	0.529
	6.0	24.00	0.529
	6.3	25.20	0.529
	8.0	31.60	0.529
	10.0	39.00	0.529
	12.0	46.30	0.529
	12.5	48.00	0.529
193.7	5.0	23.30	0.609
	6.0	27.80	0.609
	6.3	29.10	0.609
	8.0	36.60	0.609
	10.0	45.30	0.609
	12.0	53.80	0.609
	12.5	55.90	0.609
219.1	5.0	26.40	0.688
	6.0	31.50	0.688
	6.3	33.10	0.688
	8.0	41.60	0.688
	10.0	51.60	0.688
	12.0	61.30	0.688
	12.5	63.70	0.688
	16.0	80.10	0.688
244.5	5.0	29.50	0.768
	6.0	35.30	0.768
	6.3	37.00	0.768
	8.0	46.70	0.768
	10.0	57.80	0.768
	12.0	68.80	0.768
	12.5	71.50	0.768
	16.0	90.20	0.768
273.0	5.0	33.00	0.858
	6.0	39.50	0.858
	6.3	41.40	0.858
	8.0	52.30	0.858
	10.0	64.90	0.858
	12.0	77.20	0.858
	12.5	80.30	0.858
	16.0	101.00	0.858
323.9	5.0	39.30	1.02
	6.0	47.00	1.02
	6.3	49.30	1.02
	8.0	62.30	1.02
	10.0	77.40	1.02

METAL

Outside diameter (mm)	Wall thickness (mm)	Mass (kg/m)	Superficial area (m²/m)
	12.0	92.30	1.02
	12.5	96.00	1.02
	16.0	121.00	1.02
355.6	6.3	54.30	1.12
	8.0	68.60	1.12
	10.0	85.30	1.12
	12.0	102.00	1.12
	12.5	106.00	1.12
	16.0	134.00	1.12
406.4	6.3	62.20	1.28
	8.0	79.60	1.28
	10.0	97.80	1.28
	12.0	117.00	1.28
	12.5	121.00	1.28
	16.0	154.00	1.28
457.0	6.3	70.00	1.44
	8.0	88.60	1.44
	10.0	110.00	1.44
	12.0	132.00	1.44
	12.5	137.00	1.44
	16.0	174.00	1.44
508.0	6.3	77.90	1.60
	8.0	98.60	1.60
	10.0	123.00	1.60
	12.0	147.00	1.60
	12.5	153.00	1.60
	16.0	194.00	1.60

METAL

Spacing of Holes in Angles

Nominal leg length (mm)	Spacing of holes						Maximum diameter of bolt or rivet		
	A	B	C	D	E	F	A	B and C	D, E and F
200		75	75	55	55	55		30	20
150		55	55					20	
125		45	60					20	
120									
100	55						24		
90	50						24		
80	45						20		
75	45						20		
70	40						20		
65	35						20		
60	35						16		
50	28						12		
45	25								
40	23								
30	20								
25	15								

KERBS, PAVING, ETC.

KERBS/EDGINGS/CHANNELS

Precast Concrete Kerbs to BS 7263

Straight kerb units: length from 450 to 915 mm

150 mm high × 125 mm thick		
bullnosed	type BN	
half battered	type HB3	
255 mm high × 125 mm thick		
45° splayed	type SP	
half battered	type HB2	
305 mm high × 150 mm thick		
half battered	type HB1	
Quadrant kerb units		
150 mm high × 305 and 455 mm radius to match	type BN	type QBN
150 mm high × 305 and 455 mm radius to match	type HB2, HB3	type QHB
150 mm high × 305 and 455 mm radius to match	type SP	type QSP
255 mm high × 305 and 455 mm radius to match	type BN	type QBN
255 mm high × 305 and 455 mm radius to match	type HB2, HB3	type QHB
225 mm high × 305 and 455 mm radius to match	type SP	type QSP
Angle kerb units		
305 × 305 × 225 mm high × 125 mm thick		
bullnosed external angle	type XA	
splayed external angle to match type SP	type XA	
bullnosed internal angle	type IA	
splayed internal angle to match type SP	type IA	
Channels		
255 mm wide × 125 mm high flat	type CS1	
150 mm wide × 125 mm high flat type	CS2	
255 mm wide × 125 mm high dished	type CD	

KERBS, PAVING, ETC.

Transition kerb units			
from kerb type SP to HB	left handed	type TL	
	right handed	type TR	
from kerb type BN to HB	left handed	type DL1	
	right handed	type DR1	
from kerb type BN to SP	left handed	type DL2	
	right handed	type DR2	

Number of kerbs required per quarter circle (780 mm kerb lengths)

Radius (m)	Number in quarter circle
12	24
10	20
8	16
6	12
5	10
4	8
3	6
2	4
1	2

Precast Concrete Edgings

Round top type ER	Flat top type EF	Bullnosed top type EBN
150 × 50 mm	150 × 50 mm	150 × 50 mm
200 × 50 mm	200 × 50 mm	200 × 50 mm
250 × 50 mm	250 × 50 mm	250 × 50 mm

KERBS, PAVING, ETC.

BASES

Cement Bound Material for Bases and Subbases

CBM1:	very carefully graded aggregate from 37.5–75 mm, with a 7-day strength of 4.5 N/mm^2
CBM2:	same range of aggregate as CBM1 but with more tolerance in each size of aggregate with a 7-day strength of 7.0 N/mm^2
CBM3:	crushed natural aggregate or blast furnace slag, graded from 37.5–150 mm for 40 mm aggregate, and from 20–75 mm for 20 mm aggregate, with a 7-day strength of 10 N/mm^2
CBM4:	crushed natural aggregate or blast furnace slag, graded from 37.5–150 mm for 40 mm aggregate, and from 20–75 mm for 20 mm aggregate, with a 7-day strength of 15 N/mm^2

INTERLOCKING BRICK/BLOCK ROADS/PAVINGS

Sizes of Precast Concrete Paving Blocks

Type R blocks
200 × 100 × 60 mm
200 × 100 × 65 mm
200 × 100 × 80 mm
200 × 100 × 100 mm

Type S
Any shape within a 295 mm space

Sizes of clay brick pavers
200 × 100 × 50 mm
200 × 100 × 65 mm
210 × 105 × 50 mm
210 × 105 × 65 mm
215 × 102.5 × 50 mm
215 × 102.5 × 65 mm

Type PA: 3 kN
Footpaths and pedestrian areas, private driveways, car parks, light vehicle traffic and over-run

Type PB: 7 kN
Residential roads, lorry parks, factory yards, docks, petrol station forecourts, hardstandings, bus stations

KERBS, PAVING, ETC.

PAVING AND SURFACING

Weights and Sizes of Paving and Surfacing

Description of item	Size	Quantity per tonne
Paving 50 mm thick	900 × 600 mm	15
Paving 50 mm thick	750 × 600 mm	18
Paving 50 mm thick	600 × 600 mm	23
Paving 50 mm thick	450 × 600 mm	30
Paving 38 mm thick	600 × 600 mm	30
Path edging	914 × 50 × 150 mm	60
Kerb (including radius and tapers)	125 × 254 × 914 mm	15
Kerb (including radius and tapers)	125 × 150 × 914 mm	25
Square channel	125 × 254 × 914 mm	15
Dished channel	125 × 254 × 914 mm	15
Quadrants	300 × 300 × 254 mm	19
Quadrants	450 × 450 × 254 mm	12
Quadrants	300 × 300 × 150 mm	30
Internal angles	300 × 300 × 254 mm	30
Fluted pavement channel	255 × 75 × 914 mm	25
Corner stones	300 × 300 mm	80
Corner stones	360 × 360 mm	60
Cable covers	914 × 175 mm	55
Gulley kerbs	220 × 220 × 150 mm	60
Gulley kerbs	220 × 200 × 75 mm	120

KERBS, PAVING, ETC.

Weights and Sizes of Paving and Surfacing

Material	kg/m³	lb/cu yd
Tarmacadam	2306	3891
Macadam (waterbound)	2563	4325
Vermiculite (aggregate)	64–80	108–135
Terracotta	2114	3568
Cork – compressed	388	24
	kg/m²	lb/sq ft
Clay floor tiles, 12.7 mm	27.3	5.6
Pavement lights	122	25
Damp-proof course	5	1
	kg/m² per mm thickness	lb/sq ft per inch thickness
Paving slabs (stone)	2.3	12
Granite setts	2.88	15
Asphalt	2.30	12
Rubber flooring	1.68	9
Polyvinyl chloride	1.94 (avg)	10 (avg)

Coverage (m²) Per Cubic Metre of Materials Used as Subbases or Capping Layers

Consolidated thickness laid in (mm)	Square metre coverage		
	Gravel	Sand	Hardcore
50	15.80	16.50	–
75	10.50	11.00	–
100	7.92	8.20	7.42
125	6.34	6.60	5.90
150	5.28	5.50	4.95
175	–	–	4.23
200	–	–	3.71
225	–	–	3.30
300	–	–	2.47

KERBS, PAVING, ETC.

Approximate Rate of Spreads

Average thickness of course (mm)	Description	Approximate rate of spread			
		Open Textured		Dense, Medium & Fine Textured	
		(kg/m²)	(m²/t)	(kg/m²)	(m²/t)
35	14 mm open textured or dense wearing course	60–75	13–17	70–85	12–14
40	20 mm open textured or dense base course	70–85	12–14	80–100	10–12
45	20 mm open textured or dense base course	80–100	10–12	95–100	9–10
50	20 mm open textured or dense, or 28 mm dense base course	85–110	9–12	110–120	8–9
60	28 mm dense base course, 40 mm open textured of dense base course or 40 mm single course as base course		8–10	130–150	7–8
65	28 mm dense base course, 40 mm open textured or dense base course or 40 mm single course	100–135	7–10	140–160	6–7
75	40 mm single course, 40 mm open textured or dense base course, 40 mm dense roadbase	120–150	7–8	165–185	5–6
100	40 mm dense base course or roadbase	–	–	220–240	4–4.5

KERBS, PAVING, ETC.

Surface Dressing Roads: Coverage (m²) per Tonne of Material

Size in mm	Sand	Granite chips	Gravel	Limestone chips
Sand	168	–	–	–
3	–	148	152	165
6	–	130	133	144
9	–	111	114	123
13	–	85	87	95
19	–	68	71	78

Sizes of Flags

Reference	Nominal size (mm)	Thickness (mm)
A	600 × 450	50 and 63
B	600 × 600	50 and 63
C	600 × 750	50 and 63
D	600 × 900	50 and 63
E	450 × 450	50 and 70 chamfered top surface
F	400 × 400	50 and 65 chamfered top surface
G	300 × 300	50 and 60 chamfered top surface

Sizes of Natural Stone Setts

Width (mm)		Length (mm)		Depth (mm)
100	×	100	×	100
75	×	150 to 250	×	125
75	×	150 to 250	×	150
100	×	150 to 250	×	100
100	×	150 to 250	×	150

SEEDING/TURFING AND PLANTING

Topsoil Quality

Topsoil grade	Properties
Premium	Natural topsoil, high fertility, loamy texture, good soil structure, suitable for intensive cultivation.
General purpose	Natural or manufactured topsoil of lesser quality than Premium, suitable for agriculture or amenity landscape, may need fertilizer or soil structure improvement.
Economy	Selected subsoil, natural mineral deposit such as river silt or greensand. The grade comprises two subgrades; 'Low clay' and 'High clay' which is more liable to compaction in handling. This grade is suitable for low-production agricultural land and amenity woodland or conservation planting areas.

Forms of Trees

Standards:	Shall be clear with substantially straight stems. Grafted and budded trees shall have no more than a slight bend at the union. Standards shall be designated as Half, Extra light, Light, Standard, Selected standard, Heavy, and Extra heavy.
Sizes of Standards	
Heavy standard	12–14 cm girth × 3.50 to 5.00 m high
Extra Heavy standard	14–16 cm girth × 4.25 to 5.00 m high
Extra Heavy standard	16–18 cm girth × 4.25 to 6.00 m high
Extra Heavy standard	18–20 cm girth × 5.00 to 6.00 m high
Semi-mature trees:	Between 6.0 m and 12.0 m tall with a girth of 20 to 75 cm at 1.0 m above ground.
Feathered trees:	Shall have a defined upright central leader, with stem furnished with evenly spread and balanced lateral shoots down to or near the ground.
Whips:	Shall be without significant feather growth as determined by visual inspection.
Multi-stemmed trees:	Shall have two or more main stems at, near, above or below ground.

Seedlings grown from seed and not transplanted shall be specified when ordered for sale as:

1+0	one year old seedling
2+0	two year old seedling
1+1	one year seed bed, one year transplanted = two year old seedling
1+2	one year seed bed, two years transplanted = three year old seedling
2+1	two years seed bed, one year transplanted = three year old seedling
1u1	two years seed bed, undercut after 1 year = two year old seedling
2u2	four years seed bed, undercut after 2 years = four year old seedling

SEEDING/TURFING AND PLANTING

Cuttings

The age of cuttings (plants grown from shoots, stems, or roots of the mother plant) shall be specified when ordered for sale. The height of transplants and undercut seedlings/cuttings (which have been transplanted or undercut at least once) shall be stated in centimetres. The number of growing seasons before and after transplanting or undercutting shall be stated.

0 + 1	one year cutting
0 + 2	two year cutting
0 + 1 + 1	one year cutting bed, one year transplanted = two year old seedling
0 + 1 + 2	one year cutting bed, two years transplanted = three year old seedling

Grass Cutting Capacities in m² per hour

Speed mph	Width of cut in metres												
	0.5	0.7	1.0	1.2	1.5	1.7	2.0	2.0	2.1	2.5	2.8	3.0	3.4
1.0	724	1127	1529	1931	2334	2736	3138	3219	3380	4023	4506	4828	5472
1.5	1086	1690	2293	2897	3500	4104	4707	4828	5069	6035	6759	7242	8208
2.0	1448	2253	3058	3862	4667	5472	6276	6437	6759	8047	9012	9656	10944
2.5	1811	2816	3822	4828	5834	6840	7846	8047	8449	10058	11265	12070	13679
3.0	2173	3380	4587	5794	7001	8208	9415	9656	10139	12070	13518	14484	16415
3.5	2535	3943	5351	6759	8167	9576	10984	11265	11829	14082	15772	16898	19151
4.0	2897	4506	6115	7725	9334	10944	12553	12875	13518	16093	18025	19312	21887
4.5	3259	5069	6880	8690	10501	12311	14122	14484	15208	18105	20278	21726	24623
5.0	3621	5633	7644	9656	11668	13679	15691	16093	16898	20117	22531	24140	27359
5.5	3983	6196	8409	10622	12834	15047	17260	17703	18588	22128	24784	26554	30095
6.0	4345	6759	9173	11587	14001	16415	18829	19312	20278	24140	27037	28968	32831
6.5	4707	7322	9938	12553	15168	17783	20398	20921	21967	26152	29290	31382	35566
7.0	5069	7886	10702	13518	16335	19151	21967	22531	23657	28163	31543	33796	38302

Number of Plants per m² at the following offset spacings

(All plants equidistant horizontally and vertically)

Distance mm	Nr of Plants
100	114.43
200	28.22
250	18.17
300	12.35
400	6.72
500	4.29
600	3.04
700	2.16
750	1.88
900	1.26
1000	1.05
1200	0.68
1500	0.42
2000	0.23

SEEDING/TURFING AND PLANTING

Grass Clippings Wet: Based on 3.5 m³/tonne

Annual kg/100 m²		Average 20 cuts kg/100 m²			m²/tonne	m²/m³
32.0		1.6			61162.1	214067.3

Nr of cuts	22	20	18	16	12	4
kg/cut	1.45	1.60	1.78	2.00	2.67	8.00
Area capacity of 3 tonne vehicle per load						
m²	206250	187500	168750	150000	112500	37500
Load m³	**100 m² units/m³ of vehicle space**					
1	196.4	178.6	160.7	142.9	107.1	35.7
2	392.9	357.1	321.4	285.7	214.3	71.4
3	589.3	535.7	482.1	428.6	321.4	107.1
4	785.7	714.3	642.9	571.4	428.6	142.9
5	982.1	892.9	803.6	714.3	535.7	178.6

Transportation of Trees

To unload large trees a machine with the necessary lifting strength is required. The weight of the trees must therefore be known in advance. The following table gives a rough overview. The additional columns with root ball dimensions and the number of plants per trailer provide additional information, for example about preparing planting holes and calculating unloading times.

Girth in cm	Rootball diameter in cm	Ball height in cm	Weight in kg	Numbers of trees per trailer
16–18	50–60	40	150	100–120
18–20	60–70	40–50	200	80–100
20–25	60–70	40–50	270	50–70
25–30	80	50–60	350	50
30–35	90–100	60–70	500	12–18
35–40	100–110	60–70	650	10–15
40–45	110–120	60–70	850	8–12
45–50	110–120	60–70	1100	5–7
50–60	130–140	60–70	1600	1–3
60–70	150–160	60–70	2500	1
70–80	180–200	70	4000	1
80–90	200–220	70–80	5500	1
90–100	230–250	80–90	7500	1
100–120	250–270	80–90	9500	1

Data supplied by Lorenz von Ehren GmbH
The information in the table is approximate; deviations depend on soil type, genus and weather

FENCING AND GATES

Types of Preservative

Creosote (tar oil) can be 'factory' applied	by pressure to BS 144: pts 1&2 by immersion to BS 144: pt 1 by hot and cold open tank to BS 144: pts 1&2
Copper/chromium/arsenic (CCA)	by full cell process to BS 4072 pts 1&2
Organic solvent (OS)	by double vacuum (vacvac) to BS 5707 pts 1&3 by immersion to BS 5057 pts 1&3
Pentachlorophenol (PCP)	by heavy oil double vacuum to BS 5705 pts 2&3
Boron diffusion process (treated with disodium octaborate to BWPA Manual 1986)	

Note: Boron is used on green timber at source and the timber is supplied dry

Cleft Chestnut Pale Fences

Pales	Pale spacing	Wire lines	
900 mm	75 mm	2	temporary protection
1050 mm	75 or 100 mm	2	light protective fences
1200 mm	75 mm	3	perimeter fences
1350 mm	75 mm	3	perimeter fences
1500 mm	50 mm	3	narrow perimeter fences
1800 mm	50 mm	3	light security fences

Close-Boarded Fences

Close-boarded fences 1.05 to 1.8 m high
Type BCR (recessed) or BCM (morticed) with concrete posts 140 × 115 mm tapered and Type BW with timber posts

Palisade Fences

Wooden palisade fences
Type WPC with concrete posts 140 × 115 mm tapered and Type WPW with timber posts

For both types of fence:
Height of fence 1050 mm: two rails
Height of fence 1200 mm: two rails
Height of fence 1500 mm: three rails
Height of fence 1650 mm: three rails
Height of fence 1800 mm: three rails

FENCING AND GATES

Post and Rail Fences

Wooden post and rail fences
Type MPR 11/3 morticed rails and Type SPR 11/3 nailed rails
Height to top of rail 1100 mm
Rails: three rails 87 mm, 38 mm

Type MPR 11/4 morticed rails and Type SPR 11/4 nailed rails
Height to top of rail 1100 mm
Rails: four rails 87 mm, 38 mm

Type MPR 13/4 morticed rails and Type SPR 13/4 nailed rails
Height to top of rail 1300 mm
Rail spacing 250 mm, 250 mm, and 225 mm from top
Rails: four rails 87 mm, 38 mm

Steel Posts

Rolled steel angle iron posts for chain link fencing

Posts	Fence height	Strut	Straining post
1500 × 40 × 40 × 5 mm	900 mm	1500 × 40 × 40 × 5 mm	1500 × 50 × 50 × 6 mm
1800 × 40 × 40 × 5 mm	1200 mm	1800 × 40 × 40 × 5 mm	1800 × 50 × 50 × 6 mm
2000 × 45 × 45 × 5 mm	1400 mm	2000 × 45 × 45 × 5 mm	2000 × 60 × 60 × 6 mm
2600 × 45 × 45 × 5 mm	1800 mm	2600 × 45 × 45 × 5 mm	2600 × 60 × 60 × 6 mm
3000 × 50 × 50 × 6 mm with arms	1800 mm	2600 × 45 × 45 × 5 mm	3000 × 60 × 60 × 6 mm

Concrete Posts

Concrete posts for chain link fencing

Posts and straining posts	Fence height	Strut
1570 mm 100 × 100 mm	900 mm	1500 mm × 75 × 75 mm
1870 mm 125 × 125 mm	1200 mm	1830 mm × 100 × 75 mm
2070 mm 125 × 125 mm	1400 mm	1980 mm × 100 × 75 mm
2620 mm 125 × 125 mm	1800 mm	2590 mm × 100 × 85 mm
3040 mm 125 × 125 mm	1800 mm	2590 mm × 100 × 85 mm (with arms)

FENCING AND GATES

Rolled Steel Angle Posts

Rolled steel angle posts for rectangular wire mesh (field) fencing

Posts	Fence height	Strut	Straining post
1200 × 40 × 40 × 5 mm	600 mm	1200 × 75 × 75 mm	1350 × 100 × 100 mm
1400 × 40 × 40 × 5 mm	800 mm	1400 × 75 × 75 mm	1550 × 100 × 100 mm
1500 × 40 × 40 × 5 mm	900 mm	1500 × 75 × 75 mm	1650 × 100 × 100 mm
1600 × 40 × 40 × 5 mm	1000 mm	1600 × 75 × 75 mm	1750 × 100 × 100 mm
1750 × 40 × 40 × 5 mm	1150 mm	1750 × 75 × 100 mm	1900 × 125 × 125 mm

Concrete Posts

Concrete posts for rectangular wire mesh (field) fencing

Posts	Fence height	Strut	Straining post
1270 × 100 × 100 mm	600 mm	1200 × 75 × 75 mm	1420 × 100 × 100 mm
1470 × 100 × 100 mm	800 mm	1350 × 75 × 75 mm	1620 × 100 × 100 mm
1570 × 100 × 100 mm	900 mm	1500 × 75 × 75 mm	1720 × 100 × 100 mm
1670 × 100 × 100 mm	600 mm	1650 × 75 × 75 mm	1820 × 100 × 100 mm
1820 × 125 × 125 mm	1150 mm	1830 × 75 × 100 mm	1970 × 125 × 125 mm

Cleft Chestnut Pale Fences

Timber Posts

Timber posts for wire mesh and hexagonal wire netting fences

Round timber for general fences

Posts	Fence height	Strut	Straining post
1300 × 65 mm dia.	600 mm	1200 × 80 mm dia.	1450 × 100 mm dia.
1500 × 65 mm dia.	800 mm	1400 × 80 mm dia.	1650 × 100 mm dia.
1600 × 65 mm dia.	900 mm	1500 × 80 mm dia.	1750 × 100 mm dia.
1700 × 65 mm dia.	1050 mm	1600 × 80 mm dia.	1850 × 100 mm dia.
1800 × 65 mm dia.	1150 mm	1750 × 80 mm dia.	2000 × 120 mm dia.

Squared timber for general fences

Posts	Fence height	Strut	Straining post
1300 × 75 × 75 mm	600 mm	1200 × 75 × 75 mm	1450 × 100 × 100 mm
1500 × 75 × 75 mm	800 mm	1400 × 75 × 75 mm	1650 × 100 × 100 mm
1600 × 75 × 75 mm	900 mm	1500 × 75 × 75 mm	1750 × 100 × 100 mm
1700 × 75 × 75 mm	1050 mm	1600 × 75 × 75 mm	1850 × 100 × 100 mm
1800 × 75 × 75 mm	1150 mm	1750 × 75 × 75 mm	2000 × 125 × 100 mm

FENCING AND GATES

Steel Fences to BS 1722: Part 9: 1992

	Fence height	Top/bottom rails and flat posts	Vertical bars
Light	1000 mm	40 × 10 mm 450 mm in ground	12 mm dia. at 115 mm cs
	1200 mm	40 × 10 mm 550 mm in ground	12 mm dia. at 115 mm cs
	1400 mm	40 × 10 mm 550 mm in ground	12 mm dia. at 115 mm cs
Light	1000 mm	40 × 10 mm 450 mm in ground	16 mm dia. at 120 mm cs
	1200 mm	40 × 10 mm 550 mm in ground	16 mm dia. at 120 mm cs
	1400 mm	40 × 10 mm 550 mm in ground	16 mm dia. at 120 mm cs
Medium	1200 mm	50 × 10 mm 550 mm in ground	20 mm dia. at 125 mm cs
	1400 mm	50 × 10 mm 550 mm in ground	20 mm dia. at 125 mm cs
	1600 mm	50 × 10 mm 600 mm in ground	22 mm dia. at 145 mm cs
	1800 mm	50 × 10 mm 600 mm in ground	22 mm dia. at 145 mm cs
Heavy	1600 mm	50 × 10 mm 600 mm in ground	22 mm dia. at 145 mm cs
	1800 mm	50 × 10 mm 600 mm in ground	22 mm dia. at 145 mm cs
	2000 mm	50 × 10 mm 600 mm in ground	22 mm dia. at 145 mm cs
	2200 mm	50 × 10 mm 600 mm in ground	22 mm dia. at 145 mm cs

Notes: Mild steel fences: round or square verticals; flat standards and horizontals. Tops of vertical bars may be bow-top, blunt, or pointed. Round or square bar railings

Timber Field Gates to BS 3470: 1975

Gates made to this standard are designed to open one way only
All timber gates are 1100 mm high
Width over stiles 2400, 2700, 3000, 3300, 3600, and 4200 mm
Gates over 4200 mm should be made in two leaves

Steel Field Gates to BS 3470: 1975

All steel gates are 1100 mm high
Heavy duty: width over stiles 2400, 3000, 3600 and 4500 mm
Light duty: width over stiles 2400, 3000, and 3600 mm

FENCING AND GATES

Domestic Front Entrance Gates to BS 4092: Part 1: 1966

Metal gates:	Single gates are 900 mm high minimum, 900 mm, 1000 mm and 1100 mm wide

Domestic Front Entrance Gates to BS 4092: Part 2: 1966

Wooden gates:	All rails shall be tenoned into the stiles Single gates are 840 mm high minimum, 801 mm and 1020 mm wide Double gates are 840 mm high minimum, 2130, 2340 and 2640 mm wide

Timber Bridle Gates to BS 5709:1979 (Horse or Hunting Gates)

Gates open one way only Minimum width between posts Minimum height	1525 mm 1100 mm

Timber Kissing Gates to BS 5709:1979

Minimum width	700 mm
Minimum height	1000 mm
Minimum distance between shutting posts	600 mm
Minimum clearance at mid-point	600 mm

Metal Kissing Gates to BS 5709:1979

Sizes are the same as those for timber kissing gates Maximum gaps between rails 120 mm

Categories of Pedestrian Guard Rail to BS 3049:1976

Class A for normal use Class B where vandalism is expected Class C where crowd pressure is likely

DRAINAGE

Width Required for Trenches for Various Diameters of Pipes

Pipe diameter (mm)	Trench n.e. 1.50 m deep	Trench over 1.50 m deep
n.e. 100 mm	450 mm	600 mm
100–150 mm	500 mm	650 mm
150–225 mm	600 mm	750 mm
225–300 mm	650 mm	800 mm
300–400 mm	750 mm	900 mm
400–450 mm	900 mm	1050 mm
450–600 mm	1100 mm	1300 mm

Weights and Dimensions – Vitrified Clay Pipes

Product	Nominal diameter (mm)	Effective length (mm)	BS 65 limits of tolerance min (mm)	max (mm)	Crushing strength (kN/m)	Weight (kg/pipe)	(kg/m)
Supersleve	100	1600	96	105	35.00	14.71	9.19
	150	1750	146	158	35.00	29.24	16.71
Hepsleve	225	1850	221	236	28.00	84.03	45.42
	300	2500	295	313	34.00	193.05	77.22
	150	1500	146	158	22.00	37.04	24.69
Hepseal	225	1750	221	236	28.00	85.47	48.84
	300	2500	295	313	34.00	204.08	81.63
	400	2500	394	414	44.00	357.14	142.86
	450	2500	444	464	44.00	454.55	181.63
	500	2500	494	514	48.00	555.56	222.22
	600	2500	591	615	57.00	796.23	307.69
	700	3000	689	719	67.00	1111.11	370.45
	800	3000	788	822	72.00	1351.35	450.45
Hepline	100	1600	95	107	22.00	14.71	9.19
	150	1750	145	160	22.00	29.24	16.71
	225	1850	219	239	28.00	84.03	45.42
	300	1850	292	317	34.00	142.86	77.22
Hepduct (conduit)	90	1500	–	–	28.00	12.05	8.03
	100	1600	–	–	28.00	14.71	9.19
	125	1750	–	–	28.00	20.73	11.84
	150	1750	–	–	28.00	29.24	16.71
	225	1850	–	–	28.00	84.03	45.42
	300	1850	–	–	34.00	142.86	77.22

DRAINAGE

Weights and Dimensions – Vitrified Clay Pipes

Nominal internal diameter (mm)	Nominal wall thickness (mm)	Approximate weight (kg/m)
150	25	45
225	29	71
300	32	122
375	35	162
450	38	191
600	48	317
750	54	454
900	60	616
1200	76	912
1500	89	1458
1800	102	1884
2100	127	2619

Wall thickness, weights and pipe lengths vary, depending on type of pipe required
The particulars shown above represent a selection of available diameters and are applicable to strength class 1 pipes with flexible rubber ring joints
Tubes with Ogee joints are also available

Weights and Dimensions – PVC-u Pipes

	Nominal size	Mean outside diameter (mm)		Wall thickness	Weight
		min	max	(mm)	(kg/m)
Standard pipes	82.4	82.4	82.7	3.2	1.2
	110.0	110.0	110.4	3.2	1.6
	160.0	160.0	160.6	4.1	3.0
	200.0	200.0	200.6	4.9	4.6
	250.0	250.0	250.7	6.1	7.2
Perforated pipes heavy grade	As above	As above	As above	As above	As above
thin wall	82.4	82.4	82.7	1.7	–
	110.0	110.0	110.4	2.2	–
	160.0	160.0	160.6	3.2	–

Width of Trenches Required for Various Diameters of Pipes

Pipe diameter (mm)	Trench n.e. 1.5 m deep (mm)	Trench over 1.5 m deep (mm)
n.e. 100	450	600
100–150	500	650
150–225	600	750
225–300	650	800
300–400	750	900
400–450	900	1050
450–600	1100	1300

DRAINAGE

DRAINAGE BELOW GROUND AND LAND DRAINAGE

Flow of Water Which Can Be Carried by Various Sizes of Pipe

Clay or concrete pipes

	Gradient of pipeline							
	1:10	1:20	1:30	1:40	1:50	1:60	1:80	1:100
Pipe size	Flow in litres per second							
DN 100 15.0	8.5	6.8	5.8	5.2	4.7	4.0	3.5	
DN 150 28.0	19.0	16.0	14.0	12.0	11.0	9.1	8.0	
DN 225 140.0	95.0	76.0	66.0	58.0	53.0	46.0	40.0	

Plastic pipes

	Gradient of pipeline							
	1:10	1:20	1:30	1:40	1:50	1:60	1:80	1:100
Pipe size	Flow in litres per second							
82.4 mm i/dia.	12.0	8.5	6.8	5.8	5.2	4.7	4.0	3.5
110 mm i/dia.	28.0	19.0	16.0	14.0	12.0	11.0	9.1	8.0
160 mm i/dia.	76.0	53.0	43.0	37.0	33.0	29.0	25.0	22.0
200 mm i/dia.	140.0	95.0	76.0	66.0	58.0	53.0	46.0	40.0

Vitrified (Perforated) Clay Pipes and Fittings to BS En 295-5 1994

Length not specified		
75 mm bore	250 mm bore	600 mm bore
100	300	700
125	350	800
150	400	1000
200	450	1200
225	500	

Precast Concrete Pipes: Prestressed Non-pressure Pipes and Fittings: Flexible Joints to BS 5911: Pt. 103: 1994

Rationalized metric nominal sizes: 450, 500	
Length:	500–1000 by 100 increments
	1000–2200 by 200 increments
	2200–2800 by 300 increments
Angles: length:	450–600 angles 45, 22.5,11.25°
	600 or more angles 22.5, 11.25°

DRAINAGE

Precast Concrete Pipes: Unreinforced and Circular Manholes and Soakaways to BS 5911: Pt. 200: 1994

Nominal sizes:	
Shafts:	675, 900 mm
Chambers:	900, 1050, 1200, 1350, 1500, 1800, 2100, 2400, 2700, 3000 mm
Large chambers:	To have either tapered reducing rings or a flat reducing slab in order to accept the standard cover
Ring depths:	1. 300–1200 mm by 300 mm increments except for bottom slab and rings below cover slab, these are by 150 mm increments
	2. 250–1000 mm by 250 mm increments except for bottom slab and rings below cover slab, these are by 125 mm increments
Access hole:	750 × 750 mm for DN 1050 chamber
	1200 × 675 mm for DN 1350 chamber

Calculation of Soakaway Depth

The following formula determines the depth of concrete ring soakaway that would be required for draining given amounts of water.

$$h = \frac{4ar}{3\pi D^2}$$

h = depth of the chamber below the invert pipe
a = the area to be drained
r = the hourly rate of rainfall (50 mm per hour)
π = pi
D = internal diameter of the soakaway

This table shows the depth of chambers in each ring size which would be required to contain the volume of water specified. These allow a recommended storage capacity of ⅓ (one third of the hourly rainfall figure).

Table Showing Required Depth of Concrete Ring Chambers in Metres

Area m²	50	100	150	200	300	400	500
Ring size							
0.9	1.31	2.62	3.93	5.24	7.86	10.48	13.10
1.1	0.96	1.92	2.89	3.85	5.77	7.70	9.62
1.2	0.74	1.47	2.21	2.95	4.42	5.89	7.37
1.4	0.58	1.16	1.75	2.33	3.49	4.66	5.82
1.5	0.47	0.94	1.41	1.89	2.83	3.77	4.72
1.8	0.33	0.65	0.98	1.31	1.96	2.62	3.27
2.1	0.24	0.48	0.72	0.96	1.44	1.92	2.41
2.4	0.18	0.37	0.55	0.74	1.11	1.47	1.84
2.7	0.15	0.29	0.44	0.58	0.87	1.16	1.46
3.0	0.12	0.24	0.35	0.47	0.71	0.94	1.18

DRAINAGE

Precast Concrete Inspection Chambers and Gullies to BS 5911: Part 230: 1994

Nominal sizes:	375 diameter, 750, 900 mm deep
	450 diameter, 750, 900, 1050, 1200 mm deep
Depths:	from the top for trapped or untrapped units:
	centre of outlet 300 mm
	invert (bottom) of the outlet pipe 400 mm
Depth of water seal for trapped gullies:	
	85 mm, rodding eye int. dia. 100 mm
Cover slab:	65 mm min

Bedding Flexible Pipes: PVC-u Or Ductile Iron

Type 1 =	100 mm fill below pipe, 300 mm above pipe: single size material
Type 2 =	100 mm fill below pipe, 300 mm above pipe: single size or graded material
Type 3 =	100 mm fill below pipe, 75 mm above pipe with concrete protective slab over
Type 4 =	100 mm fill below pipe, fill laid level with top of pipe
Type 5 =	200 mm fill below pipe, fill laid level with top of pipe
Concrete =	25 mm sand blinding to bottom of trench, pipe supported on chocks, 100 mm concrete under the pipe, 150 mm concrete over the pipe

DRAINAGE

Bedding Rigid Pipes: Clay or Concrete
(for vitrified clay pipes the manufacturer should be consulted)

Class D:	Pipe laid on natural ground with cut-outs for joints, soil screened to remove stones over 40 mm and returned over pipe to 150 m min depth. Suitable for firm ground with trenches trimmed by hand.
Class N:	Pipe laid on 50 mm granular material of graded aggregate to Table 4 of BS 882, or 10 mm aggregate to Table 6 of BS 882, or as dug light soil (not clay) screened to remove stones over 10 mm. Suitable for machine dug trenches.
Class B:	As Class N, but with granular bedding extending half way up the pipe diameter.
Class F:	Pipe laid on 100 mm granular fill to BS 882 below pipe, minimum 150 mm granular fill above pipe: single size material. Suitable for machine dug trenches.
Class A:	Concrete 100 mm thick under the pipe extending half way up the pipe, backfilled with the appropriate class of fill. Used where there is only a very shallow fall to the drain. Class A bedding allows the pipes to be laid to an exact gradient.
Concrete surround:	25 mm sand blinding to bottom of trench, pipe supported on chocks, 100 mm concrete under the pipe, 150 mm concrete over the pipe. It is preferable to bed pipes under slabs or wall in granular material.

PIPED SUPPLY SYSTEMS

Identification of Service Tubes From Utility to Dwellings

Utility	Colour	Size	Depth
British Telecom	grey	54 mm od	450 mm
Electricity	black	38 mm od	450 mm
Gas	yellow	42 mm od rigid 60 mm od convoluted	450 mm
Water	may be blue	(normally untubed)	750 mm

ELECTRICAL SUPPLY/POWER/LIGHTING SYSTEMS

Electrical Insulation Class En 60.598 BS 4533

Class 1:	luminaires comply with class 1 (I) earthed electrical requirements
Class 2:	luminaires comply with class 2 (II) double insulated electrical requirements
Class 3:	luminaires comply with class 3 (III) electrical requirements

Protection to Light Fittings

BS EN 60529:1992 Classification for degrees of protection provided by enclosures.
(IP Code – International or ingress Protection)

1st characteristic: against ingress of solid foreign objects

The figure	2	indicates that fingers cannot enter
	3	that a 2.5 mm diameter probe cannot enter
	4	that a 1.0 mm diameter probe cannot enter
	5	the fitting is dust proof (no dust around live parts)
	6	the fitting is dust tight (no dust entry)

2nd characteristic: ingress of water with harmful effects

The figure	0	indicates unprotected
	1	vertically dripping water cannot enter
	2	water dripping 15° (tilt) cannot enter
	3	spraying water cannot enter
	4	splashing water cannot enter
	5	jetting water cannot enter
	6	powerful jetting water cannot enter
	7	proof against temporary immersion
	8	proof against continuous immersion

Optional additional codes:		A–D protects against access to hazardous parts
	H	high voltage apparatus
	M	fitting was in motion during water test
	S	fitting was static during water test
	W	protects against weather

Marking code arrangement:	(example) IPX5S = IP (International or Ingress Protection)
	X (denotes omission of first characteristic)
	5 = jetting
	S = static during water test

RAIL TRACKS

	kg/m of track	lb/ft of track
Standard gauge		
Bull-head rails, chairs, transverse timber (softwood) sleepers etc.	245	165
Main lines		
Flat-bottom rails, transverse prestressed concrete sleepers, etc.	418	280
Add for electric third rail	51	35
Add for crushed stone ballast	2600	1750
	kg/m²	**lb/sq ft**
Overall average weight – rails connections, sleepers, ballast, etc.	733	150
	kg/m of track	**lb/ft of track**
Bridge rails, longitudinal timber sleepers, etc.	112	75

RAIL TRACKS

Heavy Rails

British Standard Section No.	Rail height (mm)	Foot width (mm)	Head width (mm)	Min web thickness (mm)	Section weight (kg/m)
Flat Bottom Rails					
60 A	114.30	109.54	57.15	11.11	30.62
70 A	123.82	111.12	60.32	12.30	34.81
75 A	128.59	114.30	61.91	12.70	37.45
80 A	133.35	117.47	63.50	13.10	39.76
90 A	142.88	127.00	66.67	13.89	45.10
95 A	147.64	130.17	69.85	14.68	47.31
100 A	152.40	133.35	69.85	15.08	50.18
110 A	158.75	139.70	69.85	15.87	54.52
113 A	158.75	139.70	69.85	20.00	56.22
50 'O'	100.01	100.01	52.39	10.32	24.82
80 'O'	127.00	127.00	63.50	13.89	39.74
60R	114.30	109.54	57.15	11.11	29.85
75R	128.59	122.24	61.91	13.10	37.09
80R	133.35	127.00	63.50	13.49	39.72
90R	142.88	136.53	66.67	13.89	44.58
95R	147.64	141.29	68.26	14.29	47.21
100R	152.40	146.05	69.85	14.29	49.60
95N	147.64	139.70	69.85	13.89	47.27
Bull Head Rails					
95R BH	145.26	69.85	69.85	19.05	47.07

Light Rails

British Standard Section No.	Rail height (mm)	Foot width (mm)	Head width (mm)	Min web thickness (mm)	Section weight (kg/m)
Flat Bottom Rails					
20M	65.09	55.56	30.96	6.75	9.88
30M	75.41	69.85	38.10	9.13	14.79
35M	80.96	76.20	42.86	9.13	17.39
35R	85.73	82.55	44.45	8.33	17.40
40	88.11	80.57	45.64	12.3	19.89
Bridge Rails					
13	48.00	92	36.00	18.0	13.31
16	54.00	108	44.50	16.0	16.06
20	55.50	127	50.00	20.5	19.86
28	67.00	152	50.00	31.0	28.62
35	76.00	160	58.00	34.5	35.38
50	76.00	165	58.50	–	50.18
Crane Rails					
A65	75.00	175.00	65.00	38.0	43.10
A75	85.00	200.00	75.00	45.0	56.20
A100	95.00	200.00	100.00	60.0	74.30
A120	105.00	220.00	120.00	72.0	100.00
175CR	152.40	152.40	107.95	38.1	86.92

RAIL TRACKS

Fish Plates

British Standard Section No.	Overall plate length		Hole diameter	Finished weight per pair	
	4 Hole (mm)	6 Hole (mm)	(mm)	4 Hole (kg/pair)	6 Hole (kg/pair)
For British Standard Heavy Rails: Flat Bottom Rails					
60 A	406.40	609.60	20.64	9.87	14.76
70 A	406.40	609.60	22.22	11.15	16.65
75 A	406.40	–	23.81	11.82	17.73
80 A	406.40	609.60	23.81	13.15	19.72
90 A	457.20	685.80	25.40	17.49	26.23
100 A	508.00	–	pear	25.02	–
110 A (shallow)	507.00	–	27.00	30.11	54.64
113 A (heavy)	507.00	–	27.00	30.11	54.64
50 'O' (shallow)	406.40	–	–	6.68	10.14
80 'O' (shallow)	495.30	–	23.81	14.72	22.69
60R (shallow)	406.40	609.60	20.64	8.76	13.13
60R (angled)	406.40	609.60	20.64	11.27	16.90
75R (shallow)	406.40	–	23.81	10.94	16.42
75R (angled)	406.40	–	23.81	13.67	–
80R (shallow)	406.40	609.60	23.81	11.93	17.89
80R (angled)	406.40	609.60	23.81	14.90	22.33
For British Standard Heavy Rails: Bull head rails					
95R BH (shallow)	–	457.20	27.00	14.59	14.61
For British Standard Light Rails: Flat Bottom Rails					
30M	355.6	–	–	–	2.72
35M	355.6	–	–	–	2.83
40	355.6	–	–	3.76	–

FRACTIONS, DECIMALS AND MILLIMETRE EQUIVALENTS

FRACTIONS, DECIMALS AND MILLIMETRE EQUIVALENTS

Fractions	Decimals	(mm)		Fractions	Decimals	(mm)
1/64	0.015625	0.396875		33/64	0.515625	13.096875
1/32	0.03125	0.79375		17/32	0.53125	13.49375
3/64	0.046875	1.190625		35/64	0.546875	13.890625
1/16	0.0625	1.5875		9/16	0.5625	14.2875
5/64	0.078125	1.984375		37/64	0.578125	14.684375
3/32	0.09375	2.38125		19/32	0.59375	15.08125
7/64	0.109375	2.778125		39/64	0.609375	15.478125
1/8	0.125	3.175		5/8	0.625	15.875
9/64	0.140625	3.571875		41/64	0.640625	16.271875
5/32	0.15625	3.96875		21/32	0.65625	16.66875
11/64	0.171875	4.365625		43/64	0.671875	17.065625
3/16	0.1875	4.7625		11/16	0.6875	17.4625
13/64	0.203125	5.159375		45/64	0.703125	17.859375
7/32	0.21875	5.55625		23/32	0.71875	18.25625
15/64	0.234375	5.953125		47/64	0.734375	18.653125
1/4	0.25	6.35		3/4	0.75	19.05
17/64	0.265625	6.746875		49/64	0.765625	19.446875
9/32	0.28125	7.14375		25/32	0.78125	19.84375
19/64	0.296875	7.540625		51/64	0.796875	20.240625
5/16	0.3125	7.9375		13/16	0.8125	20.6375
21/64	0.328125	8.334375		53/64	0.828125	21.034375
11/32	0.34375	8.73125		27/32	0.84375	21.43125
23/64	0.359375	9.128125		55/64	0.859375	21.828125
3/8	0.375	9.525		7/8	0.875	22.225
25/64	0.390625	9.921875		57/64	0.890625	22.621875
13/32	0.40625	10.31875		29/32	0.90625	23.01875
27/64	0.421875	10.71563		59/64	0.921875	23.415625
7/16	0.4375	11.1125		15/16	0.9375	23.8125
29/64	0.453125	11.50938		61/64	0.953125	24.209375
15/32	0.46875	11.90625		31/32	0.96875	24.60625
31/64	0.484375	12.30313		63/64	0.984375	25.003125
1/2	0.5	12.7		1.0	1	25.4

IMPERIAL STANDARD WIRE GAUGE (SWG)

SWG No.	Diameter (inches)	Diameter (mm)	SWG No.	Diameter (inches)	Diameter (mm)
7/0	0.5	12.7	23	0.024	0.61
6/0	0.464	11.79	24	0.022	0.559
5/0	0.432	10.97	25	0.02	0.508
4/0	0.4	10.16	26	0.018	0.457
3/0	0.372	9.45	27	0.0164	0.417
2/0	0.348	8.84	28	0.0148	0.376
1/0	0.324	8.23	29	0.0136	0.345
1	0.3	7.62	30	0.0124	0.315
2	0.276	7.01	31	0.0116	0.295
3	0.252	6.4	32	0.0108	0.274
4	0.232	5.89	33	0.01	0.254
5	0.212	5.38	34	0.009	0.234
6	0.192	4.88	35	0.008	0.213
7	0.176	4.47	36	0.008	0.193
8	0.16	4.06	37	0.007	0.173
9	0.144	3.66	38	0.006	0.152
10	0.128	3.25	39	0.005	0.132
11	0.116	2.95	40	0.005	0.122
12	0.104	2.64	41	0.004	0.112
13	0.092	2.34	42	0.004	0.102
14	0.08	2.03	43	0.004	0.091
15	0.072	1.83	44	0.003	0.081
16	0.064	1.63	45	0.003	0.071
17	0.056	1.42	46	0.002	0.061
18	0.048	1.22	47	0.002	0.051
19	0.04	1.016	48	0.002	0.041
20	0.036	0.914	49	0.001	0.031
21	0.032	0.813	50	0.001	0.025
22	0.028	0.711			

PIPES, WATER, STORAGE, INSULATION

WATER PRESSURE DUE TO HEIGHT

Imperial

Head (Feet)	Pressure (lb/in²)		Head (Feet)	Pressure (lb/in²)
1	0.43		70	30.35
5	2.17		75	32.51
10	4.34		80	34.68
15	6.5		85	36.85
20	8.67		90	39.02
25	10.84		95	41.18
30	13.01		100	43.35
35	15.17		105	45.52
40	17.34		110	47.69
45	19.51		120	52.02
50	21.68		130	56.36
55	23.84		140	60.69
60	26.01		150	65.03
65	28.18			

Metric

Head (m)	Pressure (bar)		Head (m)	Pressure (bar)
0.5	0.049		18.0	1.766
1.0	0.098		19.0	1.864
1.5	0.147		20.0	1.962
2.0	0.196		21.0	2.06
3.0	0.294		22.0	2.158
4.0	0.392		23.0	2.256
5.0	0.491		24.0	2.354
6.0	0.589		25.0	2.453
7.0	0.687		26.0	2.551
8.0	0.785		27.0	2.649
9.0	0.883		28.0	2.747
10.0	0.981		29.0	2.845
11.0	1.079		30.0	2.943
12.0	1.177		32.5	3.188
13.0	1.275		35.0	3.434
14.0	1.373		37.5	3.679
15.0	1.472		40.0	3.924
16.0	1.57		42.5	4.169
17.0	1.668		45.0	4.415

1 bar	=	14.5038 lbf/in²
1 lbf/in²	=	0.06895 bar
1 metre	=	3.2808 ft or 39.3701 in
1 foot	=	0.3048 metres
1 in wg	=	2.5 mbar (249.1 N/m²)

PIPES, WATER, STORAGE, INSULATION

Dimensions and Weights of Copper Pipes to BSEN 1057, BSEN 12499, BSEN 14251

Outside diameter (mm)	Internal diameter (mm)	Weight per metre (kg)	Internal diameter (mm)	Weight per metre (kg)	Internal siameter (mm)	Weight per metre (kg)
	Formerly Table X		Formerly Table Y		Formerly Table Z	
6	4.80	0.0911	4.40	0.1170	5.00	0.0774
8	6.80	0.1246	6.40	0.1617	7.00	0.1054
10	8.80	0.1580	8.40	0.2064	9.00	0.1334
12	10.80	0.1914	10.40	0.2511	11.00	0.1612
15	13.60	0.2796	13.00	0.3923	14.00	0.2031
18	16.40	0.3852	16.00	0.4760	16.80	0.2918
22	20.22	0.5308	19.62	0.6974	20.82	0.3589
28	26.22	0.6814	25.62	0.8985	26.82	0.4594
35	32.63	1.1334	32.03	1.4085	33.63	0.6701
42	39.63	1.3675	39.03	1.6996	40.43	0.9216
54	51.63	1.7691	50.03	2.9052	52.23	1.3343
76.1	73.22	3.1287	72.22	4.1437	73.82	2.5131
108	105.12	4.4666	103.12	7.3745	105.72	3.5834
133	130.38	5.5151	–	–	130.38	5.5151
159	155.38	8.7795	–	–	156.38	6.6056

Dimensions of Stainless Steel Pipes to BS 4127

Outside siameter (mm)	Maximum outside siameter (mm)	Minimum outside diameter (mm)	Wall thickness (mm)	Working pressure (bar)
6	6.045	5.940	0.6	330
8	8.045	7.940	0.6	260
10	10.045	9.940	0.6	210
12	12.045	11.940	0.6	170
15	15.045	14.940	0.6	140
18	18.045	17.940	0.7	135
22	22.055	21.950	0.7	110
28	28.055	27.950	0.8	121
35	35.070	34.965	1.0	100
42	42.070	41.965	1.1	91
54	54.090	53.940	1.2	77

PIPES, WATER, STORAGE, INSULATION

Dimensions of Steel Pipes to BS 1387

Nominal Size	Approx. Outside Diameter	Outside diameter				Thickness		
		Light		Medium & Heavy		Light	Medium	Heavy
		Max	Min	Max	Min			
(mm)	(mm)	(mm)	(mm)	(mm)	(mm)	(mm)	(mm)	(mm)
6	10.20	10.10	9.70	10.40	9.80	1.80	2.00	2.65
8	13.50	13.60	13.20	13.90	13.30	1.80	2.35	2.90
10	17.20	17.10	16.70	17.40	16.80	1.80	2.35	2.90
15	21.30	21.40	21.00	21.70	21.10	2.00	2.65	3.25
20	26.90	26.90	26.40	27.20	26.60	2.35	2.65	3.25
25	33.70	33.80	33.20	34.20	33.40	2.65	3.25	4.05
32	42.40	42.50	41.90	42.90	42.10	2.65	3.25	4.05
40	48.30	48.40	47.80	48.80	48.00	2.90	3.25	4.05
50	60.30	60.20	59.60	60.80	59.80	2.90	3.65	4.50
65	76.10	76.00	75.20	76.60	75.40	3.25	3.65	4.50
80	88.90	88.70	87.90	89.50	88.10	3.25	4.05	4.85
100	114.30	113.90	113.00	114.90	113.30	3.65	4.50	5.40
125	139.70	–	–	140.60	138.70	–	4.85	5.40
150	165.1*	–	–	166.10	164.10	–	4.85	5.40

* 165.1 mm (6.5in) outside diameter is not generally recommended except where screwing to BS 21 is necessary
All dimensions are in accordance with ISO R65 except approximate outside diameters which are in accordance with ISO R64
Light quality is equivalent to ISO R65 Light Series II

Approximate Metres Per Tonne of Tubes to BS 1387

Nom. size	BLACK						GALVANIZED					
	Plain/screwed ends			Screwed & socketed			Plain/screwed ends			Screwed & socketed		
	L	M	H	L	M	H	L	M	H	L	M	H
(mm)	(m)	(m)	(m)	(m)	(m)	(m)	(m)	(m)	(m)	(m)	(m)	(m)
6	2765	2461	2030	2743	2443	2018	2604	2333	1948	2584	2317	1937
8	1936	1538	1300	1920	1527	1292	1826	1467	1254	1811	1458	1247
10	1483	1173	979	1471	1165	974	1400	1120	944	1386	1113	939
15	1050	817	688	1040	811	684	996	785	665	987	779	661
20	712	634	529	704	628	525	679	609	512	673	603	508
25	498	410	336	494	407	334	478	396	327	474	394	325
32	388	319	260	384	316	259	373	308	254	369	305	252
40	307	277	226	303	273	223	296	268	220	292	264	217
50	244	196	162	239	194	160	235	191	158	231	188	157
65	172	153	127	169	151	125	167	149	124	163	146	122
80	147	118	99	143	116	98	142	115	97	139	113	96
100	101	82	69	98	81	68	98	81	68	95	79	67
125	–	62	56	–	60	55	–	60	55	–	59	54
150	–	52	47	–	50	46	–	51	46	–	49	45

The figures for 'plain or screwed ends' apply also to tubes to BS 1775 of equivalent size and thickness
Key:
L – Light
M – Medium
H – Heavy

PIPES, WATER, STORAGE, INSULATION

Flange Dimension Chart to BS 4504 & BS 10

Normal Pressure Rating (PN 6) 6 Bar

Nom. size	Flange outside dia.	Table 6/2 Forged Welding Neck	Table 6/3 Plate Slip on	Table 6/4 Forged Bossed Screwed	Table 6/5 Forged Bossed Slip on	Table 6/8 Plate Blank	Raised face Dia.	T'ness	Nr. bolt hole	Size of bolt
15	80	12	12	12	12	12	40	2	4	M10 × 40
20	90	14	14	14	14	14	50	2	4	M10 × 45
25	100	14	14	14	14	14	60	2	4	M10 × 45
32	120	14	16	14	14	14	70	2	4	M12 × 45
40	130	14	16	14	14	14	80	3	4	M12 × 45
50	140	14	16	14	14	14	90	3	4	M12 × 45
65	160	14	16	14	14	14	110	3	4	M12 × 45
80	190	16	18	16	16	16	128	3	4	M16 × 55
100	210	16	18	16	16	16	148	3	4	M16 × 55
125	240	18	20	18	18	18	178	3	8	M16 × 60
150	265	18	20	18	18	18	202	3	8	M16 × 60
200	320	20	22	–	20	20	258	3	8	M16 × 60
250	375	22	24	–	22	22	312	3	12	M16 × 65
300	440	22	24	–	22	22	365	4	12	M20 × 70

Normal Pressure Rating (PN 16) 16 Bar

Nom. size	Flange outside dia.	Table 6/2 Forged Welding Neck	Table 6/3 Plate Slip on	Table 6/4 Forged Bossed Screwed	Table 6/5 Forged Bossed Slip on	Table 6/8 Plate Blank	Raised face Dia.	T'ness	Nr. bolt hole	Size of bolt
15	95	14	14	14	14	14	45	2	4	M12 × 45
20	105	16	16	16	16	16	58	2	4	M12 × 50
25	115	16	16	16	16	16	68	2	4	M12 × 50
32	140	16	16	16	16	16	78	2	4	M16 × 55
40	150	16	16	16	16	16	88	3	4	M16 × 55
50	165	18	18	18	18	18	102	3	4	M16 × 60
65	185	18	18	18	18	18	122	3	4	M16 × 60
80	200	20	20	20	20	20	138	3	8	M16 × 60
100	220	20	20	20	20	20	158	3	8	M16 × 65
125	250	22	22	22	22	22	188	3	8	M16 × 70
150	285	22	22	22	22	22	212	3	8	M20 × 70
200	340	24	24	–	24	24	268	3	12	M20 × 75
250	405	26	26	–	26	26	320	3	12	M24 × 90
300	460	28	28	–	28	28	378	4	12	M24 × 90

PIPES, WATER, STORAGE, INSULATION

Minimum Distances Between Supports/Fixings

Material	BS Nominal pipe size		Pipes – Vertical	Pipes – Horizontal on to low gradients
	(inch)	(mm)	Support distance in metres	Support distance in metres
Copper	0.50	15.00	1.90	1.30
	0.75	22.00	2.50	1.90
	1.00	28.00	2.50	1.90
	1.25	35.00	2.80	2.50
	1.50	42.00	2.80	2.50
	2.00	54.00	3.90	2.50
	2.50	67.00	3.90	2.80
	3.00	76.10	3.90	2.80
	4.00	108.00	3.90	2.80
	5.00	133.00	3.90	2.80
	6.00	159.00	3.90	2.80
muPVC	1.25	32.00	1.20	0.50
	1.50	40.00	1.20	0.50
	2.00	50.00	1.20	0.60
Polypropylene	1.25	32.00	1.20	0.50
	1.50	40.00	1.20	0.50
uPVC	–	82.40	1.20	0.50
	–	110.00	1.80	0.90
	–	160.00	1.80	1.20
Steel	0.50	15.00	2.40	1.80
	0.75	20.00	3.00	2.40
	1.00	25.00	3.00	2.40
	1.25	32.00	3.00	2.40
	1.50	40.00	3.70	2.40
	2.00	50.00	3.70	2.40
	2.50	65.00	4.60	3.00
	3.00	80.40	4.60	3.00
	4.00	100.00	4.60	3.00
	5.00	125.00	5.50	3.70
	6.00	150.00	5.50	4.50
	8.00	200.00	8.50	6.00
	10.00	250.00	9.00	6.50
	12.00	300.00	10.00	7.00
	16.00	400.00	10.00	8.25

PIPES, WATER, STORAGE, INSULATION

Litres of Water Storage Required Per Person Per Building Type

Type of building	Storage (litres)
Houses and flats (up to 4 bedrooms)	120/bedroom
Houses and flats (more than 4 bedrooms)	100/bedroom
Hostels	90/bed
Hotels	200/bed
Nurses homes and medical quarters	120/bed
Offices with canteen	45/person
Offices without canteen	40/person
Restaurants	7/meal
Boarding schools	90/person
Day schools – Primary	15/person
Day schools – Secondary	20/person

Recommended Air Conditioning Design Loads

Building type	Design loading
Computer rooms	500 W/m² of floor area
Restaurants	150 W/m² of floor area
Banks (main area)	100 W/m² of floor area
Supermarkets	25 W/m² of floor area
Large office block (exterior zone)	100 W/m² of floor area
Large office block (interior zone)	80 W/m² of floor area
Small office block (interior zone)	80 W/m² of floor area

PIPES, WATER, STORAGE, INSULATION

Capacity and Dimensions of Galvanized Mild Steel Cisterns – BS 417

Capacity (litres)	BS type (SCM)	Dimensions		
		Length (mm)	Width (mm)	Depth (mm)
18	45	457	305	305
36	70	610	305	371
54	90	610	406	371
68	110	610	432	432
86	135	610	457	482
114	180	686	508	508
159	230	736	559	559
191	270	762	584	610
227	320	914	610	584
264	360	914	660	610
327	450/1	1220	610	610
336	450/2	965	686	686
423	570	965	762	787
491	680	1090	864	736
709	910	1070	889	889

Capacity of Cold Water Polypropylene Storage Cisterns – BS 4213

Capacity (litres)	BS type (PC)	Maximum height (mm)
18	4	310
36	8	380
68	15	430
91	20	510
114	25	530
182	40	610
227	50	660
273	60	660
318	70	660
455	100	760

PIPES, WATER, STORAGE, INSULATION

Minimum Insulation Thickness to Protect Against Freezing for Domestic Cold Water Systems (8 Hour Evaluation Period)

Pipe size (mm)	Insulation thickness (mm)					
	Condition 1			Condition 2		
	λ = 0.020	λ = 0.030	λ = 0.040	λ = 0.020	λ = 0.030	λ = 0.040
Copper pipes						
15	11	20	34	12	23	41
22	6	9	13	6	10	15
28	4	6	9	4	7	10
35	3	5	7	4	5	7
42	3	4	5	8	4	6
54	2	3	4	2	3	4
76	2	2	3	2	2	3
Steel pipes						
15	9	15	24	10	18	29
20	6	9	13	6	10	15
25	4	7	9	5	7	10
32	3	5	6	3	5	7
40	3	4	5	3	4	6
50	2	3	4	2	3	4
65	2	2	3	2	3	3

Condition 1: water temperature 7°C; ambient temperature –6°C; evaluation period 8 h; permitted ice formation 50%; normal installation, i.e. inside the building and inside the envelope of the structural insulation
Condition 2: water temperature 2°C; ambient temperature –6°C; evaluation period 8 h; permitted ice formation 50%; extreme installation, i.e. inside the building but outside the envelope of the structural insulation
λ = thermal conductivity [W/(mK)]

Insulation Thickness for Chilled And Cold Water Supplies to Prevent Condensation

On a Low Emissivity Outer Surface (0.05, i.e. Bright Reinforced Aluminium Foil) with an Ambient Temperature of +25°C and a Relative Humidity of 80%

Steel pipe size (mm)	t = +10			t = +5			t = 0		
	Insulation thickness (mm)			Insulation thickness (mm)			Insulation thickness (mm)		
	λ = 0.030	λ = 0.040	λ = 0.050	λ = 0.030	λ = 0.040	λ = 0.050	λ = 0.030	λ = 0.040	λ = 0.050
15	16	20	25	22	28	34	28	36	43
25	18	24	29	25	32	39	32	41	50
50	22	28	34	30	39	47	38	49	60
100	26	34	41	36	47	57	46	60	73
150	29	38	46	40	52	64	51	67	82
250	33	43	53	46	60	74	59	77	94
Flat surfaces	39	52	65	56	75	93	73	97	122

t = temperature of contents (°C)
λ = thermal conductivity at mean temperature of insulation [W/(mK)]

PIPES, WATER, STORAGE, INSULATION

Insulation Thickness for Non-domestic Heating Installations to Control Heat Loss

Steel pipe size (mm)	t = 75			t = 100			t = 150		
	Insulation thickness (mm)			Insulation thickness (mm)			Insulation thickness (mm)		
	$\lambda = 0.030$	$\lambda = 0.040$	$\lambda = 0.050$	$\lambda = 0.030$	$\lambda = 0.040$	$\lambda = 0.050$	$\lambda = 0.030$	$\lambda = 0.040$	$\lambda = 0.050$
10	18	32	55	20	36	62	23	44	77
15	19	34	56	21	38	64	26	47	80
20	21	36	57	23	40	65	28	50	83
25	23	38	58	26	43	68	31	53	85
32	24	39	59	28	45	69	33	55	87
40	25	40	60	29	47	70	35	57	88
50	27	42	61	31	49	72	37	59	90
65	29	43	62	33	51	74	40	63	92
80	30	44	62	35	52	75	42	65	94
100	31	46	63	37	54	76	45	68	96
150	33	48	64	40	57	77	50	73	100
200	35	49	65	42	59	79	53	76	103
250	36	50	66	43	61	80	55	78	105

t = hot face temperature (°C)
λ = thermal conductivity at mean temperature of insulation [W/(mK)]

Index

Ebook Single-User Licence Agreement

We welcome you as a user of this Spon Price Book ebook and hope that you find it a useful and valuable tool. Please read this document carefully. **This is a legal agreement** between you (hereinafter referred to as the "Licensee") and Taylor and Francis Books Ltd. (the "Publisher"), which defines the terms under which you may use the Product. **By accessing and retrieving the access code on the label inside the front cover of this book you agree to these terms and conditions outlined herein. If you do not agree to these terms you must return the Product to your supplier intact, with the seal on the label unbroken and with the access code not accessed**.

1. **Definition of the Product**

    The product which is the subject of this Agreement, (the "Product") consists of online and offline access to the VitalSource ebook edition of *Spon's Mechanical & Electrical Services Price Book 2023*.

2. **Commencement and licence**

    2.1   This Agreement commences upon the breaking open of the document containing the access code by the Licensee (the "Commencement Date").

    2.2   This is a licence agreement (the "Agreement") for the use of the Product by the Licensee, and not an agreement for sale.

    2.3   The Publisher licenses the Licensee on a non-exclusive and non-transferable basis to use the Product on condition that the Licensee complies with this Agreement. The Licensee acknowledges that it is only permitted to use the Product in accordance with this Agreement.

3. **Multiple use**

    Use of the Product is not provided or allowed for more than one user or for a wide area network or consortium.

4. **Installation and Use**

    4.1   The Licensee may provide access to the Product for individual study in the following manner: The Licensee may install the Product on a secure local area network on a single site for use by one user.

    4.2   The Licensee shall be responsible for installing the Product and for the effectiveness of such installation.

    4.3   Text from the Product may be incorporated in a coursepack. Such use is only permissible with the express permission of the Publisher in writing and requires the payment of the appropriate fee as specified by the Publisher and signature of a separate licence agreement.

    4.4   The Product is a free addition to the book and the Publisher is under no obligation to provide any technical support.

5. **Permitted Activities**

    5.1   The Licensee shall be entitled to use the Product for its own internal purposes;

    5.2   The Licensee acknowledges that its rights to use the Product are strictly set out in this Agreement, and all other uses (whether expressly mentioned in Clause 6 below or not) are prohibited.

6. **Prohibited Activities**

    The following are prohibited without the express permission of the Publisher:

    6.1   The commercial exploitation of any part of the Product.

    6.2   The rental, loan, (free or for money or money's worth) or hire purchase of this product, save with the express consent of the Publisher.

    6.3   Any activity which raises the reasonable prospect of impeding the Publisher's ability or opportunities to market the Product.

    6.4   Any networking, physical or electronic distribution or dissemination of the product save as expressly permitted by this Agreement.

    6.5   Any reverse engineering, decompilation, disassembly or other alteration of the Product save in accordance with applicable national laws.

    6.6   The right to create any derivative product or service from the Product save as expressly provided for in this Agreement.

    6.7   Any alteration, amendment, modification or deletion from the Product, whether for the purposes of error correction or otherwise.

7. General Responsibilities of the License

7.1 The Licensee will take all reasonable steps to ensure that the Product is used in accordance with the terms and conditions of this Agreement.

7.2 The Licensee acknowledges that damages may not be a sufficient remedy for the Publisher in the event of breach of this Agreement by the Licensee, and that an injunction may be appropriate.

7.3 The Licensee undertakes to keep the Product safe and to use its best endeavours to ensure that the product does not fall into the hands of third parties, whether as a result of theft or otherwise.

7.4 Where information of a confidential nature relating to the product of the business affairs of the Publisher comes into the possession of the Licensee pursuant to this Agreement (or otherwise), the Licensee agrees to use such information solely for the purposes of this Agreement, and under no circumstances to disclose any element of the information to any third party save strictly as permitted under this Agreement. For the avoidance of doubt, the Licensee's obligations under this sub-clause 7.4 shall survive the termination of this Agreement.

8. Warrant and Liability

8.1 The Publisher warrants that it has the authority to enter into this agreement and that it has secured all rights and permissions necessary to enable the Licensee to use the Product in accordance with this Agreement.

8.2 The Publisher warrants that the Product as supplied on the Commencement Date shall be free of defects in materials and workmanship, and undertakes to replace any defective Product within 28 days of notice of such defect being received provided such notice is received within 30 days of such supply. As an alternative to replacement, the Publisher agrees fully to refund the Licensee in such circumstances, if the Licensee so requests, provided that the Licensee returns this copy of *Spon's Mechanical & Electrical Services Price Book 2023* to the Publisher. The provisions of this sub-clause 8.2 do not apply where the defect results from an accident or from misuse of the product by the Licensee.

8.3 Sub-clause 8.2 sets out the sole and exclusive remedy of the Licensee in relation to defects in the Product.

8.4 The Publisher and the Licensee acknowledge that the Publisher supplies the Product on an "as is" basis. The Publisher gives no warranties:

8.4.1 that the Product satisfies the individual requirements of the Licensee; or

8.4.2 that the Product is otherwise fit for the Licensee's purpose; or

8.4.3 that the Product is compatible with the Licensee's hardware equipment and software operating environment.

8.5 The Publisher hereby disclaims all warranties and conditions, express or implied, which are not stated above.

8.6 Nothing in this Clause 8 limits the Publisher's liability to the Licensee in the event of death or personal injury resulting from the Publisher's negligence.

8.7 The Publisher hereby excludes liability for loss of revenue, reputation, business, profits, or for indirect or consequential losses, irrespective of whether the Publisher was advised by the Licensee of the potential of such losses.

8.8 The Licensee acknowledges the merit of independently verifying the price book data prior to taking any decisions of material significance (commercial or otherwise) based on such data. It is agreed that the Publisher shall not be liable for any losses which result from the Licensee placing reliance on the data under any circumstances.

8.9 Subject to sub-clause 8.6 above, the Publisher's liability under this Agreement shall be limited to the purchase price.

9. Intellectual Property Rights

9.1 Nothing in this Agreement affects the ownership of copyright or other intellectual property rights in the Product.

9.2 The Licensee agrees to display the Publishers' copyright notice in the manner described in the Product.

9.3 The Licensee hereby agrees to abide by copyright and similar notice requirements required by the Publisher, details of which are as follows:

"© 2023 Taylor & Francis. All rights reserved. All materials in *Spon's Mechanical & Electrical Services Price Book 2023* are copyright protected. All rights reserved. No such materials may be used, displayed, modified, adapted, distributed, transmitted, transferred, published or otherwise reproduced in any form or by any means now or hereafter developed other than strictly in accordance with the terms of the licence agreement enclosed with *Spon's Mechanical & Electrical Services Price Book 2023*. However, text and images may be printed and copied for research and private study within the preset program limitations. Please note the copyright notice above, and that any text or images printed or copied must credit the source."

9.4   This Product contains material proprietary to and copyedited by the Publisher and others. Except for the licence granted herein, all rights, title and interest in the Product, in all languages, formats and media throughout the world, including copyrights therein, are and remain the property of the Publisher or other copyright holders identified in the Product.

10. Non-assignment

This Agreement and the licence contained within it may not be assigned to any other person or entity without the written consent of the Publisher.

11. Termination and Consequences of Termination.

11.1   The Publisher shall have the right to terminate this Agreement if:

11.1.1 the Licensee is in material breach of this Agreement and fails to remedy such breach (where capable of remedy) within 14 days of a written notice from the Publisher requiring it to do so; or

11.1.2 the Licensee becomes insolvent, becomes subject to receivership, liquidation or similar external administration; or

11.1.3 the Licensee ceases to operate in business.

11.2   The Licensee shall have the right to terminate this Agreement for any reason upon two month's written notice. The Licensee shall not be entitled to any refund for payments made under this Agreement prior to termination under this sub-clause 11.2.

11.3   Termination by either of the parties is without prejudice to any other rights or remedies under the general law to which they may be entitled, or which survive such termination (including rights of the Publisher under sub-clause 7.4 above).

11.4   Upon termination of this Agreement, or expiry of its terms, the Licensee must destroy all copies and any back up copies of the product or part thereof.

12. General

12.1   *Compliance with export provisions*
The Publisher hereby agrees to comply fully with all relevant export laws and regulations of the United Kingdom to ensure that the Product is not exported, directly or indirectly, in violation of English law.

12.2   *Force majeure*
The parties accept no responsibility for breaches of this Agreement occurring as a result of circumstances beyond their control.

12.3   *No waiver*
Any failure or delay by either party to exercise or enforce any right conferred by this Agreement shall not be deemed to be a waiver of such right.

12.4   *Entire agreement*
This Agreement represents the entire agreement between the Publisher and the Licensee concerning the Product. The terms of this Agreement supersede all prior purchase orders, written terms and conditions, written or verbal representations, advertising or statements relating in any way to the Product.

12.5   *Severability*
If any provision of this Agreement is found to be invalid or unenforceable by a court of law of competent jurisdiction, such a finding shall not affect the other provisions of this Agreement and all provisions of this Agreement unaffected by such a finding shall remain in full force and effect.

12.6   *Variations*
This agreement may only be varied in writing by means of variation signed in writing by both parties.

12.7   *Notices*
All notices to be delivered to: Spon's Price Books, Taylor & Francis Books Ltd., 4 Park Square, Milton Park, Abingdon, Oxfordshire, OX14 4RN, UK.

12.8   *Governing law*
This Agreement is governed by English law and the parties hereby agree that any dispute arising under this Agreement shall be subject to the jurisdiction of the English courts.

If you have any queries about the terms of this licence, please contact:

Spon's Price Books
Taylor & Francis Books Ltd.
4 Park Square, Milton Park, Abingdon, Oxfordshire, OX14 4RN
Tel: +44 (0) 20 7017 6000
www.sponpress.com

Spon Press
an imprint of Taylor & Francis

Ebook options

Your print copy comes with a free ebook on the VitalSource® Bookshelf platform. Further copies of the ebook are available from https://www.routledge.com/
Pick your price book and select the ebook under the 'Select Format' option.

To buy Spon's ebooks for five or more users of in your organisation please contact:

Spon's Price Books
eBooks & Online Sales
Taylor & Francis Books Ltd.
4 Park Square, Milton Park, Oxfordshire, OX14 4RN
Tel: (020) 7017 6000
onlinesales@informa.com